Reinforcement Learning and Stochastic Optimization:
A Unified Framework for Sequential Decisions

Reinforcement Learning and Stochastic Optimization

A Unified Framework for Sequential Decisions

Warren B. Powell
Princeton University
Princeton, NJ

Contents

Preface   *xxv*
Acknowledgments   *xxxi*

Part I – Introduction   *1*

1   **Sequential Decision Problems**   *3*
1.1   The Audience   *7*
1.2   The Communities of Sequential Decision Problems   *8*
1.3   Our Universal Modeling Framework   *10*
1.4   Designing Policies for Sequential Decision Problems   *15*
1.4.1   Policy Search   *15*
1.4.2   Policies Based on Lookahead Approximations   *17*
1.4.3   Mixing and Matching   *18*
1.4.4   Optimality of the Four Classes   *19*
1.4.5   Pulling it All Together   *19*
1.5   Learning   *20*
1.6   Themes   *21*
1.6.1   Blending Learning and Optimization   *21*
1.6.2   Bridging Machine Learning to Sequential Decisions   *21*
1.6.3   From Deterministic to Stochastic Optimization   *23*
1.6.4   From Single to Multiple Agents   *26*
1.7   Our Modeling Approach   *27*
1.8   How to Read this Book   *27*
1.8.1   Organization of Topics   *28*
1.8.2   How to Read Each Chapter   *31*
1.8.3   Organization of Exercises   *32*

1.9	Bibliographic Notes	*33*
	Exercises	*34*
	Bibliography	*38*

2   **Canonical Problems and Applications**   *39*

2.1	Canonical Problems	*39*
2.1.1	Stochastic Search – Derivative-based and Derivative-free	*40*
2.1.1.1	Derivative-based Stochastic Search	*42*
2.1.1.2	Derivative-free Stochastic Search	*43*
2.1.2	Decision Trees	*44*
2.1.3	Markov Decision Processes	*45*
2.1.4	Optimal Control	*47*
2.1.5	Approximate Dynamic Programming	*50*
2.1.6	Reinforcement Learning	*50*
2.1.7	Optimal Stopping	*54*
2.1.8	Stochastic Programming	*56*
2.1.9	The Multiarmed Bandit Problem	*57*
2.1.10	Simulation Optimization	*60*
2.1.11	Active Learning	*61*
2.1.12	Chance-constrained Programming	*61*
2.1.13	Model Predictive Control	*62*
2.1.14	Robust Optimization	*63*
2.2	A Universal Modeling Framework for Sequential Decision Problems	*64*
2.2.1	Our Universal Model for Sequential Decision Problems	*65*
2.2.2	A Compact Modeling Presentation	*68*
2.2.3	MDP/RL vs. Optimal Control Modeling Frameworks	*68*
2.3	Applications	*69*
2.3.1	The Newsvendor Problems	*70*
2.3.1.1	Basic Newsvendor – Final Reward	*70*
2.3.1.2	Basic Newsvendor – Cumulative Reward	*71*
2.3.1.3	Contextual Newsvendor	*71*
2.3.1.4	Multidimensional Newsvendor Problems	*72*
2.3.2	Inventory/Storage Problems	*73*
2.3.2.1	Inventory Without Lags	*73*
2.3.2.2	Inventory Planning with Forecasts	*75*
2.3.2.3	Lagged Decisions	*75*
2.3.3	Shortest Path Problems	*76*
2.3.3.1	A Deterministic Shortest Path Problem	*76*
2.3.3.2	A Stochastic Shortest Path Problem	*77*
2.3.3.3	A Dynamic Shortest Path Problem	*78*

2.3.3.4   A Robust Shortest Path Problem   *78*
2.3.4   Some Fleet Management Problems   *78*
2.3.4.1   The Nomadic Trucker   *79*
2.3.4.2   From One Driver to a Fleet   *80*
2.3.5   Pricing   *80*
2.3.6   Medical Decision Making   *81*
2.3.7   Scientific Exploration   *82*
2.3.8   Machine Learning vs. Sequential Decision Problems   *83*
2.4   Bibliographic Notes   *85*
        Exercises   *90*
        Bibliography   *93*

3   **Online Learning**   *101*
3.1   Machine Learning for Sequential Decisions   *102*
3.1.1   Observations and Data in Stochastic Optimization   *102*
3.1.2   Indexing Input x^n and Response y^{n+1} *103*
3.1.3   Functions We are Learning   *103*
3.1.4   Sequential Learning: From Very Little Data to ... More Data   *105*
3.1.5   Approximation Strategies   *106*
3.1.6   From Data Analytics to Decision Analytics   *108*
3.1.7   Batch vs. Online Learning   *109*
3.2   Adaptive Learning Using Exponential Smoothing   *110*
3.3   Lookup Tables with Frequentist Updating   *111*
3.4   Lookup Tables with Bayesian Updating   *112*
3.4.1   The Updating Equations for Independent Beliefs   *113*
3.4.2   Updating for Correlated Beliefs   *113*
3.4.3   Gaussian Process Regression   *117*
3.5   Computing Bias and Variance*   *118*
3.6   Lookup Tables and Aggregation*   *121*
3.6.1   Hierarchical Aggregation   *121*
3.6.2   Estimates of Different Levels of Aggregation   *125*
3.6.3   Combining Multiple Levels of Aggregation   *129*
3.7   Linear Parametric Models   *131*
3.7.1   Linear Regression Review   *132*
3.7.2   Sparse Additive Models and Lasso   *134*
3.8   Recursive Least Squares for Linear Models   *136*
3.8.1   Recursive Least Squares for Stationary Data   *136*
3.8.2   Recursive Least Squares for Nonstationary Data*   *138*
3.8.3   Recursive Estimation Using Multiple Observations*   *139*
3.9   Nonlinear Parametric Models   *140*
3.9.1   Maximum Likelihood Estimation   *141*

3.9.2   Sampled Belief Models   *141*

3.9.3   Neural Networks – Parametric*   *143*

3.9.4   Limitations of Neural Networks   *148*

3.10   Nonparametric Models*   *149*

3.10.1   K-Nearest Neighbor   *150*

3.10.2   Kernel Regression   *151*

3.10.3   Local Polynomial Regression   *153*

3.10.4   Deep Neural Networks   *154*

3.10.5   Support Vector Machines   *155*

3.10.6   Indexed Functions, Tree Structures, and Clustering   *156*

3.10.7   Comments on Nonparametric Models   *157*

3.11   Nonstationary Learning*   *159*

3.11.1   Nonstationary Learning I – Martingale Truth   *159*

3.11.2   Nonstationary Learning II – Transient Truth   *160*

3.11.3   Learning Processes   *161*

3.12   The Curse of Dimensionality   *162*

3.13   Designing Approximation Architectures in Adaptive Learning   *165*

3.14   Why Does It Work?**   *166*

3.14.1   Derivation of the Recursive Estimation Equations   *166*

3.14.2   The Sherman-Morrison Updating Formula   *168*

3.14.3   Correlations in Hierarchical Estimation   *169*

3.14.4   Proof of Proposition 3.14.1   *172*

3.15   Bibliographic Notes   *174*

          Exercises   *176*

          Bibliography   *180*

4   **Introduction to Stochastic Search**   *183*

4.1   Illustrations of the Basic Stochastic Optimization Problem   *185*

4.2   Deterministic Methods   *188*

4.2.1   A "Stochastic" Shortest Path Problem   *189*

4.2.2   A Newsvendor Problem with Known Distribution   *189*

4.2.3   Chance-Constrained Optimization   *190*

4.2.4   Optimal Control   *191*

4.2.5   Discrete Markov Decision Processes   *192*

4.2.6   Remarks   *192*

4.3   Sampled Models   *193*

4.3.1   Formulating a Sampled Model   *194*

4.3.1.1   A Sampled Stochastic Linear Program   *194*

4.3.1.2   Sampled Chance-Constrained Models   *195*

4.3.1.3   Sampled Parametric Models   *196*
4.3.2   Convergence   *197*
4.3.3   Creating a Sampled Model   *199*
4.3.4   Decomposition Strategies*   *200*
4.4   Adaptive Learning Algorithms   *202*
4.4.1   Modeling Adaptive Learning Problems   *202*
4.4.2   Online vs. Offline Applications   *204*
4.4.2.1   Machine Learning   *204*
4.4.2.2   Optimization   *205*
4.4.3   Objective Functions for Learning   *205*
4.4.4   Designing Policies   *209*
4.5   Closing Remarks   *210*
4.6   Bibliographic Notes   *210*
    Exercises   *212*
    Bibliography   *218*

Part II – Stochastic Search   *221*

5   **Derivative-Based Stochastic Search**   *223*
5.1   Some Sample Applications   *225*
5.2   Modeling Uncertainty   *228*
5.2.1   Training Uncertainty $W^1, \ldots, W^N$ *228*
5.2.2   Model Uncertainty S^0 *229*
5.2.3   Testing Uncertainty   *230*
5.2.4   Policy Evaluation   *231*
5.2.5   Closing Notes   *231*
5.3   Stochastic Gradient Methods   *231*
5.3.1   A Stochastic Gradient Algorithm   *232*
5.3.2   Introduction to Stepsizes   *233*
5.3.3   Evaluating a Stochastic Gradient Algorithm   *235*
5.3.4   A Note on Notation   *236*
5.4   Styles of Gradients   *237*
5.4.1   Gradient Smoothing   *237*
5.4.2   Second-Order Methods   *237*
5.4.3   Finite Differences   *238*
5.4.4   SPSA   *240*
5.4.5   Constrained Problems   *242*
5.5   Parameter Optimization for Neural Networks*   *242*
5.5.1   Computing the Gradient   *244*
5.5.2   The Stochastic Gradient Algorithm   *246*

5.6   Stochastic Gradient Algorithm as a Sequential
          Decision Problem   *247*
5.7   Empirical Issues   *248*
5.8   Transient Problems*   *249*
5.9   Theoretical Performance*   *250*
5.10   Why Does it Work?   *250*
5.10.1   Some Probabilistic Preliminaries   *251*
5.10.2   An Older Proof*   *252*
5.10.3   A More Modern Proof**   *256*
5.11   Bibliographic Notes   *263*
          Exercises   *264*
          Bibliography   *270*

6   Stepsize Policies   *273*
6.1   Deterministic Stepsize Policies   *276*
6.1.1   Properties for Convergence   *276*
6.1.2   A Collection of Deterministic Policies   *278*
6.1.2.1   Constant Stepsizes   *278*
6.1.2.2   Generalized Harmonic Stepsizes   *279*
6.1.2.3   Polynomial Learning Rates   *280*
6.1.2.4   McClain's Formula   *280*
6.1.2.5   Search-then-Converge Learning Policy   *281*
6.2   Adaptive Stepsize Policies   *282*
6.2.1   The Case for Adaptive Stepsizes   *283*
6.2.2   Convergence Conditions   *283*
6.2.3   A Collection of Stochastic Policies   *284*
6.2.3.1   Kesten's Rule   *285*
6.2.3.2   Trigg's Formula   *286*
6.2.3.3   Stochastic Gradient Adaptive Stepsize Rule   *286*
6.2.3.4   ADAM   *287*
6.2.3.5   AdaGrad   *287*
6.2.3.6   RMSProp   *288*
6.2.4   Experimental Notes   *289*
6.3   Optimal Stepsize Policies*   *289*
6.3.1   Optimal Stepsizes for Stationary Data   *291*
6.3.2   Optimal Stepsizes for Nonstationary Data – I   *293*
6.3.3   Optimal Stepsizes for Nonstationary Data – II   *294*
6.4   Optimal Stepsizes for Approximate Value Iteration*   *297*
6.5   Convergence   *300*
6.6   Guidelines for Choosing Stepsize Policies   *301*
6.7   Why Does it Work*   *303*

6.7.1   Proof of BAKF Stepsize   *303*
6.8   Bibliographic Notes   *306*
         Exercises   *307*
         Bibliography   *314*

7   Derivative-Free Stochastic Search   317
7.1   Overview of Derivative-free Stochastic Search   *319*
7.1.1   Applications and Time Scales   *319*
7.1.2   The Communities of Derivative-free Stochastic Search   *321*
7.1.3   The Multiarmed Bandit Story   *321*
7.1.4   From Passive Learning to Active Learning to Bandit Problems   *323*
7.2   Modeling Derivative-free Stochastic Search   *325*
7.2.1   The Universal Model   *325*
7.2.2   Illustration: Optimizing a Manufacturing Process   *328*
7.2.3   Major Problem Classes   *329*
7.3   Designing Policies   *330*
7.4   Policy Function Approximations   *333*
7.5   Cost Function Approximations   *335*
7.6   VFA-based Policies   *338*
7.6.1   An Optimal Policy   *338*
7.6.2   Beta-Bernoulli Belief Model   *340*
7.6.3   Backward Approximate Dynamic Programming   *342*
7.6.4   Gittins Indices for Learning in Steady State   *343*
7.7   Direct Lookahead Policies   *348*
7.7.1   When do we Need Lookahead Policies?   *349*
7.7.2   Single Period Lookahead Policies   *350*
7.7.3   Restricted Multiperiod Lookahead   *353*
7.7.4   Multiperiod Deterministic Lookahead   *355*
7.7.5   Multiperiod Stochastic Lookahead Policies   *357*
7.7.6   Hybrid Direct Lookahead   *360*
7.8   The Knowledge Gradient (Continued)*   *362*
7.8.1   The Belief Model   *363*
7.8.2   The Knowledge Gradient for Maximizing Final Reward   *364*
7.8.3   The Knowledge Gradient for Maximizing Cumulative Reward   *369*
7.8.4   The Knowledge Gradient for Sampled Belief Model*   *370*
7.8.5   Knowledge Gradient for Correlated Beliefs   *375*
7.9   Learning in Batches   *380*
7.10   Simulation Optimization*   *382*
7.10.1   An Indifference Zone Algorithm   *383*
7.10.2   Optimal Computing Budget Allocation   *383*
7.11   Evaluating Policies   *385*

7.11.1   Alternative Performance Metrics*   *386*
7.11.2   Perspectives of Optimality*   *392*
7.12   Designing Policies   *394*
7.12.1   Characteristics of a Policy   *395*
7.12.2   The Effect of Scaling   *396*
7.12.3   Tuning   *398*
7.13   Extensions*   *398*
7.13.1   Learning in Nonstationary Settings   *399*
7.13.2   Strategies for Designing Time-dependent Policies   *400*
7.13.3   A Transient Learning Model   *401*
7.13.4   The Knowledge Gradient for Transient Problems   *402*
7.13.5   Learning with Large or Continuous Choice Sets   *403*
7.13.6   Learning with Exogenous State Information – the Contextual Bandit
            Problem   *405*
7.13.7   State-dependent vs. State-independent Problems   *408*
7.14   Bibliographic Notes   *409*
            Exercises   *412*
            Bibliography   *424*

Part III – State-dependent Problems   *429*

8   **State-dependent Problems**   *431*
8.1   Graph Problems   *433*
8.1.1   A Stochastic Shortest Path Problem   *433*
8.1.2   The Nomadic Trucker   *434*
8.1.3   The Transformer Replacement Problem   *435*
8.1.4   Asset Valuation   *437*
8.2   Inventory Problems   *439*
8.2.1   A Basic Inventory Problem   *439*
8.2.2   The Inventory Problem – II   *440*
8.2.3   The Lagged Asset Acquisition Problem   *443*
8.2.4   The Batch Replenishment Problem   *444*
8.3   Complex Resource Allocation Problems   *446*
8.3.1   The Dynamic Assignment Problem   *447*
8.3.2   The Blood Management Problem   *450*
8.4   State-dependent Learning Problems   *456*
8.4.1   Medical Decision Making   *457*
8.4.2   Laboratory Experimentation   *458*
8.4.3   Bidding for Ad-clicks   *459*
8.4.4   An Information-collecting Shortest Path Problem   *459*

8.5   A Sequence of Problem Classes   *460*
8.6   Bibliographic Notes   *461*
        Exercises   *462*
        Bibliography   *466*

9   **Modeling Sequential Decision Problems**   *467*
9.1   A Simple Modeling Illustration   *471*
9.2   Notational Style   *476*
9.3   Modeling Time   *478*
9.4   The States of Our System   *481*
9.4.1   Defining the State Variable   *481*
9.4.2   The Three States of Our System   *485*
9.4.3   Initial State S_0 vs. Subsequent States S_t, $t > 0$ *488*
9.4.4   Lagged State Variables*   *490*
9.4.5   The Post-decision State Variable*   *490*
9.4.6   A Shortest Path Illustration   *493*
9.4.7   Belief States*   *495*
9.4.8   Latent Variables*   *496*
9.4.9   Rolling Forecasts*   *497*
9.4.10 Flat vs. Factored State Representations*   *498*
9.4.11 A Programmer's Perspective of State Variables   *499*
9.5   Modeling Decisions   *500*
9.5.1   Types of Decisions   *502*
9.5.2   Initial Decision x_0 vs. Subsequent Decisions x_t, $t > 0$ *502*
9.5.3   Strategic, Tactical, and Execution Decisions   *503*
9.5.4   Constraints   *504*
9.5.5   Introducing Policies   *505*
9.6   The Exogenous Information Process   *506*
9.6.1   Basic Notation for Information Processes   *506*
9.6.2   Outcomes and Scenarios   *509*
9.6.3   Lagged Information Processes*   *510*
9.6.4   Models of Information Processes*   *511*
9.6.5   Supervisory Processes*   *514*
9.7   The Transition Function   *515*
9.7.1   A General Model   *515*
9.7.2   Model-free Dynamic Programming   *516*
9.7.3   Exogenous Transitions   *518*
9.8   The Objective Function   *518*
9.8.1   The Performance Metric   *518*
9.8.2   Optimizing the Policy   *519*
9.8.3   Dependence of Optimal Policy on S_0 *520*

9.8.4   State-dependent Variations   *520*
9.8.5   Uncertainty Operators   *523*
9.9   Illustration: An Energy Storage Model   *523*
9.9.1   With a Time-series Price Model   *525*
9.9.2   With Passive Learning   *525*
9.9.3   With Active Learning   *526*
9.9.4   With Rolling Forecasts   *526*
9.10   Base Models and Lookahead Models   *528*
9.11   A Classification of Problems*   *529*
9.12   Policy Evaluation*   *532*
9.13   Advanced Probabilistic Modeling Concepts**   *534*
9.13.1   A Measure-theoretic View of Information**   *535*
9.13.2   Policies and Measurability   *538*
9.14   Looking Forward   *540*
9.15   Bibliographic Notes   *542*
          Exercises   *544*
          Bibliography   *557*

10   Uncertainty Modeling   *559*
10.1   Sources of Uncertainty   *560*
10.1.1   Observational Errors   *562*
10.1.2   Exogenous Uncertainty   *564*
10.1.3   Prognostic Uncertainty   *564*
10.1.4   Inferential (or Diagnostic) Uncertainty   *567*
10.1.5   Experimental Variability   *568*
10.1.6   Model Uncertainty   *569*
10.1.7   Transitional Uncertainty   *571*
10.1.8   Control/implementation Uncertainty   *571*
10.1.9   Communication Errors and Biases   *572*
10.1.10   Algorithmic Instability   *573*
10.1.11   Goal Uncertainty   *574*
10.1.12   Political/regulatory Uncertainty   *574*
10.1.13   Discussion   *574*
10.2   A Modeling Case Study: The COVID Pandemic   *575*
10.3   Stochastic Modeling   *575*
10.3.1   Sampling Exogenous Information   *575*
10.3.2   Types of Distributions   *577*
10.3.3   Modeling Sample Paths   *578*
10.3.4   State-action-dependent Processes   *579*
10.3.5   Modeling Correlations   *581*
10.4   Monte Carlo Simulation   *581*

10.4.1   Generating Uniform [0, 1] Random Variables   *582*

10.4.2   Uniform and Normal Random Variable   *583*

10.4.3   Generating Random Variables from Inverse Cumulative
Distributions   *585*

10.4.4   Inverse Cumulative From Quantile Distributions   *586*

10.4.5   Distributions with Uncertain Parameters   *587*

10.5   Case Study: Modeling Electricity Prices   *589*

10.5.1   Mean Reversion   *590*

10.5.2   Jump-diffusion Models   *590*

10.5.3   Quantile Distributions   *591*

10.5.4   Regime Shifting   *592*

10.5.5   Crossing Times   *593*

10.6   Sampling vs. Sampled Models   *595*

10.6.1   Iterative Sampling: A Stochastic Gradient Algorithm   *595*

10.6.2   Static Sampling: Solving a Sampled Model   *595*

10.6.3   Sampled Representation with Bayesian Updating   *596*

10.7   Closing Notes   *597*

10.8   Bibliographic Notes   *597*

Exercises   *598*

Bibliography   *601*

11   **Designing Policies**   *603*

11.1   From Optimization to Machine Learning to Sequential
Decision Problems   *605*

11.2   The Classes of Policies   *606*

11.3   Policy Function Approximations   *610*

11.4   Cost Function Approximations   *613*

11.5   Value Function Approximations   *614*

11.6   Direct Lookahead Approximations   *616*

11.6.1   The Basic Idea   *616*

11.6.2   Modeling the Lookahead Problem   *619*

11.6.3   The Policy-Within-a-Policy   *620*

11.7   Hybrid Strategies   *620*

11.7.1   Cost Function Approximation with Policy Function
Approximations   *621*

11.7.2   Lookahead Policies with Value Function Approximations   *622*

11.7.3   Lookahead Policies with Cost Function Approximations   *623*

11.7.4   Tree Search with Rollout Heuristic and a Lookup Table Policy   *623*

11.7.5   Value Function Approximation with Policy Function
Approximation   *624*

11.7.6   Fitting Value Functions Using ADP and Policy Search   *624*

11.8   Randomized Policies   *626*

11.9   Illustration: An Energy Storage Model Revisited   *627*

11.9.1   Policy Function Approximation   *628*

11.9.2   Cost Function Approximation   *628*

11.9.3   Value Function Approximation   *628*

11.9.4   Deterministic Lookahead   *629*

11.9.5   Hybrid Lookahead-Cost Function Approximation   *629*

11.9.6   Experimental Testing   *629*

11.10   Choosing the Policy Class   *631*

11.10.1   The Policy Classes   *631*

11.10.2   Policy Complexity-Computational Tradeoffs   *636*

11.10.3   Screening Questions   *638*

11.11   Policy Evaluation   *641*

11.12   Parameter Tuning   *642*

11.12.1   The Soft Issues   *644*

11.12.2   Searching Across Policy Classes   *645*

11.13   Bibliographic Notes   *646*

Exercises   *646*

Bibliography   *651*

Part IV – Policy Search   *653*

12   Policy Function Approximations and Policy Search   *655*

12.1   Policy Search as a Sequential Decision Problem   *657*

12.2   Classes of Policy Function Approximations   *658*

12.2.1   Lookup Table Policies   *659*

12.2.2   Boltzmann Policies for Discrete Actions   *659*

12.2.3   Linear Decision Rules   *660*

12.2.4   Monotone Policies   *661*

12.2.5   Nonlinear Policies   *662*

12.2.6   Nonparametric/Locally Linear Policies   *663*

12.2.7   Contextual Policies   *665*

12.3   Problem Characteristics   *665*

12.4   Flavors of Policy Search   *666*

12.5   Policy Search with Numerical Derivatives   *669*

12.6   Derivative-Free Methods for Policy Search   *670*

12.6.1   Belief Models   *671*

12.6.2   Learning Through Perturbed PFAs   *672*

12.6.3   Learning CFAs   *675*

12.6.4   DLA Using the Knowledge Gradient   *677*

12.6.5 Comments *677*

12.7 Exact Derivatives for Continuous Sequential Problems* *677*

12.8 Exact Derivatives for Discrete Dynamic Programs** *680*

12.8.1 A Stochastic Policy *681*

12.8.2 The Objective Function *683*

12.8.3 The Policy Gradient Theorem *683*

12.8.4 Computing the Policy Gradient *684*

12.9 Supervised Learning *686*

12.10 Why Does it Work? *687*

12.10.1 Derivation of the Policy Gradient Theorem *687*

12.11 Bibliographic Notes *690*

Exercises *691*

Bibliography *698*

13 **Cost Function Approximations** *701*

13.1 General Formulation for Parametric CFA *703*

13.2 Objective-Modified CFAs *704*

13.2.1 Linear Cost Function Correction *705*

13.2.2 CFAs for Dynamic Assignment Problems *705*

13.2.3 Dynamic Shortest Paths *707*

13.2.4 Dynamic Trading Policy *711*

13.2.5 Discussion *713*

13.3 Constraint-Modified CFAs *714*

13.3.1 General Formulation of Constraint-Modified CFAs *715*

13.3.2 A Blood Management Problem *715*

13.3.3 An Energy Storage Example with Rolling Forecasts *717*

13.4 Bibliographic Notes *725*

Exercises *726*

Bibliography *729*

Part V – Lookahead Policies *731*

14 **Exact Dynamic Programming** *737*

14.1 Discrete Dynamic Programming *738*

14.2 The Optimality Equations *740*

14.2.1 Bellman's Equations *741*

14.2.2 Computing the Transition Matrix *745*

14.2.3 Random Contributions *746*

14.2.4 Bellman's Equation Using Operator Notation* *746*

14.3   Finite Horizon Problems   *747*
14.4   Continuous Problems with Exact Solutions   *750*
14.4.1   The Gambling Problem   *751*
14.4.2   The Continuous Budgeting Problem   *752*
14.5   Infinite Horizon Problems*   *755*
14.6   Value Iteration for Infinite Horizon Problems*   *757*
14.6.1   A Gauss-Seidel Variation   *758*
14.6.2   Relative Value Iteration   *758*
14.6.3   Bounds and Rates of Convergence   *760*
14.7   Policy Iteration for Infinite Horizon Problems*   *762*
14.8   Hybrid Value-Policy Iteration*   *764*
14.9   Average Reward Dynamic Programming*   *765*
14.10   The Linear Programming Method for Dynamic
         Programs**   *766*
14.11   Linear Quadratic Regulation   *767*
14.12   Why Does it Work?**   *770*
14.12.1 The Optimality Equations   *770*
14.12.2 Convergence of Value Iteration   *774*
14.12.3 Monotonicity of Value Iteration   *778*
14.12.4 Bounding the Error from Value Iteration   *780*
14.12.5 Randomized Policies   *781*
14.13   Bibliographic Notes   *783*
         Exercises   *783*
         Bibliography   *793*

15   **Backward Approximate Dynamic Programming**   *795*
15.1   Backward Approximate Dynamic Programming for
         Finite Horizon Problems   *797*
15.1.1   Some Preliminaries   *797*
15.1.2   Backward ADP Using Lookup Tables   *799*
15.1.3   Backward ADP Algorithm with Continuous Approximations   *801*
15.2   Fitted Value Iteration for Infinite Horizon Problems   *804*
15.3   Value Function Approximation Strategies   *805*
15.3.1   Linear Models   *806*
15.3.2   Monotone Functions   *807*
15.3.3   Other Approximation Models   *810*
15.4   Computational Observations   *810*
15.4.1   Experimental Benchmarking of Backward ADP   *810*
15.4.2   Computational Notes   *815*
15.5   Bibliographic Notes   *816*
         Exercises   *816*

Bibliography   *821*

16   Forward ADP I: The Value of a Policy   *823*
16.1   Sampling the Value of a Policy   *824*
16.1.1   Direct Policy Evaluation for Finite Horizon Problems   *824*
16.1.2   Policy Evaluation for Infinite Horizon Problems   *826*
16.1.3   Temporal Difference Updates   *828*
16.1.4   TD(λ)   *829*
16.1.5   TD(0) and Approximate Value Iteration   *830*
16.1.6   TD Learning for Infinite Horizon Problems   *832*
16.2   Stochastic Approximation Methods   *835*
16.3   Bellman's Equation Using a Linear Model*   *837*
16.3.1   A Matrix-based Derivation**   *837*
16.3.2   A Simulation-based Implementation   *840*
16.3.3   Least Squares Temporal Difference Learning (LSTD)   *840*
16.3.4   Least Squares Policy Evaluation   *841*
16.4   Analysis of TD(0), LSTD, and LSPE Using a Single
          State*   *842*
16.4.1   Recursive Least Squares and TD(0)   *842*
16.4.2   Least Squares Policy Evaluation   *844*
16.4.3   Least Squares Temporal Difference Learning   *844*
16.4.4   Discussion   *844*
16.5   Gradient-based Methods for Approximate Value
          Iteration*   *845*
16.5.1   Approximate Value Iteration with Linear Models**   *845*
16.5.2   A Geometric View of Linear Models*   *850*
16.6   Value Function Approximations Based on Bayesian
          Learning*   *852*
16.6.1   Minimizing Bias for Infinite Horizon Problems   *852*
16.6.2   Lookup Tables with Correlated Beliefs   *853*
16.6.3   Parametric Models   *854*
16.6.4   Creating the Prior   *855*
16.7   Learning Algorithms and Atepsizes   *855*
16.7.1   Least Squares Temporal Differences   *856*
16.7.2   Least Squares Policy Evaluation   *857*
16.7.3   Recursive Least Squares   *857*
16.7.4   Bounding $1/n$ Convergence for Approximate value Iteration   *859*
16.7.5   Discussion   *860*
16.8   Bibliographic Notes   *860*
          Exercises   *862*
          Bibliography   *864*

17 **Forward ADP II: Policy Optimization** *867*
17.1 Overview of Algorithmic Strategies *869*
17.2 Approximate Value Iteration and *Q*-Learning Using Lookup Tables *871*
17.2.1 Value Iteration Using a Pre-Decision State Variable *872*
17.2.2 *Q*-Learning *873*
17.2.3 Value Iteration Using a Post-Decision State Variable *875*
17.2.4 Value Iteration Using a Backward Pass *877*
17.3 Styles of Learning *881*
17.3.1 Offline Learning *882*
17.3.2 From Offline to Online *883*
17.3.3 Evaluating Offline and Online Learning Policies *885*
17.3.4 Lookahead Policies *885*
17.4 Approximate Value Iteration Using Linear Models *886*
17.5 On-policy vs. off-policy learning and the exploration–exploitation problem *888*
17.5.1 Terminology *889*
17.5.2 Learning with Lookup Tables *890*
17.5.3 Learning with Generalized Belief Models *891*
17.6 Applications *894*
17.6.1 Pricing an American Option *894*
17.6.2 Playing "Lose Tic-Tac-Toe" *898*
17.6.3 Approximate Dynamic Programming for Deterministic Problems *900*
17.7 Approximate Policy Iteration *900*
17.7.1 Finite Horizon Problems Using Lookup Tables *901*
17.7.2 Finite Horizon Problems Using Linear Models *903*
17.7.3 LSTD for Infinite Horizon Problems Using Linear Models *903*
17.8 The Actor–Critic Paradigm *907*
17.9 Statistical Bias in the Max Operator* *909*
17.10 The Linear Programming Method Using Linear Models* *912*
17.11 Finite Horizon Approximations for Steady-State Applications *915*
17.12 Bibliographic Notes *917*
Exercises *918*
Bibliography *924*

18 **Forward ADP III: Convex Resource Allocation Problems** *927*
18.1 Resource Allocation Problems *930*

18.1.1 The Newsvendor Problem *930*

18.1.2 Two-Stage Resource Allocation Problems *931*

18.1.3 A General Multiperiod Resource Allocation Model* *933*

18.2 Values Versus Marginal Values *937*

18.3 Piecewise Linear Approximations for Scalar Functions *938*

18.3.1 The Leveling Algorithm *939*

18.3.2 The CAVE Algorithm *941*

18.4 Regression Methods *941*

18.5 Separable Piecewise Linear Approximations *944*

18.6 Benders Decomposition for Nonseparable Approximations** *946*

18.6.1 Benders' Decomposition for Two-Stage Problems *947*

18.6.2 Asymptotic Analysis of Benders with Regularization** *952*

18.6.3 Benders with Regularization *956*

18.7 Linear Approximations for High-Dimensional Applications *956*

18.8 Resource Allocation with Exogenous Information State *958*

18.9 Closing Notes *959*

18.10 Bibliographic Notes *960*

Exercises *962*

Bibliography *967*

19 Direct Lookahead Policies *971*

19.1 Optimal Policies Using Lookahead Models *974*

19.2 Creating an Approximate Lookahead Model *978*

19.2.1 Modeling the Lookahead Model *979*

19.2.2 Strategies for Approximating the Lookahead Model *980*

19.3 Modified Objectives in Lookahead Models *985*

19.3.1 Managing Risk *985*

19.3.2 Utility Functions for Multiobjective Problems *991*

19.3.3 Model Discounting *992*

19.4 Evaluating DLA Policies *992*

19.4.1 Evaluating Policies in a Simulator *994*

19.4.2 Evaluating Risk-Adjusted Policies *994*

19.4.3 Evaluating Policies in the Field *996*

19.4.4 Tuning Direct Lookahead Policies *997*

19.5 Why Use a DLA? *997*

19.6 Deterministic Lookaheads *999*

19.6.1 A Deterministic Lookahead: Shortest Path Problems *1001*

19.6.2   Parameterized Lookaheads   *1003*

19.7   A Tour of Stochastic Lookahead Policies   *1005*

19.7.1   Lookahead PFAs   *1005*

19.7.2   Lookahead CFAs   *1007*

19.7.3   Lookahead VFAs for the Lookahead Model   *1007*

19.7.4   Lookahead DLAs for the Lookahead Model   *1008*

19.7.5   Discussion   *1009*

19.8   Monte Carlo Tree Search for Discrete Decisions   *1009*

19.8.1   Basic Idea   *1010*

19.8.2   The Steps of MCTS   *1010*

19.8.3   Discussion   *1014*

19.8.4   Optimistic Monte Carlo Tree Search   *1016*

19.9   Two-Stage Stochastic Programming for Vector
          Decisions*   *1018*

19.9.1   The Basic Two-Stage Stochastic Program   *1018*

19.9.2   Two-Stage Approximation of a Sequential Problem   *1020*

19.9.3   Discussion   *1023*

19.10   Observations on DLA Policies   *1024*

19.11   Bibliographic Notes   *1025*

          Exercises   *1027*

          Bibliography   *1031*

Part VI – Multiagent Systems   *1033*

20   Multiagent Modeling and Learning   *1035*

20.1   Overview of Multiagent Systems   *1036*

20.1.1   Dimensions of a Multiagent System   *1036*

20.1.2   Communication   *1038*

20.1.3   Modeling a Multiagent System   *1040*

20.1.4   Controlling Architectures   *1043*

20.2   A Learning Problem – Flu Mitigation   *1044*

20.2.1   Model 1: A Static Model   *1045*

20.2.2   Variations of Our Flu Model   *1046*

20.2.3   Two-Agent Learning Models   *1050*

20.2.4   Transition Functions for Two-Agent Model   *1052*

20.2.5   Designing Policies for the Flu Problem   *1054*

20.3   The POMDP Perspective*   *1059*

20.4   The Two-Agent Newsvendor Problem   *1062*

20.5   Multiple Independent Agents – An HVAC Controller
          Model   *1067*

20.5.1   Model   *1067*
20.5.2   Designing Policies   *1069*
20.6   Cooperative Agents – A Spatially Distributed Blood
          Management Problem   *1070*
20.7   Closing Notes   *1074*
20.8   Why Does it Work?   *1074*
20.8.1   Derivation of the POMDP Belief Transition Function   *1074*
20.9   Bibliographic Notes   *1076*
          Exercises   *1077*
          Bibliography   *1083*

          Index   *1085*

Preface

Preface to *Reinforcement Learning and Stochastic Optimization: A unified framework for sequential decisions*

This books represents a lifetime of research into what I now call sequential decision problems, which dates to 1982 when I was introduced to the problem arising in truckload trucking (think of Uber/Lyft for trucks) where we have to choose which driver to assign to a load, and which loads to accept to move, given the high level of randomness in future customer demands, representing requests to move full truckloads of freight.

It took me 20 years to figure out a practical algorithm to solve this problem, which led to my first book (in 2007) on approximate dynamic programming, where the major breakthrough was the introduction of the post-decision state and the use of hierarchical aggregation for approximating value functions to solve these high-dimensional problems. However, I would argue today that the most important chapter in the book (and I recognized it at the time), was chapter 5 on how to model these problems, without any reference to algorithms to solve the problem. I identified five elements to a sequential decision problem, leading up to the objective function which was written

$$\max_{\pi} \mathbb{E} \left\{ \sum_{t=0}^{T} C(S_t, X^{\pi}(S_t)) | S_0 \right\}.$$

It was not until the second edition (in 2011) that I realized that approximate dynamic programming (specifically, policies that depend on value functions) was not the only way to solve these problems; rather, there were four classes of policies, and only one used value functions. The 2011 edition of the book listed three of the four classes of policies that are described in this book, but most of the book still focused on approximating value functions. It was not until a 2014 paper ("Clearing the Jungle of Stochastic Optimization") that I identified the four classes of policies I use now. Then, in 2016 I realized that the four classes of policies could be divided between two major strategies: the policy search strategy, where we search over a family of functions to find the one that works best, and the lookahead strategy, where we make good decisions by approximating the downstream impact of a decision made now.

Finally, I combined these ideas in a 2019 paper ("A Unified Framework for Stochastic Optimization" published in the *European Journal for Operational Research*) with a better appreciation of major classes of problems such as state-independent problems (the pure learning problems that include derivative-based and derivative-free stochastic search) and the more general state-dependent problems; cumulative and final reward objective functions; and the realization that any adaptive search algorithm was a sequential decision problem. The material in the 2019 paper is effectively the outline for this book.

This book builds on the 2011 edition of my approximate dynamic programming book, and includes a number of chapters (some heavily edited) from the ADP book. It would be nice to call this a third edition, but the entire framework of this book is completely different. "Approximate dynamic programming" is a term that still refers to making decisions based on the idea of approximating the downstream value of being in a state. After decades of working with this approach (which is still covered over a span of five chapters in this volume), I can now say with confidence that value function approximations, despite all the attention they have received, is a powerful methodology for a surprisingly narrow set of decision problems.

By contrast, I finally developed the confidence to claim that the four classes of policies are universal. This means that *any* method for making decisions will fall in one of these four classes, or a hybrid of two or more. This is a game changer, because it shifts the focus from an algorithm (the method for making decisions) to the model (specifically the optimization problem above, along with the state-transition function and the model of the exogenous information process). This means we write out the elements of a problem *before* we tackle the problem of designing policies to decisions. I call this:

Model first, then solve.

The communities working on sequential decision problems are very focused on methods, just as I was with my earlier work with approximate dynamic programming. The problem is that any particular method will be inherently limited to a narrow class of problems. In this book, I demonstrate how you can take a simple inventory problem, and then tweak the data to make each of the four classes work best.

This new approach has opened up an entirely new way of approaching a problem class that, in the last year of writing the book, I started calling "sequential decision analytics," which is any problem consisting of the sequence:

Decision, information, decision, information, ...

I allow decisions to range from binary (selling an asset) to discrete choices (favored in computer science) to the high-dimensional resource allocation problems popular in operations research. This approach starts with a problem, shifts to the challenging task of modeling uncertainty, and then finishes with designing policies to make decisions to optimize some metric. The approach is practical, scalable, and universally applicable.

It is exciting to be able to create a single framework that spans 15 different communities, and which represents every possible method for solving sequential decision problems. While having a common language to model any sequential decision problem, combined with the general approach of the four classes of policies, is clearly of value, this framework has been developed by standing on the shoulders of the giants who have laid the foundational work for all of these methods. I have had to make choices regarding the best notation and modeling conventions, but my framework is completely inclusive of all the methods that have been developed to solve these problems. Rather than joining the chorus of researchers promoting specific algorithmic strategies (as I once did), my goal is to raise the visibility of all methods, so that someone looking to solve a real problem is working with the biggest possible toolbox, rather than just the tools developed within a specific community.

A word needs to be said about the title of the book. As this is being written, there is a massive surge of interest in "reinforcement learning," which started as a form of approximate dynamic programming (I used to refer to ADP and RL as similar to American English and British English). However, as the RL community has grown and started working on harder problems, they encountered the same experience that I and everyone else working in ADP found: value function approximations are not a panacea. Not only is it the case that they often do not work, they usually do not work. As a result, the RL community branched out (just as I did) into other methods such as "policy gradient methods" (my "policy function approximations" or PFA), upper confidence bounding (a form of "cost function approximation" or CFA), the original Q-learning (which produces a policy based on "value function approximations" or VFA), and finally

Monte Carlo tree search (a policy based on "direct lookahead approximations" or DLA). All of these methods are found in the second edition of Sutton and Barto's landmark book *Reinforcement Learning: An introduction*, but only as specific methods rather than general classes. This book takes the next step and identifies the general classes.

This evolution from one core method to all four classes of policies is being repeated among other fields that I came to call the "jungle of stochastic optimization." Stochastic search, simulation-optimization, and bandit problems all feature methods from each of the four classes of policies. Over time, I came to realize that all these fields (including reinforcement learning) were playing catchup to the grandfather of all of this work, which is optimal control (and stochastic control). The field of optimal control was the first to introduce and seriously explore the use of value function approximations (they call these cost-to-go functions), linear decision rules (a form of PFA), and the workhorse "model predictive control" (a great name for a simple rolling horizon procedure, which is a "direct lookahead approximation" in this book). I also found that my modeling framework was closest to that used in the optimal control literature, which was the first field to introduce the concept of a transition function, a powerful modeling device that has been largely overlooked by the other communities. I make a few small tweaks such as using state S_t instead of x_t, and decision x_t (widely used in the field of math programming) instead of u_t.

Then I introduce one big change, which is to maximize over all four classes of policies. Perhaps the most important innovation of this book is to break the almost automatic link between optimizing over policies, and then assuming that we are going to compute an optimal policy from either Bellman's equation or the Hamilton-Jacobi equations. These are rarely computable for real problems, which then leads people to assume that the natural next step is to approximate these equations. This is simply false, supported by decades of research where people have developed methods that do not depend on HJB equations. I recognize this body of research developing different classes of policies by making the inclusion of all four classes of policies fundamental to the original statement of the optimization problem above.

It will take some time for people from the different communities to learn to speak this common language. More likely, there will be an adaptation of existing modeling languages to this framework. For example, the optimal control community could keep their notation, but learn to write their objective functions as I have above, recognizing that the search over policies needs to span all four classes (which, I might point out, they are already using). I would hope that the reinforcement learning community, which adopted the notation for discrete action a, might learn to use the more general x (as the bandit community has already done).

I have tried to write this book to appeal to newcomers to the field, as well as people who already have training in one or more of the subfields that deal with decisions and uncertainty; recognizing these two broad communities was easily the biggest challenge while writing this book. Not surprisingly, the book is quite long. I have tried to make it more accessible to people who are new to the field by marking many sections with an * as an indication that this section can be skipped on a first-read. I also hope that the book will appeal to people from many application domains. However, the core audience is people who are looking to solve real problems by modeling applications and implementing the work in software. The notation is designed to facilitate writing computer programs, where there should be a direct relationship between the mathematical model and the software. This is particularly important when modeling the flow of information, something that is often overlooked in mainstream reinforcement learning papers.

Warren B. Powell

Princeton, New Jersey
August, 2021

Acknowledgments

The foundation of this book is a modeling framework for sequential decision problems that involves searching over four classes of policies for making decisions. The recognition that we needed all four classes of policies came from working on a wide range of problems spanning freight transportation (almost all modes), energy, health, e-commerce, finance, and even materials science (!!).

This research required a *lot* of computational work, which was only possible through the efforts of the many students and staff that worked in CASTLE Lab. Over my 39 years of teaching at Princeton, I benefited tremendously from the interactions with 70 graduate students and post-doctoral associates, along with nine professional staff. I am deeply indebted to the contributions of this exceptionally talented group of men and women who allowed me to participate in the challenges of getting computational methods to work on such a wide range of problems. It was precisely this diversity of problem settings that led me to appreciate the motivation for the different methods for solving problems. In the process, I met people from across the jungle, and learned to speak their language not just by reading papers, but by talking to them and, often, working on their problems.

I would also like to acknowledge what I learned from supervising over 200 senior theses. While not as advanced as the graduate research, the undergraduates helped expose me to an even wider range of problems, spanning topics such as sports, health, urban transportation, social networks, agriculture, pharmaceuticals, and even optimizing Greek cargo ships. It was the undergraduates who accelerated my move into energy in 2008, allowing me to experiment with modeling and solving a variety of problems spanning microgrids, solar arrays, energy storage, demand management, and storm response. This experience exposed me to new challenges, new methods, and most important, new communities in engineering and economics.

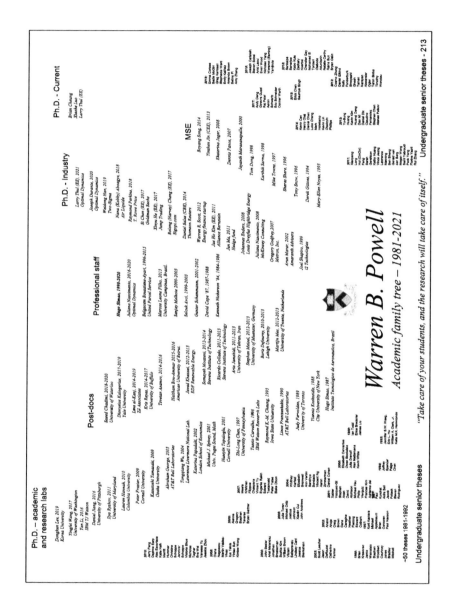

The group of students and staff that participated in CASTLE Lab is much too large to list in this acknowledgment, but I have included my academic family tree above. To everyone in this list, my warmest thanks!

I owe a special thanks to the sponsors of CASTLE Lab, which included a number of government funding agencies including the National Science Foundation, the Air Force Office of Scientific Research, DARPA, the Department of Energy (through Columbia University and the University of Delaware), and Lawrence Livermore National Laboratory (my first energy sponsor). I would particularly like to highlight the Optimization and Discrete Mathematics Program of AFOSR that provided me with almost 30 years of unbroken funding. I would like to express my appreciation to the program managers of the ODM program, including Neal Glassman (who gave me my start in this program), Donald Hearn (who introduced me to the materials science program), Fariba Fahroo (whose passion for this work played a major role in its survival at AFOSR), and Warren Adams. Over the years I came to have a deep appreciation for the critical role played by these program managers who provide a critical bridge between academic researchers and the policymakers who have to then sell the work to Congress.

I want to recognize my industrial sponsors and the members of my professional staff that made this work possible. Easily one of the most visible features of CASTLE Lab was that we did not just write academic papers and run computer simulations; our work was implemented in the field. We would work with a company, identify a problem, build a model, and then see if it worked, and it often did not. This was true research, with a process that I once documented with a booklet called "From the Laboratory to the Field, and Back." It was this back and forth process that allowed me to learn how to model and solve real problems. We had some early successes, followed by a period of frustrating failures as we tackled even harder problems, but we had two amazing successes in the early 2000s with our implementation of a locomotive optimization system at Norfolk Southern Railway using approximate dynamic programming, and our strategic fleet simulator for Schneider National (one of the largest truckload carriers in the U.S.). This software was later licensed to Optimal Dynamics which is implementing the technology in the truckload industry. My industrial sponsors received no guarantees when they funded our research, and their (sometimes misplaced) confidence in me played a critical role in our learning process.

Working with industry from a university research lab, especially for a school like Princeton, introduces administrative challenges that few appreciate. Critical to my ability to work with industry was the willingness of a particular grants officer at Princeton, John Ritter, to negotiate contracts where companies funded the research, and were then given royalty-free licenses to use the software. This was key, since it was through their use of the software that I learned what worked, and what did not. John understood that the first priority at a university is supporting the faculty and their research mission rather than maximizing royalties. I think that I can claim that my $50 million in research funding over my career paid off pretty well for Princeton.

Finally, I want to recognize the contributions of my professional staff who made these industrial projects possible. Most important is the very special role played by Hugo Simao, my first Ph.D. student who graduated, taught in Brazil, and returned in 1990 to help start CASTLE Lab. Hugo played so many roles, but most important as the lead developer on a number of major projects that anchored the lab, notably the multidecade relationship with Yellow Freight System/YRC. He was also the lead developer of our award-winning model for Schneider National that was later licensed to Optimal Dynamics, in addition to our big energy model, SMART-ISO, which simulated the PJM power grid. This is not work that can be done by graduate students, and Hugo brought his tremendous skill to the development of complex systems, starting in the 1990s when the tools were relatively primitive. Hugo also played an important role guiding students (graduate and undergraduate) with their software projects, given that I retired from programming in 1990 as the world transitioned from Fortan to C. Hugo brought talent, patience, and an unbelievable work ethic that provided the foundation in funding that made CASTLE Lab possible. Hugo was later joined by Belgacem Bouzaiene-Ayari who worked at the lab for almost 20 years and was the lead developer on another award-winning project with Norfolk Southern Railway, along with many other contributions. I cannot emphasize enough the value of the experience of working with these industrial sponsors, but this is not possible without talented research staff such as Hugo and Belgacem.

W. B. P.

Part I – Introduction

We have divided the book into 20 chapters organized into six parts. Part I includes four chapters that set the foundation for the rest of the book:

- Chapter 1 provides an introduction to the broad field that we are calling "sequential decision analytics." It introduces our universal modeling framework which reduces sequential decision problems to one of finding methods (rules) for making decisions, which we call *policies*.
- Chapter 2 introduces fifteen major canonical modeling frameworks that have been used by different communities. These communities all approach sequential decision problems under uncertainty from different perspectives, using eight different modeling systems, typically focusing on a major problem class, and featuring a particular solution method. Our modeling framework will span all of these communities.
- Chapter 3 is an introduction to online learning, where the focus is on sequential vs. batch learning. This can be viewed as an introduction to machine learning, but focusing almost exclusively on adaptive learning, which is something we are going to be doing throughout the book.
- Chapter 4 sets the stage for the rest of the book by organizing sequential decision problems into three categories: (1) problems that can be solved using deterministic mathematics, (2) problems where randomness can be reasonably approximated using a sample (and then solved using deterministic mathematics), and (3) problems that can only be solved with adaptive learning algorithms, which is the focus of the remainder of the book.

Chapter 1 provides an overview of the universal modeling framework that covers any sequential decision problem. It provides a big picture of our entire framework for modeling and solving sequential decision problems, which should be of value to any reader regardless of their background in decisions under uncertainty. It describes the scope of problems, a brief introduction to modeling sequential decision problems, and sketches the four classes of policies (methods for making decisions) that we use to solve these problems.

Chapter 2 summarizes the canonical modeling frameworks for each of the communities that address some form of sequential decision problem, using the notation of that field. Readers who are entirely new to the field might skim this chapter to get a sense of the variety of approaches that have been taken. Readers with more depth will have some level of expertise in one or more of these canonical problems, and it will help provide a bridge between that problem class and our framework.

Chapter 3 covers online learning in some depth. This chapter should be skimmed, and then used as a reference source as needed. A good starting point is to read section 3.1, and then skim the headers of the remaining sections. The book will repeatedly refer back to methods in this chapter.

Finally, chapter 4 organizes stochastic optimization problems into three categories:

(1) Stochastic optimization problems that can be solved exactly using deterministic mathematics.

(2) Stochastic optimization problems where uncertainty can be represented using a fixed sample. These can still be solved using deterministic mathematics.

(3) Stochastic optimization problems that can only be solved using sequential, adaptive learning algorithms. This will be the focus of the rest of the book.

This chapter reminds us that there are special cases of problems that can be solved exactly, possibly by replacing the original expectation with a sampled approximation. The chapter closes by setting up some basic concepts for learning problems, including making the important distinction between online and offline problems, and by identifying different strategies for designing policies for adaptive learning.

1

Sequential Decision Problems

A sequential decision problem, simply stated, consists of the sequence

decision, information, decision, information, decision, ...

As we make decisions, we incur costs or earn rewards. Our challenge is how to represent the information that will arrive in the future, and how to make decisions, both now and in the future. Modeling these problems, and making effective decisions in the presence of the uncertainty of new information, is the goal of this book.

The first step in sequential decision problems is to understand what decisions are being made. It is surprising how often it is that people faced with complex problems, which spans scientists in a lab to people trying to solve major health problems, are not able to identify the decisions they face.

We then want to find a method for making decisions. There are at least 45 words in the English language that are equivalent to "method for making a decision," but the one we have settled on is *policy*. The term *policy* is very familiar to fields such as Markov decision processes and reinforcement learning, but with a much narrower interpretation than we will use. Other fields do not use the term at all. Designing effective policies will be the focus of most of this book.

Even more subtle is identifying the different sources of uncertainty. It can be hard enough trying to identify potential decisions, but thinking about all the random events that might affect whatever it is that you are managing, whether it is reducing disease, managing inventories, or making investments, can seem like a hopeless challenge. Not only are there a wide range of sources of uncertainty, but there is also tremendous variety in how they behave.

Making decisions under uncertainty spans an exceptionally rich set of problems in analytics, arising in fields such as engineering, the sciences, business, economics, finance, psychology, health, transportation, and energy. It encompasses active learning problems, where the decision is to collect information,

Reinforcement Learning and Stochastic Optimization: A Unified Framework for Sequential Decisions, First Edition. Warren B. Powell.
© 2022 John Wiley & Sons, Inc. Published 2022 John Wiley & Sons, Inc.

that arise in the experimental sciences, medical decision making, e-commerce, and sports. It also includes iterative algorithms for stochastic search, which arises in machine learning (finding the model that best fits the data) or finding the best layout for an assembly line using a simulator. Finally, it includes two-agent games and multiagent systems. In fact, we might claim that virtually any human enterprise will include instances of sequential decision problems.

Decision making under uncertainty is a universal experience, something every human has had to manage since our first experiments trying new foods when we were two years old. Some samples of everyday problems where we have to manage uncertainty in our own lives include:

- Personal decisions – These might include how much to withdraw from an ATM machine, finding the best path to a new job, and deciding what time to leave to make an appointment.
- Food shopping – We all have to eat, and we cannot run to the store every day, so we have to make decisions of when to go shopping, and how much to stock up on different items when we do go.
- Health decisions – Examples include designing diet and exercise programs, getting annual checkups, performing mammograms and colonoscopies.
- Investment decisions – Which mutual fund should you use? How should you allocate your investments? How much should you put away for retirement? Should you rent or purchase a house?

Sequential decision problems are ubiquitous, and as a result come in many different styles and flavors. Decisions under uncertainty span virtually every major field. Table 1.1 provides a list of problem domains and a sample of questions that can arise in each of these fields. Not surprisingly, a number of different analytical fields have emerged to solve these problems, often using different notational systems, and presenting solution approaches that are suited to the characteristics of the problems in each setting.

This book will provide the analytical foundation for sequential decision problems using a "model first, then solve" philosophy. While this is standard in fields such as deterministic optimization and machine learning, it is not at all standard in the arena of making decisions under uncertainty. The communities that work on sequential decision problems tend to come up with a method for solving a problem, and then look for applications. This can come across as if we have a hammer looking for a nail.

The limitation of this approach is that the different methods that have been developed can only serve a subset of problems. Consider one of the simplest and most classical sequential decision problems: managing an inventory of product to serve demands over time. Let R_t be our inventory at time t, x_t is how much we order (that arrives instantly), to serve a demand $\hat{D}_{t+1}$ that is not known at time t. The evolution of the inventory R_t is given by

Table 1.1 A list of application domains and decisions that need to be made.

Field	Questions
Business	What products should we sell, with what features? Which supplies should you use? What price should you charge? How should we manage our fleet of delivery vehicles? Which menu attracts the most customers?
Economics	What interest rate should the Federal Reserve charge given the state of the economy? What levels of market liquidity should be provided? What guidelines should be imposed on investment banks?
Finance	What stocks should a portfolio invest in? How should a trader hedge a contract for potential downside? When should we buy or sell an asset?
Internet	What ads should we display to maximize ad-clicks? Which movies attract the most attention? When/how should mass notices be sent?
Engineering	How to design devices from aerosol cans to electric vehicles, bridges to transportation systems, transistors to computers?
Materials science	What combination of temperatures, pressures, and concentrations should we use to create a material with the highest strength?
Public health	How should we run testing to estimate the progression of a disease? How should vaccines be allocated? Which population groups should be targeted?
Medical research	What molecular configuration will produce the drug which kills the most cancer cells? What set of steps are required to produce single-walled nanotubes?
Supply chain management	When should we place an order for inventory from China? What mode of transportation should be used? Which supplier should be used?
Freight transportation	Which driver should move a load? What loads should a truckload carrier commit to move? Where should drivers be domiciled?
Information collection	Where should we send a drone to collect information on wildfires or invasive species? What drug should we test to combat a disease?
Multiagent systems	How should a large company in an oligopolistic market bid on contracts, anticipating the response of its competitors? How should a submarine behave given the presence of adversarial submarines?
Algorithms	What stepsize rule should we use in a search algorithm? How do we determine the next point to evaluate an expensive function?

$$R_{t+1} = \max\{0, R_t + x_t - \hat{D}_{t+1}\}. \tag{1.1}$$

For this problem, we might use the following policy: when the inventory falls below a value θ^{min}, order enough to bring it up to θ^{max}. All we have to do is to determine the parameter vector $\theta = (\theta^{min}, \theta^{max})$. The policy is quite simple, but finding the best value of θ can be quite challenging.

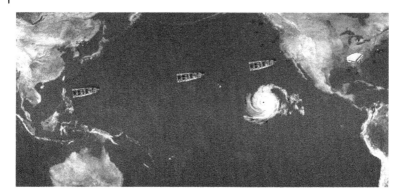

Figure 1.1 Illustration of shipments coming from China to the U.S. with a threat of a storm. *Source:* Masaqui/Wikimedia Commons/CC BY-SA 3.0

Now consider a series of inventory problems with increasing complexity, illustrated by the setting in Figure 1.1 of a warehouse in the southeastern United States ordering inventory:

(1) The inventory we order comes from China, and might take 90 to 150 days to arrive.
(2) We have to serve demand that varies seasonally (and dramatically changes around the Christmas holiday season).
(3) We are given the option to use air freight for a particular order that reduces the time by 30 days.
(4) We are selling expensive gowns, and we have to pay special attention to the risk of a stockout if there is a delay in either the production (which we can handle by using air freight) or a delay in offloading at the port.
(5) The gowns come in different styles and colors. If we run short of one color, the customer might be willing to accept a different color.
(6) We are allowed to adjust the price of the item, but we do not know precisely how the market will respond. As we adjust the price and observe the market response, we learn from this observation and use what we learn to guide future pricing decisions.

Each of these modifications would affect our decision, which means a modification of the original policy in some way.

The simple inventory problem in equation (1.1) has just a single decision, x_t, specifying how much inventory to order now. In a real problem, there is a spectrum of downstream decisions that might be considered, including:

- How much to order, and the choice of delivery commitment that determines how quickly the order arrives: rush orders, normal, relaxed.

- Pricing of current inventory while waiting for the new inventory to arrive.
- Reservations for space on cargo ships in the future.
- The speed of the cargo ship.
- Whether to rush additional inventory via air freight to fill a gap due to a delay.
- Whether to use truck or rail to move the cargo in from the port.

Then, we have to think about the different forms of uncertainty for a product that might take at least 90 days to arrive:

- The time to complete manufacturing.
- Weather delays affecting ship speeds.
- Land transportation delays.
- Product quality on arrival.
- Currency changes.
- Demand for inventory on hand between now and the arrival of new inventory.

If you set up a toy problem such as equation (1.1), you would never think about all of these different decisions and sources of uncertainty. Our presentation will feature a rich modeling framework that emphasizes our philosophy:

Model first, then solve.

We will introduce, for the first time in a textbook, a universal modeling framework for *any* sequential decision problem. We will introduce four broad classes of methods, known as policies, for making decisions that span *any* method that might be used, including anything in the academic literature or used in practice. Our goal is not to always choose the policy that performs the best, since there are multiple dimensions to evaluating a policy (computational complexity, transparency, flexibility, data requirements). However, we will always choose our policy with one eye to performance, which means the statement of an objective function will be standard. This is not the case in all communities that work on sequential decision problems.

1.1 The Audience

This book is aimed at readers who want to develop models that are practical, flexible, scalable, and implementable for sequential decision problems in the presence of different forms of uncertainty. The ultimate goal is to create software tools that can solve real problems. We use careful mathematical modeling as a necessary step for translating real problems into software. The readers who appreciate both of these goals will enjoy our presentation the most.

Given this, we have found that this material is accessible to professionals from a wide range of fields, spanning application domains (engineering, economics, and the sciences) to those with more of a methodological focus (such as machine learning, computer science, optimal control, and operations research) with a comfort level in probability and statistics, linear algebra, and, of course, computer programming.

Our presentation emphasizes modeling and computation, with minimal deviations into theory. The vast majority of the book can be read with a good course in probability and statistics, and a basic knowledge of linear algebra. Occasionally we will veer into higher dimensional applications such as resource allocation problems (e.g. managing inventories of different blood types, or trading portfolios of assets) where some familiarity with linear, integer, and/or nonlinear programming will be useful. However, these problems can all be solved using powerful solvers with limited knowledge of how these algorithms actually work.

This said, there is no shortage of algorithmic challenges and theoretical questions for the advanced Ph.D. student with a strong background in mathematics.

1.2 The Communities of Sequential Decision Problems

Figure 1.2 shows some prominent books from various methodological communities in the sequential decision-making field. These communities, which are discussed in greater depth in chapter 2, are listed in Table 1.2 in the approximate order in which the field emerged. We note that there are two distinct fields that are known as derivative-based stochastic search, and derivative-free stochastic search, that both trace their roots to separate papers published in 1951.

Each of these communities deals with some flavor of sequential decision problems, using roughly eight notational systems, and an overlapping set of algorithmic strategies. Each field is characterized by at least one book (often several), and thousands of papers (in some cases, thousands of papers each year). Each community tends to have problems that best fit the tools developed by that community, but the problem classes (and tools) are continually evolving.

The fragmentation of the communities (and their differing notational systems) disguises common approaches developed in different areas of practice, and challenges cross-fertilization of ideas. A problem that starts off simple (like the inventory problem in (1.1)) lends itself to a particular solution strategy, such as dynamic programming. As the problem grows in realism (and complexity), the original technique will no longer work, and we need to look to other communities to find a suitable method.

Figure 1.2 A sampling of major books representing different fields in stochastic optimization.

We organize all of these fields under the title of "reinforcement learning and stochastic optimization." "Stochastic optimization" refers generally to the analytical fields that address decisions under uncertainty. The inclusion of "reinforcement learning" in the title reflects the growing popularity of this community, and the use of the term to apply to a steadily expanding set of methods for solving sequential decision problems. The goal of this book is to provide a unified framework that covers *all* of the communities that work on these problems, rather than to favor any particular method. We refer to this broader field as *sequential decision analytics*.

Sequential decision analytics requires integrating tools and concepts from three core communities from the mathematical sciences:

Statistical machine learning – Here we bring together the fields of statistics, machine learning, and data sciences. Most (although not all) of our applications of these tools will involve recursive learning. We will also draw on the fields of both frequentist and Bayesian statistics, but all of this material is provided here.

Mathematical programming – This field covers the core methodologies in derivative-based and derivative-free search algorithms, which we use for purposes ranging from computing policies to optimizing the parameters of a

Table 1.2 Fields that deal with sequential decisions under uncertainty.

(1) Derivative-based stochastic search	(9) Stochastic programming
(2) Derivative-free stochastic search	(10) Multiarmed bandit problem
(3) Decision trees	(11) Simulation optimization
(4) Markov decision processes	(12) Active learning
(5) Optimal control	(13) Chance constrained programming
(6) Approximate dynamic programming	(14) Model predictive control
(7) Reinforcement learning	(15) Robust optimization
(8) Optimal stopping	

policy. Occasionally we will encounter vector-valued decision problems that require drawing on tools from linear, integer, and possibly nonlinear programming. Again, all of these methods are introduced and presented without assuming any background in stochastic optimization.

Stochastic modeling and simulation – Optimizing a problem in the presence of uncertainty often requires a careful model of the uncertain quantities that affect the performance of a process. We include a basic introduction to Monte Carlo simulation methods, but expect a background in probability and statistics, including the use of Bayes theorem.

While our presentation does not require advanced mathematics or deep preparation in any methodological area, we will be blending concepts and methods from all three of these fields. Dealing with uncertainty is inherently more subtle than deterministic problems, and requires more sophisticated modeling than arises in machine learning.

1.3 Our Universal Modeling Framework

Central to the entire book is the use of a single modeling framework, as is done in deterministic optimization and machine learning. Ours is based heavily on the one widely used in optimal control. This has proven to be the most practical and flexible, and offers a clear relationship between the mathematical model and its implementation in software. While much of our presentation will focus on modeling sequential decision problems and developing practical methods for making decisions, we also recognize the importance of developing models of the different sources of uncertainty (a topic that needs a book of its own).

Although we revisit this in more detail in chapter 9, it helps to sketch our universal modeling framework. The core elements are:

- State variables S_t – The state variable contains everything we know, and only what we need to know, to make a decision and model our problem. State variables include physical state variables R_t (the location of a drone, inventories, investments in stocks), other information I_t about parameters and quantities we know perfectly (such as current prices and weather), and beliefs B_t, in the form of probability distributions describing parameters and quantities that we do not know perfectly (this could be an estimate of how much a drug will lower the blood sugar in a new patient, or how the market will respond to price).

- Decision variables x_t – A decision variable can be binary (hold or sell), a discrete set (drugs, products, paths), a continuous variable (such as a price or dosage), and vectors of discrete and continuous variables. Decisions are subject to constraints $x_t \in \mathcal{X}_t$, and we make decisions using a method we call a policy $X^\pi(S_t)$. We introduce the notation for a policy, but we defer the design of the policy until after we complete the model. This is the basis of what we call *model first, then solve.*

- Exogenous information W_{t+1} – This is the information that we learn after we make a decision (market response to a price, patient response to a drug, the time to traverse a path), that we did not know when we made the decision. Exogenous information comes from outside whatever system we are modeling. (Decisions, on the other hand, can be thought of as an *endogenous information process* since we make decisions, a form of information, internally to the process.)

- The transition function $S^M(S_t, x_t, W_{t+1})$ which consists of the equations required to update *each* element of the state variable. This covers all the dynamics of our system, including the updating of estimates and beliefs for sequential learning problems. Transition functions are widely used in control theory using the notation $f(x, u, w)$ (for state x, control u and information w); our notation, which stands for the "state transition model" or "system model" helps us avoid using the popular letter $f(\cdot)$.

- The objective function – This first consists of the contribution (or reward, or cost, ...) we make each time period, given by $C(S_t, x_t)$, where $x_t = X^\pi(S_t)$ is determined by our policy, and S_t is our current state, which is computed by the transition function. As we are going to demonstrate later in the book, there are different ways to write the objective function, but our most common will be to maximize the cumulative contributions, which we write as

$$\max_\pi \mathbb{E}\left\{ \sum_{t=0}^{T} C(S_t, X^\pi(S_t)) | S_0 \right\}, \tag{1.2}$$

where the expectation $\mathbb{E}$ means "take an average over all types of uncertainty" which might be uncertainty about how a drug will perform, or how the market will respond to price (captured in the initial state S_0), as well as the uncertainty in the information $W_1, \dots, W_t, \dots$ that arrives over time. The maximization over policies simply means that we want to find the best method for making decisions. Most of this book is dedicated to the challenge of searching over policies.

Once we have identified these five elements, we still have two remaining steps to complete before we are done:

- Stochastic modeling (also known as uncertainty quantification) – There can be uncertainty about parameters and quantities in the state variable (including the initial state S_0), as well as our exogenous information process $W_1, W_2, \dots, W_t, \dots$ In some instances, we may avoid modeling the W_t process by observing a physical system. Otherwise, we need a mathematical model of the possible realizations of W_{t+1} given S_t and our decision x_t (either of which can influence W_{t+1}).
- Designing policies – Only after we are done with modeling do we turn to the problem of designing the policy $X^\pi(S_t)$. This is the point of departure between this book and all the books in our jungle of stochastic optimization. We do not pick policies before we develop our model; instead, once the modeling is done, we will provide a roadmap to every possible policy, with guidelines of how to choose among them.

The policy π consists of some type of function $f \in \mathcal{F}$, possibly with tunable parameters $\theta \in \Theta^f$ that are associated with the function f, where the policy maps the state to a decision. The policy will often contain an imbedded optimization problem within the function. This means that we can write (1.2) as

$$\max_{\pi = (f \in \mathcal{F}, \theta \in \Theta^f)} \mathbb{E}\left\{ \sum_{t=0}^{T} C(S_t, X^\pi(S_t)) | S_0 \right\}. \tag{1.3}$$

This leaves the question: How do we search over functions? Most of this book is dedicated to describing precisely how to do this.

Using this notation, we can revise our original characterization of a sequential decision problem, which we earlier described as *decision, information, decision, information, ...* as the sequence

$$(S_0, x_0, W_1, S_1, x_1, W_2, \dots, S_t, x_t, W_{t+1}, \dots, S_T),$$

where we now write the triplet "state, decision, new information" to capture what we know (the state variable S_t), which we use to make a decision x_t, followed by what we learn after we make a decision, the exogenous information W_{t+1}. We earn a contribution $C(S_t, x_t)$ from our decision x_t (we could say we earn a reward or incur a cost), where the decision comes from a policy $X^\pi(S_t)$.

There are many problems where it is more natural to use a counter n (the nth experiment, the nth customer arrival), in which case we would write our sequential decision problem as

$$(S^0, x^0, W^1, S^1, x^1, W^2, \dots, S^n, x^n, W^{n+1}, \dots, S^N).$$

There are even settings where we use both, as in $(S_t^n, x_t^n, W_{t+1}^n)$ to capture, for example, decisions in the nth week at hour t.

We note in passing that there are problems that consist of "decision, information, stop," "decision, information, decision, stop," "information, decision, information, decision, ...," and problems where the sequencing proceeds over an infinite horizon. We use a finite sequence as our default model.

We can illustrate our modeling framework using our simple inventory problem that we started with above.

- State variables S_t – For the simplest problem this is the inventory R_t.
- Decision variables x_t – This is how much we order at time t, and for now, we assume it arrives right away. We also introduce our policy $X^\pi(S_t)$, where $x_t = X^\pi(S_t)$, which we will design after we create our model.
- Exogenous information W_{t+1} – This would be the demand $\hat{D}_{t+1}$ that arises between t and $t+1$.
- The transition function $S^M(S_t, x_t, W_{t+1})$ – This would be the evolution of our inventory R_t, given by

$$R_{t+1} = \max\{0, R_t + x_t - \hat{D}_{t+1}\}. \tag{1.4}$$

- The objective function – This is an example of a problem where it is more natural to write the single-period contribution function *after* we observe the information W_{t+1} since this contains the demand $\hat{D}_{t+1}$ that we will serve with the inventory x_t we order in period t. For this reason, we might write our contribution function as

$$C(S_t, x_t, W_{t+1}) = p \min\{R_t + x_t, \hat{D}_{t+1}\} - cx_t$$

where p is the price at which we sell our product, and c is the cost per unit of product. Our objective function would be given by

$$\max_\pi \mathbb{E}\left\{\sum_{t=0}^{T} C(S_t, X^\pi(S_t), W_{t+1}) | S_0\right\},$$

where $x_t = X^\pi(S_t)$, and we have to be given a model of the exogenous information process $W_1, \ldots, W_T$. Since the exogenous information is random, we have to take the expectation $\mathbb{E}$ of the sum of contributions to average over all the possible outcomes of the information process.

Our next step would be to develop a mathematical model of the distribution of demand $\hat{D}_1, \hat{D}_2, \ldots, \hat{D}_t, \ldots$ which draws on tools that we introduce in chapter 10.

To design our policy $X^\pi(S_t)$, we might turn to the academic literature that shows, for this simple problem, that the policy has an order-up-to structure given by

$$
X^{Inv}(S_t|\theta) = \begin{cases} \theta^{max} - R_t & \text{if } R_t < \theta^{min}, \\ 0 & \text{otherwise.} \end{cases} \tag{1.5}
$$

This is a parameterized policy, which leaves us the challenge of finding $\theta = (\theta^{min}, \theta^{max})$ by solving

$$
\max_\theta \mathbb{E}\left\{ \sum_{t=0}^{T} C(S_t, X^{Inv}(S_t|\theta), W_{t+1})|S_0 \right\}. \tag{1.6}
$$

Here we chose a particular class of policy, and then optimized within the class.

We pause to note that using our modeling approach creates a direct relationship between our mathematical model and computer software. Each of the variables above can be translated directly to a variable name in a computer program, with the only exception that the expectation operator has to be replaced with an estimate based on simulation (we show how to do this). This relationship between mathematical model and computer software does not exist with most of the current modeling frameworks used for decisions under uncertainty, with one major exception – optimal control.

Earlier in the chapter we proposed a number of generalizations to this simple inventory problem. As we progress through the book, we will show that our five-step universal modeling framework holds up for modeling much more complex problems. In addition, we will introduce four classes of policies that will span any method that we might want to consider to solve more complex versions of our problem. In other words, not only will our modeling framework be able to model any sequential decision problem, we will outline four classes of policies that are also universal: they encompass any method that has been studied in the research literature or used in practice. The next section provides an overview of these four classes of policies.

1.4 Designing Policies for Sequential Decision Problems

What often separates one field of stochastic optimization from another is the type of policy that is used to solve a problem. Possibly the most important aspect of our unified framework in this book is how we have identified and organized different classes of policies. These are first introduced in chapter 7 in the context of derivative-free stochastic optimization (a form of pure learning problem), and then in greater depth in chapter 11 on designing policies, which sets the stage for the entire remainder of the book. In this section we are going to provide a peek at our approach for designing policies.

The entire literature on making decisions under uncertainty can be organized along two broad strategies for creating policies:

Policy search – This includes all policies where we need to search over:

- Different classes of functions $f \in \mathcal{F}$ for making decisions. For example, the order-up-to policy in equation (1.5) is a form of nonlinear parametric function.
- Any tunable parameters $\theta \in \Theta^f$ that are introduced by the function f. $\theta = (\theta^{min}, \theta^{max})$ in equation (1.5) is an example.

If we select a policy that contains parameters, then we have to find the set of parameters θ to maximize (or minimize) an objective function such as (1.6).

Lookahead approximations – These are policies formed so we make the best decision now given an approximation of the downstream impact of the decision. These are the policy classes that have attracted the most attention from the research community.

Our order-up-to policy $X^{Inv}(S_t|\theta)$ is a nice example of a policy that has to be optimized (we might say tuned). The optimization can be done using a simulator, as is implied in equation (1.6), or in the field.

Each of these two strategies produce policies that can be divided into two classes, creating four classes of policies. We describe these below.

1.4.1 Policy Search

Policies in the policy search class can be divided into two subclasses:

(1) Policy function approximations (PFAs) – These are analytical functions that map a state (which includes all the information available to us) to a decision (the order-up-to policy in equation (1.5) is a PFA). These are discussed in greater depth in chapter 12.

(2) Cost function approximations (CFAs) – CFA policies are parameterized optimization models (typically deterministic optimization models) that have been modified to help them respond better over time, and under uncertainty. CFAs have an imbedded optimization problem *within the policy*. The concept of CFAs are presented in this book for the first time as a major new class of policies. CFAs are introduced and illustrated in chapter 13.

PFAs are any analytical function that maps what we know in the state variable to a decision. These analytical functions come in three flavors:

Lookup tables – These are used when a discrete state S can be mapped to a discrete action, such as:

- If the patient is male, over 60 with high blood sugar, then prescribe metformin.
- If your car is at a particular intersection, turn left.

Parametric functions – These describe any analytical functions parameterized by a vector of parameters θ. Our order-up-to policy is a simple example. We might also write it as a linear model such as

$$X^{PFA}(S_t|\theta) = \theta_1\phi_1(S_t) + \theta_2\phi_2(S_t) + \theta_3\phi_3(S_t) + \theta_4\phi_4(S_t)$$

where $\phi_f(S_t)$ are features extracted from information in the state variable. Neural networks are another option.

Nonparametric functions – These include functions that might be locally linear approximations, or deep neural networks.

The second class of functions that can be optimized using policy search is called parametric *cost function approximations*, or CFAs, which are parameterized optimization problems. A simple CFA used in learning problems is called interval estimation and might be used to determine which ad gets the most clicks on a website. Let $\mathcal{X} = \{x_1, \ldots, x_M\}$ be the set of ads (there may be thousands of them), and let $\bar{\mu}_x^n$ be our current best estimate of the probability that ad x will be clicked on after we have run n observations (across all ads). Then let $\bar{\sigma}_x^n$ be the standard deviation of the estimate $\bar{\mu}_x^n$. Interval estimation would choose as the next ad using the policy

$$X^{CFA}(S^n|\theta) = \arg\max_{x\in\mathcal{X}} \left(\bar{\mu}_x^n + \theta\bar{\sigma}_x^n\right), \tag{1.7}$$

where "arg max$_x$" means to find the value of x that maximizes the expression in parentheses. The distinguishing features of a CFA is that it requires solving an imbedded optimization problem (the max over ads), and there is a tunable parameter θ.

Once we introduce the idea of solving an optimization problem within the policy (as we did with the policy in (1.7)), we can solve *any* parameterized optimization problem. We are no longer restricted to the idea that x has to be one of a set of discrete choices; it can be a large integer program, such as those used to plan airline schedules with schedule slack inserted to handle possible weather delays, or planning energy generation for tomorrow with reserves in case a generator fails (both of these are real instances of CFAs used in practice).

1.4.2 Policies Based on Lookahead Approximations

A natural strategy for making decisions is to consider the downstream impact of a decision you make now. There are two ways of doing this:

(3) Value function approximations (VFAs) – One popular approach for solving sequential decision problems applies the principles of a field known as *dynamic programming* (or Markov decision processes). Imagine our state variable tells us where we are on a network where we have to make a decision, or the amount of inventory we are holding. Assume that someone tells us that if we are in state S_{t+1} at time $t+1$ (that is, we are at some node in the network or will have some level of inventory), that $V_{t+1}(S_{t+1})$ is the "value" of being in state S_{t+1}, which we can think of as the cost of the shortest path to the destination, or our expected profits from time $t+1$ onward if we start with inventory S_{t+1}.

Now assume we are in a state S_t at time t and trying to determine which decision x_t we should make. After we make the decision x_t, we observe the random variable(s) W_{t+1} that take us to $S_{t+1} = S^M(S_t, x_t, W_{t+1})$ (for example, our inventory equation (1.4) in our example above). Assuming we know $V_{t+1}(S_{t+1})$, we can find the value of being in state S_t by solving

$$V_t(S_t) = \max_{x_t} \left(C(S_t, x_t) + \mathbb{E}_{W_{t+1}} \{ V_{t+1}(S_{t+1}) | S_t \} \right), \qquad (1.8)$$

where it is best to think of the expectation operator $\mathbb{E}_{W_{t+1}}$ as averaging over all outcomes of W_{t+1}. The value of x_t^* that optimizes equation (1.8) is then the optimal decision for state S_t. The first period contribution $C(S_t, x_t^*)$ plus the future contributions $\mathbb{E}_{W_{t+1}} \{ V_{t+1}(S_{t+1}) | S_t \}$ gives us the value $V_t(S_t)$ of being in state S_t now. When we know the values $V_t(S_t)$ for all time periods, and all states, we have a VFA-based policy given by

$$X_t^{VFA}(S_t) = \arg\max_{x_t} \left(C(S_t, x_t) + \mathbb{E}_{W_{t+1}} \{ V_{t+1}(S_{t+1}) | S_t \} \right), \qquad (1.9)$$

where "arg max$_{x_t}$" returns the value x_t that maximizes (1.9).

Equation (1.9) is a powerful way of computing optimal policies, but it is rarely computable in practical problems (chapter 14 presents some problem classes that can be solved exactly). For this reason, a number of communities have developed ways of approximating the value function under names such as approximate dynamic programming, adaptive dynamic programming, or, most visibly, reinforcement learning. These fields replace the exact value function $V_{t+1}(S_{t+1})$ with an approximation $\overline{V}_{t+1}(S_{t+1})$ estimated using machine learning.

VFA-based policies have attracted considerable attention from the research literature, and are possibly the most difficult of the four classes of policies. We cover approximations over four chapters (chapters 15 – 18).

(4) Direct lookahead approximations (DLAs) – The easiest example of a lookahead policy is a navigation system which plans a path to your destination, and then tells you which turn to take next. As new information arrives, the path is updated.

This is an example of a deterministic lookahead for a stochastic problem. While deterministic lookaheads are useful in some applications, there are many where we have to explicitly consider uncertainty as we make a decision, which means we have to solve a stochastic optimization problem within our direct lookahead policy! There are entire fields of research focused on specific methods for solving direct lookahead models under uncertainty. We present a general framework for modeling and solving direct lookahead policies in chapter 19.

1.4.3 Mixing and Matching

It is possible to create hybrid policies by blending strategies from multiple classes. We can create a lookahead policy H periods into the future, and then use a value function approximation to approximate the states at the end of the planning horizon. We can use a deterministic lookahead, but introduce tunable parameters to make it work better under uncertainty. We can combine a PFA (think of this as some analytical function that suggests a decision), and weight any deviation of the decision from the PFA and add it to any other optimization-based policy. When we get to stochastic lookaheads in chapter 19, we may end up using all four classes at the same time.

An example of a hybrid policy is determining both the path to drive to a destination, and the time of departure. Navigation systems use a deterministic lookahead, solving a shortest path problem using point estimates of the travel times on each link of a network. This path might produce an estimated travel time of 40 minutes, but when do you actually leave? Now you are aware of the uncertainty of traffic, so you might decide to add in a buffer. As you repeat the trip, you may adjust the buffer up or down as you evaluate the accuracy of the estimate. This is a combined direct lookahead (since it plans a path into the future) with a tunable parameter for the departure time (making it a form of PFA).

As we said, we cannot tell you how to solve any particular problem (the diversity is simply too great), but we will give you a complete toolbox, with some guidelines to help in your choice.

1.4.4 Optimality of the Four Classes

There is a widespread misconception in the academic research literature that equation (1.8) (known either as Bellman's equation, or the Hamilton-Jacobi equation) is the basis for creating optimal policies, and that any path to designing good (that is, near optimal) policies have to start with Bellman's equation. This is simply not true.

Any of the four classes of policies can contain the optimal policy for specific problem classes. The problem that arises is purely computational. For example, for the vast majority of real applications, Bellman's equation (1.8) is not computable. Trying to replace the true value function $V_{t+1}(S_{t+1})$ in equation (1.8) with some approximation $\overline{V}_{t+1}(S_{t+1})$ may work quite well, but there are many settings where it is just not going to produce effective policies. In addition, once you start talking about using approximations of the value function, you open yourself up to the possibility that any of the other three classes of policies may work just as well or (often) better. This is the reason that there are so many people making decisions over time, in the presence of new information, and who do not use (and have not even heard of) Bellman's equation.

1.4.5 Pulling it All Together

We claim that the four classes of policies (PFAs, CFAs, VFAs, and DLAs) are universal, and cover every method that has been proposed by any of the communities listed earlier, as well as anything used in practice.

Of the four classes, the academic community has focused primarily on VFAs and various forms of DLAs (both deterministic and stochastic). By contrast, our belief is that PFAs and CFAs are much more widely used in practice. CFAs in particular have been largely overlooked in the academic community, but are widely used in practice in an ad hoc way (they are typically not tuned). PFAs and CFAs (that is, the policy search classes) are preferred in practice because they are simpler, but as we will see over and over again:

The price of simplicity is tunable parameters, and tuning is hard!

1.5 Learning

A significant part of decision analytics involves learning. Traditional machine learning involves being given a dataset consisting of inputs x^n and the associated response y^n, and then finding a function $f(x|\theta)$ which might be a linear model such as

$$f(x|\theta) = \theta_0 + \theta_1 \phi_f(x) + \theta_2 \phi_f(x) + \ldots + \theta_F \phi_F(x)$$

where the functions $\phi_f(x)$ extract features from the data in x. The inputs x might be the words in a document, a patient history, weather data, or customer data such as personal data and recent buying history. We might also look at nonlinear models, hierarchical models, and even a neural network. We then have to fit the model by solving the optimization problem

$$\min_\theta \frac{1}{N} \sum_{n=1}^{N} (y^n - f(x^n|\theta))^2.$$

This is classical batch learning.

When we are making decisions sequentially, we also learn sequentially. We might have a patient arrive with medical history h^n; we then decide on treatment $x^{treat,n}$ using a policy $X^\pi(S^n)$ (where S^n includes the patient history h^n). After choosing the treatment, we wait to observe the response, which we would index by y^{n+1} for the same reason that after making decision x^n we observe W^{n+1}. The index "$n + 1$" indicates that this is new information not contained in any variable indexed by n.

Our belief state B^n (within the state variable S^n) contains all the information we need to update our estimate θ^n using the new observation y^{n+1}. All of this updating is buried in the transition

$$S^{n+1} = S^M(S^n, x^n, W^{n+1}),$$

just as y^{n+1} is contained within W^{n+1}. The methods for doing this adaptive updating are all covered in chapter 3 on online learning, which is the term the machine learning community uses for learning in a sequential, versus batch, setting.

There are a number of opportunities for using online learning in sequential decision analytics:

(1) Approximating the expectation of a function $\mathbb{E}F(x, W)$ to be maximized.
(2) Creating an approximate policy $X^\pi(S|\theta)$.
(3) Approximating the value of being in a state S_t which we typically represent by $\overline{V}_t(S_t)$.
(4) Learning any of the underlying models in a dynamic system. These include:

(4a) The *transition function* $S^M(S_t, x_t, W_{t+1})$ which might describe how a future activity depends on the past.

(4b) The cost or contribution functions which might be unknown if we are trying to replicate human behavior.

(5) Parametric cost function approximations, where we use learning to modify the objective function and/or constraints imbedded in the policy.

The tools for estimating these functions are covered in chapter 3, but we visit the specific settings of these different problems throughout the rest of the book.

1.6 Themes

Our presentation features a series of themes that run through the book. This section reviews some of these.

1.6.1 Blending Learning and Optimization

Our applications will typically involve some mixture of decisions that influence learning (directly or indirectly) and decisions (perhaps the same decisions) that influence what we learn. It helps to think of three broad classes of problems:

- Pure learning problems – In this problem class decisions only control the information that we acquire for learning. This might arise in laboratory experimentation, computer simulations, and even market tests.
- State-dependent problems without learning – We will occasionally encounter problems where decisions impact a physical system, but where there is no learning. Using a navigation system to tell us which way to turn might be an example where the decisions affect the physical system (planning the path of our car) but where there is no learning.
- Hybrid problems – We will see many settings where a decision both changes the physical system and influences information we acquire for learning. There will also be systems with multiple decisions, such as physical decisions for allocating vaccines and testing decisions that guide information collection about the spread of disease or the efficacy of a drug.

1.6.2 Bridging Machine Learning to Sequential Decisions

Finding the best policy is the same as finding the best function that achieves the lowest cost, highest profits, or best performance. Analogs to this stochastic optimization problem appear in statistics and machine learning, where a common problem is to use a dataset (x^n, y^n), where $x^n = (x_1^n, \ldots, x_K^n)$ is used to predict y^n. For example, we might specify a linear function of the form:

$$y^n = f(x^n|\theta) = \theta_0 + \theta_1 x_1^n + \dots + \theta_K^n x_K^n + \epsilon^n, \tag{1.10}$$

where ϵ^n is a random error term that is often assumed to be normally distributed with mean 0 and some variance σ^2.

We can find the parameter vector $\theta = (\theta_1, \dots, \theta_K)$ by solving

$$\min_{\theta} \frac{1}{N} \sum_{n=1}^{N} \left(y^n - f(x^n|\theta)\right)^2. \tag{1.11}$$

Our problem of fitting a model to the data, then, involves two steps. The first is to choose the function $f(x|\theta)$, which we have done by specifying the linear model in equation (1.10) (note that this model is called "linear" because it is linear in θ). The second step involves solving the optimization problem given in (1.11). The only difference is the specific choice of performance metric.

Now consider how we approach sequential decision problems. Assume we are minimizing costs $C(S^n, x^n)$ that depend on our decision x^n as well as other information that we carry in the state variable S^n. Decisions are made with a policy $x^n = X^\pi(S^n|\theta)$ parameterized by θ which is analogous to the statistical model $f(x^n|\theta)$ that is used to predict (or estimate) y^{n+1} before it becomes known. Our objective function would then be

$$\min_{\theta} \mathbb{E} \sum_{n=0}^{N-1} C(S^n, X^\pi(S^n|\theta)) \tag{1.12}$$

where $S^{n+1} = S^M(S^n, X^\pi(S^n), W^{n+1})$, and where we are given a source of the sequence $(S^0, W^1, \dots, W^N)$.

When we compare (1.11) to (1.12), we see that both are searching over a set of functions to minimize some metric. In statistical modeling, the metric requires a dataset $(x^n, y^n)_{n=1}^N$, while our decision problem just requires a contribution (or cost) function $C(S, x)$, along with the transition function $S^{n+1} = S^M(S^n, x^n, W^{n+1})$ and a source of the exogenous information process $W^1, \dots, W^N$. The tools for searching for θ to solve (1.11) or (1.12) are the same, but the input requirements (a training dataset, or a model of the physical problem) are quite different.

Our statistical model may take any of a wide range of forms, but they are all in the broad class of analytical models that might be a lookup table, parametric or nonparametric model. All of these classes of functions fall in just one of our four classes of policies that we refer to as policy function approximations.

Table 1.3 provides a quick comparison of some major problem classes in statistical learning, and corresponding problems in stochastic optimization. The first row compares the standard batch machine learning problem to our canonical stochastic optimization problem (for a state-independent problem). The second row compares online learning (where we have to adapt to data as

Table 1.3 Comparison of classical problems faced in statistics (left) versus similar problems in stochastic optimization (right).

	Statistical learning	Stochastic optimization
(1)	Batch estimation: $\min_\theta \frac{1}{N} \sum_{n=1}^N (y^n - f(x^n\|\theta))^2$	Sample average approximation: $\min_{x \in \mathcal{X}} \frac{1}{N} \sum_{n=1}^N F(x, W(\omega^n))$
(2)	Online learning: $\min_\theta \mathbb{E}F(Y - f(X\|\theta))^2$	Stochastic search: $\min_\theta \mathbb{E}F(X, W)$
(3)	Searching over functions: $\min_{f \in \mathcal{F}, \theta \in \Theta^f} \mathbb{E}F(Y - f(X\|\theta))^2$	Policy search: $\min_\pi \mathbb{E} \sum_{t=0}^T C(S_t, X^\pi(S_t))$

it arrives) to online decision making. We use expectations in both cases since the goal is to make decisions now that work well in expectation after the next observation. Finally, the third row is making clear that we are searching for functions in both machine learning and stochastic optimization, where we use the canonical expectation-based form of the objective function. As of this writing, we feel that the research community has only begun to exploit these links, so we ask the reader to be on the lookout for opportunities to help build this bridge.

1.6.3 From Deterministic to Stochastic Optimization

Our approach shows how to generalize a deterministic problem to a stochastic one. Imagine we are solving the inventory problem above, although we are going to start with a deterministic model, and we are going to use standard matrix–vector math to keep the notation as compact as possible. Since the problem is deterministic, we need to make decisions $x_0, x_1, \ldots, x_t, \ldots$ over time (x_t may be a scalar or vector). Let $C_t(x_t)$ be our contribution in period t, given by

$$C_t(x_t) = p_t x_t$$

where p_t is a (known) price at time t. We also require that the decisions x_t satisfy a set of constraints that we write generally as:

$$A_t x_t \;=\; R_t, \tag{1.13}$$

$$x_t \;\geq\; 0, \tag{1.14}$$

$$R_{t+1} \;=\; B_t x_t + \hat{R}_{t+1}. \tag{1.15}$$

We wish to solve

$$\max_{x_0,\dots,x_T} \sum_{t=0}^{T} C_t(x_t),$$ (1.16)

subject to equations (1.13)–(1.15). This is a math program that can be solved with a number of packages.

Now assume that we wish to make $\hat{R}_{t+1}$ a random variable, which means it is not known until time $t+1$. In addition, assume that the price p_t varies randomly over time, which means we do not learn p_{t+1} until time $t + 1$. These changes turn the problem into a sequential decision problem under uncertainty.

There are some simple steps to turn this deterministic optimization problem into a fully sequential one under uncertainty. To begin, we write our contribution function as

$$C_t(S_t, x_t) = p_t x_t$$

where the price p_t is random information in the state S_t. We then write the objective function as

$$\max_{\pi} \mathbb{E}\left\{\sum_{t=0}^{T} C_t(S_t, X^{\pi}(S_t))|S_0\right\},$$ (1.17)

where $X^{\pi}(S_t)$ has to produce decisions that satisfy the constraints (1.13) – (1.14). Equation (1.15) is represented by the transition function $S^M(S_t, x_t, W_{t+1})$, where W_{t+1} includes $\hat{R}_{t+1}$ and the updated price p_{t+1}. We now have a properly modeled sequential decision problem.

We made the transition from deterministic optimization to a stochastic optimization formulation by making four changes:

- We replaced each occurrence of x_t with the function (policy) $X^{\pi}(S_t)$.
- We made the contribution function $C_t(x_t)$ depend on the state S_t to capture information (such as the price p_t) that is evolving randomly over time.
- We now take the expectation of the sum of the contributions since the evolution $S_{t+1} = S^M(S_t, x_t, W_{t+1})$ depends on the random variable W_{t+1}. It is helpful to think of the expectation operator $\mathbb{E}$ as averaging all the possible outcomes of the information process $W_1, \dots, W_T$.
- We replace the $\max_{x_0,\dots,x_T}$ with $\max_{\pi}$, which means we switch from finding the best set of *decisions*, to finding the best set of *policies*.

Care has to be taken when converting constraints for deterministic problems to the format we need when there is uncertainty. For example, we might be allocating resources and have to impose a budget over time that we can write as

$$\sum_{t=0}^{T} x_t \ \leq \ B,$$

where B is a budget for how much we use over all time periods. This constraint cannot be directly used in a stochastic problem since it assumes that we "decide" the variables $x_0, x_1, \ldots, x_T$ all at the same time. When we have a sequential decision problem, these decisions have to be made sequentially, reflecting the information available at each point in time. We would have to impose budget constraints recursively, as in

$$x_t \ \leq \ B - R_t, \tag{1.18}$$

$$R_{t+1} \ = \ R_t + x_t. \tag{1.19}$$

In this case, R_t would go into our state variable, and the policy $X^\pi(S_t)$ would have to be designed to reflect the constraint (1.18), while constraint (1.19) is captured by the transition function. Each decision $x_t = X^\pi(S_t)$ has to reflect what is known (captured by S_t) at the time the decision is made.

In practice, computing the expectation is hard (typically impossible) so we resort to methods known as *Monte Carlo simulation*. We introduce these methods in chapter 10. That leaves us with the usual problem of designing the policy. For this, we return to section 1.4.

All optimization problems involve a mixture of modeling and algorithms. With integer programming, modeling is important (especially for integer problems), but modeling has always taken a back seat to the design of algorithms. A testament of the power of modern algorithms is that they generally work well (for a problem class) with modest expertise in modeling strategy.

Sequential decision problems are different.

Figure 1.3 illustrates some of the major differences between how we approach deterministic and stochastic optimization problems:

	Deterministic	Stochastic
Models	System of equations	Complex functions, numerical simulations, physical systems
Objective	Minimize cost	Performance metrics, risk measures
What we are searching for	Real-valued vectors	Functions (policies)
What is hard	Designing algorithms	(1) Modeling (2) Designing policies

Figure 1.3 Deterministic vs. stochastic optimization.

Models – Deterministic models are systems of equations. Stochastic models are often complex systems of equations, numerical simulators, or even physical systems with unknown dynamics.

Objectives – Deterministic models minimize or maximize some well-defined metric such as cost or profit. Stochastic models require that we deal with statistical performance measures and uncertainty operators such as risk. Many stochastic dynamic problems are quite complicated (think of managing supply chains, trucking companies, energy systems, hospitals, fighting diseases) and involve multiple objectives.

What we are searching for – In deterministic optimization, we are looking for a deterministic scalar or vector. In stochastic optimization, we are almost always looking for functions that we will refer to as policies.

What is hard – The challenge of deterministic optimization is designing an effective algorithm. The hardest part of stochastic optimization, by contrast, is the modeling. Designing and calibrating a stochastic model can be surprisingly difficult. Optimal policies are rare, and a policy is not optimal if the model is not correct.

1.6.4   From Single to Multiple Agents

We close the book by extending these ideas to multiagent systems. Multiagent modeling is effective for breaking up complex systems such as supply chains (where different suppliers operate independently), as well as large transportation networks such as major carriers in trucking and rail. Multiagent modeling is essential in military applications, adversarial settings such as homeland security, oligopolies that describe markets with a small number of competitors, and a host of other applications.

Multiagent modeling is important in problems involving robots, drones, and underwater vehicles, which are often used for distributed information collection. For example, a drone might be used to identify areas where wildfires are burning to guide planes and helicopters dropping fire retardant. Robots can be used to sense landmines, and underwater vehicles might be used to collect information about fish populations.

Multiagent settings almost always require learning, since there is an unavoidable compartmentalization of knowledge. This in turn introduces the dimension of communication and coordination, where coordination may be through a central agent, or where we wish to design policies that encourage agents to work together.

We use this chapter to compare our modeling strategy to the most widely used modeling and algorithmic framework for learning systems, known as partially observable Markov decision processes, or POMDPs. This is a mathematically sophisticated theory which does not lead to scalable algorithms. We are going to use our multiagent framework to clarify knowledge of the transition function, and then draw on all four classes of policies to develop practical, scalable, implementable solutions.

1.7 Our Modeling Approach

The five elements in the modeling framework (section 1.3) can be used to model *any* sequential decision problem, recognizing that there are a variety of objective functions that can be used (these will be covered later). The four classes of policies in section 1.4 cover *any* method that might be used to make decisions in a sequential decision problem.

The four classes of policies are central to our modeling framework in section 1.3. We claim that *any* method used to make decisions for a sequential decision problem (and we mean *any* sequential decision problem) will be made with one of these four classes (or a hybrid of two or more). This represents a major change compared to the approaches used by the communities listed in section 1.2, which are typically associated with a particular solution approach (sometimes more than one).

We note that our approach precisely parallels that used in deterministic optimization, where people write out an optimization model (with decision variables, constraints, and an objective) before searching for a solution. This is exactly what we are doing: we are writing out our model without specifying the policy, and then we search for effective policies. We call this approach:

Model first, then solve.

The generality of the four classes of policies is what allows us to separate the process of designing the model (in section 1.3) from the solution of the model (that is, finding an acceptable policy). We will first see this applied in the context of pure learning problems in chapter 7. Next, chapter 8 will present a much richer set of applications, followed by a greatly expanded version of the modeling framework given in chapter 9. Then, after touching on modeling uncertainty in chapter 10, chapter 11 revisits the four classes of policies in more detail. Chapters 12–19 describe each of the four classes of policies in depth before transitioning to multiagent systems in chapter 20.

1.8 How to Read this Book

The book has been carefully designed to present topics in a logical order, with a progression from simpler to more sophisticated concepts. This section provides a guide to how to approach this material.

1.8.1 Organization of Topics

The book is organized into six parts, as follows:

Part I – Introduction and foundations – We start by providing a summary of some of the most familiar canonical problems, followed by an introduction to approximation strategies which we draw on throughout the book.

- Canonical problems and applications (chapter 2) – We begin by listing a series of canonical problems that are familiar to different communities, primarily using the notation familiar to those communities. This is a chapter that can be skimmed by readers new to the general area of stochastic optimization.
- Online learning (chapter 3) – Most books on statistical learning focus on batch applications, where a model is fit to a static dataset. In our work, learning is primarily sequential, known as "online learning" in the machine learning community. Our use of online learning is purely endogenous, in that we do not need an external dataset for training.
- Introduction to stochastic search (chapter 4) – We begin with a problem we call the *basic stochastic optimization problem* which provides the foundation for most stochastic optimization problems. In this chapter we also provide examples of how some problems can be solved exactly. We then introduce the idea of solving sampled models before transitioning to adaptive learning methods, which will be the focus of the rest of the book.

Part II – State-independent problems – There is a wide range of optimization problems where the problem itself is not changing over time (for any reason). All "state-independent problems" are pure learning problems, since all that is changing as a result of our decisions is our belief about the problem. These are also known as stochastic search problems. We defer until Part III the study of more general state-dependent problems, which includes the massive class of dynamic resource allocation problems (where decisions change the allocation of resources), as well as other settings where the problem itself is evolving over time (e.g. changing weather, market prices, temperature in a room, ...).

- Derivative-based stochastic search (chapter 5) – Derivative-based algorithms represent one of the earliest adaptive methods proposed for stochastic optimization. These methods form the foundation of what is classically referred to as (derivative-based) stochastic search, or stochastic gradient algorithms.
- Stepsize policies (chapter 6) – Sampling-based algorithms need to perform smoothing between old and new estimates using what are commonly

known as stepsizes (or learning rates). Stepsize policies play a critical role in derivative-based stochastic search, where the stochastic gradient determines the direction in which we move to improve a parameter vector, but the stepsize determines how far we move in the direction of the gradient.

- Derivative-free stochastic search (chapter 7) – We then transition to derivative-free stochastic search, which encompasses a variety of fields with names such as ranking and selection (for offline learning), response surface methods, and multiarmed bandit problems (for online formulations). In this chapter that we demonstrate all four classes of policies for deciding where to next make a (typically noisy) observation of a function that we are trying to optimize.

Part III – State-dependent problems – Here we transition to the much richer class of sequential problems where the *problem* being optimized is evolving over time, which means the *problem* depends on information or parameters that are changing over time. This means the objective function and/or constraints depend on dynamic data in the state variable, where this dynamic data can depend on decisions being made (such as the inventory or location of a drone), or may just evolve exogenously (such as market prices or weather). These problems may or may not have a belief state.

- State-dependent applications (chapter 8) – We begin with a series of applications where the function is state dependent. State variables can arise in the objective function (e.g. prices), or in the constraints, which is typical of problems that involve the management of physical resources. We also illustrate problems that include evolving beliefs, which introduces the dimension of active learning (which we first encounter in chapter 7).
- Modeling sequential decision problems (chapter 9) – This chapter provides a comprehensive summary of how to model general (state-dependent) sequential decision problems. This is a substantial chapter that starts by illustrating the modeling framework in the context of a simple problem, before exposing the full depth of the modeling framework for complex problems.
- Uncertainty modeling (chapter 10) – To find good policies, you need a good model of uncertainty, which is arguably the most subtle dimension of modeling. In this chapter we identify 12 different sources of uncertainty and discuss how to model them.
- Designing policies (chapter 11) – Here we provide a more comprehensive overview of the different strategies for creating policies, leading to the four classes of policies that we first introduced in Part I for learning problems. In this chapter we also provide guidance into how to choose among the four classes for a particular problem, and present the results of a series of

experiments on variations of an energy storage problem that show that we can make *each* of the four classes of policies work best depending on the characteristics of the data.

Part IV – Policies based on policy search – These chapters describe policies in the "policy search" class that have to be tuned, either in a simulator or in the field.

- PFAs- Policy function approximations (chapter 12) – In this chapter we consider the use of parametric functions (plus some variations) which directly map from the state variable to a decision, without solving an imbedded optimization problem. This is the only class which does not solve an imbedded optimization problem. We search over a well-defined parameter space to find the policy that produces the best performance over time, in either offline or online settings. PFAs are well suited to problems with scalar action spaces, or low-dimensional continuous actions.

- CFAs- Cost function approximations (chapter 13) – This strategy spans effective policies for solving optimal learning problems (also known as multiarmed bandit problems), to policies for high-dimensional problems that require the use of solvers for linear, integer, or nonlinear programs. This policy class has been overlooked in the research literature, but is widely used (heuristically) in industry.

Part V – Policies based on lookahead approximations – Policies based on lookahead approximations are the counterpart to policies derived from policy search. Here, we design good policies by understanding the impact of a decision now on the future. We can do this by finding (usually approximately) the value of being in a state, or by planning over some horizon.

- VFAs- Policies based on value function approximations – This class covers a very rich literature that span exact methods for special cases, and an extensive literature based on approximating value functions that are described by terms such as approximate dynamic programming, adaptive (or neuro) dynamic programming, and (initially) reinforcement learning. Given the depth and breadth of the work in this area, we cover this class of policy in five chapters:
  - Exact dynamic programming (chapter 14) – There are certain classes of sequential decision problems that can be solved exactly. One of the best known is characterized by discrete states and actions (known as discrete Markov decision processes), a topic we cover in considerable depth. We also briefly cover an important problem from the optimal controls literature known as *linear quadratic regulation*, as well as some simple problems that can be solved analytically.

– Backward approximate dynamic programming (chapter 15) – Backward approximate dynamic programming parallels classical backward dynamic programming (from chapter 14), but avoids the need to enumerate states or compute expectations through Monte Carlo sampling and using machine learning to estimate value functions approximately.

– Forward approximate dynamic programming I: The value of a policy (chapter 16) – This is the first step using machine learning methods to approximate the value of policy as a function of the starting state. This is the foundation of a broad class of methods known as approximate (or adaptive) dynamic programming, or reinforcement learning.

– Forward approximate dynamic programming II: Policy optimization (chapter 17) – In this chapter we build on foundational algorithms such as Q-learning, value iteration, and policy iteration, first introduced in chapter 14, to try to find high-quality policies based on value function approximations.

– Forward approximate dynamic programming III: Convex functions (chapter 18) – This chapter focuses on convex problems, with special emphasis on stochastic linear programs with applications in dynamic resource allocation. Here we exploit convexity to build high-quality approximations of value functions.

- DLAs- Policies based on direct lookahead approximations (chapter 19) – A direct lookahead policy optimizes over a horizon, but instead of optimizing the original model, we allow ourselves to introduce a variety of approximations to make it more tractable. A standard approximation is to make the model deterministic, which can work well in some applications. For those where it does not, we revisit the entire process of solving a stochastic optimization problem, but with considerably more emphasis on computation.

Part VI – Multiagent systems and learning – We close by showing how our framework can be extended to handle multiagent systems, which inherently requires learning.

- Multiagent systems and learning (chapter 20) – We start by showing how to model learning systems as two agent problems (a controlling agent observing an environment agent), and show how this produces an alternative framework to partially observable Markov decision processes (known as POMDPs). We then extend to problems with multiple controlling agents, in particular the need to model communication.

1.8.2 How to Read Each Chapter

This book covers a lot of material, which should not be surprising given the scope of the topic. However, it has been written to "read short." In every chapter, there are sections marked by "*" – this is our indication of material that can be skipped on a first pass.

There are a few sections marked with ** which is our indication of mathematically advanced material. For mathematically sophisticated readers (especially those with a measure–theoretic probability background), there are many opportunities to approach this material using the full range of this training. This book is not designed for these readers, although we will occasionally hint at this material. We will say, however, that much of our notational style has been designed with an understanding of how probabilists (in particular) think of and approach sequential decision problems. This book will lay a proper foundation for readers who want to use this as a launching pad into more theoretical research.

Readers new to the entire topic of sequential decision problems (and by this we mean any form of dynamic programming, stochastic programming and stochastic control) should start with the relatively simpler "starter" models. It is quite easy to learn how to model the simpler problems. By contrast, complex problems can become quite rich, especially when it comes to developing stochastic models. It is important to find the problems that you are comfortable with, and then grow from there.

The book will talk at length about the four classes of policies. Of these, two are relatively simple (PFAs and CFAs) and two are much richer (VFAs and stochastic DLAs). You should not assume that you need to become an expert in all of them right away. Everyone makes decisions over time in the presence of evolving information, and the vast majority of these people have never heard of Bellman's equation (VFA-based policies). Also, while deterministic DLAs (think of navigation systems planning a path) are also relatively easy to understand, stochastic DLAs are another matter. It is much more important to get an understanding of the concept of a policy and tuning a policy (which you can do using PFAs and CFAs) than it is to jump into the more complex policies that are popular in the academic literature (VFAs and stochastic DLAs).

1.8.3 Organization of Exercises

Each chapter is accompanied by a series of exercises at the end of the chapter, divided into the following categories:

- Review questions – These are relatively simple questions drawn directly from the chapter, without any need for creative problem solving.
- Modeling questions – These will be questions that describe an application which you then have to put into the modeling framework given above.
- Computational exercises – These are exercises that require that you perform specific calculations related to methods described in the chapter.

- Theory questions – From time to time we will pose classical theory questions. Most texts on stochastic optimization emphasize these questions. This book emphasizes modeling and computation, so theory questions play a relatively minor role.
- Problem-solving questions – These questions will pose a setting and require that you go through modeling and policy design.
- Readings from *Sequential Decision Analytics and Modeling* – This is an online book that uses a teach by example style. Each chapter (except for chapters 1 and 7) illustrates how to model and solve a specific decision problem. These have been designed to bring out the features of different classes of policies. There are Python modules that go with most of these exercises that provide an opportunity to do computational work. These exercises will generally require that the reader use the Python module as a start, but where additional programming is required.
- Diary problem – This is a single problem of your choosing that you will use as a context to answer a question at the end of each chapter. It is like "keeping a diary" since you will accumulate answers that draw from the material throughout the book, but using the setting of a problem that is relevant to you.

Not all of these topics will be included in the exercises for each chapter.

1.9 Bibliographic Notes

Section 1.2 – We defer to chapter 2 for a discussion of the different communities of stochastic optimization, and review the literature there. It cannot be emphasized enough how much our universal framework draws on all these communities.

Section 1.3 – We first articulated the five elements of the universal framework in Powell (2011) (Chapter 5, which has always been available at https://tinyurl.com/PowellADP), which built on the initial model from the first edition which had six elements (Powell (2007)). Our framework draws heavily from the framework that has long been used in optimal control (there are many books, but see Lewis & Vrabie (2012) which is a popular reference in this field), but there are some differences. Our framework is compared to the optimal control framework and that used in Markov decision processes (and now reinforcement learning) in Powell (2021). Some key differences is that the optimal control framework, which is originally based on deterministic control, often optimizes over the controls $u_0, u_1, \ldots, u_T$, even when the problem is stochastic. Our notation makes it explicit that if the problem is stochastic, u_t is a function which we call a policy (the controls people will call it a control law), and we always optimizes over policies π.

Section 1.4 – Powell (2011) appears to be the first published reference to "four classes of policies" for solving dynamic programs, but it did not list the four classes used here (one class was myopic policies, and cost function approximations were overlooked). The first reference to list the four classes of policies used here was the tutorial Powell (2014) *Clearing the Jungle of Stochastic Optimization*, without recognizing that the four classes can (and should) be divided into two major strategies. The first paper to identify the two strategies of "policy search" and "lookahead policies" was given in the tutorial Powell (2016). All these ideas came together in Powell (2019) which combined the four classes of policies with the identification of state-independent and state-dependent problem classes, along with different types of objectives such as cumulative and final reward. This paper laid the foundation for this book.

Exercises

Review questions

1.1 What are the three classes of state variables?

1.2 What are the five elements of a sequential decision problem?

1.3 What is meant by "model first, then solve"?

1.4 What is the price of simplicity? Give an cxample, either from the chapter or a problem of your own choosing.

1.5 What are the two strategies for designing policies for sequential decision problems? Briefly describe the principles behind each one.

1.6 What are the four classes of policies? Briefly describe each one.

Modeling questions

1.7 Pick three examples of sequential decision problems. Provide a brief narrative describing the context, and list (a) the decision being made, (b) information that arrives after the decision is made that is likely to be relevant to the decision, and (c) at least one metric that can be used to evaluate how well the decision has performed.

1.8 For each of the three types of state variables, do the following:

(a) Give three examples of physical state variables.
(b) Give three examples of information about parameters or quantities that we know perfectly, but which would not be considered a physical state variable.
(c) Give three examples of parameters or quantities that we would not know perfectly, but could approximate with a probability distribution.

1.9 Section 1.3 shows how to model a simple inventory problem. Repeat this model assuming that we sell our product at a price p_t that changes from time period to time period according to the equation

$$p_{t+1} = p_t + \varepsilon_{t+1},$$

where ε_{t+1} is a normally distributed random variable with mean 0 and variance σ^2.

Problem-solving questions

1.10 Consider an asset-selling problem where you need to decide when to sell an asset. Let p_t be the price of the asset if it is sold at time t, and assume that you model the evolution of the price of the asset using

$$p_{t+1} = p_t + \theta(p_t - 60) + \varepsilon_{t+1}.$$

We assume that the noise terms ε_t, $t = 1, 2, \dots$ are independent and identically distributed over time, where $\varepsilon_t \sim N(0, \sigma_\varepsilon^2)$. Let

$$R_t = \begin{cases} 1 & \text{if we are still holding the asset at time } t, \\ 0 & \text{otherwise.} \end{cases}$$

Further let

$$x_t = \begin{cases} 1 & \text{if we sell the asset at time } t, \\ 0 & \text{otherwise.} \end{cases}$$

Of course, we can only sell the asset if we are still holding it. We now need a rule for deciding if we should sell the asset. We propose

$$X^\pi(S_t | \rho) = \begin{cases} 1 & \text{if } p_t \geq \bar{p}_t + \rho \text{ and } R_t = 1, \\ 0 & \text{otherwise,} \end{cases}$$

where

$$S_t \;=\; \text{the information we have available to make a decision}$$
$$\text{(we have to design this),,}$$

$$\bar{p}_t \;=\; .9\bar{p}_{t-1} + .1p_t.$$

(a) What are the elements of the state variable S_t for this problem?
(b) What is the uncertainty?
(c) Imagine running a simulation in a spreadsheet where you are given a sample realization of the noise terms over T time periods as $(\hat{\varepsilon})_{t=1}^T = (\hat{\varepsilon}_1, \hat{\varepsilon}_2, \dots, \hat{\varepsilon}_T)$. Note that we treat $\hat{\varepsilon}_t$ as a number, such as $\hat{\varepsilon}_t = 1.67$ as opposed to ε_t which is a normally distributed random variable. Write an expression for computing the value of the policy $X^\pi(S_t | \rho)$ given the sequence $(\hat{\varepsilon})_{t=1}^T$. Given this sequence, we could evaluate different values of ρ, say $\rho = 0.75, 2.35$ or 3.15 to see which performs the best.
(d) In reality, we are not going to be given the sequence $(\hat{\varepsilon})_{t=1}^T$. Assume that $T = 20$ time periods, and that

$$\sigma_\varepsilon^2 \;=\; 4^2,$$

$$p_0 \;=\; \$65,$$

$$\theta \;=\; 0.1.$$

Write out the value of the policy as an expectation (see section 1.3).
(e) Develop a spreadsheet to create 10 sample paths of the sequence $((\varepsilon_t),\ t = 1, \dots, 20)$ using the parameters above. You can generate a random observation of ε_t in a spreadsheet using the function `NORM.INV(RAND(),0,$\sigma$)`. Let the performance of our decision rule $X^\pi(S_t | \rho)$ be given by the price that it decides to sell (if it decides to sell), averaged over all 10 sample paths. Now test $\rho = 1, 2, 3, 4, \dots, 10$ and find the value of ρ that seems to work the best.
(f) Repeat (e), but this time we are going to solve the problem

$$\max_{x_0, \dots, x_T} \mathbb{E} \sum_{t=0}^{T} p_t x_t.$$

We do this by picking the time t when we are going to sell (that is, when $x_t = 1$) before seeing any information. Evaluate the solutions $x_2 = 1, x_4 = 1, \dots, x_{20} = 1$. Which is best? How does its performance compare to the performance of $X^\pi(S_t | \rho)$ for the best value of ρ?

(g) Finally, repeat (f), but now you get to see all the prices and then pick the best one. This is known as a *posterior bound* because it gets to see all the information in the future to make a decision now. How do the solutions in parts (e) and (f) compare to the posterior bound? (There is an entire field of stochastic optimization that uses this strategy as an approximation.)

(h) Classify the policies in (e), (f), and (g) (yes, (g) is a class of policy) according to the classification described in section 1.5 of the text.

1.11 The inventory problem describes a policy where an order is made if the inventory falls below θ^{min}, where we order up to θ^{max}. Which of the four classes does this represent? Write out the objective function we would have to use to find the best value of θ.

Sequential decision analytics and modeling

These exercises are drawn from the online book *Sequential Decision Analytics and Modeling* available at http://tinyurl.com/sdaexamplesprint.

1.12 Read chapter 2 on the asset selling problem (sections 2.1–2.4).

(a) Which of the four classes of policies introduced in section 1.4 are used for this problem?

(b) What tunable parameters are used in the policy?

(c) Describe the process you might use for tuning the policy using historical data.

Diary problem

The diary problem is a single problem you design that you will use for this category throughout the rest of the book.

1.13 For this chapter, you need to pick a problem context. The ideal problem is one with some richness (e.g. different types of decisions and sources of uncertainty), but the best problem is one that you are familiar with, or have a special interest in. To bring out the richness of our modeling and algorithmic framework, it would help if your sequential decision problem involved learning in some form. For now, prepare a one to two paragraph summary of the context. You will be providing additional details in later chapters.

Bibliography

Lewis, F. L. and Vrabie, D. (2012). *Design Optimal Adaptive Controllers*, 3 e. Hoboken, NJ: John Wiley & Sons.

Powell, W. B. (2007). Approximate Dynamic Programming: Solving the curses of dimensionality, John Wiley & Sons.

Powell, W. B. (2011). *Approximate Dynamic Programming: Solving the Curses of Dimensionality*, 2 e. John Wiley & Sons.

Powell, W. B. (2014). Clearing the Jungle of Stochastic Optimization. *INFORMS Tutorials in Operations Research: Bridging Data and Decisions*, pp. 109-137, November, 2014.

Powell, W. B. (2016). A Unified Framework for Optimization under Uncertainty, *in* 'Informs TutORials in Operations Research', 45–83.

Powell, W. B. (2019). A unified framework for stochastic optimization. *European Journal of Operational Research* **275** (3): 795–821.

Powell, W. B. (2021). From reinforcement learning to optimal control: A unified framework for sequential decisions. In: *Handbook on Reinforcement Learning and Optimal Control, Studies in Systems, Decision and Control*, 29–74.

2

Canonical Problems and Applications

The vast array of sequential decision problems has produced at least 15 distinct communities (which we listed in section 1.2) that have developed methods for modeling and solving these problems. Just as written and spoken languages have evolved from different roots, these communities feature roughly eight fundamentally different notational systems, in addition to what could be called dialects, with notation derived from one of the core systems.

Hidden in these different notational "languages" are methods that are sometimes truly original, while others are creative evolutions, and yet others are simply the same method with a different name. Motivating the different methods are the classes of problems that have caught the imagination of each community. Not surprisingly, individual research communities steadily move into new problems, which then motivate new methods.

This chapter provides, in section 2.1, an overview of these different communities and their modeling style. This chapter provides very brief introductions to the most important canonical models of each community, in the notation of that community. In some cases we pause to hint at how we would take a different perspective. Then, section 2.2 summarizes the universal modeling framework that we will use in this book, which can be used to model each of the canonical problems in section 2.1. Finally, section 2.3 provides a short summary of different application settings.

2.1 Canonical Problems

Each community in stochastic optimization has a canonical modeling framework that they use to illustrate their problem domain. Often, these canonical problems lend themselves to an elegant solution technique which then becomes a hammer looking for a nail. While these tools are typically limited to

Reinforcement Learning and Stochastic Optimization: A Unified Framework for Sequential Decisions, First Edition. Warren B. Powell.
© 2022 John Wiley & Sons, Inc. Published 2022 John Wiley & Sons, Inc.

a specific problem class, they often illustrate important ideas that become the foundation of powerful approximation methods. For this reason, understanding these canonical problems helps to provide a foundation for the full range of sequential decision problems under uncertainty.

For a reader new to all of these fields, these canonical problems can just be skimmed the first time through the book. It is important to realize that every one of these fields studies a form of sequential decision problem which can be modeled with the universal modeling framework that we first introduced in section 1.3, and then present in section 2.2 in more detail.

2.1.1 Stochastic Search – Derivative-based and Derivative-free

As we are going to learn, if there is a single problem that serves as an umbrella for almost all stochastic optimization problems (at least, all the ones that use an expectation), it is a problem that is often referred to as stochastic search, which is written

$$\max_x \mathbb{E}F(x, W), \tag{2.1}$$

where x is a deterministic variable, or a vector (or, as we will show, a function). The expectation is over the random variable W, which can be a vector, as well as a sequence of random variables $W_1, \dots, W_t, \dots, W_T$ that evolve over time. We are going to refer to the notational style used in the expectation, where we do not indicate what we are taking the expectation over, in equation (2.1) as the *compact form* of the expectation.

We prefer the style where we make the dependence on the random variable explicit by writing

$$\max_x \mathbb{E}_W F(x, W). \tag{2.2}$$

We refer to the style used in equation (2.2), where we indicate what random variable we are taking the expectation over, as the *expanded form* of the expectation. While probabilists frown on this habit, any notation that improves clarity should be encouraged. We are also going to introduce problems where it is useful to express the dependence on an initial state variable S^0, which might include probabilistic beliefs about uncertain parameters such as how the market might respond to a change in price. We express this dependence by writing

$$\max_x \mathbb{E}\{F(x, W)|S^0\} \quad = \quad \max_x \mathbb{E}_{S^0} \mathbb{E}_{W|S^0} F(x, W). \tag{2.3}$$

Initial state variables can express the dependence of the problem on either deterministic or probabilistic information (say, a distribution about an unknown parameter). For example, we might assume that W is normally distributed with mean μ, where μ is also uncertain (it might be uniformly distributed between 0 and 10). In this case, the first expectation in (2.3), $\mathbb{E}_{S^0}$, is over the uniform distribution for μ, while the second expectation, $\mathbb{E}_{W|S^0}$, is over the normal distribution for W given a value for the mean μ. We see that the form in equation (2.3) does a better job of communicating the uncertainties involved.

There are problems where the initial state S^0 may change each time we solve a problem. For example, S^0 might capture the medical history of a patient, after which we have to decide on a course of treatment, and then we observe medical outcomes. We will sometimes use the style in (2.1) for compactness, but we are going to use the expanded form in (2.3) as our default style (the motivation for this becomes more apparent when you start working on real applications).

This basic problem class comes in a number of flavors, depending on the following:

- Initial state S^0 – The initial state will include any deterministic parameters, as well as initial distributions of uncertain parameters. S^0 might be a fixed set of deterministic parameters (such as the temperature at which water boils), or it might change each time we solve our problem (it might include temperature and humidity in a lab), and it might include a probability distribution describing an unknown parameter (such as how a market responds to price).
- Decision x – x can be binary, discrete (and finite, and not too large), categorical (finite, but a potentially very large number of choices), continuous (scalar or vector), or a discrete vector.
- Random information W – The distribution of W may be known or unknown, and the distribution can be normal or exponential, or one with heavy tails, spikes, and rare events. W may be a single variable or vector that is realized all at once, or it can be a sequence of variables (or vectors) $W_1, \ldots, W_t, \ldots, W_T$.
- The function $F(x, W)$ may be characterized along several dimensions:
  - The cost of a function evaluation – The function $F(x, W)$ may be easy to evaluate (fractions of a second to seconds), or more expensive (minutes to hours to days to weeks).
  - Search budget – May be finite (for example, we are limited to N evaluations of the function or its gradient), or infinite (obviously this is purely for analysis purposes – real budgets are always finite). There are even problems where a rule determines when we stop, which may be exogenous or dependent on what we have learned (these are called *anytime problems*).
  - The noise level (and the nature of the noise) – There are applications where the noise in a function evaluation is minimal (or nonexistent), and others where the noise level is exceptionally high.

Problem (2.1) is the asymptotic form of the basic stochastic optimization problem where we seek an optimal solution x^* that is deterministic (there is exactly one of them). Most of this book will focus on the more practical finite budget versions where we run an algorithm (that we call π for reasons that become clear later), for N iterations, to produce a solution $x^{\pi,N}$ which is a random variable, since it depends on the observations of W along the way.

There are two flavors of this problem:

- Final reward objective – Here we run our algorithm π for N iterations, producing a solution $x^{\pi,N}$. We care only about the performance of the final solution, and not how well we do while we are performing the search. After we find $x^{\pi,N}$, we then have to evaluate it, and we introduce a random variable $\widehat{W}$ that is used for testing (as opposed to training). The final reward objective function is written (in its expanded form) as

$$\max_{\pi} \mathbb{E}_{S^0} E_{W^1,\dots,W^N|S^0} E_{\widehat{W}|S^0, x^{\pi,N}} F(x^{\pi,N}, \widehat{W}). \tag{2.4}$$

- Cumulative reward objective – In this setting we care about the total rewards while we are performing our search, which produces the objective function

$$\max_{\pi} \mathbb{E}_{S^0} \mathbb{E}_{W^1,\dots,W^N|S^0} \sum_{n=0}^{N-1} F(X^{\pi}(S^n), W^{n+1}). \tag{2.5}$$

The general problem of stochastic search has been pursued as two distinct fields that depend on the algorithmic strategy. These are known as derivative-based stochastic search, and derivative-free stochastic search. Both fields trace their roots to 1951 but have evolved independently as completely separate lines of investigation.

2.1.1.1   Derivative-based Stochastic Search

We accept the practical reality that we cannot take the derivative of an expectation, which prevents us from taking the derivative of $F(x) = \mathbb{E}F(x, W)$. However, there are many problems where we observe W, and then take the derivative of $F(x, W)$, which we write as the stochastic gradient

$$\nabla_x F(x, W(\omega)).$$

The most common way to illustrate a stochastic gradient uses the newsvendor problem

$$F(x, W) = p \min\{x, W\} - cx.$$

The stochastic gradient is easily verified to be

$$\nabla_x F(x, W) = \begin{cases} p - c & x < W, \\ -c & x > W. \end{cases}$$

As we can see, we can compute the gradient of $F(x, W)$ after we observe W. We then use this gradient in a stochastic gradient algorithm

$$x^{n+1} = x^n + \alpha_n \nabla_x F(x^n, W^{n+1}), \tag{2.6}$$

where α_n is known as a stepsize. A famous paper by Robbins and Monro published in 1951 proved that the stochastic gradient algorithm (2.6) converges asymptotically to the optimum of the objective function (2.4). This is stated formally

$$\lim_{n \to \infty} x^n = x^* = \arg\max_x \mathbb{E} F(x, W).$$

70 years later, this algorithm continues to attract considerable interest. We cover this important class in chapter 5, with an entire chapter, chapter 6, dedicated to the design of stepsize formulas for α_n.

2.1.1.2 Derivative-free Stochastic Search

While there are many problems where we can compute the stochastic gradient $\nabla_x F(x, W)$, there are far more problems where we cannot. Instead, we assume that all we can do is make random observations of the function $F(x, W)$ which we write as

$$\hat{F}^{n+1} = F(x^n, W^{n+1}),$$

where the indexing communicates that we choose x^n first, then observe W^{n+1} after which we can compute the sampled observation of the function $\hat{F}^{n+1} = F(x^n, W^{n+1})$. We then use the sampled observation $\hat{F}^{n+1}$ to update an estimate $\bar{F}^n(x)$ of $\mathbb{E} F(x, W)$ to obtain $\bar{F}^{n+1}(x)$.

Derivative-free stochastic search involves two core components:

- Creating the belief $\bar{F}^n(x)$. We do this with any of a range of machine learning tools that we review in chapter 3.
- Choosing the point to observe x^n. This is generally referred to as the algorithm, but in this book we will refer to it as a policy. This problem is addressed in considerable depth in chapter 7.

Derivative-free stochastic search is such a rich problem class that there are entire fields that pursue particular algorithmic strategies without acknowledging competing approaches.

2.1.2 Decision Trees

Decision trees are easily one of the most familiar ways to depict sequential decision problems, with or without uncertainty. Figure 2.1 illustrates a simple problem of determining whether to hold or sell an asset. If we decide to hold, we observe changes in the price of the asset and then get to make the decision of holding or selling.

Figure 2.1 illustrates the basic elements of a decision tree. Square nodes represent points where decisions are made, while circles represent points where random information is revealed. We solve the decision tree by rolling backward, calculating the value of being at each node. At an outcome node, we average across all the downstream nodes (since we do not control which node we transition to), while at decision nodes, we pick the best decision based on the one-period reward plus the downstream value.

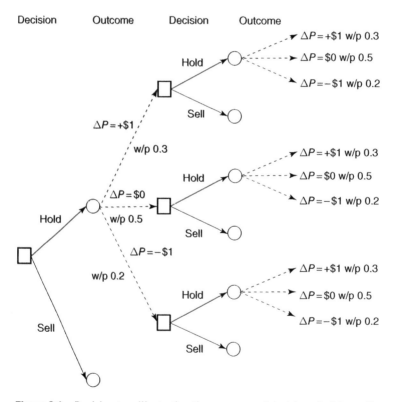

Figure 2.1 Decision tree illustrating the sequence of decisions (hold or sell an asset) and new information (price changes).

Almost any dynamic program with discrete states and actions can be modeled as a decision tree. The problem is that decision trees grow explosively, even for relatively small problems. Imagine a setting where there are three decisions (buy, sell, hold an asset), and three random outcomes (say the price changes by +1, -1 or 0). Each sequence of price change followed by decision grows the tree by a factor of 9. Now imagine a trading problem where we get to make decisions once each minute. After just one hour, our tree has grown to $9^{60} \approx 10^{57}$ branches!

2.1.3 Markov Decision Processes

A Markov decision process is modeled using a very standard framework which is outlined in Figure 2.2. Note that this is modeled without indexing time, since the standard canonical model is for a problem in steady state. Some authors also include a set of "decision epochs" which are the points in time, typically modeled as $t = 1, 2, ...$, when we compute our state variable and choose an action.

For example, if we have $s \in \mathcal{S}$ units of inventory, purchase $a \in \mathcal{A}_s$ more units, and then sell a random quantity $\hat{D}$, our updated inventory is computed using

$$s' = \max\{0, s + a - \hat{D}\}.$$

Our one-step transition matrix would be computed from

$$P(s'|s, a) = Prob[\hat{D} = \max\{0, (s + a) - s'\}].$$

State space – $\mathcal{S} = \{s_1, ..., s_{|\mathcal{S}|}\}$ is the set of (discrete states) the system may occupy.

Action space – $\mathcal{A}_s = \{a_1, ... a_M\}$ is the set of actions we can take when we are in state s.

Transition matrix – We assume we are given the one-step state transition matrix with element

$$P(s'|s, a) = \text{the probability that state } S_{t+1} = s' \text{ given that we are in}$$
$$\text{state } S_t = s \text{ and take action } a.$$

Reward function – Let $r(s, a)$ be the reward we receive when we take action a when we are in state s.

Figure 2.2 Canonical model for a Markov decision process.

Our reward function might be

$$r(s, a) = p \min\{s + a, \hat{D}\} - ca,$$

where c is the unit cost of purchasing an item of inventory and p is our sales price for meeting as much demand as we can.

If we are solving a finite horizon problem, let $V_t(S_t)$ be the optimal value of being in state S_t and behaving optimally from time t onward. If we are given $V_{t+1}(S_{t+1})$, we can compute $V_t(S_t)$ using

$$V_t(S_t) = \max_{a \in \mathcal{A}_s} \left(r(S_t, a) + \gamma \sum_{s' \in \mathcal{S}} P(s'|S_t, a)V_{t+1}(S_{t+1} = s') \right), \tag{2.7}$$

where γ is a discount factor (presumably to capture the time value of money). Note that to compute $V_t(S_t)$, we have to loop over every possible value of $S_t \in \mathcal{S}$ and then solve the maximization problem.

Equation (2.7) may seem somewhat obvious, but when first introduced it was actually quite a breakthrough, and is known as *Bellman's optimality equation* in operations research and computer science, or *Hamilton-Jacobi equations* in control theory (although this community typically writes it for continuous states and actions/controls).

Equation (2.7) is the foundation for a major class of policies that we refer to as policies based on value function approximations (or VFA policies). Specifically, if we know $V_{t+1}(S_{t+1})$, then we would make a decision at time t when we are in state S_t by solving

$$X_t^\pi(S_t) = \arg\max_{a \in \mathcal{A}_s} \left(r(S_t, a) + \gamma \sum_{s' \in \mathcal{S}} P(s'|S_t, a)V_{t+1}(S_{t+1} = s') \right).$$

If we can compute the value functions exactly using equation (2.7), then this is a rare instance of an optimal policy.

If the one-step transition matrix $P(s'|S_t, a)$ can be computed (and stored), then equation (2.7) is quite easy to compute starting at time T (when we assume $V_T(S_T)$ is given, where it is fairly common to use $V_T(S_T) = 0$) and progressing backward in time.

There has been considerable interest in this community in steady state problems, where we assume that as $t \to \infty$, that $V_t(S_t) \to V(S)$. In this case, (2.7) becomes

$$V(s) = \max_{a \in \mathcal{A}_s} \left(r(s, a) + \gamma \sum_{s' \in \mathcal{S}} P(s'|s, a)V(s') \right). \tag{2.8}$$

Now we have a system of equations that we have to solve to find $V(s)$. We review these methods in considerable depth in chapter 14.

Bellman's equation was viewed as a major computational breakthrough when it was first introduced, because it avoids the explosion of decision trees. However, people (including Bellman) quickly realized that there was a problem when the state s is a vector (even if it is still discrete). The size of the state space grows exponentially with the number of dimensions, typically limiting this method to problems where the state variable has at most three or four dimensions. This is widely known as the "curse of dimensionality."

Bellman's equation actually suffers from three curses of dimensionality. In addition to the state variable, the random information W (buried in the one-step transition $P(s'|s, a)$) might also be a vector. Finally, the action a might be a vector x. It is common for people to dismiss "dynamic programming" (but they mean discrete Markov decision processes) because of "the curse of dimensionality" (they could say because of the "curses of dimensionality"), but the real issue is the use of lookup tables. There are strategies for overcoming the curses of dimensionality, but if it were easy, this would be a much shorter book.

2.1.4 Optimal Control

The optimal control community is most familiar with the deterministic form of a control problem, which is typically written in terms of the "system model" (transition function)

$$x_{t+1} = f(x_t, u_t),$$

where x_t is the state variable and u_t is the control (or action, or decision). A typical engineering control problem might involve controlling a rocket (think of getting SpaceX to land after its takeoff), where the state x_t is the location and velocity of the rocket (each in three dimensions), while the control u_t would be the forces in all three dimensions on the rocket. How the forces affect the location and speed of the rocket (that is, its state x_t) is all contained in the transition function $f(x_t, u_t)$.

The transition function $f(x_t, u_t)$ is a particularly powerful piece of notation that we will use throughout the book (we write the transition as $S_{t+1} = S^M(S_t, x_t, W_{t+1})$). It captures the effect of a decision x_t (such as move to a location, add inventory, use a medical treatment, or apply force to a vehicle) on the state x_t. Note that the canonical MDP framework described in Figure 2.2 uses a one-step transition matrix $P(S_{t+1}|S_t, a_t)$; we show in chapter 9 that the one-step transition matrix has to be computed using the transition function. In practice, one-step transition matrices are rarely computable, while transition functions are easy to compute.

The problem is to find u_t that solves

$$\min_{u_0,...,u_T} \sum_{t=0}^{T} L(x_t, u_t) + J_T(x_T), \tag{2.9}$$

where $L(x, u)$ is a "loss function" and $J_T(x_T)$ is a terminal cost. Equation (2.9) can be stated recursively using

$$J_t(x_t) = \max_{u_t} \left(L(x_t, u_t) + J_{t+1}(x_{t+1}) \right) \tag{2.10}$$

where $x_{t+1} = f(x_t, u_t)$. Here, $J_t(x_t)$ is known as the "cost-to-go" function, which is simply different notation for the value function $V_t(S_t)$ in section 2.1.3.

A solution strategy which is so standard that it is often stated as part of the model is to view the transition $x_{t+1} = f(x_t, u_t)$ as a constraint that can be relaxed, producing the objective

$$\min_{u_0,...,u_T} \sum_{t=0}^{T} \left(L(x_t, u_t) + \lambda_t(x_{t+1} - f(x_t, u_t)) \right) + J_T(x_T), \tag{2.11}$$

where λ_t is a set of Lagrange multipliers known as "co-state variables." The function

$$H(x_0, u) = \sum_{t=0}^{T} \left(L(x_t, u_t) + \lambda_t(x_{t+1} - f(x_t, u_t)) \right) + J_T(x_T)$$

is known as the Hamiltonian.

A common form for the objective in (2.9) is an objective function that is quadratic in the state x_t and control u_t, given by

$$\min_{u_0,...,u_T} \sum_{t=0}^{T} \left((x_t)^T Q_t x_t + (u_t)^T R_t u_t \right). \tag{2.12}$$

Although it takes quite a bit of algebra, it is possible to show that the optimal solution to (2.16) can be written in the form of a function $U^{\pi}(x_t)$ which has the form

$$U^*(x_t) = -K_t x_t, \tag{2.13}$$

where K_t is a suitably dimensioned matrix that depends on the matrices $(Q_{t'}, R_{t'}), t' \leq t$.

A limitation of this theory is that it is easy to break. For example, simply adding a nonnegativity constraint $u_t \geq 0$ invalidates this result. The same is true if we make any changes to the objective function, and there are a *lot* of problems where the objective is not quadratic in the state variables and decision variables.

There are many problems where we need to model uncertainty in how our process evolves over time. The most common way to introduce uncertainty is through the transition function, which is typically written as

$$x_{t+1} = f(x_t, u_t, w_t) \tag{2.14}$$

where w_t is random at time t (this is standard notation in the optimal control literature, where it is common to model problems in continuous time). w_t might represent random demands in an inventory system, the random cost when traversing from one location to another, or the noise when measuring the presence of disease in a population. Often, w_t is modeled as additive noise which would be written

$$x_{t+1} = f(x_t, u_t) + w_t, \tag{2.15}$$

where w_t might be thought of as the wind pushing our rocket off course.

When we introduce noise, it is common to write the optimization problem as

$$\min_{u_0,\dots,u_T} \mathbb{E} \sum_{t=0}^{T} \left((x_t)^T Q_t x_t + (u_t)^T R_t u_t \right). \tag{2.16}$$

The problem with this formulation is that we have to recognize that the control u_t at time t is a random variable that depends on the state x_t, which in turn depends on the noise terms $w_0, \dots, w_{t-1}$.

To convert our original deterministic control problem to a stochastic control problem, we just have to follow the guidance we provided in section 1.6.3. We begin by introducing a *control law* (in the language of optimal control) that we represent by $U^\pi(x_t)$ (we refer to this as a policy). Now our problem is to find the best policy ("control law") that solves

$$\min_{\pi} \mathbb{E}_{w_0,\dots,w_T} \sum_{t=0}^{T} \left((x_t)^T Q_t x_t + (U_t^\pi(x_t))^T R_t U_t^\pi(x_t) \right), \tag{2.17}$$

where x_t evolves according to (2.14), and where we have to be given a model to describe the random variables w_t. A significant part of this book is focused on describing methods for finding good policies. We revisit optimal control problems in section 14.11.

The language of optimal control is widely used in engineering (mostly for deterministic problems) as well as finance, but is otherwise limited to these communities. However, it is the notation of optimal control that will form the foundation of our own modeling framework.

2.1.5 Approximate Dynamic Programming

The core idea of approximate dynamic programming is to use machine learning methods to replace the value function $V_t(S_t)$ (see equation (2.7)) with an approximation $\overline{V}_t^n(S_t|\theta)$ (assume this is after n iterations). We can use any of a variety of approximation strategies that we cover in chapter 3. Let a_t be our decision at time t (such as how much inventory to order or what drug to prescribe). Let $\bar{\theta}^n$ be our estimate of θ after n updates. Assuming we are in a state S_t^n (this might be our inventory at time t during iteration n), we could use this approximation to create a sampled observation of the value of being in state S_t^n using

$$\hat{v}_t^n = \max_{a_t} \left(C(S_t^n, a_t) + \mathbb{E}_{W_{t+1}} \{ \overline{V}_{t+1}(S_{t+1}^n | \bar{\theta}^{n-1}) | S_t^n \} \right), \tag{2.18}$$

where $S_{t+1}^n = S^M(S_t^n, a_t, W_{t+1})$, and where $\bar{\theta}^{n-1}$ is our estimate of θ after $n-1$ iterations.

We can then use $\hat{v}_t^n$ to update our estimate $\bar{\theta}_t^{n-1}$ to obtain $\bar{\theta}_t^n$. How this is done depends on how we are approximating $\overline{V}_t^n(S_t|\theta)$ (a variety of methods are described in chapter 3). There are, in addition, other ways to obtain the sampled observation $\hat{v}_t^n$ which we review in chapter 16.

Given a value function approximation $\overline{V}_{t+1}(S_{t+1}^n|\bar{\theta}^{n-1})$, we have a method for making decisions (that is, a policy) using

$$A^\pi(S_t^n) = \arg\max_{a_t} \left(C(S_t^n, a_t) + \mathbb{E}_{W_{t+1}} \{ \overline{V}_{t+1}(S_{t+1}^n | \bar{\theta}^{n-1}) | S_t^n \} \right),$$

where "$\arg\max_a$ returns the value of a that maximizes the expression. This is what we will be calling a "VFA-based policy."

The idea of using approximate value functions started with Bellman in 1959, but was then independently re-invented in the optimal control community (which used neural networks to approximate continuous value functions) in the 1970s, and computer science, where it became known as reinforcement learning, in the 1980s and 1990s. These methods are covered in considerably more depth in chapters 16 and 17. It has also been adapted to stochastic resource allocation problems where methods have been developed that exploit concavity (when maximizing) of the value function (see chapter 18).

2.1.6 Reinforcement Learning

While the controls community was developing methods for approximating value functions using neural networks, two computer scientists, Andy Barto and his student Richard Sutton, were trying to model animal behavior, as with mice trying to find their way through a maze to a reward (see Figure 2.3). Successes were learned over time by capturing the probability that a path from a particular point in the maze eventually leads to a success.

The basic idea closely parallels the methods of approximate dynamic programming, but it evolved on its own, with its own style. Instead of learning the value $V(s)$ of being in a state s, the core algorithmic strategy of reinforcement learning involves learning the value $Q(s, a)$ of being in a state s and then taking an action a. The basic algorithm, known as Q-learning, proceeds by computing

$$\hat{q}^n(s^n, a^n) = r(s^n, a^n) + \lambda \max_{a'} \bar{Q}^{n-1}(s', a'), \tag{2.19}$$

$$\bar{Q}^n(s^n, a^n) = (1 - \alpha_{n-1})\bar{Q}^{n-1}(s^n, a^n) + \alpha_{n-1}\hat{q}^n(s^n, a^n). \tag{2.20}$$

Here, λ is a discount factor, but it is different than the discount factor γ that we use elsewhere (occasionally) when solving dynamic problems (see, for example, equation (2.7)). The parameter λ is what could be called an "algorithmic discount factor" since it helps to "discount" the effect of making mistakes in the future that have the effect of reducing (incorrectly) the value of being in state s^n and taking action a^n.

The updating equation (2.21) is sometimes written

$$\bar{Q}^n(s^n, a^n) = \bar{Q}^{n-1}(s^n, a^n) + \alpha_{n-1}(\hat{q}^n(s^n, a^n) - \bar{Q}^{n-1}(s^n, a^n))$$

$$= \bar{Q}^{n-1}(s^n, a^n) + \alpha_{n-1} \underbrace{(r(s^n, a^n) + \lambda \max_{a'} \bar{Q}^{n-1}(s', a') - \bar{Q}^{n-1}(s^n, a^n))}_{\delta} \tag{2.21}$$

Figure 2.3 Finding a path through a maze.

where

$$\delta = r(s^n, a^n) + \lambda \max_{a'} \bar{Q}^{n-1}(s', a') - \bar{Q}^{n-1}(s^n, a^n)$$

is known as a "temporal difference" because it is capturing the difference between the current estimate $\bar{Q}^{n-1}(s^n, a^n)$ and the updated estimate $(r(s^n, a^n) + \lambda \max_{a'} \bar{Q}^{n-1}(s', a') - \bar{Q}^{n-1}(s^n, a^n))$ from one iteration to the next. Equation (2.21) is known as *temporal difference learning* which is performed with a fixed policy for choosing states and actions. The algorithm is referred to as "TD(λ)" (reflecting the role of the algorithmic discount factor λ) and the method is called "TD learning." In chapters 16 and 17 this will be known as approximate value iteration.

To compute (2.19), we assume we are given a state s^n, such as the location of the mouse in the maze in Figure 2.3. We use some method ("policy") to choose an action a^n, which produces a reward $r(s^n, a^n)$. Next, we choose a downstream state that might result from being in a state s^n and taking action a^n. There are two ways we can do this:

(1) Model-free learning – We assume that we have a physical system we can observe, such as a doctor making medical decisions or people choosing products off the internet.
(2) Model-based learning – Here we assume that we sample the downstream state from the one-step transition matrix $p(s'|s, a)$. In practice, what we are really doing is simulating the transition function from $s' = S^M(s^n, a^n, W^{n+1})$ where the function $S^M(\cdot)$ (using our notation) is the same as equation (2.14) from optimal control, and W^{n+1} is a random variable that we have to sample from some (known) distribution.

Computer scientists often work on problems where a system is being observed, which means they do not use an explicit model of the transition function.

Once we have our simulated downstream state s', we then find what appears to be the best action a' based on our current estimates $\bar{Q}^{n-1}(s', a')$ (known as "Q-factors"). Finally, we then update the estimates of the value of being in states s^n and action a^n. When this logic is applied to our maze in Figure 2.3, the algorithm steadily learns the state/action pairs with the highest probability of finding the exit, but it does require sampling all states and actions often enough.

There are many variations of Q-learning that reflect different rules for choosing the state s^n, choosing the action a^n, what is done with the updated estimate $\hat{q}^n(s^n, a^n)$, and how the estimates $\bar{Q}^n(s, a)$ are calculated. For example, equations (2.19)–(2.21) reflect a lookup table representation, but there is considerable ongoing research where $\bar{Q}(s, a)$ is approximated with a deep neural network.

As readers of this book will see, approximating value functions is not an algorithmic panacea. As the RL community expanded to a broader range of problems, researchers started introducing different algorithmic strategies, which will emerge in this book as samples from each of four classes of policies (policies based on value function approximations is just one). Today, "reinforcement learning" applies more to a community working on sequential decision problems using a wide range of strategies, which is how it made its way into the title of this book.

There are many people today who equate "reinforcement learning" with Q-learning, which is an algorithm, not a problem. Yet, today, leaders of the field will describe reinforcement learning as

   a) A problem class consisting of an agent acting on an environment receiving a reward.
   b) A community that identifies its work as "reinforcement learning."
   c) The set of methods developed by the community using the methods it self-identifies as "reinforcement learning" applied to the problem class.

Stated more compactly, this characterization consists of a community that self-describes its work as "reinforcement learning" consisting of any method that solves the problem class of an "agent acting on an environment receiving a reward." In effect, "reinforcement learning" is now being described as a problem class rather than a method, because there is a lot of work under the umbrella of "reinforcement learning" that does not use Q-learning (or any method for approximating value functions). The question of whether reinforcement learning is a problem or a method remains a major area of confusion at the time of the writing of this book.

We would make the argument that a more general characterization of this problem class would be a sequential decision problem, which includes any problem consisting of an agent acting on an environment, but would also include problems where an agent simply observes an environment (this is an important problem class in the RL community). In addition, rather than focusing just on VFA-based policies (such as Q-learning), we are generalizing to all four classes of policies. We note that the RL community is already working on algorithms that fall in all four classes of policies, so we would claim that our universal model characterizes not only everything the RL community is work-ing on today, but the entire problem class and methods that the RL community is likely to evolve into.

2.1.7 Optimal Stopping

A classical problem in stochastic optimization is known as the optimal stopping problem. Imagine that we have a stochastic process W_t (this might be prices of an asset) which determines a reward $f(W_t)$ if we stop at time t (the price we receive if we stop and sell the asset). Let $\omega \in \Omega$ be a sample path of $W_1, \dots, W_T$, where we are going to limit our discussion to finite horizon problems, which might represent a maturation date on a financial option. Let

$$
X_t(\omega) = \begin{cases} 1 & \text{if we stop at time } t, \\ 0 & \text{otherwise.} \end{cases}
$$

Let τ be the time t when $X_t = 1$ (we assume that $X_t = 0$ for $t > \tau$). This notation creates a problem, because ω specifies entire sample path, which seems to sug-gest that we are allowed to look into the future before making our decision at time t. Don't laugh – this this mistake is easy to make when backtesting policies using historical data. Furthermore, it is actually a fairly standard approxima-tion in the field of stochastic programming which we revisit in chapter 19 (in particular, see two-stage stochastic programming in section 19.9).

To fix this, we require that the function X_t be constructed so that it depends only on the history $W_1, \dots, W_t$. When this is the case τ is called a *stopping time*. The optimization problem can then be stated as

$$
\max_{\tau} \mathbb{E} X_\tau f(W_\tau), \tag{2.22}
$$

where we require τ to be a "stopping time." Mathematicians will often express this by requiring that τ (or equivalently, X_t) be an "$\mathcal{F}_t$-measurable function" which is just another way of saying that τ is not computed with information from points in time later than τ.

This language is familiar to students with training in measure-theoretic probability, which is *not* necessary for developing models and algorithms for stochastic optimization. Later, we are going to provide an easy introduction to these concepts in chapter 9, section 9.13, and then explain why we do not need to use this vocabulary.

More practically, the way we are going to solve the stopping problem in (2.22) is that we are going to create a function $X^\pi(S_t)$ that depends on the state of the system at time t. For example, imagine that we need a policy for selling an asset. Let $R_t = 1$ if we are holding the asset, and 0 otherwise. Assume that $p_1, p_2, \ldots, p_t$ is the history of the price process, where we receive p_t if we sell at time t. Further assume that we create a smoothed process $\bar{p}_t$ using

$$\bar{p}_t = (1 - \alpha)\bar{p}_{t-1} + \alpha p_t.$$

At time t, our state variable is $S_t = (R_t, \bar{p}_t, p_t)$. A sell policy might look like

$$X^\pi(S_t|\theta) = \begin{cases} 1 & \text{if } \bar{p}_t > \theta^{max} \text{ or } \bar{p}_t < \theta^{min}, \\ 0 & \text{otherwise.} \end{cases}$$

Finding the best policy means finding the best $\theta = (\theta^{min}, \theta^{max})$ by solving

$$\max_\theta \mathbb{E} \sum_{t=0}^{T} p_t X^\pi(S_t|\theta). \tag{2.23}$$

Our stopping time, then, is the earliest time $\tau = t$ where $X^\pi(S_t|\theta) = 1$.

Optimal stopping problems arise in a variety of settings. Some examples include:

American options – An American option gives you the right to sell the asset on or before a specified date. We provide an illustration of using approximate dynamic programming for American options in section 17.6.1. This strategy can be applied to any stopping problem.

European options – A European option on a financial asset gives you the right to sell the asset at a specified date in the future.

Machine replacement – While monitoring the status of a (typically complex) piece of machinery, we need to create a policy that tells us when to stop and repair or replace.

Homeland security – The National Security Administration collects information on many people. The NSA needs to determine when to start tracking someone, when to stop (if they feel the target is of no risk), or when to act (when they feel the target is of high risk).

Clinical trials – A drug company running a clinical trial for a drug has to know when to stop the trial and declare success or failure. For a more complete model of clinical trials, see http://tinyurl.com/sdaexamplesprint, chapter 14.

Optimal stopping may look like a disarmingly easy problem, given the simplicity of the state variable. However, in real applications there is almost always additional information that needs to be considered. For example, our asset-selling problem may depend on a basket of indices or securities that greatly expands the dimensionality of the state variable. The machine replacement problem might involve a number of measurements that are combined to make a decision. The homeland security application could easily involve a number of factors (places the person has visited, the nature of communications, and recent purchases). Finally, health decisions invariably depend on a number of factors that are unique to each patient.

2.1.8 Stochastic Programming

Imagine that we are an online retailer that has to allocate inventory to different fulfillment centers, after which it has to actually fill demands from the fulfillment centers that have inventory. Call the initial decision to allocate inventory x_0 (this is the "here and now" decision). Then we see the demand for the product D_1 and the prices p_1 that the retailer will be paid.

Let $W_1 = (D_1, p_1)$ represent this random information, and let ω refer to a sample realization of W_1, so that $W_1(\omega) = (D_1(\omega), p_1(\omega))$ is one possible realization of demands and prices. We make the decision x_1 after we see this information, so we have a decision $x_1(\omega)$ of shipping decisions for each possible realization ω of demands. The stochastic programming community usually refers to each outcome ω as a *scenario*.

Assume for the moment that $\Omega = (\omega_1, \omega_2, \ldots, \omega_K)$ is a (not too large) set of possible outcomes ("scenarios") for the demand $D_1(\omega)$ and price $p_1(\omega)$. Our second stage decisions $x_1(\omega)$ are constrained by the initial inventory decisions we made in the first stage x_0. These two constraints are written as

$$A_1 x_1(\omega) \leq x_0,$$
$$B_1 x_1(\omega) \leq D_1(\omega).$$

Let $\mathcal{X}_1(\omega)$ be the feasible region for $x_1(\omega)$ defined by these constraints. This allows us to write our problem over both stages as

$$\max_{x_0} \left(-c_0 x_0 + \sum_{\omega \in \Omega} p(\omega) \max_{x_1(\omega) \in \mathcal{X}_1(\omega)} (p_1(\omega) - c_1) x_1(\omega) \right). \tag{2.24}$$

In the language of stochastic programming, the second stage decision variables, $x_1(\omega)$, are called "recourse variables" since they represent how we may respond as new information becomes available (which is the definition of "recourse"). Two-stage stochastic programs are basically deterministic optimization problems, but they can be *very large* deterministic optimization problems, albeit ones with special structure.

For example, imagine that we allow the first stage decision x_0 to "see" the information in the second stage, in which case we would write it as $x_0(\omega)$. In this case, we obtain a series of smaller problems, one for each ω. However, now we are allowing x_0 to cheat by seeing into the future. We can overcome this by introducing a *nonanticipativity constraint* which might be written

$$x^0(\omega) - x^0 = 0. \tag{2.25}$$

Now, we have a family of first stage variables $x_0(\omega)$, one for each ω, and then a single variable x_0, where we are trying to force each $x_0(\omega)$ to be the same (at which point we would say that x_0 is "nonanticipative"). Algorithmic specialists can exploit the nonanticipacity constraint (2.25) by relaxing it, then solving a series of smaller problems (perhaps in parallel), and then introducing linking mechanisms so that the overall procedure converges toward a solution that satisfies the nonanticipativity constraint.

We would call the optimization problem in (2.24) (along with the associated constraints for time periods 0 and 1) a stochastic optimization *problem*. In practice, these applications tend to arise in the context of sequential decision problems, where we would be looking for the best decision x_t at time t that considers the uncertain future (call this $t + 1$, although it can be multiple time periods $t + 1, \dots, t + H$), giving us a *policy*

$$X_t^\pi(S_t) = \arg\max_{x_t \in \mathcal{X}_t} \left(-c_t x_t + \sum_{\omega \in \Omega} p_{t+1}(\omega) \max_{x_{t+1}(\omega) \in \mathcal{X}_{t+1}(\omega)} \left((p_{t+1}(\omega) - c_{t+1}) x_{t+1}(\omega) \right) \right). \tag{2.26}$$

The optimization problems in (2.24) and (2.26) are the same, but the goal in solving (2.26) is just to find a decision x_t to implement, after which we roll forward to time $t+1$, update the uncertain future $t+2$, and repeat the process. The decisions $x_{t+1}(\omega)$ for each of the scenarios ω are never actually implemented; we plan them only to help us improve the decision x_t that we are going to implement now. This is a policy for solving an optimization problem which is typically not modeled explicitly. We show how the objective function should be modeled in section 2.2 below.

2.1.9 The Multiarmed Bandit Problem

The classic information acquisition problem is known as the *multiarmed bandit problem* which is a colorful name for our cumulative reward problem introduced in section 2.1.1. This problem has received considerable attention since it was first introduced in the 1950s; the term appears in thousands of papers (per year!).

The bandit story proceeds as follows. Consider the situation faced by a gambler trying to choose which slot machine $x \in \mathcal{X} = \{1, 2, \dots, M\}$ to play. Now assume that the winnings may be different for each machine, but the gambler does not know the winning probabilities. The only way to obtain information is to actually play a slot machine. To formulate this problem, let

$$x^n = \text{the machine we choose tp play next after finishing the}$$
$$n^{th} \text{ trial,}$$
$$W_x^n = \text{winnings from playing slot machine } x = x^{n-1} \text{ during the}$$
$$n^{th} \text{ trial.}$$

We choose what arm to play in the nth trial after finishing the $n - 1^{st}$ trial. We let S^n be the belief state after playing n machines. For example, let

$$\mu_x = \text{a random variable giving the true expected winnings from}$$
$$\text{machine } x,$$
$$\bar{\mu}_x^n = \text{our estimate of the expected value of } \mu_x \text{ after } n \text{ trials,}$$
$$\sigma_x^{2,n} = \text{the variance of our belief about } \mu_x \text{ after } n \text{ trials.}$$

Now assume that our belief about μ is normally distributed (after n trials) with mean $\bar{\mu}_x^n$ and variance $\sigma_x^{2,n}$. We can write our belief state as

$$S^n = (\bar{\mu}_x^n, \sigma_x^{2,n})_{x \in \mathcal{X}}.$$

Our challenge is to find a policy $X^\pi(S^n)$ that determines which machine x^n to play for the $n + 1^{st}$ trial. We have to find a policy that allows us to better learn the true mean values μ_x, which means we are going to have to sometimes play a machine x^n where the estimated reward $\bar{\mu}_x^n$ is not the highest, but where we acknowledge that this estimate may not be accurate. However, we may end up playing a machine whose average reward μ_x actually is lower than the best, which means we are likely to incur lower winnings. The problem is to find the policy that maximizes winnings over time.

One way to state this problem is to maximize expected discounted winnings over an infinite horizon

$$\max_\pi \mathbb{E} \sum_{n=0}^{\infty} \gamma^n W_{x^n}^{n+1},$$

where $x^n = X^\pi(S^n)$ and where $\gamma < 1$ is a discount factor. Of course, we could also pose this as a finite horizon problem (with or without discounting).

An example of a policy that does quite well is known as the interval estimation policy, given by

$$X^{IE,n}(S^n | \theta^{IE}) = \arg \max_{x \in \mathcal{X}} \left(\bar{\mu}_x^n + \theta^{IE} \bar{\sigma}_x^{2,n} \right),$$

where $\bar{\sigma}_x^{2,n}$ is our estimate of the variance of $\bar{\mu}_x^n$, given by

$$\bar{\sigma}_x^{2,n} = \frac{\sigma_x^{2,n}}{N_x^n}.$$

where N_x^n is the number of times we test alternative x over the first n experiments. Our policy is parameterized by θ^{IE} which determines how much weight to put on the uncertainty in the estimate $\bar{\mu}_x^n$. If $\theta^{IE} = 0$, then we have a pure exploitation policy where we are simply choosing the alternative that seems best. As θ^{IE} increases, we put more emphasis on the uncertainty in the estimate. As we are going to see in chapter 7, effective learning policies have to strike a balance between exploring (trying alternatives which are uncertain) and exploiting (doing what appears to be best).

The multiarmed bandit problem is an example of an online learning problem (that is, where we have to learn by doing), where we want to maximize the cumulative rewards. Some examples of these problems are:

■ **EXAMPLE 2.1**

Consider someone who has just moved to a new city and who now has to find the best path to work. Let T_p be a random variable giving the time he will experience if he chooses path p from a predefined set of paths $\mathcal{P}$. The only way he can obtain observations of the travel time is to actually travel the path. Of course, he would like to choose the path with the shortest average time, but it may be necessary to try a longer path because it may be that he simply has a poor estimate.

■ **EXAMPLE 2.2**

A baseball manager is trying to decide which of four players makes the best designated hitter. The only way to estimate how well they hit is to put them in the batting order as the designated hitter.

■ **EXAMPLE 2.3**

A doctor is trying to determine the best blood pressure medication for a patient. Each patient responds differently to each medication, so it is necessary to try a particular medication for a while, and then switch if the doctor feels that better results can be achieved with a different medication.

Multiarmed bandit problems have a long history as a niche problem in applied probability and statistics (going back to the 1950s), computer science (starting in the mid 1980s), and engineering and the geosciences (starting in the 1990s). The bandit community has broadened to consider a much wider range of problems (for example, x could be continuous and/or a vector), and a growing range of policies. We revisit this important problem class in chapter 7 where we are going to argue that so-called "multiarmed bandit problems" are actually just derivative-free stochastic optimization problems, which can be solved with any of our four classes of policies. The difference between bandit problems and the early research into derivative-free stochastic search is that the stochastic search literature did not explicitly recognize the value of active learning: evaluating the function at x just to better learn the approximation, enabling better decisions later.

We note that derivative-free stochastic search is classically approached using a "final reward" objective function (see section 2.1.1), while the multiarmed bandit literature has been centered on cumulative reward objective, but this is not universally true. There is a version of the multiarmed bandit problem known as the "best arm" bandit problem, which uses a final reward objective.

2.1.10 Simulation Optimization

The field known as "simulation optimization" evolved originally from within the simulation community which developed Monte Carlo simulation models for simulating complex systems such as manufacturing processes. Early simulation models in the 1960s were quite slow, and they were often used to search over a series of designs, creating an interest in performing these searches efficiently.

Searching over a finite set of alternatives using noisy evaluations is an example of ranking and selection (a form of derivative free stochastic search), but these applications fostered a group of researchers within the simulation community. One of the first methodological innovations from this community was an algorithm called *optimal computing budget allocation*, or OCBA.

The general idea of an OCBA algorithm proceeds by taking an initial sample $N_x^0 = n_0$ of each alternative $x \in \mathcal{X}$, which means we use $B^0 = Mn_0$ experiments from our budget B. The algorithm then uses rules for determining how to allocate its computing budget among the different alternatives. A more detailed summary of a typical OCBA algorithm is given in section 7.10.2.

For a number of years, OCBA was closely associated with "simulation optimization," but the community has continued to evolve, tackling a wider range of problems and creating new methods to meet new challenges. Inevitably there was also some cross-over from other communities. However, similar to other communities, the scope of activities under the umbrella of "simulation optimization" has continued to broaden, encompassing other results from stochastic search (both derivative-free and derivative-based), as well as the tools for sequential decision problems such as approximate dynamic programming and reinforcement learning. Today, the simulation-optimization community would classify any search method based on Monte Carlo sampling as a form of "simulation optimization."

2.1.11 Active Learning

Classical (batch) machine learning addresses the problem of fitting a model $f(x|\theta)$ given a dataset (x^n, y^n), $n = 1, \dots, N$ to minimize some error (or loss) function $L(x, y)$. Online learning addresses the setting of fitting the model as the data is arriving in a stream. Given an estimate $\bar{\theta}^n$ based on the first n datapoints, find $\bar{\theta}^{n+1}$ given (x^{n+1}, y^{n+1}). We assume we have no control over the inputs x^n.

Active learning arises when we have partial or complete control over the inputs x^n. It might be a price, size, or concentration that we completely control. Or, we might have partial control, as occurs when we choose a treatment for a patient, but cannot control the attributes of the patient.

There are many approaches to active learning, but a popular one is to make choices where there is the greatest uncertainty. For example, imagine that we have binary outcomes (a customer does or does not purchase a product at a price x). Let x be the attributes of the customer, and let $\bar{p}(x)$ be the probability that this customer will purchase the product. We know the attributes of the customer from their login credentials. The variance of the response is given by $\bar{p}^n(x)(1 - \bar{p}^n(x))$. To minimize the variance, we would want to make an offer to a customer with attribute x that has the greatest uncertainty given by the variance $\bar{p}^n(x)(1 - \bar{p}^n(x))$. This means we would choose the x that solves

$$\max_x \bar{p}^n(x)(1 - \bar{p}^n(x)).$$

This is a very simple example of active learning.

The relationship between bandit problems and active learning is quite close. As of this writing, the term "active learning" has been increasingly replacing the artificial "multiarmed bandit problem."

2.1.12 Chance-constrained Programming

There are problems where we have to satisfy a constraint that depends on uncertain information at the time we make a decision. For example, we may wish to allocate inventory with the goal that we cover demand 80% of the time. Alternatively, we may wish to schedule a flight so that it is on time 90% of the time. We can state these problems using the general form

$$\min_x f(x),$$ (2.27)

subject to the probabilistic constraint (often referred to as a chance constraint)

$$\mathbb{P}[C(x, W) \geq 0] \leq \alpha,$$ (2.28)

where $0 \leq \alpha \leq 1$. The constraint (2.28) is often written in the equivalent form

$$\mathbb{P}[C(x, W) \leq 0] \geq 1 - \alpha.$$ (2.29)

Here, $C(x, W)$ is the amount that a constraint is violated (if positive). Using our examples, it might be the demand minus the inventory which is the lost demand if positive, or the covered demand if negative. Or, it could be the arrival time of a plane minus the scheduled time, where positive means a late arrival.

Chance-constrained programming is a method for handling a particular class of constraints that involve uncertainty, typically in the setting of a static problem: make decision, see information, stop. Chance-constrained programs convert these problems into deterministic, nonlinear programs, with the challenge of computing the probabilistic constraint within the search algorithm.

2.1.13 Model Predictive Control

There are many settings where we need to think about what is going to happen in the future in order to make a decision now. An example most familiar to all of us is the use of navigation systems that plan a path to the destination using estimated travel times on each link of the network. As we progress, these times may change and the path will be updated.

Making decisions now by optimizing (in some way) over the future is known in the optimal control literature as *model predictive control*, because we are using a (typically approximate) model of the future to make a decision now. An example of an MPC policy is

$$U^{\pi}(x_t) = \arg\min_{u_t} \left(L(x_t, u_t) + \min_{u_{t+1}, \ldots, u_{t+H}} \sum_{t'=t}^{t+H} L(x_{t'}, u_{t'}) \right)$$

$$= \arg\min_{u_t, \ldots, u_{t+H}} \sum_{t'=t}^{t+H} L(x_{t'}, u_{t'}).$$ (2.30)

The optimization problem in (2.30) requires a model over the horizon $t, \ldots, t + H$, which means we need to be able to model losses as well as the system dynamics using $x_{t+1} = f(x_t, u_t)$. A slightly more precise name for this might be "model-based predictive control," but "model predictive control" (or MPC, as it is often known) is the term that evolved in the controls community.

Model predictive control is a widely used idea, often under names such as "rolling horizon procedure" or "receding horizon procedure." Model predictive control is most often written using a deterministic model of the future, primarily because most control problems are deterministic. However, the proper use of the term refers to *any* model of the future (even an approximation) that is used to make a decision now. The two-stage stochastic programming model in section 2.1.8 is a form of model predictive control which uses a stochastic model of the future. We could even solve a full dynamic program, which is typically done when we solve an approximate stochastic model of the future. All of these are forms of "model predictive control." In this book we refer to this approach as a class of policy called "direct lookahead approximations" which we cover in chapter 19.

2.1.14 Robust Optimization

The term "robust optimization" has been applied to classical stochastic optimization problems (in particular stochastic programming), but in the mid-1990s, it became associated with problems where we need to make a decision, such as the design of a device or structure, that works under the *worst* possible settings of the uncontrollable parameters. Examples where robust optimization might arise are

---

■ **EXAMPLE 2.4**

A structural engineer has to design a tall building that minimizes cost (which might involve minimizing materials) so that it can withstand the worst storm conditions in terms of wind speed and direction.

■ **EXAMPLE 2.5**

An engineer designing wings for a large passenger jet wishes to minimize the weight of the wing, but the wing still has to withstand the stresses under the worst possible conditions.

The classical notation used in the robust optimization community is to let u be the uncertain parameters. In this book we use w, and assume that w falls within an *uncertainty set* W. The set W is designed to capture the random outcomes with some level of confidence that we can parameterize with θ, so we are going to write the uncertainty set as $W(\theta)$.

The robust optimization problem is stated as

$$\min_{x \in X} \max_{w \in W(\theta)} F(x, w). \tag{2.31}$$

Creating the uncertainty set $W(\theta)$ can be a difficult challenge. For example, if w is a vector with element w_i, one way to formulate $W(\theta)$ is the box:

$$W(\theta) = \{w | \theta_i^{lower} \leq w_i \leq \theta_i^{upper}, \forall i\},$$

where $\theta = (\theta^{lower}, \theta^{upper})$ are tunable parameters that govern the creation of the uncertainty set.

The problem is that the worst outcome in $W(\theta)$ is likely to be one of the corners of the box, where all the elements w_i are at their upper or lower bound. In practice, this is likely to be an extremely rare event. A more realistic uncertainty set captures the likelihood that a vector w may happen. There is considerable research in robust optimization focused on creating the uncertainty set $W(\theta)$.

We note that just as we formulated a two-stage stochastic programming *problem* in equation (2.24), and then pointed out that this was really a lookahead *policy* (see equation (2.26)), our robust optimization *problem* given by (2.31) can be written as a robust optimization *policy* if we write it as

$$X^{RO}(S_t) = \arg\min_{x_t \in X_t} \max_{w_{t+1} \in W_{t+1}(\theta)} F(x_t, w_{t+1}). \tag{2.32}$$

A number of papers in the robust optimization literature are doing precisely this: they formulate a robust optimization problem at time t, and then use it to make a decision x_t, after which they step forward, observe new information W_{t+1}, and repeat the process. This means that their robust optimization *problem* is actually a form of lookahead *policy*.

2.2 A Universal Modeling Framework for Sequential Decision Problems

Now that we have covered most of the major communities dealing with sequential decisions under uncertainty, it is useful to review the elements of all sequential decision problems. We are going to revisit this topic in considerably greater depth in chapter 9, but this discussion provides an introduction and a chance to compare our framework to those we reviewed above.

Our presentation focuses on sequential decision problems under uncertainty, which means new information arrives after each decision is made, but we can always ignore the new information to create a problem comparable to the deterministic control problem in section 2.1.4. We are going to assume our problem evolves over time, but there are many settings where it is more natural to use a counter (the nth experiment, the nth customer).

2.2.1   Our Universal Model for Sequential Decision Problems

These problems consist of the following elements:

The state variable – S_t – This captures all the information we need to model the system from time t onward, which means computing the cost/contribution function, constraints on decisions, and any other variables needed to model the transition of this information over time. The state S_t may consist of the physical resources R_t (such as inventories), other deterministic information I_t (price of a product, weather), and the belief state B_t which captures information about a probability distribution describing parameters or quantities that cannot be directly (and perfectly) observed. It is important to recognize that the state variable, regardless of whether it is describing physical resources, attributes of a system, or the parameters of a probability distribution, is always a form of information.

The decision variable – x_t – Decisions (which might be called actions a_t or controls u_t) represent how we control the process. Decisions are determined by decision functions known as *policies*, also known as control laws in control theory. If our decision is x_t, we will designate our policy by $X^\pi(S_t)$. Similarly, if we wish to use a_t or u_t as our decision variable, we would use $A^\pi(S_t)$ or $U^\pi(S_t)$ as our policy. If $\mathcal{X}_t$ is our feasible region (which depends on information in S_t), we assume that $X^\pi(S_t) \in \mathcal{X}_t$.

Exogenous information – W_{t+1} – This is the information that first becomes known at time $t + 1$ from an exogenous source (for example, the demand for product, the speed of the wind, the outcome of a medical treatment, the results of a laboratory experiment). W_{t+1} can be a high dimensional vector of prices (for all the different stocks) or demands for products.

The transition function – This function determines how the system evolves from the state S_t to the state S_{t+1} given the decision that was made at time t and the new information that arrived between t and $t + 1$. We designate the transition function (also known as the system model or the state transition model) by

$$S_{t+1} = S^M(S_t, x_t, W_{t+1}).$$

Note that W_{t+1} is a random variable when we make the decision x_t. Throughout, we assume that any variable indexed by t (or n) is known at time t (or after n observations).

The objective function – This function specifies the costs being minimized, the contributions/rewards being maximized, or other performance metrics. Let $C(S_t, x_t)$ be the contribution we are maximizing given the decision x_t, and given the information in S_t which may contain costs, prices, and information for constraints. A basic form of objective function might be given by

$$F^\pi(S_0) = \mathbb{E}_{S_0} \mathbb{E}_{W_1,\dots,W_T|S_0} \left\{ \sum_{t=0}^{T} C(S_t, X^\pi(S_t)) \right\}. \tag{2.33}$$

Our goal would be to find the policy that solves

$$\max_\pi F^\pi(S_0). \tag{2.34}$$

In chapters 7 and 9 we will illustrate a number of other forms of objectives.

If we are using a counter, we would represent the state by S^n, decisions by x^n, and the exogenous information by W^{n+1}. There are some problems where we need to index by both time (such as the hour within a week) and a counter (such as the nth week). We would do this using S_t^n.

We now illustrate this framework using an asset acquisition problem:

Narrative – Our asset acquisition problem involves maintaining an inventory of some resource (cash in a mutual fund, spare engines for aircraft, vaccines, ...) to meet random demands over time. We assume that purchase costs and sales prices also vary over time.

State variables – The state variable is the information we need to make a decision and compute functions that determine how the system evolves into the future. In our asset acquisition problem, we need three pieces of information. The first is R_t, the resources on hand before we make any decisions (including how much of the demand to satisfy). The second is the demand itself, denoted D_t, and the third is the price p_t. We would write our state variable as $S_t = (R_t, D_t, p_t)$.

Decision variables – We have two decisions to make. The first, denoted x_t^D, is how much of the demand D_t during time interval t that should be satisfied using available assets, which means that we require $x_t^D \leq R_t$. The second, denoted x_t^O, is how many new assets should be acquired at time t which can be used to satisfy demands during time interval $t + 1$.

Exogenous information – The exogenous information process consists of three types of information. The first is the new demands that arise between t and $t + 1$, denoted $\hat{D}_{t+1}$. The second is the change between t and $t + 1$ in the price at which we can sell our assets, denoted $\hat{p}_{t+1}$. Finally, we are going to assume that there may be exogenous changes to our available resources. These might be blood donations or cash deposits (producing positive changes), or equipment failures and cash withdrawals (producing negative changes). We denote these changes by $\hat{R}_{t+1}$. We let W_{t+1} represent all the new information that is first learned between t and $t+1$ (that is, after decision x_t is made), which for our problem would be written $W_{t+1} = (\hat{R}_{t+1}, \hat{D}_{t+1}, \hat{p}_{t+1})$.

In addition to specifying the types of exogenous information, for stochastic models we also have to specify the likelihood of a particular outcome. This might come in the form of an assumed probability distribution for $\hat{R}_{t+1}, \hat{D}_{t+1}$, and $\hat{p}_{t+1}$, or we may depend on an exogenous source for sample realizations (the actual price of the stock or the actual travel time on a path).

Transition function – The evolution of the state variables S_t is described using

$$S_{t+1} = S^M(S_t, x_t, W_{t+1}),$$

where

$$
\begin{aligned}
R_{t+1} &= R_t - x_t^D + x_t^O + \hat{R}_{t+1}, \\
D_{t+1} &= D_t - x_t^D + \hat{D}_{t+1}, \\
p_{t+1} &= p_t + \hat{p}_{t+1}.
\end{aligned}
$$

This model assumes that unsatisfied demands are held until the next time period.

Objective function – We compute our contribution $C_t(S_t, x_t)$ which might depend on our current state and the action x_t that we take at time t. For our asset acquisition problem (where the state variable is R_t), the contribution function is

$$C_t(S_t, x_t) = p_t x_t^D - c_t x_t^O.$$

In this particular model, $C_t(S_t, x_t)$ is a deterministic function of the state and action. In other applications, the contribution from action x_t depends on what happens during time $t+1$.

Our objective function is given by

$$\max_{\pi \in \Pi} \mathbb{E}\left\{ \sum_{t=0}^{T} C_t(S_t, X^{\pi}(S_t)) | S_0 \right\}.$$

Designing policies will occupy most of the rest of this volume. For an inventory problem such as this, we might use simple rules, or more complex lookahead policies, where we may look into the future with a point forecast, or while capturing the uncertainty of the future.

Chapter 9 is an entire chapter dedicated to filling in details of this basic modeling framework. When modeling a real problem, we encourage readers to describe each of these five elements in this order.

2.2.2 A Compact Modeling Presentation

For readers looking for a more compact way of writing a sequential decision problem that is perhaps more in the style of a classical deterministic math program, we suggest writing it as

$$
\max_{\pi} \mathbb{E}_{S_0} \mathbb{E}_{W_1,\ldots,W_T | S_0} \left\{ \sum_{t=0}^{T} C(S_t, X^{\pi}(S_t)) \right\}, \tag{2.35}
$$

where we assume that the policy is designed to satisfy the constraints

$$
x_t = X^{\pi}(S_t) \in \mathcal{X}_t. \tag{2.36}
$$

The transition function is given by

$$
S_{t+1} = S^M(S_t, X^{\pi}(S_t), W_{t+1}), \tag{2.37}
$$

and where we are given an exogenous information process

$$
(S_0, W_1, W_2, \ldots, W_T). \tag{2.38}
$$

Of course, this still leaves the problem of describing how to sample the exogenous information process, and how to design the policy. However, we do not feel that we need to say anything about the policy, any more than we need to say something about the decision x in a deterministic math program.

2.2.3 MDP/RL vs. Optimal Control Modeling Frameworks

It helps to pause and ask a natural question: of all the fields listed in section 2.1, do any of them match our universal framework? The answer is that there is one that comes close: optimal control (section 2.1.4).

Before describing the strengths of the optimal control modeling framework, we think it helps to start with the modeling framework that has been adopted by the reinforcement learning community, which is the most popular of all of these fields (as of the writing of this book). From its origins in the 1980s, the RL community adopted the modeling framework long used for Markov decision processes, which we presented in section 2.1.3. This framework may be mathematically elegant, but it is extremely clumsy in terms of modeling actual problems. For example, we learn nothing about a problem by defining some "state space" $\mathcal{S}$ or "action space" $\mathcal{A}$. In addition, the one-step transition matrix $P(s'|s, a)$ is almost never computable. Finally, while it is nice to specify the single-period reward function, the real problem is to sum the rewards and optimize over policies.

Now let's contrast this style with that used in optimal control. In this field, we specify state *variables* and decision/control *variables*. It is the field of optimal control that introduced the powerful construct of a transition *function*, which can seem so obvious, and yet is largely ignored by the other communities. The optimal control literature focuses predominantly on deterministic problems, but there are stochastic control problems, most often using the additive noise of equation (2.15).

The optimal control community does not use our standard format of optimizing over policies. Yet, this community has aggressively developed different classes of policies. We observe that the optimal control literature first introduced "linear control laws" (because they are optimal for linear quadratic regulation problems). It was the first to use value function approximations under a variety of names including heuristic dynamic programming, neuro-dynamic programming, and approximate/adaptive dynamic programming. Finally, it introduced (deterministic) lookahead policies (known as "model predictive control"). This spans three of our four classes of policies (PFAs, VFAs, and DLAs). We suspect that someone has used the idea of parameterized optimization models for policies (what we call CFAs), but since this strategy has not been recognized as a formal methodology, it is difficult to know if and when it has been first used.

All of the fields in section 2.1 suffer from the habit of tying a modeling framework to a solution approach. Optimal control, along with dynamic programming, assumes that the starting point is Bellman's equation (known as Hamilton-Jacobi equations in the controls community). This is our major point of departure with all of the fields listed above. In our universal modeling framework, none of the five elements provides any indication of how to design policies. Instead, we end with an objective function (equations (2.33)–(2.34)) where we state that our objective is to find an optimal policy. We defer until later the search over the four classes of policies which we first introduced in section 1.4.1, and will revisit throughout the book.

2.3 Applications

We now illustrate our modeling framework using a series of applications. These problems illustrate some of the modeling issues that can arise in actual applications. We often start from a simpler problem, and then show how details can be added. Pay attention to the growth in the dimensionality of the state variable as these complications are introduced.

2.3.1 The Newsvendor Problems

A popular problem in operations research is known as the newsvendor problem, which is described as the story of deciding how many newspapers to put out for sale to meet an unknown demand. The newsvendor problem arises in many settings where we have to choose a fixed parameter that is then evaluated in a stochastic setting. It often arises as a subproblem in a wide range of resource allocation problems (managing blood inventories, budgeting for emergencies, allocating fleets of vehicles, hiring people). It also arises in other settings, such as bidding a price for a contract (bidding too high means you may lose the contract), or allowing extra time for a trip.

The newsvendor problem is classically presented as a static final reward formulation, but we are going to keep an open mind regarding final-reward and cumulative-reward formulations.

2.3.1.1 Basic Newsvendor – Final Reward

The basic newsvendor is modeled as

$$F(x, W) = p \min\{x, W\} - cx, \tag{2.39}$$

where x is the number of "newspapers" we have to order before observing our random "demand" W. We sell our newspapers at a price p (the smaller of x and W), but we have to buy all of them at a unit cost c. The goal is to solve the problem

$$\max_x \mathbb{E}_W F(x, W). \tag{2.40}$$

In most cases, the newsvendor problem arises in settings where we can observe W, but we do not know its distribution (this is often referred to as "data driven"). When this is the case, we assume that we have to determine the amount to order x^n at the end of day n, after which we observe demand W^{n+1}, giving us a profit (at the end of day $n + 1$) of

$$\hat{F}^{n+1} = F(x^n, W^{n+1}) = p \min\{x^n, W^{n+1}\} - cx^n.$$

After each iteration, we may assume we observe W^{n+1}, although often we only observe $\min(x^n, W^{n+1})$ (which is known as censored observations) or perhaps just the realized profit

$$\hat{F}^{n+1} \;=\; p\min\{x^n, W^{n+1}\} - cx^n.$$

We can devise strategies to try to learn the distribution of W, and then use our ability to solve the problem optimally (given in exercise 4.12).

Another approach is to try to learn the function $\mathbb{E}_W F(x, W)$ directly. Either way, let S^n be our belief state (about W, or about $\mathbb{E}_W F(x, W)$) about our unknown quantities. S^n might be a point estimate, but it is often a probability distribution. For example, we might let $\mu_x = \mathbb{E}F(x, W)$ where we assume that x is discrete (say, the number of newspapers). After n iterations, we might have estimates $\bar{\mu}_x^n$ of $\mathbb{E}F(x, W)$, with standard deviation $\bar{\sigma}_x^n$ where we would then assume that $\mu_x \sim N(\bar{\mu}_x^n, \bar{\sigma}_x^{n,2})$. In this case, we would write $S^n = (\bar{\mu}^n, \bar{\sigma}^n)$ where $\bar{\mu}^n$ and $\bar{\sigma}^n$ are both vectors over all values of x.

Given our (belief) state S^n, we then have to define a policy (we might also call this a rule, or it might be a form of algorithm) that we denote by $X^\pi(S^n)$ where $x^n = X^\pi(S^n)$ is the decision we are going to use in our next trial where we either observe W^{n+1} or $\hat{F}^{n+1}$. While we would like to run this policy until $n \to \infty$, in practice we are going to be limited to N trials which then gives us a solution $x^{\pi,N}$. This solution depends on our initial state S^0, the observations $W^1, \dots, W^N$ which occurred while we were finding $x^{\pi,N}$, and then we observe $\widehat{W}$ to evaluate $x^{\pi,N}$. We want to find the policy that solves

$$\max_\pi \mathbb{E}_{S^0} \mathbb{E}_{W^1,\dots,W^N | S^0} \mathbb{E}_{\widehat{W} | S^0} F(x^{\pi,N}, \widehat{W}). \tag{2.41}$$

2.3.1.2 Basic Newsvendor – Cumulative Reward

A more realistic presentation of an actual newsvendor problem recognizes that we are accumulating profits while simultaneously learning about the demand W (or the function $\mathbb{E}_W F(x, W)$). If this is the case, then we would want to find a policy that solves

$$\max_\pi \mathbb{E}_{S_0} \mathbb{E}_{W_1,\dots,W_T | S_0} \sum_{t=0}^{T-1} F(X^\pi(S_t), W_{t+1}). \tag{2.42}$$

The cumulative reward formulation of the newsvendor problem, which captures the active learning process, appears to be new, despite being the most natural model of an actual newsvendor problem.

2.3.1.3 Contextual Newsvendor

Imagine a newsvendor problem where the price p of our product is dynamic, given by p_t, which is revealed before we have to make a decision. Our profit would be given by

$$F(x, W | S_t) = p_t \min\{x, W\} - cx. \tag{2.43}$$

As before, assume that we do not know the distribution of W, and let B_t be the state of our belief about W (or about $\mathbb{E}F(x, W)$). Our state $S_t = (p_t, B_t)$, since we have to capture both the price p_t and our state of belief B_t. We can write our problem now as

$$\max_x \mathbb{E}_W F(x, W | S_t).$$

Now, instead of finding the optimal order quantity x^*, we have to find the optimal order quantity as a function of the state S_t, which we might write as $x^*(S_t)$. While x^* is a deterministic value, $x^*(S)$ is a function of the state which represents the "context" for the decision x^*.

As we see, the "context" (a popular term in the learning community) is really just a state variable, and $x^*(S)$ is a form of policy. Finding an optimal policy will always be hard, but finding a practical, implementable policy simply involves the exercise of going through each of the four classes of policies to find one that seems promising.

2.3.1.4 Multidimensional Newsvendor Problems

Newsvendor problems can be multidimensional. One version is the *additive newsvendor problem* where there are K products to serve K demands, but using a production process that limits the total amount delivered. This would be formulated as

$$F(x_1, \dots, x_K) = E_{W_1, \dots, W_K} \sum_{k=1}^{K} p_k \min(x_k, W_k) - c_k x_k, \tag{2.44}$$

where

$$\sum_{k=1}^{K} x_k \leq U. \tag{2.45}$$

A second version arises when there are multiple products (different types/-colors of cars) trying to satisfy the same demand W. This is given by

$$F(x_1, \dots, x_K) = \mathbb{E}_W \left\{ \sum_{k=1}^{K} p_k \min \left[x_k, \left(W - \sum_{\ell=1}^{k-1} x_\ell \right)^+ \right] - \sum_{k=1}^{K} c_k x_k \right\}, \tag{2.46}$$

where $(Z)^+ = \max(0, Z)$.

2.3.2 Inventory/Storage Problems

Inventory (or storage) problems represent an astonishingly broad class of applications that span any problem where we buy/acquire (or sell) a resource to meet a demand, where excess inventory can be held to the next time period. Elementary inventory problems (with discrete quantities) appear to be the first problem to illustrate the power of a compact state space, which overcomes the exponential explosion that occurs if you try to formulate and solve these problems as decision trees. However, these elementary problems become complicated very quickly as we move into real applications.

2.3.2.1 Inventory Without Lags

The simplest problem allows us to order new product x_t at time t that arrives right away. We begin by defining the notation:

$$R_t \quad = \quad \text{Amount of inventory left over at the end of period } t.$$

$$x_t \quad = \quad \text{Amount ordered at the end of period } t \text{ that will be available at the beginning of time period } t.$$

$$\hat{D}_{t+1} \quad = \quad \text{Demand for the product that arises between } t \text{ and } t+1.$$

$$c_t \quad = \quad \text{The unit cost of order product for product ordered at time } t.$$

$$p_t \quad = \quad \text{The price we are paid when we sell a unit during the period } (t, t+1).$$

Our basic inventory process is given by

$$R_{t+1} = \max\{0, R_t + x_t - \hat{D}_{t+1}\}.$$

We add up our total contribution at the end of each period. Let y_t be the sales during time period $(t-1, t)$. Our sales are limited by the demand $\hat{D}_t$ as well as our available product $R_{t-1} + x_{t-1}$, but we are going to allow ourselves to choose how much to sell, which may be smaller than either of these. So we would write

$$y_t \quad \leq \quad R_{t-1} + x_{t-1},$$
$$y_t \quad \leq \quad \hat{D}_t.$$

We are going to assume that we determine y_t at time t after we have learned the demands D_t for the preceding time period. So, at time t, the revenues and costs are given by

$$C_t(x_t, y_t) = p_t y_t - c_t x_t.$$

If this were a deterministic problem, we would formulate it as

$$\max_{(x_t, y_t), t=0,\ldots,T} \sum_{t=0}^{T} (p_t y_t - c_t x_t).$$

However, we often want to represent the demands $\hat{D}_{t+1}$ as being random at time t. We might want to allow our prices p_t, and perhaps even our costs c_t, to vary over time with both predictable (e.g. seasonal) and stochastic (uncertain) patterns. In this case, we are going to need to define a state variable S_t that captures what we know at time t before we make our decisions x_t and y_t. Designing state variables is subtle, but for now we would assume that it would include R_t, p_t, c_t, as well as the demands D_{t+1} that have arisen during interval $(t, t+1)$.

Unlike the newsvendor problem, the inventory problem can be challenging even if the distribution of demand D_t is known. However, if it is unknown, then we may need to maintain a belief state B_t about the distribution of demand, or perhaps the expected profits when we place an order x_t.

The features of this problem allow us to create a family of problems:

Static data – If the prices p_t and costs c_t are constant (which is to say that $p_t = p$ and $c_t = c$), with a known distribution of demand, then we have a stochastic optimization problem where the state is just $S_t = R_t$.

Dynamic data – Assume the price p_t evolves randomly over time, where $p_{t+1} = p_t + \varepsilon_{t+1}$, then our state variable is $S_t = (R_t, p_t)$.

History-dependent processes – Imagine now that our price process evolves according to

$$p_{t+1} = \theta_0 p_t + \theta_1 p_{t-1} + \theta_2 p_{t-2} + \varepsilon_{t+1},$$

then we would write the sate as $S_t = \left(R_t, (p_t, p_{t-1}, p_{t-2}) \right)$.

Learning process – Now assume that we do not know the distribution of the demand. We might put in place a process to try to learn it, either from observations of demands or sales. Let B_t capture our belief about the distribution of demand, which may itself be a probability distribution. In this case, our state variable would be $S_t = (R_t, p_t, B_t)$.

Let $Y^\pi(S_t)$ be the selling policy we use to determine y_t, and let $X^\pi(S_t)$ be the buying policy we use for determining x_t, where π carries the parameters that determine both policies. We would write our objective function as

$$\max_{\pi} \mathbb{E} \sum_{t=0}^{T} (p_t Y^\pi(S_t) - c_t X^\pi(S_t)).$$

Inventory problems are quite rich. This is a problem where it is quite easy to create variations that can be solved with each of the four classes of policies introduced in section 1.4. We describe these four classes of policies in much more depth in chapter 11. In section 11.9, we illustrate an inventory problem that arises in energy storage where each of the four classes of policies may work best.

2.3.2.2 Inventory Planning with Forecasts

An important extension that arises in many real applications is where the data (demands, prices, even costs) may follow time-varying patterns which can be approximately forecasted. Let

$$f^W_{tt'} \quad = \quad \text{forecast of some activity (demands, prices, costs) made at time}$$
$$t \text{ that we think will happen at time } t'.$$

Forecasts evolve over time. They may be given to us from an exogenous source (a forecasting vendor), or we may use observed data to do our own updating of forecasts. Assuming they are provided by an external vendor, we might describe the evolution of forecasts using

$$f^W_{t+1,t'} = f^W_{tt'} + \hat{f}^W_{t+1,t'},$$

where $\hat{f}^W_{t+1,t'}$ is the (random) change in the forecasts over all future time periods t'.

When we have forecasts, the vector $f^W_t = (f^W_{tt'})_{t' \geq t}$ technically becomes part of the state variable. When forecasts are available, the standard approach is to treat these as latent variables, which means that we do not explicitly model the evolution of the forecasts, but rather just treat the forecast as a static vector. We will return to this in chapter 9, and describe a strategy for handling rolling forecasts in chapter 13.

2.3.2.3 Lagged Decisions

There are many applications where we make a decision at time t (say, ordering new inventory) that does not arrive until time t' (as a result of shipping delays). In global logistics, these lags can extend for several months. For an airline ordering new aircraft, the lags can span several years.

We can represent lags using the notation

$$x_{tt'} \quad = \quad \text{inventory ordered at time } t \text{ to arrive at time } t',$$
$$R_{tt'} \quad = \quad \text{inventory that has been ordered at some time before } t \text{ that is}$$
$$\text{going to arrive at time } t'.$$

The variable $R_{tt'}$ is how we capture the effect of previous decisions. We can roll these variables up into the vectors $x_t = (x_{tt'})_{t' \geq t}$ and $R_t = (R_{tt'})_{t' \geq t}$.

Lagged problems are particularly difficult to model. Imagine that we want to sign contracts to purchase natural gas in month t'' that might be three years into the future to serve uncertain demands. This decision has to consider the possibility that we may place an order $x_{t't''}$ at a time t' that is between now (time t) and time t''. At time t, the decision $x_{t't''}$ is a random variable that depends not just on the price of natural gas at time t', but also the decisions we might make between t and t', as well as evolving forecasts.

2.3.3 Shortest Path Problems

Shortest path problems represent a particularly elegant and powerful problem class, since a node in the network can represent any discrete state, while links out of the node can represent a discrete action.

2.3.3.1 A Deterministic Shortest Path Problem
A classical sequential decision problem is the shortest path problem. Let

$$\mathcal{J} = \text{the set of nodes (intersections) in the network,}$$

$$\mathcal{L} = \text{the set of links } (i, j) \text{ in the network,}$$

$$c_{ij} = \text{the cost (typically the time) to drive from node } i \text{ to node } j, i, j \in \mathcal{J}, (i, j) \in \mathcal{L},$$

$$\mathcal{J}_i^+ = \text{the set of nodes } j \text{ for which there is a link } (i, j) \in \mathcal{L},$$

$$\mathcal{J}_j^- = \text{the set of nodes } i \text{ for which there is a link } (i, j) \in \mathcal{L}.$$

A traveler at node i needs to choose the link (i, j) where $j \in \mathcal{J}_i^+$ is a downstream node from node i. Assume that the traveler needs to get from an origin node q to a destination node r at least cost. Let

$$v_j = \text{the minimum cost required to get from node } j \text{ to node } r.$$

We can think of v_j as the value of being in state j. At optimality, these values will satisfy

$$v_i = \min_{j \in \mathcal{J}_i^+} (c_{ij} + v_j).$$

This fundamental equation underlies all the shortest path algorithms used in navigation systems, although these have been heavily engineered to achieve the rapid response we have become accustomed to. A basic shortest path algorithm is given in Figure 2.4, although this represents just the skeleton of what a real algorithm would look like.

Step 0. Let

$$v_j^0 = \begin{cases} M & j \neq r, \\ 0 & j = r \end{cases}$$

where "M" is known as "big-M" and represents a large number. Let $n = 1$.

Step 1. Solve for all $i \in \mathcal{J}$,

$$v_i^n = \min_{j \in \mathcal{J}_i^+} \left(c_{ij} + v_j^{n-1} \right).$$

Step 2. If $v_i^n < v_i^{n-1}$ for any i, let $n = n + 1$ and return to step 1. Else stop.

Figure 2.4 A basic shortest path algorithm.

2.3.3.2 A Stochastic Shortest Path Problem

We are often interested in shortest path problems where there is uncertainty in the cost of traversing a link. For our transportation example, it is natural to view the travel time on a link as random, reflecting the variability in traffic conditions on each link.

To handle this new dimension correctly, we have to specify whether we see the outcome of the random cost on a link before or after we make the decision whether to traverse the link. If the actual cost is only realized after we traverse the link, then our decision at node x_i that we made when we are at node i would be written

$$x_i = \arg\min_{j \in \mathcal{J}_i^+} \mathbb{E}\left(\hat{c}_{ij} + v_j \right),$$

where the expectation is over the (assumed known) distribution of the random cost $\hat{c}_{ij}$. For this problem, our state variable S is simply the node at which we are located.

If we get to make our decision after we learn $\hat{c}_{ij}$, then our decision would be written

$$x_i = \arg\min_{j \in \mathcal{J}_i^+} \left(\hat{c}_{ij} + v_j \right).$$

In this setting, the state variable S is given by $S = (i, (\hat{c}_{ij})_j)$ includes both our current node, but also the costs on links emanating from node i.

2.3.3.3 A Dynamic Shortest Path Problem

Now imagine the problem being solved by any online navigation system which gets live information from the network, and updates the shortest path periodically. Assume at time t that the navigation system has estimates $\bar{c}_{tij}$ of the cost of traversing link $(i, j) \in \mathcal{L}$ where $\mathcal{L}$ is the set of all the links in the network. The system uses these estimates to solve a deterministic shortest path problem which recommends what to do right now.

Assume that the vector of estimated costs $\bar{c}_t$ is updated each time period (perhaps this is every 5 minutes), so at time $t+1$ we are given the vector of estimates $\bar{c}_{t+1}$. Let N_t be the node where the traveler is located (or is heading inbound to). The state variable is now

$$S_t = (N_t, \bar{c}_t).$$

Remembering that there is an element of $\bar{c}_t$ for each link in the network, our state variable S_t has dimensionality $|\mathcal{L}| + 1$. In chapter 19 we will describe how it is that we can solve such a complex problem using simple shortest path calculations.

2.3.3.4 A Robust Shortest Path Problem

We know that costs c_{ij} are uncertain. The navigation services can use their observations to build probability distributions for $\bar{c}_{tij}$ for the estimates of the travel times given what we know at time t. Now, imagine that, rather than taking an average, we use the θ-percentile, which we represent by $\bar{c}_{tij}(\theta)$. So, if we set $\theta = 0.90$, we would be using the 90th percentile travel time, which would discourage using links that can become highly congested.

Now let $\ell_t^\pi(\theta) \in \mathcal{L}$ be the link that is recommended when we are in state $S_t = (N_t, \bar{c}_t(\theta))$ and choose a direction by solving a deterministic shortest path problem using the link costs $\bar{c}_t(\theta)$. Let $\hat{c}_{t,\ell_t^\pi(\theta)}$ be the actual cost the traveler experiences traversing link $\ell_t^\pi(\theta) = (i, j) \in \mathcal{L}$ at time t. The problem is now to optimize across this class of policies by solving

$$\min_\theta \mathbb{E}\left\{ \sum_t \hat{c}_{t,\ell_t^\pi(\theta)} | S_0 \right\},$$

where S_0 captures the starting point of the vehicle and initial estimates of the costs. We discuss this strategy in further depth in chapter 19.

2.3.4 Some Fleet Management Problems

Fleet management problems, such as those that arise with ride hailing fleets, represent a special class of resource allocation problem. In this section we start by describing the problem faced by a single truck driver we call the "nomadic trucker," and then show how to extend the basic idea to fleets of trucks.

2.3.4.1 The Nomadic Trucker

The nomadic trucker is a problem where a single truck driver will pick up a load at A, drive it from A to B, drop it off at B, and then has to look for a new load (there are places to call in to get a list of available loads). The driver has to think about how much money he will make moving the load, but he then also has to recognize that the load will move him to a new city. His problem is to choose from a set of loads out of his location at A.

The driver is characterized at each point in time by his current or future location ℓ_t (which is a region of the country), his equipment type E_t which is the type of trailer he is pulling (which can change depending on the needs of the freight), his estimated time of arrival at ℓ_t (denoted by τ_t^{eta}), and the time τ_t^{home} that he has been away from his home. We roll these attributes into an attribute vector a_t given by

$$a_t = (\ell_t, E_t, \tau_t^{eta}, \tau_t^{home}).$$

When the driver arrives at the destination of a load, he calls a freight broker and gets a set of loads $\mathcal{L}_t$ that he can choose from. This means that his state variable (the information just before he makes a decision), is given by

$$S_t = (a_t, \mathcal{L}_t).$$

The driver has to choose among a set of actions $\mathcal{X}_t = (\mathcal{L}_t, \text{"hold"})$ that includes the loads in the set $\mathcal{L}_t$, or doing nothing. Once the driver makes this choice, the set $\mathcal{L}_t$ is no longer relevant. His state immediately after he makes his decision is called the *post-decision state* $S_t^x = a_t^x$ (the state immediately after a decision is made), which is updated to reflect the destination of the load, and the time he is expected to arrive at this location.

The natural way for a driver to choose which action to take is to balance the contribution of the action, which we write as $C(S_t, x_t)$, and the value of the driver in his post-decision state a_t^x. We might write this policy, which we call $X^\pi(S_t)$, using

$$X^\pi(S_t) = \arg\max_{x \in \mathcal{X}_t} \left(C(S_t, x) + \overline{V}_t^x(a_t^x) \right). \tag{2.47}$$

The algorithmic challenge is creating the estimates $\overline{V}_t^x(a_t^x)$, which is an example of what we will call a *value function approximation*. If the number of possible values of the driver attribute vector a_t^x was not too large, we could solve this problem using the same way we would solve the stochastic shortest path problem introduced in section 2.3.3. The hidden assumption in this problem is that the number of nodes is not too large (even a million nodes is considered manageable). When a "node" is a multidimensional vector a_t, then we may have trouble manipulating all the possible values this may take (another instance of the curse of dimensionality).

2.3.4.2 From One Driver to a Fleet

We can model a fleet of drivers by defining

$$R_{ta} = \text{the number of drivers with attribute vector } a \text{ at time } t,$$

$$R_t = (R_{ta})_{a \in \mathcal{A}},$$

where $a \in \mathcal{A}$ is in an attribute space that spans all the possible values that each element of a_t may take.

Similarly, we might describe loads by an attribute vector b that contains information such as origin, destination, scheduled pickup and delivery windows, required equipment type, and whether the load contains hazardous materials. In the United States, it is typical to aggregate the country into 100 regions, giving us 10,000 origin-destination pairs. Let

$$L_{tb} = \text{the number of loads with attribute vector } b \text{ at time } t,$$

$$L_t = (L_{tb})_{b \in \mathcal{B}}.$$

Our state variable is then given by

$$S_t = (R_t, L_t).$$

We leave it as an exercise to the reader to try to estimate the size of the state space for this problem. We show how this problem can be solved in chapter 18 using value function approximations.

2.3.5 Pricing

Imagine that we are trying to determine the price of a product, and that we feel that we can model the demand for the product using a logistics curve given by

$$D(p|\theta) = \theta_0 \frac{e^{\theta_1 - \theta_2 p}}{1 + e^{\theta_1 - \theta_2 p}}.$$

The total revenue from charging price p is given by

$$R(p|\theta) = pD(p|\theta).$$

If we knew θ, finding the optimal price would be a fairly simple exercise. But now assume that we do not know θ. Figure 2.5 illustrates a family of potential curves that might describe revenue as a function of price.

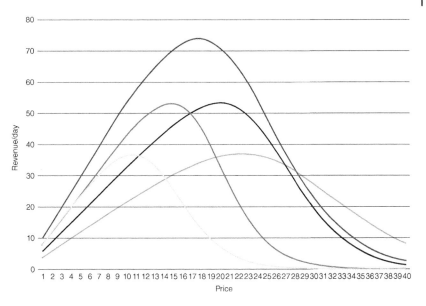

Figure 2.5 Illustration of a family of possible revenue curves.

We can approach this problem as one of learning the true value of θ. Let $\Theta = (\theta_1, \dots, \theta_K)$ be a family of possible values of θ where we assume that one of the elements of Θ is the true value. Let p_k^n be the probability that $\theta = \theta_k$ after we have made n observations. The state of our learning system, then, is $S^n = (p_k^n)_{k=1}^K$ which captures our belief about θ. We revisit this problem in chapter 7.

2.3.6 Medical Decision Making

Physicians have to make decisions about patients who arrive with some sort of complaint. The process starts by taking a medical history which consists of a series of questions about the patients history and lifestyle. Let h^n be this history, where h^n might consist of thousands of different possible characteristics (humans are complicated!). The physician might then order additional tests which produce additional information, or she might prescribe medication or request a surgical procedure. Let d^n capture these decisions. We can wrap this combination of patient history h^n and medical decisions d^n into a set of explanatory variables that we designate $x^n = (h^n, d^n)$. Also let θ be a parameter vector with the same dimensionality as x^n.

Now assume we observe an outcome y^n which for simplicity we are going to represent as binary, where $y^n = 1$ can be interpreted as "success" and $y^n = 0$ is a "failure." We are going to assume that we can model the random variable y^n (random, that is, before we observe the results of the treatment) using a logistic regression model, which is given by

$$\mathbb{P}[y^n = 1|x^n = (h^n, d^n), \theta] = \frac{e^{\theta^T x^n}}{1 + e^{\theta^T x^n}}. \tag{2.48}$$

This problem illustrates two types of uncertainty. The first is the patient history h^n, where we typically would not have a probability distribution describing these attributes. It is difficult (actually, impossible) to develop a probabilistic model of the complex characteristics captured in h^n describing a person, since a history is going to exhibit complex correlations. By contrast, the random variable y^n has a well-defined mathematical model, characterized by an unknown (and high dimensional) parameter vector θ.

We can use two different approaches for handling these different types of uncertainty. For patient attributes, we are going to use an approach that is often known as *data driven*. We might have access to a large dataset of prior attributes, decisions, and outcomes, that we might represent as $(x^n = (h^n, d^n), y^n)_{n=1}^N$. Alternatively, we may assume that we simply observe a patient h^n (this is the data-driven part), then make a decision $d^n = D^\pi(S^n)$ using a decision function $D^\pi(S^n)$ that can depend on a state variable S^n, and then observe an outcome y^n which we can describe using our probability model.

2.3.7 Scientific Exploration

Scientists looking to discover new drugs, new materials, or new designs for a wing or rocket engine, are often faced with the need to run difficult laboratory experiments looking for the inputs and processes to produce the best results. Inputs might be a choice of catalyst, the shape of a nanoparticle, or the choice of molecular compound. There might be different steps in a manufacturing process, or the choice of a machine for polishing a lens.

Then, there are the continuous decisions. Temperatures, pressures, concentrations, ratios, locations, diameters, lengths, and times are all examples of continuous parameters. In some settings these are naturally discretized, although this can be problematic if there are three or more continuous parameters we are trying to tune at the same time.

We can represent a discrete decision as choosing an element $x \in \mathcal{X} = \{x_1, \ldots, x_M\}$. Alternatively, we may have a continuous vector $x = (x_1, x_2, \ldots, x_K)$. Let x^n be our choice of x (whether it is discrete or continuous). We are going to assume that x^n is the choice we make *after* running the nth experiment that guides the $n+1^{st}$ experiment, from which we observe W^{n+1}. The outcome W^{n+1}

might be the strength of a material, the reflexivity of a surface, or the number of cancer cells killed.

We use the results of an experiment to update a belief model. If x is discrete, imagine we have an estimate $\bar{\mu}_x^n$ which is our estimate of the performance of running an experiment with choice x. If we choose $x = x^n$ and observe W^{n+1}, then we can use statistical methods (which we describe in chapter 3) to obtain updated estimates $\bar{\mu}_x^{n+1}$. In fact, we can use a property known as *correlated beliefs* that may allow us to run experiment $x = x^n$ and update estimates $\bar{\mu}_{x'}^{n+1}$ for values x' other than x.

Often, we are going to use some parametric model to predict a response. For example, we might create a linear model which can be written

$$f(x^n|\theta) = \theta_0 + \theta_1\phi_1(x^n) + \theta_2\phi_2(x^n) + \dots, \tag{2.49}$$

where $\phi_f(x^n)$ is a function that pulls out relevant pieces of information from the inputs x^n of an experiment. For example, if element x_i is the temperature, we might have $\phi_1(x^n) = x_i^n$ and $\phi_2(x^n) = (x_i^n)^2$. If x_{i+1} is the pressure, we could also have $\phi_3(x^n) = x_i^n x_{i+1}^n$ and $\phi_4(x^n) = x_i^n(x_{i+1}^n)^2$.

Equation (2.49) is known as a linear model because it is linear in the parameter vector θ. The logistic regression model in (2.48) is an example of a nonlinear model (since it is nonlinear in θ). Whether it is linear or nonlinear, parametric belief models capture the structure of a problem, reducing the uncertainty from an unknown $\bar{\mu}_x$ for each x (where the number of different values of x can number in the thousands to millions or more) down to a set of parameters θ that might number in the tens to hundreds.

2.3.8 Machine Learning vs. Sequential Decision Problems

There are close parallels between designing policies for sequential decision problems and machine learning. Let:

x^n = the data corresponding to the nth instance of a problem (the characteristics of a patient, the attributes of a document, the data for an image) that we want to use to predict an outcome y^n,

y^n = the response, which might be the response of a patient to a treatment, the categorization of a document, or the classification of an image,

$$f(x^n|\theta) \quad = \quad \text{our model which we use to predict } y^n \text{ given } x^n,$$

$$\theta \quad = \quad \text{an unknown parameter vector used to determine the model.}$$

We assume we have some metric that indicates how well our model $f(x|\theta)$ is performing. For example, we might use

$$L(x^n, y^n|\theta) \quad = \quad (y^n - f(x^n|\theta))^2.$$

The function $f(x|\theta)$ can take on many forms. The simplest is a basic linear model of the form

$$f(x|\theta) = \sum_{f \in \mathcal{F}} \theta_f \phi_f(x),$$

where $\phi_f(x)$ is known as a feature, and $\mathcal{F}$ is the set of features. There may be just a handful of features, or thousands. The statistics and machine learning communities have developed a broad array of functions, each of which is parameterized by some vector θ (sometimes designated as weights w). We review these in some depth in chapter 3.

The machine learning problem is to first pick a class of statistical model $f \in \mathcal{F}$, and then tune the parameters $\theta \in \Theta^f$ associated with that class of function. We write this as

$$\min_{f \in \mathcal{F}, \theta \in \Theta^f} \frac{1}{N} \sum_{n=1}^{N} (y^n - f(x^n|\theta))^2. \tag{2.50}$$

When we are solving a sequential decision problem, we need to find the best policy. We can think of a policy π as consisting of choosing a function $f \in \mathcal{F}$ along with tunable parameters $\theta \subset \Theta^f$. When we write our problem of optimizing over policies, we typically use

$$\max_{\pi = (f \in \mathcal{F}, \theta \in \Theta^f)} \mathbb{E} \left\{ \sum_{t=0}^{T} C(S_t, X^\pi(S_t|\theta))|S_0 \right\}. \tag{2.51}$$

When we compare the machine learning problem (2.50) with the sequential decision problem (2.51), we see that both are searching over classes of functions. We argue in chapter 3 that there are three (overlapping) classes of functions used for machine learning: lookup tables, parametric and nonparametric functions. Then we are going to argue in chapter 11 that there are four classes of policies (that is, four sets of functions in $\mathcal{F}$ when we are designing policies), where one of them, policy function approximations, includes all the functions that we might use in machine learning. The other three are all forms of optimization problems.

2.4   Bibliographic Notes

Section 2.1.1 – The field of stochastic search traces its roots to two papers: Robbins and Monro (1951) for derivative-based stochastic search, and Box and Wilson (1951) for derivative-free methods. Some early papers include the work on unconstrained stochastic search including Wolfowitz (1952) (using numerical derivatives), Blum (1954) (extending to multidimensional problems), and Dvoretzky (1956), which contributed theoretical research. A separate line of research focused on constrained problems under the umbrella of "stochastic quasi-gradient" methods, with seminal contributions from Ermoliev (1988), Shor (1979), Pflug (1988), Kushner and Clark (1978), Shapiro and Wardi (1996), and Kushner and Yin (2003). As with other fields, this field broadened over the years. The best modern review of the field (under this name) is Spall (2003), which was the first book to pull together the field of stochastic search as it was understood at that time. Bartlett et al. (2007) approaches this topic from the perspective of online algorithms, which refers to stochastic gradient methods where samples are provided by an exogenous source.

    The derivative-free version of stochastic search with discrete alternatives has been widely studied as the *ranking and selection* problem. Ranking and selection enjoys a long history dating back to the 1950s, with an excellent treatment of this early research given by the classic DeGroot (1970), with a more up-to-date review in Kim and Nelson (2007). Recent research has focused on parallel computing (Luo et al. (2015), Ni et al. (2016)) and handling unknown correlation structures (Qu et al., 2012). However, ranking and selection is just another name for derivative-free stochastic search, and has been widely studied under this umbrella (Spall, 2003). The field has attracted considerable attention from the simulation-optimization community, reviewed next.

Section 2.1.2 – Decision trees represent the simplest approach to modeling and, for simple settings, solving sequential decision problems. They lend themselves to complex decision problems in health (should a patient receive an MRI?), business (should a business enter a new market?), and policy (should the military pursue a new strategy?). (Skinner, 1999) is one of many books on decision trees, and there are literally dozens of survey articles addressing the use of decision trees in different application areas.

Section 2.1.3 – The field of Markov decision processes was introduced, initially in the form of deterministic dynamic programs, by Bellman (1952), leading to his classic reference (Bellman, 1957) (see also (Bellman, 1954) and (Bellman et al., 1955)), but this work was continued by a long stream of books including Howard (1960) (another classic), Nemhauser (1966), Denardo (1982),

Heyman and Sobel (1984), leading up to Puterman (2005) (this first appeared in 1994). Puterman's book represents the last but best in a long series of books on Markov decision processes, and now represents the major reference in what is a largely theoretical field, since the core of the field depends on one-step transition matrices which are rarely computable, and only for extremely small problems. More recently, Bertsekas (2017) provides an in-depth summary of the field of dynamic programming and Markov decision processes using a style that is a hybrid of notation from optimal control, with the principles of Markov decision processes, while also covering many of the concepts from approximate dynamic programming and reinforcement learning (covered below).

Section 2.1.4 – There is a long history in the development of optimal control dating to the 1950s, summarized by many books including Kirk (2012), Stengel (1986), Sontag (1998), Sethi (2019), and Lewis and Vrabie (2012). The canonical control problem is continuous, low-dimensional, and unconstrained, which leads to an analytical solution. Of course, applications evolved past this canonical problem, leading to the use of numerical methods. Deterministic optimal control is widely used in engineering, whereas stochastic optimal control has tended to involve much more sophisticated mathematics. Some of the most prominent books include Astrom (1970), Kushner and Kleinman (1971), Bertsekas and Shreve (1978), Yong and Zhou (1999), Nisio (2014), and Bertsekas (2017) (note that some of the books on deterministic controls touch on the stochastic case).

As a general problem, stochastic control covers any sequential decision problem, so the separation between stochastic control and other forms of sequential stochastic optimization tends to be more one of vocabulary and notation (Bertsekas (2017) is a good example of a book that bridges these vocabularies). Control-theoretic thinking has been widely adopted in inventory theory and supply chain management (e.g. Ivanov and Sokolov (2013) and Protopappa-Sieke and Seifert (2010)), finance (Yu et al., 2010), and health services (Ramirez-Nafarrate et al., 2014), to name a few.

There is considerable overlap between the fields of dynamic programming (including Markov decision processes) and optimal control (including stochastic control), but the two fields have evolved largely independently, using different notation, and motivated by very different applications. However, there are numerous parallels in the development of numerical methods for solving problems in both fields. Both fields start from the same foundation, known as Bellman's equations in dynamic programming, and Hamilton-Jacobi equations in optimal control (leading some to refer to them as Hamilton-Jacobi-Bellman (or HJB) equations).

Section 2.1.5 – Approximate dynamic programming (also referred to as adaptive dynamic programming and, for a period, neuro-dynamic programming) has been studied since Bellman first recognized that discrete dynamic programming suffered from the curse of dimensionality (see Bellman and Dreyfus (1959) and Bellman et al. (1963)), but the operations research community then seemed to drop any further research in approximation methods until the 1980s. As computers improved, researchers began tackling Bellman's equation using numerical approximation methods, with the most comprehensive presentation in Judd (1998) which summarized almost a decade of research (see also Chen et al. (1999)).

A completely separate line of research in approximations evolved in the control theory community with the work of Paul Werbos (Werbos (1974)) who recognized that the "cost-to-go function" (the same as the value function in dynamic programming) could be approximated using various techniques. Werbos helped develop this area through a series of papers (examples include Werbos (1989), Werbos (1990), Werbos (1992) and Werbos (1994)). Important references are the edited volumes (White and Sofge, 1992) and (Si et al., 2004) which highlighted what had already become a popular approach using neural networks to approximate both policies ("actor nets") and value functions ("critic nets"). Si et al. (2004) contains a nice review of the field as of 2002. Tsitsiklis (1994) and Jaakkola et al. (1994) were the first to recognize that the basic algorithms being developed under the umbrella of reinforcement learning represented generalizations of the early stochastic gradient algorithms of Robbins and Monro (1951). Bertsekas and Tsitsiklis (1996) laid the foundation for adaptive learning algorithms in dynamic programming, using the name "neuro-dynamic programming." Werbos, (e.g. Werbos (1992)), had been using the term "approximate dynamic programming," which became the title of Powell (2007) (with a major update in Powell (2011)), a book that also merged math programming and value function approximations to solve high-dimensional, convex stochastic optimization problems (but, see the developments under stochastic programming below). Later, the engineering controls community reverted to "adaptive dynamic programming" as the operations research community adopted "approximate dynamic programming."

Section 2.1.6 – A third line of research into approximation methods started in the 1980s in the computer science community under the umbrella of "reinforcement learning" with the work of Richard Sutton and Andy Barto into Q-learning. The field took off with the appearance of their now widely cited book (Sutton and Barto, 2018), although by this time the field was quite active (see the review Kaelbling et al. (1996)). Research under the umbrella of "reinforcement learning" has evolved to include other algorithmic strategies under names such as policy search and Monte Carlo tree search. Other references from the reinforcement learning community include Busoniu et al. (2010) and Szepesvári (2010). In 2017, Bertsekas published the fourth edition of his optimal control book (Bertsekas (2017)), which covers a range of topics spanning classical Markov decision processes and the approximate algorithms associated with approximate dynamic programming and optimal control, but using the notation of optimal control and constructs from Markov decision processes (such as one-step transition matrices). Bertsekas' book easily has the most comprehensive review of the ADP/RL literature, and we recommend this book for readers looking for a comprehensive bibliography of these fields (as of 2017). In 2018, Sutton and Barto came out with a greatly expanded second edition of their classic *Reinforcement Learning* book (Sutton and Barto (2018)) which features methods that move far behind the basic Q-learning algorithms of the first edition. In the language of this book, readers comparing the first and second editions of *Reinforcement Learning* will see the transition from policies based on value functions alone (Q-learning in the RL community), to examples from all four classes of policies.

The characterization of "reinforcement learning" along the lines of the three features (e.g. "agent acting on the environment receiving a reward") was provided at a workshop by Professor Benjamin van Roy, a leader in the RL community.

Section 2.1.7 – Optimal stopping is an old and classic topic. An elegant presentation is given in Cinlar (1975) with a more recent discussion in Cinlar (2011) where it is used to illustrate filtrations. DeGroot (1970) provides a nice summary of the early literature. One of the earliest books dedicated to the topic is Shiryaev (1978) (originally in Russian). Moustakides (1986) describes an application to identifying when a stochastic process has changed, such as the increase of incidence in a disease or a drop in quality on a production line. Feng and Gallego (1995) uses optimal stopping to determine when to start end-of-season sales on seasonal items. There are numerous uses of optimal stopping in finance (Azevedo and Paxson, 2014), energy (Boomsma et al., 2012), and technology adoption (Hagspiel et al., 2015), to name just a few.

Section 2.1.8 – There is an extensive literature exploiting the natural convexity of $Q(x_0, W_1)$ in x_0, starting with Van Slyke and Wets (1969), followed by the seminal papers on stochastic decomposition (Higle and Sen, 1991) and the stochastic dual dynamic programming (SDDP) (Pereira and Pinto, 1991). A substantial literature has unfolded around this work, including Shapiro (2011) who provides a careful analysis of SDDP, and its extension to handle risk measures (Shapiro et al. (2013), Philpott et al. (2013)). A number of papers have been written on convergence proofs for Benders-based solution methods, but the best is Girardeau et al. (2014). Kall and Wallace (2009) and Birge and Louveaux (2011) are excellent introductions to the field of stochastic programming. King and Wallace (2012) is a nice presentation on the process of modeling problems as stochastic programs. A modern overview of the field is given by Shapiro et al. (2014).

Section 2.1.9 – Active learning problems have been studied as "multiarmed bandit problems" since 1960 in the applied probability community. DeGroot (1970) was the first to show that an optimal policy for the multiarmed bandit problem could be formulated (if not solved) using Bellman's equation (this is true of *any* learning problem, regardless of whether we are maximizing final or cumulative rewards). The first real breakthrough occurred in Gittins and Jones (1974) (the first and most famous paper), followed by Gittins (1979). The theory of Gittins indices was described thoroughly in his first book (Gittins, 1989), but the "second edition" (Gittins et al., 2011), which was a complete rewrite of the first edition, represents the best introduction to the field of Gittins indices, which now features hundreds of papers. However, the field is mathematically demanding, with index policies that are difficult to compute.

A parallel line of research started in the computer science community with the work of Lai and Robbins (1985) who showed that a simple policy known as *upper confidence bounding* possessed the property that the number of times we test the wrong arm can be bounded (although it continues to grow with n). The ease of computation, combined with these theoretical properties, made this line of research extremely attractive, and has produced an explosion of research. While no books on this topic have appeared as yet, an excellent monograph is Bubeck and Cesa-Bianchi (2012).

These same ideas have been applied to bandit problems using a terminal reward objective using the label the "best arm" bandit problem (see Audibert and Bubeck (2010), Kaufmann et al. (2016), Gabillon et al. (2012)).

Section 2.1.10 – The original work on optimal computing budget allocation was developed by Chun-Hung Chen in Chen (1995), followed by a series of articles (Chen (1996), Chen et al. (1997), Chen et al. (1998), Chen et al. (2003), Chen et al. (2008)), leading up to the book Chen and Lee (2011) that provides a thorough overview of this field. The field has focused primarily on discrete alternatives (e.g. different designs of a manufacturing system), but has also included work on continuous alternatives (e.g. Hong and Nelson (2006)). An important recent result by Ryzhov (2016) shows the asymptotic equivalence of OCBA and expected improvement policies which maximize the value of information. When the number of alternatives is much larger (say, 10,000), techniques such as simulated annealing, genetic algorithms and tabu search (adapted for stochastic environments) have been brought to bear. Swisher et al. (2000) contains a nice review of this literature. Other reviews include Andradóttir (1998*a*), Andradóttir (1998*b*), Azadivar (1999),

Fu (2002), and Kim and Nelson (2007). The recent review Chau et al. (2014) focuses on gradient-based methods.

The scope of problems and methods studied under the umbrella of "simulation-optimization" has steadily grown (a pattern similar to other communities in stochastic optimization). The best evidence of this is Michael Fu's *Handbook of Simulation Optimization* (Fu (2014)) which is a superb reference for many of the tools in this field.

Section 2.1.11 – Active learning is a field that emerged from within the machine learning community; parallels the bandit community in that an agent could control (or influence) the inputs x^n to a learning process that produces observations y^n. The field emerged primarily in the 1990s (see in particular Cohn et al. (1996) and Cohn et al. (1994)). The book Settles (2010) provides a nice introduction to the field which indicates a strong awareness of the parallels between active learning and multiarmed bandit problems. A recent tutorial is given by Krempl et al. (2016).

Section 2.1.12 – Chance-constrained optimization was first introduced by Charnes et al. (1959), followed by Charnes and Cooper (1963), for handling constraints that involve uncertainty. It has also been studied as "probabilistic constrained programming" (Prekopa (1971), Prekopa (2010)) and continues to attract hundreds of papers each year. Chance-constrained programming is standard in many books on stochastic optimization (see, for example, Shapiro et al. (2014)).

Section 2.1.13 – This is a subfield of optimal control, but it evolved into a field of its own, with popular books such as Camacho and Bordons (2003) and thousands of articles (see Lee (2011) for a 30-year review). As of this writing, there are over 50 review articles feature modeling predictive control since 2010.

Section 2.1.14 – A thorough review of the field of robust optimization is contained in Ben-Tal et al. (2009) and Bertsimas et al. (2011), with a more recent review given in Gabrel et al. (2014). Bertsimas and Sim (2004) studies the price of robustness and describes a number of important properties. Robust optimization is attracting interest in a variety of application areas including supply chain management (Bertsimas and Thiele (2006), Keyvanshokooh et al. (2016)), energy (Zugno and Conejo, 2015), and finance (Fliege and Werner, 2014).

Exercises

Review questions

2.1 What is meant by the *compact form* and *expanded form* of the expectation operator? Give an illustration of each.

2.2 Write out the objective functions that we would use when maximizing the cumulative reward or maximizing the final reward.

2.3 Compare the Markov decision process model in section 2.1.3 to the optimal control model in section 2.1.4 by creating a table showing how each approach models the following:

- State variables.
- Decision/control variables.
- The transition function (use the version in the optimal control formulation that includes the randomness w_t).
- The value of being in a state at time t.
- How this value can be used to find the best decision given the state x_t (otherwise known as the policy).

2.4 From the very brief presentation in this chapter, what is the difference between approximate dynamic programming and reinforcement learning (using Q-learning).

2.5 Write out an optimal stopping problem as an optimal control problem. Would the optimal policy take the form in equation (2.13)? Justify your answer.

2.6 Does solving the optimization problem in (2.23) produce an optimal policy? Discuss why or why not.

2.7 In the stochastic programming model in section 2.24, what is meant by "ω"? Use the setting of allocating inventories to warehouses at time 0 (this decision is given by x_0), after which we see demands, and then determine which warehouse should satisfy each demand.

2.8 For the multiarmed bandit problem, write out the objective function for finding the best interval estimation policy.

2.9 Describe in words the decision that is being optimized over using the OCBA algorithm in simulation optimization. Contrast how OCBA operates (in general terms) compared to interval estimation for the multi-armed bandit problem.

2.10 What objective is being optimized in active learning? Could you solve this same problem using interval estimation?

2.11 What is the core computational challenge that arises in chance-constrained programming?

2.12 Compare model predictive control to using stochastic programming as a policy.

2.13 Describe in words, using an example, the core idea of robust optimization. Just as the two-stage stochastic program in (2.24) could be written as a policy (as we do in equation (2.26)) show how robust optimization can also be written as a policy.

2.14 From section 2.3.8, what is the difference between a machine learning problem, and a sequential decision problem?

Modeling questions

2.15 Provide three examples of:

(a) Problems where we would want to maximize the cumulative reward (or minimize cumulative cost).
(b) Problems where we would want to maximize the final reward (or minimize the final cost).

2.16 Show how to write solve a decision tree (section 2.1.2) as a Markov decision process (section 2.1.3) using Bellman's equation (2.7).

2.17 Put the contextual newsvendor problem in section 2.3.1 into the format of the universal modeling framework in section 2.2. Introduce and define any additional notation you may need.

2.18 Put the inventory planning problem with forecasts in section 2.3.2 into the format of the universal modeling framework in section 2.2. Introduce and define any additional notation you may need.

2.19 Put the dynamic shortest path problem in section 2.3.3 into the format of the universal modeling framework in section 2.2. Introduce and define any additional notation you may need.

2.20 Put the robust shortest path problem in section 2.3.3 into the format of the universal modeling framework in section 2.2. Introduce and define any additional notation you may need.

2.21 Put the nomadic trucker problem in section 2.3.4.1 into the format of the universal modeling framework in section 2.2. The state variable $S_t = (a_t, \mathcal{L}_t)$ given in the section is incomplete. What is missing? Introduce and define any additional notation you may need. [Hint: Carefully review the definition of the state variable given in section 2.2. Now look at the policy in (2.47), and see if there is any statistic that will be changing over time that is needed to make a decision (which means it has to go into the state variable).]

2.22 Put the pricing problem in section 2.3.5 into the format of the universal modeling framework in section 2.2. Introduce and define any additional notation you may need.

2.23 Put the medical decision-making problem in section 2.3.6 into the format of the universal modeling framework in section 2.2. Introduce and define any additional notation you may need.

2.24 Put the scientific exploration problem in section 2.3.7 into the format of the universal modeling framework in section 2.2. Introduce and define any additional notation you may need.

Diary problem

The diary problem is a single problem you chose (see chapter 1 for guidelines). Answer the following for your diary problem.

2.25 Which of the canonical problems (you may name more than one) seem to use the language that best fits your diary problem. Give examples from your diary problem that seem to fit a particular canonical problem.

Bibliography

Andradóttir, S. (1998a). A review of simulation optimization techniques. *1998 Winter Simulation Conference. Proceedings* 1 (0): 151–158.

Andradóttir, S. (1998b). Simulation Optimimzation. In: *Handbook of simulation* (ed. J. Banks), 307–333. Hoboken, NJ: John Wiley & Sons. chapter 9.

Astrom, K.J. (1970). *Introduction to Stochastic Control Theory*. Mineola, NY: Dover Publications.

Audibert, J.-y. and Bubeck, S. (2010). Best Arm Identification in Multi-Armed Bandits. *CoLT*. 13.

Azadivar, F. (1999). Simulation optimization methodologies. In: *Proceedings of the 1999 Winter Simulation Confer-ence* (eds. P. Farrington, H. Nemb-hard, D. Sturrock and G. Evans), 93–100. IEEE.

Azevedo, A. and Paxson, D. (2014). Developing real option game models. *European Journal of Operational Research* 237 (3): 909–920.

Bartlett, P. L., Hazan, E. & Rakhlin, A. (2007). Adaptive Online Gradient Descent. *Advances in neural information processing systems* pp. 1–8.

Bellman, R. (1952). On the theory of dynamic programming. *Proceedings of the National Academy of Sciences* 38 (8): 716–719.

Bellman, R. E. (1954). The Theory of Dynamic Programming. *Bulletin of the American Mathematical Society* 60: 503–516.

Bellman, R.E. (1957). *Dynamic Programming.* Princeton, N.J.: Princeton University Press.

Bellman, R.E. and Dreyfus, S.E. (1959). Functional approximations and dynamic programming. *Mathematical Tables and Other Aids to Computation* 13: 247–251.

Bellman, R.E., Glicksberg, I., and Gross, O. (1955). On the optimal inventory equation. *Management Science* 1: 83–104.

Bellman, R., Kalaba, R. and Kotkin, B. (1963). Polynomial approximationI a new computational technique in dynamic programming: Allocation processes. *Mathematics of Computation* 17: 155–161.

Ben-Tal, A., El Ghaoui, L., and Nemirovski, A. (2009). Robust optimization. *Princeton University Press* 53 (3): 464–501.

Bertsekas, D.P. (2017). *Dynamic Programming and Optimal Control: Approximate Dy-namic Programming*, 4 e. Belmont, MA.: Athena Scientific.

Bertsekas, D.P. and Shreve, S.E. (1978). *Stochastic Optimal Control: The Discrete Time Case*, Vol. 0. Academic Press.

Bertsekas, D.P. and Tsitsiklis, J.N. (1996). *Neuro-Dynamic Programming*. Belmont, MA: Athena Scientific.

Bertsimas, D., Iancu, D.A., and Parrilo, P.A. (2011). A hierarchy of nearoptimal policies for multistage adaptive optimization. *IEEE Transactions on Automatic Control* 56 (12): 2809–2824.

Bertsimas, D.J. and Sim, M. (2004). The price of robustness. *Operations Research* 52 (1): 35–53.

Bertsimas, D.J. and Thiele, A. (2006). A robust optimization approach to inventory theory. *Operations Research* 54 (1): 150–168.

Birge, J.R. and Louveaux, F. (2011). *Introduction to Stochastic Programming*, 2e. New York: Springer.

Blum, J. (1954). Multidimensional stochastic approximation methods. *Annals of Mathematical Statistics* 25: 737–74462.

Boomsma, T.K., Meade, N., and Fleten, S.E. (2012). Renewable energy investments under different support schemes: A real options approach. *European Journal of Operational Research* 220 (1): 225–237.

Box, G.E.P. and Wilson, K.B. (1951). On the experimental attainment of optimum conditions. *Journal of the Royal Statistical Society Series B* 13 (1): 1–45.

Bubeck, S. and Cesa-Bianchi, N. (2012). Regret analysis of stochastic and nonstochastic multi-armed bandit problems. *Foundations and Trends in Machine Learning* 5 (1): 1–122.

Busoniu, L., Babuska, R., De Schutter, B., and Ernst, D. (2010). *Reinforcement Learning and Dynamic Programming using Function Approximators*. New York: CRC Press

Camacho, E. and Bordons, C. (2003). *Model Predictive Control*. London: Springer.

Charnes, A. and Cooper, W.W. (1963). Deterministic equivalents for optimizing and satisficing under chance constraints. *Operations Research* 11: 18–39.

Charnes, A., Cooper, W.W., and Cooper, A.A. (1959). Chance constrained programming. *Management Science* 5: 73–79.

Chau, M., Fu, M.C., Qu, H., and Ryzhov, I.O. (2014). Simulation optimization: A tutorial overview and recent developments in gradient-based Methods, In: *Winter Simulation Conference* (eds. A. Tolk, S. Diallo, I. Ryzhov, L. Yilmaz, S. Buckley and J. Miller), 21–35. Informs.

Chen, C.H. (1995). An effective approach to smartly allocate computing budget for discrete event simulation. In: *34th IEEE Conference on Decision and Control*, Vol. 34, New Orleans, LA, 2598–2603.

Chen, C.H. (1996). A lower bound for the correct subsetselection probability and its application to discrete event system simulations. *IEEE Transactions on Automatic Control* 41 (8): 1227–1231.

Chen, C.H. and Lee, L.H. (2011). *Stochastic Simulation Optimization.* Hackensack, N.J.: World Scientific Publishing Co.

Chen, C.H., Donohue, K., Yücesan, E., and Lin, J. (2003). Optimal computing budget allocation for Monte Carlo simulation with application to product design. *Simulation Modelling Practice and Theory* 11: 57–74.

Chen, C.H., He, D., Fu, M.C., and Lee, L. H. (2008). Efficient simulation budget allocation for selecting an optimal subset. *INFORMS Journal on Computing* 20 (4): 579–595.

Chen, C.H., Yuan, Y., Chen, H.C., Yücesan, E., and Dai, L. (1998). Computing budget allocation for simulation experiments with different system structure. In: *Proceedings of the 30th conference on Winter simulation.* 735–742.

Chen, H.C., Chen, C.H., Dai, L., and Yucesan, E. (1997). A gradient approach for smartly allocating computing budget for discrete event simulation. In: *Proceedings of the 1996 Winter Simulation Conference* (eds. J. Charnes, D. Morrice, D. Brunner and J. Swain), 398–405. Piscataway, NJ, USA: IEEE Press.

Chen, V.C.P., Ruppert, D., and Shoemaker, C.A. (1999). Applying experimental design and regression splines to high-dimensional continuous-state stochastic dynamic programming. *Operations Research* 47 (1): 38–53.

Cinlar, E. (1975). *Introduction to Stochastic Processes.* Upper Saddle River, NJ: Prentice Hall.

Cinlar, E. (2011). *Probability and Stochastics.* New York: Springer.

Cohn, D.A., Ghahramani, Z., and Jordan, M.I. (1996). Active learning with statistical models. *Learning* 4: 129–145.

Cohn, D., Atlas, L., and Ladner, R. (1994). Improving generalization with active learning. *Machine Learning* 5 (2201): 221.

DeGroot, M.H. (1970). *Optimal Statistical Decisions.* John Wiley and Sons.

Denardo, E.V. (1982). *Dynamic Programming*, Englewood Cliffs, NJ: PrenticeHall.

Dvoretzky, A. (1956). On stochastic approximation. In: *Proceedings 3rd Berkeley Symposium on Mathematical Statistics and Probability* (ed. J. Neyman), 39–55. University of California Press.

Ermoliev, Y. (1988). Stochastic quasigradient methods. In: *Numerical Techniques for Stochastic Optimization* (eds Y. Ermoliev and R. Wets). Berlin: SpringerVerlag.

Feng, Y. and Gallego, G. (1995). Optimal starting times for end-of-season sales and optimal stopping times for promotional fares. *Management Science* 41 (8): 1371–1391.

Fliege, J. and Werner, R. (2014). Robust multiobjective optimization and applications in portfolio optimization. *European Journal of Operational Research* 234 (2): 422–433.

Fu, M.C. (2002). Optimization for simulation: Theory vs. practice. *Informs Journal on Computing* 14 (3): 192–215.

Fu, M.C. (2014). *Handbook of Simulation Optimization*. New York: Springer.

Gabillon, V., Ghavamzadeh, M., and Lazaric, A. (2012). Best arm identification: A unified approach to fixed budget and fixed confidence. *Nips*. 1–9.

Gabrel, V., Murat, C., and Thiele, A. (2014). Recent advances in robust optimization: An overview. *European Journal of Operational Research* 235 (3): 471–483.

Girardeau, P., Leclere, V., and Philpott, A.B. (2014). On the convergence of decomposition methods for multistage stochastic convex programs. *Mathematics of Operations Research* 40 (1): 130–145.

Gittins, J. (1979). Bandit processes and dynamic allocation indices. *Journal of the Royal Statistical Society. Series B (Methodological)* 41 (2): 148–177.

Gittins, J. (1989). *Multiarmed Bandit Allocation Indices*. New York: Wiley and Sons.

Gittins, J. and Jones, D. (1974). A dynamic allocation index for the sequential design of experiments. In: *Progress in statistics* (ed. J. Gani), 241–266. Amsterdam: North Holland.

Gittins, J., Glazebrook, K.D., and Weber, R.R. (2011). *MultiArmed Bandit Allocation Indices*. New York: John Wiley & Sons.

Hagspiel, V., Huisman, K.J., and Nunes, C. (2015). Optimal technology adoption when the arrival rate of new technologies changes. *European Journal of Operational Research* 243 (3): 897–911.

Heyman, D.P. and Sobel, M. (1984). *Stochastic Models in Operations Research, Volume II: Stochastic Optimization*. New York: McGraw Hill.

Higle, J.L. and Sen, S. (1991). Stochastic decomposition: An algorithm for twostage linear programs with recourse. *Mathematics of Operations Research* 16 (3): 650–669.

Hong, J. and Nelson, B. L. (2006). Discrete optimization via simulation using Compass. *Operations Research* 54 (1): 115–129.

Howard, R.A. (1960). *Dynamic programming and Markov processes*. Cambridge, MA: MIT Press.

Ivanov, D. and Sokolov, B. (2013). Control and systemtheoretic identification of the supply chain dynamics domain for planning, analysis and adaptation of performance under uncertainty. *European Journal of Operational Research* 224 (2): 313–323.

Jaakkola, T., Jordan, M.I., and Singh, S.P. (1994). On the convergence of stochastic iterative dynamic programming algorithms. *Neural Computation* 6 (6): 1185–1201.

Judd, K.L. (1998). *Numerical Methods in Economics*. MIT Press.

Kaelbling, L.P., Littman, M.L., and Moore, A.W. (1996). Reinforcement learning: a survey. *Journal of Artificial Intelligence Research* 4: 237–285.

Kall, P. and Wallace, S.W. (2009). *Stochastic Programming*, Vol. 10. Hoboken, NJ: John Wiley & Sons.

Kaufmann, E., Cappé, O., and Garivier, A. (2016). On the complexity of best-arm identification in multi-armed bandit models. *Journal of Machine Learning Research* 17: 1–42.

Keyvanshokooh, E., Ryan, S. M., and Kabir, E. (2016). Hybrid robust and stochastic optimization for closedloop supply chain network design using accelerated Benders decomposition. *European Journal of Operational Research* 249 (1): 76–92

Kim, S.-H. and Nelson, B.L. (2007). *Recent advances in ranking and selection*, 162–172. Piscataway, NJ, USA: IEEE Press.

King, A.J. and Wallace, S.W. (2012). *Modeling with Stochastic Programming*, New York: Springer Verlag.

Kirk, D.E. (2012). *Optimal Control Theory: An introduction*. New York: Dover.

Krempl, G., Lemaire, V., Lughofer, E., and Kottke, D. (2016). Active learning: Applications, foundations and emerging trends (tutorial). *CEUR Workshop Proceedings* 1707: 1–2.

Kushner, H.J. and Clark, S. (1978). *Stochastic Approximation Methods for Constrained and Unconstrained Systems*. New York: SpringerVerlag.

Kushner, H.J. and Kleinman, A.J. (1971). Accelerated procedures for the solution of discrete Markov control problems. *IEEE Transactions on Automatic Control* 16: 2147–152.

Kushner, H.J. and Yin, G.G. (2003). *Stochastic Approximation and Recursive Algorithms and Applications*. New York: Springer.

Lai, T.L. and Robbins, H. (1985). Asymptotically efficient adaptive allocation rules. *Advances in Applied Mathematics* 6: 4–22.

Lee, J.H. (2011). Model predictive control: Review of the three decades of development. *International Journal of Control, Automation and Systems* 9 (3): 415–424.

Lewis, F.L. and Vrabie, D. (2012). *Design Optimal Adaptive Controllers*, 3e. Hoboken, NJ: JohnWiley & Sons.

Luo, J., Hong, L.J., Nelson, B.L., and Wu, Y. (2015). Fully sequential procedures for large-scale ranking-and-selection problems in parallel computing environments. *Operations Research* 63 (5): 1177–1194.

Moustakides, G.V. (1986). Optimal stopping times for detecting changes in distributions. *Annals of Statistics* 14 (4): 1379–1387.

Nemhauser, G.L. (1966). *Introduction to Dynamic Programming*. New York: JohnWiley & Sons.

Ni, E.C., Henderson, S.G., and Hunter, S.R. (2016). Efficient ranking and selection in parallel computing environments. *Operations Research* 65 (3): 821–836.

Nisio, M. (2014). *Stochastic Control Theory: Dynamic Programming Principle*. New York: Springer.

Pereira, M.F. and Pinto, L.M.V.G. (1991). Multistage stochastic optimization applied to energy planning. *Mathematical Programming* 52: 359–375.

Pflug, G. (1988). Stepsize rules, stopping times and their implementation in stochastic quasigradient algorithms. In: *Numerical Techniques for Stochastic Optimization*, 353–372. New York: SpringerVerlag.

Philpott, A.B., De Matos, V., and Finardi, E. (2013). On solving multistage stochastic programs with coherent risk measures. *Operations Research* 51 (4): 957–970.

Powell, W.B. (2007). *Approximate Dynamic Programming: Solving the Curses of Dimensionality*, John Wiley & Sons.

Powell, W.B. (2011). *Approximate Dynamic Programming: Solving the Curses of Dimensionality*, 2e. John Wiley & Sons.

Prekopa, A. (1971). On probabilistic constrained programming. In: *Proceedings of the Princeton Symposium on Mathematical Programming*, 113–123. Princeton NJ, Princeton University Press.

Prekopa, A. (2010). *Stochastic Programming*. Dordrecht, The Netherlands: Kluwer Academic Publishers.

Protopappa-Sieke, M. and Seifert, R.W. (2010). Interrelating operational and financial performance measurements in inventory control. *European Journal of Operational Research* 204 (3): 439–448.

Puterman, M.L. (2005). *Markov Decision Processes*, 2e., Hoboken, NJ: John Wiley and Sons.

Qu, H., Ryzhov, I.O., and Fu, M.C. (2012). Ranking and selection with unknown correlation structures. In: *Proceedings Winter Simulation Conference* (ed. A.U.C. Laroque, J. Himmelspach, R. Pasupathy, and O. Rose). number 1995.

Ramirez-Nafarrate, A., Baykal Hafizoglu, A., Gel, E.S., and Fowler, J.W. (2014). Optimal control policies for ambulance diversion. *European Journal of Operational Research* 236 (1): 298–312.

Robbins, H. and Monro, S. (1951). A stochastic approximation method. *The Annals of Mathematical Statistics* 22 (3): 400–407.

Ryzhov, I.O. (2016). On the convergence rates of expected improvement methods. *Operations Research* 64 (6): 1515–1528.

Sethi, S.P. (2019). *Optimal Control Theory: Applications to Management Science and Economics*, 3 e. Boston: SpringerVerlag.

Settles, B. (2010). *Active Learning*. New York: Sciences.

Shapiro, A. (2011). Analysis of stochastic dual dynamic programming method. *European Journal of Operational Research* 209 (1): 63–72.

Shapiro, A. and Wardi, Y. (1996). Convergence analysis of stochastic algorithms. *Mathematics of Operations Research* 21: 615–628.

Shapiro, A., Dentcheva, D., and Ruszczyński, A. (2014). *Lectures on Stochastic Programming: Modeling and theory*, 2 e. Philadelphia: SIAM.

Shapiro, A., Tekaya,W., Da Costa, J.P., and Soares, M.P. (2013). Risk neutral and risk averse stochastic dual dynamic programming method. *European Journal of Operational Research* 224 (2): 375–391.

Shiryaev, A.N. (1978). *Optimal Stopping Rules*. Moscow: Springer.

Shor, N.K. (1979). *The Methods of Nondifferentiable Op[timization and their Applications*. Kiev: Naukova Dumka.

Si, J., Barto, A.G., Powell, W.B., and Wunsch, D. (eds.) (2004). *Learning and Approximate Dynamic Programming: Scaling up to the Real World*. New York: John Wiley and Sons.

Skinner, D.C. (1999). *Introduction to Decision Analysis*. Gainesville, Fl: Probabilistic Publishing.

Sontag, E. (1998). *Mathematical Control Theory*, 2ed., 1–544. Springer.

Spall, J.C. (2003). *Introduction to Stochastic Search and Optimization: Estimation, simulation and control*. Hoboken, NJ: John Wiley & Sons.

Stengel, R.F. (1986). *Stochastic Optimal Control: Theory and Application*. Hoboken, NJ: John Wiley & Sons.

Sutton, R.S. and Barto, A.G. (2018). *Reinforcement Learning: An Introduction*, 2e. Cambridge, MA: MIT Press.

Swisher, J.R., Hyden, P.D., and Schruben, L.W. (2000). A survey of simulation optimization techniques and procedures. In: *Simulation Conference Proceedings, 2000. Winter*, 119–128.

Szepesvári, C. (2010). Algorithms for reinforcement learning. *Synthesis Lectures on Artificial Intelligence and Machine Learning* 4 (1): 1–103.

Tsitsiklis, J.N. (1994). Asynchronous stochastic approximation and Q-learning. *Machine Learning* 16: 185–202.

Van Slyke, R.M. and Wets, R.J.-B. (1969). Lshaped linear programs with applications to optimal control and stochastic programming. *SIAM Journal of Applied Mathematics* 17: 638–663.

Werbos, P.J. (1974). Beyond regression: new tools for prediction and analysis in the behavioral sciences, PhD thesis, Harvard University.

Werbos, P.J. (1989). Backpropagation and neurocontrol: A review and prospectus. In: *IJCNN, International Joint Conference on Neural Networks*, 209–216.

Werbos, P.J. (1990). Backpropagation Through Time: What It Does and How to Do It. *Proceedings of the IEEE* 78 (10): 1550–1560.

Werbos, P.J. (1992). Approximate dynamic programming for real-time control and neural modelling. In: *Handbook of Intelligent Control: Neural, Fuzzy, and Adaptive Approaches* (eds. D.J. White and D.A. Sofge), 493–525. Van Nostrand.

Werbos, P.J. (1994). *The Roots of Backpropagation: From Ordered Derivatives to Neural Networks and Political Forecasting*. New York: John Wiley & Sons.

White, D. and Sofge, D. (1992). *Handbook of intelligent control: Neural, fuzzy, and adaptive approaches*. New York: Van Nostrand Reinhold Company.

Wolfowitz, J. (1952). On the stochastic approximation method of Robbins and Monro. *Annals of Mathematical Statistics* 23: 457–461.

Yong, J. and Zhou, X.Y. (1999). *Stochastic Controls: Hamiltonian Systems and HJB Equations*. New York: Springer.

Yu, M., Takahashi, S., Inoue, H., and Wang, S. (2010). Dynamic portfolio optimization with risk control for absolute deviation model. *European Journal of Operational Research* 201 (2): 349–364.

Zugno, M. and Conejo, A.J. (2015). A robust optimization approach to energy and reserve dispatch in electricity markets. *European Journal of Operational Research* 247 (2): 659–671.

3

Online Learning

There is a massive community that has evolved under names such as statistics, statistical learning, machine learning, and data sciences. The vast majority of this work, known as *supervised learning*, involves taking a dataset (x^n, y^n), $n = 1, \ldots, N$ of input data x^n and corresponding observations (sometimes called "labels") y^n and using this to design a statistical model $f(x|\theta)$ that produces the best fit between $f(x^n|\theta)$ and the associated observation (or label) y^n. This is the world of big data.

This book is on the topic of making decisions (that we call x). So why do we need a chapter on learning? The simple explanation is that machine learning arises throughout the process of helping computers make decisions. Classical machine learning is focused on learning something about an exogenous process: forecasting weather, predicting demand, estimating the performance of a drug or material. In this book, we need exogenous learning for the same reason everyone else does, but most of the time we will focus on *endogenous learning*, where we are learning about value functions, policies, and response surfaces, which are learning problems that arise in the context of methods for making decisions.

We open this chapter with an overview of the role of machine learning in the context of sequential decision making. The remainder of the chapter is an introduction to machine learning, with an emphasis on learning over time, a topic known as online learning, since this will dominate the applications of machine learning for sequential decisions.

As elsewhere, the sections marked with an * can easily be skipped on an initial pass through this chapter. Readers should understand the information that is available in this chapter, but otherwise should view it as a reference that is turned to on an as needed basis (and there will be many references to this chapter in the rest of the book).

Reinforcement Learning and Stochastic Optimization: A Unified Framework for Sequential Decisions, First Edition. Warren B. Powell.
© 2022 John Wiley & Sons, Inc. Published 2022 John Wiley & Sons, Inc.

3.1 Machine Learning for Sequential Decisions

It is useful to begin our discussion of statistical learning by describing the learning issues that arise in the context of sequential decisions. This section provides an overview of the following dimensions of learning problems:

- Observations and data in sequential decisions – While classical statistical learning problems consist of datasets composed of input (or independent) variables x and output (or dependent) variables y, in sequential decision making the input variables x^n are decisions that we control (at least partially).
- Indexing data – When we do batch learning, we use a dataset (x^n, y^n), $n = 1, \dots, N$ where y^n is the response associated with the input data x^n. In the context of sequential decisions, we pick x^n and then observe y^{n+1}.
- Functions we are learning – There are a half dozen different classes of functions that we may need to approximate in different stochastic optimization contexts.
- Sequential learning – Most of our applications involve starting with little or no data, and then successively acquiring more data. This often means we have to transition from low-dimensional models (which can be fitted with little data) to higher-dimensional models.
- Approximation strategies – Here we summarize the three major classes of approximation strategies from the statistical learning literature. The rest of this chapter summarizes these strategies.
- Objectives – Sometimes we are trying to fit a function to data which minimizes errors, and sometimes we are finding a function to maximize contributions or minimize costs. Either way, learning functions is always its own optimization problem, sometimes buried within a larger stochastic optimization problem.
- Batch vs. recursive learning – Most of the statistical learning literature focuses on using a given dataset (and of late, these are very large datasets) to fit complex statistical models. In the context of sequential decision problems, we primarily depend on adaptive (or online) learning, so this chapter describes recursive learning algorithms.

3.1.1 Observations and Data in Stochastic Optimization

Before we present our overview of statistical techniques, we need to say a word about the data we are using to estimate functions. In statistical learning, it is typically assumed that we are given input data x, after which we observe a response y. Some examples include:

- We may observe the characteristics x of a patient to predict the likelihood y of responding to a treatment regime.

- We wish to predict the weather y based on meteorological conditions x that we observe now.
- We observe the pricing behavior of nearby hotels along with the price of rooms in our hotel, which we represent by x, to predict the response y of whether a customer books a room, y.

In these settings, we obtain a dataset where we associate the response y^n with the observations x^n, which gives us a dataset $(x^n, y^n)_{n=1}^N$.

In the context of sequential decision problems, x may be a decision, such as a choice of drug treatment, the price of a product, the inventory of vaccines, or the choice of movies to display on a user's internet account. In many settings, x may consist of a mixture of controllable elements (such as a drug dosage), and uncontrollable elements (the characteristics of the patient). We can always view machine learning as taking information that is known, x, to predict or estimate something that is unknown, which we call y.

3.1.2 Indexing Input x^n and Response y^{n+1}

Most work in machine learning uses a batch dataset that we can describe by (x^n, y^n), $n = 1, \ldots, N$, where x^n is the input, or independent, variables, and y^n is the associated response (sometimes called a label).

In the context of sequential decisions, we are going to find it more convenient to pick a decision $x^n = X^\pi(S^n)$ based on what we know, given by S^n, and some rule or policy $X^\pi(S^n)$. The decision x^n is based on our history of observations $y^1, \ldots, y^n$ that are used to create our state variable S^n. We then observe y^{n+1}, which gives us an updated state S^{n+1}. Note that we start with $n = 0$, where x^0 is the first decision, which we have to make before seeing any observations.

This style of indexing is consistent with how we index time, where $x_t = S^\pi(S_t)$, after which we observe W_{t+1} which is the information that arrives between t and $t + 1$. It can, however, create unnatural labeling. Imagine a medical setting where we have treated n patients. We use what we know from the first n patients, captured in S^n, to decide the treatment for the $n + 1^{st}$ patient, after which we observe the response by the $n + 1^{st}$ patient as y^{n+1} (or W^{n+1} if we use our "W" notation). This can seem unnatural. It is important, however, to keep to the principle that if a variable is indexed by n, it depends only on information from the first n observations.

3.1.3 Functions We are Learning

The need to approximate functions arises in a number of settings in stochastic optimization. Some of the most important include:

1) Approximating the expectation of a function $\mathbb{E}F(x, W)$ to be maximized, where we assume that we have access to unbiased observations $\hat{F} = F(x, W)$ for a given decision x, which draws on a major branch of statistical learning known as supervised learning.

2) Creating an approximate policy $X^\pi(S|\theta)$. We may fit these functions using one of two ways. We may assume that we have an exogenous source of decisions x that we can use to fit our policy $X^\pi(S|\theta)$ (this would be supervised learning). More frequently, we are tuning the policy to maximize a contribution (or minimize a cost), which is sometimes referred to as a kind of reinforcement learning.

3) Approximating the value of being in a state S, given by $V_t(S_t)$. We wish to find an approximation $\overline{V}_t(S_t)$ that will give us an estimate even when one or more of the elements of S_t is continuous, and/or when S_t is multidimensional. The difference between approximating $\mathbb{E}F(x, W)$ vs. $V_t(S_t)$ is that we can get unbiased observations of $\mathbb{E}F(x, W)$, whereas observations of $V_t(S_t)$ depend on simulations using suboptimal policies to make decisions over $t + 1, t + 2, ...$, which introduces the bias.

4) Learning any of the underlying models in a dynamic system. These include:

4a) The *transition function* that describes how the system evolves over time, which we will write as $S^M(S_t, x_t, W_{t+1})$ which is used to compute the next state S_{t+1}. This arises in complex environments where the dynamics are not known, such as modeling how much water is retained in a reservoir, which depends in a complex way on rainfall and temperature. We might approximate losses using a parametric model that has to be estimated.

4b) The cost or contribution functions (also known as rewards, gains, losses). This might be unknown if a human is making a decision to maximize an unknown utility, which we might represent as a linear model with parameters to be determined from observed behaviors.

4c) The evolution of exogenous quantities such as wind or prices, where we might model an observation W_{t+1} as a function of the history $W_t, W_{t-1}, W_{t-2}, ...$, where we have to fit our model from past observations.

There are three strategies we can use to approach the learning problems in this category:

Exogenous learning – An example of a transition function is a time series model of wind speeds w_t which we might write as

$$w_{t+1} = \bar{\theta}_{t0}w_t + \bar{\theta}_{t1}w_{t-1} + \bar{\theta}_{t2}w_{t-2} + \varepsilon_{t+1},$$

where the input $x_t = (w_t, w_{t-1}, w_{t-2})$ and the response $y_{t+1} = w_{t+1}$ allows us to update our estimate of the parameter vector $\bar{\theta}_t$. The response y_{t+1} comes from outside the system.

Endogenous learning – We may have an estimate of a value function

$$\overline{V}_t^n(S_t|\bar{\theta}_t) = \sum_{f \in \mathcal{F}} \bar{\theta}_{tf}^n \phi_f(S_t).$$

We can then generate a sampled observation $\hat{v}_t^n$ using

$$\hat{v}_t^n = \max_{a_t} \left(C(S_t^n, a_t) + \mathbb{E}_{W_{t+1}} \{ \overline{V}_{t+1}(S_{t+1}^n | \bar{\theta}^{n-1}) | S_t^n \} \right),$$

to update our parameters $\bar{\theta}_t^n$. The sampled estimate $\hat{v}_t^n$ is created endogenously.

Inverse optimization – Imagine that we are watching a human make decisions (playing a game, managing a robot, dispatching a truck, deciding on a medical treatment) where we do not have a well-defined contribution function $C(S_t, x_t)$. Assume that we can come up with a parameterized contribution function $C(S_t, x_t | \theta^{cont})$. We do not have exogenous observations of contributions, and we also do not have endogenous calculations such as $\hat{v}_t$ that provide noisy estimates of the contribution. However, we are given a history of actual decisions x_t. Assume that we are using a policy $X^\pi(S_t | \theta^{cont})$ that depends on $C(S_t, x_t | \theta^{cont})$ (and therefore depends on θ^{cont}). In this case, the policy $X^\pi(S_t | \theta^{cont})$ plays a role exactly analogous to a statistical model, where we choose θ^{cont} to get the best fit between our policy $X^\pi(S_t | \theta^{cont})$ and the observed decisions. Of course this is a form of exogenous learning, but the decisions only hint at what the contribution function should be.

5) Later we will introduce a class of policies that we call *parametric cost function approximations* where we have to learn two types of functions:

   5a) Parametric modifications of cost functions (for example, a penalty for not serving a demand now but instead holding it for the future). This is not the same as estimating the reward function (see bullet 4) from observed decisions.

   5b) Parametric modifications of constraints (for example, inserting schedule slack into an airline schedule to handle uncertainty in travel times).

   Each of these parametric modifications have to be tuned (which is a form of function estimation) to produce the best results over time.

3.1.4 Sequential Learning: From Very Little Data to ... More Data

A common theme in learning problems in the context of sequential decision problems is that the learning has to be done adaptively. This typically means that instead of fitting just one model, we have to transition from models with relatively few parameters (we might call these low-dimensional architectures) to higher-dimensional architectures.

There has been considerable attention to the online updating of parameter estimates. This is particularly easy in the case of linear models, although more challenging with nonlinear models like neural networks. However, there has been much less attention given to the updating of the structure of the model itself in an online setting.

3.1.5 Approximation Strategies

Our tour of statistical learning makes a progression through the following classes of approximation strategies:

Lookup tables – Here we estimate a function $f(x)$ where x falls in a discrete region $\mathcal{X}$ given by a set of points $x_1, x_2, \dots, x_M$. A point x_m could be the characteristics of a person, a type of material, or a movie. Or it could be a point in a discretized, continuous region. As long as x is some discrete element, $f(x)$ is a function where we pick x, and then "look up" its value $f(x)$. Some authors call these "tabular" representations.

In most applications, lookup tables work well in one or two dimensions, then become difficult (but feasible) in three or four dimensions, and then quickly become impractical starting at four or five dimensions. This is the classical "curse of dimensionality." Our presentation focuses on using aggregation, and especially hierarchical aggregation, both to handle the curse of dimensionality, as well as to manage the transition in recursive estimation from initial estimates with very little data, to produce better estimates as more data becomes available.

Parametric models – There are many problems where we can approximate a function using an analytical model in terms of some unknown parameters. These come in two broad categories:

Linear models – The simplest parametric model is linear in the parameters, which we might write

$$f(x|\theta) = \theta_0 + \theta_1 \phi_1(x) + \theta_2 \phi_2(x) + \dots, \tag{3.1}$$

where $(\phi_f(x))_{f \in \mathcal{F}}$ are *features* that extract possibly useful information from x which could be a vector, or the data describing a movie or ad. Equation (3.1) is called a linear model because it is linear in θ (it may be highly nonlinear in x). Alternatively, we may have a nonlinear model such as

$$f(x|\theta) = e^{\sum_{f \in \mathcal{F}} \theta_f \phi_f(x)}.$$

Parametric models may be low-dimensional (1-100 parameters), or high-dimensional (e.g. several hundred to thousands of parameters).

Nonlinear models – Nonlinear parametric models are usually chosen with a particular form motivated by the problem. Some examples are step functions (useful in asset buying and selling or inventory problems)

$$f(x|\theta) = \begin{cases} -1 & x \le \theta^{low}, \\ 0 & \theta^{low} < x < \theta^{high}, \\ +1 & x \ge \theta^{high}, \end{cases} \tag{3.2}$$

or logistic regression (useful for pricing and recommendation problems)

$$f(x|\theta) = \frac{1}{1 + e^{\theta_0 + \theta_1 x_1 + \dots}}. \tag{3.3}$$

There are models such as neural networks whose primary advantage is that they do not impose any structure, which means they can approximate almost anything (especially the very large instances known as deep neural networks). These models can feature tens of thousands to as many as hundreds of millions of parameters. Not surprisingly, they require very large datasets to determine these parameters.

Nonparametric models – Nonparametric models create estimates by building a structure directly from the data. A simple example is where we estimate $f(x)$ from a weighted combination of nearby observations drawn from a set (f^n, x^n), $n = 1, \dots, N$. We can also construct approximations through locally linear approximations.

The three categories of statistical models – lookup tables, parametric, and nonparametric – are best thought of as overlapping sets, as illustrated in Figure 3.1. For example, neural networks, which we describe below, can be classified as either parametric models (for simpler neural networks) or nonparametric models (for deep neural networks). Other methods are effectively hybrids, such as those based on tree regression which might create a linear approximation (parametric) around specific regions of the input data (the definitions of the regions are lookup table).

Notably missing from this chapter is approximation methods for convex functions. There are many applications where $F(x, W)$ is convex in x. This function is so special that we defer handling this problem class until chapter 5 (and especially chapter 18) when we address stochastic convex (or concave) stochastic optimization problems such as linear programs with random data.

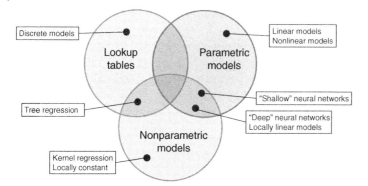

Figure 3.1 Illustration of the overlap between lookup table, parametric, and nonparametric statistical models.

We begin our presentation with lookup tables, which are the simplest way to represent a function without assuming any structure. We begin by presenting lookup tables from both frequentist and Bayesian perspectives. In sequential decision problems, we need both belief models. As a general rule, Bayesian models are best when we have access to some prior, and where function evaluations are expensive.

3.1.6 From Data Analytics to Decision Analytics

Learning in the context of sequential decision problems can be approached from the perspective of two broad objectives:

- Learning a function – We might want to learn an approximation of a function such as an objective function $\mathbb{E}F(x, W)$, or a value function $V(s)$ or perhaps even the transition function $S^M(s, x, W)$. In these settings, we assume we have a source of observations of our function that may be noisy, and even biased. For example, we might have access to y^{n+1} which is a noisy observation of $\mathbb{E}F(x^n, W^{n+1})$ that we are going to approximate with some function $f(x|\theta)$. If we collect a dataset $(x^0, y^1, x^1, y^2, ..., x^{n-1}, y^n)$, we would look to find θ that minimizes the error between the observations y and $f(x|\theta)$ using

$$\min_{\theta} \frac{1}{N} \sum_{n=0}^{N-1} \left(y^{n+1} - f(x^n|\theta) \right)^2. \tag{3.4}$$

- Maximizing rewards (or minimizing costs) – We can search for a policy $X^\pi(S|\theta)$ that maximizes a contribution function $C(S, x)$ using

$$\max_\theta \mathbb{E} C(S, X^\pi(S)) \approx \frac{1}{N} \sum_{n=0}^{N-1} C(S^n, X^\pi(S^n|\theta)), \qquad (3.5)$$

where the states evolve according to a known transition function $S^{n+1} = S^M(S^n, x^n, W^{n+1})$.

The objective function in (3.4) is characteristic of classical machine learning, which we put under the umbrella of "data analytics." There are different ways to express objectives (for example, we might want to use $|y^{n+1} - f(x^n|\theta)|$, but they always involve predictions from a model, $f(x|\theta)$, and observations y.

The objective function in (3.5) is characteristic of optimization problems, which we put under the umbrella of "decision analytics." It assumes some form of pre-defined performance metric (cost, contribution, reward, utility), and notably does not require an exogenous dataset $(y^n)_{n=1}^N$.

3.1.7 Batch vs. Online Learning

Equation (3.4) (or (3.5)) is the standard problem that arises in batch learning problems, where we use a fixed dataset (possibly a very large one in the modern era of "big data") to fit a model (increasingly high-dimensional models such as neural networks which we introduce below).

While batch learning can arise in stochastic optimization, the most common learning problems are adaptive, which means updating estimates as new data arrives as happens in online applications. Imagine that after n iterations (or samples), we have the sequence

$$(x^0, W^1, y^1, x^1, W^2, y^2, x^2, \dots, W^n, y^n).$$

Assume that we use this data to obtain an estimate of our function that we call $\bar{F}^n(x)$. Now assume we use this estimate to make a decision x^n, after which we experience exogenous information W^{n+1} and then the response y^{n+1}. We need to use our prior estimate $\bar{F}^n(x)$ along with the new information (W^{n+1}, y^{n+1}) to produce a new estimate $\bar{F}^{n+1}(x)$.

We could, of course, just solve a new batch problem with one more observation. This can be computationally expensive, and it also puts equal weight on the entire history. There are some settings where the more recent observations are more important.

3.2 Adaptive Learning Using Exponential Smoothing

The most common method we will use for adaptive learning is known by various names, but is popularly referred to as *exponential smoothing*. Assume we have a sequence of observations of some quantity, which might be the number of people booking a room, the response of a patient to a particular drug, or the travel time on a path. Let μ be the unknown truth, which could be the average number of people booking a room at a particular price, or the probability a patient responds to a drug, or the average travel time of our path. We want to estimate the average from a sequence of observations.

Let W^n be the nth observation of the quantity we are trying to estimate, and let $\bar{\mu}^n$ be our estimate of the true mean μ after n observations. The most widely used method for computing $\bar{\mu}^{n+1}$ given $\bar{\mu}^n$ and a new observation W^{n+1} is given by

$$\bar{\mu}^{n+1} = (1 - \alpha_n)\bar{\mu}^n + \alpha_n W^{n+1}. \tag{3.6}$$

In chapter 5 we are going to motivate (3.6) using an algorithmic strategy known as stochastic gradient algorithms for solving a specific optimization problem. For now, it is enough to say that this basic equation will arise frequently in a variety of online learning problems.

Not surprisingly, the biggest challenge with this method is choosing α_n. The variable α_n is known variously as a learning rate, smoothing factor or (in this book), a stepsize (we will see the motivation for the term stepsize in chapter 5). This topic is so rich that we dedicate an entire chapter (chapter 6) to this topic. For now, we can hint at some simple strategies:

- Constant stepsizes – Easily the simplest strategy is one that is actually widely used, which is to simply set $\alpha_n = \bar{\alpha}$ where $\bar{\alpha}$ is a constant chosen in advance.
- Harmonic stepsize – This is an arithmetically declining sequence

$$\alpha_n = \frac{\theta^{step}}{\theta^{step} + n - 1}.$$

If $\theta^{step} = 1$, this gives us $\alpha_n = 1/n$ (we show in chapter 6 that this produces a simple average). Often this stepsize declines too quickly. Increasing θ^{step} slows the decline in the stepsize which can accelerate learning. It is also possible to have a declining sequence that approaches a limit point.
- In chapter 6 we also introduce a family of adaptive stepsizes that respond to the data.

3.3   Lookup Tables with Frequentist Updating

The frequentist view is arguably the approach that is most familiar to people with an introductory course in statistics. Assume we are trying to estimate the mean μ of a random variable W which might be the performance of a device or policy. Let W^n be the nth sample observation, such as the sales of a product or the blood sugar reduction achieved by a particular medication. Also let $\bar{\mu}^n$ be our estimate of μ, and $\hat{\sigma}^{2,n}$ be our estimate of the variance of W. We know from elementary statistics that we can write $\bar{\mu}^n$ and $\hat{\sigma}^{2,n}$ using

$$\bar{\mu}^n \quad = \quad \frac{1}{n} \sum_{m=1}^{n} W^m, \tag{3.7}$$

$$\hat{\sigma}^{2,n} \quad = \quad \frac{1}{n-1} \sum_{m=1}^{n} (W^m - \bar{\mu}^n)^2. \tag{3.8}$$

The estimate $\bar{\mu}^n$ is a random variable (in the frequentist view) because it is computed from other random variables, namely $W^1, W^2, \ldots, W^n$. Imagine if we had 100 people each choose a sample of n observations of W. We would obtain 100 different estimates of $\bar{\mu}^n$, reflecting the variation in our observations of W. The best estimate of the variance of the estimator $\bar{\mu}^n$ is given by

$$\bar{\sigma}^{2,n} = \frac{1}{n} \hat{\sigma}^{2,n}.$$

Note that as $n \to \infty$, $\bar{\sigma}^{2,n} \to 0$, but $\hat{\sigma}^{2,n} \to \sigma^2$ where σ^2 is the true variance of W. If σ^2 is known, there would be no need to compute $\hat{\sigma}^{2,n}$ and $\bar{\sigma}^{2,n}$ would be given as above with $\hat{\sigma}^{2,n} = \sigma^2$.

We can write these expressions recursively using

$$\bar{\mu}^n \quad = \quad \left(1 - \frac{1}{n}\right) \bar{\mu}^{n-1} + \frac{1}{n} W^n, \tag{3.9}$$

$$\hat{\sigma}^{2,n} \quad = \quad \frac{n-2}{n-1} \hat{\sigma}^{2,n-1} + \frac{1}{n}(W^n - \bar{\mu}^{n-1})^2, \quad n \geq 2. \tag{3.10}$$

We will often speak of our belief state which captures what we know about the parameters we are trying to estimate. Given our observations, we would write our belief state as

$$B^n = \left(\bar{\mu}^n, \hat{\sigma}^{2,n}\right).$$

Equations (3.9) and (3.10) describe how our belief state evolves over time.

3.4   Lookup Tables with Bayesian Updating

The Bayesian perspective casts a different interpretation on the statistics we compute which is particularly useful in the context of learning when observations are expensive (imagine having to run expensive simulations or field experiments). In the frequentist perspective, we do not start with any knowledge about the system before we have collected any data. It is easy to verify from equations (3.9) and (3.10) that we never use $\bar{\mu}^0$ or $\hat{\sigma}^{2,0}$.

By contrast, in the Bayesian perspective we assume that we begin with a prior distribution of belief about the unknown parameter μ. In other words, any number whose value we do not know is interpreted as a random variable, and the distribution of this random variable represents our belief about how likely μ is to take on certain values. So if μ is the true but unknown mean of W, we might say that while we do not know what this mean is, we think it is normally distributed around θ^0 with standard deviation σ^0.

Thus, the true mean μ is treated as a random variable with a known mean and variance, but we are willing to adjust our estimates of the mean and variance as we collect additional information. If we add a distributional assumption such as the normal distribution, we would say that this is our initial distribution of belief, known generally as the Bayesian prior.

The Bayesian perspective is well suited to problems where we are collecting information about a process where observations are expensive. This might arise when trying to price a book on the internet, or plan an expensive laboratory experiment. In both cases, we can be expected to have some prior information about the right price for a book, or the behavior of an experiment using our knowledge of physics and chemistry.

We note a subtle change in notation from the frequentist perspective, where $\bar{\mu}^n$ was our statistic giving our estimate of μ. In the Bayesian view, we let $\bar{\mu}^n$ be our estimate of the mean of the random variable μ after we have made n observations. It is important to remember that μ is a random variable whose distribution reflects our prior belief about μ. The parameter $\bar{\mu}^0$ is not a random variable. This is our initial estimate of the mean of our prior distribution. After n observations, $\bar{\mu}^n$ is our updated estimate of the mean of the random variable μ (the true mean).

Below we first use some simple expressions from probability to illustrate the effect of collecting information. We then give the Bayesian version of (3.9) and (3.10) for the case of independent beliefs, where observations of one choice do not influence our beliefs about other choices. We follow this discussion by giving the updating equations for correlated beliefs, where an observation of μ_x for alternative x tells us something about $\mu_{x'}$. We round out our presentation by touching on other important types of distributions.

3.4.1 The Updating Equations for Independent Beliefs

We begin by assuming (as we do through most of our presentation) that our random variable W is normally distributed. Let σ_W^2 be the variance of W, which captures the noise in our ability to observe the true value. To simplify the algebra, we define the *precision* of W as

$$\beta^W = \frac{1}{\sigma_W^2}.$$

Precision has an intuitive meaning: smaller variance means that the observations will be closer to the unknown mean, that is, they will be more precise.

Now let $\bar{\mu}^n$ be our estimate of the true mean μ after n observations, and let β^n be the precision of this estimate. If we observe W^{n+1}, $\bar{\mu}^n$ and β^n are updated according to

$$\bar{\mu}^{n+1} = \frac{\beta^n \bar{\mu}^n + \beta^W W^{n+1}}{\beta^n + \beta^W}, \tag{3.11}$$

$$\beta^{n+1} = \beta^n + \beta^W. \tag{3.12}$$

Equations (7.26) and (7.27) are the Bayesian counterparts of (3.9) and (3.10), although we have simplified the problem a bit by assuming that the variance of W is known. The belief state in the Bayesian view (with normally distributed beliefs) is given by the belief state

$$B^n = (\bar{\mu}^n, \beta^n).$$

If our prior distribution of belief about μ is normal, and if the observation W is normal, then the posterior distribution is also normal. It turns out that after a few observations (perhaps five to ten), the distribution of belief about μ will be approximately normal due to the law of large numbers for almost any distribution of W. For the same reason, the posterior distribution is also approximately normal regardless of the distribution of W! So, our updating equations (7.26) and (7.27) produce the mean and precision of a normal distribution for almost all problems!

3.4.2 Updating for Correlated Beliefs

We are now going to make the transition that instead of one number μ, we now have a vector $\mu_{x_1}, \mu_{x_2}, \ldots, \mu_{x_M}$ where $\mathcal{X} = \{x_1, \ldots, x_M\}$ is our set we are choosing among. We can think of an element of μ as μ_x, which might be our estimate of a function $\mathbb{E}F(x, W)$ at x. Often, μ_x and $\mu_{x'}$ are correlated, as might happen when x is continuous, and x and x' are close to each other. There are a number of examples that exhibit what we call *correlated beliefs*:

■ **EXAMPLE 3.1**

We are interested in finding the price of a product that maximizes total revenue. We believe that the function $R(p)$ that relates revenue to price is continuous. Assume that we set a price p^n and observe revenue R^{n+1} that is higher than we had expected. If we raise our estimate of the function $R(p)$ at the price p^n, our beliefs about the revenue at nearby prices should be higher.

■ **EXAMPLE 3.2**

We choose five people for the starting lineup of our basketball team and observe total scoring for one period. We are trying to decide if this group of five people is better than another lineup that includes three from the same group with two different people. If the scoring of these five people is higher than we had expected, we would probably raise our belief about the other group, since there are three people in common.

■ **EXAMPLE 3.3**

A physician is trying to treat diabetes using a treatment of three drugs, where she observes the drop in blood sugar from a course of a particular treatment. If one treatment produces a better-than-expected response, this would also increase our belief of the response from other treatments that have one or two drugs in common.

■ **EXAMPLE 3.4**

We are trying to find the highest concentration of a virus in the population. If the concentration of one group of people is higher than expected, our belief about other groups that are close (either geographically, or due to other relationships) would also be higher.

Correlated beliefs are a particularly powerful device in learning functions, allowing us to generalize the results of a single observation to other alternatives that we have not directly measured.

Let $\bar{\mu}_x^n$ be our belief about alternative x after n measurements. Now let

$$Cov^n(\mu_x, \mu_y) = \text{the covariance in our belief about } \mu_x \text{ and } \mu_y$$

$$\text{given the first } n \text{ observations.}$$

We let Σ^n be the covariance matrix, with element $\Sigma^n_{xy} = Cov^n(\mu_x, \mu_y)$. Just as we defined the precision β^n_x to be the reciprocal of the variance, we are going to define the precision matrix M^n to be

$$M^n = (\Sigma^n)^{-1}.$$

Let e_x be a column vector of zeroes with a 1 for element x, and as before we let W^{n+1} be the (scalar) observation when we decide to measure alternative x. We could label W^{n+1} as W^{n+1}_x to make the dependence on the alternative more explicit. For this discussion, we are going to use the notation that we choose to measure x^n and the resulting observation is W^{n+1}.

If we choose to measure x^n, we can also interpret the observation as a column vector given by $W^{n+1}e_{x^n}$. Keeping in mind that $\bar{\mu}^n$ is a column vector of our beliefs about the expectation of μ, the Bayesian equation for updating this vector in the presence of correlated beliefs is given by

$$\bar{\mu}^{n+1} \quad = \quad (M^{n+1})^{-1}\left(M^n\bar{\mu}^n + \beta^W W^{n+1}e_{x^n}\right), \tag{3.13}$$

where M^{n+1} is given by

$$M^{n+1} \quad = \quad (M^n + \beta^W e_{x^n}(e_{x^n})^T). \tag{3.14}$$

Note that $e_x(e_x)^T$ is a matrix of zeroes with a one in row x, column x, whereas β^W is a scalar giving the precision of our measurement W.

It is possible to perform these updates without having to deal with the inverse of the covariance matrix. This is done using a result known as the Sherman-Morrison formula. If A is an invertible matrix (such as Σ^n) and u is a column vector (such as e_x), the Sherman-Morrison formula is

$$[A + uu^T]^{-1} = A^{-1} - \frac{A^{-1}uu^T A^{-1}}{1 + u^T A^{-1}u}. \tag{3.15}$$

See section 3.14.2 for the derivation of this formula.

Using the Sherman-Morrison formula, and letting $x = x^n$, we can rewrite the updating equations as

$$\bar{\mu}^{n+1}(x) \quad = \quad \bar{\mu}^n + \frac{W^{n+1} - \bar{\mu}^n_x}{\sigma^2_W + \Sigma^n_{xx}}\Sigma^n e_x, \tag{3.16}$$

$$\Sigma^{n+1}(x) \quad = \quad \Sigma^n - \frac{\Sigma^n e_x(e_x)^T\Sigma^n}{\sigma^2_W + \Sigma^n_{xx}}, \tag{3.17}$$

where we express the dependence of $\bar{\mu}^{n+1}(x)$ and $\Sigma^{n+1}(x)$ on the alternative x which we have chosen to measure.

To illustrate, assume that we have three alternatives with mean vector

$$
\bar{\mu}^n = \begin{bmatrix} 20 \\ 16 \\ 22 \end{bmatrix}.
$$

Assume that $\sigma_W^2 = 9$ and that our covariance matrix Σ^n is given by

$$
\Sigma^n = \begin{bmatrix} 12 & 6 & 3 \\ 6 & 7 & 4 \\ 3 & 4 & 15 \end{bmatrix}.
$$

Assume that we choose to measure $x = 3$ and observe $W^{n+1} = W_3^{n+1} = 19$. Applying equation (3.16), we update the means of our beliefs using

$$
\begin{aligned}
\bar{\mu}^{n+1}(3) &= \begin{bmatrix} 20 \\ 16 \\ 22 \end{bmatrix} + \frac{19 - 22}{9 + 15} \begin{bmatrix} 12 & 6 & 3 \\ 6 & 7 & 4 \\ 3 & 4 & 15 \end{bmatrix} \begin{bmatrix} 0 \\ 0 \\ 1 \end{bmatrix} \\
&= \begin{bmatrix} 20 \\ 16 \\ 22 \end{bmatrix} + \frac{-3}{24} \begin{bmatrix} 3 \\ 4 \\ 15 \end{bmatrix} \\
&= \begin{bmatrix} 19.625 \\ 15.500 \\ 20.125 \end{bmatrix}.
\end{aligned}
$$

The update of the covariance matrix is computed using

$$
\begin{aligned}
\Sigma^{n+1}(3) &= \begin{bmatrix} 12 & 6 & 3 \\ 6 & 7 & 4 \\ 3 & 4 & 15 \end{bmatrix} - \frac{\begin{bmatrix} 12 & 6 & 3 \\ 6 & 7 & 4 \\ 3 & 4 & 15 \end{bmatrix} \begin{bmatrix} 0 \\ 0 \\ 1 \end{bmatrix} [0\ 0\ 1] \begin{bmatrix} 12 & 6 & 3 \\ 6 & 7 & 4 \\ 3 & 4 & 15 \end{bmatrix}}{9 + 15} \\
&= \begin{bmatrix} 12 & 6 & 3 \\ 6 & 7 & 4 \\ 3 & 4 & 15 \end{bmatrix} - \frac{1}{24} \begin{bmatrix} 3 \\ 4 \\ 15 \end{bmatrix} [3\ 4\ 15] \\
&= \begin{bmatrix} 12 & 6 & 3 \\ 6 & 7 & 4 \\ 3 & 4 & 15 \end{bmatrix} - \frac{1}{24} \begin{bmatrix} 9 & 12 & 45 \\ 12 & 16 & 60 \\ 45 & 60 & 225 \end{bmatrix} \\
&= \begin{bmatrix} 12 & 6 & 3 \\ 6 & 7 & 4 \\ 3 & 4 & 15 \end{bmatrix} - \begin{bmatrix} 0.375 & 0.500 & 1.875 \\ 0.500 & 0.667 & 2.500 \\ 1.875 & 2.500 & 9.375 \end{bmatrix} \\
&= \begin{bmatrix} 11.625 & 5.500 & 1.125 \\ 5.500 & 6.333 & 1.500 \\ 1.125 & 1.500 & 5.625 \end{bmatrix}.
\end{aligned}
$$

These calculations are fairly easy, which means we can execute them even if we have thousands of alternatives. However, the method starts to become impractical if the number of alternatives is in the range of 10^5 or more, which arises when we consider problems where an alternative x is itself a multidimensional vector.

3.4.3 Gaussian Process Regression

A common strategy for approximating continuous functions is to discretize them, and then capture continuity by noting that the value of nearby points will be correlated, simply because of continuity. This is known as *Gaussian process regression*.

Assume that we have an unknown function $f(x)$ that is continuous in x which for the moment we will assume is a scalar that is discretized into the values $(x_1, x_2, \ldots, x_M)$. Let $\bar{\mu}^n(x)$ be our estimate of $f(x)$ over our discrete set. Let $\mu(x)$ be the true value of $f(x)$ which, with our Bayesian hat on, we will interpret as a normally distributed random variable with mean $\bar{\mu}_x^0$ and variance $(\sigma_x^0)^2$ (this is our prior). We will further assume that μ_x and $\mu_{x'}$ are correlated with covariance

$$Cov(\mu_x, \mu_{x'}) = (\sigma^0)^2 e^{\alpha \|x - x'\|}, \tag{3.18}$$

where $\|x - x'\|$ is some distance metric such as $|x - x'|$ or $(x - x')^2$ (if x is a scalar) or $\sqrt{\sum_{i=1}^{I}(x_i - x_i')^2}$ if x is a vector. If $x = x'$ then we just pick up the variance in our belief about μ_x. The parameter α captures the degree to which x and x' are related as they get further apart.

Figure 3.2 illustrates a series of curves randomly generated from a belief model using the covariance function given in equation (3.18) for different values of α. Smaller values of α produce smoother curves with fewer undulations, because a smaller α translates to a higher covariance between more distant values of x and x'. As α increases, the covariance drops off and two different points on the curve become more independent.

Gaussian process regression (often shortened to just "GPR") is a powerful approach for approximating smooth functions that are continuous but otherwise have no specific structure. We present GPR here as a generalization of lookup table belief models, but it can also be characterized as a form of nonparametric statistics which we discuss below. In chapter 7 we will show how using GPR as a belief model can dramatically accelerate optimizing functions of continuous parameters such as drug dosages for medical applications, or the choice of temperature, pressure, and concentration in a laboratory science application.

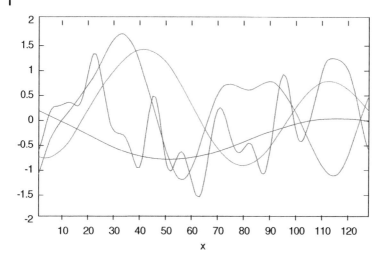

Figure 3.2 Illustration of a series of functions generated using Gaussian process regression (correlated beliefs) for different values of α.

3.5 Computing Bias and Variance*

A powerful strategy for estimating functions of multidimensional vectors using lookup tables is hierarchical aggregation, where we estimate the function at different levels of aggregation. To lay the foundation for this approach, we are going to need some basic results on bias and variance in statistical estimation.

Assume we are trying to estimate a true but unknown parameter μ which we can observe, but we have to deal with both bias β and noise ε, which we write as

$$\hat{\mu}^n = \mu + \beta + \varepsilon^n. \tag{3.19}$$

Both μ and β are unknown, but we are going to assume that we have some way to make a noisy estimate of the bias that we are going to call $\hat{\beta}^n$. Later we are going to provide examples of how to get estimates of β.

Now let $\bar{\mu}^n$ be our estimate of μ after n observations. We will use the following recursive formula for $\bar{\mu}^n$

$$\bar{\mu}^n = (1 - \alpha_{n-1})\bar{\mu}^{n-1} + \alpha_{n-1}\hat{\mu}^n.$$

We are interested in estimating the variance of $\bar{\mu}^n$ and its bias $\bar{\beta}^n$. We start by computing the variance of $\bar{\mu}^n$. We assume that our observations of μ can be represented using equation (3.19), where $\mathbb{E}\varepsilon^n = 0$ and $Var[\varepsilon^n] = \sigma^2$. With this model, we can compute the variance of $\bar{\mu}^n$ using

$$Var[\bar{\mu}^n] = \lambda^n \sigma^2, \tag{3.20}$$

where λ^n (this is λ at iteration n, not raised to the nth power) can be computed from the simple recursion

$$\lambda^n = \begin{cases} \alpha_{n-1}^2, & n = 1, \\ (1 - \alpha_{n-1})^2 \lambda^{n-1} + \alpha_{n-1}^2, & n > 1. \end{cases} \tag{3.21}$$

To see this, we start with $n = 1$. For a given (deterministic) initial estimate $\bar{\mu}^0$, we first observe that the variance of $\bar{\mu}^1$ is given by

$$\begin{aligned} Var[\bar{\mu}^1] &= Var[(1 - \alpha_0)\bar{\mu}^0 + \alpha_0 \hat{\mu}^1] \\ &= \alpha_0^2 Var[\hat{\mu}^1] \\ &= \alpha_0^2 \sigma^2. \end{aligned}$$

For $\bar{\mu}^n$ for $n > 1$, we use a proof by induction. Assume that $Var[\bar{\mu}^{n-1}] = \lambda^{n-1}\sigma^2$. Then, since $\bar{\mu}^{n-1}$ and $\hat{\mu}^n$ are independent, we find

$$\begin{aligned} Var[\bar{\mu}^n] &= Var\left[(1 - \alpha_{n-1})\bar{\mu}^{n-1} + \alpha_{n-1}\hat{\mu}^n\right] \\ &= (1 - \alpha_{n-1})^2 Var\left[\bar{\mu}^{n-1}\right] + \alpha_{n-1}^2 Var[\hat{\mu}^n] \\ &= (1 - \alpha_{n-1})^2 \lambda^{n-1}\sigma^2 + \alpha_{n-1}^2\sigma^2 \tag{3.22} \\ &= \lambda^n \sigma^2. \tag{3.23} \end{aligned}$$

Equation (3.22) is true by assumption (in our induction proof), while equation (3.23) establishes the recursion in equation (3.21). This gives us the variance, assuming of course that σ^2 is known.

Using our assumption that we have access to a noisy estimate of the bias given by β^n, we can compute the mean-squared error using

$$\mathbb{E}\left[\left(\bar{\mu}^{n-1} - \bar{\mu}^n\right)^2\right] = \lambda^{n-1}\sigma^2 + \beta^{2,n}. \tag{3.24}$$

See exercise 3.11 to prove this. This formula gives the variance around the known mean, $\bar{\mu}^n$. For our purposes, it is also useful to have the variance around the observations $\hat{\mu}^n$. Let

$$\nu^n = \mathbb{E}\left[\left(\bar{\mu}^{n-1} - \hat{\mu}^n\right)^2\right]$$

be the mean squared error (including noise and bias) between the current estimate $\bar{\mu}^{n-1}$ and the observation $\hat{\mu}^n$. It is possible to show that (see exercise 3.12)

$$\nu^n = (1 + \lambda^{n-1})\sigma^2 + \beta^{2,n},$$ (3.25)

where λ^n is computed using (3.21).

In practice, we do not know σ^2, and we certainly do not know the bias β. As a result, we have to estimate both parameters from our data. We begin by providing an estimate of the bias using

$$\bar{\beta}^n = (1 - \eta_{n-1})\bar{\beta}^{n-1} + \eta_{n-1}\beta^n,$$

where η_{n-1} is a (typically simple) stepsize rule used for estimating the bias and variance. As a general rule, we should pick a stepsize for η_{n-1} which produces larger stepsizes than α_{n-1} because we are more interested in tracking the true signal than producing an estimate with a low variance. We have found that a constant stepsize such as .10 works quite well on a wide range of problems, but if precise convergence is needed, it is necessary to use a rule where the stepsize goes to zero such as the harmonic stepsize rule (equation (6.15)).

To estimate the variance, we begin by finding an estimate of the total variation ν^n. Let $\bar{\nu}^n$ be the estimate of the total variance which we might compute using

$$\bar{\nu}^n = (1 - \eta_{n-1})\bar{\nu}^{n-1} + \eta_{n-1}(\bar{\mu}^{n-1} - \hat{\mu}^n)^2.$$

Using $\bar{\nu}^n$ as our estimate of the total variance, we can compute an estimate of σ^2 using

$$\bar{\sigma}^{2,n} = \frac{\bar{\nu}^n - \bar{\beta}^{2,n}}{1 + \lambda^{n-1}}.$$

We can use $(\bar{\sigma}^n)^2$ in equation (3.20) to obtain an estimate of the variance of $\bar{\mu}^n$.

If we are doing true averaging (as would occur if we use a stepsize of $1/n$), we can get a more precise estimate of the variance for small samples by using the recursive form of the small sample formula for the variance

$$\hat{\sigma}^{2,n} = \frac{n-2}{n-1}\hat{\sigma}^{2,n-1} + \frac{1}{n}(\bar{\mu}^{n-1} - \hat{\mu}^n)^2.$$ (3.26)

The quantity $\hat{\sigma}^{2,n}$ is an estimate of the variance of $\hat{\mu}^n$. The variance of our estimate $\bar{\mu}^n$ is computed using

$$\bar{\sigma}^{2,n} = \frac{1}{n}\hat{\sigma}^{2,n}.$$

We are going to draw on these results in two settings, which are both distinguished by how estimates of the bias β^n are computed:

- Hierarchical aggregation – We are going to estimate a function at different levels of aggregation. We can assume that the estimate of the function at the most disaggregate level is noisy but unbiased, and then let the difference between the function at some level of aggregation and the function at the most disaggregate level as an estimate of the bias.

- Transient functions – Later, we are going to use these results to approximate value functions. It is the nature of algorithms for estimating value functions that the underlying process varies over time (we see this most clearly in chapter 14). In this setting, we are making observations from a truth that is changing over time, which introduces a bias.

3.6 Lookup Tables and Aggregation*

Lookup table representations are the simplest and most general way to represent a function. If we are trying to model a function $f(x) = \mathbb{E}F(x, W)$, or perhaps a value function $V_t(S_t)$, assume that our function is defined over a discrete set of values $x_1, \dots, x_M$ (or discrete states $\mathcal{S} = \{1, 2, \dots, |\mathcal{S}|\}$). We wish to use observations of our function, whether they be $f^n = F(x^n, W^{n+1})$ (or $\hat{v}_t^n$, derived from simulations of the value of being in a state S_t), to create an estimate $\bar{F}^{n+1}$ (or $\bar{V}_t^{n+1}(S_t)$).

The problem with lookup table representations is that if our variable x (or state S) is a vector, then the number of possible values grows exponentially with the number of dimensions. This is the classic curse of dimensionality. One strategy for overcoming the curse of dimensionality is to use aggregation, but picking a single level of aggregation is generally never satisfactory. In particular, we typically have to start with no data, and steadily build up an estimate of a function.

We can accomplish this transition from little to no data, to increasing numbers of observations, by using hierarchical aggregation. Instead of picking a single level of aggregation, we work with a family of aggregations which are hierarchically structured.

3.6.1 Hierarchical Aggregation

Lookup table representations of functions often represent the first strategy we consider because it does not require that we assume any structural form. The problem is that lookup tables suffer from the curse of dimensionality. A powerful strategy that makes it possible to extend lookup tables is the use of hierarchical aggregation. Rather than simply aggregating a state space into a smaller space, we pose a family of aggregations, and then combine these based on the statistics of our estimates at each level of aggregation. This is not a panacea (nothing is), and should not be viewed as a method that "solves" the curse of dimensionality, but it does represent a powerful addition to our toolbox of approximation strategies. As we will see, this is particularly useful when being applied in the context of sequential decision problems.

We can illustrate hierarchical aggregation using our nomadic trucker example that we first introduced in section 2.3.4.1. In this setting, we are managing a truck driver who is picking up and dropping off loads (imagine taxicabs for freight), where the driver has to choose loads based on both how much money he will make moving the load, and the value of landing at the destination of the load. Complicating the problem is that the driver is described by a multi-dimensional attribute vector $a = (a_1, a_2, ..., a_d)$ which includes attributes such as the location of a truck (which means location in a region), his equipment type, and his home location (again, a region).

If our nomadic trucker is described by the state vector $S_t = a_t$ which we act on with an action x_t (moving one of the available loads), the transition function $S_{t+1} = S^M(S_t, x_t, W_{t+1})$ may represent the state vector at a high level of detail (some values may be continuous). But the decision problem

$$\max_{x_t \in \mathcal{X}} \left(C(S_t, x_t) + \mathbb{E}\{\overline{V}_{t+1}(G(S_{t+1}))|S_t\} \right) \tag{3.27}$$

uses a value function $\overline{V}_{t+1}(G(S_{t+1}))$, where $G(\cdot)$ is an aggregation function that maps the original (and very detailed) state S into something much simpler. The aggregation function G may ignore a dimension, discretize it, or use any of a variety of ways to reduce the number of possible values of a state vector. This also reduces the number of parameters we have to estimate. In what follows, we drop the explicit reference of the aggregation function G and simply use $\overline{V}_{t+1}(S_{t+1})$. The aggregation is implicit in the value function approximation.

Some major characteristics that can be used for aggregation are:

- Spatial – A transportation company is interested in estimating the value of truck drivers at a particular location. Locations may be calculated at the level of a five-digit zip code (there are about 55,000 in the United States), three-digit zip code (about 1,000), or the state level (48 contiguous states).
- Temporal – A bank may be interested in estimating the value of holding an asset at a point in time. Time may be measured by the day, week, month, or quarter.
- Continuous parameters – The state of an aircraft may be its fuel level; the state of a traveling salesman may be how long he has been away from home; the state of a water reservoir may be the depth of the water; the state of the cash reserve of a mutual fund is the amount of cash on hand at the end of the day. These are examples of systems with at least one dimension of the state that is at least approximately continuous. The variables may all be discretized into intervals of varying lengths.

- Hierarchical classification – A portfolio problem may need to estimate the value of investing money in the stock of a particular company. It may be useful to aggregate companies by industry segment (for example, a particular company might be in the chemical industry, and it might be further aggregated based on whether it is viewed as a domestic or multinational company). Similarly, problems of managing large inventories of parts (for cars, for example) may benefit by organizing parts into part families (transmission parts, engine parts, dashboard parts).

The examples below provide additional illustrations.

■ EXAMPLE 3.5

The state of a jet aircraft may be characterized by multiple attributes which include spatial and temporal dimensions (location and flying time since the last maintenance check), as well other attributes. A continuous parameter could be the fuel level, an attribute that lends itself to hierarchical aggregation might be the specific type of aircraft. We can reduce the number of states (attributes) of this resource by aggregating each dimension into a smaller number of potential outcomes.

■ EXAMPLE 3.6

The state of a portfolio might consist of the number of bonds which are characterized by the source of the bond (a company, a municipality or the federal government), the maturity (6 months, 12 months, 24 months), when it was purchased, and its rating by bond agencies. Companies can be aggregated up by industry segment. Bonds can be further aggregated by their bond rating.

■ EXAMPLE 3.7

Blood stored in blood banks can be characterized by type, the source (which might indicate risks for diseases), age (it can be stored for up to 42 days), and the current location where it is being stored. A national blood management agency might want to aggregate the state space by ignoring the source (ignoring a dimension is a form of aggregation), discretizing the age from days into weeks, and aggregating locations into more aggregate regions.

■ **EXAMPLE 3.8**

The value of an asset is determined by its current price, which is continuous. We can estimate the asset using a price discretized to the nearest dollar.

There are many applications where aggregation is naturally hierarchical. For example, in our nomadic trucker problem we might want to estimate the value of a truck based on three attributes: location, home domicile, and fleet type. The first two represent geographical locations, which can be represented (for this example) at three levels of aggregation: 400 sub-regions, 100 regions, and 10 zones. Table 3.1 illustrates five levels of aggregation that might be used. In this example, each higher level can be represented as an aggregation of the previous level.

Aggregation is also useful for continuous variables. Assume that our state variable is the amount of cash we have on hand, a number that might be as large as $10 million dollars. We might discretize our state space in units of $1 million, $100 thousand, $10 thousand, $1,000, $100, and $10. This discretization produces a natural hierarchy since 10 segments at one level of aggregation naturally group into one segment at the next level of aggregation.

Hierarchical aggregation is a natural way to generate a family of estimates, but in most cases there is no reason to assume that the structure is hierarchical. In fact, we may even use overlapping aggregations (sometimes known as "soft" aggregation), where the same state s aggregates into multiple elements in $\mathcal{S}^g$. For example, assume that s represents an (x, y) coordinate in a continuous space which has been discretized into the set of points $(x_i, y_i)_{i \in \mathcal{I}}$. Further assume that we have a distance metric $\rho((x, y), (x_i, y_i))$ that measures the distance from any point (x, y) to every aggregated point (x_i, y_i), $i \in \mathcal{I}$. We might

Table 3.1 Examples of aggregations of the state space for the nomadic trucker problem. ‫ indicates that the particular dimension is ignored.

Aggregation level	Location	Fleet type	Domicile	Size of state space
0	Sub-region	Fleet	Region	$400 \times 5 \times 100 = 200,000$
1	Region	Fleet	Region	$100 \times 5 \times 100 = 50,000$
2	Region	Fleet	Zone	$100 \times 5 \times 10 = 5,000$
3	Region	Fleet	-	$100 \times 5 \times 1 = 500$
4	Zone	-	-	$10 \times 1 \times 1 = 10$

use an observation at the point (x, y) to update estimates at each (x_i, y_i) with a weight that declines with $\rho((x, y), (x_i, y_i))$.

3.6.2 Estimates of Different Levels of Aggregation

Assume we are trying to approximate a function $f(x)$, $x \in \mathcal{X}$. We begin by defining a family of aggregation functions

$$G^g : \mathcal{X} \to \mathcal{X}^{(g)}.$$

$\mathcal{X}^{(g)}$ represents the gth level of aggregation of the domain $\mathcal{X}$. Let

$$\mathcal{G} = \text{the set of indices corresponding to the levels of aggregation.}$$

In this section, we assume we have a single aggregation function G that maps the disaggregate state $x \in \mathcal{X} = \mathcal{X}^{(0)}$ into an aggregated space $\mathcal{X}^{(g)}$. In section 3.6.3, we let $g \in \mathcal{G} = \{0, 1, 2, ...\}$ and we work with all levels of aggregation at the same time.

To begin our study of aggregation, we first need to characterize how we sample values x at the disaggregate level. For this discussion, we assume we have two exogenous processes: At iteration n, the first process chooses a value to sample (which we denote by x^n), and the second produces an observation of the value of being in state

$$\hat{f}^n(x^n) = f(x^n) + \varepsilon^n.$$

Later, we are going to assume that x^n is determined by some policy, but for now, we can treat this as purely exogenous.

We need to characterize the errors that arise in our estimate of the function. Let

$$f_x^{(g)} = \text{the true estimate of the } g^{th} \text{ aggregation}$$
$$\text{of the original function } f(x).$$

We assume that $f^{(0)}(x) = f(x)$, which means that the zeroth level of aggregation is the true function.

Let

$$\bar{f}_x^{(g,n)} = \text{the estimate of the value of } f(x) \text{ at the } g^{th} \text{ level}$$
$$\text{of aggregation after } n \text{ observations.}$$

Throughout our discussion, a bar over a variable means it was computed from sample observations. A hat means the variable was an exogenous observation.

When we are working at the most disaggregate level ($g = 0$), the state s that we measure is the observed state $s = \hat{s}^n$. For $g > 0$, the subscript x in $\bar{f}_x^{(g,n)}$

refers to $G^g(x^n)$, or the gth level of aggregation of $f(x)$ at $x = x^n$. Given an observation $(x^n, \hat{f}^n(x^n))$, we would update our estimate of the $f^{(g)}(x)$ using

$$\bar{f}_x^{(g,n)} = (1 - \alpha_{x,n-1}^{(g)})\bar{f}_x^{(g,n-1)} + \alpha_{x,n-1}^{(g)}\hat{f}^n(x).$$

Here, we have written the stepsize $\alpha_{x,n-1}^{(g)}$ to explicitly represent the dependence on the decision x and level of aggregation g. Implicit is that this is also a function of the number of times that we have updated $\bar{f}_x^{(g,n)}$ by iteration n, rather than a function of n itself.

To illustrate, imagine that our nomadic trucker is described by the vector $x =$ (Loc, Equip, Home, DOThrs, Days), where "Loc" is location, "Equip" denotes the type of trailer (long, short, refrigerated), "Home" is the location of where he lives, "DOThrs" is a vector giving the number of hours the driver has worked on each of the last eight days, and "Days" is the number of days the driver has been away from home. We are going to estimate the value $f(x)$ for different levels of aggregation of x, where we aggregate purely by ignoring certain dimensions of s. We start with our original disaggregate observation $\hat{f}(x)$, which we are going to write as

$$\hat{f}\begin{pmatrix} \text{Loc} \\ \text{Equip} \\ \text{Home} \\ \text{DOThrs} \\ \text{Days} \end{pmatrix} = f(x) + \varepsilon.$$

We now wish to use this estimate of the value of a driver with attribute x to produce value functions at different levels of aggregation. We can do this by simply smoothing this disaggregate estimate in with estimates at different levels of aggregation, as in

$$\bar{f}^{(1,n)}\begin{pmatrix} \text{Loc} \\ \text{Equip} \\ \text{Home} \end{pmatrix} = (1 - \alpha_{x,n-1}^{(1)})\bar{f}^{(1,n-1)}\begin{pmatrix} \text{Loc} \\ \text{Equip} \\ \text{Home} \end{pmatrix} + \alpha_{x,n-1}^{(1)}\hat{f}\begin{pmatrix} \text{Loc} \\ \text{Equip} \\ \text{Home} \\ \text{DOThrs} \\ \text{Days} \end{pmatrix},$$

$$\bar{f}^{(2,n)}\begin{pmatrix} \text{Loc} \\ \text{Equip} \end{pmatrix} = (1 - \alpha_{x,n-1}^{(2)})\bar{f}^{(2,n-1)}\begin{pmatrix} \text{Loc} \\ \text{Equip} \end{pmatrix} + \alpha_{x,n-1}^{(2)}\hat{f}\begin{pmatrix} \text{Loc} \\ \text{Equip} \\ \text{Home} \\ \text{DOThrs} \\ \text{Days} \end{pmatrix},$$

$$\bar{f}^{(3,n)}(\text{Loc}) = (1 - \alpha_{x,n-1}^{(3)})\bar{f}^{(3,n-1)}(\text{Loc}) + \alpha_{x,n-1}^{(3)}\hat{v}\begin{pmatrix} \text{Loc} \\ \text{Equip} \\ \text{Home} \\ \text{DOThrs} \\ \text{Days} \end{pmatrix}.$$

In the first equation, we are smoothing the value of a driver based on a five-dimensional state vector, given by x, in with an approximation indexed by a three-dimensional state vector. The second equation does the same using value function approximation indexed by a two-dimensional state vector, while the third equation does the same with a one-dimensional state vector. It is very important to keep in mind that the stepsize must reflect the number of times a state has been updated.

We need to estimate the variance of $\bar{f}_x^{(g,n)}$. Let

$$(s_x^2)^{(g,n)} \quad = \quad \text{The estimate of the variance of observations made}$$
of the function at x, using data from aggregation level g, after n observations.

$(s_x^2)^{(g,n)}$ is the estimate of the variance of the observations $\hat{f}$ when we observe the function at $x = x^n$ which aggregates to x (that is, $G^g(x^n) = x$). We are really interested in the variance of our estimate of the mean, $\bar{f}_x^{(g,n)}$. In section 3.5, we showed that

$$(\bar{\sigma}_x^2)^{(g,n)} \quad = \quad Var[\bar{f}_x^{(g,n)}]$$
$$= \quad \lambda_x^{(g,n)}(s_x^2)^{(g,n)}, \tag{3.28}$$

where $(s_x^2)^{(g,n)}$ is an estimate of the variance of the observations $\hat{f}^n$ at the gth level of aggregation (computed below), and $\lambda_s^{(g,n)}$ can be computed from the recursion

$$\lambda_x^{(g,n)} = \begin{cases} (\alpha_{x,n-1}^{(g)})^2, & n = 1, \\ (1 - \alpha_{x,n-1}^{(g)})^2 \lambda_x^{(g,n-1)} + (\alpha_{x,n-1}^{(g)})^2, & n > 1. \end{cases}$$

Note that if the stepsize $\alpha_{x,n-1}^{(g)}$ goes to zero, then $\lambda_x^{(g,n)}$ will also go to zero, as will $(\bar{\sigma}_x^2)^{(g,n)}$. We now need to compute $(s_x^2)^{(g,n)}$ which is the estimate of the variance of observations $\hat{f}^n$ at points x^n for which $G^g(x^n) = x$ (the observations of states that aggregate up to x). Let $\bar{\nu}_x^{(g,n)}$ be the total variation, given by

$$\bar{\nu}_x^{(g,n)} = (1 - \eta_{n-1})\bar{\nu}_x^{(g,n-1)} + \eta_{n-1}(\bar{f}_x^{(g,n-1)} - \hat{f}_x^n)^2,$$

where η_{n-1} follows some stepsize rule (which may be just a constant). We refer to $\bar{\nu}_x^{(g,n)}$ as the total variation because it captures deviations that arise both due to measurement noise (the randomness when we compute $\hat{f}^n(x)$) and bias (since $\bar{f}_x^{(g,n-1)}$ is a biased estimate of the mean of $\hat{f}^n(x)$).

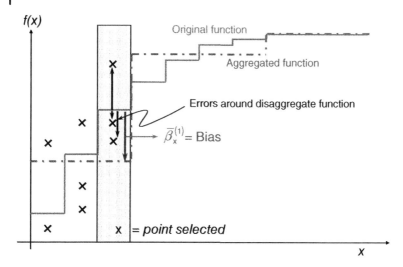

Figure 3.3 Illustration of a disaggregate function, an aggregated approximation, and a set of samples. For a particular state s, we show the estimate and the bias.

We finally need an estimate of the bias from aggregation which we find by computing

$$\bar{\beta}_x^{(g,n)} = \bar{f}_x^{(g,n)} - \bar{f}_x^{(0,n)}. \tag{3.29}$$

We can separate out the effect of bias to obtain an estimate of the variance of the error using

$$(s_x^2)^{(g,n)} = \frac{\bar{\nu}_x^{(g,n)} - (\bar{\beta}_x^{(g,n)})^2}{1 + \lambda^{n-1}}. \tag{3.30}$$

In the next section, we put the estimate of aggregation bias, $\bar{\beta}_x^{(g,n)}$, to work.

The relationships are illustrated in Figure 3.3, which shows a simple function defined over a single, continuous state (for example, the price of an asset). If we select a particular state s, we find we have only two observations for that state, versus seven for that section of the function. If we use an aggregate approximation, we would produce a single number over that range of the function, creating a bias between the true function and the aggregated estimate. As the illustration shows, the size of the bias depends on the shape of the function in that region.

One method for choosing the best level of aggregation is to choose the level that minimizes $(\bar{\sigma}_s^2)^{(g,n)} + (\bar{\beta}_s^{(g,n)})^2$, which captures both bias and variance. In the next section, we use the bias and variance to develop a method that uses estimates at all levels of aggregation at the same time.

3.6.3 Combining Multiple Levels of Aggregation

Rather than try to pick the best level of aggregation, it is intuitively appealing to use a weighted sum of estimates at different levels of aggregation. The simplest strategy is to use

$$\bar{f}_x^n = \sum_{g \in \mathcal{G}} w^{(g)} \bar{f}_x^{(g)}, \tag{3.31}$$

where $w^{(g)}$ is the weight applied to the gth level of aggregation. We would expect the weights to be positive and sum to one, but we can also view these simply as coefficients in a regression function. In such a setting, we would normally write the regression as

$$\bar{F}(x|\theta) = \theta_0 + \sum_{g \in \mathcal{G}} \theta_g \bar{f}_x^{(g)},$$

(see section 3.7 for a presentation of linear models). The problem with this strategy is that the weight does not depend on the value of x. Intuitively, it makes sense to put a higher weight on points x which have more observations, or where the estimated variance is lower. This behavior is lost if the weight does not depend on x.

In practice, we will generally observe some states much more frequently than others, suggesting that the weights should depend on x. To accomplish this, we need to use

$$\bar{f}_x^n = \sum_{g \in \mathcal{G}} w_x^{(g)} \bar{f}_x^{(g,n)}.$$

Now the weight depends on the point being estimated, allowing us to put a higher weight on the disaggregate estimates when we have a lot of observations. This is clearly the most natural, but when the domain $\mathcal{X}$ is large, we face the challenge of computing thousands (perhaps hundreds of thousands) of weights. If we are going to go this route, we need a fairly simple method to compute the weights.

We can view the estimates $(\bar{f}^{(g,n)})_{g \in \mathcal{G}}$ as different ways of estimating the same quantity. There is an extensive statistics literature on this problem. For example, it is well known that the weights that minimize the variance of $\bar{f}_x^n$ in equation (3.31) are given by

$$w_x^{(g)} \propto \left((\bar{\sigma}_x^2)^{(g,n)} \right)^{-1}.$$

Since the weights should sum to one, we obtain

$$w_x^{(g)} = \left(\frac{1}{(\bar{\sigma}_x^2)^{(g,n)}} \right) \left(\sum_{g \in \mathcal{G}} \frac{1}{(\bar{\sigma}_x^2)^{(g,n)}} \right)^{-1}. \tag{3.32}$$

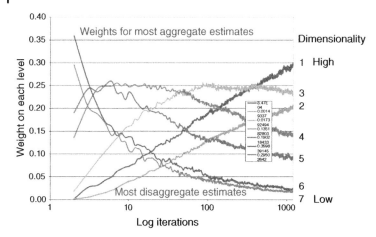

Figure 3.4 Average weight (across all states) for each level of aggregation using equation (3.33).

These weights work if the estimates are unbiased, which is clearly not the case. This is easily fixed by using the total variation (variance plus the square of the bias), producing the weights

$$
w_x^{(g,n)} = \frac{1}{\left((\bar{\sigma}_x^2)^{(g,n)} + \left(\bar{\beta}_x^{(g,n)} \right)^2 \right)} \left(\sum_{g' \in \mathcal{G}} \frac{1}{\left((\bar{\sigma}_x^2)^{(g',n)} + \left(\bar{\beta}_x^{(g',n)} \right)^2 \right)} \right)^{-1}. \quad (3.33)
$$

These are computed for each level of aggregation $g \in \mathcal{G}$. Furthermore, we compute a different set of weights for each point x. $(\bar{\sigma}_x^2)^{(g,n)}$ and $\bar{\beta}_x^{(g,n)}$ are easily computed recursively using equations (3.28) and (3.29), which makes the approach well suited to large-scale applications. Note that if the stepsize used to smooth $\bar{f}^n$ goes to zero, then the variance $(\bar{\sigma}_x^2)^{(g,n)}$ will also go to zero as $n \to \infty$. However, the bias $\bar{\beta}_x^{(g,n)}$ will in general not go to zero.

Figure 3.4 shows the average weight put on each level of aggregation (when averaged over all the inputs x) for a particular application. The behavior illustrates the intuitive property that the weights on the aggregate level are highest when there are only a few observations, with a shift to the more disaggregate level as the algorithm progresses. This is a very important behavior when approximating functions recursively. It is simply not possible to produce good function approximations with only a few data points, so it is important to use simple functions (with only a few parameters).

3.7   Linear Parametric Models

Up to now, we have focused on lookup-table representations of functions, where if we are at a point x (or state s), we compute an approximation $\bar{F}(x)$ (or $\bar{V}(s)$) that is an estimate of the function at x (or state s). Using aggregation (even mixtures of estimates at different levels of aggregation) is still a form of look-up table (we are just using a simpler lookup-table). Lookup tables offer tremendous flexibility, but generally do not scale to higher dimensional variables (x or s), and do not allow you to take advantage of structural relationships.

There has been considerable interest in estimating functions using regression methods. A classical presentation of linear regression poses the problem of estimating a parameter vector θ to fit a model that predicts a variable y using a set of observations (known as covariates in the machine learning community) $(x_i)_{i \in \mathcal{I}}$, where we assume a model of the form

$$y \;=\; \theta_0 + \sum_{i=1}^{I} \theta_i x_i + \varepsilon. \tag{3.34}$$

The variables x_i might be called independent variables, explanatory variables, or covariates, depending on the community. In dynamic programming where we want to estimate a value function $V^\pi(S_t)$, we might write

$$\bar{V}(S|\theta) = \sum_{f \in \mathcal{F}} \theta_f \phi_f(S),$$

where $(\phi_f(S))_{f \in \mathcal{F}}$ are known variously as *basis functions* or *features*, but are also referred to by names such as *covariates* or simply "independent variables." We might use this vocabulary regardless of whether we are approximating a value function or the policy itself. In fact, if we write our policy using

$$X^\pi(S_t|\theta) = \sum_{f \in \mathcal{F}} \theta_f \phi_f(S_t),$$

we would refer to $X^\pi(S_t|\theta)$ as a *linear decision rule* or, alternatively, as an *affine policy* ("affine" is just a fancy name for linear, by which we mean linear in θ).

Linear models are arguably the most popular approximation strategy for complex problems because they handle high-dimensionality by imposing a linear structure (which also means separable and additive). Using this language, instead of an independent variable x_i, we would have a basis function $\phi_f(S)$, where $f \in \mathcal{F}$ is a *feature*. $\phi_f(S)$ might be an indicator variable (e.g., 1 if we have an 'X' in the center square of our tic-tac-toe board), a discrete number (the number of X's in the corners of our tic-tac-toe board), or a continuous quantity (the price of an asset, the amount of oil in our inventories, the amount of $AB-$ blood on hand at the hospital). Some problems might have fewer than 10 features; others may have dozens; and some may have hundreds of thousands. In general, however, we would write our value function in the form

$$\overline{V}(S|\theta) = \sum_{f \in \mathcal{F}} \theta_f \phi_f(S).$$

In a time-dependent model, the parameter vector θ would typically also be indexed by time, which can dramatically increase the number of parameters we have to estimate.

In the remainder of this section, we provide a brief review of linear regression, followed by some examples of regression models. We close with a more advanced presentation that provides insights into the geometry of basis functions (including a better understanding of why they are called "basis functions"). Given the tremendous amount of attention this class of approximations has received in the literature, we defer to chapter 16 a full description of how to approximate value functions.

3.7.1 Linear Regression Review

Let y^n be the nth observation of our dependent variable (what we are trying to predict) based on the observation $(x_1^n, x_2^n, \ldots, x_I^n)$ of our independent (or explanatory) variables (the x_i are equivalent to the basis functions we used earlier). Our goal is to estimate a parameter vector θ that solves

$$\min_{\theta} \sum_{m=1}^{n} \left(y^m - \left(\theta_0 + \sum_{i=1}^{I} \theta_i x_i^m \right) \right)^2. \tag{3.35}$$

This is the standard linear regression problem.

Throughout this section, we assume that the underlying process from which the observations y^n are drawn is stationary (an assumption that is often not the case in the context of sequential decision problems).

If we define $x_0 = 1$, we let

$$x^n = \begin{pmatrix} x_0^n \\ x_1^n \\ \vdots \\ x_I^n \end{pmatrix}$$

be an $I+1$-dimensional column vector of observations. Throughout this section, and unlike the rest of the book, we use traditional vector operations, where $x^T x$ is an inner product (producing a scalar) while xx^T is an outer product, producing a matrix of cross terms.

Letting θ be the column vector of parameters, we can write our model as

$$y = \theta^T x + \varepsilon.$$

We assume that the errors $(\varepsilon^1, \ldots, \varepsilon^n)$ are independent and identically distributed. We do not know the parameter vector θ, so we replace it with an estimate $\bar{\theta}$ which gives us the predictive formula

$$\bar{y}^n = (\bar{\theta})^T x^n,$$

where $\bar{y}^n$ is our predictor of y^{n+1}. Our prediction error is

$$\hat{\varepsilon}^n = y^n - (\bar{\theta})^T x^n.$$

Our goal is to choose θ to minimize the mean squared error

$$\min_{\theta} \sum_{m=1}^{n} (y^m - \theta^T x^m)^2. \tag{3.36}$$

It is well known that this can be solved very simply. Let X^n be the n by $I + 1$ matrix

$$X^n = \begin{pmatrix} x_0^1 & x_1^1 & & x_I^1 \\ x_0^2 & x_1^2 & \cdots & x_I^2 \\ \vdots & \vdots & & \vdots \\ x_0^n & x_1^n & & x_I^n \end{pmatrix}.$$

Next, denote the vector of observations of the dependent variable as

$$Y^n = \begin{pmatrix} y^1 \\ y^2 \\ \vdots \\ y^n \end{pmatrix}.$$

The optimal parameter vector $\bar{\theta}$ (after n observations) is given by

$$\bar{\theta} = [(X^n)^T X^n]^{-1} (X^n)^T Y^n. \tag{3.37}$$

These are known as the *normal equations*.

Solving a static optimization problem such as (3.36), which produces the elegant equations for the optimal parameter vector in (3.37), is the most common approach taken by the statistics community. It has little direct application in the context of our sequential decision problems since our applications tend to be recursive in nature, reflecting the fact that at each iteration we obtain new observations, which require updates to the parameter vector. In addition, our observations tend to be notoriously nonstationary. Later, we show how to overcome this problem using the methods of recursive statistics.

3.7.2 Sparse Additive Models and Lasso

It is not hard to create models where there are a large number of explanatory variables. Some examples include:

■ **EXAMPLE 3.9**

A physician is trying to choose the best medical treatment for a patient, which may be described by thousands of different characteristics. It is unlikely that all of these characteristics have strong explanatory power.

■ **EXAMPLE 3.10**

A scientist is trying to design probes to identify the structure of RNA molecules. There are hundreds of locations where a probe can be attached. The challenge is to design probes to learn a statistical model that has hundreds of parameters (corresponding to each location).

■ **EXAMPLE 3.11**

An internet provider is trying to maximize ad-clicks, where each ad is characterized by an entire dataset consisting of all the text and graphics. A model can be created by generating hundreds (perhaps thousands) of features based on word patterns within the ad. The problem is to learn which features are most important by carefully selecting ads.

In these settings, we are trying to approximate a function $f(S)$ where S is our "state variable" consisting of all the data (describing patients, the RNA molecule, or the features within an ad). $f(S)$ might be the response (medical successes or costs, or clicks on ads), which we approximate using

$$\bar{F}(S|\theta) = \sum_{f \in \mathcal{F}} \theta_f \phi_f(S). \tag{3.38}$$

Now imagine that there are hundreds of features in the set $\mathcal{F}$, but we anticipate that $\theta_f = 0$ for many of these. In this case, we would view equation (3.38) as a *sparse additive* model, where the challenge is to identify a model with the highest explanatory power which means excluding the parameters which do not contribute very much.

Imagine we have a dataset consisting of $(f^n, S^n)_{n=1}^N$ where f^n is the observed response corresponding to the information in S^n. If we use this data to fit (3.38), virtually every fitted value of θ_f will be nonzero, producing a huge model with little explanatory power. To overcome this, we introduce what is known as a *regularization* term where we penalize nonzero values of θ. We would write the optimization problem as

$$\min_{\theta} \left(\sum_{n=1}^{N} (f^n - \bar{F}(S^n|\theta))^2 + \lambda \sum_{f \in \mathcal{F}} \|\theta_f\|_1 \right), \tag{3.39}$$

where $\|\theta_f\|_1$ represents what is known as "L_1" regularization, which is the same as taking the absolute value $|\theta_f|$. L_2 regularization would use θ_f^2, which means that there is almost no penalty for values of θ_f that are close to zero. This means we are assessing a penalty when $\theta_f \neq 0$, and the marginal penalty is the same for any value of θ_f other than zero.

We refer to $\lambda \sum_f \|\theta_f\|_1$ as a *regularization* term. As we increase λ, we put a higher penalty for allowing θ_f to be in the model. It is necessary to increase λ, take the resulting model, and then test it on an out-of-sample dataset. Typically, this is done repeatedly (five times is typical) where the out-of-sample observations are drawn from a different 20% of the data (this process is known as *cross-validation*). We can plot the error from this testing for each value of λ, and find the best value of λ.

This procedure is known as Lasso, for "Least absolute shrinkage and selection operator." The procedure is inherently batch, although there is a recursive form that has been developed. The method works best when we assume there is access to an initial testing dataset that can be used to help identify the best set of features.

A challenge with regularization is that it requires determining the best value of λ. It should not be surprising that you will get the best fit if you set $\lambda = 0$, creating a model with a large number of parameters. The problem is that these models do not offer the best predictive power, because many of the fitted parameters $\theta_f > 0$ reflect spurious noise rather than the identification of truly important features.

The way to overcome this is to use cross-validation, which works as follows. Imagine fitting the model on an 80% sample of the data, and then evaluating the model on the remaining 20%. Now, repeat this five times by rotating through the dataset, using different portions of the data for testing. Finally, repeat this entire process for different values of λ to find the value of λ that produces the lowest error.

Regularization is sometimes referred to as modern statistical learning. While not an issue for very low dimensional models where all the variables are clearly important, regularization is arguably one of the most powerful tools for modern models which feature large numbers of variables. Regularization can be introduced into virtually any statistical model, including nonlinear models and neural networks.

3.8 Recursive Least Squares for Linear Models

Perhaps one of the most appealing features of linear regression is the ease with which models can be updated recursively. Recursive methods are well known in the statistics and machine learning communities, but these communities often focus on batch methods. Recursive statistics is especially valuable in stochastic optimization because they are well suited to any adaptive algorithm.

We start with a basic linear model

$$y = \theta^T x + \varepsilon,$$

where $\theta = (\theta_1, \dots, \theta_I)^T$ is a vector of regression coefficients. We let X^n be the $n \times I$ matrix of observations (where n is the number of observations). Using batch statistics, we can estimate θ from the normal equation

$$\theta \ = \ [(X^n)^T X^n]^{-1}(X^n)^T Y^n. \tag{3.40}$$

We note in passing that equation (3.40) represents an optimal solution of a statistical model using a sampled dataset, one of the major solution strategies that we are going to describe in chapter 4 (stay tuned!).

We now make the conversion to the vocabulary where instead of a feature x_i, we are going to let x be our data and let $\phi_f(x)$ be a feature (also known as basis functions), where $f \in \mathcal{F}$ is our set of features. We let $\phi(x)$ be a column vector of the features, where $\phi^n = \phi(x^n)$ replaces x^n. We also write our function approximation using

$$\bar{F}(x|\theta) = \sum_{f \in \mathcal{F}} \theta_f \phi_f(x) = \phi(x)^T \theta.$$

Throughout our presentation, we assume that we have access to an observation $\hat{f}^n$ of our function $F(x, W)$.

3.8.1 Recursive Least Squares for Stationary Data

In the setting of adaptive algorithms in stochastic optimization, estimating the coefficient vector θ using batch methods such as equation (3.40) would be very expensive. Fortunately, it is possible to compute these formulas recursively. The updating equation for θ is

$$\theta^n = \theta^{n-1} - H^n \phi^n \hat{\varepsilon}^n, \tag{3.41}$$

where H^n is a matrix computed using

$$H^n = \frac{1}{\gamma^n} M^{n-1}. \tag{3.42}$$

The error $\hat{\varepsilon}^n$ is computed using

$$\hat{\varepsilon}^n = \bar{F}(x|\theta^{n-1}) - \hat{y}^n. \tag{3.43}$$

Note that it is common in statistics to compute the error in a regression using "actual minus predicted" while we are using "predicted minus actual" (see equation (3.43) above). Our sign convention is motivated by the derivation from first principles of optimization, which we cover in more depth in chapter 5.

Now let M^n be the $|\mathcal{F}| \times |\mathcal{F}|$ matrix given by

$$M^n = [(X^n)^T X^n]^{-1}.$$

Rather than do the matrix inversion, we can compute M^n recursively using

$$M^n = M^{n-1} - \frac{1}{\gamma^n}(M^{n-1}\phi^n(\phi^n)^T M^{n-1}), \tag{3.44}$$

where γ^n is a scalar computed using

$$\gamma^n = 1 + (\phi^n)^T M^{n-1}\phi^n. \tag{3.45}$$

The derivation of equations (3.41)-(3.45) is given in section 3.14.1.

It is possible in any regression problem that the matrix $(X^n)^T X^n$ (in equation (3.40)) is non-invertible. If this is the case, then our recursive formulas are not going to overcome this problem. When this happens, we will observe $\gamma^n = 0$. Alternatively, the matrix may be invertible, but unstable, which occurs when γ^n is very small (say, $\gamma^n < \epsilon$ for some small ϵ). When this occurs, the problem can be circumvented by using

$$\bar{\gamma}^n = \gamma^n + \delta,$$

where δ is a suitably chosen small perturbation that is large enough to avoid instabilities. Some experimentation is likely to be necessary, since the right value depends on the scale of the parameters being estimated.

The only missing step in our algorithm is initializing M^0. One strategy is to collect a sample of m observations where m is large enough to compute M^m using full inversion. Once we have M^m, we use it to initialize M^0 and then we can proceed to update it using the formula above. A second strategy is to use $M^0 = \epsilon I$, where I is the identity matrix and ϵ is a "small constant." This strategy is not guaranteed to give the exact values, but should work well if the number of observations is relatively large.

In our stochastic optimization applications, the observations $\hat{f}^n$ will represent observations of the value of a function, or estimates of the value of being in a state, or even decisions we should make given a state. Our data can be a decision x (or possibly the decision x and initial state S_0), or a state S. The updating equations assume implicitly that the estimates come from a stationary series.

There are many problems where the number of basis functions can be extremely large. In these cases, even the efficient recursive expressions in this section cannot avoid the fact that we are still updating a matrix where the number of rows and columns may be large. If we are only estimating a few dozen or a few hundred parameters, this can be fine. If the number of parameters extends into the thousands, even this strategy would probably bog down. It is very important to work out the approximate dimensionality of the matrices before using these methods.

3.8.2 Recursive Least Squares for Nonstationary Data*

It is generally the case in approximate dynamic programming that our observations $\hat{f}^n$ (typically, updates to an estimate of a value function) come from a nonstationary process. This is true even when we are estimating the value of a fixed policy if we use TD learning, but it is always true when we introduce the dimension of optimizing over policies. Recursive least squares puts equal weight on all prior observations, whereas we would prefer to put more weight on more recent observations.

Instead of minimizing total errors (as we do in equation (3.35)) it makes sense to minimize a geometrically weighted sum of errors

$$\min_{\theta} \sum_{m=1}^{n} \lambda^{n-m} \left(f^m - \left(\theta_0 + \sum_{i=1}^{I} \theta_i \phi_i^m \right) \right)^2 , \tag{3.46}$$

where λ is a discount factor that we use to discount older observations. If we repeat the derivation in section 3.8.1, the only changes we have to make are in the updating formula for M^n, which is now given by

$$M^n = \frac{1}{\lambda} \left(M^{n-1} - \frac{1}{\gamma^n} (M^{n-1} \phi^n (\phi^n)^T M^{n-1}) \right) , \tag{3.47}$$

and the expression for γ^n, which is now given by

$$\gamma^n = \lambda + (\phi^n)^T M^{n-1} \phi^n . \tag{3.48}$$

λ works in a way similar to a stepsize, although in the opposite direction. Setting $\lambda = 1$ means we are putting an equal weight on all observations, while smaller values of λ puts more weight on more recent observations. In this way, λ plays a role similar to our use of λ in $TD(\lambda)$.

We could use this logic and view λ as a tunable parameter. Of course, a constant goal in the design of algorithms is to avoid the need to tune yet another parameter. For the special case where our regression model is just a constant (in which case $\phi^n = 1$), we can develop a simple relationship between α_n and the discount factor (which we now compute at each iteration, so we write it as λ_n). Let $G^n = (H^n)^{-1}$, which means that our updating equation is now given by

$$\theta^n = \theta^{n-1} - (G^n)^{-1}\phi^n\hat{\varepsilon}^n.$$

Recall that we compute the error ε^n as predicted minus actual as given in equation (3.43). This is required if we are going to derive our optimization algorithm based on first principles, which means that we are minimizing a stochastic function. The matrix G^n is updated recursively using

$$G^n = \lambda_n G^{n-1} + \phi^n(\phi^n)^T, \tag{3.49}$$

with $G^0 = 0$. For the case where $\phi^n = 1$ (in which case G^n is also a scalar), $(G^n)^{-1}\phi^n = (G^n)^{-1}$ plays the role of our stepsize, so we would like to write $\alpha_n = G^n$. Assume that $\alpha_{n-1} = (G^{n-1})^{-1}$. Equation (3.49) implies that

$$\begin{aligned} \alpha_n &= (\lambda_n G^{n-1} + 1)^{-1} \\ &= \left(\frac{\lambda_n}{\alpha_{n-1}} + 1\right)^{-1}. \end{aligned}$$

Solving for λ_n gives

$$\lambda_n = \alpha_{n-1}\left(\frac{1 - \alpha_n}{\alpha_n}\right). \tag{3.50}$$

Note that if $\lambda_n = 1$, then we want to put equal weight on all the observations (which would be optimal if we have stationary data). We know that in this setting, the best stepsize is $\alpha_n = 1/n$. Substituting this stepsize into equation (3.50) verifies this identity.

The value of equation (3.50) is that it allows us to relate the discounting produced by λ_n to the choice of stepsize rule, which has to be chosen to reflect the nonstationary of the observations. In chapter 6, we introduce a much broader range of stepsize rules, some of which have tunable parameters. Using (3.50) allows us to avoid introducing yet another tunable parameter.

3.8.3 Recursive Estimation Using Multiple Observations*

The previous methods assume that we get one observation and use it to update the parameters. Another strategy is to sample several paths and solve a classical least-squares problem for estimating the parameters. In the simplest implementation, we would choose a set of realizations $\hat{\Omega}^n$ (rather than a single sample ω^n) and follow all of them, producing a set of estimates $(f(\omega))_{\omega \in \hat{\Omega}^n}$ that we can use to update our estimate of the function $\bar{F}(s|\theta)$.

If we have a set of observations, we then face the classical problem of finding a vector of parameters $\hat{\theta}^n$ that best match all of these function estimates. Thus, we want to solve

$$\hat{\theta}^n = \arg\min_\theta \frac{1}{|\hat{\Omega}^n|} \sum_{\omega \in \hat{\Omega}^n} (\bar{F}(s|\theta) - f(\omega))^2.$$

This is the standard parameter estimation problem faced in the statistical estimation community. If $\bar{F}(s|\theta)$ is linear in θ, then we can use the usual formulas for linear regression. If the function is more general, we would typically resort to nonlinear programming algorithms to solve the problem. In either case, $\hat{\theta}^n$ is still an update that needs to be smoothed in with the previous estimate θ^{n-1}, which we would do using

$$\theta^n = (1 - \alpha_{n-1})\theta^{n-1} + \alpha_{n-1}\hat{\theta}^n. \tag{3.51}$$

One advantage of this strategy is that in contrast with the updates that depend on the gradient of the value function, updates of the form given in equation (3.51) do not encounter a scaling problem, and therefore we return to our more familiar territory where $0 < \alpha_n \leq 1$. Of course, as the sample size $\hat{\Omega}$ increases, the stepsize should also be increased because there is more information in $\hat{\theta}^n$. Using stepsizes based on the Kalman filter (see sections 6.3.2 and 6.3.3) will automatically adjust to the amount of noise in the estimate.

The usefulness of this particular strategy will be very problem-dependent. In many applications, the computational burden of producing multiple estimates $\hat{v}^n(\omega), \omega \in \hat{\Omega}^n$ before producing a parameter update will simply be too costly.

3.9 Nonlinear Parametric Models

While linear models are exceptionally powerful (recall that "linear" means linear in the parameters), it is inevitable that some problems will require models that are nonlinear in the parameters. We might want to model the nonlinear response of price, dosage, or temperature. Nonlinear models introduce challenges in model estimation as well as learning in stochastic optimization problems.

We begin with a presentation on maximum likelihood estimation, one of the most widely used estimation methods for nonlinear models. We then introduce the idea of a sampled nonlinear model, which is a simple way of overcoming the complexity of a nonlinear model. We close with an introduction to neural networks, a powerful approximation architecture that has proven to be useful in machine learning as well as dynamic programs arising in engineering control problems.

3.9.1 Maximum Likelihood Estimation

The most general method for estimating nonlinear models is known as maximum likelihood estimation. Let $f(x|\theta)$ the function given θ, and assume that we observe

$$y = f(x|\theta) + \epsilon$$

where $\epsilon \sim N(0, \sigma^2)$ is the error with density

$$f^\epsilon(w) = \frac{1}{\sqrt{2\pi\sigma}} \exp \frac{w^2}{2\sigma^2}.$$

Now imagine that we have a set of observations $(y^n, x^n)_{n=1}^N$. The likelihood of observing $(y^n)_{n=1}^N$ is given by

$$L(y|x, \theta) = \Pi_{n=1}^N \exp \frac{(y^n - f(x^n|\theta))^2}{2\sigma^2}.$$

It is common to use the log likelihood $\mathcal{L}(y|x, \theta) = \log L(y|x, \theta)$, which gives us

$$\mathcal{L}(y|x, \theta) = \sum_{n=1}^N \frac{1}{\sqrt{2\pi\sigma}} (y^n - f(x^n|\theta))^2, \tag{3.52}$$

where we can, of course, drop the leading constant $\frac{1}{\sqrt{2\pi\sigma}}$ when maximizing $\mathcal{L}(y|x, \theta)$.

Equation (3.52) can be used by nonlinear programming algorithms to estimate the parameter vector θ. This assumes that we have a batch dataset $(y^n, x^n)_{n=1}^N$, which is not our typical setting. In addition, the log likelihood $\mathcal{L}(y|x, \theta)$ can be nonconvex when $f(x|\theta)$ is nonlinear in θ, which further complicates the optimization challenge.

The next section describes a method for handling nonlinear models in a recursive setting.

3.9.2 Sampled Belief Models

A powerful strategy for estimating models that are nonlinear in the parameters assumes that the unknown parameter θ can only take on one of a finite set

$\theta_1, \theta_2, \dots, \theta_K$. Let θ be a random variable representing the true value of θ, where θ takes on one of the values $\Theta = (\theta_k)_{k=1}^K$.

Assume we start with a prior set of probabilities $p_k^0 = \mathbb{P}[\theta = \theta_k]$, and let $p^n = (p_k^n), k = 1, \dots K$ be the probabilities after n experiments. This is framework we use when we adopt a Bayesian perspective: we view the true value of θ as a random variable θ, with a prior distribution of belief p^0 (which might be uniform).

We refer to $B^n = (p^n, \Theta)$ as a *sampled belief model*. Sampled belief models are powerful ways for representing the uncertainty in a nonlinear belief model. The process of generating the set Θ (which actually can change with iterations) makes it possible for a user to ensure that each member of the sample is reasonable (for example, we can ensure that some coefficients are positive). Updating the probability vector p^n can be done fairly simply using Bayes theorem, as we show below.

What we are now going to do is to use observations of the random variable Y to update our probability distribution. To illustrate this, assume that we are observing successes and failures, so $Y \in \{0, 1\}$, as might happen with medical outcomes. In this setting, the vector x would consist of information about a patient as well as medical decisions. Assume that the probability that $Y = 1$ is given by a logistic regression, given by

$$f(y|x, \theta) = \mathbb{P}[Y = 1|x, \theta] \tag{3.53}$$

$$= \frac{\exp^{U(x|\theta)}}{1 + \exp^{U(x|\theta)}}, \tag{3.54}$$

where $U(x|\theta)$ is a linear model given by

$$U(x|\theta) = \theta_0 + \theta_1 x_1 + \theta_2 x_2 + \dots + \theta_M.$$

We assume that θ is one of the elements $(\theta_k)_{k=1}^K$, where θ_k is a vector of elements $(\theta_{km})_{m=1}^M$. Let $H^n = (y^1, \dots, y^n)$ be our history of observations of the random outcome Y. Now assume that $p_k^n = \mathbb{P}[\theta = \theta_k|H^n]$, and that we next choose x^n and observe $Y = y^{n+1}$ (later, we are going to talk about how to choose x^n). We can update our probabilities using Bayes theorem

$$p_k^{n+1} = \frac{\mathbb{P}[Y = y^{n+1}|x^n, \theta_k, H^n]\mathbb{P}[\theta = \theta_k|x^n, H^n]}{\mathbb{P}[Y = y^{n+1}|x^n, H^n]}. \tag{3.55}$$

We start by observing that $p_k^n = \mathbb{P}[\theta = \theta_k|x^n, H^n] = \mathbb{P}[\theta = \theta_k|H^n]$. The conditional probability $\mathbb{P}[Y = y^{n+1}|x^n, \theta_k, H^n]$ comes from our logistic regression in (3.54):

$$\mathbb{P}[Y = y^{n+1}|x^n, \theta_k, H^n] = \begin{cases} f(x^n|\theta^n) & \text{if } y^{n+1} = 1, \\ 1 - f(x^n|\theta^n) & \text{if } y^{n+1} = 0. \end{cases}$$

Finally, we compute the denominator using

$$\mathbb{P}[Y = y^{n+1}|x^n, H^n] = \sum_{k=1}^{K} \mathbb{P}[Y = y^{n+1}|x^n, \theta_k, H^n] p_k^n.$$

This idea can be extended to a wide range of distributions for Y. Its only limitation (which may be significant) is the assumption that θ can be only one of a finite set of discrete values. A strategy for overcoming this limitation is to periodically generate new possible values of θ, use the past history of observations to obtain updated probabilities, and then drop the values with the lowest probability.

3.9.3   Neural Networks – Parametric*

Neural networks represent an unusually powerful and general class of approximation strategies that has been widely used in optimal control and statistical learning. There are a number of excellent textbooks on the topic along with widely available software packages, so our presentation is designed only to introduce the basic idea and encourage readers to experiment with this technology if simpler models are not effective.

In this section, we restrict our attention to low-dimensional neural networks, although these "low-dimensional" neural networks may still have thousands of parameters. Neural networks in this class have been very popular for many years in the engineering controls community, where they are used for approximating both policies and value functions for deterministic control problems.

We return to neural networks in section 3.10.4 where we discuss the transition to "deep" neural networks, which are extremely high-dimensional functions allowing them to approximate almost anything, earning them the classification as a nonparametric model.

In this section we describe the core algorithmic steps for performing estimation with neural networks. We defer until chapter 5 the description of how we optimize the parameters, since we are going to use the methods of derivative-based stochastic optimization which are covered in that chapter.

Up to now, we have considered approximation functions of the form

$$\bar{F}(x|\theta) = \sum_{f \in \mathcal{F}} \theta_f \phi_f(x),$$

where $\mathcal{F}$ is our set of features, and $(\phi_f(x))_{f \in \mathcal{F}}$ are the basis functions which extract what are felt to be the important characteristics of the state variable which explain the value of being in a state. We have seen that when we use an approximation that is linear in the parameters, we can estimate the parameters

θ recursively using standard methods from linear regression. For example, if x^n is the nth input with element x_i^n, our approximation might look like

$$\bar{F}(x^n|\theta) = \sum_{i \in \mathcal{I}} \left(\theta_{1i} x_i^n + \theta_{2i} (x_i^n)^2 \right).$$

Now assume that we feel that the best function might not be quadratic in R_i, but we are not sure of the precise form. We might want to estimate a function of the form

$$\bar{F}(x^n|\theta) = \sum_{i \in \mathcal{I}} \left(\theta_{1i} x_i^n + \theta_{2i} (x_i^n)^{\theta_3} \right).$$

Now we have a function that is nonlinear in the parameter vector $(\theta_1, \theta_2, \theta_3)$, where θ_1 and θ_2 are vectors and θ_3 is a scalar. If we have a training dataset of state-value observations, $(\hat{f}^n, R^n)_{n=1}^N$, we can find θ by solving

$$\min_{\theta} F(\theta) = \sum_{n=1}^{N} \left(\hat{f}^n - \bar{F}(x^n|\theta) \right)^2, \tag{3.56}$$

which generally requires the use of nonlinear programming algorithms. One challenge is that nonlinear optimization problems do not lend themselves to the simple recursive updating equations that we obtained for linear (in the parameters) functions. More problematic is that we have to experiment with various functional forms to find the one that fits best.

Neural networks are, ultimately, a form of nonlinear model which can be used to approximate the function $\mathbb{E} f(x, W)$ (or a policy $X^{\pi}(S)$, or a value function $V(S)$). We will have an input x (or S), and we are using a neural network to predict an output $\hat{f}$ (or a decision x^n, or a value v^n). Using the traditional notation of statistics, let x^n be a vector of inputs which could be features $\phi_f(S^n)$ for $f \in \mathcal{F}$. If we were using a linear model, we would write

$$f(x^n|\theta) = \theta_0 + \sum_{i=1}^{I} \theta_i x_i^n.$$

In the language of neural networks, we have I inputs (we have $I+1$ parameters since we also include a constant term), which we wish to use to estimate a single output f^{n+1} (a random observations of our function). The relationships are illustrated in Figure 3.5 where we show the I inputs which are then "flowed" along the links to produce $f(x^n|\theta)$. After this, we then learn the sample realization $\hat{f}^{n+1}$ that we were trying to predict, which allows us to compute the error $\epsilon^{n+1} = \hat{f}^{n+1} - f(x^n|\theta)$.

Define the random variable X to describe a set of inputs (where x^n is the value of X at the nth iteration), and let $\hat{f}$ be the random variable giving the response from input X. We would like to find a vector θ that solves

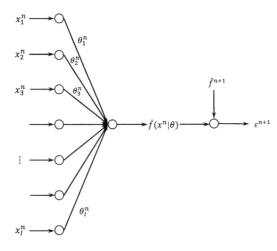

Figure 3.5 Neural networks with a single layer.

$$\min_\theta \mathbb{E}\frac{1}{2}(f(X|\theta) - \hat{f})^2.$$

Let $F(\theta) = \mathbb{E}\big(0.5(f(X|\theta) - \hat{f})^2\big)$, and let $F(\theta, \hat{f}) = 0.5(f(X|\theta) - \hat{f})^2$ where $\hat{f}$ is a sample realization of our function. As before, we can solve this iteratively using the algorithm we first introduced in section 3.2 which gives us the updating equation

$$\theta^{n+1} = \theta^n - \alpha_n \nabla_\theta F(\theta^n, \hat{f}^{n+1}), \tag{3.57}$$

where $\nabla_\theta F(\theta^n, \hat{f}^{n+1}) = \varepsilon^{n+1} = (f(x^n|\theta) - \hat{f}^{n+1})$ for a given input $X = x^n$ and observed response $\hat{f}^{n+1}$.

We illustrated our linear model by assuming that the inputs were the individual dimensions of the control variable which we denoted x_i^n. We may not feel that this is the best way to represent the state of the system (imagine representing the states of a Connect-4 game board). We may feel it is more effective (and certainly more compact) if we have access to a set of basis functions $\phi_f(X)$, $f \in \mathcal{F}$, where $\phi_f(X)$ captures a relevant feature of our system given the inputs X. In this case, we would be using our standard basis function representation, where each basis function provides one of the inputs to our neural network.

This was a simple illustration, but it shows that if we have a linear model, we get the same basic class of algorithms that we have already used. A richer model, given in Figure 3.6, illustrates a more classical neural network. Here, the "input signal" x^n (this can be the state variable or the set of basis functions) is communicated through several layers. Let $x^{(1,n)} = x^n$ be the input to the first layer (recall that x_i^n might be the ith dimension of the state variable itself, or a

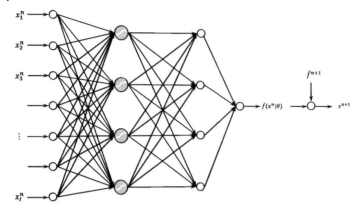

Figure 3.6 A three-layer neural network.

basis function). Let $\mathcal{J}^{(1)}$ be the set of inputs to the first layer (for example, the set of basis functions).

Here, the first linear layer produces $|\mathcal{J}^{(2)}|$ outputs given by

$$y_j^{(2,n)} = \sum_{i \in \mathcal{J}^{(1)}} \theta_{ij}^{(1)} x_i^{(1,n)}, \quad j \in \mathcal{J}^{(2)}.$$

$x_j^{(2,n)}$ becomes the input to a nonlinear *perceptron* node which is characterized by a nonlinear function that may dampen or magnify the input. A typical functional form for a perceptron node is the logistics function given by

$$\sigma(y) = \frac{1}{1 + e^{-\beta y}}, \tag{3.58}$$

where β is a scaling coefficient. The function $\sigma(y)$ is illustrated in Figure 3.7. The sigmoid function $\sigma(x)$ introduces nonlinear behavior into the communication of the "signal" x^n. In addition let

$$\sigma'(y) = \frac{\partial \sigma(y)}{\partial y}.$$

We next calculate

$$x_i^{(2,n)} = \sigma(y_i^{(2,n)}), \quad i \in \mathcal{J}^{(2)}$$

and use $x_i^{(2,n)}$ as the input to the second linear layer. We then compute

$$y_j^{(3,n)} = \sum_{i \in \mathcal{J}^{(2)}} \theta_{ij}^{(2)} x_i^{(2,n)}, \quad j \in \mathcal{J}^{(3)}$$

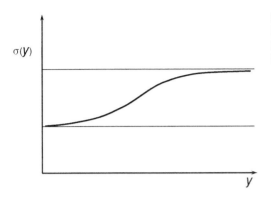

Figure 3.7 Illustrative logistics function for introducing nonlinear behavior into neural networks.

and then calculate the input to layer 3

$$x_i^{(3,n)} = \sigma(y_i^{(3,n)}), \ i \in \mathcal{J}^{(3)}.$$

Finally, we compute the single output using

$$\bar{f}^n(x^n|\theta) = \sum_{i \in \mathcal{J}^{(3)}} \theta_i^{(3)} x_i^{(3,n)}.$$

As before, f^n is our estimate of the response from input x^n. This is our function approximation $\bar{F}^n(s|\theta)$ which we update using the observation $\hat{f}^{n+1}$. Now that we know how to produce estimates using a neural network given the vector θ, the next step is optimize θ.

We update the parameter vector $\theta = (\theta^{(1)}, \theta^{(2)}, \theta^{(3)})$ using the stochastic gradient algorithm given in equation (3.57). The only difference is that the derivatives have to capture the fact that changing $\theta^{(1)}$, for example, impacts the "flows" through the rest of the network. There are standard packages for fitting neural networks to data using gradient algorithms, but for readers interested in the algorithmic side, we defer until section 5.5 the presentation of this algorithm since it builds on the methods of derivative-based stochastic search.

This presentation should be viewed as a simple illustration of an extremely rich field. The advantage of neural networks is that they offer a much richer class of nonlinear functions ("nonlinear architectures" in the language of machine learning) which can be trained in an iterative way. Calculations involving neural networks exploit the layered structure, and naturally come in two forms: feed forward propagation, where we step forward through the layers "simulating" the evolution of the input variables to the outputs, and backpropagation, which is used to compute derivatives so we can calculate the marginal impact of changes in the parameters (shown in section 5.5).

3.9.4 Limitations of Neural Networks

Neural networks offer an extremely flexible architecture, which reduces the need for designing and testing different nonlinear (parametric) models. They have been particularly successful in the context of deterministic problems such as optimal control of engineering systems, and the familiar voice and image recognition tools that have been so visible. There is a price, however, to this flexibility:

- To fit models with large numbers of parameters, you need large datasets. This is problematic in the context of sequential decision problems since we are often starting with little or no data, and then generating a series of inputs that allow us to create increasingly more accurate estimates of whatever function we are approximating.
- The flexibility of a neural network also means that, when applied to problems with noise, the network may just be fitting the noise (this is classic overfitting, well known to the statistical learning community). When the underlying problem exhibits noise (and many of the sequential decision problems in this book exhibit high levels of noise), the data requirements grow dramatically. Sadly, this is often overlooked in the neural network community, where it is not unusual to fit neural networks to datasets where there are more parameters in the neural network than data points.
- Neural networks struggle to replicate structure. There are many problems in business, engineering, economics, and the sciences that exhibit structure: monotonicity (the higher the price, the lower the demand); concavity (very common in resource allocation problems); unimodularity (there is an optimum response to dosage, which declines when the dosage is too high or too low).

The problem of handling noise, and not being able to capture structure, is illustrated in Figure 3.8, where we sampled data from the newsvendor problem

$$F(x, W) = 10 \min\{x, W\} - 8x,$$

where W is distributed according to the density

$$f^W(w) = .1e^{-.1w}.$$

We sampled 1000 observations of the demand W and the profit $F(x, W)$, for values of x that were drawn uniformly between 0 and 40. This data was then fitted with a neural network.

The expected profit $F(x) = \mathbb{E}_W F(x, W)$ is shown as the concave red line. The fitted neural network does not come close to capturing this structure. 1000 observations is a lot of data for approximating a one-dimensional function.

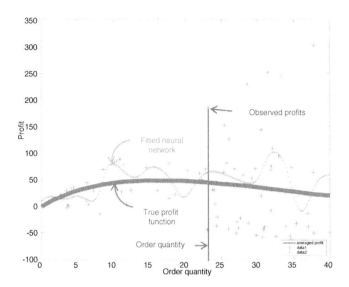

Figure 3.8 A neural network fitted to sampled data from a newsvendor problem, demonstrating the tendency of neural networks to overfit noisy data without capturing problem structure (such as concavity).

We urge readers to use caution when using neural networks in the context of noisy but structured applications since these arise frequently in the application domains discussed in the opening of chapter 1.

3.10 Nonparametric Models*

The power of parametric models is matched by their fundamental weakness: they are only effective if you can find the right structure, and this remains a frustrating art. For this reason, nonparametric statistics have attracted recent attention. They avoid the art of specifying a parametric model, but introduce other complications. Nonparametric methods work primarily by building local approximations to functions using observations rather than depending on functional approximations.

Nonparametric models are characterized by the property that as the number of observations $N \to \infty$, we can approximate any function with arbitrary accuracy. This means that the working definition of a nonparametric model is that with enough data, they can approximate any function. However, the price of such flexibility is that you need very large datasets.

There is an extensive literature on the use of approximation methods for continuous functions. These problems, which arise in many applications in engineering and economics, require the use of approximation methods that can adapt to a wide range of functions. Interpolation techniques, orthogonal polynomials, Fourier approximations, and splines are just some of the most popular techniques. Often, these methods are used to closely approximate the expectation using a variety of numerical approximation techniques.

We note that lookup tables are, technically, a form of nonparametric approximation methods, although these can also be expressed as parametric models by using indicator variables (this is the reason why the three classes of statistical models are illustrated as overlapping functions in Figure 3.1). For example, assume that $\mathcal{X} = \{x_1, x_2, \ldots, x_M\}$ is a set of discrete inputs, and let

$$\mathbb{1}_{\{X=x\}} = \begin{cases} 1 & \text{if } X = x \in \mathcal{X}, \\ 0 & \text{otherwise} \end{cases}$$

be an indicator variable that tells us when X takes on a particular value. We can write our function as

$$f(X|\theta) = \sum_{x \in \mathcal{X}} \theta_x \mathbb{1}_{\{X=x\}}.$$

This means that we need to estimate a parameter θ_x for each $x \in \mathcal{X}$. In principle, this is a parametric representation, but the parameter vector θ has the same dimensionality as the input vector x. However, the working definition of a nonparametric model is one that, given an infinite dataset, will produce a perfect representation of the true function, a property that our lookup table model clearly satisfies. It is precisely for this reason that we treat lookup tables as a special case since parametric models are always used for settings where the parameter vector θ is much lower dimensional than the size of $\mathcal{X}$.

In this section, we review some of the nonparametric methods that have received the most attention within the approximate dynamic programming community. This is an active area of research which offers potential as an approximation strategy, but significant hurdles remain before this approach can be widely adopted. We start with the simplest methods, closing with a powerful class of nonparametric methods known as support vector machines.

3.10.1 K-Nearest Neighbor

Perhaps the simplest form of nonparametric regression forms estimates of functions by using a weighted average of the k-nearest neighbors. As above, we assume we have a response y^n corresponding to a measurement $x^n = (x_1^n, x_2^n, \ldots, x_I^n)$. Let $\rho(x, x^n)$ be a distance metric between a query point x (in dynamic programming, this would be a state) and an observation x^n. Then let $\mathcal{N}^n(x)$ be the set of the k-nearest points to the query point x, where clearly we require $k \leq n$. Finally let $\bar{Y}^n(x)$ be the response function, which is our best estimate of the true function $Y(x)$ given the observations $x^1, \ldots, x^n$. When we use a k-nearest neighbor model, this is given by

$$\bar{Y}^n(x) = \frac{1}{k} \sum_{n \in \mathcal{N}^n(x)} y^n. \tag{3.59}$$

Thus, our best estimate of the function $Y(x)$ is made by averaging the k points nearest to the query point x.

Using a k-nearest neighbor model requires, of course, choosing k. Not surprisingly, we obtain a perfect fit of the data by using $k = 1$ if we base our error on the training dataset.

A weakness of this logic is that the estimate $\bar{Y}^n(x)$ can change abruptly as x changes continuously, as the set of nearest neighbors changes. An effective way of avoiding this behavior is using kernel regression, which uses a weighted sum of all data points.

3.10.2 Kernel Regression

Kernel regression has attracted considerable attention in the statistical learning literature. As with k-nearest neighbor, kernel regression forms an estimate $\bar{Y}(x)$ by using a weighted sum of prior observations which we can write generally as

$$\bar{Y}^n(x) = \frac{\sum_{m=1}^{n} K_h(x, x^m) y^m}{\sum_{m=1}^{n} K_h(x, x^m)} \tag{3.60}$$

where $K_h(x, x^m)$ is a weighting function that declines with the distance between the query point x and the measurement x^m. h is referred to as the *bandwidth* which plays an important scaling role. There are many possible choices for the weighting function $K_h(x, x^m)$. One of the most popular is the Gaussian kernel, given by

$$K_h(x, x^m) = e^{-\left(\frac{\|x - x^m\|}{h}\right)^2}$$

where $\| \cdot \|$ is the Euclidean norm. Here, h plays the role of the standard deviation. Note that the bandwidth h is a tunable parameter that captures the range of influence of a measurement x^m. The Gaussian kernel, often referred to as *radial basis functions*, provides a smooth, continuous estimate $\bar{Y}^n(x)$. Another popular choice of kernel function is the symmetric Beta family, given by

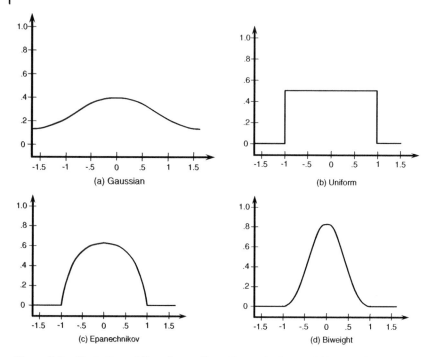

Figure 3.9 Illustration of Gaussian, uniform, Epanechnikov, and biweight kernel weighting functions.

$$K_h(x, x^m) = \max(0, (1 - \|x - x^m\|)^2)^h.$$

Here, h is a nonnegative integer. $h = 1$ gives the uniform kernel; $h = 2$ gives the Epanechnikov kernel; and $h = 3$ gives the biweight kernel. Figure 3.9 illustrates each of these four kernel functions.

We pause to briefly discuss some issues surrounding k-nearest neighbors and kernel regression. First, it is fairly common to see k-nearest neighbors and kernel regression being treated as a form of aggregation. The process of giving a set of states that are aggregated together has a certain commonality with k-nearest neighbor and kernel regression, where points near each other will produce estimates of $Y(x)$ that are similar. But this is where the resemblance ends. Simple aggregation is actually a form of parametric regression using dummy variables, and it offers neither the continuous approximations, nor the asymptotic unbiasedness of kernel regression.

Kernel regression is a method of approximation that is fundamentally different from linear regression and other parametric models. Parametric models use an explicit estimation step, where each observation results in an update to a vector of parameters. At any point in time, our approximation consists of the pre-specified parametric model, along with the current estimates of the regression parameters. With kernel regression, all we do is store data until we need an estimate of the function at a query point. Only then do we trigger the approximation method, which requires looping over all previous observation, a step that clearly can become expensive as the number of observations grow.

Kernel regression enjoys an important property from an old result known as Mercer's theorem. The result states that there exists a set of basis functions $\phi_f(S)$, $f \in \mathcal{F}$, possibly of very high dimensionality, where

$$K_h(S, S') = \phi(S)^T \phi(S'),$$

as long as the kernel function $K_h(S, S')$ satisfies some basic properties (satisfied by the kernels listed above). In effect this means that using appropriately designed kernels is equivalent to finding potentially very high-dimensional basis functions, without having to actually create them.

Unfortunately, the news is not all good. First, there is the annoying dimension of bandwidth selection, although this can be mediated partially by scaling the explanatory variables. More seriously, kernel regression (and this includes k-nearest neighbors) cannot be immediately applied to problems with more than about five dimensions (and even this can be a stretch). The problem is that these methods are basically trying to aggregate points in a multidimensional space. As the number of dimensions grows, the density of points in the d-dimensional space becomes quite sparse, making it very difficult to use "nearby" points to form an estimate of the function. A strategy for high-dimensional applications is to use separable approximations. These methods have received considerable attention in the broader machine learning community, but have not been widely tested in an ADP setting.

3.10.3 Local Polynomial Regression

Classical kernel regression uses a weighted sum of responses y^n to form an estimate of $Y(x)$. An obvious generalization is to estimate locally linear regression models around each point x^n by solving a least squares problem which minimizes a weighted sum of least squares. Let $\bar{Y}^n(x|x^i)$ be a linear model around the point x^k, formed by minimizing the weighted sum of squares given by

$$\min_\theta \left(\sum_{m=1}^{n} K_h(x^k, x^m) \left(y^m - \sum_{i=1}^{I} \theta_i x_i^m \right)^2 \right). \tag{3.61}$$

Thus, we are solving a classical linear regression problem, but we do this for each point x^k, and we fit the regression using all the other points (y^m, x^m), $m = 1, \ldots, n$. However, we weight deviations between the fitted model and each observation y^m by the kernel weighting factor $K_h(x^k, x^m)$ which is centered on the point x^k.

Local polynomial regression offers significant advantages in modeling accuracy, but with a significant increase in complexity.

3.10.4 Deep Neural Networks

Low-dimensional (basically finite) neural networks are a form of parametric regression. Once you have specified the number of layers and the nodes per layer, all that is left are the weights in the network, which represent the parameters. However, there is a class of high-dimensional neural networks known as deep learners, which typically have four or more layers (see Figure 3.10). These behave as if they have an unlimited number of layers and nodes per layer.

Deep learners have shown tremendous power in terms of their ability to capture complex patterns in language and images. It is well known that they require notoriously large datasets for training, but there are settings where massive amounts of data are available such as the results of internet searches, images of people, and text searches. In the context of algorithms for sequential decision problems, there are settings (such as the algorithms used for playing video games) where it is possible to run the algorithm for millions of iterations.

As of this writing, it is not yet clear if deep learners will prove useful in stochastic optimization, partly because our data comes from the iterations of an algorithm, and partly because the high-dimensional capabilities of neural networks raise the risk of overfitting in the context of stochastic optimization problems. Deep neural networks are very high-dimensional architectures, which means that they tend to fit noise, as we illustrated in Figure 3.8. In

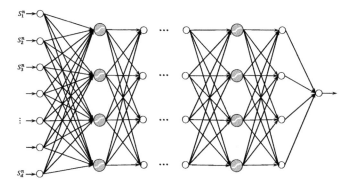

Figure 3.10 Illustration of a deep neural network.

addition, they are not very good at imposing structure such as monotonicity (although this has been a topic of research).

3.10.5 Support Vector Machines

Support vector machines (for classification) and support vector regression (for continuous problems) have attracted considerable interest in the machine learning community. For the purpose of fitting value function approximations, we are primarily interested in support vector regression, but we can also use regression to fit policy function approximations, and if we have discrete actions, we may be interested in classification. For the moment, we focus on fitting continuous functions.

Support vector regression, in its most basic form, is linear regression with a different objective than simply minimizing the sum of the squares of the errors. With support vector regression, we consider two goals. First, we wish to minimize the absolute sum of deviations that are larger than a set amount ξ. Second, we wish to minimize the regression parameters themselves, to push as many as possible close to zero.

As before, we let our predictive model be given by

$$y = \theta x + \epsilon.$$

Let $\epsilon^i = y^i - \theta x^i$ be the error. We then choose θ by solving the following optimization problem

$$\min_{\theta}\left(\frac{\eta}{2}\|\theta\|^2 + \sum_{i=1}^{n}\max\{0, |\epsilon^i| - \xi\}\right). \tag{3.62}$$

The first term penalizes positive values of θ, encouraging the model to minimize values of θ unless they contribute in a significant way to producing a better model. The second term penalizes errors that are greater than ξ. The parameters η and ξ are both tunable parameters. The error ϵ^i and error margin ξ are illustrated in Figure 3.11.

It can be shown by solving the dual that the optimal value of θ and the best fit $\bar{Y}(x)$ have the form

$$\theta = \sum_{i=1}^{n}(\bar{\beta}^i - \bar{\alpha}^i)x^i,$$

$$\bar{Y}(x) = \sum_{i=1}^{n}(\bar{\beta}^i - \bar{\alpha}^i)(x^i)^T x^i.$$

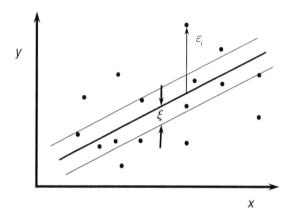

Figure 3.11 Illustration of penalty structure for support vector regression. Deviations within the gray area are assessed a value of zero. Deviations outside the gray area are measured based on their distance to the gray area.

Here, $\bar{\beta}^i$ and $\bar{\alpha}^i$ are scalars found by solving

$$\min_{\bar{\beta}^i,\bar{\alpha}^i} \xi \sum_{i=1}^{n}(\bar{\beta}^i + \bar{\alpha}^i) - \sum_{i=1}^{n} y^i(\bar{\beta}^i + \bar{\alpha}^i) + \frac{1}{2}\sum_{i=1}^{n}\sum_{i'=1}^{n}(\bar{\beta}^i + \bar{\alpha}^i)(\bar{\beta}^{i'} + \bar{\alpha}^{i'})(x^i)^T x^{i'},$$

subject to the constraints

$$0 \le \bar{\alpha}^i, \bar{\beta}^i \;\le\; 1/\eta,$$

$$\sum_{i=1}^{n}(\bar{\beta}^i - \bar{\alpha}^i) \;=\; 0,$$

$$\bar{\alpha}^i\bar{\beta}^i \;=\; 0.$$

3.10.6 Indexed Functions, Tree Structures, and Clustering

There are many problems where we feel comfortable specifying a simple set of basis functions for some of the parameters, but we do not have a good feel for the nature of the contribution of other parameters. For example, we may wish to plan how much energy to hold in storage over the course of the day. Let R_t be the amount of energy stored at time t, and let H_t be the hour of the day. Our state variable might be $S_t = (R_t, H_t)$. We feel that the value of energy in storage is a concave function in R_t, but this value depends in a complex way on the hour of day. It would not make sense, for example, to specify a value function approximation using

$$\overline{V}(S_t) = \theta_0 + \theta_1 R_t + \theta_2 R_t^2 + \theta_3 H_t + \theta_4 H_t^2.$$

There is no reason to believe that the hour of day will be related to the value of energy storage in any convenient way. Instead, we can estimate a function $\overline{V}(S_t|H_t)$ given by

$$\overline{V}(S_t|h) = \theta_0(h) + \theta_1(h)R_t + \theta_2(h)R_t^2.$$

What we are doing here is estimating a linear regression model for each value of $h = H_t$. This is simply a form of lookup table using regression given a particular value of the complex variables. Imagine that we can divide our state variable S_t into two sets: the first set, f_t, contains variables where we feel comfortable capturing the relationship using linear regression. The second set, g_t, includes more complex variables whose contribution is not as easily approximated. If g_t is a discrete scalar (such as hour of day), we can consider estimating a regression model for each value of g_t. However, if g_t is a vector (possibly with continuous dimensions), then there will be too many values.

When the vector g_t cannot be enumerated, we can resort to various clustering strategies. These fall under names such as regression trees and local polynomial regression (a form of kernel regression). These methods cluster g_t (or possibly the entire state S_t) and then fit simple regression models over subsets of data. In this case, we would create a set of clusters $\mathcal{C}^n$ based on n observations of states and values. We then fit a regression function $\overline{V}(S_t|c)$ for each cluster $c \in \mathcal{C}^n$. In traditional batch statistics, this process proceeds in two stages: clustering and then fitting. In approximate dynamic programming, we have to deal with the fact that we may change our clusters as we collect additional data.

A much more complex strategy is based on a concept known as Dirichlet process mixtures. This is a fairly sophisticated technique, but the essential idea is that you form clusters that produce good fits around local polynomial regressions. However, unlike traditional cluster-then-fit methods, the idea with Dirichlet process mixtures is that membership in a cluster is probabilistic, where the probabilities depend on the query point (e.g., the state whose value we are trying to estimate).

3.10.7 Comments on Nonparametric Models

Nonparametric models are extremely flexible, but two characteristics make them hard to work with:

- They need a *lot* of data.
- Due to their ability to closely fit data, nonparametric models are susceptible to overfitting when used to fit functions where observations are subject

to noise (which describes almost everything in this book). Figure 3.12 illustrates observations of revenue as we vary price. We are expecting a smooth, concave function. A kernel regression model closely fits the data, producing a behavior that does not seem realistic. By contrast, we might fit a quadratic model that captures the structure that we are expecting.

• Nonparametric models can be very clumsy to store. Kernel regression models effectively need the entire dataset. Deep neural networks may involve hundreds of thousands or even millions of parameters.

Neural networks have attracted considerable attention in recent years as they have demonstrated their ability to recognize faces and voices. These are problems that do not have a well-known structure matching bitmapped images to the identity of a person. We also note that the right answer is deterministic, which helps with training.

We anticipate that parametric models will remain popular for problems which have known structure. A difficulty with parametric models is that they are generally accurate only over some region. This is not a problem if we are searching for a unique point on a function, such as the best price, the best dosage of a drug, or the right temperature for running an experiment. However, there are problems such as estimating the value $V_t(S_t)$ of being in a state S_t at time t, which is a random variable that depends on the history up to time t. If we want to develop an approximate $\overline{V}_t(S_t) \approx V_t(S_t)$, then it has to be accurate over the range of states that we are likely to visit (and of course we may not know this).

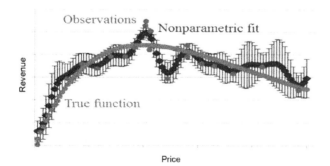

Figure 3.12 Fitting noisy data of revenue as a function of price using kernel regression versus a quadratic function.

3.11 Nonstationary Learning*

There are a number of settings where the true mean varies over time. We begin with the simplest setting where the mean may evolve up or down, but on average stays the same. We then consider the situation where the signal is steadily improving up to some unknown limit.

In chapter 7 we are going to use this in the context of optimizing functions of nonstationary random variables, or time-dependent functions of (typically) stationary random variables.

3.11.1 Nonstationary Learning I – Martingale Truth

In the stationary case, we might write observations as

$$W_{t+1} = \mu + \varepsilon_{t+1},$$

where $\varepsilon \sim N(0, \sigma_\varepsilon^2)$. This means that $\mathbb{E}W_{t+1} = \mu$, which is an unchanging truth that we are trying to learn. We refer to this as the stationary case because the distribution of W_t does not depend on time.

Now assume that the true mean μ is also changing over time. We write the dynamics of the mean using

$$\mu_{t+1} = \mu_t + \varepsilon_{t+1}^\mu,$$

where ε^μ is a random variable with distribution $N(0, \sigma_\mu^2)$. This means that $\mathbb{E}\{\mu_{t+1}|\mu_t\} = \mu_t$, which is the definition of a martingale process. This means that on average, the true mean μ_{t+1} at time $t + 1$ will be the same as at time t, although the actual may be different. Our observations are then made from

$$W_{t+1} = \mu_{t+1} + \varepsilon_{t+1}.$$

Typically, the variability of the mean process $\mu_0, \mu_1, \dots, \mu_t, \dots$ is much lower than the variance of the noise of an observation W of μ.

Now assume that μ_t is a vector with element μ_{tx}, where x will allow us to capture the performance of different drugs, paths through a network, people doing a job, or the price of a product. Let $\bar{\mu}_{tx}$ be the estimate of μ_{tx} at time t. Let Σ_t be the covariance matrix at time t, with element $\Sigma_{txx'} = Cov^n(\mu_{tx}, \mu_{tx'})$. This means we can write the distribution of μ_t as

$$\mu_t \sim N(\bar{\mu}_t, \Sigma_t).$$

This is the posterior distribution of μ_t, which is to say the distribution of μ_t given our prior observations $W_1, \dots, W_t$, and our prior $N(\bar{\mu}_0, \sigma_0)$. Let Σ^μ be the covariance matrix for the random variable ε^μ describing the evolution of μ. The *predictive distribution* is the distribution of μ_{t+1} given μ_t, which we write as

$$\mu_{t+1}|\mu_t \sim N(\bar{\mu}_t, \tilde{\Sigma}_t^\mu),$$

where

$$\tilde{\Sigma}_t^\mu = \Sigma_t + \Sigma^\mu.$$

Let e_{t+1} be the error in a vector of observations W_{t+1} given by

$$e_{t+1} = W_{t+1} - \bar{\mu}_t.$$

Let Σ^ε be the covariance matrix for e_{t+1}. The updated mean and covariance is computed using

$$\begin{aligned}
\bar{\mu}_{t+1} &= \bar{\mu}_t + \tilde{\Sigma}_t^\mu (\Sigma^\varepsilon + \tilde{\Sigma}_t^\mu)^{-1} e_{t+1}, \\
\Sigma_{t+1} &= \tilde{\Sigma}_t^\mu - \tilde{\Sigma}_t^\mu (\Sigma^\varepsilon + \tilde{\Sigma}_t^\mu) \tilde{\Sigma}_t^\mu.
\end{aligned}$$

3.11.2 Nonstationary Learning II – Transient Truth

A more general, but slightly more complex model, allows for predictable changes in θ_t. For example, we may know that θ_t is growing over time (perhaps θ_t is related to age or the population size), or we may be modeling variations in solar energy and have to capture the rising and setting of the sun.

We assume that μ_t is a vector with element x. Now assume we have a diagonal matrix M_t with factors that govern the predictable change in μ_t, allowing us to write the evolution of μ_t as

$$\mu_{t+1} = M_t \mu_t + \delta_{t+1}.$$

The evolution of the covariance matrix Σ_t becomes

$$\tilde{\Sigma}_t = M_t \Sigma_t M_t + \Sigma^\delta.$$

Now the evolution of the estimates of the mean and covariance matrix $\bar{\mu}_t$ and Σ_t are given by

$$\begin{aligned}
\bar{\mu}_{t+1} &= M_t \bar{\mu}_t + \tilde{\Sigma}_t (\Sigma^\varepsilon + \tilde{\Sigma}_t)^{-1} e_{t+1}, \\
\Sigma_{t+1} &= \tilde{\Sigma}_t - \tilde{\Sigma}_t (\Sigma^\varepsilon + \tilde{\Sigma}_t) \tilde{\Sigma}_t.
\end{aligned}$$

Note there is no change in the formula for Σ_{t+1} since M_t is built into $\tilde{\Sigma}_t$.

3.11.3 Learning Processes

There are many settings where we know that a process is improving over time up to an unknown limit. We refer to these as *learning processes* since we are modeling a process that learns as it progresses. Examples of learning processes are:

■ **EXAMPLE 3.12**

We have to choose a new basketball player x and then watch him improve as he gains playing time.

■ **EXAMPLE 3.13**

We observe the reduction in blood sugar due to diabetes medication x for a patient who has to adapt to the drug.

■ **EXAMPLE 3.14**

We are testing an algorithm where x are the parameters of the algorithm. The algorithm may be quite slow, so we have to project how good the final solution will be.

We model our process by assuming that observations come from

$$W_x^n = \mu_x^n + \varepsilon^n, \tag{3.63}$$

where the true mean μ_x^n rises according to

$$\mu_x^n(\theta) = \theta_x^s + [\theta_x^\ell - \theta_x^s][1 - e^{-n\theta_x^r}]. \tag{3.64}$$

Here, θ_x^s is the expected starting point at $n = 0$, while θ_x^ℓ is the limiting value as $n \to \infty$. The parameter θ_x^r controls the rate at which the mean approaches θ_x^ℓ. Let $\theta = (\theta^s, \theta^\ell, \theta^r)$ be the vector of unknown parameters.

If we fix θ^r, then $\mu_x^n(\theta)$ is linear in θ^s and θ^ℓ, allowing us to use our equations for recursive least squares for linear models that we presented in section 3.8. This will produce estimates $\bar{\theta}^{s,n}(\theta^r)$ and $\bar{\theta}^{\ell,n}(\theta^r)$ for each possible value of θ^r.

To handle the one nonlinear parameter θ^r, assume that we discretize this parameter into the values $\theta_1^r, \dots, \theta_K^r$. Let $p_k^{r,n}$ be the probability that $\theta^r = \theta_k^r$, which can be shown to be given by

$$p_k^{r,n} = \frac{L_k^n}{\sum_{k'=1}^{K} L_{k'}^n}$$

where L_k^n is the likelihood that $\theta^r = \theta_k^r$ which is given by

$$L_k^n \propto e^{-\left(\frac{W^{n+1} - \mu_x^n}{\sigma_\varepsilon}\right)^2},$$

where σ_ε^2 is the variance of ε. This now allows us to write

$$\bar{\mu}_x^n(\theta) = \sum_{k=1}^{K} p_k^{r,n} \bar{\mu}_x^n(\theta|\theta^r).$$

This approach provides us with conditional point estimates and variances of $\bar{\theta}^{s,n}(\theta^r), \bar{\theta}^{\ell,n}(\theta^r)$ for each θ^r, along with the distribution $p^{r,n}$ for θ^r.

3.12   The Curse of Dimensionality

There are many applications where state variables have multiple, possibly continuous dimensions. In some applications, the number of dimensions can number in the millions or larger (see section 2.3.4.2). Some examples are

---

■ **EXAMPLE 3.15**

An unmanned aerial vehicle may be described by location (three dimensions), velocity (three dimensions), in addition to fuel level. All dimensions are continuous.

■ **EXAMPLE 3.16**

A utility is trying to plan the amount of energy that should be put in storage as a function of the wind history (six hourly measurements), the history of electricity spot prices (six measurements), and the demand history (six measurements).

■ **EXAMPLE 3.17**

A trader is designing a policy for selling an asset that is priced against a basket of 20 securities, creating a 20-dimensional state variable.

■ **EXAMPLE 3.18**

A medical patient can be described by several thousand characteristics, beginning with basic information such as age, weight, gender, but extending to lifestyle variables (diet, smoking, exercise) to an extensive array of variables describing someone's medical history.

Each of these problems has a multi-dimensional state vector, and in all but the last example the dimensions are continuous. If we have 10 dimensions, and discretize each dimension into 100 elements, our input vector x (which might be a state) is $100^{10} = 10^{20}$ which is clearly a very large number. A reasonable strategy might be to aggregate. Instead of discretizing each dimension into 100 elements, what if we discretize into 5 elements? Now our state space is $5^{10} = 9.76 \times 10^6$, or almost 10 million states. Much smaller, but still quite large. Figure 3.13 illustrates the growth in the state space with the number of dimensions.

Each of our examples explode with the number of dimensions because we are using a lookup table representation for our function. It is important to realize that the curse of dimensionality is tied to the use of lookup tables. The other approximation architectures avoid the curse, but they do so by assuming structure such as a parametric form (linear or nonlinear).

Approximating high-dimensional functions is fundamentally intractable without exploiting structure. Beware of anyone claiming to "solve the curse of dimensionality." Pure lookup tables (which make no structural assumptions) are typically limited to four or five dimensions (depending on the number of

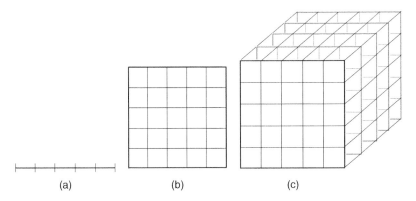

(a)   (b)   (c)

Figure 3.13 Illustration of the effect of higher dimensions on the number of grids in an aggregated state space.

values each dimension can take). However, we can handle thousands, even millions, of dimensions if we are willing to live with a linear model with separable, additive basis functions.

We can improve the accuracy of a linear model by adding features (basis functions) to our model. For example, if we use a second-order parametric representation, we might approximate a two-dimensional function using

$$F(x) \approx \theta_0 + \theta_1 x_1 + \theta_2 x_2 + \theta_{11} x_1^2 + \theta_{22} x_2^2 + \theta_{12} x_1 x_2.$$

If we have N dimensions, the approximation would look like

$$F(x) \approx \theta_0 + \sum_{i=1}^{N} \theta_i x_i + \sum_{i=1}^{N} \sum_{j=1}^{N} \theta_{ij} x_i x_j,$$

which means we have to estimate $1 + N + N^2$ parameters. As N grows, this grows very quickly, and this is only a second-order approximation. If we allow N^{th} order interactions, the approximation would look like

$$F(x) \approx \theta_0 + \sum_{i=1}^{N} \theta_i x_i + \sum_{i_1=1}^{N} \sum_{i_2=1}^{N} \theta_{i_1 i_2} x_{i_1} x_{i_2} + \sum_{i_1=1}^{N} \sum_{i_2=1}^{N} \cdots \sum_{i_N=1}^{N} \theta_{i_1, i_2, \dots, i_N} x_{i_1} x_{i_2} \cdots x_{i_N}.$$

The number of parameters we now have to estimate is given by $1 + N + N^2 + N^3 + \dots + N^N$. Not surprisingly, this becomes intractable even for relatively small values of N.

The problem follows us if we were to use kernel regression, where an estimate of a function at a point s can be estimated from a series of observations $(\hat{f}^i, x^i)_{i=1}^{N}$ using

$$F(x) \approx \frac{\sum_{i=1}^{N} \hat{f}^i k(x, x^i)}{\sum_{i=1}^{N} k(x, x^i)}$$

where $k(x, x^i)$ might be the Gaussian kernel

$$k(x, x^i) = e^{-\frac{\|x - x^i\|^2}{b}}$$

where b is a bandwidth. Kernel regression is effectively a soft form of the aggregation depicted in Figure 3.13(c). The problem is that we would have to choose a bandwidth that covers most of the data to get a statistically reliable estimate of a single point.

To see this, imagine that our observations are uniformly distributed in an N-dimensional cube that measures 1.0 on each side, which means it has a volume of 1.0. If we carve out an N-dimensional cube that measures .5 on a side, then this would capture 12.5% of the observations in a 3-dimensional cube, and 0.1 percent of the observations in a 10-dimensional cube. If we would like to choose a cube that captures $\eta = .1$ of our cube, we would need a cube that measures $r = \eta^{1/N} = .1^{1/10} = .794$, which means that our cube is covering almost 80% of the range of each input dimension.

The problem is that we have a multidimensional function, and we are trying to capture the joint behavior of all N dimensions. If we are willing to live with separable approximations, then we can scale to very large number of dimensions. For example, the approximation

$$F(x) \approx \theta_0 + \sum_{i=1}^{N} \theta_{1i} x_i + \sum_{i=1}^{N} \theta_{2i} x_i^2,$$

captures quadratic behavior but without any cross terms. The number of parameters is $1 + 2N$, which means we may be able to handle very high-dimensional problems. However, we lose the ability to handle interactions between different dimensions.

Kernel regression, along with essentially all nonparametric methods, is basically a fancy form of lookup table. Since these methods do not assume any underlying structure, they depend on capturing the local behavior of a function. The concept of "local," however, breaks down in high dimensions, where by "high" we typically mean four or five or more.

3.13 Designing Approximation Architectures in Adaptive Learning

Most solution methods in stochastic optimization are adaptive, which means that the data is arriving over time as a sequence of inputs x^n and observations $\hat{f}^{n+1}$. With each observation, we have to update our estimate of whatever function we are approximating, which might be the objective function $\mathbb{E}F(x, W)$, a value function $V(s)$, a policy $X^\pi(s)$, or a transition function $S^M(s, x, W)$. This entire chapter has focused on adaptive learning, but in the context where we used a fixed model and just adapt the parameters to produce the best fit.

Adaptive learning means that we have to start with small datasets (sometimes no data at all), and then adapt as new decisions and observations arrive. This raises a challenge we have not addressed above: we need to do more than just update a parameter vector θ^n with new data to produce θ^{n+1}. Instead, we need to update the architecture of the function we are trying to estimate. Said differently, the dimensionality of θ^n (or at least the set of nonzero elements of θ^n) will need to change as we acquire more data.

A key challenge with any statistical learning problem is designing a function that strikes the right tradeoff between the dimensionality of the function and the amount of data available for approximating the function. For a batch problem, we can use powerful tools such as regularization (see equation (3.39)) for identifying models that have the right number of variables given the available data. But this only works for batch estimation, where the size of the dataset is fixed.

As of this writing, additional research is needed to create the tools that can help to identify not just the best parameter vector θ^n, but the structure of the function itself. One technique that does this is hierarchical aggregation which we presented in the context of lookup tables in section 3.6. This is a powerful methodology that adaptively adjusts from a low-dimensional representation (that is, estimates of the function at a high level of aggregation) to higher dimensional representations, which is accomplished by putting higher weights on the more disaggregate estimates. However, lookup table belief models are limited to relatively low-dimensional problems.

3.14 Why Does It Work?**

3.14.1 Derivation of the Recursive Estimation Equations

Here we derive the recursive estimation equations given by equations (3.41)-(3.45). To begin, we note that the matrix $(X^n)^T X^n$ is an $I + 1$ by $I + 1$ matrix where the element for row i, column j is given by

$$[(X^n)^T X^n]_{i,j} = \sum_{m=1}^{n} x_i^m x_j^m.$$

This term can be computed recursively using

$$[(X^n)^T X^n]_{i,j} = \sum_{m=1}^{n-1} (x_i^m x_j^m) + x_i^n x_j^n.$$

In matrix form, this can be written

$$[(X^n)^T X^n] = [(X^{n-1})^T X^{n-1}] + x^n (x^n)^T.$$

Keeping in mind that x^n is a column vector, $x^n (x^n)^T$ is an $I + 1$ by $I + 1$ matrix formed by the cross products of the elements of x^n. We now use the Sherman-Morrison formula (see section 3.14.2 for a derivation) for updating the inverse of a matrix

$$[A + uu^T]^{-1} = A^{-1} - \frac{A^{-1} uu^T A^{-1}}{1 + u^T A^{-1} u},$$

where A is an invertible $n \times n$ matrix, and u is an n-dimensional column vector. Applying this formula to our problem, we obtain

$$
\begin{aligned}
[(X^n)^T X^n]^{-1} &= [(X^{n-1})^T X^{n-1} + x^n(x^n)^T]^{-1} \\
&= [(X^{n-1})^T X^{n-1}]^{-1} \\
&\quad - \frac{[(X^{n-1})^T X^{n-1}]^{-1} x^n (x^n)^T [(X^{n-1})^T X^{n-1}]^{-1}}{1 + (x^n)^T [(X^{n-1})^T X^{n-1}]^{-1} x^n}.
\end{aligned}
\tag{3.65}
$$

The term $(X^n)^T Y^n$ can also be updated recursively using

$$
(X^n)^T Y^n = (X^{n-1})^T Y^{n-1} + x^n(y^n).
\tag{3.66}
$$

To simplify the notation, let

$$
\begin{aligned}
M^n &= [(X^n)^T X^n]^{-1}, \\
\gamma^n &= 1 + (x^n)^T [(X^{n-1})^T X^{n-1}]^{-1} x^n.
\end{aligned}
$$

This simplifies our inverse updating equation (3.65) to

$$
M^n = M^{n-1} - \frac{1}{\gamma^n}(M^{n-1} x^n (x^n)^T M^{n-1}).
$$

Recall that

$$
\bar{\theta}^n = [(X^n)^T X^n]^{-1}(X^n)^T Y^n.
\tag{3.67}
$$

Combining (3.67) with (3.65) and (3.66) gives

$$
\begin{aligned}
\bar{\theta}^n &= [(X^n)^T X^n]^{-1}(X^n)^T Y^n \\
&= \left(M^{n-1} - \frac{1}{\gamma^n}(M^{n-1} x^n (x^n)^T M^{n-1})\right)\left((X^{n-1})^T Y^{n-1} + x^n y^n\right), \\
&= M^{n-1}(X^{n-1})^T Y^{n-1} \\
&\quad - \frac{1}{\gamma^n} M^{n-1} x^n (x^n)^T M^{n-1} \left[(X^{n-1})^T Y^{n-1} + x^n y^n\right] + M^{n-1} x^n y^n.
\end{aligned}
$$

We can start to simplify by using $\bar{\theta}^{n-1} = M^{n-1}(X^{n-1})^T Y^{n-1}$. We are also going to bring the term $x^n M^{n-1}$ inside the square brackets. Finally, we are going to bring the last term $M^{n-1} x^n y^n$ inside the brackets by taking the coefficient $M^{n-1} x^n$ outside the brackets and multiplying the remaining y^n by the scalar $\gamma^n = 1 +$

$(x^n)^T M^{n-1} x^n$, giving us

$$
\begin{aligned}
\bar{\theta}^n \;=\; & \bar{\theta}^{n-1} - \frac{1}{\gamma^n} M^{n-1} x^n \left[(x^n)^T (M^{n-1}(X^{n-1})^T Y^{n-1}) \right. \\
& \left. + (x^n)^T M^{n-1} x^n y^n - (1 + (x^n)^T M^{n-1} x^n) y^n \right].
\end{aligned}
$$

Again, we use $\bar{\theta}^{n-1} = M^{n-1}(X^{n-1})^T Y^{n-1}$ and observe that there are two terms $(x^n)^T M^{n-1} x^n y^n$ that cancel, leaving

$$
\bar{\theta}^n \;=\; \bar{\theta}^{n-1} - \frac{1}{\gamma^n} M^{n-1} x^n \left((x^n)^T \bar{\theta}^{n-1} - y^n \right).
$$

We note that $(\bar{\theta}^{n-1})^T x^n$ is our prediction of y^n using the parameter vector from iteration $n-1$ and the explanatory variables x^n. y^n is, of course, the actual observation, so our error is given by

$$
\hat{\varepsilon}^n = y^n - (\bar{\theta}^{n-1})^T x^n.
$$

Let

$$
H^n = -\frac{1}{\gamma^n} M^{n-1}.
$$

We can now write our updating equation using

$$
\bar{\theta}^n = \bar{\theta}^{n-1} - H^n x^n \hat{\varepsilon}^n. \tag{3.68}
$$

3.14.2 The Sherman-Morrison Updating Formula

The Sherman-Morrison matrix updating formula (also known as the Woodbury formula or the Sherman-Morrison-Woodbury formula) assumes that we have a matrix A and that we are going to update it with the outer product of the column vector u to produce the matrix B, given by

$$
B = A + uu^T. \tag{3.69}
$$

Pre-multiply by B^{-1} and post-multiply by A^{-1}, giving

$$
A^{-1} = B^{-1} + B^{-1} uu^T A^{-1}. \tag{3.70}
$$

Post-multiply by u

$$
\begin{aligned}
A^{-1} u \;=\; & B^{-1} u + B^{-1} uu^T A^{-1} u \\
\;=\; & B^{-1} u \left(1 + u^T A^{-1} u \right).
\end{aligned}
$$

Note that $u^T A^{-1} u$ is a scalar. Divide through by $\left(1 + u^T A^{-1} u \right)$

$$
\frac{A^{-1} u}{(1 + u^T A^{-1} u)} \;=\; B^{-1} u.
$$

Now post-multiply by $u^T A^{-1}$

$$\frac{A^{-1}uu^T A^{-1}}{(1 + u^T A^{-1}u)} = B^{-1}uu^T A^{-1}. \tag{3.71}$$

Equation (3.70) gives us

$$B^{-1}uu^T A^{-1} = A^{-1} - B^{-1}. \tag{3.72}$$

Substituting (3.72) into (3.71) gives

$$\frac{A^{-1}uu^T A^{-1}}{(1 + u^T A^{-1}u)} = A^{-1} - B^{-1}. \tag{3.73}$$

Solving for B^{-1} gives us

$$\begin{aligned} B^{-1} &= [A + uu^T]^{-1} \\ &= A^{-1} - \frac{A^{-1}uu^T A^{-1}}{(1 + u^T A^{-1}u)}, \end{aligned}$$

which is the desired formula.

3.14.3 Correlations in Hierarchical Estimation

It is possible to derive the optimal weights for the case where the statistics $\bar{v}_s^{(g)}$ are not independent. In general, if we are using a hierarchical strategy and have $g' > g$ (which means that aggregation g' is more aggregate than g), then the statistic $\bar{v}_s^{(g',n)}$ is computed using observations $\hat{v}_s^n$ that are also used to compute $\bar{v}_s^{(g,n)}$.

We begin by defining

$$\begin{aligned} \mathcal{N}_s^{(g,n)} &= \text{The set of iterations } n \text{ where } G^g(\hat{s}^n) = G^g(s) \text{ (that is,} \\ &\quad \hat{s}^n \text{ aggregates to the same state as } s). \\ N_s^{(g,n)} &= |\mathcal{N}_s^{(g,n)}| \\ \bar{\varepsilon}_s^{(g,n)} &= \text{An estimate of the average error when observing} \\ &\quad \text{state } s = G(\hat{s}^n). \\ &= \frac{1}{N_s^{(g,n)}} \sum_{n \in \mathcal{N}_s^{(g,n)}} \hat{\varepsilon}_s^{(g,n)}. \end{aligned}$$

The average error $\bar{\varepsilon}_s^{(g,n)}$ can be written

$$
\begin{aligned}
\bar{\varepsilon}_s^{(g,n)} &= \frac{1}{N_s^{(g,n)}} \left(\sum_{n \in \mathcal{N}_s^{(0,n)}} \varepsilon^n + \sum_{n \in \mathcal{N}_s^{(g,n)} \setminus \mathcal{N}_s^{(0,n)}} \varepsilon^n \right) \\
&= \frac{N_s^{(0,n)}}{N_s^{(g,n)}} \bar{\varepsilon}_s^{(0)} + \frac{1}{N_s^{(g,n)}} \sum_{n \in \mathcal{N}_s^{(g,n)} \setminus \mathcal{N}_s^{(0,n)}} \varepsilon^n.
\end{aligned}
\tag{3.74}
$$

This relationship shows us that we can write the error term at the higher level of aggregation g' as a sum of a term involving the errors at the lower level of aggregation g (for the same state s) and a term involving errors from other states s'' where $G^{g'}(s'') = G^{g'}(s)$, given by

$$
\begin{aligned}
\bar{\varepsilon}_s^{(g',n)} &= \frac{1}{N_s^{(g',n)}} \left(\sum_{n \in \mathcal{N}_s^{(g,n)}} \varepsilon^n + \sum_{n \in \mathcal{N}_s^{(g',n)} \setminus \mathcal{N}_s^{(g,n)}} \varepsilon^n \right) \\
&= \frac{1}{N_s^{(g',n)}} \left(N_s^{(g,n)} \frac{\sum_{n \in \mathcal{N}_s^{(g,n)}} \varepsilon^n}{N_s^{(g,n)}} + \sum_{n \in \mathcal{N}_s^{(g',n)} \setminus \mathcal{N}_s^{(g,n)}} \varepsilon^n \right) \\
&= \frac{N_s^{(g,n)}}{N_s^{(g',n)}} \bar{\varepsilon}_s^{(g,n)} + \frac{1}{N_s^{(g',n)}} \sum_{n \in \mathcal{N}_s^{(g',n)} \setminus \mathcal{N}_s^{(g,n)}} \varepsilon^n.
\end{aligned}
\tag{3.75}
$$

We can overcome this problem by rederiving the expression for the optimal weights. For a given (disaggregate) state s, the problem of finding the optimal weights $(w_s^{(g,n)})_{g \in \mathcal{G}}$ is stated by

$$
\min_{w_s^{(g,n)}, g \in \mathcal{G}} \mathbb{E} \left[\frac{1}{2} \left(\sum_{g \in \mathcal{G}} w_s^{(g,n)} \cdot \bar{v}_s^{(g,n)} - v_s^{(g,n)} \right)^2 \right]
\tag{3.76}
$$

subject to

$$
\sum_{g \in \mathcal{G}} w_s^{(g,n)} = 1
\tag{3.77}
$$

$$
w_s^{(g,n)} \geq 0, \quad g \in \mathcal{G}.
\tag{3.78}
$$

Let

$$
\begin{aligned}
\bar{\delta}_s^{(g,n)} &= \text{The error in the estimate } \bar{v}_s^{(g,n)} \text{ from the true value} \\
&\quad \text{associated with attribute vector } s. \\
&= \bar{v}_s^{(g,n)} - v_s.
\end{aligned}
$$

The optimal weights are computed using the following theorem:

Theorem 3.14.1. For a given attribute vector, s, the optimal weights, $w_s^{(g,n)}$, $g \in \mathcal{G}$, where the individual estimates are correlated by way of a tree structure, are given by solving the following system of linear equations in (w, λ):

$$\sum_{g \in \mathcal{G}} w_s^{(g,n)} \mathbb{E}\left[\bar{\delta}_s^{(g,n)} \bar{\delta}_s^{(g',n)}\right] - \lambda = 0 \quad \forall \ g' \in \mathcal{G} \tag{3.79}$$

$$\sum_{g \in \mathcal{G}} w_s^{(g,n)} = 1 \tag{3.80}$$

$$w_s^{(g,n)} \geq 0 \quad \forall \ g \in \mathcal{G}. \tag{3.81}$$

Proof: The proof is not too difficult and it illustrates how we obtain the optimal weights. We start by formulating the Lagrangian for the problem formulated in (3.76)-(3.78), which gives us

$$\begin{aligned} L(w, \lambda) &= \mathbb{E}\left[\frac{1}{2}\left(\sum_{g \in \mathcal{G}} w_s^{(g,n)} \cdot \bar{v}_s^{(g,n)} - \nu_s^{(g,n)}\right)^2\right] + \lambda\left(1 - \sum_{g \in \mathcal{G}} w_s^{(g,n)}\right) \\ &= \mathbb{E}\left[\frac{1}{2}\left(\sum_{g \in \mathcal{G}} w_s^{(g,n)}\left(\bar{v}_s^{(g,n)} - \nu_s^{(g,n)}\right)\right)^2\right] + \lambda\left(1 - \sum_{g \in \mathcal{G}} w_s^{(g,n)}\right). \end{aligned}$$

The first-order optimality conditions are

$$\mathbb{E}\left[\sum_{g \in \mathcal{G}} w_s^{(g,n)}\left(\bar{v}_s^{(g,n)} - \nu_s^{(g,n)}\right)\left(\bar{v}_s^{(g',n)} - \nu_s^{(g,n)}\right)\right] - \lambda = 0 \quad \forall \ g' \in \mathcal{G} \tag{3.82}$$

$$\sum_{g \in \mathcal{G}} w_s^{(g,n)} - 1 = 0. \tag{3.83}$$

To simplify equation (3.82), we note that,

$$\begin{aligned} \mathbb{E}\left[\sum_{g \in \mathcal{G}} w_s^{(g,n)}\left(\bar{v}_s^{(g,n)} - \nu_s^{(g,n)}\right)\left(\bar{v}_s^{(g',n)} - \nu_s^{(g,n)}\right)\right] &= \mathbb{E}\left[\sum_{g \in \mathcal{G}} w_s^{(g,n)} \bar{\delta}_s^{(g,n)} \bar{\delta}_s^{(g',n)}\right] \\ &= \sum_{g \in \mathcal{G}} w_s^{(g,n)} \mathbb{E}\left[\bar{\delta}_s^{(g,n)} \bar{\delta}_s^{(g',n)}\right]. \end{aligned}$$

$$\tag{3.84}$$

Combining equations (3.82) and (3.84) gives us equation (3.79) which completes the proof. ☐

Finding the optimal weights that handle the correlations between the statistics at different levels of aggregation requires finding $\mathbb{E}\left[\bar{\delta}_s^{(g,n)} \bar{\delta}_s^{(g',n)}\right]$. We are going to compute this expectation by conditioning on the set of attributes $\hat{s}^n$ that are sampled. This means that our expectation is defined over the outcome space Ω^ε. Let $N_s^{(g,n)}$ be the number of observations of state s at aggregation level g. The expectation is computed using:

Proposition 3.14.1. The coefficients of the weights in equation (3.80) can be expressed as follows:

$$\mathbb{E}\left[\bar{\delta}_s^{(g,n)}\bar{\delta}_s^{(g',n)}\right] = \mathbb{E}\left[\bar{\beta}_s^{(g,n)}\bar{\beta}_s^{(g',n)}\right] + \frac{N_s^{(g,n)}}{N_s^{(g',n)}}\mathbb{E}\left[\bar{\varepsilon}_s^{(g,n)^2}\right] \quad \forall g \le g' \text{ and } g,g' \in \mathcal{G}.$$

$$(3.85)$$

The proof is given in section 3.14.4.

Now consider what happens when we make the assumption that the measurement error ε^n is independent of the attribute being sampled, $\hat{s}^n$. We do this by assuming that the variance of the measurement error is a constant given by σ_ε^2. This gives us the following result:

Corollary 3.14.1. For the special case where the statistical noise in the measurement of the values is independent of the attribute vector sampled, equation (3.85) reduces to

$$\mathbb{E}\left[\bar{\delta}_s^{(g,n)}\bar{\delta}_s^{(g',n)}\right] = \mathbb{E}\left[\bar{\beta}_s^{(g,n)}\bar{\beta}_s^{(g',n)}\right] + \frac{\sigma_\varepsilon^2}{N_s^{(g',n)}}. \tag{3.86}$$

For the case where $g = 0$ (the most disaggregate level), we assume that $\beta_s^{(0)} = 0$ which gives us

$$\mathbb{E}\left[\bar{\beta}_s^{(0,n)}\bar{\beta}_s^{(g',n)}\right] = 0.$$

This allows us to further simplify (3.86) to obtain

$$\mathbb{E}\left[\bar{\delta}_s^{(0,n)}\bar{\delta}_s^{(g',n)}\right] = \frac{\sigma_\varepsilon^2}{N_s^{(g',n)}}. \tag{3.87}$$

3.14.4 Proof of Proposition 3.14.1

We start by defining

$$\bar{\delta}_s^{(g,n)} = \bar{\beta}_s^{(g,n)} + \bar{\varepsilon}_s^{(g,n)}. \tag{3.88}$$

Equation (3.88) gives us

$$
\begin{aligned}
\mathbb{E}\left[\bar{\delta}_s^{(g,n)}\bar{\delta}_s^{(g',n)}\right] &= \mathbb{E}\left[(\bar{\beta}_s^{(g,n)}+\bar{\varepsilon}_s^{(g,n)})(\bar{\beta}_s^{(g',n)}+\bar{\varepsilon}_s^{(g',n)})\right] \\
&= \mathbb{E}\left[\bar{\beta}_s^{(g,n)}\bar{\beta}_s^{(g',n)}+\bar{\beta}_s^{(g',n)}\bar{\varepsilon}_s^{(g,n)}+\bar{\beta}_s^{(g,n)}\bar{\varepsilon}_s^{(g',n)}+\bar{\varepsilon}_s^{(g,n)}\bar{\varepsilon}_s^{(g',n)}\right] \\
&= \mathbb{E}\left[\bar{\beta}_s^{(g,n)}\bar{\beta}_s^{(g',n)}\right]+\mathbb{E}\left[\bar{\beta}_s^{(g',n)}\bar{\varepsilon}_s^{(g,n)}\right]+\mathbb{E}\left[\bar{\beta}_s^{(g,n)}\bar{\varepsilon}_s^{(g',n)}\right] \\
&\quad +\mathbb{E}\left[\bar{\varepsilon}_s^{(g,n)}\bar{\varepsilon}_s^{(g',n)}\right].
\end{aligned}
\tag{3.89}
$$

We note that

$$
\mathbb{E}\left[\bar{\beta}_s^{(g',n)}\bar{\varepsilon}_s^{(g,n)}\right]=\bar{\beta}_s^{(g',n)}\mathbb{E}\left[\bar{\varepsilon}_s^{(g,n)}\right]=0.
$$

Similarly

$$
\mathbb{E}\left[\bar{\beta}_s^{(g,n)}\bar{\varepsilon}_s^{(g',n)}\right]=0.
$$

This allows us to write equation (3.89) as

$$
\mathbb{E}\left[\bar{\delta}_s^{(g,n)}\bar{\delta}_s^{(g',n)}\right] = \mathbb{E}\left[\bar{\beta}_s^{(g,n)}\bar{\beta}_s^{(g',n)}\right]+\mathbb{E}\left[\bar{\varepsilon}_s^{(g,n)}\bar{\varepsilon}_s^{(g',n)}\right].
\tag{3.90}
$$

We start with the second term on the right-hand side of equation (3.90). This term can be written as

$$
\begin{aligned}
\mathbb{E}\left[\bar{\varepsilon}_s^{(g,n)}\bar{\varepsilon}_s^{(g',n)}\right] &= \mathbb{E}\left[\bar{\varepsilon}_s^{(g,n)}\cdot\frac{N_s^{(g,n)}}{N_s^{(g')}}\bar{\varepsilon}_s^{(g,n)}\right]+\mathbb{E}\left[\bar{\varepsilon}_s^{(g,n)}\cdot\frac{1}{N_s^{(g')}}\sum_{n\in\mathcal{N}_s^{(g',n)}\setminus\mathcal{N}_s^{(g,n)}}\varepsilon^n\right] \\
&= \frac{N_s^{(g,n)}}{N_s^{(g')}}\mathbb{E}\left[\bar{\varepsilon}_s^{(g,n)}\bar{\varepsilon}_s^{(g,n)}\right]+\frac{1}{N_s^{(g')}}\underbrace{\mathbb{E}\left[\bar{\varepsilon}_s^{(g,n)}\cdot\sum_{n\in\mathcal{N}_s^{(g',n)}\setminus\mathcal{N}_s^{(g,n)}}\varepsilon^n\right]}_{I}.
\end{aligned}
$$

The term I can be rewritten using

$$
\mathbb{E}\left[\bar{\varepsilon}_s^{(g,n)}\cdot\sum_{n\in\mathcal{N}_s^{(g',n)}\setminus\mathcal{N}_s^{(g,n)}}\varepsilon^n\right] = \mathbb{E}\left[\bar{\varepsilon}_s^{(g,n)}\right]\mathbb{E}\left[\sum_{n\in\mathcal{N}_s^{(g',n)}\setminus\mathcal{N}_s^{(g,n)}}\varepsilon^n\right],
$$
$$
= 0
$$

which means

$$
\mathbb{E}\left[\bar{\varepsilon}_s^{(g,n)}\bar{\varepsilon}_s^{(g',n)}\right] = \frac{N_s^{(g,n)}}{N_s^{(g')}}\mathbb{E}\left[\bar{\varepsilon}_s^{(g)^2}\right].
\tag{3.91}
$$

Combining (3.90) and (3.91) proves the proposition. $\qquad\square$

The second term on the right-hand side of equation (3.91) can be further simplified using,

$$
\begin{aligned}
\mathbb{E}\left[\bar{\varepsilon}_s^{(g)^2}\right] &= \mathbb{E}\left[\left(\frac{1}{N_s^{(g,n)}} \sum_{n \in \mathcal{N}_s^{(g,n)}} \varepsilon^n\right)^2\right], \quad \forall \ g' \in \mathcal{G} \\
&= \frac{1}{\left(N_s^{(g,n)}\right)^2} \sum_{m \in \mathcal{N}_s^{(g,n)}} \sum_{n \in \mathcal{N}_s^{(g,n)}} \mathbb{E}\left[\varepsilon^m \varepsilon^n\right] \\
&= \frac{1}{\left(N_s^{(g,n)}\right)^2} \sum_{n \in \mathcal{N}_s^{(g,n)}} \mathbb{E}\left[(\varepsilon^n)^2\right] \\
&= \frac{1}{\left(N_s^{(g,n)}\right)^2} N_s^{(g,n)} \sigma_\varepsilon^2 \\
&= \frac{\sigma_\varepsilon^2}{N_s^{(g,n)}}. \quad\quad\quad (3.92)
\end{aligned}
$$

Combining equations (3.85), (3.91), and (3.92) gives us the result in equation (3.86). □

3.15 Bibliographic Notes

This chapter is primarily a tutorial into online (adaptive) learning. Readers looking to do serious algorithmic work should obtain a good statistical reference such as Bishop (2006) or Hastie et al. (2009). The second reference can be downloaded from

```
http://www-stat.stanford.edu/~tibs/ElemStatLearn/.
```

Note that classical references in statistical learning tend to focus on batch learning, while we are primarily interested in online (or adaptive) learning.

Section 3.6 – Aggregation has been a widely used technique in dynamic programming as a method to overcome the curse of dimensionality. Early work focused on picking a fixed level of aggregation (Whitt (1978), Bean et al. (1987)), or using adaptive techniques that change the level of aggregation as the sampling process progresses (Bertsekas and Castanon (1989), Mendelssohn (1982), Bertsekas and Tsitsiklis (1996)), but which still use a fixed level of aggregation at any given time. Much of the literature on aggregation has focused on deriving error bounds (Zipkin (1980)). For a good discussion of aggregation as a general technique in modeling, see

Rogers et al. (1991). The material in section 3.6.3 is based on George et al. (2008) and Powell and George (2006). LeBlanc and Tibshirani (1996) and Yang (2001) provide excellent discussions of mixing estimates from different sources. For a discussion of soft state aggregation, see Singh et al. (1995). Section 3.5 on bias and variance is based on Powell and George (2006).

Section 3.7 – Basis functions have their roots in the modeling of physical processes. A good introduction to the field from this setting is Heuberger et al. (2005). Schweitzer and Seidmann (1985) describes generalized polynomial approximations for Markov decision processes for use in value iteration, policy iteration, and the linear programming method. Menache et al. (2005) discusses basis function adaptations in the context of reinforcement learning. For a very nice discussion of the use of basis functions in approximate dynamic programming, see Tsitsiklis and Roy (1996) and Van Roy (2001). Tsitsiklis and Van Roy (1997) proves convergence of iterative stochastic algorithms for fitting the parameters of a regression model when the policy is held fixed. For section 17.6.1, the first use of approximate dynamic programming for evaluating an American call option is given in Longstaff and Schwartz (2001), but the topic has been studied for decades (see Taylor (1967)). Tsitsiklis and Van Roy (2001) also provide an alternative ADP algorithm for American call options. Clement et al. (2002) provides formal convergence results for regression models used to price American options. This presentation on the geometric view of basis functions is based on Tsitsiklis and Van Roy (1997).

Section 3.10 – An excellent introduction to continuous approximation techniques is given in Judd (1998) in the context of economic systems and computational dynamic programming. Ormoneit and Sen (2002) and Ormoneit and Glynn (2002) discuss the use of kernel-based regression methods in an approximate dynamic programming setting, providing convergence proofs for specific algorithmic strategies. For a thorough introduction to locally polynomial regression methods, see Fan and Gijbels (1996). An excellent discussion of a broad range of statistical learning methods can be found in Hastie et al. (2009). Bertsekas and Tsitsiklis (1996) provides an excellent discussion of neural networks in the context of approximate dynamic programming. Haykin (1999) presents a much more in-depth presentation of neural networks, including a chapter on approximate dynamic programming using neural networks. A very rich field of study has evolved around support vector machines and support vector regression. For a thorough tutorial, see Smola and Schölkopf (2004). A shorter and more readable introduction is contained in chapter 12 of Hastie et al. (2009). Note that SVR does not lend itself readily to recursive updating, which we suspect will limit its usefulness in approximate dynamic programming.

Figure 3.8 was created by Larry Thul.

Section 3.12 – See Hastie et al. (2009), section 2.5, for a very nice discussion of the challenges of approximating high-dimensional functions.

Section 3.14.2 – The Sherman-Morrison updating formulas are given in a number of references, such as L. and Soderstrom (1983) and Golub and Loan (1996).

Exercises

Review questions

3.1 What are the five classes of approximations that may arise in sequential decision problems?

3.2 When using lookup table models with independent observations, what are the belief state variables for frequentist and Bayesian beliefs?

3.3 What is the belief state for lookup tables with correlated beliefs, when using a Bayesian belief model?

3.4 This chapter is organized around three major classes of approximation architectures: lookup table, parametric, and nonparametric, but some have argued that there should only be two classes: parametric and nonparametric. Justify your answer by presenting an argument why a lookup table can be properly modeled as a parametric model, and then a counter argument why a lookup table is more similar to a nonparametric model. [Hint: What is the defining characteristic of a nonparametric model? – see section 3.10.]

3.5 What is the belief state if you are doing recursive updating of a linear model?

3.6 A deep neural network is just a bigger neural network. So why are deep neural networks considered nonparametric models? After all they are just a nonlinear model with a very large number of parameters. How many parameters does a neural network have with four layers and 100 nodes per layer.?

Computational exercises

3.7 Use equations (3.16) and (3.17) to update the mean vector with prior

$$\bar{\mu}^0 = \begin{bmatrix} 10 \\ 18 \\ 12 \end{bmatrix}.$$

Assume that we test alternative 3 and observe $W = 19$ and that our prior covariance matrix Σ^0 is given by

$$\Sigma^0 = \begin{bmatrix} 12 & 4 & 2 \\ 4 & 8 & 3 \\ 2 & 3 & 10 \end{bmatrix}.$$

Assume that $\lambda^W = 4$. Give $\bar{\mu}^1$ and Σ^1.

3.8 In a spreadsheet, create a 4×4 grid where the cells are numbered 1, 2, ..., 16 starting with the upper left-hand corner and moving left to right, as shown below.

1	2	3	4
5	6	7	8
9	10	11	12
13	14	15	16

We are going to treat each number in the cell as the mean of the observations drawn from that cell. Now assume that if we observe a cell, we observe the mean plus a random variable that is uniformly distributed between -1 and $+1$. Next define a series of aggregations where aggregation 0 is the disaggregate level, aggregation 1 divides the grid into four 2×2 cells, and aggregation 2 aggregates everything into a single cell. After n iterations, let $\bar{f}_s^{(g,n)}$ be the estimate of cell "s" at the nth level of aggregation, and let

$$\bar{f}_s^n = \sum_{g \in \mathcal{G}} w_s^{(g)} \bar{f}_s^{(g,n)}$$

be your best estimate of cell s using a weighted aggregation scheme. Compute an overall error measure using

$$(\bar{\sigma}^2)^n = \sum_{s \in \mathcal{S}} (\bar{f}_s^n - v_s)^2,$$

where v_s is the true value (taken from your grid) of being in cell s. Also let $w^{(g,n)}$ be the average weight after n iterations given to the aggregation level g when averaged over all cells at that level of aggregation (for example, there is only one cell for $w^{(2,n)}$). Perform 1000 iterations where at each iteration you randomly sample a cell and measure it with noise. Update your estimates at each level of aggregation, and compute the variance of your estimate with and without the bias correction.

(a) Plot $w^{(g,n)}$ for each of the three levels of aggregation at each iteration. Do the weights behave as you would expect? Explain.

(b) For each level of aggregation, set the weight given to that level equal to one (in other words, we are using a single level of aggregation) and plot the overall error as a function of the number of iterations.

(c) Add to your plot the average error when you use a weighted average, where the weights are determined by equation (3.32) without the bias correction.

(d) Finally add to your plot the average error when you used a weighted average, but now determine the weights by equation (3.33), which uses the bias correction.

(e) Repeat the above assuming that the noise is uniformly distributed between -5 and $+5$.

3.9 In this exercise you will use the equations in section 3.8.1 to update a linear model. Assume you have an estimate of a linear model given by

$$\bar{F}(x|\theta^0) = \theta_0 + \theta_1\phi_1(x) + \theta_2\phi_2(x)$$
$$= -12 + 5.2\phi_1 + 2.8\phi_2.$$

Assume that the matrix B^0 is a 3×3 identity matrix. Assume the vector $\phi = (\phi_0 \ \phi_1 \ \phi_2) = (5 \ 15 \ 22)$ and that you observe $\hat{f}^1 = 90$. Give the updated regression vector θ^1.

Theory questions

3.10 Show that

$$\sigma_s^2 = (\sigma_s^2)^{(g)} + (\beta_s^{(g)})^2 \tag{3.93}$$

which breaks down the total variation in an estimate at a level of aggregation is the sum of the variation of the observation error plus the bias squared.

3.11 Show that $\mathbb{E}\left[\left(\bar{\mu}^{n-1} - \mu(n)\right)^2\right] = \lambda^{n-1}\sigma^2 + (\beta^n)^2$ (which proves equation (3.24)). [Hint: Add and subtract $\mathbb{E}\bar{\mu}^{n-1}$ inside the expectation and expand.]

3.12 Show that $\mathbb{E}\left[\left(\bar{\theta}^{n-1} - \hat{\theta}^n\right)^2\right] = (1 + \lambda^{n-1})\sigma^2 + (\beta^n)^2$ (which proves equation 3.25). [Hint: See previous exercise.]

3.13 Derive the small sample form of the recursive equation for the variance given in (3.26). Recall that if

$$\bar{\mu}^n = \frac{1}{n}\sum_{m=1}^{n}\hat{\mu}^m$$

then an estimate of the variance of $\hat{\theta}$ is

$$Var[\hat{\mu}] = \frac{1}{n-1} \sum_{m=1}^{n} (\hat{\mu}^m - \bar{\mu}^n)^2.$$

Problem-solving questions

3.14 Consider the problem where you are observing the number of arrivals Y^{n+1} which you believe are coming from a Poisson distribution with mean λ which is given by

$$Prob[Y^{n+1} = y|\lambda] = \frac{\lambda^y e^{-\lambda}}{\lambda!},$$

where we assume $y = 0, 1, 2, ...$ Your problem is that you do not know what λ is, but you think it is one of $\{\lambda_1, \lambda_2, ..., \lambda_K\}$. Assume that after n observations of the number of arrivals Y, we have estimated the probability

$$p_k^n = Prob[\lambda = \lambda_k|Y^1, ..., Y^n].$$

Using the methods of section 3.9.2 for sampled belief models, write the expression for p_k^{n+1} given the observation Y^{n+1}. Your expression has to be in terms of p_k^n and the Poisson distribution above.

3.15 Bayes' theorem comes from the identity $P(A|B)P(B) = P(B|A)P(A)$ where A and B are probabilistic events. From this, we can

$$P(B|A) = \frac{P(A|B)P(B)}{P(A)}.$$

Use this identity to derive equation (3.55) used for updating beliefs for sampled belief models. Clearly identify events A and B. [Hint: an equivalent form of Bayes theorem involves conditionining everything on a third event C, as in

$$P(B|A,C) = \frac{P(A|B,C)P(B|C)}{P(A|C)}.$$

What is the event C in equation (3.55)?]

Diary problem

The diary problem is a single problem you chose (see chapter 1 for guidelines). Answer the following for your diary problem.

3.16 Review the different classes of approximations described in section 3.1.3, and identify examples of as many of these that may arise in your approximation.

Bibliography

Bean, J.C., Birge, J.R., and Smith, R.L. (1987). Aggregation in dynamic programming. *Operations Research* 35: 215–220.

Bertsekas, D.P. and Castanon, D.A. (1989). Adaptive aggregation methods for infinite horizon dynamic programming. *IEEE Transactions on Automatic Control* 34: 589–598.

Bertsekas, D.P. and Tsitsiklis, J.N. (1996). *Neuro-Dynamic Programming*. Belmont, MA: Athena Scientific.

Bishop, C.M. (2006). *Pattern Recognition and Machine Learning*. New York: Springer.

Clement, E., Lamberton, D., and Protter, P. (2002). An analysis of a least squares regression method for American option pricing. *Finance and Stochastics* 17: 448–471.

Fan, J. and Gijbels, I. (1996). *Local Polynomial Modelling and Its Applications*. London: Chapman and Hall.

George, A., Powell, W.B., and Kulkarni, S. (2008). Value function approximation using multiple aggregation for multiattribute resource management. *Journal of Machine Learning Research*. 2079–2111.

Golub, G.H. and Loan, C.F.V. (1996). *Matrix Computations*. Baltimore, MD: John Hopkins University Press.

Hastie, T.J., Tibshirani, R.J., and Friedman, J.H. (2009). *The Elements of Statistical Learning: Data Mining, Inference, and Prediction*. New York: Springer.

Haykin, S. (1999). *Neural Networks: A comprehensive foundation*. Englewood Cliffs, N.J: Prentice Hall.

Heuberger, P.S.C., den Hov, P.M.J.V., and Wahlberg, B. (eds) (2005). *Modeling and Identification with Rational Orthogonal Basis Functions*. New York: Springer.

Judd, K.L. (1998). *Numerical Methods in Economics*. MIT Press.

Ljung, l. and Soderstrom, T. (1983). *Theory and Practice of Recursive Identification*. Cambridge, MA: MIT Press.

LeBlanc, M. and Tibshirani, R. (1996). Combining estimates in regression and classification. *Journal of the American Statistical Association* 91: 1641–1650.

Longstaff, F.A. and Schwartz, E.S. (2001). Valuing American options by simulation: A simple least squares approach. *The Review of Financial Studies* 14 (1): 113–147.

Menache, I., Mannor, S., and Shimkin, N. (2005). Basis function adaptation in temporal difference reinforcement learning. *Annals of Operations Research* 134 (1): 215–238.

Mendelssohn, R. (1982). An iterative aggregation procedure for Markov decision processes. *Operations Research* 30: 62–73.

Ormoneit, D. and Glynn, P. W. (2002). Kernelbased reinforcement learning averagecost problems. In: *IEEE Transactions on Automatic Control*. 1624–1636.

Ormoneit, D. and Sen, Ś. (2002), Kernelbased reinforcement learning. *Machine Learning*.

Powell, W.B. and George, A.P. (2006). Adaptive stepsizes for recursive estimation with applications in approximate dynamic programming. *Journal of Machine Learning* 65 (1): 167–198.

Rogers, D., Plante, R., Wong, R., and Evans, J. (1991). Aggregation and disaggregation techniques and methodology in optimization. *Operations Research* 39: 553–582.

Schweitzer, P. and Seidmann, A. (1985). Generalized polynomial approximations in Markovian decision processes. *Journal of Mathematical Analysis and Applications* 110 (6): 568–582.

Singh, S.P., Jaakkola, T., and Jordan, M.I. (1995). Reinforcement learning with soft state aggregation. *Advances in Neural Information Processing Systems* 7: 361–368. MIT Press.

Smola, A. J. and Schölkopf, B. (2004). A tutorial on support vector regression. *Statistics and Computing* 14 (3): 199–222.

Taylor, H. (1967). Evaluating a call option and optimal timing strategy in the stock market. *Management Science* 12: 111–120.

Tsitsiklis, J.N. and Roy, B.V. (1996). Feature-based methods for large scale dynamic programming. *Machine Learning* 22: 59–94.

Tsitsiklis, J.N. and Van Roy, B. (1997). An analysis of temporal difference learning with function approximation. *IEEE Transactions on Automatic Control* 42 (5): 674–690.

Tsitsiklis, J. N. and Van Roy, B. (2001). Regression methods for pricing complex American-style options. *IEEE Transactions on Neural Networks* 12: 694–703.

Van Roy, B. (2001). Neuro-dynamic programming: Overview and recent trends. In: *Handbook of Markov Decision Processes: Methods and Applications* (eds. E. Feinberg and A. Shwartz), 431–460. Boston: Kluwer.

Whitt, W. (1978). Approximations of dynamic programs I. *Mathematics of Operations Research.* 231–243.

Yang, Y. (2001). Adaptive regression by mixing. *Journal of the American Statistical Association.*

Zipkin, P. (1980). Bounds on the effect of aggregating variables in linear programming. *Operations Research* 28: 155–177.

4

Introduction to Stochastic Search

Our most basic optimization problem can be written

$$\max_{x \in \mathcal{X}} \mathbb{E}_W F(x, W), \tag{4.1}$$

where x is our decision and W is any form of random variable. A simple example of this problem is the newsvendor problem which we might write

$$\max_{x \in \mathcal{X}} \mathbb{E}_W \big(p \min(x, W) - cx \big),$$

where x is a quantity of product we order at cost c, W is the demand, and we sell the smaller of x and W to the market at a price p.

This problem is the one most identified with the field that goes under the name of "stochastic search." It is typically presented as a "static" stochastic optimization problem because it consists of making a single decision x, then observing an outcome W allowing us to assess the performance $F(x, W)$, at which point we stop. However, this all depends on how we interpret "$F(x, W)$," "x," and "W."

For example, we can use $F(x, W)$ to represent the results of running a simulation, a set of laboratory experiments, or the profits from managing a fleet of trucks. The input x could be the set of controllable inputs that govern the behavior of the simulator, the materials used in the laboratory experiments, or the size of our fleet of trucks. In addition, x could also be the parameters of a policy for making decisions, such as the order-up-to parameters $\theta = (\theta^{min}, \theta^{max})$ in the inventory problem we introduced in section 1.3 (see equation (1.5)).

At the same time, the variable W could be the sequence $W = (W^1, W^2, \dots, W^N)$ representing the events within the simulator, the outcomes of individual laboratory experiments, or the loads that arrive while dispatching our fleet of trucks. Finally, $F(x, W)$ could be the performance of the simulation or set of experiments or our fleet of trucks over a week. This

Reinforcement Learning and Stochastic Optimization: A Unified Framework for Sequential Decisions, First Edition. Warren B. Powell.
© 2022 John Wiley & Sons, Inc. Published 2022 John Wiley & Sons, Inc.

means that we could write $F(x, W)$ as

$$F(x, W) = \sum_{t=0}^{T} C(S_t, X^\pi(S_t))$$

where "x" is our policy π and our state variable evolves according to $S_{t+1} = S^M(S_t, X^\pi(S_t), W_{t+1})$ given the sequence $W = (W_1, \dots, W_T)$.

While equation (4.1) is the most standard way of writing this problem, we are going to use the expanded form as our default statement of the problem, which is written

$$\max_{x \in \mathcal{X}} \mathbb{E}_{S^0} \mathbb{E}_{W|S^0} \{F(x, W)|S^0\}, \tag{4.2}$$

which allows us to express the expectation on information in an initial state S^0, which can include deterministic parameters as well as probabilistic information (which we need when we use Bayesian belief models). For example, our problem may depend on an unknown physical parameter θ which we believe may be one of a set $\theta_1, \dots, \theta_K$ with probability $p_k^0 = \mathbb{P}[\theta = \theta_k]$.

There are three core strategies for solving our basic stochastic optimization problem (4.2):

Deterministic methods – There are some problems with sufficient structure that allows us to compute any expectations exactly, which reduces a stochastic problem to a deterministic one (more precisely, it reduces a stochastic problem to one that can be solved using deterministic mathematics). In some cases problems can be solved analytically, while others require the use of deterministic optimization algorithms.

Sampled approximations – This is a powerful and widely used approach for turning computationally intractable expectations into tractable ones. We note that sampled problems, while solvable, may not be easily solvable, and as a result have attracted considerable research interest, especially for problems where x is high-dimensional and possibly integer. However, we will also make the argument that a sampled stochastic problem is, fundamentally, a problem that can be solved with deterministic mathematics, although the analysis of the properties of the resulting solution may require stochastic tools.

Adaptive learning methods – The vast majority of stochastic optimization problems will end up requiring adaptive learning methods, which are fundamentally stochastic, and require stochastic tools. These are the approaches that will attract most of our attention in this volume. We will be particularly interested in the performance of these methods using finite learning budgets.

We begin our presentation by discussing different perspectives of our basic stochastic optimization problem, which encompasses fully sequential problems when we interpret "x" as a policy π. We then observe that there are examples of stochastic optimization problems that can be solved using standard deterministic methods, either by directly exploiting the structure of the uncertainty (which allows us to compute the expectation directly), or by using the powerful idea of sampled models.

We then close by setting up some preliminary discussions about adaptive learning methods, which are then discussed in more detail in chapters 5, 6, and 7. As we point out below, adaptive learning methods represent a form of sequential decision problem where the state variable S^n captures only what we know (or believe). There is no other physical process (such as inventory) or informational process (such as a time series) which links decisions over time. We defer until Part III of the book the handling of these more complex problems.

The perspectives presented in this chapter appear to be new, and set up the approach we use throughout the rest of the book.

4.1 Illustrations of the Basic Stochastic Optimization Problem

There is no shortage of applications of our basic stochastic optimization problem. Some examples that illustrate applications in different settings include:

■ **EXAMPLE 4.1**

Engineering design – Here x is the design of an airplane wing where we have to create a design that minimizes costs over a range of different conditions. We can learn from numerical simulations, laboratory strength tests, and examining actual aircraft for stress fractures.

■ **EXAMPLE 4.2**

Let $(y^n, x^n)_{n=1}^N$ be a set of explanatory variables x^n and response variables y^n. We would like to fit a statistical model (this might be a linear parametric model, or a neural network) where θ is the parameters (or weights) that characterize the model. We want to find θ that solves

$$\min_\theta \frac{1}{N} \sum_{n=1}^N (y^n - f(x^n|\theta))^2.$$

This problem, which is very familiar in statistics, is a sampled approximation of

$$\min_{\theta} \mathbb{E}(Y - f(X|\theta))^2,$$

where X is a random input and Y is the associated random response.

■ **EXAMPLE 4.3**

We would like to design an energy system where R is a vector of energy investments (in wind farms, solar fields, battery storage, gas turbines), which we have to solve subject to random realizations of energy from wind and solar (which we represent using the vector W) defined over a year. Let $C^{cap}(R)$ be the capital cost of these investments, and let $C^{op}(R, W)$ be the net operating revenue given W (computed from a numerical simulator). Now we want to solve

$$\max_{R} \mathbb{E}(-C^{cap}(R) + C^{op}(R, W)).$$

■ **EXAMPLE 4.4**

A bank uses a policy $X^\pi(S|\theta)$ that covers how much to move into or out of cash given the state S which describes how much cash is on hand, the forward price/earnings ratio of the S&P 500 (an important index of the stock market), and current 10-year bond rates. The vector θ captures upper and lower limits on each variable that triggers decisions to move money into or out of cash. If $C(S_t, X^\pi(S_t|\theta), W_{t+1})$ is the cash flow given the current state S_t and the next-period returns W_{t+1}, then we want to find the policy control parameters θ that solves

$$\max_{\theta} \mathbb{E} \sum_{t=0}^{T} e^{-rt} C(S_t, X^\pi(S_t|\theta), W_{t+1}).$$

Each of these examples involve making some decision: The design of the airplane wing, the model parameter θ, the energy investment R, or the parameters θ of a cash transfer policy. In each case, we have to choose a design either to optimize a deterministic function, a sampled approximation of a stochastic problem, or by adaptive learning (either from a simulator, laboratory experiments or field observations).

While there are some settings where we can solve (4.2) directly (possibly with an approximation of the expectation), most of the time we are going to turn to iterative learning algorithms. We will start with a state S^n that captures our belief state about the function $F(x) = \mathbb{E}\{F(x, W)|S^0\}$ after n experiments (or observations). We then use this knowledge to make a decision x^n after which we observe W^{n+1} which leads us to a new belief state S^{n+1}. The problem is designing a good rule (or policy) that we call $X^\pi(S^n)$ that determines x^n. For example, we might want to find the best answer that we can with a budget of N iterations.

We pose this as one of finding the best policy to determine a solution $x^{\pi,N}$, which is a random variable that might depend on any initial distributions S^0 (if necessary), and the sequence of observations $(W^1, ..., W^N)$ that, combined with our policy (algorithm) π produces $x^{\pi,N}$. We can think of $(W^1, ..., W^N)$ as the training observations. Then, we let $\widehat{W}$ be observations we make to perform testing of $x^{\pi,N}$. This can all be written (using our expanded form of the expectation) as

$$\max_\pi \mathbb{E}_{S^0}\mathbb{E}_{W^1,...,W^N|S^0}\mathbb{E}_{\widehat{W}|S^0}\{F(x^{\pi,N}, \widehat{W})|S^0\}. \tag{4.3}$$

We ask the reader to contrast our original version of this problem in equation (4.1) with (4.3). The version in (4.1) can be found throughout the research literature. But the version in (4.3) is the problem we are actually solving in practice.

The formulations in (4.1), (4.2), and (4.3) all focus on finding the best decision (or design) to maximize some function. We refer to these as *final reward* formulations. This distinction is important when we use adaptive learning policies $X^\pi(S)$, since this involves optimizing using intelligent trial and error.

When we use adaptive learning (which is a widely used strategy), then we have to think about our attitude toward the intermediate decisions x^n for $n < N$. If we have to "count" the results of these intermediate experiments, then we would write our objective as

$$\max_\pi \mathbb{E}_{S^0}\mathbb{E}_{W^1,...,W^N|S^0}\left\{\sum_{n=0}^{N-1} F(X^\pi(S^n), W^{n+1})|S^0\right\}. \tag{4.4}$$

When we are using an adaptive learning strategy, we are going to refer to (4.3) as the *final reward* formulation, while the objective function in (4.4) is the *cumulative reward* formulation.

It is not by accident that the function $F(x, W)$ does not depend on our evolving state variable S^n (or S_t), while the policy $X^\pi(S^n)$ does. We are assuming here that our function $F(x, W)$ itself is not evolving over time; all that is changing are the inputs x and W. When we want our performance to depend on the state, we will use $C(S, x)$ to indicate this dependence.

The number of applications that fit the basic model given in equation (4.2) is limitless. For discussion purposes, it is helpful to recognize some of the major problem classes that arise in this setting:

- Discrete problems, where $\mathcal{X} = \{x_1, \dots, x_M\}$. Examples might be where x_m is a set of features for a product, catalysts for a type of material, drug cocktails, or even paths over a network.
- Concave problems, where $F(x, W)$ is concave in x (often x is a vector in this case).
- Linear programs, where $F(x, W)$ is a linear cost function and $\mathcal{X}$ is a set of linear constraints.
- Continuous, nonconcave problems, where x is continuous.
- Expensive functions – There are many settings where computing $F(x, W)$ involves running time-consuming computer simulations or laboratory experience that may take hours to days to weeks, or field experiments that may take weeks or months.
- Noisy functions – There are many problems where the measurement or observation errors in the function are extremely high, which introduces the need to develop methods that manage this level of noise.

For these problems, the decision x may be finite, continuous scalar, or a vector (that may be discrete or continuous).

As we progress, we are going to see many instances of (4.1) (or (4.2)) where we sequentially guess at a decision x^n, then observe W^{n+1}, and use this information to make a better guess x^{n+1}, with the goal of solving (4.1). In fact, before we are done, we are going to show that we can reduce our formulations of fully sequential problems such as our inventory problem to the same form as in (4.3) (or (4.4)). For this reason, we have come to refer to (4.2) as the basic stochastic optimization model.

4.2 Deterministic Methods

There are a handful of stochastic optimization problems that can be solved to optimality using purely deterministic methods. We are going to provide a brief illustration of some examples as an illustration, but in practice, exact solutions of stochastic problems will be quite rare. The discussion in this section is relatively advanced, but the point is important since the research community often overlooks that there are a number of so-called "stochastic optimization problems" that are solved using purely deterministic mathematics.

4.2.1 A "Stochastic" Shortest Path Problem

In section 2.3.3, we introduced a stochastic shortest path problem where a traveler arriving to node i would see the sample realizations of the random costs C_{ij} to each node j that can be reached from i. Assume that on the nth day we arrive to node i and observe the sample realization $\hat{c}_{ij}^n$ of the random variable C_{ij}. We would then get a sampled observation of the value of being at node i from

$$\hat{v}_i^n = \min_{j \in \mathcal{I}_i^+} \left(\hat{c}_{ij}^n + \overline{V}_j^{n-1} \right),$$

where $\mathcal{I}_i^+$ is the set of all nodes that we can reach from node i. Now assume that we do not see the sample realization of the random variable C_{ij} before we make our decision. Assume we have to make the decision before we see the realization. In this case, we have to use the expected value $\bar{c}_{ij} = \mathbb{E}C_{ij}$, which means we are solving

$$
\begin{aligned}
\hat{v}_i^n &= \min_{j \in \mathcal{I}_i^+} \mathbb{E}\left(C_{ij} + \overline{V}_j^{n-1} \right), \\
&= \min_{j \in \mathcal{I}_i^+} \left(\bar{c}_{ij} + \overline{V}_j^{n-1} \right),
\end{aligned}
$$

which is just what we would solve if we had a deterministic shortest path problem. In other words, when we have a linear objective, if we have to make decisions before we see information, then the resulting problem reduces to a deterministic optimization problem which can (generally) be solved exactly.

The key difference between this "stochastic" shortest path problem and the one in section 2.3.3.2 is how information is revealed. The problem in section 2.3.3.2 is harder (and more interesting) because information is revealed just before we make the decision of the next link to traverse. Here, information is revealed after we make a decision, which means decisions have to be made using distributional information. Since the problem is linear in the costs, then all we need are the means, turning our stochastic problem into a deterministic problem.

4.2.2 A Newsvendor Problem with Known Distribution

We next consider one of the oldest stochastic optimization problems, known as the newsvendor problem, which is given by

$$\max_x EF(x, W) = \mathbb{E}\left(p \min\{x, W\} - cx \right). \tag{4.5}$$

Assume that we know the cumulative distribution $F^W(w) = \mathbb{P}[W \leq w]$ of the demand W. We begin by computing the stochastic gradient, given by

$$
\nabla_x F(x, W) = \begin{cases} p - c & \text{if } x \leq W, \\ -c & \text{if } x > W. \end{cases} \tag{4.6}
$$

We next observe that if $x = x^*$, the optimal solution, then the expectation of the gradient should be zero. This means

$$
\begin{aligned}
\mathbb{E}\nabla_x F(x, W) &= (p - c)\mathbb{P}[x^* \leq W] - c\mathbb{P}[x^* > W], \\
&= (p - c)\mathbb{P}[x^* \leq W] - c(1 - \mathbb{P}[x^* \leq W]), \\
&= 0.
\end{aligned}
$$

Solving for $\mathbb{P}[x^* \leq W]$ gives

$$
\mathbb{P}[x^* \leq W] = \frac{c}{p}. \tag{4.7}
$$

Under the (reasonable) assumption that the unit purchase cost c is less than the sales price p, we see that the optimal solution x^* corresponds to the point where the probability that x^* is less than the demand W is the ratio of the cost over the price. Thus if the cost is low, the probability that the demand is greater than the supply (which means we lose sales) should be low.

Equation (4.7) gives the optimal solution of the newsvendor problem. It requires that we know the distribution of demand, and also requires that we be able to take the expectation of the gradient and solve for the optimal probability analytically. Not surprisingly, these conditions are rarely met in practice.

4.2.3 Chance-Constrained Optimization

There are some problems where we can compute the expectation exactly, but the result is (typically) a nonlinear problem that can only be solved numerically. A good illustration of this is a method known as chance-constrained programming, which is itself a rich area of study. A classical formulation (which we first saw in section 2.1.12) poses the problem

$$
\min_x f(x), \tag{4.8}
$$

subject to the constraint

$$
p(x) \leq \alpha, \tag{4.9}
$$

where

$$
p(x) = \mathbb{P}[C(x, W) \geq 0] \tag{4.10}
$$

is the probability that a constraint violation, captured by $C(x, W)$, is violated. Thus, $C(x, W)$ might be the uncovered demand for energy, or the degree to which two driverless cars get closer than an allowed tolerance. If we can compute $p(x)$ (analytically or numerically), we can draw on powerful nonlinear programming algorithms to solve (4.8) directly.

4.2.4 Optimal Control

In section 2.1.4, we formulated an optimal control problem of the form

$$\min_{u_0,\dots,u_T} \sum_{t=0}^{T} L_t(x_t, u_t).$$

where states evolve according to $x_{t+1} = f(x_t, u_t)$. We may introduce a stochastic noise term giving us the state transition equation

$$x_{t+1} = f(x_t, u_t) + w_t,$$

where (following the standard convention of the controls community) w_t is random at time t. The historical basis for this notational convention is the roots of optimal control in continuous time, where w_t would represent the noise between t and $t + dt$. In the presence of noise, we need to introduce a policy $U^\pi(x_t)$. We would now write our objective function as

$$\min_{\pi} \mathbb{E} \sum_{t=0}^{T} L_t(x_t, U_t^\pi(x_t)). \tag{4.11}$$

Now assume that the loss function has the quadratic form

$$L_t(x_t, u_t) = (x_t)^T Q_t x_t + (u_t)^T R_t u_t.$$

After quite a bit of algebra, it is possible to show that the optimal policy has the form

$$U_t^\pi(x_t) = K_t x_t, \tag{4.12}$$

where K_t is a complex matrix that depends on the matrices Q_t and R_t.

This solution depends on three critical features of this problem:

- The objective function is quadratic in the state x_t and the control u_t.
- The control u_t is unconstrained.
- The noise term w_t is additive in the transition function.

Despite these limitations, this result has proved quite important for many problems in engineering.

4.2.5 Discrete Markov Decision Processes

As with the field of stochastic control, there is an incredibly rich body of litera-ture that has grown up around the basic problem of discrete dynamic programs, a problem that we first introduced in section 2.1.3, but address in much more depth in chapter 14. Imagine that we have a contribution $C(s, x)$ when we are in state $s \in \mathcal{S}$ and take discrete action $x \in \mathcal{X} = \{x_1, \ldots, x_M\}$, and a one-step transition matrix $P(s'|s, x)$ which gives the probability that we evolve to state $S_{t+1} = s'$ given that we are in state $S_t = s$ and take action x. It is possible to show that the value of being in a state $S_t = s$ at time t is given by

$$V_t(S_t) = \max_{x \in \mathcal{X}} \left(C(S_t, x) + \sum_{s' \in \mathcal{S}} P(s'|S_t, x)V_{t+1}(s') \right). \tag{4.13}$$

We can compute (4.13) if we start at time T with some initial value, say $V_T(s) = 0$, and then step backward in time. This produces the optimal policy $X_t^*(S_t)$ given by

$$X_t^*(S_t) = \arg\max_{x \in \mathcal{X}} \left(C(S_t, x) + \sum_{s' \in \mathcal{S}} P(s'|S_t, x)V_{t+1}(s') \right). \tag{4.14}$$

Again, we have found our optimal policy purely using deterministic mathemat-ics. The critical element of this formulation is the assumption that the one-step transition matrix $P(s'|s, x)$ is known (and computable). This requirement also requires that the state space $\mathcal{S}$ and action space $\mathcal{X}$ be discrete and not too large.

4.2.6 Remarks

These are a representative list of the very small handful of stochastic opti-mization problems that can be solved either analytically or numerically using deterministic methods. While we have not covered every problem that can be solved this way, the list is not long. This is not to minimize the importance of these results, which sometimes serve as the foundation for algorithms for more general problems.

Often, the most difficult aspect of a stochastic optimization problem is the expectation (or other operators such as risk metrics to deal with uncertainty). It should not be surprising, then, that the techniques used to solve more gen-eral stochastic optimization problems tend to focus on simplifying or breaking down the representation of uncertainty. The next section introduces the con-cept of sampled models, a powerful strategy that is widely used in stochastic optimization. We then transition to a discussion of adaptive sampling-based methods that is the focus of most of the rest of this book.

4.3 Sampled Models

One of the most powerful and widely used methods in stochastic optimization is to replace the expectation in the original model in equation (4.1), which is typically computationally intractable, with a sampled model. For example, we might represent the possible values of W (which might be a vector) using the set $\hat{W} = \{w^1, \ldots, w^N\}$. Assume that each w^n can happen with equal probability. We can then approximate the expectation in equation (4.1) using

$$\mathbb{E}F(x, W) \approx \bar{F}(x) = \frac{1}{N} \sum_{n=1}^{N} F(x, w^n).$$

The use of samples can transform intractable expectations into relatively easy calculations. More difficult is understanding the properties of the resulting approximation $\bar{F}(x)$, and the effect of sampling errors on the solution of

$$\max_{x} \bar{F}(x). \tag{4.15}$$

These questions have been addressed under the umbrella of a method called the *sample average approximation*, but the idea has been applied in a variety of settings.

Our newsvendor problem is a nice example of a stochastic optimization problem where the uncertain random variable is a scalar, but real applications can feature random inputs W that are very high dimensional. A few examples illustrate how large random variables can be:

■ **EXAMPLE 4.5**

A blood management problem requires managing eight blood types, which can be anywhere from 0 to 5 weeks old, and may or may not be frozen, creating $6 \times 8 \times 2 = 96$ blood types. Patients needing blood create demands for eight different types of blood. Each week there are random supplies (96 dimensions) and random demands (8 dimensions), creating an exogenous information variable W_t with 104 dimensions.

■ **EXAMPLE 4.6**

A freight company is moving parcels among 1,000 different terminals. Since each parcel has an origin and destination, the vector of new demands has 1,000,000 dimensions.

■ **EXAMPLE 4.7**

Patients arriving to a doctor's office may exhibit as many as 300 different characteristics. Since each patient may or may not have any of these characteristics, there are as many as $2^{300} \sim 2 \times 10^{90}$ different types of patients (far more than the population of planet Earth!)

This section provides a brief introduction to what has evolved into an incredibly rich literature. We start by addressing the following questions:

- How do we formulate a sampled model?
- How good is the quality of the sampled solution (and how fast does it approach the optimal as K is increased)?
- For large problems (high dimensional x), what are strategies for solving (4.15)?
- Again for large problems, what are the best ways of creating the sample $w^1, \dots, w^N$?

We are going to return to sampled models from time to time since they represent such a powerful strategy for handling expectations.

4.3.1 Formulating a Sampled Model

Assume that W is one of these multidimensional (and possibly very high dimensional) random variables. Further assume that we have some way of generating a set of samples $w^1, \dots, w^N$. These may be generated from a known probability distribution, or perhaps from a historical sample. We can replace our original stochastic optimization problem (4.1) with

$$\max_x \frac{1}{N} \sum_{n=1}^{N} F(x, w^n). \tag{4.16}$$

Solving (4.16) as an approximation of the original problem in (4.1) is known as the *sample average approximation*. It is important to realize that both our original stochastic optimization problem (4.1) and the sampled problem (4.16) are deterministic optimization problems. The challenge is computation.

Below we illustrate several uses of sampled models.

4.3.1.1 A Sampled Stochastic Linear Program

As with W, the decision variable x can be a scalar, or a very high-dimensional vector. For example, we might have a linear program where we are optimizing the flows of freight x_{ij} from location i to location j by solving

$$\min_x F(x, W) = \sum_{i,j \in \mathcal{J}} c_{ij} x_{ij},$$

subject to a set of linear constraints

$$\begin{aligned} Ax &= b, \\ x &\geq 0. \end{aligned}$$

A common application of this model arises when making a decision to allo-cate a resource such as blood inventories from central blood banks to hospitals, before knowing the results of weekly donations of blood, and the schedule of operations that need blood, at each hospital for the following week.

Now assume that the random information is the cost vector c (which might reflect the types of surgeries that require blood transfusions), the coefficient matrix A (which might capture travel times between inventory locations and hospitals), and the vector b (which captures blood donations and surgeries). Thus, $W = (A, b, c)$.

If we have one sample of W, then we have a straightforward linear program which may not be too hard to solve. But now imagine that we have $N = 100$ samples of the data, given by $(A^n, b^n, c^n)_{n=1}^N$. We could then solve

$$\min_x \frac{1}{N} \sum_{n=1}^{N} c_{ij}^n x_{ij},$$

subject to, for $n = 1, \dots, 100$,

$$\begin{aligned} A^n x &= b^n, \\ x &\geq 0. \end{aligned}$$

If we choose a sample of $N = 100$ outcomes, then our sampled problem in (4.16) becomes a linear program that is 100 times larger (remember we have just one vector x, but 100 samples of A, b and c). This may be computationally difficult (in fact, coming up with a single vector x that is feasible for all 100 samples of the data (A, b, c) may not even be possible).

4.3.1.2 Sampled Chance-Constrained Models

We can use our idea of sampling to solve chance-constrained programs. We begin by noting that a probability is like an expectation. Let $\mathbb{1}_{\{E\}} = 1$ if event E

is true. Then we can write our probability as

$$\mathbb{P}[C(x, W) \leq 0] = \mathbb{E}_W \mathbb{1}_{\{C(x,W) \leq 0\}}.$$

We can replace the chance constraint in (4.10) with a sampled version, where we basically average the random indicator variable to obtain

$$\mathbb{P}[C(x, W) \leq 0] \approx \frac{1}{N} \sum_{n=1}^{N} \mathbb{1}_{\{C(x,w^n) \leq 0\}}.$$

If x is discrete, then each $\mathbb{1}_{\{C(x,w^n)\}}$ can be calculated in advance for each w^n. If x is continuous, then it is likely that these indicator functions can be written as linear constraints.

4.3.1.3 Sampled Parametric Models

Sampled models may take other forms. Imagine that we wish to model demand as a function of price using a logistic function

$$D(p|\theta) = D^0 \frac{e^{\theta_0 - \theta_1 p}}{1 + e^{\theta_0 - \theta_1 p}}.$$

We want to pick a price that maximizes revenue using

$$R(p|\theta) = pD(p|\theta).$$

Our problem is that we do not know θ. We might assume that our vector θ follows a multivariate normal distribution, in which case we would want to solve

$$\max_p \mathbb{E}_\theta pD(p|\theta), \tag{4.17}$$

but computing the expectation may be hard. However, perhaps we are willing to say that θ may take on one of a set of values $\theta^1, \ldots, \theta^N$, each with probability q^n. Now we can solve

$$\max_p \sum_{n=1}^{N} pD(p|\theta^n)q^n. \tag{4.18}$$

Whereas equation (4.17) may be intractable, (4.18) may be much easier.

Both (4.16) and (4.18) are examples of sampled models. However, the representation in (4.16) is used in settings where $(w^1, \ldots, w^N)$ is a sample drawn from a typically large (often infinite) set of potential outcomes. The model in (4.18) is used when we have an uncertain belief about parameters, and are using the set $\theta^1, \ldots, \theta^N$, with a probability vector q that may evolve over time.

4.3.2 Convergence

The first question that arises with sampled models concerns how large N needs to be. Fortunately, the sample average approximation enjoys some nice convergence properties. We start by defining

$$F(x) = \mathbb{E}F(x, W),$$

$$\bar{F}^N(x) = \frac{1}{N}\sum_{n=1}^{N} F(x, w^n).$$

The simplest (and most intuitive) result is that we get closer to the optimal solution as the sample size grows. We write this by saying

$$\lim_{N\to\infty} \bar{F}^N(x) \to \mathbb{E}F(x, W).$$

Let x^N be the optimal solution of the approximate function, which is to say

$$x^N = \arg\max_{x\in\mathcal{X}} \bar{F}^N(x).$$

The asymptotic convergence means that we will eventually achieve the optimum solution, a result we state by writing

$$\lim_{N\to\infty} \bar{F}^N(x^N) \to F(x^*).$$

These results tell us that we will eventually achieve the best possible objective function (note that there may be more than one optimal solution). The most interesting and important result is the rate at which we achieve this result. We start by assuming that our feasible region $\mathcal{X}$ is a set of discrete alternatives $x_1, \ldots, x_M$. This might be a set of discrete choices (e.g. different product configurations or different drug cocktails), or a discretized continuous parameter such as a price or concentration. Or, it could be a random sample of a large set of possibly vector-valued decisions.

Now, let ϵ be some small value (whatever that means). The amazing result is that as N increases, the probability that the optimal solution to the approximate problem, X^N, is more than ϵ from the optimal shrinks at an *exponential rate*. We can write this statement mathematically as

$$\mathbb{P}[F(x^N) < F(x^*) - \epsilon] < |\mathcal{X}|e^{-\eta N}, \tag{4.19}$$

for some constant $\eta > 0$. What equation (4.19) is saying is that the probability that the quality of our estimated solution x^N, given by $F(x^N)$, is more than ϵ away from the optimal $F(x^*)$, decreases at an exponential rate $e^{-\eta N}$ with a constant, $|\mathcal{X}|$, that depends on the size of the feasible region. The coefficient $\mathcal{X}$ is quite large, of course, and we have no idea of the magnitude of η. However, the result suggests that the probability that we do worse than $F(x^*) - \epsilon$ (remember that we are maximizing) declines exponentially with N, which is comforting.

A similar but stronger result is available when x is continuous and $f(x, W)$ is concave, and the feasible region $\mathcal{X}$ might be specified by a set of linear inequalities. In this case, the convergence is given by

$$\mathbb{P}[F(x^N) < F(x^*) - \epsilon] < Ce^{-\eta N}, \qquad (4.20)$$

for given constants $C > 0$ and $\eta > 0$. Note that unlike (4.19), equation (4.20) does not depend on the size of the feasible region, although the practical effect of this property is unclear.

The convergence rate results (4.19) (for discrete decisions) or (4.20) (for convex functions) tell us that as we allow our sample size N to increase, the optimal objective function $F(x^N)$ approaches the optimal solution $F(x^*)$ at an exponential rate, which is a very encouraging result. Of course, we never know the parameters η, or C and η, so we have to depend on empirical testing to get a sense of the actual convergence rate. However, knowing that the rate of convergence is exponential (regardless of the values of C and η) is exceptionally important. We would also note that while solving a sampled model is fundamentally deterministic (since the sample gives us an approximate expectation that can be calculated exactly), the analysis of the rate of convergence with respect to the sample size N is pure stochastic analysis.

The exponential convergence rates are encouraging, but there are problems such as linear (or especially integer) programs that are computationally challenging even when $N = 1$. We are going to see these later in the context of models where we use sampling to look into the future. There are two computational issues that will need to be addressed:

Sampling – Rather than just doing random sampling to obtain $W^1, \ldots, W^N$, it is possible to choose these samples more carefully so that a smaller sample can be used to produce a more realistic representation of the underlying sources of uncertainty.

Decomposition – The sampled problem (4.16) can still be quite large (it is N times bigger than the problem we would obtain if we just used expectations for uncertain quantities), but the sampled problem has structure we can exploit using decomposition algorithms.

We defer until chapter 10 a more complete description of sampling methods to represent uncertainty. We then wait until chapter 19 to show how decomposition methods can be used in the setting of lookahead policies.

4.3.3 Creating a Sampled Model

A particularly important problem with large-scale applications is the design of the sample $W^1, \ldots, W^N$. The most popular methods for generating a sample are:

- From history – We may not have a formal probability model for W, but we can draw samples from history. For example, W^n might be a sample of wind speeds over a week, or currency fluctuations over a year.
- Monte Carlo simulation – There is a powerful set of tools on the computer known as Monte Carlo simulation which allow us to create samples of random variables as long as we know the underlying distribution (we cover this in more detail in chapter 10).

In some instances we have an interest in creating a reasonable representation of the underlying uncertainty with the smallest possible sample. For example, imagine that we are replacing the original problem $max_x \mathbb{E}F(x, W)$ with a sampled representation

$$\max_x \frac{1}{N} \sum_{n=1}^{N} F(x, W^n).$$

Now imagine that x is a (possibly large) vector of integer variables, which might arise if we are trying to schedule aircraft for an airline, or to design the location of warehouses for a large logistics network. In such settings, even a deterministic version of the problem might be challenging, whereas we are now trying to solve a problem that is N times as large. Instead of solving the problem over an entire sample $W^1, \ldots, W^N$, we may be interested in using a good representative subset (W^j), $j \in \mathcal{J}$. Assume that W^n is a vector with elements $W^n = (W_1^n, \ldots, W_k^n, \ldots, W_K^n)$. One way to compute such a subset is to compute a distance metric $d^1(n, n')$ between W^n and $W^{n'}$ which we might do using

$$d^1(n, n') = \sum_{k=1}^{K} |W_k^n - W_k^{n'}|.$$

This would be called an "L_1-norm" because it is measuring distances by the absolute value of the distances between each of the elements. We could also use an "L_2-norm" by computing

$$d^2(n, n') = \sqrt{\left(\sum_{k=1}^{K} (W_k^n - W_k^{n'})^2 \right)}.$$

The L_2-norm puts more weight on large deviations in an individual element, rather than a number of small deviations spread over many dimensions. We can generalize this metric using

$$d^p(n, n') = \left(\sum_{k=1}^{K} (W_k^n - W_k^{n'})^p \right)^{\frac{1}{p}}.$$

However, other than the L_1 and L_2 metrics, the only other metric that is normally interesting is the L_∞-norm, which is the same as setting $d^\infty(n, n')$ equal to the absolute value of the largest difference across all the dimensions.

Using the distance metric $d^p(n, n')$, we choose a number of clusters J and then organize the original set of observations $W^1, \ldots, W^n$ into J clusters. This can be done using a popular family of algorithms that go under names such as k-means clustering or k-nearest neighbor clustering. There are different variations of these algorithms which can be found in standard libraries. The core idea in these procedures can be roughly described as:

Step 0 – Use some rule to pick J centroids. This might be suggested by problem structure, or you can pick J elements out of the set $W^1, \ldots, W^N$ at random.

Step 1 – Now step through each $W^1, \ldots, W^N$ and assign each one to the centroid that minimizes the distance $d^p(n, j)$ over all centroids $j \in \mathcal{J}$.

Step 2 – Find the centroids of each of the clusters and return to Step 1 until you find that your clusters are the same as the previous iteration (or you hit some limit).

A nice feature of this approach is that it can be applied to high-dimensional random variables W, as might arise when W represents observations (wind speed, prices) over many time periods, or if it represents observations of the attributes of groups of people (such as medical patients).

The challenge of representing uncertain events using well-designed samples is growing into a mature literature. We refer the reader to the bibliographic notes for some guidance as of the time that this book is being written.

4.3.4 Decomposition Strategies*

Let $\overline{W} = \mathbb{E}W$ be a point estimate of our random variable W. From time to time, we encounter problems where the deterministic problem

$$\max_{x \in \mathcal{X}} F(x, \overline{W}),$$

is reasonably difficult to solve. For example, it might be a large integer program which might arise when scheduling airlines or planning when energy generators should turn on and off. In this case, $F(x, \overline{W})$ would be the contribution function and $\mathcal{X}$ would contain all the constraints, including integrality. Imagine that we can solve the deterministic problem, but it might not be that easy (integer programs might have 100,000 integer variables). If we want to capture the uncertainty of W using a sample of, say, 20 different values of W, then we create an integer program that is 20 times larger. Even modern solvers on today's computers have difficulty with this.

Now imagine that we decompose the problem so that there is a different solution for each possible value of W. Assume we have N sample outcomes $\omega^1, \omega^2, \dots, \omega^N$ where $W^n = W(\omega^n)$ is the set of sample realizations of W corresponding to outcome ω^n. Let $x(\omega^n)$ be the optimal solution corresponding to this outcome.

We might start by rewriting our sampled stochastic optimization problem (4.16) as

$$\max_{x(\omega^1),\dots,x(\omega^N)} \frac{1}{N} \sum_{n=1}^{N} F(x(\omega^n), W(\omega^n)). \tag{4.21}$$

We can solve this problem by creating N parallel problems and obtaining a different solution $x^*(\omega^n)$ for each ω. That is,

$$x^*(\omega^n) = \arg\max_{x(\omega^n) \in \mathcal{X}} F(x(\omega^n), W(\omega^n)).$$

This is a much smaller problem, but it also means choosing x assuming you know the outcome W. This would be like allowing an aircraft to arrive late to an airport because we already knew that the crew for the next leg was also going to be late.

The good news is that this is a starting point. What we really want is a solution where all the $x(\omega)$ are the same. We can introduce a constraint, often known as a *nonanticipativity constraint*, that looks like

$$x(\omega^n) - \bar{x} \;=\; 0, \quad n = 1, \dots, N. \tag{4.22}$$

If we introduce this constraint, we are just back to our original (and very large) problem. But what if we relax this constraint and add it to the objective function with a penalty λ. This produces the relaxed problem

$$\max_{x(\omega^1),\dots,x(\omega^N)} \frac{1}{N} \sum_{n=1}^{N} \big(F(x(\omega^n), W(\omega^n)) + \lambda^n(x(\omega^n) - \bar{x}) \big). \tag{4.23}$$

What is nice about this new objective function is that, just as with the problem in (4.21), it decomposes into N problems, which makes the overall problem solvable. Now the difficulty is that we have to coordinate the different subproblems by tuning the vector $\lambda^1, \dots, \lambda^N$ until our nonanticipativity constraint (4.22) is satisfied.. We are not going to address this problem in detail, but this hints at a path for solving large scale problems using sampled means.

4.4 Adaptive Learning Algorithms

When we cannot calculate the expectation exactly, either through structure or resorting to a sampled model, we have to turn to adaptive learning algorithms. This transition fundamentally changes how we approach stochastic optimization problems, since any adaptive algorithm can be modeled as a sequential decision problem, otherwise known as a dynamic program.

We separate our discussion of adaptive learning algorithms between derivative-based algorithms, discussed in chapter 5, and derivative-free algorithms, presented in chapter 7. In between, chapter 6 discusses the problem of adaptively learning a signal, a problem that introduces the annoying but persistent problem of stepsizes that we first encountered in chapter 5, but which pervades the design of adaptive learning algorithms.

We begin by offering a general model of adaptive learning problems, which are basically a simpler example of the dynamic programs that we consider later in the book. As we illustrate in chapters 5 and 7, adaptive learning methods can be viewed as sequential decision problems (dynamic programs) where the state variable captures only what we know about the state of the search algorithm. This gives us an opportunity to introduce some of the core ideas of sequential decision problems, without all the richness and complexity that come with this problem class.

Below, we sketch the core elements of any sequential decision problem, and then outline the fundamental class of policies (or algorithms) that are used to solve them.

4.4.1 Modeling Adaptive Learning Problems

Whether we are solving a derivative-based or derivative-free problem, any adaptive learning algorithm is going to have the structure of a sequential decision problem, which has five core components:

State S^n – This will capture the current point in the search, and other information required by the algorithm. The nature of the state variable depends heavily on how we are structuring our search process. The state variable may capture beliefs about the function (this is a major issue in derivative-free stochastic search), as well as the state of the algorithm itself. In chapter 9, we tackle the problem of modeling general dynamic programs which include states that are directly controllable (most often, these are physical problems).

Decision x^n – While this is sometimes x^n, the precise "decision" being made within an adaptive learning algorithm depends on the nature of the algorithm, as we see in chapter 5. Depending on the setting, decisions are made by a decision rule, an algorithm, or (the term we primarily use), a policy. If x is our decision, we designate $X^\pi(S)$ as the policy (or algorithm).

Exogenous information W^{n+1} – This is the new information that is sampled during the nth iteration (but after making decision x^n), either from a Monte Carlo simulation or observations from an exogenous process (which could be a computer simulation, or the real world).

Transition function – The transition function includes the equations that govern the evolution from S^n to S^{n+1}. Our default notation used throughout this volume is to write

$$S^{n+1} = S^M(S^n, x^n, W^{n+1}).$$

Objective function – This is how we evaluate how well the policy is performing. The notation depends on the setting. We may have a problem where we make a decision x^n at the end of iteration n, then observe information W^{n+1} in iteration $n + 1$, from which we can evaluate our performance using $F(x^n, W^{n+1})$. This is going to be our default notation for learning problems.

When we make the transition to more complex problems with a physical state, we are going to encounter problems where the contribution (cost if minimizing) depends on the state S^n and decision x^n, which we would write as $C(S^n, x^n)$, but there are other variations. We return to the objective function below.

We are going to be able to model any sequential learning algorithm as a sequential decision process that can be modeled as the sequence

$$(S^0, x^0 = X^\pi(S^0), W^1, S^1, x^1 = X^\pi(S^1), W^2, ...).$$

Thus, all sequential learning algorithms, for *any* stochastic optimization problem, can ultimately be reduced to a sequential decision problem.

For now (which is to say, chapters 5 and 7), we are going to limit our attention to where decisions only affect what we learn about the function we are optimizing. In chapter 8, we are going to introduce the complex dimension of controllable physical states. Mathematically, there is no difference in how we formulate a problem where the state consists only of what we know about a function, versus problems where the state captures the locations of people, equipment, and inventory. However, pure learning problems are much simpler, and represent a good starting point for modeling and solving stochastic optimization problems using sequential (adaptive) methods. In addition, we will be using these methods throughout the remainder of the book. For example, policy search methods (chapters 12 and 13) both require that we solve stochastic search problems, which we may approach using either derivative-based or derivative-free methods.

4.4.2 Online vs. Offline Applications

The terms "online" and "offline" are terms that are widely used in both machine learning and stochastic optimization settings, but they take on different interpretations which can be quite important, and which have created considerable confusion in the literature. Below we explain the terms in the context of these two communities, and then describe how these terms are used in this volume.

4.4.2.1 Machine Learning

Machine learning is an optimization problem that involves minimizing the error between a proposed model (typically parametric) and a dataset. We can represent the model by $f(x|\theta)$ where the model may be linear or nonlinear in θ (see chapter 3). The most traditional representation is to assume that we have a set of input variables $x^1, \ldots, x^n$ with a corresponding set of observations $y^1, \ldots, y^n$, to which we are going to fit our model by solving

$$\min_\theta \sum_{i=1}^n (y^i - f(x^i|\theta))^2, \tag{4.24}$$

where we might represent the optimal solution to (4.24) by θ^*. This problem is solved as a batch optimization problem using any of a set of deterministic optimization algorithms. This process is classically known as offline learning in the machine learning. Once we find θ^*, we would presumably use our model $f(x|\theta^*)$ to make an estimate of something, such as a forecast of the future, or a product recommendation.

In online learning, we assume that data is arriving sequentially over time. In this case, we are going to assume that we see x^n and then observe y^{n+1}, where the use of $n+1$ is our way of showing that y^{n+1} is observed after seeing $x^0, \ldots, x^n$. Let D^n be our dataset at time n where

$$D^n = \{x^0, y^1, x^1, y^2, \ldots, x^{n-1}, y^n\}.$$

We need to estimate a new value of θ, which we call θ^n, for each new piece of information which includes (x^{n-1}, y^n). We would call any method we use to compute θ^n a *learning policy*, but one obvious example would be

$$\theta^n = \arg\min_\theta \sum_{i=0}^{n-1} (y^{i+1} - f(x^i|\theta))^2. \tag{4.25}$$

More generally, we could write our learning policy as $\theta^n = \Theta^\pi(D^n)$. As our dataset evolves $D^1, D^2, \ldots, D^n, D^{n+1}, \ldots$, we update our estimate θ^n sequentially.

In the eyes of the machine learning community, the difference between the offline problem in equation (4.24) and the online learning problem in (4.25) is that the first is a single, batch optimization problem, while the second is implemented sequentially.

4.4.2.2 Optimization

Imagine that we are trying to design a new material to maximize the conversion of solar energy to electricity. We will go through a series of experiments testing different materials, as well as continuous parameters such as the thickness of a layer of a material. We wish to sequence our experiments to try to create a surface that maximizes energy conversion within our experimental budget. What we care about is how well we do in the end; trying a design that does not work is not a problem as long as the final design works well.

Now consider the problem of actively tilting solar panels to maximize the energy production over the course of the day, where we have to handle not just the changing angle of the sun during the day (and over seasons), but also with changes in cloud cover. Again, we may have to experiment with different angles, but now we need to maximize the total energy created while we are trying to learn the best angle.

We would treat the first problem as an offline problem since we are learning in the lab, while the second is an online problem since we are optimizing in the field. When we are in the lab, we do not mind failed experiments as long as we get the best result in the end, which means we would maximize final reward. By contrast, when we are learning in the field we want to optimize the cumulative reward. Note that both problems are fully sequential, which means the machine learning community would view both as forms of online learning.

We show how to write out the objective functions for our offline and online settings next.

4.4.3 Objective Functions for Learning

In contrast with the exact methods for solving stochastic optimization problems, there are different ways to formulate the objective function for adaptive learning problems. For learning problems, we are going to let $F(x, W)$ be the function that captures our performance objective when we make decision x and then observe random information W. In an iterative setting, we will write $F(x^n, W^{n+1})$; in a temporal setting, we will write $F(x_t, W_{t+1})$. Our choice $x^n = X^\pi(S^n)$ will be made by a policy that depends on the state, but otherwise the contribution $F(x, W)$ depends only on the action and random information.

The function $\mathbb{E}F(x, W)$ captures the performance of our implementation decision x. To make a good decision, we need to design an algorithm, or more precisely, a learning policy $X^\pi(S)$, that allows us to find the best x. There are different objective functions for capturing the performance of a learning policy:

Final reward – Let $x^{\pi,n} = X^{\pi}(S^n)$ be our solution at iteration n while following policy π. We may analyze the policy π in two ways:

Finite time analysis – Here, we want to solve

$$\max_{\pi} \mathbb{E}\{F(x^{\pi,N}, W)|S^0\} = \mathbb{E}_{S^0}\mathbb{E}_{W^1,...,W^N|S^0}\mathbb{E}_{\widehat{W}|S^0}F(x^{\pi,N}, \widehat{W}) \qquad (4.26)$$

where:

- S^0 might include a distribution of belief about unknown parameters (such as whether a patient is allergic to a drug),
- $W^1, ..., W^N$ are the observations we make while running our search policy π for N iterations (these are the training iterations),
- $\widehat{W}$ is the sampling done to test the performance of the final design $x^{\pi,N}$.

Asymptotic analysis – In this setting, we are trying to establish that

$$\lim_{N\to\infty} x^{\pi,N} \to x^*$$

where x^* solves $\max_x \mathbb{E}F(x, W)$. In both of these settings, we are only interested in the quality of the final solution, whether it is $x^{\pi,N}$ or x^*. We do not care about the solutions obtained along the way.

Cumulative reward – Cumulative reward objectives arise when we are interested not just in the performance after we have finished learning the best asymptotic design x^*, or the best design in a finite budget N, $x^{\pi,N}$, or finite time T, x_T^{π}. We divide these problems into two broad classes:

Deterministic policy – The most common setting is where we want to design a single policy that optimizes the cumulative reward over some horizon. We can further divide deterministic policies into two classes:

Stationary policy – This is the simplest setting, where we wish to find a single policy $X^{\pi}(S_t)$ to solve:

$$\max_{\pi} \mathbb{E} \sum_{t=0}^{T-1} F(X^{\pi}(S_t), W_{t+1}), \qquad (4.27)$$

within a finite time horizon T. We may write this in either a discounted objective,

$$\max_{\pi} \mathbb{E} \sum_{t=0}^{T} \gamma^t C(S_t, X^{\pi}(S_t)), \qquad (4.28)$$

or average reward,

$$\max_{\pi} \mathbb{E} \frac{1}{T} \sum_{t=0}^{T} C(S_t, X^{\pi}(S_t)). \tag{4.29}$$

Both (4.28) and (4.29) can be extended to infinite horizon, where we would replace (4.29) with

$$\max_{\pi} \lim_{T \to \infty} \mathbb{E} \frac{1}{T} \sum_{t=0}^{T} C(S_t, X^{\pi}(S_t)). \tag{4.30}$$

Time-dependent policy – There are many problems where we need a time-dependent policy $X_t^{\pi}(S_t)$, either because the behavior needs to vary by time of day, or because we need different behaviors based on how close the decisions are to the end of horizon. We denote the policy by time t as $X_t^{\pi}(S_t)$, but let π_t refer to the choices (type of function, parameters) we need to make for each time period. These problems would be formulated

$$\max_{\pi_0,\dots,\pi_{T-1}} \mathbb{E} \sum_{t=0}^{T-1} F(X_t^{\pi}(S_t), W_{t+1}). \tag{4.31}$$

Although the policies are time dependent, they are in the class of static policies because they are designed before we start the process of making observations.

Adaptive policy – Now we allow our policies to learn over time, as would often happen in an online setting. Modeling this is a bit subtle, and it helps to use an example. Imagine that our policy is of the form

$$X^{\pi}(S_t|\theta) = \theta_0 + \theta_1 S_t + \theta_2 S_t^2.$$

This would be an example of a stationary policy parameterized by $\theta = (\theta_0, \theta_1, \theta_2)$. Now imagine that θ is a function of time, so we would write our policy as

$$X_t^{\pi}(S_t|\theta_t) = \theta_{t0} + \theta_{t1} S_t + \theta_{t2} S_t^2,$$

where we have now written the *policy* $X_t^{\pi}(S_t)$ as being time dependent, since the function depends on time (through the parameter vector θ_t). Finally, imagine that we have an adaptive policy that updates θ_t after computing $x_t = X^{\pi}(S_t|\theta_t)$ and observing W_{t+1}. Just as we have to make a decision x_t, we have to "decide" on how to set θ_{t+1} given S_{t+1} (which depends on S_t, x_t, and W_{t+1}). In this case, θ_t becomes a part of the state variable (along with any other statistics needed to compute θ_{t+1} given what we know at time t).

We refer to the policy for learning θ as our *learning policy*, which we designate $\Theta^{\pi^{lrn}}$, where we would write

$$\theta_t = \Theta^{\pi^{lrn}}(S_t).$$

We refer to $\Theta^{\pi^{lrn}}(S_t)$ as the *learning policy* (also known as the "behavior policy") while $X^{\pi^{imp}}(S_t|\theta_t)$ is the *implementation policy*, which is the policy that makes the decisions that are implemented (this is also known as the "target policy"). This problem is formulated as

$$\max_{\pi^{imp}} \max_{\pi^{lrn}} \mathbb{E} \sum_{t=0}^{T-1} F(X^{\pi^{imp}}(S_t|\theta_t), W_{t+1}).$$

For learning problems (problems where the *function* $F(x, W)$ does not depend on the state), we are going to use (4.26) (the final reward) or (4.27) (the cumulative reward for stationary policies) as our default notation for the objective function.

It is common, especially in the machine learning community, to focus on *regret* rather than the total reward, cost, or contribution. Regret is simply a measure of how well you do relative to how well you could have done (but recognize that there are different ways of defining the best we could have done). For example, imagine that our learning policy has produced the approximation $\bar{F}^{\pi,N}(x)$ of the function $\mathbb{E}F(x, W)$ by following policy π after N samples, and let

$$x^{\pi,N} = \arg\max_x \bar{F}^{\pi,N}(x)$$

be the best solution based on the approximation. The regret $\mathcal{R}^{\pi,N}$ would be given by

$$\mathcal{R}^{\pi,N} = \max_x \mathbb{E}F(x, W) - \mathbb{E}F(x^{\pi,N}, W). \tag{4.32}$$

Of course, we cannot compute the regret in a practical application, but we can study the performance of algorithms in a setting where we assume we know the true function (that is, $\mathbb{E}F(x, W)$), and then compare policies to try to discover this true value. Regret is popular in theoretical research (for example, computing bounds on the performance of policies), but it can also be used in computer simulations comparing the performance of different policies.

4.4.4 Designing Policies

Now that we have presented a framework for modeling our learning problems, we need to address the problem of designing policies (we will sometimes refer to these as algorithms), especially in chapter 7 when we deal with derivative-free optimization.

We originally introduced the different classes of policies in section 1.4. As a brief reminder, there are two fundamental strategies for designing policies, each of which break down into two subclasses, creating four classes of policies:

Policy search – These are functions that are tuned to work well over time without directly modeling the effect of a decision now on the future. Policies designed using policy search fall into two styles:

Policy function approximations (PFAs) – PFAs are analytical functions that map directly from state to a decision.

Cost function approximations (CFAs) – CFAs involve maximizing (or minimizing) a parameterized optimization problem that returns a decision.

Lookahead policies – These are policies that are designed by estimating, directly or indirectly, the impact of a decision now on the future. There are again two ways of creating these policies:

Value function approximations (VFAs) – If we are in a state S^n, make a decision x^n, that leads (with the introduction of new information) to a new state S^{n+1}, assume we have a function $V^{n+1}(S^{n+1})$ that estimates (exactly or, more often, approximately) the value of being in state S^{n+1}. The value function $V^{n+1}(S^{n+1})$ captures the downstream impact of decision x^n, and can be used to help us make the best decision now.

Direct lookahead policies (DLAs) – These are policies where we model the downstream trajectory of each decision, and the optimizing across decisions now as well as decisions in the future (which may have to incorporate uncertainty).

The importance of each of these four classes depends on the characteristics of the problem. We are going to see all four of these classes used in the setting of derivative-free optimization in chapter 7. By contrast, derivative-based search strategies reviewed in chapter 5 have historically been more limited, although this perspective potentially introduces new strategies that might be pursued. When we transition to problems with physical states starting in chapter 8, we are going to see that we will need to draw on all four classes. For this reason, we discuss these four classes in more depth in chapter 11.

4.5 Closing Remarks

This chapter offers three fundamental perspectives of stochastic optimization problems. Section 4.2 is basically a reminder that any stochastic optimization problem can be solved as a deterministic optimization problem if we are able to compute the expectation exactly. While this will not happen very often, we offer this section as a reminder to readers not to overlook this path.

Section 4.3 then introduces the powerful approach of using sampled models, where we overcome the complexity of computing an expectation by replacing the underlying uncertainty model with a small sampled set, which is much easier to model. This strategy should always be in your toolbox, even when it will not solve the entire problem.

When all else fails (which is most of the time), we are going to need to turn to adaptive learning strategies, which are increasingly being grouped under the umbrella known as *reinforcement learning*. These approaches have evolved into substantial areas of research, which we divide into derivative-based methods in chapter 5, and derivative-free methods in chapter 7. In chapter 5, we are going to see that we need a device called "stepsizes" (which we cover in chapter 6), which can be viewed as a type of decision, where different stepsize rules are actually types of policies.

4.6 Bibliographic Notes

- Section 4.2.2 – The newsvendor problem where the distribution of W is known can be found in any standard textbook on inventory theory (see, e.g., Porteus (2002)), and is also a standard canonical problem in many books on stochastic optimization (see, e.g., Shapiro et al. (2014)).
- Section 4.2.3 – See the bibliographic notes for section 2.1.12 for references on chance-constrained programming.
- Section 4.2.4 – See the bibliographic notes for section 2.1.4 for references on optimal control.
- Section 4.2.5 – We address Markov decision processes in detail in chapter 14 and the references cited there.
- Section 4.3.1 – See the references for section 2.1.8 for some references on stochastic programming.

- Section 4.3.2 – The convergence rate results given in equations (4.19) and (4.20) are presented in Shapiro et al. (2014), based on work in Shapiro and Wardi (1996) and Shapiro and Homem-de Mello (2000). An excellent presentation of sampled methods and the convergence rates is given in Kim, Pasupathy, and Henderson's chapter in Fu (2014) [Chapter 8], as well as Ghadimi and Lan's chapter on finite time convergence properties [Chapter 7].
- Section 4.3.4 – The decomposition of stochastic programs was exploited in Rockafellar and Wets (1991) using a technique called "progressive hedging." Mulvey et al. (1995) implemented the method and performed numerical testing.
- Section 4.3.3 – The use of scenarios to approximate the future can make what are already large problems much larger, so there has been considerable attention given to the process of sampling efficient scenarios; see Dupacova et al. (2003) and Heitsch and Romisch (2009) for early, but important, contributions to this field.
- Section 4.4.1 – Every adaptive problem, whether it be a sequential decision problem or a stochastic algorithm, can be modeled using the five elements listed here. This structure was first presented in this style in Powell (2011). This framework follows the style of deterministic math programs, which consist of three core elements: decision variables, constraints, and the objective function. Our framework builds off the modeling framework used in stochastic, optimal control (see, for example, Kirk (2012), Stengel (1986), Sontag (1998), Sethi (2019), and Lewis and Vrabie (2012)). Powell (2021) contrasts the modeling framework used in this volume to the modeling style of Markov decision processes (which has been adopted in reinforcement learning) to that used in optimal control.
- Section 4.4.3 – We first used the finite time formulation of the stochastic search problem given in equation (4.3) in Powell (2019); we have not seen this formulation used elsewhere, since the asymptotic formulation in (4.1) is so standard in the stochastic optimization literature.
- Section 4.4.4 – Powell (2011)[Chapter 6] is the first reference to discuss different classes of policies, but overlooked cost function approximations. Powell (2014) was the first time the four classes of policies (as listed here) were given, without recognizing that they belonged in two classes. Powell (2016) presented the four classes of policies, divided between the two strategies: policy search and lookahead policies. Powell (2019) summarized these again, introducing additional modeling insights such as final and cumulative reward (equation (4.3) is written as final reward, but it can also be stated in a cumulative reward format, as we will do in chapter 7). This book is the first to present these ideas formally.

Exercises

Review questions

4.1 In your own words, explain why $\min_x \mathbb{E}_W F(x, W)$ is properly viewed as a deterministic optimization problem if we can compute the expectation W.

4.2 How would we compute $\mathbb{E}_W F(x, W)$ using a sampled approximation? Does this meet the conditions for a deterministic optimization problem? Explain (briefly!).

4.3 Assume we take a sample $\{w^1, \ldots, w^N\}$ and then solve the sampled representation

$$\max_x \frac{1}{N} \sum_{n=1}^{N} F(x, w^n)$$

to obtain an optimal solution x^N. Let x^* solve $\max_x \mathbb{E} F(x, W)$ (if we could compute this). What is the rate that $F(x^N)$ approaches $F(x^*)$? When $F(x)$ is concave?

4.4 What is the difference between offline and online learning in the machine learning community?

4.5 Write out the objective functions for final reward and cumulative reward? Be sure to use the expanded form of the expectation, which means you need to indicate what random variables each expectation is being taken over.

Modeling questions

4.6 Our basic newsvendor problem

$$F(x, W) = p \max\{0, x - W\} - cx,$$

can be written as different forms of optimization problems:

a) Write out the asymptotic form of the optimization problem to maximize the final reward.

b) Write out the final reward version of the newsvendor problem assuming we can only perform N observations of the newsvendor problem.

c) Assume that we have to perform our learning in the field, which means we need to maximize the sum of the rewards over N observations. Write out the objective function for this problem.

4.7 We illustrated $F(x, W)$ above using our basic newsvendor problem

$$F(x, W) = p \max\{0, x - W\} - cx,$$

but this is general notation that can be used to represent an entire range of sequential decision problems. Imagine that we have an asset selling problem, where we are determining when to sell an asset. Let W be a sequence of prices $p_1, p_2, \dots, p_t, \dots, p_T$. Assume we are going to sell our stock when $p_t \geq x$, which means that "x" defines a policy. Write out what $F(x, W)$ means for this problem, and formulate the objective function to optimize over policies.

Problem-solving questions

4.8 In a flexible spending account (FSA), a family is allowed to allocate x pretax dollars to an escrow account maintained by the employer. These funds can be used for medical expenses in the following year. Funds remaining in the account at the end of the following year are given back to the employer. Assume that you are in a 35% tax bracket (sounds nice, and the arithmetic is a bit easier).

Let W be the random variable representing total medical expenses in the upcoming year, and let $P^W(S) = Prob[W \leq w]$ be the cumulative distribution function of the random variable W.

a) Write out the objective function $F(x)$ that we would want to solve to find x to minimize the total cost (in pretax dollars) of covering your medical expenses next year.

b) If x^* is the optimal solution and $\nabla_x F(x)$ is the gradient of your objective function if you allocate x to the FSA, use the property that $\nabla_x F(x) = 0$ to derive the critical ratio that gives the relationship between $x *$ and the cumulative distribution function $P^W(w)$.

c) Given your 35% tax bracket, what percentage of the time should you have funds left over at the end of the year?

4.9 Consider the problem faced by a mutual fund manager who has to decide how much to keep in liquid assets versus investing to receive market returns. Assume he has R_t dollars to invest at the end of day t, and needs to determine the quantity x_t to put in cash at the end of day t to meet the demand $\hat{D}_{t+1}$ for cash in day $t + 1$. The remainder, $R_t - x_t$, is to be invested and will receive a market return of $\hat{\rho}_{t+1}$ (for example, we might have $\hat{\rho}_{t+1} = 1.0002$, implying a dollar invested is worth 1.0002 tomorrow). Assume there is nothing earned for the amount held in cash.

If $\hat{D}_t > x_{t-1}$, the fund manager has to redeem stocks. Not only is there a transaction cost of 0.20% (redeeming $1000 costs $2.00), the manager also has to pay capital gains. His fund pays taxes on the average gain of the total assets he is holding (rather than the gain on the money that was just invested). At the moment, selling assets generates a tax commitment of 10% which is deducted and held in escrow. Thus, selling $1000 produces net proceeds of 0.9(1000–2). As a result, if he needs to cover a cash request of $10,000, he will need to sell enough assets to cover both the transaction costs (which are tax deductible) and the taxes, leaving $10,000 net proceeds to cover the cash request.

a) Formulate the problem of determining the amount of money to hold in cash as a stochastic optimization problem. Formulate the objective function $F(x)$ giving the expected return when holding x dollars in cash.

b) Give an expression for the stochastic gradient $\nabla_x F(x)$.

c) Find the optimal fraction of the time that you have to liquidate assets to cover cash redemption. For example, if you manage the fund for 100 days, how many days would you expect to liquidate assets to cover cash redemptions?

4.10 Independent system operators (ISOs) are companies that manage our power grid by matching generators (which create the energy) with customers. Electricity can be generated via steam, which takes time, or gas turbines which are fast but expensive. Steam generation has to be committed in the day-ahead market, while gas turbines can be brought on line with very little advance notification.

Let x_t be the amount of steam generation capacity (measured in megawatt-hours) that is requested on day t to be available on day $t + 1$. Let $p_{t,t+1}^{steam}$ be the price of steam on day $t + 1$ that is bid on day t (which is known on day t). Let D_{t+1} be the demand for electricity (also measured in megawatt-hours) on day $t + 1$, which depends on temperature and other factors that cannot be perfectly forecasted. However, we do know the cumulative distribution function of D_{t+1}, given by $F^D(d) = Prob[D_{t+1} < d]$. If the demand exceeds the energy available from steam (planned on day t), then the balance has to be generated from gas turbines. These are bid at the last minute, and therefore we have to pay a random price p_{t+1}^{GT}. At the same time, we are not able to store energy; there is no inventory held over if $D_{t+1} < x_t$. Assume that the demand D_{t+1} and the price of electricity from gas turbines p_{t+1}^{GT} are independent.

a) Formulate the objective function $F(x)$ to determine x_t as an optimization problem.

b) Compute the stochastic gradient of your objective function $F(x)$ with respect to x_t. Identify which variables are known at time t, and which only become known at time $t + 1$.

c) Find an expression that characterizes the optimal value of x_t in terms of the cumulative probability distribution $F^D(d)$ of the demand D_T.

4.11 We are going to illustrate the difference between

$$\max_{x} \mathbb{E}F(x, W) \tag{4.33}$$

and

$$\max_{x} F(x, \mathbb{E}W) \tag{4.34}$$

using a sampled belief model. Assume we are trying to price a product where the demand function is given by

$$D(p|\theta) = \theta^0 \frac{e^{U(p|\theta)}}{1 + e^{U(p|\theta)}}, \tag{4.35}$$

where

$$U(p|\theta) = \theta_1 + \theta_2 p.$$

Our goal is to find the price that maximizes total revenue given by

$$R(p|\theta) = pD(p|\theta). \tag{4.36}$$

Here, our random variable W is the vector of coefficients $\theta = (\theta_0, \theta_1, \theta_2)$ which can take one of four possible values of θ given by the set $\Theta = \{\theta^1, \theta^2, \theta^3, \theta^4\}$.

a) Find the price $p(\theta)$ that maximizes

$$\max_{p} R(p|\theta), \tag{4.37}$$

for each of the four values of θ. You may do this analytically, or to the nearest integer (the relevant range of prices is between 0 and

Table 4.1 Data for exercise 4.11.

θ	$P(\theta)$	θ_0	θ_1	θ_2
θ^1	0.20	50	4	-0.2
θ^2	0.35	65	4	-0.3
θ^3	0.30	75	4	-0.4
θ^4	0.15	35	7	-0.25

40). Either way, it is a good idea to plot the curves (they are carefully chosen). Let $p^*(\theta)$ be the optimal price for each value of θ and compute

$$R^1 = \mathbb{E}_\theta \max_{p(\theta)} R(p^*(\theta)|\theta). \tag{4.38}$$

b) Find the price p that maximizes

$$R^2 = \max_p \mathbb{E}_\theta R(p|\theta), \tag{4.39}$$

where $R(p|\theta)$ is given by equation (4.36).

c) Now find the price p that maximizes

$$R^3 = \max_p R(p|\mathbb{E}\theta).$$

d) Compare the optimal prices and the optimal objective functions R^1, R^2, and R^3 produced by solving (4.37), (4.39), and (4.40). Use the relationships among the revenue functions to explain as much as possible about the relevant revenues and prices.

Theory questions

4.12 Recall our newsvendor problem

$$\max_x \mathbb{E}_W F(x, W)$$

where $F(x, W) = p \min(x, W) - cx$. Assume that W is given by a known distribution $f^W(w)$ with cumulative distribution

$$F^W(w) = \mathbb{P}[W \leq w].$$

You are going to show that the optimal solution x^* satisfies

$$F^W(x^*) = \frac{p - c}{p}. \tag{4.40}$$

Do this by first finding the stochastic gradient $\nabla_x F(x, W)$ which will give you a gradient that depends on whether $W < x$ or $W > x$. Now take the expected value of this gradient and set it equal to zero, and use this to show (4.40).

4.13 The newsvendor problem is given by

$$\max_x F(x) = \mathbb{E}_W F(x, W),$$

where

$$F(x, W) = p \min\{x, W\} - cx,$$

where we assume that our sales price p is strictly greater than the purchase cost c. An important property of the newsvendor problem is that $F(x)$ is concave in x. This means, for example, that

$$\lambda F(x_1) + (1 - \lambda)F(x_2) \leq F(\lambda x_1 + (1 - \lambda)x_2), \tag{4.41}$$

for $0 \leq \lambda \leq 1$, and where $x_1 \leq x_2$. This property is illustrated in Figure 4.1.

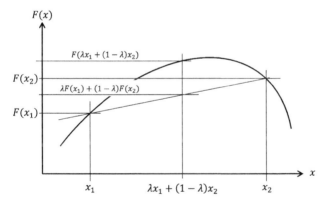

Figure 4.1 Concave function, showing that
$\lambda F(x_1) + (1 - \lambda)F(x_2) \leq F(\lambda x_1 + (1 - \lambda)x_2)$.

a) To begin, fix the random variable W, and show that $F(x, W)$ is concave (this should be apparent just by plotting the graph of $F(x, W)$ for a fixed W).

b) Now assume that W can only take on a fixed set of values $w^1, \ldots, w^N$, where each occurs with probability $p^n = Prob[W = w^n]$. Let $F(x) =$

$\sum_{n=1}^{N} p^n F(x, w^n)$. Substitute this into equation (4.41) to show that the weighted sum of concave functions is concave.

c) Finally argue that (b) implies that the newsvendor problem is concave in x.

Diary problem

The diary problem is a single problem you chose (see chapter 1 for guidelines). Answer the following for your diary problem.

4.14 For your diary problem:

a) Do you think you could reduce your problem to a deterministic problem as described in section 4.2? If not, could you approximate it using the sampled methods described in section 4.3? For each of these two approaches, explain why or why not. Assuming neither of these work, can you sketch how an adaptive search algorithm might work?

b) Identify whether you would formulate your decision problem using a final reward or cumulative reward objective?

Bibliography

Dupacova, J., GroweKuska, N., and Romisch, W. (2003). Scenario reduction in stochastic programming: An approach using probability metrics. Mathematical Programming, Sereis A 95: 493–511.

Fu, M.C. (2014). *Handbook of Simulation Optimization*. New York: Springer.

Heitsch, H. and Romisch, W. (2009). Scenario tree modeling for multistage stochastic programs. *Mathematical Programming* 118: 371–406.

Kirk, D.E. (2012). *Optimal Control Theory: An introduction*. New York: Dover.

Lewis, F.L. and Vrabie, D. (2012). *Design Optimal Adaptive Controllers*, 3e. Hoboken, NJ: JohnWiley & Sons.

Mulvey, J.M., Vanderbei, R.J., and Zenios, S.A. (1995). Robust optimization of large-scale systems. *Operations Research* 43 (2): 264–281.

Porteus, E.L. (2002). *Foundations of Stochastic Inventory Theory*. Stanford: Stanford University Press.

Powell, W.B. (2011). *Approximate Dynamic Programming: Solving the Curses of Dimensionality*, 2e. John Wiley & Sons.

Powell, W.B. (2014). Clearing the jungle of stochastic optimization. *Informs TutORials in Operations Research 2014*.

Powell, W.B. (2016). A unified framework for optimization under uncertainty. In: *Informs TutORials in Operations Research*. 45–83.

Powell, W.B. (2019). A unified framework for stochastic optimization. *European Journal of Operational Research* 275 (3): 795–821.

Powell, W.B. (2021). From reinforcement learning to optimal control: A unified framework for sequential decisions. In: *Handbook on Reinforcement Learning and Optimal Control, Studies in Systems, Decision and Control* , 29–74.

Rockafellar, R.T. and Wets, R.J.-B. (1991). Scenarios and policy aggregation in optimization under uncertainty. *Mathematics of Operations Research* 16 (1): 119–147.

Sethi, S.P. (2019). *Optimal Control Theory: Applications to Management Science and Economics*, 3e. Boston: SpringerVerlag.

Shapiro, A. and Homem-de Mello, T. (2000). On the rate of convergence of optimal solutions of Monte Carlo approximations of stochastic programs. *SIAM Journal on Optimization* 11: 70–86.

Shapiro, A. and Wardi, Y. (1996). Convergence analysis of stochastic algorithms. *Mathematics of Operations Research* 21: 615–628.

Shapiro, A., Dentcheva, D., and Ruszczyński, A. (2014). *Lectures on Stochastic Programming: Modeling and theory*, 2e. Philadelphia: SIAM.

Sontag, E. (1998). *Mathematical Control Theory*, 2e., 1–544. Springer.

Stengel, R.F. (1986). *Stochastic optimal control: theory and application*. Hoboken, NJ: John Wiley & Sons.

Part II – Stochastic Search

Stochastic search covers a broad class of problems that are typically grouped under names such as stochastic approximation methods (derivative-based stochastic search), ranking and selection (derivative-free stochastic search), simulation-optimization, and multiarmed bandit problems. We include in this part problems that are often solved using iterative algorithms, where the only information carried from one iteration to the next is what we have learned about the function. This is the defining characteristic of a learning problem.

Chapter 5 begins with derivative-based algorithms, where we describe the difference between asymptotic and finite-time analysis. This chapter identifies the importance of stepsizes, which are actually "decisions" in derivative-based methods. Chapter 6 provides an in-depth discussion of stepsize policies.

We then transition to derivative-free problems in chapter 7, where there is a much richer tradition of designing policies compared to derivative-based methods. This will be the first time we fully explore our canonical framework and the four classes of policies. Derivative-free stochastic search is a sequential decision problem characterized by a pure belief state which captures our approximation of the underlying problem. This allows us to build a bridge to the multiarmed bandit community. We also introduce the idea of *active learning*, where we make decisions specifically to improve our knowledge of the function we are optimizing.

By the end of Part II, we will have laid the foundation for the much richer class of sequential decision problems that involve controllable physical states that link decisions and dynamics from one time period to the next. However, we will use the tools of these three chapters throughout the rest of the book, especially in the context of tuning parameters for policies.

5

Derivative-Based Stochastic Search

We begin our discussion of adaptive learning methods in stochastic optimization by addressing problems where we have access to derivatives (or gradients, if x is a vector) of our function $F(x, W)$. It is common to start with the asymptotic form of our basic stochastic optimization problem

$$\max_{x \in \mathcal{X}} \mathbb{E}\{F(x, W)|S^0\}, \tag{5.1}$$

but soon we are going to shift attention to finding the best algorithm (or policy) for finding the best solution within a finite budget. We are going to show that with any adaptive learning algorithm, we can define a state S^n that captures what we know after n iterations. We can represent any algorithm as a "policy" $X^\pi(S^n)$ which tells us the next point $x^n = X^\pi(S^n)$ given what we know, S^n, after n iterations. Eventually we complete our budget of N iterations, and produce a solution that we call $x^{\pi,N}$ to indicate that the solution was found with policy (algorithm) π after N iterations.

After we choose x^n, we observe a random variable W^{n+1} that is not known when we chose x^n. We then evaluate the performance through a function $F(x^n, W^{n+1})$ which can serve as a placeholder for a number of settings, including the results of a computer simulation, how a product works in the market, the response of a patient to medication, or the strength of a material produced in a lab. The initial state S^0 might contain fixed parameters (say the boiling point of a material), the attributes of a patient, the starting point of an algorithm, and beliefs about any uncertain parameters.

When we focus on this finite-budget setting, the problem in (5.1) becomes

$$\max_{\pi} \mathbb{E}\{F(x^{\pi,N}, W)|S^0\}, \tag{5.2}$$

but this way of writing the problem hides what is actually happening. Starting with what we know in S^0, we are going to apply our policy $X^\pi(S^n)$ while we generate the sequence

Reinforcement Learning and Stochastic Optimization: A Unified Framework for Sequential Decisions, First Edition. Warren B. Powell.
© 2022 John Wiley & Sons, Inc. Published 2022 John Wiley & Sons, Inc.

$$(S^0, x^0, W^1, S^1, \ldots, S^n, x^n, W^{n+1}, \ldots, S^N)$$

where the observations $W^1, \ldots, W^N$ might be called *training data* to produce the solution $x^{\pi,N}$. Once we have $x^{\pi,N}$, we evaluate it using a new random variable that we denote by $\widehat{W}$ which is what we use for testing. We then use $\widehat{W}$ to evaluate the performance of $x^{\pi,N}$ which is computed using

$$\bar{F}^{\pi,N} = \mathbb{E}_{\widehat{W}} F(x^{\pi,N}, \widehat{W}). \tag{5.3}$$

We are almost there. The problem with $\bar{F}^{\pi,N}$ is that it is a random variable that depends on the specific sequence $W^1, \ldots, W^N$, as well as any distributional information in S^0 (we return to this issue later). We have potentially three sources of uncertainty:

The initial state S^0 – The initial state S^0 might include a probability distribution describing our belief (say) of the mean of a random variable.

The training sequence $W^1, \ldots, W^N$ – These are our observations while we are computing $x^{\pi,N}$.

The testing process – Finally, we are going to repeatedly sample from W, using a random variable we call $\widehat{W}$ to make the distinction with the random variable W that we use for training $x^{\pi,N}$.

The value F^π of our policy (algorithm) $X^\pi(S)$ can now be written as (using our expanded form of the expectation)

$$F^\pi = \mathbb{E}_{S^0} \mathbb{E}_{W^1, \ldots, W^N | S^0} \mathbb{E}_{\widehat{W} | S^0} \{ F(x^{\pi,N}, \widehat{W}) | S^0 \}. \tag{5.4}$$

These expectations can be a little frightening. In practice we are going to simulate them, but we defer this to later in the chapter.

The objective in (5.2) would be the natural finite-budget version of (5.1) (which we also call the *final reward* objective), but we should keep an open mind and recognize that we may also be interested in the *cumulative reward* formulation given by

$$\max_{\pi} \mathbb{E}_{S^0} \mathbb{E}_{W^1, \ldots, W^N | S^0} \left\{ \sum_{n=0}^{N-1} F(x^n, W^{n+1}) | S^0 \right\} \tag{5.5}$$

where $x^n = X^\pi(S^n)$ is our search policy (typically known as an "algorithm"). Note that when we maximize cumulative reward, we add up our performance as we go, so we do not have that final training step with $\widehat{W}$ that we did above with our final reward objective.

The transition from searching for a solution x to finding a function π is one of the central differences between deterministic and stochastic optimization problems. We are moving from looking for the best *solution* x to finding the best *algorithm* (or policy) π.

In this chapter, we assume that we can compute the gradient $\nabla F(x, W)$ once the random information W becomes known. This is most easily illustrated using the newsvendor problem. Let x be the number of newspapers placed in a bin, with unit cost c. Let W be the random demand for newspapers (which we learn after choosing x), which are sold at price p. We wish to find x that solves

$$\max_x F(x) = \mathbb{E}F(x, W) = \mathbb{E}\big(p \min\{x, W\} - cx\big). \tag{5.6}$$

We can use the fact that we can compute *stochastic gradients*, which are gradients that we compute only after we observe the demand W, given by

$$\nabla_x F(x, W) = \begin{cases} p - c & \text{if } x \leq W, \\ -c & \text{if } x > W. \end{cases} \tag{5.7}$$

The gradient $\nabla_x F(x, W)$ is known as a stochastic gradient because it depends on the random demand W, which is to say that we calculate it after we have observed W.

We are going to show how to design simple algorithms that exploit our ability to compute gradients after the random information becomes known. Even when we do not have direct access to gradients, we may be able to estimate them using finite differences. We are also going to see that the core ideas of stochastic gradient methods pervade a wide range of adaptive learning algorithms.

We start by summarizing a variety of applications.

5.1 Some Sample Applications

Derivative-based problems exploit our ability to use the derivative after the random information has been observed (but remember that our decision x must be made before we have observed this information). These derivatives, known as stochastic gradients, require that we understand the underlying dynamics of the problem. When this is available, we have access to some powerful algorithmic strategies that have been developed since these ideas were first invented in 1951 by Robbins and Monro.

Some examples of problems where derivatives can be computed directly are:

- Cost-minimizing newsvendor problem – A different way of expressing the newsvendor problem is one of minimizing overage and underage costs. Using the same notation as above, our objective function would be written

$$\min_x \mathbb{E}F(x, W) = \mathbb{E}[c^o \max\{0, x - W\} + c^u \max\{0, W - x\}]. \qquad (5.8)$$

We can compute the derivative of $F(x, \hat{D})$ with respect to x after W becomes known using

$$\nabla_x F(x, W) = \begin{cases} c^0 & \text{if } x > W, \\ -c^u & \text{if } x \le W. \end{cases}$$

- Nested newsvendor – This hints at a multidimensional problem which would be hard to solve even if we knew the demand distribution. Here there is a single random demand D that we can satisfy with products $1, \dots, K$ where we use the supply of products $1, \dots, k - 1$ before using product k. The profit-maximizing version is given by

$$\max_{x_1,\dots,x_K} = \sum_{k=1}^{K} p_k \mathbb{E} \min\left\{ x_k, \left(D - \sum_{j=1}^{k-1} x_j\right)^+ \right\} - \sum_{k=1}^{K} c_k x_k. \qquad (5.9)$$

Although more complicated than the scalar newsvendor, it is still fairly straightforward to find the gradient with respect to the vector x once the demand becomes known.

- Statistical learning – Let $f(x|\theta)$ be a statistical model which might be of the form

$$f(x|\theta) = \theta_0 + \theta_1 \phi_1(x) + \theta_2 \phi_2(x) + \dots .$$

Imagine we have a dataset of input variables $x^1, \dots, x^N$ and corresponding response variables $y^1, \dots, y^N$. We would like to find θ to solve

$$\min_\theta \frac{1}{N} \sum_{n=1}^{N} (y^n - f(x^n|\theta))^2.$$

- Finding the best inventory policy – Let R_t be the inventory at time t. Assume we place an order x_t according to the rule

$$X^\pi(R_t|\theta) = \begin{cases} \theta^{max} - R_t & \text{If } R_t < \theta^{min} \\ 0 & \text{Otherwise.} \end{cases}$$

Our inventory evolves according to

$$R_{t+1} = \max\{0, R_t + x_t - D_{t+1}\}.$$

Assume that we earn a contribution $C(R_t, x_t, D_{t+1})$

$$C(R_t, x_t, D_{t+1}) = p \min\{R_t + x_t, D_{t+1}\} - cx_t.$$

We then want to choose θ to maximize

$$\max_{\theta} \mathbb{E} \sum_{t=0}^{T} C(R_t, X^{\pi}(R_t|\theta), D_{t+1}).$$

If we let $F(x, W) = \sum_{t=0}^{T-1} C(R_t, X^{\pi}(R_t|\theta), D_{t+1})$ where $x = (\theta^{min}, \theta^{max})$ and $W = D_1, D_2, \dots, D_T$, then we have the same problem as our newsvendor problem in equation (5.6). In this setting, we simulate our policy, and then look back and determine how the results would have changed if θ is perturbed for the same sample path. It is sometimes possible to compute the derivative analytically, but if not, we can also do a numerical derivative (but using the same sequence of demands).

- Maximizing e-commerce revenue – Assume that demand for a product is given by

$$D(p|\theta) = \theta_0 - \theta_1 p + \theta_2 p^2.$$

Now, find the price p to maximize the revenue $R(p) = pD(p|\theta)$ where θ is unknown.

- Optimizing engineering design – An engineering team has to tune the timing of a combustion engine to maximize fuel efficiency while minimizing emissions. Assume the design parameters x include the pressure used to inject fuel, the timing of the beginning of the injection, and the length of the injection. From this the engineers observe the gas consumption $G(x)$ for a particular engine speed, and the emissions $E(x)$, which are combined into a utility function $U(x) = U(E(x), G(x))$ which combines emissions and mileage into a single metric. $U(x)$ is unknown, so the goal is to find an estimate $\bar{U}(x)$ that approximates $U(x)$, and then maximize it.

- Derivatives of simulations – In the previous section we illustrated a stochastic gradient algorithm in the context of a simple newsvendor problem. Now imagine that we have a multiperiod simulation, such as we might encounter when simulating flows of jobs around a manufacturing center. Perhaps we use a simple rule to govern how jobs are assigned to machines once they have finished a particular step (such as being drilled or painted). However, these rules have to reflect physical constraints such as the size of buffers for holding jobs before a machine can start working on them. If the buffer for a downstream machine is full, the rule might specify that a job be routed to a different machine or to a special holding queue.

   This is an example of a policy that is governed by static variables such as the size of the buffer. We would let x be the vector of buffer sizes. It would be helpful, then, if we could do more than simply run a simulation for a fixed vector x. What if we could compute the derivative with respect to each element of x, so that after running a simulation, we obtain all the derivatives?

Computing these derivatives from simulations is the focus of an entire branch of the simulation community. A class of algorithms called *infinitesimal perturbation analysis* was developed specifically for this purpose. It is beyond the scope of our presentation to describe these methods in any detail, but it is important for readers to be aware that the field exists.

5.2 Modeling Uncertainty

Before we progress too far, we need to pause and say a few words about how we are modeling uncertainty, and the meaning of what is perhaps the most dangerous piece of notation in stochastic optimization, the expectation operator $\mathbb{E}$.

We are going to talk about uncertainty from three perspectives. The first is the random variable W that arises when we evaluate a solution, which we refer to as *training uncertainty*. The second is the initial state S^0, where we express *model uncertainty*, typically in the form of uncertainty about parameters (but sometimes in the structure of the model itself). The third addresses testing uncertainty. In final-reward problems, we use the random variable $\widehat{W}$ for testing. In cumulative-reward settings, we test as we proceed.

5.2.1 Training Uncertainty $W^1, \ldots, W^N$

Consider an adaptive algorithm (which we first introduced in chapter 4) that proceeds by guessing x^n and then observing W^{n+1} which leads to x^{n+1} and so on (we give examples of these procedures in this chapter). If we limit the algorithm to N iterations, our sequence will look like

$$(x^0, W^1, x^1, W^2, x^2, \ldots, x^n, W^{n+1}, \ldots, x^N).$$

Table 5.1 illustrates six sample paths for the sequence $W^1, \ldots, W^{10}$. We often let ω to represent an outcome of a random variable, or an entire sample path (as we would here). We might let Ω be the set of all the sample paths, which for this problem we would write as

$$\Omega = (\omega_1, \omega_2, \omega_3, \omega_4, \omega_5, \omega_6).$$

We could then let $W_t(\omega)$ be the outcome of the random variable W_t at time t for sample path ω. Thus, $W_5(\omega_2) = 7$. If we are following sample path ω using policy π, we obtain the final design $x^{\pi,N}(\omega)$. By running policy π for each

Table 5.1 Illustration of six sample paths for the random variable W.

ω	W^1	W^2	W^3	W^4	W^5	W^6	W^7	W^8	W^9	W^{10}
1	0	1	6	3	6	1	6	0	2	4
2	3	2	2	1	7	5	4	6	5	4
3	5	2	3	2	3	4	2	7	7	5
4	6	3	7	3	2	3	4	7	3	4
5	3	1	4	5	2	4	3	4	3	1
6	3	4	4	3	3	3	2	2	6	1

outcome $\omega \in \Omega$, we would generate a population of designs $x^{\pi,N}$ which provide a nice way to represent $x^{\pi,N}$ as a random variable.

5.2.2 Model Uncertainty S^0

We illustrate model uncertainty using our newsvendor problem, where we make a decision x, then observe a random demand $W = \hat{D}$, after which we calculate our profit using equation (5.6). Imagine that our demand follows a Poisson distribution given by

$$\mathbb{P}[W = w] = \frac{\mu^w e^{-\mu}}{w!},$$

where $w = 0, 1, 2, ...$ In this setting, our expectation would be over the possible outcomes of W, so we could write the optimization problem in equation (5.6) as

$$F(x|\mu) = \sum_{w=0}^{\infty} \frac{\mu^w e^{-\mu}}{w!} (p \min\{x, w\} - cx).$$

This does not look too hard, but what happens if we do not know μ? This parameter would be carried by our initial state S^0. If we are uncertain about μ, we may feel that we can describe it using an exponential distribution given by

$$\mu \sim \lambda e^{-\lambda u},$$

where the parameter λ is known as a hyperparameter, which is to say it is a parameter that determines a distribution that describes the uncertainty of a problem parameter. The assumption is that even if we do not know λ precisely, it still does a good job of describing the uncertainty in the mean demand μ. In this case, S^0 would include both λ and the assumption that μ is described by an exponential distribution.

We would now write our expectation of $F(x, W)$ as

$$F(x) = \mathbb{E}\{F(x, W)|S^0\},$$
$$= \mathbb{E}_{S^0}\mathbb{E}_{W|S^0}\{F(x, W)|S^0\}.$$

For our example, this would be translated as

$$F(x|\lambda) = \mathbb{E}_{\mu|\lambda}\mathbb{E}_{W|\mu}\{F(x, W)|\mu\}.$$

The notation $\mathbb{E}_{W|\mu}$ means the conditional expectation of W given μ. Using our distributions where the random demand W follows a Poisson distribution with mean μ which is itself random with an exponential distribution with mean λ, we would write the expectation as

$$F(x|\lambda) = \int_{u=0}^{\infty} \lambda e^{-\lambda u} \sum_{w=0}^{\infty} \frac{u^w e^{-u}}{w!} (p \min(x, w) - cx) du.$$

In practice, we are rarely using explicit probability distributions. One reason is that we may not know the distribution, but we may have an exogenous source for generating random outcomes. The other is that we may have a distribution, but it might be multidimensional and impossible to compute.

5.2.3 Testing Uncertainty

When we finally obtain our solution $x^{\pi,N}$, we then have to evaluate the quality of the solution. For the moment, let's fix $x^{\pi,N}$. We let $\widehat{W}$ denote the random observations we use when testing the performance of our final solution $x^{\pi,N}$. We use $\widehat{W}$ to represent the random observations while testing to avoid confusion with the random observations W we use while training.

We write the value of the solution $x^{\pi,N}$ using

$$F(x^{\pi,N}) = \mathbb{E}_{\widehat{W}}\{F(x^{\pi,N}, \widehat{W})|S^0\}. \tag{5.10}$$

In practice we will typically evaluate the expectation using Monte Carlo simulation. Assume that we have a set of outcomes of $\widehat{W}$ that we call $\hat{\Omega}$, where $\omega \in \hat{\Omega}$ is one outcome of $\widehat{W}$ which we represent using $\widehat{W}(\omega)$. Once again assume that we have taken a random sample to create $\hat{\Omega}$ where every outcome is equally likely. Then we could evaluate our solution $x^{\pi,N}$ using

$$\bar{F}(x^{\pi,N}) = \frac{1}{|\hat{\Omega}|} \sum_{\omega \in \hat{\Omega}} F(x^{\pi,N}, \widehat{W}(\omega)).$$

The estimate $\bar{F}(x^{\pi,N})$ evaluates a single decision $x^{\pi,N}$, which hints to the performance of the learning policy π.

5.2.4 Policy Evaluation

If we wish to evaluate a policy $X^\pi(S^n)$, we have to combine all three types of uncertainty. This is done by computing

$$F^\pi = \mathbb{E}_{S^0} E_{W^1,\dots,W^N|S^0} E_{\widehat{W}|S^0} F(x^{\pi,N}, \widehat{W}).$$

In practice, we can replace each expectation by a sample over whatever is random. Furthermore, these samples can be (a) sampled from a probability distribution, (b) represented by a large, batch dataset, or (c) observed from an exogenous process (which involves online learning).

5.2.5 Closing Notes

This section is hardly a comprehensive treatment of modeling uncertainty. Given the richness of this topic, chapter 10 is dedicated to describing the process of modeling uncertainty. The discussion here was to bring out the basic forms of uncertainty when evaluating algorithms for stochastic search.

We mention only in passing the growing interest in replacing the expectation $\mathbb{E}$ with some form of risk measure that recognizes that the possibility of extreme outcomes is more important that is represented by their likelihood (which may be low). Expectations average over all outcomes, so if extreme events occur with low probability, they do not have much effect on the solution. Also, expectations may have the effect of letting high outcomes cancel low outcomes, when in fact one tail is much more important than the other. We discuss risk in more detail in section 9.8.5. Replacing the expectation operator with some form of risk measure does not change the core steps when evaluating a policy.

5.3 Stochastic Gradient Methods

One of the oldest and earliest methods for solving our basic stochastic optimization problem

$$\max_x \mathbb{E}F(x, W), \tag{5.11}$$

uses the fact that we can often compute the gradient of $F(x, W)$ with respect to x after the random variable W becomes known. For example, assume that we are trying to solve a newsvendor problem, where we wish to allocate a quantity x of resources ("newspapers") before we know the demand W. The optimization problem is given by

$$\max_{x} F(x) = \mathbb{E}p \min\{x, W\} - cx. \tag{5.12}$$

If we could compute $F(x)$ exactly (that is, analytically), and its derivative, then we could find x^* by taking its derivative and setting it equal to zero as we did in section 4.2.2. If this is not possible, we could still use a classical steepest ascent algorithm

$$x^{n+1} \;=\; x^n + \alpha_n \nabla_x F(x^n), \tag{5.13}$$

where α_n is a stepsize. For deterministic problems, we typically choose the best stepsize by solving the one-dimensional optimization problem

$$\alpha^n = \arg\max_{\alpha \geq 0} F\big(x^n + \alpha \nabla F(x^n)\big). \tag{5.14}$$

For stochastic problems, we would have to be able to compute $F(x) = \mathbb{E}F(x, W)$ in (5.14), which is computationally intractable (otherwise we return to the techniques in chapter 4). This means that we cannot solve the one-dimensional search for the best stepsize in equation (5.14).

Instead, we resort to an algorithmic strategy known as stochastic gradients, where we use the gradient $\nabla_x F(x^n, W^{n+1})$, which means we wait until we observe W^{n+1} and then take the gradient of the function. This is not possible for all problems (hence the reason for chapter 7), but for problems where we can find the gradient, this overcomes the issues associated with computing the derivative of an expectation. The idea that we are allowed to wait until *after* we observe W^{n+1} before computing the gradient is the magic of stochastic gradient algorithms.

5.3.1 A Stochastic Gradient Algorithm

For our stochastic problem, we assume that we either cannot compute $F(x)$, or we cannot compute the gradient exactly. However, there are many problems where, if we fix $W = W(\omega)$, we can find the derivative of $F(x, W(\omega))$ with respect to x. Then, instead of using the deterministic updating formula in (5.13), we would instead use

$$x^{n+1} \;=\; x^n + \alpha_n \nabla_x F(x^n, W^{n+1}). \tag{5.15}$$

Here, $\nabla_x F(x^n, W^{n+1})$ is called a *stochastic gradient* because it depends on a sample realization of W^{n+1}.

It is important to note our indexing. A variable such as x^n or α_n that is indexed by n is assumed to be a function of the observations $W^1, W^2, \ldots, W^n$, but not W^{n+1}. Thus, our stochastic gradient $\nabla_x F(x^n, W^{n+1})$ depends on our current solution x^n and the next observation W^{n+1}.

To illustrate, consider the simple newsvendor problem with the profit maximizing objective

$$F(x, W) \quad = \quad p \min\{x, W\} - cx.$$

In this problem, we order a quantity $x = x^n$ (determined at the end of day n), and then observe a random demand W^{n+1} that was observed the next day $n + 1$. We earn a revenue given by $p \min\{x^n, W^{n+1}\}$ (we cannot sell more than we bought, or more than the demand), but we had to pay for our order, producing a negative cost $-cx$. Let $\nabla F(x^n, W^{n+1})$ be the sample gradient, taken when $W = W^{n+1}$. In our example, this is given by

$$\frac{\partial F(x^n, W^{n+1})}{\partial x} = \begin{cases} p - c & \text{If } x^n < W^{n+1}, \\ -c & \text{If } x^n > W^{n+1}. \end{cases} \tag{5.16}$$

The quantity x^n is the estimate of x computed from the previous iteration (using the sample realization ω^n), while W^{n+1} is the sample realization in iteration $n + 1$ (the indexing tells us that x^n was computed without knowing W^{n+1}). When the function is deterministic, we would choose the stepsize by solving the one-dimensional optimization problem determined by (5.14).

5.3.2 Introduction to Stepsizes

Now we face the problem of finding the stepsize α_n when we have to work with the stochastic gradient $\nabla F(x^n, W^{n+1})$. Unlike our deterministic algorithm, we cannot solve a one-dimensional search (as we did in (5.14)) to find the best stepsize after seeing W^{n+1}, simply because we cannot compute the expectation.

We overcome our inability to compute the expectation by working with stochastic gradients. While the computational advantages are tremendous, it means that the gradient is now a random variable. This means that the stochastic gradient can even point away from the optimal solution such that any positive stepsize actually makes the solution worse. Figure 5.1 compares the behavior of a deterministic search algorithm, where the solution improves at each iteration, and a stochastic gradient algorithm.

This behavior is easily illustrated using our newsvendor problem. It might be that the optimal order quantity is 15. However, even if we order $x = 20$, it is possible that the demand is 24 on a particular day, pushing us to move our order quantity to a number larger than 20, which is even further from the optimum.

The major challenge when using stochastic gradients is the stepsize; we can no longer use the one-dimensional search as we did with our deterministic application in equation (5.14). Interestingly, when we are working on stochastic

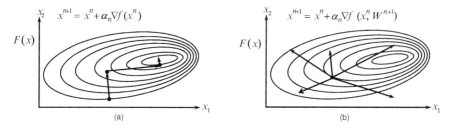

Figure 5.1 Illustration of gradient ascent for a deterministic problem (a), and stochastic gradients (b).

problems, we overcome our inability to solve the one-dimensional search problem by using relatively simple stepsize rules that we are going to call stepsize policies. For example, a classic formula is nothing more than

$$\alpha_n = \frac{1}{n+1} \tag{5.17}$$

for $n = 0, 1, \dots$. With this formula, we can show that

$$\lim_{n\to\infty} x^n \to x^*, \tag{5.18}$$

where x^* is the optimal solution of our original optimization problem (5.1). But note – we did not promise that convergence was fast, only that it would converge (eventually). (See section 5.10 for proofs of this convergence.) There is a *very* large literature that proves asymptotic convergence, but then runs the algorithm for a finite number of iterations and just assumes the resulting solution is good.

There are many applications where the units of the gradient, and the units of the decision variable, are different. This happens with our newsvendor example, where the gradient is in units of dollars, while the decision variable x is in units of newspapers. This is a significant problem that causes headaches in practice.

A problem where we avoid this issue arises if we are trying to learn the mean of a random variable W. We can formulate this task as a stochastic optimization problem using

$$\min_x \mathbb{E} \frac{1}{2}(x - W)^2. \tag{5.19}$$

Here, our function $F(x, W) = \frac{1}{2}(x - W)^2$, and it is not hard to see that the value of x that minimizes this function is $x = \mathbb{E}W$. Now assume that we want to produce a sequence of estimates of $\mathbb{E}W$ by solving this problem using a sequential (online) stochastic gradient algorithm, which looks like

$$
\begin{aligned}
x^{n+1} &= x^n - \alpha_n \nabla F_x(x^n, W^{n+1}), & (5.20)\\
&= x^n - \alpha_n(x^n - W^{n+1}),\\
&= (1 - \alpha_n)x^n + \alpha_n W^{n+1}. & (5.21)
\end{aligned}
$$

Equation (5.20) illustrates α_n as the stepsize in a stochastic gradient algorithm, while equation (5.21) is exponential smoothing (see section 3.2). In this context, α_n is widely known as a smoothing factor or "learning rate."

There are going to be problems where our "one over n" stepsize formula (5.17) is very slow. However, for the problem of estimating the mean of a random variable, we are going to show in chapter 6 that "one over n" is actually *the optimal stepsize formula*!! That is, no other stepsize formula will give faster convergence. This is just a hint of the richness we are going to encounter with stepsize rules.

There are problems where we may start with a prior estimate of $\mathbb{E}W$ which we can express as x^0. In this case, we would want to use an initial stepsize $\alpha^0 < 1$. However, we often start with no information, in which case an initial stepsize $\alpha^0 = 1$ gives us

$$
\begin{aligned}
x^1 &= (1 - \alpha_0)x^0 + \alpha_0 W^1\\
&= W^1,
\end{aligned}
$$

which means we do not need the initial estimate for x^0. Smaller initial stepsizes would only make sense if we had access to a reliable initial guess, and in this case, the stepsize should reflect the confidence in our original estimate (for example, we might be warm starting an algorithm from a previous iteration).

This section is just a peek into stepsizes. We cover this rich topic in considerably more detail in chapter 6.

5.3.3 Evaluating a Stochastic Gradient Algorithm

In section 5.10 we are going to provide two proofs of asymptotic optimality. The problem is that we never run these algorithms to the limit, which means we are only interested in our finite time performance. If we are only interested in the quality of our final solution $x^{\pi,N}$, then we want to use the final reward objective given by (5.4), but this raises the issue: How do we compute this? The answer is that we have to simulate it.

Let ω^ℓ be a sample realization of our random variables $W^1(\omega^\ell), \dots, W^N(\omega^\ell)$ that we use for training (estimating) $x^{\pi,N}(\omega^\ell)$ for $\ell = 1, 2, \dots, L$. Then let ψ^k be a sample realization of our testing information $\widehat{W}(\psi^k)$, for $k = 1, 2, \dots, K$.

Assume that there is no probabilistic information in S^0. We can estimate the performance of our algorithm π using

$$\bar{F}^\pi = \frac{1}{L}\sum_{\ell=1}^{L}\left(\frac{1}{K}\sum_{k=1}^{K}F(x^{\pi,N}(\omega^\ell),\widehat{W}(\psi^k))\right),\tag{5.22}$$

where $x^n(\omega^\ell) = X^\pi(S^n(\omega^\ell))$ is determined by our stochastic gradient formula (5.20) and α_n comes from our stepsize formula (say, equation (5.17)). For this problem our state variable $S^n = x^n$, which means that our state transition equation $S^{n+1}(\omega^\ell) = S^M(S^n(\omega^\ell),x^n(\omega^\ell),W^{n+1}(\omega^\ell))$ is just the stochastic gradient (5.20). We then let $x^{\pi,N} = x^N$ be the ending point.

The final reward objective in (5.22) is easily the most classical way of evaluating a stochastic search algorithm, but there are several arguments to be made for using the cumulative reward, which we would simulate using

$$\bar{F}^\pi = \frac{1}{L}\sum_{\ell=1}^{L}\left(\sum_{n=0}^{N-1}F(x^n(\omega^\ell),W^{n+1}(\omega^\ell))\right).\tag{5.23}$$

It is possible that we have to apply this algorithm in a field situation such as a real newsvendor problem, where we have to live with the results of each solution x^n. However, we may simply be interested in the overall rate of convergence, which would be better captured by (5.23).

5.3.4   A Note on Notation

Throughout this book, we index variables (whether we are indexing by iterations or time) to clearly identify the *information content* of each variable. Thus, x^n is the decision made after W^n becomes known. When we compute our stochastic gradient $\nabla_x F(x^n,W^{n+1})$, we use x^n which was determined after observing W^n. If the iteration counter refers to an experiment, then it means that x^n is determined after we finish the nth experiment. If we are solving a newsvendor problem where n indexes days, then it is like determining the amount of newspapers to order for day $n+1$ after observing the sales for day n. If we are performing a laboratory experiment, we use the information up through the first n experiments to choose x^n, which specifies the design settings for the $n+1^{st}$ experiment. This indexing makes sense when you realize that the index n reflects the information content, not when it is being implemented.

In chapter 6, we are going to present a number of formulas to determine stepsizes. Some of these are deterministic, such as $\alpha_n = 1/n$, and some are stochastic, adapting to information as it arrives. Our stochastic gradient formula in equation (5.15) communicates the property that the stepsize α_n that is multiplied times the gradient $\nabla_x F(x^n, W^{n+1})$ is allowed to see W^n and x^n, but not W^{n+1}.

We return to this issue in chapter 9, but we urge readers to adopt this notational system.

5.4   Styles of Gradients

There are a few variants of the basic stochastic gradient method. Below we introduce the idea of gradient smoothing and describe a method for approximating a second-order algorithm.

5.4.1   Gradient Smoothing

In practice, stochastic gradients can be *highly* stochastic, which is the reason why we have to use stepsizes. However, it is possible to mitigate some of the variability by smoothing the gradient itself. If $\nabla F(x^n, W^{n+1})$ is our stochastic gradient, computed after the $n + 1^{\text{st}}$ experiment, we could then smooth this using

$$g^{n+1} = (1 - \eta)g^n + \eta \nabla F(x^n, W^{n+1}),$$

where η is a smoothing factor where $0 < \eta \leq 1$. We could replace this with a declining sequence η_n, although common practice is to keep this process as simple as possible. Regardless of the strategy, gradient smoothing has the effect of introducing at least one more tunable parameter. The open empirical question is whether gradient smoothing adds anything beyond the smoothing produced by the stepsize policy used for updating x^n.

5.4.2   Second-Order Methods

Second-order methods for deterministic optimization have proven to be particularly attractive. For smooth, differentiable functions, the basic update step looks like

$$x^{n+1} = x^n + (H^n)^{-1} \nabla_x f(x^n), \tag{5.24}$$

where H^n is the Hessian, which is the matrix of second derivatives. That is,

$$H^n_{xx'} = \left. \frac{\partial^2 f(x)}{\partial x \partial x'} \right|_{x=x^n}.$$

The attraction of the update in equation (5.24) is that there is no stepsize. The reason (and this requires that $f(x)$ be smooth with continuous first derivatives) is that the inverse Hessian solves the problem of scaling. In fact, if $f(x)$ is quadratic, then equation (5.24) takes us to the optimal solution in one step!

Since functions are not always as nice as we would like, it is sometimes useful to introduce a constant "stepsize" α, giving us

$$x^{n+1} = x^n + \alpha(H^n)^{-1}\nabla_x f(x^n),$$

where $0 < \alpha \leq 1$. Note that this smoothing factor does not have to solve any scaling problems (again, this is solved by the Hessian).

If we have access to second derivatives (which is not always the case), then our only challenge is inverting the Hessian. This is not a problem with a few dozen or even a few hundred variables, but there are problems with thousands to tens of thousands of variables. For large problems, we can strike a compromise and just use the diagonal of the Hessian. This is both much easier to compute, as well as being easy to invert. Of course, we lose some of the fast convergence (and scaling).

There are many problems (including all stochastic optimization problems) where we do not have access to Hessians. One strategy to overcome this is to construct an approximation of the Hessian using what are known as rank-one updates. Let $\bar{H}^n$ be our approximate Hessian which is computed using

$$\bar{H}^{n+1} = \bar{H}^n + \nabla f(x^n)(\nabla f(x^n))^T. \tag{5.25}$$

Recall that $\nabla f(x^n)$ is a column vector, so $\nabla f(x^n)(\nabla f(x^n))^T$ is a matrix with the dimensionality of x. Since it is made up of an outer product of two vectors, this matrix has rank 1.

This methodology could be applied to a stochastic problem. As of this writing, we are not aware of any empirical study showing that these methods work, although there has been recent interest in second-order methods for online machine learning.

5.4.3 Finite Differences

It is often the case that we do not have direct access to a derivative. Instead, we can approximate gradients using finite differences which requires running the simulation multiple times with perturbed inputs.

Assume that x is a P-dimensional vector, and let e_p be a P-dimensional column vector of zeroes with a 1 in the pth position. Let $W_p^{n+1,+}$ and $W_p^{n+1,-}$ be sequences of random variables that are generated when we run each simulation, which would be run in the $n+1^{st}$ iteration. The subscript p only indicates that these are the random variables for the pth run.

Now assume that we can run two simulations for each dimension, $F(x^n + \delta x^n e_p, W_p^{n+1,+})$ and $F(x^n - \delta x^n e_p, W_p^{n+1,-})$ where $\delta x^n e_p$ is the change in x^n, multiplied by e_p so that we are only changing the pth dimension. Think of $F(x^n + \delta x^n e_p, W_p^{n+1,+})$ and $F(x^n - \delta x^n e_p, W_p^{n+1,-})$ as calls to a black-box simulator where we start with a set of parameters x^n, and then perturb it to $x^n + \delta x^n e_p$ and $x^n - \delta x^n e_p$ and run two separate, independent simulations. We then have to do this for each dimension p, allowing us to compute

$$g_p^n(x^n, W^{n+1,+}, W^{n+1,-}) = \frac{F(x^n + \delta x^n e_p, W_p^{n+1,+}) - F(x^n - \delta x^n e_p, W_p^{n+1,-})}{2\delta x_p^n}, \qquad (5.26)$$

where we divide the difference by the width of the change, given by $2\delta x_p^n$, to get the slope.

The calculation of the derivative (for one dimension) is illustrated in Figure 5.2. We see from Figure 5.2 that shrinking δx can introduce a lot of noise in the estimate of the gradient. At the same time, as we increase δx, we introduce bias, which we see in the difference between the dashed line showing $\mathbb{E}g^n(x^n, W^{n+1,+}, W^{n+1,-})$, and the dotted line that depicts $\partial \mathbb{E}F(x^n, W^{n+1})/\partial x^n$. If we want an algorithm that converges asymptotically in the limit, we need δx^n decreasing, but in practice it is often set to a constant δx, which is then handled as a tunable parameter.

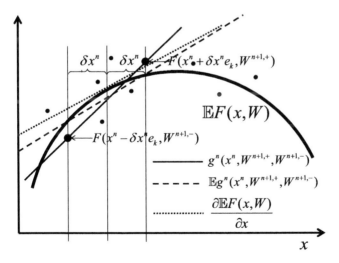

Figure 5.2 Different estimates of the gradient of $F(x, W)$ with the stochastic gradient $g^n(x^n, W^{n+1,+}, W^{n+1,-})$ (solid line), the expected finite difference $\mathbb{E}g^n(x^n, W^{n+1,+}, W^{n+1,-})$ (dashed line), and the exact slope at x^n, $\partial \mathbb{E}F(x^n, W^{n+1})/\partial x^n$.

Finite differences can be expensive. Running a function evaluation can require seconds to minutes, but there are computer models that can take hours or days (or more) to run. Equation (5.26) requires $2P$ function evaluations, which can be especially problematic when $F(x, W)$ is an expensive simulation, as well as when the number of dimensions P is large. Fortunately, these simulations can often be run in parallel. In the next section we introduce a strategy for handling multidimensional parameter vectors.

5.4.4 SPSA

A powerful method for handling higher-dimensional parameter vectors is *simultaneous perturbation stochastic approximation* (or SPSA). SPSA computes gradients in the following way. Let $Z_p, p = 1, \dots, P$ be a vector of zero-mean random variables, and let Z^n be a sample of this vector at iteration n. We approximate the gradient by perturbing x^n by the vector Z using $x^n + \eta^n Z^n$ and $x^n - \eta^n Z^n$, where η^n is a scaling parameter that may be a constant over iterations, or may vary (typically it will decline). Now let $W^{n+1,+}$ and $W^{n+1,-}$ represent two different samples of the random variables driving the simulation (these can be generated in advance or on the fly). We then run our simulation twice: once to find $F(x^n + \eta^n Z^n, W^{n+1,+})$, and once to find $F(x^n - \eta^n Z^n, W^{n+1,-})$. The estimate of the gradient is then given by

$$
g^n(x^n, W^{n+1,+}, W^{n+1,-}) = \begin{bmatrix} \dfrac{F(x^n+\eta^n Z^n, W^{n+1,+})-F(x^n-\eta^n Z^n, W^{n+1,-})}{2\eta^n Z_1^n} \\[6pt] \dfrac{F(x^n+\eta^n Z^n, W^{n+1,+})-F(x^n-\eta^n Z^n, W^{n+1,-})}{2\eta^n Z_2^n} \\ \vdots \\ \dfrac{F(x^n+\eta^n Z^n, W^{n+1,+})-F(x^n-\eta^n Z^n, W^{n+1,-})}{2\eta^n Z_P^n} \end{bmatrix}. \quad (5.27)
$$

Note that the numerator of each element of g^n in equation (5.27) is the same, which means we only need two function evaluations: $F(x^n + \eta^n Z^n, W^{n+1,+})$ and $F(x^n - \eta^n Z^n, W^{n+1,-})$. The only difference is the Z_p^n in the denominator for each dimension p.

The real power of SPSA arises in applications where simulations are noisy, and these can be *very* noisy in many settings. A way to overcome this is with the use of "mini-batches" where the simulations to compute $F(x^n + \eta^n Z^n, W^{n+1,+})$ and $F(x^n - \eta^n Z^n, W^{n+1,-})$ are run, say, M times and averaged. Keep in mind that these can be done in parallel; this does not mean they are free, but if you have access to parallel processing capability (which is quite common), it means that repeated simulations may not add to the completion time for your algorithm.

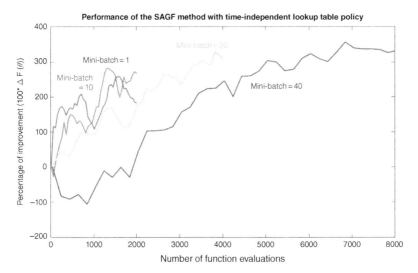

Figure 5.3 Convergence of SPSA for different mini-batch sizes, showing the slower convergence to a better solution with larger mini-batches.

Figure 5.3 illustrates the effect of mini-batches; larger mini-batches produce slower initial performance, but better performance over more iterations. Note that figure shows performance in terms of function evaluations, not CPU time, so the benefits of parallel computing are ignored. This graphic suggests a strategy of using increasing mini-batch sizes. Smaller mini-batches work well in the beginning, while larger mini-batches help as the algorithm progresses.

SPSA seems like magic: we are getting a P-dimensional gradient from just two function evaluations, regardless of the value of P. The open question is the rate of convergence, which will depend very much on the characteristics of the problem at hand. A reader will naturally ask: "Does it work?" The unqualified answer is: "It *can* work," but you will need to spend time understanding the characteristics of your problem, and tuning the algorithmic choices of SPSA, notably:

- Choice of stepsize formula, and tuning of any stepsize parameters (there is always at least one). Be careful with tuning, as it may depend on the starting point of your algorithm x^0, as well as other problem characteristics.
- The choice of mini-batch size. SPSA is trying to get a lot of information from just two function evaluations, so there is going to be a price to be paid in terms of convergence rates. A key issue here is whether you have access to parallel computing resources.

- You may also experiment with gradient smoothing, which is another way to stabilize the algorithm but without the price of repeated simulations required by mini-batches. This introduces the additional dimension of tuning the smoothing factor for gradient smoothing.
- Don't forget that all gradient-based methods are designed for maximizing concave functions (minimizing convex functions), but your function may not be concave. For complex problems, it is not necessarily easy (or even possible) to verify the behavior of the function, especially for higher dimensional problems (three or more).

5.4.5 Constrained Problems

There are problems where x has to stay in a feasible region $\mathcal{X}$, which might be described by a system of linear equations such as

$$\mathcal{X} = \{x | Ax = b, x \geq 0\}.$$

When we have constraints, we would first compute

$$y^{n+1} \quad = \quad x^n + \alpha_n \nabla_x F(x^n, W^{n+1}),$$

which might produce a solution y^{n+1} which does not satisfy the constraints. To handle this we project y^{n+1} using a projection step that we write using

$$x^{n+1} \leftarrow \Pi_{\mathcal{X}}[y^{n+1}].$$

The definition of the projection operator $\Pi_{\mathcal{X}}[\cdot]$ is given by

$$\Pi_{\mathcal{X}}[y] = \arg\min_{x \in \mathcal{X}} \|x - y\|_2, \tag{5.28}$$

where $\|x - y\|_2$ is the "L_2 norm" defined by

$$\|x - y\|_2 = \sum_i (x_i - y_i)^2.$$

The projection operator $\Pi_{\mathcal{X}}[\cdot]$ can often be solved easily by taking advantage of the structure of a problem. For example, we may have box constraints of the form $0 \leq x_i \leq u_i$. In this case, any element x_i falling outside of this range is just mapped back to the nearest boundary (0 or u_i).

5.5 Parameter Optimization for Neural Networks*

In section 3.9.3 we descibed how to produce an estimate from a neural network given the set of parameters. Now we are going to show how to estimate these parameters using the stochastic gradient concepts that we presented in this chapter.

We are going to show how to derive the gradient for the three-layer network in Figure 5.4. We will use the following relationships that we first derived in section 3.9.3 from the forward pass:

$$\bar{f}(x^n | \theta) = \sum_{i \in \mathcal{J}^{(1)}} \sum_{j \in \mathcal{J}^{(2)}} \theta_{ij}^{(1)} x_i^{(1,n)}, \tag{5.29}$$

$$y_j^{(2,n)} = \sum_{i \in \mathcal{J}^{(1)}} \theta_{ij}^{(1)} x_i^{(1,n)}, \quad j \in \mathcal{J}^{(2)}, \tag{5.30}$$

$$x_i^{(2,n)} = \sigma(y_i^{(2,n)}), \quad i \in \mathcal{J}^{(2)}, \tag{5.31}$$

$$y_j^{(3,n)} = \sum_{i \in \mathcal{J}^{(2)}} \theta_{ij}^{(2)} x_i^{(2,n)}, \quad j \in \mathcal{J}^{(3)}, \tag{5.32}$$

$$x_i^{(3,n)} = \sigma(y_i^{(3,n)}), \quad i \in \mathcal{J}^{(3)}. \tag{5.33}$$

Recall that $\sigma(y)$ is the sigmoid function

$$\sigma(y) = \frac{1}{1 + e^{-\beta y}}, \tag{5.34}$$

and that $\sigma'(y) = \frac{\partial \sigma(y)}{\partial y}$.

We are going to start by showing how to compute the gradient. Then, we will present the stochastic gradient algorithm and discuss some issues that arise in the context of neural networks.

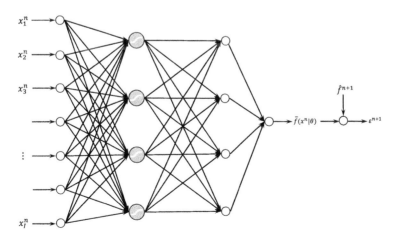

Figure 5.4 A three-layer neural network.

5.5.1 Computing the Gradient

We compute the stochastic gradient $\nabla_\theta F(\theta)$ for a given input x^n and observed response $\hat{f}^{n+1}$. $\hat{f}^{n+1}$ plays the role of W^{n+1} in our original derivation.

Assume that we are given an input x^n. If we follow the forward instructions above, our final estimate is produced by

$$\bar{f}(x^n|\theta) = \sum_{i \in \mathcal{J}^{(3)}} \theta_i^{(3)} x_i^{(3,n)}. \tag{5.35}$$

The goal is to find θ that solves

$$\min_\theta F(\theta) = \mathbb{E}\frac{1}{2}\sum_{n=1}^{N-1}(\bar{f}(x^n|\theta) - \hat{f}^{n+1})^2. \tag{5.36}$$

We want to understand the effect of a change in θ given an input x^n and response $\hat{f}^{n+1}$. Specifically, we want the gradient $\nabla_\theta F(\theta)$. First, since we cannot compute the expectation, we are going to compute a *stochastic gradient* which means we are going to replace $F(\theta)$ with the function evaluated for a particular x^n, given the response $\hat{f}^{n+1}$ which we write as

$$F(x^n, \hat{f}^{n+1}|\theta) = \frac{1}{2}(\bar{f}(x^n|\theta) - \hat{f}^{n+1})^2.$$

Computing $\nabla_\theta F(x^n, \hat{f}^{n+1}|\theta)$ will prove to be a nice exercise in applying the chain rule. We are going to compute the gradient by stepping backward through the neural network in Figure 5.4. Our hope is that by illustrating how to do it for this network, the process of extending this to other neural networks will be apparent.

We start with the derivative with respect to $\theta^{(3)}$:

$$\frac{\partial F(\theta|x^n, \hat{f}^{n+1})}{\partial \theta_i^{(3)}} = (\bar{f}(x^n|\theta) - \hat{f}^{n+1})\frac{\partial \bar{f}(x^n|\theta)}{\partial \theta_i^{(3)}}, \tag{5.37}$$

$$= (\bar{f}(x^n|\theta) - \hat{f}^{n+1})x_i^{(3,n)}, \tag{5.38}$$

where (5.38) comes from differentiating (5.29). The derivation of the gradient with respect to $\theta^{(2)}$ is given by

$$\frac{\partial F(\theta|x^n, \hat{f}^{n+1})}{\partial \theta_{ij}^{(2)}} = (\bar{f}(x^n|\theta) - \hat{f}^{n+1})\frac{\partial \bar{f}(x^n|\theta)}{\partial \theta_{ij}^{(2)}}, \tag{5.39}$$

$$\frac{\partial \bar{f}(x^n|\theta)}{\partial \theta_{ij}^{(2)}} = \frac{\partial \bar{f}(x^n|\theta)}{\partial x_j^{(3,n)}}\frac{\partial x_j^{(3,n)}}{\partial y_j^{(3)}}\frac{\partial y_j^{(3,n)}}{\partial \theta_{ij}^{(2)}}, \tag{5.40}$$

$$= \theta_j^{(3)}\sigma'(y_j^{(3,n)})x_i^{(2,n)}. \tag{5.41}$$

Remember that $\sigma'(y)$ is the derivative of our sigmoid function (equation (3.58)) with respect to y.

Finally, the gradient with respect to $\theta^{(1)}$ is found using

$$\frac{\partial F(\theta|x^n, \hat{f}^{n+1})}{\partial \theta_{ij}^{(1)}} = (\bar{f}(x^n|\theta) - \hat{f}^{n+1})\frac{\partial \bar{f}(x^n|\theta)}{\partial \theta_{ij}^{(1)}}, \tag{5.42}$$

$$\frac{\partial \bar{f}(x^n|\theta)}{\partial \theta_{ij}^{(1)}} = \sum_k \frac{\partial \bar{f}(x^n|\theta)}{\partial x_k^{(3)}}\frac{\partial x_k^{(3,n)}}{\partial y_k^{(3)}}\frac{\partial y_k^{(3)}}{\partial \theta_{ij}^{(1)}}, \tag{5.43}$$

$$= \sum_k \theta_k^{(3)}\sigma'(y_k^{(3)})\frac{\partial y_k^{(3)}}{\partial \theta_{ij}^{(1)}}, \tag{5.44}$$

$$\frac{\partial y_k^{(3)}}{\partial \theta_{ij}^{(1)}} = \frac{\partial y_k^{(3)}}{\partial x_j^{(2)}}\frac{\partial x_j^{(2)}}{\partial \theta_{ij}^{(1)}}, \tag{5.45}$$

$$= \frac{\partial y_k^{(3)}}{\partial x_j^{(2)}}\frac{\partial x_j^{(2)}}{\partial y_j^{(2)}}\frac{\partial y_j^{(2)}}{\partial \theta_{ij}^{(1)}}, \tag{5.46}$$

$$= \theta_{jk}^{(2)}\sigma'(y_j^{(2)})x_i^{(1)}. \tag{5.47}$$

Combining the above gives us

$$\frac{\partial F(\theta|x^n, \hat{f}^{n+1})}{\partial \theta_i^{(1)}} = (\bar{f}(x^n|\theta) - \hat{f}^{n+1})\left(\sum_k \theta_k^{(3)}\sigma'(y_k^{(3)})\theta_{jk}^{(2)}\right)\sigma'(y_j^{(2)})x_i^{(1)},$$

$$\frac{\partial F(\theta|x^n, \hat{f}^{n+1})}{\partial \theta_i^{(2)}} = (\bar{f}(x^n|\theta) - \hat{f}^{n+1})\theta_j^{(3)}\sigma'(y_j^{(3,n)})x_i^{(2,n)},$$

$$\frac{\partial F(\theta|x^n, \hat{f}^{n+1})}{\partial \theta_i^{(3)}} = (\bar{f}(x^n|\theta) - \hat{f}^{n+1})x_i^{(3,n)}.$$

The complete stochastic gradient is then given by

$$\nabla_\theta F(\theta|x^n, \hat{f}^{n+1}|\theta) = \begin{pmatrix} \nabla_{\theta^{(1)}} F(x^n, \hat{f}^{n+1}|\theta) \\ \nabla_{\theta^{(2)}} F(x^n, \hat{f}^{n+1}|\theta) \\ \nabla_{\theta^{(3)}} F(x^n, \hat{f}^{n+1}|\theta) \end{pmatrix}.$$

We are now ready to execute our parameter search using a stochastic gradient algorithm.

5.5.2 The Stochastic Gradient Algorithm

The search for θ is done using a basic stochastic gradient algorithm given by

$$\theta^{n+1} = \theta^n - \alpha_n \nabla_\theta F(\theta^n, \hat{f}^{n+1}). \tag{5.48}$$

We return to this method in considerably more detail in chapter 5. In particular, we have an entire chapter devoted to the design of the stepsize α_n, although for now we note that we could use a formula as simple as

$$\alpha_n = \frac{\theta^{step}}{\theta^{step} + n - 1}.$$

For now, we are going to focus on the properties of the function $F(\theta)$ in equation (5.36). In particular, readers need to be aware that the function $F(\theta)$ is highly nonconvex, as illustrated in Figure 5.5(a) for a two-dimensional problem. Figure 5.5(b) shows that when we start from two different starting points, we can end up at two different local minima. This behavior is typical of nonlinear models, but is especially true of neural networks.

The lack of convexity in the objective function $F(\theta)$ is well known to the neural network community. One strategy is to try a number of different starting points, and then use the best of the optimized values of θ. The real issue, of course, is not which θ produces the lowest error for a particular dataset, but which θ produces the best performance with new data.

This behavior also complicates using neural networks in an online setting. If we have fitted a neural network and then add one more data point, there is not a natural process for incrementally updating the estimate of θ. Simply doing one iteration of the gradient update in (5.48) accomplishes very little, since we are never truly at the optimum.

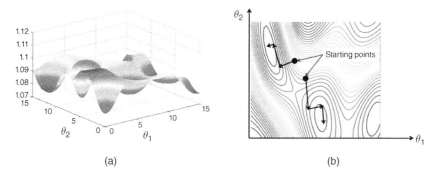

(a)   (b)

Figure 5.5 (a) Illustration of nonconvex behavior of the response surface for $F(\theta)$; (b) the path to local minima from two different starting points.

5.6   Stochastic Gradient Algorithm as a Sequential Decision Problem

We think of stochastic gradient algorithms as methods for solving a problem such as our basic optimization problem in (5.1). However, designing a stochastic gradient algorithm can itself be formulated as a sequential decision problem and modeled using the canonical framework we presented in section 2.2.

We start by restating our stochastic gradient algorithm

$$x^{n+1} \quad = \quad x^n + \alpha_n \nabla_x F(x^n).$$

In practice we have to tune the stepsize formula. While there are many rules we could use, we will illustrate the key idea using a simple adaptive stepsize policy known as Kesten's rule given by

$$\alpha_n(\theta^{kest}) = \frac{\theta^{kest}}{\theta^{kest} + N^n}, \qquad (5.49)$$

where θ^{kest} is a tunable parameter and N^n counts the number of times that the gradient $\nabla_x F(x^n)$ has changed sign, which means that

$$(\nabla_x F(x^{n-1}))^T \nabla_x F(x^n) < 0.$$

When this inner product is negative, it means that the algorithm is starting to criss-cross which is an indication that it is in the vicinity of the optimum.

The stochastic gradient algorithm is stated as a sequential decision problem in Figure 5.6. Formulating the stochastic gradient algorithm as a sequential decision problem produces an optimization problem that looks for the best *algorithm*. Here, we have limited ourselves to (a) the use of a stochastic gradient algorithm, and (b) the use of Kesten's stepsize policy. This means that we are just optimizing over the tunable parameter θ^{kest} which implies that we are optimizing within a class of algorithms, which is quite common.

There is a substantial literature on stochastic optimization algorithms which prove asymptotic convergence, and then examine the rate of convergence empirically. The goal stated in (5.50) of finding an *optimal algorithm* is aspirational, but formalizes what people are trying to achieve in practice. In chapter 6, we are going to review a variety of stepsize policies. We could search over all of these, although this is never done.

Even if we limit ourselves to a single class of stepsize policy, it is easy to overlook that tuning θ^{kest} depends on problem parameters such as the starting point x^0. It is easy to overlook the dependence of tuned parameters on problem data. This issue has been largely overlooked by the research community.

In section 1.4, we introduced four classes of policies that will describe any search over policies. Virtually all stochastic gradient algorithms, however, use

State variables $S^n = (x^n, N^n)$.

Decision variables The stepsize α_n. This is determined by Kesten's stepsize policy in equation (5.49) which is parameterized by θ.

Exogenous information W^{n+1}, which depends on the motivating problem. This could involve observing the demand in a newsvendor problem, or it may involve running a simulator and observing $\hat{F}^{n+1} = F(x^n, W^{n+1})$.

Transition function These consist of equations for each state variable:

$$x^{n+1} = x^n + \alpha_n \nabla_x F(x^n),$$

$$N^{n+1} = \begin{cases} N^n + 1 & \text{if } (\nabla_x F(x^{n-1}))^T \nabla_x F(x^n) < 0, \\ N^n & \text{otherwise.} \end{cases}$$

Note that $\nabla_x F(x^n)$ may be approximated using numerical derivatives such as SPSA.

Objective function We wish to maximize the performance of the final solution that we call $x^{\pi,N}$, which means we wish to optimize:

$$\max_\pi \mathbb{E}_{S^0} \mathbb{E}_{W^1,\ldots,W^N|S^0} \mathbb{E}_{\widehat{W}|S^0} F(x^{\pi,N}, \widehat{W}). \tag{5.50}$$

If we limit ourselves to Kesten's rule, then we can write the objective in terms of optimizing over θ^{kest}. More generally, we might want to search over different classes of stepsize rules, each of which is likely to have its own tunable parameter.

Figure 5.6 A stochastic gradient algorithm as a sequential decision problem.

policies from the first class, policy function approximations. This raises the question whether any of the other three classes might work well. As of this writing, we are not aware of any work to explore these options.

5.7 Empirical Issues

Invariably, the process of actually implementing these algorithms raises issues that are often ignored when describing the algorithms. To help mitigate this transition, below are some of the challenges an experimentalist is likely to encounter.

Tunable parameters – Arguably one of the most frustrating aspects of any algorithm is the need to tune parameters. For gradient-based algorithms, this typically refers to the tunable parameters in the stepsize policy, but could include a smoothing factor for gradient smoothing. These tunable parameters are a direct result of the use of first-order algorithms, which are easy to compute but which exploit very little information about the underlying function. Particularly frustrating is that this tuning really matters. A poorly tuned stepsize algorithm may decrease too quickly, creating apparent convergence. It is completely possible that a poorly tuned stepsize policy can result in a conclusion that an algorithm is not working. Stepsizes that are too large can introduce too much noise.

Scaling – In most (but not all) applications, the units of the gradient $\nabla F(x, W)$ are different than the units of x. A rough rule is that the initial stepsize should be chosen so that the initial change in x is on the order of 30% to 50% of the starting value.

Benchmarking – Whenever possible, it helps to run an algorithm on a simpler problem where the optimal solution can be found using other means, either analytically or numerically. For example, it might be possible to apply the stochastic gradient algorithm on a deterministic sequence that can be solved using deterministic algorithms.

Robustness – A desirable property of any algorithm is that it work reliably, on any problem instance (that is, within a problem class). For example, tuning parameters in the stepsize policy is annoying, but bearable if it only has to be done once.

5.8 Transient Problems*

There are many applications where we are trying to solve our basic stochastic optimization problem in an online setting, where the random variable W comes from field observations. In these settings, it is not unusual to find that the underlying distribution describing W is changing over time. For example, the demands in our newsvendor application may be changing as the purchasing patterns of the market change.

We tend to design algorithms so they exhibit asymptotic convergence. For example, we would insist that the stepsize α_n decline to zero as the algorithm progresses. In a transient setting, this is problematic because it means we are putting decreasing emphasis on the latest information, which is more important than older information. Over time, as α_n approaches zero, the algorithm will stop responding to new information. If we use a stepsize such as $\alpha_n = 1/n$, it is possibly to show that the algorithm will eventually adapt to new information, but the rate of adaptation is so slow that the results are not useful.

Practitioners avoid this problem by either choosing a constant stepsize, or one that starts large but converges to a constant greater than zero. If we do this, the algorithm will start bouncing around the optimum. While this behavior may seem undesirable, in practice this is preferable, partly because the optimum of stochastic optimization problems tend to be smooth, but mostly because it means the algorithm is still adapting to new information, which makes it responsive to a changing signal.

5.9 Theoretical Performance*

In practice, finding and tuning search algorithms tends to be ad hoc. Formal analysis of search algorithms tends to fall in one of three categories:

Asymptotic convergence – Probably the most standard result for an algorithm is a proof that the solution will asymptotically approach the optimal solution (that is, as the number of iterations $N \rightarrow \infty$). The criticism of asymptotic convergence is that it says nothing about rate of convergence, which means it is not telling us anything about the quality of the solution after N iterations. See sections 5.10.2 and 5.10.3 in the appendix for samples of asymptotic convergence proofs.

Finite-time bounds – These are results that suggest that the quality of the solution after N iterations is within some limit. These bounds tend to be quite weak, and almost always feature unknown coefficients.

Asymptotic rate of convergence – It is often possible to provide high-quality estimates of the rate of convergence, but only when the solution is in the vicinity of the optimal.

The holy grail of theoretical analysis of algorithms is tight bounds for the performance after n itcrations. These are rare, and are limited to very simple problems. For this reason, empirical analysis of algorithms remains an important part of the design and analysis of search algorithms. Frustratingly, the performance of a search algorithm on one dataset may not guarantee good performance on a different dataset, even for the same problem class. We anticipate that this is typically due to a failure to properly tune the algorithm.

5.10 Why Does it Work?

Stochastic approximation methods have a rich history starting with the seminal paper Robbins and Monro (1951) and followed by Blum (1954*b*) and Dvoretzky (1956). The serious reader should see Kushner and Yin (1997) for a modern treatment of the subject. Wasan (1969) is also a useful reference for fundamental results on stochastic convergence theory. A separate line of investigation was undertaken by researchers in eastern European community focusing on constrained stochastic optimization problems (Gaivoronski (1988), Ermoliev (1988), Ruszczyński (1980), Ruszczyński (1987)). This work is critical to our fundamental understanding of Monte Carlo-based stochastic learning methods.

The theory behind these proofs is fairly deep and requires some mathematical maturity. For pedagogical reasons, we start in section 5.10.1 with some probabilistic preliminaries, after which section 5.10.2 presents one of the original proofs, which is relatively more accessible and which provides the basis for the universal requirements that stepsizes must satisfy for theoretical proofs. Section 5.10.3 provides a more modern proof based on the theory of martingales.

5.10.1 Some Probabilistic Preliminaries

The goal in this section is to prove that these algorithms work. But what does this mean? The solution $\bar{x}^n$ at iteration n is a random variable. Its value depends on the sequence of sample realizations of the random variables over iterations 1 to n. If $\omega = (W^1, W^2, \dots, W^n, \dots)$ represents the sample path that we are following, we can ask what is happening to the limit $\lim_{n \to \infty} \bar{x}^n(\omega)$. If the limit is x^*, does x^* depend on the sample path ω?

In the proofs below, we show that the algorithms converge *almost surely*. What this means is that

$$\lim_{n \to \infty} \bar{x}^n(\omega) = x^*$$

for all $\omega \in \Omega$ that can occur with positive measure. This is the same as saying that we reach x^* with probability 1. Here, x^* is a deterministic quantity that does not depend on the sample path. Because of the restriction $p(\omega) > 0$, we accept that in theory, there could exist a sample outcome that can never occur that would produce a path that converges to some other point. As a result, we say that the convergence is "almost sure," which is universally abbreviated as "*a.s.*" Almost sure convergence establishes the core theoretical property that the algorithm will eventually settle in on a single point. This is an important property for an algorithm, but it says nothing about the rate of convergence (an important issue in approximate dynamic programming).

Let $x \in \mathfrak{R}^n$. At each iteration n, we sample some random variables to compute the function (and its gradient). The sample realizations are denoted by W^n. We let $\omega = (W^1, W^2, \dots,)$ be a realization of all the random variables over all iterations. Let Ω be the set of all possible realizations of ω, and let $\mathfrak{F}$ be the σ-algebra on Ω (that is to say, the set of all possible events that can be defined using Ω). We need the concept of the history up through iteration n. Let

H^n = a random variable giving the history of all random variables up through iteration n.

A sample realization of H^n would be

$$h^n \;=\; H^n(\omega)$$
$$\;=\; (W^1, W^2, ..., W^n).$$

We could then let W^n be the set of all outcomes of the history (that is, $h^n \in H^n$) and let $\mathcal{H}^n$ be the σ-algebra on W^n (which is the set of all events, including their complements and unions, defined using the outcomes in W^n). Although we could do this, this is not the convention followed in the probability community. Instead, we define a sequence of σ-algebras $\mathfrak{F}^1, \mathfrak{F}^2, ..., \mathfrak{F}^n$ as the sequence of σ-algebras on Ω that can be generated as we have access to the information through the first $1, 2, ..., n$ iterations, respectively. What does this mean? Consider two outcomes $\omega \neq \omega'$ for which $H^n(\omega) = H^n(\omega')$. If this is the case, then any event in $\mathfrak{F}^n$ that includes ω must also include ω'. If we say that a function is $\mathfrak{F}^n$-measurable, then this means that it must be defined in terms of the events in $\mathfrak{F}^n$, which is in turn equivalent to saying that we cannot be using any information from iterations $n + 1, n + 2, ...$

We would say, then, that we have a standard probability space $(\Omega, \mathfrak{F}, \mathcal{P})$ where $\omega \in \Omega$ represents an elementary outcome, $\mathfrak{F}$ is the σ-algebra on $\mathfrak{F}$, and $\mathcal{P}$ is a probability measure on Ω. Since our information is revealed iteration by iteration, we would also then say that we have an increasing set of σ-algebras $\mathfrak{F}^1 \subseteq \mathfrak{F}^2 \subseteq ... \subseteq \mathfrak{F}^n$ (which is the same as saying that $\mathcal{F}^n$ is a filtration).

5.10.2 An Older Proof*

Enough with probabilistic preliminaries. We wish to solve the unconstrained problem

$$\max_x \mathbb{E}F(x, \omega) \tag{5.51}$$

with x^* being the optimal solution. Let $g(x, \omega)$ be a stochastic ascent vector that satisfies

$$g(x, \omega)^T \nabla F(x, \omega) \geq 0. \tag{5.52}$$

For many problems, the most natural ascent vector is the gradient itself

$$g(x, \omega) \;=\; \nabla F(x, \omega) \tag{5.53}$$

which clearly satisfies (5.52).

We assume that $F(x) = \mathbb{E}F(x, \omega)$ is continuously differentiable and concave, with bounded first and second derivatives so that for finite M

$$-M \leq g(x, \omega)^T \nabla^2 F(x)g(x, \omega) \leq M. \tag{5.54}$$

A stochastic gradient algorithm (sometimes called a stochastic approximation method) is given by

$$\bar{x}^n = \bar{x}^{n-1} + \alpha_{n-1}g(\bar{x}^{n-1}, \omega). \tag{5.55}$$

We first prove our result using the proof technique of Blum (1954*b*) that generalized the original stochastic approximation procedure proposed by Robbins and Monro (1951) to multidimensional problems. This approach does not depend on more advanced concepts such as martingales and, as a result, is accessible to a broader audience. This proof helps the reader understand the basis for the conditions $\sum_{n=0}^{\infty} \alpha_n = \infty$ and $\sum_{n=0}^{\infty} (\alpha_n)^2 < \infty$ that are required of all stochastic approximation algorithms.

We make the following (standard) assumptions on stepsizes

$$\alpha_n > 0 \text{ for all } n \geq 0, \tag{5.56}$$

$$\sum_{n=0}^{\infty} \alpha_n = \infty, \tag{5.57}$$

$$\sum_{n=0}^{\infty} (\alpha_n)^2 < \infty. \tag{5.58}$$

We want to show that under suitable assumptions, the sequence generated by (5.55) converges to an optimal solution. That is, we want to show that

$$\lim_{n \to \infty} x^n = x^* \text{ a.s.} \tag{5.59}$$

We now use Taylor's theorem (remember Taylor's theorem from freshman calculus?), which says that for any continuously differentiable convex function $F(x)$, there exists a parameter $0 \leq \eta \leq 1$ that satisfies for a given x and x^0

$$F(x) = F(x^0) + \nabla F(x^0 + \eta(x - x^0))(x - x^0). \tag{5.60}$$

This is the first-order version of Taylor's theorem. The second-order version takes the form

$$F(x) = F(x^0) + \nabla F(x^0)(x - x^0) + \frac{1}{2}(x - x^0)^T \nabla^2 F(x^0 + \eta(x - x^0))(x - x^0) \tag{5.61}$$

for some $0 \leq \eta \leq 1$. We use the second-order version. In addition, since our problem is stochastic, we will replace $F(x)$ with $F(x, \omega)$ where ω tells us what sample path we are on, which in turn tells us the value of W.

To simplify our notation, we are going to replace x^0 with x^{n-1}, x with x^n, and finally we will use

$$g^n = g(x^{n-1}, \omega). \tag{5.62}$$

This means that, by definition of our algorithm,

$$
\begin{aligned}
x - x^0 &= x^n - x^{n-1} \\
&= (x^{n-1} + \alpha_{n-1} g^n) - x^{n-1} \\
&= \alpha_{n-1} g^n.
\end{aligned}
$$

From our stochastic gradient algorithm (5.55), we may write

$$
\begin{aligned}
F(x^n, \omega) &= F(x^{n-1} + \alpha_{n-1} g^n, \omega) \\
&= F(x^{n-1}, \omega) + \nabla F(x^{n-1}, \omega)(\alpha_{n-1} g^n) \\
&\quad + \frac{1}{2}(\alpha_{n-1} g^n)^T \nabla^2 F(x^{n-1} + \eta \alpha_{n-1} g^n, \omega)(\alpha_{n-1} g^n). \tag{5.63}
\end{aligned}
$$

It is now time to use a *standard mathematician's trick*. We sum both sides of (5.63) to get

$$
\sum_{n=1}^{N} F(x^n, \omega) = \sum_{n=1}^{N} F(x^{n-1}, \omega) + \sum_{n=1}^{N} \nabla F(x^{n-1}, \omega)(\alpha_{n-1} g^n) +
$$

$$
\frac{1}{2} \sum_{n=1}^{N} (\alpha_{n-1} g^n)^T \nabla^2 F\left(x^{n-1} + \eta \alpha_{n-1} g^n, \omega\right)(\alpha_{n-1} g^n). \tag{5.64}
$$

Note that the terms $F(x^n)$, $n = 2, 3, \ldots, N$ appear on both sides of (5.64). We can cancel these. We then use our lower bound on the quadratic term (5.54) to write

$$
F(x^N, \omega) \geq F(x^0, \omega) + \sum_{n=1}^{N} \nabla F(x^{n-1}, \omega)(\alpha_{n-1} g^n) + \frac{1}{2} \sum_{n=1}^{N} (\alpha_{n-1})^2 (-M)(5.65)
$$

We now want to take the limit of both sides of (5.65) as $N \to \infty$. In doing so, we want to show that everything must be bounded. We know that $F(x^N)$ is bounded (*almost surely*) because we assumed that the original function was bounded. We next use the assumption (5.58) that the infinite sum of the squares of the stepsizes is also bounded to conclude that the rightmost term in (5.65) is bounded. Finally, we use (5.52) to claim that all the terms in the remaining summation $(\sum_{n=1}^{N} \nabla F(x^{n-1})(\alpha_{n-1} g^n))$ are positive. That means that this term is also bounded (from both above and below).

What do we get with all this boundedness? Well, if

$$\sum_{n=1}^{\infty} \alpha_{n-1} \nabla F(x^n, \omega) g^n < \infty \quad \text{for all } \omega \tag{5.66}$$

and (from (5.57))

$$\sum_{n=1}^{\infty} \alpha_{n-1} = \infty. \tag{5.67}$$

We can conclude that

$$\sum_{n=1}^{\infty} \nabla F(x^{n-1}, \omega) g^n < \infty. \tag{5.68}$$

Since all the terms in (5.68) are positive, they must go to zero. (Remember, everything here is true *almost surely*; after a while, it gets a little boring to keep saying *almost surely* every time. It is a little like reading Chinese fortune cookies and adding the automatic phrase "under the sheets" at the end of every fortune.)

We are basically done except for some relatively difficult (albeit important if you are ever going to do your own proofs) technical points to really prove convergence. At this point, we would use technical conditions on the properties of our ascent vector g^n to argue that if $\nabla F(x^n, \omega) g^n \to 0$ then $\nabla F(x^n, \omega) \to 0$, (it is okay if g^n goes to zero as $F(x^n, \omega)$ goes to zero, but it cannot go to zero too quickly).

This proof was first proposed in the early 1950s by Robbins and Monro and became the basis of a large area of investigation under the heading of stochastic approximation methods. A separate community, growing out of the Soviet literature in the 1960s, addressed these problems under the name of stochastic gradient (or stochastic quasi-gradient) methods. More modern proofs are based on the use of martingale processes, which do not start with Taylor's formula and do not (always) need the continuity conditions that this approach needs.

Our presentation does, however, help to present several key ideas that are present in most proofs of this type. First, concepts of almost sure convergence are virtually standard. Second, it is common to set up equations such as (5.63) and then take a finite sum as in (5.64) using the alternating terms in the sum to cancel all but the first and last elements of the sequence of some function (in our case, $F(x^{n-1}, \omega)$). We then establish the boundedness of this expression as $N \to \infty$, which will require the assumption that $\sum_{n=1}^{\infty} (\alpha_{n-1})^2 < \infty$. Then, the assumption $\sum_{n=1}^{\infty} \alpha_{n-1} = \infty$ is used to show that if the remaining sum is bounded, then its terms must go to zero.

More modern proofs will use functions other than $F(x)$. Popular is the introduction of so-called Lyapunov functions, which are artificial functions that provide a measure of optimality. These functions are constructed for the purpose of the proof and play no role in the algorithm itself. For example, we might let $T^n = ||x^n - x^*||$ be the distance between our current solution x^n and the optimal solution. We will then try to show that T^n is suitably reduced to prove convergence. Since we do not know x^*, this is not a function we can actually measure, but it can be a useful device for proving that the algorithm actually converges.

It is important to realize that stochastic gradient algorithms of all forms do not guarantee an improvement in the objective function from one iteration to the next. First, a sample gradient g^n may represent an appropriate ascent vector for a sample of the function $F(x^n, \omega)$ but not for its expectation. In other words, randomness means that we may go in the wrong direction at any point in time. Second, our use of a nonoptimizing stepsize, such as $\alpha_{n-1} = 1/n$, means that even with a good ascent vector, we may step too far and actually end up with a lower value.

5.10.3 A More Modern Proof**

Since the original work by Robbins and Monro, more powerful proof techniques have evolved. Below we illustrate a basic martingale proof of convergence. The concepts are somewhat more advanced, but the proof is more elegant and requires milder conditions. A significant generalization is that we no longer require that our function be differentiable (which our first proof required). For large classes of resource allocation problems, this is a significant improvement.

First, just what is a martingale? Let $\omega_1, \omega_2, \dots, \omega_t$ be a set of exogenous random outcomes, and let $h_t = H_t(\omega) = (\omega_1, \omega_2, \dots, \omega_t)$ represent the history of the process up to time t. We also let $\mathfrak{F}_t$ be the σ-algebra on Ω generated by H_t. Further, let U_t be a function that depends on h_t (we would say that U_t is a $\mathfrak{F}_t$-measurable function), and bounded ($\mathbb{E}|U_t| < \infty, \forall t \geq 0$). This means that if we know h_t, then we know U_t deterministically (needless to say, if we only know h_t, then U_{t+1} is still a random variable). We further assume that our function satisfies

$$\mathbb{E}[U_{t+1}|\mathfrak{F}_t] = U_t.$$

If this is the case, then we say that U_t is a *martingale*. Alternatively, if

$$\mathbb{E}[U_{t+1}|\mathfrak{F}_t] \leq U_t \tag{5.69}$$

then we say that U_t is a *supermartingale*. If U_t is a supermartingale, then it has the property that it drifts downward, usually to some limit point U^*. What is important is that it only drifts downward in expectation. That is, it could easily be the case that $U_{t+1} > U_t$ for specific outcomes. This captures the behavior of stochastic approximation algorithms. Properly designed, they provide solutions that improve on average, but where from one iteration to another the results can actually get worse.

Finally, assume that $U_t \geq 0$. If this is the case, we have a sequence U_t that drifts downward but which cannot go below zero. Not surprisingly, we obtain the following key result:

Theorem 5.10.1. *Let U_t be a positive supermartingale. Then, U_t converges to a finite random variable U^* almost surely.*

Note that "almost surely" (which is typically abbreviated "a.s.") means "for all (or every) ω." Mathematicians like to recognize every possibility, so they will add "every ω that might happen with some probability," which means that we are allowing for the possibility that U_t might not converge for some sample realization ω that would never actually happen (that is, where $p(\omega) > 0$). This also means that it converges with probability one.

So what does this mean for us? We assume that we are still solving a problem of the form

$$\max_x \mathbb{E}F(x, \omega), \tag{5.70}$$

where we assume that $F(x, \omega)$ is continuous and concave (but we do not require differentiability). Let $\bar{x}^n$ be our estimate of x at iteration n (remember that $\bar{x}^n$ is a random variable). Instead of watching the evolution of a process of time, we are studying the behavior of an algorithm over iterations. Let $F^n = \mathbb{E}F(\bar{x}^n)$ be our objective function at iteration n and let F^* be the optimal value of the objective function. If we are maximizing, we know that $F^n \leq F^*$. If we let $U^n = F^* - F^n$, then we know that $U^n \geq 0$ (this assumes that we can find the true expectation, rather than some approximation of it). A stochastic algorithm will not guarantee that $F^n \geq F^{n-1}$, but if we have a good algorithm, then we may be able to show that U^n is a supermartingale, which at least tells us that in the limit, U^n will approach some limit $\bar{U}$. With additional work, we might be able to show that $\bar{U} = 0$, which means that we have found the optimal solution.

A common strategy is to define U^n as the distance between $\bar{x}^n$ and the optimal solution, which is to say

$$U^n = (\bar{x}^n - x^*)^2. \tag{5.71}$$

Of course, we do not know x^*, so we cannot actually compute U^n, but that is not really a problem for us (we are just trying to prove convergence). Note that we immediately get $U^n \geq 0$ (without an expectation). If we can show that U^n is a supermartingale, then we get the result that U^n converges to a random variable U^* (which means the algorithm converges). Showing that $U^* = 0$ means that our algorithm will (eventually) produce the optimal solution. We are going to study the convergence of our algorithm for maximizing $\mathbb{E}F(x, W)$ by studying the behavior of U^n.

We are solving this problem using a stochastic gradient algorithm

$$\bar{x}^n = \bar{x}^{n-1} + \alpha_{n-1} g^n, \tag{5.72}$$

where g^n is our stochastic gradient. If F is differentiable, we would write

$$g^n = \nabla_x F(\bar{x}^{n-1}, W^n).$$

But in general, F may be nondifferentiable, in which case we may have multiple gradients at a point $\bar{x}^{n-1}$ (for a single sample realization). In this case, we write

$$g^n \in \partial_x F(\bar{x}^{n-1}, W^n),$$

where $\partial_x F(\bar{x}^{n-1}, W^n)$ refers to the set of subgradients at $\bar{x}^{n-1}$. We assume our problem is unconstrained, so $\nabla_x F(\bar{x}^*, W^n) = 0$ if F is differentiable. If it is nondifferentiable, we would assume that $0 \in \partial_x F(\bar{x}^*, W^n)$.

Throughout our presentation, we assume that x (and hence g^n) is a scalar (exercise 6.17 provides an opportunity to redo this section using vector notation). In contrast with the previous section, we are now going to allow our stepsizes to be stochastic. For this reason, we need to slightly revise our original assumptions about stepsizes (equations (5.56) to (5.58)) by assuming

$$\alpha_n > 0 \quad a.s., \tag{5.73}$$

$$\sum_{n=0}^{\infty} \alpha_n = \infty \quad a.s., \tag{5.74}$$

$$\mathbb{E}\left[\sum_{n=0}^{\infty} (\alpha_n)^2\right] < \infty. \tag{5.75}$$

The requirement that α_n be nonnegative "almost surely" (a.s.) recognizes that α_n is a random variable. We can write $\alpha_n(\omega)$ as a sample realization of the stepsize (that is, this is the stepsize at iteration n if we are following sample path ω). When we require that $\alpha_n \geq 0$ "almost surely" we mean that $\alpha_n(\omega) \geq 0$ for all ω where the probability (more precisely, probability measure) of ω, $p(\omega)$, is greater than zero (said differently, this means that the probability that

$\mathbb{P}[\alpha_n \geq 0] = 1$). The same reasoning applies to the sum of the stepsizes given in equation (5.74). As the proof unfolds, we will see the reason for needing the conditions (and why they are stated as they are).

We next need to assume some properties of the stochastic gradient g^n. Specifically, we need to assume the following:

Assumption 1 – $\mathbb{E}[g^{n+1}(\bar{x}^n - x^*)|\mathfrak{F}^n] \geq 0$,
Assumption 2 – $|g^n| \leq B_g$,
Assumption 3 – For any x where $|x - x^*| > \delta$, $\delta > 0$, there exists $\epsilon > 0$ such that $\mathbb{E}[g^{n+1}|\mathfrak{F}^n] > \epsilon$.

Assumption 1 assumes that on average, the gradient g^n points toward the optimal solution x^*. This is easy to prove for deterministic, differentiable functions. While this may be harder to establish for stochastic problems or problems where $F(x)$ is nondifferentiable, we do not have to assume that $F(x)$ is differentiable. Nor do we assume that a particular gradient g^{n+1} moves toward the optimal solution (for a particular sample realization, it is entirely possible that we are going to move away from the optimal solution). Assumption 2 assumes that the gradient is bounded. Assumption 3 requires that the expected gradient cannot vanish at a nonoptimal value of x. This assumption will be satisfied for any concave function.

To show that U^n is a supermartingale, we start with

$$U^{n+1} - U^n = (\bar{x}^{n+1} - x^*)^2 - (\bar{x}^n - x^*)^2$$
$$= \left((\bar{x}^n - \alpha_n g^{n+1}) - x^*\right)^2 - (\bar{x}^n - x^*)^2$$
$$= \left((\bar{x}^n - x^*)^2 - 2\alpha_n g^{n+1}(\bar{x}^n - x^*) + (\alpha_n g^{n+1})^2\right) - (\bar{x}^n - x^*)^2$$
$$= (\alpha_n g^{n+1})^2 - 2\alpha_n g^{n+1}(\bar{x}^n - x^*). \tag{5.76}$$

Taking conditional expectations on both sides gives

$$\mathbb{E}[U^{n+1}|\mathfrak{F}^n] - \mathbb{E}[U^n|\mathfrak{F}^n] = \mathbb{E}[(\alpha_n g^{n+1})^2|\mathfrak{F}^n] - 2\mathbb{E}[\alpha_n g^{n+1}(\bar{x}^n - x^*)|\mathfrak{F}^n]. \tag{5.77}$$

We note that

$$\mathbb{E}[\alpha_n g^{n+1}(\bar{x}^n - x^*)|\mathfrak{F}^n] = \alpha_n \mathbb{E}[g^{n+1}(\bar{x}^n - x^*)|\mathfrak{F}^n] \tag{5.78}$$
$$\geq 0. \tag{5.79}$$

Equation (5.78) is subtle but important, as it explains a critical piece of notation in this book. Keep in mind that we may be using a stochastic stepsize formula, which means that α_n is a random variable. We assume that α_n is $\mathfrak{F}^n$-measurable, which means that we are not allowed to use information from iteration $n + 1$ to compute it. This is why we use α_{n-1} in updating equations such as equation (5.13) instead of α_n. When we condition on $\mathfrak{F}^n$ in equation (5.78), α_n is deterministic, allowing us to take it outside the expectation. This allows us to write the conditional expectation of the product of α_n and g^{n+1} as the product of the expectations. Equation (5.79) comes from Assumption 1 and the nonnegativity of the stepsizes.

Recognizing that $\mathbb{E}[U^n|\mathfrak{F}^n] = U^n$ (given $\mathfrak{F}^n$), we may rewrite (5.77) as

$$\mathbb{E}[U^{n+1}|\mathfrak{F}^n] = U^n + \mathbb{E}[(\alpha_n g^{n+1})^2|\mathfrak{F}^n] - 2\mathbb{E}[\alpha_n g^{n+1}(\bar{x}^n - x^*)|\mathfrak{F}^n]$$

$$\leq U^n + \mathbb{E}[(\alpha_n g^{n+1})^2|\mathfrak{F}^n]. \tag{5.80}$$

Because of the positive term on the right-hand side of (5.80), we cannot directly get the result that U^n is a supermartingale. But hope is not lost. We appeal to a neat little trick that works as follows. Let

$$W^n = \mathbb{E}[U^n + \sum_{m=n}^{\infty}(\alpha_m g^{m+1})^2|\mathfrak{F}^n]. \tag{5.81}$$

We are going to show that W^n is a supermartingale. From its definition, we obtain

$$W^n = \mathbb{E}[W^{n+1} + U^n - U^{n+1} + (\alpha_n g^{n+1})^2|\mathfrak{F}^n],$$

$$= \mathbb{E}[W^{n+1}|\mathfrak{F}^n] + U^n - \mathbb{E}[U^{n+1}|\mathfrak{F}^n] + \mathbb{E}[(\alpha_n g^{n+1})^2|\mathfrak{F}^n]$$

which is the same as

$$\mathbb{E}[W^{n+1}|\mathfrak{F}^n] = W^n - \underbrace{(U^n + \mathbb{E}[(\alpha_n g^{n+1})^2|\mathfrak{F}^n] - \mathbb{E}[U^{n+1}|\mathfrak{F}^n])}_{I}.$$

We see from equation (5.80) that $I \geq 0$. Removing this term gives us the inequality

$$\mathbb{E}[W^{n+1}|\mathfrak{F}^n] \leq W^n. \tag{5.82}$$

This means that W^n is a supermartingale. It turns out that this is all we really need because $\lim_{n\to\infty} W^n = \lim_{n\to\infty} U^n$. This means that

$$\lim_{n\to\infty} U^n \to U^* \quad a.s. \tag{5.83}$$

Now that we have the basic convergence of our algorithm, we have to ask: but what is it converging to? For this result, we return to equation (5.76) and sum it over the values $n = 0$ up to some number N, giving us

$$\sum_{n=0}^{N}(U^{n+1} - U^n) = \sum_{n=0}^{N}(\alpha_n g^{n+1})^2 - 2\sum_{n=0}^{N} \alpha_n g^{n+1}(\bar{x}^n - x^*). \qquad (5.84)$$

The left-hand side of (5.84) is an alternating sum (sometimes referred to as a telescoping sum), which means that every element cancels out except the first and the last, giving us

$$U^{N+1} - U^0 = \sum_{n=0}^{N}(\alpha_n g^{n+1})^2 - 2\sum_{n=0}^{N} \alpha_n g^{n+1}(\bar{x}^n - x^*).$$

Taking expectations of both sides gives

$$\mathbb{E}[U^{N+1} - U^0] = \mathbb{E}\left[\sum_{n=0}^{N}(\alpha_n g^{n+1})^2\right] - 2\mathbb{E}\left[\sum_{n=0}^{N} \alpha_n g^{n+1}(\bar{x}^n - x^*)\right]. \qquad (5.85)$$

We want to take the limit of both sides as N goes to infinity. To do this, we have to appeal to the *Dominated Convergence Theorem* (DCT), which tells us that

$$\lim_{N\to\infty} \int_x f^n(x)dx = \int_x \left(\lim_{N\to\infty} f^n(x)\right) dx$$

if $|f^n(x)| \leq g(x)$ for some function $g(x)$ where

$$\int_x g(x)dx < \infty.$$

For our application, the integral represents the expectation (we would use a summation instead of the integral if x were discrete), which means that the DCT gives us the conditions needed to exchange the limit and the expectation. Above, we showed that $\mathbb{E}[U^{n+1}|\mathfrak{F}^n]$ is bounded (from (5.80) and the boundedness of U^0 and the gradient). This means that the right-hand side of (5.85) is also bounded for all n. The DCT then allows us to take the limit as N goes to infinity inside the expectations, giving us

$$U^* - U^0 = \mathbb{E}\left[\sum_{n=0}^{\infty}(\alpha_n g^{n+1})^2\right] - 2\mathbb{E}\left[\sum_{n=0}^{\infty} \alpha_n g^{n+1}(\bar{x}^n - x^*)\right].$$

We can rewrite the first term on the right-hand side as

$$\mathbb{E}\left[\sum_{n=0}^{\infty}(\alpha_n g^{n+1})^2\right] \leq \mathbb{E}\left[\sum_{n=0}^{\infty}(\alpha_n)^2(B)^2\right] \tag{5.86}$$

$$= B^2 \mathbb{E}\left[\sum_{n=0}^{\infty}(\alpha_n)^2\right] \tag{5.87}$$

$$< \infty. \tag{5.88}$$

Equation (5.86) comes from Assumption 2 which requires that $|g^n|$ be bounded by B, which immediately gives us Equation (5.87). The requirement that $\mathbb{E}\sum_{n=0}^{\infty}(\alpha_n)^2 < \infty$ (equation (5.58)) gives us (5.88), which means that the first summation on the right-hand side of (5.85) is bounded. Since the left-hand side of (5.85) is bounded, we can conclude that the second term on the right-hand side of (5.85) is also bounded.

Now let

$$\beta^n = \mathbb{E}\left[g^{n+1}(\bar{x}^n - x^*)\right]$$

$$= \mathbb{E}\left[\mathbb{E}\left[g^{n+1}(\bar{x}^n - x^*)|\mathfrak{F}^n\right]\right]$$

$$\geq 0,$$

since $\mathbb{E}[g^{n+1}(\bar{x}^n - x^*)|\mathfrak{F}^n] \geq 0$ from Assumption 1. This means that

$$\sum_{n=0}^{\infty}\alpha_n\beta^n < \infty \text{ a.s.} \tag{5.89}$$

But, we have required that $\sum_{n=0}^{\infty}\alpha_n = \infty$ a.s. (equation (5.74)). Since $\alpha_n > 0$ and $\beta^n \geq 0$ (a.s.), we conclude that

$$\lim_{n\to\infty}\beta^n \to 0 \text{ a.s.} \tag{5.90}$$

If $\beta^n \to 0$, then $\mathbb{E}[g^{n+1}(\bar{x}^n - x^*)] \to 0$, which allows us to conclude that $\mathbb{E}[g^{n+1}(\bar{x}^n - x^*)|\mathfrak{F}^n] \to 0$ (the expectation of a nonnegative random variable cannot be zero unless the random variable is always zero). But what does this tell us about the behavior of $\bar{x}^n$? Knowing that $\beta^n \to 0$ does not necessarily imply that $g^{n+1} \to 0$ or $\bar{x}^n \to x^*$. There are three scenarios:

1) $\bar{x}^n \to x^*$ for all n, and of course all sample paths ω. If this were the case, we are done.
2) $\bar{x}^{n_k} \to x^*$ for a subsequence $n_1, n_2, \ldots, n_k, \ldots$. For example, it might be that the sequence $\bar{x}^1, \bar{x}^3, \bar{x}^5, \ldots \to x^*$, while $\mathbb{E}[g^2|\mathfrak{F}^1], \mathbb{E}[g^4|\mathfrak{F}^3], \ldots, \to 0$. This would mean that for the subsequence n_k, $U^{n_k} \to 0$. But we already know that $U^n \to$

U^* where U^* is the unique limit point, which means that $U^* = 0$. But if this is the case, then this is the limit point for every sequence of $\bar{x}^n$.

3) There is no subsequence $\bar{x}^{n_k}$ which has $\bar{x}^*$ as its limit point. This means that $\mathbb{E}[g^{n+1}|\mathfrak{F}^n] \to 0$. However, assumption 3 tells us that the expected gradient cannot vanish at a nonoptimal value of x. This means that this case cannot happen.

This completes the proof. □

5.11 Bibliographic Notes

Section 5.3 – The theoretical foundation for estimating value functions from Monte Carlo estimates has its roots in stochastic approximation theory, originated by Robbins and Monro (1951), with important early contributions made by Kiefer and Wolfowitz (1952), Blum (1954*a*) and Dvoretzky (1956). For thorough theoretical treatments of stochastic approximation theory, see Wasan (1969), Kushner and Clark (1978), and Kushner and Yin (1997). Very readable treatments of stochastic optimization can be found in Pflug (1996) and Spall (2003) (Spall's book is a modern classic on stochastic approximation methods). More modern treatments of stochastic gradient methods are given in Fu (2014) and Shapiro et al. (2014).

Section 5.4 – There are a number of ways to compute gradients, including numerical derivatives (when exact gradients are not available), gradient smoothing, mini-batches (averages of sampled gradients). Excellent modern treatments can be found in Michael Fu's edited volume Fu (2014), including Fu's chapter on stochastic gradient estimation [Chapter 5], and Chau and Fu's chapter on stochastic approximation methods and finite-difference methods [Chapter 6].

Section 5.4.4 – The simultaneous perturbation stochastic approximation (SPSA) method, which provides a practical strategy for estimating numerical gradients for higher-dimensional problems, is due to Spall (see Spall (2003)). Figure 5.3 was prepared by Saeed Ghadimi.

Section 5.6 – The formulation of a stochastic gradient algorithm as a sequential decision problem was first described in Powell (2019). However, mention should be made of the work of Harold Kushner (see Kushner and Yin (2003) for a summary) which viewed algorithms as dynamical systems. Our work viewing algorithms as *controlled* dynamical systems appears to be new, although this is hard to verify.

Section 5.10.2 – This proof is based on Blum (1954*b*), which generalized the original paper by Robbins and Monro (1951).

Section 5.10.3 – The proof in section 5.10.3 uses standard techniques drawn from several sources, notably Wasan (1969), Chong (1991), Kushner and Yin (1997), and, for this author, Powell and Cheung (2000).

Exercises

Review questions

5.1 Write out a basic stochastic gradient algorithm for iteration n, and explain why W^{n+1} is indexed by $n + 1$ instead of n. Write out the stochastic gradient for the newsvendor problem.

5.2 There are potentially three forms of uncertainty that arise in the use of stochastic gradient algorithms. Give the notation for each and explain with an example (you may use an example from the chapter).

5.3 A gradient for a continuous, deterministic function points in the direction of steepest ascent. Is this true for stochastic gradients? Illustrate this for the problem of estimating the mean of a random variable.

5.4 Consider the newsvendor problem

$$F(x, W) = 10 \max\{x, W\} - 8x.$$

For $x = 9$ and $W = 10$ compute the numerical derivative $\nabla_x F(x, W)$ using the increment $\delta = 1$. What if you use $\delta = 4$?

Modeling questions

5.5 Consider a function $F(x, W)$ that depends on a decision $x = x^n$ after which we observe a random outcome W^{n+1}. Assume that we can compute the gradient $\nabla_x F(x^n, W^{n+1})$. We would like to optimize this problem using a standard stochastic gradient algorithm:

$$x^{n+1} = x^n + \alpha_n \nabla_x F(x^n, W^{n+1}).$$

Our goal is to find the best answer we can after N iterations.

a) Assume that we are using a stepsize policy of

$$\alpha_n = \frac{\theta}{\theta + n - 1}.$$

Model the problem of finding the best stepsize policy as a stochastic optimization problem. Give the state variable(s), the decision variable, the exogenous information, the transition function, and the objective function. Please use precise notation.

b) How does your model change if you switch to Kesten's stepsize rule which uses

$$\alpha_n = \frac{\theta}{\theta + N^n - 1},$$

where N^n is the number of times that the gradient has changed signs, which is computed using

$$N^{n+1} = \begin{cases} N^n + 1 & \text{if } \nabla_x F(x^{n-1}, W^n) \nabla_x F(x^n, W^{n+1}) < 0 \\ N^n & \text{otherwise.} \end{cases}$$

5.6 A customer is required by her phone company to pay for a minimum number of minutes per month for her cell phone. She pays 12 cents per minute of guaranteed minutes, and 30 cents per minute that she goes over her minimum. Let x be the number of minutes she commits to each month, and let M be the random variable representing the number of minutes she uses each month, where M is normally distributed with mean 300 minutes and a standard deviation of 60 minutes.

(a) Write down the objective function in the form $\min_x \mathbb{E} f(x, M)$.

(b) Derive the stochastic gradient for this function.

(c) Let $x^0 = 0$ and choose as a stepsize $\alpha_{n-1} = 10/n$. Use 100 iterations to determine the optimum number of minutes the customer should commit to each month.

5.7 An oil company covers the annual demand for oil using a combination of futures and oil purchased on the spot market. Orders are placed at the end of year $t - 1$ for futures that can be exercised to cover demands in year t. If too little oil is purchased this way, the company can cover the remaining demand using the spot market. If too much oil is purchased with futures, then the excess is sold at 70% of the spot market price (it is not held to the following year – oil is too valuable and too expensive to store).

To write down the problem, model the exogenous information using

$$\hat{D}_t = \text{demand for oil during year } t,$$

$$\hat{p}_t^s = \text{spot price paid for oil purchased in year } t,$$

$$\hat{p}_{t,t+1}^f = \text{futures price paid in year } t \text{ for oil to be used in year } t + 1.$$

The demand (in millions of barrels) is normally distributed with mean 600 and standard deviation of 50. The decision variables are given by

$$\bar{\theta}^f_{t,t+1} = \text{number of futures to be purchased at the end of year } t \text{ to be used in year } t + 1,$$

$$\bar{\theta}^s_t = \text{spot purchases made in year } t.$$

(a) Set up the objective function to minimize the expected total amount paid for oil to cover demand in a year $t + 1$ as a function of $\bar{\theta}^f_t$. List the variables in your expression that are not known when you have to make a decision at time t.

(b) Give an expression for the stochastic gradient of your objective function. That is, what is the derivative of your function for a particular sample realization of demands and prices (in year $t + 1$)?

(c) Generate 100 years of random spot and futures prices as follows:

$$\hat{p}^f_t = 0.80 + 0.10 U^f_t,$$

$$\hat{p}^s_{t,t+1} = \hat{p}^f_t + 0.20 + 0.10 U^s_t,$$

where U^f_t and U^s_t are random variables uniformly distributed between 0 and 1. Run 100 iterations of a stochastic gradient algorithm to determine the number of futures to be purchased at the end of each year. Use $\bar{\theta}^f_0 = 30$ as your initial order quantity, and use as your stepsize $\alpha_t = 20/t$. Compare your solution after 100 years to your solution after 10 years. Do you think you have a good solution after 10 years of iterating?

Computational exercises

5.8 We want to compute a numerical derivative of the newsvendor problem

$$F(x, W) = 10 \min\{x, W\} - 8x.$$

Assume that we have generated a random sample of $W = 12$, and that we want to generate a numerical derivative to estimate the gradient $\nabla_x F(x, W)$ for $x = 8$ and $W = 12$.

a) Compute a right-biased numerical derivative using $\delta = 1.0$. Show how to perform the computation and given the resulting estimate.

b) Compute a balanced numerical derivative centered on $x = 8$, but using estimates perturbed by $+\delta$ and $-\delta$.

c) Write the software using any environment to optimize $F(x, W)$ using numerical derivatives, assuming $W \in Uniform[5, 20]$. Carefully specify any assumptions you make. Run your algorithm for 20 iterations.

5.9 Below is a form of two-dimensional newsvendor problem, where we allocate two types of resource, x_1 and x_2, to meet a common demand W:

$$F(x_1, x_2, W) = 10 \min\{x_1, W\} + 14 \min\{x_2, (\max\{0, W - x_1\})\} - 8x_1 - 10x_2.$$

We are going to pretend that our vector x might have a dozen or more dimensions, but use this two-dimensional version to perform a detailed numerical example of the SPSA method for estimating gradients.

a) Use the SPSA algorithm to compute an estimate of the gradient $\nabla_x F(x_1, x_2, W)$ using two-function evaluations around the point $x_1 = 8$, $x_2 = 10$. Show all the detailed calculations and the resulting gradient. Show how you handled the sampling of W.

b) Write the software using any environment to optimize $F(x_1, x_2, W)$ using the SPSA algorithm, assuming $W \in Uniform[5, 20]$. Carefully specify any assumptions you make. Run your algorithm for 20 iterations.

Theory questions

5.10 The proof in section 5.10.3 was performed assuming that x is a scalar. Repeat the proof assuming that x is a vector. You will need to make adjustments such as replacing Assumption 2 with $\|g^n\| < B$. You will also need to use the triangle inequality which states that $\|a + b\| \leq \|a\| + \|b\|$.

Problem solving questions

5.11 Write out the stochastic gradient for the nested newsvendor problem given in equation (5.9).

5.12 In a flexible spending account (FSA), a family is allowed to allocate x pretax dollars to an escrow account maintained by the employer. These funds can be used for medical expenses in the following year. Funds remaining in the account at the end of the following year revert back to the employer. Assume that you are in a 40% tax bracket (sounds nice, and the arithmetic is a bit easier). Let M be the random variable representing total medical expenses in the upcoming year, and let $F(x) = Prob[M \leq x]$ be the cumulative distribution function of the random variable M.

a) Write out the objective function that we would want to solve to find x to minimize the total cost (in pretax dollars) of covering your medical expenses next year.

b) If x^* is the optimal solution and $g(x)$ is the gradient of your objective function if you allocate x to the FSA, use the property that $g(x^*) = 0$ to derive (you must show the derivation) the critical ratio that gives the relationship between x^* and the cumulative distribution function $F(x)$.

c) If you are in a 35% tax bracket, what percentage of the time should you have funds left over at the end of the year?

5.13 We are going to solve a classic stochastic optimization problem using the newsvendor problem. Assume we have to order x assets after which we try to satisfy a random demand D for these assets, where D is randomly distributed between 100 and 200. If $x > D$, we have ordered too much and we pay $5(x - D)$. If $x < D$, we have an underage, and we have to pay $20(D - x)$.

(a) Write down the objective function in the form $\min_x \mathbb{E}f(x, D)$.

(b) Derive the stochastic gradient for this function.

(c) Find the optimal solution analytically [Hint: take the expectation of the stochastic gradient, set it equal to zero and solve for the quantity $\mathbb{P}(D \leq x^*)$. From this, find x^*.]

(d) Since the gradient is in units of dollars while x is in units of the quantity of the asset being ordered, we encounter a scaling problem. Choose as a stepsize $\alpha_{n-1} = \alpha_0/n$ where α_0 is a parameter that has to be chosen. Use $x^0 = 100$ as an initial solution. Plot x^n for 1000 iterations for $\alpha_0 = 1, 5, 10, 20$. Which value of α_0 seems to produce the best behavior?

(e) Repeat the algorithm (1000 iterations) 10 times. Let $\omega = (1, \dots, 10)$ represent the 10 sample paths for the algorithm, and let $x^n(\omega)$ be the solution at iteration n for sample path ω. Let $Var(x^n)$ be the variance of the random variable x^n where

$$\overline{V}(x^n) = \frac{1}{10} \sum_{\omega=1}^{10} (x^n(\omega) - x^*)^2.$$

Plot the standard deviation as a function of n for $1 \leq n \leq 1000$.

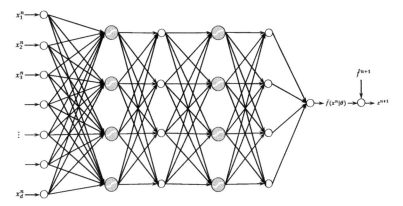

Figure 5.7 A four-layer neural network for exercise 5.14.

5.14 Following the methods of section 5.5.1, compute the gradient $\nabla_\theta F(x^n, \hat{f}^{n+1}|\theta)$ for the network depicted in Figure 5.7.

Sequential decision analytics and modeling

These exercises are drawn from the online book *Sequential Decision Analytics and Modeling* available at `http://tinyurl.com/sdaexamplesprint`.

5.15 Read chapter 3, sections 3.1-3.4 on the adaptive market planning problem. The presentation provides a classical derivation of the optimal order quantity for a newsvendor problem, but then presents a version of the problem where the objective is the cumulative reward.

a) When we are optimizing a single-period newsvendor problem, we are looking for the best x. What are we searching for when we are optimizing a multiperiod newsvendor problem where we are maximizing cumulative rewards (as is done in the book)?
b) Compute the gradient of the cumulative reward objective function with respect to the stepsize parameter θ^{step} in the stepsize rule.
c) Describe a stochastic gradient algorithm for optimizing θ^{step} when using a cumulative reward objective.

Diary problem

The diary problem is a single problem you chose (see chapter 1 for guidelines). Answer the following for your diary problem.

5.16 Your diary problem may have a decision x which is continuous (a quantity, a temperature, a price, a dosage). If not, you may wish to use a parameterized procedure $X^\pi(S_t|\theta)$ for determining x_t when you are in state S_t, where you are going to face the need to tune θ. At this point in the book, you have not developed the background to model and solve your problem, but discuss any continuous decisions (or tunable parameters) that you think might arise when modeling and solving your problem. Then, describe any new information that you might learn after you fix your continuous variable (this would play the role of W in the canonical problem for this chapter).

Bibliography

Blum, J. (1954a). Multidimensional stochastic approximation methods. *Annals of Mathematical Statistics* 25: 737–74462.

Blum, J.R. (1954b). Approximation methods which converge with probability one. *Annals of Mathematical Statistics* 25: 382–386.

Chong, E.K.P. (1991). On-Line Stochastic Optimization of Queueing Systems.

Dvoretzky, A. (1956). On stochastic approximation. In: *Proceedings 3rd Berkeley Symposium on Mathematical Statistics and Probability* (ed. J. Neyman), 39–55. University of California Press.

Ermoliev, Y. (1988). Stochastic quasigradient methods. In: *Numerical Techniques for Stochastic Optimization* (eds. Y. Ermoliev and R. Wets). Berlin: SpringerVerlag.

Fu, M.C. (2014). *Handbook of Simulation Optimization*. New York: Springer.

Gaivoronski, A. (1988). Stochastic quasigradient methods and their implementation. In: *Numerical Techniques for Stochastic Optimization* (eds. Y. Ermoliev and R. Wets). Berlin: SpringerVerlag.

Kiefer, J. and Wolfowitz, J. (1952). Stochastic estimation of the maximum of a regression function *Annals of Mathematical Statistics* 23: 462–466.

Kushner, H.J. and Clark, S. (1978). *Stochastic Approximation Methods for Constrained and Unconstrained Systems*. New York: SpringerVerlag.

Kushner, H.J. and Yin, G.G. (1997). *Stochastic Approximation Algorithms and Applications*. New York: SpringerVerlag.

Kushner, H.J. and Yin, G.G. (2003). *Stochastic Approximation and Recursive Algorithms and Applications*. New York: Springer.

Pflug, G. (1996). *Optimization of Stochastic Models: The Interface Between Simulation and Optimization, Kluwer International Series in Engineering and Computer Science: Discrete Event Dynamic Systems*. Boston: Kluwer Academic Publishers.

Powell, W.B. (2019). A unified framework for stochastic optimization. *European Journal of Operational Research* 275 (3): 795–821.

Powell, W.B. and Cheung, R.K.M. (2000). SHAPE: A Stochastic Hybrid Approximation Procedure for TwoStage Stochastic Programs. *Operations Research* 48: 73–79.

Robbins, H. and Monro, S. (1951). A stochastic approximation method. *The Annals of Mathematical Statistics* 22 (3): 400–407.

Ruszczyński, A. (1980). Feasible direction methods for stochastic programming problems. *Mathematical Programming* 19: 220–229.

Ruszczyński, A. (1987). A linearization method for nonsmooth stochastic programming problems. *Mathematics of Operations Research* 12: 32–49.

Shapiro, A., Dentcheva, D., and Ruszczyński, A. (2014), *Lectures on Stochastic Programming: Modeling and Theory*, 2e. Philadelphia: SIAM.

Spall, J.C. (2003). *Introduction to Stochastic Search and Optimization: Estimation, simulation and control*. Hoboken, NJ: John Wiley & Sons.

Wasan, M.T. (1969). *Stochastic approximation*. Cambridge: Cambridge University Press.

6

Stepsize Policies

There is a wide range of adaptive learning problems that depend on an iteration of the form we first saw in chapter 5 that looks like

$$x^{n+1} \;=\; x^n + \alpha_n \nabla_x F(x^n, W^{n+1}). \tag{6.1}$$

The stochastic gradient $\nabla_x F(x^n, W^{n+1})$ tells us what direction to go in, but we need the stepsize α_n to tell us how far we should move.

There are two important settings where this formula is used. The first is where we are maximizing some metric such as contributions, utility, or performance. In these settings, the units of $\nabla_x F(x^n, W^{n+1})$ and the decision variable x are different, so the stepsize has to perform the scaling so that the size of $\alpha_n \nabla_x F(x^n, W^{n+1})$ is not too large or too small relative to x^n.

A second and very important setting arises in what is known as *supervised learning*. In this context, we are trying to estimate some function $f(x|\theta)$ using observations $y = f(x|\theta) + \varepsilon$. In this context, $f(x|\theta)$ and y have the same scale. We encounter these problems in three settings:

- Approximating the function $\mathbb{E}F(x, W)$ to create an estimate $\bar{F}(x)$ that can be optimized.
- Approximating the value $V_t(S_t)$ of being in a state S_t and then following some policy (we encounter this problem starting in chapters 16 and 17 when we introduce approximate dynamic programming).
- Creating a parameterized policy $X^\pi(S|\theta)$ to fit observed decisions. Here, we assume we have access to some method of creating a decision x and then we use this to create a parameterized policy $X^\pi(S|\theta)$. One source of decisions x is watching human behavior (for example, the choices made by a physician), but we could use any of our four classes of policies.

Reinforcement Learning and Stochastic Optimization: A Unified Framework for Sequential Decisions, First Edition. Warren B. Powell.
© 2022 John Wiley & Sons, Inc. Published 2022 John Wiley & Sons, Inc.

In chapter 3, we saw a range of methods for approximating functions. Imagine that we face the simplest problem of estimating the mean of a random variable W, which we can show (see exercise 6.21) solves the following stochastic optimization problem

$$\min_{x} \mathbb{E}\frac{1}{2}(x - W)^2. \tag{6.2}$$

Let $F(x, W) = \frac{1}{2}(x - W)^2$. The stochastic gradient of $F(x, W)$ with respect to x is

$$\nabla_x F(x, W) = (x - W).$$

We can optimize (6.2) using a stochastic gradient algorithm which we would write (remember that we are minimizing):

$$
\begin{aligned}
x^{n+1} &= x^n - \alpha_n \nabla F(x^n, W^{n+1}) & (6.3) \\
&= x^n - \alpha_n(x^n - W^{n+1}) & (6.4) \\
&= (1 - \alpha_n)x^n + \alpha_n W^{n+1}. & (6.5)
\end{aligned}
$$

Equation (6.5) will be familiar to many readers as exponential smoothing (also known as a *linear filter* in signal processing). The important observation is that in this setting, the stepsize α_n needs to be between 0 and 1 since x and W are the same scale.

One of the challenges in Monte Carlo methods is finding the stepsize α_n. We refer to a method for choosing a stepsize as a *stepsize policy*, although popular terms include stepsize rule or learning rate schedules. To illustrate, we begin by rewriting the optimization problem (6.2) in terms of finding the estimate $\bar{\mu}$ of μ which is the true mean of the random variable W which we write as

$$\min_{\bar{\mu}} \mathbb{E}\frac{1}{2}(\bar{\mu} - W)^2. \tag{6.6}$$

This switch in notation will allow us to later make decisions about how to estimate $\mu_x = \mathbb{E}_W F(x, W)$ where we observe $\hat{F} = F(x, W)$. For now, we just want to focus on a simple estimation problem.

Our stochastic gradient updating equation (6.4) becomes

$$\bar{\mu}^{n+1} = \bar{\mu}^n - \alpha_n(\bar{\mu}^n - W^{n+1}). \tag{6.7}$$

With a properly designed stepsize rule (such as $\alpha_n = 1/n$), we can guarantee that

$$\lim_{n \to \infty} \bar{\mu}^n \to \mu,$$

but our interest is doing the best we can within a budget of N iterations which means we are trying to solve

$$\max_{\pi} \mathbb{E}_{S^0} \mathbb{E}_{W^1,...,W^N|S^0} \mathbb{E}_{\widehat{W}|S^0} F(x^{\pi,N}, \widehat{W}), \tag{6.8}$$

where π refers to our stepsize rule, covering both the type of rule and any tunable parameters. We note that in this chapter, we do not care if we are solving the final-reward objective (6.8), or the cumulative-reward objective given by

$$\max_{\pi} \mathbb{E}_{S^0} \mathbb{E}_{W^1,...,W^N|S^0} \sum_{n=0}^{N} F(x^n, W^{n+1}), \tag{6.9}$$

where $x^n = X^{\pi}(S^n)$. Our goal is to search for the best stepsize formula (and the best within a class) regardless of the objective.

There are two issues when designing a good stepsize rule. The first is the question of whether the stepsize produces some theoretical guarantee, such as asymptotic convergence or a finite time bound. While this is primarily of theoretical interest, these conditions do provide important guidelines to follow to produce good behavior. The second issue is whether the rule produces good empirical performance.

We divide our presentation of stepsize rules into three classes:

Deterministic policies – These are stepsize policies that are deterministic functions of the iteration counter n. This means that we know before we even start running our algorithm what the stepsize α_n will be.

Adaptive policies – These are policies where the stepsize at iteration n depends on the statistics computed from the trajectory of the algorithm. These are also known as *stochastic stepsize rules.*

Optimal policies – Our deterministic and adaptive stepsize policies may have provable guarantees of asymptotic convergence, but were not derived using a formal optimization model. A byproduct of this heritage is that they require tuning one or more parameters. Optimal policies are derived from a formal model which is typically a simplified problem. These policies tend to be more complex, but eliminate or at least minimize the need for parameter tuning.

The deterministic and stochastic rules presented in section 6.1 and section 6.2 are, for the most part, designed to achieve good rates of convergence, but are not supported by any theory that they will produce the best rate of convergence. Some of these stepsize rules are, however, supported by asymptotic proofs of convergence and/or regret bounds.

In section 6.3 we provide a theory for choosing stepsizes that produce the fastest possible rate of convergence when estimating value functions based on policy evaluation. Finally, section 6.4 presents an optimal stepsize rule designed specifically for approximate value iteration.

6.1 Deterministic Stepsize Policies

Deterministic stepsize policies are the simplest to implement. Properly tuned, they can provide very good results. We begin by presenting some basic properties that a stepsize rule has to satisfy to ensure asymptotic convergence. While we are going to be exclusively interested in performance in finite time, these rules provide guidelines that are useful regardless of the experimental budget. After this, we present a variety of recipes for deterministic stepsize policies.

6.1.1 Properties for Convergence

The theory for proving convergence of stochastic gradient algorithms was first developed in the early 1950s and has matured considerably since then (see section 5.10). However, all the proofs require three basic conditions:

$$\alpha_n > 0, \quad n = 0, 1, ..., \tag{6.10}$$

$$\sum_{n=0}^{\infty} \alpha_n = \infty, \tag{6.11}$$

$$\sum_{n=0}^{\infty} (\alpha_n)^2 < \infty. \tag{6.12}$$

Equation (6.10) requires that the stepsizes be strictly positive (we cannot allow stepsizes equal to zero). The most important requirement is (6.11), which states that the infinite sum of stepsizes must be infinite. If this condition did not hold, the algorithm might stall prematurely. Finally, condition (6.12) requires that the infinite sum of the squares of the stepsizes be finite. This condition, in effect, requires that the stepsize sequence converge "reasonably quickly."

An intuitive justification for condition (6.12) is that it guarantees that the *variance* of our estimate of the optimal solution goes to zero in the limit. Sections 5.10.2 and 5.10.3 illustrate two proof techniques that both lead to these requirements on the stepsize. However, it is possible under certain conditions to replace equation (6.12) with the weaker requirement that $\lim_{n \to \infty} \alpha_n = 0$.

Condition (6.11) effectively requires that the stepsizes decline according to an arithmetic sequence such as

$$\alpha_{n-1} = \frac{1}{n}. \tag{6.13}$$

This rule has an interesting property. Exercise 6.21 asks you to show that a step-size of $1/n$ produces an estimate $\bar{\mu}^n$ that is simply an average of all previous observations, which is to say

$$\bar{\mu}^n \quad = \quad \frac{1}{n} \sum_{m=1}^{n} W^m. \tag{6.14}$$

Of course, we have a nice name for equation (6.14): it is called a sample average. And we are all aware that in general (some modest technical conditions are required) as $n \to \infty$, $\bar{\mu}^n$ will converge (in some sense) to the mean of our random variable W.

The issue of the rate at which the stepsizes decrease is of considerable practical importance. Consider, for example, the stepsize sequence

$$\alpha_n \quad = \quad .5\alpha_{n-1},$$

which is a geometrically decreasing progression. This stepsize formula violates condition (6.11). More intuitively, the problem is that the stepsizes would decrease so quickly that the algorithm would stall prematurely. Even if the gradient pointed in the right direction at each iteration, we likely would never reach the optimum.

There are settings where the "$1/n$" stepsize formula is the best that we can do (as in finding the mean of a random variable), while in other situations it can perform extremely poorly because it can decline to zero too quickly. One situation where it works poorly arises when we are estimating a function that is changing over time (or iterations). For example, the algorithmic strategy called Q-learning (which we first saw in section 2.1.6) involves two steps:

$$\hat{q}^n(s^n, a^n) \quad = \quad r(s^n, a^n) + \gamma \max_{a'} \bar{Q}^{n-1}(s', a'),$$
$$\bar{Q}^n(s^n, a^n) \quad = \quad (1 - \alpha_{n-1})\bar{Q}^{n-1}(s^n, a^n) + \alpha_{n-1}\hat{q}^n(s^n, a^n).$$

Here, we create a sampled observation $\hat{q}^n(s^n, a^n)$ of being in a state s^n and taking an action a^n, which we compute using the one period reward $r(s^n, a^n)$ plus an estimate of the downstream value, computed by sampling a downstream state s' given the current state s^n and action a^n, and then choosing the best action a' based on our current estimate of the value of different state-action pairs $\bar{Q}^{n-1}(s', a')$. We then smooth $\hat{q}^n(s^n, a^n)$ using our stepsize α_{n-1} to obtain updated estimates $\bar{Q}^n(s^n, a^n)$ of the value of the state-action pair s^n and a^n.

Figure 6.1 illustrates the behavior of using $1/n$ in this setting, which shows that we are significantly underestimating the values. Below, we fix this by generalizing $1/n$ using a tunable parameter. Later, we are going to present stepsize formulas that help to mitigate this behavior.

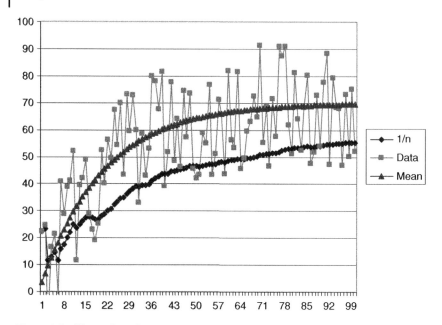

Figure 6.1 Illustration of poor convergence of the $1/n$ stepsize rule in the presence of transient data.

6.1.2 A Collection of Deterministic Policies

The remainder of this section presents a series of deterministic stepsize formulas designed to overcome this problem. These rules are the simplest to implement and are typically a good starting point when implementing adaptive learning algorithms.

6.1.2.1 Constant Stepsizes
A constant stepsize rule is simply

$$\alpha_{n-1} = \begin{cases} 1 & \text{if } n = 1, \\ \bar{\alpha} & \text{otherwise,} \end{cases}$$

where $\bar{\alpha}$ is a stepsize that we have chosen. It is common to start with a stepsize of 1 so that we do not need an initial value $\bar{\mu}^0$ for our statistic.

Constant stepsizes are popular when we are estimating not one but many parameters (for large-scale applications, these can easily number in the thousands or millions). In these cases, no single rule is going to be right for all of the parameters and there is enough noise that any reasonable stepsize rule will work well.

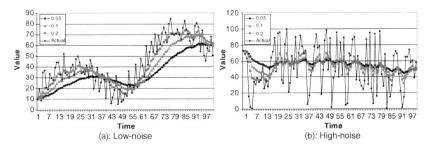

Figure 6.2 Illustration of the effects of smoothing using constant stepsizes. Case (a) represents a low-noise dataset, with an underlying nonstationary structure; case (b) is a high-noise dataset from a stationary process.

Constant stepsizes are easy to code (no memory requirements) and, in particular, easy to tune (there is only one parameter). Perhaps the biggest point in their favor is that we simply may not know the rate of convergence, which means that we run the risk with a declining stepsize rule of allowing the stepsize to decline too quickly, producing a behavior we refer to as "apparent convergence."

In dynamic programming, we are typically trying to estimate the value of being in a state using observations that are not only random, but which are also changing systematically as we try to find the best policy. As a general rule, as the noise in the observations of the values increases, the best stepsize decreases. But if the values are increasing rapidly, we want a larger stepsize.

Choosing the best stepsize requires striking a balance between stabilizing the noise and responding to the changing mean. Figure 6.2 illustrates observations that are coming from a process with relatively low noise but where the mean is changing quickly (6.2a), and observations that are very noisy but where the mean is not changing at all (6.2b). For the first, the ideal stepsize is relatively large, while for the second, the best stepsize is quite small.

6.1.2.2 Generalized Harmonic Stepsizes

A generalization of the $1/n$ rule is the generalized harmonic sequence given by

$$\alpha_{n-1} = \frac{\theta}{\theta + n - 1}. \tag{6.15}$$

This rule satisfies the conditions for convergence, but produces larger stepsizes for $\theta > 1$ than the $1/n$ rule. Increasing θ slows the rate at which the stepsize drops to zero, as illustrated in Figure 6.3. In practice, it seems that despite theoretical convergence proofs to the contrary, the stepsize $1/n$ can decrease to zero far too quickly, resulting in "apparent convergence" when in fact the solution is far from the best that can be obtained.

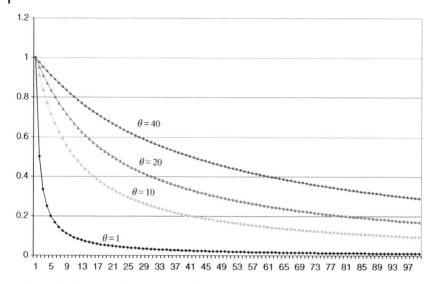

Figure 6.3 Stepsizes for $a/(a+n)$ while varying a.

6.1.2.3 Polynomial Learning Rates

An extension of the basic harmonic sequence is the stepsize

$$\alpha_{n-1} = \frac{1}{(n)^\beta},\tag{6.16}$$

where $\beta \in (\frac{1}{2}, 1]$. Smaller values of β slow the rate at which the stepsizes decline, which improves the responsiveness in the presence of initial transient conditions. The best value of β depends on the degree to which the initial data is transient, and as such is a parameter that needs to be tuned.

6.1.2.4 McClain's Formula

McClain's formula is an elegant way of obtaining $1/n$ behavior initially but approaching a specified constant in the limit. The formula is given by

$$\alpha_n = \frac{\alpha_{n-1}}{1 + \alpha_{n-1} - \bar{\alpha}},\tag{6.17}$$

where $\bar{\alpha}$ is a specified parameter. Note that steps generated by this model satisfy the following properties

$$\alpha_n > \alpha_{n+1} > \bar{\alpha} \quad \text{if} \quad \alpha > \bar{\alpha},$$
$$\alpha_n < \alpha_{n+1} < \bar{\alpha} \quad \text{if} \quad \alpha < \bar{\alpha}.$$

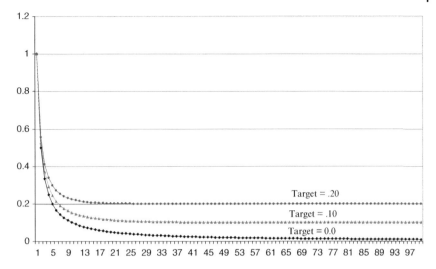

Figure 6.4   The McClain stepsize rule with varying targets.

McClain's rule, illustrated in Figure 6.4, combines the features of the "$1/n$" rule which is ideal for stationary data, and constant stepsizes for nonstationary data. If we set $\bar{\alpha} = 0$, then it is easy to verify that McClain's rule produces $\alpha_{n-1} = 1/n$. In the limit, $\alpha_n \rightarrow \bar{\alpha}$. The value of the rule is that the $1/n$ averaging generally works quite well in the very first iterations (this is a major weakness of constant stepsize rules), but avoids going to zero. The rule can be effective when you are not sure how many iterations are required to start converging, and it can also work well in nonstationary environments.

6.1.2.5   Search-then-Converge Learning Policy

The search-then-converge (STC) stepsize rule is a variation on the harmonic stepsize rule that produces delayed learning. The rule can be written as

$$\alpha_{n-1} = \alpha_0 \frac{\left(\dfrac{b}{n} + a\right)}{\left(\dfrac{b}{n} + a + n^\beta\right)}. \tag{6.18}$$

If $\beta = 1$, then this formula is similar to the STC policy. In addition, if $b = 0$, then it is the same as the harmonic stepsize policy $\theta/(\theta + n)$. The addition of the term b/n to the numerator and the denominator can be viewed as a kind of harmonic stepsize policy where a is very large but declines with n. The effect of the b/n term, then, is to keep the stepsize larger for a longer period of time, as illustrated in Figure 6.5(a). This can help algorithms that have to go through an

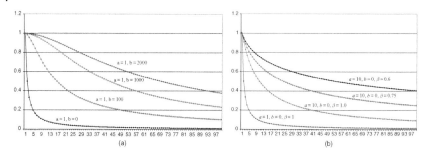

Figure 6.5 The search-then-converge rule while (a) varying b, and (b) varying β.

extended learning phase when the values being estimated are relatively unstable. The relative magnitude of b depends on the number of iterations which are expected to be run, which can range from several dozen to several million.

This class of stepsize rules is termed "search-then-converge" because they provide for a period of high stepsizes (while searching is taking place) after which the stepsize declines (to achieve convergence). The degree of delayed learning is controlled by the parameter b, which can be viewed as playing the same role as the parameter a but which declines as the algorithm progresses. The rule is designed for approximate dynamic programming methods applied to the setting of playing games with a delayed reward (there is no reward until you win or lose the game).

The exponent β in the denominator has the effect of increasing the stepsize in later iterations (see Figure 6.5(b)). With this parameter, it is possible to accelerate the reduction of the stepsize in the early iterations (by using a smaller a) but then slow the descent in later iterations (to sustain the learning process). This may be useful for problems where there is an extended transient phase requiring a larger stepsize for a larger number of iterations.

6.2 Adaptive Stepsize Policies

There is considerable appeal to the idea that the stepsize should depend on the actual trajectory of the algorithm. For example, if we are consistently observing that our estimate $\bar{\mu}^{n-1}$ is smaller (or larger) than the observations W^n, then it suggests that we are trending upward (or downward). When this happens, we typically would like to use a larger stepsize to increase the speed at which we reach a good estimate. When the stepsizes depend on the observations W^n, then we say that we are using a *adaptive stepsize*. This means, however, that we have to recognize that it is a random variable (some refer to these as stochastic stepsize rules).

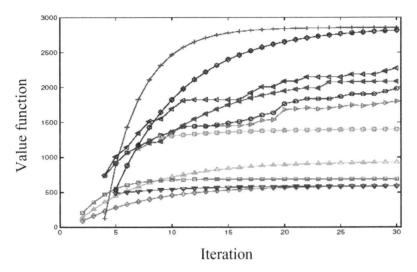

Figure 6.6 Different parameters can undergo significantly different initial rates.

In this section, we first review the case for adaptive stepsizes, then present the revised theoretical conditions for convergence, and finally outline a series of heuristic recipes that have been suggested in the literature. After this, we present some stepsize rules that are optimal until special conditions.

6.2.1 The Case for Adaptive Stepsizes

Assume that our estimates are consistently under or consistently over the actual observations. This can easily happen during early iterations due to either a poor initial starting point or the use of biased estimates (which is common in dynamic programming) during the early iterations. For large problems, it is possible that we have to estimate thousands of parameters. It seems unlikely that all the parameters will approach their true value at the same rate. Figure 6.6 shows the change in estimates of the value of being in different states, illustrating the wide variation in learning rates that can occur within the same dynamic program.

Adaptive stepsizes try to adjust to the data in a way that keeps the stepsize larger while the parameter being estimated is still changing quickly. Balancing noise against the change in the underlying signal, particularly when both of these are unknown, is a difficult challenge.

6.2.2 Convergence Conditions

When the stepsize depends on the history of the process, the stepsize itself becomes a random variable, which means we could replace the stepsize α_n with

$\alpha_n(\omega)$ to express its dependence on the sample path ω that we are following. This change requires some subtle modifications to our requirements for convergence (equations (6.11) and (6.12)). For technical reasons, our convergence criteria change to

$$\alpha_n > 0, \text{ almost surely,} \tag{6.19}$$

$$\sum_{n=0}^{\infty} \alpha_n = \infty, \text{ almost surely,} \tag{6.20}$$

$$\mathbb{E}\left\{\sum_{n=0}^{\infty}(\alpha_n)^2\right\} < \infty. \tag{6.21}$$

The condition "almost surely" (universally abbreviated "a.s.") means that equations (6.19)–(6.20) holds for every sample path ω, and not just on average. For example we could replace equation (6.20) with

$$\sum_{n=0}^{\infty} \alpha_n(\omega) = \infty, \text{ for all } \omega, p(\omega) > 0. \tag{6.22}$$

More precisely, we mean every sample path ω that might actually happen, which is why we introduced the condition $p(\omega) > 0$. We exclude sample paths where the probability that the sample path would happen is zero, which is something that mathematicians stress over. Note that almost surely is *not* the same as requiring

$$\mathbb{E}\left\{\sum_{n=0}^{\infty} \alpha_n\right\} = \infty, \tag{6.23}$$

which requires that this condition be satisfied on average but would allow it to fail for specific sample paths. This is a much weaker condition, and would *not* guarantee convergence every time we run the algorithm. Note that the condition (6.21) does, in fact, use an expectation, which hints that this is a weaker condition.

For the reasons behind these conditions, go to our "Why does it work" section (5.10). Note that while the theoretical conditions provide broad guidance, there are significant empirical differences between policies that satisfy the conditions for asymptotic optimality.

6.2.3 A Collection of Stochastic Policies

The desire to find stepsize policies that adapt to the data has become a cottage industry which has produced a variety of formulas with varying degrees of sophistication and convergence guarantees. This section provides a brief sample of some popular policies, some (such as AdaGrad) with strong performance guarantees. Later, we present some optimal policies for specialized problems.

To present our adaptive stepsize formulas, we need to define a few quantities. Recall that our basic updating expression is given by

$$\bar{\mu}^n = (1 - \alpha_{n-1})\bar{\mu}^{n-1} + \alpha_{n-1}W^n.$$

$\bar{\mu}^{n-1}$ is an estimate of whatever value we are estimating. Note that we may be estimating a function $\mu(x) = EF(x, W)$ (for discrete x), or we may be estimating a continuous function where smoothing is required. We can compute an error by comparing the difference between our current estimate $\bar{\mu}^{n-1}$ and our latest observation W^n which we write as

$$\varepsilon^n = \bar{\mu}^{n-1} - W^n.$$

Some formulas depend on tracking changes in the sign of the error. This can be done using the indicator function

$$\mathbb{1}_{\{X\}} = \begin{cases} 1 & \text{if the logical condition } X \text{ is true,} \\ 0 & \text{otherwise.} \end{cases}$$

Thus, $\mathbb{1}_{\{\varepsilon^n \varepsilon^{n-1} < 0\}}$ indicates if the sign of the error has changed in the last iteration.

Below, we first summarize three classic rules. Kesten's rule is the oldest and is perhaps the simplest illustration of an adaptive stepsize rule. Trigg's formula is a simple rule widely used in the demand-forecasting community. The stochastic gradient adaptive stepsize rule enjoys a theoretical convergence proof, but is controlled by several tunable parameters that complicate its use in practice. Then, we present three more modern rules: ADAM, AdaGrad, and RMSProp are rules that were developed by the machine learning community for fitting neural networks to data.

6.2.3.1 Kesten's Rule

Kesten's rule was one of the earliest stepsize rules which took advantage of a simple principle. If we are far from the optimal, the gradients $\nabla_x F(x^n, W^{n+1})$ tend to point in the same direction. As we get closer to the optimum, the gradients start to switch directions. Exploiting this simple observation, Kesten proposed the simple rule

$$\alpha_{n-1} = \frac{\theta}{\theta + K^n - 1},$$

(6.24)

where θ is a parameter to be calibrated. K^n counts the number of times that the sign of the error has changed, where we use

$$K^n = \begin{cases} n & \text{if } n = 1, 2, \\ K^{n-1} + \mathbb{1}_{\{(\nabla_x F(x^{n-1}, W^n))^T \nabla_x F(x^n, W^{n+1}) < 0\}} & \text{if } n > 2. \end{cases}$$

(6.25)

Kesten's rule is particularly well suited to initialization problems. It slows the reduction in the stepsize as long as successive gradients generally point in the same direction. They decline when the gradients begin to alternate sign, indicating that we are moving around the optimum.

6.2.3.2 Trigg's Formula

Let $S(\cdot)$ be the smoothed estimate of errors calculated using

$$S(\varepsilon^n) = (1 - \beta)S(\varepsilon^{n-1}) + \beta \varepsilon^n.$$

Trigg's formula is given by

$$\alpha_n = \frac{|S(\varepsilon^n)|}{S(|\varepsilon^n|)}.$$

(6.26)

The formula takes advantage of the simple property that smoothing on the absolute value of the errors is greater than or equal to the absolute value of the smoothed errors. If there is a series of errors with the same sign, that can be taken as an indication that there is a significant difference between the true mean and our estimate of the mean, which means we would like larger stepsizes.

6.2.3.3 Stochastic Gradient Adaptive Stepsize Rule

This class of rules uses stochastic gradient logic to update the stepsize. We first compute

$$\psi^n = (1 - \alpha_{n-1})\psi^{n-1} + \varepsilon^n.$$

(6.27)

The stepsize is then given by

$$\alpha_n = \left[\alpha_{n-1} + \nu \psi^{n-1} \varepsilon^n\right]_{\alpha_-}^{\alpha_+},$$

(6.28)

where α_+ and α_- are, respectively, upper and lower limits on the stepsize. $[\cdot]_{\alpha_-}^{\alpha_+}$ represents a projection back into the interval $[\alpha_-, \alpha_+]$, and ν is a scaling factor.

$\psi^{n-1}\varepsilon^n$ is a stochastic gradient that indicates how we should change the step-size to improve the error. Since the stochastic gradient has units that are the square of the units of the error, while the stepsize is unitless, ν has to perform an important scaling function. The equation $\alpha_{n-1} + \nu\psi^{n-1}\varepsilon^n$ can easily produce stepsizes that are larger than 1 or smaller than 0, so it is customary to spec-ify an allowable interval (which is generally smaller than (0,1)). This rule has provable convergence, but in practice, ν, α_+ and α_- all have to be tuned.

6.2.3.4 ADAM

ADAM (Adaptive Moment Estimation) is another stepsize policy that has attracted attention in recent years. As above, let $g^n = \nabla_x F(x^{n-1}, W^n)$ be our gra-dient, and let g_i^n be the ith element. ADAM proceeds by adaptively computing means and variances according to

$$m_i^n = \beta_1 m_i^{n-1} + (1 - \beta_1)g_i^n, \tag{6.29}$$

$$v_i^n = \beta_2 v_i^{n-1} + (1 - \beta_2)(g_i^n)^2. \tag{6.30}$$

These updating equations introduce biases when the data is nonstationary, which is typically the case in stochastic optimization. ADAM compensates for these biases using

$$\bar{m}_i^n = \frac{m_i^n}{1 - \beta_1},$$

$$\bar{v}_i^n = \frac{v_i^n}{1 - \beta_2}.$$

The stochastic gradient equation for ADAM is then given by

$$x_i^{n+1} = x_i^n + \frac{\eta}{\sqrt{\bar{v}_i^n} + \epsilon}\bar{m}_i^n. \tag{6.31}$$

6.2.3.5 AdaGrad

AdaGrad ("adaptive gradient") is a relatively recent stepsize policy that has attracted considerable attention in the machine learning literature which not only enjoys nice theoretical performance guarantees, but has also become quite popular because it seems to work quite well in practice.

Assume that we are trying to solve our standard problem

$$\max_x \mathbb{E}_W F(x, W),$$

where we make the assumption that not only is x a vector, but also that the scaling for each dimension might be different (an issue we have ignored so far). To simplify the notation a bit, let the stochastic gradient with respect to x_i, $i = 1, \ldots, I$ be given by

$$g_i^n = \nabla_{x_i} F(x^{n-1}, W^n).$$

Now create a $I \times I$ diagonal matrix G^n where the $(i, i)^{th}$ element G_{ii}^n is given by

$$G_{ii}^n = \sum_{m=1}^n (g_i^n)^2.$$

We then set a stepsize for the ith dimension using

$$\alpha_{ni} = \frac{\eta}{(G_{ii}^n)^2 + \epsilon}, \tag{6.32}$$

where ϵ is a small number (e.g. 10^{-8}) to avoid the possibility of dividing by zero. This can be written in matrix form using

$$\alpha_n = \frac{\eta}{\sqrt{G^n} + \epsilon} \otimes g_t, \tag{6.33}$$

where α_n is an I-dimensional matrix.

AdaGrad does an unusually good job of adapting to the behavior of a function. It also adapts to potentially different behaviors of each dimension. For example, we might be solving a machine learning problem to learn a parameter vector θ (this would be the decision variable instead of x) for a linear model of the form

$$y = \theta_0 + \theta_1 X_1 + \theta_2 X_2 + \ldots.$$

The explanatory variables $X_1, X_2, \ldots$ can take on values in completely different ranges. In a medical setting, X_1 might be blood sugar with values between 5 and 8, while X_2 might be the weight of a patient that could range between 100 and 300 pounds. The coefficients θ_1 and θ_2 would be scaled according to the inverse of the scales of the explanatory variables.

6.2.3.6 RMSProp

RMSProp (Root Mean Squared Propagation) was designed to address the empirical observation that AdaGrad declines too quickly. We continue to let

$g^n = \nabla_x F(x^n, W^{n+1})$ be our stochastic gradient. Let $\bar{g}^n$ be a smoothed version of the inner product $(g^n)^T g^n$ given by

$$\bar{g}^n = (1 - \beta)\bar{g}^n + \beta\|g^n\|^2. \tag{6.34}$$

We then compute our stepsize using

$$\alpha_n = \frac{\eta}{\sqrt{\bar{g}^n}}. \tag{6.35}$$

Suggested parameter values are $\beta = 0.1$ and $\eta = 0.001$, but we always suggest performing some exploration with tunable parameters.

6.2.4 Experimental Notes

A word of caution is offered when testing out stepsize rules. It is quite easy to test out these ideas in a controlled way in a simple spreadsheet on randomly generated data, but there is a big gap between showing a stepsize that works well in a spreadsheet and one that works well in specific applications. Adaptive stepsize rules work best in the presence of transient data where the degree of noise is not too large compared to the change in the signal (the mean). As the variance of the data increases, adaptive stepsize rules begin to suffer and simpler deterministic rules tend to work better.

6.3 Optimal Stepsize Policies*

Given the variety of stepsize formulas we can choose from, it seems natural to ask whether there is an optimal stepsize rule. Before we can answer such a question, we have to define exactly what we mean by it. Assume that we are trying to estimate a parameter that we denote by μ that may be static, or evolving over time (perhaps as a result of learning behavior), in which case we will write it as μ^n.

At iteration n, assume we are trying to track a time-varying process μ^n. For example, when we are estimating approximate value functions $\bar{V}^n(s)$, we will use algorithms where the estimate $\bar{V}^n(s)$ tends to rise (or perhaps fall) with the iteration n. We will use a learning policy π, so we are going to designate our estimate $\bar{\mu}^{\pi,n}$ to make the dependence on the learning policy explicit. At time n, we would like to choose a stepsize policy to minimize

$$\min_{\pi} \mathbb{E}(\bar{\mu}^{\pi,n} - \mu^n)^2. \tag{6.36}$$

Here, the expectation is over the entire history of the algorithm (note that it is not conditioned on anything, although the conditioning on S^0 is implicit) and requires (in principle) knowing the true value of the parameter being estimated.

The best way to think of this is to first imagine that we have a stepsize policy such as the harmonic stepsize rule

$$\alpha_n(\theta) = \frac{\theta}{\theta + n - 1},$$

which means that optimizing over π is the same (for this stepsize policy) as optimizing over θ. Assume that we observe our process with error ε, which is to say

$$W^{n+1} = \mu^n + \varepsilon^{n+1}.$$

Our estimate of $\bar{\mu}^{\pi,n}$ is given by

$$\bar{\mu}^{\pi,n+1} = (1 - \alpha_n(\theta))\bar{\mu}^{\pi,n} + \alpha_n(\theta)W^{n+1}.$$

Now imagine that we create a series of sample paths ω of observations of $(\varepsilon^n)_{n=1}^N$. If we follow a particular sample realization of observation errors $(\varepsilon^n(\omega))_{n=1}^N$, then this gives us a sequence of observations $(W^n(\omega))_{n=1}^N$, which will then produce, for a given stepsize policy π, a sequence of estimates $(\bar{\mu}^{\pi,n}(\omega))_{n=1}^N$. We can now write our optimization problem as

$$\min_\theta \frac{1}{N} \sum_{n=1}^N (\bar{\mu}^{\pi,n}(\omega^n) - \mu^n)^2. \tag{6.37}$$

The optimization problem in (6.37) illustrates how we might go through the steps of optimizing stepsize policies. Of course, we will want to do more than just tune the parameter of a particular policy. We are going to want to compare different stepsize policies, such as those listed in section 6.2.

We begin our discussion of optimal stepsizes in section 6.3.1 by addressing the case of estimating a constant parameter which we observe with noise. Section 6.3.2 considers the case where we are estimating a parameter that is changing over time, but where the changes have mean zero. Finally, section 6.3.3 addresses the case where the mean may be drifting up or down with nonzero mean, a situation that we typically face when approximating a value function.

6.3.1 Optimal Stepsizes for Stationary Data

Assume that we observe W^n at iteration n and that the observations W^n can be described by

$$W^n \;=\; \mu + \varepsilon^n$$

where μ is an unknown constant and ε^n is a stationary sequence of independent and identically distributed random deviations with mean 0 and variance σ_ε^2. We can approach the problem of estimating μ from two perspectives: choosing the best stepsize and choosing the best linear combination of the estimates. That is, we may choose to write our estimate $\bar{\mu}^n$ after n observations in the form

$$\bar{\mu}^n = \sum_{m=1}^{n} a_m^n W^m.$$

For our discussion, we will fix n and work to determine the coefficients of the vector $a_1, \ldots, a_n$ (where we suppress the iteration counter n to simplify notation). We would like our statistic to have two properties: It should be unbiased, and it should have minimum variance (that is, it should solve (6.36)). To be unbiased, it should satisfy

$$\mathbb{E}\left[\sum_{m=1}^{n} a_m W^m\right] \;=\; \sum_{m=1}^{n} a_m \mathbb{E} W^m$$

$$=\; \sum_{m=1}^{n} a_m \mu$$

$$=\; \mu,$$

which implies that we must satisfy

$$\sum_{m=1}^{n} a_m = 1.$$

The variance of our estimator is given by

$$Var(\bar{\mu}^n) \;=\; Var\left[\sum_{m=1}^{n} a_m W^m\right].$$

We use our assumption that the random deviations are independent, which allows us to write

$$
Var(\bar{\mu}^n) = \sum_{m=1}^{n} Var[a_m W^m]
$$

$$
= \sum_{m=1}^{n} a_m^2 Var[W^m]
$$

$$
= \sigma_\varepsilon^2 \sum_{m=1}^{n} a_m^2. \tag{6.38}
$$

Now we face the problem of finding $a_1, \dots, a_n$ to minimize (6.38) subject to the requirement that $\sum_m a_m = 1$. This problem is easily solved using the Lagrange multiplier method. We start with the nonlinear programming problem

$$
\min_{\{a_1,\dots,a_n\}} \sum_{m=1}^{n} a_m^2,
$$

subject to

$$
\sum_{m=1}^{n} a_m = 1, \tag{6.39}
$$

$$
a_m \geq 0. \tag{6.40}
$$

We relax constraint (6.39) and add it to the objective function

$$
\min_{\{a_m\}} L(a, \lambda) = \sum_{m=1}^{n} a_m^2 - \lambda \left(\sum_{m=1}^{n} a_m - 1 \right),
$$

subject to (6.40). We are now going to try to solve $L(a, \lambda)$ (known as the "Lagrangian") and hope that the coefficients a are all nonnegative. If this is true, we can take derivatives and set them equal to zero

$$
\frac{\partial L(a, \lambda)}{\partial a_m} = 2a_m - \lambda. \tag{6.41}
$$

The optimal solution (a^*, λ^*) would then satisfy

$$
\frac{\partial L(a, \lambda)}{\partial a_m} = 0.
$$

This means that at optimality

$$
a_m = \lambda/2,
$$

which tells us that the coefficients a_m are all equal. Combining this result with the requirement that they sum to one gives the expected result:

$$a_m = \frac{1}{n}.$$

In other words, our best estimate is a sample average. From this (somewhat obvious) result, we can obtain the optimal stepsize, since we already know that $\alpha_{n-1} = 1/n$ is the same as using a sample average.

This result tells us that if the underlying data is stationary, and we have no prior information about the sample mean, then the best stepsize rule is the basic $1/n$ rule. Using any other rule requires that there be some violation in our basic assumptions. In practice, the most common violation is that the observations are not stationary because they are derived from a process where we are searching for the best solution.

6.3.2 Optimal Stepsizes for Nonstationary Data – I

Assume now that our parameter evolves over time (iterations) according to the process

$$\mu^n = \mu^{n-1} + \xi^n, \tag{6.42}$$

where $\mathbb{E}\xi^n = 0$ is a zero mean drift term with variance σ_ξ^2. As before, we measure μ^n with an error according to

$$W^{n+1} = \mu^n + \varepsilon^{n+1}.$$

We want to choose a stepsize so that we minimize the mean squared error. This problem can be solved using a method known as the *Kalman filter*. The Kalman filter is a powerful recursive regression technique, but we adapt it here for the problem of estimating a single parameter. Typical applications of the Kalman filter assume that the variance of ξ^n, given by σ_ξ^2, and the variance of the measurement error, ε^n, given by σ_ε^2, are known. In this case, the Kalman filter would compute a stepsize (generally referred to as the gain) using

$$\alpha_n = \frac{\sigma_\xi^2}{\nu^n + \sigma_\varepsilon^2}, \tag{6.43}$$

where ν^n is computed recursively using

$$\nu^n = (1 - \alpha_{n-1})\nu^{n-1} + \sigma_\xi^2. \tag{6.44}$$

Remember that $\alpha_0 = 1$, so we do not need a value of ν^0. For our application, we do not know the variances so these have to be estimated from data. We first estimate the bias using

$$\bar{\beta}^n \ = \ (1 - \eta_{n-1})\bar{\beta}^{n-1} + \eta_{n-1}\left(\bar{\mu}^{n-1} - W^n\right), \tag{6.45}$$

where η_{n-1} is a simple stepsize rule such as the harmonic stepsize rule or McClain's formula. We then estimate the total error sum of squares using

$$\bar{\nu}^n \ = \ (1 - \eta_{n-1})\bar{\nu}^{n-1} + \eta_{n-1}\left(\bar{\mu}^{n-1} - W^n\right)^2. \tag{6.46}$$

Finally, we estimate the variance of the error using

$$(\bar{\sigma}_\varepsilon^{2,n}) \ = \ \frac{\bar{\nu}^n - (\bar{\beta}^n)^2}{1 + \bar{\lambda}^{n-1}}, \tag{6.47}$$

where $\bar{\lambda}^{n-1}$ is computed using

$$\lambda^n = \begin{cases} (\alpha_{n-1})^2, & n = 1, \\ (1 - \alpha_{n-1})^2\lambda^{n-1} + (\alpha_{n-1})^2, & n > 1. \end{cases}$$

We use $(\bar{\sigma}_\varepsilon^{2,n})$ as our estimate of σ_ε^2. We then propose to use $(\bar{\beta}^n)^2$ as our estimate of σ_ξ^2. This is purely an approximation, but experimental work suggests that it performs quite well, and it is relatively easy to implement.

6.3.3 Optimal Stepsizes for Nonstationary Data – II

In dynamic programming, we are trying to estimate the value of being in a state (call it v) by $\bar{v}$ which is estimated from a sequence of random observations $\hat{v}$. The problem we encounter is that $\hat{v}$ might depend on a value function approximation which is steadily increasing (or decreasing), which means that the observations $\hat{v}$ are nonstationary. Furthermore, unlike the assumption made by the Kalman filter that the mean of $\hat{v}$ is varying in a zero-mean way, our observations of $\hat{v}$ might be steadily increasing. This would be the same as assuming that $\mathbb{E}\xi = \mu > 0$ in the section above. In this section, we derive the Kalman filter learning rate for biased estimates.

Our challenge is to devise a stepsize that strikes a balance between minimizing error (which prefers a smaller stepsize) and responding to the nonstationary data (which works better with a large stepsize). We return to our basic model

$$W^{n+1} = \mu^n + \varepsilon^{n+1},$$

where μ^n varies over time, but it might be steadily increasing or decreasing. This would be similar to the model in the previous section (equation (6.42)) but where ξ^n has a nonzero mean. As before we assume that $\{\varepsilon^n\}_{n=1,2,\dots}$ are independent and identically distributed with mean value of zero and variance, σ^2.

We perform the usual stochastic gradient update to obtain our estimates of the mean

$$\bar{\mu}^n(\alpha_{n-1}) \quad = \quad (1 - \alpha_{n-1})\bar{\mu}^{n-1}(\alpha_{n-1}) + \alpha_{n-1}W^n. \tag{6.48}$$

We wish to find α_{n-1} that solves

$$\min_{\alpha_{n-1}} F(\alpha_{n-1}) = \mathbb{E}\left[\left(\bar{\mu}^n(\alpha_{n-1}) - \mu^n\right)^2\right]. \tag{6.49}$$

It is important to realize that we are trying to choose α_{n-1} to minimize the *unconditional* expectation of the error between $\bar{\mu}^n$ and the true value μ^n. For this reason, our stepsize rule will be deterministic, since we are not allowing it to depend on the information obtained up through iteration n.

We assume that the observation at iteration n is unbiased, which is to say

$$\mathbb{E}\left[W^{n+1}\right] \quad = \quad \mu^n. \tag{6.50}$$

But the smoothed estimate is biased because we are using simple smoothing on nonstationary data. We denote this bias as

$$\beta^{n-1} \quad = \quad \mathbb{E}\left[\bar{\mu}^{n-1} - \mu^n\right]$$
$$= \quad \mathbb{E}\left[\bar{\mu}^{n-1}\right] - \mu^n. \tag{6.51}$$

We note that β^{n-1} is the bias computed after iteration $n - 1$ (that is, after we have computed $\bar{\mu}^{n-1}$). β^{n-1} is the bias when we use $\bar{\mu}^{n-1}$ as an estimate of μ^n.

The variance of the observation W^n is computed as follows:

$$Var[W^n] \quad = \quad \mathbb{E}\left[(W^n - \mu^n)^2\right]$$
$$= \quad \mathbb{E}\left[(\varepsilon^n)^2\right]$$
$$= \quad \sigma_\varepsilon^2. \tag{6.52}$$

It can be shown (see section 6.7.1) that the optimal stepsize is given by

$$\alpha_{n-1} \quad = \quad 1 - \frac{\sigma_\varepsilon^2}{(1 + \lambda^{n-1})\sigma_\varepsilon^2 + (\beta^{n-1})^2}, \tag{6.53}$$

where λ is computed recursively using

$$\lambda^n = \begin{cases} (\alpha_{n-1})^2, & n = 1, \\ (1 - \alpha_{n-1})^2\lambda^{n-1} + (\alpha_{n-1})^2, & n > 1. \end{cases} \tag{6.54}$$

We refer to the stepsize rule in equation (6.53) as the *bias adjusted Kalman filter*, or BAKF. The BAKF stepsize formula enjoys several nice properties:

Stationary data For a sequence with a static mean, the optimal stepsizes are given by

$$\alpha_{n-1} \;=\; \frac{1}{n} \quad \forall\, n = 1, 2, \ldots \ . \tag{6.55}$$

This is the optimal stepsize for stationary data.

No noise For the case where there is no noise ($\sigma^2 = 0$), we have the following:

$$\alpha_{n-1} \;=\; 1 \quad \forall\, n = 1, 2, \ldots \ . \tag{6.56}$$

This is ideal for nonstationary data with no noise.

Bounded by $1/n$ At all times, the stepsize obeys

$$\alpha_{n-1} \geq \frac{1}{n} \quad \forall\, n = 1, 2, \ldots \ .$$

This is important since it guarantees asymptotic convergence.

These are particularly nice properties since we typically have to do parameter tuning to get this behavior. The properties are particularly when estimating value functions, since sampled estimates of the value of being in a state tends to be transient.

The problem with using the stepsize formula in equation (6.53) is that it assumes that the variance σ^2 and the bias $(\beta^n)^2$ are known. This can be problematic in real instances, especially the assumption of knowing the bias, since computing this basically requires knowing the real function. If we have this information, we do not need this algorithm.

As an alternative, we can try to estimate these quantities from data. Let

$$\bar{\sigma}_\varepsilon^{2,n} \;=\; \text{estimate of the variance of the error after iteration } n,$$

$$\bar{\beta}^n \;=\; \text{estimate of the bias after iteration } n,$$

$$\bar{\nu}^n \;=\; \text{estimate of the variance of the bias after iteration } n.$$

To make these estimates, we need to smooth new observations with our current best estimate, something that requires the use of a stepsize formula. We could attempt to find an optimal stepsize for this purpose, but it is likely that a reasonably chosen deterministic formula will work fine. One possibility is McClain's formula (equation (6.17)):

$$\eta_n \;=\; \frac{\eta_{n-1}}{1 + \eta_{n-1} - \bar{\eta}}.$$

A limit point such as $\bar{\eta} \in (0.05, 0.10)$ appears to work well across a broad range of functional behaviors. The property of this stepsize that $\eta_n \to \bar{\eta}$ can be a strength, but it does mean that the algorithm will not tend to converge in the limit, which requires a stepsize that goes to zero. If this is needed, we suggest a harmonic stepsize rule:

$$\eta_{n-1} = \frac{a}{a + n - 1},$$

where a in the range between 5 and 10 seems to work quite well for many dynamic programming applications.

Care needs to be used in the early iterations. For example, if we let $\alpha_0 = 1$, then we do not need an initial estimate for $\bar{\mu}^0$ (a trick we have used throughout). However, since the formulas depend on an estimate of the variance, we still have problems in the second iteration. For this reason, we recommend forcing η_1 to equal 1 (in addition to using $\eta_0 = 1$). We also recommend using $\alpha_n = 1/(n + 1)$ for the first few iterations, since the estimates of $(\bar{\sigma}^2)^n, \bar{\beta}^n$ and $\bar{\nu}^n$ are likely to be very unreliable in the very beginning.

Figure 6.7 summarizes the entire algorithm. Note that the estimates have been constructed so that α_n is a function of information available up through iteration n.

Figure 6.8 illustrates the behavior of the bias-adjusted Kalman filter stepsize rule for two signals: very low noise (Figure 6.8a) and with higher noise (Figure 6.8b). For both cases, the signal starts small and rises toward an upper limit of 1.0 (on average). In both figures, we also show the stepsize $1/n$. For the low-noise case, the stepsize stays quite large. For the high-noise case, the stepsize roughly tracks $1/n$ (note that it never goes below $1/n$).

6.4   Optimal Stepsizes for Approximate Value Iteration*

All the stepsize rules that we have presented so far are designed to estimate the mean of a nonstationary series. In this section, we develop a stepsize rule that is specifically designed for approximate value iteration, which is an algorithm we are going to see in chapters 16 and 17. Another application is Q-learning, which we first saw in section 2.1.6.

We use as our foundation a dynamic program with a single state and single action. We use the same theoretical foundation that we used in section 6.3. However, given the complexity of the derivation, we simply provide the expression for the optimal stepsize, which generalizes the BAKF stepsize rule given in equation (6.53).

We start with the basic relationship for our single state problem

$$v^n(\alpha_{n-1}) = (1 - (1 - \gamma)\alpha_{n-1})v^{n-1} + \alpha_{n-1}\hat{C}^n. \qquad (6.57)$$

Let $c = \hat{C}$ be the expected one-period contribution for our problem, and let $Var(\hat{C}) = \sigma^2$. For the moment, we assume c and σ^2 are known. We next define the iterative formulas for two series, λ^n and δ^n, as follows:

$$\lambda^n = \begin{cases} \alpha_0^2 & n = 1 \\ \alpha_{n-1}^2 + (1 - (1 - \gamma)\alpha_{n-1})^2 \lambda^{n-1} & n > 1. \end{cases}$$

$$\delta^n = \begin{cases} \alpha_0 & n = 1 \\ \alpha_{n-1} + (1 - (1 - \gamma)\alpha_{n-1})\delta^{n-1} & n > 1. \end{cases}$$

Step 0. Initialization:

    Step 0a. Set the baseline to its initial value, $\bar{\mu}_0$.
    Step 0b. Initialize the parameters – $\bar{\beta}_0$, $\bar{\nu}_0$ and $\bar{\lambda}_0$.
    Step 0c. Set initial stepsizes $\alpha_0 = \eta_0 = 1$, and specify the stepsize rule for η.
    Step 0d. Set the iteration counter, $n = 1$.

Step 1. Obtain the new observation, W^n.
Step 2. Smooth the baseline estimate.

$$\bar{\mu}^n = (1 - \alpha_{n-1})\bar{\mu}^{n-1} + \alpha_{n-1} W^n.$$

Step 3. Update the following parameters:

$$\varepsilon^n = \bar{\mu}^{n-1} - W^n,$$
$$\bar{\beta}^n = (1 - \eta_{n-1})\bar{\beta}^{n-1} + \eta_{n-1}\varepsilon^n,$$
$$\bar{\nu}^n = (1 - \eta_{n-1})\bar{\nu}^{n-1} + \eta_{n-1}(\varepsilon^n)^2,$$
$$(\bar{\sigma}^2)^n = \frac{\bar{\nu}^n - (\bar{\beta}^n)^2}{1 + \lambda^{n-1}}.$$

Step 4. Evaluate the stepsizes for the next iteration.

$$\alpha_n = \begin{cases} 1/(n+1) & n = 1, 2, \\ 1 - \frac{(\bar{\sigma}^2)^n}{\bar{\nu}^n}, & n > 2, \end{cases}$$

$$\eta_n = \frac{a}{a+n-1}. \quad \text{Note that this gives us } \eta_1 = 1.$$

Step 5. Compute the coefficient for the variance of the smoothed estimate of the baseline.

$$\bar{\lambda}^n = (1 - \alpha_{n-1})^2 \bar{\lambda}^{n-1} + (\alpha_{n-1})^2.$$

Step 6. If $n < N$, then $n = n + 1$ and go to Step 1, else stop.

Figure 6.7 The bias-adjusted Kalman filter stepsize rule.

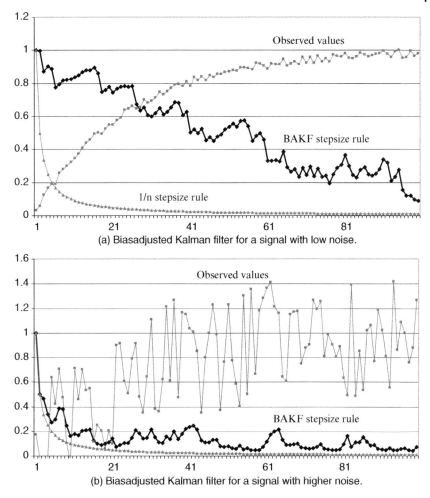

Figure 6.8 The BAKF stepsize rule for low-noise (a) and high-noise (b). Each figure shows the signal, the BAKF stepsizes, and the stepsizes produced by the $1/n$ stepsize rule.

It is possible to then show that

$$\mathbb{E}(v^n) = \delta^n c,$$
$$Var(v^n) = \lambda^n \sigma^2.$$

Let $v^n(\alpha_{n-1})$ be defined as in equation (6.57). Our goal is to solve the optimization problem

$$\min_{\alpha_{n-1}} \mathbb{E}\left[\left(v^n(\alpha_{n-1}) - \mathbb{E}\hat{v}^n\right)^2\right].$$ (6.58)

The optimal solution can be shown to be given by

$$\alpha_{n-1} = \frac{(1-\gamma)\lambda^{n-1}\sigma^2 + (1-(1-\gamma)\delta^{n-1})^2 c^2}{(1-\gamma)^2\lambda^{n-1}\sigma^2 + (1-(1-\gamma)\delta^{n-1})^2 c^2 + \sigma^2}. \tag{6.59}$$

We refer to equation (6.59) as the *optimal stepsize for approximate value iteration* (OSAVI). Of course, it is only optimal for our single state problem, and it assumes that we know the expected contribution per time period c, and the variance in the contribution $\hat{C}$, σ^2.

OSAVI has some desirable properties. If $\sigma^2 = 0$, then $\alpha_{n-1} = 1$. Also, if $\gamma = 0$, then $\alpha_{n-1} = 1/n$. It is also possible to show that $\alpha_{n-1} \geq (1-\gamma)/n$ for any sample path.

All that remains is adapting the formula to more general dynamic programs with multiple states and where we are searching for optimal policies. We suggest the following adaptation. We propose to estimate a single constant $\bar{c}$ representing the average contribution per period, averaged over all states. If $\hat{C}^n$ is the contribution earned in period n, let

$$\bar{c}^n = (1 - \nu_{n-1})\bar{c}^{n-1} + \nu_{n-1}\hat{C}^n,$$
$$(\bar{\sigma}^n)^2 = (1 - \nu_{n-1})(\bar{\sigma}^{n-1})^2 + \nu_{n-1}(\bar{c}^n - \hat{C}^n)^2.$$

Here, ν_{n-1} is a separate stepsize rule. Our experimental work suggests that a constant stepsize works well, and that the results are quite robust with respect to the value of ν_{n-1}. We suggest a value of $\nu_{n-1} = 0.2$. Now let $\bar{c}^n$ be our estimate of c, and let $(\bar{\sigma}^n)^2$ be our estimate of σ^2.

We could also consider estimating $\bar{c}^n(s)$ and $(\bar{\sigma}^n)^2(s)$ for each state, so that we can estimate a state-dependent stepsize $\alpha_{n-1}(s)$. There is not enough experimental work to support the value of this strategy, and lacking this we favor simplicity over complexity.

6.5 Convergence

A practical issue that arises with all stochastic approximation algorithms is that we simply do not have reliable, implementable stopping rules. Proofs of convergence in the limit are an important theoretical property, but they provide no guidelines or guarantees in practice.

A good illustration of the issue is given in Figure 6.9. Figure 6.9a shows the objective function for a dynamic program over 100 iterations (in this application, a single iteration required approximately 20 minutes of CPU time). The figure shows the objective function for an ADP algorithm which was run 100 iterations, at which point it appeared to be flattening out (evidence of convergence). Figure 6.9b is the objective function for the same algorithm run

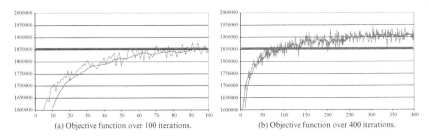

(a) Objective function over 100 iterations. (b) Objective function over 400 iterations.

Figure 6.9   The objective function, plotted over 100 iterations (a) displays "apparent convergence." The same algorithm, continued over 400 iterations (b) shows significant improvement.

for 400 iterations, which shows that there remained considerable room for improvement after 100 iterations.

We refer to this behavior as "apparent convergence," and it is particularly problematic on large-scale problems where run times are long. Typically, the number of iterations needed before the algorithm "converges" requires a level of subjective judgment. When the run times are long, wishful thinking can interfere with this process.

Complicating the analysis of convergence in stochastic search is the behavior in some problems to go through periods of stability which are simply a precursor to breaking through to new plateaus. During periods of exploration, a stochastic gradient algorithm might discover a strategy that opens up new opportunities, moving the performance of the algorithm to an entirely new level.

Special care has to be made in the choice of stepsize rule. In any algorithm using a declining stepsize, it is possible to show a stabilizing objective function simply because the stepsize is decreasing. This problem is exacerbated when using algorithms based on value iteration, where updates to the value of being in a state depend on estimates of the values of future states, which can be biased. We recommend that initial testing of a stochastic gradient algorithm start with inflated stepsizes. After getting a sense for the number of iterations needed for the algorithm to stabilize, decrease the stepsize (keeping in mind that the number of iterations required to convergence may increase) to find the right tradeoff between noise and rate of convergence.

6.6   Guidelines for Choosing Stepsize Policies

Given the plethora of strategies for computing stepsizes, it is perhaps not surprising that there is a need for general guidance when choosing a stepsize

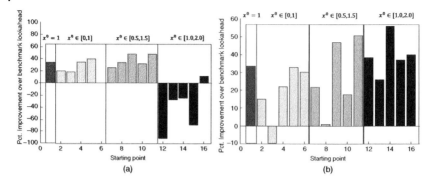

Figure 6.10 Performance of stochastic gradient algorithm using starting point $x^0 = 1$, $x^0 \in [0,1]$, $x^0 \in [0.5, 1.5]$, and $x^0 \in [1.0, 2.0]$ using two different tuned values of the stepsize parameter θ.

formula. Strategies for stepsizes are problem-dependent, and as a result any advice reflects the experience of the individual giving the advice.

An issue that is often overlooked is the role of tuning the stepsize policy. If a stepsize is not performing well, is it because you are not using an effective stepsize policy? Or is it because you have not properly tuned the one that you are using? Even more problematic is when you feel that you have tuned your stepsize policy as well as it can be tuned, but then you change something in your problem. For example, the distance from starting point to optimal solution matters. Changing your starting point, or modifying problem parameters so that the optimal solution moves, can change the optimal tuning of your stepsize policy.

This helps to emphasize the importance of our formulation which poses stochastic search algorithms as optimization problems *searching for the best algorithm*. Since parameter tuning for stepsizes is a manual process, people tend to overlook it, or minimize it. Figure 6.10 illustrates the risk of failing to recognize the point of tuning.

Figure 6.10(a) shows the performance of a stochastic gradient algorithm using a "tuned" stepsize, for four sets of starting points for x^0: $x^0 = 1$, $x^0 \in [0, 1.0]$, $x^0 \in [0.5, 1.5]$, and $x^0 \in [1.0, 2.0]$. Note the poor performance when the starting point was chosen in the range $x^0 \in [1.0, 2.0]$. Figure 6.10(b) shows the same algorithm after the stepsize was re-tuned for the range $x^0 \in [1.0, 2.0]$ (the same stepsize was used for all four ranges).

With this in mind, we offer the following general strategies for choosing stepsizes:

Step 1 Start with a constant stepsize α and test out different values. Problems with a relatively high amount of noise will require smaller stepsizes.

Periodically stop the search and test the quality of your solution (this will require running multiple simulations of $F(x, \widehat{W})$ and averaging). Plot the results to see roughly how many iterations are needed before your results stop improving.

Step 2 Now try the harmonic stepsize $\theta/(\theta + n - 1)$. $\theta = 1$ produces the $1/n$ stepsize rule that is provably convergent, but is likely to decline too quickly. To choose θ, look at how many iterations seemed to be needed when using a constant stepsize. If 100 iterations appears to be enough for a stepsize of 0.1, then try $\theta \approx 10$, as it produces a stepsize of roughly .1 after 100 iterations. If you need 10,000 iterations, choose $\theta \approx 1000$. But you will need to tune θ. An alternative rule is the polynomial stepsize rule $\alpha_n = 1/n^\beta$ with $\beta \in (0.5, 1]$ (we suggest 0.7 as a good starting point).

Step 3 Now start experimenting with the adaptive stepsize policies. RMSProp has become popular as of this writing for stationary stochastic search. For nonstationary settings, we suggest the BAKF stepsize rule (section 6.3.3). We will encounter an important class of nonstationary applications when we are estimating value function approximations in chapters 16 and 17.

There is always the temptation to do something simple. A constant stepsize, or a harmonic rule, are both extremely simple to implement. Keep in mind that both have a tunable parameter, and that the constant stepsize rule will not converge to anything (although the final solution may be quite acceptable). A major issue is that the best tuning of a stepsize not only depends on a problem, but also on the parameters of a problem such as the discount factor.

BAKF and OSAVI are more difficult to implement, but are more robust to the setting of the single, tunable parameter. Tunable parameters can be a major headache in the design of algorithms, and it is good strategy to absolutely minimize the number of tunable parameters your algorithm needs. Stepsize rules should be something you code once and forget about, but keep in mind the lesson of Figure 6.10.

6.7 Why Does it Work*

6.7.1 Proof of BAKF Stepsize

We now have what we need to derive an optimal stepsize for nonstationary data with a mean that is steadily increasing (or decreasing). We refer to this as the *bias-adjusted Kalman filter* stepsize rule (or BAKF), in recognition of its close relationship to the Kalman filter learning rate. We state the formula in the following theorem:

Theorem 6.7.1. *The optimal stepsizes* $(\alpha_m)_{m=0}^n$ *that minimize the objective function in equation (6.49) can be computed using the expression*

$$\alpha_{n-1} = 1 - \frac{\sigma^2}{(1 + \lambda^{n-1})\sigma^2 + (\beta^{n-1})^2}, \tag{6.60}$$

where λ *is computed recursively using*

$$\lambda^n = \begin{cases} (\alpha_{n-1})^2, & n = 1 \\ (1 - \alpha_{n-1})^2 \lambda^{n-1} + (\alpha_{n-1})^2, & n > 1. \end{cases} \tag{6.61}$$

Proof: We present the proof of this result because it brings out some properties of the solution that we exploit later when we handle the case where the variance and bias are unknown. Let $F(\alpha_{n-1})$ denote the objective function from the problem stated in (6.49).

$$F(\alpha_{n-1}) = \mathbb{E}\left[(\bar{\mu}^n(\alpha_{n-1}) - \mu^n)^2\right] \tag{6.62}$$

$$= \mathbb{E}\left[((1 - \alpha_{n-1})\bar{\mu}^{n-1} + \alpha_{n-1}W^n - \mu^n)^2\right] \tag{6.63}$$

$$= \mathbb{E}\left[((1 - \alpha_{n-1})(\bar{\mu}^{n-1} - \mu^n) + \alpha_{n-1}(W^n - \mu^n))^2\right] \tag{6.64}$$

$$= (1 - \alpha_{n-1})^2 \mathbb{E}\left[(\bar{\mu}^{n-1} - \mu^n)^2\right] + (\alpha_{n-1})^2 \mathbb{E}\left[(W^n - \mu^n)^2\right]$$
$$+ 2\alpha_{n-1}(1 - \alpha_{n-1})\underbrace{\mathbb{E}\left[(\bar{\mu}^{n-1} - \mu^n)(W^n - \mu^n)\right]}_{I}. \tag{6.65}$$

Equation (6.62) is true by definition, while (6.63) is true by definition of the updating equation for $\bar{\mu}^n$. We obtain (6.64) by adding and subtracting $\alpha_{n-1}\mu^n$. To obtain (6.65), we expand the quadratic term and then use the fact that the stepsize rule, α_{n-1}, is deterministic, which allows us to pull it outside the expectations. Then, the expected value of the cross-product term, I, vanishes under the assumption of independence of the observations and the objective function reduces to the following form

$$F(\alpha_{n-1}) = (1 - \alpha_{n-1})^2 \mathbb{E}\left[(\bar{\mu}^{n-1} - \mu^n)^2\right] + (\alpha_{n-1})^2 \mathbb{E}\left[(W^n - \mu^n)^2\right] \tag{6.66}$$

In order to find the optimal stepsize, α_{n-1}^*, that minimizes this function, we obtain the first-order optimality condition by setting $\frac{\partial F(\alpha_{n-1})}{\partial \alpha_{n-1}} = 0$, which gives us

$$-2\left(1-\alpha_{n-1}^{*}\right)\mathbb{E}\left[\left(\bar{\mu}^{n-1}-\mu^{n}\right)^{2}\right]+2\alpha_{n-1}^{*}\mathbb{E}\left[\left(W^{n}-\mu^{n}\right)^{2}\right]=0. \tag{6.67}$$

Solving this for α_{n-1}^{*} gives us the following result

$$\alpha_{n-1}^{*}=\frac{\mathbb{E}\left[\left(\bar{\mu}^{n-1}-\mu^{n}\right)^{2}\right]}{\mathbb{E}\left[\left(\bar{\mu}^{n-1}-\mu^{n}\right)^{2}\right]+\mathbb{E}\left[\left(W^{n}-\mu^{n}\right)^{2}\right]}. \tag{6.68}$$

Recall that we can write $(\bar{\mu}^{n-1}-\mu^{n})^{2}$ as the sum of the variance plus the bias squared using

$$\mathbb{E}\left[\left(\bar{\mu}^{n-1}-\mu^{n}\right)^{2}\right]=\lambda^{n-1}\sigma^{2}+\left(\beta^{n-1}\right)^{2}. \tag{6.69}$$

Using (6.69) and $\mathbb{E}\left[\left(W^{n}-\mu^{n}\right)^{2}\right]=\sigma^{2}$ in (6.68) gives us

$$\begin{aligned}\alpha_{n-1}&=\frac{\lambda^{n-1}\sigma^{2}+(\beta^{n-1})^{2}}{\lambda^{n-1}\sigma^{2}+(\beta^{n-1})^{2}+\sigma^{2}}\\&=1-\frac{\sigma^{2}}{(1+\lambda^{n-1})\sigma^{2}+(\beta^{n-1})^{2}},\end{aligned}$$

which is our desired result (equation (6.60)). □

From this result, we can next establish several properties through the following corollaries.

Corollary 6.7.1. *For a sequence with a static mean, the optimal stepsizes are given by*

$$\alpha_{n-1}=\frac{1}{n}\quad \forall\, n=1,2,\dots\ . \tag{6.70}$$

Proof: In this case, the mean $\mu^{n}=\mu$ is a constant. Therefore, the estimates of the mean are unbiased, which means $\beta^{n}=0\quad \forall t=2,\dots,.$ This allows us to write the optimal stepsize as

$$\alpha_{n-1}=\frac{\lambda^{n-1}}{1+\lambda^{n-1}}. \tag{6.71}$$

Substituting (6.71) into (6.54) gives us

$$\alpha_{n}=\frac{\alpha_{n-1}}{1+\alpha_{n-1}}. \tag{6.72}$$

If $\alpha_{0}=1$, it is easy to verify (6.70). □

For the case where there is no noise ($\sigma^{2}=0$), we have the following:

Corollary 6.7.2. *For a sequence with zero noise, the optimal stepsizes are given by*

$$\alpha_{n-1}=1\quad \forall\, n=1,2,\dots\ . \tag{6.73}$$

The corollary is proved by simply setting $\sigma^2 = 0$ in equation (6.53). As a final result, we obtain

Corollary 6.7.3. *In general,*

$$\alpha_{n-1} \geq \frac{1}{n} \quad \forall\, n = 1, 2, \ldots .$$

Proof: We leave this more interesting proof as an exercise to the reader (see exercise 6.17).

Corollary 6.7.3 is significant since it establishes one of the conditions needed for convergence of a stochastic approximation method, namely that $\sum_{n=1}^{\infty} \alpha_n = \infty$. An open theoretical question, as of this writing, is whether the BAKF stepsize rule also satisfies the requirement that $\sum_{n=1}^{\infty} (\alpha_n)^2 < \infty$.

6.8 Bibliographic Notes

Sections 6.1–6.2 A number of different communities have studied the problem of "stepsizes," including the business forecasting community (Brown (1959) 1963), Gardner (1983), Giffin (1971), Holt et al. (1960), Trigg (1964), artificial intelligence Darken and Moody (1991), Darken et al. (1992), Jaakkola et al. (1994), Sutton and Singh (1994), stochastic programming Kesten (1958), Mirozahmedov and Uryasev (1983), Pflug (1988), Ruszczyński and Syski (1986) and signal processing (Douglas and Mathews (1995)), Goodwin and Sin (1984). The neural network community refers to "learning rate schedules"; see Haykin (1999). Even-dar and Mansour (2003) provides a thorough analysis of convergence rates for certain types of stepsize formulas, including $1/n$ and the polynomial learning rate $1/n^\beta$, for Q-learning problems. These sections are based on the presentation in Powell and George (2006) Broadie et al. (2011) revisits the stepsize conditions (6.19)–(6.19).

Section 6.3.1 – The optimality of averaging for stationary data is well known. Our presentation was based on Kushner and Yin (2003)[pp. 1892–185].

Section 6.3.2 – This result for nonstationary data is a classic result from Kalman filter theory (see, for example, Meinhold and Singpurwalla (2007)).

Section 6.3.3 – The BAKF stepsize formula was developed by Powell and George (2006), where it was initially called the "optimal stepsize algorithm" (or OSA).

Section 6.4 – The OSAVI stepsize formula for approximate value iteration was developed in Ryzhov et al. (2015).

Section 6.6 – Figure 6.10 was prepared by Saeed Ghadimi.

Exercises

Review questions

6.1 What is a harmonic stepsize policy? Show that a stepsize $\alpha_n = 1/n$ is the same as simple averaging.

6.2 What three conditions have to be satisfied for convergence of a deterministic stepsize policy.

6.3 Describe Kesten's rule and provide an intuitive explanation for the design of this policy.

6.4 Assume that the stepsize α_n is an adaptive (that is, stochastic) stepsize policy. What do we mean when we require

$$\sum_{n=0}^{\infty} \alpha_n = \infty$$

to be true *almost surely*. Why is this not equivalent to requiring

$$\mathbb{E}\left\{\sum_{n=0}^{\infty} \alpha_n\right\} = \infty?$$

What is the practical implication of requiring the condition to be true "almost surely."

6.5 Explain why $1/n$ is the *optimal* stepsize policy when estimating the mean of a random variable from observations that are stationary over the iterations.

6.6 Give the underlying stochastic model assumed by the Kalman filter. What is the optimal policy for this model?

Computational exercises

6.7 Let U be a uniform $[0, 1]$ random variable, and let

$$\mu^n = 1 - \exp(-\theta_1 n).$$

Now let $\hat{R}^n = \mu^n + \theta_2(U^n - .5)$. We wish to try to estimate μ^n using

$$\bar{R}^n = (1 - \alpha_{n-1})\bar{R}^{n-1} + \alpha_{n-1}\hat{R}^n.$$

In the exercises below, estimate the mean (using $\bar{R}^n$) and compute the standard deviation of $\bar{R}^n$ for $n = 1, 2, \ldots, 100$, for each of the following stepsize rules:

- $\alpha_{n-1} = 0.10$.
- $\alpha_{n-1} = a/(a + n - 1)$ for $a = 1, 10$.
- Kesten's rule.
- The bias-adjusted Kalman filter stepsize rule.

For each of the parameter settings below, compare the rules based on the average error (1) over all 100 iterations and (2) in terms of the standard deviation of $\bar{R}^{100}$.

(a) $\theta_1 = 0, \theta_2 = 10$.
(b) $\theta_1 = 0.05, \theta_2 = 0$.
(c) $\theta_1 = 0.05, \theta_2 = 0.2$.
(d) $\theta_1 = 0.05, \theta_2 = 0.5$.
(e) Now pick the single stepsize that works the best on all four of the above exercises.

6.8 Consider a random variable given by $R = 10U$ (which would be uniformly distributed between 0 and 10). We wish to use a stochastic gradient algorithm to estimate the mean of R using the iteration $\bar{\theta}^n = \bar{\theta}^{n-1} - \alpha_{n-1}(R^n - \bar{\theta}^{n-1})$, where R^n is a Monte Carlo sample of R in the nth iteration. For each of the stepsize rules below, use the mean squared error

$$MSE = \sqrt{\frac{1}{N}\sum_{n=1}^{N}(R^n - \bar{\theta}^{n-1})^2} \qquad (6.74)$$

to measure the performance of the stepsize rule to determine which works best, and compute an estimate of the bias and variance at each iteration. If the stepsize rule requires choosing a parameter, justify the choice you make (you may have to perform some test runs).

(a) $\alpha_{n-1} = 1/n$.
(b) Fixed stepsizes of $\alpha_n = .05, .10$ and $.20$.
(c) The stochastic gradient adaptive stepsize rule (equations 6.27)–(6.28)).
(d) The Kalman filter (equations (6.43)–(6.47)).
(e) The optimal stepsize rule (algorithm 6.7).

6.9 Repeat exercise 6.8 using

$$R^n = 10(1 - e^{-0.1n}) + 6(U - 0.5).$$

6.10 Repeat exercise 6.8 using

$$R^n = \left(10/(1 + e^{-0.1(50-n)})\right) + 6(U - 0.5).$$

6.11 Use a stochastic gradient algorithm to solve the problem

$$\min_x \frac{1}{2}(X - x)^2,$$

where X is a random variable. Use a harmonic stepsize rule (equation (6.15)) with parameter $\theta = 5$. Perform 1000 iterations assuming that you observe $X^1 = 6, X^2 = 2, X^3 = 5$ (this can be done in a spreadsheet). Use a starting initial value of $x^0 = 10$. What is the best possible value for θ for this problem?

6.12 Consider a random variable given by $R = 10U$ (which would be uniformly distributed between 0 and 10). We wish to use a stochastic gradient algorithm to estimate the mean of R using the iteration $\bar{\mu}^n = \bar{\mu}^{n-1} - \alpha_{n-1}(R^n - \bar{\mu}^{n-1})$, where R^n is a Monte Carlo sample of R in the nth iteration. For each of the stepsize rules below, use equation (6.74) (see exercise 6.8) to measure the performance of the stepsize rule to determine which works best, and compute an estimate of the bias and variance at each iteration. If the stepsize rule requires choosing a parameter, justify the choice you make (you may have to perform some test runs).

(a) $\alpha_{n-1} = 1/n$.
(b) Fixed stepsizes of $\alpha_n = .05, .10$ and .20.
(c) The stochastic gradient adaptive stepsize rule (equations (6.27)–(6.28)).
(d) The Kalman filter (equations (6.43)–(6.47)).
(e) The optimal stepsize rule (algorithm 6.7).

6.13 Repeat exercise 6.8 using

$$R^n = 10(1 - e^{-0.1n}) + 6(U - 0.5).$$

6.14 Repeat exercise 6.8 using

$$R^n = \left(10/(1 + e^{-0.1(50-n)})\right) + 6(U - 0.5).$$

6.15   Let U be a uniform $[0, 1]$ random variable, and let

$$\mu^n = 1 - \exp(-\theta_1 n).$$

Now let $\hat{R}^n = \mu^n + \theta_2(U^n - .5)$. We wish to try to estimate μ^n using

$$\bar{R}^n = (1 - \alpha_{n-1})\bar{R}^{n-1} + \alpha_{n-1}\hat{R}^n.$$

In the exercises below, estimate the mean (using $\bar{R}^n$) and compute the standard deviation of $\bar{R}^n$ for $n = 1, 2, \dots, 100$, for each of the following stepsize rules:

- $\alpha_{n-1} = 0.10$.
- $\alpha_{n-1} = \theta/(\theta + n - 1)$ for $a = 1, 10$.
- Kesten's rule.
- The bias-adjusted Kalman filter stepsize rule.

For each of the parameter settings below, compare the rules based on the average error (1) over all 100 iterations and (2) in terms of the standard deviation of $\bar{R}^{100}$.

(a) $\theta_1 = 0, \theta_2 = 10$.
(b) $\theta_1 = 0.05, \theta_2 = 0$.
(c) $\theta_1 = 0.05, \theta_2 = 0.2$.
(d) $\theta_1 = 0.05, \theta_2 = 0.5$.
(e) Now pick the single stepsize that works the best on all four of the above exercises.

Theory questions

6.16   Show that if we use a stepsize rule $\alpha_{n-1} = 1/n$, then $\bar{\mu}^n$ is a simple average of $W^1, W^2, \dots, W^n$ (thus proving equation 6.14). Use this result to argue that any solution of equation (6.7) produces the mean of W.

6.17   Prove corollary 6.7.3.

6.18   The bias adjusted Kalman filter (BAKF) stepsize rule (equation (6.53)), is given by

$$\alpha_{n-1} = 1 - \frac{\sigma_\varepsilon^2}{(1 + \lambda^{n-1})\sigma_\varepsilon^2 + (\beta^{n-1})^2},$$

where λ is computed recursively using

$$\lambda^n = \begin{cases} (\alpha_{n-1})^2, & n = 1 \\ (1 - \alpha_{n-1})^2 \lambda^{n-1} + (\alpha_{n-1})^2, & n > 1. \end{cases}$$

Show that for a stationary data series, where the bias $\beta^n = 0$, produces stepsizes that satisfy

$$\alpha_{n-1} = \frac{1}{n} \quad \forall n = 1, 2, \dots .$$

6.19 An important property of the BAKF stepsize policy (equation (6.53)) satisfies the property that $\alpha_n \geq 1/n$.

(a) Why is this important?
(b) Prove that this result holds.

Problem-solving questions

6.20 Assume we have to order x assets after which we try to satisfy a random demand D for these assets, where D is randomly distributed between 100 and 200. If $x > D$, we have ordered too much and we pay $5(x - D)$. If $x < D$, we have an underage, and we have to pay $20(D - x)$.

(a) Write down the objective function in the form $\min_x \mathbb{E} f(x, D)$.
(b) Derive the stochastic gradient for this function.
(c) Find the optimal solution analytically [Hint: take the expectation of the stochastic gradient, set it equal to zero and solve for the quantity $\mathbb{P}(D \leq x^*)$. From this, find x^*.]
(d) Since the gradient is in units of dollars while x is in units of the quantity of the asset being ordered, we encounter a scaling problem. Choose as a stepsize $\alpha_{n-1} = \alpha_0/n$ where α_0 is a parameter that has to be chosen. Use $x^0 = 100$ as an initial solution. Plot x^n for 1000 iterations for $\alpha_0 = 1, 5, 10, 20$. Which value of α_0 seems to produce the best behavior?
(e) Repeat the algorithm (1000 iterations) 10 times. Let $\omega = (1, \dots, 10)$ represent the 10 sample paths for the algorithm, and let $x^n(\omega)$ be the solution at iteration n for sample path ω. Let $Var(x^n)$ be the variance of the random variable x^n where

$$\overline{V}(x^n) = \frac{1}{10} \sum_{\omega=1}^{10} (x^n(\omega) - x^*)^2.$$

Plot the standard deviation as a function of n for $1 \leq n \leq 1000$.

6.21 Show that if we use a stepsize rule $\alpha_{n-1} = 1/n$, then $\bar{\mu}^n$ is a simple average of $W^1, W^2, \ldots, W^n$ (thus proving equation 6.14).

6.22 A customer is required by her phone company to pay for a minimum number of minutes per month for her cell phone. She pays 12 cents per minute of guaranteed minutes, and 30 cents per minute that she goes over her minimum. Let x be the number of minutes she commits to each month, and let M be the random variable representing the number of minutes she uses each month, where M is normally distributed with mean 300 minutes and a standard deviation of 60 minutes.

(a) Write down the objective function in the form $\min_x \mathbb{E} f(x, M)$.
(b) Derive the stochastic gradient for this function.
(c) Let $x^0 = 0$ and choose as a stepsize $\alpha_{n-1} = 10/n$. Use 100 iterations to determine the optimum number of minutes the customer should commit to each month.

6.23 An oil company covers the annual demand for oil using a combination of futures and oil purchased on the spot market. Orders are placed at the end of year $t - 1$ for futures that can be exercised to cover demands in year t. If too little oil is purchased this way, the company can cover the remaining demand using the spot market. If too much oil is purchased with futures, then the excess is sold at 70% of the spot market price (it is not held to the following year – oil is too valuable and too expensive to store).

To write down the problem, model the exogenous information using

$$\hat{D}_t \quad = \quad \text{Demand for oil during year } t,$$

$$\hat{p}_t^s \quad = \quad \text{Spot price paid for oil purchased in year } t,$$

$$\hat{p}_{t,t+1}^f \quad = \quad \text{Futures price paid in year } t \text{ for oil to be used in year } t + 1.$$

The demand (in millions of barrels) is normally distributed with mean 600 and standard deviation of 50. The decision variables are given by

$$\bar{\mu}_{t,t+1}^f \quad = \quad \text{Number of futures to be purchased at the end of year } t \text{ to be used in year } t + 1.$$

$$\bar{\mu}_t^s \quad = \quad \text{Spot purchases made in year } t.$$

(a) Set up the objective function to minimize the expected total amount paid for oil to cover demand in a year $t + 1$ as a function of $\bar{\mu}_t^f$. List the variables in your expression that are not known when you have to make a decision at time t.

(b) Give an expression for the stochastic gradient of your objective function. That is, what is the derivative of your function for a particular sample realization of demands and prices (in year $t + 1$)?

(c) Generate 100 years of random spot and futures prices as follows:

$$\hat{p}_t^f = 0.80 + 0.10U_t^f,$$
$$\hat{p}_{t,t+1}^s = \hat{p}_t^f + 0.20 + 0.10U_t^s,$$

where U_t^f and U_t^s are random variables uniformly distributed between 0 and 1. Run 100 iterations of a stochastic gradient algorithm to determine the number of futures to be purchased at the end of each year. Use $\bar{\mu}_0^f = 30$ as your initial order quantity, and use as your stepsize $\alpha_t = 20/t$. Compare your solution after 100 years to your solution after 10 years. Do you think you have a good solution after 10 years of iterating?

Sequential decision analytics and modeling

These exercises are drawn from the online book *Sequential Decision Analytics and Modeling* available at http://tinyurl.com/sdaexamplesprint.

6.24 Read sections 5.1–5.6 on the static shortest path problem. We are going to focus on the extension in section 5.6, where the traveler gets to see the actual link cost $\hat{c}_{ij}$ before traversing the link.

(a) Write out the five elements of this dynamic model. Use our style of representing the policy as $X^\pi(S_t)$ without specifying the policy.

(b) We are going to use a VFA-based policy which requires estimating the function:

$$\overline{V}_t^{x,n}(i) = (1 - \alpha_n)\overline{V}_t^{x,n-1}(i) + \alpha_n \hat{v}_t^n(i).$$

We cover value function approximations in much greater depth later, but at the moment, we are interested in the stepsize α_n, which has a major impact on the performance of the system. The ADP algorithm has been implemented in Python, which can be downloaded from http://tinyurl.com/sdagithub using the module "StochasticShortestPath_Static." The code currently uses the harmonic stepsize rule

$$\alpha_n = \frac{\theta^\alpha}{\theta^\alpha + n - 1},$$

where θ^α is a tunable parameter. Run the code for 50 iterations using $\theta^\alpha = 1, 2, 5, 10, 20, 50$ and report on the performance.

(c) Implement the stepsize rule RMSProp (described in section 6.2.3) (which has its own tunable parameter), and compare your best implementation of RMSProp with your best version of the harmonic stepsize.

Diary problem

The diary problem is a single problem you chose (see chapter 1 for guidelines). Answer the following for your diary problem.

6.25 Try to identify at least one, but more if possible, parameters (or functions) that you would have to adaptively estimate in an online fashion, either from a flow of real data, or from an iterative search algorithm. For each case, answer the following:

(a) Describe the characteristics of the observations in terms of the degree of stationary or nonstationary behavior, the amount of noise, and whether the series might undergo sudden shifts (this would only be the case for data coming from live observations).

(b) Suggest one deterministic stepsize policy, and one adaptive stepsize policy, for each data series, and explain your choice. Then compare these to the BAKF policy and discuss strengths and weaknesses.

Bibliography

Broadie, M., Cicek, D., and Zeevi, A. (2011). General bounds and finite-time improvement for the Kiefer-Wolfowitz stochastic approximation algorithm. *Operations Research* 59 (5): 1211–1224.

Brown, R.G. (1959). *Statistical Forecasting for Inventory Control*. New York: McGrawHill.

Brown, R.G. (1963). *Smoothing, Forecasting and Prediction of Discrete Time Series*. Englewood Cliffs, N.J: PrenticeHall.

Darken, C. and Moody, J. (1991). Note on learning rate schedules for stochastic optimization. In: *Advances in Neural Information Processing Systems 3* (eds. R.P. Lippmann, J. Moody and D.S. Touretzky), 1009–1016.

Darken, C., Chang, J., and Moody, J. (1992). Learning rate schedules for faster stochastic gradient search. In: *Neural Networks for Signal Processing 2 Proceedings of the 1992 IEEE Workshop.*

Douglas, S.C. and Mathews, V.J. (1995). Stochastic gradient adaptive step size algorithms for adaptive filtering. *Proc. International Conference on Digital Signal Processing*, Limassol, Cyprus 1: 142–147.

Evendar, E. and Mansour, Y. (2003). Learning rates for Q-learning. *Journal of Machine Learning Research* 5: 1–25.

Gardner, E.S. (1983). Automatic monitoring of forecast errors. *Journal of Forecasting* 2: 1–21.

Giffin, W.C. (1971). *Introduction to Operations Engineering.* Homewood, IL: R. D. Irwin, Inc.

Goodwin, G.C. and Sin, K.S. (1984). *Adaptive Filtering and Control.* Englewood Cliffs, NJ: PrenticeHall.

Haykin, S. (1999). *Neural Networks: A comprehensive foundation.* Englewood Cliffs, N.J: Prentice Hall.

Holt, C.C., Modigliani, F., Muth, J., and Simon, H. (1960). *Planning, Production, Inventories and Work Force.* Englewood Cliffs, NJ: PrenticeHall.

Jaakkola, T., Singh, S.P., and Jordan, M.I. (1994). Reinforcement learning algorithm for partially observable Markov decision problems. *Advances in Neural Information Processing Systems* 7: 345.

Kesten, H. (1958). Accelerated stochastic approximation. *The Annals of Mathematical Statistics* 29: 41–59.

Kushner, H.J. and Yin, G.G. (2003). *Stochastic Approximation and Recursive Algorithms and Applications*, New York: Springer.

Meinhold, R.J. and Singpurwalla, N.D. (2007). Understanding the Kalman Filter. *The American Statistician* 37 (2): 123–127.

Mirozahmedov, F. and Uryasev, S. (1983). Adaptive Stepsize regulation for stochastic optimization algorithm. Zurnal vicisl. mat. i. mat. fiz. 23 (6): 1314–1325.

Pflug, G. (1988). Stepsize rules, stopping times and their implementation in stochastic quasigradient algorithms. In: *Numerical Techniques for Stochastic Optimization*, 353–372. New York: SpringerVerlag.

Powell, W.B. and George, A.P. (2006). Adaptive stepsizes for recursive estimation with applications in approximate dynamic programming. *Journal of Machine Learning* 65 (1): 167–198.

Ruszczyński, A. and Syski, W. (1986). A method of aggregate stochastic subgradients with online stepsize rules for convex stochastic programming problems. *Mathematical Programming Study* 28: 113–131.

Ryzhov, I.O., Frazier, P.I. and Powell, W.B. (2015). A newoptimal stepsize for approximate dynamic programming. *IEEE Transactions on Automatic Control* 60 (3): 743–758.

Sutton, R.S. and Singh, S.P. (1994). On step-size and bias in temporal-difference learning. In: *Eight Yale Workshop on Adaptive and Learning Systems* (ed. C. for System Science), 91–96.

Yale University. Trigg, D.W. (1964). Monitoring a forecasting system. *Operations Research Quarterly* 15: 271–274.

7

Derivative-Free Stochastic Search

There are many settings where we wish to solve

$$\max_{x \in \mathcal{X}} \mathbb{E}\{F(x, W)|S^0\}, \tag{7.1}$$

which is the same problem that we introduced in the beginning of chapter 5. When we are using derivative-free stochastic search, we assume that we can choose a point x^n according to some policy that uses a belief about the function that we can represent by $\bar{F}^n(x) \approx \mathbb{E}F(x, W)$ (as we show below, there is more to the belief than a simple estimate of the function). Then, we observe the performance $\hat{F}^{n+1} = F(x^n, W^{n+1})$. Random outcomes can be the response of a patient to a drug, the number of ad-clicks from displaying a particular ad, the strength of a material from a mixture of inputs and how the material is prepared, or the time required to complete a path over a network. After we run our experiment, we use the observed performance $\hat{F}^{n+1}$ to obtain an updated belief about the function, $\bar{F}^{n+1}(x)$.

We may use derivative-free stochastic search because we do not have access to the derivative (or gradient) $\nabla F(x, W)$, or even a numerical approximation of the derivative. The most obvious examples arise when x is a member of a discrete set $\mathcal{X} = \{x_1, \dots, x_M\}$, such as a set of drugs or materials, or perhaps different choices of websites. In addition, x may be continuous, and yet we cannot even approximate a derivative. For example, we may want to test a drug dosage on a patient, but we can only do this by trying different dosages and observing the patient for a month.

There may also be problems which can be solved using a stochastic gradient algorithm (possibly using numerical derivatives). It is not clear that a gradient-based solution is necessarily better. We suspect that if stochastic gradients can be calculated directly (without using numerical derivatives), that this is likely going to be the best approach for high-dimensional problems (fitting neural networks are a good example). But there are going to be problems where both methods may apply, and it simply will not be obvious which is the best approach.

We are going to approach our problem by designing a policy (or algorithm) $X^\pi(S^n)$ that chooses $x^n = X^\pi(S^n)$ given what we know about $\mathbb{E}\{F(x, W)|S^0\}$ as captured by our approximation

$$\bar{F}^n \approx \mathbb{E}\{F(x, W)|S^0\}.$$

For example, if we are using a Bayesian belief for discrete $x \in \mathcal{X} = \{x_1, \dots, x_M\}$, our belief B^n would consist of a set of estimates $\bar{\mu}_x^n$ and precisions β_x^n for each $x \in \mathcal{X}$. Our belief state is then $B^n = (\bar{\mu}_x^n, \beta_x^n)_{x \in \mathcal{X}}$ which is updated using, for $x = x^n$,

$$\bar{\mu}_x^{n+1} = \frac{\beta_x^n \bar{\mu}_x^n + \beta^W W^{n+1}}{\beta_x^n + \beta^W},$$

$$\beta_x^{n+1} = \beta_x^n + \beta^W.$$

We first saw these equations in chapter 3. Alternatively, we might use a linear model $f(x|\theta)$ which we would write

$$f(x|\bar{\theta}^n) = \bar{\theta}_0^n + \bar{\theta}_1^n \phi_1(x) + \bar{\theta}_2^n \phi_2(x) + \bar{\theta}_2^n \phi_2(x) + \dots,$$

where $\phi_f(x)$ is a feature drawn from the input x, which could include data from a website, a movie, or a patient (or patient type). The coefficient vector $\bar{\theta}^n$ would be updated using the equations for recursive least squares (see section 3.8) where the belief state B^n consists of the estimates of the coefficients $\bar{\theta}^n$ and a matrix M^n.

After choosing $x^n = X^\pi(S^n)$, then observing a response $\hat{F}^{n+1} = F(x^n, W^{n+1})$, we update our approximation to obtain $\bar{F}^{n+1}$ which we capture in our belief state S^{n+1} using the methods we presented in chapter 3. We represent the updating of beliefs using

$$S^{n+1} = S^M(S^n, x^n, W^{n+1}).$$

This could be done using any of the updating methods described in chapter 3. This process produces a sequence of states, decisions, and information that we will typically write as

$$(S^0, x^0 = X^\pi(S^0), W^1, S^1, x^1 = X^\pi(S^1), W^2, \dots, S^n, x^n = X^\pi(S^n), W^{n+1}, \dots).$$

In real applications, we have to stop at some finite N. This changes our optimization problem from the asymptotic formulation in (7.1) to the problem (which we now state using the expanded form):

$$\max_{\pi} \mathbb{E}_{S^0}\mathbb{E}_{W^1,\dots,W^N|S^0}\mathbb{E}_{\widehat{W}|S^0}\{F(x^{\pi,N},\widehat{W})|S^0\} \tag{7.2}$$

where $x^{\pi,N}$ depends on the sequence $W^1,\dots,W^N$.

This is the final-reward formulation that we discussed in chapter 4. We can also consider a cumulative reward objective given by

$$\max_{\pi} \mathbb{E}_{S^0}\mathbb{E}_{W^1,\dots,W^N|S^0}\left\{\sum_{n=0}^{N-1} F(X^{\pi}(S^n),W^{n+1})|S^0\right\}. \tag{7.3}$$

For example, we might use (7.2) when we are running laboratory experiments to design a new solar panel, or running computer simulations of a manufacturing process that produces the strongest material. By contrast, we would use (7.3) if we want to find the price that maximizes the revenue from selling a product on the internet, since we have to maximize revenues over time while we are experimenting. We note here the importance of using the expanded form for expectations when comparing (7.2) and (7.3).

An entire book could be written on derivative-free stochastic search. In fact, entire books and monographs have been written on specific versions of the problem, as well as specific classes of solution strategies. This chapter is going to be a brief tour of this rich field.

Our goal will be to provide a unified view that covers not only a range of different formulations (such as final reward and cumulative reward), but also the different classes of policies that we can use. This will be the first chapter where we do a full pass over all four classes of policies that we first introduced in chapter 1. We will now see them all put to work in the context of pure learning problems. We note that in the research literature, each of the four classes of policies are drawn from completely different fields. This is the first time that all four are illustrated at the same time.

7.1 Overview of Derivative-free Stochastic Search

There are a number of dimensions to the rich problem class known as derivative-free stochastic search. This section is designed as an introduction to this challenging field.

7.1.1 Applications and Time Scales

Examples of applications that arise frequently include:

- Computer simulations – We may have a simulator of a manufacturing system or logistics network that models inventories for a global supply chain. The simulation may take anywhere from several seconds to several days to run. In fact, we can put in this category any setting that involves the computer to evaluate a complex function.

- Internet applications – We might want to find the ad that produces the most ad-clicks, or the features of a website that produce the best response.
- Transportation – Choosing the best path over a network – After taking a new position and renting a new apartment, you use the internet to identify a set of K paths – many overlapping, but covering modes such as walking, transit, cycling, Uber, and mixtures of these. Each day you get to try a different path x to try to learn the time required μ_x to traverse path x.
- Sports – Identifying the best team of basketball players – A coach has 15 players on a basketball team, and has to choose a subset of five for his starting lineup. The players vary in terms of shooting, rebounding and defensive skills.
- Laboratory experiments – We may be trying to find the catalyst that produces a material of the highest strength. This may also depend on other experimental choices such as the temperature at which a material is baked, or the amount of time it is exposed to the catalyst in a bath.
- Medical decision making – A physician may wish to try different diabetes medications on a patient, where it may take several weeks to know how a patient is responding to a drug.
- Field experiments – We may test different products in a market, which can take a month or more to evaluate the product. Alternatively, we may experiment with different prices for the product, where we may wait several weeks to assess the market response. Finally, a university may admit students from a high-school to learn how many accept the offer of admission; the university cannot use this information until the next year.
- Policy search – We have to decide when to store energy from a solar array, when to buy from or sell to the grid, and how to manage storage to meet the time varying loads of a building. The rules may depend on the price of energy from the grid, the availability of energy from the solar array, and the demand for energy in the building. Policy search is typically performed in a simulator, but may also be done in the field.

These examples bring out the range of time scales that can arise in derivative-free learning:

- Fractions of a second to seconds – Running simple computer simulations, or assessing the response to posting a popular news article.
- Minutes – Running more expensive computer simulations, testing the effect of temperature on the toxicity of a drug.
- Hour – Assessing the effect of bids for internet ads.

- Hours to days – Running expensive computer simulations, assessing the effect of a drug on reducing fevers, evaluating the effect of a catalyst on materials strength.
- Weeks – Test marketing new products and testing prices.
- Year – Evaluating the performance of people hired from a particular university, observing matriculation of seniors from a high school.

7.1.2 The Communities of Derivative-free Stochastic Search

Derivative-free search arises in so many settings that the literature has evolved in a number of communities. It helps to understand the diversity of perspectives.

Statistics The earliest paper on derivative-free stochastic search appeared in 1951, which interestingly appeared in the same year as the original paper for derivative-based stochastic search.

Applied probability The 1950s saw the first papers on "one-armed" and "two-armed" bandits laying the foundation for the multiarmed bandit literature that has emerged as one of the most visible communities in this field (see below).

Simulation In the 1970s the simulation community was challenged with the problem of designing manufacturing systems. Simulation models were slow, and the challenge was finding the best configuration given limited computing resources. This work became known as "simulation optimization."

Geosciences Out in the field, geoscientists were searching for oil and faced the problem of deciding where to dig test wells, introducing the dimension of evaluating surfaces that were continuous but otherwise poorly structured.

Operations research Early work in operations research on derivative-free search focused more on optimizing complex deterministic functions. The OR community provided a home for the simulation community and their work on ranking and selection.

Computer science The computer science community stumbled into the multiarmed bandit problem in the 1980s, and developed methods that were much simpler than those developed by the applied probability community. This has produced an extensive literature on upper confidence bounding.

7.1.3 The Multiarmed Bandit Story

We would not be doing justice to the learning literature if we did not acknowledge the contribution of a substantial body of research that addresses what is known as the *multiarmed bandit problem*. The term comes from the common

Figure 7.1 A set of slot machines.

description (in the United States) that a slot machine (in American English), which is sometimes known as a "fruit machine" (in British English), is a "one-armed bandit" since each time you pull the arm on the slot machine you are likely to lose money (see Figure 7.1).

Now imagine that you have to choose which out of a group of slot machines to play (a surprising fiction since winning probabilities on slot machines are carefully calibrated). Imagine (and this is a stretch) that each slot machine has a different winning probability, and that the only way to learn about the winning probability is to play the machine and observe the winnings. This may mean playing a machine where your estimate of winnings is low, but you acknowledge that your estimate may be wrong, and that you have to try playing the machine to improve your knowledge.

This classic problem has several notable characteristics. The first and most important is the tradeoff between exploration (trying an arm that does not seem to be the best in order to learn more about it) and exploitation (trying arms with higher estimated winnings in order to maximize winnings over time), where winnings are accumulated over time. Other distinguishing characteristics of the basic bandit problem include: discrete choices (that is, slot machines, generally known as "arms"), lookup table belief models (there is a belief about each individual machine), and an underlying process that is stationary (the distribution of winnings does not change over time). Over time, the bandit community has steadily generalized the basic problem.

Multiarmed bandit problems first attracted the attention of the applied probability community in the 1950s, initially in the context of the simpler two-armed problem. It was first formulated in 1970 as a dynamic program that characterized the optimal policy, but it could not be computed. The multiarmed problem resisted computational solution until the development in 1974 by J.C. Gittins who identified a novel decomposition that led to what are known as *index policies* which involves computing a value ("index") for each arm, and then choosing the arm with the greatest index. While "Gittins indices" (as they came to be known) remain computationally difficult to compute, the elegant simplicity of index policies has guided research into an array of policies that are quite practical.

In 1985, a second breakthrough came from the computer science community, when it was found that a very simple class of policies known as *upper confidence bound* (or UCB) policies (also described below) enjoyed nice theoretical properties in the form of bounds on the number of times that the wrong arm would be visited. The ease with which these policies can be computed (they are a form of index policy) has made them particularly popular in high-speed settings such as the internet where there are many situations where it is necessary to make good choices, such as which ad to post to maximize the value of an array of services.

Today, the literature on "bandit problems" has expanded far from its original roots to include any sequential learning problem (which means the state S^n includes a belief state about the function $\mathbb{E}F(x, W)$) where we control the decisions of where to evaluate $F(x, W)$. However, bandit problems now include many problem variations, such as

- Maximizing the final reward rather than just cumulative rewards.
- "Arms" no longer have to be discrete; x may be continuous and vector-valued.
- Instead of one belief about each arm, a belief might be in the form of a linear model that depends on features drawn from x.
- The set of available "arms" to play may change from one round to the next.

The bandit community has fostered a culture of creating problem variations, and then deriving index policies and proving properties (such as regret bounds) that characterize the performance of the policy. While the actual performance of the UCB policies requires careful experimentation and tuning, the culture of creating problem variations is a distinguishing feature of this community. Table 7.1 lists a sampling of these bandit problems, with the original multiarmed bandit problem at the top.

7.1.4 From Passive Learning to Active Learning to Bandit Problems

Chapter 3 describes recursive (or adaptive) learning methods that can be described as a sequence of inputs x^n followed by an observed response y^{n+1}.

Table 7.1 A sample of the growing population of "bandit" problems.

Bandit problem	Description
Multiarmed bandits	Basic problem with discrete alternatives, online (cumulative regret) learning, lookup table belief model with independent beliefs
Best-arm bandits	Identify the optimal arm with the largest confidence given a fixed budget
Restless bandits	Truth evolves exogenously over time
Adversarial bandits	Distributions from which rewards are being sampled can be set arbitrarily by an adversary
Continuum-armed bandits	Arms are continuous
X-armed bandits	Arms are a general topological space
Contextual bandits	Exogenous state is revealed which affects the distribution of rewards
Dueling bandits	The agent gets a relative feedback of the arms as opposed to absolute feedback
Arm-acquiring bandits	New machines arrive over time
Intermittent bandits	Arms are not always available
Response surface bandits	Belief model is a response surface (typically a linear model)
Linear bandits	Belief is a linear model
Dependent bandits	A form of correlated beliefs
Finite horizon bandits	Finite-horizon form of the classical infinite horizon multiarmed bandit problem
Parametric bandits	Beliefs about arms are described by a parametric belief model
Nonparametric bandits	Bandits with nonparametric belief models
Graph-structured bandits	Feedback from neighbors on graph instead of single arm
Extreme bandits	Optimize the maximum of recieved rewards
Quantile-based bandits	The arms are evaluated in terms of a specified quantile
Preference-based bandits	Find the correct ordering of arms

If we have no control over the inputs x^n, then we would describe this as *passive learning*.

In this chapter, the inputs x^n are the results of decisions that we make, where it is convenient that the standard notation for the inputs to a statistical model, and decisions for an optimization model, both use x. When we directly control the inputs (that is, we choose x^n), or when decisions influence the inputs, then we would refer to this as *active learning*. Derivative-free stochastic search can always be described as a form of active learning, since we control (directly or indirectly) the inputs which updates a belief model.

At this point you should be asking: what is the difference between derivative-free stochastic search (or as we now know it, active learning) and multiarmed bandit problems? At this stage, we think it is safe to say that the following problem classes are equivalent:

(a) Sequential decision problems with (a) a dynamic belief state and (b) where decisions influence the observations used to update beliefs.
(b) Derivative-free stochastic search problems.
(c) Active learning problems.
(d) Multiarmed bandit problems.

Our position is that problem class (a) is the clearest description of these problems. We note that we are not excluding derivative-based stochastic search in principle. Our presentation of derivative-based stochastic search in chapter 5 did not include any algorithms with a belief state, but we suspect that this will happen in the near future.

A working definition of a bandit problem could be any active learning problem that has been given a label "[adjective]-bandit problem." We claim that *any* sequential decision problem with a dynamic belief state, and where decisions influence the evolution of the belief state, is either a form of bandit problem, or waiting to be labeled as such.

7.2 Modeling Derivative-free Stochastic Search

As with all sequential decision problems, derivative-free stochastic search can be modeled using the five core elements: state variables, decision variables, exogenous information, transition function, and objective function. We first describe each of these five elements in a bit more detail, and then illustrate the model using the context of a problem that involves designing a manufacturing process.

7.2.1 The Universal Model

Our universal model of any sequential decision problem consists of five elements: state variables, decision variables, exogenous information, the transition function, and the objective function. Below we describe these elements in slightly more detail for the specific context of derivative-free stochastic optimization.

State variables – For derivative-free stochastic optimization, our state variable S^n after n experiments consists purely of the belief state B^n about the function $\mathbb{E}F(x, W)$. In chapter 8 we will introduce problems where we have a physical state R^n such as our budget for making experiments, or the location of a drone collecting information, in which case our state would be $S^n = (R^n, B^n)$. We might have the attributes of a patient in addition to the belief how the patient will respond to a treatment, which gives us a state $S^n = (I^n, B^n)$ (these are often called "contextual problems"), in addition to all three classes of state variables, giving us $S^n = (R^n, I^n, B^n)$. However, this chapter will focus almost exclusively on problems where $S^n = B^n$.

The belief B^0 will contain initial estimates of unknown parameters of our belief models. Often, we will have a prior distribution of belief about parameters, in which case B^0 will contain the parameters describing this distribution.

If we do not have any prior information, we will likely have to do some initial exploration, which tends to be guided by some understanding of the problem (especially scaling).

Decision variables – The decision x^n, made after n experiments (which means using the information from S^n), may be binary (do we accept web site A or B), discrete (one of a finite set of choices), continuous (scalar or vector), integer (scalar or vector), and categorical (e.g. the choice of patient type characterized by age, gender, weight, smoker, and medical history).

Decisions are typically made subject to a constraint $x^n \in \mathcal{X}^n$, using a policy that we denote $X^\pi(S^n)$. Here, "π" carries information about the type of function and any tunable parameters. If we run N experiments using policy $X^\pi(S^n)$, we let $x^{\pi,N}$ be the final design. In some cases the policy will be time dependent, in which case we would write it as $X^{\pi,n}(S^n)$.

Most of the time we are going to assume that our decision is to run a single, discrete experiment that returns an observation $W^{n+1}_{x^n}$ or $\hat{F}^{n+1} = F(x^n, W^{n+1})$, but there will be times where x^n_a represents the number of times we run an experiment on "arm" a.

Exogenous information – We let W^{n+1} be the new information that arrives after we choose to run experiment x^n. Often, W^{n+1} is the performance of an experiment, which we would write $W^{n+1}_{x^n}$. More generally, we will write our response function as $F(x^n, W^{n+1})$, in which case W^{n+1} represents observations made that allow us to compute $F(x, W)$ given the decision x. In some cases we may use $F(x^n, W^{n+1})$ to represent the process of running an experiment, where we observe a response $\hat{F}^{n+1} = F(x^n, W^{n+1})$.

Transition function – We denote the transition function by

$$S^{n+1} = S^M(S^n, x^n, W^{n+1}). \tag{7.4}$$

In derivative-free search where S^n is typically the belief about the unknown function $\mathbb{E}F(x, W)$, the transition function represents the recursive updating of statistical model using the methods described in chapter 3. The nature of the updating equations will depend on the nature of the belief model (e.g. lookup table, parametric, neural networks) and whether we are using frequentist or Bayesian belief models.

Objective functions – There are a number of ways to write objective functions in sequential decision problems. Our default notation for derivative-free stochastic search is to let

$$
\begin{aligned}
F(x, W) \quad = \quad &\text{the response (could be a contribution or cost, or any} \\
&\text{performance metric) of running experiment } x^n \text{ (said} \\
&\text{differently, running an experiment with design} \\
&\text{parameters } x^n).
\end{aligned}
$$

Note that $F(x, W)$ is not a function of S_t; we deal with those problems starting in chapter 8.

If we are running a series of experiments in a computer or laboratory setting, we are typically interested in the final design $x^{\pi,N}$, which is a random variable that depends on the initial state S^0 (that may contain a prior distribution of belief B^0) and the experiments $W^1_{x^0}, W^2_{x^1}, \dots, W^N_{x^{N-1}}$. This means that $x^{\pi,N} = X^{\pi}(S^N)$ is a random variable. We can evaluate this random variable by running it through a series of tests that we capture with the random variable $\widehat{W}$, which gives us the final-reward objective function

$$
\max_{\pi} \mathbb{E}\{F(x^{\pi,N}, W)|S^0\} = \mathbb{E}_{S^0}\mathbb{E}_{W^1,\dots,W^N|S^0}\mathbb{E}_{\widehat{W}|S^0}F(x^{\pi,N}, \widehat{W}) \tag{7.5}
$$

where $S^0 = B^0$ which is our initial belief about the function.

There are settings where we are running the experiments in the field, and we care about the performance of each of the experiments. In this case, our objective would be the cumulative reward given by

$$
\max_{\pi} \mathbb{E}_{S^0}\mathbb{E}_{W^1,\dots,W^N|S^0}\left\{\sum_{n=0}^{N-1} F(x^n, W^{n+1})|S^0\right\} \tag{7.6}
$$

where $x^n = X^{\pi}(S^n)$ and where S^0 includes in B^0 anything we know (or believe) about the function before we start.

There are many flavors of performance metrics. We list a few more in section 7.11.1.

We encourage readers to write out all five elements any time you need to represent a sequential decision problem. We refer to this problem as the *base model*. We need this term because later we are going to introduce the idea of a *lookahead model* where approximations are introduced to simplify calculations.

Our challenge, then, is to design effective policies that work well in our base model. We first illustrate this in the context of a classical problem of optimizing a simulation of a manufacturing system.

7.2.2   Illustration: Optimizing a Manufacturing Process

Assume that $x \in \mathcal{X} = \{x_1, \dots, x_M\}$ represents different configurations for manufacturing a new model of electric vehicle which we are going to evaluate using a simulator. Let $\mu_x = \mathbb{E}_W F(x, W)$ be the expected performance if we could run an infinitely long simulation. We assume that a single simulation (of reasonable duration) produces the performance

$$\hat{F}_x = \mu_x + \varepsilon,$$

where $\varepsilon \sim N(0, \sigma_W^2)$ is the noise from running a single simulation.

Assume we use a Bayesian model (we could do the entire exercise with a frequentist model), where our prior on the truth μ_x is given by $\mu_x \sim N(\bar{\mu}_x^0, \bar{\sigma}_x^{2,0})$. Assume that we have performed n simulations, and that $\mu_x \sim N(\bar{\mu}_x^n, \bar{\sigma}_x^{2,n})$. Our belief B^n about μ_x after n simulations is then given by

$$B^n = (\bar{\mu}_x^n, \bar{\sigma}_x^{2,n})_{x \in \mathcal{X}}. \tag{7.7}$$

For convenience, we are going to define the *precision* of an experiment as $\beta^W = 1/\sigma_W^2$, and the precision of our belief about the performance of configuration x as $\beta_x^n = 1/\bar{\sigma}_x^{2,n}$.

If we choose to try configuration x^n and then run the $n + 1^{st}$ simulation and observe $\hat{F}^{n+1} = F(x^n, W^{n+1})$, we update our beliefs using

$$\bar{\mu}_x^{n+1} = \frac{\beta_x^n \bar{\mu}_x^n + \beta^W \hat{F}_x^{n+1}}{\beta_x^n + \beta^W}, \tag{7.8}$$

$$\beta_x^{n+1} = \beta_x^n + \beta^W, \tag{7.9}$$

if $x = x^n$; otherwise, $\bar{\mu}_x^{n+1} = \bar{\mu}_x^n$ and $\beta_x^{n+1} = \beta_x^n$. These updating equations assume that beliefs are independent; it is a minor extension to allow for correlated beliefs.

We are now ready to state our model using the canonical framework:

State variables The state variable is the belief $S^n = B^n$ given by equation (7.7).
Decision variables The decision variable is the configuration $x \in \mathcal{X}$ that we wish to test next, which will be determined by a policy $X^\pi(S^n)$.

Exogenous information This is the simulated performance given by
$$\hat{F}^{n+1}(x^n) = F(x^n, W^{n+1}).$$

Transition function These are given by equations (7.8)–(7.9) for updating the beliefs.

Objective function We have a budget to run N simulations of different configurations. When the budget is exhausted, we choose the best design according to

$$x^{\pi,N} = \arg\max_{x \in \mathcal{X}} \bar{\mu}_x^N,$$

where we introduce the policy π because $\bar{\mu}_x^N$ has been estimated by running experiments using experimentation policy $X^\pi(S^n)$. The performance of a policy $X^\pi(S^n)$ is given by

$$F^\pi(S^0) = \mathbb{E}_{S^0} \mathbb{E}_{W^1,\dots,W^N|S^0} \mathbb{E}_{\widehat{W}|S^0} F(x^{\pi,N}, \widehat{W}).$$

Our goal is to then solve

$$\max_\pi F^\pi(S^0).$$

This problem called for an objective that optimized the performance of the final design $x^{\pi,N}$, which we call the final reward objective. However, we could change the story to one that involved learning in the field, where we want to optimize as we learn, in which case we would want to optimize the cumulative reward. The choice of objective does not change the analysis approach, but it will change the choice of policy that works best.

7.2.3 Major Problem Classes

There is a wide range of applications that fall under the umbrella of derivative-free stochastic search. Some of the most important features from the perspective of design policies (which we address next) are:

- Characteristics of the design x – The design variable x may be binary, finite, continuous scalar, vector (discrete or continuous), and multiattribute.
- Noise level – This captures the variability in the outcomes from one experiment to the next. Experiments may exhibit little to no noise, up to tremendously high noise levels, where the noise greatly exceeds the variations among μ_x.
- Time required for an experiment – Experiments can take fractions of a second, seconds, minutes up to hours, weeks, and months.
- Learning budget – Closely related to the time required for an experiment is the budget we have for completing a series of experiments and choosing

a design. There are problems where we have a budget of 5,000 observations of ad-clicks to learn the best of 1,000 ads, or a budget of 30 laboratory experiments to learn the best compound out of 30,000.

- Belief model – It helps when we can exploit underlying structural properties when developing belief models. Beliefs may be correlated, continuous (for continuous x), concave (or convex) in x, monotone (outcomes increase or decrease with x). Beliefs may also be Bayesian or frequentist.
- Steady state or transient – It is standard to assume we are observing a process that is not changing over time, but this is not always true.
- Hidden variables – There are many settings where the response depends on variables that we either cannot observe, or simply are not aware of (this may be revealed as a transient process).

The range of problems motivates our need to take a general approach toward designing policies.

7.3 Designing Policies

We now turn to the problem of designing policies for either the final reward objective (7.2) or the cumulative reward (7.3). There are two strategies for designing policies, each of which can be further divided into two classes, producing four classes of policies. We provide a brief sketch of these here, and then use the rest of the chapter to give more in-depth examples. It will not be apparent at first, but all four classes of policies will be useful for particular instances of derivative-free stochastic search problems.

Most of the time through this book we use t as our time index, as in x_t and S_t. With derivative-free stochastic search, the most natural indexing is the counter n, as in the nth experiment, observation or iteration. We index the counter n in the superscript (as we first described in chapter 1), which means we have the decision x^n (this is our decision *after* we run our n^{th} experiment), and S^n, which is the information we use to make the decision x^n.

The four classes of polices are given by:

Policy search – Here we use any of the objective functions such as (7.5) or (7.6) to search within a family of functions to find the policy that works best. Policies in the policy-search class can be further divided into two classes:

   Policy function approximations (PFAs) – PFAs are analytical functions that map states to actions. They can be lookup tables, or linear models which might be of the form

$$X^{PFA}(S^n|\theta) = \sum_{f \in \mathcal{F}} \theta_f \phi_f(S^n).$$

PFAs can also be nonlinear models such as a neural network, although these can require an extremely large number of training iterations.

Cost function approximations (CFAs) – CFAs are parameterized optimization models. A simple one that is widely used in pure learning problems, called interval estimation, is given by

$$X^{CFA-IE}(S^n|\theta^{IE}) = \arg \max_{x \in \mathcal{X}} (\bar{\mu}_x^n + \theta^{IE} \bar{\sigma}_x^n) \tag{7.10}$$

where $\bar{\sigma}_x^n$ is the standard deviation of $\bar{\mu}_x^n$ which declines as the number of times we observe alternative x grows.

A CFA could be a simple sort, as arises with the interval estimation policy in (7.10), but it could also be a linear, nonlinear, or integer program, which makes it possible for x to be a large vector instead of one of a discrete set. We can state this generally as

$$X^{CFA}(S^n|\theta) = \arg \max_{x \in \mathcal{X}^\pi(\theta)} \bar{C}^\pi(S^n, x|\theta),$$

where $\bar{C}^\pi(S^n, x|\theta)$ might be a parametrically modified objective function (e.g. with penalties), while $\mathcal{X}^\pi(\theta)$ might be parametrically modified constraints.

Lookahead approximations – An optimal policy can be written as

$$X^{*,n}(S^n) = \arg \max_{x^n} \Bigg(C(S^n, x^n) + $$
$$\mathbb{E} \Bigg\{ \max_\pi \mathbb{E} \Bigg\{ \sum_{m=n+1}^{N} C(S^m, X^{\pi,m}(S^m)) \Bigg| S^{n+1} \Bigg\} \Bigg| S^n, x^n \Bigg\} \Bigg) \tag{7.11}$$

Remember that $S^{n+1} = S^M(S^n, x^n, W^{n+1})$, where there are two potential sources of uncertainty: the exogenous information W^{n+1}, as well as uncertainty about parameters that would be captured in S^n. Remember that for derivative-free stochastic search, the state S^n is our belief state after n observations, which typically consists of continuous parameters (in some cases, vectors of continuous parameters, such as the presence of diseases across countries).

In practice, equation (7.11) cannot be computed, so we have to use approximations. There are two approaches for creating these approximations:

Value function approximations (VFAs) – The ideal VFA policy involves solving Bellman's equation

$$V^n(S^n) = \max_x \left(C(S^n, x) + \mathbb{E}\{V^{n+1}(S^{n+1}) | S^n, x\} \right), \qquad (7.12)$$

where

$$V^{n+1}(S^{n+1}) = \max_\pi \mathbb{E} \left\{ \left. \sum_{m=n+1}^N C(S^m, X^{\pi,m}(S^m)) \right| S^{n+1} \right\}.$$

If we could compute this, our optimal policy would be given by

$$X^{*,n}(S^n) = \arg\max_{x \in \mathcal{X}^n} \left(C(S^n, x) + \mathbb{E}\{V^{n+1}(S^{n+1}) | S^n, x\} \right). \qquad (7.13)$$

Typically we cannot compute $V^{n+1}(S^{n+1})$ exactly. A popular strategy known as "approximate dynamic programming" involves replacing the value function with an approximation $\overline{V}^{n+1}(S^{n+1})$ which gives us

$$X^{VFA,n}(S^n) = \arg\max_{x \in \mathcal{X}^n} \left(C(S^n, x) + \mathbb{E}\{\overline{V}^{n+1}(S^{n+1} | \theta) | S^n, x\} \right). \qquad (7.14)$$

Since expectations can be impossible to compute (and approximations are computationally expensive), we often use a value function approximation around the post-decision state, which eliminates the expectation:

$$X^{VFA,n}(S^n) = \arg\max_{x \in \mathcal{X}^n} \left(C(S^n, x) + \overline{V}^{x,n}(S^{x,n} | \theta) \right). \qquad (7.15)$$

Direct lookaheads (DLAs) – The second approach is to create an *approximate lookahead model*. If we are making a decision at time t, we represent our lookahead model using the same notation as the base model, but replace the state S^n with $\tilde{S}^{n,m}$, the decision x^n with $\tilde{x}^{n,m}$ which is determined with policy $\tilde{X}^{\tilde{\pi}}(\tilde{S}^{n,m})$, and the exogenous information W^n with $\tilde{W}^{n,m}$. This creates a *lookahead model* that can be written

$$(S^n, x^n, \tilde{W}^{n,n+1}, \tilde{S}^{n,n+1}, \tilde{x}^{n,n+1}, \tilde{W}^{n,n+2}, \dots, \tilde{S}^{n,m}, \tilde{x}^{n,m}, \tilde{W}^{n,m+1}, \dots).$$

We are allowed to introduce any approximations that we think are appropriate for a lookahead model. For example, we may change the belief model, or we may simplify the different types of uncertainty. This gives us an approximate lookahead policy

$$X^{DLA,n}(S^n) = \arg\max_x \left(C(S^n, x) + \right.$$

$$\left. \tilde{\mathbb{E}} \left\{ \max_{\tilde{\pi}} \tilde{\mathbb{E}} \left\{ \left. \sum_{m=n+1}^N C(\tilde{S}^{n,m}, \tilde{X}^{\tilde{\pi}}(\tilde{S}^{n,m})) | \tilde{S}^{n,n+1} \right\} | S^n, x \right\} \right). \qquad (7.16)$$

We emphasize that the lookahead model may be deterministic, but in learning problems the lookahead model has to capture uncertainty. These can be hard to solve, which is why we create a lookahead model that is distinct from the base model which is used to evaluate the policy. We return to lookahead models below.

There are communities in derivative-free stochastic search that focus on *each* of these four classes of policies, so we urge caution before jumping to any conclusions about which class seems best. We emphasize that these are four meta-classes. There are numerous variations within each of the four classes.

We make the claim (backed up by considerable empirical work) that it is important to understand all four classes of policies, given the tremendous variety of problems that we highlighted in section 7.2.3. Finding the best compound out of 3,000 possible choices, with experiments that take 2-4 days to complete, with a budget of 60 days (this is a real problem), is very different than finding the best ads to display to maximize ad-clicks, when we might test 2,000 different ads each day, with millions of views from users. It is inconceivable that we could solve both settings with the same policy.

The next four sections cover each of the four classes of policies:

- Section 7.4 – Policy function approximations
- Section 7.5 – Cost function approximations
- Section 7.6 – Policies based on value function approximations
- Section 7.7 – Policies based on direct lookahead models

After these, sections 7.8, 7.9 and 7.10 provide additional background into two important classes of policies. Section 7.11 discusses evaluating policies, followed by section 7.12 provides some guidance in choosing a policy. We close with a discussion of a series of extensions to our basic model.

7.4 Policy Function Approximations

A PFA is any function that maps directly from a state to an action without solving an imbedded optimization problem. PFAs may be any of the function classes we covered in chapter 3, but for pure learning problems, they are more likely to be a parametric function. Some examples include

- An excitation policy – Imagine that demand as a function of price is given by

$$D(p) = \theta_0 - \theta_1 p.$$

We might want to maximize revenue $R(p) = pD(p) = \theta_0 p - \theta_1 p^2$, where we do not know θ_0 and θ_1. Imagine that we have estimates $\bar{\theta}^n = (\bar{\theta}_0^n, \bar{\theta}_1^n)$ after n experiments. Given $\bar{\theta}^n$, the price that optimizes revenue is

$$p^n = \frac{\bar{\theta}_0^n}{2\bar{\theta}_1^n}.$$

After we post price p^n, we observe demand $\hat{D}^{n+1}$, and then use this to update our estimate $\bar{\theta}^n$ using recursive least squares (see section 3.8).

We can learn more effectively if we introduce some noise, which we can do using

$$p^n = \frac{\bar{\theta}_0^n}{2\bar{\theta}_1^n} + \varepsilon^{n+1} \tag{7.17}$$

where $\varepsilon \sim N(0, \sigma_\varepsilon^2)$, and where the exploration variance σ_ε^2 is a tunable parameter. Let $P^{exc}(S^n | \sigma_\varepsilon)$ represent the excitation policy that determines the price p^n in equation (7.17), parameterized by σ_ε. Also let

$$\hat{R}(p^n, \hat{D}^{n+1}) \quad = \quad \text{the revenue we earn when we charge price } p^n$$
$$\text{and then observe demand } \hat{D}_{t+1}.$$

We tune σ_ε by solving

$$\max_{\sigma_\varepsilon} F(\sigma_\varepsilon) = \mathbb{E} \sum_{n=0}^{N-1} \hat{R}(P^{exc}(S^n | \sigma_\varepsilon), \hat{D}^{n+1}).$$

Excitation policies are quite popular in engineering for learning parametric models. They are ideally suited for online learning, because they favor trying points near the optimum.

- For our pricing problem we derived an optimal price given the belief about demand response, but we could simply pose a linear function of the form

$$X^\pi(S^n | \theta) = \sum_{f \in \mathcal{F}} \theta_f \phi_f(S^n). \tag{7.18}$$

Recall that the recursive formulas provided in section 3.8 imply a state variable given by $S^n = B^n = (\bar{\theta}^n, M^n)$. We now determine θ by solving

$$\max_{\theta} F(\theta) = \mathbb{E} \sum_{n=0}^{N-1} F(X^\pi(S^n | \theta), W^{n+1}). \tag{7.19}$$

We note that $F(\theta)$ is typically highly nonconcave in θ. Algorithms for solving (7.19) remain an active area of research. We revisit this in chapter 12 when we consider policy function approximations for state-dependent problems.

- Neural networks – While neural networks are growing in popularity as policies, as of this writing we are not aware of their use in pure learning problems as a policy, but this might be an area of research. For example, it is not obvious how to design features $\phi_f(S)$ when the state variables are given by $S^n = (\bar{\theta}^n, M^n)$. A neural network would be able to handle this type of nonlinear response.

If $X^\pi(S^n|\theta)$ is the neural network and θ is the weight vector (note that θ might have thousands of dimensions), the challenge would be to optimize the weights using equation (7.19). Note that the price of this generality is that it would require many iterations to find a good weight vector.

We note in passing that if there is an imbedded optimization problem (which is usually the case) then the policy is technically a form of cost function approximation.

7.5 Cost Function Approximations

Cost function approximations represent what is today one of the most visible and popular classes of learning policies. CFAs describe policies where we have to maximize (or minimize) something to find the alternative to try next, and where we do not make any effort at approximating the impact of a decision now on the future. CFAs cover a wide range of practical, and surprisingly powerful, policies.

Simple greedy policies – We use the term "simple greedy policy" to refer to a policy which chooses an action which maximizes the expected reward given current beliefs, which would be given by

$$X^{SG}(S^n) = \arg\max_x \bar{\mu}_x^n.$$

Now imagine that we have a nonlinear function $F(x, \theta)$ where θ is an unknown parameter where, after n experiments, might be normally distributed with distribution $N(\theta^n, \sigma^{2,n})$. Our simple greedy policy would solve

$$\begin{aligned} X^{SG}(S^n) &= \arg\max_x F(x, \theta^n), \\ &= \arg\max_x F(x, \mathbb{E}(\theta|S^n)). \end{aligned}$$

This describes a classical approach known under the umbrella as *response surface methods* where we pick the best action based on our latest statistical approximation of a function. We can then add a noise term as we did in our excitation policy in equation (7.17), which introduces a tunable parameter σ_ε.

Bayes greedy – Bayes greedy is just a greedy policy where the expectation is kept on the outside of the function (where it belongs), which would be written

$$X^{BG}(S^n) = \arg\max_x \mathbb{E}_\theta\{F(x,\theta)|S^n\}.$$

When the function $F(x,\theta)$ is nonlinear in θ, this expectation can be tricky to compute. One strategy is to use a sampled belief model and assume that $\theta \in \{\theta_1, \dots, \theta_K\}$, and let $p_k^n = Prob[\theta = \theta_k]$ after n iterations. We would then write our policy as

$$X^{BG}(S^n) = \arg\max_x \sum_{k=1}^K p_k^n F(x,\theta_k).$$

Finally, we can add a noise term $\varepsilon \sim N(0,\sigma_\varepsilon^2)$, which would then have to be tuned.

Upper confidence bounding – UCB policies, which are very popular in computer science, come in many flavors, but they all share a form that follows one of the earliest UCB policies given by

$$\nu_x^{UCB,n} = \bar{\mu}_x^n + 4\sigma^W \sqrt{\frac{\log n}{N_x^n}}, \tag{7.20}$$

where $\bar{\mu}_x^n$ is our estimate of the value of alternative x, and N_x^n is the number of times we evaluate alternative x within the first n iterations. The coefficient $4\sigma^W$ has a theoretical basis, but is typically replaced with a tunable parameter θ^{UCB} which we might write as

$$\nu_x^{UCB,n}(\theta^{UCB}) = \bar{\mu}_x^n + \theta^{UCB} \sqrt{\frac{\log n}{N_x^n}}. \tag{7.21}$$

The UCB policy, then, would be

$$X^{UCB}(S^n|\theta^{UCB}) = \arg\max_x \nu_x^{UCB,n}(\theta^{UCB}), \tag{7.22}$$

where θ^{UCB} would be tuned using an optimization formulation such as that given in (7.19).

UCB policies all use an index composed of a current estimate of the value of alternative ("arm" in the language of the bandit-oriented UCB community), given by $\bar{\mu}_x^n$, plus a term that encourages exploration, sometimes called an "uncertainty bonus." As the number of observations grows, $\log n$ also grows (but logarithmically), while N_x^n counts how many times we have sampled alternative x. Note that since initially $N_x^0 = 0$, the UCB policy assumes that we have a budget to try every alternative at least once. When the number of alternatives exceeds the budget, we either need a prior, or to move away from a lookup table belief model.

Interval estimation – Interval estimation is a class of UCB policy, with the difference that the uncertainty bonus is given by the standard deviation $\bar{\sigma}^n_x$ of the estimate $\bar{\mu}^n_x$ of the value of alternative x. The interval estimation policy is then given by

$$X^{IE}(S^n|\theta^{IE}) = \arg\max_x \left(\bar{\mu}^n_x + \theta^{IE} \bar{\sigma}^n_x \right). \qquad (7.23)$$

Here, $\bar{\sigma}^n_x$ is our estimate of the standard deviation of $\bar{\mu}^n_x$. As the number of times we observe action x goes to infinity, $\bar{\sigma}^n_x$ goes to zero. The parameter θ^{IE} is a tunable parameter, which we would tune using equation (7.19).

Thompson sampling – Thompson sampling works by sampling from the current belief about $\mu_x \sim N(\bar{\mu}^n_x, \bar{\sigma}^{n,2}_x)$, which can be viewed as the prior distribution for experiment $n+1$. Now choose a sample $\hat{\mu}^n_x$ from the distribution $N(\bar{\mu}^n_x, \bar{\sigma}^{n,2}_x)$. The Thompson sampling policy is then given by

$$X^{TS}(S^n) = \arg\max_x \hat{\mu}^n_x.$$

Thompson sampling is more likely to choose the alternative x with the largest $\bar{\mu}^n_x$, but because we sample from the distribution, we may also choose other alternatives, but are unlikely to choose alternatives where the estimate $\bar{\mu}^n_x$ is low relative to the others.

Note that we can create a tunable version of Thompson sampling by choosing $\hat{\mu}^n_x \sim N(\bar{\mu}^n_x, (\theta^{TS}\bar{\sigma}^n_x)^2)$, in which case we would write our policy as $X^{TS}(S^n|\theta^{TS})$. Now we just have to tune θ^{TS}.

Boltzmann exploration – A different form of maximizing over actions involves computing a probability that we pick an action x, given an estimate $\bar{\mu}^n_x$ of the reward from this action. This is typically computed using

$$p^n(x|\theta) = \frac{e^{\theta \bar{\mu}^n_x}}{\sum_{x'} e^{\theta \bar{\mu}^n_{x'}}}. \qquad (7.24)$$

Now pick x^n at random according to the distribution $p^n(x|\theta)$. Boltzmann exploration is sometimes referred to as "soft max" since it is performing a maximization in a probabilistic sense.

Both PFAs and CFAs require tuning a parameter θ (which is often a scalar but may be a vector), where the tuning can be used to maximize either a final reward or cumulative reward objective function. We note that searching for θ is its own sequential decision problem, which requires a policy that we would call a *learning policy* (finding θ) that then produces a good *implementation policy* $X^\pi(S^n|\theta)$.

7.6 VFA-based Policies

A powerful algorithmic strategy for some problem classes is based on Bellman's equation, which we introduced briefly in section 2.1.3. This approach has received less attention in the context of learning problems, with the notable exception of one community that is centered on the idea known as "Gittins indices" which we review below. Gittins indices, popular in the applied probability community, are virtually unheard of in computer science which focuses on upper confidence bounding. Gittins indices are much harder to compute (as we show below), but it was Gittins indices that introduced the idea of using an index policy that was the original inspiration for UCB policies (this connection occurred in 1984, 10 years after the first paper on Gittins indices).

In this section, we are going to begin in section 7.6.1 by introducing the general idea of using Bellman's equation for pure learning problems. Section 7.6.2 will illustrate the ideas in the context of a simple problem where we are testing a new drug, where observations are all 0 or 1. Then, section 7.6.3 will introduce a powerful approximation strategy based on the idea of approximating value functions. We close in section 7.6.4 by covering the rich history and theory behind Gittins indices, which laid the foundation for modern research in pure learning problems.

7.6.1 An Optimal Policy

Consider the graph in Figure 7.2(a). Imagine that our state (that is, the node where we are located) is $S^n = 2$, and we are considering a decision $x_s = 5$ that puts us in state $S^{n+1} = 5$. Let $\mathcal{X}^n$ be the states (nodes) we can reach from state (node) S^n, and assume that we have a value $V^{n+1}(s')$ for each $s \in \mathcal{X}^n$. Then we can write the value of being in state S^n using Bellman's equation, which gives us

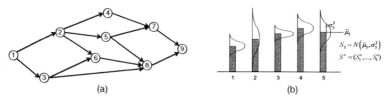

Figure 7.2 (a) Optimizing over a graph, where we are considering the transition from node (state) 2 to node (state) 5. (b) Optimizing a learning problem, where we are considering evaluating alternative 5, which would change our belief (state) $S_5^n = (\bar{\mu}_x^n, \sigma_5^{2,n})$ to belief (state) $S_5^{n+1} = (\bar{\mu}_x^{n+1}, \sigma_5^{2,n+1})$.

$$V^n(S^n) = \max_{s' \in \mathcal{X}^n} \left(C(S^n, s') + V^{n+1}(s') \right). \tag{7.25}$$

Bellman's equation, as given in equation (7.25), is fairly intuitive. In fact, this is the foundation of every shortest path algorithm that is used in modern navigation systems.

Now consider the learning problem in Figure 7.2(b). Our state is our belief about the true performance of each of the five alternatives. Recall that the precision $\beta_x^n = 1/\sigma_x^{2,n}$. We can express our state as $S^n = (\bar{\mu}_x^n, \beta_x^n)_{x \in \mathcal{X}}$. Now imagine we decide to experiment with the 5th alternative, which will give us an observation W_5^{n+1}. The effect of our observation W_5^{n+1} will take us to state

$$\bar{\mu}_5^{n+1} = \frac{\beta_5^n \bar{\mu}_5^n + \beta^W W_5^{n+1}}{\beta_5^n + \beta^W}, \tag{7.26}$$

$$\beta_5^{n+1} = \beta_5^n + \beta^W. \tag{7.27}$$

The values for $\bar{\mu}_x^n$ and β_x^n for x other than 5 are unchanged.

The only differences between our graph problem and our learning problem are:

(a) The decision to move from state (node) 2 to state 5 in the graph problem is a deterministic transition.
(b) The states in our graph problem are discrete, while the state variables in the learning problem are continuous and vector valued.

In our learning problem, we make a decision $x^n = 5$ to test alternative 5, but the outcome W_5^{n+1} is random, so we do not know what state the experiment will take us to. However, we can fix this by inserting an expectation in Bellman's equation, giving us

$$V^n(S^n) = \max_{x \in \mathcal{X}} \left(C(S^n, x) + \mathbb{E}_{S^n} \mathbb{E}_{W|S^n} \{ V^{n+1}(S^{n+1}) | S^n, x \} \right), \tag{7.28}$$

where the first expectation $\mathbb{E}_{S^n}$ handles the uncertainty in the true value μ_x given the belief in S^n, while the second expectation $\mathbb{E}_{W|S^n}$ handles the noise in the observation $W_x^{n+1} = \mu_x + \varepsilon^{n+1}$ of our unknown truth μ_x. Note that if we are using a frequentist belief model, we would just use $\mathbb{E}_W$.

Other than the expectation in equation (7.28), equations (7.25) and (7.28) are basically the same. The point is that we can use Bellman's equation regardless of whether the state is a node in a network, or the belief about the performance of a set of alternatives. State variables are state variables, regardless of their interpretation.

If we could solve equation (7.28), we would have an optimal policy given by

$$X^*(S^n) = \arg\max_{x \in \mathcal{X}} \left(C(S^n, x) + \mathbb{E}_{S^n} \mathbb{E}_{W|S^n}\{V^{n+1}(S^{n+1})|S^n, x\}\right). \quad (7.29)$$

Equation (7.28) is set up for problems which maximize undiscounted cumulative rewards. If we want to solve a final reward problem, we just have to ignore contributions until the final evaluation, which means we write Bellman's equation as

$$V^n(S^n) = \max_{x \in \mathcal{X}} \begin{cases} \left(0 + \mathbb{E}_{S^n} \mathbb{E}_{W|S^n}\{V^{n+1}(S^{n+1})|S^n, x\}\right) & n < N, \\ C(S^N, x) & n = N. \end{cases} \quad (7.30)$$

We could also change our objective to a discounted, infinite horizon model by simply adding a discount factor $\gamma < 1$, which changes equation (7.28) to

$$V^n(S^n) = \max_{x \in \mathcal{X}} \left(C(S^n, x) + \gamma \mathbb{E}_W\{V^{n+1}(S^{n+1})|S^n, x\}\right). \quad (7.31)$$

The problem with Bellman's equation is that while it is not hard finding the value of being at each node in a graph (even if there are 100,000 nodes), handling a belief state is much harder. If we have a problem with just 20 alternatives, the state $S^n = (\bar{\mu}_x^n, \beta_x^n)_{x \in \mathcal{X}}$ would have 40 continuous dimensions, which is an extremely difficult estimation problem given noisy measurements and reasonable computing budgets. There are real applications with thousands of alternatives (or more), as would occur when finding the best out of thousands of compounds to fight a disease, or the best news article to display on the website of a news organization.

7.6.2 Beta-Bernoulli Belief Model

There is an important class of learning problem that can be solved exactly. Imagine that we are trying to learn whether a new drug is successful. We are testing patients, where each test yields the outcome of success ($W^n = 1$) or failure ($W^n = 0$). Our only decision is the set

$$\mathcal{X} = \{continue, patent, cancel\}.$$

The decision to "patent" means to stop the trial and file for a patent on the drug as the step before going to market. The decision to "cancel" means to stop the trial and cancel the drug. We maintain a state variable R^n where

$$R^n = \begin{cases} 1 & \text{if we are still testing,} \\ 0 & \text{if we have stopped testing.} \end{cases}$$

The evolution of R^n is given by

$$R^{n+1} = \begin{cases} 1 & R^n = 1 \text{ and } x^n = \text{"continue",} \\ 0 & \text{otherwise.} \end{cases}$$

As the experiments progress, we are going to keep track of successes α^n and failures β^n using

$$\alpha^{n+1} = \alpha^n + W^{n+1},$$
$$\beta^{n+1} = \beta^n + (1 - W^{n+1}).$$

We can then estimate the probability of success of the drug using

$$\rho^n = \frac{\alpha^n}{\alpha^n + \beta^n}.$$

The state variable is

$$S^n = (R^n, \alpha^n, \beta^n).$$

We can create a belief about the true probability of success, ρ, by assuming that it is given by a beta distribution with parameters (α^n, β^n) which is given by

$$f(\rho|\alpha, \beta) = \frac{\Gamma(\alpha)}{\Gamma(\alpha) + \Gamma(\beta)} \rho^{\alpha-1}(1 - \rho)^{\beta-1}$$

where $\Gamma(k) = k!$. The beta distribution is illustrated in Figure 7.3. Given we are in state S^n, we then assume that $\rho^{n+1} \sim beta(\alpha^n, \beta^n)$.

This is a model that allows us to compute Bellman's equation in (7.28) exactly (for finite horizon problems) since R^n, α^n, and β^n are all discrete. We have to specify the negative cost of continuing (that is, the cost of running the study), along with the contribution of stopping to patent, versus canceling the drug.

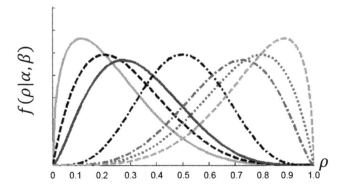

Figure 7.3 A family of densities from a beta distribution.

7.6.3 Backward Approximate Dynamic Programming

Relatively little attention has been directed to learning problems from the communities that use the concepts of approximate dynamic programming. Readers will not see these ideas in this book until chapters 15, 16, and 17. To avoid duplication we are just going to sketch how we might use what is becoming known as backward approximate dynamic programming, which we introduce in greater depth in chapter 15.

Assume for the moment that we have a standard discrete learning model where we are trying to learn the true values μ_x for each $x \in X = \{x_1, \ldots, x_M\}$. Assume our beliefs are normally distributed, which means that after n experiments,

$$\mu_x \sim N(\bar{\mu}_x^n, \bar{\sigma}_x^{2,n}).$$

Our belief state, then, would be

$$S^n = (\bar{\mu}_x^n, \bar{\sigma}_x^n)_{x \in X}.$$

What we are going to do is to replace our value function $V^n(S^n)$ with a linear model of the form

$$V^n(S^n | \theta^n) \approx \bar{V}^n(S^n) = \sum_{f \in \mathcal{F}} \theta_f^n \phi_f(S^n). \tag{7.32}$$

Note that we are making a point of modeling our problem as a finite horizon problem, which means that our value function approximation $\bar{V}^n(S^n)$ depends on the index n, which is why we had to index the vector $\theta^n = (\theta_f^n)_{f \in \mathcal{F}}$.

The features $(\phi_f)_{f \in \mathcal{F}}$ (which do not depend on n) are features that we have designed from our state variable S^n. For example, imagine for our features that we sort the alternatives based on the basis of the index v_x^n given by

$$v_x^n = \bar{\mu}_x^n + 2\bar{\sigma}_x^n.$$

Now assume the alternatives are sorted so that $v_{x_1}^n \geq v_{x_2}^n \geq \ldots \geq v_{x_F}^n$, where F might be the top 20 alternatives (this sorting is very important – the sorting has to be done at every iteration). Next create features such as

$$\begin{aligned}
\phi_{x,1} &= \bar{\mu}_x^n, \\
\phi_{x,2} &= (\bar{\mu}_x^n)^2, \\
\phi_{x,3} &= \bar{\sigma}_x^n, \\
\phi_{x,4} &= \bar{\mu}_x^n \bar{\sigma}_x^n, \\
\phi_{x,5} &= \bar{\mu}_x^n + 2\bar{\sigma}_x^n.
\end{aligned}$$

Now we have five features per alternative, times 20 alternatives which gives us a model with 100 features (and 101 parameters, when we include a constant term).

Backward approximate dynamic programming works roughly as follows:

Step 0 Set $\theta^{N+1} = 0$, giving us $\overline{V}^{N+1}(S^{N+1}) = 0$. Set $n = N$.

Step 1 Sample K states from the set of possible values of the state S. For each sampled state $\hat{s}_k^n$, compute an estimated value $\hat{v}_k^n$ from

$$\hat{v}_k^n = \max_x \left(C(\hat{s}_k^n, x) + \mathbb{E}\{\overline{V}^{n+1}(S^{n+1}|\theta^{n+1})|\hat{s}_k^n, x\} \right), \tag{7.33}$$

where S^{n+1} is computed by sampling what we might observe W_x^{n+1} from choosing to test alternative x, and where $\overline{V}^{n+1}(S^{n+1}|\theta^{n+1})$ is given by our approximation in equation (7.32). The expectation has to average over these outcomes.

Step 2 Take the set of values $(\hat{s}_k^n, \hat{v}_k^n)_{k=1}^K$ and fit a new linear model $\overline{V}^n(s|\theta^n)$ using batch linear regression (see section 3.7.1).

Step 3 Set $n \leftarrow n - 1$. If $n \geq 0$, return to Step 1.

This approach is relatively immune to the dimensionality of the state variable or exogenous information variable. In fact, we can even use these ideas if the belief is expressed using a parametric model, although we would have to design a new set of features. Finally, the logic does not even require that the alternatives x be discrete, since we can think of solving equation (7.33) as a nonlinear programming problem. However, this idea has seen only minimal experimentation.

We suspect that a VFA-based policy is best suited for problems with small learning budgets. Note that the resulting policy is nonstationary (it depends on n), in contrast with the CFA-based policies that are so popular in some communities for their simplicity.

We emphasize the idea of using value function approximations for learning models is quite young, and at this stage we offer no guarantees on the performance of this particular approximation strategy. We present this logic to illustrate the idea that we may be able to solve the so-called "curse of dimensionality" of dynamic programming using statistical models.

7.6.4 Gittins Indices for Learning in Steady State*

We are going to finally turn to what was viewed as a breakthrough result in the 1970s. "Bandit problems" were initially proposed in the 1950s, with some theoretical interest in the 1970s. The idea of characterizing an optimal policy using Bellman's equation (as in equation (7.28)) first emerged during this time. The idea of using the belief as a state variable was a central insight. However, solving Bellman's equation looked completely intractable.

Basic Idea

In 1974, John Gittins introduced the idea of decomposing bandit problems using what is known as "Lagrangian relaxation." In a nutshell, the bandit problem requires that we observe one, and only one, alternative at a time. If we write our decision as $x_i^n = 1$ if we choose to test alternative i, then we would introduce a constraint

$$\sum_{i=1}^{M} x_i^n = 1. \tag{7.34}$$

Now imagine solving our optimization problem where we relax the constraint (7.34). Once we do this, the problem decomposes into M dynamic programs (one for each arm). However, we put a price, call it ν^n, on the constraint (7.34), which means we would write our optimization problem as

$$\min_{(\nu^n)_{n=1}^{N}} \max_{(\pi^n)_{n=1}^{N}} \mathbb{E}\left\{\sum_{n=0}^{N}\left(W_{x^n}^{n+1} + \nu^n\left(\sum_{i=1}^{M} x_i^n - 1\right)\right) | S^0\right\}, \tag{7.35}$$

where $x^n = X^{\pi^n}(S^n)$ is the policy at iteration n.

This problem looks complicated, because we have to find a policy $X^{\pi^n}(S^n)$ for each iteration n, and we also have to optimize the penalty (also known as a dual variable or shadow price) ν^n for each n. But what if we solve the steady state version of the problem, given by

$$\min_{\nu} \max_{\pi} \mathbb{E}\left\{\sum_{n=0}^{\infty} \gamma^n\left(W_{x^n}^{n+1} + \nu\left(\sum_{i=1}^{M} x_i^n - 1\right)\right) | S^0\right\} \tag{7.36}$$

where $x^n = X^{\pi}(S^n)$ is our now stationary policy, and there is a single penalty ν. For the infinite horizon problem we have to introduce a discount factor γ (we could do this for the finite horizon version, but this is generally not necessary).

Now the problem decomposes by alternative. For these problems, we now choose between continuing to test, or stopping. When we continue testing alternative x, we not only receive W_x^{n+1}, we also pay a "penalty" ν. A more natural way to formulate the problem is to assume that if you continue to play, you receive the reward W_x^{n+1}, whereas if you stop you receive a reward $r = -\nu$.

We now have the following dynamic program for each arm where the decision is only whether to "Stop" or "Continue," which gives us

$$V_x(S|r) = \max\{\underbrace{r + \gamma V_x(S|r)}_{Stop}, \underbrace{\mathbb{E}_W\{W_x + \gamma V_x(S'|r)|S^n\}}_{Continue}\}, \tag{7.37}$$

where S' is the updated state given the random observation W whose distribution is given by the belief in S^n. For example, if we have binomial (0/1) outcomes, S^n would be the probability ρ^n that $W = 1$ (as we did in section 7.6.2). For normally distributed rewards, we would have $\mathbb{E}_{W|S^n} = \bar{\mu}^n$. If we stop, we do not learn anything so the state S stays the same.

It can be shown that if we choose to stop sampling in iteration n and accept the fixed payment ρ, then that is the optimal strategy for all future rounds. This means that starting at iteration n, our optimal future payoff (once we have decided to accept the fixed payment) is

$$\begin{aligned} V(S|r) &= r + \gamma r + \gamma^2 r + \cdots \\ &= \frac{r}{1-\gamma}, \end{aligned}$$

which means that we can write our optimality recursion in the form

$$V(S^n|r) = \max\left[\frac{r}{1-\gamma}, \bar{\mu}^n + \gamma\mathbb{E}\{V(S^{n+1}|r)\big|S^n\}\right]. \tag{7.38}$$

Now for the magic of Gittins indices. Let Γ be the value of r which makes the two terms in the brackets in (7.38) equal (the choice of Γ is in honor of Gittins). That is,

$$\frac{\Gamma}{1-\gamma} = \mu + \gamma\mathbb{E}\{V(S|\Gamma)\big|S\}. \tag{7.39}$$

The hard part of Gittins indices is that we have to iteratively solve Bellman's equation for different values of Γ until we find one where equation (7.39) is true. The reader should conclude from this that Gittins indices are computable (this is the breakthrough), but computing them is not easy.

We assume that W is random with a known variance σ_W^2. Let $\Gamma^{Gitt}(\mu, \sigma, \sigma_W, \gamma)$ be the solution of (7.39). The optimal solution depends on the current estimate of the mean, μ, its variance σ^2, the variance of our measurements σ_W^2, and the discount factor γ. For notational simplicity, we are assuming that the experimental noise σ_W^2 is independent of the action x, but this assumption is easily relaxed.

Next assume that we have a set of alternatives $\mathcal{X}$, and let $\Gamma_x^{Gitt,n}(\bar{\mu}_x^n, \bar{\sigma}_x^n, \sigma_W, \gamma)$ be the value of Γ that we compute for each alternative $x \in \mathcal{X}$ given state $S^n = (\bar{\mu}_x^n, \bar{\sigma}_x^n)_{x \in \mathcal{X}}$. An optimal policy for selecting the alternative x is to choose the one with the highest value for $\Gamma_x^{Gitt,n}(\bar{\mu}_x^n, \bar{\sigma}_x^n, \sigma_W, \gamma)$. That is, we would make our choice using

$$\max_x \Gamma_x^{Gitt,n}(\bar{\mu}_x^n, \bar{\sigma}_x^n, \sigma_W, \gamma).$$

Such policies are known as *index policies*, which refer to the property that the parameter $\Gamma_x^{Gitt,n}(\bar{\mu}_x^n, \bar{\sigma}_x^n, \sigma_W, \gamma)$ for alternative x depends only on the characteristics of alternative x. For this problem, the parameters $\Gamma_x^{Gitt,n}(\bar{\mu}_x^n, \bar{\sigma}_x^n, \sigma_W, \gamma)$ are called Gittins indices. While Gittins indices have attracted little attention outside the probability community (given the computational complexity), the concept of index policies captured the attention of the research community in 1984 in the first paper that introduced upper confidence bounding (in our CFA class). So, the real contribution of Gittins index policies is the simple idea of an index policy.

We next provide some specialized results when our belief is normally distributed.

Gittins Indices for Normally Distributed Rewards

Students learn in their first statistics course that normally distributed random variables enjoy a nice property. If Z is normally distributed with mean 0 and variance 1 and if

$$X = \mu + \sigma Z$$

then X is normally distributed with mean μ and variance σ^2. This property simplifies what are otherwise difficult calculations about probabilities of events.

The same property applies to Gittins indices. Although the proof requires some development, it is possible to show that

$$\Gamma^{Gitt,n}(\bar{\mu}^n, \bar{\sigma}^n, \sigma_W, \gamma) = \mu + \Gamma(\frac{\bar{\sigma}^n}{\sigma_W}, \gamma)\sigma_W,$$

where

$$\Gamma(\frac{\bar{\sigma}^n}{\sigma_W}, \gamma) = \Gamma^{Gitt,n}(0, \sigma, 1, \gamma)$$

is a "standard normal Gittins index" for problems with mean 0 and variance 1. Note that $\bar{\sigma}^n/\sigma_W$ decreases with n, and that $\Gamma(\frac{\bar{\sigma}^n}{\sigma_W}, \gamma)$ decreases toward zero as $\bar{\sigma}^n/\sigma_W$ decreases. As $n \to \infty$, $\Gamma^{Gitt,n}(\bar{\mu}^n, \bar{\sigma}^n, \sigma_W, \gamma) \to \bar{\mu}^n$.

Unfortunately, as of this writing, there do not exist easy-to-use software utilities for computing standard Gittins indices. Table 7.2 is exactly such a table for Gittins indices. The table gives indices for both the variance-known and

Table 7.2 Gittins indices $\Gamma(\frac{\sigma^n}{\sigma_w}, \gamma)$ for the case of observations that are normally distributed with mean 0, variance 1, and where $\frac{\sigma^n}{\sigma_w} = \frac{1}{n}$ Adapted from Gittins (1989), 'Multiarmed Bandit Allocation Indices', Wiley and Sons: New York.

	Discount factor			
	Known variance		Unknown variance	
Observations	0.95	0.99	0.95	0.99
1	0.9956	1.5758	-	-
2	0.6343	1.0415	10.1410	39.3343
3	0.4781	0.8061	1.1656	3.1020
4	0.3878	0.6677	0.6193	1.3428
5	0.3281	0.5747	0.4478	0.9052
6	0.2853	0.5072	0.3590	0.7054
7	0.2528	0.4554	0.3035	0.5901
8	0.2274	0.4144	0.2645	0.5123
9	0.2069	0.3808	0.2353	0.4556
10	0.1899	0.3528	0.2123	0.4119
20	0.1058	0.2094	0.1109	0.2230
30	0.0739	0.1520	0.0761	0.1579
40	0.0570	0.1202	0.0582	0.1235
50	0.0464	0.0998	0.0472	0.1019
60	0.0392	0.0855	0.0397	0.0870
70	0.0339	0.0749	0.0343	0.0760
80	0.0299	0.0667	0.0302	0.0675
90	0.0267	0.0602	0.0269	0.0608
100	0.0242	0.0549	0.0244	0.0554

variance-unknown cases, but only for the case where $\frac{\sigma^n}{\sigma_w} = \frac{1}{n}$. In the variance-known case, we assume that σ^2 is given, which allows us to calculate the variance of the estimate for a particular slot machine just by dividing by the number of observations.

Lacking standard software libraries for computing Gittins indices, researchers have developed simple approximations. As of this writing, the most recent of these works as follows. First, it is possible to show that

$$\Gamma(s, \gamma) = \sqrt{-\log \gamma} \cdot b\left(-\frac{s^2}{\log \gamma}\right). \tag{7.40}$$

A good approximation of $b(s)$, which we denote by $\tilde{b}(s)$, is given by

$$
\tilde{b}(s) = \begin{cases}
\frac{s}{\sqrt{2}} & s \leq \frac{1}{7}, \\
e^{-0.02645(\log s)^2 + 0.89106 \log s - 0.4873} & \frac{1}{7} < s \leq 100, \\
\sqrt{s} \,(2 \log s - \log \log s - \log 16\pi)^{\frac{1}{2}} & s > 100.
\end{cases}
$$

Thus, the approximate version of (7.40) is

$$
\Gamma^{Gitt,n}(\mu, \sigma, \sigma_W, \gamma) \approx \bar{\mu}^n + \sigma_W \sqrt{-\log \gamma} \cdot \tilde{b}\left(-\frac{\bar{\sigma}^{2,n}}{\sigma_W^2 \log \gamma}\right). \tag{7.41}
$$

Comments

While Gittins indices were considered a major breakthrough, it has largely remained an area of theoretical interest in the applied probability community. Some issues that need to be kept in mind when using Gittins indices are:

- While Gittins indices were viewed as a computational breakthrough, they are not, themselves, easy to compute.
- Gittins index theory only works for infinite horizon, discounted, cumulative reward problems. Gittins indices are not optimal for finite horizon problems which is what we always encounter in practice, but Gittins indices may still be a useful approximation.
- Gittins theory is limited to lookup table belief models (that is, discrete arms/alternatives) with independent beliefs. This is a major restriction in real applications.

We note that Gittins indices are not widely used in practice, but it was the development of Gittins indices that established the idea of using index policies, which laid the foundation for all the work on upper confidence bounding, first developed in 1984, but which has seen explosive growth post 2000, largely driven by search algorithms on the internet.

7.7 Direct Lookahead Policies

There are certain classes of learning problems that require that we actually plan into the future, just as a navigation package will plan a path to the destination. The difference is that while navigation systems can get away with solving a deterministic approximation, recognizing and modeling uncertainty is central to learning problems.

We are going to begin in section 7.7.1 with a discussion of the types of learning problems where a lookahead policy is likely to add value. Then, we are going to describe a series of lookahead strategies that progress in stages before going to a full stochastic lookahead. These will be presented as follows:

- Section 7.7.2 discusses the powerful idea of using one-step lookaheads, which is very useful for certain problem classes where experiments are expensive.
- When one-step lookaheads do not work, a useful strategy is to do a restricted, multistep lookahead, which we describe in section 7.7.3.
- Section 7.7.4 proposes a full multiperiod, deterministic lookahead.
- Section 7.7.5 illustrates a full, multiperiod stochastic lookaheads, although a proper discussion is given in chapter 19.
- Section 7.7.6 describes a class of hybrid policies.

7.7.1   When do we Need Lookahead Policies?

Lookahead policies are very important when we are managing a physical resource. That is why navigation systems have to plan a path to the destination in order to figure out what to do now. By contrast, the most popular policies for pure learning problems are in the CFA class such as upper confidence bounding, Thompson sampling, and interval estimation.

   However, there are pure learning problems where different classes of lookahead policies are particularly useful. As we progress through our direct lookahead policies, the following problem characteristics will prove to be important:

- Complex belief models – Imagine testing the market response to charging a price $p = \$100$ for a book, and we find that sales are higher than expected. Then we would expect that prices $p = \$95$ and $p = \$105$ would also be higher than expected. This reflects correlated beliefs. Direct lookahead models can capture these interactions, while pure index policies such as upper confidence bounding and Thompson sampling cannot (although they pick up these effects indirectly through the tuning process).
- Expensive experiments/small budgets – There are many settings where experiments are expensive, which means that we will have budgets (typically for both time and money) that limit how many we can do. With a limited budget, we will be more interested in exploring in the early experiments than toward the end. This is particularly pronounced with cumulative rewards, but it is also true with final rewards.
- Noisy experiments/S-curve value of information – There are problems that enjoy the intuitive behavior where the value of repeating the same experiment multiple times provides increasing value, but with decreasing marginal returns, as illustrated in Figure 7.4(a). When the noise from running an experiment is high enough, the marginal value of running an experiment can actually grow, as illustrated in Figure 7.4(b).

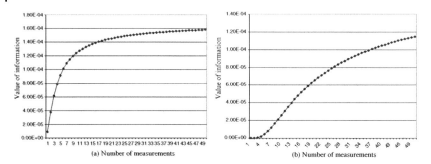

Figure 7.4 Value of making n observations. In (a), the value of information is concave, while in (b) the value of information follows an S-curve.

The S-curve behavior in Figure 7.4(b) arises when experiments are noisy, which means that a single experiment contributes little information. This behavior is actually quite common, especially when the outcome of an experiment is a success or failure (perhaps indicated by 1 or 0).

When we have an S-curve value of information, this means we have to think about how many times we can evaluate an alternative x. It may take 10 repetitions before we are learning anything. If we have 100 alternatives and a budget of 50, we are simply not going to be able to evaluate each alternative 10 times, which means we are going to have to completely ignore a number of alternatives. However, making this decision requires planning into the future given the experimental budget.

- Large number of alternatives relative to the budget – There are problems where the number of alternatives to test is much larger than our budget. This means that we are simply not going to be able to do a good job evaluating alternatives. We would need to plan into the future to know if it is worth trying even one experiment. This is most pronounced when the value of information follows an S-curve, but arises even when the value of information is concave.

7.7.2 Single Period Lookahead Policies

A single period lookahead would never work well if we had to deal with a physical state (imagine solving a shortest path problem over a graph with a single period lookahead). However, they often work exceptionally well in the setting of learning problems. Some approaches for performing single-period lookaheads are given below:

Knowledge gradient for final reward – The most common form of single-period lookahead policy are value-of-information policies, which maximize the value of information from a single experiment. Let $S^n = (\bar{\mu}_x^n, \beta_x^n)_{x \in \mathcal{X}}$ be our belief state now where $\bar{\mu}_x^n$ is our estimate of the performance of design x, and β_x^n is the precision (one over the variance).

Imagine that we are trying to find the design x that maximizes μ_x, where μ_x is unknown. Let $\bar{\mu}_x^n$ be our best estimate of μ given our state of knowledge (captured by $S^n = B^n$). If we stopped now, we would choose our design x^n by solving

$$x^n = \arg\max_{x'} \bar{\mu}_{x'}^n.$$

Now imagine running experiment $x = x^n$, where we will make a noisy observation

$$W_{x^n}^{n+1} = \mu_{x^n} + \varepsilon^{n+1},$$

where $\varepsilon^{n+1} \sim N(0, \sigma_W^2)$. This will produce an updated estimate $\bar{\mu}_{x'}^{n+1}(x^n)$ for $x' = x^n$ which we compute using

$$\bar{\mu}_{x^n}^{n+1} = \frac{\beta_{x^n}^n \bar{\mu}_{x^n}^n + \beta^W W_{x^n}^{n+1}}{\beta_{x^n}^n + \beta^W}, \tag{7.42}$$

$$\beta_{x^n}^{n+1} = \beta_{x^n}^n + \beta^W. \tag{7.43}$$

For $x' \neq x^n$, $\bar{\mu}_{x'}^{n+1}$ and $\beta_{x'}^{n+1}$ are unchanged. This gives us an updated state $S^{n+1}(x) = (\bar{\mu}_{x'}^{n+1}, \beta_{x'}^{n+1})_{x' \in \mathcal{X}}$ which is random because we have not yet observed W_x^{n+1} (we are still trying to decide if we should run experiment x). The value of our solution after this experiment (given what we know at time n) is given by

$$\mathbb{E}_{S^n} \mathbb{E}_{W|S^n} \{\max_{x'} \bar{\mu}^{n+1}(x)) | S^n\}.$$

We can expect that our experiment using parameters (or design) x would improve our solution, so we can evaluate this improvement using

$$\nu^{KG}(x) = \mathbb{E}_{S^n} \mathbb{E}_{W|S^n} \{\max_{x'} \bar{\mu}^{n+1}(x)) | S^n\} - \max_{x'} \bar{\mu}_{x'}^n. \tag{7.44}$$

The quantity $\nu^{KG}(x)$ is known as the *knowledge gradient*, and it gives the expected value of the information from experiment x. This calculation is made by looking one experiment into the future. We cover knowledge gradient policies in considerably greater depth in section 7.8.

Expected improvement – Known as EI in the literature, expected improvement is a close relative of the knowledge gradient, given by the formula

$$v_x^{EI,n} = \mathbb{E}\left[\max\left\{ 0, \mu_x - \max_{x'} \bar{\mu}_{x'}^n \right\} \middle| S^n, x = x^n \right].$$ (7.45)

Unlike the knowledge gradient, EI does not explicitly capture the value of an experiment, which requires evaluating the ability of an experiment to change the final design decision. Rather, it measures the degree to which an alternative x *might* be better. It does this by capturing the degree to which the random truth μ_x *might* be greater than the current best estimate $\max_{x'} \bar{\mu}_{x'}^n$.

Sequential kriging – This is a methodology developed in the geosciences to guide the investigation of geological conditions, which are inherently continuous and two- or three-dimensional. Kriging evolved in the setting of geo-spatial problems where x is continuous (representing a spatial location, or even a location underground in three dimensions). For this reason, we let the truth be the function $\mu(x)$, rather than μ_x (the notation we used when x was discrete).

Kriging uses a form of meta-modeling where the surface is assumed to be represented by a linear model, a bias model and a noise term which can be written as

$$\mu(x) = \sum_{f \in \mathcal{F}} \theta_f \phi_f(x) + Z(x) + \varepsilon,$$

where $Z(x)$ is the bias function and $(\phi_f(x))_{f \in \mathcal{F}}$ are a set of features extracted from data associated with x. Given the (assumed) continuity of the surface, it is natural to assume that $Z(x)$ and $Z(x')$ are correlated with covariance

$$Cov(Z(x), Z(x')) = \beta \exp\left[-\sum_{i=1}^{d} \alpha_i (x_i - x_i')^2 \right],$$

where β is the variance of $Z(x)$ while the parameters α_i perform scaling for each dimension.

The best linear model, which we denote $\bar{Y}^n(x)$, of our surface $\mu(x)$, is given by

$$\bar{Y}^n(x) = \sum_{f \in \mathcal{F}} \theta_f^n \phi_f(x) +$$

$$\sum_{i=1}^{n} Cov(Z(x_i), Z(x)) \sum_{j=1}^{n} Cov(Z(x_j), Z(x))(\hat{y}_i - \sum_{f \in \mathcal{F}} \theta_f^n \phi_f(x)),$$

where θ^n is the least squares estimator of the regression parameters, given the n observations $\hat{y}^1, \dots, \hat{y}^n$.

Kriging starts with the expected improvement in equation (7.45), with a heuristic modification to handle the uncertainty in an experiment (ignored in (7.45)). This gives an adjusted EI of

$$\mathbb{E}^n I(x) = \mathbb{E}^n \left[\max(\bar{Y}^n(x^{**}) - \mu(x), 0) \right] \left(1 - \frac{\sigma_\varepsilon}{\sqrt{\sigma^{2,n}(x) + \sigma_\varepsilon^2}} \right), \quad (7.46)$$

where x^{**} is a point chosen to maximize a utility that might be given by

$$u^n(x) = -(\bar{Y}^n(x) + \sigma^n(x)).$$

Since x is continuous, maximizing $u^n(x)$ over x can be hard, so we typically limit our search to previously observed points

$$x^{**} = \arg \max_{x \in \{x^1, \dots, x^n\}} u^n(x).$$

The expectation in (7.46) can be calculated analytically using

$$\mathbb{E}^n \left[\max(\bar{Y}^n(x^{**}) - \mu(x), 0) \right] = (\bar{Y}^n(x^{**}) - \bar{Y}^n(x)) \Phi \left(\frac{\bar{Y}^n(x^{**}) - \bar{Y}^n(x)}{\sigma^n(x)} \right)$$

$$+ \sigma^n(x) \phi \left(\frac{\bar{Y}^n(x^{**}) - \bar{Y}^n(x)}{\sigma^n(x)} \right),$$

where $\phi(z)$ is the standard normal density, and $\Phi(z)$ is the cumulative density function for the normal distribution.

Value of information policies are well-suited to problems where information is expensive, since they focus on running the experiments with the highest value of information. These policies are particularly effective when the value of information is concave, which means that the marginal value of each additional experiment is lower than the previous one. This property is not always true, especially when experiments are noisy, as we discussed above.

A lookahead policy such as the knowledge gradient is able to take advantage of more complex belief models than simple lookup table beliefs. As we show below, beliefs may be correlated, or may even be parametric models. This is because the knowledge gradient for, say, alternative x has to consider the beliefs for all $x' \in \mathcal{X}$. This is in contrast with the other index policies (such as upper confidence bounding) where the value associated with alternative x has nothing to do with the beliefs for the other alternatives.

7.7.3 Restricted Multiperiod Lookahead

The knowledge gradient, which only looks one step ahead, has been found to be particularly effective when the value of information is concave, as we previously depicted in Figure 7.4(a). However, when the value of information follows an S-curve, as in Figure 7.4(b), the value of doing a single experiment can be almost zero, and provides no guidance toward identifying the best experiments to run.

A common problem with one-step lookahead policies such as the knowledge gradient arises when experiments are very noisy, which means the value of information from a single experiment is quite low. This problem almost always arises when outcomes are 0/1, such as happens when you advertise a product and wait to see if a customer clicks on the ad. Needless to say, no-one would ever make a choice about which alternative is best based on 0/1 outcomes; we would perform multiple experiments and take an average. This is what we call a restricted lookahead, since we restrict our evaluation of the future to using one alternative at a time.

We can formalize this notion of performing repeated experiments. Imagine that instead of doing a single experiment, that we can repeat our evaluation of alternative x n_x times. If we are using a lookup table belief model, this means the updated precision is

$$\beta_x^{n+1}(n_x) = \beta_x^n + n_x \beta^W,$$

where as before, β^W is the precision of a single experiment. Then, we compute the knowledge gradient as given in equation (7.44) (the details are given in section 7.8), but using a precision of $n_x \beta^W$ rather than just β^W.

This leaves the question of deciding how to choose n_x. One way is to use the KG(*) algorithm, where we find the value of n_x that produces the highest *average* value of information. We first compute n_x^* from

$$n_x^* = \arg\max_{n_x > 0} \frac{\nu_x(n_x)}{n_x}. \qquad (7.47)$$

This is illustrated in Figure 7.5. We do this for each x, and then run the experiment with the highest value of $\frac{\nu_x(n_x)}{n_x}$. Note that we are not requiring that each experiment be repeated n_x^* times; we only use this to produce a new index (the maximum average value of information), which we use to identify the next single experiment. This is why we call this strategy a *restricted lookahead policy*; we are looking ahead n_x steps, but only considering doing the same experiment x multiple times.

A more general policy uses the concept of *posterior reshaping*. The idea is quite simple. Introduce a repetition parameter θ^{KGLA} where we let the precision of an experiment be given by

$$\beta_x^{n+1}(\theta^{KGLA}) = \beta_x^n + \theta^{KGLA} \beta^W.$$

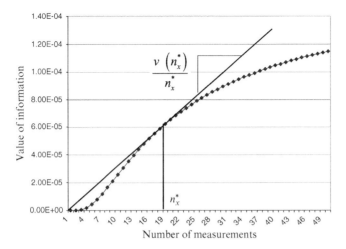

Figure 7.5 The KG(*) policy, which maximizes the average value of a series of experiments testing a single alternative.

Now let $v_x^{KG,n}(\theta^{KGLA})$ be the knowledge gradient when we use repetition factor θ^{KGLA}. Our knowledge gradient policy would be still given by

$$X^{KG}(S^n|\theta^{KGLA}) = \arg\max_x v_x^{KG,n}(\theta^{KGLA}).$$

We now have a tunable parameter, but this is the price of managing this complexity. The value of using a tunable parameter is that the tuning process implicitly captures the effect of the experimental budget N.

7.7.4 Multiperiod Deterministic Lookahead

Assume we have a setting where experiments are noisy, which means we are likely to face an S-shaped value-of-information curve as depicted in Figure 7.4(b). Instead of using θ^{KGLA} as the repeat factor for the knowledge gradient, let y_x be the number of times we plan on repeating the experiment for alternative x given our prior distribution of belief about each alternative. Then let $v_x^{KG,0}(y_x)$ be the value-of-information curve for alternative x, using our prior (time 0) belief, if we plan on running the experiment y_x times.

Assume we start with a budget of R^0 experiments. Previously, we assumed we had a budget of N experiments, but this notation will give us a bit more flexibility.

We can determine the vector $y = (y_x)_{x \in \mathcal{X}}$ by solving the optimization problem

$$\max_y \sum_{x \in \mathcal{X}} v_x^{KG,0}(y_x), \tag{7.48}$$

subject to the constraints:

$$\sum_{x \in \mathcal{X}} y_x \leq R^0, \tag{7.49}$$

$$y_x \geq 0, \quad x \in \mathcal{X}. \tag{7.50}$$

The optimization problem described by equations (7.48)-(7.50) is a non-concave integer programming problem. The good news is that it is very easy to solve optimally using a simple dynamic programming recursion.

Assume that $\mathcal{X} = \{1, 2, ..., M\}$, so that x is an integer between 1 and M. We are going to solve a dynamic program over the alternatives, starting with $x = M$. Let R_x^0 be the number of experiments that we have remaining to allocate over alternatives $x, x + 1, ..., M$. We start with the last alternative where we need to solve

$$\max_{y_M \leq R_M^0} v_M^{KG,0}(y_M). \tag{7.51}$$

Since $v_M^{KG,0}(y_M)$ is strictly increasing, the optimal solution would be $y_M = R_M^0$. Now let

$$V_M(R_M) = v_M^{KG,0}(R_M),$$

which you obtain for $R_M = 0, 1, ..., R^0$ by solving equation (7.51) for each value of R_M^0 (note that you do not have to really "solve" (7.51) at this point, since the solution is just $v_M^{KG,0}(y_M)$).

Now that we have $V_M(R_M)$, we step backward through the alternatives using Bellman's recursion (see chapter 14 for more details):

$$V_x(R_x^0) = \max_{y_x \leq R_x^0} \left(v_x^{KG,0}(y_x) + V_{x+1}(R_{x+1}^0) \right)$$

$$= \max_{y_x \leq R_x^0} \left(v_x^{KG,0}(y_x) + V_{x+1}(R_x^0 - y_x) \right), \tag{7.52}$$

where equation (7.53) has to be solved for $R_x^0 = 0, 1, ..., R^0$. This equation is solved for $x = M - 1, M - 2, ..., 1$.

After we obtain $V_x(R_x^0)$ for each $x \in \mathcal{X}$ and all $0 \le R_x^0 \le R^0$, we can then find an optimal allocation y^0 from

$$y_x^0 = \arg\max_{y_x \le R_x^0} \left(\nu_x^{KG,0}(y_x) + V_{x+1}(R_x^0 - y_x) \right). \tag{7.53}$$

Given the allocation vector $y^0 = (y_x^0)_{x \in \mathcal{X}}$, we now have to decide how to implement this solution. If we can only do one experiment at a time, a reasonable strategy might be to choose the experiment x for which y_x is largest. This would give us a policy that we can write as

$$X^{DLA,n}(S^n) = \arg\max_{x \in \mathcal{X}} y_x^n, \tag{7.54}$$

where we replace y^0 with y^n for iteration n in the calculations above. At iteration n, we would replace R^0 with R^n. After implementing the decision to perform experiment $x^n = X^{DLA,n}(S^n)$, we update $R^{n+1} = R^n - 1$ (assuming we are only performing one experiment at a time). We then observe W^{n+1}, and update the beliefs using our transition function $S^{n+1} = S^M(S^n, x^n, W^{n+1})$ using equations (7.42)–(7.43).

7.7.5 Multiperiod Stochastic Lookahead Policies

A full multiperiod lookahead policy considers making different decisions as we step into the future. We illustrate a full multiperiod lookahead policy for learning using the setting of trying to identify the best hitter on a baseball team. The only way to collect information is to put the hitter into the lineup and observe what happens. We have an estimate of the probability that the player will get a hit, but we are going to update this estimate as we make observations (this is the essence of learning).

Assume that we have three candidates for the position. The information we have on each hitter from previous games is given in Table 7.3. If we choose player A, we have to balance the likelihood of getting a hit, and the value of the information we gain about his true hitting ability, since we will use the event of whether or not he gets a hit to update our assessment of his probability of

Table 7.3 History of hitting performance for three candidates.

Player	No. hits	No. at-bats	Average
A	36	100	0.360
B	1	3	0.333
C	7	22	0.318

getting a hit. We are going to again use Bayes' theorem to update our belief about the probability of getting a hit. Fortunately, this model produces some very intuitive updating equations. Let H^n be the number of hits a player has made in n at-bats. Let $\hat{H}^{n+1} = 1$ if a hitter gets a hit in his $(n + 1)$st at-bat. Our prior probability of getting a hit after n at-bats is

$$\mathbb{P}[\hat{H}^{n+1} = 1 | H^n, n] = \frac{H^n}{n}.$$

Once we observe $\hat{H}^{n+1}$, it is possible to show that the posterior probability is

$$\mathbb{P}[\hat{H}^{n+2} = 1 | H^n, n, \hat{H}^{n+1}] = \frac{H^n + \hat{H}^{n+1}}{n + 1}.$$

In other words, all we are doing is computing the batting average (hits over at-bats).

Our challenge is to determine whether we should try player A, B, or C right now. At the moment, A has the best batting average of 0.360, based on a history of 36 hits out of 100 at-bats. Why would we try player B, whose average is only 0.333? We easily see that this statistic is based on only three at-bats, which would suggest that we have a lot of uncertainty in this average.

We can study this formally by setting up the decision tree shown in Figure 7.6. For practical reasons, we can only study a problem that spans two at-bats. We show the current prior probability of a hit, or no hit, in the first at-bat. For the second at-bat, we show only the probability of getting a hit, to keep the figure from becoming too cluttered.

Figure 7.7 shows the calculations as we roll back the tree. Figure 7.7(c) shows the expected value of playing each hitter for exactly one more at-bat using the information obtained from our first decision. It is important to emphasize that after the first decision, only one hitter has had an at-bat, so the batting averages only change for that hitter. Figure 7.7(b) reflects our ability to choose what we think is the best hitter, and Figure 7.7(a) shows the expected value of each hitter before any at-bats have occurred. We use as our reward function the expected number of total hits over the two at-bats. Let R_x be our reward if batter x is allowed to hit, and let H_{1x} and H_{2x} be the number of hits that batter x gets over his two at-bats. Then

$$R_x = H_{1x} + H_{2x}.$$

Taking expectations gives us

$$\mathbb{E}R_x = \mathbb{E}H_{1x} + \mathbb{E}H_{2x}.$$

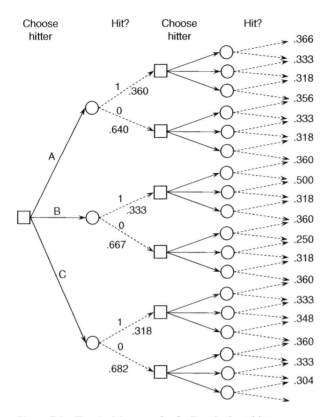

Figure 7.6 The decision tree for finding the best hitter.

So, if we choose batter A, the expected number of hits is

$$\mathbb{E}R_A = .360(1 + .366) + .640(0 + .356)$$
$$= .720$$

where 0.360 is our prior belief about his probability of getting a hit; .366 is the expected number of hits in his second at-bat (the same as the probability of getting a hit) given that he got a hit in his first at-bat. If player A did not get a hit in his first at-bat, his updated probability of getting a hit, 0.356, is still higher than any other player. This means that if we have only one more at-bat, we would still pick player A even if he did not get a hit in his first at-bat.

Although player A initially has the highest batting average, our analysis says that we should try player B for the first at-bat. Why is this? On further exami-nation, we realize that it has a lot to do with the fact that player B has had only three at-bats. If this player gets a hit, our estimate of his probability of getting a

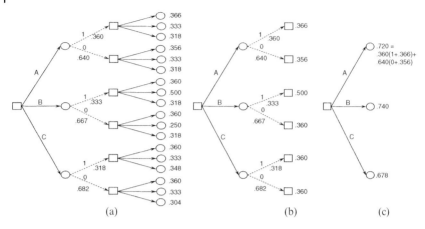

Figure 7.7 (a) Expected value of a hit in the second at-bat; (b) Value of best hitter after one at-bat; (c) Expected value of each hitter before first at-bat.

hit jumps to 0.500, although it drops to 0.250 if he does not get a hit. If player A gets a hit, his batting average moves from 0.360 to 0.366, reflecting the weight of his much longer record. This is our first hint that it can be useful to collect information about choices where there is the greatest uncertainty.

This example illustrates a setting where observations change our beliefs, which we build into the tree. We could have built our tree where all probabilities remain static, which is typical in decision trees. Imbedding the process of updating probabilities within the decision tree is what distinguishes classical decision trees from the use of decision trees in a learning setting.

Decision trees are actually a powerful strategy for learning, although they have not attracted much attention in the learning literature. One reason is simply that they are computationally more difficult, and for most applications, they do not actually work better. Another is that they are harder to analyze, which makes them less interesting in the research communities that analyze algorithms.

7.7.6 Hybrid Direct Lookahead

When we first introduced direct lookahead policies, we described a strategy of doing a full lookahead using an approximate model. That is, if

$$S^0, x^0, W^1, \dots, S^n, x^n, W^{n+1}, \dots$$

represents our base model, the lookahead model might use approximate states $\tilde{S}^{n,m}$, simplified decisions $\tilde{x}^{n,m}$ and/or a sampled information process $\tilde{W}^{n,m}$.

In addition, we might use a simplified policy $\tilde{X}^{\tilde{\pi}}(\tilde{S}^{n,m})$, as illustrated in the lookahead policy which we first presented in equation (11.24), but which we replicate here:

$$
X^{DLA,n}(S^n) = \arg\max_{x^n} \Bigg(C(S^n, x^n) +
$$

$$
\tilde{E}\left\{\max_{\tilde{\pi}} \tilde{E}\left\{\sum_{m=n+1}^{N} C(\tilde{S}^{n,m}, \tilde{X}^{\tilde{\pi}}(\tilde{S}^{n,m}))|\tilde{S}^{n,n+1}\right\}|S^n, x^n\right\}\Bigg). \tag{7.55}
$$

To illustrate a type of hybrid policy, we are not going to approximate the model in any way (the state variables, decisions, and observations will all be just as they are in the base model). But we are going to suggest using a simple UCB or interval estimation policy for $\tilde{X}^{\tilde{\pi}}(\tilde{S}^{n,m})$. For example, we might use

$$
\tilde{X}^{\tilde{\pi}}(\tilde{S}^{n,m}|\tilde{\theta}^{IE}) = \arg\max_{x \in \mathcal{X}} \left(\tilde{\mu}^{n,m} + \tilde{\theta}^{IE}\tilde{\sigma}^{n,m}\right).
$$

Let's see how this works. First, the imbedded optimization over policies in equation (20.24) is replaced with a search over $\tilde{\theta}^{IE}$, which we write as

$$
X^{DLA,n}(S^n) = \arg\max_{x^n} \Bigg(C(S^n, x^n) +
$$

$$
\tilde{E}\left\{\max_{\tilde{\theta}^{IE}} \tilde{E}\left\{\sum_{m=n+1}^{N} C(\tilde{S}^{n,m}, \tilde{X}^{\tilde{\pi}}(\tilde{S}^{n,m}|\tilde{\theta}^{IE}))|\tilde{S}^{n,n+1}\right\}|S^n, x^n\right\}\Bigg). \tag{7.56}
$$

This seems easier, but we have to understand what is meant by that max operator imbedded in the policy. What is happening is that as we make a decision x^n for observation n, we then sample an outcome W^{n+1} given x^n which brings us to a (stochastic) state $\tilde{S}^{n,n+1}$. It is only then that we are supposed to optimize over $\tilde{\theta}^{IE}$, which means we are actually trying to find a function $\tilde{\theta}^{IE}(\tilde{S}_{n,n+1})$. Wow!

Clearly, no-one is going to do this, so we are just going to fix a single parameter θ^{IE}. Note that we no longer have a tilde over it, because it is now a part of the base policy, not the lookahead policy. This means that it is tuned as part of the base policy, so the lookahead is now written

$$
X^{DLA,n}(S^n|\theta^{IE}) = \arg\max_{x^n} \Bigg(C(S^n, x^n) +
$$

$$
\tilde{E}\left\{\tilde{E}\left\{\sum_{m=n+1}^{N} C(\tilde{S}^{n,m}, \tilde{X}^{\tilde{\pi}}(\tilde{S}^{n,m}|\tilde{\theta}^{IE}))|\tilde{S}^{n,n+1}\right\}|S^n, x^n\right\}\Bigg). \tag{7.57}
$$

Now we have gotten rid of the imbedded max operator entirely. We now have a parameterized policy $X^{DLA,n}(S^n|\theta^{IE})$ where θ^{IE} has to be tuned. However, this seems much more manageable.

So, how does our policy $X^{DLA,n}(S^n|\theta^{IE})$ actually work? Assume that our decision x^n is a scalar that we can enumerate. What we can do is for each value of x^n, we can simulate our interval estimation lookahead policy some number of times, and take an average.

We mention this idea primarily to illustrate how we can use a simpler policy inside a lookahead model. Of course, there has to be a reason why we are using a lookahead policy in the first place, so why would we expect a simple IE policy to work? Part of the reason is that approximations within a lookahead policy do not introduce the same errors as if we tried to use this policy instead of the lookahead policy.

7.8 The Knowledge Gradient (Continued)*

The knowledge gradient belongs to the class of value-of-information policies which choose alternatives based on the improvement in the quality of the objective from better decisions that arise from a better understanding of the problem. The knowledge gradient works from a Bayesian belief model where our belief about the truth is represented by a probability distribution of possible truths. The basic knowledge gradient calculates the value of a single experiment, but this can be used as a foundation for variations that allow for repeated experiments.

The knowledge gradient was originally developed for offline (final reward) settings, so we begin with this problem class. Our experience is that the knowledge gradient is particularly well suited for settings where experiments (or observations) are expensive. For example:

- An airline wants to know the effect of allowing additional schedule slack, which can only be evaluated by running dozens of simulations to capture the variability due to weather. Each simulation may take several hours to run.
- A scientist needs to evaluate the effect of increasing the temperature of a chemical reaction or the strength of a material. A single experiment may take several hours, and needs to be repeated to reduce the effect of the noise in each experiment.
- A drug company is running clinical trials on a new drug, where it is necessary to test the drug at different dosages for toxicity. It takes several days to assess the effect of the drug at a particular dosage.

After developing the knowledge gradient for offline (final reward) settings, we show how to compute the knowledge gradient for online (cumulative reward) problems. We begin by discussing belief models, but devote the rest of this section to handling the special case of independent beliefs. Section 7.8.4 extends the knowledge gradient to a general class of nonlinear parametric belief models.

7.8.1 The Belief Model

The knowledge gradient uses a Bayesian belief model where we begin with a prior on $\mu_x = \mathbb{E}F(x, W)$ for $x \in \{x_1, \dots, x_M\}$. We are going to illustrate the key ideas using a lookup table belief model (which is to say, we have an estimate for each value of x), where we initially assume the beliefs are independent. This means that anything we learn about some alternative x does not teach us anything about an alternative x'.

We assume that we believe that the true value of μ_x is described by a normal distribution $N(\bar{\mu}_x^0, \bar{\sigma}_x^{2,0})$, known as the prior. This may be based on prior experience (such as past experience with the revenue from charging a price x for a new book), some initial data, or from an understanding of the physics of a problem (such as the effect of temperature on the conductivity of a metal).

It is possible to extend the knowledge gradient to a variety of belief models. A brief overview of these is:

Correlated beliefs Alternatives x may be related, perhaps because they are discretizations of a continuous parameter (such as temperature or price), so that μ_x and μ_{x+1} are close to each other. Trying x then teaches us something about μ_{x+1}. Alternatively, x and x' may be two drugs in the same class, or a product with slightly different features. We capture these relationships with a covariance matrix Σ^0 where $\Sigma_{xx'}^0 = Cov(\mu_x, \mu_{x'})$. We show how to handle correlated beliefs below.

Parametric linear models We may derive a series of features $\phi_f(x)$, for $f \in \mathcal{F}$. Assume that we represent our belief using

$$f(x|\theta) = \sum_{f \in \mathcal{F}} \theta_f \phi_f(x),$$

where $f(x|\theta) \approx \mathbb{E}F(x, W)$ is our estimate of $\mathbb{E}F(x, W)$. We now treat θ as the unknown parameter, where we might assume that the vector θ is described by a multivariate normal distribution $N(\theta^0, \Sigma^{\theta,0})$, although coming up with these priors (in the parameter space) can be tricky.

Parametric nonlinear models Our belief model might be nonlinear in θ. For example, we might use a logistic regression

$$f(x|\theta) = \frac{e^{U(x|\theta)}}{1 + e^{U(x|\theta)}}, \tag{7.58}$$

where $U(x|\theta)$ is a linear model given by

$$U(x|\theta) = \theta_0 + \theta_1 x_1 + \theta_2 x_2 + \dots + \theta_K x_K$$

where $(x_1, \dots, x_K)$ are the features of a decision x.

Belief models that are nonlinear in the parameters can cause some difficulty, but we can circumvent this by using a sampled belief model, where we assume the uncertain θ is one of the set $\{\theta_1, \dots, \theta_K\}$. Let p_k^n be the probability that $\theta = \theta_k$, which means that $p^n = (p_k^n)$, $k = 1, \dots, K$ is our belief at time n. See section 3.9.2 for more information.

Nonparametric models Simpler nonparametric models are primarily local approximations, so we could use constant, linear, or nonlinear models defined over local regions. More advanced models include neural networks (the kind known as "deep learners") or support vector machines, both of which were introduced in chapter 3.

Below we show how to calculate the knowledge gradient for each of these belief models, with the exception of the nonparametric models (listed for completeness).

7.8.2 The Knowledge Gradient for Maximizing Final Reward

The knowledge gradient seeks to learn about the value of different actions by maximizing the value of information from a single observation. Let S^n be our belief state about the value of each action x. The knowledge gradient uses a Bayesian model, so

$$S^n = (\bar{\mu}_x^n, \sigma_x^{2,n})_{x \in \mathcal{X}},$$

captures the mean and variance of our belief about the true value $\mu_x = \mathbb{E}F(x, W)$, where we also assume that $\mu_x \sim N(\bar{\mu}_x^n, \sigma_x^{2,n})$.

The value of being in belief state S^n is given by

$$V^n(S^n) = \mu_{x^n},$$

where x^n is the choice that appears to be best given what we know after n experiments, calculated using

$$x^n = \arg\max_{x' \in \mathcal{X}} \bar{\mu}_{x'}^n.$$

If we choose action x^n, we then observe $W_{x^n}^{n+1}$ which we then use to update our estimate of our belief about μ_x using our Bayesian updating equations (7.42)–(7.43).

The value of state $S^{n+1}(x)$ when we try action x is given by

$$V^{n+1}(S^{n+1}(x)) = \max_{x' \in \mathcal{X}} \bar{\mu}_{x'}^{n+1}(x)$$

where $\bar{\mu}_{x'}^{n+1}(x)$ is the updated estimate of $\mathbb{E}\mu$ given S^n (that is, our estimate of the distribution of μ after n experiments), and the result of implementing x

and observing W_x^{n+1}. We have to decide which experiment to run after the nth observation, so we have to work with the expected value of running experiment x, given by

$$\mathbb{E}\{V^{n+1}(S^{n+1}(x))|S^n\} = \mathbb{E}\{\max_{x' \in \mathcal{X}} \bar{\mu}_{x'}^{n+1}(x)|S^n\}.$$

The knowledge gradient is then given by

$$v_x^{KG,n} = \mathbb{E}\{V^{n+1}(S^M(S^n, x, W^{n+1}))|S^n, x\} - V^n(S^n),$$

which is equivalent to

$$v^{KG}(x) = \mathbb{E}\{\max_{x'} \bar{\mu}_{x'}^{n+1}(x)|S^n\} - \max_{x'} \bar{\mu}_{x'}^n. \tag{7.59}$$

Here, $\bar{\mu}^{n+1}(x)$ is the updated value of $\bar{\mu}^n$ after running an experiment with setting $x = x^n$, after which we observe W_x^{n+1}. Since we have not yet run the experiment, W_x^{n+1} is a random variable, which means that $\bar{\mu}^{n+1}(x)$ is random. In fact, $\bar{\mu}^{n+1}(x)$ is random for two reasons. To see this, we note that when we run experiment x, we observe an updated value from

$$W_x^{n+1} = \mu_x + \varepsilon_x^{n+1},$$

where $\mu_x = \mathbb{E}F(x, W)$ is the true value, while ε_x^{n+1} is the noise in the observation. This introduces two forms of uncertainty: the unknown truth μ_x, and the noise ε_x^{n+1}. Thus, it would be more accurate to write equation (7.59) as

$$v^{KG}(x) = \mathbb{E}_\mu\{\mathbb{E}_{W|\mu} \max_{x'} \bar{\mu}_{x'}^{n+1}(x)|S^n\} - \max_{x'} \bar{\mu}_{x'}^n \tag{7.60}$$

where the first expectation $\mathbb{E}_\mu$ is conditioned on our belief state S^n, while the second expectation $\mathbb{E}_{W|\mu}$ is over the experimental noise W given our distribution of belief about the truth μ.

To illustrate how equation (7.60) is calculated, imagine that μ takes on values $\{\mu_1, \ldots, \mu_K\}$, and that p_k^μ is the probability that $\mu = \mu_k$. Assume that μ is the mean of a Poisson distribution describing the number of customers W that click on a website and assume that

$$P^W[W = \ell | \mu = \mu_k] = \frac{\mu_k^\ell e^{-\mu_k}}{\ell!}.$$

We would then compute the expectation in equation (7.60) using

$$v^{KG}(x) = \sum_{k=1}^K \left(\sum_{\ell=0}^\infty \left(\max_{x'} \bar{\mu}_{x'}^{n+1}(x|W = \ell) \right) P^W[W = \ell | \mu = \mu_k] \right) p_k^\mu - \max_{x'} \bar{\mu}_{x'}^n,$$

where $\bar{\mu}_{x'}^{n+1}(x|W = \ell)$ is the updated estimate of $\bar{\mu}_{x'}^n$ if we run experiment x (which might be a price or design of a website) and we then observe $W = \ell$. The updating would be done using any of the recursive updating equations described in chapter 3.

We now want to capture how well we can solve our optimization problem, which means solving $\max_{x'} \bar{\mu}_{x'}^{n+1}(x)$. Since $\bar{\mu}_{x'}^{n+1}(x)$ is random (since we have to pick x before we know W^{n+1}), then $\max_{x'} \bar{\mu}_{x'}^{n+1}(x)$ is random. This is why we have to take the expectation, which is conditioned on S^n which captures what we know now.

Computing a knowledge gradient policy for independent beliefs is extremely easy. We assume that all rewards are normally distributed, and that we start with an initial estimate of the mean and variance of the value of decision x, given by

$$
\begin{aligned}
\bar{\mu}_x^0 &= \text{the initial estimate of the expected reward from making decision } x, \\
\bar{\sigma}_x^0 &= \text{the initial estimate of the standard deviation of our belief about } \mu.
\end{aligned}
$$

Each time we make a decision we receive a reward given by

$$
W_x^{n+1} = \mu_x + \varepsilon^{n+1},
$$

where μ_x is the true expected reward from action x (which is unknown) and ε is the experimental error with standard deviation σ_W (which we assume is known).

The estimates $(\bar{\mu}_x^n, \bar{\sigma}_x^{2,n})$ are the mean and variance of our belief about μ_x after n observations. We are going to find that it is more convenient to use the idea of *precision* (as we did in chapter 3) which is the inverse of the variance. So, we define the precision of our belief and the precision of the experimental noise as

$$
\begin{aligned}
\beta_x^n &= 1/\bar{\sigma}_x^{2,n}, \\
\beta^W &= 1/\sigma_W^2.
\end{aligned}
$$

If we take action x and observe a reward W_x^{n+1}, we can use Bayesian updating to obtain new estimates of the mean and variance for action x, following the steps we first introduced in section 3.4. To illustrate, imagine that we try an action x where $\beta_x^n = 1/(20^2) = 0.0025$, and $\beta^W = 1/(40^2) = .000625$. Assume $\bar{\mu}_x^n = 200$ and that we observe $W_x^{n+1} = 250$. The updated mean and precision are given by

$$
\begin{aligned}
\bar{\mu}_x^{n+1} &= \frac{\beta_x^n \bar{\mu}_x^n + \beta^W W_x^{n+1}}{\beta_x^n + \beta^W} \\
&= \frac{(.0025)(200) + (.000625)(250)}{.0025 + .000625} \\
&= 210.
\end{aligned}
$$

$$\beta_x^{n+1} = \beta_x^n + \beta^W$$
$$= .0025 + .000625$$
$$= .003125.$$

We next find the variance of the *change* in our estimate of μ_x assuming we choose to sample action x in iteration n. For this we define

$$\tilde{\sigma}_x^{2,n} = Var[\bar{\mu}_x^{n+1} - \bar{\mu}_x^n | S^n] \tag{7.61}$$
$$= Var[\bar{\mu}_x^{n+1} | S^n]. \tag{7.62}$$

We use the form of equation (7.61) to highlight the definition of $\tilde{\sigma}_x^{2,n}$ as the change in the variance given what we know at time n, but when we condition on what we know (captured by S^n) it means that $Var[\bar{\mu}_x^n | S^n] = 0$ since $\bar{\mu}_x^n$ is just a number at time n.

With a little work, we can write $\tilde{\sigma}_x^{2,n}$ in different ways, including

$$\tilde{\sigma}_x^{2,n} = \bar{\sigma}_x^{2,n} - \bar{\sigma}_x^{2,n+1}, \tag{7.63}$$
$$= \frac{(\bar{\sigma}_x^{2,n})}{1 + \sigma_W^2 / \bar{\sigma}_x^{2,n}}. \tag{7.64}$$

Equation (7.63) expresses the (perhaps unexpected) result that $\tilde{\sigma}_x^{2,n}$ measures the change in the estimate of the standard deviation of the reward from decision x from iteration $n-1$ to n. Using our numerical example, equations (7.63) and (7.64) both produce the result

$$\tilde{\sigma}_x^{2,n} = 400 - 320 = 80$$
$$= \frac{40^2}{1 + \frac{10^2}{40^2}} = 80.$$

Finally, we compute

$$\zeta_x^n = - \left| \frac{\bar{\mu}_x^n - \max_{x' \neq x} \bar{\mu}_{x'}^n}{\tilde{\sigma}_x^n} \right|.$$

ζ_x^n is called the *normalized influence* of decision x. It gives the number of standard deviations from the current estimate of the value of decision x, given by $\bar{\mu}_x^n$, and the best alternative other than decision x. We then find

$$f(\zeta) = \zeta \Phi(\zeta) + \phi(\zeta),$$

where $\Phi(\zeta)$ and $\phi(\zeta)$ are, respectively, the cumulative standard normal distribution and the standard normal density. Thus, if Z is normally distributed with mean 0, variance 1, $\Phi(\zeta) = \mathbb{P}[Z \leq \zeta]$ while

Table 7.4 The calculations behind the knowledge gradient algorithm.

Decision	$\bar{\mu}$	$\bar{\sigma}$	$\tilde{\sigma}$	ζ	$f(z)$	KG index
1	1.0	2.5	1.569	-1.275	0.048	0.075
2	1.5	2.5	1.569	-0.956	0.090	0.142
3	2.0	2.5	1.569	-0.637	0.159	0.249
4	2.0	2.0	1.400	-0.714	0.139	0.195
5	3.0	1.0	0.981	-1.020	0.080	0.079

$$\phi(\zeta) = \frac{1}{\sqrt{2\pi}} \exp\left(-\frac{\zeta^2}{2}\right).$$

The knowledge gradient algorithm chooses the decision x with the largest value of $v_x^{KG,n}$ given by

$$v_x^{KG,n} = \tilde{\sigma}_x^n f(\zeta_x^n).$$

The knowledge gradient algorithm is quite simple to implement. Table 7.4 illustrates a set of calculations for a problem with five options. $\bar{\mu}$ represents the current estimate of the value of each action, while $\bar{\sigma}$ is the current standard deviation of μ. Options 1, 2 and 3 have the same value for $\bar{\sigma}$, but with increasing values of $\bar{\mu}$.

The table illustrates that when the variance is the same, the knowledge gradient prefers the decisions that appear to be the best. Decisions 3 and 4 have the same value of $\bar{\mu}$, but decreasing values of $\bar{\sigma}$, illustrating that the knowledge gradient prefers decisions with the highest variance. Finally, decision 5 appears to be the best of all the decisions, but has the lowest variance (meaning that we have the highest confidence in this decision). The knowledge gradient is the smallest for this decision out of all of them.

The knowledge gradient trades off how well an alternative is expected to perform, and how uncertain we are about this estimate. Figure 7.8 illustrates this tradeoff. Figure 7.8(a) shows five alternatives, where the estimates are the same across all three alternatives, but with increasing standard deviations. Holding the mean constant, the knowledge gradient increases with standard deviation of the estimate of the mean. Figure 7.8(b) repeats this exercise, but now holding the standard deviation the same, with increasing means, showing that the knowledge gradient increases with the estimate of the mean. Finally, Figure 7.8(c) varies the estimates of the mean and standard deviation so that the knowledge gradient stays constant, illustrating the tradeoff between the estimated mean and its uncertainty.

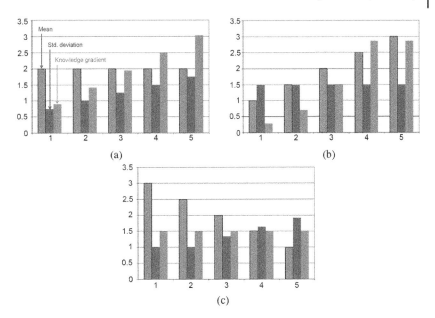

Figure 7.8 The knowledge gradient for lookup table with independent beliefs with equal means (a), equal variances (b), and adjusting means and variances so that the KG is equal (c).

This tradeoff between the expected performance of a design, and the uncertainty about its performance, is a feature that runs through well designed policies. However, all the other policies with this property (interval estimation, upper confidence bounding, Gittins indices), achieve this with indices that consist of the sum of the expected performance and a term that reflects the uncertainty of an alternative.

The knowledge gradient, however, achieves this behavior without the structure of the expected reward plus an uncertainty term with a tunable parameter. In fact, this brings out a major feature of the knowledge gradient, which is that it does not have a tunable parameter.

7.8.3 The Knowledge Gradient for Maximizing Cumulative Reward

There are many online applications of dynamic programming where there is an operational system which we would like to optimize in the field. In these settings, we have to live with the rewards from each experiment. As a result, we have to strike a balance between the value of an action and the information we gain that may improve our choice of actions in the future. This is precisely the tradeoff that is made by Gittins indices for the multiarmed bandit problem.

It turns out that the knowledge gradient is easily adapted for online problems. As before, let $v_x^{KG,n}$ be the offline knowledge gradient, giving the value of observing action x, measured in terms of the improvement in a single decision. Now imagine that we have a budget of N decisions. After having made n decisions (which means, n observations of the value of different actions), if we observe $x = x^n$ which allows us to observe W_x^{n+1}, we received an expected reward of $\mathbb{E}^n W_x^{n+1} = \bar{\mu}_x^n$, and obtain information that improves the contribution from a single decision by $v_x^{KG,n}$. However, we have $N - n$ more decisions to make. Assume that we learn from the observation of W_x^{n+1} by choosing $x^n = x$, but we do not allow ourselves to learn anything from future decisions. This means that the remaining $N - n$ decisions have access to the same information.

From this analysis, the knowledge gradient for online applications consists of the expected value of the single-period contribution of the experiment, plus the improvement in all the remaining decisions in our horizon. This implies

$$v_x^{OLKG,n} = \bar{\mu}_x^n + (N - n)v_x^{KG,n}. \tag{7.65}$$

This is a general formula that extends any calculation of the offline knowledge gradient to online (cumulative reward) learning problems. Note that we now obtain the same structure we previously saw in UCB and IE policies (as well as Gittins indices) where the index is a sum of a one-period reward, plus a bonus term for learning.

It is also important to recognize that the online policy is time-dependent, because of the presence of the $(N - n)$ coefficient. When n is small, $(N - n)$ is large, and the policy will emphasize exploration. As we progress, $(N - n)$ shrinks and we place more emphasis on maximizing the immediate reward.

7.8.4 The Knowledge Gradient for Sampled Belief Model*

There are many settings where our belief model is nonlinear in the parameters. For example, imagine that we are modeling the probability that a customer will click on an ad when we bid x, where higher bids improve our ability of getting attractive placements that increase the number of clicks. Assume that the response is a logistic regression given by

$$P^{purchase}(x|\theta) = \frac{e^{U(x|\theta)}}{1 + e^{U(x|\theta)}} \tag{7.66}$$

where $U(x|\theta) = \theta_0 + \theta_1 x$. We do not know θ, so we are going to represent it as a random variable with some distribution. We might represent it as a multivariate normal distribution, but this makes computing the knowledge gradient very complicated.

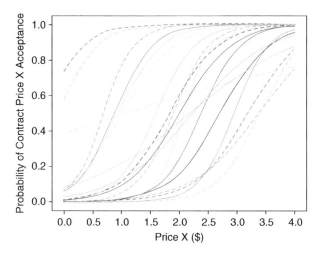

Figure 7.9 Sampled set of bid response curves.

A very practical strategy is to use a sampled belief model which we first introduced in section 3.9.2. Using this approach we assume that θ takes an outcome in the set $\{\theta_1, \dots, \theta_K\}$. Let $p_k^n = Prob[\theta = \theta_k]$ after we have run n experiments. Note that these probabilities represent our belief state, which means that

$$S^n = (p_k^n)_{k=1}^K.$$

We might use an initial distribution $p_k^0 = 1/K$. A sampled belief for a logistic curve is illustrated in Figure 7.9.

Let W_x^{n+1} be the observation when we run experiment $x^n = x$, and let θ^{n+1} be the random variable representing θ after we have run experiment x and observed W_x^{n+1}. When we used a lookup table belief model, we wrote the knowledge gradient as (see (7.59))

$$\nu^{KG,n}(x) = \mathbb{E}\{\max_{x'} \bar{\mu}_{x'}^{n+1}(x)|S^n, x^n = x\} - \max_{x'} \bar{\mu}_{x'}^n. \tag{7.67}$$

For our nonlinear model, we let $\mu_x = f(x|\theta)$ where we assume we know the function $f(x|\theta)$ but we do not know θ. The knowledge gradient would then be written

$$\nu^{KG,n}(x) = \mathbb{E}\{\max_{x'} \mathbb{E}\{f(x', \theta^{n+1}(x))|S^{n+1}\}|S^n, x^n = x\}$$

$$- \max_{x'} \mathbb{E}_\theta\{f(x', \theta)|S^n\}. \tag{7.68}$$

We are going to step through this expression more carefully since we are going to have to compute it directly. Readers just interested in computing the knowledge gradient can jump right to equation (7.76) which can be directly implemented. The derivation that we provide next will provide insights into the principles of the knowledge gradient as a concept.

First, we need to realize that $\bar{\mu}_x^n = \mathbb{E}\{\mu_x|S^n\}$ is the expectation of μ_x given what we know after n iterations, which is captured in S^n. For our nonlinear model, this would be written

$$
\begin{aligned}
\mathbb{E}_\theta\{f(x',\theta)|S^n\} &= \sum_{k=1}^{K} p_k^n f(x',\theta_k) \\
&= \bar{f}^n(x').
\end{aligned}
$$

Next, $\bar{\mu}_{x'}^{n+1}(x)$ in equation (7.67) is our estimate of μ_x after running experiment $x^n = x$ and observing W_x^{n+1}. For the lookup table model, we would write this as

$$
\bar{\mu}_{x'}^{n+1}(x) = \mathbb{E}_\mu\{\mu_{x'}|S^{n+1}\} \tag{7.69}
$$

where $S^{n+1} = S^M(S^n, x^n, W^{n+1})$. This means that (7.69) can also be written

$$
\bar{\mu}_{x'}^{n+1}(x) = \mathbb{E}_\mu\{\mu_{x'}|S^n, x^n, W^{n+1}\}. \tag{7.70}
$$

For our sampled belief model, we use p_k^n instead of $\bar{\mu}^n$, and we use the updated probabilities $p_k^{n+1}(S^n, x^n = s, W_x^{n+1} = W)$ instead of $\bar{\mu}^{n+1}(x)$ where

$$
p_k^{n+1}(S^n, x^n = x, W_x^{n+1} = W) = Prob[\theta = \theta_k|S^n, x^n = x, W_x^{n+1} = W].
$$

We express the dependence of $p_k^{n+1}(S^n, x^n = s, W_x^{n+1} = W)$ on the prior state S^n, decision x^n and experimental outcome W^{n+1} explicitly to make these dependencies clear. The random variable $W = W_x^{n+1}$ depends on θ since

$$
W_x^{n+1} = f(x|\theta) + \varepsilon^{n+1}.
$$

Our belief about θ depends on when we are taking the expectation, which is captured by conditioning on S^n (or later, $\mathbb{E}^n... = \mathbb{E}...|S^n$). To emphasize the dependence on S^n, we are going to write $\mathbb{E}^n\{\cdot|S^n\}$ to emphasize when we are conditioning on S^n. This will help when we have to use nested expectations, conditioning on both S^n and S^{n+1} in the same equation.

The expectation inside the max operator is

$$
\begin{aligned}
\mathbb{E}_\theta^{n+1}\{f(x',\theta^{n+1}(x))|S^{n+1}\} &= \mathbb{E}_\theta^{n+1}\{f(x',\theta^{n+1}(x))|S^n, x^n = x, W_x^{n+1} = W\} \\
&= \sum_{k=1}^{K} f(x',\theta_k) p_k^{n+1}(S^n, x^n = x, W_x^{n+1} = W).
\end{aligned}
$$

Note that we are only taking the expectation over θ, since W^{n+1} is known at this point. We take the expectation over θ given the posterior p_k^{n+1} because even after we complete the $n + 1^{st}$ experiment, we still have to make a decision (that is, choosing x') without knowing the true value of θ.

We now have to compute $p_k^{n+1}(S^n, x^n = x, W_x^{n+1} = W)$. We first assume that we know the distribution of W_x^{n+1} given θ (that is, we know the distribution of W_x^{n+1} if we know θ). For our ad-click problem, this would just have outcomes 0 or 1 where $Prob[W = 1]$ is given by our logistic curve in equation (7.66). For more general problems, we are going to assume we have the distribution.

$$f^W(w|x, \theta_k) \quad = \quad \mathbb{P}[W^{n+1} = w|x, \theta = \theta_k].$$

We compute $p_k^{n+1}(S^n, x^n = x, W_x^{n+1} = W)$ using Bayes theorem by first writing

$$p_k^{n+1}(S^n, x^n = x, W_x^{n+1} = w) = Prob[\theta = \theta_k|S^n, x^n = x, W_x^{n+1} = w]$$

$$= \frac{Prob[W_x^{n+1} = w|\theta = \theta_k, S^n, x^n = x]Prob[\theta = \theta_k|S^n, x^n = x]}{Prob[W_x^{n+1} = w|S^n, x^n = x]}$$

$$= \frac{f^W(W^{n+1} = w|x^n, \theta_k)p_k^n}{C(w)}, \tag{7.71}$$

where $C(w)$ is the normalizing constant given $W^{n+1} = w$, which is calculated using

$$C(w) = \sum_{k=1}^{K} f^W(W^{n+1} = w|x^n, \theta_k)p_k^n.$$

Below, we are going to treat $C(W)$ (with capital W) as a random variable with realization $C(w)$. Note that we condition $p_k^{n+1}(S^n, x^n = x, W_x^{n+1} = w)$ on S^n since this gives us the prior p_k^n which we use in Bayes theorem. However, once we have computed $p_k^{n+1}(S^n, x^n = x, W_x^{n+1} = w)$ we write the posterior probability as $p_k^{n+1})(w)$ since we no longer need to remember $x^n = x$ or the prior distribution $S^n = (p_k^n)_{k=1}^K$, but we do need to express the dependence on the outcome $W^{n+1} = w$. We will write the posterior distribution as $p^n(W)$ when we want to express the outcome as a random variable.

We are now ready to compute the knowledge gradient in equation (7.68). We begin by writing it with expanded expectations as

$$\nu^{KG,n}(x) = \mathbb{E}_\theta^n \mathbb{E}_{W^{n+1}|\theta}\{\max_{x'} \mathbb{E}_\theta^{n+1}\{f(x', \theta^{n+1})|S^{n+1}\}|S^n, x^n = x\}$$

$$- \max_{x'} \mathbb{E}_\theta^n\{f(x', \theta)|S^n\}. \tag{7.72}$$

We have to take the expectations $\mathbb{E}_\theta^n \mathbb{E}_{W^{n+1}|\theta}$ because when we are trying to decide which experiment x to run, we do not know the outcome W^{n+1}, and we do not know the true value of θ on which W^{n+1} depends.

The posterior distribution of belief allows us to write $\mathbb{E}_\theta^{n+1} f(x', \theta^{n+1}) | S^{n+1}\}$ using

$$\mathbb{E}_\theta^{n+1}\{f(x', \theta^{n+1}) | S^{n+1}\} = \sum_{k=1}^{K} f(x', \theta_k) p_k^{n+1}(W^{n+1}).$$

Substituting this into equation (7.72) gives us

$$\nu^{KG,n}(x) = \mathbb{E}_\theta^n \mathbb{E}_{W^{n+1}|\theta} \left\{ \max_{x'} \sum_{k=1}^{K} f(x', \theta_k) p_k^{n+1}(W^{n+1}) \middle| S^n, x^n = x \right\}$$

$$- \max_{x'} \bar{f}^n(x'). \tag{7.73}$$

We now focus on computing the first term of the knowledge gradient. Substituting $p_k^{n+1}(W^{n+1})$ from equation (7.71) into (7.73) gives us

$$\mathbb{E}_\theta^n \mathbb{E}_{W^{n+1}|\theta} \left\{ \max_{x'} \sum_{k=1}^{K} f(x', \theta_k) p_k^{n+1}(W^{n+1}) \middle| S^n \right\}$$

$$= \mathbb{E}_\theta^n \mathbb{E}_{W^{n+1}|\theta} \left\{ \max_{x'} \sum_{k=1}^{K} f(x', \theta_k) \left(\frac{f^W(W^{n+1}|x^n, \theta_k) p_k^n}{C(W^{n+1})} \right) \middle| S^n, x = x^n \right\}.$$

Keeping in mind that the entire expression is a function of x, the expectation can be written

$$\mathbb{E}_\theta^n \mathbb{E}_{W^{n+1}|\theta} \left\{ \max_{x'} \frac{1}{C(W)} \sum_{k=1}^{K} f(x', \theta)(f^W(W^{n+1}|x^n, \theta_k) p_k^n) | S^n, x = x^n \right\}$$

$$= \mathbb{E}_\theta^n \mathbb{E}_{W|\theta} \frac{1}{C(W)} \left\{ \max_{x'} \sum_{k=1}^{K} f(x', \theta)(f^W(W^{n+1}|x^n, \theta_k) p_k^n) | S^n, x = x^n \right\}$$

$$= \sum_{j=1}^{K} \left(\sum_{\ell=1}^{L} \frac{1}{C(w_\ell)} \{A_\ell\} f^W(W^{n+1} = w_\ell | x, \theta_j) \right) p_j^n, \tag{7.74}$$

where

$$A_\ell = \max_{x'} \sum_{k=1}^{K} f(x', \theta_k)(f^W(W^{n+1} = w_\ell | x^n, \theta_k) p_k^n).$$

We pause to note that the density $f^W(w, x, \theta)$ appears twice in equation (7.74): once as $f^W(W^{n+1} = w_\ell | x^n, \theta_k)$, and once as $f^W(W^{n+1} = w_\ell | x, \theta_j)$.

The first one entered the equation as part of the use of Bayes' theorem to find $p_x^{n+1}(W)$. This calculation is done inside the max operator after W^{n+1} has been observed. The second one arises because when we are deciding the experiment x^n, we do not yet know W^{n+1} and we have to take the expectation over all possible outcomes. Note that if we have binary outcomes (1 if the customer clicks on the ad, 0 otherwise), then the summation over w_ℓ is only over those two values.

We can further simplify this expression by noticing that the terms $f^W(W = w_\ell | x, \theta_j)$ and p_j^n are not a function of x' or k, which means we can take them outside of the max operator. We can then reverse the order of the other sums over k and w_ℓ, giving us

$$
\mathbb{E}_\theta \mathbb{E}_{W|\theta} \left\{ \max_{x'} \frac{1}{C(W)} \sum_{k=1}^{K} f(x', \theta_k) f^W(W|x^n, \theta_k)) p_k^n | S^n, x = x^n \right\}
$$

$$
= \sum_{\ell=1}^{L} \sum_{j=1}^{K} \left(\frac{f^W(W = w_\ell | x, \theta_j) p_j^n}{C(w_\ell)} \right) \left\{ \max_{x'} \sum_{k=1}^{K} f(x', \theta_k) f^W(W = w_\ell | x^n, \theta_k) p_k^n | S^n, x = x^n \right\}. \quad (7.75)
$$

Using the definition of the normalizing constant $C(w)$ we can write

$$
\sum_{j=1}^{K} \left(\frac{f^W(W = w_\ell | x, \theta_j) p_j^n}{C(w_\ell)} \right) = \left(\frac{\sum_{j=1}^{K} f^W(W = w_\ell | x, \theta_j) p_j^n}{C(w_\ell)} \right)
$$

$$
= \left(\frac{\sum_{j=1}^{K} f^W(W = w_\ell | x, \theta_j) p_j^n}{\sum_{k=1}^{K} f^W(W = w_\ell | x, \theta_k) p_k^n} \right)
$$

$$
= 1.
$$

We just simplified the problem by cancelling two summations over the K values of θ. This is a significant simplification, since these sums were nested. This allows us to write (7.75) as

$$
\mathbb{E}_\theta \mathbb{E}_{W|\theta} \left\{ \max_{x'} \frac{1}{C(W)} \sum_{k=1}^{K} p_k^n f^W(W|x^n, \theta_k) f(x', \theta_k) | S^n, x = x^n \right\}
$$

$$
= \sum_{\ell=1}^{L} \left\{ \max_{x'} \sum_{k=1}^{K} p_k^n f^W(W = w_\ell | x^n, \theta_k) f(x', \theta_k) | S^n, x = x^n \right\}. \quad (7.76)
$$

This is surprisingly powerful logic, since it works with any nonlinear belief model.

7.8.5 Knowledge Gradient for Correlated Beliefs

A particularly important feature of the knowledge gradient is that it can be adapted to handle the important problem of correlated beliefs. In fact, the vast majority of real applications exhibit some form of correlated beliefs. Some examples are given below.

■ **EXAMPLE 7.1**

Correlated beliefs can arise when we are maximizing a continuous surface (nearby points will be correlated) or choosing subsets (such as the location of a set of facilities) which produce correlations when subsets share common elements. If we are trying to estimate a continuous function, we might assume that the covariance matrix satisfies

$$Cov(x, x') \propto e^{-\rho \|x-x'\|},$$

where ρ captures the relationship between neighboring points. If x is a vector of $0's$ and $1's$ indicating elements in a subset, the covariance might be proportional to the number of 1's that are in common between two choices.

■ **EXAMPLE 7.2**

There are about two dozen drugs for reducing blood sugar, divided among four major classes. Trying a drug in one class can provide an indication of how a patient will respond to other drugs in that class.

■ **EXAMPLE 7.3**

A materials scientist is testing different catalysts in a process to design a material with maximum conductivity. Prior to running any experiment, the scientist is able to estimate the likely relationship in the performance of different catalysts, shown in Table 7.5. The catalysts that share an Fe (iron) or Ni (nickel) molecule show higher correlations.

Constructing the covariance matrix involves incorporating the structure of the problem. This may be relatively easy, as with the covariance between discretized choices of a continuous surface.

There is a more compact way of updating our estimate of $\bar{\mu}^n$ in the presence of correlated beliefs. Let $\lambda^W = \sigma_W^2 = 1/\beta^W$ (this is basically a trick to get rid of that nasty square). Let $\Sigma^{n+1}(x)$ be the updated covariance matrix given that we have chosen to evaluate alternative x, and let $\tilde{\Sigma}^n(x)$ be the change in the covariance matrix due to evaluating x, which is given by

$$\tilde{\Sigma}^n(x) = \Sigma^n - \Sigma^{n+1},$$
$$= \frac{\Sigma^n e_x (e_x)^T \Sigma^n}{\Sigma_{xx}^n + \lambda^W},$$

Table 7.5 Correlation matrix describing the relationship between estimated performance of different catalysts, as estimated by an expert.

	1.4nmFe	1nmFe	2nmFe	10nm-Fe	2nmNi	Ni0.6nm	10nm-Ni
1.4nmFe	1.0	0.7	0.7	0.6	0.4	0.4	0.2
1nmFe	0.7	1.0	0.7	0.6	0.4	0.4	0.2
2nmFe	0.7	0.7	1.0	0.6	0.4	0.4	0.2
10nmFe	0.6	0.6	0.6	1.0	0.4	0.3	0.0
2nmNi	0.4	0.4	0.4	0.4	1.0	0.7	0.6
Ni0.6nm	0.4	0.4	0.4	0.3	0.7	1.0	0.6
10nmNi	0.2	0.2	0.2	0.0	0.6	0.6	1.0

where e_x is a vector of 0s with a 1 in the position corresponding to alternative x. Now define the vector $\tilde{\sigma}^n(x)$, which gives the square root of the change in the variance due to measuring x, which is given by

$$\tilde{\sigma}^n(x) = \frac{\Sigma^n e_x}{\sqrt{\Sigma^n_{xx} + \lambda^W}}. \tag{7.77}$$

Let $\tilde{\sigma}_i(\Sigma, x)$ be the component $(e_i)^T \tilde{\sigma}(x)$ of the vector $\tilde{\sigma}(x)$, and let $Var^n(\cdot)$ be the variance given what we know after n experiments. We note that if we evaluate alternative x^n, then

$$Var^n \left[W^{n+1} - \bar{\mu}^n_{x^n} \right] = Var^n \left[\mu_{x^n} + \varepsilon^{n+1} \right]$$
$$= \Sigma^n_{x^n x^n} + \lambda^W. \tag{7.78}$$

Next define the random variable

$$Z^{n+1} = (W^{n+1} - \bar{\mu}^n_{x^n})/\sqrt{Var^n \left[W^{n+1} - \bar{\mu}^n_{x^n} \right]}.$$

We can now rewrite our expression which we first saw in chapter 3, equation (7.26) for updating our beliefs about the mean as

$$\bar{\mu}^{n+1} = \bar{\mu}^n + \tilde{\sigma}(x^n)Z^{n+1}. \tag{7.79}$$

Note that $\bar{\mu}^{n+1}$ and $\bar{\mu}^n$ are vectors giving beliefs for all alternatives, not just the alternative x^n that we tested. The knowledge gradient policy for correlated beliefs is computed using

$$X^{KG}(s) = \arg\max_x \mathbb{E} \left[\max_i \mu_i^{n+1} \mid S^n = s \right] \tag{7.80}$$
$$= \arg\max_x \mathbb{E} \left[\max_i \left(\bar{\mu}_i^n + \tilde{\sigma}_i(x^n)Z^{n+1} \right) \mid S^n, x \right]$$

where Z is a scalar, standard normal random variable. The problem with this expression is that the expectation is harder to compute, but a simple algorithm can be used to compute the expectation exactly. We start by defining

$$h(\bar{\mu}^n, \tilde{\sigma}(x)) = \mathbb{E}\left[\max_i\left(\bar{\mu}_i^n + \tilde{\sigma}_i(x^n)Z^{n+1}\right) \mid S^n, x = x^n\right]. \tag{7.81}$$

Substituting (7.81) into (7.80) gives us

$$X^{KG}(s) = \arg\max_x h(\bar{\mu}^n, \tilde{\sigma}(x)). \tag{7.82}$$

Let $a_i = \bar{\mu}_i^n$, $b_i = \tilde{\sigma}_i(\Sigma^n, x^n)$, and let Z be our standard normal deviate. Now define the function $h(a, b)$ as

$$h(a, b) = \mathbb{E}\max_i\left(a_i + b_i Z\right). \tag{7.83}$$

Both a and b are M-dimensional vectors. Sort the elements b_i so that $b_1 \leq b_2 \leq \dots$ so that we get a sequence of lines with increasing slopes, as depicted in Figure 7.10. There are ranges for z over a particular line may dominate the other lines, and some lines may be dominated all the time (such as alternative 3).

We need to identify and eliminate the dominated alternatives. To do this we start by finding the points where the lines intersect. The lines $a_i + b_i z$ and $a_{i+1} + b_{i+1} z$ intersect at

$$z = c_i = \frac{a_i - a_{i+1}}{b_{i+1} - b_i}.$$

For the moment, we are going to assume that $b_{i+1} > b_i$. If $c_{i-1} < c_i < c_{i+1}$, then we can find a range for z over which a particular choice dominates, as depicted in Figure 7.10. A line is dominated when $c_{i+1} < c_i$, at which point they are dropped from the set. Once the sequence c_i has been found, we can compute (7.80) using

$$h(a, b) = \sum_{i=1}^{M}(b_{i+1} - b_i)f(-|c_i|),$$

where as before, $f(z) = z\Phi(z) + \phi(z)$. Of course, the summation has to be adjusted to skip any choices i that were found to be dominated.

It is important to recognize that there is more to incorporating correlated beliefs than simply using the covariances when we update our beliefs after an experiment. With this procedure, we anticipate the updating before we even perform an experiment.

The ability to handle correlated beliefs in the choice of what experiment to perform is an important feature that has been overlooked in other procedures. It makes it possible to make sensible choices when our experimental budget is much smaller than the number of potential choices we have to evaluate. There

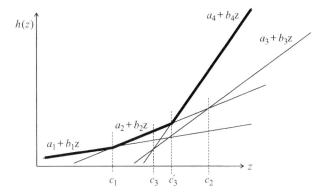

Figure 7.10 Regions of z over which different choices dominate. Choice 3 is always dominated.

are, of course, computational implications. It is relatively easy to handle dozens or hundreds of alternatives, but as a result of the matrix calculations, it becomes expensive to handle problems where the number of potential choices is in the thousands. If this is the case, it is likely the problem has special structure. For example, we might be discretizing a p-dimensional parameter surface, which suggests using a parametric model for the belief model.

A reasonable question to ask is: given that the correlated KG is considerably more complex than the knowledge gradient policy with independent beliefs, what is the value of using correlated KG? Figure 7.11(a) shows the sampling pattern when learning a quadratic function, starting with a uniform prior, when using the knowledge gradient with independent beliefs for the learning policy, but using correlated beliefs to update beliefs after an experiment has been run.

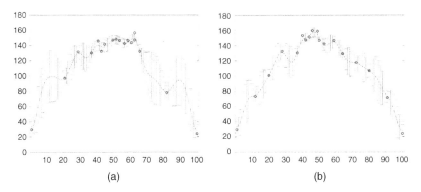

Figure 7.11 (a) Sampling pattern from knowledge gradient using independent beliefs; (b) sampling pattern from knowledge gradient using correlated beliefs.

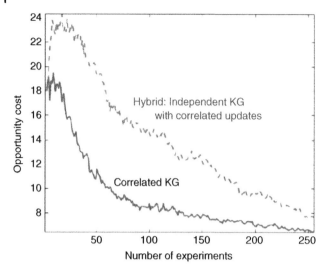

Figure 7.12 Comparison of correlated KG policy against a KG policy with independent beliefs, but using correlated updates, showing the improvement when using the correlated KG policy.

This policy tends to produce sampling that is more clustered in the region near the optimum. Figure 7.11(b) shows the sampling pattern for the knowledge gradient policy with correlated beliefs, showing a more uniform pattern that shows a better spread of experiments.

So, the correlated KG logic seems to do a better job of exploring, but how well does it work? Figure 7.12 shows the opportunity cost for each policy, where smaller is better. For this example, the correlated KG works quite a bit better, probably due to the tendency of the correlated KG policy to do explore more efficiently.

While these experiments suggest strong support for the correlated KG policy when we have correlated beliefs, we need to also note that tunable CFA-based policies such as interval estimation or the UCB policies can also be tuned in the context of problems with correlated beliefs. The tradeoff is that the correlated KG policy does not require tuning, but is more difficult to implement. A tuned CFA policy requires tuning (which can be a challenge) but is otherwise trivial to implement. This is the classic tradeoff between a CFA policy (in the policy search class) and a DLA policy (in the lookahead class).

7.9 Learning in Batches

There are many settings where it is possible to do simultaneous observations, effectively learning in batch. Some examples are:

- If learning is being done via computer simulation, different runs can be run in parallel.
- An automotive manufacturer looking to tune its robots can try out different ideas at different plants.
- A company can perform local test marketing in different cities, or by using ads targeted for different types of people shopping online.
- A materials scientist looking for new materials can divide a plate into 25 squares and perform 25 experiments in batch.

When parallel testing is possible, the natural question is then: how do we determine the set of tests before we know the outcomes of the other tests? We cannot simply apply a policy repeatedly, as it might just end up choosing the same point (unless there is some form of forced randomization).

A simple strategy is to simulate a sequential learning process. That is, use some policy to determine the first test, and then either use the expected outcome, or a simulated outcome, to update the beliefs from running the test, and then repeat the process. The key is to update the beliefs after each simulated outcome. If you can perform K experiments in parallel, repeat this process K times.

Figure 7.13 shows the effect of running an experiment when using correlated beliefs when using the knowledge gradient, although the principle applies to a number of learning policies. On the left is the knowledge gradient before running the indicated experiment, and the right shows the knowledge gradient after running the experiment. As a result of using correlated beliefs, the knowledge gradient drops in the region around the first experiment, discouraging the choice of another experiment nearby. Note that what is important is where you

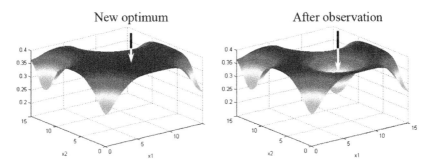

Figure 7.13 The knowledge gradient with correlated beliefs before running an experiment (left) and after (right), showing the drop in the knowledge gradient both at the point of the test, and the neighboring region, after running the experiment.

are planning on doing the first experiment, not the outcome. This is particularly true of the knowledge gradient.

7.10 Simulation Optimization*

A subcommunity within the larger stochastic search community goes by the name *simulation optimization*. This community also works on problems that can be described in the form of $\max_x \mathbb{E}F(x, W)$, but the context typically arises when x represents the design of a physical system, which is then evaluated (noisily) using discrete-event simulation. The number of potential designs $\mathcal{X}$ is typically in the range of 5 to perhaps 100. The standard approach in simulation optimization is to use a frequentist belief model, where it is generally assumed that our experimental budget is large enough for us to run some initial testing of each of the alternatives to build an initial belief.

The field of simulation-optimization has its roots in the analysis of designs, such as the layout of a manufacturing system, where we can get better results if we run a discrete event simulation model for a longer time. We can evaluate a design x more accurately by increasing the run length n_x of the simulation, where n_x might be the number of time periods, the CPU time, or the number of discrete events (e.g. customer arrivals). We assume that we have a global budget N, and we need to find n_x for each x so that

$$\sum_{x \in \mathcal{X}} n_x = N.$$

For our purposes, there is no difference between a potential design of a physical system and a policy. Searching for the best design and searching for the best policy is, algorithmically speaking, identical as long as the set of policies is not too large.

We can tackle this problem using the strategies described above (such as the knowledge gradient) if we break up the problem into a series of short simulations (say, 1 time step or 1 unit of CPU time). Then, at each iteration we have to decide which design x to evaluate, contributing to our estimate θ_x^n for design x. The problem with this strategy is that it ignores the startup time for a simulation. It is much easier to set a run length n_x for each design x, and then run the entire simulation to obtain an estimate of θ_x.

The simulation-optimization problem is traditionally formulated in a frequentist framework, reflecting the lack of prior information about the alternatives. A standard strategy is to run the experiments in two stages. In the first stage, a sample n^0 is collected for each design. The information from this first stage is used to develop an estimate of the value of each design. We might learn, for example, that certain designs seem to lack any promise at all, while other designs may seem more interesting. Rather than spreading our budget across all the designs, we can use this information to focus our computing budget across the designs that offer the greatest potential.

7.10.1 An Indifference Zone Algorithm

There are a number of algorithms that have been suggested to search for the best design using the indifference zone criterion, which is one of the most popular in the simulation-optimization community. The algorithm in Figure 7.14 summarizes a method which successively reduces a set of candidates at each iteration, focusing the evaluation effort on a smaller and smaller set of alternatives. The method (under some assumptions) using a user-specified indifference zone of δ. Of course, as δ is decreased, the computational requirements increase.

7.10.2 Optimal Computing Budget Allocation

The value of the indifference zone strategy is that it focuses on achieving a specific level of solution quality, being constrained by a specific budget. However, it is often the case that we are trying to do the best we can within a specific computing budget. For this purpose, a line of research has evolved under the name *optimal computing budget allocation*, or OCBA.

Figure 7.15 illustrates a typical version of an OCBA algorithm. The algorithm proceeds by taking an initial sample $N_x^0 = n_0$ of each alternative $x \in \mathcal{X}$, which means we use $B^0 = Mn_0$ experiments from our budget B. Letting $M = |\mathcal{X}|$, we divide the remaining budget of experiments $B - B^0$ into equal increments of size Δ, so that we do $N = (B - Mn_0)\Delta$ iterations.

After n iterations, assume that we have tested alternative x N_x^n times, and let W_x^m be the mth observation of x, for $m = 1, \dots, N_x^n$. The updated estimate of the value of each alternative x is given by

$$\theta_x^n = \frac{1}{N_x^n} \sum_{m=1}^{N_x^n} W_x^m.$$

Let $x^n = \arg\max \theta_x^n$ be the current best option.

After using Mn_0 observations from our budget, at each iteration we increase our allowed budget by $B^n = B^{n-1} + \Delta$ until we reach $B^N = B$. After each increment, the allocation N_x^n, $x \in \mathcal{X}$ is recomputed using

$$\frac{N_x^{n+1}}{N_{x'}^{n+1}} = \frac{\hat{\sigma}_x^{2,n}/(\theta_{x^n}^n - \theta_{x'}^n)^2}{\hat{\sigma}_{x'}^{2,n}/(\theta_{x^n}^n - \theta_{x'}^n)^2} \qquad x \neq x' \neq x^n, \tag{7.84}$$

Step 0. Initialization:

Step 0a. Select the probability of correct selection $1 - \alpha$, indifference zone parameter δ and initial sample size $n_0 \geq 2$.

Step 0b. Compute

$$\eta = \frac{1}{2} \left[\left(\frac{2\alpha}{k-1} \right)^{-2/(n_0-1)} - 1 \right].$$

Step 0c. Set $h^2 = 2\eta(n_0 - 1)$.

Step 0d. Set $\mathcal{X}^0 = \mathcal{X}$ as the set of systems in contention.

Step 0e. Obtain samples W_x^m, $m = 1, \ldots, n_0$ of each $x \in \mathcal{X}^0$ and let θ_x^0 be the resulting sample means for each alternative computing using

$$\theta_x^0 = \frac{1}{n_0} \sum_{m=1}^{n_0} W_x^m.$$

Compute the sample variances for each pair using

$$\hat{\sigma}_{xx'}^2 = \frac{1}{n_0 - 1} \sum_{m=1}^{n_0} \left[W_x^m - W_{x'}^m - \left(\theta_x^0 - \theta_{x'}^0 \right) \right]^2.$$

Set $r = n_0$.

Step 0f. Set $n = 1$.

Step 1. Compute

$$W_{xx'}(r) = \max \left\{ 0, \frac{\delta}{2r} \left(\frac{h^2 \hat{\sigma}_{xx'}^2}{\delta^2} - r \right) \right\}.$$

Step 2. Refine the eligible set using

$$\mathcal{X}^n = \left\{ x : x \in \mathcal{X}^{n-1} \text{ and } \theta_x^n \geq \theta_{x'}^n - W_{xx'}(r), x' \neq x \right\}.$$

Step 3. If $|\mathcal{X}^n| = 1$, stop and select the element in $\mathcal{X}^n$. Otherwise, perform an additional sample W_x^{n+1} of each $x \in \mathcal{X}^n$, set $r = r + 1$ and return to step 1.

Figure 7.14 Policy search algorithm using the indifference zone criterion. Adapted from Nelson and Kim (2001), 'A fully sequential procedure for indifference zone selection in simulation', ACM Trans. Model. Comput. Simul. 11(3), 251–273.

$$N_{x^n}^{n+1} = \hat{\sigma}_{x^n}^n \sqrt{\sum_{i=1, i \neq x^n}^{M} \left(\frac{N_x^{n+1}}{\hat{\sigma}_i^n} \right)^2}. \tag{7.85}$$

We use equations (7.84)–(7.85) to produce an allocation N_x^n such that $\sum_x N_x^n = B^n$. Note that after increasing the budget, it is not guaranteed that $N_x^n \geq N_x^{n-1}$

Step 0. Initialization:

> **Step 0a.** Given a computing budget B, let n^0 be the initial sample size for each of the $M = |\mathcal{X}|$ alternatives. Divide the remaining budget $T - Mn_0$ into increments so that $N = (T - Mn_0)/\delta$ is an integer.
>
> **Step 0b.** Obtain samples W_x^m, $m = 1, \dots, n_0$ samples of each $x \in \mathcal{X}$.
>
> **Step 0c.** Initialize $N_x^1 = n_0$ for all $x \in \mathcal{X}$.
>
> **Step 0d.** Initialize $n = 1$.

Step 1. Compute

$$\theta_x^n = \frac{1}{N_x^n} \sum_{m=1}^{N_x^n} W_x^m.$$

Compute the sample variances for each pair using

$$\hat{\sigma}_x^{2,n} = \frac{1}{N_x^n - 1} \sum_{m=1}^{N_x^n} (W_x^m - \theta_x^n)^2.$$

Step 2. Let $x^n = \arg\max_{x \in \mathcal{X}} \theta_x^n$.

Step 3. Increase the computing budget by Δ and calculate the new allocation $N_1^{n+1}, \dots, N_M^{n+1}$ so that

$$\frac{N_x^{n+1}}{N_{x'}^{n+1}} = \frac{\hat{\sigma}_x^{2,n}/(\theta_{x^n}^n - \theta_{x'}^n)^2}{\hat{\sigma}_{x'}^{2,n}/(\theta_{x^n}^n - \theta_{x'}^n)^2} \quad x \neq x' \neq x^n,$$

$$N_{x^n}^{n+1} = \hat{\sigma}_{x^n}^n \sqrt{\sum_{i=1, i \neq x^n}^{M} \left(\frac{N_x^{n+1}}{\hat{\sigma}_i^n}\right)^2}.$$

Step 4. Perform $\max\left(N_x^{n+1} - N_x^n, 0\right)$ additional simulations for each alternative x.

Step 5. Set $n = n + 1$. If $\sum_{x \in \mathcal{X}} N_x^n < B$, go to step 1.

Step 6. Return $x^n \arg\max_{x \in \mathcal{X}} \theta_x^n$.

Figure 7.15 Optimal computing budget allocation procedure.

for some x. If this is the case, we would not evaluate these alternatives at all in the next iteration. We can solve these equations by writing each N_x^n in terms of some fixed alternative (other than x^n), such as N_1^n (assuming $x^n \neq 1$). After writing N_x^n as a function of N_1^n for all x, we then determine N_1^n so that $\sum N_x^n \approx B^n$ (within rounding).

The complete algorithm is summarized in Figure 7.15.

7.11 Evaluating Policies*

There are a number of ways to approach evaluating policies in the context of derivative-free stochastic search. We start by presenting alternative performance metrics and close with a discussion of alternative perspectives of optimality.

7.11.1 Alternative Performance Metrics*

Up to now we have evaluated the performance of a policy based on the expected value of the final or cumulative reward. However, there are many ways to evaluate a policy. Below is a list of metrics that have been drawn from different communities.

Empirical Performance

We might simulate a policy K times, where each repetition involves making N observations of W. We let ω^k represent a full sample realization of these N observations, which we would denote by $W^1(\omega^k), \dots, W^N(\omega^k)$. Each sequence ω^k creates a design decision $x^{\pi,N}(\omega^k)$.

It is useful to separate the random variable W that we observe while learning from the random variable we use to evaluate a design, so we are going to let W be the random variable we observe while learning, and we are going to let $\widehat{W}$ be the random variable we use for evaluating a design. Most of the time these are the same random variable with the same distribution, but it opens the door to allowing them to be different.

Once we obtain a design $x^{\pi,N}(\omega^k)$, we then have to evaluate it by taking, say, L observations of $\widehat{W}$, which we designate by $\widehat{W}^1, \dots, \widehat{W}^\ell, \dots, \widehat{W}^L$. Using this notation, we would approximate the performance of a design $x^{\pi,N}(\omega^k)$ using

$$\bar{F}^\pi(\omega^k) = \frac{1}{L} \sum_{\ell=1}^{L} F(x^{\pi,N}(\omega^k), \widehat{W}^\ell).$$

We then average over all ω^k using

$$\bar{F}^\pi = \frac{1}{K} \sum_{k=1}^{K} \bar{F}^\pi(\omega^k).$$

Quantiles

Instead of evaluating the average performance, we may wish to evaluate a policy based on some quantile. For example, if we are maximizing performance, we might be interested in the 10th percentile, since a policy that produces good average performance may work very poorly some of the time.

Let $Q_\alpha(R)$ be the α quantile of a random variable R. Let $F^\pi = F(x^{\pi,N}, W)$ be the random variable describing the performance of policy π, recognizing that we may have uncertainty about the model (captured by S^0), uncertainty in the experiments $W^1, \dots, W^N$ that go into the final design $x^{\pi,N}$, and then uncertainty in how well we do when we implement $x^{\pi,N}$ due to $\widehat{W}$. Now, instead of taking an expectation of F^π as we did before, we let

$$V_\alpha^\pi = Q_\alpha F(x^{\pi,N}, \widehat{W}).$$

We anticipate that there are many settings where the α quantile is more interesting than an expectation. However, we have to caution that optimizing the α quantile is much harder than optimizing an expectation.

Static Regret – Deterministic Setting

We illustrate static regret for deterministic problems using the context of machine learning where our decision is to choose a parameter θ that fits a model $f(x|\theta)$ to observations y. Here, "x" plays the role of data rather than a decision, although later we will get to "decide" what data to collect (confused yet?).

The machine learning community likes to evaluate the performance of a machine learning algorithm (known as a "learner") which is searching for the best parameters θ to fit some model $f(x|\theta)$ to predict a response y. Imagine a dataset $x^1, \dots, x^n, \dots, x^N$ and let $L^n(\theta)$ be the loss function that captures how well our function $f(x^n|\theta^n)$ predicts the response y^{n+1}, where θ^n is our estimate of θ based on the first n observations. Our loss function might be

$$L^{n+1}(x^n, y^{n+1}|\theta^n) = (y^{n+1} - f(x^n|\theta^n))^2.$$

Assume now that we have an algorithm (or policy) for updating our estimate of θ that we designate $\Theta^\pi(S^n)$, where S^n captures whatever the algorithm (or policy) needs to update θ^{n-1} to θ^n. One example of a policy is to optimize over the first n data points, so we would write

$$\Theta^\pi(S^n) = \arg\min_\theta \sum_{m=0}^{n-1} L^{m+1}(x^m, y^{m+1}|\theta).$$

Alternatively, we could use one of the gradient-based algorithms presented in chapter 5. If we fix this policy, our total loss would be

$$L^\pi = \sum_{n=0}^{N-1} L^{n+1}(x^n, y^{n+1}|\Theta^\pi(S^n)).$$

Now imagine that we pick the best value of θ, which we call θ^*, based on all the data. This requires solving

$$L^{static,*} = \min_\theta \sum_{n=0}^{N-1} L^{n+1}(x^n, y^{n+1}|\theta).$$

We now compare the performance of our policy, L^π, to our static bound, $L^{static,*}$. The difference is known as the *static regret* in the machine learning community, or the *opportunity cost* in other fields. The regret (or opportunity cost) is given by

$$R^{static,\pi} = L^\pi - L^{static,*}.$$ (7.86)

Static Regret – Stochastic Setting

Returning to the setting where we have to decide which alternative x to try, we now illustrate static regret in a stochastic setting, where we seek to maximize rewards ("winnings") W_x^n by trying alternative x in the nth trial. Let $X^\pi(S^n)$ be a policy that determines the alternative x^n to evaluate given what we know after n experiments (captured by our state variable S^n). Imagine that we can generate the entire sequence of winnings W_x^n for all alternatives x, and all iterations n. If we evaluate our policy on a single dataset (as we did in the machine learning setting), we would evaluate our regret (also known as *static regret*) as

$$R^{\pi,n} = \max_x \sum_{m=1}^n W_x^m - \sum_{m=1}^n W_{X^\pi(S^m)}^m.$$ (7.87)

Alternatively, we could write our optimal solution at time n as

$$x^n = \arg\max_x \sum_{m=1}^n W_x^m,$$

and then write the regret as

$$R^{\pi,n} = \sum_{m=1}^n W_{x^n}^m - \sum_{m=1}^n W_{X^\pi(S^m)}^m.$$

The regret (for a deterministic problems) $R^{\pi,n}$ is comparing the best decision at time n assuming we know all the values W_x^m, $x \in X$ for $m = 1, \dots, n$, against what our policy $X^\pi(S^m)$ would have chosen given just what we know at time m (please pay special attention to the indexing). This is an instance of static regret for a deterministic problem.

In practice, W_x^m is a random variable. Let $W_x^m(\omega)$ be one sample realization for a sample path $\omega \in \Omega$ (we can think of regret for a deterministic problem as the regret for a single sample path). Here, ω represents a set of all possible realizations of W over all alternatives x, and all iterations n. Think of specifying ω as pre-generating all the observations of W that we *might* experience over all experiments. However, when we make a decision $X^\pi(S^m)$ at time m, we are not allowed to see any of the information that might arrive at times after m.

When we introduce uncertainty, there are now two ways of evaluating regret. The first is to assume that we are going to first observe the outcomes $W_x^m(\omega)$ for all the alternatives and the entire history $m = 1, \dots, n$, and compare this to what our policy $X^\pi(S^m)$ would have done at each time m knowing only what has happened up to time m. The result is the regret for a single sample path ω

$$R^{\pi,n}(\omega) = \max_{x(\omega)} \sum_{m=1}^{n} W_{x(\omega)}^{m}(\omega) - \sum_{m=1}^{n} W_{X^{\pi}(S^{m})}^{m}(\omega). \tag{7.88}$$

As we did above, we can also write our optimal decision for the stochastic case as

$$x^{n}(\omega) = \arg\max_{x \in \mathcal{X}} \sum_{m=1}^{n} W_{x}^{m}(\omega).$$

We would then write our regret for sample path ω as

$$R^{\pi,n}(\omega) = \sum_{m=1}^{n} W_{x^{n}(\omega)}^{m}(\omega) - \sum_{m=1}^{n} W_{X^{\pi}(S^{m})}^{m}(\omega).$$

Think of $x^{n}(\omega)$ as the best answer if we actually did know $W_{x}^{m}(\omega)$ for $m = 1, \ldots, n$, which in practice would never be true.

If we use our machine learning setting, the sample ω would be a single dataset used to fit our model. In machine learning, we typically have a single dataset, which is like working with a single ω. This is typically what is meant by a deterministic problem (think about it). Here, we are trying to design policies that will work well across many datasets.

In the language of probability, we would say that $R^{\pi,n}$ is a random variable (since we would get a different answer each time we run the simulation), while $R^{\pi,n}(\omega)$ is a sample realization. It helps when we write the argument (ω) because it tells us what is random, but $R^{\pi,n}(\omega)$ and $x^{n}(\omega)$ are sample realizations, while $R^{\pi,n}$ and x^{n} are considered random variables (the notation does not tell you that they are random – you just have to know it). We can "average" over all the outcomes by taking an expectation, which would be written

$$\mathbb{E} R^{\pi,n} = \mathbb{E} \left\{ W_{x^{n}}^{n} - \sum_{m=1}^{n} W_{X^{\pi}(S^{m})}^{m} \right\}.$$

Expectations are mathematically pretty, but we can rarely actually compute them, so we run simulations and take an average. Assume we have a set of sample realizations $\omega \in \hat{\Omega} = \{\omega^{1}, \ldots, \omega^{\ell}, \ldots, \omega^{L}\}$. We can compute an average regret (approximating expected regret) using

$$\mathbb{E} R^{\pi,n} \approx \frac{1}{L} \sum_{\ell=1}^{L} R^{\pi,n}(\omega^{\ell}).$$

Classical static regret assumes that we are allowed to find a solution $x^{n}(\omega)$ for each sample path. There are many settings where we have to find solutions before we see any data, that works well, on average, over all sample paths. This produces a different form of regret known in the computer science community as *pseudo-regret* which compares a policy $X^\pi(S^n)$ to the solution x^* that works best *on average* over all possible sample paths. This is written

$$\bar{R}^{\pi,n} = \max_x \mathbb{E}\left\{\sum_{m=1}^{n} W_x^n\right\} - \mathbb{E}\left\{\sum_{m=1}^{n} W_{X^\pi(S^n)}^n(\omega)\right\}. \tag{7.89}$$

Again, we will typically need to approximate the expectation using a set of sample paths $\hat{\Omega}$.

Dynamic Regret

A criticism of static regret is that we are comparing our policy to the best decision x^* (or best parameter θ^* in a learning problem) for an entire dataset, but made after the fact with perfect information. In online settings, it is necessary to make decisions x^n (or update our parameter θ^n) using only the information available up through iteration n.

Dynamic regret raises the bar by choosing the best value θ^n that minimizes $L^n(x^{n-1}, y^n|\theta)$, which is to say

$$\theta^{*,n} = \arg\min_\theta L^n(x^{n-1}, y^n|\theta), \tag{7.90}$$

$$= \arg\min_\theta (y^n - f(x^{n-1}|\theta))^2. \tag{7.91}$$

The dynamic loss function is then

$$L^{dynamic,*} = \sum_{n=0}^{N-1} L^{n+1}(x^n, y^{n+1}|\theta^{*,n}).$$

More generally, we could create a policy Θ^π for adaptively evolving θ (equation (7.91) is an example of one such policy). In this case we would compute θ using $\theta^n = \Theta^\pi(S^n)$, where S^n is our belief state at time n (this could be current estimates, or the entire history of data). We might then write our dynamic loss problem in terms of finding the best policy Θ^π for adaptively searching for θ as

$$L^{dynamic,*} = \min_{\Theta^\pi} \sum_{n=0}^{N-1} L^{n+1}(x^n, y^{n+1}|\Theta^\pi(S^n)).$$

We then define dynamic regret using

$$R^{dynamic,\pi} = L^\pi - L^{dynamic,*}.$$

Dynamic regret is simply a performance metric using a more aggressive benchmark. It has attracted recent attention in the machine learning community as a way of developing theoretical benchmarks for evaluating learning policies.

Opportunity Cost (Stochastic)

Opportunity cost is a term used in the learning community that is the same as regret, but often used to evaluate policies in a stochastic setting. Let $\mu_x = \mathbb{E}F(x, \theta)$ be the true value of design x, let

$$x^* = \arg\max_x \mu_x,$$

$$x^\pi = \arg\max_x \mu_{x^{\pi,N}}.$$

So, x^* is the best design if we knew the truth, while $x^{\pi,N}$ is the design we obtained using learning policy π after exhausting our budget of N experiments. In this setting, μ_x is treated deterministically (think of this as a known truth), but $x^{\pi,N}$ is random because it depends on a noisy experimentation process. The expected regret, or opportunity cost, of policy π is given by

$$R^\pi = \mu_{x^*} - \mathbb{E}\mu_{x^{\pi,N}}. \tag{7.92}$$

Competitive Analysis

A strategy that is popular in the field known as *online computation* (which has nothing to do with "online learning") likes to compare the performance of a policy to the best that could have been achieved. There are two ways to measure "best." The most common is to assume we know the future. Assume we are making decisions $x^0, x^1, \ldots, x^T$ over our horizon $0, \ldots, T$. Let ω represent a sample path $W^1(\omega), \ldots, W^N(\omega)$, and let $x^{*,t}(\omega)$ be the best decision given that we know that all random outcomes (over the entire horizon) are known (and specified by ω). Finally, let $F(x^n, W^{n+1}(\omega))$ be the performance that we observe at time $t + 1$. We can then create a perfect foresight (PF) policy using

$$X^{PF,n}(\omega) = \arg\max_{x^n(\omega)} \left(c^n x^n(\omega) + \max_{x^{n+1}(\omega),\ldots,x^N(\omega)} \sum_{m=n+1}^N c^m x^m(\omega) \right).$$

Unlike every other policy that we consider in this volume, this policy is allowed to see into the future, producing decisions that are better than anything we could achieve without this ability. Now consider some $X^\pi(S^n)$ policy that is only allowed to see the state at time S^n. We can compare policy $X^\pi(S)$ to our perfect foresight using the *competitive ratio* given by

$$\rho^\pi = \mathbb{E}\frac{\sum_{n=0}^{N-1} F(X^{\pi,n}(\omega), W^{n+1}(\omega))}{\sum_{n=0}^{N-1} F(X^{PF,n}(\omega), W^{n+1}(\omega))}$$

where the expectation is over all sample paths ω (competitive analysis is often performed for a single sample path). Researchers like to prove bounds on the competitive ratio, although these bounds are never tight.

Indifference Zone Selection

A variant of the goal of choosing the best alternative $x^* = \arg\max_x \mu_x$ is to maximize the likelihood that we make a choice $x^{\pi,N}$ that is almost as good as x^*. Assume we are equally happy with any outcome within δ of the best, by which we mean

$$\mu_{x^*} - \mu_{x^{\pi,N}} \le \delta.$$

The region $(\mu_{x^*} - \delta, \mu_{x^*})$ is referred to as the *indifference zone*. Let $V^{n,\pi}$ be the value of our solution after n experiments. We require $\mathbb{P}^\pi\{\mu_{d^*} = \bar{\mu}^* | \mu\} > 1 - \alpha$ for all μ where $\mu_{[1]} - \mu_{[2]} > \delta$, and where $\mu_{[1]}$ and $\mu_{[2]}$ represent, respectively, the best and second best choices.

We might like to maximize the likelihood that we fall within the indifference zone, which we can express using

$$P^{IZ,\pi} = \mathbb{P}^\pi(V^{\pi,n} > \mu^* - \delta).$$

As before, the probability has to be computed with the appropriate Bayesian or frequentist distribution.

7.11.2 Perspectives of Optimality*

In this section we review different perspectives of optimality in sequential search procedures.

Asymptotic Convergence for Final Reward

While in practice we need to evaluate how an algorithm does in a finite budget, there is a long tradition in the analysis of algorithms to study the asymptotic performance of algorithms when using a final-reward criterion. In particular, if x^* is the solution to our asymptotic formulation in equation (7.1), we would like to know if our policy that produces a solution $x^{\pi,N}$ after N evaluations would eventually converge to x^*. That is, we would like to know if

$$\lim_{N\to\infty} x^{\pi,N} \to x^*.$$

Researchers will often begin by proving that an algorithm is asymptotically convergent (as we did in chapter 5), and then evaluate the performance in a finite budget N empirically. Asymptotic analysis generally only makes sense when using a final-reward objective.

Finite Time Bounds on Choosing the Wrong Alternative

There is a body of research that seeks to bound the number of times a policy chooses a suboptimal alternative (where alternatives are often referred to as "arms" for a multiarmed bandit problem). Let μ_x be the (unknown) expected reward for alternative x, and let $W_x^n = \mu_x + \epsilon_x^n$ be the observed random reward from trying x. Let x^* be the optimal alternative, where

$$x^* = \arg\max_x \mu_x.$$

For these problems, we would define our loss function as

$$L^n(x^n) = \begin{cases} 1 & \text{if } x^n \neq x^*, \\ 0 & \text{otherwise.} \end{cases}$$

Imagine that we are trying to minimize the cumulative reward, which means the total number of times that we do not choose the best alternative. We can compare a policy that chooses $x^n = X^\pi(S^n)$ against a perfect policy that chooses x^* each time. The regret for this setting is then simply

$$R^{\pi,n} = \sum_{m=1}^{n} L^n(X^\pi(S^n)).$$

Not surprisingly, R^π grows monotonically in n, since good policies have to be constantly experimenting with different alternatives. An important research goal is to design bounds on $R^{\pi,n}$, which is called a finite-time bound, since it applies to $R^{\pi,n}$ for finite n.

Probability of Correct Selection

A different perspective is to focus on the probability that we have selected the best out of a set X alternatives. In this setting, it is typically the case that the number of alternatives is not too large, say 10 to 100, and certainly not 100,000. Assume that

$$x^* = \arg\max_{x \in X} \mu_x$$

is the best decision (for simplicity, we are going to ignore the presence of ties). After n samples, we would make the choice

$$x^n = \arg\max_{x \in X} \bar{\mu}_x^n.$$

This is true regardless of whether we are using a frequentist or Bayesian estimate.

We have made the correct selection if $x^n = x^*$, but even the best policy cannot guarantee that we will make the best selection every time. Let $\mathbb{1}_{\{\mathcal{E}\}} = 1$ if the event $\mathcal{E}$ is true, 0 otherwise. We write the probability of correct selection as

$$P^{CS,\pi} \quad = \quad \text{probability we choose the best alternative}$$

$$= \quad \mathbb{E}\mathbb{1}_{\{x^n=x^*\}},$$

where the underlying probability distribution depends on our experimental policy π. The probability is computed using the appropriate distribution, depending on whether we are using Bayesian or frequentist perspectives. This may be written in the language of loss functions. We would define the loss function as

$$L^{CS,\pi} \quad = \quad \mathbb{1}_{\{x^n \neq x^*\}}.$$

Although we use $L^{CS,\pi}$ to be consistent with our other notation, this is more commonly represented as L_{0-1} for "0-1 loss."

Note that we write this in terms of the negative outcome so that we wish to minimize the loss, which means that we have not found the best selection. In this case, we would write the probability of correct selection as

$$P^{CS,\pi} = 1 - \mathbb{E}L^{CS,\pi}.$$

Subset Selection –
Ultimately our goal is to pick the best design. Imagine that we are willing to choose a subset of designs S, and we would like to ensure that $P(x^* \in S) \geq 1-\alpha$, where $1/|\mathcal{X}| < 1-\alpha < 1$. Of course, it would be idea if $|S| = 1$ or, failing this, as small as possible. Let $\bar{\mu}_x^n$ be our estimate of the value of x after n experiments, and assume that all experiments have a constant and known variance σ. We include x in the subset if

$$\bar{\mu}_x^n \geq \max_{x' \neq x} \bar{\mu}_{x'}^n - h\sigma\sqrt{\frac{2}{n}}.$$

The parameter h is the $1-\alpha$ quantile of the random variable $\max_i Z_i^n$ where Z_i^n is given by

$$Z_i^n = \frac{(\bar{\mu}_i^n - \bar{\mu}_x^n) - (\mu_i - \mu_x)}{\sigma\sqrt{2/n}}.$$

7.12 Designing Policies

By now we have reviewed a number of solution approaches organized by our four classes of policies:

PFAs – Policy function approximations, which are analytical functions such as a linear decision rule (that has to be tuned), or the setting of the optimal price to which we add noise (the excitation policy).

CFAs – Cost function approximations, which are probably the most popular class of policy for these problems. A good example is the family of upper confidence bounding policies, such as

$$X^{UCB}(S^n|\theta) = \arg\max_{x^n \in \mathcal{X}} \left(\bar{\mu}_x^n + \theta \bar{\sigma}_x^n \right).$$

VFAs – Policies based on value function approximation, such as using Gittins indices or backward ADP to estimate the value of information.

DLAs – Policies based on direct lookaheads such as the knowledge gradient (a one-step lookahead) or kriging.

It is easy to assume that the policy we want is the policy that performs the best. This is simply not the case. A representative from a large tech company that used active learning policies extensively stated their criteria very simply:

We will use the best policy that can be computed in under 50 milliseconds.

This hints that there is more to using a policy than just its performance. We begin our discussion with a list of characteristics of good learning policies. We then raise the issue of scaling for tunable parameters, and close with a discussion of the whole process of tuning.

7.12.1 Characteristics of a Policy

Our standard approach to evaluating policies is to look for the policy that performs the best (on average) according to some performance metric. In practice, the choice of policy tends to consider the following characteristics:

Performance This is our objective function which is typically written as

$$\max_{\pi} \mathbb{E}_{S^0} \mathbb{E}_{W^1,\ldots,W^N|S^0} \mathbb{E}_{\widehat{W}|S^0} \{ F(x^{\pi,N}, \widehat{W}) | S^0 \},$$

or

$$\max_{\pi} \mathbb{E}_{S^0} \mathbb{E}_{W^1,\ldots,W^N|S^0} \sum_{n=0}^{N-1} F(X^{\pi}(S^n), W^{n+1}).$$

Computational complexity CPU times matter. The tech company above required that we be able to compute a policy in 50 milliseconds, while an energy company faced a limit of 4 hours.

Robustness Is the policy reliable? Does it produce consistently reliable solutions under a wide range of data inputs? This might be important in the
setting of recommending prices for hotel rooms, where the policy involves
learning in the field. A hotel would not want a system that recommends
unrealistic prices.

Tuning The less tuning that is required, the better.

Transparency A bank might need a system that recommends whether a loan
should be approved. There are consumer protection laws that protect against
bias that require a level of transparency in the reasons that a loan is turned
down.

Implementation complexity How hard is it to code? How likely is it that a
coding error will affect the results?

The tradeoff between simplicity and complexity is particularly important. As of
this writing, CFA-based policies such as upper confidence bounding are receiving tremendous attention in the tech sector due in large part to their simplicity,
as well as their effectiveness, but always at the price of introducing tunable
parameters.

The problem of tuning is almost uniformly overlooked by the theory community that focuses on theoretical performance metrics such as regret bounds.
Practitioners, on the other hand, are aware of the importance of tuning, but
tuning has historically been an ad hoc activity, and far too often it is simply
overlooked! Tuning is best done in a simulator, but simulators are only approximations of the real world, and they can be expensive to build. We need more
research into online tuning.

Lookahead policies may have no tuning, such as the knowledge gradient in
the presence of concave value of information or the deterministic direct lookahead, or some tuning such as the parameter θ^{KGLA} in section 7.7.3. Either
way, a lookahead policy requires a lookahead model, which introduces its own
approximations. So, there is no free lunch.

7.12.2 The Effect of Scaling

Consider the case of two policies. The first is interval estimation, given by

$$X^{IE}(S^n|\theta^{IE}) = \arg\max_x \left(\bar{\mu}_x^n + \theta^{IE}\sigma_x^n\right),$$

which exhibits a unitless tunable parameter θ^{IE}. The second policy is a type of
upper confidence bounding policy known in the literature as UCB-E, given by

Table 7.6 Optimal tuned parameters for interval estimation IE, and UCBE. Adapted from Wang, Y., Wang, C., Powell, W. B. and Edu, P. P. (2016), The Knowledge Gradient for Sequential Decision Making with Stochastic Binary Feedbacks, in 'ICML2016', Vol. 48.

Problem	IE	UCBE
Goldstein	0.0099	2571
AUF_HNoise	0.0150	0.319
AUF_MNoise	0.0187	1.591
AUF_LNoise	0.0109	6.835
Branin	0.2694	.000366
Ackley	1.1970	1.329
HyperEllipsoid	0.8991	21.21
Pinter	0.9989	0.000164
Rastrigin	0.2086	0.001476

$$X^{UCB-E,n}(S^n|\theta^{UCB-E}) = \arg\max_x \left(\bar{\mu}_x^n + \sqrt{\frac{\theta^{UCB-E}}{N_x^n}} \right),$$

where N_x^n is the number of times that we have evaluated alternative x. We note that unlike the interval estimation policy, the tunable parameter θ^{UCB-E} has units, which means that we have to search over a much wider range than we would when optimizing θ^{IE}.

Each of these parameters were tuned on a series of benchmark learning problems using the testing system called MOLTE, with the results reported in Table 7.6. We see that the optimal value of θ^{IE} ranges from around 0.01 to 1.2. By contrast, θ^{UCB-E} ranges from 0.0001 to 2500.

These results illustrate the effect of units on tunable parameters. The UCB-E policy enjoys finite time bounds on its regret, but would never produce reasonable results without tuning. By contrast, the optimal values of θ^{IE} for interval estimation vary over a narrower range, although conventional wisdom for this parameter is that it should range between around 1 and 3. If θ^{IE} is small, then the IE policy is basically a pure exploitation policy.

Parameter tuning can be difficult in practice. Imagine, for example, an actual setting where an experiment is expensive. How would tuning be done? This issue is typically ignored in the research literature where standard practice is to focus on provable qualities. We argue that despite the presence of provable properties, the need for parameter tuning is the hallmark of a heuristic.

If tuning cannot be done, the actual empirical performance of a policy may be quite poor.

Bayesian policies such as the knowledge gradient do not have tunable parameters, but do require the use of priors. Just as we do not have any real theory to characterize the behavior of algorithms that have (or have not) been tuned, we do not have any theory to describe the effect of incorrect priors.

7.12.3 Tuning

An issue that will keep coming back in the design of algorithms is tuning. We will keep repeating the mantra:

The price of simplicity is tunable parameters... and tuning is hard!

We are designing policies to solve any of a wide range of stochastic search problems, but when our policy involves tunable parameters, we are creating a stochastic search problem (tuning the parameters) to solve our stochastic search problem. The hope, of course, is that the problem of tuning the parameters of our policy is easier than the search problem that our problem is solving.

What readers need to be aware of is that the performance of their stochastic search policy can be very dependent on the tuning of the parameters of the policy. In addition, the best value of these tunable parameters can depend on anything from the characteristics of a problem to the starting point of the algorithm. This is easily the most frustrating aspect of tuning of policy parameters, since you have to know when to stop and revisit the settings of your parameters.

7.13 Extensions*

This section covers a series of extensions to our basic learning problem:

- Learning in nonstationary settings
- Strategies for designing policies
- A transient learning model
- The knowledge gradient for transient problems
- Learning with large or continuous choice sets
- Learning with exogenous state information
- State-dependent vs. state-independent problems

7.13.1 Learning in Nonstationary Settings

Our classic "bandit" problem involves learning the value of μ_x for $x \in \mathcal{X} = \{x_1, \dots, x_M\}$ using observations where we choose the alternative to evaluate $x^n = X^{\pi}(S^n)$ from which we observe

$$W^{n+1} = \mu_{x^n} + \varepsilon^{n+1}.$$

In this setting, we are trying to learn a static set of parameters μ_x, $x \in \mathcal{X}$, using a stationary policy $X^{\pi}(S_t)$. An example of a stationary policy for learning is upper confidence bounding, given by

$$X^{UCB}(S^n|\theta^{UCB}) = \arg\max_x \left(\bar{\mu}_x^n + \theta^{UCB} \sqrt{\frac{\log n}{N_x^n}} \right), \tag{7.93}$$

where N_x^n is the number of times we have tried alternative x over the first n experiments.

It is natural to search for a stationary policy $X^{\pi}(S^n)$ (that is, a policy where the *function* does not depend on time t) by optimizing an infinite horizon, discounted objective such as

$$\max_{\pi} \mathbb{E} \sum_{n=0}^{\infty} \gamma^n F(X^{\pi}(S^n), W^{n+1}). \tag{7.94}$$

In practice, truly stationary problems are rare. Nonstationarity can arise in a number of ways:

Finite-horizon problems – Here we are trying to optimize performance over a finite horizon $(0, N)$ for a problem where the exogenous information W_t comes from a stationary process. The objective would be given by

$$\max_{\pi} \mathbb{E} \sum_{n=0}^{N} F(X^{\pi}(S^n), W^{n+1}).$$

Note that we may use a stationary policy such as upper confidence bounding to solve this problem, but it would not be optimal. The knowledge gradient policy for cumulative rewards (equation (7.65)).

Learning processes – x might be an athlete who gets better as she plays, or a company might get better at making a complex component.

Exogenous nonstationarities – Field experiments might be affected by weather which is continually changing.

Adversarial response – x might be a choice of an ad to display, but the market response depends on the behavior of other players who are changing their strategies. This problem class is known as "restless bandits" in the bandit community.

Availability of choices – We may wish to try different people for a job, but they may not be available on any given day. This problem is known as "intermittent bandits."

7.13.2 Strategies for Designing Time-dependent Policies

There are two strategies for handling time-dependencies:

Time-dependent policies A time-dependent policy is simply a policy that depends on time. We already saw one instance of a nonstationary policy when we derived the knowledge gradient for cumulative rewards, which produced the policy

$$X^{OLKG,n}(S^n) = \arg\max_{x \in \mathcal{X}} \left(\bar{\mu}_x^n + (N - n)\bar{\sigma}_x^n \right). \tag{7.95}$$

Here we see that not only is the state $S^n = (\bar{\mu}^n, \bar{\sigma}^n)$ time-dependent, the policy itself is time-dependent because of the coefficient $(N - n)$. The same would be true if we used a UCB policy with coefficient $\theta^{UCB,n}$, but this means that instead of learning one parameter θ^{UCB}, we have to learn $(\theta_0^{UCB}, \theta_1^{UCB}, ..., \theta_N^{UCB})$.

Note that a time-dependent policy is designed in advance, before any observations have been made. This can be expressed mathematically as solving the optimization problem

$$\max_{\pi^0, ..., \pi^N} \mathbb{E} \sum_{n=0}^{N} F(X^{\pi^n}(S^n), W^{n+1}). \tag{7.96}$$

Adaptive policies These are policies which adapt to the data, which means the function itself is changing over time. This is easiest to understand if we assume we have a parameterized policy $X^\pi(S_t|\theta)$ (such as interval estimation – see equation (12.46)). Now imagine that the market has shifted which means we would like to increase how much exploration we are doing.

We can do this by allowing the parameter θ to vary over time, which means we would write our decision policy as $X^\pi(S^n|\theta^n)$. We need logic to adjust θ^n which we depict using $\theta^{n+1} = \Theta^{\pi^\theta}(S^n)$. The function $\Theta^{\pi^\theta}(S^n)$ can be viewed as a policy (some would call it an algorithm) to adjust θ^n. Think of this as a "policy to tune a policy."

For a given policy $X^\pi(S_t|\theta^n)$, the problem of tuning the π^θ-policy would be written as

$$\max_{\pi^\theta} \mathbb{E} \sum_{n=0}^{N} F(X^\pi(S^n|\Theta^\pi(S^n)), W^{n+1}).$$

We still have to choose the best implementation policy $X^\pi(S^n|\theta^n)$. We could write the combined problem as

$$\max_{\pi^\theta} \max_{\pi} \mathbb{E} \sum_{n=0}^{N} F(X^\pi(S^n|\Theta^\pi(S^n)), W^{n+1}).$$

Both policies π^θ, which determines $\Theta^{\pi^\theta}(S^n)$, and π, which determines $X^\pi(S^n|\theta^n)$ have to be determined offline, but the decision policy is being tuned adaptively while in the field (that is, "online").

7.13.3 A Transient Learning Model

We first introduced this model in section 3.11 where the true mean varies over time. It is most natural to talk about nonstationary problems in terms of varying over time t, but we will stay with our counter index n for consistency.

When we have a transient process, we update our beliefs according to the model

$$\mu^{n+1} = M^n \mu^n + \varepsilon^{\mu,n+1},$$

where $\varepsilon^{\mu,n+1}$ is a random variable with distribution $N(0, \sigma_\mu^2)$, which means that $\mathbb{E}\{\mu^{n+1}|\mu^n\} = M^n\mu^n$. The matrix M^n is a diagonal matrix that captures predictable changes (e.g. where the means are increasing or decreasing predictably). If we let M^n be the identity matrix, then we have the simpler problem where the changes in the means have mean 0 which means that we expect $\mu^{n+1} = \mu^n$. However, there are problems where there can be a predictable drift, such as estimating the level of a reservoir changing due to stochastic rainfall and predictable evaporation. We then make noisy observations of μ^n using

$$W^n = M^n \mu^n + \varepsilon^n.$$

It used to be that if we did not observe an alternative x' that our belief $\bar{\mu}_{x'}^n$ did not change (and of course, nor did the truth). Now, the truth may be changing, and to the extent that there is predictable variation (that is, M^n is not the identity matrix), then even our beliefs may change.

The updating equation for the mean vector is given by

$$\bar{\mu}_x^{n+1} = \begin{cases} M_x^n \bar{\mu}_x^n + \frac{W^{n+1} - M_x^n \bar{\mu}_x^n}{\sigma_\varepsilon^2 + \Sigma_{xx}^n} \Sigma_{xx}^n & \text{if } x^n = x, \\ M_x^n \bar{\mu}_x^n & \text{otherwise.} \end{cases} \tag{7.97}$$

To describe the updating of Σ^n, let Σ_x^n be the column associated with alternative x, and let e_x be a vector of 0's with a 1 in the position corresponding to alternative x. The updating equation for Σ^n can then be written

$$\Sigma_x^{n+1} = \begin{cases} \Sigma_x^n - \frac{(\Sigma_x^n)^T \Sigma_x^n}{\sigma_\varepsilon^2 + \Sigma_{xx}^n} e_x & \text{if } x^n = x, \\ \Sigma_x^n & \text{otherwise.} \end{cases} \tag{7.98}$$

These updating equations can play two roles in the design of learning policies. First, they can be used in a lookahead policy, as we illustrate next with the knowledge gradient (a one-step lookahead policy). Alternatively, they can be used in a simulator for the purpose of doing policy search for the best PFA or CFA.

7.13.4 The Knowledge Gradient for Transient Problems

To compute the knowledge gradient, we first compute

$$\tilde{\sigma}_x^{2,n} = \text{The conditional change in the variance of } \bar{\mu}_x^{n+1}$$
$$\text{given what we know now,}$$
$$= Var(\bar{\mu}_x^{n+1}|\bar{\mu}^n) - Var(\bar{\mu}^n),$$
$$= Var(\bar{\mu}_x^{n+1}|\bar{\mu}^n),$$
$$= \tilde{\Sigma}_{xx}^n.$$

We can use $\tilde{\sigma}_x^n$ to write the updating equation for $\bar{\mu}^n$ using

$$\bar{\mu}^{n+1} = M^n \bar{\mu}^n + \tilde{\sigma}_x^n Z^{n+1} e_p,$$

where $Z^{n+1} \sim N(0,1)$ is a scalar, standard normal random variable.

We now present some calculations that parallel the original knowledge gradient calculations. First, we define ζ_{tx} as we did before

$$\zeta_x^n = -\left| \frac{\bar{\mu}_x^n - \max_{x' \neq x} \bar{\mu}_{x'}^n}{\tilde{\sigma}_x^n} \right|.$$

This is defined for our stationary problem. We now define a modified version that we call ζ_x^M that is given by

$$\zeta_x^{M,n} = M^n \zeta_x^n.$$

We can now compute the knowledge gradient for nonstationary truths using a form that closely parallels the original knowledge gradient,

$$\nu_x^{KG-NS,n} = \bar{\sigma}_x^n \left(\zeta_x^{M,n} \Phi(\zeta_x^{M,n}) + \phi(\zeta_x^{M,n}) \right) \tag{7.99}$$

$$= \bar{\sigma}_x^n \left(M^n \zeta_x^n \Phi(M^n \zeta_x^n) + \phi(M^n \zeta_x^n) \right). \tag{7.100}$$

It is useful to compare this version of the knowledge gradient to the knowledge gradient for our original problem with static truths. If M^n is the identity matrix, then this means that the truths μ^n are not changing in a predictable way; they might increase or decrease, but on average μ^{n+1} is the same as μ^n. When this happens, the knowledge gradient for the transient problem is the same as the knowledge gradient when the truths are not changing at all.

So, does this mean that the problem where the truths are changing is the same as the one where they remain constant? Not at all. The difference arises in the updating equations, where the precision of alternatives x' that are not tested decrease, which will make them more attractive from the perspective of information collection.

7.13.5 Learning with Large or Continuous Choice Sets

There are many problems where our choice set $\mathcal{X}$ is either extremely large or continuous (which means the number of possible values is infinite). For example:

■ **EXAMPLE 7.4**

A website advertising movies has the space to show 10 suggestions out of hundreds of movies within a particular genre. The website has to choose from all possible combinations of 10 movies out of the population.

■ **EXAMPLE 7.5**

A scientist is trying to choose the best from a set of over 1000 different materials, but has a budget to only test 20.

■ **EXAMPLE 7.6**

A bakery chef for a food producer has to find the best proportions of flour, milk, yeast, and salt.

■ **EXAMPLE 7.7**

A basketball coach has to choose the best five starting players from a team of 12. It takes approximately half a game to draw conclusions about the performance of how well five players work together.

Each of these examples exhibit large choice sets, particularly when evaluated relative to the budget for running experiments. Such situations are surprisingly common. We can handle these situations using a combination of strategies:

Generalized learning The first step in handling large choice sets is using a belief model that provides for a high level of generalization. This can be done using correlated beliefs for lookup table models, and parametric models, where we only have to learn a relatively small number of parameters (which we hope is smaller than our learning budget).

Sampled actions Whether we have continuous actions or large (often multidimensional) actions, we can create smaller problems by just using a sampled set of actions, just as we earlier used sampled beliefs about a parameter vector θ.

Action sampling is simply another use of Monte Carlo simulation to reduce a large set to a small one, just as we have been doing when we use Monte Carlo sampling to reduce large (often infinite) sets of outcomes of random variables to smaller, discrete sets. Thus, we might start with the optimization problem

$$F^* = \max_{x \in \mathcal{X}} \mathbb{E}_W F(x, W).$$

Often the expectation cannot be computed, so we replace the typically large set of outcomes of W, represented by some set Ω, with a sampled set of outcomes $\hat{\Omega} = \{w_1, w_2, \ldots, w_K\}$, giving us

$$\bar{F}^K = \max_{x \in \mathcal{X}} \frac{1}{K} \sum_{k=1}^{K} F(x, w_k).$$

When $\mathcal{X}$ is too large, we can play the same game and replace it with a random sample $\hat{\mathcal{X}} = \{x_1, \ldots, x_L\}$, giving us the problem

$$W^{K,L} = \max_{x \in \hat{\mathcal{X}}} \frac{1}{K} \sum_{k=1}^{K} F(x, w_k). \tag{7.101}$$

Section 4.3.2 provides results that demonstrate that the approximation $\bar{F}^K$ converges quite quickly to F^* as K increases. We might expect a similar result from

$W^{K,L}$ as L increases, although there are problems where it is not possible to grow L past a certain amount. For example, see equation (7.76) for our sampled belief model, which becomes computationally challenging if the number of sampled values of θ is too large.

A strategy for overcoming this limitation is to periodically drop, say, $L/2$ elements of $\mathcal{X}$ (based on the probabilities p_k^n), and then go through a process of randomly generating new values and adding them to the set until we again have L elements. We may even be able to obtain an estimate of the value of each of the new alternatives before running any new experiments. This can be done using the following:

- If we have a parametric belief model, we can estimate a value of x using our current estimate of θ. This could be a point estimate, or distribution $(p_k^n)_{k=1}^K$ over a set of possible values $\theta_1, \dots, \theta_K$.
- If we are using lookup tables with correlated beliefs, and assuming we have access to a correlation function that gives us $Cov(F(x), F(x'))$ for any pair x and x', we can construct a belief from experiments we have run up to now. We just have to rerun the correlated belief model from chapter 3 including the new alternative, but without running any new experiments.
- We can always use nonparametric methods (such as kernel regression) to estimate the value of any x from the observations we have made so far, simply by smoothing over the new point. Nonparametric methods can be quite powerful (hierarchical aggregation is an example, even though we present it alongside lookup table models in chapter 3), but they assume no structure and as a result need more observations.

Using these estimates, we might require that any newly generated alternative x be at least as good as any of the estimates of values in the current set. This process might stop if we cannot add any new alternatives after testing some number M.

7.13.6 Learning with Exogenous State Information – the Contextual Bandit Problem

The original statement of our basic stochastic optimization problem (in its asymptotic form),

$$\max_x \mathbb{E}F(x, W)$$

is looking for a solution in the form of a deterministic decision x^*. We then proposed that a better form was

$$\max_x \mathbb{E}\{F(x, W)|S^0\}. \tag{7.102}$$

Again, we assume that we are looking for a single decision x^*, although now we have to recognize that technically, this decision is a function of the initial state S^0.

Now consider an adaptive learning process where a new initial state S^0 is revealed each time we try to evaluate $F(x, W)$. This changes the learning process, since each time we observe $F(x, W)$ for some x and a sampled W, what we learn has to reflect that it is in the context of the initial state S^0. Some illustrations of this setting are:

■ **EXAMPLE 7.8**

Consider a newsvendor problem where S^0 is the weather forecast for tomorrow. We know that if it is raining or very cold, that sales will be lower. We need to find an optimal order decision that reflects the weather forecast. Given the forecast, we make a decision of how many newspapers to stock, and then observe the sales.

■ **EXAMPLE 7.9**

A patient arrives to a hospital with a complaint, and a doctor has to make treatment decisions. The attributes of the patient represent initial information that the patient provides in the form of a medical history, then a decision is made, followed by a random outcome (the success of the treatment).

In both of these examples, we have to make our decision given advance information (the weather, or the attributes of the patient). Instead of finding a single optimal solution x^*, we need to find a function $x^*(S^0)$. This function is a form of policy (since it is a mapping of state to action).

This problem was first studied as a type of multiarmed bandit problems, which we first introduced in chapter 2. In this community, these are known as *contextual bandit problems*, but as we show here, when properly modeled this problem is simply an instance of a *state dependent* sequential decision problem.

We propose the following model of contextual problems. First, we let B_t be our belief state at time t that captures our belief about the function $F(x) = \mathbb{E}F(x, W)$ (keep in mind that this is distributional information). We then model two types of exogenous information:

Exogenous information – W_t^e This is information that arrives before we make a decision (this would be the weather in our newsvendor problem, or the attributes of the patient before making the medical decision).

Outcome W_t^o This is the information that arrives as a result of a decision, such as how the patient responds to a drug.

Using this notation, the sequencing of information, belief states and decisions is

$$(B^0, W^{e,0}, x^0, W^{o,1}, B^1, W^{e,1}, x^1, W^{o,2}, B^2, ...).$$

We have written the sequence $(W^{o,n}, B^n, W^{e,n})$ to reflect the logical progression where we first learn the outcome of a decision $W^{o,n}$, then update our belief state producing B^n, and then observe the new exogenous information $W^{e,n}$ before making decision x^n. However, we can write $W^n = (W^{o,n}, W^{e,n})$ as the exogenous information, which leads to a new state $S^n = (B^n, W^{e,n})$.

This change of variables, along with defining $S^0 = (B^0, W^{e,0})$, gives us our usual sequence of states, actions, and new information that we can write as

$$(S^0 = (B^0, W^{e,0}), x^0, W^{o,1}, B^1 = B^M(B^0, x^0, W^1 = (W^{e,1}, W^{o,1})), S^1 = (B^1, W^{e,1}), x^1,$$
$$W^{o,2}, B^2 = B^M(B^1, x^1, W^2 = (W^{e,2}, W^{o,2})), ...).$$

This, then, is the same as our basic sequence

$$(S^0, x^0, W^1, S^1, x^1, S^2, ..., S^n, x^n, W^{n+1}, ...).$$

Our policy $X^{\pi,n}(S^n)$ will now depend on both our belief state B^n about $\mathbb{E}F(x, W)$, as well as the new exogenous information $W^{e,n}$.

So why is this an issue? Simply put, pure learning problems are easier than state-dependent problems. In particular, consider one of the popular CFA policies such as upper confidence bounding or interval estimation. Instead of learning $\bar{\mu}_x^n$, we have to learn $\bar{\mu}_x^n(W^e)$. For example, if $\bar{\mu}_x^n$ describes the reduction in blood sugar from using drug x, we now have to learn the reduction in blood sugar for drug x for a patient with attributes $W^{e,n}$.

In other words, exogenous state information makes the learning more complex. If we are solving a problem where the exogenous information is weather, we might be able to describe weather using a handful of states (cold/hot, dry/rainy). However, if the exogenous information is the attributes of a patient, then it could have many dimensions. This is problematic if we are using a lookup table representation (as we might with weather), but perhaps we are just using a parametric model.

As an illustration, assume that we are deciding on the bid for an ad. The probability that a customer clicks on the ad depends on our bid b, and is given by the logistics curve:

$$p(b|\theta) \quad = \quad \frac{e^{U(b|\theta)}}{1 + e^{U(b|\theta)}},$$ (7.103)

where $U(b|\theta)$ is a linear model given by

$$U(b|\theta) = \theta_0 + \theta_1 b.$$

Now assume we are given additional information that arrives in W_t^e that provides attributes of the consumer as well as attributes of the ad. Let a_t capture this vector of attributes (this means that $W_t^e = a_t$). Then this has the effect of changing our utility function to

$$U(b|a, \theta) = \theta_0 + \theta_1 b + \theta_2 a_1 + \theta_3 a_2 + \dots .$$

As we can see, if we are using a parametric model, the additional attributes expands the number of features in $U(b|a, \theta)$, which would increase the number of observations required to estimate the vector of coefficients θ. The number of observations needed depends on the number of parameters, and the level of noise in the data.

7.13.7 State-dependent vs. State-independent Problems

We are going to spend the rest of this book on what we call "state-dependent problems," which refers to settings where the *problem* depends on the state variable. To illustrate, consider a simple newsvendor problem

$$\max_{x} F(x) = \mathbb{E}_W(p \min\{x, W\} - cx).$$ (7.104)

Assume we do not know the distribution of W, but we can collect information by choosing x^n, then observing

$$\hat{F}^{n+1} = p \min\{x^n, W^{n+1}\} - cx^n.$$

We can then use the observation $\hat{F}^{n+1}$ to produce an updated estimate $\bar{F}^{n+1}(x)$. The parameters describing the approximation $\bar{F}^n(x)$ make up our belief state B^n, which for this problem represents the only state variables. The goal is to explore different values of x to develop a good approximation $\bar{F}^n(x)$ to help choose the best value of x.

Now assume that the prices change each period, and that we are given the price p^n just before we make our choice x^n. The price p^n is a form of exogenous information, which means that instead of trying to find the best x, we are trying to find the best function $x(p)$. Now we have to decide what type of function we want to use to represent $x(p)$ (lookup table? a parametric function of p?).

Finally, assume that we have to choose product from inventory to satisfy the demand W, where R^n is our inventory. Assume that we have to observe $x^n \leq R^n$, and that the inventory is updated according to

$$R^{n+1} = R^n - \min\{x^n, W^{n+1}\} + \max\{0, x^n - W^{n+1}\}.$$

Now our decision x^n at time n affects our state R^{n+1}. For this problem, our state variable is given by

$$S^n = (R^n, p^n, B^n).$$

A special case of a state-dependent problem was the learning problem we saw in section 7.13.6, since the problem depends on the exogenous information $W^{e,n}$. This is a type of state-dependent problem, but decisions only affect the belief; the exogenous information $W^{e,n+1}$ is not affected by the decisions x^n. This quality means that this is closer to a learning problem than the broader class of state-dependent problems.

State-dependent problems may or may not involve a belief state, but will involve information other than a belief (which is what makes them state-dependent problems). A major problem class includes problems that involve the management of resources. A simple example involves managing a vehicle moving over a graph, where the decision changes the location of the vehicle.

We will show in the remainder of the book that we can approach these more complex problems with the same five-element modeling framework that we first introduced in chapter 2, and again in this chapter. Also, we will design policies using the same four classes of policies that were covered here. What changes is the choice of which policies work best.

7.14 Bibliographic Notes

Section 7.1 – The earliest paper on derivative-free stochastic search is the seminal paper (Box and Wilson, 1951), which interestingly appeared in the same year as the original paper for derivative-based stochastic search (Robbins and Monro, 1951).

Section 7.2.1 – Our formulation of derivative-free stochastic search was first suggested in Powell (2019). Of particular value is writing out the objective function for evauating policies in an explicit way; perhaps surprisingly, this is often (although not always) overlooked. We are not aware of another reference formulating stochastic search problems as formal optimization problems searching for optimal policies.

Section 7.1.4 – This is the first time in writing that the equivalence of these four classes of problems have been observed.

Section 7.3 – The idea of using all four classes of policies for pure learning problems was first suggested in Powell (2019), but this book is the first to illustrate this idea in a comprehensive way.

Section 7.5 – There is by now an extensive literature in the reinforcement learning community using what are generally referred to as "upper confidence bounding" policies, which we classify under the heading of parametric cost function approximations. A nice introduction to these learning strategies is contained in Kaelbling (1993) and Sutton and Barto (2018). Thrun (1992) contains a good discussion of exploration in the learning process, which is achieved by the "uncertainty bonus" in UCB policies. The discussion of Boltzmann exploration and epsilon-greedy exploration is based on Singh et al. (2000). The upper confidence bound is due to Lai and Robbins (1985). We use the version of the UCB rule given in Lai (1987). The UCB1 policy is given in Auer et al. (2002). Analysis of UCB policies are given in Lai and Robbins (1985), as well as Chang et al. (2007).

For a nice review of Bayesian optimization, see Frazier (2018).

Interval estimation is due to Kaelbling (1993) (interval estimation today is viewed (correctly) as just another form of upper confidence bounding).

See Russo et al. (2017) for a nice tutorial on Thompson sampling, which was first introduced in 1933 (Thompson (1933)).

Section 7.6 – DeGroot (1970) was the first to express pure learning problems (known at the time as multiarmed bandit problems) using Bellman's optimality equation, although it was computationally intractable. Gittins and Jones (1974) was the first to propose a decomposition of discounted infinite horizon learning problems into dynamic programs for each arm (hence of much lower dimensionality). This result produced an explosion of research into what became known as "Gittins indices" (or simply "index policies"). See Gittins (1979), Gittins (1981), and Gittins (1989). Whittle (1983) and Ross (1983) provide very clear tutorials on Gittins indices, helping to launch an extensive literature on the topic (see, for example, Lai and Robbins (1985), Berry and Fristedt (1985), and Weber (1992)). The work on approximating Gittins indices is due to Brezzi and Lai (2002), Yao (2006), and Chick and Gans (2009). In 2011 Gittins' former student, Kevin Glazebrook, came out with a "second edition" of Gittins' original book (Gittins et al. (2011)). The book is actually entirely new.

Index policies are limited to discounted, infinite horizon problems since the "index," which is related to the Lagrange multiplier on the coupling constraint requiring that we try at most one arm, needs to be independent of time. It is possible, however, to use the tools of approximate dynamic programming (in particular backward dynamic programming, described in chapter 15) to approximate the value functions around the belief state. This idea was developed by a former student (Weidong Han), but never published.

Section 7.7.2 – There are a variety of strategies based on the idea of approximating the value of one or more experiments. There is by now an extensive line of research based on the principle of the knowledge gradient, which we review in section 7.8 (see the bibliographic notes below). Sequential kriging optimization was proposed by Huang et al. (2006). Stein (1999) provides a thorough introduction to the field of kriging, which evolved from the field of spatial statistics.

An example of a restricted lookahead policy is the KG(*) policy proposed in Frazier and Powell (2010) to overcome the potential nonconcavity in the value of information.

The deterministic multiperiod lookahead was work performed jointly with graduate student Ahmet Duzgun, but was never published. It is presented here just to illustrate the range of different policies that can be tried.

The idea of using a decision tree to evaluate the value of information is standard material in the decision sciences (see, for example, Skinner (1999).

Section 7.7.5 – The hitting example in section was taken from Powell and Ryzhov (2012).

Section 7.8 – The knowledge gradient policy for normally distributed rewards and independent beliefs was introduced by Gupta and Miescke (1996), and subsequently analyzed in greater depth by Frazier et al. (2008). The knowledge gradient for correlated beliefs was introduced by Frazier et al. (2009). The adaptation of the knowledge gradient for online problems is due to Ryzhov and Powell (2009). A fairly thorough introduction to the knowledge gradient policy is given in Powell and Ryzhov (2012) (as of this writing, a partially finished second edition is available for download from https://tinyurl.com/optimallearningcourse). Portions of this section are adapted from material in Powell and Ryzhov (2012).

Section 7.10 – There is an advanced field of research within the simulation community that has addressed the problem of using simulation (in particular, discrete event simulation) to find the best setting of a set of parameters that controls the behavior of the simulation. An early survey is given by Bechhofer et al. (1995); a more recent survey can be found in Fu et al. (2007). Kim et al. (2005) provides a nice tutorial overview of methods based on ordinal optimization. Other important contributions in this line include Hong and Nelson (2006) and Hong and Nelson (2007). Most of this literature considers problems where the number of potential alternatives is not too large. Nelson et al. (2001) considers the case when the number of designs is large. Ankenman et al. (2009) discusses the use of a technique called kriging, which is useful when the parameter vector x is continuous. The literature on optimal computing budget allocation is based on a series of articles originating with Chen (1995), and including Chen et al. (1997, 1998), and Chen et al. (2000). Chick et al. (2001) introduces the $LL(B)$ strategy which maximizes the linear loss with measurement budget B. He et al. (2007) introduce an

OCBA procedure for optimizing the expected value of a chosen design, using the Bonferroni inequality to approximate the objective function for a single stage. A common strategy in simulation is to test different parameters using the same set of random numbers to reduce the variance of the comparisons. Fu et al. (2007) apply the OCBA concept to measurements using common random numbers. The field of simulation-optimization continues to evolve. For a more modern overview of the scope of activities, see Fu (2014).

Section 7.11.1 – The list of different objective functions is taken from Powell and Ryzhov (2012)[Chapter 6].

Exercises

Review questions

7.1 Explain in words each of the three nested expectations in equation (7.2).

7.2 Why do we go from maximizing over x in our original stochastic search problem in equation (7.1) to maximizing over policies π in equation (7.2)?

7.3 What is the meaning of "bandit" and "arms" in multi-armed bandit problems?

7.4 What is meant by passive learning and active learning? Why is derivative-free stochastic search an active learning problem?

7.5 State in words the information that would be needed in the state variable when describing a search algorithm for derivative-free stochastic search.

7.6 Which of the four classes of policies are used in the derivative-based stochastic search algorithms that we described in chapter 5? Which of the four classes of policies are described in this chapter for derivative-free stochastic search? Can you explain why there is the difference between derivative-based and derivative-free settings?

7.7 Give an example of a PFA-based policy for derivative-free stochastic search.

7.8 Give an example of a CFA-based policy for stochastic search.

7.9 State mathematically the definition of the knowledge gradient, and state in words what it is doing.

7.10 The knowledge gradient policy is a one-step lookahead that finds the value of one more experiment. Under what conditions does this approach fail?

7.11 What is meant by a restricted multi-step lookahead?

7.12 Give both the final-reward and cumulative reward objectives for learning problems.

7.13 Define the objective function that minimizes expected static regret.

7.14 What is meant by the indifference zone?

Modeling questions

7.15 Consider the problem of finding the best in a set of discrete choices $\mathcal{X} = \{x_1, ..., x_M\}$. Assume that for each alternative you maintain a lookup table belief model, where $\bar{\mu}_x^n$ is your estimate of the true mean μ_x, with precision β_x^n. Assume that your belief about μ_x is Gaussian, and let $X^\pi(S^n)$ be a policy that specifies the experiment $x^n = X^\pi(S^n)$ that you will run next, where you will learn $W_{x^n}^{n+1}$ which you will use to update your beliefs.

   (a) Formulate this learning problem as a stochastic optimization problem. Define your state variable, decision variable, exogenous information, transition function, and objective function.

   (b) Specify three possible policies, with no two from the same policy class (PFA, CFA, VFA, and DLA).

7.16 Section 7.3 introduces four classes of policies for derivative-free stochastic search, a concept that was not discussed when we introduced derivative-based stochastic search in chapter 5. In which of the four classes of policies would you classify a stochastic gradient algorithm? Explain and describe a key step in the design of stochastic gradient algorithms that is explained by your choice of policy class.

7.17 A newsvendor problem where the demand distribution W is known is a static problem. When we use learning, it is a fully sequential problem. Assume we are using a derivative-based stochastic gradient algorithm from chapter 5 with a deterministic, harmonic stepsize rule. Model this system as a fully sequential problem assuming you are limited to N iterations.

7.18 Assume we are using a quadratic approximation to approximate the expected profit of a newsvendor problem:

$$F(x_t) = \mathbb{E}\{p\min\{x_t, W_{t+1}\} - cx_t\}.$$

Table 7.7 Priors for exercise 7.19

Choice	$\bar{\mu}^n$	σ^n
1	3.0	8.0
2	4.0	8.0
3	5.0	8.0
4	5.0	9.0
5	5.0	10.0

Assume you are going to be using recursive least squares to update your quadratic belief model

$$\bar{F}_t(x|\bar{\theta}_t) = \bar{\theta}_{t0} + \bar{\theta}_{t1}x + \bar{\theta}_{t2}x_t^2.$$

Further assume that you are going to choose your decision using an excitation policy of the form

$$X^{\pi}(S_t|\bar{\theta}_t) = \arg\max_{x_t} \bar{F}_t(x|\bar{\theta}_t) + \varepsilon_{t+1},$$

where $\varepsilon_{t+1} \sim N(0, \sigma_\varepsilon^2)$. Model this learning problem as a sequential decision problem. What class of policy are you using? What are the tunable parameters?

Computational exercises

7.19 Table 7.7 shows the priors $\bar{\mu}^n$ and the standard deviations σ^n for five alternatives.

(a) Three of the alternatives have the same standard deviation, but with increasing priors. Three have the same prior, but with increasing standard deviations. Using only this information, state any relationships that you can between the knowledge gradients for each alternative. Note that you will not be able to completely rank all the alternatives.

(b) Compute the knowledge gradient for each alternative assuming that $\sigma^W = 4$.

7.20 You have to find the best of five alternatives. After n experiments, you have the data given in the table below. Assume that the precision of the experiment is $\beta^W = 0.6$.

Choice	θ^n	β^n	β^{n+1}	$\tilde{\sigma}$	$\max_{x'\neq x}\theta^n_{x'}$	ζ	$f(\zeta)$	ν^{KG}_x
1	3.0	0.444	1.044	1.248	6	-2.404	0.003	0.003
2	5.0	0.160	0.760	2.321	6	-0.431	0.220	0.511
3	6.0	0.207	0.807	2.003	5	-0.499	0.198	0.397
4	4.0	0.077	?	?	?	?	?	?
5	2.0	0.052	0.652	4.291	6	-0.932	0.095	0.406

(a) Give the definition of the knowledge gradient, first in plain English and second using mathematics.

(b) Fill in the missing entries for alternative 4 in the Table above. Be sure to clearly write out each expression and then perform the calculation. For the knowledge gradient ν^{KG}_x, you will need to use a spreadsheet (or MATLAB) to compute the normal distribution.

(c) Now assume that we have an online learning problem. We have a budget of 20 experiments, and the data in the table above shows what we have learned after three experiments. Assuming no discounting, what is the online knowledge gradient for alternative 2? Give both the formula and the number.

7.21 You have to find the best of five alternatives. After n experiments, you have the data given in the Table below. Assume that the precision of the experiment is $\beta^W = 0.6$.

Alternative	$\bar{\mu}^n$	$\bar{\sigma}^n$	$\tilde{\sigma}$	ζ	$f(\zeta)$	KG index
1	4.0	2.5	2.321	-0.215	0.300	0.696
2	4.5	3.0	?	?	?	?
3	4.0	3.5	3.365	-0.149	0.329	1.107
4	4.2	4.0	3.881	-0.077	0.361	1.401
5	3.7	3.0	2.846	-0.281	0.274	0.780

(a) Give the definition of the knowledge gradient, first in plain English and second using mathematics.

(b) Fill in the missing entries for alternative 2 in the table above. Be sure to clearly write out each expression and then perform the calculation. For the knowledge gradient ν^{KG}_x, you will need to use a spreadsheet (or programming environment) to compute the normal distribution.

(c) Now assume that we have an online learning problem. We have a budget of 20 experiments, and the data in the table above

Table 7.8 Three observations, for three alternatives, given a normally distributed belief, and assuming normally distributed observations.

Iteration	A	B	C
Prior (μ_x^0, β_x^0)	(32,0.2)	(24,0.2)	(27,0.2)
1	36	-	-
2	-	-	23
3	-	22	-

shows what we have learned after three experiments. Assuming no discounting, what is the online knowledge gradient for alternative 2? Give both the formula and the number.

7.22 You have three alternatives, with priors (mean and precision) as given in the first line of Table 7.8. You then observe each of the alternatives in three successive experiments, with outcomes shown in the table. All observations are made with precision $\beta^W = 0.2$. Assume that beliefs are independent.

(a) Give the objective function (algebraically) for offline learning (maximizing final reward) if you have a budget of three experiments, and where you evaluate the policy using the truth (as you would do in a simulator).

(b) Give the numerical value of the policy that was used to generate the choices that created Table 7.8, using our ability to use the simulated truth (as you have done in your homeworks). This requires minimal calculations (which can be done without a calculator).

(c) Now assume that you need to run experiments in an online (cumulative reward) setting. Give the objective function (algebraically) to find the optimal policy for online learning (maximizing cumulative reward) if you have three experiments. Using the numbers in the table, give the performance of the policy that generated the choices that were made. (This again requires minimal calculations.)

7.23 There are four paths you can take to get to your new job. On the map, they all seem reasonable, and as far as you can tell, they all take 20 minutes, but the actual times vary quite a bit. The value of taking a path is your current estimate of the travel time on that path. In the table below, we show the travel time on each path if you had travelled that path. Start with an initial estimate of each value function of 20 minutes with your tie-breaking rule to use the lowest numbered path. At each iteration, take the path with the best estimated value, and update your estimate of the value of the path based on your experience. After 10 iterations, compare your estimates of each path to the estimate you obtain by averaging the "observations" for each path over all 10 days. Use a constant stepsize of 0.20. How well did you do?

| | **Paths** | | | |
Day	1	2	3	4
1	37	29	17	23
2	32	32	23	17
3	35	26	28	17
4	30	35	19	32
5	28	25	21	26
6	24	19	25	31
7	26	37	33	30
8	28	22	28	27
9	24	28	31	30
10	33	29	17	29

7.24 Assume you are considering five options. The actual value μ_d, the initial estimate $\bar{\mu}_d^0$, and the initial standard deviation $\bar{\sigma}_d^0$ of each $\bar{\mu}_d^0$ are given in Table 7.9. Perform 20 iterations of each of the following algorithms:

(a) Interval estimation using $\theta^{IE} = 2$.
(b) The upper confidence bound algorithm using $\theta^{UCB} = 6$.
(c) The knowledge gradient algorithm.
(d) A pure exploitation policy.
(e) A pure exploration policy.

Each time you sample a decision, randomly generate an observation $W_d = \mu_d + \sigma^\varepsilon Z$ where $\sigma^\varepsilon = 1$ and Z is normally distributed with mean 0 and variance 1. [Hint: You can generate random observations of Z in Excel by using $=NORM.INV(RAND())$.]

7.25 Repeat exercise 7.24 using the data in Table 7.10, with $\sigma^\varepsilon = 10$.

7.26 Repeat exercise 7.24 using the data in Table 7.11, with $\sigma^\varepsilon = 20$.

Table 7.9 Data for exercise 7.24.

Decision	μ	$\bar{\theta}^0$	$\bar{\sigma}^0$
1	1.4	1.0	2.5
2	1.2	1.2	2.5
3	1.0	1.4	2.5
4	1.5	1.0	1.5
5	1.5	1.0	1.0

Table 7.10 Data for exercise 7.25.

Decision	μ	$\bar{\theta}^0$	$\bar{\sigma}^0$
1	100	100	20
2	80	100	20
3	120	100	20
4	110	100	10
5	60	100	30

Table 7.11 Data for exercise 7.26.

Decision	μ	$\bar{\theta}^0$	$\bar{\sigma}^0$
1	120	100	30
2	110	105	30
3	100	110	30
4	90	115	30
5	80	120	30

Theory questions

7.27 Assume that we have a standard normal prior about a true parameter μ which we assume is normally distributed with mean $\bar{\mu}^0$ and variance $(\sigma^0)^2$.

    (a) Given the observations $W^1, \dots, W^n$, is $\bar{\mu}^n$ deterministic or random?

    (b) Given the observations $W^1, \dots, W^n$, what is $\mathbb{E}(\mu|W^1, \dots, W^n)$ (where μ is our truth)? Why is μ random given the first n experiments?

Table 7.12 Priors.

Choice	$\bar{\mu}^n$	σ^n
1	5.0	9.0
2	3.0	8.0
3	5.0	10.0
4	4.5	12.0
5	5.0	8.0
6	5.5	6.0
7	4.0	8.0

(c) Given the observations $W^1, \dots, W^n$, what is the mean and variance of $\bar{\mu}^{n+1}$? Why is $\bar{\mu}^{n+1}$ random?

7.28 What is the relationship between the deterministic regret $R^{static,\pi}$ (recall that this was done for a machine learning problem where the "decision" is to choose a parameter θ) in equation (7.86) and the regret $R^{\pi,n}(\omega)$ for a single sample path ω in equation (7.88)? Write the regret $R^{\pi,n}(\omega)$ in equation (7.88) in the context of a learning problem and explain what is meant by a sample ω.

7.29 What is the relationship between the expected regret $\mathbb{E}R^{\pi,n}$ in equation (7.89) and the pseudo-regret $\bar{R}^{\pi,n}$ in equation (7.89)? Is one always at least as large as the other? Describe a setting under which each would be appropriate.

Problem-solving questions

7.30 There are seven alternatives with normally distributed priors on μ_x for $x \in \{1, 2, 3, 4, 5, 6, 7\}$ given in table 7.12.
Without doing any calculations, state any relationships between the alternatives based on the knowledge gradient. For example, $1 < 2 < 3$ means 3 has a higher knowledge gradient than 2 which is better than 1 (if this was the case, you do not have to separately say that $1 < 3$).

7.31 Figure 7.16 shows the belief about an unknown function as three possible curves, where one of the three curves is the true function. Our goal is to find the point x^* that maximizes the function. Without doing any computation (or math), create a graph and draw the general shape of the knowledge gradient for each possible experiment x. [Hint: the

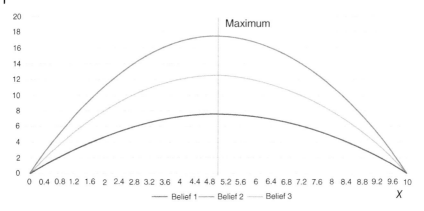

Figure 7.16 Use to plot the shape of the knowledge gradient for all x.

knowledge gradient captures your ability to make a better decision using more information.]

7.32 Assume you are trying to find the best of five alternatives. The actual value μ_x, the initial estimate $\bar{\mu}_x^0$ and the initial standard deviation $\bar{\sigma}_x^0$ of each $\bar{\mu}_d^0$ are given in Table 7.13. [This exercise does not require any numerical work.]

(a) Consider the following learning policies:
  (1) Pure exploitation.
  (2) Interval estimation.
  (3) The upper confidence bounding (pick any variant).
  (4) Thompson sampling.
  (5) The knowledge gradient.
  Write out each policy and identify any tunable parameters. How would you go about tuning the parameters?

(b) Classify each of the policies above as a (i) Policy function approximation (PFA), (ii) Cost function approximation (CFA), (iii) Policy based on a value function approximation (VFA), or (iv) Direct lookahead approximation (DLA).

(c) Set up the optimization formulation that can serve as a basis for evaluating these policies in an online (cumulative reward) setting (just one general formulation is needed – not one for each policy).

7.33 Joe Torre, former manager of the great Yankees, had to struggle with the constant game of guessing who his best hitters are. The problem is that he can only observe a hitter if he puts him in the order. He has four batters that he is looking at. The table below shows their actual

Table 7.13 Prior beliefs for learning exercise.

Alternative	μ	$\bar{\mu}^0$	$\bar{\sigma}^0$
1	1.4	1.0	2.5
2	1.2	1.2	2.5
3	1.0	1.4	2.5
4	1.5	1.0	1.5
5	1.5	1.0	1.0

batting averages (that is to say, batter 1 will produce hits 30% of the time, batter 2 will get hits 32% of the time, and so on). Unfortunately, Joe does not know these numbers. As far as he is concerned, these are all .300 hitters.

For each at-bat, Joe has to pick one of these hitters to hit. Table 7.14 below shows what would have happened if each batter were given a chance to hit (1 = hit, 0 = out). Again, Joe does not get to see all these numbers. He only gets to observe the outcome of the hitter who gets to hit.

Assume that Joe always lets the batter hit with the best batting average. Assume that he uses an initial batting average of .300 for each hitter (in case of a tie, use batter 1 over batter 2 over batter 3 over batter 4). Whenever a batter gets to hit, calculate a new batting average by putting an 80% weight on your previous estimate of his average plus a 20% weight on how he did for his at-bat. So, according to this logic, you would choose batter 1 first. Since he does not get a hit, his updated average would be $0.80(.200) + .20(0) = .240$. For the next at-bat, you would choose batter 2 because your estimate of his average is still .300, while your estimate for batter 1 is now .240.

After 10 at-bats, who would you conclude is your best batter? Comment on the limitations of this way of choosing the best batter. Do you have a better idea? (It would be nice if it were practical.)

7.34 In section 7.13.3, we showed for the transient learning problem that if M_t is the identity matrix, that the knowledge gradient for a transient truth was the same as the knowledge gradient for a stationary environment. Does this mean that the knowledge gradient produces the same behavior in both environments?

7.35 Describe the state variable S^n for a problem where $X = \{x_1, ..., x_M\}$ is a set of discrete actions (also known as "arms") using a Bayesian belief model where $\bar{\mu}_x^n$ is the belief about alternative x and β_x^n is the

Table 7.14 Data for problem 7.33.

Day	Actual batting average			
	0.300	0.320	0.280	0.260
	Batter			
	A	B	C	D
1	0	1	1	1
2	1	0	0	0
3	0	0	0	0
4	1	1	1	1
5	1	1	0	0
6	0	0	0	0
7	0	0	1	0
8	1	0	0	0
9	0	1	0	0
10	0	1	0	1

precision. Now set up Bellman's equation and characterize an optimal policy (assume we have a budget of N experiments) and answer the following:

(a) What makes this equation so hard to solve?
(b) What is different about the approach used for Gittins indices that makes this approach tractable? This approach requires a certain decomposition; how is the problem decomposed?

Sequential decision analytics and modeling

These exercises are drawn from the online book *Sequential Decision Analytics and Modeling* available at http://tinyurl.com/sdaexamplesprint.

7.36 Read chapter 4, sections 4.1–4.4, on learning the best diabetes medication.

(a) This is a sequential decision problem. What is the state variable?
(b) Which of the four classes of policies are presented as a solution for this problem?
(c) The problem of learning how a patient responds to different medications has to be resolved through field testing. What is the appropriate objective function for these problems?

(d) The policy has a tunable parameter. Formulate the problem of tuning the parameter as a sequential decision problem. Assume that this is being done off-line in a simulator. Take care when formulating the objective function for optimizing the policy.

7.37 Read chapter 12, sections 12.1–12.4 (but only section 12.4.2), on ad-click optimization.

(a) Section 12.4.2 presents an excitation policy. Which of the four classes of policies does this fall in?

(b) The excitation policy has a tunable parameter ρ. One way to search for the best ρ is to discretize it to create a set of possible values $\{\rho_1, \rho_2, \dots, \rho_K\}$. Describe belief models using:

(i) Independent beliefs.

(ii) Correlated beliefs.

Describe a CFA policy for finding the best value of ρ within this set using either belief model.

7.38 Read chapter 12, sections 12.1–12.4 on ad-click optimization. We are going to focus on section 12.4.3 which proposes a knowledge gradient policy.

(a) Describe in detail how to implement a knowledge gradient policy for this problem.

(b) When observations are binary (the customer did or did not click on the ad), the noise in a single observation $W_{t+1,x}$ of ad x can be very noisy, which means the value of information from a single experiment can be quite low. A way to handle this is to use a lookahead model that looks forward τ time periods. Describe how to calculate the knowledge gradient when looking forward τ time periods (instead of just one time period).

(c) How would you go about selecting τ?

(d) There are versions of the knowledge gradient for offline learning (maximizing final reward) and online learning (maximizing cumulative reward). Give the expressions for the knowledge gradient for both offline and online learning.

7.39 Continuing the exercise for chapter 4, assume that we have to tune the policy in the field rather than in the simulator. Model this problem as a sequential decision problem. Note that you will need a "policy" (some would call this an algorithm) for updating the tunable parameter θ that is separate from the policy for choosing the medication.

Diary problem

The diary problem is a single problem you chose (see chapter 1 for guidelines). Answer the following for your diary problem.

7.40 Pick one of the learning problems that arises in your diary problem, where you would need to respond adaptively to new information. Is the information process stationary or nonstationary? What discuss the pros and cons of:

(a) A deterministic stepsize policy (identify which one you are considering).

(b) A stochastic stepsize policy (identify which one you are considering).

(c) An optimal stepsize policy (identify which one you are considering).

Bibliography

Ankenman, B., Nelson, B.L., and Staum, J. (2009). Stochastic Kriging for simulation metamodeling. *Operations Research* 58 (2): 371–382.

Auer, P., Cesabianchi, N., and Fischer, P. (2002). Finitetime analysis of the multiarmed bandit problem. *Machine Learning* 47 (2): 235–256.

Bechhofer, R.E., Santner, T.J., and Goldsman, D.M. (1995). *Design and Analysis of Experiments for Statistical Selection, Screening, and Multiple Comparisons*. New York: John Wiley & Sons.

Berry, D.A. and Fristedt, B. (1985). *Bandit Problems*. London: Chapman and Hall.

Box, G.E.P. and Wilson, K.B. (1951). On the experimental attainment of optimum conditions. *Journal of the Royal Statistical Society Series B* 13 (1): 1–45.

Brezzi, M. and Lai, T.L. (2002). Optimal learning and experimentation in bandit problems. *Journal of Economic Dynamics and Control* 27: 87–108.

Chang, H.S.: Fu, M.C.: Hu, J., and Marcus, S.I. (2007., *Simulationbased Algorithms for Markov Decision Processes*. Berlin: Springer.

Chen, C.H. (1995). An effective approach to smartly allocate computing budget for discrete event simulation. In *34th IEEE Conference on Decision and Control*, Vol. 34, 2598–2603, New Orleans, LA.

Chen, C.H., Yuan, Y., Chen, H.C., Yücesan, E., and Dai, L. (1998). Computing budget allocation for simulation experiments with different system structure. In: *Proceedings of the 30th conference on Winter simulation*, 735–742.

Chen, H.C., Chen, C.H., Dai, L., and Yucesan, E. (1997). A gradient approach for smartly allocating computing budget for discrete event simulation. In: *Proceedings of the 1996 Winter Simulation Conference* (eds. J. Charnes, D. Morrice, D. Brunner and J. Swain), 398–405. Piscataway, NJ, USA: IEEE Press.

Chen, H.C., Chen, C.H., Yucesan, E., and Yücesan, E. (2000). Computing efforts allocation for ordinal optimization and discrete event simulation. *IEEE Transactions on Automatic Control* 45 (5): 960–964.

Chick, S.E. and Gans, N. (2009). Economic analysis of simulation selection problems. *Management Science* 55 (3): 421–437.

Chick, S.E. and Inoue, K. (2001). New two-stage and sequential procedures for selecting the best simulated system. *Operations Research* 49 (5): 732—743.

DeGroot, M.H. (1970). *Optimal Statistical Decisions*. John Wiley and Sons.

Frazier, P.I. (2018). A Tutorial on Bayesian Optimization, Technical report, Cornell University, Ithaca NY.

Frazier, P.I. and Powell, W.B. (2010). Paradoxes in learning and the marginal value of information. *Decision Analysis* 7 (4): 378–403.

Frazier, P.I., Powell, W.B., and Dayanik, S. (2009). The knowledge-gradient policy for correlated normal beliefs. *INFORMS Journal on Computing* 21 (4): 599–613.

Frazier, P.I., Powell, W.B., and Dayanik, S.E. (2008). A knowledge-gradient policy for sequential information collection. *SIAM Journal on Control and Optimization* 47 (5): 2410–2439.

Fu, M.C. (2014). *Handbook of Simulation Optimization*. New York: Springer.

Fu, M.C., Hu, J.Q., Chen, C.H., and Xiong, X. (2007). Simulation allocation for determining the best design in the presence of correlated sampling. INFORMS Journal on Computing 19: 101–111.

Gittins, J. (1979). Bandit processes and dynamic allocation indices. *Journal of the Royal Statistical Society. Series B (Methodological)* 41 (2): 148–177.

Gittins, J. (1981). Multiserver scheduling of jobs with increasing completion times. *Journal of Applied Probability* 16: 321–324.

Gittins, J. (1989). *Multiarmed Bandit Allocation Indices*. New York: Wiley and Sons.

Gittins, J. and Jones, D. (1974). A dynamic allocation index for the sequential design of experiments. In: *Progress in statistics* (ed. J. Gani), 241—266. North Holland, Amsterdam.

Gittins, J., Glazebrook, K.D., and Weber, R.R. (2011). *Multi-Armed Bandit Allocation Indices*. New York: John Wiley & Sons.

Gupta, S.S. and Miescke, K.J. (1996). Bayesian look ahead one-stage sampling allocations for selection of the best population. *Journal of statistical planning and inference* 54 (2): 229—244.

He, D., Chick, S.E., and Chen, C.-H. (2007). Opportunity cost and OCBA selection procedures in ordinal optimization for a fixed number of alternative systems. *IEEE Transactions on Systems Man and Cybernetics Part CApplications and Reviews* 37 (5): 951–961.

Hong, J. and Nelson, B.L. (2006). Discrete optimization via simulation using COMPASS. *Operations Research* 54 (1): 115–129.

Hong, L. and Nelson, B. L. (2007). A framework for locally convergent randomsearch algorithms for discrete optimization via simulation. *ACM Transactions on Modeling and Computer Simulation* 17 (4): 1–22.

Huang, D., Allen, T.T., Notz, W.I., and Zeng, N. (2006). Global optimization of stochastic black-box systems via sequential Kriging meta-models. *Journal of Global Optimization* 34 (3): 441–466.

Kaelbling, L.P. (1993). *Learning in embedded systems*, Cambridge, MA: MIT Press.

Kim, S.-H., Nelson, B. L., and Sciences, M. (2005). On the asymptotic validity of fully sequential selection procedures for steady-state simulation. *Industrial Engineering*, 1–37.

Lai, T.L. (1987). Adaptive treatment allocation and the multi-armed bandit problem. *Annals of Statistics* 15 (3): 1091–1114.

Lai, T.L. and Robbins, H. (1985). Asymptotically efficient adaptive allocation rules. *Advances in Applied Mathematics* 6: 4–22.

Nelson, B.L. and Kim, S.H. (2001). A fully sequential procedure for indifferencezone selection in simulation. *ACM Transactions on Modeling and Computer Simulation* 11 (3): 251–273.

Nelson, B.L., Swann, J., Goldsman, D., and Song, W. (2001). Simple procedures for selecting the best simulated system when the number of alternatives is large. *Operations Research* 49: 950–963.

Powell, W.B. (2019). A unified framework for stochastic optimization. *European Journal of Operational Research* 275 (3): 795–821.

Powell, W.B. and Ryzhov, I.O. (2012). *Optimal Learning*. Hoboken, NJ: John Wiley & Sons.

Robbins, H. and Monro, S. (1951). A stochastic approximation method. *The Annals of Mathematical Statistics* 22 (3): 400–407.

Ross, S.M. (1983). *Introduction to Stochastic Dynamic Programming*. New York: Academic Press.

Russo, D., Van Roy, B., Kazerouni, A., Osband, I., and Wen, Z. (2017). A tutorial on thompson sampling. 11 (1): 1–96.

Ryzhov, I.O. and Powell, W.B. (2009). A monte Carlo knowledge gradient method for learning abatement potential of emissions reduction technologies. In: *Proceedings of the 2009 Winter Simulation Conference* (eds. M. D. Rossetti, R. R. Hill, B. Johansson, A. Dunkin and R. G. Ingalls), 1492–1502.

Winter Simulation Conference', pp. 1492–1502. Singh, S., Jaakkola, T., Littman, M., and Szepesvari, C. (2000). Convergence results for single-step on-policy reinforcement-learning algorithms. *Machine Learning* 38 (3): 287—308.

Skinner, D.C. (1999). *Introduction to Decision Analysis*. Gainesville, Fl: Probabilistic Publishing.

Stein, M.L. (1999). *Interpolation of spatial data: Some theory for kriging*. New York: Springer Verlag.

Sutton, R.S. and Barto, A.G. (2018). *Reinforcement Learning: An Introduction*, 2e. Cambridge, MA: MIT Press.

Thompson, W.R. (1933). On the likelihood that one unknown probability exceeds another in view of the evidence of two samples. *Biometrika* 25 (3/4): 285–294.

Thrun, S.B. (1992). The role of exploration in learning control. In White, D.A., and Sofge, D.A.

Wang, Y., Wang, C., Powell, W.B., and Edu, P.P. (2016), The knowledge gradient for sequential decision making with stochastic binary feedbacks. In: *ICML2016,* Vol. 48. New York.

Weber, R.R. (1992) On the gittins index for multiarmed bandits. *The Annals of Applied Probability* 2 (4): 1024–1033.

Whittle, P. (1983). *Optimization Over Time: Dynamic Programming and Stochastic Control Volumes I and II, Wiley Series in Probability and Statistics: Probability and Statistics.* New York: John Wiley & Sons.

Yao, Y. (2006). Some results on the Gittins index for a normal reward process. In: *Time Series and Related Topics: In Memory of Ching-Zong-Wei* (eds H. Ho, C. Ing and T. Lai), 284–294. Beachwood, OH, USA: Institute of Mathematical Statistics.

Part III – State-dependent Problems

We now transition to a much richer class of dynamic problems where some aspect of the *problem* depends on dynamic information. This might arise in three ways:

- The objective function depends on dynamic information, such as a cost or price.
- The constraints may depend on the availability of resources (that are being controlled dynamically), or other information in constraints such as the travel time in a graph or the rate at which water is evaporating.
- The distribution of a random variable such as weather, or the distribution of demand, may be varying over time, which means the parameters of the distribution are in the state variable.

When we worked on state-independent problems, we often wrote the function being maximized as $F(x, W)$ to express the dependence on the decision x or random information W, but not on any information in our state S_t (or S^n). As we move to our state-dependent world, we are going to write our cost or contribution function as $C(S_t, x_t)$ or, in some cases, $C(S_t, x_t, W_{t+1})$, to capture the possible dependence of the objective function on dynamic information in S_t. In addition, our decision x_t might be constrained by $x_t \in \mathcal{X}_t$, where the constraints $\mathcal{X}_t$ may depend on dynamic data such as inventories, travel times, or conversion rates.

Finally, our random information W may itself depend on known information in the state variable S_t, or possibly on hidden information that we cannot observe, but have beliefs about (these beliefs would also be captured in the state variable). For example, W might be the number of clicks on an ad which is described by some probability distribution whose parameters (e.g. the mean) is also uncertain. Thus, at time t (or time n), we may find ourselves solving a problem that looks like

$$\max_{x_t \in \mathcal{X}_t} \mathbb{E}_{S_t} \mathbb{E}_{W|S_t} \{C(S_t, x_t, W_{t+1})|S_t\}.$$

If the cost/contribution function $C(S_t, x_t, W_{t+1})$, and/or the constraints $\mathcal{X}_t$, and/or the expectation depends on time-dependent data, then we have an instance of a state-dependent problem.

We are not trying to say that all state-dependent problems are the same, but we do claim that state-dependent problems represent an important transition from state-independent problems, where the only state is the belief B_t about our function. This is why we also refer to state-independent problems as learning problems.

We lay the foundation for state-dependent problems with the following chapters:

- State-dependent applications (chapter 8) – We begin our presentation with a series of applications of problems where the function is state dependent. State variables can arise in the objective function (e.g. prices), but in most of the applications the state arises in the constraints, which is typical of problems that involve the management of physical resources.
- Modeling general sequential decision problems (chapter 9) – This chapter provides a comprehensive summary of how to model general (state-dependent) sequential decision problems in all of their glory.
- Modeling uncertainty (chapter 10) – To find good policies (to make good decisions), you need a good model, and this means an accurate model of uncertainty. In this chapter we identify different sources of uncertainty and discuss how to model them.
- Designing policies (chapter 11) – Here we provide a much more comprehensive overview of the different strategies for creating policies, leading to the four classes of policies that we first introduce in part I for learning problems. If you have a particular problem you are trying to solve (rather than just building your toolbox), this chapter should guide you to the policies that seem most relevant to your problem.

After these chapters, the remainder of the book is a tour through the four classes of policies which we illustrated in chapter 7 in the context of derivative-free stochastic optimization.

8

State-dependent Problems

In chapters 5 and 7, we introduced sequential decision problems in which the state variable consisted only of the state of the algorithm (chapter 5) or the state of our belief about an unknown function $\mathbb{E}\{F(x, W)|S_0\}$ (chapter 7). These problems cover a very important class of applications that involve maximizing or minimizing functions that can represent anything from complex analytical functions and black-box simulators to laboratory and field experiments.

The distinguishing feature of state-dependent problems is that the *problem* being optimized now depends on our state variable, where the "problem" might be the function $F(x, W)$, the expectation (e.g. the distribution of W), or the feasible region $\mathcal{X}$. The state variable may be changing purely exogenously (where decisions do not impact the state of the system), purely endogenously (the state variable only changes as a result of decisions), or both (which is more typical).

There is a genuinely vast range of problems where the performance metric (costs or contributions), the distributions of random variables W, and/or the constraints, depend on information that is changing over time, either exogenously or as a result of decisions (or both). When information changes over time, it is captured in the state variable S_t (or S^n if we are counting events with n).

Examples of state variables that affect the problem itself include:

- Physical state variables, which might include inventories, the location of a vehicle on a graph, the medical condition of a patient, the speed and location of a robot, and the condition of an aircraft engine. Physical state variables are typically expressed through the constraints.
- Informational state variables, such as prices, a patient's medical history, the humidity in a lab, or attributes of someone logging into the internet. These variables might affect the objective function (costs and contributions), or the constraints. This information may evolve exogenously (e.g. the weather), or

might be directly controlled (e.g. setting the price of a product), or influenced by decisions (selling energy into the grid may lower electricity prices).
- Distributional information capturing beliefs about unknown parameters or quantities, such as information about the how a patient might respond to a drug, or the state of the materials in a jet engine, or how the market might respond to the price of a product.

While physical resource management problems are perhaps the easiest to envision, state-dependent problems can include any problem where the function being minimized depends on dynamic information, either in the objective function itself, or the constraints, or the equations that govern how the system evolves over time (the transition function).

For our state-independent problems, we wrote the objective function as $F(x, W)$, since the function itself did not depend on the state variable. For state-dependent problems, we will usually write our single-period contribution (or cost) function as $C(S_t, x_t)$, although there are settings where it is more natural to write it as $C(S_t, x_t, W_{t+1})$ or, in some settings, $C(S_t, x_t, S_{t+1})$.

We will communicate the dependence of expectations on the state variables through conditioning, by writing $\mathbb{E}\{F(\cdot)|S_t\}$ (or $\mathbb{E}\{F(\cdot)|S^n\}$). We will express the dependence of constraints on dynamic state information by writing $x \in \mathcal{X}_t$. Note that writing $C(S_t, x_t)$ means that the contribution function depends on dynamic information such as

$$C(S_t, x_t) = p_t x_t,$$

where the price p_t evolves randomly over time.

At this point, it is useful to highlight what is probably the biggest class of state-dependent problems, which is those that involve the management of physical resources. Generally known as *dynamic resource allocation problems*, these problems are the basis of the largest and most difficult problems that we will encounter. These problems are typically high-dimensional, often with complex dynamics and types of uncertainty.

In this chapter we present four classes of examples:

- Graph problems – These are problems where we are modeling a single resource that is controlled using a discrete set of actions moving over a discrete set of states.
- Inventory problems – This is a classical problem in dynamic programming which comes in a virtually unlimited set of variations.
- Information acquisition problems – These are state-dependent active learning problems that we touched on at the end of chapter 7, but now we imbed them in more complex settings.

- Complex resource allocation problems – Here we put our toe in the water and describe some high-dimensional applications.

These illustrations are designed to teach by example. The careful reader will pick up subtle modeling choices, in particular the indexing with respect to time. We suggest that readers skim these problems, selecting examples that are of interest. In chapter 9, we are going to present a very general modeling framework, and it helps to have a sense of the complexity of applications that may arise.

Finally, we forewarn the reader that this chapter just presents models, not solutions. This fits with our "model first, then solve" approach. We do not even use our universal modeling framework. The idea is to introduce applications with notation. We present our universal modeling framework in detail for these more complex problems in chapter 9. Then, after introducing the rich challenges of modeling uncertainty in chapter 10, we turn to the problem of designing policies in chapter 11. All we can say at this point is: There are four classes of policies, and any approach we may choose will come from one of these four classes (or a hybrid). What we will not do is assume that we can solve it with a particular strategy, such as approximate dynamic programming.

8.1 Graph Problems

A popular class of stochastic optimization problems involve managing a single physical asset moving over a graph, where the nodes of the graph capture the physical state.

8.1.1 A Stochastic Shortest Path Problem

We are often interested in shortest path problems where there is uncertainty in the cost of traversing a link. For our transportation example, it is natural to view the travel time on a link as random, reflecting the variability in traffic conditions on each link. There are two ways we can handle this uncertainty. The simplest is to assume that our driver has to make a decision before seeing the travel time over the link. In this case, our updating equation would look like

$$v_i^n = \min_{j \in \mathcal{I}_i^+} \mathbb{E}\{\hat{c}_{ij} + v_j^{n-1}\},$$

where $\hat{c}_{ij}$ is a random variable describing the cost of traversing i to j. If $\bar{c}_{ij} = \mathbb{E}\hat{c}_{ij}$, then our problem reduces to

$$v_i^n = \min_{j \in \mathcal{J}_i^+} \left(\bar{c}_{ij} + v_j^{n-1} \right),$$

which is a simple deterministic problem.

An alternative model is to assume that we know the cost on a link from i to j as soon as we arrive at node i. In this case, we would have to solve

$$v_i^n = \mathbb{E} \left\{ \min_{j \in \mathcal{J}_i^+} \left(\hat{c}_{ij} + v_j^{n-1} \right) \right\}.$$

Here, the expectation is outside of the min operator that chooses the best decision, capturing the fact that now the decision itself is random.

Note that our notation is ambiguous, in that with the same notation, we have two very different models. In chapter 9, we are going to refine our notation so that it will be immediately apparent when a decision "sees" the random information and when the decision has to be made before the information becomes available.

8.1.2 The Nomadic Trucker

A nice illustration of sequential decisions is a problem we are going to call the nomadic trucker, depicted in Figure 8.1. In this problem, our trucker has to move loads of freight (which fill his truck) from one city to the next. When he arrives in a city i ("Texas" in Figure 8.1), he is offered a set of loads to different destinations, and has to choose one. Once he makes his choice (in the figure, he chooses the load to New Jersey), he moves the load to its destination, delivers the freight, and then the problem repeats itself. The other loads are offered to other drivers, so if he returns to node i at a later time, he is offered an entirely new set of loads (that are entirely random).

We model the state of our nomadic trucker by letting R_t be his location. From a location, our trucker is able to choose from a set of demands $\hat{D}_t$. Thus, our state variable is $S = (R_t, \hat{D}_t)$, where R_t is a scalar (the location) while $\hat{D}_t$ is a vector giving the number of loads from R_t to each possible destination. A decision $x_t \in \mathcal{X}_t$ represents the decision to accept a load in $\hat{D}_t$ and go to the destination of that load.

Let $C(S_t, x_t)$ be the contribution earned from being in location R_t (this is contained in S_t) and taking decision x_t. Any demands not covered in $\hat{D}_t$ at time t are lost. After implementing decision x_t, the driver either stays in his current location (if he does nothing), or moves to a location that corresponds to the destination of the load the driver selected in the set $\hat{D}_t$.

Let R_t^x be the location that decision x_t sends the driver to. We will later call this the *post-decision state*, which is the state after we have made a decision but before any new information has arrived. The post-decision state variable $S_t^x = R_t^x$ is the location the truck will move to, but before any demands have

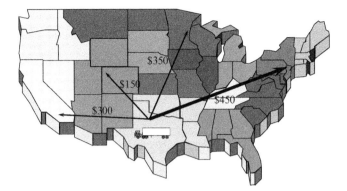

Figure 8.1 Illustration of a nomadic trucker in location "Texas" with the choice of four loads to move.

been revealed. We assume that the decision x_t determines the downstream destination, so $R_{t+1} = R_t^x$.

The driver makes his decision by solving

$$\hat{v}_t = \max_{x \in \hat{D}_t} \left(C(S_t, x) + \overline{V}_t^x(R_t^x) \right),$$

where R_t^x is the downstream location (the "post-decision state"), and $\overline{V}_t^x(R_t^x)$ is our current estimate (as of time t) of the value of the truck being in the destination R_t^x. Let x_t be the best decision given the downstream values $\overline{V}_t^x(R_t^x)$. Noting that R_t is the current location of the truck, we update the value of our previous, post-decision state using

$$\overline{V}_{t-1}^x(R_{t-1}^x) \leftarrow (1 - \alpha)\overline{V}_{t-1}^x(R_{t-1}^x) + \alpha\hat{v}_t.$$

Note that we are smoothing $\hat{v}_t$, which is the value of being in the pre-decision state S_t, with the current estimate $\overline{V}_{t-1}^x(R_{t-1}^x)$ of the previous post-decision state.

8.1.3 The Transformer Replacement Problem

The electric power industry uses equipment known as transformers to convert the high-voltage electricity that comes out of power generating plants into currents with successively lower voltage, finally delivering the current we can use in our homes and businesses. The largest of these transformers can weigh 200 tons, might cost millions of dollars to replace and may require a year or more to build and deliver. Failure rates are difficult to estimate (the most powerful transformers were first installed in the 1960s and have yet to reach the end of their natural lifetimes). Actual failures can be very difficult to predict, as they often depend on heat, power surges, and the level of use.

We are going to build an aggregate replacement model where we only capture the age of the transformers. Let

$$a \quad = \quad \text{the age of a transformer (in units of time periods) at time } t,$$

$$R_{ta} \quad = \quad \text{the number of active transformers of age } a \text{ at time } t.$$

Here and elsewhere, we need to model the attributes of a resource (in this case, the age).

For our model, we assume that age is the best predictor of the probability that a transformer will fail. Let

$$\hat{R}_{t+1,a}^{fail} \quad = \quad \text{the number of transformers of age } a \text{ that fail between } t \text{ and } t+1,$$

$$p_a \quad = \quad \text{the probability a transformer of age } a \text{ will fail between } t \text{ and } t+1.$$

Of course, $\hat{R}_{t+1,a}^{fail}$ depends on R_{ta} since transformers can only fail if we own them.

It can take a year or two to acquire a new transformer. Assume that we are measuring time in quarters (three-month periods). Normally it can take about six quarters from the time of purchase before a transformer is installed in the network. However, we may pay extra and get a new transformer in as little as three quarters. If we purchase a transformer that arrives in six time periods, then we might say that we have acquired a transformer that is $a = -6$ time periods old. Paying extra gets us a transformer that is $a = -3$ time periods old. Of course, the transformer is not productive until it is at least $a = 0$ time periods old. Let

$$x_{ta} \quad = \quad \text{the number of transformers of age } a \text{ that we purchase at time } t.$$

The transition function is given by

$$R_{t+1,a} \quad = \quad R_{t,a-1} + x_{t,a-1} - \hat{R}_{t+1,a}^{fail}.$$

If we have too few transformers, then we incur what are known as "congestion costs," which represent the cost of purchasing power from more expensive utilities because of bottlenecks in the network. To capture this, let

$$\bar{R} \quad = \quad \text{target number of transformers that we should have available,}$$

$$R_t^A \quad = \quad \text{actual number of transformers that are available at time } t,$$

$$= \sum_{a \geq 0} R_{ta},$$

c_a = the cost of purchasing a transformer of age a,

$C_t(R_t^A, \bar{R})$ = expected congestion costs if R_t^A transformers are available,

$$= c_0 \left(\frac{\bar{R}}{R_t^A} \right)^{\beta}.$$

The function $C_t(R_t^A, \bar{R})$ captures the behavior that as R_t^A falls below $\bar{R}$, the congestion costs rise quickly.

The total cost function is then given by

$$C(S_t, x_t) = C_t(R_t^A, \bar{R}) + c_a x_t.$$

For this application, our state variable R_t might have as many as 100 dimensions. If we have, say, 200 transformers, each of which might be as many as 100 years old, then the number of possible values of R_t could be 100^{200}. It is not unusual for modelers to count the size of the state space, although this is an issue only for particular solution methods that depend on lookup table representations of the value of being in a state, or the action we should take given that we are in a state.

8.1.4 Asset Valuation

Imagine you are holding an asset that you can sell at a price that fluctuates randomly. In this problem we want to determine the best time to sell the asset, and from this, infer the value of the asset. For this reason, this type of problem arises frequently in the context of asset valuation and pricing.

Let p_t be the price at which we can sell our asset at time t, at which point you have to make a decision

$$x_t = \begin{cases} 1 & \text{sell}, \\ 0 & \text{hold}. \end{cases}$$

For our simple model, we assume that p_t is independent of prior prices (a more typical model would assume that the *change* in price is independent of prior history). With this assumption, our system has two physical states that we denote by R_t, where

$$R_t = \begin{cases} 1 & \text{we are holding the asset}, \\ 0 & \text{we have sold the asset}. \end{cases}$$

Our state variable is then given by

$$S_t = (R_t, p_t).$$

Let

τ = the time at which we sell our asset.

τ is known as a *stopping time* (recall the discussion in section 2.1.7), which means it can only depend on information that has arrived on or before time t. By definition, $x_\tau = 1$ indicates the decision to sell at time $t = \tau$. It is common to think of τ as the decision variable, where we wish to solve

$$\max_\tau \mathbb{E} p_\tau. \tag{8.1}$$

Equation (8.1) is a little tricky to interpret. Clearly, the choice of when to stop is a random variable since it depends on the price p_t. We cannot optimally choose a random variable, so what is meant by (8.1) is that we wish to choose a *function* (or *policy*) that determines when we are going to sell. For example, we would expect that we might use a rule that says

$$X_t^{PFA}(S_t | \theta^{\text{sell}}) = \begin{cases} 1 & \text{if } p_t \geq \theta^{\text{sell}} \text{ and } S_t = 1, \\ 0 & \text{otherwise.} \end{cases} \tag{8.2}$$

In this case, we have a function parameterized by θ^{sell} which allows us to write our problem in the form

$$\max_{\theta^{\text{sell}}} \mathbb{E} \left\{ \sum_{t=0}^\infty \gamma^t p_t X_t^{PFA}(S_t | \theta^{\text{sell}}) \right\}, \tag{8.3}$$

where $\gamma < 1$ is a discount factor. This formulation raises two questions. First, while it seems very intuitive that our policy would take the form given in equation (8.2), there is the theoretical question of whether this in fact is the structure of an optimal policy.

The second question is how to find the best policy within this class. For this problem, that means finding the parameter θ^{sell}. This is precisely the type of problem that we addressed in our stochastic search chapters 5 and 7. However, this is not the only policy we might use. Another is to define the function

$$V_t(S_t) \quad = \quad \text{the value of being in state } S_t \text{ at time } t \text{ and then making}$$
$$\text{optimal decisions from time } t \text{ onward.}$$

More practically, let $V^\pi(S_t)$ be the value of being in state S_t and then following policy π from time t onward. This is given by

$$V_t^\pi(S_t) = \mathbb{E}\left\{\sum_{t'=t}^{\infty} \gamma^{t'-t} p_{t'} X_{t'}^\pi(S_{t'}|\theta^{\text{sell}})\right\}.$$

Of course, it would be nice if we could find an optimal policy since this would maximize $V_t^\pi(S_t)$. More often, we need to use some approximation that we call $\overline{V}_t(S_t)$. In this case, we might define a policy

$$X^{VFA}(S_t) = \arg\max_{x_t} \left(p_t x_t + \gamma \mathbb{E}\{\overline{V}_{t+1}(S_{t+1})|S_t, x_t\}\right). \tag{8.4}$$

We have just illustrated two styles of policies: X^{PFA} and X^{VFA}. These are two of the four classes we first visited in chapter 7, called policy function approximation and value function approximation. We will again review all four classes of policies in chapter 11, which we will discuss in depth in chapters 12–19.

8.2 Inventory Problems

Another popular class of problems involving managing a quantity of resources that are held in some sort of inventory. The inventory can be money, products, blood, people, water in a reservoir or energy in a battery. The decisions govern the quantity of resource moving into and out of the inventory.

8.2.1 A Basic Inventory Problem

A basic inventory problem arises in applications where we purchase product at time t to be used during time interval $t+1$. We are going to encounter this problem again, sometimes as discrete problems, but often as continuous problems, and sometimes as vector valued problems (when we have to acquire different types of assets).

We can model the problem using

R_t = the inventory on hand at time t before we make a new ordering decision, and before we have satisfied any demands arising in time interval t,

x_t = the amount of product purchased at time t which we assume arrives immediately,

D_t = the demand known at time t that we have to satisfy.

We have chosen to model R_t as the resources on hand in period t before demands have been satisfied. Our definition here makes it easier to introduce (in the next section) the decision of how much demand we should satisfy. In our most basic problem, the state variable S_t is given by

$$S_t = (R_t, D_t).$$

Our inventory R_t is described using the equation

$$R_{t+1} = R_t - \min\{R_t, D_t\} + x_t.$$

Let

$$\hat{D}_{t+1} \quad = \quad \text{new demands that we learn about during time}$$
$$\text{interval } (t, t+1).$$

We assume that any unsatisfied demands are lost. This means that D_t evolves according to

$$D_{t+1} = \hat{D}_{t+1}.$$

Here we are assuming that D_{t+1} is revealed to us through the new information $\hat{D}_{t+1}$. Below we are going to introduce the ability to backlog unsatisfied demands.

We assume we purchase new assets at a fixed price p^{buy} and sell them at a fixed price p^{sell}. The amount we earn between $t-1$ and t (satisfying the demand D_t that becomes known by time t), including the decision we make at time t, is given by

$$C(S_t, x_t) = p^{\text{sell}} \min\{R_t, D_t\} - p^{\text{buy}} x_t.$$

An alternative formulation of this problem is to write the contribution based on what we will receive between t and $t+1$. In this case, we would write the contribution as

$$C(S_t, x_t, \hat{D}_{t+1}) = p^{\text{sell}} \min\{(R_t - \min\{R_t, D_t\} + x_t), \hat{D}_{t+1}\} - p^{\text{buy}} x_t. \quad (8.5)$$

It is because of problems like this that we sometimes write our contribution function as $C(S_t, x_t, W_{t+1})$.

8.2.2 The Inventory Problem – II

Many inventory problems introduce additional sources of uncertainty. The inventory we are managing could be stocks, planes, energy commodities such as oil, consumer goods, and blood. In addition to the need to satisfy random demands (the only source of uncertainty we considered in our basic inventory problem), we may also have randomness in the prices at which we buy and sell assets. We may also include exogenous changes to the inventory on hand due to additions (cash deposits, blood donations, energy discoveries) and subtractions (cash withdrawals, equipment failures, theft of product).

We can model the problem using

$$
\begin{aligned}
x_t^{\text{buy}} &= \text{inventory purchased at time } t \text{ to be used during time} \\
&\quad \text{interval } t + 1, \\
x_t^{\text{sell}} &= \text{amount of inventory sold to satisfy demands during time} \\
&\quad \text{interval } t, \\
x_t &= (x_t^{\text{buy}}, x_t^{\text{sell}}), \\
R_t &= \text{inventory level at time } t \text{ before any decisions are made,} \\
D_t &= \text{demands waiting to be served at time } t.
\end{aligned}
$$

Of course, we are going to require that $x_t^{\text{sell}} \leq \min\{R_t, D_t\}$, since we cannot sell what we do not have, and we cannot sell more than the market demand. We are also going to assume that we buy and sell our inventory at market prices that fluctuate over time. These are described using

$$
\begin{aligned}
p_t^{\text{buy}} &= \text{market price for purchasing inventory at time } t, \\
p_t^{\text{sell}} &= \text{market price for selling inventory at time } t, \\
p_t &= (p_{t+1}^{\text{sell}}, p_{t+1}^{\text{buy}}).
\end{aligned}
$$

Our system evolves according to several types of exogenous information processes that include random changes to the supplies (inventory on hand), demands, and prices. We model these using

$$
\begin{aligned}
\hat{R}_{t+1} &= \text{exogenous changes to the inventory on hand that occur} \\
&\quad \text{during time interval } (t, t + 1) \text{ (e.g. rainfall adding} \\
&\quad \text{water to a reservoir, deposits/withdrawals of cash} \\
&\quad \text{to a mutual fund, or blood donations),} \\
\hat{D}_{t+1} &= \text{new demands for inventory that arises during time} \\
&\quad \text{interval } (t, t + 1), \\
\hat{p}_{t+1}^{\text{buy}} &= \text{change in the purchase price that occurs during time} \\
&\quad \text{interval } (t, t + 1), \\
\hat{p}_{t+1}^{\text{sell}} &= \text{change in the selling price that occurs during time} \\
&\quad \text{interval } (t, t + 1), \\
\hat{p}_{t+1} &= (\hat{p}_{t+1}^{\text{buy}}, \hat{p}_{t+1}^{\text{sell}}).
\end{aligned}
$$

We assume that the exogenous changes to inventory, $\hat{R}_t$, occur before we satisfy demands at time t.

For more complex problems such as this, it is convenient to have a generic variable for exogenous information. We use the notation W_{t+1} to represent all the information that first arrives between t and $t + 1$, where for this problem, we would have

$$W_{t+1} = (\hat{R}_{t+1}, \hat{D}_{t+1}, \hat{p}_{t+1}).$$

The state of our system is described by

$$S_t = (R_t, D_t, p_t).$$

The state variables evolve according to

$$
\begin{aligned}
R_{t+1} &= R_t - x_t^{\text{sell}} + x_t^{\text{buy}} + \hat{R}_{t+1}, \\
D_{t+1} &= D_t - x_t^{\text{sell}} + \hat{D}_{t+1}, \\
p_{t+1}^{\text{buy}} &= p_t^{\text{buy}} + \hat{p}_{t+1}^{\text{buy}}, \\
p_{t+1}^{\text{sell}} &= p_t^{\text{sell}} + \hat{p}_{t+1}^{\text{sell}}.
\end{aligned}
$$

We can add an additional twist if we assume the market price, for instance, follows a time-series model

$$p_{t+1}^{\text{sell}} = \theta_0 p_t^{\text{sell}} + \theta_1 p_{t-1}^{\text{sell}} + \theta_2 p_{t-2}^{\text{sell}} + \varepsilon_{t+1},$$

where $\varepsilon_{t+1} \sim N(0, \sigma_\varepsilon^2)$. In this case, the state of our price process is captured by $(p_t^{\text{sell}}, p_{t-1}^{\text{sell}}, p_{t-2}^{\text{sell}})$ which means our state variable would be given by

$$S_t = (R_t, D_t, (p_t, p_{t-1}, p_{t-2})).$$

Note that if we did not allow backlogging, then we would update demands with just

$$D_{t+1} = \hat{D}_{t+1}. \tag{8.6}$$

Contrast this with our updating of the prices p_{t+1} which depends on either p_t or even p_{t-1} and p_{t-2}. To model the evolution of prices, we have an explicit mathematical model, including an assumed error such as ε_{t+1} where we assumed $\varepsilon_{t+1} \sim N(0, \sigma_\varepsilon^2)$. When we simply observe the updated value of demand (as we are doing in (8.6)), then we describe the process as "data driven." We would need a source of data from which to draw the observations $\hat{D}_1, \hat{D}_2, \dots, \hat{D}_t, \dots$. We revisit this concept in more depth in chapter 10.

The one-period contribution function is

$$C_t(S_t, x_t) = p_t^{\text{sell}} x_t^{\text{sell}} - p_t^{\text{buy}} x_t.$$

8.2.3 The Lagged Asset Acquisition Problem

A variation of the basic asset acquisition problem we introduced in section 8.2.1 arises when we can purchase assets now to be used in the future. For example, a hotel might book rooms at time t for a date t' in the future. A travel agent might purchase space on a flight or a cruise line at various points in time before the trip actually happens. An airline might purchase contracts to buy fuel in the future. In all of these cases, it will generally be the case that assets purchased farther in advance are cheaper, although prices may fluctuate.

For this problem, we are going to assume that selling prices are

$$
\begin{aligned}
x_{tt'} \;=\; & \text{resources purchased at time } t \text{ to be used to satisfy demands} \\
& \text{that become known during time interval between } t'-1 \\
& \text{and } t',
\end{aligned}
$$

$$
\begin{aligned}
x_t \;&=\; (x_{t,t+1}, x_{t,t+2}, \dots,), \\
&=\; (x_{tt'})_{t'>t},
\end{aligned}
$$

$$
\begin{aligned}
D_{tt'} \;&=\; \text{total demand known at time } t \text{ to be served at time } t', \\
D_t \;&=\; (D_{tt'})_{t'\ge t},
\end{aligned}
$$

$$
\begin{aligned}
R_{tt'} \;=\; & \text{inventory acquired on or before time } t \text{ that may be used to} \\
& \text{satisfy demands that become known between } t'-1 \text{ and } t',
\end{aligned}
$$

$$
R_t \;=\; (R_{tt'})_{t'\ge t}.
$$

Now, R_{tt} is the resources on hand in period t that can be used to satisfy demands D_t that become known during time interval t. In this formulation, we do not allow x_{tt}, which would represent purchases on the spot market. If this were allowed, purchases at time t could be used to satisfy unsatisfied demands arising during time interval between $t-1$ and t.

After we make our decisions x_t, we observe new demands

$$
\begin{aligned}
\hat{D}_{t+1,t'} \;=\; & \text{new demands for the resources that become known} \\
& \text{during time interval } (t, t+1) \text{ to be served at time } t'.
\end{aligned}
$$

The state variable for this problem would be

$$
S_t = (R_t, D_t),
$$

where R_t is the vector capturing inventory that will arrive in the future.

The transition equation for R_t is given by

$$
R_{t+1,t'} = \begin{cases} (R_{t,t} - \min(R_{tt}, D_{tt})) + x_{t,t+1} + R_{t,t+1}, & t' = t+1, \\ R_{tt'} + x_{tt'}, & t' > t+1. \end{cases}
$$

The transition equation for D_t is given by

$$D_{t+1,t'} = \begin{cases} (D_{tt} - \min(R_{tt}, D_{tt})) + \hat{D}_{t,t+1} + D_{t,t+1}, & t' = t+1, \\ D_{tt'} + \hat{D}_{t+1,t'}, & t' > t+1. \end{cases}$$

To compute profits, let

p_t^{sell} = the sales price, which varies stochastically over time as it did earlier,

$p_{t,t'-t}^{\text{buy}}$ = the purchase price, which depends on both time t as well as how far into the future we are purchasing.

The one-period contribution function (measuring forward in time) is

$$C_t(S_t, x_t) = p_t^{\text{sell}} \min(R_{tt}, D_{tt}) - \sum_{t'>t} p_{t,t'-t}^{\text{buy}} x_{tt'}.$$

Note that we index the contribution function $C_t(S_t, x_t)$ by time t. This is not because the prices p_t^{sell} and $p_{t,\tau}^{\text{buy}}$ depend on time. This information is captured in the state variable S_t. Rather, it is because of the sum $\sum_{t'>t}$ which depends on t.

8.2.4 The Batch Replenishment Problem

One of the classical problems in operations research is one that we refer to here as the batch replenishment problem. To illustrate the basic problem, assume that we have a single type of resource that is consumed over time. As the reserves of the resource run low, it is necessary to replenish the resources. In many problems, there are economies of scale in this process. It is more economical to increase the level of resources in one jump (see examples).

■ **EXAMPLE 8.1**

An oil company maintains an aggregate level of oil reserves. As these are depleted, it will undertake exploration expeditions to identify new oil fields, which will produce jumps in the total reserves under the company's control.

■ **EXAMPLE 8.2**

A startup company has to maintain adequate reserves of operating capital to fund product development and marketing. As the cash is depleted, the finance officer has to go to the markets to raise additional capital. There are fixed costs of raising capital, so this tends to be done in batches.

■ **EXAMPLE 8.3**

A delivery vehicle for an e-commerce food delivery company would like to do several deliveries at the same time. As orders come in, it has to decide whether to continue waiting or to leave with the orders that it has already accumulated.

To introduce the core elements, let

$$D_t = \text{demand waiting to be served at time } t,$$
$$R_t = \text{resource level at time } t,$$
$$x_t = \text{additional resources acquired at time } t \text{ to be used during time interval } t + 1.$$

Our state variable is

$$S_t = (R_t, D_t).$$

After we make our decision x_t of how much new product to order, we observe new demands

$$\hat{D}_{t+1} = \text{new demands that arrive during the interval } (t, t + 1).$$

The transition function is given by

$$R_{t+1} = \max\{0, (R_t + x_t - D_t)\},$$
$$D_{t+1} = D_t - \min\{R_t + x_t, D_t\} + \hat{D}_{t+1}.$$

Our one-period cost function (which we wish to minimize) is given by

$$
\begin{aligned}
C(S_t, x_t, \hat{D}_{t+1}) &= \text{total cost of acquiring } x_t \text{ units of the resource} \\
&= c^f I_{\{x_t > 0\}} + c^p x_t + c^h R^M_{t+1}(R_t, x_t, \hat{D}_{t+1}),
\end{aligned}
$$

where

$$c^f = \text{the fixed cost of placing an order,}$$
$$c^p = \text{the unit purchase cost,}$$
$$c^h = \text{the unit holding cost.}$$

For our purposes, $C(S_t, x_t, \hat{D}_{t+1})$ could be any nonconvex function; this is a simple example of one. Since the cost function is nonconvex, it helps to order larger quantities at the same time.

Assume that we have a family of decision functions $X^{\pi}(R_t)$, $\pi \in \Pi$, for determining x_t. For example, we might use a decision rule such as

$$X^{\pi}(R_t|\theta) = \begin{cases} \theta^{max} - R_t & \text{if } R_t < \theta^{min}, \\ 0 & \text{if } R_t \geq \theta^{min} \end{cases}$$

where $\theta = (\theta^{min}, \theta^{max})$ are specified parameters. In the language of sequential decision problems, a decision rule such as $X^{\pi}(S_t)$ is known as a *policy* (literally, a rule for making decisions). We index policies by π, and denote the set of policies by Π. In this example, a combination $(\theta^{min}, \theta^{max})$ represents an instance of our order-up-to policy, and Θ would represent all the possible values of θ^{min} and θ^{max} (this would be the set of policies in this class).

Our goal is to solve

$$\min_{\theta \in \Theta} \mathbb{E} \left\{ \sum_{t=0}^{T} \gamma^t C(S_t, X^{\pi}(R_t|\theta), \hat{D}_{t+1}) \right\}.$$

This means that we want to search over all possible values of θ^{min} and θ^{max} to find the best performance (on average).

The basic batch replenishment problem, where R_t and x_t are scalars, is quite easy (if we know things like the distribution of demand). But there are many real problems where these are vectors because there are different types of resources. The vectors may be small (different types of fuel or blood) or extremely large (hiring different types of people for a consulting firm or the military; maintaining spare parts inventories).

8.3 Complex Resource Allocation Problems

Problems involving the management of physical resources can become quite complex. Below we illustrate a dynamic assignment problem that arises in the context of assigning fleets of drivers (and cars) to riders requesting trips over time, and a problem involving the modeling inventories of different types of blood.

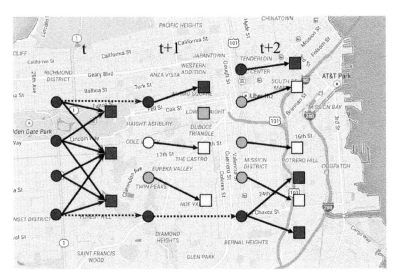

Figure 8.2 Illustration of the dynamic assignment of drivers (circles) to riders (squares).

8.3.1 The Dynamic Assignment Problem

Consider the challenge of matching drivers (or perhaps driverless electric vehicles) to customers calling in dynamically over time, illustrated in Figure 8.2. We have to think about which driver to assign to which rider based on the characteristics of the driver (or car), such as where the driver lives (or how much energy is in the car's battery), along with the characteristics of the trip (origin, destination, length).

We describe drivers (and cars) using

$$
a_t = \begin{pmatrix} a_1 \\ a_2 \\ a_3 \end{pmatrix} = \begin{pmatrix} \text{The location of the car} \\ \text{The type of car} \\ \text{Hours the driver has been on duty} \end{pmatrix}.
$$

We can model our fleet of drivers and cars using

$$
\begin{aligned}
\mathcal{A} &= \text{the set of all possible attribute vectors,} \\
R_{ta} &= \text{the number of cars with attribute } a \in \mathcal{A} \text{ at time } t, \\
R_t &= (R_{ta})_{a \in \mathcal{A}}.
\end{aligned}
$$

We note that R_t can be very high dimensional, since the attribute a is a vector. In practice, we never generate the vector R_t, since it is more practical to just create a list of drivers and cars. The notation R_t is used just for modeling purposes.

Demands for trips arise over time, which we can model using

$$b \quad = \quad \text{the characteristics of a trip (origin, destination, car type requested),}$$

$$\mathcal{B} \quad = \quad \text{the set of all possible values of the vector } b,$$

$$\hat{D}_{tb} \quad = \quad \text{the number of new customer requests with attribute } b \text{ that were first learned at time } t,$$

$$\hat{D}_t \quad = \quad (\hat{D}_{tb})_{b \in \mathcal{B}},$$

$$D_{tb} \quad = \quad \text{the total number of unserved trips with attribute } b \text{ waiting at time } t,$$

$$D_t \quad = \quad (D_{tb})_{b \in \mathcal{B}}.$$

We next have to model the decisions that we have to make. Assume that at any point in time, we can either assign a driver to handle a customer, or send her home. Let

$$\mathcal{D}^H \quad = \quad \text{the set of decisions representing sending a driver to her home location,}$$

$$\mathcal{D}^D \quad = \quad \text{the set of decisions to assign a driver to a rider, where } d \in \mathcal{D}^D \text{ represents a decision to serve a demand of type } b_d,$$

$$d^{\phi} \quad = \quad \text{the decision to "do nothing,"}$$

$$\mathcal{D} \quad = \quad \mathcal{D}^H \cup \mathcal{D}^D \cup d^{\phi}.$$

A decision has the effect of changing the attributes of a driver, as well as possibly satisfying a demand. The impact on the resource attribute vector of a driver is captured using the attribute transition function, represented using

$$a_{t+1} = a^M(a_t, d).$$

For algebraic purposes, it is useful to define the indicator function

$$\delta_{a'}(a_t, d) = \begin{cases} 1 & \text{for } a^M(a_t, d) = a', \\ 0 & \text{otherwise.} \end{cases}$$

A decision $d \in \mathcal{D}^D$ means that we are serving a customer described by an attribute vector b_d. This is only possible, of course, if $D_{tb} > 0$. Typically, D_{tb} will be 0 or 1, although our model allows for multiple trips with the same attributes.

We indicate which decisions we have made using

$$x_{tad} = \text{the number of times we apply a decision of type } d \text{ to trip}$$
$$\text{with attribute } a,$$

$$x_t = (x_{tad})_{a \in A, d \in \mathcal{D}}.$$

Similarly, we define the cost of a decision to be

$$c_{tad} = \text{the cost of applying a decision of type } d \text{ to driver with}$$
$$\text{attribute } a,$$

$$c_t = (c_{tad})_{a \in A, d \in \mathcal{D}}.$$

We could solve this problem myopically by making what appears to be the best decisions now, ignoring their impact on the future. We would do this by solving

$$\min_{x_t} \sum_{a \in A} \sum_{d \in \mathcal{D}} c_{tad} x_{tad}, \tag{8.7}$$

subject to

$$\sum_{d \in \mathcal{D}} x_{tad} = R_{ta}, \tag{8.8}$$

$$\sum_{a \in A} x_{tad} \le D_{tb_d}, \quad d \in \mathcal{D}^D, \tag{8.9}$$

$$x_{tad} \ge 0. \tag{8.10}$$

Equation (8.8) says that we either have to send a driver home, or assign her to serve a customer. Equation (8.9) says that we can only assign the driver to a job of type b_d if there is in fact a job of type b_d. Said differently, we cannot assign more than one driver per passenger. However, we do not have to cover every trip.

The problem posed by equations (8.7)–(8.10) is a linear program. Real problems may involve managing hundreds or even thousands of individual entities. The decision vector $x_t = (x_{tad})_{a \in A, d \in \mathcal{D}}$ may have over ten thousand dimensions (variables in the language of linear programming). However, commercial linear programming packages handle problems of this size quite easily.

If we make decisions by solving (8.7)–(8.10), we say that we are using a *myopic policy* since we are using only what we know now, and we are ignoring the impact of decisions now on the future. For example, we may decide to send a driver home rather than have her sit in a hotel room waiting for a job, but this ignores the likelihood that another job may suddenly arise close to the driver's current location.

Given a decision vector, the dynamics of our system can be described using

$$R_{t+1,a} = \sum_{a' \in \mathcal{A}} \sum_{d \in \mathcal{D}} x_{ta'd} \delta_a(a', d), \tag{8.11}$$

$$D_{t+1,b_d} = D_{t,b_d} - \sum_{a \in \mathcal{A}} x_{tad} + \hat{D}_{t+1,b_d}, \quad d \in \mathcal{D}^D. \tag{8.12}$$

Equation (8.11) captures the effect of all decisions (including serving demands) on the attributes of the drivers. This is easiest to visualize if we assume that all tasks are completed within one time period. If this is not the case, then we simply have to augment the state vector to capture the attribute that we have partially completed a task. Equation (8.12) subtracts from the list of available demands any of type b_d that are served by a decision $d \in \mathcal{D}^D$ (recall that each element of $\mathcal{D}^D$ corresponds to a type of trip, which we denote b_d).

The state of our system is given by

$$S_t = (R_t, D_t).$$

The evolution of our state variable over time is determined by equations (8.11) and (8.12). We can now set up an optimality recursion to determine the decisions that minimize costs over time using

$$V_t(S_t) = \min_{x_t \in \mathcal{X}_t} \left(C_t(S_t, x_t) + \gamma \mathbb{E} V_{t+1}(S_{t+1}) \right),$$

where S_{t+1} is the state at time $t + 1$ given that we are in state S_t and action x_t. S_{t+1} is random because at time t, we do not know $\hat{D}_{t+1}$. The feasible region $\mathcal{X}_t$ is defined by equations (8.8)–(8.10).

Needless to say, the state-variable for this problem is quite large. The dimensionality of R_t is determined by the number of attributes of our driver, while the dimensionality of D_t is determined by the relevant attributes of a demand. In real applications, these attributes can become fairly detailed. Fortunately, this problem has a lot of structure which we exploit in chapter 18.

8.3.2 The Blood Management Problem

The problem of managing blood inventories serves as a particularly elegant illustration of a resource allocation problem. We are going to start by assuming that we are managing inventories at a single hospital, where each week we have to decide which of our blood inventories should be used for the demands that need to be served in the upcoming week.

We have to start with a bit of background about blood. For the purposes of managing blood inventories, we care primarily about blood type and age. Although there is a vast range of differences in the blood of two individuals,

Table 8.1 Allowable blood substitutions for most operations, 'X' means a substitution is allowed. Adapted from Cant, L. (2006), 'Life Saving Decisions: A Model for Optimal Blood Inventory Management'.

Donor	Recipient							
	AB+	AB−	A+	A−	B+	B−	O+	O−
AB+	X							
AB−	X	X						
A+	X		X					
A−	X	X	X	X				
B+	X				X			
B−	X	X			X	X		
O+	X		X		X		X	
O−	X	X	X	X	X	X	X	X

for most purposes doctors focus on the eight major blood types: $A+$ (" A positive"), $A-$ ("A negative"), $B+$, $B-$, $AB+$, $AB-$, $O+$, and $O-$. While the ability to substitute different blood types can depend on the nature of the operation, for most purposes blood can be substituted according to Table 8.1.

A second important characteristic of blood is its age. The storage of blood is limited to six weeks, after which it has to be discarded. Hospitals need to anticipate if they think they can use blood before it hits this limit, as it can be transferred to blood centers which monitor inventories at different hospitals within a region. It helps if a hospital can identify blood it will not need as soon as possible so that the blood can be transferred to locations that are running short.

One mechanism for extending the shelf-life of blood is to freeze it. Frozen blood can be stored up to 10 years, but it takes at least an hour to thaw, limiting its use in emergency situations or operations where the amount of blood needed is highly uncertain. In addition, once frozen blood is thawed it must be used within 24 hours.

We can model the blood problem as a heterogeneous resource allocation problem. We are going to start with a fairly basic model which can be easily extended with almost no notational changes. We begin by describing the attributes of a unit of stored blood using

$$ a = \begin{pmatrix} a_1 \\ a_2 \end{pmatrix} = \begin{pmatrix} \text{Blood type } (A+, A-, \ldots) \\ \text{Age (in weeks)} \end{pmatrix}, $$

$\mathcal{B}$ = Set of all attribute types.

We will limit the age to the range $0 \leq a_2 \leq 6$. Blood with $a_2 = 6$ (which means blood that is already six weeks old) is no longer usable. We assume that decision epochs are made in one-week increments.

Blood inventories, and blood donations, are represented using

$$
\begin{aligned}
R_{ta} & = \text{units of blood of type } a \text{ available to be assigned or held} \\
& \quad \text{at time } t, \\
R_t & = (R_{ta})_{a \in A}, \\
\hat{R}_{ta} & = \text{number of new units of blood of type } a \text{ donated between} \\
& \quad t - 1 \text{ and } t, \\
\hat{R}_t & = (\hat{R}_{ta})_{a \in A}.
\end{aligned}
$$

The attributes of demand for blood are given by

$$
d = \begin{pmatrix} d_1 \\ d_2 \\ d_3 \end{pmatrix} = \begin{pmatrix} \text{Blood type of patient} \\ \text{Surgery type: urgent or elective} \\ \text{Is substitution allowed?} \end{pmatrix},
$$

$$
\begin{aligned}
d^\phi & = \text{decision to hold blood in inventory (``do nothing''),} \\
\mathcal{D} & = \text{set of all demand types } d \text{ plus } d^\phi.
\end{aligned}
$$

The attribute d_3 captures the fact that there are some operations where a doctor will not allow any substitution. One example is childbirth, since infants may not be able to handle a different blood type, even if it is an allowable substitute. For our basic model, we do not allow unserved demand in one week to be held to a later week. As a result, we need only model new demands, which we accomplish with

$$
\begin{aligned}
\hat{D}_{td} & = \text{units of demand with attribute } d \text{ that arose between} \\
& \quad t - 1 \text{ and } t, \\
\hat{D}_t & = (\hat{D}_{td})_{d \in \mathcal{D}}.
\end{aligned}
$$

We act on blood resources with decisions given by

$$
\begin{aligned}
x_{tad} & = \text{number of units of blood with attribute } a \text{ that we assign} \\
& \quad \text{to a demand of type } d, \\
x_t & = (x_{tad})_{a \in A, d \in \mathcal{D}}.
\end{aligned}
$$

The feasible region X_t is defined by the following constraints:

$$\sum_{d \in \mathcal{D}} x_{tad} = R_{ta}, \tag{8.13}$$

$$\sum_{a \in \mathcal{A}} x_{tad} \leq \hat{D}_{td}, \quad d \in \mathcal{D}, \tag{8.14}$$

$$x_{tad} \geq 0. \tag{8.15}$$

Blood that is held simply ages one week, but we limit the age to six weeks. Blood that is assigned to satisfy a demand can be modeled as being moved to a blood-type sink, denoted, perhaps, using $a_{t,1} = \phi$ (the null blood type). The blood attribute transition function $a^M(a_t, d_t)$ is given by

$$a_{t+1} = \begin{pmatrix} a_{t+1,1} \\ a_{t+1,2} \end{pmatrix} = \begin{cases} \begin{pmatrix} a_{t,1} \\ \min\{6, a_{t,2} + 1\} \end{pmatrix}, & d_t = d^\phi, \\[2ex] \begin{pmatrix} \phi \\ - \end{pmatrix}, & d_t \in \mathcal{D}. \end{cases}$$

To represent the transition function, it is useful to define

$$\delta_{a'}(a, d) = \begin{cases} 1 & a_t^x = a' = a^M(a_t, d_t), \\ 0 & \text{otherwise}, \end{cases}$$

$$\Delta = \text{matrix with } \delta_{a'}(a, d) \text{ in row } a' \text{ and column } (a, d).$$

We note that the attribute transition function is deterministic. A random element would arise, for example, if inspections of the blood resulted in blood that was less than six weeks old being judged to have expired. The resource transition function can now be written

$$R_{ta'}^x = \sum_{a \in \mathcal{A}} \sum_{d \in \mathcal{D}} \delta_{a'}(a, d) x_{tad},$$

$$R_{t+1,a'} = R_{ta'}^x + \hat{R}_{t+1,a'}.$$

In matrix form, these would be written

$$R_t^x = \Delta x_t, \tag{8.16}$$

$$R_{t+1} = R_t^x + \hat{R}_{t+1}. \tag{8.17}$$

Figure 8.3 illustrates the transitions that are occurring in week t. We either have to decide which type of blood to use to satisfy a demand (Figure 8.3a), or to hold the blood until the following week. If we use blood to satisfy a demand, it is assumed lost from the system. If we hold the blood until the following week,

Table 8.2 Contributions for different types of blood and decisions

Condition	Description	Value
if $d = d^\phi$	Holding	0
if $a_1 = a_1$ when $d \in \mathcal{D}$	No substitution	0
if $a_1 \neq a_1$ when $d \in \mathcal{D}$	Substitution	-10
if $a_1 = O-$ when $d \in \mathcal{D}$	O- substitution	5
if $d_2 =$ Urgent	Filling urgent demand	40
if $d_2 =$ Elective	Filling elective demand	20

it is transformed into blood that is one week older. Blood that is six weeks old may not be used to satisfy any demands, so we can view the bucket of blood that is six weeks old as a sink for unusable blood (the value of this blood would be zero). Note that blood donations are assumed to arrive with an age of 0. The pre- and post-decision state variables are given by

$$S_t = (R_t, \hat{D}_t),$$
$$S_t^x = (R_t^x).$$

There is no real "cost" to assigning blood of one type to demand of another type (we are not considering steps such as spending money to encourage additional donations, or transporting inventories from one hospital to another). Instead, we use the contribution function to capture the preferences of the doctor. We would like to capture the natural preference that it is generally better not to substitute, and that satisfying an urgent demand is more important than an elective demand. For example, we might use the contributions described in Table 8.2. Thus, if we use $O-$ blood to satisfy the needs for an elective patient with $A+$ blood, we would pick up a -\$10 contribution (penalty since it is negative) for substituting blood, a +\$5 for using $O-$ blood (something the hospitals like to encourage), and a +\$20 contribution for serving an elective demand, for a total contribution of +\$15.

The total contribution (at time t) is finally given by

$$C_t(S_t, x_t) = \sum_{a \in \mathcal{A}} \sum_{d \in \mathcal{D}} c_{tad} x_{tad}.$$

As before, let $X_t^\pi(S_t)$ be a policy (some sort of decision rule) that determines $x_t \in \mathcal{X}_t$ given S_t. We wish to find the best policy by solving

$$\max_{\pi \in \Pi} \mathbb{E} \sum_{t=0}^{T} C(S_t, X^\pi(S_t)). \tag{8.18}$$

R_t^x

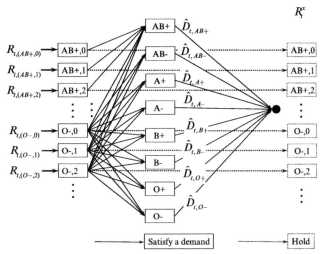

(a) - Assigning blood supplies to demands in week t. Solid lines represent assigning blood to a demand, dotted lines represent holding blood.

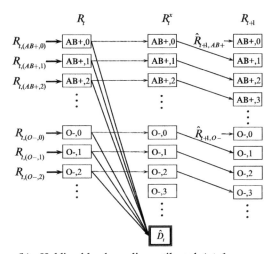

(b) - Holding blood supplies until week $t + 1$.

Figure 8.3 (a) The assignment of different blood types (and ages) to known demands in week t, and (b) holding blood until the following week.

The most obvious way to solve this problem is with a simple myopic policy, where we maximize the contribution at each point in time without regard to the effect of our decisions on the future. We can obtain a family of myopic policies by adjusting the one-period contributions. For example, our bonus of \$5 for using $O-$ blood (in Table 8.2), is actually a type of myopic policy. We encourage using $O-$ blood since it is generally more available than other blood types. By changing this bonus, we obtain different types of myopic policies that we can represent by the set Π^M, where for $\pi \in \Pi^M$ our decision function would be given by

$$X_t^\pi(S_t) = \arg\max_{x_t \in \mathcal{X}_t} \sum_{a \in \mathcal{A}} \sum_{d \in \mathcal{D}} c_{tad} x_{tad}. \qquad (8.19)$$

The optimization problem in (8.19) is a simple linear program (known as a "transportation problem"). Solving the optimization problem given by equation (8.18) for the set $\pi \in \Pi^M$ means searching over different values of the bonus for using $O-$ blood.

In chapter 13 we will introduce a way of improving this policy through simple parameterizations, using a class of policy we call a *cost function approximation*. We then revisit the same problem in chapter 18 when we develop a powerful strategy based on approximate dynamic programming, where we exploit the natural concavity of the value function. Finally, we touch on this problem one last time in chapter 20 when we show how to optimize the management of blood over many hospitals using a multiagent formulation.

8.4 State-dependent Learning Problems

Information acquisition is an important problem in many applications where we face uncertainty about the value of an action, but the only way to obtain better estimates of the value is to take the action. For example, a baseball manager may not know how well a particular player will perform at the plate. The only way to find out is to put him in the lineup and let him hit. The only way a mutual fund can learn how well a manager will perform may be to let her manage a portion of the portfolio. A pharmaceutical company does not know how the market will respond to a particular pricing strategy. The only way to learn is to offer the drug at different prices in test markets.

We have already seen information acquisition problems in chapter 7 where we called them active learning problems. Here we are going to pick up this thread but in the context of more complex problems that combine a hybrid of physical and informational states, in addition to belief states (which is what makes them learning problems).

Information acquisition plays a particularly important role in sequential decison problems, but the presence of physical states can complicate the learning process. Imagine that we are managing a medical team that is currently in zone i testing for the presence of a communicable disease. We are thinking about moving the team to zone j to do more testing there. We already have estimates of the presence of the disease in different zones, but visiting zone j not only improves our estimate for zone j, but also other zones through correlated beliefs.

Information acquisition problems are examples of dynamic optimization problems with belief states. These have not received much attention in the research literature, but we suspect that they arise in many practical applications that combine decisions under uncertainty with field observations.

8.4.1 Medical Decision Making

Patients arrive at a doctor's office for treatment. They begin by providing a medical history, which we capture as a set of attributes $a_1, a_2, \ldots$ which includes patient characteristics (gender, age, weight), habits (smoking, diet, exercise patterns), results from a blood test, and medical history (e.g. prior diseases). Finally, the patient may have some health issue (fever, knee pain, elevated blood sugar, ...) that is the reason for the visit. This attribute vector can have hundreds of elements.

Assume that our patient is dealing with elevated blood sugar. The doctor might prescribe lifestyle changes (diet and exercise), or a form of medication (along with the dosage), where we can represent the choice as $d \in \mathcal{D}$. Let t index visits to the doctor, and let

$$
x_{td} = \begin{cases} 1 & \text{if the physician chooses medication } d \in \mathcal{D}, \\ 0 & \text{otherwise.} \end{cases}
$$

After the physician makes a decision x_t, we observe the change in blood sugar levels by $\hat{y}_{t+1}$ which we assume is learned at the next visit.

Let $U(a, x|\theta)$ be a linear model of patient attributes and medical decisions which we write using

$$
U(a, x|\theta) = \sum_{f \in \mathcal{F}} \theta_f \phi_f(a, x),
$$

where $\phi_f(a, x)$ for $f \in \mathcal{F}$ represents features that we design from a combination of the patient attributes a (which are given) and the medical decision x. We believe that we can predict the patient response $\hat{y}$ using the logistic function

$$
\hat{y}|\theta \sim \frac{e^{U(a,x|\theta)}}{1 + e^{U(a,x|\theta)}}. \tag{8.20}
$$

Of course, we do not know what θ is. We can use data across a wide range of patients to get a population estimate $\bar{\theta}_t^{pop}$ that is updated every time we treat a patient and then observe an outcome. A challenge with medical decision-making is that every patient responds to treatments differently. Ideally, we would like to estimate $\bar{\theta}_{ta}$ that depends on the attribute a of an individual patient.

This is a classical learning problem similar to the derivative-free problems we saw in chapter 7, with one difference: we are given the attribute a of a patient, after which we make a decision. Our state, then, consists of our estimate of $\bar{\theta}_t^{pop}$ (or $\bar{\theta}_{ta}$), which goes into our belief state, along with the patient attribute a_t, which affects the response function itself. The attributes a_t represents the dynamic information for the problem.

8.4.2 Laboratory Experimentation

We may be trying to design a new material. For the nth experiment we have to choose

$$x_1^n \quad = \quad \text{the temperature at which the experiment will be run,}$$

$$x_2^n \quad = \quad \text{the length of time that the material will be heated,}$$

$$x_3^n \quad = \quad \text{the concentration of oxygen in the chamber,}$$

$$x_4^n \quad = \quad \text{the water concentration.}$$

When the experiment is complete, we then test the strength of the resulting material, which we model as

$$W^{n+1} \quad = \quad \text{the strength of the material produced by the experiment.}$$

We use this observation to update our estimate of the relationship between the inputs x^n and the resulting strength W^{n+1} which we represent using

$$f(x|\bar{\theta}^n) \quad = \quad \text{statistical estimate of the strength of the material} \\ \text{produced by inputs } x^n.$$

Our belief state B^n would consist of the estimate $\bar{\theta}^n$ as well as any other information needed to perform recursive updating (you have to pick your favorite method from chapter 3). The transition function for B^n would be the recursive updating equations for $\bar{\theta}^n$ from the method you chose from chapter 3.

For our laboratory experiment, we often have physical state variables R^n that might capture inventories of the materials needed for an experiment, and we might capture in our information state I^n information such as the temperature and humidity of the room where the experiment is being run. This gives us a state variable that consists of our physical state R^n, additional information I^n, and our belief B^n.

8.4.3 Bidding for Ad-clicks

Companies looking to advertise their product on the internet need to first choose a series of keywords, such as "hotels in New York," "pet-friendly hotel," "luxury hotel in New York," that will attract the people who they feel will find their offering most attractive. Then, they have to decide what to bid to have their ad featured on sites such as Google and Facebook. Assume that the probability that we win one of the limited "sponsored" slots on either platform, when we bid p, is given by

$$P^{\text{click}}(p|\theta) = \frac{e^{\theta_0 + \theta_1 p}}{1 + e^{\theta_0 + \theta_1 p}}. \tag{8.21}$$

Our problem is that we do not know $\theta = (\theta_0, \theta_1)$, but we think it is one of the family $\Theta = \{\theta_1, \dots, \theta_K\}$.

Assume that after n trials, we have a belief $p_k^n = Prob[\theta = \theta_k]$ that θ_k is the true value of θ. Now, we have a budget R^n that starts at R^0 at the beginning of each week. We have to learn how to place our bids, given our budget constraint, so that we learn how to maximize ad-clicks. Our state variable consists of our remaining budget R^n and our belief vector $p^n = (p_1^n, \dots, p_K^n)$.

8.4.4 An Information-collecting Shortest Path Problem

Assume that we have to choose a path through a network, but we do not know the actual travel time on any of the links of the network. In fact, we do not even know the mean or variance (we might be willing to assume that the probability distribution is normal).

To get a sense of some of the complexity when learning while moving around a graph, imagine that you have to find the best path from your apartment in New York City to your new job. You start by having to decide whether to walk to a subway station, take the subway to a station close to your workplace, and then walk. Or you can walk to a major thoroughfare and wait for a taxi, or you can make the decision to call Uber or Lyft if the wait seems long. Finally, you can call an Uber or Lyft from your apartment, which involves waiting for the car at your apartment and takes you right to your office.

Each decision involves collecting information by making a decision and observing the time required for each leg of the trip. Collecting information requires participating in the process, and changes your location. Also, observing the wait for an Uber or Lyft at your apartment hints at how long you might have to wait if you call one while waiting for a taxi. Your location on the graph (along with other information that might be available, such as weather) represents the dynamic information. In contrast to the medical decision-making example where we have no control over the patient attributes a_t, in our dynamic shortest path problem, our current location is directly a result of decisions we have made in the past.

Information-collecting shortest path problems arise in any information collection problem where the decision now affects not only the information you collect, but also the decisions you can make in the future. While we can solve basic bandit problems optimally, this broader problem class remains unsolved.

8.5 A Sequence of Problem Classes

Eventually, we are going to show that most stochastic optimization problems can be formulated using a common framework. However, this seems to suggest that all stochastic optimization problems are the same, which is hardly the case. It helps to identify major problem classes.

- Deterministically solvable problems – These are optimization problems where the uncertainty has enough structure that we can solve the problem exactly using deterministic methods. This covers an important class of problems, but we are going to group these together for now. All remaining problem classes require some form of adaptive learning.
- Pure learning problems – We make a decision x^n (or x_t), then observe new information W^{n+1} (or W_{t+1}), after which we update our knowledge to make a new decision. In pure learning problems, the only information passed from iteration n to $n+1$ (or from time t to time $t+1$) is updated knowledge, while in other problems, there may be a physical state (such as inventory) linking decisions.
- Stochastic problems with a physical state – Here we are managing resources, which arise in a vast range of problems where the resource might be people, equipment, or inventory of different products. Resources might also be money or different types of financial assets. There are a wide range of physical state problems depending on the nature of the setting. Some major problem classes include

  Stopping problems – The state is 1 (process continues) or 0 (process has been stopped). This arises in asset selling, where 1 means we are still holding the asset, and 0 means it has been sold.

  Inventory problems – We hold a quantity of resource to meet demands, where leftover inventory is held to the next period. Two important subclasses include:

  Inventory problems with static attributes – A static attribute might reflect the type of equipment or resource which does not change.

> Inventory problems with dynamic attributes – A dynamic attribute might be spatial location, age, or deterioration.

Multiattribute resource allocation – Resources might have static and dynamic attributes, and may be re-used over time (such as people or equipment).

Discrete resource allocation – This includes dynamic transportation problems, vehicle routing problems, and dynamic assignment problems.

- Physical state problems with an exogenous information state – While managing resources, we may also have access to exogenous information such as prices, weather, past history, or information about the climate or economy. Information states come in three flavors:
  - Memoryless – The information I_t at time t does not depend on past history, and is "forgotten" after a decision is made.
  - First-order exogenous process – I_t depends on I_{t-1}, but not on previous decisions.
  - State-dependent exogenous process – I_t depends on S_{t-1} and possibly x_{t-1}.
- Physical state with a belief state – Here, we are both managing resources while learning at the same time.

This list provides a sequence of problems of increasing complexity. However, each problem class can be approached with any of the four classes of policies.

8.6 Bibliographic Notes

All of the problems in this chapter are popular topics in the operations research literature. Most of the work in this chapter is based on work with former students.

Section 8.1.1 – The stochastic shortest path problem is a classic problem in operations research (see, for example, Bertsekas et al. (1991)). We use it to illustrate modeling strategy when we make different assumptions about what a traveler sees while traversing the network.

Section 8.1.2 – The "nomadic trucker" problem was first introduced in Powell (2011).

Section 8.1.3 – Equipment replacement problems are a popular topic. This section was based on the work of Johannes Enders (Enders et al. (2010).

Section 8.2.4 – Batch replenishment problems are a popular topic in operations research, often arising in the context of bulk service queues. This section was based on the work of Katerina Padaki (Powell and Papadaki (2002)), but see also Puterman (2005).

Section 8.3.1 – The material on the dynamic assignment problem is based on the work of Michael Spivey (Spivey and Powell (2004)).

Section 8.3.2 – This model of the blood management problem is based on the undergraduate senior thesis research of Lindsey Cant (Cant (2006)).

Section 8.4.4 – The work on the information collecting shortest path problem is based on (Ryzhov and Powell (2011)).

Section 8.4.3 – This section is based on (Han and Powell (2020)).

Exercises

Review questions

8.1   What is meant by a "state-dependent" problem? Give three examples.

8.2   You are moving over a static graph. At each time period, you arrive at another node. Why is this a "state-dependent problem"?

8.3   What are the essential differences between a shortest path problem with random costs, and an inventory problem with random demands (and deterministic prices and costs)?

8.4   Consider a discrete inventory problem where you can order at most 10 items at a point in time, but where you can order them to arrive in 1 day, 2 days, ... 5 days. Give the state variable, and compute how many states we may have for this (very simple) problem.

8.5   For the dynamic assignment problem in section 8.3.1, assume that space has been divided into 200 zones, there are three types of cars, and drivers may be on duty up to 10 hours (treat hours as an integer starting at 1). To understand the complexity of the problem, answer the following:

   (a) What is the dimensionality of the state variable?
   (b) What is the dimensionality of the decision vector?
   (c) What is the dimensionality of the exogenous information vector?

8.6   For the blood management problem in section 8.3.2, answer the following:

   (a) What is the dimensionality of the state variable?
   (b) What is the dimensionality of the decision vector?
   (c) What is the dimensionality of the exogenous information vector?

Modeling questions

8.7 Consider a discrete inventory problem with deterministic demands, but (possibly) random costs. Starting with an inventory of 0, sketch a few time periods of this problem assuming you cannot order more than 2 items per time period, and show that this can be modeled as a dynamic shortest path problem.

8.8 What is the distinguishing characteristic of a state-dependent *problem*, as opposed to the state-independent problems we considered in chapters 5 and 7? Contrast what we mean by a solution to a stochastic optimization problem with a state-independent function, versus what we mean by a solution to a stochastic optimization problem with a state-dependent function?

8.9 Repeat the gambling problem assuming that the value of ending up with S^N dollars is $\sqrt{S^N}$.

8.10 Section 8.2.1 describes an inventory problem that uses a contribution function $C(S_t, x_t, W_{t+1})$, and shows that it can also be modeled so the single-period contribution is written $C(S_t, x_t)$. Show how to convert any problem where you are given the contribution function in the form of $C(S_t, x_t, W_{t+1})$ into a problem where the single period contribution is given by $C(S_t, x_t)$ without changing sum of the contributions over time. This result allows us to write $C(S_t, x_t)$ without loss of generality, but there will be problems (such as the inventory problem in section 8.2.1), where it will be more natural to write $C(S_t, x_t, W_{t+1})$. The choice is up to the modeler.

8.11 Rewrite the transition function for the asset acquisition problem II (section 8.2.2) assuming that R_t is the resources on hand after we satisfy the demands.

8.12 Write out the transition equations for the lagged asset acquisition problem in section 8.2.3 when we allow spot purchases, which means that we may have $x_{tt} > 0$. x_{tt} refers to purchases that are made at time t which can be used to serve unsatisfied demands D_t that occur during time interval t.

8.13 Model the sequence of states, decisions, and information for the medical decision making problem in section 8.4.1 using the notation described in section 7.13.6.

Theory questions

8.14   Consider three variations of a shortest path problem:

Case I – All costs are known in advance. Here, we assume that we have a real-time network tracking system that allows us to see the cost on each link of the network before we start our trip. We also assume that the costs do not change during the time from which we start the trip to when we arrive at the link.

Case II – Costs are learned as the trip progresses. In this case, we assume that we see the actual link costs for links out of node i when we arrive at node i.

Case III – Costs are learned after the fact. In this setting, we only learn the cost on each link after the trip is finished.

Let v_i^I be the expected cost to get from node i to the destination for Case I. Similarly, let v_i^{II} and v_i^{III} be the expected costs for cases II and III. Show that $v_i^I \leq v_i^{II} \leq v_i^{III}$.

Problem-solving questions

8.15   We are now going to do a budgeting problem where the reward function does not have any particular properties. It may have jumps, as well as being a mixture of convex and concave functions. But this time we will assume that $R = 30$ dollars and that the allocations x_t must be in integers between 0 and 30. Assume that we have $T = 5$ products, with a contribution function $C_t(x_t) = cf(x_t)$ where $c = (c_1, \ldots, c_5) = (3, 1, 4, 2, 5)$ and where $f(x)$ is given by

$$f(x) = \begin{cases} 0, & x \leq 5, \\ 5, & x = 6, \\ 7, & x = 7, \\ 10, & x = 8, \\ 12, & x \geq 9. \end{cases}$$

Find the optimal allocation of resources over the five products.

8.16   You suddenly realize toward the end of the semester that you have three courses that have assigned a term project instead of a final exam. You quickly estimate how much each one will take to get 100 points (equivalent to an A+) on the project. You then guess that if you invest t hours in a project, which you estimated would need T hours to get 100 points, then for $t < T$ your score will be

$$R = 100\sqrt{t/T}.$$

That is, there are declining marginal returns to putting more work into a project. So, if a project is projected to take 40 hours and you only invest 10, you estimate that your score will be 50 points (100 times the square root of 10 over 40). You decide that you cannot spend more than a total of 30 hours on the projects, and you want to choose a value of t for each project that is a multiple of 5 hours. You also feel that you need to spend at least 5 hours on each project (that is, you cannot completely ignore a project). The time you estimate to get full score on each of the three projects is given by

Project	Completion time T
1	20
2	15
3	10

Show how to solve this problem as a decision tree. Assume you have to decide how many hours to allocate to each project, in increments of 5 hours. Set up your tree so you enumerate the decisions for project 1 (5, 10, 15, 20), then project 2, then project 3. There are 12 possible decisions over the first two projects (not all of them feasible). For each combination, look at the time remaining for the third project and find the optimal time allocation for the third project. Work backward to find the optimal allocation over all three projects.

Diary problem

The diary problem is a single problem you chose (see chapter 1 for guidelines). Answer the following for your diary problem.

8.17 It is quite likely that your diary problem falls in the "state-dependent problem" class. Describe some of the key state variables that characterize your problem, using the dimensions of physical states, other information, and belief states. Indicate in each case whether the state variables evolve from decisions, exogenous sources, or both.

Bibliography

Bertsekas, D.P., Tsitsiklis, J.N., and An. (1991). Analysis of stochastic shortest path problems. *Mathematics of Operations Research* 16 (3): 580–595.

Cant, L. (2006). Life saving decisions: A model for optimal blood inventory management.

Enders, J., Powell,W.B., and Egan, D. (2010). A dynamic model for the failure replacement of aging high-voltage transformers. *Energy Systems*.

Han, W. and Powell, W. B. (2020). Optimal online learning for nonlinear belief models using discrete priors. *Operations Research*.

Powell, W.B. (2011). *Approximate Dynamic Programming: Solving the Curses of Dimensionality*, 2e. John Wiley & Sons.

Powell, W.B. and Papadaki, K.P. (2002). Exploiting structure in adaptive dynamic programming algorithms for a stochastic batch service problem. *European Journal Of Operational Research* 142: 108–127.

Puterman, M.L. (2005). *Markov Decision Processes*, 2e. Hoboken, NJ: John Wiley and Sons.

Ryzhov, I.O. and Powell, W.B. (2011). Information collection on a graph. *Operations Research* 59 (1): 188–201.

Spivey, M.Z. and Powell, W.B. (2004). The dynamic assignment problem. *Transportation Science* 38 (4): 399–419.

9

Modeling Sequential Decision Problems

Perhaps one of the most important skills to develop when solving sequential decision problems is the ability to write down a mathematical model of the problem. As illustrated in Figure 9.1, the path from a real application to doing computational work on the computer has to pass through the process of mathematical modeling. Unlike fields such as deterministic optimization and machine learning, there is not a standard modeling framework for decisions under uncertainty. This chapter will develop, in much greater detail, our universal modeling framework for any sequential decision problem. Although we have introduced this framework in earlier chapters, this chapter is dedicated to modeling, bringing out the incredible richness of sequential decision problems. This chapter is written to stand alone, so there is some repetition of elements of our universal model.

While the problem domain of sequential decision problems is astonishingly rich, we can write any sequential decision problem as the sequence:

$$(decision, information, decision, information, ...).$$

Let x_t be the decision we make at time t, and let W_{t+1} be the new information that arrives between t (that is, after the decision has been made), and $t+1$ (when we have to make the next decision). We will find it convenient to represent what we know when we make a decision. We refer to this information as the "state" variable S_t (think of this as our "state of knowledge"). Using this bit of notation, we can write our sequential decision problem as

$$(S_0, x_0, W_1, S_1, ..., S_t, x_t, W_{t+1}, S_{t+1}, ..., S_T). \tag{9.1}$$

There are many applications where it is more natural to use a counter n, which might be the nth arrival of a customer, the nth experiment, or the nth iteration of an algorithm, in which case we would write our sequence as

Reinforcement Learning and Stochastic Optimization: A Unified Framework for Sequential Decisions, First Edition. Warren B. Powell.
© 2022 John Wiley & Sons, Inc. Published 2022 John Wiley & Sons, Inc.

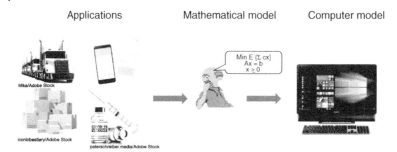

Figure 9.1 The path from applications to computation passes through the need for a mathematical model.

$$(S^0, x^0, W^1, S^1, \dots, S^n, x^n, W^{n+1}, S^{n+1}, \dots, S^N, x^N). \tag{9.2}$$

Note that the nth arrival might occur in continuous time, where we might let τ^n be the time at which the nth event occurs. This notation allows us to model systems in continuous time (we can let t^n be the time of the nth decision event).

There are problems where we might repeatedly simulate over time, in which case we would let write our sequence as

$$(S_0^1, x_0^1, W_1^1, \dots, S_t^1, x_t^1, W_{t+1}^1, \dots, S_0^n, x_0^n, W_1^n, \dots, S_t^n, x_t^n, W_{t+1}^n, \dots),$$

where we assume that our first pass is viewed as iteration $n = 1$. For the remainder of this chapter, we will assume our underlying physical process evolves over time, but any search for better policies (or parameters) will use iteration n.

After each decision, we evaluate our performance using a metric such as a contribution (there are many terms we might use) that we typically will write as $C(S_t, x_t)$ or, in some situations, $C(S_t, x_t, W_{t+1})$ (again, there are other styles that we discuss shortly). Decisions x_t are determined using a function we call a policy and denote by $X^\pi(S_t)$. Our ultimate goal is finding the best policy that optimizes the contribution function in some way.

The sequence in (9.1) (or (9.2) if we are using counts) can be used to describe virtually *any* sequential decision problem, but it requires that the problem be modeled correctly, and this means properly using one of the most misunderstood concepts in sequential decision problems: the state variable. The remainder of this chapter develops this basic model in considerably more depth.

Up to now, we have avoided discussing some important subtleties that arise in the modeling of sequential decision systems. We intentionally overlooked trying to define a state variable, which we have viewed as simply S_t. We have avoided discussions of how to properly model time or more complex information processes. We have also ignored the richness of modeling all the different sources of uncertainty for which we have a dedicated chapter (chapter 10). This style has facilitated introducing some basic ideas in dynamic programming, but would severely limit our ability to apply these methods to real problems.

There are five elements to any sequential decision problem, consisting of the following:

State variables – The state variables describe what we need to know (from history) to model the system forward in time. The initial state S_0 is also where we specify fixed parameters, the initial values of parameters (or quantities) that vary over time, as well as our distribution of belief about parameters we do not know perfectly.

Decision/action/control variables – These are the variables we control. Choosing these variables ("making decisions") represents the central challenge in sequential decision problems. This is where we describe constraints that limit what decisions we can make. Here is where we introduce the concept of a policy, but do not describe how to design the policy.

Exogenous information variables – These variables describe information that arrives to us exogenously, representing what we learn after we make each decision. Modeling exogenous information processes can be a significant challenge for many applications.

Transition function – This is the function that describes how each state variable evolves from one point in time to another. We may have explicit equations relating the next state to the current state, decision, and the exogenous information we learn after making the decision, for some, all, or none of the state variables.

Objective function – We assume we are trying to maximize or minimize some metric that is specified. This function describes how well we are doing at a point in time, and represents the foundation for evaluating policies.

An important point to make about the modeling framework is that there will at all times be a direct relationship to the software implementation. The mathematical model can be translated directly into software, and it will be possible to translate changes in the software back to the mathematical model.

We are going to start by illustrating these elements in the context of a simple energy storage problem in section 9.1. This is a nice starter problem because the state variable is fairly obvious. However, in section 9.9, we demonstrate how simple problems become complicated quickly using extensions of the initial energy storage application in section 9.1. The variations in section 9.9 introduces modeling issues that have never been addressed in the academic literature.

A reader can actually skip the entire rest of the chapter after this illustration if they are new to the field and just getting started. The remainder of the chapter will lay the foundation for modeling (and then solving) an exceptionally wide range of problems, including all the application areas covered by the fields presented in chapter 2, and all the applications sketched in chapter 8. We demonstrate these concepts when we show how to model the energy storage variations in section 9.9.

Even for more determined readers who are willing to read past section 9.1, we have still marked a number of sections with * that can be skipped on a first read.

This entire chapter is predicated on our "model first, then solve" style because we are going to present the model without addressing how we might solve it (hint: we will use one or more of our four classes of policies). This contrasts with the standard style used in the literature on sequential decision problems which is to present a method (see, for example, the introduction to reinforcement learning in section 2.1.6), but we also include so-called models where it is clear that the next step is to use Bellman's equation.

The rest of this chapter is organized as follows. We begin by describing the principles of good notation in section 9.2, followed by section 9.3 which addresses the subtleties of modeling time. These two sections lay the critical foundation for notation that is used throughout the book. Notation is not as critical for simple problems, as long as it is precise and consistent. But what seems like benign notational decisions for a simple problem can cause unnecessary difficulties, possibly producing a model that simply does not capture the real problem, or making the model completely intractable.

The five elements of a dynamic model are covered in the following sections:

- State variables – section 9.4.
- Decision variables – section 9.5
- Exogenous information variables – section 9.6
- Transition function – section 9.7
- Objective function – section 9.8

We then present a more complex energy storage problem in section 9.9.

Having laid this foundation, we transition to a series of topics that can be skipped on a first pass, but which help to expand the readers' appreciation of modeling of dynamic systems. These include:

Base models vs. lookahead models – Section 9.10 introduces the concept of base models (which is what we describe in this chapter) and lookahead models, which represent one of our classes of policies (described in much greater detail in chapter 19).

Problem classification – Section 9.11 describes four fundamental problem classes differentiated based on whether we have state-independent or state-dependent problems, and whether we are working in an offline setting (maximizing the final reward) or an online setting (maximizing cumulative reward).

Policy evaluation – Section 9.12 describes how to evaluate a policy using Monte Carlo simulation. This can actually be somewhat subtle. We have found that a good test of whether you understand an expectation is that you know how to estimate it using Monte Carlo simulation.

Advanced probabilistic modeling concepts – For readers who enjoy bridging to more advanced concepts in probability theory, section 9.13 provides an introduction to the vocabulary of measure-theoretic concepts and probability modeling. This discussion is designed for readers who do not have any formal training in this area, but would like to understand some of the language (and concepts) that measure-theoretic probability brings to this area.

Once we have laid out the five core elements of the model, we still have two components that we deal in much greater depth in subsequent chapters:

Uncertainty modeling – Chapter 10 deals with the exceptionally rich area of modeling uncertainty, which enters our model through the initial state S_0, which can capture uncertainties about parameters and quantities, and the exogenous information process $W_1, \dots, W_T$. We recommend that when providing the basic model of our problem, the discussion of "exogenous information" should be limited to just listing the variables, without delving into how we model the uncertainties.

Designing policies – Chapter 11 describes in more detail the four classes of policies, which are the topic of chapters 12-19. We believe strongly that the design of policies can only occur *after* we have developed our model.

This chapter describes modeling in considerable depth, and as a result it is quite long. Sections marked with a '*' can be skipped on a first read. The section on more advanced probabilistic modeling is marked with a '**' to indicate that this is more difficult material.

9.1 A Simple Modeling Illustration

We are going to first describe a simple energy storage problem in an unstructured way, then we are going to pull the problem together into the five dimensions of a sequential decision problem. We do this by beginning with a plain English narrative (which we recommend for any problem).

Narrative: We have a single battery connected to the grid, where we can either buy energy from the grid or sell it back to the grid. Electricity prices are highly volatile, and may jump from an average of around $20 per megawatt-hour (MWh) to over $1000 per MWh (in some areas of the country, prices can exceed $10,000 per MWh for brief periods). Prices change every 5 minutes, and we will make the assumption that we can observe the price and then decide if we want to buy or sell. We can only buy or sell for an entire 5-minute interval at a maximum rate of 10 kilowatts (0.01 megawatts). The capacity of our battery storage is 100 kilowatts, which means it may take 10 hours of continuous charging to charge an empty battery.

To model our problem, we introduce the following notation:

$$x_t \;=\; \text{The rate at which we buy from the grid to charge the battery}$$
$(x_t > 0)$ or sell back to the grid to discharge the battery $(x_t < 0)$.

$$u \;=\; \text{The maximum charge or discharge rate for the battery}$$
(the power rating, which is 10 kwh).

$$p_t \;=\; \text{The price of electricity on the grid at time } t.$$

$$R_t \;=\; \text{The charge level of the battery.}$$

$$R^{max} \;=\; \text{The capacity of the battery.}$$

Assume for the purposes of this simple model that the prices p_t are random and independent over time.

If the prices were known in advance (which makes this a deterministic problem), we might formulate the problem as a linear program using

$$\max_{x_0,\dots,x_T} \sum_{t=0}^{T} -p_t x_t \qquad (9.3)$$

subject to:

$$R_{t+1} \;=\; R_t + x_t,$$
$$x_t \;\le\; u,$$
$$x_t \;\le\; R^{max} - R_t,$$
$$x_t \;\ge\; 0.$$

Now let's introduce the assumption that prices p_t are random, and independent across time. Our first step is to replace the deterministic decision x_t with a policy $X^\pi(S_t)$ that depends on the state (that is, what we know), which we need to define (we will do this in a minute).

Next, imagine that we are going to run a simulation. Assume that we have compiled from history a series of sample paths that we are going to index by the Greek letter ω. If we have 20 sample paths, think of having 20 values of ω where each ω implies a sequence of prices $p_0(\omega), p_1(\omega), \ldots, p_T(\omega)$. Let Ω be the entire set of sample paths (you can think of this as the numbers $1, 2, \ldots, 20$ if we have 20 sample paths of prices). We assume each sample path occurs with equal likelihood.

We haven't yet described how we are going to design our policy (that is for later), but let's say we have a policy. We can simulate the policy for sample path ω and get the sample value of the policy using

$$F^\pi(\omega) = \sum_{t=0}^{T} -p_t(\omega)X^\pi(S_t(\omega)),$$

where the notation $S_t(\omega)$ for our as-yet undefined state variable indicates that, as we would expect, depends on the sample path ω (technically, it also depends on the policy π that we have been following). Next, we want to average over all the sample paths, so we compute an average

$$\bar{F}^\pi = \frac{1}{|\Omega|} \sum_{\omega \in \Omega} F^\pi(\omega).$$

This averaging is an approximation of an expectation. If we could enumerate every sample path, we could write

$$F^\pi = \mathbb{E} \sum_{t=0}^{T} -p_t X^\pi(S_t). \tag{9.4}$$

Here, we drop the dependence on ω, but need to remember that prices p_t, as well as the state variables S_t, are random variables because they do depend on the sample path. There will be many times that we will write the objective function using the expectation as in equation (9.4). Whenever you see this, keep in mind that we are virtually always assuming that we will approximate the expectation using an average as in equation (9.4).

Our final step is to find the best policy. We would write this objective using

$$\max_{\pi} \mathbb{E} \sum_{t=0}^{T} -p_t X^\pi(S_t). \tag{9.5}$$

So, we now have our objective function (equation (9.5)), which is frustratingly stated in terms of optimizing over policies, but we have yet to provide any indication how we do this! This is what we call "model first, then solve." In this chapter, we will only get as far as the objective function. This parallels what is done in every single paper on deterministic optimization or optimal control, as well as any paper on machine learning. They all present a model (which includes an objective function) and only then do they set about solving it (which in our setting means coming up with a policy).

At this point, we have two tasks remaining:

(1) Uncertainty quantification. We need to develop a model of any sources of uncertainty such as our price process. In a real problem, we would not be able to assume that the prices are independent. In fact, this is a fairly difficult problem.
(2) Designing policies. We need to design effective policies for buying and selling that depend only on the information in the state variable S_t.

Now that we have described the problem, presented our notation, and described the two remaining tasks, we are going to step back and describe the problem in terms of the five elements described earlier:

State variables – This has to capture all the information we need at time t. We can see that we need to know how much is stored in the battery R_t, and the grid price p_t, so

$$S_t = (R_t, p_t).$$

Decision variables – Clearly this is x_t. We then need to express the constraints on x_t which are given by

$$x_t \leq u,$$
$$x_t \leq R^{max} - R_t,$$
$$x_t \geq 0.$$

Finally, when we define the decision variables we also introduce the policy $X^\pi(S_t)$ to be designed later. We introduce it now because we need it to present the objective function. Note that we have defined the state variable S_t, so this is given.

Exogenous information variables – This is where we model any information that becomes available *after* we make the decision x_t. For our simple problem, this would be the updated price, so

$$W_{t+1} = p_{t+1}.$$

Transition function – These are the equations that govern how the state variables evolve over time. We have two variables in the state variables. R_t evolves according to

$$R_{t+1} = R_t + x_t.$$

The price process evolves according to

$$p_{t+1} = W_{t+1}.$$

In other words, we just observe the next price rather than derive it from an equation. This is an example of "model free dynamic programming." There are many problems where we observe S_{t+1} rather than compute it; here we have an instance where we compute R_{t+1} but observe p_{t+1}. As an alternative model, we might assume that we model the *change* in the price using

$$\hat{p}_{t+1} = \text{the change in the price from } t \text{ to } t+1,$$

which means that $W_{t+1} = \hat{p}_{t+1}$ and our transition function becomes

$$p_{t+1} = p_t + \hat{p}_{t+1}.$$

Objective function – Our contribution function at time t is given by

$$C(S_t, x_t) = -p_t x_t,$$

where p_t is pulled from the state variable S_t. We would then write our objective function as

$$\max_{\pi} \mathbb{E} \sum_{t=0}^{T} C(S_t, X^{\pi}(S_t)). \tag{9.6}$$

Now we have modeled our problem using the five dimensions outlined earlier. This follows the same style used for deterministic optimization, but adapted for sequential decision problems. As problems become more complicated, the state variable will become more complex, as will the transition function (which requires an equation for each state variable). It would be nice if we could write a model from top to bottom, starting with state variables, but that is not how it works in practice. Modeling is iterative.

We repeat: this is a good stopping point for readers new to the field. Sections 9.2 on notational style and 9.3 on modeling time are useful to develop notational skills. The remainder of the chapter is primarily useful if you want a solid foundation for more complex problems (we give a hint of this in the expanded energy storage problem in section 9.9). In particular, section 9.4 is an in-depth introduction to the notion of state variables, which get complicated very quickly, even with relatively modest extensions of a simple problem. We demonstrate precisely this process in section 9.9 which presents a series of seemingly modest extensions of this energy problem, focusing on how to model the state variable as we add details to the problem.

9.2 Notational Style

Good modeling begins with good notation. The choice of notation has to balance traditional style with the needs of a particular problem class. Notation is easier to learn if it is mnemonic (the letters look like what they mean) and compact (avoiding a profusion of symbols). Notation also helps to bridge communities. Notation is a language: the simpler the language, the easier it is to understand the problem.

As a start, it is useful to adopt notational conventions to simplify the style of our presentation. For this reason, we adopt the following notational conventions:

Variables – Variables are *always* a single letter. We would never use, for example, CH for "cost of holding inventory."

Modeling time – We always use t to represent a point in time, while we use τ to represent an interval over time. When we need to represent different points in time, we might use $t, t', \bar{t}, t^{max}$, and so on. Time is *always* represented as a subscript such as S_t.

Indexing time – If we are modeling activities in discrete time, then t is an index and should be put in the subscript. So x_t would be an activity at time t, with the vector $x = (x_0, x_1, \ldots, x_t, \ldots, x_T)$ giving us all the activities over time. When modeling problems in continuous time, it is more common to write t as an argument, as in $x(t)$. x_t is notationally more compact (try writing a complex equation full of variables written as $x(t)$ instead of x_t).

Indexing vectors – Vectors are almost always indexed in the subscript, as in x_{ij}. Since we use discrete time models throughout, an activity at time t can be viewed as an element of a vector. When there are multiple indices, they should be ordered from outside in the general order over which they might be summed (think of the outermost index as the most detailed information). So, if x_{tij} is the flow from i to j at time t with cost c_{tij}, we might sum up the total cost using $\sum_t \sum_i \sum_j c_{tij} x_{tij}$. Dropping one or more indices creates a vector over the elements of the missing indices to the right. So, $x_t = (x_{tij})_{\forall i, \forall j}$ is the vector of all flows occurring at time t. Time, when present, is always the innermost index.

Temporal indexing of functions – A common notational error is to index a function by time t when in fact the function itself does not depend on time, but depends on inputs that do depend on time. For example, imagine that we have a stochastic price process where the state $S_t = p_t$ which is the price of the asset, and x_t is how much we sell $x_t > 0$ or buy $x_t < 0$. We might want to write our contribution as

$$C_t(S_t, x) = p_t x_t.$$

However, in this case the *function* does not depend on time t; it only depends on data $S_t = p_t$ that depends on time. So the proper way to write this would be

$$C(S_t, x) = p_t x_t.$$

Now imagine that our contribution function is given by

$$C_t(S_t, x_t) = \sum_{t'=t}^{t+H} p_{tt'} x_{tt'}.$$

Here, the *function* depends on time because the summation runs from t to $t + H$.

Flavors of variables – It is often the case that we need to indicate different flavors of variables, such as holding costs and order costs. These are always indicated as superscripts, where we might write c^h or c^{hold} as the holding cost. Note that while variables must be a single letter, superscripts may be words (although this should be used sparingly). We think of a variable like "c^h" as a single piece of notation. It is better to write c^h as the holding cost and c^p as the purchasing cost than to use h as the holding cost and p as the purchasing cost (the first approach uses a single letter c for cost, while the second approach uses up two letters – the roman alphabet is a scarce resource). Other ways of indicating flavors is hats ($\hat{x}$), bars ($\bar{x}$), tildes ($\tilde{x}$), and primes (x').

Iteration counters – There are problems where it is more natural to count events such as customer arrivals, experiments, observations, or iterations of an algorithm, rather than representing the actual time at which a decision is being made.

We place iteration counters in the superscript, since we view it as indicating the value of a single variable at iteration n, as opposed to the nth element of a vector. So, x^n is our activity at iteration n, while x^{n+1} is the value of x at iteration $n + 1$. If we are using a descriptive superscript, we might write $x^{h,n}$ to represent x^h at iteration n. Sometimes algorithms require inner and outer iterations. In this case, we use n to index the outer iteration and m for the inner iteration.

While this will prove to be the most natural way to index iterations, there is potential for confusion where it may not be clear if the superscript n is an index (as we view it) or raising a variable to the nth power. One notable exception to this convention is indexing stepsizes which we first saw in chapter 5. If we write α^n, it looks like we are raising α to the nth power, so we use α_n.

Sets are represented using capital letters in a calligraphic font, such as $\mathcal{X}, \mathcal{F}$, or $\mathcal{J}$. We generally use the lowercase roman letter as an element of a set, as in $x \in \mathcal{X}$ or $i \in \mathcal{J}$.

Exogenous information – Information that first becomes available (from outside the system) at time t is denoted using hats, for example, $\hat{D}_t$ or $\hat{p}_t$. Our only exception to this rule is W_t which is our generic notation for exogenous information (since W_t *always* refers to exogenous information, we do not use a hat).

Statistics – Statistics computed using exogenous information are generally indicated using bars, for example $\bar{x}_t$ or $\overline{V}_t$. Since these are functions of random variables, they are also random. We do not use hats, because we have reserved "hat" variables for exogenous information.

Index variables – Throughout, i, j, k, l, m and n are always scalar indices.

Superscripts/subscripts on superscripts/subscripts – As a general rule, avoid superscripts on superscripts (and so forth). For example, it is tempting to think of x_{b_t} as saying that x is a function of time t, when in fact this means it is a function of b which itself depends on time.

For example, x might be the number of clicks when the bid at time t is b_t, but what this notation is saying is that the number of clicks just depends on the bid, and not on time. If we want to capture the effect of both the bid and time, we have to write $x_{b,t}$.

Similarly, the notation F_{T^D} cannot be used as the forecast of the demand D at time T. To do this, you should write F_T^D. The notation F_{T^D} is just a forecast at a time $t = T^D$ that might correspond to the time, say, at which a demand occurs. But if you also write F_{T^p} where it just happens that $T^D = T^p$, you cannot refer to these as different forecasts because one is indexed by T^D while the other is indexed by T^p.

Of course, there are exceptions to every rule, and you have to keep an eye on standard notational conventions within pocket research communities.

9.3 Modeling Time

There are two strategies for modeling "time" in a sequential decision problem:

- Counters – There are many settings where we make decisions corresponding to discrete events, such as running an experiment, the arrival of a customer, or iterations of an algorithm. We generally let n the variable we use for counting, and we place it in the superscript, as in X^n or $\bar{f}^n(x)$. $n = 1$ corresponds to the first event, while $n = 0$ means no events have happened. However, our first

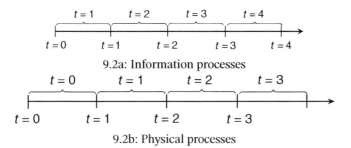

Figure 9.2 Relationship between discrete and continuous time for information processes (9.2a) and physical processes (9.2b).

decision occurs at $n = 0$ since we generally have to make a decision before anything has happened.

- Time – We may wish to directly model time. If time is continuous, we would write a function as $f(t)$, but all of the problems in this book are modeled in discrete time $t = 0, 1, 2, \ldots$ If we wish to model the time of the arrival of the nth customer, we would write t^n. However, we would write X^n for a variable that depends on the nth arrival rather than X_{t^n}.

Our style of indexing counters in the superscripts and time in subscripts helps when we are modeling simulations where we have to run a simulation multiple times. Thus, we might write X_t^n as information at time t in the nth iteration of our simulation.

The confusion over modeling time arises in part because there are two processes that we have to capture: the flow of information, and the flow of physical and financial resources. For example, a buyer may purchase an option now (an information event) to buy a commodity in the future (the physical event). Customers may call an airline (the information event) to fly on a future flight (the physical event). An electric power company has to purchase equipment now to be used one or two years in the future. All of these problems represent examples of *lagged information processes* and force us to explicitly model the informational and physical events.

Notation can easily become confused when an author starts by writing down a deterministic model of a physical process, and then adds uncertainty. The problem arises because the proper convention for modeling time for information processes is different than what should be used for physical processes.

We begin by establishing the relationship between discrete and continuous time. All of the models in this book assume that decisions are made in discrete time (sometimes referred to as *decision epochs*). However, the flow of information is best viewed in continuous time.

The relationship of our discrete time approximation to the real flow of information and physical resources is depicted in Figure 9.2. Above the line, "t" refers to a time interval while below the line, "t" refers to a point in time. When we are modeling information, time $t = 0$ is special; it represents "here and now" with the information that is available at the moment. The discrete time t refers to the time interval from $t - 1$ to t (illustrated in Figure 9.2a). This means that the first new information arrives during time interval 1.

This notational style means that any variable indexed by t, say S_t or x_t, is assumed to have access to the information that arrived up to time t, which means up through time interval t. This property will dramatically simplify our notation in the future. For example, assume that f_t is our forecast of the demand for electricity. If $\hat{D}_t$ is the observed demand during time interval t, we would write our updating equation for the forecast using

$$f_{t+1} = (1 - \alpha)f_t + \alpha\hat{D}_{t+1}. \tag{9.7}$$

We refer to this form as the *informational representation*. Note that the forecast f_{t+1} is written as a function of the information that became available during time interval $(t, t + 1)$, given by the demand $\hat{D}_{t+1}$.

When we are modeling a physical process, it is more natural to adopt a different convention (illustrated in Figure 9.2b): discrete time t refers to the time interval between t and $t + 1$. This convention arises because it is most natural in deterministic models to use time to represent when something is happening or when a resource can be used. For example, let R_t be our cash on hand that we can use during day t (implicitly, this means that we are measuring it at the beginning of the day). Let $\hat{D}_t$ be the demand for cash during the day, and let x_t represent additional cash that we have decided to add to our balance (to be used during day t). We can model our cash on hand using

$$R_{t+1} = R_t + x_t - \hat{D}_t. \tag{9.8}$$

We refer to this form as the *physical representation*. Note that the left-hand side is indexed by $t + 1$, while all the quantities on the right-hand side are indexed by t.

Throughout this book, we are going to use the informational representation as indicated in equation (9.7). We first saw this in our presentation of stochastic gradients in chapter 5, when we wrote the updates from a stochastic gradient using

$$x^{n+1} = x^n + \alpha_n \nabla_x F(x^n, W^{n+1}),$$

where here we are using iteration n instead of time t.

9.4   The States of Our System

The most important quantity in any sequential decision process is the state variable. This is the set of variables that captures everything that we know, and need to know, to model our system. Without question this is the most subtle, and poorly understood, dimension of modeling sequential decision problems.

9.4.1   Defining the State Variable

Surprisingly, other presentations of dynamic programming spend little time defining a state variable. Bellman's seminal text [Bellman (1957), p. 81] says "... we have a physical system characterized at any stage by a small set of parameters, the *state variables*." In a much more modern treatment, Puterman first introduces a state variable by saying [Puterman (2005), p. 18] "At each decision epoch, the system occupies a *state*." In both cases, the italics are in the original manuscript, indicating that the term "state" is being introduced. In effect, both authors are saying that given a system, the state variable will be apparent from the context.

Interestingly, different communities appear to interpret state variables in slightly different ways. We adopt an interpretation that is fairly common in the control theory community, which effectively models the state variable S_t as all the information needed to model the system from time t onward. We agree with this definition, but it does not provide much guidance in terms of actually translating real applications into a formal model. We suggest the following definitions:

Definition 9.4.1. *A* **state variable** *is:*

(a) **Policy-dependent version** A function of history that, combined with the exogenous information (and a policy), is necessary and sufficient to compute the cost/contribution function, the decision function (the policy), and any information required by the transition function to model the information needed for the cost/contribution and decision functions.

(b) **Optimization version** A function of history that is necessary and sufficient to compute the cost/contribution function, the constraints, and any information required by the transition function to model the information needed for the cost/contribution function and the constraints.

Some remarks are in order:

(i) The policy-dependent definition defines the state variable in terms of the information needed to compute the core model information (cost/contribution function, and the policy (or decision function)), and any other information needed to model the evolution of the core information over time (that is, the transition function). Note that constraints (at a point in time t) are assumed to be captured by the policy. Since the policy can be any function, it could potentially be a function that includes information that does not seem relevant to the problem, and which would never be used in an optimal policy. For example, a policy that says "turn left if the sun is shining" with an objective to minimize travel time would put whether or not the sun is shining in the state variable, although this does not contribute to minimizing travel times.

(ii) The optimization version defines a state variable in terms of the information needed to compute the core model information (costs/contributions and constraints), and any other information needed to model the evolution of the core information over time (their transition function). This definition limits the state variable to information needed by the optimization problem, and cannot include information that is irrelevant to the core model.

(iii) Both definitions include any information that might be needed to compute the evolution of core model information, as well as information needed to model the evolution of this information over time. This includes information needed to represent the stochastic behavior, which includes distributional information needed to compute or approximate expectations. In section 9.9.4, we present an example of how rolling forecasts enter the state variable because they are needed in the transition function.

(iv) Both definitions imply that the state variable includes the information needed to compute the transition function for core model information. For example, if we model a price process using

$$p_{t+1} = \theta_0 p_t + \theta_1 p_{t-1} + \theta_2 p_{t-2} + \varepsilon^p_{t+1}, \tag{9.9}$$

then the state variable for this price process would be $S_t = (p_t, p_{t-1}, p_{t-2})$. At time t, the prices p_{t-1} and p_{t-2} are not needed to compute the cost/-contribution function or constraints, but they are needed to model the evolution of p_t, which is part of the cost/contribution function.

(v) The qualifier "necessary and sufficient" is intended to eliminate irrelevant information. For example, with our lagged price model shown earlier, we need p_t, p_{t-1} and p_{t-2} but not p_{t-3}, p_{t-4}. A similar term used in the statistics literature is "sufficient statistic," which means it contains all the information needed for any future calculations.

(vi) A byproduct of our definitions is the observation that *all* properly modeled dynamic systems are Markovian, by construction. It is surprisingly common for people to make a distinction between "Markovian" and "history-dependent" processes. For example, if our price process evolves according to equation (9.9), many would call this a history-dependent process, but consider what happens when we define

$$\bar{p}_t = \begin{pmatrix} p_t \\ p_{t-1} \\ p_{t-2} \end{pmatrix}$$

and let

$$\bar{\theta}_t = \begin{pmatrix} \theta_0 \\ \theta_1 \\ \theta_2 \end{pmatrix}$$

which means we can write

$$p_{t+1} = \bar{\theta}^T \bar{p}_t + \varepsilon_{t+1}. \tag{9.10}$$

Here we see that $\bar{p}_t$ is a vector known at time t (who cares when the information first became known?). We would say that equation (9.10) describes a Markov process with state $S_t = (p_t, p_{t-1}, p_{t-2})$.

(vii) There is an issue of missing information and/or incorrect models. For example, we may assume that our price process evolves according to the model in equation (9.9), but this is really just an approximation of a much more complex process that is not known to us. As a simple illustration, assume that the true model is given by

$$\begin{aligned} p_{t+1} &= \theta_0 p_t + \theta_1 p_{t-1} + \theta_2 p_{t-1}^2 + \theta_3 p_{t-2} + \theta_4 p_{t-2}^2 \\ &+ \theta_5 p_{t-1} p_{t-2} + \varepsilon_{t+1}^p. \end{aligned} \tag{9.11}$$

We use equation (9.9) because it is simpler. Even if we tried equation (9.11), the noise in the data may lead us to conclude that θ_2, θ_4 and θ_5 are statistically indistinguishable from zero. If we had enough data, we might realize that the model (9.9) violates the assumptions that the error term ε_t is independent across time with the same distribution. If we knew that (9.11) was the true model (perhaps because we coded it into a simulator we are trying to optimize), we might say that the model in equation (9.9) is non-Markovian. For this issue, we turn to the famous quote by G.E.P. Box who noted: "All models are wrong, and some are useful," which is a way of saying there are errors in all models. The model in equation (9.9) is Markovian because we *assume* it to be Markovian.

There will be problems where we know that we do not know a parameter or quantity, but in these cases, the solution is to introduce a belief about these values. This belief is added to the state variable, which then produces a Markov model. If someone claims that a model is non-Markovian, then either it is missing known information that should be added, or we should add beliefs about unknown parameters and quantities.

These definitions provide a very quick test of the validity of a state variable. If there is a piece of data in either the decision function (policy), the transition function, or the contribution function which is not in the state variable, then we do not have a complete state variable. Similarly, if there is information in the state variable that is never needed in any of these three functions, then we can drop it and still have a valid state variable.

We use the term "necessary and sufficient" so that our state variable is as compact as possible. For example, we could argue that we need the entire history of events up to time t to model future dynamics, but in practice, this is rarely the case. As we start doing computational work, we are going to want S_t to be as compact as possible. Furthermore, there are many problems where we simply do not need to know the entire history. It might be enough to know the status of all our resources at time t (the resource variable R_t). But there are examples where this is not enough.

Assume, for example, that we need to use our history to forecast the price of a stock. Our history of prices is given by $(\hat{p}_1, \hat{p}_2, \ldots, \hat{p}_t)$. If we use a simple exponential smoothing model, our estimate of the mean price $\bar{p}_t$ can be computed using

$$\bar{p}_t = (1 - \alpha)\bar{p}_{t-1} + \alpha\hat{p}_t,$$

where α is a stepsize satisfying $0 \leq \alpha \leq 1$. With this forecasting mechanism, we do not need to retain the history of prices, but rather only the latest estimate $\bar{p}_t$. As a result, $\bar{p}_t$ is called a *sufficient statistic*, which is a statistic that captures all relevant information needed to compute any additional statistics from new information. A state variable, according to our definition, is always a sufficient statistic.

Consider what happens when we switch from exponential smoothing to an N-period moving average. Our forecast of future prices is now given by

$$\bar{p}_t = \frac{1}{N} \sum_{\tau=0}^{N-1} \hat{p}_{t-\tau}.$$

Now, we have to retain the N-period rolling set of prices $(\hat{p}_t, \hat{p}_{t-1}, \ldots, \hat{p}_{t-N+1})$ in order to compute the price estimate in the next time period. With exponential smoothing, we could write

$$S_t = \bar{p}_t.$$

If we use the moving average, our state variable would be

$$S_t = (\hat{p}_t, \hat{p}_{t-1}, \dots, \hat{p}_{t-N+1}). \tag{9.12}$$

We discuss latent variables (state variables that we *choose* to approximate as deterministic, but which really are changing stochastically over time), and unobservable state variables (which are also changing stochastically, but which we cannot observe).

9.4.2 The Three States of Our System

To set up our discussion, assume that we are interested in solving a relatively complex resource management problem, one that involves multiple (possibly many) different types of resources which can be modified in various ways (changing their attributes). For such a problem, it is necessary to work with three types of state variables:

The physical state R_t – This is a snapshot of the status of the physical resources we are managing and their attributes. This might include the amount of water in a reservoir, the price of a stock or the location of a sensor on a network. It could also refer to the location and speed of a robot.

The information state I_t – This encompasses any other information we need to make a decision, compute the transition or compute the objective function. We can think of I_t as information about quantities and parameters that we know perfectly, but which do not seem to belong in the physical state R_t which typically captures resources we are managing.

The belief (or knowledge) state B_t – The belief state is information specifying a probability distribution describing an unknown quantity or parameter. The type of distribution (e.g. binomial, normal, or exponential) is typically specified in the initial state S_0, although there are exceptions to this. The belief state B_t is information just like R_t and I_t, except that it is information specifying a probability distribution (such as the mean and variance of a normal distribution), or the statistics characterizing a frequentist model (see sections 3.3 and 3.4).

We then pull these together to create our state variable

$$S_t = (R_t, I_t, B_t).$$

Mathematically, the information state I_t should include information about resources R_t, since R_t is, after all, a form of information. The distinction between I_t (such as wind speed, temperature or the stock market), and R_t (how much energy is in the battery, water in a reservoir or money invested in the

The state variable

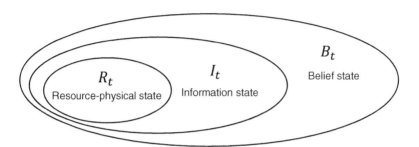

Figure 9.3 Illustration of the growing sets of state variables, where information state includes physical state variables, while the belief state includes everything.

stock market) is not important. We separate the variables simply because there are so many problems that involve managing physical or financial resources, and it is often the case that decisions impact only the physical resources. At the same time, B_t includes probabilistic information about parameters that we do not know perfectly. Knowing a parameter perfectly, as is the case with R_t and I_t, is just a special case of a probability distribution.

A proper representation of the relationship between B_t, I_t and R_t is illustrated in Figure 9.3. However, we find it more useful to make a distinction (even if it is subjective) of what constitutes a variable that describes part of the physical state R_t, and then let I_t be all remaining variables that describe quantities that are known perfectly. Then, we let B_t consist entirely of probability distributions that describe parameters that we do not know perfectly.

State variables take on different flavors depending on the mixture of physical, informational and knowledge states, as well as the relationship between the state of the system now, and the states in the past.

- Physical state – There are three important variations that involve a physical state:
  - Pure physical state – There are many problems which involve only a physical state which is typically some sort of resource being managed. There are problems where R_t is a vector, a low-dimensional vector (as in $R_t = (R_{ti})_{i \in \mathcal{I}}$ where i might be a blood type, or a type of piece of equipment), or a high-dimensional vector (as in $R_t = (R_{ta})_{a \in \mathcal{A}}$ where a is a multidimensional attribute vector).
  - Physical state with information – We may be managing the water in a reservoir (captured by R_t) given temperature and wind speed (which affects evaporation) captured by I_t.

- Physical state, information state, and belief state – We need the cash on hand in a mutual fund, R_t, information I_t about interest rates, and a probability model B_t describing, say, our belief about whether the stock market is going up or down.

- Information state – In most applications information evolves exogenously, although there are exceptions. The evolution of information comes in several flavors:

  - Memoryless – The information I_{t+1} does not depend on I_t. For example, we may feel that the characteristics of a patient arriving at time $t + 1$ to a doctor's office is independent of the patient arriving at time t. We may also believe that rainfall in month $t+1$ is independent of the rainfall in month t.

  - First-order Markov – Here we assume that I_{t+1} depends on I_t. For example, we may feel that the spot market price of oil, the wind speed, or temperature and humidity at $t+1$ depend on the value at time t. We might also insist that a decision x_{t+1} not deviate more than a certain amount from the decision x_t at time t.

  - Higher-order Markov – We may feel that the price of a stock p_{t+1} depends on p_t, p_{t-1}, and p_{t-2}. However, we can create a variable $\bar{p}_t = (p_t, p_{t-1}, p_{t-2})$ and convert such a system to a first-order Markov system, so we really only have to deal with memoryless and first-order Markov systems.

  - Full history dependent – This arises when the evolution of the information I_{t+1} depends on the full history, as might happen when modeling the progress of currency prices or the progression of a disease. This type of model is typically used when we are not comfortable with a compact state variable (and there are methods designed to handle these problems – see section 19.9).

- Belief state – Belief states capture beliefs we have about uncertain quantities or parameters that are (typically) evolving over time, often as a direct or indirect results of a decision. Uncertainty in the belief state can arise in three ways:

  - Uncertainty about a static parameter – For example, we may not know the impact of price on demand, or the sales of a laptop with specific features. The nature of the unknown parameter depends on the type of belief model: the features of the laptop correspond to a lookup table, while the demand-price tradeoff represents the parameter of a parametric model. These problems are broadly known under the umbrella of optimal learning, but are often associated with the literature on multiarmed bandit problems.

  - Uncertainty about a dynamic (uncontrollable) parameter – The sales of a laptop with a specific set of features may change over time. This may occur because of unobservable variables. For example, the demand elasticity of a product (such as housing) may depend on other market characteristics (such as the growth of industry in the area).

– Uncertainty about a dynamic, controllable parameter – Imagine that we control the inventory of a product that we cannot observe perfectly. We may control purchases that replenish inventory which is then used to complete sales, but our ability to track sales is imperfect, giving us an imprecise estimate of the inventory. These problems are typically referred to as partially observable Markov decision processes (POMDPs).

There has been a tendency in the literature to treat the belief state as if it were somehow different than "the" state variable. It is not. The state variable is all the information that describes the system at time t, whether that information is the amount of inventory, the location of a vehicle, the current weather or interest rates, or the parameters of a distribution describing some unknown quantity. If the decision maker only has a belief about an uncertain parameter, then for that decision problem, the belief is very much a part of the state variable.

We believe we resolve this unique point of confusion in chapter 20 (Multiagent modeling and learning) by offering a two-agent model (the environment and the controlling agent), which means there are two state variables: one for the environment, and one for the controlling agent. When making a decision, the controlling agent only has access to what is in their state variable, and if this is a belief about an uncertain quantity, then we work with this, just as we did in chapter 7 (think of the interval estimation policy).

We can use S_t to be the state of a single resource (if this is all we are managing), or let $S_t = R_t$ be the state of all the resources we are managing. There are many problems where the state of the system consists only of R_t. We suggest using S_t as a generic state variable when it is not important to be specific, but it must be used when we may wish to include other forms of information. For example, we might be managing resources (consumer products, equipment, people) to serve customer demands $\hat{D}_t$ that become known at time t. If R_t describes the state of the resources we are managing, our state variable would consist of $S_t = (R_t, \hat{D}_t)$, where $\hat{D}_t$ represents additional information we need to solve the problem.

9.4.3 Initial State S_0 vs. Subsequent States S_t, $t > 0$

It is important to distinguish between the initial state S_0 and subsequent states S_t, $t > 0$, as we explain:

The Initial State S_0

The initial state plays a special role in the modeling of a sequential decision problem. It stores any data that is an input to the system, which may include:

- Any deterministic parameters – This might include the deterministic data describing a graph (for example), or any problem parameters that never change.
- Initial values of parameters that evolve over time – For example, this could be the initial inventory, the starting location of a robot, or the initial speed of wind at a wind farm.
- The distribution of belief about uncertain parameters – This is known as the *prior* distribution of belief about anything that is not known perfectly. We emphasize that this prior can be a Bayesian prior, or the initial statistics of a frequentist model.

The Subsequent States S_t, $t > 0$

By convention, the dynamic state S_t (for $t > 0$) only contains the information that changes over time. Thus, if we were solving a shortest path problem over a deterministic graph, S_t would tell us the node which we currently occupy, but would not include, for example, the deterministic data describing the graph which is not changing (by assumption) as we move over the graph. Similarly, it would not include any deterministic parameters such as the maximum speed of our vehicle.

As our system evolves, we drop any deterministic parameters that do not change. These become *latent* (or hidden) variables, since our problem depends on them, but we drop them from S_t for $t > 0$. However, it is important to recognize that these values may change each time we solve an instance of the problem. Examples of these random starting states include:

■ **EXAMPLE 9.1**

We wish to optimize the management of a fleet of trucks. We fix the number of trucks in our fleet, but this is a parameter that we specify, and we may change the fleet size from one instance of the problem to another.

■ **EXAMPLE 9.2**

We wish to optimize the amount of energy to store in a battery given a forecast of clouds over a 24-hour planning horizon. Let $f_{0t'}$ is the forecast of energy at time t' which is given to us at time 0, the vector of forecasts $f_0 = (f_{0t'})_{t'=0}^{24}$ (which does not evolve over time) is part of the initial state. However, each time we optimize our problem, we are given a new forecast.

■ **EXAMPLE 9.3**

We are designing an optimal policy for finding the best medication for type II diabetes, but the policy depends on the attributes of the patient (age, weight, gender, ethnicity, and medical history), which do not change over the course of the treatment.

9.4.4 Lagged State Variables*

There are a number of settings where our state variable is actually telling us information about the future. The simplest example arises in resource allocation problems, where resources (trucks/trains/planes enroute to a destination, inbound inventory, people undergoing training) are known now, but will not be available to be used until some point in the future. We would capture this using

$$R_{tt'} = \text{The resources on hand at time } t \text{ that cannot be used until time } t',$$

$$R_t = (R_{tt'})_{t' \geq t}.$$

Another example would be customer orders being made at time t to be served in the future. For example, we might have

$$D_{tt'} = \text{The number of reservations to fly on an airplane at time } t' \text{ that we know about at time } t,$$

$$D_t = (D_{tt'})_{t' \geq t}.$$

Both R_t and D_t would be considered part of our state S_t.

9.4.5 The Post-decision State Variable*

Our standard strategy is to model the state variable S_t as all the information we need to make a decision (as well as computing costs, constraints and the transition function). This allows us to write the sequence of state, decision, information as

$$(S_0, x_0, W_1, S_1, x_1, W_2, S_2, x_2, \ldots, x_{t-1}, W_t, S_t). \tag{9.13}$$

Since the state S_t is what we know just before we make a decision, we might also refer to it as the *pre-decision state*. There are settings where we will find it useful to model the state immediately after we make a decision. We model this as S_t^x to indicate that it is still being observed at time t, but immediately after we make the decision x (hence the superscript). We refer to S_t^x as the *post-decision state*. Our information sequence (9.13) becomes

$$(S_0, x_0, S_0^x, W_1, S_1, x_1, S_1^x, W_2, S_2, x_2, S_2^x, \ldots, x_{t-1}, S_{t-1}^x, W_t, S_t). \qquad (9.14)$$

Since there is no new exogenous information between making the decision x_t and the observation of the post-decision state S_t^x, the post-decision state is a deterministic function of the pre-decision state S_t and x_t.

The examples given provide some illustrations of pre- and post-decision states.

■ **EXAMPLE 9.4**

A traveler is driving through a network, where the travel time on each link of the network is random. As she arrives at node i, she is allowed to see the travel times on each of the links out of node i, which we represent by $\hat{\tau}_i = (\hat{\tau}_{ij})_j$. As she arrives at node i, her pre-decision state is $S_t = (i, \hat{\tau}_i)$. Assume she decides to move from i to k. Her post-decision state is $S_t^x = (k)$. Note that she is still at node i; the post-decision state captures the fact that she will next be at node k, and we no longer have to include the travel times on the links out of node i.

■ **EXAMPLE 9.5**

The nomadic trucker revisited. Let $R_{ta} = 1$ if the trucker has attribute vector a at time t and 0 otherwise. Now let D_{tb} be the number of customer demands (loads of freight) of type b available to be moved at time t. The pre-decision state variable for the trucker is $S_t = (R_t, D_t)$, which tells us the state of the trucker and the demands available to be moved. Assume that once the trucker makes a decision, all the unserved demands in D_t are lost, and new demands become available at time $t + 1$. The post-decision state variable is given by $S_t^x = R_t^x$ where $R_{ta}^x = 1$ if the trucker has attribute vector r after a decision has been made.

■ **EXAMPLE 9.6**

Imagine playing backgammon where R_{ti} is the number of your pieces on the ith "point" on the backgammon board (there are 24 points on a board). The transition from S_t to S_{t+1} depends on the player's decision x_t, the play of the opposing player, and the next roll of the dice. The post-decision state variable is simply the state of the board after a player moves but before his opponent has moved.

The post-decision state can be particularly valuable in the context of dynamic programming, which we are going to address in depth in chapters 16 and 17. There are three ways of finding a post-decision state variable:

Decomposing Decisions and Information

There are many problems where we can create functions $S^{M,x}(\cdot)$ and $S^{M,W}(\cdot)$ from which we can compute

$$S_t^x = S^{M,x}(S_t, x_t), \tag{9.15}$$

$$S_{t+1} = S^{M,W}(S_t^x, W_{t+1}). \tag{9.16}$$

The structure of these functions is highly problem-dependent. However, there are sometimes significant computational benefits, primarily when we face the problem of making a decision when we are in state S_t, and would like to know the value of the state the decision takes us to. The post-decision state is a deterministic function of the pre-decision state S_t and the decision x_t, which can be computationally very convenient (see chapters 15 and 16).

State-decision Pairs

A very generic way of representing a post-decision state is to simply write

$$S_t^x = (S_t, x_t).$$

Figure 9.4 provides a nice illustration using our tic-tac-toe example. Figure 9.4a shows a tic-tac-toe board just before player O makes his move. Figure 9.4b shows the augmented state-decision pair, where the decision (O decides to place his move in the upper right hand corner) is distinct from the state. Finally, Figure 9.4c shows the post-decision state. For this example, the pre- and post-decision state spaces are the same, while the augmented state-decision pair is nine times larger.

The augmented state (S_t, x_t) is closely related to the post-decision state S_t^x (not surprising, since we can compute S_t^x deterministically from S_t and x_t). But computationally, the difference is significant. If $\mathcal{S}$ is the set of possible values of S_t, and $\mathcal{X}$ is the set of possible values of x_t, then our augmented state space has size $|\mathcal{S}| \times |\mathcal{X}|$, which is obviously much larger (especially if x is a vector!).

The augmented state variable is used in a popular class of algorithms known as Q-learning (which we first introduced in chapter 2), where the challenge is to statistically estimate Q-factors which give the value of being in state S_t *and*

Pre-decision
S_t

State-decision
(S_t, x_t)

Post-decision
(S_t^x)

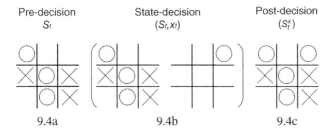

9.4a   9.4b   9.4c

Figure 9.4 Pre-decision state, augmented state-decision, and post-decision state for tic-tac-toe.

taking decision x_t. The Q-factors are written $Q(S_t, x_t)$, in contrast with value functions $V_t(S_t)$ which provide the value of being in a state. This allows us to directly find the best decision by solving $\min_x Q(S_t, x_t)$. This is the essence of Q-learning, but the price of this algorithmic step is that we have to estimate $Q(S_t, x_t)$ for each S_t and x_t. It is not possible to determine x_t by optimizing a function of S_t^x alone, since we generally cannot determine which decision x_t brought us to S_t^x.

The Post-decision as a Point Estimate

Assume that we have a problem where we can compute a point estimate of future information. Let $\overline{W}_{t,t+1}$ be a point estimate, computed at time t, of the outcome of W_{t+1}. If W_{t+1} is a numerical quantity, we might use $\overline{W}_{t,t+1} = \mathbb{E}(W_{t+1}|S_t)$ or $\overline{W}_{t,t+1} = 0$.

If we can create a reasonable estimate $\overline{W}_{t,t+1}$, we can compute post- and pre-decision state variables using

$$S_t^x = S^M(S_t, x_t, \overline{W}_{t,t+1}),$$
$$S_{t+1} = S^M(S_t, x_t, W_{t+1}).$$

Measured this way, we can think of S_t^x as a point estimate of S_{t+1}, but this does not mean that S_t^x is necessarily an approximation of the expected value of S_{t+1}.

9.4.6 A Shortest Path Illustration

We are going to use a simple shortest-path problem to illustrate the process of defining a state variable. We start with a deterministic graph shown in Figure 9.5, where we are interested in finding the best path from node 1 to node 11. Let t be the number of links we have traversed, and let N_t be the node number were we are located after $t = 2$ transitions. What state are we in?

Most people answer this with

$$S_t = N_t = 6.$$

This answer hints at two conventions that we use when defining a state variable. First, we exclude any information that is not changing, which in this case is any information about our deterministic graph. It also excludes the prior nodes in our path (1 and 3) since these are not needed for any future decisions.

Now assume that the travel times are random, but where we know the probability distribution of travel times over each link (and these distributions are not changing over time). This graph is depicted in Figure 9.6. We are going to assume, however, that when a traveler arrives at node i, she is able to see the actual cost $\hat{c}_{ij}$ for the link (i, j) out of node i (if this is the link that is chosen now). Now, what is our state variable?

Obviously, we still need to know our current node $N_t = 6$. However, the revealed link costs also matter. If the cost of moving from node 6 to node 9 changes from 9.7 to 2.3 or 18.4, our decision may change. This means that these costs are very much a part of our state of information. Thus, we would write our state as

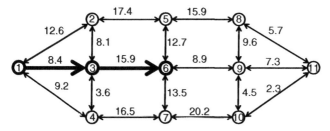

Figure 9.5 A deterministic network for a traveler moving from node 1 to node 11 with known arc costs.

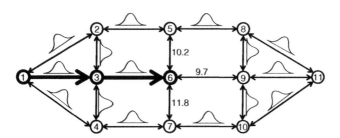

Figure 9.6 A stochastic network, where arc costs are revealed as the traveler arrives to a node.

$$S_t = (\underbrace{N_t}_{R_t}, \underbrace{(\hat{c}_{N_t,\cdot})}_{I_t}) = (\underbrace{6}_{R_t}, \underbrace{(10.2, 9.7, 11.8)}_{I_t}),$$

where $(\hat{c}_{N_t,\cdot})$ represents the costs on all the links out of node N_t. Thus, we see an illustration of both a physical state $R_t = N_t$, and information $I_t = (10.2, 9.7, 11.8)$.

For our last example, we introduce the problem of left-hand turn penalties. If our turn from node 6 to node 5 is a left hand turn, we are going to add a penalty of .7 minutes. Now what is our state variable?

The left-hand turn penalty requires that we know if the move from 6 to 5 is a left hand turn. This calculation requires knowing where we are coming from. Thus, we now need to include our previous node, N_{t-1} in our state variable, giving us

$$S_t = (\underbrace{N_t}_{R_t}, \underbrace{(\hat{c}_{N_t,\cdot}), N_{t-1}}_{I_t}) = (\underbrace{6}_{R_t}, \underbrace{(10.2, 9.7, 11.8), 3}_{I_t}).$$

Now, N_t is our physical state, but N_{t-1} is a piece of information required to compute the cost function.

9.4.7 Belief States*

There are many applications where we are not able to observe (or measure) the state of the system precisely. Instead, we will maintain a probabilistic belief about the unknown parameter or quantity. Some examples include:

■ **EXAMPLE 9.7**

A patient may have cancer in the colon which might be indicated by the presence of polyps (small growths in the colon). The number of polyps is not directly observable. There are different methods for testing for the presence of polyps that allow us to infer how many there may be, but these are imperfect.

■ **EXAMPLE 9.8**

The military has to make decisions about sending out aircraft to remove important military targets that may have been damaged in previous raids. These decisions typically have to be made without knowing the precise state of the targets.

■ **EXAMPLE 9.9**

Policy makers have to decide how much to reduce CO_2 emissions, and would like to plan a policy over 200 years that strikes a balance between costs and the rise in global temperatures. Scientists cannot measure temperatures perfectly (in large part because of natural variations), and the impact of CO_2 on temperature is unknown and not directly observable.

For each of these examples, we have a system with quantities or parameters that cannot be observed, along with variables that can be observed. When this happens, we handle these values using our belief state B_t.

There is an extensive literature on what are known as "partially observable Markov decision processes" (or "POMDPs") which are sequential decision problems with quantities or parameters that are not known perfectly. The POMDP literature is both mathematically sophisticated as well as computationally challenging. In other words, once you figure out the math, you do not end up with tools that can solve real problems.

Our belief is that the POMDP literature is not modeling these problems correctly. We believe they should be modeled as multiagent systems where there is an "environment agent" and a "controlling agent" that cannot observe the environment perfectly. When models of each agent are formulated with our modeling framework, problems with belief states become much more practical. We defer this discussion to our chapter on multiagent problems in chapter 20.

9.4.8 Latent Variables*

One of the more subtle dimensions of any dynamic model is the presence of information that is not explicitly captured in the state variable S_t. Remember that we do not model in the dynamic state S_t any information in S_0 that does not change over time. We may have many static parameters in S_0. While these are used in the model, they are not in S_t. An optimal policy will depend on this information, but the dependence is not explicit.

Here are some examples of observable latent variables:

■ **EXAMPLE 9.10**

We are solving a shortest path from origin node r to destination node s, and let $i, j \in \mathcal{J}$ be intermediate nodes. If we are currently at node i, we would model our "state" as being at node i. We use a shortest path algorithm to find the best path from each $i \in \mathcal{I}$, which would give us the "value" $V(i)$ (really the cost) of being at each node i and following the optimal path to node s. The destination s is actually a latent variable, since it is not captured in our state. If we included s in our state variable, we would have to compute the optimal value $V(i, s)$ for every combination of i and s, which is much harder.

■ **EXAMPLE 9.11**

A company is optimizing inventories at different distribution centers. Each DC optimizes its own inventory given the demands that it sees. However, since orders can be satisfied from any DC (not always the closest), the optimal inventory at any single DC depends on the inventories at the other DCs, which become latent variables in the planning of each DC.

■ **EXAMPLE 9.12**

A drug company is determining the optimal dosage given the weight of a patient. The right dosage also depends on age, as well as medical conditions such as diabetes. A dosage table that depends on weight, age, and blood sugar would be too complicated, but a patient's physician would need to consider these variables. This means that the physician needs to optimize the dosage for each patient, starting with the dosage table from the drug company which ignores variables other than weight.

When deciding whether to model a variable explicitly (which means modeling how it evolves over time) or as a latent variable (which means holding it constant) introduces an important tradeoff: including a dynamically varying parameter in the state variable produces a more complex, higher dimensional state variable, but one which does not have to be reoptimized when the parameter changes. By contrast, treating a parameter as a latent variable simplifies the model, but requires that the model be reoptimized when the parameter changes.

9.4.9 Rolling Forecasts*

An important dimension of any dynamic model is the probability distribution of random activities that will happen in the future. For example, imagine that we are planning the commitment of energy resources given a forecast $f_{tt'}^W$ made at time t of the energy $W_{t'}$ that will be generated from wind at some time t' in the future. We might now assume that the wind in the future is given by

$$W_{t'} = f^W_{tt'} + \varepsilon_{tt'},$$

where $\varepsilon_{tt'} \sim N(0, (t' - t)\sigma^2)$.

For this simple model, where the variance of the error depends only on how far into the future we are projecting, the forecast determines the probability distribution of the energy from wind. In the vast majority of models, we treat the forecast $(f^W_{tt'})^{t+H}_{t'=t}$ as fixed. This means that our forecast is a form of latent variable. In practice, the forecast evolves over time as new information arrives. We might model this evolution using

$$f^W_{t+1,t'} = f^W_{tt'} + \hat{f}^W_{t+1,t'}, \tag{9.17}$$

where $\hat{f}^W_{t+1,t'} \sim N(0, \sigma^2_W)$ represents the exogenous change in the forecast for time t'. Equation (9.17) is known as a *martingale model of forecast evolution*, or MMFE, in the inventory literature, which means that the expected value of the forecast in the future, $f^W_{t'',t'}$, is equal to the forecast $f^W_{t,t'}$ now. [A "martingale" is a stochastic process that might evolve up or down from one time period to the next, but on average stays the same.] This means that if we are given $f^W_{tt'}, t' = t, \dots, T$, then this vector of forecasts is a part of the state variable, since all of this information is needed to model the evolution of W_t.

For many problems, however, forecasts are not modeled as a dynamically evolving stochastic process; instead, they are viewed as static, which means that they are not part of the state variable. In this case they would be *latent variables*, and are not explicitly being modeled. As we see in chapter 19 on direct lookahead models, holding forecasts constant is a common approximation in direct lookahead models. This is what navigation systems are doing when planning a path over the network. At a point in time, the system fixes the estimate of the travel time over each link and plans a path. A few minutes later, the times are updated and the path is recalculated, but the logic for finding the path does not explicitly model the possible changes in the estimates of link times.

Classical dynamic programming models seem to almost universally ignore the role of forecasts in the modeling of a dynamic optimization problem, which means that they are being treated as latent variables. This in turn means that the problem has to be re-optimized from scratch when the forecasts are updated. By contrast, forecasts are easily handled in lookahead policies, as we see later. We describe methods for handling forecasts in chapter 19; for now, we just want to show how the presence of rolling forecasts can affect the state variable.

9.4.10 Flat vs. Factored State Representations*

It is very common in the dynamic programming literature to define a discrete set of states $S = (1, 2, \dots, |S|)$, where $s \in S$ indexes a particular state. For example, consider an inventory problem where S_t is the number of items we have in inventory (where S_t is a scalar). Here, our state space $\mathcal{S}$ is the set of integers, and $s \in \mathcal{S}$ tells us how many products are in inventory.

Now assume that we are managing a set of K product types. The state of our system might be given by $S_t = (S_{t1}, S_{t2}, \ldots, S_{tk}, \ldots)$ where S_{tk} is the number of items of type k in inventory at time t. Assume that $S_{tk} \leq M$. Our state space $\mathcal{S}$ would consist of all possible values of S_t, which could be as large as K^M. A state $s \in \mathcal{S}$ corresponds to a particular vector of quantities $(S_{tk})_{k=1}^K$.

Modeling each state with a single scalar index is known as a flat or unstructured representation. Such a representation is simple and elegant, and produces mathematically compact models that have been popular in communities like operations research and computer science. We first saw this used in section 2.1.3, and we will return to this in chapter 14 in much more depth. However, the use of a single index completely disguises the structure of the state variable, and often produces intractably large state spaces.

In the design of algorithms, it is often essential that we exploit the structure of a state variable. For this reason, we generally find it necessary to use what is known as a *factored* representation, where each factor represents a *feature* of the state variable. For example, in our inventory example we have K factors (or features). It is possible to build approximations that exploit the structure that each dimension of the state variable is a particular quantity.

In section 8.3.1, we solved a problem of managing resources (people, equipment) which were described by an attribute vector $a \in \mathcal{A}$, where we assumed that the attribute space $\mathcal{A}$ was discrete. This is an example of a flat representation. Each element a_i of an attribute vector represents a particular feature of the entity. This representation allowed us to model the resource vector as $R_t = (R_{ta})_{a \in \mathcal{A}}$, where R_t is now a vector with element $a \in \mathcal{A}$.

9.4.11 A Programmer's Perspective of State Variables

State variables are easily one of the least understood concepts in dynamic modeling, as evidenced by the large number of books on dynamic programs that do not even define a state variable (at least not properly). For a different perspective, imagine that you are programming a simulator of a dynamic system where decisions are made over time. You are going to create a set of variables to model your system (for many problems, this can be a lot of variables). We can divide the variables into four broad categories:

Category 1 – These are all the variables that are set initially (either hard coded into the program, or read in from an external data source). We can divide these into several subcategories:

(1a) – Fixed parameters (such as the boiling point of water, or the maximum speed of a vehicle) that never change.

(1b) – Initial values of variables that evolve over time (whether due to decisions and/or exogenous inputs).

(1c) – Initial beliefs about parameters and quantities that are not known perfectly. These beliefs may or may not evolve over the course of the simulation.

Category 2 – Variables that change over the course of the simulation, either due to decisions or exogenous inputs (and we exclude decisions and exogenous information which we put in categories 3 and 4). These can include:

(2a) – Variables that describe quantities and parameters that are known perfectly.

(2b) – Variables that describe probability distributions that evolve over time. These variables might describe the parameters of parametric distributions, probabilities, or sufficient statistics.

Category 3 – Variables that represent decisions that are determined by some policy.

Category 4 – Information that enters our system exogenously. This information may be used to make a decision and discarded, or may become included in a category 2 variable.

All the variables that fall in Category 1 are what we put in our initial state variable S_0. All the variables that fall in Category 2 are what we put in our initial state variable S_t for $t > 0$. Category 3 refers to variables that we control, known as control variables, actions or, in this book, decisions that we call x_t (see section 9.5). Finally, Category 4 refers to new information that arrives from outside our system, which we have modeled as W_{t+1}. We may use this to make a decision and then discard it. However, it may be blended into one of the variables in Category 2.

From this discussion, we see that "state variables" are what a programmer would call a "variable," although we exclude decision variables and exogenous information variables, unless these are retained for future use (in which case they become included in Category 2). Also, a programmer may retain a lot of information in "variables" for reporting purposes, whereas we restrict our definition of state variables to information that we actually need to model our system.

9.5 Modeling Decisions

There are a number of words in English that can mean "decision," as illustrated in Table 9.1. The optimization literature assumes that the types of decisions are known in advance, overlooking what can be one of the most subtle dimensions of modeling. Arguably one of the most challenging dimensions of optimization is recognizing exactly what decisions need to be optimized!

It should not be surprising that even the optimization communities use different words (and notation) to mean decision. The classical literature on Markov decision process talks about choosing an action $a \in \mathcal{A}$ (or $a \in \mathcal{A}_s$, where $\mathcal{A}_s$ is the set of actions available when we are in state s). The optimal control community chooses a control $u \in \mathcal{U}_x$ when the system is in state x. The math programming community wants to choose a decision represented by the vector x. We have also noticed that the bandit community in computer science has also adopted "x" as its notation for a decision which is typically discrete. In this book, we use "x" as our default notation, although we occasionally slip back to using action a (in particular, see chapter 14) when we are using methods where the action must be discrete.

When we model decisions in a sequential decision problem, we recommend introducing the following elements:

- The types of decisions, and notation for whether a decision is made (and if appropriate, how much).
- Constraints on decisions made at time t.

Table 9.1 Sample of words in English that represent a decision. The second column describes decisions in the context of collecting information, as in choice of experiment to run, or what to listen to or observe.

General terms	Collecting information
Action	Examine
Acquire/buy/purchase	Experiment
Choice	Listen
Control	Observe
Decision	Probe
Design	Research
Intervention (medical)	Sample
Option	Sense
Move	Test
Response	View
Task	Scan
Trade (finance)	

- The notation for a policy (or method) for making decisions, but without specifying the policy.

9.5.1 Types of Decisions

Decisions come in many forms. We illustrate this using our notation x which tends to be the notation of choice for more complex problems. Examples of different types of decisions are

- Binary, where x can be 0 or 1.
- Discrete set, where $x \in \{x_1, \dots, x_M\}$.
- Continuous scalar, where $x \in [a, b]$.
- Continuous vector, where $x \in \Re^n$.
- Integer vector, where $x \in \mathbb{Z}^n$.
- Subset selection, where x is a vector of 0's and 1's, indicating which members are in the set.
- Multidimensional categorical, where $x_a = 1$ if we make a choice described by an attribute $a = (a_1, \dots, a_K)$. For example, a could be the attributes of a drug or patient, or the features of a movie.

There are many applications where a decision is either continuous or vector-valued. For example, in chapter 8 we describe applications where a decision at time t involves the assignment of resources to tasks. Let $x = (x_d)_{d \in \mathcal{D}}$ be the vector of decisions, where $d \in \mathcal{D}$ is a *type* of decision, such as assigning resource i to task j, or purchasing a particular type of equipment. It is not hard to create problems with hundreds, thousands, and even tens of thousands of *dimensions*. These high-dimensional decision vectors arise frequently in the types of resource allocation problems addressed in operations research.

This discussion makes it clear that the complexity of the space of decisions (or actions or controls) can vary considerably across applications. There are entire communities dedicated to problems with a specific class of decisions. For example, optimal stopping problems feature binary actions (hold or sell). The entire field of Markov decision processes, as well as all the problems described in chapter 7 for derivative-free stochastic optimization, assume discrete sets. Derivative-based stochastic optimization, as well as the field of stochastic programming, assumes that x is a vector, usually continuous.

9.5.2 Initial Decision x_0 vs. Subsequent Decisions x_t, $t > 0$

Just as we distinguished between the initial state S_0 and subsequent states S_t, $t \geq 0$, it is useful to distinguish between the first decision x_0 and ongoing decisions x_t, $t > 0$:

Initial Decision x_0

The first decision x_0 is a mixture of initial design decisions that are only made once, and the first instances of ongoing control decisions. Examples of initial design decisions include:

- Location and capacity of fixed facilities.
- The configuration of a manufacturing system or network.
- The design of a robot or other machines.
- The people who are hired to staff the system.
- The initial location and quantities of resources (robots, trucks, nurses) that will be managed over the course of a simulation.
- The parameters that govern the behavior of policies.

All of these are parameters that can be viewed as design variables to be optimized. Particularly important is recognizing that the design of the policy is no different than any of the other decisions that affect the design of a system.

Subsequent Decisions x_t, $t > 0$

The decisions x_t represents the decisions that are controlling the system that are made on an ongoing basis. The array of controlling decisions is much too long to list, but we can characterize them in broad categories:

- Decisions which manage physical resources: people, robots, machinery, inventories (of any product), water, energy.
- Decisions which manage financial resources: investments, contracts.
- Decisions that affect the performance of a process: prices, speeds, temperatures.
- Information collection decisions from computer simulations, laboratory experiments, field experiments.
- Decisions to communicate or share information: ads, marketing, promotions.

Feel free to jump back to Table 1.1 in chapter 1 for a hint at the diversity of control decisions.

9.5.3 Strategic, Tactical, and Execution Decisions

It is important to recognize that there are often lags between when a decision is made (which determines its information content) and when it is implemented (which is the point at which it impacts our system). To handle lagged decision processes we define

$$x_{tt'} \quad = \quad \text{a decision made at time } t \text{ to be implemented at time } t' \geq t.$$

We now describe three classes of decisions based on the lag:

- Strategic planning – x_0 refers to all decisions made at time $t = 0$. These are our design decisions discussed earlier.
- Tactical planning – $x_{tt'}$ where $t' > t$ – These are decisions now that impact the future, which means we have to model exogenous information $W_{t+1}, \ldots, W_{t'}$, as well as the decisions $x_{t+1}, \ldots, x_{t'}$ that we make between t and t'.
- Execution – x_{tt} – These are decisions that we implement at time t.

Each of these decisions require simulating other decisions. For example

- Strategic planning – We will need to simulate decisions $x_1, x_2, \ldots, x_T$ in order to evaluate the performance of the design decisions x_0.
- Tactical planning – Here we are making a decision $x_{tt'}$ at time t to implement at time t', which means we need to simulate the decisions $x_t, x_{t+1}, \ldots, x_{t'-1}$ to anticipate the state that we will be in at time t' when we make a decision at time t.
- Execution – To help us make the decision x_{tt} that we are going to implement now (at time t), we will often need to simulate the downstream impact of this decision, which means simulating the decisions $x_{t+1}, x_{t+2}, \ldots, x_T$.

9.5.4 Constraints

When we make decisions at time t, we often have to specify constraints on the decisions. The simplest type of "constraint" is to specify a set of possible (discrete) decisions $\mathcal{D}_s$ given that we are in state s. Often the set of possible types of decisions $\mathcal{D}$ is static, but if it depends on the state (which can vary over time), we would write

$$\mathcal{D}_t \quad = \quad \text{the set of types of decisions given that we are in state } S_t \text{ at time } t. \text{ The dependence on the state } S_t \text{ is implicit through our indexing the set by time,}$$

$$x_{td} \quad = \quad \text{the number of times we execute decision } d \in \mathcal{D}_t \text{ at time } t.$$

An example can be assigning drivers to loads at time t, where $\mathcal{D}_t$ is the set of loads available at time t.

If we have a vector of decisions x_{td} for $d \in \mathcal{D}$, we may easily have constraints on the vector x_t. For example, x_{td} might be the amount we invest in stock d, but we have to limit our investments to the amount of cash R_t we have on hand, so we would write:

$$\sum_{d \in \mathcal{D}} x_{td} \leq R_t,$$

$$x_{td} \geq 0.$$

We can write constraints like this in the general format

$$A_t x_t = R_t,$$

$$x_t \geq 0.$$

Even more general is to write

$$x_t \in \mathcal{X}_t,$$

where $\mathcal{X}_t$ may be a discrete set such as $\{x_1, \dots, x_M\}$, or the solution to our system of linear equations. When we index a set (or variable) by t as in $\mathcal{X}_t$, this means it depends on information in the state S_t. We do not write it as $\mathcal{X}(S_t)$ just to keep the notation compact.

9.5.5 Introducing Policies

The challenge of any optimization problem (including stochastic optimization) is making decisions. In a sequential (stochastic) decision problem, the decision x_t depends on the information available at time t, which is captured by S_t. This means we need a decision x_t for each S_t, which means we need a function $x_t(S_t)$. This function is known as a *policy*, often designated by π. While many authors use $\pi(S_t)$ to represent the policy, we use π to carry the information that describes the function, and designate the function as $X^\pi(S_t)$. If we are using action a_t, we would designate our policy as $A^\pi(S_t)$, or $U^\pi(S_t)$ if we are finding control u_t. Policies may be stationary (as we have written them), or time-dependent, in which case we would write $X_t^\pi(S_t)$.

We introduce the notation for the policy, such as $X^\pi(S_t)$ when we introduce decisions in our model, but we do not make any effort at choosing the policy. This is at the heart of our philosophy:

Model first, then solve.

The choice of policy depends not only on the structure of the problem, but it may even depend on the nature of the data for a particular problem. In chapter 11, we are going to describe an energy storage problem (which we model in section 9.9 below) where we show that each of four classes of policies (plus a fifth hybrid) can work best depending on the specific characteristics of a dataset.

Starting in chapter 11, we are going to spend the rest of the book identifying different classes of policies that are suited to problems with different characteristics. Note that it is not an accident that we address the design of policies after we discuss modeling uncertainty in chapter 10, which we "model first, then solve."

9.6 The Exogenous Information Process

An important dimension of many of the problems that we address is the arrival of exogenous information, which changes the state of our system. Modeling the flow of exogenous information represents, along with states, the most subtle dimension of modeling a stochastic optimization problem. We sketch the basic notation for modeling exogenous information here, and defer to chapter 10 a more complete discussion of uncertainty.

We begin by noting that this section only addresses the exogenous information that arrives at times $t > 0$. This ignores the initial state S_0 which is an entirely different source of information (which technically is exogenous).

9.6.1 Basic Notation for Information Processes

Consider a problem of tracking the value of an asset. Assume the price evolves according to

$$p_{t+1} = p_t + \hat{p}_{t+1}.$$

Here, $\hat{p}_{t+1}$ is an exogenous random variable representing the change in the price during time interval $t + 1$. At time t, p_t is a number, while (at time t) p_{t+1} is random.

We might assume that $\hat{p}_{t+1}$ comes from some probability distribution such as a normal distribution with mean 0 and variance σ^2. However, rather than work with a random variable described by some probability distribution, we are going to primarily work with sample realizations. Table 9.2 shows 10 sample realizations of a price process that starts with $p_0 = 29.80$ but then evolves according to the sample realization. These samples might come from a mathematical model, or observations from history.

Following standard convention, we index each path by the Greek letter ω (in the example, ω runs from 1 to 10). At time $t = 0$, p_t and $\hat{p}_t$ is a random variable (for $t \geq 1$), while $p_t(\omega)$ and $\hat{p}_t(\omega)$ are *sample realizations*. We refer to the sequence

$$p_1(\omega), \ p_2(\omega), \ p_3(\omega), \ ..., p_T(\omega)$$

as a *sample path* for the prices p_t.

We are going to use "ω" notation throughout this volume, so it is important to understand what it means. As a rule, we will primarily index exogenous random variables such as $\hat{p}_t$ using ω, as in $\hat{p}_t(\omega)$. $\hat{p}_{t'}$ is a random variable if we are sitting at a point in time $t < t'$. $\hat{p}_t(\omega)$ is not a random variable; it is a sample realization. For example, if $\omega = 5$ and $t = 2$, then $\hat{p}_t(\omega) = -0.73$. We are going to create randomness by choosing ω at random. To make this more specific, we need to define

$$\Omega \ = \ \text{the set of all possible sample realizations (with } \omega \in \Omega),$$

$$p(\omega) \ = \ \text{the probability that outcome } \omega \text{ will occur.}$$

A word of caution is needed here. We will often work with continuous random variables, in which case we have to think of ω as being continuous. In this case, we cannot say $p(\omega)$ is the "probability of outcome ω." However, in all of our work, we will use discrete samples. For this purpose, we can define

$$\hat{\Omega} \ = \ \text{a set of discrete sample observations of } \omega \in \Omega.$$

Table 9.2 A set of sample realizations of prices (p_t) and the changes in prices ($\hat{p}_t$).

Sample path	$t = 0$	$t = 1$		$t = 2$		$t = 3$	
ω	p_0	$\hat{p}_1$	p_1	$\hat{p}_2$	p_2	$\hat{p}_3$	p_3
1	29.80	2.44	32.24	1.71	33.95	−1.65	32.30
2	29.80	−1.96	27.84	0.47	28.30	1.88	30.18
3	29.80	−1.05	28.75	−0.77	27.98	1.64	29.61
4	29.80	2.35	32.15	1.43	33.58	−0.71	32.87
5	29.80	0.50	30.30	−0.56	29.74	−0.73	29.01
6	29.80	−1.82	27.98	−0.78	27.20	0.29	27.48
7	29.80	−1.63	28.17	0.00	28.17	−1.99	26.18
8	29.80	−0.47	29.33	−1.02	28.31	−1.44	26.87
9	29.80	−0.24	29.56	2.25	31.81	1.48	33.29
10	29.80	−2.45	27.35	2.06	29.41	−0.62	28.80

In this case, we can talk about $p(\omega)$ being the probability that we sample ω from within the set $\hat{\Omega}$. Often, we will assume that each element of $\hat{\Omega}$ occurs with equal probability:

$$p(\omega) = \frac{1}{|\hat{\Omega}|}.$$

For more complex problems, we may have an entire family of random variables. In such cases, it is useful to have a generic "information variable" that represents all the information that arrives during time interval t. For this purpose, we define

$W_{t+1} =$ the exogenous information becoming available during time interval $(t, t+1)$.

We might also say that W_{t+1} is the information that first becomes known by time $t + 1$, which means it is not known when we make the decision x_t.

W_t may be a single variable, or a collection of variables (travel times, equipment failures, customer demands). We note that while we use the convention of putting hats on variables representing exogenous information ($\hat{D}_t, \hat{p}_t$), we do not use a hat for W_t since this is our only use for this variable, whereas D_t and p_t have other meanings. We always think of information as arriving in continuous time, hence W_t is the information arriving during time interval t, rather than at time t. This eliminates the ambiguity over the information available when we make a decision at time t.

We sometimes need to refer to the *history* of our process, for which we define

$\begin{aligned} h_t &= \text{the history of the process, consisting of all the information} \\ &\quad \text{known through time } t, \\ &= (W_1, W_2, \ldots, W_t), \\ \mathcal{H}_t &= \text{the set of all possible histories through time } t, \\ &= \{h_t(\omega) | \omega \in \Omega\}, \\ \Omega_t(h_t) &= \text{the set of all sample paths that correspond to history } h_t, \\ &= \{\omega \in \Omega | h_t(\omega) = h_t\}. \end{aligned}$

In some applications, we might refer to h_t as the state of our system, but this is usually a very clumsy representation. However, we will use the history of the process for a specific modeling and algorithmic strategy.

9.6.2 Outcomes and Scenarios

Some communities prefer to use the term *scenario* to refer to a sample realization of random information. For most purposes, "outcome," "sample path," and "scenario" can be used interchangeably (although sample path refers to a sequence of outcomes over time). There are many, however, who use the term "scenario" to represent a major event. For example, a company may launch a new product that may receive a market response that can be described as strong, medium or weak. For each of these scenarios, there are still going to be daily fluctuations in sales. We prefer to use "scenario" to refer to the market response (that is, the major event), and "outcome" to capture the variations around the market response.

We recommend denoting the set of scenarios by Ψ, with $\psi \in \Psi$ representing an individual scenario. Then, for a particular scenario ψ, we might have a set of outcomes $\omega \in \Omega$ (or $\Omega(\psi)$) representing various minor events (daily sales volume).

Two examples illustrate this notation:

■ **EXAMPLE 9.13**

Planning spare transformers – In the electric power sector, a certain type of transformer was invented in the 1960's. As of this writing, the industry does not really know the failure rate curve for these units (is their lifetime roughly 50 years? 60 years?). Let ψ be the scenario that the failure curve has a particular shape (for example, where failures begin happening at a higher rate around 50 years). For a given scenario ψ (the failure rate curve), ω represents a sample outcome of failures (transformers can fail at any time, although the likelihood they will fail depends on ψ).

■ **EXAMPLE 9.14**

Long term contracts for electricity – The price of electricity today depends largely on the price of natural gas. Electricity prices on an hourly basis can be highly volatile, but they average a price that reflects the price of natural gas. This relationship may depend on (a) the aggregate production of natural gas (which can depend on government policy) and (b) the availability of renewables. We can describe the relative supplies of energy from natural gas and renewables as a scenario ψ, and then model hourly variations as a sample path ω.

9.6.3 Lagged Information Processes*

There are many settings where the information about a new arrival comes before the new arrival itself, as we saw earlier in state variables. These also happen in exogenous information processes, as illustrated in the following examples:

■ **EXAMPLE 9.15**

A customer may make a reservation at time t to be served at time t'.

■ **EXAMPLE 9.16**

An orange juice products company may purchase futures for frozen concentrated orange juice at time t that can be exercised at time t'.

■ **EXAMPLE 9.17**

A programmer may start working on a piece of coding at time t with the expectation that it will be finished at time t'.

We handle these problems using two time indices, a form that we refer to as the "(t, t')" notation.

Lagged information processes are surprisingly common. Let $\hat{D}_{tt'}$ be the number of customers calling in at time t to book a hotel room at time t'. We can write our set of orders arriving on day t as

$$
\begin{aligned}
\hat{D}_{tt'} \;&=\; \text{the demands that first become known during time} \\
&\quad\ \text{interval } t \text{ to be served during time interval } t', \\
\hat{D}_t \;&=\; (\hat{D}_{tt'})_{t' \geq t}.
\end{aligned}
$$

Then, $\hat{D}_1, \hat{D}_2, \dots, \hat{D}_t, \dots$ is the sequence of orders, where each $\hat{D}_t$ can be orders being called in for different times into the future.

An important class of lagged processes are forecasts. Let

$$
\begin{aligned}
f^D_{tt'} \;&=\; \text{the forecast of the demand } \hat{D}_{t'} \text{ during time interval } t' \text{ made} \\
f^D_{tt'} \;&=\; \text{using the information available up through time } t, \\
f^D_t \;&=\; (f^D_{tt'})_{t' \geq t}.
\end{aligned}
$$

An important special case of each of these variables is when $t' = t$. We would describe this version of each of the variables as follows:

$$\hat{D}_{tt} = \text{The actual demand during time } t,$$

$$f_{tt}^{D} = \text{this is another way of writing } \hat{D}_{tt},$$

$$R_{tt} = \text{the resources we know about at time } t \text{ that we can use at}$$
$$\text{time } t.$$

Note that these variables are now written in terms of the information content. For example, $\hat{D}_{tt'}$ are the demands we know about at time t that will need to be served at time t'. The first time index specifies when the information becomes known.

9.6.4 Models of Information Processes*

Information processes come in varying degrees of complexity. Needless to say, the structure of the information process plays a major role in the models and algorithms used to solve the problem. We describe information processes in increasing levels of complexity.

State-independent Processes

Information might be generated by independent, unintelligent, exogenous processes such as weather, markets, biological processes, chemical reactions, and complex simulators, where the information is independent of the state S_t or decision x_t.

■ **EXAMPLE 9.18**

A publicly traded index fund has a price process that can be described (in discrete time) as $p_{t+1} = p_t + \sigma\delta$, where δ is normally distributed with mean μ, variance 1, and σ is the standard deviation of the change over the length of the time interval.

■ **EXAMPLE 9.19**

Requests for credit card confirmations arrive according to a Poisson process with rate λ. This means that the number of arrivals during a period of length Δt is given by a Poisson distribution with mean $\lambda\Delta t$, which is independent of the history of the system.

The practical challenge we typically face in these applications is that we do not know the parameters of the system. In our price process, the price may be trending upward or downward, as determined by the parameter μ. In our customer arrival process, we need to know the rate λ (which can also be a function of time).

State-independent information processes are attractive because they can be generated and stored in advance, simplifying the process of testing policies. In chapter 19, we will describe an algorithmic strategy based on the use of *scenario trees* which have to be created in advance.

State/action-dependent Information Processes

There are many problems where the exogenous information W_{t+1} depends on the state S_t and/or the decision x_t. Some illustrations include:

■ **EXAMPLE 9.20**

The change in the speed of wind at a wind farm depends on the current speed. If the current speed is low, the change is likely to be an increase. If it is high, the change is likely to be a decrease.

■ **EXAMPLE 9.21**

A market with limited information may respond to price changes. If the price drops over the course of a day, the market may interpret the change as a downward movement, increasing sales, and putting further downward pressure on the price. The market may also respond to decisions by mutual funds to sell large amounts of stock.

■ **EXAMPLE 9.22**

Customers arriving to a bank are served by a group of tellers, where the number of tellers on duty are controlled by a bank manager. The arrival rate of customers depend on the length of the queue (which is the state of our system), which depends on the decisions (made hourly) of how many people to have on duty.

State/action-dependent information processes make it impossible to pre-generate sample outcomes when testing policies. While not a major issue, it complicates comparing policies since we cannot fix the sample outcomes.

State-dependent information processes introduce a subtle notational complication. Following standard convention, the notation ω almost universally refers to a sample path. Thus, $W_t(\omega)$ represents the exogenous information arriving between $t - 1$ and t when we are following sample path ω. If we write $S_t(\omega)$, we mean the state we are in at time t when we are following sample path ω, but now we have to make it clear what policy we are following to get there. For example, we might write $S_{t+1}^\pi = S^M(S_t^\pi, X_t^\pi(S_t), W_{t+1}^\pi(\omega))$, where it is clear that we are using policy π to get from S_t^π to S_{t+1}^π.

Multiagent Sytems

The exogenous information may come from the decisions made by another agent. We can make the argument that W_{t+1}, which is really the decisions of another agent, would be a random variable that depends on some observable system state variables (such as the state of a game board), and the decision x_t made by the first agent. However, with enough training, the behavior of each agent tends to become predictable (this is typical of experts playing against each other), which means deterministic (although one strategy in an adversarial game is to introduce noise to keep the opponent from learning your strategies).

We cover the topic of multiagent systems in chapter 20.

More Complex Information Processes

Now consider the problem of modeling currency exchange rates. The change in the exchange rate between one pair of currencies is usually followed quickly by changes in others. If the Japanese yen rises relative to the US dollar, it is likely that the Euro will also rise relative to it, although not necessarily proportionally. As a result, we have a vector of information processes that are correlated.

In addition to correlations between information processes, we can also have correlations over time. An upward push in the exchange rate between two currencies in one day is likely to be followed by similar changes for several days while the market responds to new information. Sometimes the changes reflect long-term problems iin the economy of a country. Such processes may be modeled using advanced statistical models which capture correlations between processes as well as over time.

An *information model* is a mathematical model of the underlying information process. This falls under the broad umbrella of uncertainty modeling or uncertainty quantification, which we cover in chapter 10. In some cases with complex information models, it is possible to proceed without any model at all. Instead, we can use realizations drawn from history. For example, we may take samples of changes in exchange rates from different periods in history and assume that these are representative of changes that may happen in the future. The value of using samples from history is that they capture all of the properties of the real system. This is an example of planning a system without a model of an information process.

Deterministic Models

While listing different types of exogenous information processes, we cannot ignore the possibility that we do not have an exogenous information process, as would be the case with any deterministic system. We note that a large majority of the work in optimal control performed (primarily) in engineering applications is deterministic.

9.6.5 Supervisory Processes*

We are sometimes trying to control systems where we have access to a set of decisions from an exogenous source. These may be decisions from history, or they may come from a knowledgeable expert. Either way, this produces a dataset of states $(S^m)^n_{m=1}$ and decisions $(x^m)^n_{m=1}$. In some cases, we can use this information to fit a statistical model which we use to try to predict the decision that would have been made given a state.

The nature of such a statistical model depends very much on the context, as illustrated in the examples:

■ **EXAMPLE 9.23**

We can capture data on patient histories and complaints, along with the treatment decisions by physicians. We can use this history to train a neural network to recommend a treatment given the characteristics of a patient.

■ **EXAMPLE 9.24**

We can use the history of decisions when playing games (notably video games, but also games such as chess and computer Go), to train a statistical model what decision to make given the state of the game.

We can use supervisory processes to statistically estimate a decision function that forms an initial policy. We can then use this policy in the context of methods to create even better policies using the principles of policy search. The supervisory process helps provide an initial policy that may not be perfect, but at least is reasonable.

9.7 The Transition Function

The next step in modeling a dynamic system is the specification of the *transition function* which is a concept that is widely used in the optimal control community. This function describes how the system evolves from one state to another as a result of decisions and information. If you have ever written a simulator of a dynamic system, you have written a transition function, since this is nothing more than the equations that describe how variables evolve over time.

We begin our discussion of system dynamics by introducing some general mathematical notation. While useful, this generic notation does not provide much guidance into how specific problems should be modeled. We then describe how to model the dynamics of some simple problems, followed by a more general model for complex resources.

9.7.1 A General Model

The dynamics of our system are represented by a function that describes how the state evolves as new information arrives and decisions are made. The optimal control community will usually write the transition function (using controls notation) as

$$x_{t+1} = f(x_t, u_t, w_t)$$

where x_t is their notation for state, u_t is the decision or control, and w_t is the exogenous information which is random at time t (there is a long history behind this). The function $f(\cdot)$ goes by different names such as "plant model" (literally, the model of a physical production plant), "plant equation," "law of motion," "transfer function," "system dynamics," "system model," "state equations," "transition law," as well as "transition function."

When modeling complex problems, the letters f, g, and h are widely used for "functions," where f in particular is popular for being used in many ways. To avoid taking this valuable piece of real estate in the alphabet, we use the notation

$$S_{t+1} = S^M(S_t, x_t, W_{t+1}). \tag{9.18}$$

We use the notation $S^M(\cdot)$ since it hints at "state model" or "state transition model." This style avoids using another letter from the alphabet.

For real-world problems, the transition function often hides tremendous complexity in the modeling of the dynamics of a system. A transition function can easily consist of hundreds or thousands of lines of code. Of course, we started with a simple example in section 9.1 that required only two equations.

This is a very general way of representing the dynamics of a system. Assuming we have a proper state variable S_t that captures all the information we need to model the system from time t onward, the information W_{t+1} arriving during time interval $(t, t+1)$ depends on the state S_t at the end of time interval t (and possibly the decision x_t). In this case, we can store the system dynamics in the form of a one-step transition matrix using

$$P(s'|s,x) \quad = \quad \text{the probability that } S_{t+1} = s' \text{ given } S_t = s \text{ and } X^\pi(S_t) = x.$$

The one-step transition matrix is the foundation of a field known as discrete Markov decision processes, which we cover in chapter 14. There is a simple relationship between the transition function and the one-step transition matrix. Define the indicator function

$$\mathbb{1}_X = \begin{cases} 1 & \text{if } X \text{ is true,} \\ 0 & \text{otherwise.} \end{cases}$$

Assuming that the set of outcomes of $W_{t+1} = w \in \Omega^W$ is discrete, the one-step transition matrix can be computed using

$$P(s'|s,x) = \mathbb{E}_{W_{t+1}}\{\mathbb{1}_{\{s'=S^M(S_t=s,x_t=x,W_{t+1})\}}|S_t = s, x_t = x\}$$
$$= \sum_{w \in \Omega^W} P(W_{t+1} = w|S_t = s, x_t = x)\mathbb{1}_{\{s'=S^M(S_t=s,x_t=x,w)\}}. \quad (9.19)$$

We now have two ways of representing the dynamics of our system: the transition function $S^M(S_t, x_t, W_{t+1})$, and the one-step transition matrix $P(s'|s,x)$. The controls community (which is substantial) uses the transition function, while the community that works with Markov decision processes (which was adopted by the reinforcement learning community within computer science) uses the one-step transition matrix $P(s'|s,x)$. Given the derivation in equation (9.19), it seems clear that you need the one-step transition function in order to compute the one-step transition matrix. Yet, the MDP community will often treat the one-step transition matrix as input data.

In this book we exclusively use the one-step transition *function*, since this is trivially computable, even when the state variable S_t is high-dimensional (and even continuous). It is literally the equations you would use to simulate the system. By contrast, the one-step transition matrix is a powerful theoretical device, but it is utterly incomputable for all but the most trivial problems.

9.7.2 Model-free Dynamic Programming

There are many complex operational problems where we simply do not have a transition function. Some examples include

---

■ **EXAMPLE 9.25**

We are trying to find an effective policy to tax carbon to reduce CO_2 emissions. We may try increasing the carbon tax, but the dynamics of climate change are so complex that the best we can do is wait a year and then repeat our measurements.

■ **EXAMPLE 9.26**

A ride hailing service encourages drivers to go on duty by raising prices (surge pricing). Since it is impossible to predict how drivers will behave, it is necessary to simply raise the price and observe how many drivers come on duty (or go off duty).

■ **EXAMPLE 9.27**

A utility managing a water reservoir can observe the level of the reservoir and control the release of water, but the level is also affected by rainfall, river inflows, and exchanges with ground water, which are unobservable.

These examples illustrate problems where we do not know the dynamics, where the system reflects the unknown utility function of Uber drivers, and unobservable exogenous information. As a result, we either do not know the transition function itself, or there are decisions that we cannot model, or exogenous information we cannot simulate. In all three cases, we cannot compute the transition $S_{t+1} = S^M(S_t, x_t, W_{t+1})$.

In such settings (which are surprisingly common), we assume that given the state S_t, we take an action x_t and then simply observe the next state S_{t+1}. We can put this in the format of our original model by letting W_{t+1} be the new state, and writing our transition function as

$$S_{t+1} = W_{t+1}.$$

However, it is more natural (and compact) to simply assume that our system evolves according to

$$S_0 \rightarrow x_0 \rightarrow S_1 \rightarrow x_1 \rightarrow S_2 \rightarrow \dots .$$

We note that in many systems, there may be state variables where we do know the transition equation(s) (such as in an inventory problem), while there are other state variables where we do not know the transition, such as demands and prices.

9.7.3 Exogenous Transitions

There are many problems where some of the state variables evolve exogenously over time: rainfall, a stock price (assuming we cannot influence the price), the travel time on a congested road network, and equipment failures. There are two ways of modeling these processes.

The first models the change in the variable. If our state variable is a price p_t, we might let $\hat{p}_{t+1}$ be the change in the price between t and $t + 1$, giving us the transition function

$$p_{t+1} = p_t + \hat{p}_{t+1}.$$

This has the advantage of giving us a clean transition function that describes how the price evolves over time. With this notation, we would write $W_{t+1} = (\hat{p}_{t+1})$, so that the exogenous information is distinct from the state variable.

Alternatively, we could simply assume that the new state p_{t+1} is the exogenous information, which means we would write $W_{t+1} = p_{t+1}$. This requires that we have a process we are observing that gives us p_{t+1} without telling us how we transitioned from p_t to p_{t+1}.

9.8 The Objective Function

The final dimension of our model is the objective function. We divide our discussion between creating performance metrics for evaluating a decision x_t, and evaluating the policy $X^\pi(S_t)$.

9.8.1 The Performance Metric

Performance metrics are described using a variety of terms such as

1. Rewards, profits, revenues, costs (business)
2. Gains, losses (engineering)
3. Strength, conductivity, diffusivity (materials science)
4. Tolerance, toxicity, effectiveness (health)
5. Stability, reliability (engineering)
6. Risk, volatility (finance)
7. Utility (economics)
8. Errors (machine learning)
9. Time (to complete a task)

These differ primarily in terms of units and whether we are minimizing or maximizing. These are modeled using a variety of notation systems such as c for cost, r for revenue or reward, g for gain, L or ℓ for loss, U for utility, and $\rho(X)$ as a risk measure for a random variable X.

There are many problems where there are multiple metrics. There are three strategies we can use to handle these:

(1) Utility functions – We can combine different metrics into a single utility, which requires specifying weights on each metric.
(2) We maximize one metric subject to constraints on the other metrics.
(3) Multiobjective programming – Here we capture different objectives at the same time (such as expected profit and risk), and then let a decision-maker make an appropriate tradeoff.

Both methods (1) and (2) produce a single performance metric. These are the approaches we use in this book, since they make it possible for a computer to identify a single best decision.

9.8.2 Optimizing the Policy

We close our first pass through modeling by giving the objective function for finding the best policy. Our default objective function for state-dependent problems (that is, where the contribution function and/or constraints depend on the state S_t) can be written

$$\max_{\pi \in \Pi} \mathbb{E}_{S_0} \mathbb{E}_{W_1,\dots,W_T | S_0} \left\{ \sum_{t=0}^{T} C_t(S_t, X_t^\pi(S_t)) | S_0 \right\}. \tag{9.20}$$

Once we get used to what we have to take the expectation over, we may just use the compact form of the expectation

$$\max_{\pi \in \Pi} \mathbb{E} \left\{ \sum_{t=0}^{T} C_t(S_t, X_t^\pi(S_t)) | S_0 \right\}. \tag{9.21}$$

As we did in chapter 7, we write our expectation in nested form to express the possible presence of a probabilistic initial state S_0 (where we might have a distribution of belief about some information), and the observations $W_1, \dots, W_T$. We explicitly express the dependence on S_0, even if it does not contain any probabilistic beliefs, to communicate the dependence on any static data (which could include latent variables).

The objective (9.21) is written using cumulative rewards, but there will be settings where we should use a final-reward objective. We return to this issue shortly.

9.8.3 Dependence of Optimal Policy on S_0

Our notation for the objective function (9.20) captures the dependence of the optimal policy on S_0 which is always present, but generally overlooked in the optimization literature. Specifically, if we find an optimal policy $X^*(S_t)$, it really should be written $X^*(S_t|S_0)$. Yes, this means that if we change the initial state, we may change the optimal policy, possibly significantly.

We already saw this in section 6.6 when we discussed tuning the stepsize policy, and then demonstrated how poorly it could work when we changed the starting point of the algorithm (this would be captured by S^0) if we did not retune the stepsize. The problem is vividly demonstrated in Figure 6.10(a) when we picked a starting point in the region $[1,2]$. In practice, reoptimizing the policy when we change S_0 can quickly become impractical. We simply make the point: reoptimizing the policy each time we change S_0 *is* impractical, but this does not mean we can pretend it is not an issue. The dependence is higly problem-dependent, but something that any algorithmic reseacher needs to be aware of.

We are going to see this issue again when designing policies in the context of stochastic lookahead models. Section 19.7.1 proposes (ahem!) that we ignore the dependence of tunable parameters on the starting state, but makes the argument that we can tolerate approximations such as this when the policy is just be used to simulate the downstream impact of a decision. Needless to say, there are a lot of unanswered questions here.

9.8.4 State-dependent Variations

Depending on the setting, we might use any of the following ways of expressing our contribution function:

$$F(x, W) = \text{A general performance metric (to be minimized}$$
$$\text{or maximized) that depends only on the decision } x$$
$$\text{and information } W \text{ that is revealed after we choose } x.$$

$$C(S_t, x_t) = \text{A cost/contribution function that depends on}$$
$$C(S_t, x_t) = \text{the state } S_t \text{ and decision } x_t.$$

$$C(S_t, x_t, W_{t+1}) = \text{A cost/contribution function that depends on}$$
$$\text{the state } S_t \text{ and the decision } x_t, \text{ and the}$$
$$\text{information } W_{t+1} \text{ that is revealed after } x_t \text{ is}$$
$$\text{determined.}$$

$$C(S_t, x_t, S_{t+1}) = \text{A cost/contribution function that depends on}$$
$$\text{the state } S_t \text{ and the decision } x_t, \text{ after which}$$
$$\text{we observe the subsequent state } S_{t+1}. \text{ This format}$$
$$\text{is used in model-free settings where we do not}$$
$$\text{know the transition function.}$$

$$C_t(S_t, x_t) = \text{The cost/contribution function when the function}$$
$$\text{itself depends on time } t.$$

We have used the notation $F(x, W)$ (as we did in chapters 5 and 7) when our problem does not depend on the state. However, as we transition to state-dependent problems, we use $C(S_t, x_t)$ (or $C(S_t, x_t, W_{t+1})$ or $C(S_t, x_t, S_{t+1})$) to communicate that the objective function (or constraints or expectation) depend on the state. Readers may choose to use any notation such as $r(\cdot)$ for reward, $g(\cdot)$ for gain, $L(\cdot)$ for loss, or $U(\cdot)$ for utility.

The state-dependent representations all depend on the state S_t (or S^n if we wish), but it is useful to say what this means. When we make a decision, we need to work with a cost function and possibly constraints where we express the dependence on S_t by writing the feasible region X_t as depending on t (the notation $X(S_t)$ seems clumsy). For example, we might move money in a mutual fund to or from cash, buying or selling an index that is at price p_t. Let R_t be the amount of available cash, which evolves as people make deposits or withdrawals. The amount of cash could be defined by

$$R_{t+1}^{cash} = R_t^{cash} + x_t + \hat{R}_{t+1}, \tag{9.22}$$
$$R_{t+1}^{index} = R_t^{index} - x_t. \tag{9.23}$$

where $x_t > 0$ is the amount of money moved into cash by selling the index fund, while $x_t < 0$ represents money from from cash into the index fund. We have to observe the constraints

$$x_t \leq R_t^{index},$$
$$-x_t \leq R_t^{cash}.$$

The money we make is based on what we receive from buying or selling the index fund, which we would write as

$$C(S_t, x_t) = p_t x_t,$$

where the price evolves according to the model

$$p_{t+1} = \theta_0 p_t + \theta_1 p_{t-1} + \varepsilon_{t+1}.$$

For this problem, our state variable would be $S_t = (R_t, p_t, p_{t-1})$. For this example, the contribution function itself depends on the state through the prices, while the constraints (R_t^{index} and R_t^{cash}) also vary dynamically and are part of the state.

Now imagine that we have to make the decision to buy or sell shares of our index fund, but the price we get is based on the closing price, which is not known when we make our decision. In this case, we would write our contribution function as

$$C(S_t, x_t, W_{t+1}) = p_{t+1} x_t,$$

where $W_{t+1} = \hat{p}_{t+1} = p_{t+1} - p_t$. We note that our policy $X^\pi(S_t)$ for making the decision x_t is not allowed to use W_{t+1}; rather, we have to wait until time $t + 1$ before evaluating the quality of the decision.

Finally, consider a model of a hydroelectric reservoir where we have to manage the inventory in the reservoir, but where the dynamics describing its evolution is much more complicated than equations such as (9.22) and (9.23). In this setting, we can observe the reservoir level R_t, then make a decision of how much water to release out of the reservoir x_t, after which we observe the updated reservoir level R_{t+1}. This is similar to observing an updated price p_{t+1}. For these problems, we might let W_{t+1} be the new state, in which case our "transition equations" are just

$$S_{t+1} = W_{t+1}.$$

Alternatively, we may find it more natural to write the contribution function $C(S_t, x_t, S_{t+1})$, which is fairly common, but there are settings where we have transition equations for some variables but not others.

We use $C(S_t, x_t)$ as our standard notation (in some settings we will index the contribution function by time, as in $C_t(S_t, x_t)$). If we find ourselves writing the contribution in a form that needs $C(S_t, x_t, W_{t+1})$ as we illustrated, we can always break the contribution into the parts that can be computed at time t, and the parts that cannot be computed until time $t + 1$. We can easily write this as

$$C_t(S_t, x_t, W_{t+1}) = C_t^1(S_t, x_t) + C_{t+1}^2(S_t, x_t, W_{t+1}).$$

where $C_t^1(S_t, x_t) = -cx_t$ captures the components of the contribution function that can be computed at time t, and $C_{t+1}^2(S_t, x_t, W_{t+1}) = p \min\{S_t + x_t, W_{t+1}\}$ captures the components that cannot be computed until time $t + 1$.

Next create the contribution function

$$\tilde{C}_t(S_t, x_t) = C_t^2(S_{t-1}, x_{t-1}, W_t) + C_t^1(S_t, x_t).$$

Now optimize the sum of contribution functions $\tilde{C}_t(S_t, x_t)$ over the horizon. This strategy may seem unintuitive (or unappealing) since $C_{t-1}^2(S_{t-1}, x_{t-1}, W_t)$ does not depend on x_t, and we are not capturing the impact of x_t on revenue. However, these are simply cosmetic issues. Simply moving the contributions that depend on W_{t+1} to the next time period will not change the overall performance of any optimizing policy that we propose in chapter 11 (or develop in the rest of the book).

9.8.5   Uncertainty Operators

An important issue when optimizing under uncertainty is that we have to decide how to evaluate the distribution of the objective function for a policy. Some choices we can use are:

- The expectation operator $\mathbb{E}\{\cdot|S_0\}$ – We use this as our default operator, since it is easily the one that is most commonly used.
- The risk operator $\rho(\cdot)$ – This is actually a family of operators that are designed to capture the tails or spread of the distribution of outcomes. Some examples are:
    - Value at risk $F_\alpha^\pi = VaR_\alpha(F^\pi)$ – This is the value F_α^π of the α-quantile of a random variable F^π giving the performance of the policy $X^\pi(S)$. If we are maximizing, we might use the 10th percentile to protect ourselves from doing poorly.
    - Conditional value at risk $CVaR_\alpha(Z)$ – Also known as the average value at risk or expected shortfall, this is the expectation of $Z = \max\{0, F_\alpha - F^\pi\}$ (if we are maximizing).
    - There is a host of potential other measures, such as the worst performance over the horizon, the α-percentile over all the time periods, and so on.
- Robust optimization, where we would use the worst possible outcome which we can write

$$\min_{\omega \in \Omega} F^\pi(\omega),$$

where $F^\pi(\omega)$ is the performance of the policy for sample path ω. This means that our optimization problem is

$$\max_\pi \min_{\omega \in \Omega} F^\pi(\omega).$$

Our default operator is the expectation, which is often used even when a risk measure is used in a stochastic lookahead model. For example, there is a substantial community called "robust optimization" (see section 2.1.14) which might use a stochastic lookahead policy with a robust objective, but which then evaluates the "robust" policy by simulating it many times and taking an average (which means using an expectation to evaluate the policy). We revisit this in chapter 19.

9.9   Illustration: An Energy Storage Model

In section 9.1 we presented a very simple energy storage problem where we have to determine when to buy energy from the grid, or sell it back to the grid.

We are going to expand on this model, first by introducing the ability to draw energy from the grid or a wind farm which is stored in a battery, from which we draw energy to meet a demand D_t. Then, we are going to make the price process into a simple first-order process.

The decision variables are given by

$$x_t^G = \text{the energy we purchase from the grid } (x_t^G > 0) \text{ or sell back}$$
$$\text{to the grid } (x_t^G < 0) \text{ which moves to or from the battery,}$$

$$x_t^E = \text{the energy generated from a wind farm at time } t \text{ to the battery,}$$

$$x_t^D = \text{the energy moved from the battery to meet the demand } D_t.$$

We then define the exogenous inputs

$$E_t = \text{the energy available from the wind farm at time } t,$$

$$D_t = \text{the demand for energy at time } t.$$

The flows have to satisfy the constraints

$$x_t^E \leq E_t, \tag{9.24}$$
$$x_t^G + x_t^E \leq R^{max} - R_t, \tag{9.25}$$
$$x_t^D \leq R_t, \tag{9.26}$$
$$x_t^D \leq D_t, \tag{9.27}$$
$$-x_t^G \leq R_t. \tag{9.28}$$

Equation (9.24) limits the energy we store in the battery from the wind farm to the available wind in the wind farm. Equation (9.25) limits the total energy from the grid and the wind farm to the available capacity in the battery. Equation (9.26) limits the amount we use from the battery to serve the demand to the amount in the battery, while equation (9.27) limits the energy sent to meet the demand to the demand itself. Equation (9.28) limits the amount of energy sent back to the grid (this is where $x_t^G < 0$) to the amount in the battery.

The transition equations are given by

$$R_{t+1} = R_t + x_t,$$
$$p_{t+1} = p_t + \varepsilon_{t+1},$$

where $\varepsilon_{t+1} \sim N(0, \sigma^2)$ (before we had assumed that we just observed p_{t+1}). We assume that the changes in prices $\hat{p}_t$ are independent across time. We assume that the energy E_t from the wind farm and the demand D_t is observed without models of their evolution. We address some modeling issues related to forecasting E_t.

For this basic system, the state variable would be

$$S_t = ((R_t, E_t, D_t), p_t).$$

We are now going to step through a series of variations where we modify the price process, and then describe the effect of the change on the state variable.

9.9.1 With a Time-series Price Model

We begin by replacing our simple price process in equation (9.29) with a time series model given by

$$p_{t+1} = \theta_0 p_t + \theta_1 p_{t-1} + \theta_2 p_{t-2} + \varepsilon_{t+1}. \tag{9.29}$$

It is surprisingly common for people to say that p_t is the "state" of the price process, and then insist that it is no longer Markovian (it would be called "history dependent"), but "it can be made Markovian by expanding the state variable," which would be done by including p_{t-1} and p_{t-2}. Using our definition, the state is all the information needed to model the process from time t onward, which means that the state of our price process is (p_t, p_{t-1}, p_{t-2}). This means our system state variable is now

$$S_t = ((R_t, E_t, D_t), (p_t, p_{t-1}, p_{t-2})).$$

We then have to modify our transition function so that the "price state variable" at time $t + 1$ becomes (p_{t+1}, p_t, p_{t-1}).

9.9.2 With Passive Learning

The price model in equation (9.29) assumed the coefficients $\theta = (\theta_0, \theta_1, \theta_2)$ were known. Now assume that the coefficients are unknown and have to be learned along the way, as in

$$p_{t+1} = \bar{\theta}_{t0} p_t + \bar{\theta}_{t1} p_{t-1} + \bar{\theta}_{t2} p_{t-2} + \varepsilon_{t+1}. \tag{9.30}$$

Here, we have to recursively update our estimate $\bar{\theta}_t$ which we can do using recursive least squares which we introduced in section 3.8. To do this, let

$$\bar{p}_t = (p_t, p_{t-2}, p_{t-2})^T,$$
$$\bar{F}_t(\bar{p}_t | \bar{\theta}_t) = (\bar{p}_t)^T \bar{\theta}_t.$$

The updating equations for $\bar{\theta}_t$ are given by

$$\bar{\theta}_{t+1} = \bar{\theta}_t + \frac{1}{\gamma_t} M_t \bar{p}_t \varepsilon_{t+1}, \tag{9.31}$$

$$\varepsilon_{t+1} = \bar{F}_t(\bar{p}_t|\bar{\theta}_t) - p_{t+1}, \tag{9.32}$$

$$M_{t+1} = M_t - \frac{1}{\gamma_t} M_t (\bar{p}_t)(\bar{p}_t)^T M_t, \tag{9.33}$$

$$\gamma_t = 1 - (\bar{p}_t)^T M_t \bar{p}_t. \tag{9.34}$$

To compute these equations, we need the three-element vector $\bar{\theta}_t$ and the 3×3 matrix M_t. These then need to be added to our state variable, giving us

$$S_t = \big((R_t, E_t, D_t), (p_t, p_{t-1}, p_{t-2}), (\bar{\theta}_t, M_t)\big),$$

which has 18 continuous dimensions. We then have to include equations (9.31)–(9.34) in our transition function.

9.9.3   With Active Learning

There are many settings where the decisions we make either directly affect or at least influence what we observe. We are going to assume that our decision x_t^{GB} to buy or sell energy from or to the grid can have an impact on prices. We might propose a modified price model given by

$$p_{t+1} = \bar{\theta}_{t0} p_t + \bar{\theta}_{t1} p_{t-1} + \bar{\theta}_{t2} p_{t-2} + \bar{\theta}_{t3} x_t^{GB} + \varepsilon_{t+1}. \tag{9.35}$$

Now, buying or selling large quantities from or to the grid can push prices higher or lower, allowing us to explore different regions of the model. This is known as *active learning,* a topic we introduced in chapter 7 for both offline and online settings.

This change in our price model does not affect the state variable from the previous model, aside from adding one more element to $\bar{\theta}_t$, with the required changes to the matrix M_t. The change will, however, have an impact on the policy. It is easier to learn θ_{t3} by varying x_t^{GB} over a wide range, which means trying values of x_t^{GB} that do not appear to be optimal given our current estimate of the vector $\bar{\theta}_t$. Making decisions partly just to learn (to make better decisions in the future) is the essence of *active learning,* best known in the field of multiarmed bandit problems.

9.9.4   With Rolling Forecasts

Forecasting is such a routine activity in operational problems, it may come as a surprise that we have been modelling these problems incorrectly.

Assume we have a forecast $f^E_{t,t+1}$ of the energy E_{t+1} from the wind farm, which means

$$E_{t+1} = f^E_{t,t+1} + \varepsilon_{t+1,1}, \tag{9.36}$$

where $\varepsilon_{t+1,1} \sim N(0, \sigma^2_\varepsilon)$ is the random variable capturing the one-period-ahead error in the forecast.

Equation (9.36) introduces a new variable, the forecast $f^E_{t,t+1}$, which must now be added to the state variable. This means we now need a transition equation to describe how $f^E_{t,t+1}$ evolves over time. We do this by using a two-period-ahead forecast, $f^E_{t,t+2}$, which is basically a forecast of $f^E_{t+1,t+2}$, plus an error, giving us

$$f^E_{t+1,t+2} = f^E_{t,t+2} + \varepsilon_{t+1,2}, \tag{9.37}$$

where $\varepsilon_{t+1,2} \sim N(0, \sigma^2_\varepsilon)$ is the two-period-ahead error (we are assuming that the variance in a forecast increases linearly with time). Now we have to put $f^E_{t,t+2}$ in the state variable, which generates a new transition equation. This generalizes to

$$f^E_{t+1,t'} = f^E_{t,t'} + \varepsilon_{t+1,t'-t}, \tag{9.38}$$

where $\varepsilon_{t+1,t'-t} \sim N(0, \sigma^2_\varepsilon)$.

This stops, of course, when we hit the planning horizon H. This means that we now have to add

$$f^E_t = (f^E_{tt'})^{t+H}_{t'=t+1}$$

to the state variable, with the transition equations (9.38) for $t' = t+1, \ldots, t+H$. Combined with the learning statistics, our state variable is now

$$S_t = \big((R_t, E_t, D_t), (p_t, p_{t-1}, p_{t-2}), (\bar{\theta}_t, M_t), f^E_t\big).$$

It is useful to note that we have a nice illustration of the three elements of our state variable:

$$(R_t, E_t, D_t) = \text{the physical state variables (energy in the battery,}$$
$$\text{energy available from the wind farm, current}$$
$$\text{demand for energy),}$$

$$(p_t, p_{t-1}, p_{t-2}) = \text{other information (recent prices),}$$

$$((\bar{\theta}_t, M_t), f^E_t) = \text{the belief state, since these parameters determine}$$
$$\text{the distribution of belief about variables that}$$
$$\text{are not known perfectly.}$$

This state variable has 42 dimensions: three for the physical states, three for prices, 12 for the endogenous forecasts, and 24 for the rolling forecasts.

9.10 Base Models and Lookahead Models

There is a subtle but critical distinction between a "model" of a real problem, and what we will come to know as a "lookahead model," which is an approximation that is used to peek into the future (typically with various convenient approximations) for the purpose of making a decision now. We are going to describe lookahead models in far greater depth in chapter 19, but we feel that it is useful to make the distinction now.

Using the framework presented in this chapter, we can write almost any sequential decision process in the compact form

$$
max_{\pi \in \Pi} \mathbb{E} \left\{ \sum_{t=0}^{T} C_t(S_t, X_t^{\pi}(S_t)) | S_0 \right\},
\tag{9.39}
$$

where $S_{t+1} = S^M(S_t, X_t^{\pi}(S_t), W_{t+1})$. Of course, we have to specify our model for $(W_t)_{t=0}^{T+1}$ in addition to defining the state variable (later we will address the issue of identify our class of policies).

For the moment, we view (9.39) (along with the transition function) as "the problem" that we are trying to solve. If we find an effective policy, we assume we have solved "the problem." However, we are going to learn that in dynamic systems, we are often solving a problem at some time t over a horizon $(t, \ldots, t+H)$, where we simply set $t = 0$ and number time periods accordingly. The question is: are we interested in the solution over the entire planning horizon, or just the decision in the first time period?

Given the widespread use of lookahead models, we need a term to identify when we are presenting a model of a problem we wish to solve. We might use the term "real model" to communicate that this is our model of the real world. Statisticians use the term "true model," but this seems to assume that we have somehow perfectly modeled a real problem, which is never the case. Some authors use the term "nominal model," but we feel that this is not sufficiently descriptive.

In this book, we use the term *base model* since we feel that this communicates the idea that this is the model we wish to solve. We take the position that regardless of any modeling approximations that have been introduced (either for reasons of tractability or availability of data), this is "the" model we are trying to solve.

Later, we are going to introduce approximations of our base model, which may still be quite difficult to solve. Most important will be the use of lookahead models, which we discuss in depth in chapter 19.

9.11   A Classification of Problems*

It is useful to contrast problems based on two key dimensions: First, whether the objective function is final-reward or cumulative-reward, and second, whether the objective function is state-independent (learning problems, which we covered in chapters 5 and 7) or state-dependent (traditional dynamic programs), which we began treating in chapter 8, and which will be the focus of the remaining chapters.

This produces four problem classes which are depicted in Table 9.3. We have numbered the classes in increasing order of complexity, with the warning that class 4 is particularly difficult to parse. In this section, we are going to write out the objectives in expectation form, but in the section that follows, we are going to show how to simulate the expectations, which we feel will make the expectations easier to understand. It may help to flip forward to section 9.12 to peek at the simulated version of each expression.

(Class 1) State-independent, final reward – This describes classical search problems where we are trying to find the best algorithm (which we call a policy π) for finding the best solution $x^{\pi,N}$ within our budget N. After n experiments the state S^n captures only our belief state about the function $\mathbb{E}F(x,W)$, and our decisions are made with a policy (or algorithm) $x^n = X^\pi(S^n)$. We can write this problem as

$$\max_\pi \mathbb{E}\{F(x^{\pi,N}, \widehat{W})|S^0\} = \mathbb{E}_{S^0}\mathbb{E}_{W^1,...,W^N|S^0}\mathbb{E}_{\widehat{W}|S^0} F(x^{\pi,N}, \widehat{W}), \qquad (9.40)$$

where $W^1, ..., W^N$ are the observations of W while learning the function $\mathbb{E}F(x,W)$, and $\widehat{W}$ is the random variable used for testing the final design $x^{\pi,N}$. The distinguishing characteristics of this problem are (a) that the function $F(x,W)$ depends only on x and W, and not on the state S^n, and (b) that we evaluate our policy $X^\pi(S)$ only after we have exhausted our budget of N experiments. We do allow the function $F(x,W)$, the observations $W^1, ..., W^N$

Table 9.3   Comparison of formulations for state-independent (learning) vs. state-dependent problems, and offline (final reward) and online (cumulative reward).

	Offline Final reward	Online Cumulative reward
State-independent problems	$\max_\pi \mathbb{E}\{F(x^{\pi,N}, W)|S_0\}$ (1) Stochastic search	$\max_\pi \mathbb{E}\{\sum_{n=0}^{N-1} F(X^\pi(S^n), W^{n+1})|S_0\}$ (2) Multiarmed bandit problem
State-dependent problems	$\max_{\pi^{lrn}} \mathbb{E}\{C(S, X^{\pi^{imp}}(S|\theta^{imp}), W)|S_0\}$ (4) Offline dynamic programming	$\max_\pi \mathbb{E}\{\sum_{t=0}^{T} C(S_t, X^\pi(S_t), W_{t+1})|S_0\}$ (3) Online dynamic programming

and the random variable $\widehat{W}$ to depend on the initial state S_0, which includes any deterministic parameters, as well as probabilistic information (such as a Bayesian prior) that describes any unknown parameters (such as how the market responds to price).

(Class 2) State-independent, cumulative reward – Here we are looking for the best policy that learns while it optimizes. This means that we are trying to maximize the sum of the rewards received within our budget. This is the classic multiarmed bandit problem that we first saw in chapter 7 if the decisions x were discrete and we did not have access to derivatives (but we are not insisting on these limitations). We can write the problem as

$$\max_{\pi} \mathbb{E}\left\{\sum_{n=0}^{N-1} F(X^\pi(S^n), W^{n+1})|S^0\right\} = \mathbb{E}_{S^0}\mathbb{E}_{W^1,...,W^N|S^0}\sum_{n=0}^{N-1} F(X^\pi(S^n), W^{n+1}). \quad (9.41)$$

(Class 3) State-dependent, cumulative reward – We now transition to problems where we are maximizing contributions that depend on the state variable, the decision, and possibly (but not always) random information that arrives after we make a decision (if it arrived before, it would be included in the state variable). For this reason, we are going to switch from our notation $F(x, W)$ to our notation $C(S, x, W)$ (or, in a time-indexed environment, $C(S_t, x_t, W_{t+1})$). As with the multiarmed bandit problem (or more generally, Class (2) problems), we want to find a policy that learns while implementing. These problems can be written as

$$\max_{\pi} \mathbb{E}\left\{\sum_{t=0}^{T} C(S_t, X^\pi(S_t), W_{t+1})|S_0\right\} = \mathbb{E}_{S_0}\mathbb{E}_{W_1,...,W_T|S_0}\left\{\sum_{t=0}^{T} C(S_t, X^\pi(S_t), W_{t+1})|S_0\right\}. \quad (9.42)$$

State variables in this problem class may include any of the following:

- Variables that are controlled (or influenced) by decisions (such as inventory or the location of a sensor on a graph). These variables directly affect the contribution function (such as price) or the constraints (such as the inventory).
- Variables that evolve exogenously (such as the wind speed or price of an asset).
- Variables that capture our belief about a parameter that are only used by the policy.

When we consider that our state S_t may include a controllable physical state R_t, exogenous information I_t and/or a belief state B_t, we see that this covers a very broad range of problems. The key feature here is that our policy has to maximize cumulative contributions as we progress, which may include learning (if there is a belief state).

(Class 4) State-dependent, final reward – For our state-independent function $F(x, W)$ we were looking for the best policy to learn the decision $x^{\pi,N}$ to be implemented. In this setting, we can think of the policy as a *learning policy*, while $x^{\pi,N}$ is the *implementation decision*. In the state-dependent case, the implementation decision becomes one that depends on the state (at least, part of the state), which is a function we call the *implementation policy*. We designate the implementation policy by $X^{\pi^{imp}}(S|\theta^{imp})$, which we write as depending on a set of parameters θ^{imp} which have to be learned. We designate the learning policy for learning θ^{imp} by $\Theta^{\pi^{lrn}}(S|\theta^{lrn})$ which proceeds by giving us parameters $\theta^{imp,n} = \Theta^{\pi^{lrn}}(S^n|\theta^{lrn})$. The problem can be written as

$$\max_{\pi^{lrn}} \mathbb{E}\{C(S, X^{\pi^{imp}}(S|\theta^{imp}), \widehat{W})|S^0\} =$$

$$\mathbb{E}_{S^0} \mathbb{E}^{\pi^{lrn}}_{W^1,\ldots,W^N|S^0} \mathbb{E}^{\pi^{imp}}_{S|S^0} \mathbb{E}_{\widehat{W}|S^0} C(S, X^{\pi^{imp}}(S|\theta^{imp}), \widehat{W}). \tag{9.43}$$

where $W^1, \ldots, W^N$ represents the observations made while using our budget of N experiments to learn a policy, and $\widehat{W}$ is the random variable observed when evaluating the policy at the end. We use the expectation operator $\mathbb{E}^{\pi^{lrn}}$ indexed by the learning policy when the expectation is over a random variable whose distribution is affected by the learning policy.

The learning policy could be a stochastic gradient algorithm to learn the parameters θ^{imp}, or it could be one of our derivative-free methods such as interval estimation or upper confidence bounding. The learning policy could be algorithms for learning value functions such as Q-learning (see equations (2.19)–(2.21) in chapter 2), or the parameters of any of the derivative-free search algorithms in chapter 7.

We typically cannot compute the expectation $\mathbb{E}^{\pi^{imp}}_S$ since it depends on the implementation policy which in turn depends on the learning policy. As an alternative, we can run a simulation over a horizon $t = 0, \ldots, T$ and then divide by T to get an average contribution per unit time. This simulation is performed using our testing random variable $\widehat{W}_t$, since we are evaluating the policy after we have learned the implementation policy. Let $\widehat{W}^n = (\widehat{W}^n_1, \ldots, \widehat{W}^n_T)$ be a simulation over our horizon. This allows us to write our learning problem as

$$\max_{\pi^{lrn}} \mathbb{E}_{S^0} \mathbb{E}^{\pi^{imp}}_{((w^n_t)^T_{t=0})^N_{n=0}|S^0} \left(\mathbb{E}^{\pi^{imp}}_{(\widehat{W}_t)^T_{t=0}|S^0} \frac{1}{T} \sum_{t=0}^{T-1} C(S_t, X^{\pi^{imp}}(S_t|\theta^{imp}), \widehat{W}_{t+1}) \right). \tag{9.44}$$

This parallels class (1) problems. We are searching over learning policies that determine the implementation policy through $\theta^{imp} = \Theta^{\pi^{lrn}}(S|\theta^{lrn})$, where the simulation over time replaces $F(x, W)$ in the state-independent formulation. The sequence $(W_t^n)_{t=0}^T, n = 1, \ldots, N$ replaces the sequence $W^1, \ldots, W^N$ for the state-independent case, where we start at state $S_0 = S^0$. We then do our final evaluation by taking an expectation over $(\widehat{W}_t)_{t=0}^T$, where we again assume we start our simulations at $S^0 = S_0$.

9.12 Policy Evaluation*

While it is certainly useful to characterize these four problem classes, it is an entirely different matter to compute the expectations in equations (9.40)–(9.44). The best way to approach this task (in fact, the best way to actually understand the expectations) is to simulate them. In this section we describe how to approximate each expectation using simulation.

We begin by fixing a policy $X^\pi(S_t|\theta)$ parameterized by some vector θ, which can be anything including a learning policy such as Thompson sampling, a stochastic gradient algorithm with a particular stepsize policy, or a direct lookahead policy. In problem class (1), $X^\pi(S_t|\theta)$ is a pure learning policy which learns an implementation decision $x^{\pi,N}(\theta)$. In classes (2) and (3), it is a policy where we learn as we implement. In class (4), we use a learning policy $\Theta^{\pi^{lrn}}(S_t|\theta^{lrn})$ to learn the parameter θ^{imp} of an implementation policy $X^{\pi^{imp}}(S_t|\theta^{imp})$ where $\theta^{imp} = \Theta^{\pi^{lrn}}(\theta^{lrn})$ depends on the learning policy.

Throughout, we are going to use θ (or θ^{lrn}) as a (possibly vector-valued) parameter that controls our learning policy (for classes (1) and (4)) or the implementation policy (possibly with learning) for classes (2) and (3). The vector θ (or θ^{lrn}) might be the parameters governing the behavior of any adaptive learning algorithm.

We now need to evaluate how well this policy works. We start with state S^0 if we are in problem classes (1) or (4), or S_0 if we are in problem class (3), and S^0 or S_0 if we are in problem class (2). From the initial state, we pick initial values of any parameters, either because they are fixed, or by drawing them from an assumed distribution (that is, a Bayesian prior).

We next address the process of simulating a policy for each of the four problem classes.

(Class 1) State-independent, final reward – From an initial state S^0, we use our (learning) policy to make decision $x^0 = X^\pi(S^0|\theta)$, and then observe outcome W^1, producing an updated state S^1 (in this problem, S^n is a pure knowledge state). The parameter θ controls the behavior of our learning policy. We repeat this until our budget is depleted, during which we observe the sequence $W^1(\omega), \ldots, W^N(\omega)$, where we let ω represent a particular sample path. At the end we learn state S^N, from which we find our best solution (the final design) $x^{\pi,N}$, which we write as $x^{\pi,N}(\theta|\omega)$ to express its dependence on the learning policy π (parameterized by θ) and the sample path ω.

We then evaluate $x^{\pi,N}(\theta|\omega)$ by simulating $\mathbb{E}_{\widehat{W}}F(x^{\pi,N}(\theta|\omega), \widehat{W})$ by repeatedly sampling from $\widehat{W}$ to get sampled estimates of $\mathbb{E}_{\widehat{W}}F(x^{\pi,N}(\theta|\omega), \widehat{W})$. Let $\widehat{W}(\psi)$ be a particular realization of $\widehat{W}$. A sampled estimate of the policy π (which we assume is parameterized by θ) is given by

$$F^{\pi}(\theta|\omega, \psi) = F(x^{\pi,N}(\theta|\omega), \widehat{W}(\psi)). \tag{9.45}$$

We now average over a set of K samples of ω, and L samples of ψ, giving us

$$\bar{F}^{\pi}(\theta) = \frac{1}{K}\frac{1}{L}\sum_{k=1}^{K}\sum_{\ell=1}^{L} F^{\pi}(\theta|\omega^k, \psi^\ell). \tag{9.46}$$

(Class 2) State-independent, cumulative reward – This problem can be interpreted in two ways. As the cumulative reward version of problem class (1), we simulate our policy for N iterations, giving us the sequence $(S^0, x^0, W^1, \dots, x^{N-1}, W^N, S^N)$. Here, we accumulate our rewards, producing a sampled estimate

$$F^{\pi}(\theta|\omega) = \sum_{n=0}^{N-1} F(X^{\pi}(S^n|\theta), W^{n+1}(\omega)). \tag{9.47}$$

Unlike class (1), we evaluate our policy as we go, avoiding the need for the final step at the end. We would then compute an average using

$$\bar{F}^{\pi}(\theta) = \frac{1}{K}\sum_{k=1}^{K} F^{\pi}(\theta|\omega^k), \tag{9.48}$$

over a sample of K observations.

We can also recast this problem as simulating over time, where we just replace W^n with W_t and S^n with S_t.

(Class 3) State-dependent, cumulative reward – This is the state-dependent version of problem class (2), which we model as evolving over time. Starting in state S_0, we simulate the policy much as we did in equation (9.47) which is given by

$$F^{\pi}(\theta|\omega) = \sum_{t=0}^{T-1} C(S_t(\omega), X^{\pi}(S_t(\omega)|\theta), W_{t+1}(\omega)). \tag{9.49}$$

We then average over sample paths to obtain

$$\bar{F}^{\pi}(\theta) = \frac{1}{K}\sum_{k=1}^{K} F^{\pi}(\theta|\omega^k). \tag{9.50}$$

(Class 4) State-dependent, final reward – We now have a hybrid of problem classes (1) and (3), where we use a learning policy $\Theta^{\pi^{lrn}}(S|\theta^{lrn})$ to learn the parameters of an implementation policy $X^{\pi^{imp}}(S_t|\theta^{imp})$, where the parameter $\theta^{imp} = \Theta^{\pi^{lrn}}(\theta^{lrn})$ that determines the behavior of the implementation policy depends on the learning policy π^{lrn} and its tunable parameters θ^{lrn}. We then have to evaluate the implementation policy, just as we evaluated the final design $x^{\pi,N}(\theta)$ in class (1), where $x^{\pi,N}(\theta)$ is the implementation decision that depends on the learning policy π and its parameters θ.

In class (1), we evaluated the implementation decision $x^{\pi,N}(\theta)$ by simulating $\widehat{W}$ to obtain estimates of $F(x^{\pi,N}, \widehat{W})$. Now we have to take an expectation over the state S which we do by simulating our implementation policy $X^{\pi^{imp}}(S_t|\theta^{imp})$ starting in state S_0 until the end of our horizon S_T. One simulation from 0 to T is comparable to an evaluation of $F(x, W)$. This means that a sample path ω, which in (1) was one observation of $W_1, \ldots, W_T$, is an observation of $(W_t^n, t = 1, \ldots, T), n = 0, \ldots, N$. This observation then produces the implementation policy $X^{\pi^{imp}}(S_t|\theta^{imp})$ (whereas in class (1) problems it produced the implementation decision $x^{\pi,N}(\theta|\omega)$).

To simulate the value of the policy, we simulate one last set of observations $\widehat{W}_1(\psi), \ldots, \widehat{W}_T(\psi)$ which, combined with our implementation policy which we write as $X^{\pi^{imp},N}(S_t|\theta^{imp}, \omega)$ produces a sequence of states $S_t(\psi)$, giving us the estimate

$$F^\pi(\theta^{lrn}|\omega, \psi) = \frac{1}{T} \sum_{t=0}^{T} C(S_t(\psi), X^{\pi^{imp}}(S_t(\psi)|\theta^{imp}, \omega), \widehat{W}_{t+1}(\psi)), \quad (9.51)$$

where we need to remember that $\theta^{imp} = \Theta^{\pi^{lrn}}(\theta^{lrn})$. We finally average over a set of K samples of ω, and L samples of ψ, giving us

$$\bar{F}^\pi(\theta^{lrn}) = \frac{1}{K}\frac{1}{L} \sum_{k=1}^{K} \sum_{\ell=1}^{L} F^\pi(\theta^{lrn}|\omega^k, \psi^\ell), \quad (9.52)$$

We now have a way of computing the performance of a policy $\bar{F}^\pi(\theta)$, which may be a learning policy for classes (1) and (4), or an implementation (and learning) policy for classes (2) and (3).

9.13 Advanced Probabilistic Modeling Concepts**

Sequential decision problems introduce some very subtle issues when bridging with classical probability theory. This material is not important for readers who just want to focus on models and algorithms. However, understanding how the probability community thinks of stochastic dynamic programs provides a fresh perspective that brings a deep pool of theory from the probability community.

Section 9.13.1 provides a beginners introduction to what is known as a measure-theoretic view of information, which provides some basic concepts that are used throughout advanced research papers in stochastic optimization. Then, section 9.13.2 provides a short primer of terms that are widely used throughout stochastic optimization papers which represent what is arguably the most common uses of the measure-theoretic terminology presented in section 9.13.1. We emphasize that while these concepts are widely used in the mathematical research literature, they are not necessary for modeling and solving real problems.

9.13.1 A Measure-theoretic View of Information**

For readers interested in proving theorems or reading theoretical research articles, it is useful to have a more fundamental understanding of information.

When we work with random information processes and uncertainty, it is standard in the probability community to define a probability space, which consists of three elements. The first is the set of outcomes Ω, which is generally assumed to represent all possible outcomes of the information process (actually, Ω can include outcomes that can never happen). If these outcomes are discrete, then all we would need is the probability of each outcome $p(\omega)$.

It is nice to have a terminology that allows for continuous quantities. We want to define the probabilities of our events, but if ω is continuous, we cannot talk about the probability of an outcome ω. However we can talk about a set of outcomes $\mathcal{E}$ that represent some specific event (if our information is a price, the event $\mathcal{E}$ could be all the prices that constitute the event that the price is greater than some number). In this case, we can define the probability of an outcome $\mathcal{E}$ by integrating the density function $p(\omega)$ over all ω in the event $\mathcal{E}$.

Probabilists handle continuous outcomes by defining a set of events $\mathfrak{F}$, which is literally a "set of sets" because each element in $\mathfrak{F}$ is itself a set of outcomes in Ω. This is the reason we resort to the script font $\mathfrak{F}$ as opposed to our calligraphic font for sets; it is easy to read $\mathcal{E}$ as "calligraphic E" and $\mathfrak{F}$ as "script F." The set $\mathfrak{F}$ has the property that if an event $\mathcal{E}$ is in $\mathfrak{F}$, then its complement $\Omega \setminus \mathcal{E}$ is in $\mathfrak{F}$, and the union of any two events $\mathcal{E}_X \cup \mathcal{E}_Y$ in $\mathfrak{F}$ is also in $\mathfrak{F}$.

$\mathfrak{F}$ is called a "sigma-algebra" (which may be written "σ-algebra"), and is a countable union of events in Ω. An understanding of sigma-algebras is not important for computational work, but can be useful in certain types of proofs (the proof in section 5.10.3 is a good example). Sigma-algebras are without question one of the more arcane devices used by the probability community, but once they are mastered, they are a powerful theoretical tool (but useless for modeling or computation, which is the reason why we do not use them elsewhere).

Finally, it is required that we specify a probability measure denoted $\mathcal{P}$, which gives the probability (or density) of an outcome ω, which can then be used to compute the probability of an event in $\mathfrak{F}$.

We can now define a formal probability space for our exogenous information process as $(\Omega, \mathfrak{F}, \mathcal{P})$, sometimes known as the "holy trinity" in probability. If we wish to take an expectation of some quantity that depends on the information, say $Ef(W)$, then we would sum (or integrate) over the set $\mathcal{E} \in \mathfrak{F}$ multiplied by the probability (or density) $\mathcal{P}$.

This notation is especially powerful for "static" problems where there are two points in time: before we see the random variable W, and after. This creates a challenge when we have sequential problems where information evolves over time. Probabilists have adapted the original concept of probability spaces $(\Omega, \mathfrak{F}, \mathcal{P})$ by manipulating the set of events $\mathfrak{F}$, as we show next.

It is important to emphasize that ω represents *all* the information that will become available, over all time periods. As a rule, we are solving a problem at time t, which means we do not have the information that will become available after time t. To handle this, we let $\mathfrak{F}_t$ be the sigma-algebra representing events that can be created using only the information up to time t. To illustrate, consider an information process W_t consisting of a single 0 or 1 in each time period. W_t may be the information that a customer purchases a jet aircraft, or the event that an expensive component in an electrical network fails. If we look over three time periods, there are eight possible outcomes, as shown in Table 9.4.

Let $\mathcal{E}_{\{W_1\}}$ be the set of outcomes ω that satisfy some logical condition on W_1. If we are at time $t = 1$, we only see W_1. The event $W_1 = 0$ would be written

$$\mathcal{E}_{\{W_1=0\}} = \{\omega | W_1 = 0\} = \{1, 2, 3, 4\}.$$

The sigma-algebra $\mathfrak{F}_1$ would consist of the events

$$\{\mathcal{E}_{\{W_1=0\}}, \mathcal{E}_{\{W_1=1\}}, \mathcal{E}_{\{W_1 \in \{0,1\}\}}, \mathcal{E}_{\{W_1 \notin \{0,1\}\}}\}.$$

Now assume that we are at time $t = 2$ and have access to W_1 and W_2. With this information, we are able to divide our outcomes Ω into finer subsets. Our history H_2 consists of the elementary events $\mathcal{H}_2 = \{(0,0), (0,1), (1,0), (1,1)\}$. Let $h_2 = (0,1)$ be an element of H_2. The event $\mathcal{E}_{\{h_2=(0,1)\}} = \{3, 4\}$. At time $t = 1$, we could not tell the difference between outcomes 1, 2, 3, and 4; now that we are at time 2, we can differentiate between $\omega \in \{1, 2\}$ and $\omega \in \{3, 4\}$. The sigma-algebra $\mathfrak{F}_2$ consists of all the events $\mathcal{E}_{h_2}, h_2 \in \mathcal{H}_2$, along with all possible unions and complements.

Table 9.4 Set of demand outcomes.

Outcome	Time period		
ω	1	2	3
1	0	0	0
2	0	0	1
3	0	1	0
4	0	1	1
5	1	0	0
6	1	0	1
7	1	1	0
8	1	1	1

Another event in $\mathfrak{F}_2$ is $\{\omega|(W_1, W_2) = (0,0)\} = \{1,2\}$. A third event in $\mathfrak{F}_2$ is the union of these two events, which consists of $\omega = \{1,2,3,4\}$ which, of course, is one of the events in $\mathfrak{F}_1$. In fact, every event in $\mathfrak{F}_1$ is an event in $\mathfrak{F}_2$, but not the other way around. The reason is that the additional information from the second time period allows us to divide $\mathfrak{F}$ into finer set of subsets. Since $\mathfrak{F}_2$ consists of all unions (and complements), we can always take the union of events, which is the same as ignoring a piece of information.

By contrast, we cannot divide $\mathfrak{F}_1$ into a finer subsets. The extra information in $\mathfrak{F}_2$ allows us to filter Ω into a finer set of subsets than was possible when we only had the information through the first time period. If we are in time period 3, $\mathfrak{F}$ will consist of each of the individual elements in Ω as well as all the unions needed to create the same events in $\mathfrak{F}_2$ and $\mathfrak{F}_1$.

From this example, we see that more information (that is, the ability to see more elements of $W_1, W_2, ...$) allows us to divide Ω into finer-grained subsets. For this reason, we can always write $\mathfrak{F}_{t-1} \subseteq \mathfrak{F}_t$. $\mathfrak{F}_t$ always consists of every event in $\mathfrak{F}_{t-1}$ in addition to other finer events. As a result of this property, $\mathfrak{F}_t$ is termed a *filtration*. It is because of this interpretation that the sigma-algebras are typically represented using the script letter F (which literally stands for filtration) rather the more natural letter H (which stands for history). The fancy font used to denote a sigma-algebra is used to designate that it is a set of sets (rather than just a set).

It is *always* assumed that information processes satisfy $\mathfrak{F}_{t-1} \subseteq \mathfrak{F}_t$. Interestingly, this is not always the case in practice. The property that information forms a filtration requires that we never "forget" anything. In real applications, this is not always true. Assume, for example, that we are doing forecasting using a moving average. This means that our forecast f_t might be written as $f_t = (1/T) \sum_{t'=1}^{T} \hat{D}_{t-t'}$. Such a forecasting process "forgets" information that is older than T time periods.

By far the most widespread use of the notation $\mathfrak{F}_t$ is to represent the information we know at time t. For example, let W_{t+1} be the information that we will learn at time $t + 1$. If we are sitting at time t, we might use a forecast $f_{t,t+1}^W$ which would be written

$$f_{t,t+1}^W = \mathbb{E}\{W_{t+1}|\mathfrak{F}_t\}. \tag{9.53}$$

Conditioning on $\mathfrak{F}_t$ means conditioning on what we know at time t which some authors will write as

$$f_{t,t+1}^W = \mathbb{E}_t W_{t+1}. \tag{9.54}$$

Equations (9.53) and (9.54) are equivalent, and both would be read "the conditional expectation of W_{t+1} given what we know at time t."

If we do not include this conditioning, then this is the same as an expectation we would make at time 0, which we could write

$$
\begin{aligned}
f_{0,t+1}^W &= \mathbb{E}W_{t+1} \\
&= \mathbb{E}\{W_{t+1}|\mathfrak{F}_0\}.
\end{aligned}
$$

There are numerous textbooks on measure theory. For a nice introduction to measure-theoretic thinking (and in particular the value of measure-theoretic thinking), see Pollard (2002) for an introduction to measure-theoretic probability, or the advanced text Cinlar (2011). For an illustration of mathematics using this notation, see the "More modern proof" of convergence for stochastic gradient algorithms in section 5.10.3.

9.13.2 Policies and Measurability

An immediate use of our new measure-theory vocabulary is to communicate the relatively simple concept that decisions have to be made without using information from the future. What this section will do is to allow you to talk like a trained stochastic optimizer, but you will also learn a simpler, and perhaps more accurate, way of communicating this simple idea.

As before, let x_t be a decision at time t. The decision x_t, made at time t, depends on the information that has arrived up to time t. The standard mathematical style is to express this dependence by writing the decision as $x_t(\omega)$, where ω represents the sample path as described in section 9.6 and illustrated in Table 9.2. It is important to remember that when we use ω, we are specifying the *entire* sample path over the horizon $0, \dots, T$. This means that we are allowing x_t to "see" not only the entire history, but the entire future!

The probability community has learned how to fix this problem. The following statements all mean the decision x_t depends only on information available up to and including time t:

- "x_t is $\mathcal{F}_t$ measurable." – The fast translation of this statement is that "x_t only uses information that is known at time t." Given our tutorial in the previous subsection, we can provide a little more background. Recall (from earlier) that $\mathfrak{F}_t$ is a set of sets, where one of the sets in $\mathfrak{F}_t$ will consist of all the sample paths ω that have the same history $h_t(\omega) = (W_1, \dots, W_t)$ but without regard to the outcomes of $W_{t+1}, \dots, W_T$. Let $\mathcal{E}_t(h_t)$ be the elementary event that includes all ω where $h_t(\omega) = h_t$ (remember that $\mathfrak{F}_t$ consists of all unions and complements, which means that $\mathfrak{F}_t$ will include events of all ω where $h_{t-1}(\omega) = h_{t-1}$).

  Now, any sample path belonging to $\mathcal{E}_t(h_t)$ should produce the same decision x_t. So, for each elementary event $\mathcal{E}_t(h_t)$ (remember that there is one h_t for each ω) there is a decision, which means we can create a set of decisions that we will call $\mathfrak{X}_t$, where there is a one-to-one correspondence between sets in $\mathfrak{F}_t$ and sets in $\mathfrak{X}_t$.

  Assuming that the sample paths ω are discrete (this would be the case whenever we use a sampled set of ω's), we assume that we have a probability $p(\omega)$ for each ω (probabilists refer to $p^W(\omega)$ as a *measure*). We can find the probability that each set of decisions in $\mathfrak{X}_t$ occur by finding the corresponding set of ω's in $\mathfrak{F}_t$. So if $\mathcal{E}_t$ is an elementary set in $\mathfrak{F}_t$, we can compute its probability using

  $$P(\mathcal{E}_t) = \sum_{\omega \in \mathcal{E}_t} p^W(\omega).$$

  Then, for each elementary event $\mathcal{E}_t$ there is a single decision $x_t(\mathcal{E}_t)$ which occurs with probability $P(\mathcal{E}_t)$. From this thinking, we can compute the probability of each event in $\mathfrak{X}_t$. So the measure on the sets in $\mathfrak{X}_t$ are computed from the probabilities we already computed from the probabilities in $\mathfrak{F}_t$.

  This is what is meant by saying that a decision x_t is $\mathfrak{F}_t$-measurable.

- "x_t is nonanticipative" – We first encountered "nonanticipativity" in chapter 2 where we introduced "nonanticipativity constraints" in section 2.1.8 when we introduced two-stage stochastic programs (see in particular equation (2.25)). This is really just another way of saying that x_t cannot depend on the actual outcome of $W_{t'}$ for any $t' > t$.

- "x_t is an adapted policy." – This is nothing more than another way of saying that x_t can only depend on what we know up to time t (or that x_t is $\mathfrak{F}_t$-measurable), which in turn means that x_t "adapts" to new information. As we move forward in time, the decision "adapts" to the new information.

- "τ is a stopping time," – For optimal stopping problems, where we are looking to sell an asset at time τ that depends on the price process, we say that "τ is a stopping time," which means that the decision to sell at time $\tau = t$ must be $\mathcal{F}_t$ measurable.

All of these statements require some mathematical sophistication to understand, and they all mean:

$$x_t = X^\pi(S_t) \text{ is a function of the state } S_t.$$

By constructing our policy as depending on the state S_t, we guarantee that the decision does not have access to any information from the future. This follows immediately from our transition function

$$S_t = S^M(S_{t-1}, x_{t-1}, W_t)$$

which tells us that S_t is only a function of W_t, as well as S_{t-1} and x_{t-1}. By repeating this, we see that S_t is only a function of $S_0, W_1, \dots, W_t$.

This (much simpler) discussion also brings out that we are not actually interested in the entire history $h_t = (S_0, W_1, \dots, W_t)$. We really only need the state S_t. For example, in an inventory problem, we only care about how much inventory we have at time t, but if we want to compute the probability of a decision x_t, we need the probability of being in the state S_t, which means we need to know the set of outcomes ω that led us to state S_t. Not surprisingly, this also depends on the prior decisions $x_0, \dots, x_{t-1}$, which depends on the policy that produced these decisions. Sounds complicated, but we will never actually need to compute the probability of a decision. What we do need is the expected performance of the policy, which we estimate using simulation.

We can conclude from this discussion that you do not need to understand "x_t is $\mathfrak{F}_t$-measurable," beyond understanding that it just means that x_t only has access to information that has arrived on or before time t. All you really need to understand is that x_t depends only on the state S_t, but this means you need to understand what a state variable is. Every theoretician working in stochastic optimization understands "$\mathfrak{F}_t$-measurable," but there are many who do not know what a state variable is.

9.14 Looking Forward

We are not quite done with modeling. Chapter 10 addresses the rich area of modeling uncertainty which comes in a number of forms. For some applications, it can easily be argued that modeling uncertainty is more important than pursuing optimal policies. However, even books have to limit what they can cover.

After we provide a basic introduction to uncertainty modeling, the rest of the book focuses on designing policies. This material is organized as follows:

Designing policies (chapter 11) – This chapter describes four fundamental (meta) classes of policies, called policy function approximations (PFAs), cost function approximations (CFAs), policies based on value function approximations (VFAs), and direct lookahead policies (DLAs).

Policy function approximations (chapter 12) – The simplest class of policies are policy function approximations, which is where we describe a policy as some sort of analytical function (lookup tables, parametric or nonparametric functions)

Cost function approximations (chapter 13) – Here we find approximations of cost functions which we then minimize (possibly subject to a set of constraints, which we might also modify).

Value function approximations (chapters 14–18) – These chapters develop policies based on value functions. Given the richness of this general approach, we present this material in a series of chapters as follows:

Exact dynamic programming (chapter 14) – This is the classical material on dynamic programs with discrete states, discrete actions, and randomness that is simple enough that we can take expectations.

Backward approximate dynamic programming (chapter 15) – This is the first of a series of chapters that present iterative methods for learning approximations of value functions. In this chapter, we introduce a technique we call *backward approximate dynamic programming* since it builds on classical "backward" methods of Markov decision processes presented in chapter 14. The rest of the material on approximate value functions focuses on "forward" methods.

Forward approximate dynamic programming I (chapter 16) – We begin with a presentation of methods for approximating value functions using forward methods. In this chapter, the policy is fixed.

Forward approximate dynamic programming II (chapter 17) – We build on the tools in chapter 16 but now we use our approximate value functions to define our policy.

Forward approximate dynamic programming III (chapter 18) – This chapter focuses on the important special case where the value function is convex in the state variable. This arises in a applications that involve the allocation of resources.

Direct lookahead approximations (chapter 19) – The last class of policies optimizes an approximate lookahead model. We deal with two important problem classes: where decisions are discrete (or discretized), and it is possible to enumerate all actions, and where the decision x is a vector, making it impossible to enumerate all actions.

Multiagent modeling and learning (chapter 20) – We close by addressing the important topic of multiagent modeling, which arises in a wide array of applications, from controlling a fleet of drones or robots, modeling teams of medical technicians or soldiers, or modeling a global supply chain. Multiagent modeling introduces the need for modeling communication, an issue that does not arise in this chapter. We start by modeling basic learning problems as a two-agent system.

9.15   Bibliographic Notes

This chapter is a revised version of Chapter 5 from Powell (2011). To our knowledge, this book (and its predecessor in Powell (2011)) are the only books to clearly articulate the five elements of sequential decision problems in this way. However, as we review in Powell (2021) (available on arXiv), our framework closely follows the general style used throughout the optimal control community, with a few minor tweaks, and some major ones. We view the minor tweaks as consisting of:

- We switch from the standard notation of the controls community that uses state x_t and "control" u_t to reflect the substantial community in math programming that uses x for decisions, and we adopt the standard (and more mnemonic) S_t for state (we use a capital letter following the standard style of the applied probability community).
- We use $S^M(s, x, w)$ for the transition function rather than $f(s, x, w)$ for the simple reason that "$f(\cdot)$" is too popular for modeling a wide range of functions. $S^M(\cdot)$ has the mnemonic "state model" or "system model."
- The controls community often writes

$$x_{t+1} = f(x_t, u_t, w_t)$$

where w_t is random at time t (see, for example, Bertsekas (2017). This notation is inherited from continuous time models. We use

$$S_{t+1} = S^M(S_t, x_t, W_{t+1})$$

where W_{t+1} is random at time t, but known at $t + 1$. This notation allows us to keep to the convention that any variable indexed by t is known at time t.

It is surprisingly common in the controls literature to see people writing the objective function (for stochastic problems) as

$$\min_{u_0,\ldots,u_T} \mathbb{E} \sum_{t=0}^{T} p_t u_t,$$

where (for this example) the prices p_t vary randomly over time (there may be other random elements in the constraints). The problem is that writing $\min_{u_0,\ldots,u_T}$ does not recognize that u_t is a random variable. Mathematically sophisticated authors understand that u_t is random, and can be written $u_t(\omega)$ where ω is a sample path of any random information. It is important to require that "u_t be $\mathcal{F}_t$-measurable," which recognizes that u_t is a function, but it does not provide any indication of how to construct the policy. Often, authors are simply assuming that we will find an optimal policy by solving the Hamilton-Jacobi-Bellman equations, without recognizing that this is often not possible (even approximately).

Our modeling style would write the objective function as

$$\max_{\pi} \mathbb{E}\left\{ \sum_{t=0}^{T} C(S_t, X^\pi(S_t))|S_0 \right\}$$

where we explicitly search over policies (the switch from min to max is simple preference). Then, we identify four specific classes of policies, which brings transparency to all the approaches that might be used.

Our modeling approach, then, clearly separates the model (which requires searching over policies) from how we solve the model, which we do by designing policies from the four classes.

Section 9.3 – Figure 9.2 which describes the mapping from continuous to discrete time was outlined for me by Erhan Cinlar.

Section 9.4 – The definition of states is amazingly confused in the literature on sequential decision problems. The first recognition of the difference between the physical state and the belief state appears to be in Bellman and Kalaba (1959) which used the term "hyperstate" to refer to the belief state, making the distinction from "physical states" which, even today, are equated by many authors with "state variable."

The control literature has long used state to represent a sufficient statistic (see, for example, Kirk (2012)), representing the information needed to model the system forward in time. For an introduction to partially observable Markov decision processes, see White (1991). An excellent description of the modeling of Markov decision processes from an AI perspective is given in Boutilier et al. (1999), including a very nice discussion of factored representations of state variables. See also Guestrin et al. (2003) for an application of the concept of factored state spaces to a Markov decision process.

The definition of a state variable here refines the definition introduced in Powell (2011).

Section 9.5 – Our notation for decisions represents an effort to bring together the fields of dynamic programming and math programming. We believe this notation was first used in Powell et al. (2001). For a classical treatment of decisions from the perspective of Markov decision processes, see Puterman (2005). For examples of decisions from the perspective of the optimal control community, see Kirk (2012) and Lewis and Vrabie (2012). For examples of treatments of dynamic programming in economics, see Stokey and R. E. Lucas (1989) and Chow (1997).

Section 9.6 – Our representation of information follows classical styles in the probability literature (see, for example, Chung (1974)). Considerable attention has been given to the topic of supervisory control (see, for example, Werbos (1992)).

Section 9.7 – The concept of a "transition function" (which has been given a number of different names) is absolutely standard in the controls community, and yet surprisingly absent throughout the other communities (and in particular Markov decision processes, which seems to insist on using one-step transition matrices). See any of the books on optimal control (Kirk (2012), Stengel (1986), Sontag (1998), Sethi (2019), and Lewis and Vrabie (2012)). Bertsekas (2017) opens his book by stating the transition function, but then switches to using transition matrices (or kernels for continuous states), which require taking the expectation of the transition function.

Section 9.11 – Our identification of the four classes of objectives (final reward and cumulative reward, state-independent problems and state-dependent problems) was first presented in Powell (2019), although the material in section 9.12 is new.

Section 9.9 – The energy storage example is taken from Powell (2021) (also available on arXiv).

Exercises

Review questions

9.1 What is the difference between the *history* of a process, and the state of a process?

9.2 What is meant by a "martingale model of forecast evolution?"

9.3 What are the five components of a sequential decision problem?

9.4 What are the three types of state variables? Give an example of each.

9.5 What may the exogenous information W_{t+1} depend on?

9.6 Assuming the state S_t is discrete, how do you compute the one-step probability transition matrix from the transition function, assuming you know the probability distribution of W_{t+1}?

9.7 Write out, and explain, the objective functions for the following four cases:

    (1) State-independent problems, final reward.
    (2) State-independent problems, cumulative reward.
    (3) State-dependent problems, cumulative reward.
    (4) State-dependent problems, final reward.

Modeling questions

9.8 A traveler needs to traverse the graph shown in Figure 9.7 from node 1 to node 11 where the goal is to find the path that minimizes the sum of the costs over the path. To solve this problem we are going to use the deterministic version of Bellman's optimality equation that states

$$V(s) = \min_{a \in \mathcal{A}_s} \left(c(s, a) + V(s'(s, a)) \right) \qquad (9.55)$$

where $s'(s, a)$ is the state we transition to when we are in state s and take action $a \in \mathcal{A}_s$. The set $\mathcal{A}_s$ is the set of actions (in this case, traversing over a link) available when we are in state s.

To solve this problem, answer the following questions:

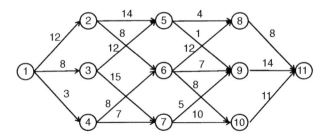

Figure 9.7 Deterministic shortest path problem.

(a) Describe an appropriate state variable for this problem (with notation).
(b) If the traveler is at node 6 by following the path 1-2-6, what is her state?
(c) Find the path that minimizes the sum of the costs on the links traversed by the traveler. Using Bellman's equation (14.2), work backward from node 11 and find the best path from each node to node 11, ultimately finding the best path from node 1 to node 11. Show your solution by drawing the graph with the links that fall on an optimal path from some node to node 11 drawn in bold.

9.9 A traveler needs to traverse the graph shown in Figure 9.8 from node 1 to node 11, where the goal is to find the path that minimizes the *largest* cost of all the links on the path. To solve this problem, answer the following questions:

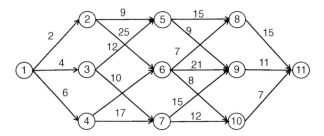

Figure 9.8 A path problem minimizing the largest cost of all the links on a path.

(a) Describe an appropriate state variable for this problem (with notation).
(b) If the traveler is at node 6 by following the path 1-2-6, what is her state?
(c) Using Bellman's equation (14.2), find the path (or paths) that minimizes the largest costs on the links traversed by the traveler. For each decision point (the nodes in the graph), give the value of the state variable corresponding to the optimal path to that decision point, and the value of being in that state (that is, the cost if we start in that state and then follow the optimal solution).

9.10 Repeat exercise 9.9, but this time minimize the *second largest* arc cost on a path.

9.11   A traveler needs to traverse the graph shown in Figure 9.9 from node 1 to node 11, where the goal is to find the path that minimizes the *second largest* link cost along the path. To solve this problem, answer the following questions:

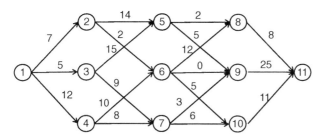

Figure 9.9   A path problem minimizing the product of the costs.

(a) Describe an appropriate state variable for this problem (with notation).

(b) If the traveler is at node 6 by following the path 1-2-6, what is her state?

(c) If the traveler is at node 10 by following the path 1-2-6-10, what is her state?

(d) Using Bellman's equation (14.2), find the path (or paths) that minimizes the product of the costs on the links traversed by the traveler. For each decision point (the nodes in the graph), give the value of the state variable corresponding to the optimal path to that decision point, and the value of being in that state (that is, the cost if we start in that state and then follow the optimal solution).

9.12   Consider our basic newsvendor problem

$$\max_{x} \mathbb{E}_D F(x, D) = \mathbb{E}_D \big(p \min\{x, D\} - cx \big). \tag{9.56}$$

Show how the following variations of this problem can be modeled using the universal modeling framework:

(a) The final reward formulation of the basic newsvendor problem.

(b) The cumulative reward formulation of the basic newsvendor problem.

(c) The asymptotic formulation of the newsvendor problem. What are the differences between the asymptotic formulation and the final reward formulation?

9.13 Now consider a dynamic version of our newsvendor problem where a decision x_t is made at time t by solving

$$\max_x \mathbb{E}_D F(x, D) = \mathbb{E}_D(p_t \min\{x, D_{t+1}\} - cx). \tag{9.57}$$

Assume that the price p_t is independent of prior history.

(a) Model the cumulative reward version of the newsvendor problem in (9.57).

(b) How does your model change if we are instead solving

$$\max_x \mathbb{E}_D F(x, D) = \mathbb{E}_D(p_{t+1} \min\{x, D_{t+1}\} - cx). \tag{9.58}$$

where we continue to assume that the price, which is now p_{t+1}, is independent of prior history.

(c) How does your model of (9.58) change if

$$p_{t+1} = \theta_0 p_t + \theta_1 p_{t-1} + \varepsilon_{t+1}, \tag{9.59}$$

where ε_{t+1} is a zero-mean noise term, independent of the state of the system.

9.14 We continue the newsvendor problem in exercise 9.13, but now assume that (θ_0, θ_1) in equation (9.59) are unknown. At time t, we have estimates $\bar{\theta}_t = (\bar{\theta}_{t0}, \bar{\theta}_{t1})$. Assume the true θ is now a random variable that follows a multivariate normal distribution with mean $\mathbb{E}_t \theta = \bar{\theta}_t$ which we initialize to

$$\bar{\theta}_0 = \begin{pmatrix} 20 \\ 40 \end{pmatrix},$$

and covariance matrix Σ_t^θ which we initialize to

$$\Sigma_0^\theta = \begin{pmatrix} \sigma_{00}^2 & \sigma_{01}^2 \\ \sigma_{10}^2 & \sigma_{11}^2 \end{pmatrix}.$$

$$= \begin{pmatrix} 36 & 16 \\ 16 & 25 \end{pmatrix}.$$

Drawing on the updating equations in section 3.4.2, give a full model of this problem using a cumulative reward objective function (that is, give the state, decision and exogenous information variables, transition function and objective unction).

9.15 See the series of variants of our familiar newsvendor (or inventory) problem. In each, describe the pre- and post-decision states, decision and exogenous information in the form:

$$(S_0, x_0, S_0^x, W_1, S_1, x_1, S_1^x, W_2, ...)$$

Specify S_t, S_t^x, x_t and W_t in terms of the variables of the problem.

(a) The basic newsvendor problem where we wish to find x that solves

$$\max_x \mathbb{E}\{p \min(x, \hat{D}) - cx\} \tag{9.60}$$

where the distribution of $\hat{D}$ is unknown.

(b) The same as (a), but now we are given a price p_t at time t and asked to solve (9.60) using this information. Note that p_t is unrelated to any prior history or decisions.

(c) Repeat (b), but now $p_{t+1} = p_t + \hat{p}_{t+1}$.

(d) Repeat (c), but now leftover inventory is held to the next time period.

(e) Of the problems mentioned, which (if any) are *not* dynamic programs? Explain.

(f) Of the problems mentioned, which would be classified as solving state-dependent vs. state-independent functions.

9.16 In this exercise you are going to model an energy storage problem, which is a problem class that arises in many settings (how much cash to keep on hand, how much inventory on a store shelf, how many units of blood to hold, how many milligrams of a drug to keep in a pharmacy, ...). We will begin by describing the problem in English with a smattering of notation. Your job will be to develop it into a formal dynamic model. Our problem is to decide how much energy to purchase from the electric power grid at a price p_t. Let x_t^{gs} be the amount of power we buy (if $x^{gs} > 0$) or sell (if $x^{gs} < 0$). We then have to decide how much energy to move from storage to meet the demand D_t in a commercial building, where $x_t^{sb} \geq 0$ is the amount we move to the building to meet the demand D_t. Unsatisfied demand is penalized at a price c per unit of energy. Assume that prices evolve according to a time-series model given by

$$p_{t+1} = \theta_0 p_t + \theta_1 p_{t-1} + \theta_2 p_{t-2} + \varepsilon_{t+1}, \tag{9.61}$$

where ε_{t+1} is a random variable with mean 0 that is independent of the price process. We do not know the coefficients θ_i for $i = 0, 1, 2$, so instead we use estimates $\bar{\theta}_{ti}$. As we observe p_{t+1}, we can update the vector $\bar{\theta}_t$ using the recursive formulas for updating linear models as described in chapter 3, section 3.8 (you will need to review this section to answer parts of this question).

Every time period we are given a forecast $f_{tt'}^D$ of the demand $D_{t'}$ at time t' in the future, where $t' = t, t+1, t+H$. We can think of $f_{tt}^D = D_t$ as the actual demand. We can also think of the forecasts $f_{t+1,t'}^D$ as the "new information" or define a "change in the forecast" $\hat{f}^D_{t+1,t'}$ in which case we would write

$$f^D_{t+1,t'} = f^D_{tt'} + \hat{f}^D_{t+1,t'}.$$

(a) What are the elements of the state variable S_t (we suggest filling in the other elements of the model to help identify the information needed in S_t). Define both the pre- and post-decision states.

(b) What are the elements of the decision variable x_t? What are the constraints (these are the equations that describe the limits on the decisions). Finally introduce a function $X^\pi(S_t)$ which will be our policy for making decisions to be designed later (but we need it in the objective function that will be explained subsequently).

(c) What are the elements of the exogenous information variable W_{t+1} that become known at time $t+1$ but which were not known at time t.

(d) Write out the transition function $S_{t+1} = S^M(S_t, x_t, W_{t+1})$, which are the equations that describe how each element of the state variable S_t evolves over time. There needs to be one equation for each state variable.

(e) Write out the objective function by writing:
The contribution function $C(S_t, x_t)$.
The objective function where you maximize expected profits over some general set of policies (to be defined later – not in this exercise).

9.17 Patients arrive at a doctor's office, each of whom are described by a vector of attribute $a = (a_1, a_2, \ldots, a_K)$ where a might describe age, gender, height, weight, whether the patient smokes, and so on. Let a^n be the attribute vector describing the nth patient. For each patient, the doctor makes a decision x^n (surgery, drug regimens, rehabilitation), and then observes an outcome y^n for patient n. From y^n, we obtain an updated estimate θ^n for the parameters of a nonlinear model $f(x|\theta)$ that helps us to predict y for other patients.

(a) Give the five elements of this decision problem. Be sure to model the state after a patient arrives (this would be the pre-decision state), S^n, after a decision is made (this would be the post-decision state), $S^{x,n}$ and after the outcome of a decision becomes known, $S^{y,n}$.

(b) The value of being in a state S^n can be computed using Bellman's equation

$$V^n(S^n) = \max_{x \in \mathcal{X}} \left((C(S^n, x) + E_W\{V^{n+1}(S^{n+1})|S^n, x\} \right). \tag{9.62}$$

Define the value of being in the state (i) after a patient arrives, (ii) after a decision is made, and (iii) before a patient arrives. Call these $V(S)$, $V^x(S^x)$, and $V^y(S^y)$. Write $V(S^n)$ as a function of $V^x(S^{x,n})$ and write $V^x(S^{x,n})$ as a function of $V^y(S^{y,n})$.

9.18 Consider the problem of controlling the amount of cash a mutual fund keeps on hand. Let R_t be the cash on hand at time t. Let $\hat{R}_{t+1}$ be the net deposits (if $\hat{R}_{t+1} > 0$) or withdrawals (if $\hat{R}_{t+1} < 0$), where we assume that $\hat{R}_{t+1}$ is independent of $\hat{R}_t$. Let M_t be the stock market index at time t, where the evolution of the stock market is given by $M_{t+1} = M_t + \hat{M}_{t+1}$ where $\hat{M}_{t+1}$ is independent of M_t. Let x_t be the amount of money moved from the stock market into cash ($x_t > 0$) or from cash into the stock market ($x_t < 0$).

(a) Give a complete model of the problem, including both pre-decision and post-decision state variables.

(b) Suggest a simple parametric policy function approximation, and give the objective function as an online learning problem.

9.19 A college student must plan what courses she takes over each of eight semesters. To graduate, she needs 34 total courses, while taking no more than five and no less than three courses in any semester. She also needs two language courses, one science course, eight departmental courses in her major and two math courses.

(a) Formulate the state variable for this problem in the most compact way possible.

(b) Give the transition function for our college student assuming that she successfully passes any course she takes. You will need to introduce variables representing her decisions.

(c) Give the transition function for our college student, but now allow for the random outcome that she may not pass every course.

9.20 A broker is working in thinly traded stocks. He must make sure that he does not buy or sell in quantities that would move the price and he feels that if he works in quantities that are no more than 10 percent of the average sales volume, he should be safe. He tracks the average sales volume of a particular stock over time. Let $\hat{v}_t$ be the sales volume on day t, and assume that he estimates the average demand f_t using $f_t = (1 - \alpha)f_{t-1} + \alpha\hat{v}_t$. He then uses f_t as his estimate of the sales volume for the next day. Assuming he started tracking demands on day $t = 1$, what information would constitute his state variable?

9.21 How would your previous answer change if our broker used a 10-day moving average to estimate his demand? That is, he would use $f_t = 0.10 \sum_{i=1}^{10} \hat{v}_{t-i+1}$ as his estimate of the demand.

9.22 The pharmaceutical industry spends millions managing a sales force to push the industry's latest and greatest drugs. Assume one of these salesmen must move between a set $\mathcal{J}$ of customers in his district. He decides which customer to visit next only after he completes a visit. For this exercise, assume that his decision does not depend on his prior history of visits (that is, he may return to a customer he has visited previously). Let S_n be his state immediately after completing his n^{th} visit that day.

(a) Assume that it takes exactly one time period to get from any customer to any other customer. Write out the definition of a state variable, and argue that his state is only his current location.

(b) Now assume that τ_{ij} is the (deterministic and integer) time required to move from location i to location j. What is the state of our salesman at any time t? Be sure to consider both the possibility that he is at a location (having just finished with a customer) or between locations.

(c) Finally assume that the travel time τ_{ij} follows a discrete uniform distribution between a_{ij} and b_{ij} (where a_{ij} and b_{ij} are integers)?

9.23 Consider a simple asset acquisition problem where x_t is the quantity purchased at the end of time period t to be used during time interval $t + 1$. Let D_t be the demand for the assets during time interval t. Let R_t be the pre-decision state variable (the amount on hand before you have ordered x_t) and R_t^x be the post-decision state variable.

(a) Write the transition function so that R_{t+1} is a function of R_t, x_t, and D_{t+1}.

(b) Write the transition function so that R_t^x is a function of R_{t-1}^x, D_t, and x_t.

(c) Write R_t^x as a function of R_t, and write R_{t+1} as a function of R_t^x.

9.24 As a buyer for an orange juice products company, you are responsible for buying futures for frozen concentrate. Let $x_{tt'}$ be the number of futures you purchase in year t that can be exercised during year t'.

(a) What is your state variable in year t?

(b) Write out the transition function.

9.25 A classical inventory problem works as follows. Assume that our state variable R_t is the amount of product on hand at the end of time period t and that D_t is a random variable giving the demand during time interval $(t-1, t)$ with distribution $p_d = P(D_t = d)$. The demand in time interval t must be satisfied with the product on hand at the beginning of the period. We can then order a quantity x_t at the end of period t that can be used to replenish the inventory in period $t + 1$. Give the transition function that relates R_{t+1} to R_t.

9.26 Many problems involve the movement of resources over networks. The definition of the state of a single resource, however, can be complicated by different assumptions for the probability distribution for the time required to traverse a link. For each example, give the state of the resource:

(a) You have a deterministic, static network, and you want to find the shortest path from an origin node q to a destination node r. There is a known cost c_{ij} for traversing each link (i, j).

(b) Next assume that the cost c_{ij} is a random variable with an unknown distribution. Each time you traverse a link (i, j), you observe the cost $\hat{c}_{ij}$, which allows you to update your estimate $\bar{c}_{ij}$ of the mean of c_{ij}.

(c) Finally assume that when the traveler arrives at node i he sees $\hat{c}_{ij}$ for each link (i, j) out of node i.

(d) A taxicab is moving people in a set of cities $\mathcal{C}$. After dropping a passenger off at city i, the dispatcher may have to decide to reposition the cab from i to j, $(i, j) \in \mathcal{C}$. The travel time from i to j is τ_{ij}, which is a random variable with a discrete uniform distribution (that is, the probability that $\tau_{ij} = t$ is $1/T$, for $t = 1, 2, \dots, T$). Assume that the travel time is known before the trip starts.

(e) Same as (d), but now the travel times are random with a geometric distribution (that is, the probability that $\tau_{ij} = t$ is $(1 - \theta)\theta^{t-1}$, for $t = 1, 2, 3, \dots$).

9.27 As the purchasing manager for a major citrus juice company, you have the responsibility of maintaining sufficient reserves of oranges for sale or conversion to orange juice products. Let x_{ti} be the amount of oranges that you decide to purchase from supplier i in week t to be used in week $t+1$. Each week, you can purchase up to $\hat{q}_{ti}$ oranges (that is, $x_{ti} \leq \hat{q}_{ti}$) at a price $\hat{p}_{ti}$ from supplier $i \in \mathcal{I}$, where the price/quantity pairs $(\hat{p}_{ti}, \hat{q}_{ti})_{i \in \mathcal{I}}$ fluctuate from week to week. Let s_0 be your total initial inventory of oranges, and let D_t be the number of oranges that the company needs for production during week t (this is our demand). If we are unable to meet demand, the company must purchase additional oranges on the spot market at a spot price $\hat{p}_{ti}^{spot}$.

(a) What is the exogenous stochastic process for this system?
(b) What are the decisions you can make to influence the system?
(c) What would be the state variable for your problem?
(d) Write out the transition equations.
(e) What is the one-period contribution function?
(f) Propose a reasonable structure for a decision rule for this problem, and call it X^{π}. Your decision rule should be in the form of a function that determines how much to purchase in period t.
(g) Carefully and precisely, write out the objective function for this problem in terms of the exogenous stochastic process. Clearly identify what you are optimizing over.
(h) For your decision rule, what do we mean by the space of policies?

9.28 Customers call in to a service center according to a (nonstationary) Poisson process. Let $\mathcal{E}$ be the set of events representing phone calls, where $t_e, e \in \mathcal{E}$ is the time that the call is made. Each customer makes a request that will require time τ_e to complete and will pay a reward r_e to the service center. The calls are initially handled by a receptionist who determines τ_e and r_e. The service center does not have to handle all calls and obviously favors calls with a high ratio of reward per time unit required (r_e/τ_e). For this reason, the company adopts a policy that the call will be refused if $(r_e/\tau_e) < \gamma$. If the call is accepted, it is placed in a queue to wait for one of the available service representatives. Assume that the probability law driving the process is known, where we would like to find the right value of γ.

(a) This process is driven by an underlying exogenous stochastic process with element $\omega \in \Omega$. What is an instance of ω?
(b) What are the decision epochs?
(c) What is the state variable for this system? What is the transition function?
(d) What is the action space for this system?
(e) Give the one-period reward function.
(f) Give a full statement of the objective function that defines the Markov decision process. Clearly define the probability space over which the expectation is defined, and what you are optimizing over.

9.29 A major oil company is looking to build up its storage tank reserves, anticipating a surge in prices. It can acquire 20 million barrels of oil, and it would like to purchase this quantity over the next 10 weeks (starting in week 1). At the beginning of the week, the company contacts its usual sources, and each source $j \in \mathcal{J}$ is willing to provide $\hat{q}_{tj}$ million barrels at a price $\hat{p}_{tj}$. The price/quantity pairs $(\hat{p}_{tj}, \hat{q}_{tj})$ fluctuate from week to week. The company would like to purchase (in discrete units of millions of barrels) x_{tj} million barrels (where x_{tj} is discrete) from source j in week $t \in \{1, 2, \dots, 10\}$. Your goal is to acquire 20 million barrels while spending the least amount possible.

(a) What is the exogenous stochastic process for this system?
(b) What would be the state variable for your problem? Give an equation(s) for the system dynamics.
(c) Propose a structure for a decision rule for this problem and call it X^{π}.
(d) For your decision rule, what do we mean by the space of policies? Give examples of two different decision rules.
(e) Write out the objective function for this problem using an expectation over the exogenous stochastic process.
(f) You are given a budget of \$300 million to purchase the oil, but you absolutely must end up with 20 million barrels at the end of the 10 weeks. If you exceed the initial budget of \$300 million, you may get additional funds, but each additional \$1 million will cost you \$1.5 million. How does this affect your formulation of the problem?

9.30 You own a mutual fund where at the end of each week t you must decide whether to sell the asset or hold it for an additional week. Let $\hat{r}_t$ be the one-week return (e.g. $\hat{r}_t = 1.05$ means the asset gained five percent in the previous week), and let p_t be the price of the asset if you were to sell it in week t (so $p_{t+1} = p_t \hat{r}_{t+1}$). We assume that the returns $\hat{r}_t$ are independent and identically distributed. You are investing this asset for eventual use in your college education, which will occur in 100 periods. If you sell the asset at the end of time period t, then it will earn a money market rate q for each time period until time period 100, at which point you need the cash to pay for college.

(a) What is the state space for our problem?
(b) What is the action space?
(c) What is the exogenous stochastic process that drives this system? Give a five time period example. What is the history of this process at time t?

(d) You adopt a policy that you will sell if the asset falls below a price $\bar{p}$ (which we are requiring to be independent of time). Given this policy, write out the objective function for the problem. Clearly identify exactly what you are optimizing over.

Theory questions

9.31 Assume that we have N discrete resources to manage, where R_a is the number of resources of type $a \in \mathcal{A}$ and $N = \sum_{a \in \mathcal{A}} R_a$. Let $\mathcal{R}$ be the set of possible values of the vector R. Show that

$$|\mathcal{R}| = \binom{N + |\mathcal{A}| - 1}{|\mathcal{A}| - 1},$$

where

$$\binom{X}{Y} = \frac{X!}{Y!(X - Y)!}$$

is the number of combinations of X items taken Y at a time.

Diary problem

The diary problem is a single problem you chose (see chapter 1 for guidelines). Answer the following for your diary problem.

9.32 Now you are finally going to model your diary problem, in its full detail (but you will not attempt to design a policy).

(a) Define each of the elements of the state variable. Note that this is an iterative process; you generally need to define the state variable as you identify the information you need at time t to model the system from time t onward. Do you have a belief state? If not, try to introduce one. All you need is some parameter that you can model as being unknown, but which you can estimate as data arrives to the system. The most interesting problems are where your decisions influence what you observe.

(b) What are the decisions? Describe in words, and then introduce notation for each decision. Now describe the constraints or the set of allowable decisions at time t. Add any information that you need at time t (that may change as we step forward in time) to the state variable. Introduce notation for the policy, although we will design the policy after we complete the model. The policy may introduce additional information that will have to be added to the state variable, but we will handle this after we start to design the policy.

(c) What is the exogenous information that arrives after you make the decision? (Note that you may have a deterministic problem, which means you do not have any exogenous information.) If your exogenous information depends on what you know at time t, then this information must be in the state variable.

(d) Define the transition function, which describes how each state variable evolves over time (not that we may not be done with the state variable). The level of detail here will depend on the complexity of your problem. Feel free to use both model-based transitions (where the equation governing the transition is known) and model-free transitions (where you simply observe the updated value of the variable).

(e) Write out the one-period contribution function, which may introduce additional information that you will need to add to the state variable (with corresponding additions to the transition function). Now write out the value of a policy, and write the objective of maximizing over policies (or classes of policies).

Bibliography

Bellman, R.E. (1957). *Dynamic Programming*. Princeton, N.J.: Princeton University Press.

Bellman, R.E. and Kalaba, R. (1959). On adaptive control processes. *IRE Transactions on Automatic Control* 4: 1–9.

Bertsekas, D.P. (2017). *Dynamic Programming and Optimal Control: Approximate Dynamic Programming*, 4e. Belmont, MA: Athena Scientific.

Boutilier, C., Dean, T., and Hanks, S. (1999). Decision-theoretic planning: Structural assumptions and computational leverage. *Journal of Artificial Intelligence Research*, 11: 1–94.

Chow, G. (1997). *Dynamic Economics*. New York: Oxford University Press.

Chung, K.L. (1974). *A Course in Probability Theory*. New York: Academic Press.

Cinlar, E. (2011). *Probability and Stochastics*. New York: Springer.

Guestrin, C., Koller, D., and Parr, R. (2003). Efficient solution algorithms for factored MDPs. *Journal of Artificial Intelligence Research* 19: 399–468.

Kirk, D.E. (2012). *Optimal Control Theory: An introduction*. New York: Dover.

Lewis, F.L. and Vrabie, D. (2012). *Design Optimal Adaptive Controllers*, 3e., Hoboken, NJ: JohnWiley & Sons.

Pollard, D. (2002). *A User's Guide to Measure Theoretic Probability*. Cambridge: Cambridge University Press.

Powell,W.B. (2011). *Approximate Dynamic Programming: Solving the Curses of Dimensionality*, 2e. John Wiley & Sons.

Powell, W.B. (2019). A unified framework for stochastic optimization. *European Journal of Operational Research* 275 (3): 795–821.

Powell, W.B. (2021). From reinforcement learning to optimal control: A unified framework for sequential decisions. *Handbook on Reinforcement Learning and Optimal Control, Studies in Systems, Decision and Control*. 29–74.

Powell, W.B., Simao, H.P., and Shapiro, J.A. (2001). A representational paradigm for dynamic resource transformation problems. In: *Annals of Operations Research* (eds. F.C. Coullard and J.H. Owens), 231–279. J.C. Baltzer AG.

Puterman, M.L. (2005). *Markov Decision Processes*, 2e. Hoboken, NJ: John Wiley and Sons.

Sethi, S.P. (2019). *Optimal Control Theory: Applications to Management Science and Economics*, 3e. Boston: Springer-Verlag.

Sontag, E. (1998). *Mathematical Control Theory*, 2e., 1–544. Springer.

Stengel, R.F. (1986). *Stochastic optimal control: theory and application*. Hoboken, NJ: John Wiley & Sons.

Stokey, N.L. and Lucas, R.E. (1989). *Recursive Methods in Dynamic Economics*. Cambridge, MA: Harvard University Press.

Werbos, P.J. (1992). Neurocontrol and supervised learning: An overview and evaluation. In: *Handbook of Intelligent Control* (eds. D.A. White and D.A. Sofge), 65–86. New York: Von Nostrand Reinhold.

White, C.C. (1991). A survey of solution techniques for the partially observable Markov decision process. *Annals of operations research* 32: 215–230.

10

Uncertainty Modeling

We cannot find an effective policy unless we are modeling the problem properly. In the realm of sequential decision problems, this means accurately modeling uncertainty. The importance of modeling uncertainty has been underrepresented in the stochastic optimization literature, although practitioners working on real problems have long been aware of both the importance and the challenges of modeling uncertainty.

Fortunately, there is a substantial body of research focused on the modeling of uncertainty and stochastic processes that has evolved in the communities working on Monte Carlo simulation and uncertainty quantification. We use uncertainty modeling as the broader term that describes the process of identifying and modeling uncertainty, while simulation refers to the vast array of tools that break down complex stochastic processes using the computational tools of Monte Carlo simulation.

It helps to remind ourselves of the two information processes that drive any sequential stochastic optimization problem: decisions, and exogenous information. Assume that we can pick some policy $X_t^\pi(S_t)$. We need to be able to simulate a sample realization of the policy, which will look like

$$S_0 \to x_0 = X_0^\pi(S_0) \to W_1 \to S_1 \to x_1 = X_1^\pi(S_1) \to W_2 \to S_3 \to$$

Given our policy, this simulation assumes that we have access to a transition function

$$S_{t+1} = S^M(S_t, X_t^\pi(S_t), W_{t+1}). \tag{10.1}$$

We can execute equation (10.1) if we are given a policy $X_t^\pi(S_t)$ and if we have access to the following:

Reinforcement Learning and Stochastic Optimization: A Unified Framework for Sequential Decisions, First Edition. Warren B. Powell.
© 2022 John Wiley & Sons, Inc. Published 2022 John Wiley & Sons, Inc.

S_0 = The initial state – This is where we place information about initial estimates (or priors) of parameters, as well as assumptions about probability distributions and functions.

W_t = Exogenous information that enters our system for the first time between $t-1$ and t for $t = 1, 2, \ldots, T$.

In this chapter, we focus on the often challenging problem of simulating the exogenous sequence $(W_t)_{t=0}^{T}$. We assume that the initial state S_0 is given, but recognize that it may include a probabilistic belief about unknown and unobservable parameters. The process of converting the characteristics of a stochastic process into a mathematical model is broadly known as *uncertainty quantification*. Since it is easy to overlook sources of uncertainty when building a model, we place considerable attention on identifying the different sources of uncertainty that we have encountered in our applied work, keeping in mind that S_0 and W_t are the only variables our modeling framework provides for representing uncertainty.

After reviewing different sources of uncertainty, we then provide a basic introduction to a powerful set of techniques known as Monte Carlo simulation, which allows us to replicate stochastic processes on the computer. Given the rich array of different types of stochastic processes, our discussion here provides little more than a taste of the tools that are available to replicate stochastic processes.

10.1 Sources of Uncertainty

Uncertainty arises in different forms. Some of the major forms that we have encountered are

- Observational errors – This arises from uncertainty in observing or measuring the state of the system. Observational errors arise when we have unknown state variables that cannot be observed directly (and accurately).
- Exogenous uncertainty – This describes the exogenous arrival of information to the system, which might be weather, demands, prices, the response of a patient to medication or the reaction of the market to a product.
- Prognostic uncertainty – We often have access to a forecast $f_{tt'}^{W}$ of the information $W_{t'}$. Prognostic uncertainty captures the deviation of the actual $W_{t'}$ from the forecast $f_{tt'}^{W}$. If we think of $W_t = f_{tt}^{W}$ as the actual value of W_t, then we can think of the realization of W_t (the exogenous information described above) as just an update to a forecast.

- Inferential (or diagnostic) uncertainty – Inferential uncertainty arises when we use observations (from field or physical measurements, or computer simulations) to draw inferences about another set of parameters. It arises from our lack of understanding of the precise properties or behavior of a system, which introduces errors in our ability to estimate parameters, partly from noise in the observations, and partly from errors in our modeling of the underlying system.
- Experimental variability – Sometimes equated with observational uncertainty, experimental variability refers to differences between the results of experiments run under similar conditions. An experiment might be a computer simulation, a laboratory experiment, or a field implementation. Even if we can perfectly measure the results of an experiment, there is variation from one experiment to the next.
- Model uncertainty – We may not know the structure of the transition function $S_{t+1} = S^M(S_t, x_t, W_{t+1})$, or the parameters that are imbedded in the function. Model uncertainty is often attributed to the transition function, but it may also apply to the model of the stochastic process W_t since we often do not know the precise structure.
- Transitional uncertainty – This arises when we have a perfect model of how a system should evolve, but exogenous shocks (wind buffeting an aircraft, rainfall affecting reservoir levels) can introduce uncertainty in how an otherwise deterministic system will evolve. Transitional uncertainty is often represented as

$$S_{t+1} = S^M(S_t, x_t) + \varepsilon_{t+1}.$$

- Control/implementation uncertainty – This is where we choose a control u_t (such as a temperature or speed), but what happens is $\hat{u}_t = u_t + \delta u_t$ where δu_t is a random perturbation.
- Communication errors and biases – Communication from an agent q about his state S_{qt} to an agent q' where errors may introduced, either accidentally or purposely.
- Algorithmic instability – Very minor changes in the input data for a problem, or small adjustments in parameters guiding an algorithm (which exist in virtually all algorithms), can completely change the path of the algorithm, introducing variability in the results.
- Goal uncertainty – Uncertainty in the desired goal of a solution, as might arise when a single model has to produce results acceptable to different people or users.
- Political/regulatory uncertainty – Uncertainty about taxes, rules, and requirements that affect costs and constraints (for example, tax energy credits, automotive mileage standards). These can be viewed as a form of

systematic uncertainty, but this is a particularly important source of uncertainty with its own behaviors.

Below we provide more detailed discussions of each type of uncertainty. One challenge is modeling each source of uncertainty, since we have only two mechanisms for introducing exogenous information into our model: the initial state S_0, and the exogenous information process $W_1, W_2, \ldots$. Thus, the different types of uncertainty may look similar mathematically, but it is important to characterize the mechanisms by which uncertainty enters our model.

10.1.1 Observational Errors

Observational (or measurement) uncertainty reflects errors in our ability to observe (or measure) the state of the system directly. Some examples include:

■ **EXAMPLE 10.1**

Different people may measure the gases in the oil of a high-voltage transformer, producing different measurements (possibly due to variations in equipment, the temperature at which the transformer was observed, or variations in the oil surrounding the coils).

■ **EXAMPLE 10.2**

The Center for Disease Control and Prevention estimates the number of mosquitoes carrying a disease by setting traps and counting how many mosquitoes are caught that are found with the disease. From day to day the number of infected mosquitoes that are caught can vary considerably.

■ **EXAMPLE 10.3**

A company may be selling a product at a price p_t which is being varied to find the best price. However, the sales (at a fixed price) will be random from one time period to the next.

■ **EXAMPLE 10.4**

Different doctors, seeing the same patient for the first time, may elicit different information about the characteristics of the patient.

Partially observable systems arise in any application where we cannot directly observe parameters. A simple example arises in pricing, where we may feel that demand varies linearly with price according to

$$D(p) = \theta_0 - \theta_1 p.$$

At time t, our best estimate of the demand function is given by

$$D(p) = \bar{\theta}_0 - \bar{\theta}_1 p.$$

We observe sales, which would be given by

$$\hat{D}_{t+1} = \theta_0 - \theta_1 p_t + \varepsilon_{t+1}.$$

We do not know (θ_0, θ_1), but we can use observations to create updated estimates. If $(\bar{\theta}_{t0}, \bar{\theta}_{t1})$ is our estimate as of time t, we can use our observation $\hat{D}_{t+1}$ of sales between t and $t + 1$ to obtain updated estimates $(\bar{\theta}_{t+1,0}, \bar{\theta}_{t+1,1})$. In this model, we would view $\bar{\theta}_t = (\bar{\theta}_{t0}, \bar{\theta}_{t1})$ as our state variable, which is our estimate of the static parameter θ. Since θ is a fixed parameter, we do not include it in the state variable, but rather treat it as a *latent variable*.

The presence of states that cannot be perfectly observable gives rise to what are widely known as *partially observable Markov decision processes*, or POMDP's. To model this, let $\check{S}_t$ be the true (but possibly unobservable) state of the system at time t, while S_t is the observable state. One way of writing our dynamics might be

$$S_{t+1} = \check{S}^M(\check{S}_t, x_t) + \varepsilon_{t+1},$$

which captures our inability to directly observe $\check{S}_t$. These systems are most often motivated by problems such as those in engineering where we cannot directly observe the state of charge of a battery, the location and velocity of an aircraft, or the number of truck trailers sitting at a terminal (terminal managers tend to hide trailers to keep up their inventories).

We can represent our unobservable state as a probability distribution. This might be a continuous distribution (perhaps the normal or multivariate normal distribution), or perhaps more simply as a discrete distribution where q_{ti}^k is the probability that the state variable S_{ti} takes on outcome k (or perhaps a parameter θ^k) at time t. Then, the vector $q_{ti} = (q_{ti}^k), k = 1, \ldots, K$ is the distribution capturing our belief about the unobservable state. We then include q_t (for each uncertain state dimension) as part of our state variable (this is where our belief state comes in).

10.1.2 Exogenous Uncertainty

Exogenous uncertainty represents the information that we typically model through the process W_t represent new information about supplies and demands, costs and prices, and physical parameters that can appear in either the objective function or constraints. Exogenous uncertainty can arise in different styles, including:

- Fine-grained time-scale uncertainty – Sometimes referred to as *aleatoric uncertainty*, fine time-scale uncertainty refers to uncertainty that varies from time-step to time-step which is assumed to reflect the dynamics of the problem. Whether a time step is minutes, hours, days or weeks, fine time-scale uncertainty means that information from one time-step to the next is either uncorrelated, or where correlations drop off fairly quickly.
- Coarse-grained time-scale uncertainty – Referred to in different settings as systematic uncertainty or *epistemic* uncertainty (popular in the medical community), coarse time-scale uncertainty reflects uncertainty in an environment which occurs over long time scales. This might reflects new technology, changes in market patterns, the introduction of a new disease, or an unobserved fault in machinery for a process.
- Distributional uncertainty – If we represent the exogenous information W_t, or the initial state S_0, as a probability distribution, there may be uncertainty in either the type of distribution or the parameters of a distribution.
- Adversarial uncertainty – The exogenous information process $W_1, \ldots, W_T$ may come from another agent who is choosing W_t in a way to make us perform poorly. We cannot be sure how the adversary may behave.

10.1.3 Prognostic Uncertainty

Prognostic uncertainty reflects errors in our ability to forecast activities in the future. Typically these are written as $f_{tt'}$ to represent the forecast of some quantity at time t', given what we know at time t (represented by our state variable S_t). Examples include:

■ **EXAMPLE 10.5**

A company may create a forecast of demand D_t for its product. If $f_{tt'}^D$ is the forecast of the demand $D_{t'}$ given what we know at time t, then the difference between $f_{tt'}^D$ and $D_{t'}$ is the uncertainty in our forecast.

■ **EXAMPLE 10.6**

A utility is interested in forecasting the price of electricity 10 years from now. Electricity prices are well approximated by the intersection of the load (the amount of electricity needed at a point in time) and the "supply stack" which is the cost of energy as a function of the total supply (typically an increasing function). The supply stack reflects the cost of different fuels (nuclear, coal, natural gas) and generators (different technologies, and different ages, affect operating costs). We have to forecast the prices of these different sources (one form of uncertainty) along with the load (a different form of uncertainty).

■ **EXAMPLE 10.7**

We might be interested in forecasting energy from wind $E_{t'}^W$ at time t. This might require that we first generate a meteorological forecast of weather systems (high and low pressure systems), as well as capturing the movement of the atmosphere (wind speed and direction).

If $W_{t'}$ is some form of random information in the future, we might be able to create a forecast $f_{tt'}^W$ using what we know at time t. We typically assume that our forecasts are unbiased, which means we can write

$$f_{tt'}^W = \mathbb{E}\{W_{t'}|S_t\}.$$

Forecasts can come from two sources. An *endogenous forecast* is obtained from a model that is created endogenously from data. For example, we might be forecasting demand using the model

$$f_{tt'}^D = \theta_{t0} + \theta_{t1}(t' - t).$$

Now assume we observe the demand D_{t+1}. We might use any of a range of algorithms to update our parameter estimates to obtain

$$f_{t+1,t'}^D = \theta_{t+1,0} + \theta_{t+1,1}(t' - (t + 1)).$$

The parameter vector θ_t can be updated recursively from observations W_{t+1}. If θ_t is our current estimate of $(\theta_{t0}, \theta_{t1})$, let Σ_t be our estimate of the covariance between the random variables θ_0 and θ_1 (these are the true values of the parameters). Let $\beta^W = 1/(\sigma_W^2)$ be the precision of an observation W_{t+1} (the precision is the inverse of the variance), and assume we can form the precision matrix given by $M_t = [(X_t)^T X_t]^{-1}$, where X_t is a matrix where each row consists of the vector of independent variables. In the case of our demand example, the design variables for time t would be $x_t = (1 \quad p_t)^T$. We can update θ_t and Σ_t (or M_t) recursively using

$$\theta_{t+1} = \theta_t - \frac{1}{\gamma_{t+1}} M_t x_{t+1} \varepsilon_{t+1}, \tag{10.2}$$

where ε_{t+1} is the error given by

$$\varepsilon_{t+1} = W_{t+1} - \theta_t x_t. \tag{10.3}$$

The matrix $M_{t+1} = [(X_{t+1})^T X_{t+1}]^{-1}$. This can be updated recursively without computing an explicit inverse using

$$M_{t+1} = M_t - \frac{1}{\gamma_{t+1}} (M_t x_{t+1} (x_{t+1})^T M_t). \tag{10.4}$$

The parameter γ_{t+1} is a scalar computed using

$$\gamma_{t+1} = 1 + (x_{t+1})^T M_t x_{t+1}. \tag{10.5}$$

Note that if we multiply (10.4) through by σ_ε^2 we obtain

$$\Sigma_{t+1}^\theta = \Sigma_t^\theta - \frac{1}{\gamma_{t+1}} (\Sigma_t^\theta x_{t+1} (x_{t+1})^T \Sigma_t^\theta), \tag{10.6}$$

where we scale γ_{t+1} by σ_ε^2, giving us

$$\gamma_{t+1} = \sigma_\varepsilon^2 + (x_{t+1})^T \Sigma_t^\theta x_{t+1}. \tag{10.7}$$

Equations (10.2)–(10.7) represent the transition function for updating θ_t.

The second source of a forecast is exogenous, where the forecast might be supplied by a vendor. In this case, we might view the updated set of forecasts $(f_{tt'})_{t' \geq t}$ as exogenous information. Alternatively, we could think of the change in forecasts as the exogenous information. If we let $\hat{f}_{t+1,t'}$ be the change between t and $t+1$ in the forecast for activities at time t', we would then write

$$f_{t+1,t'} = f_{tt'} + \hat{f}_{t+1,t'}.$$

From a modeling perspective, these forecasts differ in terms of how they are represented in the state variable. In the case of our endogenous forecast, the state variable would be captured by (θ_t, Σ_t), with the corresponding transition equations given by (10.2)–(10.7). With our exogenous forecast, the state variable would be simply $(f_{tt'})_{t'=t}^T$.

Regardless of whether the forecast is exogenous or endogenous, the new information (the exogenous observation or the updated forecast) would be modeled as a part of the exogenous information process W_t.

10.1.4 Inferential (or Diagnostic) Uncertainty

It is often the case that we cannot directly observe a parameter. Instead, we have to use (possibly imperfect) observations of one or more parameters to infer variables or parameters that we cannot directly observe. Some examples include:

■ **EXAMPLE 10.8**

We might not be able to directly observe the presence of heart disease, but we may use blood pressure as an indicator. Measuring blood pressure introduces observational error, but there is also error in making the inference that a patient suffers from heart disease from blood pressure alone.

■ **EXAMPLE 10.9**

We observe (possibly imperfectly) the sales of a product. From these sales we wish to estimate the elasticity of demand with respect to price.

■ **EXAMPLE 10.10**

Power companies generally do not know the precise location of a tree falling that creates a power outage. Rather, a falling tree can create a short circuit that will trip a circuit breaker higher in the tree (rooted at a substation), producing many outages, including to customers who may be far from the fallen tree. Diagnostic uncertainty refers to errors in our ability to precisely describe where a tree might have fallen purely from phone calls.

■ **EXAMPLE 10.11**

Sensors may detect an increase in carbon monoxide in the exhaust of a car. This information may indicate several possible causes, such as an aging catalytic converter, the improper timing of the cylinders, or an incorrect air-fuel mixture (which might hint at a problem in a different sensor).

Inferential uncertainty can be described as uncertainty in the parameters of a model. In our example involving the detection of carbon monoxide, we might use this information to update the probability that the real cause is due to each of three or four different mechanical problems. This would represent an instance of using a (possibly noisy) observation to update a lookup table model of where failures are located. By contrast, when we use sales data to update our demand elasticity, that would be an example of using noisy observational data to update a parametric model.

In some settings the term *diagnostic uncertainty* is used instead of inferential uncertainty. We feel that this term reflects the context of identifying a problem (a failed component, presence of a disease) that we are not able to observe directly. However, both inferential uncertainty and diagnostic uncertainty reflect uncertainty in parameters that have been estimated (inferred) from indirect observations.

Inferential uncertainty is a form of derived uncertainty that arises when we estimate a parameter $\bar{\theta}$ from data (simulated or observed). The raw uncertainty is contained in the sequence W_t (or W^n). We then have to derive the distribution of our estimate $\bar{\theta}$ resulting from the exogenous noise, which we contain in our belief state B_t.

10.1.5 Experimental Variability

Experimental variability reflects changes in the results of experiments run under the same conditions. Experimental settings include

Laboratory experiments – We include here physical experiments run in a laboratory setting, encompassing chemical, biological, mechanical, and even human testing.
Numerical simulations – Large simulators describing complex physical systems, ranging from models of businesses to models of physical processes, can exhibit variability from one run to the next, often reflecting minor variations in input data and parameters.
Field testing – This can range from observing sales of a product to testing of new drugs.

Experimental uncertainty arises from possibly minor variations in the dynamics of a system (simulated or physical) which introduce variability when running experiments. Experimental uncertainty typically reflects our inability to perfectly estimate parameters that drive the system, or errors in our ability to understand (or model) the system.

Some sources equate observational and experimental uncertainty, and often they are handled in the same way. However, we feel it is useful to distinguish between pure measurement (observational) errors, which might be reduced with better technology, and experimental errors, which have more to do with the process and which are not reduced through better measurement technologies.

Experimental noise might be attributed as a byproduct of the exogenous information process W_t. For example, for a given policy $X^\pi(S_t)$, an experiment might consist of evaluating

$$\hat{F}^\pi = F^\pi(\omega) = \sum_{t=0}^{T} C(S_t(\omega), X^\pi(S_t(\omega))).$$

Here, the noise is due to the variation in W_t. However, imagine that we are running a series of experiments. Let $\hat{F}^n(\theta^n)$ be the observation of the outcome of an experiment run with parameters $\theta = \theta^n$. Let $f(\theta) = \mathbb{E}\hat{F}^n(\theta)$ be the exact (but unobservable) value of running the experiment with parameter setting θ. We can write

$$\hat{F}^n(\theta^n) = f(\theta^n) + \varepsilon^n.$$

In this case, the sequence ε^n would be the exogenous information W^n.

10.1.6 Model Uncertainty

"Model uncertainty" is a bit of a catch-all phrase that often refers to the transition function, but not always. Model uncertainty comes in two forms. The first is errors in estimates of parameters of a parametric model. If we are estimating these parameters over time from observations, we would refer to this as inferential uncertainty. Now imagine that we characterize our model using a set of fixed parameters that are being updated. We are not estimating these parameters over time, but rather we are just using assumed values which are uncertain.

The second is errors in the structure of the model itself (economists refer to this as *specification errors*). Some examples include:

■ **EXAMPLE 10.12**

We may approximate demand as a function of price as a linear function, a logistics curve, or a quadratic function. We will use observational data to estimate the parameters of each function, but we may not directly address the errors introduced by assuming a particular type of function.

■ **EXAMPLE 10.13**

We may describe the diffusion of chemicals in a liquid using a first-order set of differential equations, which we fit to observational data. However, the real process may be better described by a second- (or higher) order set of differential equations. Our first-order model may be at best a good local approximation.

◼ **EXAMPLE 10.14**

Grid operators often model the supply curve of a power generator using a convex function, which is easier to solve. However, a more detailed model might capture complex relationships that reflect the fact that costs may rise in steps as different components of the generator come on (e.g. heat recovery).

Model uncertainty for dynamic problems can be found in four different parts of the model:

- Costs or rewards – Measuring the cost of a grid outage on the community may require estimating the impact of a loss of power on homes and businesses.
- Constraints – Constraints can often be written in the form $A_t x_t = R_t$. There are many applications where dynamic uncertainty enters through the right-hand side R_t; this is how we would model the supply or demand of blood which would be a more typical form of dynamic uncertainty. Model uncertainty often arises in the matrix A_t, which is where we might capture the assumed speed of an aircraft, or the efficiency of a manufacturing process.
- Stochastic modeling – If we are using a model of the exogenous information W_t, then there may be errors in this model.
- Dynamics – This is where we are uncertain about the function $S^M(S_t, x_t, W_{t+1})$ which describes how the system evolves over time.

The transition function $S^M(\cdot)$ captures all the physics of a problem, and there are many problems where we simply do not understand the physics. For example, we might be trying to explain how a person or market might respond to a price, or how global warming might respond to a change in CO_2 concentrations.

Some policies make decisions using nothing more than the current state, allowing them to be used in settings where the underlying dynamics have not been modeled. By contrast, an entire class of policies based on lookahead models (which we cover in chapter 19) depend on at least an approximate model of the problem. See 9.7.2 for a discussion of model-free dynamic programming.

Whether we are dealing with costs, constraints, or the dynamics, our model can be described in terms of the choice of the model structure, and any parameters that characterize the model. Let $m \in \mathcal{M}$ represent the structure of the model, and let $\theta \in \Theta^m$ be the parameters that characterize a model with structure m. As a general rule, the model structure m is fixed in advance (for example, we might assume that a particular relationship is linear) but with uncertain parameters.

An alternative approach is to associate a prior q_0^m that gives the probability that we believe that model m is correct. Similarly, we might start with an initial estimate θ_0^m for the parameter vector θ^m. We might even assume that we start by assuming that θ^m is described by a multivariate normal distribution with mean θ_0^m and covariance matrix Σ_0^m.

As we might expect, prior information about the model (whether it is the probability q_0 that a type of model is correct, or the prior distribution on θ^m) is communicated through the initial state S_0. If this belief is updated over time, then this would also be part of the dynamic state S_t.

10.1.7 Transitional Uncertainty

There are many problems where the dynamics of the system are modeled deterministically. This is often the case in engineering applications where we apply a control u_t (such as a force) to a dynamic system. Simple physics might describe how the control affects our system, which we would then write

$$S_{t+1} = S^M(S_t, u_t).$$

However, exogenous noise might interfere with these dynamics. For example, we might be predicting the speed and location of an aircraft after applying forces u_t. Variations in the atmosphere might interfere with our equations, so we introduce a noise term ε_{t+1}, giving us

$$S_{t+1} = S^M(S_t, u_t) + \varepsilon_{t+1}.$$

We note that despite the noise, we assume that we can observe (measure) the state perfectly.

10.1.8 Control/implementation Uncertainty

There are many problems where we cannot precisely control a process. Some examples include:

■ **EXAMPLE 10.15**

An experimentalist has requested that a rat be fed a diet with x_t grams of fat. However, variability in the preparation of the meals, and the choice the rat makes of what to eat, introduces variability in the amount of fat that is consumed.

■ **EXAMPLE 10.16**

A publisher chooses to sell a book at a wholesale price p_t^W at time t and then observes sales. However, the publisher has no control over the retail price offered to the purchasing public.

■ **EXAMPLE 10.17**

The operator of a power grid may request that a generator come online and generate x_t megawatts of power. However, this may not happen either because of a technical malfunction or human implementation errors.

Control uncertainty is widely overlooked in the dynamic programming literature, but is well known in the econometrics community as the "errors in variable" model.

We might model errors in the implementation of a decision using a simple additive model

$$\hat{x}_t = x_t + \varepsilon_t^x,$$

where $\hat{x}_t$ is the decision that is actually implemented, and ε_t^x captures the difference between what was requested, x_t, versus what was implemented, $\hat{x}_t$. We note that ε_t^x would be modeled as an element of W_t, although in practice it is not always observable.

It is important to distinguish between uncertainty in how a decision (or control) is implemented from other sources of uncertainty because of potential nonlinearities in how the decision affects the results.

10.1.9 Communication Errors and Biases

In a multiagent system, one agent might communicate location or status to another agent, but this information can contain errors (a drone might not know its exact location) or biases (a fleet driver might report being on the road for fewer hours in order to be allowed to drive longer). In supply chain management, an engine manufacturer may send inflated production targets to suppliers to encourage suppliers to have enough inventory to handle problems, say, in the quality of parts that require more returns.

10.1.10 Algorithmic Instability

A more subtle form of uncertainty is one that we refer to as algorithmic uncertainty. We use this category to describe uncertainty that is introduced by the algorithm used to solve a problem, which may also be partly attributable to the model itself. Three examples of how algorithmic uncertainty arises are:

- Algorithms that depend on Monte Carlo sampling.
- Algorithms that exhibit sensitivity to small changes in the input data.
- Algorithms that produce different results even when run on exactly the same data, possibly due to variations in run times for a parallel implementation of an algorithm.
- Optimization algorithms for nonconvex problems where the optimal solution is highly dependent on the starting point(s), which may be randomly generated.

The stochastic gradient algorithm introduced in chapter 5, which we write using

$$x^{n+1} = x^n + \alpha_n \nabla_x F(x^n, W^{n+1}),$$

is a nice example of an algorithm that depends on Monte Carlo sampling, which is how we generated the observation W^{n+1}. These algorithms depend on carefully tuned stepsize policies for α_n to mitigate the effects of the noise.

The second type of algorithmic uncertainty arises due to the sensitivity that many deterministic optimization algorithms exhibit (in particular integer programs and nonlinear programs). Small changes in the input data can produce wide swings in the solution, although often there may be little or no change in the objective function. Thus, we may solve an optimization problem (perhaps this might be a linear program) that depends on a parameter θ. Let $F(\theta)$ be the optimal objective function and let $x(\theta)$ be the optimal solution. Small changes in θ can produce large (and unpredictable) changes in $x(\theta)$, which introduces a very real form of uncertainty.

The third type of uncertainty arises primarily with complex problems such as large integer programs that might take advantage of parallel processing. The behavior of these algorithms depends on the performance of the parallel processors, which can be affected by the presence of other jobs on the system. As a result, we can observe variability in the results, even when applied to exactly the same problem with the same data.

Algorithmic uncertainty is in the same class as experimental uncertainty, thus we defer to the discussion there for a description of how to model it.

10.1.11 Goal Uncertainty

Many problems involve balancing multiple, competing objectives, such as putting different priorities on cost versus service, profits versus risk. One way to model this is to assume a linear utility function of the form

$$U(S, x) = \sum_{\ell \in \mathcal{L}} \theta_\ell \phi_\ell(S, x),$$

where S is our state variable, x is a decision, and $(\phi_\ell(S, x))_{\ell \in \mathcal{L}}$ is a set of features that capture the different metrics we use to evaluate a system such as cost, service, productivity, and total profits. The vector $(\theta_\ell)_{\ell \in \mathcal{L}}$ captures the weight we put on each feature. One way to model goal uncertainty is to represent θ as being uncertain (it may even vary from one decision-maker to another).

Another form of uncertainty might arise when we do not know all the features $\phi(S, x)$. For example, we may not even be aware that a reason to assign a particular driver to move a customer is that the customer is going to a location near the home of the driver. A human dispatcher might know this through personal interactions with the driver, but a computer might not. The result could then be a disagreement between a computer recommendation and what a human wants to do.

10.1.12 Political/regulatory Uncertainty

For problems that involve long-term planning, changes in laws and regulations can introduce a significant source of uncertainty. Supply chain relationships with China, for example, can introduce the dimension of changes in tariffs. Planning energy investments bring in the dimension of the potential of a carbon tax. Manpower planning in many countries can depend on immigration policies, in industries ranging from agriculture to software to manufacturing.

10.1.13 Discussion

Careful readers will notice some overlap between these different types of uncertainty. Observational uncertainty, which refers specifically to errors in the direct observation of a parameter, and inferential uncertainty, which refers to errors in our ability to make inferences about models and parameters indirectly from data, represents one example, but we feel that it is useful to highlight the distinction. Model uncertainty is a term that resonates with many people, but it is an umbrella for several types of uncertainty.

What matters is if this list helps people identify as many sources of uncertainty as possible. We test this idea in a brief case study next.

10.2 A Modeling Case Study: The COVID Pandemic

A particularly rich application for modeling uncertainty arises when planning the vaccination response to the COVID pandemic, which was unfolding as this book was being written. Table 10.1 lists each of the different sources of uncertainty, and provides a few examples of each.

For a problem as complex as planning the vaccination process for COVID, there are many sources of uncertainty. Working from our list of different types of uncertainty helps to highlight forms of uncertainty that might be overlooked. Keep in mind that any model of a complex problem requires simplifications, but it helps to list as many sources of uncertainty as possible so that any simplifications are conscious ones, as opposed to simply overlooking a source of uncertainty.

10.3 Stochastic Modeling

Once we have identified sources of uncertainty, the next step is generating sequences of random outcomes that represent samples of observations of exogenous information. This exercise can be relatively straightforward, or not. There are many problems where the stochastic modeling of the different sources of uncertainty is much harder, and much more important, than designing a policy.

10.3.1 Sampling Exogenous Information

Somewhere in stochastic modeling we usually end up needing to compute an expectation, as we found in chapter 9 when we formulated our objective function as

$$\min_\pi \mathbb{E} \sum_{t=0}^{T} C(S_t, X_t^\pi(S_t)).$$

With rare exceptions, we will not be able to compute the expectation, and instead we have to resort to sampling, which can be accomplished in one of several ways:

- Mathematical models – Here we develop probability distributions to describe the frequency of different outcomes. We then use the methods of Monte Carlo simulation (described below) to sample from these distributions. This approach requires the highest mathematical sophistication to generate samples that mimic actual behavior.

Table 10.1 Illustration of different types of uncertainty arising in the vaccination response to the COVID pandemic.

Type of uncertainty	Description
Observational errors	Sample error observing people with symptoms
	Errors classifying people with symptoms as having COVID
Exogenous	Reports of new cases, deaths
uncertainty	Availability of ICUs, personal protective equipment
	Actual production of vaccines
Prognostic	Projection of cases, hospital admissions
uncertainty	Estimates of future performance of vaccines
	Projections of population response to vaccines
	Projections of vaccine production
Inferential	Estimates of infection rates
uncertainty	Estimates of effectiveness of vaccines
Experimental	Uncertainty in how a drug will perform in a clinical trial
uncertainty	Uncertainty in how many people will agree to be vaccinated
Model uncertainty	Uncertainty in the structure of the transmission model used for forecasting
	Uncertainty in the geographical spread of infections
Transitional uncertainty	Additions/withdrawals to/from vaccine inventories, with noise from refrigeration failures
Control uncertainty	Which population groups were vaccinated given the planned prioritzation
	How vaccines were allocated relative to the plan
Implementation uncertainty	Deviations when vaccines are not given to the correct people
Communication	Reporting errors from the field
errors	Failure to notify people when they should be vaccinated
Goal uncertainty	Disagreements in prioritizing who should be vaccinated first
Political/regulatory	If/when a vaccine will be approved
uncertainty	Allocation of vaccines to different states, countries

- Historical data – A common strategy is to simply run a process over historical data. This is widely used to test trading strategies in finance, for example, where this is known as "back testing."
- Observational sampling – This is where we use observations from an exogenous process, most commonly referred to as the "real world," to generate sample realizations.
- Numerical simulations – We may have a (typically large) computer model of a complex process. The simulation may be of a physical system such as a supply chain or an asset allocation model. Some simulation models can require extensive calculations (a single sample realization could take hours or days on a computer). We can use such simulations as a source of observations similar to observations from real-world environments.
- Contingencies – We use the term "contingency" to refer to outcomes that *may* happen, and we have to plan for the possibility that they may happen, without building a probability model or estimating the frequency of these events. For example, companies managing power grids are required to plan for the event that their largest generator may fail. Some will use the term "scenario" to refer to a contingency, but "scenarios" are often used to refer to samples of a set of random variables, which are used to represent a sample of a probability distribution.

Often, we create simulated versions of the real world in order to test algorithms, with the understanding that the simulated source of observations will be replaced with exogenous observations. It is important to understand whether this is the eventual plan, since some policies depend on having access to an underlying model.

10.3.2   Types of Distributions

While it is easy to represent random information as a single variable such as W_t, it is important to realize that random variables can exhibit very different behaviors. The major classes of distributions that we have encountered in our work include:

- Exponential (or geometric) families of random variables – These include the continuous distributions such as normal (or Gaussian) distributions, log normal, exponential and gamma distributions, and discrete distributions such as the Poisson distribution, geometric distribution, and the negative binomial distribution. We also include in this class the uniform distribution (continuous or discrete).
- Heavy-tailed distributions – Price processes are a good example of variability that tends to exhibit very high standard deviations. An extreme example is

the Cauchy distribution which has infinite variance, but there can be less extreme distributions with heavy tails.

- Spikes – These are infrequent but extreme observations. For example, electricity prices periodically spike from typical prices in the range of 20 to 50 dollars per megawatt, to prices of 300 to as much as 10,000 dollars per megawatt for very short intervals (perhaps 5 to 10 minutes).
- Bursts – Bursts describe processes such as snow or rain, power outages due to extreme weather, or sales of a product where a new product, advertising or price reduction can produce a rise in sales over a period of time. Bursts are characterized by a sequence of observations over a short period of time.
- Rare events – Rare events are similar to spikes, but are characterized not by extreme values but rather by events that may happen, but happen rarely. For example, failures of jet engines are quite rare, but they happen, requiring that the manufacturer hold spares.
- Regime shifting – A data series may move from one regime to another as the world changes. For example, the discovery of fracking created a new supply of natural gas which resulted in electricity prices dropping from around $50 per megawatt-hour to around $20 per megawatt-hour.
- Hybrid/compound distributions – There are problems where a random variable is drawn from a distribution with a mean which is itself a random variable. The mean of a Poisson distribution, perhaps representing people clicking on an ad, might have a mean which itself is a random variable reflecting the behavior of competing ads.

10.3.3 Modeling Sample Paths

In chapter 9, section 9.8.2, we showed that we could write the value of a policy as

$$F^\pi = \mathbb{E} \sum_{t=0}^{T} C(S_t, X_t^\pi(S_t)). \tag{10.8}$$

We then wrote this as a simulation using

$$F^\pi(\omega) = \sum_{t=0}^{T} C(S_t(\omega), X_t^\pi(S_t(\omega))), \tag{10.9}$$

where the states are generated according to $S_{t+1}(\omega) = S^M(S_t(\omega), X_t^\pi(S_t(\omega)), W_{t+1}(\omega))$. In this section, we illustrate our notation for representing sample paths more carefully.

We start by assuming that we have constructed 10 potential realizations of price paths p_t, $t = 1, 2, \ldots, 8$, which we have shown in Table 10.2. Each sample

Table 10.2 Illustration of a set of sample paths for prices all starting at $45.00.

| ω^n | $t = 1$ | $t = 2$ | $t = 3$ | $t = 4$ | $t = 5$ | $t = 6$ | $t = 7$ | $t = 8$ |
	p_1	p_2	p_3	p_4	p_5	p_6	p_7	p_8
ω^1	45.00	45.53	47.07	47.56	47.80	48.43	46.93	46.57
ω^2	45.00	43.15	42.51	40.51	41.50	41.00	39.16	41.11
ω^3	45.00	45.16	45.37	44.30	45.35	47.23	47.35	46.30
ω^4	45.00	45.67	46.18	46.22	45.69	44.24	43.77	43.57
ω^5	45.00	46.32	46.14	46.53	44.84	45.17	44.92	46.09
ω^6	45.00	44.70	43.05	43.77	42.61	44.32	44.16	45.29
ω^7	45.00	43.67	43.14	44.78	43.12	42.36	41.60	40.83
ω^8	45.00	44.98	44.53	45.42	46.43	47.67	47.68	49.03
ω^9	45.00	44.57	45.99	47.38	45.51	46.27	46.02	45.09
ω^{10}	45.00	45.01	46.73	46.08	47.40	49.14	49.03	48.74

path is a particular set of outcomes of the p_t for all time periods. We index each potential set of outcomes by ω, and let Ω be the set of all sample paths where, for our example, $\Omega = \{1, 2, \ldots, 10\}$. Thus, $p_t(\omega^n)$ would be the price for sample path ω^n at time t. For example, referring to the table we see that $p_2(\omega^4) = 45.67$.

One reason that we may generate information on the fly is that it is easier to implement in software. For example, it avoids generating and storing an entire sample path of observations. However, another reason is that random information may depend on the current state, a setting we address next.

10.3.4 State-action-dependent Processes

Imagine that we are looking to optimize an energy system in the presence of increasing contributions from wind and solar energy. It is reasonable to assume that the available energy from wind or solar, which we represent generically as W_t, is not affected by any decision we make. We could create a series of sample paths of wind, which we could denote by $\hat{\omega} \in \hat{\Omega}$, where each sequence $\hat{\omega}$ is a set of outcomes of $W_1(\hat{\omega}), \ldots, W_T(\hat{\omega})$. These sample paths could be stored in a dataset and used over and over.

There are a number of examples where exogenous information depends on the state of the system. Some examples include:

■ **EXAMPLE 10.18**

A drone is monitoring a forest for evidence of fires. What the drone observes (the exogenous information) depends on its location (its state).

■ **EXAMPLE 10.19**

Imagine the setting where a patient is being given a cholesterol-lowering drug. We have to decide the dosage (10mg, 20mg, …), and then we observe blood pressure and whether the patient experiences any heart irregularities. The observations represent the random information, but these observations are influenced by the prior dosage decisions.

■ **EXAMPLE 10.20**

The price of oil reflects oil inventories. As inventories rise, the market recognizes the presence of surplus inventories which depresses prices. Decisions about how much oil to store affects the exogenous changes in market prices.

In some cases, the random information depends on the decision being made at time t. For example, imagine that we are a large investment bank buying and selling stock. Large buy and sell orders will influence the price. Imagine that we place a (large) order to sell x_t shares of stock, which will clear the market at a random price

$$p_{t+1}(x_t) = p_t - \theta x_t + \varepsilon_{t+1},$$

where θ captures the impact of the order on the market price. We are not able to directly observe this effect, so we create a single random variable $\hat{p}_{t+1}$ that captures the entire change in price, given by

$$\hat{p}_{t+1} = -\theta x_t + \varepsilon_{t+1}.$$

Thus, our random variable $\hat{p}_{t+1}$ depends on the decision x_t.

We can model problems where the exogenous information W_{t+1} depends on the action x_t as if it were depending on the post-decision state $S_t^x = (S_t, x_t)$. However, since it is the sales x_t itself that influences the change in price, it is important that x_t be captured explicitly in the post-decision state.

Whether the exogenous information depends on the state or the action, it depends on the policy, since the state at time t reflects prior decisions.

10.3.5   Modeling Correlations

One of the most difficult problems in stochastic modeling is capturing correlations. Some examples of types of correlations include:

- Correlations over time – Activities from one time period to the next can be positively correlated (increased demand suggests that the demand in the next time period may be even higher) or negatively correlated (above average observations will be followed by below average observations).
- Correlation over space – There are many problems that exhibit strong spatial correlations. Some examples include:
  - Weather – Temperature, wind speed, and rainfall will tend to show strong positive correlations with distance.
  - Presence of disease – Since diseases spread from one person (or animal) to another, the result is spatial pockets of disease that tend to grow.
  - Purchasing behavior – Word of mouth about a product may produce spatial pockets of similar buying behavior.
- Correlation based on characteristics or features – We might see similarities in how people respond to a type of medication based on gender, genetic markers, or smoking history. We might be modeling market demands for similar products.

One of the challenges when generating random samples when there are correlations is that we may have to capture these correlations at different levels of aggregation. We note that the hierarchical aggregation methodology presented in section 3.6.1 accomplishes this automatically.

10.4   Monte Carlo Simulation

We now address the problem of generating random variables from known probability distributions using a process known as Monte Carlo simulation. Although most software tools come with functions to generate observations from major distributions, it is often necessary to customize tools to handle more general distributions.

There is an entire field that focuses on developing and using tools based on the idea of Monte Carlo simulation, and our discussion should be viewed as little more than a brief introduction.

10.4.1 Generating Uniform $[0, 1]$ Random Variables

Arguably the most powerful tool in the Monte Carlo toolbox is the ability to use the computer to generate random numbers that are uniformly distributed between 0 and 1. This is so important that most computer languages and computing environments have a built-in tool for generating uniform $[0, 1]$ random variables, as well as random variables from other distributions. While we strongly recommend using these tools, it is useful to understand how they work. It starts with a simple recursion that looks like

$$R^{n+1} \leftarrow (a + bR^n) \bmod (m),$$

where a and b are very large numbers, while m might be a number such as $2^{64} - 1$ (for a 64 bit computer), or perhaps $m = 999,999,999$. For example, we might use

$$R^{n+1} \leftarrow (593845395 + 2817593R^n) \quad \bmod (999999999).$$

This process simulates randomness because the arithmetic operation $(a + bR)$ creates a number much larger than m, which means we are taking the low order digits, which move in a very random way.

We have to initialize this with some starting variable R^0 called the *random number seed*. If we fix R^0 to some number (say, 123456), then every sequence $R^1, R^2, ...$ will be exactly the same (some computers use an internal clock to keep this from happening, but sometimes this is a desirable feature). If a and b are chosen carefully, R^n and R^{n+1} will appear (even under careful statistical testing) to be independent.

Due to the mod function, all the values of R^n will be between 0 and 999999999. This is convenient because it means if we divide each of them by 999999999, we get a sequence of numbers between 0 and 1. Thus, let

$$U^n = \frac{R^n}{m}.$$

While this process looks easy, we caution readers to use built-in functions for generating random variables, because they will have been carefully designed to produce the required independence properties. Every programming language comes with this function built in. For example, in Excel, the function `Rand()` will generate a random number between 0 and 1 which is both uniformly distributed over this interval, as well as being independent (a critical feature).

Below, we are going to exploit our ability to generate a sequence of uniform $[0, 1]$ random variables to generate a variety of random variables which we denote $W^1, ..., W^n, ...$ We refer to the sequence W^n as a Monte Carlo sample, while modeling using this sample is referred to as Monte Carlo simulation.

There is a wide range of probability distributions that we may draw on to simulate different types of random phenomena, so we are not even going to attempt to provide a comprehensive list of probability distributions. However, we are going to give a summary of some major classes of distributions, primarily as a way to illustrate different methods for generating random observations.

10.4.2 Uniform and Normal Random Variable

Now that we can generate random numbers between 0 and 1, we can quickly generate random numbers that are uniform between a and b using

$$X = a + (b - a)U.$$

Below we are going to show how we can use our ability to generate $(0,1)$ random variables to generate random variables from many other distributions. However, one important exception is that we cannot easily use this capability to generate random variables that are normally distributed.

For this reason, programming languages also come with the ability to generate random variables Z that are normally distributed with mean 0 and variance 1. With this capability, we can generate random variables that are normally distributed with mean μ and variance σ^2 using the sample transformation

$$X = \mu + \sigma Z.$$

We can take one more step. While we will derive tremendous value from our ability to generate a sequence of *independent* random variables that are uniformly distributed on $[0, 1]$, we often have a need to generate a sequence of *correlated* random variables that are normally distributed. Imagine that we need a vector X

$$X = \begin{pmatrix} X_1 \\ X_2 \\ \vdots \\ X_N \end{pmatrix}.$$

Now assume that we are given a covariance matrix Σ where $\Sigma_{ij} = Cov(X_i, X_j)$. Just as we use σ above (the square root of the variance σ^2), we are going to take the "square root" of Σ by taking its Cholesky decomposition, which produces an upper right-triangular matrix. In Python (using the `numpy` package), this can be done using

```
C = numpy.linalg.cholesky(Σ).
```

The matrix C satisfies

$$\Sigma = CC^T,$$

which is why it is sometimes viewed as the square root of Σ.

Now assume that we generate a column vector Z of N independent, normally distributed random variables with mean 0 and variance 1. Let μ be a column vector of $\mu_1, \ldots, \mu_N$ which are the means of our vector of random variables. We can generate a vector of N random variables X with mean μ and covariance matrix Σ using

$$\begin{pmatrix} X_1 \\ X_2 \\ \vdots \\ X_N \end{pmatrix} = \begin{pmatrix} \mu_1 \\ \mu_2 \\ \vdots \\ \mu_N \end{pmatrix} + C \begin{pmatrix} Z_1 \\ Z_2 \\ \vdots \\ Z_N \end{pmatrix}.$$

To illustrate, assume our vector of means is given by

$$\mu = \begin{bmatrix} 10 \\ 3 \\ 7 \end{bmatrix}.$$

Assume our covariance matrix is given by

$$\Sigma = \begin{bmatrix} 9 & 3.31 & 0.1648 \\ 3.31 & 9 & 3.3109 \\ 0.1648 & 3.3109 & 9 \end{bmatrix}.$$

The Cholesky decomposition computed in Python using C = `numpy.linalg.cholesky`(Σ) is

$$C = \begin{bmatrix} 3 & 1.1033 & 0.0549 \\ 0 & 3 & 1.1651 \\ 0 & 0 & 3 \end{bmatrix}.$$

Imagine that we generate a vector Z of independent standard normal deviates

$$Z = \begin{bmatrix} 1.1 \\ -0.57 \\ 0.98 \end{bmatrix}.$$

Using this set of sample realizations of Z, a sample realization u would be

$$u = \begin{bmatrix} 10.7249 \\ 2.4318 \\ 9.9400 \end{bmatrix}.$$

10.4.3 Generating Random Variables from Inverse Cumulative Distributions

Assume we have a distribution with density $f_X(x)$ and cumulative distribution $F_X(x)$, and let $F_X^{-1}(u)$ be the inverse, which means that $x = F_X^{-1}(u)$ is the value of x such that the probability that $X \leq x$ is equal to u (it helps if $0 \leq u \leq 1$). There are some distributions where $F_X^{-1}(u)$ can be found analytically, but computing this numerically can also be quite practical. We now use the following trick from probability. Let U be a random variable that is uniform over the interval $[0, 1]$. Then $X = F_X^{-1}(U)$ is a random variable that has the distribution $X \sim f_X(x)$.

A simple example of this result is the case of an exponential density function $\lambda e^{-\lambda x}$ with cumulative distribution function $1 - e^{-\lambda x}$. Setting $U = 1 - e^{-\lambda x}$ and solving for x gives

$$X = -\frac{1}{\lambda} \ln(1 - U).$$

Since $1 - U$ is also uniformly distributed between 0 and 1, we can use

$$X = -\frac{1}{\lambda} \ln(U).$$

We can generate outputs from a gamma distribution given by

$$f(x|k, \theta) = \frac{x^{k-1} e^{-\frac{x}{\theta}}}{\theta^k \Gamma(k)}.$$

$\Gamma(k)$ is the gamma function, with $\Gamma(k) = (k - 1)!$ if k is integer. The gamma distribution is created by summing k exponential distributions, each with mean $(k\lambda)^{-1}$. This can be simulated by simply generating k random variables with an exponential distribution and adding them together.

A special case of this result allows us to generate binomial random variables. First sample U which is uniform on $[0,1]$, and compute

$$R = \begin{cases} 1 & \text{if } U < p \\ 0 & \text{otherwise.} \end{cases},$$

R will have a binomial distribution with probability p. The same idea can be used to generate a geometric distribution, which is given by (for $x = 0, 1, ...$)

$$\mathbb{P}(X \leq x) = 1 - (1 - p)^{k+1}.$$

Now generate U and find the largest k such that $1 - (1 - p)^{k+1} \leq U$.

Figure 10.1 illustrates using the inverse cumulative-distribution method to generate both uniformly distributed and exponentially distributed random

Figure 10.1 Generating uniformly and exponentially distributed random variables using the inverse cumulative distribution method.

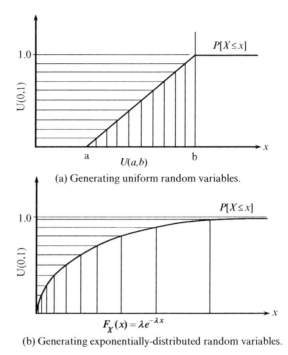

(a) Generating uniform random variables.

(b) Generating exponentially-distributed random variables.

numbers. After generating a uniformly distributed random number in the interval $[0,1]$ (denoted $U(0,1)$ in the figure), we then map this number from the vertical axis to the horizontal axis. If we want to find a random number that is uniformly distributed between a and b, the cumulative distribution simply stretches (or compresses) the uniform $(0,1)$ distribution over the range (a, b).

10.4.4 Inverse Cumulative From Quantile Distributions

This same idea can be used with a quantile distribution (which is a form of nonparametric distribution). Imagine that we compile our cumulative distribution from data. For example, we might be interested in a distribution of wind speeds. Imagine that we collect a large sample of observations $X_1, \dots, X_n, \dots, X_N$, and further assume that they are sorted so that $X_n \leq X_{n+1}$. We would then let $F_X(x)$ be the percentage of observations that are less than or equal to x. The inverse cumulative is computed by simply associating $f_n = F_X(x_n)$ with each observation x_n. Now, if we choose a uniform random number U, we simply find the smallest value of n such that $f_n \leq U$, and then output X_n as our generated random variable.

10.4.5 Distributions with Uncertain Parameters

Imagine that we have the problem of optimizing the price charged for an airline or hotel given the random requests from the market. It is reasonable to assume that the arrival process is described by a Poisson arrival process with rate λ customers per day. However, in most settings we do not know λ.

One approach is to assume that λ is described by yet another probability distribution. For example, we might assume that λ follows a gamma-distribution, which is parameterized by (k, θ). Now, instead of having to know λ, we just need to choose (k, θ), which are referred to as *hyperparameters*. Introducing a belief on unknown parameters introduces more parameters for fitting a distribution. For example, if λ is the expected number of arrivals per day, then the variance of the number of arrivals is also λ, but it is quite likely that the variance is much higher. We can tune the hyperparameters (k, θ) so that we still match the mean but produce a variance closer to what we actually observe.

Consider, for example, the problem of sampling Poisson arrivals describing the process of booking rooms for a hotel for a particular date. For simplicity, we are going to assume that the booking rate is a constant λ over the interval $[0, T]$ where T is the date where people would actually stay in the room (in reality, this rate would vary over time). If N_t is the number of customers booking rooms on day t, the probability distribution of N_t would be given by

$$\mathbb{P}[N_t = i] = \frac{\lambda^i e^{-\lambda}}{i!}.$$

We can generate random samples from this distribution using the methods presented earlier.

Now assume that we are uncertain about λ. We might assume that it has a beta distribution which is given by

$$f(x : \alpha, \beta) = \frac{\Gamma(\alpha + \beta)}{\Gamma(\alpha)\Gamma(\beta)} x^{\alpha-1}(1 - x)^{\beta-1},$$

where $\Gamma(k) = (k - 1)!$ (if k is integer). The beta distribution takes on a variety of shapes over the domain $0 \leq x \leq 1$ (check out the shapes on Wikipedia). Assume that when we observe bookings, we find that N_t has a mean μ and variance σ^2. If the arrival rate λ were known, we would have $\mu = \sigma^2 = \lambda$. However, in practice we often find that $\sigma^2 > \mu$, in which case we can view λ as a random variable.

To find the mean and variance of λ, we start by observing that

$$\mathbb{E}N_t = \mathbb{E}\{\mathbb{E}\{N_t | \lambda\}\} = \mathbb{E}\lambda = \mu.$$

Finding the variance of λ is a bit harder. We start with the identity

$$
\begin{aligned}
VarN_t &= \sigma^2 \\
&= \mathbb{E}N_t^2 - (\mathbb{E}N_t)^2.
\end{aligned}
\tag{10.10}
$$

This allows us to write

$$
\begin{aligned}
\mathbb{E}N_t^2 &= VarN_t + (\mathbb{E}N_t)^2 \\
&= \sigma^2 + \mu^2.
\end{aligned}
$$

We then use

$$
\begin{aligned}
\mathbb{E}N_t &= \mathbb{E}\{\mathbb{E}\{N_t|\lambda\}\} \\
&= \mathbb{E}\lambda, \\
&= \mu. \\
\mathbb{E}N_t^2 &= \mathbb{E}\{\mathbb{E}\{N_t^2|\lambda\}\} \\
&= \mathbb{E}\{\lambda + \lambda^2\} \\
&= \mu + (Var\lambda + \mu^2).
\end{aligned}
$$

We can now write

$$
\begin{aligned}
\sigma^2 + \mu^2 &= \mu + (Var\lambda + \mu^2), \\
Var\lambda &= \sigma^2 - \mu.
\end{aligned}
$$

So, given the mean μ and variance σ^2 of N_t, we can find the mean and variance of λ.

The next challenge is to find the parameters α and β of our beta distribution, which has mean and variance

$$
\mathbb{E}X = \frac{\alpha}{\alpha + \beta},
$$

$$
VarX = \frac{\alpha\beta}{(\alpha + \beta)^2(\alpha + \beta + 1)}.
$$

We are going to leave as an exercise to the reader to decide how to pick α and β so that the moments of our beta-distributed random variable X match the moments of λ.

The parameters α and β are called *hyperparameters* as they are distributional parameters that describe the uncertainty in the arrival rate parameter λ. α and β should be chosen so that the mean of the beta distribution closely matches the observed mean μ (which would be the mean of λ). Less critical is matching the variance, but it is important to reasonably replicate the variance σ^2 of N_t.

Once we have fit the beta distribution, we can run simulations by first simulating a value of λ from the beta distribution. Then, given our sampled value of λ (call it $\hat{\lambda}$), we would sample from our Poisson distribution using arrival rate $\hat{\lambda}$.

10.5   Case Study: Modeling Electricity Prices

With the emphasis on renewables, there has been considerable interest in modeling the stochastic processes that arise in this setting. In this section, we will look at challenges that arise when modeling the price of electricity purchased from the grid, and the energy from a wind farm.

We begin with the problem of modeling real-time electricity prices, shown in Figure 10.2. These prices, taken from the grid operated by PJM Interconnections, which operates the grid serving the mid-Atlantic states in the United States. The prices are from February, 2015, and illustrate the well-known heavy-tailed behavior of electricity prices.

The most elementary model for prices is a basic random walk, given by

$$p_{t+1} = p_t + \varepsilon_{t+1}, \tag{10.11}$$

where we typically assume that $\varepsilon_{t+1} \sim N(0, \sigma_\varepsilon^2)$, which is estimated from the sequence of observations of $p_{t+1} - p_t$.

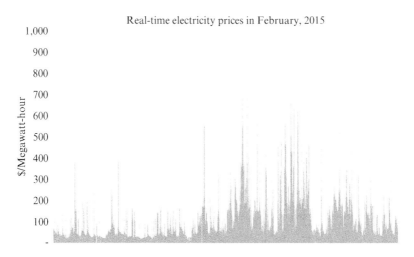

Figure 10.2   Electricity spot prices at 5-minute intervals in February 2015 for PJM Interconnections.

There are a number of problems with this model for applications such as electricity prices. The remainder of this section will suggest methods to improve the performance of this basic model.

10.5.1 Mean Reversion

The most popular stochastic model for prices is known most simply as a mean-reversion model, or, if you enjoy using jargon, the *Ornstein-Uhlenbeck process*. We start by tracking the mean of the process using a simple exponential smoothing model

$$\bar{\mu}_t = (1 - \eta)\bar{\mu}_{t-1} + \eta p_t,$$

where η is a stepsize (or smoothing factor, or learning rate) that smooths the price signal, which is typically a number in the range $[.01, 0.10]$.

Given this estimate of the mean, the mean-reversion model is given by

$$p_{t+1} = p_t + \kappa(\bar{\mu}_t - p_t) + \varepsilon_{t+1}, \tag{10.12}$$

where κ is another smoothing coefficient that has to be calibrated to produce the best fit of estimated and actual prices. If p_t is greater than the estimate of the mean $\bar{\mu}_t$, the next price is pushed down. The noise term ε_{t+1} is typically assumed to be normally distributed with distribution $N(0, \sigma_\varepsilon^2)$, where σ_ε^2 is calculated from the differences between the estimated price $\bar{p}_t$ given by

$$\bar{p}_t = p_t + \kappa(\bar{\mu}_t - p_t),$$

and the actual price p_{t+1}.

10.5.2 Jump-diffusion Models

A limitation of a basic mean-reversion model is that the distribution of p_{t+1} may not be well-described by a normal distribution (given p_t). A simple fix is to use a "jump-diffusion" model, which uses two noise terms that we will call $\varepsilon^{\text{base}}$ and $\varepsilon^{\text{jump}}$. We only add the jump term for a small percentage of the time periods, given by ρ^{jump}. We accomplish this by introducing the indicator variable

$$I_t^{\text{jump}} = \begin{cases} 1 & \text{with probability } \rho^{\text{jump}}, \\ 0 & \text{otherwise.} \end{cases}$$

We can now write our jump diffusion model as

$$p_{t+1} = p_t + \kappa(\bar{\mu}_t - p_t) + \varepsilon_{t+1}^{\text{base}} + I_{t+1}^{\text{jump}}\varepsilon_{t+1}^{\text{jump}}. \tag{10.13}$$

We estimate ρ^{jump} by starting with the basic mean-reversion model in equation (10.12), fitting the model, then estimating the variance σ_ε^2. Then, we pick a tolerance (say three standard deviations), and classify any price p_{t+1} that differs from the predicted price $\bar{p}_t$ by more than three standard deviations as falling outside of the base model. The fraction of prices falling in this range then gives us an initial estimate of ρ^{jump}.

We then estimate the distribution of the jump noise $\varepsilon_{t+1}^{\text{jump}}$ by just using those points that fall outside of the three-sigma range, and also re-estimate the distribution of $\varepsilon_{t+1}^{\text{base}}$ using only the prices that fall within the three-sigma range. Of course, the variance of the error distribution for this subset will be smaller than before. As a result, we would normally repeat the process using the jump diffusion model (10.13). This process might be repeated several times until the estimates stabilize.

The jump diffusion model produces an error

$$\varepsilon_{t+1} = \varepsilon_{t+1}^{\text{base}} + I_{t+1}^{\text{jump}} \varepsilon_{t+1}^{\text{jump}}$$

that will better approximate heavy-tailed behavior than a simple normal distribution.

10.5.3 Quantile Distributions

A common problem in many applications (such as electricity prices) is asymmetric distributions. The largest prices are much larger relative to the mean than the smallest prices. The same is also true with energy from wind, since gusts of wind can be much larger relative to the mean than zero, which is the smallest wind speed. In addition, choosing a parametric distribution that fits either of these processes is challenging.

An alternative approach to using parametric distributions such as the normal is to compile the cumulative distribution of errors directly from the data. A quantile distribution is shown in Figure 10.3, which illustrates its ability to capture asymmetric, heavy-tailed behavior. This is a form of nonparametric distribution (which is also a lookup table), since you have to store $F_X(x)$ for each possible value of x. So, if prices range from 0 to \$1,000, and we want to store the cumulative distribution in increments of 0.10, we need a table with possibly 10,000 different values (although we only have to store the cumulative distribution for prices we actually observe).

It is relatively easy to store the cumulative distribution in a more compact way. The bigger problem is when the distribution depends on other variables such as temperature and humidity. If we divide temperature into 10 ranges, and humidity into 10 ranges, then we have 100 combinations of temperature and humidity, and we would need to compute a cumulative distribution for each of these 100 combinations (this is a classic curse-of-dimensionality since we are using a lookup table for temperature and humidity). Parametric distributions

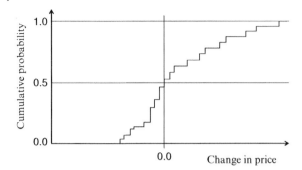

Figure 10.3 Illustration of a quantile distribution for changes in prices.

may offer more compact strategies for incorporating additional dependent variables, but this is generally not possible when using lookup tables (that is, the quantile distributions).

All the methods above focus on getting a better representation of the distribution of errors, but ignore the correlations across time. High electricity prices tend to come in bursts, correlated with hot days where temperatures stay high for periods of time. We propose two approaches, regime shifting and crossing times, that address this issue.

10.5.4 Regime Shifting

A powerful strategy is to identify "regimes" that describe different ranges of our random variable (such as price). For example, we may divide prices into five ranges, or regimes. Each regime might be associated with combinations of temperature and humidity (or any other exogenous variable), or it can be ranges of prices.

First, we compute the distributions we are interested in (such as the change of price) indexed by regime. So, rather than enumerating 100 combinations of temperature and humidity, we could group these into 5 or 10 buckets that we think best explain prices. Number the regimes $s_1, \dots, s_K$ and let S^{regime} be the set of regimes. Then, we have two tasks:

- Compute error distributions (and any other quantities) using any methodology indexed by the regime you are in. These distributions may be parametric or nonparametric (e.g. quantiles), using any of the modeling strategies described above.
- Add up the number of times f_{s_k, s_ℓ} that you transition from regime s_k to regime s_ℓ, and then normalize these to obtain the transition probabilities

$$P_{s_k,s_\ell}^{\text{regime}} = Prob[S_{t+1}^{\text{regime}} = s_\ell | S_t^{\text{regime}} = s_k]. \tag{10.14}$$

Both of these sets of calculations are performed while stepping forward in time through historical data. If your regimes depend on other variables (such as humidity and temperature), then you will need this historical data as well.

Regime shifting is a form of indexed modeling, which can be thought of as a nonparametric modeling strategy. It depends on being able to identify a reasonably small number of regimes, and simplifies modeling by allowing us to fit models that work for individual regimes, rather than globally over the entire dataset.

Regime shifting also gives us another critical feature. Our jump diffusion model, for example, assumed that the jump indicator variable I_t^{jump} was independent across time periods. However, looking at the price plot in Figure 10.2, we see that there are bursts of higher prices. We can capture these bursts, to a degree, since $P_{s_k,s_k}^{\text{regime}}$ will be the probability that we stay in regime s_k, allowing us to capture a certain level of persistence.

10.5.5 Crossing Times

An important characteristic when modeling stochastic processes is not just capturing whether the forecast is above or below the actual, but how long it *stays* above or below. This is important in many settings. For example, if the price of electricity stays high for a period of time, then a utility that has to pay this price may run out of cash reserves.

We are going to use the context of simulating energy from wind. Figure 10.4 shows a sample path of actual energy from wind, along side the forecast (made, say, at noon the day before). The figure shows two time intervals: the first where the actual is below the forecast, called a "down-crossing time," and the second where the forecast is above the forecast, called the "up-crossing time." These "crossing times" are periods where the actual is continuously below or above the forecast.

We are going to build on the methods we have described above in this section to develop a more sophisticated method for replicating how long stochastic processes stay at higher or lower levels, although this time we are going to use energy from a wind farm, where we are trying to model the errors relative to a forecast of the wind energy.

Our modeling strategy uses the following ideas:

(a) As we step forward in time through the historical dataset, each time the actual crosses from above (below) the forecast to below (above) the forecast, we are going to compute the time that the actual was above or below the forecast, and classify that into a set of ranges (say three, for short (S),

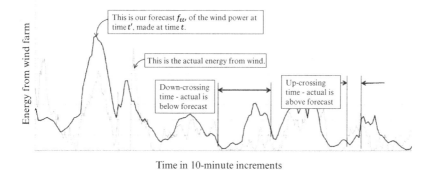

Figure 10.4 Actual vs. predicted energy from wind, showing up- and down-crossings.

    medium (M) or long (L)), that we will treat like regimes which consist of whether the actual was above or below (call this A or B), and then which of the time ranges that the length of the interval falls in (S, M, or L). If there are three time ranges, then there are six regimes, giving us $S^{\text{regime}} = \{A - S, A - M, A - L, B - S, B - M, B - L\}$.

(b) Each time we determine we have come to the end of a crossing time, we update (a) our frequency counter f_{s_k, s_ℓ} for $s_k, s_\ell \in S^{\text{regime}}$ and (b) the frequency distribution for how long the interval lasted given the regime it was in. After normalization we obtain the regime transition matrix $P_{s_k, s_\ell}^{\text{regime}}$ and the distribution of the length of each crossing time given the regime S^{regime}.

(c) For each time period where we know S_t^{regime}, aggregate the energy from wind, E_t, into a set of ranges (say five), and call this set $\mathcal{E}$.

(d) Given the aggregated energy from wind at time t, $E_t^g \in \mathcal{E}$, and the crossing state $i \in \mathcal{I}$, observe the energy E_{t+1} (which is not aggregated) and compile a cumulative distribution of E_{t+1} given the crossing state i and aggregated wind state E_t^g. This distribution will look like the distribution in Figure 10.3. So, if we have six crossing states and five aggregated wind speeds, this gives us 30 states, which means we are creating 30 wind speed cumulative distributions. The result of this calculation is the distribution

$$P_w^W(e, s) = Prob[W_{t+1} | E_t^g = e, S_t^{\text{regime}} = s].$$

    This logic is quite powerful. We now have the ability to explicitly model the distribution of *how long* the actual data stream (wind speed in this case) stays above or below a baseline (the wind speed forecast). In other applications, the baseline could be just an average.

10.6 Sampling vs. Sampled Models

Monte Carlo sampling is without question the most powerful tool in our toolbox for dealing with uncertainty. In this section, we illustrate three ways of performing Monte Carlo sampling: (1) iterative sampling, (2) solving a static, sampled model, and (3) sequentially solving a sampled model with adaptive learning.

10.6.1 Iterative Sampling: A Stochastic Gradient Algorithm

Imagine that we are interested in solving the problem

$$
\begin{aligned}
F(x) &= \mathbb{E}F(x, W) & (10.15) \\
&= \mathbb{E}\{p \min\{x, \hat{D}\} - cx\}, & (10.16)
\end{aligned}
$$

where $W = \hat{D}(\omega)$ is a sample realization of the demand $\hat{D}$, drawn from a full set of outcomes Ω. We could search for the best x using a classical stochastic gradient algorithm such as

$$
x^{n+1} = x^n + \alpha_n \nabla_x F(x^n, \hat{D}(\omega^{n+1})), \tag{10.17}
$$

where

$$
\nabla_x F(x, \hat{D}) = \begin{cases} p - c & x > \hat{D}, \\ -c & x \le \hat{D}. \end{cases} \tag{10.18}
$$

$\nabla_x F(x, \hat{D})$ is called a *stochastic gradient* because it depends on the random variable $\hat{D}$. Under some conditions (for example, the stepsize α_n needs to go to zero, but not too quickly), we can prove that this algorithm will asymptotically converge to the optimal solution.

10.6.2 Static Sampling: Solving a Sampled Model

A sampled version of this problem, on the other hand, involves picking a sample $\hat{\Omega} = \{\omega^1, \ldots, \omega^N\}$. We then solve

$$
\bar{\theta}^N = \arg \min_\theta \frac{1}{N} \sum_{n=1}^N F(\theta | \omega^n). \tag{10.19}
$$

This is actually a deterministic problem (known in some communities as the *sample average approximation*), although one that is much larger than the original stochastic problem (see section 4.3 for a more complete discussion). For many applications, equation (10.19) can be solved using a deterministic solver, although the problem may be quite large. The stochastic gradient update

(10.17) can be much easier to compute than solving the sampled problem (10.19).

The quality of the solution to (10.19) compared to the optimal solution of the original problem (10.15) depends on the application, but as we saw in section 4.3.2, the rate of convergence of $\bar{\theta}^N$ to the optimal θ (for an infinite sample) is actually quite fast.

In practice, stochastic gradient algorithms require tuning the stepsize sequence α_n which can be quite frustrating. On the other hand, stochastic gradient algorithms can be implemented in an online fashion (e.g. through field observations) while the objective (10.19) is a strictly offline approach. There is a rich theory showing that the optimal solution of (10.19), x^N, asymptotically approaches the true optimal (that is, the solution of the original problem (10.15)) as N goes to infinity, but the algorithm is always applied to a static sample $\hat{\Omega}$. Unlike our stochastic gradient algorithm in the previous section, there is no notion of asymptotic convergence (although in practice we will typically stop our stochastic gradient algorithm after a fixed number of iterations).

10.6.3 Sampled Representation with Bayesian Updating

We close our discussion with an illustration of using a sampled model where we are uncertain about the parameters of the model. We then run experiments sequentially and update our belief about the probability that each sampled parameter value is correct.

Imagine, for example, that we we are solving a stochastic revenue management problem for airlines where we assume that the customers arrive according to a Poisson process with rate λ. The problem is that we are not sure of the arrival rate λ. We assume that the true arrival rate is one of a set of values $\lambda_t^1, \dots, \lambda_t^K$, where each is true with probability q_t^k. The vector q_t captures our belief about the true parameters, and can be updated using a simple application of Bayes theorem.

Now let $N(\lambda)$ be a Poisson random variable with mean λ, and let N_{t+1} be the observed number of arrivals between t and $t+1$. We can update q_t using

$$q_{t+1}^k = \frac{\mathbb{P}(N(\lambda) = N_{t+1}|\lambda = \lambda^k)q_t^k}{\sum_{\ell=1}^{K} q_t^\ell \mathbb{P}(N(\lambda) = N_{t+1}|\lambda = \lambda^\ell)},$$

where

$$\mathbb{P}(N(\lambda) = N_{t+1}|\lambda = \lambda^\ell) = \frac{(\lambda^\ell)^{N_{t+1}} e^{\lambda^\ell}}{N_{t+1}!}.$$

The idea of using a sampled set of parameters is quite powerful, and extends to higher dimensional distributions. However, identifying an appropriate sample of parameters becomes harder as the number of parameters increases.

10.7   Closing Notes

We could have dedicated this entire book to methods for modeling stochastic systems without any reference to decisions or optimization. The study of stochastic systems can be found under names including Monte Carlo simulation and uncertainty quantification, with significant contributions from communities that include statistics, stochastic search, simulation optimization, and stochastic programming. This chapter is designed only to provide an indication of some of the topics that a reader will encounter when developing a sequential decision model.

There are a wide range of problems in energy, supply chain management, engineering and health where the process of designing a stochastic model of different sources of uncertainty is quite likely going to be harder than designing an effective policy (although this is not to minimize the importance of effective policies). As we now transition to chapter 11 on designing policies, we encourage the reader to think of developing a stochastic model and an associated policy as an iterative process. The four classes of policies are of increasing complexity, and you may want to get a simpler policy working for the purpose of testing your software while you are building a more sophisticated uncertainty model.

10.8   Bibliographic Notes

Section 10.1 – Our identification of the different sources of uncertainty from the perspective of a model is new.

Section 10.3 – Stochastic modeling is a rich and mature field of study with a long history. For example, there is a field called *uncertainty quantification*; see Smith (2014) and Sullivan (2015) for modern introductions. Stochastic modeling is a term that is often associated with Monte Carlo simulation (see the next section).

Section 10.4 – Monte Carlo simulation is a field with a deep and rich history, starting with the basic idea of using a computer to generate seemingly random numbers. The field has matured to address all the dimensions of modeling stochastic systems. Some examples of excellent introductions are Nelson (2013), Carsey and Harden (2014), Law (2007), and Thomopoulos (2013). For a rigorous treatment of the mathematics of simulation can be found in Asmussen and Glynn (2007). There are a number of books describing these methods in the context of specific fields. For example, Glasserman (2004) and McLeish (2005) describe simulation methods for finance, while Carsey and Harden (2014) presents the methods in the context of the social sciences.

Exercises

Review Questions

10.1 Section 10.5 describes a series of models: mean reversion, jump diffusion, quantile distributions, regime shifting, and crossing times. Very briefly summarize the *specific feature* that each of these strategies contributes relative to the most basic random walk model

$$p_{t+1} = p_t + \varepsilon_{t+1}$$

where $\varepsilon_{t+1} \sim N(0, \sigma_\varepsilon^2)$.

10.2 Section 10.5.5 models "crossing times" for a stochastic process.

(a) Describe what is meant by a "crossing time."
(b) The methodology is described as a form of regime shifting. What is the set of regimes introduced for the problem of modeling wind energy?

Modeling Questions

10.3 For each of the forms of uncertainty below, list the category (or categories) from section 10.1 that best describe the form of uncertainty:

(a) The response of a patient to a new drug.
(b) The energy that will be generated by a wind farm over the next hour, E_{t+1}, given the observation of wind over each of the previous six hours, E_t, E_{t-1}, E_{t-5}, and the fitted linear model:

$$E_{t+1} = \theta_0 E_t + ... + \theta_5 E_{t-5} + \varepsilon_{t+1}.$$

(c) The number of people who say they will vote for a candidate running for office in a telephone poll of 100 people.
(d) The estimated location of a ship calculated using a radar signal, which might incur distortions from weather.
(e) The performance of a dispatcher for a trucking company assigning drivers to loads.

(f) The tariffs to be paid for parts imported from another country next year.

(g) The number of units of inventory transferred from one store to another as instructed by a central manager.

(h) The performance of each member in a team managing a portfolio of physical assets.

(i) The change in market price when a large mutual fund decides to sell a large number of shares in a stock (enough to affect the market).

Computational Exercises

Exercises 10.4 to 10.10 all use the electricity price data that can be downloaded from the supplementary materials website http://tinyurl. com/RLSOsupplementary, "Spreadsheet of electricity price data" (under Chapter 10). Use the tab for the February price data.

10.4 Electricity prices tend to be very random, with very large spikes. Start by assuming that electricity prices p_t (where t steps forward in 5-minute increments) are coming from an exponential distribution, which means we can write

$$p_t \sim \lambda e^{-\lambda y}.$$

Assume that p_t is independent of p_{t+1}. There are 288 five-minute time periods in a day.

(a) Use the computed average price $\bar{p}$ (given in the spreadsheet) to compute $\lambda = 1/\bar{p}$. Then, use the cumulative distribution to compute the expected number of prices (out of the 8064 time periods in February) should be above 100, 200, ..., 500. Compare this to the actual number of prices above each of these values (use the yellow highlighted cell to enter these values to get both the expected number of prices that are over these values, and the actual number). What pattern do you see?

(b) Show how to perform a sample realization from an exponential distribution using the ability of a computer to generate a random variable U that is uniformly distributed between 0 and 1.

(c) Simulate 8064 observations of prices, and plot them as we have plotted the actual prices. How do the two graphs compare?

10.5 Using the spreadsheet for electricity prices, fit a random walk model (equation (10.11)), where you will have to estimate the variance of ε_{t+1}

from the 8064 prices. Generate a sample of 8064 prices using this model, and compare to the actual historical prices. How would you characterize the similarities, and differences, between the two sets of prices?

10.6 Again using the spreadsheet for electricity prices, fit a mean reversion model, where you will have to tune κ (do this using trial and error) to find them model that fits the best. Use $\eta = 0.10$ in your smoothing model for $\bar{\mu}_t$. You will also need to use the model to estimate the variance of ε_{t+1}. Finally, generate another sample of 8064 prices and compare the results to the actual prices.

10.7 Follow the instructions in section 10.5.2 to fit a jump diffusion model, and compare the results to the historical data.

10.8 Use the basic random walk model in equation (10.11) to compute the errors, and then fit a quantile distribution using price increments of $1. Again, simulate the 8064 prices from this model, and compare the patterns with the historical model, as well as the prices from the random walk model (and other methods that you may have implemented above).

10.9 Divide the range of prices into five ranges of your choosing (these may be of equal size, but you may wish to experiment with different sizes, given the wide range of prices). Compute the regime shifting probability distribution $P_{S_k,S_\ell}^{\text{regime}}$ defined in equation (10.14). Now fit a normal distribution for the change in prices for each region. Finally, simulate the evolution of regimes, and then draw a random price for the random distribution in each regime. Compare your results to the historical prices.

10.10 Use the steps described in section 10.5.5 to estimate the regime transition probabilities and the conditional wind distributions $Prob[W_{t+1}|E_t^g = e, S_t^{\text{regime}} = s]$. Finally, use these distributions to simulate electricity prices, and compare the resulting sample (over the 8064 time periods) to history.

Theory Questions

10.11 Let X be a random variable (any random variable with finite variance) and let $F_X(x)$ be the cumulative distribution, which means $F_X(x) = Prob[X \le x]$. Let $F^{-1}(u)$, where $0 \le u \le 1$, be the inverse cumulative distribution, where $u = Prob[X \le F^{-1}(u)]$. Show that the random variable U where $U = F^{-1}(X)$ is uniformly distributed between 0 and 1.

Sequential Decision Analytics and Modeling

These exercises are drawn from the online book *Sequential Decision Analytics and Modeling* available at http://tinyurl.com/sdaexamplesprint.

10.12 Read chapter 8, sections 8.1–8.4, but our focus will be on the uncertainty modeling in section 8.3, which describes three ways of modeling uncertainty in the forecast. Describe each method in detail, and discuss the strengths and weaknesses of each method.

10.13 Read chapter 9, sections 9.1–9.4, but our focus will be on the uncertainty modeling in section 9.3, which describes two ways of modeling uncertainty in the forecast. Describe each method in detail, and discuss the strengths and weaknesses of each method.

Diary Problem

The diary problem is a single problem you chose (see chapter 1 for guidelines). Answer the following for your diary problem.

10.14 Create your own version of Table 10.1 by listing the different categories of uncertainty, and then list the types of uncertainty in your diary problem (if any) that belong to each category. You may feel that a type of uncertainty in your problem can be listed in more than one category.

Bibliography

Asmussen, S. and Glynn, P.W. (2007). *Stochastic Simulation: Algorithms and Analysis*. Springer Science & Business Media.

Carsey, T.M. and Harden, J.J. (2014). *Monte Carlo Simulation and Resampling Methods for Social Science*. Sage Publications.

Glasserman, P. (2004). *Monte Carlo Methods in Financial Engineering*. New York: SpringerVerlag.

Law, A.M. (2007). *Simulation Modeling and Analysis*. New York: McGraw-Hill.

McLeish, D.L. (2005). *Monte Carlo Simulation and Finance*. New York: John Wiley & Sons.

Nelson, B.L. (2013). *Foundations and Methods of Stochastic Simulation: A first course*. New York: Springer.

Smith, R.C. (2014). *Uncertainty Quantification: Theory, Implementation, and Applications*. Philadelphia: SIAM.

Sullivan, T. (2015). *Introduction to Uncertainty Quantification*. New York: Springer.

Thomopoulos, N. (2013). *Essentials of Monte Carlo Simulation: Statistical methods for building simulation models*. New York: Springer.

11

Designing Policies

Now that we have learned how to model a sequential decision problem and simulate an exogenous process $W_1, \ldots, W_t, \ldots$, we return to the challenge of finding a policy that solves our objective function from chapter 9

$$\max_{\pi \in \Pi} \mathbb{E} \left\{ \sum_{t=0}^{T} C_t(S_t, X_t^{\pi}(S_t)) | S_0 \right\}. \tag{11.1}$$

This objective function has been the basis of our "model first, then solve" approach. But now it is time to solve. This leaves us with the question: How in the world do we search over some arbitrary class of policies?

This is precisely the reason that this form of the objective function is popular with mathematicians who do not care about computation, or in communities where it is already clear what type of policy is being used. However, equation (11.1) is not widely used, and we believe the reason is that there has not been a natural path to computation. In fact, entire fields have emerged which focus on particular classes of policies.

In this chapter, we address the problem of searching over policies in a general way. Our approach is quite practical in that we organize our search using classes of policies that are widely used either in practice or in the research literature. Instead of focusing on a particular hammer looking for a nail, we cover all four classes of policies, with the knowledge that when you settle on an approach, it will come from one of the four classes, or possibly a hybrid of two (or more).

We start by clarifying one area of confusion, which is the precise meaning of the term "policy" which is popular only in certain communities. A simple definition of a policy is:

Definition 11.0.1. A **policy** is a method that determines a decision given the information in state S_t... any method.

Reinforcement Learning and Stochastic Optimization: A Unified Framework for Sequential Decisions, First Edition. Warren B. Powell.
© 2022 John Wiley & Sons, Inc. Published 2022 John Wiley & Sons, Inc.

The "any method" is included to counteract the assumption by many that "policy" refers specifically and narrowly to analytical functions, which is just one of our four classes of policies.

"Policies" arise in so many settings in human behavior that it should not be surprising that there are many words that have the same meaning. Table 11.1 provides 45 different examples from the English language.

The problem with the concept of a policy is that it refers to *any* method for determining a decision given a state, and as a result it covers a wide range of algorithmic strategies, each suited to different problems with different computational requirements. Chapter 7 was the first time that we actually saw all four classes of policies applied in the context of derivative-free stochastic optimization, where there are entire research fields dedicated to each of the four classes. The reason that one of these four classes has not emerged as the best reflects the diversity of problems even within this specific problem class. As we move to the much larger class of state-dependent problems, the diversity of applications becomes even broader.

In this chapter, we are going to revisit the four classes of policies (which we first saw in chapters 1, 4, and 7) in greater depth. The hope is that after finishing this chapter, a reader looking to solve a particular problem might have an idea of which one (or two) classes of policies might be best suited for a particular

Table 11.1 Words from the English language that describe methods for making decisions.

Algorithm	Format	Prejudice
Behavior	Formula	Principle
Belief	Grammar	Procedure
Bias	Habit	Process
Canon	Laws/bylaws	Protocols
Code	Manner	Recipe
Commandment	Method	Ritual
Conduct	Mode	Rule
Control law	Mores	Style
Convention	Orthodoxy	Syntax
Culture	Patterns	Technique
Customs	Plans	Template
Duty	Policies	Tenet
Etiquette	Practice	Tradition
Fashion	Precedent	Way of life

problem. For readers looking to simply build up their toolbox of methods, this chapter will serve as an introduction to the four classes, with some guidance how to choose among them. Then we are going to spend chapters 12–19 looking into the four classes in even more detail.

We are going to start by describing a spectrum of problems ranging from (deterministic) optimization to machine learning, and then we are going to contrast our problem of searching for the best policy to the search problems that these other problem areas pose.

11.1 From Optimization to Machine Learning to Sequential Decision Problems

If we have a linear programming problem, anyone with training in deterministic optimization would write down a model that looks like

$$\min_{x} c^T x$$

subject to

$$Ax = b,$$
$$x \geq 0.$$

In real applications, the challenge is creating the A-matrix, but this process is well understood, and there are computer packages that can take these models and solve them, even when x is a vector with thousands, even hundreds of thousands, of variables (or dimensions). Formal training in linear programming is no longer a prerequisite; the users manuals for popular computer packages such as Gurobi and Cplex are sufficient to get you started.

Just as popular is the format used for deterministic optimal control, where we have to manage a system over time by choosing a set of controls $u_0, u_1, \dots, u_T$ (imagine the forces on a vehicle such as landing a SpaceX rocket) to minimize a loss function $L(x_t, u_t)$ when the system is in "state" x_t (for example, the location and velocity of our rocket). The canonical control problem would be written

$$\min_{u_0,\dots,u_T} \sum_{t=0}^{T} L(x_t, u_t), \tag{11.2}$$

where the state x_t (this is standard notation in this community) evolves according to a *transition function* which is written

$$x_{t+1} = f(x_t, u_t). \tag{11.3}$$

The controls may be subject to constraints. Again, there are standard packages for solving versions of this problem.

A different problem that is very relevant to our work arises in machine learning, where we want to find a function (typically called a "statistical model") $f(x|\theta)$, that minimizes the error between observed inputs x^n and the corresponding output y^n for a training dataset $(x^n, y^n), n = 1, \dots, N$. For example, a linear model would be written

$$y = \theta_0 + \theta_1 \phi_1(x) + \theta_2 \phi_2(x) + \dots + \varepsilon, \tag{11.4}$$

where $\phi_f(x)$ is a feature of the input data x. Let $f \in \mathcal{F}$ be a family of functions (models), where f might specify the structure (such as the linear model in (11.4)) and the features ($\phi_f(x)$). Next let $\theta \in \Theta^f$ be the tunable parameters associated with model f. Our optimization problem is to find the best function (model), and the best parameters θ associated with the function, a problem we write as

$$\min_{f \in \mathcal{F}, \theta \in \Theta^f} \sum_{n=1}^{N} (y^n - f(x^n|\theta))^2. \tag{11.5}$$

Here we see an optimization problem written in terms of optimizing over functions, along with any parameters for that function. For machine learning applications, $\mathcal{F}$ covers lookup tables, parametric models, and nonparametric models, and all the choices within these sets (as we covered in chapter 3).

These models are very standard. Readers trained in any of these fields would recognize these models, and would have access to software libraries designed to solve them. These modeling languages are spoken around the world.

The optimization for sequential decision problems, given by equation (11.1), involves searching over policies, which parallels the search over functions in machine learning ("policies" are all examples of functions). However, policies span a much wider range of functions. For example, we are going to see that the first of our four classes of policies include every class of function that we might consider in machine learning.

11.2   The Classes of Policies

There are two fundamental strategies for creating policies, each of which can be further divided into two classes, creating our four classes of policies. The two strategies are given by:

Policy search – Here we are using equation (11.1) directly to search over (a) classes of functions and (b) parameters that characterize a particular class of function.

Lookahead approximations – These are policies that approximate (sometimes exactly) the downstream value of an action taken now.

Both of these can lead to optimal policies under certain circumstances, but only in special cases where we can exploit structure. Since these are relatively rare, a variety of approximation strategies have evolved.

Policy search is based on the principle of assuming that the policy $X^\pi(S_t|\theta)$ belongs to some class of functions, which are typically parametric, but may be nonparametric (that is, locally parametric). Let the set $f \in \mathcal{F}$ capture the structure of the function, and let $\theta \in \Theta^f$ be the tunable parameters associated with each function. The design of the set $\mathcal{F}$ and the choice of $f \in \mathcal{F}$ is often (not always) more art than science. We let $\pi = (f \in \mathcal{F}, \theta \in \Theta^f)$ describe both the type of function and the parameters.

The policy search problem can be written generally as

$$\max_{\pi=(f\in\mathcal{F},\theta\in\Theta^f)} \mathbb{E}_{S_0}\mathbb{E}_{W_1,...,W_T|S_0}\left\{\sum_{t=0}^{T} C(S_t, X^\pi(S_t|\theta))|S_0\right\}. \tag{11.6}$$

Note that we can use any of our family of objective functions (cumulative reward, final reward) and uncertainty operators (described in section 9.8.5) such as expectation (the most common), max-min (robust optimization), or any of the risk measures that emphasize the tails of the distribution.

There are two class of policies within the policy search class:

Policy function approximations (PFAs) – These are analytical functions that map a state to a feasible action. These functions can be any of the three classes of functions we introduced in chapter 3:

Lookup tables – Also referred to as tabular functions, lookup tables mean that we have a discrete decision $X^\pi(S)$ for each discrete state S.

Parametric representations – These are explicit, analytical functions for $X^\pi(S)$ which generally involve a vector of parameters that we typically represent by θ. Thus, we might write our policy as

$$X(S|\theta) = \sum_{f\in\mathcal{F}} \theta_f \phi_f(S)$$

where $\phi_f(S)$, $f \in \mathcal{F}$ is a set of features tuned for approximating the value function or the policy. Neural networks are a class of parametric functions (see section 3.9.3) that are popular in the engineering controls community, where they may be used to approximate either the policy or the value function.

Nonparametric representations – Nonparametric representations offer a more general way of representing functions, but at a price of greater complexity.

PFAs are typically limited to discrete actions, or low-dimensional (and typically continuous) vectors. Note that PFAs include all the classes of statistical models such as those reviewed in chapter 3. PFAs are presented in chapter 12.

Cost function approximations (CFAs) – These are parameterized optimization models where we may use a parameterized modification of the objective function, subject to a (possibly parameterized) approximation of the constraints. CFAs are optimization problems which could be a simple sort (such as the UCB policies introduced in chapter 7), or it could involve solving large linear or integer programs such as scheduling an airline or planning a supply chain. CFAs have the general form

$$X^{CFA}(S_t|\theta) = \arg\max_{x \in \mathcal{X}_t(\theta)} \bar{C}_t(S_t, x|\theta),$$

where $\bar{C}_t(S_t, x|\theta)$ is a parametrically modified cost function, subject to a parametrically modified set of constraints. CFAs are covered in chapter 13.

Lookahead policies are based on trying to solve what will at first look like a rather frightening expression:

$$X_t^*(S_t)=\arg\max_{x_t}\left(C(S_t, x_t) + \mathbb{E}\left\{\max_{\pi}\mathbb{E}\left\{\sum_{t'=t+1}^{T} C(S_{t'}, X_{t'}^\pi(S_{t'}))\bigg|S_{t+1}\right\}\bigg|S_t, x_t\right\}\right). \qquad (11.7)$$

It should not come as a surprise that we cannot compute this, so we turn to approximations. There are two broad classes of approximation strategies, which are given by:

Value function approximations (VFAs) – These are policies based on an approximation of the value of being in a state. These have the general form

$$X^{VFA}(S_t|\theta) = \arg\max_{x \in \mathcal{X}_t} \left(C(S_t, x) + \mathbb{E}\{\overline{V}_{t+1}(S_{t+1}|\theta)|S_t, x_t\}\right) \qquad (11.8)$$

where $\overline{V}_{t+1})(S_{t+1})$ is an approximation of the value of being in state S_{t+1}. VFAs represent a rich and challenging algorithmic strategy that we cover in chapters 14–18.

Direct lookahead policies (DLAs) – This last class of policies directly solves an approximate version of the lookahead policy in equation (11.6). There are a variety of strategies for creating an approximate lookahead model. The most common approximation is to use a deterministic lookahead, but there are many applications where this would be too strong of an approximation. Stochastic lookaheads are such a rich problem class that there are entire fields dedicated to specific strategies for solving even approximate versions of stochastic lookaheads. Direct lookahead policies are covered in chapter 19.

Combined, these create four classes of policies (more precisely, these are meta-classes) that encompass every algorithmic strategy that has been proposed for any sequential stochastic optimization problem. We claim that these classes cover any heuristic methods already used in practice, as well as everything covered in the research literature.

Some observations:

- The first three classes of policies (PFAs, CFAs, and VFAs) introduce four different types of functions we might approximate (we first saw these in chapter 3). These include (1) approximating the function we are maximizing $\mathbb{E}F(x, W)$, (2) the policy $X^\pi(S)$, (3) the objective function or constraints, or (4) the downstream value of being in a state $V_t(S_t)$. Function approximation plays an important role in stochastic optimization, and this brings in the disciplines of statistics and machine learning.
- The class of functions in the PFA class is precisely the set of three classes of approximating architectures from machine learning: lookup tables, parametric, and nonparametric. The only difference between machine learning and searching for the best PFA policy is the objective function. Machine learning uses a training dataset $(x^n, y^n), n = 1, ..., N$ to solve

$$\min_{f \in \mathcal{F}, \theta \in \Theta^f} \sum_{n=1}^{N} (y^n - f(x^n|\theta))^2,$$

which requires a training dataset. Policy search requires a performance metric $C(S, x)$, and a model (the transition function $S^M(s, x, W)$) to create the objective function in equation (11.1).
- The last three classes of policies (CFAs, VFAs, and DLAs) all use an imbedded arg max (or arg min) which means we have to solve a maximization problem as a step in computing the policy. This maximization (or minimization) problem may be fairly trivial (for example, sorting the value of a set of choices), or quite complex (some applications require solving large integer programs).
- It is possible to get very high-quality results from relatively simple policies if we are allowed to tune them (these would fall under policy search). However, this opens the door to using relatively simple lookahead policies

(for example, using a deterministic lookahead) which has been modified by tunable parameters for helping to manage uncertainty.

These four classes of policies encompass all the disciplines that we reviewed in chapter 2. We started to hint at the full range of policies in chapter 7 when we addressed derivative-free stochastic optimization. We are going to cover these policies in considerably more depth over chapters 12–19. Our goal is to provide a foundation for designing effective policies for the full modeling framework we introduced in chapter 9.

In the remainder of this chapter, we describe these policies in somewhat more depth, but defer to later chapters for complete descriptions. Reading this chapter is the best way to get a sense of all four classes of policies. We use an energy storage application in section 11.9 to demonstrate that each of these four classes may work best on the same problem class, depending on the specific characteristics of the data.

11.3 Policy Function Approximations

It is often the case that we have a very good idea of how to make a decision, and we can design a function (which is to say a policy) that returns a decision which captures the structure of the problem. For example:

■ **EXAMPLE 11.1**

A policeman would like to give tickets to maximize the revenue from the citations he writes. Stopping a car requires about 15 minutes to write up the citation, and the fines on violations within 10 miles per hour of the speed limit are fairly small. Violations of 20 miles per hour over the speed limit are significant, but relatively few drivers fall in this range. It is clear that the best policy will be to choose a speed, say θ^{speed}, above which he writes out a citation. The problem is choosing θ^{speed}.

■ **EXAMPLE 11.2**

A utility wants to maximize the profits earned by storing energy in a battery when prices are lowest during the day, and releasing the energy when prices are highest. There is a fairly regular daily pattern to prices. The optimal policy can be found by solving a dynamic program or stochastic lookahead policy, but it is fairly apparent that the policy is to charge the battery at one time during the day, and discharge it at another. The problem is identifying these times.

◼ **EXAMPLE 11.3**

A trader likes to invest in IPOs, wait a few days and then sell, hoping for a quick bump. She wants to use a rule of waiting d days at which point she sells. The problem is to determine d.

◼ **EXAMPLE 11.4**

A drone can be controlled using a series of actuators that govern the force applied in each of three directions to control acceleration, speed, and location (in that order). The logic for specifying the force in each direction can be controlled by a neural network which has to be trained to produce the best results.

◼ **EXAMPLE 11.5**

We are holding a stock, and would like to sell it when it goes over a price θ^{sell}. How should we determine θ^{sell}?

◼ **EXAMPLE 11.6**

In an inventory policy, we will order new product when the inventory S_t falls below θ^{min}. When this happens, we place an order $x_t = \theta^{max} - S_t$, which means we "order up to" θ^{max}. We need to determine $\theta = (\theta^{min}, \theta^{max})$.

◼ **EXAMPLE 11.7**

We might choose to set the output x_t from a water reservoir, as a function of the state (the level of the water) S_t of the reservoir, using a linear function of the form $x_t = \theta_0 + \theta_1 S_t$. Or we might desire a nonlinear relationship with the water level, and use a basis function $\phi(S_t)$ to produce a policy $x_t = \theta_0 + \theta_1 \phi(S_t)$.

The most common type of policy function approximation is some sort of parametric model. Imagine a policy that is linear in a set of basis functions $\phi_f(S_t)$, $f \in \mathcal{F}$. For example, if S_t is a scalar, we might use $\phi_1(S_t) = S_t$ and

$\phi_2(S_t) = S_t^2$. We might also create a constant basis function $\phi_0(S_t) = 1$. Let $\mathcal{F} = \{0, 1, 2\}$ be the set of three basis functions. Assume that we feel that we can write our policy in the form

$$X^\pi(S_t|\theta) = \theta_0\phi_0(S_t) + \theta_1\phi_1(S_t) + \theta_2\phi_2(S_t). \tag{11.9}$$

Here, the index "π" carries the information that the function is linear in a set of basis functions, the set of basis functions, and the parameter vector θ. Policies with this structure are known as *linear decision rules* or, if you want to sound fancy, *affine policies*, because they are linear in the parameter vector θ.

The art is coming up with the structure of the policy. The science is in choosing θ, which we do by solving the stochastic optimization problem

$$\max_\theta F^\pi(\theta) = \mathbb{E} \sum_{t=0}^T C(S_t, X^\pi(S_t|\theta)). \tag{11.10}$$

Here, we write $\max_\theta$ because we have fixed the class of policies (that is, the search $f \in \mathcal{F}$), and we are now searching within a well-defined space. If we were to write $\max_\pi$ ..., a proper interpretation would be that we would be searching over different functions (e.g. different sets of basis functions), or perhaps even different classes, in addition to searching for whatever parameters θ are associated with that class. Note that we will let π be both the class of policy as well as its parameter vector θ, but we still write $F^\pi(\theta)$ explicitly as a function of θ.

The major challenge we face is that we cannot compute $F^\pi(\theta)$ in any compact form, primarily because we cannot compute the expectation. Instead, we have to depend on Monte Carlo samples. Fortunately, we draw on the field of stochastic search to help us with this process. We describe these algorithms in more detail in chapter 12, but the work all draws on derivative-based stochastic optimization (chapter 5) and derivative-free stochastic search (chapter 7).

Parametric policies are popular because of their compact form, but are largely restricted to stationary problems where the policy is not a function of time. Imagine, for example, a situation where the parameter vector in our policy (11.9) is time dependent, giving us a policy of the form

$$X_t^\pi(S_t|\theta) = \sum_{f \in \mathcal{F}} \theta_{tf}\phi_f(S_t). \tag{11.11}$$

Now, our parameter vector is $\theta = (\theta_t)_{t=0}^T$, which is generally dramatically larger than the stationary problem. Solving equation (11.10) for such a large parameter vector (which would easily have hundreds or thousands of dimensions) becomes intractable unless we can compute derivatives of $F^\pi(\theta)$ with respect to θ.

We cover policy function approximations, and how to optimize them, in much greater depth in chapter 12.

11.4   Cost Function Approximations

Cost function approximations represent a class of policy that has been largely overlooked in the academic literature, yet it is widely used in industry (but in an ad-hoc way). In a nutshell, CFAs involve solving a deterministic optimization problem that has been modified so that it works well over time, under uncertainty.

To illustrate, we might start with a myopic policy of the form

$$X_t^{Myopic}(S_t) = \arg\max_{x \in \mathcal{X}_t} C(S_t, x), \tag{11.12}$$

where $\mathcal{X}_t$ captures the set of constraints. We emphasize that x may be high-dimensional, with a linear cost function such as $C(S_t, x) = c_t x$, subject to a set of linear constraints:

$$
\begin{aligned}
A_t x_t &= b_t, \\
x_t &\leq u_t, \\
x_t &\geq 0.
\end{aligned}
$$

This hints at the difference in the type of problems we can consider with CFAs. A sample application might involve assigning resources (people, machines) to jobs (tasks, orders) over time. Let c_{trj} be the cost (or contribution) of assigning resource r to job j at time t, where c_t is the vector of all assignment costs. Also let $x_{trj} = 1$ if we assign resource r to job j at time t, 0 otherwise. Our myopic policy, which assigns resources to jobs to minimize costs now, may perform reasonably well. Now assume that we would like to see if we could make it work a little better.

We can sometimes improve on a myopic policy by solving a problem with a modified objective function.

$$X_t^{CFA}(S_t|\theta) = \arg\max_{x \in \mathcal{X}_t} \Big(C(S_t, x) + \underbrace{\sum_{f \in \mathcal{F}} \theta_f \phi_f(S_t, x)}_{\text{Cost function correction term}} \Big). \tag{11.13}$$

The new term in the objective is called a "cost function correction term." We note that the cost function correction term is *not* a value function approximation, even if it is in the same place, and might even have the same analytic form. The difference is how the coefficient vector θ is computed.

More often (in our experience) we work by modifying the constraints. We might use

$$A_t x_t = \theta^1 \otimes b_t + \theta^2,$$
$$x_t \leq u_t - \theta^3,$$
$$x_t \geq \theta^4.$$

Here, the operator $\otimes$ means that we multiply the ith element of θ^1 times the ith element of b_t. θ^1 and θ^2 are assumed to be the same dimension as the vector b_t, while θ^3 and θ^4 are each assumed to be the same dimension as x_t. The parameter θ^3 can be used to shrink the capacity of storage batteries so we have spare capacity to store a burst of energy from wind, while θ^4 might be used to ensure safety stocks in a supply chain problem.

There will be times when we might scale the matrix A_t. For example, airlines have to insert schedule slack to handle possible weather delays. Instead of using the average flight time between two cities, an airline may use the 80th percentile, which of course is a tunable parameter.

In chapter 13, we discuss a wider range of approximation strategies, including modified constraints and hybrid lookahead policies.

11.5 Value Function Approximations

The next class of policy is based on approximating the value of being in a state resulting from an action we take now. The core idea starts with Bellman's optimality equation (that we first saw in chapter 2 but study in much greater depth in chapter 14), which is written

$$V_t(S_t) = \max_{x \in \mathcal{X}_t} \left(C(S_t, x) + \gamma \mathbb{E}\{V_{t+1}(S_{t+1}) | S_t\} \right) \tag{11.14}$$

where $S_{t+1} = S^M(S_t, x, W_{t+1})$. If we use the post-decision state variable S_t^x,

$$V_t(S_t) = \max_{x \in \mathcal{X}_t} \left(C(S_t, x) + V_t^x(S_t^x) \right), \tag{11.15}$$

where $V_t^x(S_t^x)$ is the (optimal) value of being in post-decision state S_t^x at time t. Chapters 15–18 introduce methods for approximating the value function when it cannot be computed exactly, producing the policy

$$X_t^{VFA-pre}(S_t) = \arg\max_x \left(C(S_t, x) + \gamma \mathbb{E}\{\overline{V}_{t+1}(S_{t+1}|\theta)|S_t\} \right), \tag{11.16}$$

where $\overline{V}_{t+1}(S_{t+1}|\theta)$ approximates the term (from the optimal policy given by equation (11.7))

$$\overline{V}_{t+1}(S_{t+1}|\theta) \approx \max_{\pi} \mathbb{E}\left\{\sum_{t'=t+1}^{T} C(S_{t'}, X_{t'}^{\pi}(S_{t'})) \,\middle|\, S_{t+1}\right\}.$$

The expectation in (11.16) can be computationally problematic within the $\arg\max_x$, so a way to avoid it is to use the post-decision version of the policy (introduced in section 9.4.5), given by

$$X^{VFA}(S_t|\theta) = \arg\max_{x\in\mathcal{X}_t}\left(C(S_t, x) + \overline{V}_t^x(S_t^x|\theta)\right) \tag{11.17}$$

where

$$\overline{V}_t^x(S_t^x, x|\theta) \approx \mathbb{E}\left\{\max_{\pi} \mathbb{E}\left\{\sum_{t'=t+1}^{T} C(S_{t'}, X_{t'}^{\pi}(S_{t'})) \,\middle|\, S_{t+1}\right\} \,\middle|\, S_t, x_t\right\}.$$

The $\arg\max_x$ in (11.17) is now a deterministic optimization problem, which is much more convenient to use, and opens the door to allowing x to be a potentially high-dimensional vector.

Although dynamic programming is most often used in settings with discrete actions, we can handle vector-valued decisions x_t for problems where the contribution function $C(S_t, x_t)$ is concave in x_t, which produces concave value functions. Chapter 18 shows how to create value function approximations that exploit this property, making it possible to solve very high-dimensional resource allocation problems.

A closely related policy, developed under the umbrella of reinforcement learning within computer science, is to use Q-factors which approximate the value of being in a state S_t and taking discrete action a_t (the strategy only works for discrete actions). Let $\overline{Q}^n(s, a)$ be our approximate value of being in state s and taking action a after n iterations. Q-learning uses some rule to choose a state s^n and action a^n, and then uses some process to simulate a subsequent downstream state s' (which might be observed from a physical system). It then proceeds by computing

$$\hat{q}^n(s^n, a^n) = C(s^n, a^n) + \max_{a'} \overline{Q}^{n-1}(s', a'), \tag{11.18}$$

$$\overline{Q}^n(s^n, a^n) = (1 - \alpha)\overline{Q}^{n-1}(s^n, a^n) + \alpha\hat{q}^n(s^n, a^n). \tag{11.19}$$

Given a set of Q-factors $\overline{Q}^n(s, a)$, the policy is given by

$$A^{\pi}(S_t) = \arg\max_{a} \overline{Q}^n(S_t, a). \tag{11.20}$$

Q-learning became quite popular largely because of its simplicity, but there is a big gap between coding the basic updates in (11.18)–(11.19), and getting it to actually work. There are a number of algorithmic choices that have to be made, such as how to choose the state s^n and action a^n during the learning process, and how to approximate $Q(s, a)$ when the state space is large (which it always is).

Developing effective policies through the process of approximating value functions is a powerful solution approach, but it is no panacea, and getting it to work well can be quite challenging. It has attracted considerable attention from the academic literature, which is one reason that we need five chapters (chapters 14–18).

11.6 Direct Lookahead Approximations

We save direct lookahead policies for last because this is the most brute-force approach among the four classes of policies. A good description for DLA policies is that they are the class you turn to when all else fails, and all else often fails.

11.6.1 The Basic Idea

Imagine that we are in a state S_t. We would like to choose an action x_t that maximizes the contribution $C(S_t, x_t)$ now, plus the value of the state that our action takes us to. Given S_t and x_t, we will generally experience some randomness W_{t+1} that then takes us to state S_{t+1}. The value of being in state S_{t+1} is given by

$$V_{t+1}^*(S_{t+1}) = \max_\pi \mathbb{E}\left\{ \sum_{t'=t+1}^{T} C(S_{t'}, X_{t'}^\pi(S_{t'}))|S_{t+1} \right\}$$

$$= \mathbb{E}\left\{ \sum_{t'=t+1}^{T} C(S_{t'}, X_{t'}^*(S_{t'}))|S_{t+1} \right\}. \tag{11.21}$$

We could write our optimal policy just as we did above in equation (11.14)

$$X^*(S_t) = \arg\max_{x_t} \left(C(S_t, x_t) + \mathbb{E}\{V_{t+1}^*(S_{t+1})|S_t, x_t\} \right),$$

but now we are going to recognize that we generally cannot compute the optimal value function $V_{t+1}^*(S_{t+1})$. Rather than try to approximate this function, we are going to substitute in the definition of $V_{t+1}^*(S_{t+1})$ from (11.21), which gives us

$$X_t^*(S_t) = \arg\max_{x_t} \left(C(S_t, x_t) + \mathbb{E}\left\{ \mathbb{E}\left\{ \sum_{t'=t+1}^{T} C(S_{t'}, X_{t'}^*(S_{t'})) \middle| S_{t+1} \right\} \middle| S_t, x_t \right\} \right)$$

(11.22)

Another way of writing (11.22) is to explicitly imbed the search for the optimal policy in the lookahead portion, giving us

$$X_t^*(S_t) = \arg\max_{x_t} \left(C(S_t, x_t) + \mathbb{E}\left\{ \max_{\pi} \mathbb{E}\left\{ \sum_{t'=t+1}^{T} C(S_{t'}, X_{t'}^{\pi}(S_{t'})) \middle| S_{t+1} \right\} \middle| S_t, x_t \right\} \right)$$

(11.23)

Equation (11.23) can look particularly daunting, until we realize that this is exactly what we are doing when we solve a decision tree (exercise 11.10 provides a numerical example) which is illustrated in Figure 11.1. Remember that a "decision node" in a decision tree (the squares) corresponds to the state S_t (if we are referring to the first node), or the states $S_{t'}$ for the later nodes.

We could use some generic rule $X_{t'}^{\pi}(S_{t'})$ for making a decision, or we can solve the decision tree by stepping backward through the tree to find the optimal

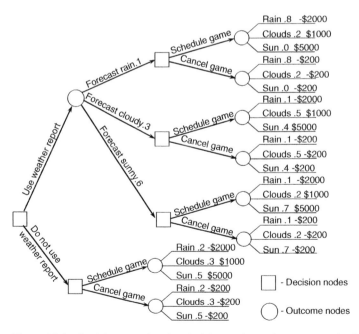

Figure 11.1 Decision tree showing decision nodes and outcome nodes for the setting of deciding whether to schedule a baseball game.

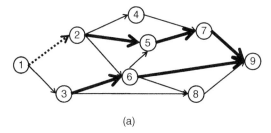

(a)

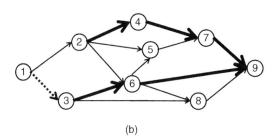

(b)

Figure 11.2 (a) Decision to go (1,2) given the path 2-5-7-9. (b) Decision to go (1,3) when path out of node 2 changes.

action $x_{t'}^*$ for each discrete state $S_{t'}$, which is a lookup table representation for the optimal policy $X_{t'}^*(S_{t'})$. We just have to recognize that $X_{t'}^\pi(S_{t'})$ refers to *some* rule for choosing an action out of node $S_{t'}$, while $X_{t'}^*(S_{t'})$ is the best action out of node $S_{t'}$.

To parse equation (11.23), the first expectation, which is conditioned on state S_t and action x_t, is over the first set of random outcomes out of the circle nodes. The inner $\max_\pi$ refers generally to the process of finding the best action out of *each* of the remaining decision nodes, before knowing the downstream random outcomes. We then evaluate this policy by taking the expectation over all outcomes.

Another way to help understand equation (11.22) (or (11.23)) is to think about a deterministic shortest path problem. Consider the networks shown in Figure 11.2. If we know that we would use the path 2-5-7-9 to get from 2 to 9, we would choose to go from 1 to 2 to take advantage of this path. But if we elect to use a different path out of node 2 (a costlier path), then our decision from node 1 might be to go to node 3. The decisions we are thinking about making downstream can affect the decision we make now.

This key insight translates to the world of uncertainty with a twist: the decision we make now depends on the *policy* we use to make decisions in future. While we dream about using optimal policies, in practice that is just a dream. Chapter 19 explores the idea of using simpler policies in our lookahead model to help streamline computation.

11.6.2 Modeling the Lookahead Problem

This hints at one of the most popular ways of approximating the future for a stochastic problem, which is simply to use a deterministic approximation of the future. We can create what we are going to call a *deterministic lookahead model*, where we act as if we are optimizing in the future, but only for an approximate model.

So we do not confuse the lookahead model with the model we are trying to solve, we are going to introduce two notational devices. First, we are going to use tilde's for state, decision variables and exogenous information variables. Second, we are going to index them by t and t', where t refers to the time at which we are making a decision, and t' indexes time within our lookahead model. Thus, a deterministic lookahead model over a horizon $t, \ldots, t+H$, would be formulated as

$$
X_t^{DLA-Det}(S_t|\theta) = \arg \max_{x_t,(\tilde{x}_{t,t+1},\ldots,\tilde{x}_{t,t+H})} \left(C(S_t, x_t) + \sum_{t'=t+1}^{t+H} C(\tilde{S}_{tt'}, \tilde{x}_{tt'}) \right)
$$

Here, we have replaced the model of the problem from time $t + 1$ to the end of horizon T with a deterministic approximation that goes out to some truncated horizon $t + H$.

There are special cases where we can solve a stochastic lookahead model. One is problems with small numbers of discrete actions, and relatively simple forms of uncertainty. In this case, we can represent our problem using a decision tree such as the one we illustrated in Figure 11.1. A decision tree allows us to find the best decision for each node (that is, each state), which is a form of lookup table policy. The problem is that decision trees explode in size for most problems, limiting their usefulness. In chapter 19, we describe methods for formulating and solving stochastic lookahead models using Monte Carlo methods.

While this is the simplest type of lookahead policy, it illustrates the basic idea. We cannot solve the true problem in (11.23), so we introduced a variety of approximations. Deterministic lookahead models tend to be relatively easy to solve (but not always). However, using a deterministic approximation of the future means that we may make decisions now that do not properly prepare us for random events that may happen in the future. Thus, there is considerable interest in solving a lookahead model that recognizes that the future is uncertain.

The design of lookahead models is as much art as science, but we can use some science to guide the art. We can simplify the lookahead model using strategies such as limiting the horizon, using a sample of random outcomes, discretization, ignoring the updating of selected variables, and using simplified policies for the lookahead model. We note that there are entire books dedicated to specific ways of approximating lookahead models.

Direct lookahead policies are covered in considerably greater depth in chapter 19. For now, we are going to provide a peek inside the design of policies in the lookahead model which we sometimes call the "policy-within-a-policy."

11.6.3 The Policy-Within-a-Policy

As a hint of what we mean by designing the "policy-with-a-policy," we might choose to use a linear decision rule that we could write as

$$\tilde{X}_t^{Lin}(\tilde{S}_{tt'}|\theta_t) = \theta_{t0} + \theta_{t1}\phi_1(\tilde{S}_{tt'}) + \theta_{t2}\phi_2(\tilde{S}_{tt'}).$$

We could then write our stochastic lookahead policy as

$$X_t^{DLA-Stoch}(S_t) = \arg\max_{x_t}\left(C(S_t, x_t) + \right.$$

$$\left. \tilde{E}\left\{ \max_{\hat{\theta}_t}\tilde{E}\left\{ \sum_{t'=t+1}^{T} C(\tilde{S}_{tt'}, \tilde{X}_t^{Lin}(\tilde{S}_{tt'}|\tilde{\theta}_t))|\tilde{S}_{t,t+1} \right\} |S_t, x_t \right\}\right). \quad (11.24)$$

Keep in mind that when you see the expectation operator $\tilde{E}$, this means taking an expectation over the approximate stochastic model. In fact, it almost always means taking an expectation over a sampled model, so just think of taking a sample of whatever random variable is involved. The first $\tilde{E}$ is taking an expectation over the first set of exogenous outcomes, $\tilde{W}_{t,t+1}$, while the second $\tilde{E}$ requires sampling the entire sequence $\tilde{W}_{t,t+2}, \ldots, \tilde{W}_{t,t+H}$. Given this sample, we now simulate our lookahead policy $\tilde{X}_t^{Lin}(\tilde{S}_{tt'}|\theta_t)$ (often called a *rollout policy*) to help us estimate the downstream effects from making decision x_t now.

Designing the policy-within-a-policy is a true art, guided only by the four classes of policies. We need to capture the essential behavior of the problem, but with a policy that is easy to compute. Since the decisions recommended by the policy-within-a-policy are not actually being implemented, we can tolerate approximations in the policy for the purpose of streamlining computation. Needless to say, this tradeoff requires considerable insight into the behavior of the problem.

11.7 Hybrid Strategies

Now that we have identified the four major (meta)classes of policies, we need to recognize that we can also create hybrids by mixing the different classes.

The set of (possibly tunable) policy function approximations, parametric cost function approximations, value function approximations, and direct lookahead policies represent the core tools in the arsenal for finding effective policies for sequential decision problems. Given the richness of applications, it perhaps should not be surprising that we often turn to mixtures of these strategies.

11.7.1 Cost Function Approximation with Policy Function Approximations

A major strength of a deterministic lookahead policy is that we can use powerful math programming solvers to solve high-dimensional deterministic models. A challenge is handling uncertainty in this framework. Policy function approximations, on the other hand, are best suited for relatively simple decisions, and are able to handle uncertainty by capturing structural properties (when they can be clearly identified). PFAs can be integrated into high-dimensional models as nonlinear penalty terms acting on individual (scalar) variables.

As an example, consider the problem of assigning resources (imagine we are managing blood supplies) to tasks, where each resource is described by an attribute vector a (the blood type and age) while each task is described by an attribute vector b (the blood type of a patient, along with other attributes such as whether the patient is an infant or has immune disorders). Let c_{ab} be the contribution we assign if we assign a resource of type a to a patient with blood type b. Let R_{ta} be the number of units of blood type a available at time t, and let D_{tb} be the demand for blood b. Finally let x_{tab} be the number of resources of type a assigned to a task of type b. A myopic policy (a form of cost function approximation) would be to solve

$$X^{CFA}(S_t) = \arg\max_{x_t} \sum_{a \in \mathcal{A}} \sum_{b \in \mathcal{B}} c_{ab} x_{tab} \tag{11.25}$$

subject to

$$\sum_{b \in \mathcal{B}} x_{tab} \leq R_{ta}, \tag{11.26}$$

$$\sum_{a \in \mathcal{A}} x_{tab} \leq D_{tb}, \tag{11.27}$$

$$x_{tab} \geq 0. \tag{11.28}$$

This policy would maximize the total contribution for all blood assignments, but might ignore issues such as a doctor's preference to avoid using blood that is not a perfect match for infants or patients with certain immune disorders.

A doctor's preferences might be expressed through a set of patterns ρ_{ab} which gives the fraction of demand of type b to be satisfied with blood of type a, where

$\sum_a \rho_{ab} = 1$. The vector $\rho_{.b} = (\rho_{ab})_{a\in\mathcal{A}}$ can be viewed as a probabilistic policy describing how to satisfy a demand for a unit of blood of type b (it is a form of PFA).

A natural question would be: why do we need the optimization model? Why can't we just use the patterns ρ_{ab}? The reason is that our patterns might specify how much demand of type b should be supplied with blood with attribute a (we could turn these probabilities around), but in reality we have to balance across all the blood types and demands, which is a much higher dimensionality problem.

The optimization problem described by equations (11.25)–(11.28) easily handles very high-dimensional problems. In fact, we can include blood attributes of blood type (8 types), age (6 types), whether it is frozen or not and, if we like, the location of the blood (this could number in the hundreds to many thousands). This means that our number of blood attributes could range from 100 to 1 million. Problems of this size easily fall in the scope of modern solvers.

We can combine the high-dimensional capabilities of the optimization model given by equations (11.25)–(11.28) with the low-dimensional patterns ρ_{ab} that capture the behavior of the blood management system. These can be combined in a hybrid that would be written

$$X^{CFA-PFA}(S_t|\theta) = \arg\max_{x_t} \sum_{a\in\mathcal{A}} \sum_{b\in\mathcal{B}} \left(c_{ab} x_{tab} + \theta(x_{tab} - D_{tb}\rho_{ab})^2 \right),$$

where θ is a tunable parameter that controls the weight placed on the PFA. This can now be optimized using policy search methods.

11.7.2 Lookahead Policies with Value Function Approximations

Deterministic rolling horizon procedures offer the advantage that we can solve them optimally, and if we have vector-valued decisions, we can use commercial solvers. Limitations of this approach are (a) they require that we use a deterministic view of the future and (b) they can be computationally expensive to solve (pushing us to use shorter horizons). By contrast, a major limitation of value function approximations is that we may not be able to capture the complex interactions that are taking place within our optimization of the future.

An obvious strategy is to combine the two approaches. For low-dimensional action spaces, we can use tree search or a roll-out heuristics for H periods, and then use a value function approximation. If we are using a rolling horizon procedure for vector-valued decisions, we might solve

$$X^\pi(S_t) = \arg\max_{x_t,\dots,x_{t+H}} \sum_{t'=t}^{t+H-1} C(S_{t'}, x_{t'}) + \overline{V}_{t+H}(S_{t+H}),$$

where S_{t+H} is determined by X_{t+H}. In this setting, $\overline{V}_{t+H}(S_{t+H})$ would have to be some convenient analytical form (linear, piecewise linear, nonlinear in S_{t+H}) in order to be used in an appropriate solver.

The hybrid strategy makes it possible to capture the future in a very precise way for a few time periods, while minimizing truncation errors by terminating the tree with an approximate value function. This is a popular strategy in computerized chess games, where a decision tree captures all the complex interactions for a few moves into the future. Then, a simple point system capturing the pieces lost is used to reduce the effect of a finite horizon.

We touch on this stategy only briefly here, but it is arguably one of the most powerful new algorithmic technologies to emerge in stochastic optimization for problems that call for a lookahead policy (of which there are many). We revisit this strategy in more depth in chapter 13.

We note that recent breakthroughs in the use of computers to solve chess or the Chinese game of Go use a hybrid strategy that mixes lookahead policies (using tree search methods we describe in chapter 19), PFAs (basically rules of how to behave based on patterns derived from looking at past games), and VFAs.

11.7.3   Lookahead Policies with Cost Function Approximations

A rolling horizon procedure using a deterministic forecast is, of course, vulnerable to the use of a point forecast of the future. For example, we might be planning inventories for our supply chain for iPhones, but a point forecast might allow inventories to drop to zero if this still allows us to satisfy our forecasts of demand. This strategy would leave the supply chain vulnerable if demands are higher than expected, or if there are delivery delays.

This limitation will not be solved by introducing value function approximations at the end of the horizon. It is possible, however, to perturb the forecasts of demands to account for uncertainty. For example, we could inflate the forecasts of demand to encourage holding inventory. We could multiply the forecast of demand $f_{tt'}^D$ at time t' made at time t by a factor $\theta_{t'-t}^D$. This gives us a vector of tunable parameters $\theta_1^D, \dots, \theta_H^D$ over a planning horizon of length H. Now we just need to tune this parameter vector to achieve good results over many sample paths.

We demonstrate this strategy in chapter 13 using an energy storage setting.

11.7.4   Tree Search with Rollout Heuristic and a Lookup Table Policy

A surprisingly powerful heuristic algorithm that has received considerable success in the context of designing computer algorithms to play games has evolved under the name "Monte Carlo tree search." MCTS uses a limited tree search, which is then augmented by a rollout heuristic assisted by a user-defined lookup table policy. In other words, this is a direct lookahead policy on a stochastic model that mimics solving the original problem, with the restriction that it is only for decision problems with discrete actions.

For example, a computer might evaluate all the options for a chess game for the next four moves, at which point the tree grows explosively. After four moves, the algorithm might resort to a rollout heuristic (which is a general term implying a simple policy-within-a-policy), assisted by rules derived from thousands of chess games (a form of PFA, similar to our patterns ρ_{ab} above). These rules are encapsulated in an aggregated form of lookup table policy that guides the search for a number of additional moves into the future.

11.7.5 Value Function Approximation with Policy Function Approximation

Assume we are given a policy $\bar{X}(S_t)$, which might be in the form of a lookup table or a parameterized policy function approximation. This policy might reflect the experience of a domain expert, or it might be derived from a large database of past decisions. For example, we might have access to the decisions of people playing online poker, or it might be the historical patterns of a company. We can think of $\bar{X}(S_t)$ as the decision of the domain expert or the decision made in the field. If the action is continuous, we could incorporate it into our decision function using

$$X^{\pi}(S_t|\theta) = \arg\max_{x} \left(C(S_t, x) + \overline{V}(S^{M,x}(S_t, x)) - \theta(\bar{X}(S_t) - x)^2 \right).$$

The term $\theta(\bar{X}(S_t) - x)^2$ can be viewed as a penalty for choosing actions that deviate from the external domain expert. The parameter θ controls how important this term is. We note that this penalty term can be set up to handle decisions at some level of aggregation.

11.7.6 Fitting Value Functions Using ADP and Policy Search

Consider any application of approximate dynamic programming to a problem where we are using a parameterized value function approximation – linear, nonlinear parametric, or a neural network. We might be playing games, pricing an option, managing energy storage, or solving a high-dimensional resource allocation problem.

We can estimate a VFA-like term in two stages. Assume we start with a pure VFA policy using a value function approximation $\overline{V}_t^x(S_t^x|\theta^{VFA})$ around the post-decision state S_t^x using the linear model

$$\overline{V}_t^x(S_t^x|\theta^{VFA}) = \sum_{f\in\mathcal{F}} \theta_f^{VFA}\phi_f(S_t^x), \tag{11.29}$$

where $(\phi_f(S_t^x))_{f\in\mathcal{F}}$ is a user-defined set of features and θ^{VFA} is a set of parameters chosen using approximate dynamic programming algorithms. This gives us a VFA policy that we can write as

$$X_t^{VFA}(S_t|\theta^{VFA}) = \arg\max_x \left(C(S_t,x) + \overline{V}_t^x(S_t^x|\theta^{VFA})\right). \tag{11.30}$$

Chapters 15–17 cover strategies for approximating value functions in much greater depth under the umbrella of approximate dynamic programming (ADP). These methods can produce good solutions, but classical ADP techniques are hardly perfect, especially when using parameterized approximations such as the linear model in equation (11.29). This is the first stage of this hybrid strategy.

For the second stage, we can take our VFA policy $X_t^{VFA}(S_t|\theta^{VFA})$ and, starting with $\theta = \theta^{VFA}$, further tune θ using policy search techniques by solving

$$\max_\theta F(\theta) = \mathbb{E}\sum_{t=0}^T C(S_t, X^{VFA}(S_t|\theta)). \tag{11.31}$$

This will typically require the use of one of the algorithms that we introduced in chapters 5 or 7. Let θ^{CFA} be the optimal solution to (11.31). When we use θ^{CFA} in our policy in equation (11.30), it gives us the policy

$$X_t^{CFA}(S_t|\theta^{CFA}) = \arg\max_x \left(C(S_t,x) + \sum_{f\in\mathcal{F}} \theta_f^{CFA}\phi_f(S_t^x)\right).$$

The policy $X_t^{CFA}(S_t|\theta^{CFA})$ is no longer a VFA policy, since there is no reason for $\sum_{f\in\mathcal{F}} \theta_f^{CFA}\phi_f(S_t^x)$ to approximate a value function at this point. The reason is that choosing θ to optimize (11.31) completely loses the objective to make $\sum_{f\in\mathcal{F}} \theta_f^{CFA}\phi_f(S_t^x)$ approximate a value function.

We note in passing that in theory, the CFA-based policy $X_t^{CFA}(S_t|\theta^{CFA})$ should always outperform the VFA-based policy $X_t^{VFA}(S_t|\theta^{VFA})$ since both have the exact same architecture, but the CFA-based policy is tuned specifically to optimize the objective function. There are two reasons why this may not be the case:

- Solving the policy search problem (11.31) introduces noise. The function $F(\theta)$ is often nonconcave, and if the search algorithm is not properly tuned, it can actually end up with a solution that is worse than the starting point.
- The VFA-based policy can easily handle a time-dependent problem, producing a time-dependent policy (in which case we would write θ_t^{VFA} as dependent on time t). The parameters in the CFA-based policy, on the other hand, are assumed to be stationary (that is, they do not depend on time). If they did depend on time, the parameter vector θ_t is now *much* bigger than the stationary parameter vector θ.

Despite these concerns, we believe that starting the stochastic search for the optimization problem in (11.31) using θ^{VFA} as a starting point is likely to produce better results than if we had to use some randomly chosen starting point.

11.8 Randomized Policies

There are several situations where it is useful to randomize a policy:

Exploration-exploitation – This is easily the most common use of randomized policies. Three popular examples of exploration-exploitation policies are:

Epsilon-greedy exploration – This is a popular policy for balancing exploration and exploitation, and can be used for any problem with discrete actions, where the policy has an imbedded $\arg\max_a$ to choose the best discrete action within a discrete set $\mathcal{X} = \{x_1, \dots, x_M\}$. Let $C(s, x)$ be the contribution from being in state s and taking action x, which might include a value function or a lookahead model. The epsilon-greedy policy chooses an action $x \in \mathcal{X}$ at random with probability ϵ, and chooses the action $\arg\max_{x \in \mathcal{X}} C(s, x)$ with probability $1 - \epsilon$.

Boltzmann exploration – Let $\bar{Q}^n(s, x)$ be the current estimate of the value of being in state s and making decision $x \in \mathcal{X} = \{x_1, \dots, x_M\}$. Now compute the probability of choosing action a according to the Boltzmann distribution

$$P(x|s, \theta) = \frac{e^{\theta \bar{Q}^n(s,x)}}{\sum_{x' \in \mathcal{X}} e^{\theta \bar{Q}^n(s,x')}}.$$

The parameter θ is a tunable parameter, where $\theta = 0$ produces a pure exploration policy, while as θ increases, the policy becomes greedy (choosing the action that appears to be best), which is a pure exploitation policy. The Boltzmann policy chooses what appears to be the best action with the highest probability, but any action may be chosen. This is the reason it is often called a *soft max* operator.

Excitation – Assume that the control x is continuous (and possibly vector-valued). Let Z be a similarly-dimensioned vector of normally distributed random variables with mean 0 and variance 1. An excitation policy perturbs the policy $X^{\pi}(S_t)$ by adding a noise term such as

$$x_t = X^{\pi}(S_t) + \sigma Z,$$

where σ is an assumed level of noise.

Thompson sampling – As we saw in chapter 7, Thompson sampling uses a prior on the value of $\mu_x = \mathbb{E}F(x, W)$ is $\mu_x \sim N(\bar{\mu}_x^n, \sigma_x^{2,n})$. Now draw $\hat{\mu}_x^n$ from the distribution $N(\bar{\mu}_x^n, \sigma_x^{2,n})$ for each x, and then choose

$$X^{TS}(S^n) = \arg \max_x \hat{\mu}_x^n.$$

Modeling unpredictable behavior – We may be trying to model the behavior of a system with human input. The policy $X^{\pi}(S_t)$ may reflect perfectly rational behavior, but a human may behave erratically.

Disguising the state – In a multiagent system, a decision can reveal private information. Randomization can help to disguise private information.

Any of the four classes of policies can be randomized, either by perturbing the decision after it comes out of the policy, or by randomizing inputs such as costs or constraints.

It is possible to convert any randomized policy into a deterministic one by including a uniformly-distributed random variable U_t (or the normally-distributed variable Z) to the exogenous information process W_t so that it becomes a part of the state variable S_t. This random variable can then be used to provide the additional information to make $X^{\pi}(S_t)$ a deterministic function of the (now expanded) state S_t. However, it is standard to refer to the policies above as "random."

11.9 Illustration: An Energy Storage Model Revisited

In section 9.9, we presented a model of an energy storage problem. We are going to return to this problem and create samples of all four classes of policies, along with a hybrid. We are going to further show that *each* of these policies may work best depending on the data. We recommend reviewing the model since we are going to use the same notation.

11.9.1 Policy Function Approximation

Our policy function approximation is given by

$$
X_t^{PFA}(S_t|\theta) = \begin{cases}
x_t^{EL} &=& \min\{L_t, E_t\}, \\[4pt]
x_t^{BL} &=& \begin{cases} h_t & \text{If } p_t > \theta^U \\ 0 & \text{If } p_t < \theta^U \end{cases} \\[10pt]
x_t^{GL} &=& L_t - x_t^{EL} - x_t^{BL}, \\[4pt]
x_t^{EB} &=& \min\{E_t - x_t^{EL}, \rho^{chrg}\}, \\[4pt]
x_t^{GB} &=& \begin{cases} \rho^{chrg} - x_t^{EB} & \text{If } p_t < \theta^L \\ 0 & \text{If } p_t > \theta^L \end{cases}
\end{cases}
$$

where $h_t = \min\{L_t - x_t^{EL}, \min\{R_t, \rho^{chrg}\}\}$. This policy is parameterized by (θ^L, θ^U) which determine the price points at which we charge or discharge.

11.9.2 Cost Function Approximation

The cost function approximation minimizes a one-period cost plus a tunable error correction term:

$$
X^{CFA-EC}(S_t|\theta) = \arg \min_{x_t \in \mathcal{X}_t} \left(C(S_t, x_t) + \theta(x_t^{GB} + x_t^{EB} + x_t^{BL}) \right), \qquad (11.32)
$$

where $\mathcal{X}_t$ captures the constraints on the flows (equations (9.24)–(9.28) are from the model given in section 9.9). We use a linear correction term for simplicity which is parameterized by the scalar θ.

11.9.3 Value Function Approximation

Our VFA policy uses an approximate value function approximation, which we write as

$$
X^{VFA}(S_t) = \arg \min_{x_t \in \mathcal{X}_t} \left(C(S_t, x_t) + \overline{V}_t^x(R_t^x) \right), \qquad (11.33)
$$

where $\overline{V}_t^x(R_t^x)$ is a piecewise linear function approximating the marginal value of the post-decision resource state. We use methods described in chapter 18 to compute the value function approximation which exploits the natural convexity of the problem. For now, we simply note that the approximation is quite good.

11.9.4 Deterministic Lookahead

The next policy is a deterministic lookahead over a horizon H which has access to a forecast of wind energy.

$$X_t^{DLA-DET}(S_t) = \arg\min_{(x_t, \tilde{x}_{t,t+1}, \dots, \tilde{x}_{t,t+H})} \left(C(S_t, x_t) + \sum_{t'=t+1}^{t+H} C(\tilde{S}_{tt'}, \tilde{x}_{tt'}) \right) \qquad (11.34)$$

subject to, for $t' = t, \dots, T$:

$$\tilde{x}_{tt'}^{EL} + \tilde{x}_{tt'}^{EB} \leq f_{tt'}^E, \qquad (11.35)$$

$$(\tilde{x}_{tt'}^{GL} + \tilde{x}_{tt'}^{EL} + \tilde{x}_{tt'}^{BL}) = f_{tt'}^L, \qquad (11.36)$$

$$\tilde{x}_{tt'}^{BL} \leq \tilde{R}_{tt'}, \qquad (11.37)$$

$$\tilde{x}_{tt'} \geq 0 \qquad (11.38)$$

where $f_{tt'}^E$ is the forecast of energy from a wind farm at time t', made at time t, and $f_{tt'}^L$ is a forecast of load (demand) for power. We use tilde's on variables in our lookahead model so they are not confused with the same variable in the base model. The variables are also indexed by t, which is when the lookahead model is formed, and t', which is the time period within the lookahead horizon.

11.9.5 Hybrid Lookahead-Cost Function Approximation

Our last policy, $X_t^{DLA-CFA}(S_t | \theta^L, \theta^U)$, is a hybrid lookahead with a form of cost function approximation in the form of two additional constraints for $t' = t + 1, \dots, T$:

$$\tilde{R}_{tt'} \geq \theta^L, \qquad (11.39)$$

$$\tilde{R}_{tt'} \leq \theta^U. \qquad (11.40)$$

These constraints provide buffers to ensure that we do not plan on the energy level getting too close to the lower or upper limits, allowing us to anticipate that there will be times when the energy from a renewable source is lower, or higher, than we planned. We note that a CFA-lookahead policy is actually a hybrid policy, combining a deterministic lookahead with a cost function approximation (where the approximation is in the modification of the constraints).

11.9.6 Experimental Testing

To test our policies, we created five problem variations:

(a) A stationary problem with heavy-tailed prices, relatively low noise, moderately accurate forecasts, and a reasonably fast storage device.

(b) A time-dependent problem with daily load patterns, no seasonalities in energy and price, relatively low noise, less accurate forecasts, and a very fast storage device.

(c) A time-dependent problem with daily load, energy and price patterns, relatively high noise, less accurate forecasts using time series (errors grow with the horizon), and a reasonably fast storage device.

(d) A time-dependent problem with daily load, energy and price patterns, relatively low noise, very accurate forecasts, and a reasonably fast storage device.

(e) Same as (c), but the forecast errors are stationary over the planning horizon.

Each problem variation was designed specifically to take advantage of the characteristics of each of our five policies. We tested all five policies on all five problems. In each case, we evaluated the policy by solving the problem using perfect information (this is known as a posterior bound), and then evaluating the policy as a fraction of this posterior bound. The results are shown in Table 11.2, where the bold entries (in the diagonal) indicates the policy that worked best on that problem class.

The table shows that *each* of the five policies works best on one of the five problems. Of course, the problems were designed so that this was the case, but this illustrates that any of the policies can be best, even on a single problem class, just by modifying the data. For example, a deterministic lookahead works best when the forecast is quite good. A VFA-based strategy works best on problems that are very time-dependent, with a high degree of uncertainty (that is, the forecasts are poor). The hybrid CFA-based policy works best when the forecast is uncertain, but adds value.

Table 11.2 Performance of each class of policy on each problem, relative to the optimal posterior solution (from Powell and Meisel (2016)). Bold indicates the best performer.

Problem:	PFA	CFA-EC	VFA	LA-DET	LA-CFA
A	**0.959**	0.839	0.936	0.887	0.887
B	0.714	**0.752**	0.712	0.746	0.746
C	0.865	0.590	**0.914**	0.886	0.886
D	0.962	0.749	0.971	**0.997**	0.997
E	0.865	0.590	0.914	0.922	**0.934**

11.10   Choosing the Policy Class

Given the choice of policies, the question naturally arises, how do we design a policy that is best for a particular problem? Not surprisingly, it depends on the characteristics of the problem, constraints on computation time, and the complexity of the algorithm. This is the art of policy design, but we feel that we have done as much as we can to guide the art, and make any choices well-informed.

Below we summarize different types of problems, and provide a sample of a policy that appears to be well suited to the application, largely based on our own experiences with real applications.

11.10.1   The Policy Classes

We begin our discussion by reviewing the characteristics of each of our four meta-classes of policies.

Policy function approximations

A utility would like to know the value of a battery that can store electricity when prices are low and release them when prices are high. The price process is highly volatile, with a modest daily cycle. The utility needs a simple policy that is easy to implement in software. The utility chose a policy where we fix two prices, and store when prices are below the lower level and release when prices are above the higher level. This requires optimizing these two price points. A different policy might involve storing at a certain time of day, and releasing at another time of day, to capture the daily cycle.

The PFA is a natural choice because we understand the structure of the policy. It seems clear (and supporting research proves that this is the case) that a "buy low, sell high" policy is optimal. In many cases, the structure of a PFA seems apparent, but lacks any proof of optimality, and may not be optimal, but likely works quite well.

An exception to this guidance is the use of neural networks which have attracted considerable attention for controlling robots, and for playing computer games that provide an environment for collecting large numbers of observations. Neural networks can handle complex inputs such as the characteristics of a player and the state of the game. The weaknesses of neural networks are:

- They require a *lot* of training iterations.
- They do a poor job of capturing structure (e.g. realizing that you should charge a higher price for a hotel room if your competitors are charging higher prices).

- They struggle with noise, and easily run the risk of overfitting (see Figure 3.8 in section 3.9.4).

Neural networks appear to work best in low-noise environments, and where you can run large numbers of repetitions to train the typically large number of parameters that make up a neural network.

Even when the structure of the policy seems apparent, there are several problem characteristics that limit the usefulness of PFAs:

- Time dependency – It may easily be the case that the parameters of our PFA (e.g. the points at which we buy and sell electricity) are time dependent. It is relatively easy to optimize over two parameters. If there are 100 time periods, it is an entirely different matter to optimize over 200 parameters.
- State dependency – Our policy may depend on other state variables such as weather (in our energy storage attribute). In a health application, we may be able to design a PFA to determine the dosage of a medication to lower blood sugar. For example, we may be able to design a simple linear (or piecewise linear) function relating the dosage to the level of blood sugar. But the choice of drug (there are dozens) may depend on patient attributes (of which there are hundreds), and we may need a different PFA for each set of patient attributes.
- Decision dimensionality – PFAs are not well suited to problems where the decision x_t is a vector. If your decision is a vector, that is a quick hint that you are going to need one of the three classes (CFAs, VFAs, and DLAs) that have an imbedded optimization problem which allows us to draw on all the tools of mathematical programming.

Cost function approximation

Cost function approximations may easily be the most widely used class of policy in real applications, although as a class they have been largely ignored by the research literature. CFAs are often used when there is a natural deterministic approximation that can be solved using standard methods. The idea is to introduce parameters that make the policy work better under uncertainty. Of course, this means that, just as with PFAs, there has to be enough structure that we can design an effective parameterization. However, rather than building a policy from scratch, we are starting with a deterministic approximation.

We first saw CFAs used very effectively in pure learning problems in chapter 7. For example, the interval estimation policy

$$X^{IE}(S^n|\theta^{IE}) = \arg\max_x \left(\bar{\mu}_x^n + \theta^{IE}\bar{\sigma}_x^n\right),$$

which trades off exploitation (by maximizing over $\bar{\mu}_x^n$ which is our estimate of how well choice x might work) and exploration (by maximizing over $\bar{\sigma}_x^n$ which is the standard deviation of our estimate $\bar{\mu}_x^n$). The weight that we put on $\bar{\sigma}_x^n$ relative to $\bar{\mu}_x^n$, given by θ^{IE}, has to be tuned.

CFAs are useful when there is a reasonable deterministic approximation that can be optimized, and where we have an intuitive idea of how to handle uncertainty. Consider the problem of deciding on a time to leave for work for your job in a dense city. Your navigation system tells you that the trip will take 37 minutes, so you add 10 minutes to be safe. After following this strategy for a week, you arrive late one day because of an unexpected delay, so you increase your buffer to 15 minutes. This is a form of CFA which is searching for the best path, and then adding a tunable buffer to account for uncertainty, where the buffer is tuned in the field.

CFAs are also well-suited to complex, high-dimensional problems such as scheduling an airline. In this setting, we would solve a large, deterministic integer program to schedule planes and crews, but we have to deal with the uncertainty of flight times due to congestion and weather delays. The airline adds a buffer which may depend on both the origin and destination, but also the time of day. This buffer might be based on a dataset where the airline chooses a buffer so that the flight should be on-time θ percent of the time. The airline will then monitor network-wide on-time performance and feedback from customers to help it tune θ.

Value function approximations

Value function approximations tend to be used for problems where we need to capture the impact of a decision now on the future, *and* where this value can be captured in a well-defined function. Since policies based on VFAs are much easier to use than policies based on DLAs (but notably when we need a stochastic lookahead), the first question should be for problems in this class: Why aren't you using VFAs?

This is where you have to look at your problem and ask how complicated the value function needs to be. Note that dimensionality is not an issue. If you feel that you can reasonably approximate the future using a function that is linear or concave (if maximizing, convex if minimizing) in the state variable, these can be estimated for very high dimensions. These can often be found in large resource allocation problems.

Some examples where VFAs seem to be relatively easy to approximate are:

A blood management problem – Consider the blood management problem presented in section 8.3.2. We can use approximate dynamic programming to solve high-dimensional, spatially distributed versions of this problem using the methods we will describe in chapter 18.

Inventory problems – There are many problems where R_t is a scalar describing the inventory of product for sale, blood supplies, energy in a battery, or cash in a mutual fund.

Routing on a graph – We are at a node i and need to determine which link (i, j) to go to, where traversing a link incurs a random cost $\hat{c}_{ij}$ which is revealed after we move from i to j. We need to learn the value $\bar{v}_i$ of being at each node to make the best decision. Note that this representation is using a lookup table version of the value functions, which means the number of nodes cannot be too large.

We can easily tweak these problems to create examples where the value functions would be quite difficult to approximate:

Blood management with backlogging – Take our blood management problem from section 8.3.2, and add the simple twist that there are elective surgeries which do not have to be satisfied right away. If we do not cover a surgery now, we can perform it at a later time. This "backlogging" introduces interactions between the amount of blood on hand, and the backlogged surgeries, which makes the structure of the value function much more complicated.

Contextual inventory problems – Imagine that while managing our inventory R_t we have to consider other dynamic data. For example, if R_t is how much energy is in the battery, we might also have to keep track of the current and previous prices of energy, the temperature, and the demand for energy. This additional data is sometimes known as a "context," and it complicates the problem because the value of inventory typically does not have structural properties that we can exploit.

Routing on a dynamic graph – Imagine we face the situation of planning a path through a real network with travel times that are constantly being updated. Although often overlooked, the state variable for this problem is a combination of the node where a vehicle has to make a decision, and the current estimates of the link times *for every link in the network!*.

There are problems where policies based on value function approximations represent an amazing breakthrough, but we imagine that they are a very small percentage of real sequential decision problems (which are ubiquitous). We note that the number of papers in the academic literature focusing on the use of value functions, as a percentage of all papers dealing with decisions under uncertainty, far exceeds the percentage of real-world problems that are actually solved using value functions.

Direct lookahead policies

There are many problems which just naturally seem to require that we plan over a horizon to make a decision now. An easy example is a navigation system that plans a path all the way to the destination to determine whether to turn right or left at the next intersection. This problem could never be solved with a PFA or CFA. One can argue that it can be solved with a VFA because deterministic shortest path problems are, in fact, dynamic programs that are solved using value functions, but this is only after we have translated the problem to a deterministic approximation (that is, a deterministic DLA), ignoring dynamic updating of travel time estimates (which is a form of rolling forecast).

There are three important strategies in the DLA class that are quite practical for many applications:

Deterministic lookahead – Sometimes known as model predictive control (MPC) or a rolling/receding horizon procedure, a deterministic lookahead is often the first policy that many will try when faced with a problem which needs a lookahead policy. There is not a simple formula that determines this, but readers should think about their problem and ask to what extent downstream decisions might affect a decision that you need to make now. Also important when choosing between a DLA and VFA is to what extent information is treated as a latent variable in the VFA. For example, it is very common for forecasts to be modeled as latent variables when using VFA-based policies, which means the VFAs have to be recomputed each time the forecasts are updated. By contrast, deterministic DLAs have the forecast built right into the model, which is often (but not always) relatively easy to solve.

    We have found that some quantities are easier to approximate deterministically than others. For example, we would never obtain an approximation of a buy-low, sell-high type of policy if we model uncertain prices with a deterministic forecast. On the other hand, we seem to be quite comfortable planning the best path over a network using point estimates of travel times.

    As a general pattern, as uncertainty increases, VFAs tend to work better, since they provide a natural mechanism for smoothing over variations. Higher levels of uncertainty also tend to make value functions smoother and easier to approximate.

    Deterministic lookaheads can often be good approximations even in the presence of uncertainty. For example, it works quite well in planning paths to a destination even though travel times over each leg of the network are random.

Sampled lookaheads – When we need to handle uncertainty in the future, we are going to return to our approximate, stochastic lookahead policy which we write as

$$X_t^{DLA-Stoch}(S_t) = \arg\max_{x_t} \Bigg(C(S_t, x_t) +$$

$$\tilde{E}\left\{\max_{\tilde{\pi}} \tilde{E}\left\{\sum_{t'=t+1}^{T} C(\tilde{S}_{tt'}, \tilde{X}_t^{\tilde{\pi}}(\tilde{S}_{tt'}|\tilde{\theta}_t))|\tilde{S}_{t,t+1}\right\}|S_t, x_t\right\}\Bigg). \quad (11.41)$$

While equation (11.41) can look frightening, we are going to break it down in chapter 19. In a nutshell, each $\tilde{E}$ is approximated by simulating the underlying sequence $\tilde{W}_{t,t+1}$ (for the first $\tilde{E}$) and then $\tilde{W}_{t,t+2}, \tilde{W}_{t,t+3}, \dots, \tilde{W}_{t,t+H}$ (for the second $\tilde{E}$). The challenge is designing the lookahead policy $\tilde{X}^{\tilde{\pi}}(\tilde{S}_{tt'})$. While of course we want the best possible decisions, in a lookahead model we can get away with less than the best policy, focusing more attention on computational complexity.

Parameterized lookaheads – A strategy that is widely used in practice, but largely ignored by the research community, is the idea of using a deterministic lookahead, but then introduce parameters that modify the deterministic model so that it works better under uncertainty. For example, imagine that we want to find the best path over a network with uncertain link times. Instead of using the estimate of the average link time, we might use the 80th percentile. Now turn this percentile into a tunable parameter θ and simulating use shortest paths based on these travel times. We demonstrate this idea in chapter 13.

11.10.2 Policy Complexity-Computational Tradeoffs

There is a simple tradeoff when choosing policies. Simply put: the more work you put into computing your policy, the less work you have to put into designing and tuning it.

For example, PFAs are the simplest functions, but they only work well for simple problems. Inventory problems are a nice example. Standard inventory policies are characterized by a lower inventory θ^{min} that triggers an inventory order, and an order-up-to amount θ^{max}, creating a tunable parameter vector $\theta = (\theta^{min}, \theta^{max})$. This sounds so deceptively simple that the inventory literature has not progressed past this elementary policy in 60 years.

In fact, the beginning of chapter 1 started with an inventory problem of goods crossing the Pacific to a warehouse in the Southeastern U.S. (see Figure 1.1) which introduces a sequence of complications. We have created a sample of these complications in Table 11.3 where we have listed three types of issues that our policy would have to consider: additional state variables, future information, and future decisions.

Table 11.3 Illustration of complicating state variables, future information and future decisions for our decision problem.

State variables	Future information	Future decisions
Cargo ships will arrive in 6 and 20 weeks	A ship due in 40 days may be delayed 0 to 7 days	We can send rush order via air freight
A storm will hit the port creating a 1-week delay	Demand for produce may shift up by 15%	We can raise prices
A surge in demand will occur in 2 weeks	Forecasted transportation capacity cannot meet the surge	Outside transportation capacity has to be arranged
A commodity price just jumped 20%	Commodity shortages may arise	We can change suppliers

The complicating state variables mean that our state is no longer our inventory R_t, but a host of other information such as the timing of previously ordered inventory (the ships arriving in 6 and 20 weeks), the storm about to hit the port and the change in commodity prices. We can roll all this into our state S_t, but how does this change our inventory policy?

Now our order-up-to parameter vector θ becomes a function $\theta(S_t)$, but what does this function look like? Most likely this will involve additional experimentation and more parameters that have to be tuned. The PFA may be simple, but the state-dependent parameter vector $\theta(S_t)$ represents a major complication. In addition, the challenge of nonstationary behavior typically means that $\theta(S_t)$ becomes $\theta_t(S_t)$, which means θ itself (or the function) is now time dependent.

Additional complications arise in the information that *might* arrive in the future, and then the decisions we *might* make in response. However, the fundamental structure of an order-up-to policy is built around a fairly simple model that can not handle the richness of a hybrid decision structure that adapts to new information in different ways.

The other classes of policies are better suited at handling these complexities. VFAs are better suited at handling time dependencies and high levels of uncertainties, while DLAs (which are often parameterized), which require solving an approximate lookahead, remove much of the guesswork of how a policy should behave by building the complicating issues directly into the model. The price is additional computation (and possibly quite a bit more).

At the risk of oversimplifying this issue, Figure 11.3 depicts the trade-off between the complexity of creating a policy and the cost of computing it. We have divided direct lookaheads between deterministic and stochastic

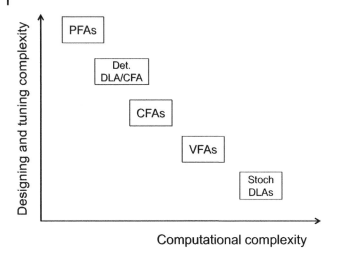

Figure 11.3 Illustration of the tradeoff between the complexity of creating a policy, and the cost of computing a policy, for each of the major policy classes.

lookaheads, since they are dramatically different. The point is to recognize that the complexities of building a policy and computing it are important issues that have to be considered when designing a policy.

11.10.3   Screening Questions

The following screening questions may help to assess which policy class may be appropriate. Keep in mind that the choice of policy is heavily dependent on the context of specific applications which you are working on.

(1) Does the problem have structure that suggests a simple and natural decision rule? If there is an "obvious" policy (e.g. replenish inventory when it gets too low), then more sophisticated algorithms based on value function approximations are likely to struggle. Exploiting structure always helps, but always remember: simple rules imply tunable parameters, and tuning can be hard.

(2) Would a greedy (that is, myopic) policy work reasonably well? This would open the door to solving relatively easy optimization problems for high-dimensional resource allocation problems (such as assigning resources to tasks).

(3) Is the problem fairly stationary, or highly nonstationary? Nonstationary problems (e.g. responding to hourly demand or daily water levels) mean that you need a policy that depends on time. Rolling horizon problems

can work well if the level of uncertainty is low relative to the predictable variability. It is hard to produce policy function approximations where the parameters vary by time period.

(4) Do you have a strong sense that decisions that you might make in the future will affect what you are going to do now? An easy example is a vehicle navigation system that plans the path all the way to the destination, but planning investments to meet major obligations (college, retirement) is another example. If this is the case, we then need to ask questions about uncertainty:

(a) Do you have a well-defined goal you have to reach by a point in time?

(b) Could a deterministic approximation of the future be a reasonable starting point? Since deterministic lookahead policies are so popular (there is an entire field of optimal control, called model predictive control, dedicated to this approach), we have to think carefully about "why not a deterministic lookahead." However, there are many problems where deterministic lookaheads would not be effective. Some examples are:

  * Asset selling problems with stochastic selling prices – Optimal policies depend very much on policies that recognize price variations and exploit them (e.g. sell when price goes above some point).

  * Managing a single resource serving discrete demands (see the nomadic trucker example in section 2.3.4.1).

  * There are actually dynamic problems where a deterministic lookahead model is too large to be solved quickly enough in a dynamic environment.

(c) How much uncertainty is in the future? Value function approximations are especially valuable when uncertainty is high, and can make the value of being in a state easier to approximate.

(5) Does the value of the most important state variables (in particular physical states or states that are being directly controlled) appear to have natural structure that can be exploited in the design of a value function approximation?

A guiding principle when working with the four classes of policies is to start with the simplest policies and then work up.

Table 11.4 lists each of the scenarios above with a suggested starting strategy. Given the massive diversity in problem classes, it is exceptionally difficult to provide precise advice, but we emphasize: these are only intended as starting suggestions.

Unless you are pursuing an algorithm as an intellectual exercise, it is best to focus on your problem and choose the method that is best suited to the application. For more complex problems, be prepared to use a hybrid strategy.

Table 11.4 Suggested starting strategies for the different scenarios in the text.

Scenario	Recommended strategy:
(1)	Clear choice for a PFA.
(2)	Likely choice for a CFA since this likely requires an imbedded optimization problem. Look for parameterizations to improve performance.
(3)	A deterministic lookahead (with imbedded forecast) can turn a nonstationary problem into a stationary one.
(4)	At this point you are headed down the class of direct lookahead policies; just have to figure out which one.
(4a)	If you can live with a deterministic lookahead, then this is your first step. If you need to reach a specific target by a specific time under uncertainty, then you are looking at a technically challenging policy.
(4b)	This suggests a deterministic direct lookahead is your natural starting point.
(4c)	If you have significant uncertainty, then deterministic lookaheads start to struggle, and a value function starts to become more attractive.
(5)	If the value of the state variables that are directly being controlled (even some relatively simple systems can have auxiliary variables that evolve exogenously), then try designing an architecture for a value function approximation. At this point, you need consider the updating mechanisms that are described in chapters 15–18.

For example, rolling horizon procedures may be combined with adjustments that depend on tunable parameters (a form of policy function approximation). You might use a lookahead policy using a decision tree combined with a simple value function approximation to help reduce the size of the tree.

CFAs incorporate more structure in the optimization problem, which makes tuning the coefficients easier. A pure CFA does not attempt to approximate the future, while VFA-based policies are approximating the future, inside an optimization problem (there is an imbedded $\arg\max_x$ within the VFA policy), which tends to further simplify coming up with both the architecture of the VFA, as well as the tuning.

Keep in mind that any approximation can be compensated with tunable parameters. This is how PFAs and (myopic) CFAs (such as UCB policies) can be effective. However, approximate DLAs can also be compensated using well-designed parameterizations, as we are going to demonstrate in chapter 13. Also, tuning a parameterized DLA is simplified because it already has a considerable amount of the problem structure built into policy, as we are going to demonstrate in chapter 13.

But ultimately, it depends on the characteristics of your particular application.

11.11   Policy Evaluation

The choice of the best policy class, and in particular any tuning within a policy class, requires that we perform policy evaluation.

We first have to decide: are we maximizing cumulative reward, as would typically happen in an online setting? In this case we would use

$$\max_{\pi} F^{\pi} \;=\; \mathbb{E}\left\{\sum_{t=0}^{T} C(S_t, X^{\pi}(S_t), W_{t+1})|S_0\right\}$$

$$=\; \mathbb{E}_{S_0}\mathbb{E}_{W_1,\dots,W_T|S_0}\left\{\sum_{t=0}^{T} C(S_t, X^{\pi}(S_t))|S_0\right\}.$$

We then simulate F^{π} using

$$F^{\pi}(\theta|\omega) = \sum_{t=0}^{T-1} C(S_t(\omega), X^{\pi}(S_t(\omega)|\theta)).$$

Finally we average over sample paths to obtain

$$\bar{F}^{\pi}(\theta) = \frac{1}{K}\sum_{k=1}^{K} F^{\pi}(\theta|\omega^k).$$

Otherwise, we are optimizing a final design, which for state-dependent problems, means we are evaluating our policy using

$$\max_{\pi^{lrn}} F^{\pi^{lrn}} = \mathbb{E}\{C(S, X^{\pi^{imp}}(S|\theta^{imp}), \widehat{W})|S^0\}$$

$$= \mathbb{F}_{S^0}\mathbb{E}^{\pi^{imp}}_{((W_t^n)_{t=0}^T)_{n=0}^N|S^0}\left(\mathbb{E}^{\pi^{imp}}_{(\widehat{W}_t)_{t=0}^T|S^0}\frac{1}{T}\sum_{t=0}^{T-1} C(S_t, X^{\pi^{imp}}(S_t|\theta^{imp}), \widehat{W}_{t+1})\right).$$

We then simulate F^{π} using

$$F^{\pi}(\theta^{lrn}|\omega, \psi) = \frac{1}{T}\sum_{t=0}^{T-1} C(S_t(\omega), X^{\pi^{imp}}(S_t(\omega)|\theta^{imp}), \widehat{W}_{t+1}(\psi)).$$

Finally we average over sample paths to obtain

$$\bar{F}^{\pi}(\theta^{lrn}) = \frac{1}{K}\frac{1}{L}\sum_{k=1}^{K}\sum_{\ell=1}^{L} \bar{F}^{\pi}(\theta^{lrn}|\omega^{k},\psi^{\ell}).$$

An important part of the evaluation is designing the observations of the testing samples represented by $\widehat{W}$.

A separate choice is the handling of risk. We use an expectation operator $\mathbb{E}$ as our default metric for averaging across outcomes, but it is entirely possible that risk is an important issue. We might be interested in worst-case performance, 10th-percentile, or one of the VaR or CVaR risk measures discussed in section 9.8.5 are relevant.

11.12 Parameter Tuning

Parameter tuning in policy search is its own stochastic optimization problem to find a policy (or algorithm) to solve a stochastic optimization problem which we can write as

$$\max_{\theta} F^{\pi}(\theta). \tag{11.42}$$

Since $F^{\pi}(\theta)$ involves an expectation we cannot compute we typically are solving

$$\max_{\theta} \bar{F}^{\pi}(\theta). \tag{11.43}$$

Independent of which class of problem produces our function $F^{\pi}(\theta)$ (or $\bar{F}^{\pi}(\theta)$), we need to find a (possibly vector-valued) parameter θ that controls our implementation policy (or how we find our implementation policy).

There are two broad strategies for performing parameter tuning: derivative-based stochastic search (which we covered in chapter 5) and derivative-free stochastic search (covered in chapter 7). Remember that we can use numerical derivatives for problems where gradients are not directly available (which is most of the time). The SPSA algorithm (see section 5.4.4) is well suited for optimizing vector-valued parameters θ even when derivatives are not available.

The process of parameter tuning will need to consider the following issues:

Simulators vs. field experiments – It has been our experience that the vast majority of formal parameter tuning is done using simulators, but building a well-calibrated simulator can be a major project. There are many sequential decision problems that need to be solved, but which do not justify the resources required to build a simulator. If this is the case, the only alternative is to use online learning in the field, which eliminates the possibility of using any derivative-based algorithm. The techniques of chapter 7 using a cumulative-reward objective should be applied here.

Tunable parameters – Choosing the best policy is going to require balancing computational complexity against the simplicity of parameterized policies. The simpler policies in the PFA and CFA classes will look appealing because of their simplicity and easy of development, but as you gain experience in this area, you will start to appreciate the line:

> "The price of simplicity is tunable parameters, and tuning is hard!"

The lookahead policies typically have a much lower burden of parameter tuning (and when there are parameters, they are easier to tune), but you trade off the computational cost of executing these policies in the field.

Latent variables – Further complicating the process of parameter tuning is the presence of "latent variables." Latent variables are, by definition, hidden, which means that if they change, their effect is not being modeled explicitly. A latent variable can be as simple as the starting point of an algorithm. If you tune the parameters of a stepsize rule for a particular starting point, the resulting stepsize rule may easily fail with starting points that are much closer to or farther from the optimal solution. Latent variables can also be the noise in an experiment, or problem features that affect the shape of the response surface.

Expensive experiments – There are many settings where experiments are time consuming (and possibly expensive). Any experiments in the field face the problem that it takes a day to observe a day. However, there are problems that require expensive computer simulations spanning hours to days for a single observation. Laboratory experiments are typically much worse. As of this writing, the research on parameter tuning with small budgets is quite limited. The key in such problem settings is exploiting as much structure and domain knowledge as possible.

Throughout the parameter tuning process, remember that parameter tuning is a sequential decision problem to solve a sequential decision problem. To get a good solution to your real application, you have to do a good job with the parameter search. We recommend testing your search procedure on some benchmark application that allows you to get an accurate measure of how well the procedure is working. Of course, you want to design a benchmark that matches the general behavior of your real application.

Just as a weak algorithm for a deterministic optimization problem can produce a poor solution, a weak search algorithm ("learning policy") can produce a poor implementation policy. In fact, the results can be quite poor. Just because you run an algorithm many iterations does not mean that you have produced a high-quality (or even good) solution. The best way to protect yourself is to design competing solution approaches (perhaps using two or more classes of policies, but this even applies within a class) and choose the one that works best.

11.12.1   The Soft Issues

If the number of classes being tested is small, a reasonable strategy is to analyze each of the policy classes and choose the best one. Of course, we can do better, since this is basically a search over discrete choices.

Rather than evaluate each policy class in depth (which is impractical), we can do a partial evaluation, just as we would examine an unknown function. This introduces the issue of having to optimize over a set of parameters in order to evaluate a particular search policy/algorithm. If this is easy, then finding the best search policy/algorithm may not be as critical. However, imagine finding the best search policy for a problem where derivatives are not available, and function evaluations take several hours (or a day). Choosing the policy is not as obvious as it may seem, given our focus on finding optimal policies. Unlike deterministic optimization, where we want the *best* solution x (lowest cost, highest profit, ...), how well the policy $X^{\pi}(S_t|\theta)$ performs is only one of a number of factors to be considered.

There are parallels with machine learning where we want the value of θ so that our model $f(x|\theta)$ produces the best fit to the data (the training dataset). However, choosing the best model $f(x|\theta)$, which requires searching over functions $f \in \mathcal{F}$, is more complicated. The goal is to work well in the field, and while producing good estimates (or predictions) is always important, issues such as transparency and robustness are also important.

Choosing the best policy depends on the context, but a list of important issues that can and will be important in the final choice include:

- Solution quality – Of course we would like solutions that perform as well as possible, especially in higher volume transactions with clear economic consequences.
- Computational tractability – A representative from Google once made a statement that they wanted the best policy for choosing what ads to display, but it could not take more than 50 milliseconds. A major grid operator has four hours to determine their plan for generation for tomorrow, but they are being asked to implement stochastic lookaheads (which we address in chapter 19) that require significantly more computational effort.

- Robustness – Is the procedure consistently reliable, across a wide range of conditions?
- Methodological complexity – If the method is captured in a black box package, then we only care if the package works, and how well. But there are very few general purpose packages, which means companies (or their consultants) have to develop the logic on their own. A company (more precisely the team doing the work) has to feel confident that the method can be implemented correctly, with good results, on time, and on budget.
- Transparency/diagnosability – We may need to understand *why* a decision is made. If an automated system turns down a loan application from a minority applicant, laws may require that this be documented. However, we may also wonder why a driver is moving a long empty move to pick up a load: Did the load have to be moved? Could it be moved later? Since data may not be perfect, it may be necessary to understand what data is having an impact on the decision. If we do not like a decision, can we trace the reasons behind the recommendation so that we can either understand it, or fix it?
- Data requirements – We need to understand what data is required, and how reliable it is.

11.12.2   Searching Across Policy Classes

The previous section focused on tuning the parameters of a particular policy class. What about searching across policy classes? We need to remember that the four "classes" of policies are really meta-classes; picking a class such as PFA or VFA still involves a lot of work identifying the best functional approximation (for the policy or value function), and then doing all the work of tuning or fitting these approximations. It is not unusual to spend several months developing a particular policy. Doing this for each of the policy classes is generally going to be impractical.

This is where it will be necessary to think of the issues raised in this chapter. Soft issues may dominate the choice of policy class. How much time do you have to develop and test a policy? How important is computational complexity, or transparency? There is nothing wrong with letting these dimensions steer the choice of policy class. For the reader who is using this book to solve a specific problem (rather than gaining general knowledge of the field), our hope is that the discussion in this chapter might guide them to the chapter that will best fit the needs of your problem.

11.13 Bibliographic Notes

Section 11.2 – The identification that there are specific classes of policies was first proposed in Powell (2011)[Chapter 6], but this discussion failed to identify cost function approximations as a specific class. The four classes of policies as they are identified in this book were first formalized in Powell (2014). Powell (2016) divided the four classes into the two core strategies: "policy search policies," and "lookahead policies." Finally, Powell (2019) introduced the concepts of state-independent problems (pure learning problems) and state-dependent problems, along with final-reward and cumulative-reward objectives.

This chapter gives a quick overview of all four classes of policies, but this is just to lay the foundation for the rest of the book. Each of these four classes have been studied in depth, and will be covered in chapters 12–19. Please look at the bibliographic notes in these chapters for more complete summaries of references.

Section 11.9 – This work was taken from Powell and Meisel (2016).

Exercises

Review questions

11.1 What is a policy?

11.2 What are the two strategies for designing policies? What distinguishes them?

11.3 Each of the two strategies consists of two classes of policies. Name them, and describe the distinguishing characteristics of each of the four classes that separates them from the other three.

11.4 For each of the four classes of policies, describe the characteristic(s) that are most difficult about that class.

11.5 What is the central message of the energy storage problem described in section 11.9?

11.6 What is meant by the "policy-within-a-policy"?

11.7 Describe what is meant by a randomized policy? Give an example of a randomized policy for (a) continuous decisions and (b) discrete decisions.

Modeling questions

11.8 What is the difference between a stationary policy, a deterministic nonstationary policy, and an adaptive policy?

11.9 Below is a list of problems with a proposed method for making decisions. Classify each method based on the four classes of policies (you may decide that a method is a hybrid of more than one class).

(a) You use Google maps to find the best path to your destination.

(b) You are managing a shuttle service between the mainland and a small resort island. You decide to dispatch the shuttle as soon as you reach a minimum number of people, or when the wait time of the first person to board exceeds a particular amount.

(c) An airline optimizes its schedule over a month using schedule slack to protect against potential delays.

(d) Upper confidence bounding policies for performing sequential learning (these were introduced in chapter 7).

(e) A computer program for playing chess using a point system to evaluate the value of each piece that has not yet been captured. Assume it chooses the move that leaves it with the highest number of points after one move.

(f) Imagine an improved computer program that enumerates all possible chess moves after three moves, and then applies its point system.

(g) Thompson sampling for sequential learning (also introduced in chapter 7).

11.10 You are the owner of a racing team, and you have to decide whether to keep going with your current driver or to stop and consider a new driver. The decision after each race is to stay with your driver or stop (and switch). The only outcome you care about is whether your driver won or not.

(a) Formulate the problem as a decision tree over three races (we index these races as 0, 1, and 2).

(b) In equation (11.23), we write our optimal policy as

$$X_t^*(S_t) = \arg\max_{x_t}\left(C(S_t, x_t) + \mathbb{E}\left\{\max_\pi \mathbb{E}\left\{\sum_{t'=t+1}^{T} C(S_{t'}, X_{t'}^\pi(S_{t'})) \,\middle|\, S_{t+1}\right\} \,\middle|\, S_t, x_t\right\}\right).$$

(11.44)

Letting $t = 0$ where we face one of two actions (stay with current driver or replace), fully enumerate all the policies we may consider for $t = 1, 2$.

(c) The outer expectation $\mathbb{E}$ in (11.44) is over which random variable(s)?

(d) The inner expectation $\mathbb{E}$ in (11.44) is over which random variable(s)?

Problem-solving questions

11.11   Following is a list of how decisions are made in specific situations. For each, classify the decision function in terms of which of the four fundamental classes of policies are being used. If a policy function approximation or value function approximation is used, identify which functional class is being used:

(a) If the temperature is below 40 degrees F when I wake up, I put on a winter coat. If it is above 40 but less than 55, I will wear a light jacket. Above 55, I do not wear any jacket.

(b) When I get in my car, I use the navigation system to compute the path I should use to get to my destination.

(c) To determine which coal plants, natural gas plants and nuclear power plants to use tomorrow, a grid operator solves an integer program that plans over the next 24 hours which generators should be turned on or off, and when. This plan is then used to notify the plants who will be in operation tomorrow.

(d) A chess player makes a move based on her prior experience of the probability of winning from a particular board position.

(e) A stock broker is watching a stock rise from $22 per share up to $36 per share. After hitting $36, the broker decides to hold on to the stock for a few more days because of the feeling that the stock might still go up.

11.12   Repeat exercise 11.11 for the following decision situations:

(a) A utility has to plan water flows from one reservoir to the next, while ensuring that a host of legal restrictions will be satisfied. The problem can be formulated as a linear program which enforces these constraints. The utility uses a forecast of rainfalls over the next 12 months to determine what it should do right now.

(b) The utility now decides to capture uncertainties in the rainfall by modeling 20 different scenarios of what the rainfall might be on a month-by-month basis over the next year.

(c) A mutual fund has to decide how much cash to keep on hand. The mutual fund uses the rule of keeping enough cash to cover total redemptions over the last 5 days.

(d) A company is planning sales of TVs over the Christmas season. It produces a projection of the demand on a week-by-week basis, but does not want to end the season with zero inventories, so the company adds a function that provides positive value for up to 20 TVs.

(e) A wind farm has to make commitments of how much energy it can provide tomorrow. The wind farm creates a forecast, including an estimate of the expected amount of wind and the standard deviation of the error. The operator then makes an energy commitment so that there is an 80% probability that he will be able to make the commitment.

11.13 Consider two policies:

$$X^{\pi^A}(S_t|\theta) = \arg\max_{x_t}\left(C(S_t, x_t) + \sum_{f\in\mathcal{F}} \theta_f \phi_f(S_t)\right), \qquad (11.45)$$

and

$$X^{\pi^B}(S_t|\theta) = \arg\max_{x_t}\left(C(S_t, x_t) + \sum_{f\in\mathcal{F}} \theta_f \phi_f(S_t)\right). \qquad (11.46)$$

In the case of the policy π^A in equation (11.45), we search for the parameter vector θ by solving

$$\max_{\theta} \mathbb{E}\sum_{t=0}^{T} C(S_t, X^{\pi^A}(S_t|\theta)). \qquad (11.47)$$

In the case of policy π^B, we wish to find θ so that

$$\sum_{f\in\mathcal{F}} \theta_f \phi_f(S_t) \approx \mathbb{E}\sum_{t'=t}^{T} C(S_t, X^{\pi^B}(S_t|\theta)). \qquad (11.48)$$

(a) Classify policies π^A and π^B among the four classes of policies.

(b) Can we expect that the value θ^A that optimizes (11.47) would be approximately equal to the value θ^B that solves equation (11.48)?

(c) Assuming that we can solve the policy search problem in (11.47) to optimality, can we make a statement about which of the two policies might be better? Explain.

11.14 Earlier we considered the problem of assigning a resource i to a task j. If the task is not covered at time t, we hold it in the hopes that we can complete it in the future. We would like to give tasks that have been delayed more higher priority, so instead of just just maximizing the contribution c_{ij}, we add in a bonus that increases with how long the task has been delayed, giving us the modified contribution

$$c_{tij}^{\pi}(\theta) = c_{ij} + \theta_0 e^{-\theta_1(\tau_j - t)}.$$

Now imagine using this contribution function, but optimizing over a time horizon T using forecasts of tasks that might arrive in the future.

(a) Write out the objective function for optimizing θ offline in a simulator.

(b) Would solving this problem, using $c_{tij}^{\pi}(\theta)$ as the contribution for covering task j using resource i at time t, give you the behavior that you want?

Sequential decision analytics and modeling

These exercises are drawn from the online book *Sequential Decision Analytics and Modeling* available at `http://tinyurl.com/sdaexamplesprint`.

11.15 Briefly summarize the policy (there may be more than one) used in the "Designing policies" section for chapters 2, 3, 4, 5, and 6. Classify each policy in terms of the four classes of policies (PFA, CFA, VFA, DLA). If it is a DLA policy, what policy was suggested for the policy-within-a-policy?

11.16 Briefly summarize the policy (there may be more than one) used in the "Designing policies" section for chapters 8 (for sections 8.4.1–8.4.5), 9 (for sections 9.4.1 and 9.4.2), and 10 (for sections 10.4.1 and 10.4.2). Classify each policy in terms of the four classes of policies (PFA, CFA, VFA, DLA). If it is a DLA policy, what policy was suggested for the policy-within-a-policy?

11.17 Briefly summarize the policy (there may be more than one) used in the "Designing policies" section for chapters 11 (for section 11.4), 12 (for sections 12.4.1–12.4.3), and 13 (for section 13.4). Classify each policy in terms of the four classes of policies (PFA, CFA, VFA, DLA). If it is a DLA policy, what policy was suggested for the policy-within-a-policy?

Diary problem

The diary problem is a single problem you chose (see chapter 1 for guidelines). Answer the following for your diary problem.

11.18 List all the decisions that arise in the context of your diary problem (there may be only one, but if your problem is sufficiently rich, you can probably find several). Suggest the class of policy you think is most promising for each type of decision. If possible, try to identify a second choice, and discuss why you feel that the first choice is better.

11.19 Discuss the soft issues (section 11.12.1) that you anticipate would be relevant to at least one of the decisions in your diary problem?

Bibliography

Powell, W.B. (2011). *Approximate Dynamic Programming: Solving the Curses of Dimensionality*, 2e. John Wiley & Sons.

Powell, W.B. (2014). Clearing the Jungle of Stochastic Optimization. *Informs TutORials in Operations Research 2014*.

Powell, W.B. (2016). A unified framework for optimization under uncertainty. In: *Informs TutORials in Operations Research*, 45–83.

Powell, W.B. (2019). A unified framework for stochastic optimization. *European Journal of Operational Research* 275 (3): 795–821.

Powell, W.B. and Meisel, S. (2016). Tutorial on stochastic optimization in energy part II: An energy storage illustration. *IEEE Transactions on Power Systems*.

Part IV – Policy Search

Policy search is a strategy where we define a class of functions that determine a decision, and then search for the best function within that class. Policies in the policy search class can be divided into two subclasses:

Policy function approximations (PFAs) – PFAs are analytical functions that relate information in the state variables to decisions. PFAs come in three (overlapping) forms: lookup tables, parametric models, and nonparametric (or locally parametric) models, which are the same classes of functions used in machine learning. PFAs are typically limited to scalar actions or low-dimensional controls.

PFAs are covered in chapter 12, along with a general discussion of methods for policy search.

Cost function approximations (CFAs) – Parametric CFAs are parameterized optimization problems, where the parameterization guides the optimization problem to produce decisions that work well (a) over time and (b) under uncertainty. We first saw a parametric CFA in chapter 7 in the form of policies for multiarmed bandit problems such as an interval estimation policy

$$X^{\pi}(S_t|\theta) = \arg\max_{x \in \mathcal{X}} \left(\bar{\mu}_x^n + \theta \bar{\sigma}_x^n \right)$$

where $\mathcal{X} = \{x_1, \dots, x_M\}$ is a discrete set of alternatives (ads, drugs) and where $\bar{\mu}_x^n$ is our current estimate of the performance of alternative x after n experiments, and $\bar{\sigma}_x^n$ is the standard deviation of $\bar{\mu}_x^n$. The parameter θ has to be tuned to optimize the policy.

The presence of the "arg max" operator opens the door to using optimization solvers which means the modified optimization problem can be a large linear, nonlinear, or integer program. Now, x can be a high-dimensional vector, with thousands, even hundreds of thousands, of variables. An example is scheduling flights for an airline where we have to introduce schedule slack for weather delays, or the scheduling of energy generators for the power grid, where schedules have to be set given the possibility of outages.

CFAs are covered in chapter 13.

Policy search applied to finding analytical policy function approximations has been widely studied in the academic literature. There are close parallels between policy search and classical machine learning: machine learning minimizes some distance metric between a model $f(x^n|\theta)$ and the corresponding observation y^n and requires a training dataset $(x^n, y^n), n = 1, \ldots, N$, while policy search requires a performance metric $C(S_t, x_t)$ and a model of the system given by the transition function $S_{t+1} = S^M(S_t, x_t, W_{t+1})$ and a model of the exogenous information process.

Parametric cost function approximations, on the other hand, represent a powerful strategy that has been widely used in practice (usually in an ad hoc manner), but almost completely ignored by the research literature, where it is viewed as a "deterministic heuristic." Our position is that it is just as valid as any parametric model used in machine learning. This book is the first to treat this approach as a valid algorithmic strategy for certain classes of stochastic optimization problems.

The policy search class of policies are simpler than the lookahead classes, and as a result they are quite popular. The academic literature places far more attention on the lookahead classes, but the policy search class is much more widely used in practice. The problem is that the price of simplicity is tunable parameters, and tuning is hard.

12

Policy Function Approximations and Policy Search

A policy function approximation (PFA) is any analytical function mapping a state to an action. These "analytical functions" come in three broad (and overlapping) flavors:

Lookup tables – These consist of discrete inputs, and produce a discrete output. Examples are: "If the chess board is in this state, I take this move" or "If this is a male patient, over 50, never smoked, high blood sugar, then take this medication."

Parametric functions – These can be linear or nonlinear models, including neural networks. The user has to specify the structure of the model which is assumed to be governed by a vector of parameters θ, and then algorithms search for the best values of the parameters.

Nonparametric functions – Nonparametric functions might be locally constant approximations, locally linear defined over regions, or high-dimensional nonlinear functions such as deep neural networks.

What distinguishes policy function approximations from the other classes of policies we introduce later in the book is that each of the remaining classes has an imbedded optimization problem within the policy. As a result, PFAs are the simplest class of policies and the easiest to compute, but require a human (typically) to specify the architecture. Not surprisingly, given the wide range of decisions that we encounter throughout life, most decisions are made with simple rules that can be characterized as PFAs, so PFAs are arguably the most widely used class of policy in day-to-day decision making.

Most of our attention will be devoted to parametric functions that are characterized by a set of parameters which we denote by θ. Some examples are listed below.

Reinforcement Learning and Stochastic Optimization: A Unified Framework for Sequential Decisions, First Edition. Warren B. Powell.

■ **EXAMPLE 12.1**

A basic inventory policy is to order product when the inventory goes below some value θ^{min} where we order up to some upper value θ^{max}. If S_t is the inventory level, this policy might be written

$$X^\pi(S_t|\theta) = \begin{cases} \theta^{max} - S_t & \text{if } S_t < \theta^{min}, \\ 0 & \text{otherwise.} \end{cases}$$

■ **EXAMPLE 12.2**

If S_t is a scalar variable giving, for example, the rainfall over the last week, we might set a policy for releasing water from a reservoir using

$$X^\pi(S_t|\theta) = \theta_0 + \theta_1 S_t + \theta_2 S_t^2.$$

■ **EXAMPLE 12.3**

A popular strategy in the engineering community is to train a policy $U^\pi(S_t|\theta)$ for controlling a robot (or a rocket like SpaceX) using a neural network which is characterized by a set of layers and a set of weights that are captured by θ (we provided a brief description of neural networks in section 3.9.3) which takes as input a state variable S_t and outputs a control u_t.

Each of these examples involves a policy parameterized by a parameter vector θ. In principle, we can represent a lookup table using this notation where there is a parameter θ_s for each discrete state s. However, most problems exhibit a large (potentially infinite) number of states, which translates to an equally large (and potentially infinite) number of parameters. There are techniques for optimizing over high-dimensional parameter vectors as long as we can compute gradients exactly (which we develop later in this chapter). However, most applications will be lower-dimensional, and can be optimized using the methods of chapters 5 and 7.

We begin by describing different classes of policies where we focus on policies that have attracted some attention in the literature. Afterward, we turn our attention to the much harder task of optimizing these parameters. The foundation of this process starts with one of our objective functions such as

$$\max_{\theta \in \Theta^\pi} \mathbb{E}\left\{\sum_{t=0}^{T} C(S_t, X^\pi(S_t|\theta))|S_0\right\}, \tag{12.1}$$

where $S_{t+1} = S^M(S_t, X^\pi(S_t|\theta), W_{t+1})$, and where the expectation is over the beliefs in S_0 (if applicable) and the different possible sequences $W_1, \ldots, W_T$. The search is over some space Θ^π that corresponds to the class of policy we have chosen. As we show, this disarmingly simple formulation can be quite hard to solve. However, we have to remember that PFAs are likely the most widely used class of policy in the vast range of sequential decision problems.

12.1 Policy Search as a Sequential Decision Problem

All policy search methods start from the basic idea of simulating a sample path ω giving us a performance metric such as

$$\hat{F}^\pi(\theta, \omega) = \sum_{t=0}^{T} C(S_t(\omega), X^\pi(S_t(\omega)|\theta)), \tag{12.2}$$

where $S_{t+1}(\omega) = S^M(S_t(\omega), X^\pi(S_t(\omega)|\theta), W_{t+1}(\omega))$, where we follow a sample path $W_1(\omega), \ldots, W_T(\omega)$. If we let $W = (W_1, \ldots, W_T)$ represent the entire sequence of random variables (dropping the index ω), we can write this problem using the standard form of a stochastic search problem given by

$$\max_{\theta} F^\pi(\theta) = \mathbb{E}F^\pi(\theta, W). \tag{12.3}$$

Of course, we only work with simulations of $\hat{F}^\pi(\theta, \omega)$, but the form in equation (12.3) is the standard way of writing stochastic search problems.

The objective function in equation (12.3) describes a sequential decision problem characterized by our five elements: (1) the state S_t, (2) the policy $X^\pi(S_t|\theta)$, (3) the exogenous information process W_t, (4) the transition function $S_{t+1} = S^M(S_t, X^\pi(S_t|\theta), W_{t+1})$, and (5) the objective function (12.3), just as we outlined in chapter 9.

The problem of searching for θ is its own sequential decision problem, which consists of the same five components:

(1) The state of the algorithm $S^{\theta,n}$, which includes our belief B^n about the function $F^\pi(\theta)$.
(2) The decision θ^n which is determined by the θ-policy $\theta^n = \Theta^\pi(S^{\theta,n})$.
(3) The exogenous information, which would be the outcome of a simulation of the policy $\hat{F}^\pi(\theta, \omega)$ from equation (12.3).

(4) The transition function

$$S^{\theta,n+1} = S^{\theta,M}(S^{\theta,n}, \theta^n, \hat{F}^\pi(\theta, \omega^{n+1})),$$

which is the equation for updating the belief B^n given the point θ^n at which we observed the function, and the observed performance $\hat{F}^\pi(\theta^n, \omega^{n+1})$, where ω^{n+1} is the sample path used for the $n + 1^{st}$ simulation.

(5) The objective function, where we use the terminal performance of our learning policy π^{lrn} for learning θ after N iterations:

$$\max_{\pi^{lrn}} \mathbb{E}_{S^{\theta,0}} \mathbb{E}_{W^1,\dots,W^N|S^0} \mathbb{E}_{\widehat{W}|S^{\theta,0}} \{F(\theta^{\pi,N}, \widehat{W})|S^{\theta,0}\}.$$

See equation (7.5) in chapter 7 for an in-depth discussion of this objective function.

Now we face the same issues as we do designing an implementation policy $X^\pi(S_t|\theta)$. This is the challenge we address in this chapter by reviewing both derivative-based and derivative-free methods of performing parameter tuning for any PFA-based policy.

12.2   Classes of Policy Function Approximations

A policy function approximation can, quite simply, use any of the strategies used in machine learning that we reviewed in chapter 3: lookup tables, parametric functions (which includes neural networks), and nonparametric functions (including deep neural networks), as well as any hybrids. The only difference between machine learning and policy function approximations is the objective function, as well as the data requirements. The reader is encouraged to flip back to section 1.6.2 where we made this connection. The main point is that machine learning involves solving the search problem over functions $f \in \mathcal{F}, \theta \in \Theta^f$ which we write as

$$\min_{\theta=(f\in\mathcal{F},\theta\in\Theta^f)} \frac{1}{N} \sum_{n=1}^{N} (y^n - f(x^n|\theta))^2,$$

where we need the training dataset (x^n, y^n), $n = 1, \dots, N$. By contrast, policy search involves solving

$$\min_{\theta=(f\in\mathcal{F},\theta\in\Theta^f)} \mathbb{E} \sum_{t=0}^{t} C(S_t, X^\pi(S_t|\theta)),$$

where we do not need a training dataset, but we do need the system model $S_{t+1} = S^M(S_t, X^\pi(S_t|\theta), W_{t+1})$ and the model of the exogenous information process $S_0, W_1, \dots, W_T$. Otherwise, both are searching over the same classes of functions $f \in \mathcal{F}$ which includes lookup tables, parametric and nonparametric functions, and any associated parameters $\theta \in \Theta^f$.

12.2.1 Lookup Table Policies

A lookup table policy is a function where for a particular discrete state s we return a discrete action $x = X^\pi(s)$. This means we have one parameter (an action) for each state. We exclude from this class any policies that can be parameterized by a smaller number of parameters.

Lookup tables are relatively common in practice, since they are easy to understand. Some examples are:

- The Transportation Safety Administration (TSA) has specific rules that determine when and how a passenger should be searched.
- Call-in centers use specific rules to govern how a call should be routed.
- Expert chess players are able to look at a board (in the initial stages of a game) and know exactly what move to make.
- A doctor will often take a set of symptoms and patient characteristics to determine the right treatment.

Lookup tables are easy to understand, and easy to enforce. But in practice, they can be very hard to optimize since there is a value (the action) for each state. So, if we have $|\mathcal{S}| = 1000$ states, searching directly for the best policy would mean searching over a 1000-dimensional parameter space (the action to be taken in each state).

One attraction of lookup table policies is that they are very easy to compute in production; imagine a real-time setting where decisions have to be made with exceptional speed. In business, lookup table policies are widely used where they are known as business rules, although these rules may often be parameterized. In practice these rules are not optimized using formal methods; this chapter will indicate how to do this.

12.2.2 Boltzmann Policies for Discrete Actions

A Boltzmann policy chooses a discrete action $x \in \mathcal{X}_s$ according to the probability distribution

$$f(x|s, \theta) = \frac{e^{\theta \bar{C}(s,x)}}{\sum_{x' \in \mathcal{X}} e^{\theta \bar{C}(s,x)}},$$

where $\bar{C}(s, x)$ is some sort of contribution to be maximized. This could be our estimate of a function $\mathbb{E}F(x, W)$ as we did in chapter 7, or an estimate of the one-step contribution plus a downstream value, as in

$$\bar{C}(S^n, x) = C(S^n, x) + \mathbb{E}\{\bar{V}^n(S^{n+1})|S^n, x\},$$

where $\bar{V}^n(S)$ is our current estimate of the value of being in state S.

Let $F(x|S^n, \theta)$ be the cumulative distribution of our probabilities

$$F(x|s, \theta) = \sum_{x' \leq x} f(x'|s, \theta).$$

Let $U \in [0, 1]$ be a uniformly distributed random number. Our policy $X^\pi(s|\theta)$ could be written as

$$X^\pi(s|\theta) = \arg\max_x\{F(x|s, \theta)|F(x|s, \theta) \leq U\}.$$

This is an example of a so-called stochastic policy, but we handle it just as we would any other policy.

Boltzmann policies are often referred to as "soft-max" because the actions with the highest estimated value are given the highest probability of being accepted. As θ increases, the probability of choosing the decision x with the highest $\bar{C}(s, x)$ quickly approaches 1.0. The purpose of using values of θ so that there is a reasonable probability of choosing actions with less attractive values is that we can observe how well the decision performs, and update our estimate of $\bar{C}(s, x)$.

12.2.3 Linear Decision Rules

A linear decision rules (also known as an "affine policy") is any policy that is linear in the unknown parameters. Thus, a linear decision rule policy might be of the form

$$X^\pi(S_t|\theta) = \theta_0 + \theta_1\phi_1(S_t) + \theta_2\phi_2(S_t).$$

A simple illustration might be a rule for setting the insulin dosage x of a drug given the blood sugar h_t of a patient. We might propose a dosing strategy given by

$$X^\pi(S_t|\theta) = \theta_0 + \theta_1 h_t + \theta_2 h_t^2 + \theta_3 h_t^3.$$

Now the challenge is determining the vector θ that keeps blood sugar within a specified range.

We first saw linear decision rules in chapter 4 when we presented the linear quadratic control problem which, in our notation, is given by

$$\min_{\theta} \mathbb{E} \sum_{t=0}^{T} \left((S_t)^T Q_t S_t + (X^{\pi}(S_t|\theta))^T R_t X^{\pi}(S_t|\theta) \right). \tag{12.4}$$

After considerable algebra, it is possible to show that the optimal policy $X_t^*(S_t)$ is given by

$$X_t^*(S_t) = -K_t S_t,$$

where K_t is a suitably dimensioned matrix that is a function of the matrices Q_t and R_t. Of course, we assume that S_t and x_t are continuous vectors. Thus, $X^*(S_t)$ is a linear function of S_t with coefficients determined by the matrix K_t. See section 14.11 for more details.

This result requires that the objective function be quadratic (or a mixture of quadratic and linear) functions of the state S_t and control x_t. It also requires that the problem be unconstrained, which can be a reasonable starting point for many problems in robotic controls where forces x_t can be positive or negative, and where some constraints (such as the maximum force) would simply not be binding.

Linear decision rules have been applied to other problems, but care has to be used. Linear approximations of functions can be quite useful in a particular region of the function, but a policy $X^{\pi}(S_t)$ has to work well over the entire range of states S_t that we might actually encounter. Low-dimensional linear models (such as a quadratic approximation) can incur fitting errors, while higher-dimensional models are harder to fit, especially when experiments are expensive.

12.2.4 Monotone Policies

There are a number of problems where the decision increases, or decreases, with the state variable. If the state variable is multidimensional, then the decision (which we assume is scalar) increases, or decreases, with *each* dimension of the state variable. Policies with this structure are known as *monotone policies*. Some examples include the following:

- There are a number of problems with binary actions that can be modeled as $x \in \{0, 1\}$. For example:
  - We may hold a stock ($x_t = 0$) or sell ($x_t = 1$) if the price p_t falls below a smoothed estimate $\bar{p}_t$ which we compute using

$$\bar{p}_t = (1 - \alpha)\bar{p}_{t-1} + \alpha p_t.$$

Our policy is then given by

$$X^\pi(S_t|\theta) = \begin{cases} 1 & \text{if } p_t \le \bar{p}_t - \theta, \\ 0 & \text{otherwise.} \end{cases}$$

The function $X^\pi(S_t|\theta)$ decreases monotonically in p_t (as p_t increases, $X^\pi(S_t|\theta)$ goes from 1 to 0).

 – A shuttle bus waits until there are at least R_t customers on the bus, or it has waited τ_t. The decision to dispatch goes from $x_t = 0$ (hold the bus) to $x_t = 1$ (dispatch the bus) as R_t exceeds a threshold θ^R or as τ_t exceeds θ^τ, which means the policy $X^\pi(S_t|\theta)$ increases monotonically in both state variables $S_t = (R_t, \tau_t)$.

• A battery is being used to buy power from the grid when electricity prices p_t fall below a lower limit θ^{min}, or sell when the price goes above θ^{max}. The battery does nothing when $\theta^{min} < p_t < \theta^{max}$. We write the policy as

$$X^\pi(S_t|\theta) = \begin{cases} -1 & \text{if } p_t \le \theta^{min}, \\ 0 & \text{if } \theta^{min} < p_t < \theta^{max}, \\ 1 & \text{if } p_t \ge \theta^{max}. \end{cases} \tag{12.5}$$

We see that $X^\pi(S_t|\theta)$ increases monotonically in the state $S_t = p_t$.

• Dosages for blood sugar control increase with both the weight of the patient, and with the patient's glycemic index. The policy is in the form of a lookup table, with different dosages for each range of weight and glycemic index.

Each of these policies is controlled by a relatively small number of parameters, although this is not always the case. For example, if we use a fine discretization of the patient's weight and glycemic index, we could find that we need to specify hundreds of dosages. However, monotonicity can dramatically reduce the search process.

12.2.5 Nonlinear Policies

The term "nonlinear policy" pretty much covers any policy that has a single parametric form, which is not linear in the parameters θ that can be tuned. This includes the following:

• There are many problems that have specific structure. Our decision might be a continuous quantity such as the amount of water to apply to a wildfire, or the dosage of a drug to be given to a patient. We might feel that the policy will have an S-curve behavior with respect to a variable such as the intensity of the wildfire, or the weight of a patient, which can be described by

$$X^{\pi}(S_t|\theta) = \frac{1}{1 + e^{\theta_0 + \theta_1 \phi_1(S_t) + ... + \theta_F \theta_F \phi_F(S_t)}}.$$

The term $\phi_1(S_t)$ might capture the intensity of the fire or weight of the patient, while the other terms might capture other variables that shift the S-curve.

- A "buy low, sell high" policy such as the one in equation (12.5) is a kind of nonlinear policy. It is not smooth, since the function increases in steps as the price increases.

- Neural networks – A neural network (even a small neural network) is a high-dimensional nonlinear model that can have thousands to millions of parameters. The advantage of neural networks is that they can fit virtually any functional form, which seems to suggest that we do not have to know the form. Neural networks have actually been used for decades in primarily deterministic engineering control problems where the decision might be a three-dimensional force on a device.

    Neural networks have three weaknesses:

    - Neural networks are very high-dimensional architectures, which means they need a lot of data. This problem is magnified when there is noise (most uses of neural networks are applied to deterministic problems such as pattern recognition or robotic control).

    - Neural networks are very flexible (they can fit virtually any function) which means they can overfit, which means that they struggle with noisy data, as can easily happen when simulating a policy.

    - It is hard to make neural networks reflect structure such as monotonicity (the higher the price, the lower the demand).

As of this writing, neural networks have attracted considerable attention from the computer science community (and they have been used for a long time in engineering control problems), but care has to be used given the issues listed here. They have attracted considerable attention in the context of optimizing games, which are low noise (you just have the behavior of your opponent) and it is possible to run millions of simulated games to train the policies.

12.2.6 Nonparametric/Locally Linear Policies

The problem with parametric models is that sometimes functions are simply too complex to fit with low-order parametric models. For example, imagine that our policy looks like the function shown in Figure 12.1. Simple quadratic fits will not work, and higher-order polynomials will struggle due to overfitting unless the number of observations is extremely large.

We could handle very general functions if we could use lookup tables (which may require that we discretize any continuous parameters). However, lookup

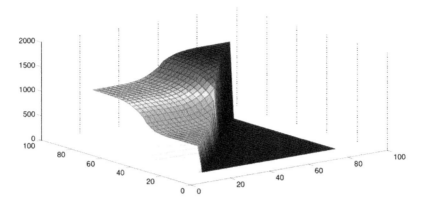

Figure 12.1 Illustration of a complex nonlinear (monotone) function.

tables can become extremely large when we have three or more dimensions in our state variable. Even three dimensional lookup tables quickly grow to thousands to millions of elements. The problem is compounded when the search algorithm has to evaluate actions for each state many times to handle noise.

A surprisingly powerful strategy for many problems with continuous states and actions is to assume locally linear responses. For example, S_t may capture the level of a reservoir, or the current speed and altitude of a helicopter. The control x_t could be the rate at which water is released from the reservoir, or the forces applied to the helicopter. Assume that we use our understanding of the problem to create a family of regions $S_1, \ldots, S_I$, which are most likely going to be a set of rectangular regions (or intervals if there is only one dimension). We might then create a family of linear (affine) policies of the form

$$X_i^\pi(S_t | \theta) = \theta_{i0} + \theta_{i1}\phi_1(S_t) + \theta_{i2}\phi_2(S_t),$$

for $S_t \in S_i$ where S_i is a user-defined region of the state space (there are only a few of these).

This approach has been found to be very effective in some classes of control problems. In practice, the regions S_i are designed by someone with an understanding of the physics of the problem. Further, instead of tuning one vector θ, we have to tune $\theta_1, \ldots, \theta_I$. While this involves considerable testing and tuning, the approach can work quite well and offers the important feature that the resulting policy can be computed extremely quickly.

12.2.7 Contextual Policies

Imagine that we have designed a policy $X^\pi(S_t|\theta)$ that depends on S_t and is parameterized by θ. This policy is actually the solution to the problem

$$
\max_{\pi=(f\in\mathcal{F},\theta\in\Theta^f)} \mathbb{E}\left\{\sum_{t=0}^{T} C(S_t, X^\pi(S_t|\theta))|S_0\right\}. \tag{12.6}
$$

Recall that $f \in \mathcal{F}$ reflects the search over policy classes (and we may have fixed this to one class), while $\theta \in \Theta^F$ captures the search over any tunable parameters for that class (and we always have this search). For now, let's assume that we have also fixed the policy to some class $f \in \mathcal{F}$, so that we are only optimizing over θ.

The attentive reader will note that we always write our objective function in this way, which means we express the explicit dependence on the initial state S_0, which captures deterministic parameters, initial values of dynamic parameters, and prior beliefs. This means that our optimized θ actually should be written $\theta(S_0)$. In other words, if we change our initial conditions, we may have to retune θ.

Some communities refer to the initial state S_0 as the "context" of the problem. If S_0 captures stable, static parameters, then it is unlikely to change very much. However, it could happen that we re-tune our policy every quarter (as happens in financial settings), so that the policy picks up current market conditions, which can be very complex. If our sequential decision problem is a search algorithm, then S_0 might be the starting point of the algorithm, while θ governs the behavior of the stepsize policy.

12.3 Problem Characteristics

When designing a policy search method, it is important to understand the characteristics of how the system responds to the parameterized policy. Some important dimensions of problem characteristics include:

Computational complexity – How you approach policy search will be quite different if a single simulation of a policy is a fraction of a second, or hours, or days, or longer. Methods based on viewing the simulator as a black-box tend to require more function evaluations.

Level of noise – Policy simulations can be reasonably stable, especially for very large systems where aggregate behavior is more stable, but they can also be extremely random.

The response surface – It may be concave, smooth but only unimodular, non-concave with local maxima, and it may feature jumps (think about buy low, sell high policies).

Parameter dimensionality – We can divide parameters into three classes:

- Scalar models – There are a number of applications with a single scalar parameter (think of our Boltzmann policy).
- Low-dimensional continuous models – In many applications the number of tunable parameters is less than five or 10, which may mean that it is too large to do a full grid search (discretizing each dimension and searching over all values), but it may simplify computing numerical derivatives.
- High-dimensional continuous models – It is easy to create policies with tens to hundreds of parameters, or even hundreds of thousands. A good example of a high-dimensional policy is a neural network, but it could also arise with a linear model with a high-dimensional state variable (as might arise in the management of complex resources).

Stationarity – The process we are controlling may be:

- Stationary, which means the parameters of the underlying process are not changing over time.
- Periodic (such as time of day patterns).
- Nonstationary, which also comes in different forms. For example, imagine we are controlling a basic inventory problem. This problem may feature:
  - Smooth transitions as demand for product steadily increases or decreases, or exhibits smooth seasonal transitions.
  - Bursts, when a product suddenly gets popular for a period of time.
  - Shifts, such as a sudden increase in demand following an advertising campaign or change in price.
  - Spikes, such as a spike in electricity prices which encourages sudden selling.

12.4 Flavors of Policy Search

Given a parametric (or locally parametric) function parameterized by θ (typically a vector, but not always), we now face the challenge of finding the best value of θ. There are different dimensions to the policy search problem:

Derivative-based vs. derivative-free – These include:

Derivative-based methods – Derivative-based methods are attractive when optimizing vectors of continuous parameters and where we feel comfortable that $\mathbb{E}F^{\pi}(\theta, W)$ is smooth (note that the expectation often helps considerably to smooth functions). The vast majority of derivative-based methods use the classical first-order, stochastic gradient algorithm described in chapter 5:

$$\theta^{n+1} = \theta^n + \alpha_n \nabla_\theta F^\pi(\theta^n, W^{n+1}). \tag{12.7}$$

We can divide derivative-based methods into two broad categories:

- Numerical derivatives – Numerical derivatives are estimates of derivatives which use only the simulations $F^\pi(\theta, \omega)$, without requiring any actual derivative information of $\nabla_\theta F^\pi(\theta, \omega)$. These methods have been described in chapter 5, but we will review methods based on numerical derivatives.
- Exact derivatives – These methods exploit the underlying structure of a sequential decision problem to compute derivatives exactly, avoiding the need for expensive numerical derivatives.

Derivative-free methods – These methods all view the policy simulator as a black box, and use the methods described in chapter 7. Let $S^{\theta,n}$ be what we know about $\mathbb{E}F^\pi(\theta, W)$ (not to be confused with S_t within our sequential decision problem), and let $\Theta^\pi(S^{\theta,n})$ be our policy for choosing the parameter vector θ^n based on what we know in $S^{\theta,n}$. The update rule $\Theta^\pi(S^{\theta,n})$ can be any of the four classes of policies described in chapter 7.

Online vs. offline learning – In online learning, we are learning in an environment where updates come to us. As a rule, we have to live with the performance of our policy, which means we are maximizing the cumulative reward. Most policy search uses some form of adaptive algorithm, although this can be done in a laboratory where we use one policy, the *learning policy*, to find the best policy to implement, called the *implementation policy*. Some in the reinforcement learning community refer to the learning policy as the *behavior policy* while the implementation policy is the *target policy*. Many refer to the learning policy as an algorithm; we think the relationship to policies creates a bridge to our entire framework with the four classes of policies.

Performance-based vs. supervised learning – Most policy search uses as a goal to maximize the total reward (either the final reward or cumulative reward), but there are settings where we have an "expert" (the supervisor) who will specify what to do, allowing us to fit our policies to the choices of the supervisor. The expert decisions could come from a physician making decisions, a financial trader making traders, or a dispatcher assigning drivers to loads. This turns a policy search problem (maximizing a performance metric) into a machine learning (effectively "predicting" what the supervisor will do), or a hybrid, where we balance a performance metric against matching the decisions of an exogenous decision-maker.

As we review the methods, there will be a clear tradeoff between efficiency and complexity. We are going to start with the simplest methods, which are those which treat $F^\pi(\theta)$ as a black box and, as a result, do not exploit any structural properties of the underlying problem. This includes derivative-free methods, and derivative-based methods using numerical derivatives.

We then progress to derivative-based methods where we work with analytical derivatives of the gradient which exploit the structure of the underlying problem. For our presentation, these come in two flavors:

Discrete dynamic programs – These are problems where we are at a node (state) s, choose a discrete action a, and then transition to a node s' with probability $P(s'|s, a)$ (which we represent but generally cannot compute). An important subclass of graph problems are those where actions are chosen at random (known as a stochastic policy), but transitions are made deterministically. Here, we wish to optimize a parameterized policy $A^\pi(s|\theta)$, where action $a_t = A^\pi(S_t|\theta)$ is discrete.

Continuous control problems – In this setting we choose a continuous control x_t (which may be vector-valued) that impacts the state S_{t+1} in a continuous way through a known (and differentiable) transition function.

Both problem classes have attracted considerable attention, and illustrate different methods for computing gradients.

Our presentation will proceed from the simplest methods to the most sophisticated:

- Derivative-based policy search using numerical derivatives – section 12.5.
- Derivative-free policy search – section 12.6.
- Derivative-based with exact derivatives: continuous dynamic programs – section 12.7.
- Derivative-based with exact derivatives: discrete dynamic programs – section 12.8.

The first two methods treat the policy simulator as a black box, and make virtually no assumptions about the internal structure of the problem. These methods are simplest, but the price you pay is that you will have to deal with the potentially high noise of a policy simulator. Also, while simulating a policy can be quite fast, there are many applications where this is computationally intensive, requiring several minutes to hours to days or more for numerical simulations.

In addition, there are settings where we do not have access to a simulator, and have to do our policy search in the field, where it takes a day to observe a day.

The third method derives an explicit formula for the gradient, which requires knowing specific relationships within the model. The derivatives required to compute the gradient are all computed for a specific sample path, and therefore avoid any complexities associated with expectations.

The fourth method is designed for discrete dynamic programs, and works directly from the expectation-based form of the objective function. This is a mathematically advanced presentation for readers with a strong probability background (which is the reason that it is marked with a **).

12.5 Policy Search with Numerical Derivatives

Any "black box" model starts with our assumption that we can perform a simulation of the policy $X^\pi(S_t|\theta)$ by simulating a sample path to get an estimate of

$$\hat{F}(\theta, \omega) = \sum_{t=0}^{T} C(S_t(\omega), X^\pi(S_t(\omega)|\theta)). \tag{12.8}$$

While there are different ways of estimating derivatives numerically, we are going to focus on the SPSA algorithm ("simultaneous perturbation stochastic approximation") which is designed for settings where θ is a vector, which we first presented in section 5.4.4. In theory SPSA can produce estimates of the gradient $\nabla_\theta F(\theta, \omega)$, regardless of the dimension of θ, with just two simulations. In practice, these estimates can be quite noisy, motivating using multiple simulations and averaging.

The method works as follows:

(1) Let $Z_k, k = 1, \dots, K$ be a vector of zero-mean random variables, and let Z^n be a sample of this vector at iteration n.
(2) Create perturbed values of θ^n using $\theta^{n+} = \theta^n + \eta^n Z^n$ and $\theta^{n-} = \theta^n - \eta^n Z^n$, where η^n is a scaling sequence (it is typically chosen as a constant that does not vary with n).
(3) Let $W^{n+1,+}$ and $W^{n+1,-}$ represent two different samples of the random variables driving the simulation (these can be generated in advance or on the fly). There is no meaning to the $+$ and $-$ in the superscript other than to indicate that these are the samples that are run to evaluate θ^{n+} and θ^{n-}.
(4) Run the simulation twice, once to find $\hat{F}^{n+} = F(\theta^{n+}, W^{n+1,+})$, and once to find $\hat{F}^{n-} = F(\theta^{n-}, W^{n+1,-})$.

(5) It is common that we have to run multiple simulations and take an average. Let $W_m^{n+1,+}$ be m^{th} sample of the random information series and let

$$\hat{F}_m^{n+1,+}(\theta^{n+1,+}) = F(\theta^{n+1,+}, W_m^{n+1,+}),$$

represent the performance of the mth simulation which we run m^{batch} times (this is called a "mini-batch"). Let $\hat{F}_m^{n+1,-}(\theta^{n+1,-})$ be parallel sets of runs. We then take an average

$$\bar{F}^{n+1,+}(\theta^{n+1,+}) = \frac{1}{m^{batch}} \sum_{m=1}^{m^{batch}} \hat{F}_m^{n+1,+}(\theta^{n+1,+}).$$

$\bar{F}^{n+1,-}(\theta^{n+1,-})$ is computed similarly.

(6) Compute the estimate of the gradient using

$$g^{n+1}(\theta^n) = \begin{bmatrix} \dfrac{\bar{F}^{n+1,+}(\theta^{n+1,+}) - \bar{F}^{n+1,-}(\theta^{n+1,-})}{2\eta^n Z_1^n} \\ \dfrac{\bar{F}^{n+1,+}(\theta^{n+1,+}) - \bar{F}^{n+1,-}(\theta^{n+1,-})}{2\eta^n Z_2^n} \\ \vdots \\ \dfrac{\bar{F}^{n+1,+}(\theta^{n+1,+}) - \bar{F}^{n+1,-}(\theta^{n+1,-})}{2\eta^n Z_P^n} \end{bmatrix}. \tag{12.9}$$

We then use this in our stochastic gradient algorithm

$$\theta^{n+1} = \theta^n + \alpha_n g^{n+1}(\theta^n). \tag{12.10}$$

While the basic gradient updating formula (12.10) is disarmingly simple (hence the reason we presented it first), it hides the need to experiment with stepsize formulas (covered in chapter 6), tuning parameters required by the stepsize formula, and tuning the size of the mini-batch (which may need to vary by iteration).

Stochastic gradients can be effective and easy to implement, but be prepared to spend some time tuning the algorithm to get good results.

12.6   Derivative-Free Methods for Policy Search

In this section we provide a tour through chapter 7 on methods that only require that we be able to perform simulations of a policy. We remind the reader of the four classes of policies that can be used to perform derivative-free stochastic search:

Policy function approximation (PFA) – section 7.4 – These are simple rules, and below we suggest one that accelerates a simple statistical learning method. Of course, the price of simplicity is (yet another) tunable parameter.

Cost function approximation (CFA) – section 7.5 – Simple CFAs include upper confidence bounding and interval estimation for problems with discrete alternatives. Below we suggest a strategy for applying these ideas to policy search.

Value function approximation (VFA) – section 7.6 – VFA-based policies are relatively complex and have not yet been demonstrated to significantly outperform simpler methods. For this reason, we do not cover these methods here.

Direct lookahead (DLA) – section 7.7 – The knowledge gradient is a one-step lookahead (easily modified to be a restricted multistep lookahead) which has proven useful in the context of expensive function evaluations requiring smaller budgets.

12.6.1   Belief Models

We can draw on a number of the different belief structures presented in chapter 3. Some that are likely to be useful in the representation of continuous vectors for the parameter vector θ include:

- Lookup table with correlated beliefs – Also known as Gaussian process regression (technically this is one form of GPR), this could work well for vectors θ with one to three dimensions. GPR does not impose any structural assumptions other than smoothness, but this also means that it is not able to produce functions that are known to be concave, convex or unimodular.
- Low-dimensional linear models (e.g. quadratic) – Low dimensional linear models can be used in a number of settings, spanning anywhere from one to dozens of variables. Particularly useful are methods that work to fit a low-dimensional model in the vicinity of the optimum (which, of course, we are trying to find).
- Sparse linear models – These models extend the linear models to the domain of high-dimensional vectors, but where we think that many of the elements of θ may be zero.
- Sampled belief models – There are problems with special structure that suggest a particular type of nonlinear model, such as logistic regression for a pricing or recommendation system. If the nonlinear function $f(x|\theta)$ is parameterized by an unknown vector θ, we might represent the uncertainty in our belief by a family of possible values $\theta \in \{\theta^1, \dots, \theta^K\}$.
- Neural networks – We have described gradient-based search models using neural network policies (which can be very high dimensional), but the biggest strength of neural networks, which is their flexibility to replicate any

functions, is also their biggest weakness, since this flexibility requires very large datasets. Their flexibility also means that they can overfit noisy data.

It is helpful, even important, to represent not only our best estimate of the belief, but also the uncertainty in the belief. We can do this for lookup tables (including with correlated beliefs) and linear models. We can do this for non-linear models using the technique of using a sampled belief model, where we maintain a population of possible values of the unknown parameter vector θ and the probability that each is the true value. However, we are not able to do this with neural networks.

We encourage the reader to review the different policies in chapter 7, but we provide some simple illustrations that have proven useful.

12.6.2 Learning Through Perturbed PFAs

One of the most popular heuristics for optimizing an unknown function is to use the first n observations, $(x^0, y^1), (x^1, y^2), \ldots, (x^{n-1}, y^n)$ to create a belief $\bar{f}^n(x|\bar{\theta}^n)$ using any of the methods in chapter 7. Then, we could compute

$$x^n = \arg\max_x \bar{f}^n(x|\bar{\theta}^n), \tag{12.11}$$

which we then use to run a simulation to obtain the updated sample

$$y^{n+1} = F(x^n, W^{n+1}).$$

It turns out that this simple idea is surprisingly ineffective, as illustrated in Figure 12.2 for the setting of learning how sales responds to price, where we need to learn the relationship between sales and price, while maximizing revenue. Figure 12.2(a) shows three different possible sales response curves, where we are making the simplistic assumption that this relationship is linear in price (remember that *any* parametric model is at best going to be locally accurate).

Now assume that we use our best estimate of the sales response to create a best estimate of revenue as a function of price, and then set the price to maximize the revenue (as we would if we used equation (12.11) to determine the next point to observe). The problem with this is that we end up testing prices near the apparent optimum, as illustrated in Figure 12.2(b). The problem with these observations is that it requires that we learn the sales response from a series of noisy observations that are clustered together, which makes it virtually impossible to get a reliable estimate of the sales curve.

The best way to learn the sales curve is to make observations that are as far from the center as possible, as shown in Figure 12.2(c). There are two problems with this strategy. First, our sales response model is only an approximation; in this example we assume it is linear in price, which is clearly accurate only near

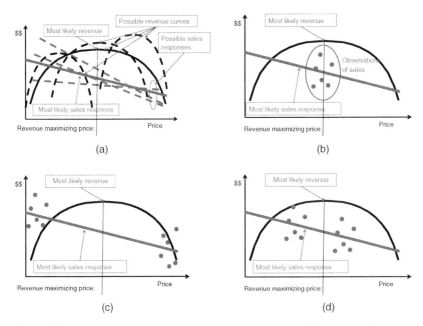

Figure 12.2   Actively learning a demand response function: (a) Three possible sales response lines and corresponding revenue curves, (b) Observed price-sales combinations if we use prices that appear to maximize revenue, (c) Observing extreme prices (high and low) to improve learning of sales response, and (d) Balancing learning (observing away from the middle) and earning (observing prices near the middle).

the middle. The second problem is that if we are learning in the field, these would be points where we perform poorly (that is, we would expect to receive very low revenue).

The most effective strategy is illustrated in Figure 12.2(d), showing observations that are not too close to the optimum, but not too far. This is known as "sampling the shoulders" of the function.

The idea of sampling a function in a region *around* the optimum, rather than the optimum itself, is supported by an analysis of the value of information from sampling each point. Figure 12.3 shows the value of information for a scalar function, $\bar{f}^n(x|\bar{\theta}^n)$ computed using the knowledge gradient (see sections 7.7.2 and 7.8), which shows that there are peaks to the value of information that is some distance from the optimum.

This raises the question: how to find this peak? The calculations used to compute the knowledge gradient are more complex, and still require knowing something about the behavior of the true function, which would never be true in practice. For this reason, an interesting strategy is to take this insight

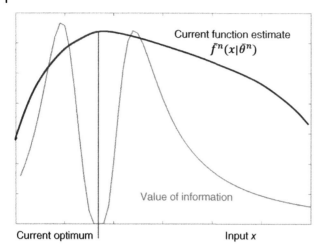

Figure 12.3 Plot of the value of information from sampling x over a range, showing the highest value of information that is some distance from the current apparent optimum.

and design a simple policy (which falls in the PFA class). In fact, we suggest two policies:

An optimum-deviation policy – The idea here is to pick a point x^n that is a distance ρ from the optimum $\bar{x}^n = \arg\max_x \bar{f}^n(x|\bar{\theta}^n)$. If x is a k–dimensional vector, this deviation can be created by sampling k normally distributed random variables $Z_1, \dots, Z_K$, each with mean 0 and variance 1, and then normalizing them so that

$$\sqrt{\sum_{k=1}^{K} Z_k^2} = \rho.$$

Let $\bar{Z}^n$ be the resulting k–dimensional vector. Now compute the sampling point using

$$x_k^n = \bar{x}_k^n + \bar{Z}_k^n.$$

Note that in one dimension, we would have $\bar{Z}^n = \pm\rho$.

An excitation policy – Here we again generate a k–dimensional perturbation vector Z^n, where each element has mean 0 and variance 1, and then set

$$x_k^n = \bar{X}_k^n + \rho Z_k^n.$$

While the optimum-deviation policy forces x^n to be a distance ρ from the optimum $\bar{x}^n$, an excitation policy simply introduces a random perturbation with mean 0, which means the most likely point to sample is the optimum of $\bar{f}^n(x|\bar{\theta}^n)$.

The excitation policy is more natural in a setting where we are learning in the field using a cumulative reward objective, providing an additional incentive to sample in the vicinity of the apparent optimum, while still forcing some exploration. The optimum-deviation policy will produce faster learning, but at a price to how well we do while we are learning, which is best if we are using a final-reward objective.

We have to remind ourselves that these policies are designed for tuning the parameter vector θ of an implementation policy, but we now have a new tunable parameter, ρ. Fortunately, we may be able to pick a reasonable value of ρ a-priori. We first note that virtually any search algorithm benefits from an assumption that the data x can be scaled. For example, we may assume that we can scale each dimension of x to be between 0 and 1, or normally distributed with mean 0 and variance 1. When we do this, we might feel that ρ will likely be between 0.1 and 0.5.

12.6.3 Learning CFAs

Section 7.5 describes a number of CFA policies for derivative-free stochastic search that can all be used for parameter search. We illustrate two policies that work through the same mechanism which highlights an important characteristic of active learning policies (which describes *any* policy where decisions affect a belief about unknown parameters).

Interval estimation – Start by assuming that we can represent the feasible region for x by a finite (or sampled) set $\mathcal{X} = \{x^1, \ldots, x^K\}$. Let $\bar{\mu}_x^n$ be our estimate of $f(x) = \mathbb{E}F(x, W)$ for $x \in \mathcal{X}$ after n experiments. Since $f(x)$ is a continuous surface, it makes sense to use correlated beliefs (also known as Gaussian process regression) which we introduced in section 3.4.2. Recall that we would maintain a covariance matrix Σ^n.
A basic interval estimation policy is given by

$$X^{IE}(S^n|\theta^{IE}) = \arg\max_{x \in \mathcal{X}} \left(\bar{\mu}_x^n + \theta^{IE} \bar{\sigma}_x^n \right), \tag{12.12}$$

where $\bar{\sigma}_x^n = \sqrt{\Sigma_{xx}^n}$. Note that our state S^n is our belief $B^n = (\bar{\mu}^n, \Sigma^n)$.
Sampled θ-percentile – A policy closely related to interval estimation is to explicitly capture the θ-percentile. Figure 12.4 shows a sampled belief model

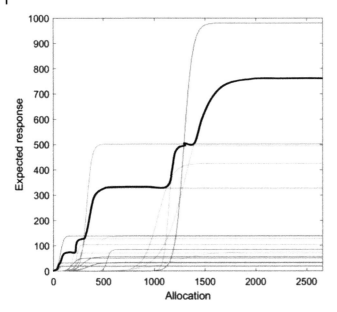

Figure 12.4 Sampled belief model, with 95th percentile highlighted (second highest belief).

with 20 possible beliefs. If we set $\theta = 0.95$, this means taking the second-highest belief, which is shown as the solid black line. As above, we still have to tune θ.

Both the interval estimation policy and the sampled θ-percentile policies make recommendations based on an optimistic estimate of the estimated function. With interval estimation, the random variable $\bar{\mu}_x^n$ will be normally distributed (from the central limit theorem), so if we pick $\theta^{IE} = 2$, for example, we will be making our choices based on the 95th percentile of the function.

Similarly, if we use $K = 20$ samples in our sampled belief model, we could use the 19th highest sample (as we did in Figure 12.4) and again obtain a 95th percentile estimate. Of course, the percentile we use is a tunable parameter that depends on the size of our experimental budget, and whether we are doing offline (final reward) or online (cumulative reward) learning. For expensive functions (and small learning budgets), the best value of θ will likely be a declining function of the number of experiments that have been completed.

There is a substantial literature that analyzes policies based on the principle of using optimistic estimates using the broad term of *upper confidence bounding*. The idea is that learning improves when using optimistic estimates of the function, since the current estimate may, as a result of experimental noise, underestimate the true function.

12.6.4 DLA Using the Knowledge Gradient

A form of direct lookahead is the knowledge gradient which we first introduced in section 7.7.2 (see also section 7.8) which is a one-step lookahead. The knowledge gradient has been found to be particularly useful for functions that are relatively expensive to evaluate, which limits the size of our experimental budget.

Figure 12.5 illustrates the knowledge gradient on a two-dimensional surface which is estimated using correlated beliefs (see section 7.8.5 for a summary of how to compute the knowledge gradient with correlated beliefs, and section 3.4.2 for the updating equations for correlated beliefs). Note that the knowledge gradient (on the right) is highest in regions of the function farthest from prior measurements, while the knowledge gradient is smallest at points that have just been evaluated (which minimizes uncertainty).

12.6.5 Comments

There is a general theme that runs through these policies (and throughout the literature on active learning problems), which is that you want to perform function evaluations that strike a balance between maximizing uncertainty, while simultaneously maximizing the *possibility* that the point in the function may prove to be best. This means that it is not enough to maintain a belief about the function $\bar{f}^n(x|\bar{\theta}^n)$; we also have to maintain a belief about our *uncertainty* in the function at each point. This section highlights the methods that we have found to be most effective in our own work.

12.7 Exact Derivatives for Continuous Sequential Problems*

We are now going to derive an exact gradient (technically, a stochastic gradient in the style of chapter 5) of the performance of a policy with respect to the parameters θ that govern the performance of the policy. This section will focus on problems where the state S_{t+1} is a differentiable function with respect to the state S_t and decision x_t, as might arise when we are managing resources (water, blood, money).

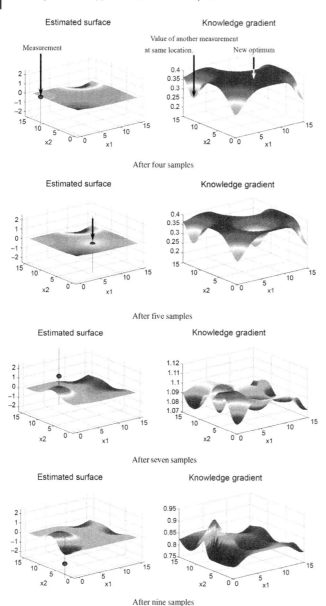

Figure 12.5 The knowledge gradient with correlated beliefs being applied to a two-dimensional surface. The plots on the left are the beliefs after n samples, while the plots on the right plot the knowledge gradient at each point.

We return again to our basic sequential optimization problem

$$F^\pi(\theta) = \mathbb{E}\left\{\sum_{t=0}^{T} C(S_t, X_t^\pi(S_t|\theta))|S_0\right\},\qquad(12.13)$$

where our dynamics evolve (as before) according to

$$S_{t+1} = S^M(S_t, x_t, W_{t+1}),$$

where we are given an initial state S_0 and access to observations of the sequence $W = (W_1, \ldots, W_T)$. Our goal in this section is to find the gradient $\nabla_\theta F^\pi(\theta, \omega)$ exactly for a particular sample path ω (rather than using a numerical derivative).

We have written our policy $X_t^\pi(S_t)$ in a time-dependent form for generality, but this means estimating time-dependent parameters θ_t that characterize the policy. In most applications we would use the stationary version $X^\pi(S_t)$, with a single set of parameters θ. However, when we can compute the gradient exactly, we can handle high-dimensional parameters much more efficiently than methods based on numerical derivatives can (SPSA may seem like magic, but it isn't!).

Continuous sequential problems are distinguished from discrete dynamic programs specifically because we assume we can compute $\partial S_{t+1}/\partial x_t$. With discrete dynamic programs, we assumed the actions a were categorical (e.g. left-/right or red/green/blue). In that setting, we had to consider the downstream impact of a decision made now by capturing the effect of changing the policy parameter θ on the probability of which state we would visit. Now we can capture this impact directly.

There are two approaches for minimizing $F^\pi(\theta)$ over the parameter vector θ:

Batch learning – Here we replace (12.13) with an average over N samples, giving us

$$\bar{F}^\pi(\theta) = \frac{1}{N}\sum_{n=1}^{N}\sum_{t=0}^{T} C(S_t(\omega^n), X_t^\pi(S_t(\omega^n)|\theta)),\qquad(12.14)$$

where $S_{t+1}(\omega^n) = S^M(S_t(\omega^n), X^\pi(S_t(\omega^n)), W_{t+1}(\omega^n))$ is the sequence of states generated following sample path ω^n. This is a classical statistical estimation problem.

Adaptive learning – Rather than solving a single (possibly very large) batch problem, we can use our standard stochastic gradient updating logic (from chapter 7)

$$\theta^{n+1} = \theta^n + \alpha_n \nabla_\theta F^\pi(\theta^n, W^{n+1}).$$

This update is executed following each forward pass through the simulation.

Both approaches depend on computing the gradient $\nabla_\theta F^\pi(\theta, \omega)$ for a given simple path ω from which we generate a sequence of state $S_{t+1} = S^M(S_t, x_t, W_{t+1}(\omega))$ where $x_t = X^\pi(S_t)$. Normally we would write $S_t(\omega)$ or $x_t(\omega)$ to indicate the dependence on sample path ω, but we suppress this here for notational compactness.

We find the gradient by differentiating (12.13) with respect to θ, which requires a meticulous application of the chain rule, recognizing that the contribution $C(S_t, x_t)$ is a function of both S_t and x_t, the policy $X^\pi(S_t|\theta)$ is a function of both the state S_t and the parameter θ, and the state S_t is a function of the previous state S_{t-1}, the previous control x_{t-1}, and the most recent exogenous information W_t, which is assumed to be independent of the control, although this could be handled. This gives us

$$\nabla_\theta F^\pi(\theta, \omega) = \left(\frac{\partial C_0(S_0, x_0)}{\partial x_0}\right)\left(\frac{\partial X_0^\pi(S_0|\theta)}{\partial \theta}\right) + \sum_{t'=1}^{T}\left[\left(\frac{\partial C_{t'}(S_{t'}, X_{t'}^\pi(S_{t'}|\theta))}{\partial S_{t'}}\frac{\partial S_{t'}}{\partial \theta}\right)\right.$$
$$\left. + \frac{\partial C_{t'}(S_{t'}, x_{t'})}{\partial x_{t'}}\left(\frac{\partial X_{t'}^\pi(S_{t'}|\theta)}{\partial S_{t'}}\frac{\partial S_{t'}}{\partial \theta} + \frac{\partial X_{t'}^\pi(S_{t'}|\theta)}{\partial \theta}\right)\right], \quad (12.15)$$

where

$$\frac{\partial S_{t'}}{\partial \theta} = \frac{\partial S_{t'}}{\partial S_{t'-1}}\frac{\partial S_{t'-1}}{\partial \theta} + \frac{\partial S_{t'}}{\partial x_{t'-1}}\left[\frac{\partial X_{t'-1}^\pi(S_{t'-1}|\theta)}{\partial S_{t'-1}}\frac{\partial S_{t'-1}}{\partial \theta} + \frac{\partial X_{t'-1}^\pi(S_{t'-1}|\theta)}{\partial \theta}\right].$$
$$(12.16)$$

The derivatives $\partial S_{t'}/\partial \theta$ are computed using (12.16) by starting at $t' = 0$ where

$$\frac{\partial S_0}{\partial \theta} = 0,$$

and stepping forward in time.

Equations (12.15) and (12.16) require that we be able to take derivatives of the cost function, the policy, and the transition function. We assume this is possible, although the complexity of these derivatives is highly problem dependent.

12.8   Exact Derivatives for Discrete Dynamic Programs**

This section is going to take a significant step up in complexity in its derivation of analytical derivatives for policy parameters in the context of discrete dynamic programs. These are problems where decisions are categorical: left-right, color, or a product recommendation. For the advanced (and determined) reader, this presentation provides a different perspective into the mathematics of sequential decision problems that works directly with expectations rather than the simulation-based strategy used in section 12.7.

To emphasize the use of discrete actions, we are going to switch notation to action a rather than our usual decision x. We assume that we are going to maximize the single-period expected reward in steady state. We use the following notation:

$$
\begin{aligned}
r(s, a) \quad &= \quad \text{Reward if we are in state } s \in \mathcal{S} \text{ and take action } a \in \mathcal{A}_s. \\
A^\pi(s|\theta) \quad &= \quad \text{Policy that determines the action } a \text{ given that we are in} \\
&\qquad \text{state } s, \text{ which is parameterized by } \theta. \\
P_t(s'|s, a) \quad &= \quad \text{Probability of transitioning to state } s' \text{ given that we} \\
&\qquad \text{are in state } s \text{ and take action } a \text{ at time } t \text{ (we use} \\
&\qquad P(s'|s, a) \text{ if the underlying dynamics are stationary).} \\
d_t^\pi(s|\theta) \quad &= \quad \text{Probability of being in state } s \text{ at time } t \text{ while} \\
&\qquad \text{following policy } \pi.
\end{aligned}
$$

This notation reflects the classical notation of the reinforcement learning community, which has adopted the notation from Markov decision processes (we will see this in much more detail in chapter 14). Normally we would use a transition function, but here we are using the one-step transition matrix (we showed how to calculate the one-step transition matrix from the transition function in section 9.7). Also, we are using for the first time the probability of being in a state while following policy π, given by $d_t^\pi(s|\theta)$, although we previously used the idea of computing the expectation over the states in section 9.11 (look at the objective function for class 4).

We first introduce a parameterized stochastic policy which is typically required for problems where decisions are discrete with no particular structure (e.g. red-green-blue). We note that the parameters that we are optimizing are primarily controlling the balance of exploration and exploitation. We then present the objective function (there is more than one way to write this, as we show later). Finally, we describe a computable method for taking the gradient of this objective function.

12.8.1 A Stochastic Policy

We follow the standard practice in the literature of using what is called a stochastic policy, where an action a is chosen probabilistically. We represent our policy using

$$p_t^\pi(a|s,\theta) \quad = \quad \text{the probability of choosing action } a \text{ at time } t, \text{ given}$$
that we are in state s, where θ is a tunable parameter (possibly a vector).

Most of the time we will use a stationary policy that we denote $\bar{p}^\pi(a|s,\theta)$ which can be viewed as a time-averaged version of our policy $p_t^\pi(a|s,\theta)$ which we might compute using

$$\bar{p}^\pi(a|s,\theta) = \lim_{T\to\infty} \frac{1}{T} \sum_{t=1}^T p_t^\pi(a|s,\theta).$$

A particularly popular policy (especially in computer science) assumes that actions are chosen at random according to a Boltzmann distribution (also known as Gibbs sampling). Assume at time t that we have

$$\bar{Q}_t(s,a) \quad = \quad \text{estimated value at time } t \text{ of being in state } s \text{ and}$$
taking action a.

Next define the probabilities (using our familiar Boltzmann distribution)

$$p_t^\pi(a|s,\theta) = \frac{e^{\theta \bar{Q}_t(s,a)}}{\sum_{a'\in\mathcal{A}_s} e^{\theta \bar{Q}_t(s,a')}}. \tag{12.17}$$

We can compute the values $\bar{Q}_t(s,a)$ using $\bar{Q}_t(s,a) = r(s,a)$, although this means choosing actions based on immediate rewards. Alternatively, we might use

$$\bar{Q}_t(s,a) \quad = \quad r(s,a) + \max_{a'} \bar{Q}_{t+1}(s',a'),$$

where s' is chosen randomly from simulating the next step (or sampling from the transition matrix $P_t(s'|s,a)$ if this is available). We first saw methods for computing Q-values under the umbrella of reinforcement learning in section 2.1.6.

If we are modeling a stationary problem, it is natural to transition to a stationary policy. Let $\bar{p}^\pi(a|s,\theta)$ be our stationary action probabilities where we replace the time-dependent values $\bar{Q}_t(s,a)$ with stationary values $\bar{Q}(s,a)$ computed using

$$\bar{Q}^\pi(s,a|\theta) \quad = \quad r(s,a) + \mathbb{E}\left\{ \sum_{t'=1}^T r(S_{t'}, A^\pi(S_{t'}|\theta))|S_0 = s, a_0 = a \right\}. \tag{12.18}$$

This is the total reward over the horizon from starting in state s and taking action a (note that we could use average or discounted rewards, over finite or infinite horizons). We remind the reader we are never going to actually compute these expectations. Using these values, we can create a stationary distribution for choosing actions using

$$\bar{p}^\pi(a|s,\theta) = \frac{e^{\theta\bar{Q}^\pi(s,a|\theta)}}{\sum_{a'\in A_s} e^{\theta\bar{Q}^\pi(s,a'|\theta)}}.$$ (12.19)

Finally, our policy $A^\pi(s|\theta)$ is to choose action a with probability given by $p_t^\pi(a|s,\theta)$. The development shown in section 12.8.2 does not require that we use the Boltzmann policy, but it helps to have an example in mind.

12.8.2 The Objective Function

To develop the gradient, we have to start by writing out our objective function which is to maximize the average reward over time, given by

$$F^\pi(\theta) = \lim_{T\to\infty} \frac{1}{T} \left\{ \sum_{t=0}^T \sum_{s\in S} \left(d_t^\pi(s|\theta) \sum_{a\in A_s} r(s,a)p_t^\pi(a|s,\theta) \right) \right\}.$$ (12.20)

A more compact form involves replacing the time-dependent state probabilities with their time averages (since we are taking the limit). Let

$$\bar{d}^\pi(s|\theta) = \lim_{T\to\infty} \frac{1}{T} \sum_{t=0}^T d_t^\pi(s|\theta).$$

We can then write our average reward per time period as

$$F^\pi(\theta) = \sum_{s\in S} \bar{d}^\pi(s|\theta) \sum_{a\in A_s} r(s,a)\bar{p}^\pi(a|s,\theta).$$ (12.21)

12.8.3 The Policy Gradient Theorem

We are now ready to take derivatives. Differentiating both sides of (12.21) and applying the chain rule gives us

$$\nabla_\theta F^\pi(\theta)$$

$$= \sum_{s\in S} \left(\nabla_\theta \bar{d}^\pi(s|\theta) \sum_{a\in A_s} r(s,a)\bar{p}^\pi(a|s,\theta) + \bar{d}^\pi(s|\theta) \sum_{a\in A_s} r(s,a)\nabla_\theta \bar{p}^\pi(a|s,\theta) \right).$$

(12.22)

While we cannot compute probabilities such as $d^\pi(s)$, we can simulate them (we show this in the next few lines). We also assume we can compute $\nabla_\theta \bar{p}^\pi(a|s,\theta)$ by differentiating our probability distribution in equation (12.19). Derivatives of probabilities such as $\nabla_\theta \bar{d}^\pi(s|\theta)$, however, are another matter.

This is where the development known as the *policy gradient theorem* helps us. This theorem tells us that we can calculate the gradient of $F^\pi(\theta)$ with respect to θ using

$$\frac{\partial F^{\pi}(\theta)}{\partial \theta} = \sum_{s} d^{\pi}(s|\theta) \sum_{a} \frac{\partial \bar{p}^{\pi}(a|s,\theta)}{\partial \theta} Q^{\pi}(s,a), \tag{12.23}$$

where $Q^{\pi}(s,a)$ is given by

$$Q^{\pi}(s,a|\theta) = \sum_{t=1}^{\infty} \mathbb{E}\{r(s_t, a_t) - F^{\pi}(\theta)|s_0 = s, a_0 = a\}.$$

This is the expected difference between rewards earned each time period from a starting state, and the expected reward (given by $F^{\pi}(\theta)$) earned each period when we are in steady state. We will not be able to compute this derivative exactly, but we show below that we can produce an unbiased estimate without too much difficulty. What is most important is that, unlike equation (12.22), we do not have to compute (or even approximate) $\nabla_{\theta} d^{\pi}(s|\theta)$. We pick this derivation up in the appendix in section 12.10.1. If you are willing to trust that equation (12.23) is true, read on!

12.8.4 Computing the Policy Gradient

As is always the case in stochastic optimization, the challenge boils down to computation. To help the discussion, we repeat the policy gradient result:

$$\frac{\partial F^{\pi}(\theta)}{\partial \theta} = \sum_{s} d^{\pi}(s|\theta) \sum_{a} \frac{\partial \bar{p}^{\pi}(a|s,\theta)}{\partial \theta} Q^{\pi}(s,a). \tag{12.24}$$

We start by assuming that we have some analytical form for the policy which allows us to compute $\partial \bar{p}^{\pi}(a|s,\theta)/\partial \theta$ (which is the case when we use our Boltzmann distribution). This leaves the stationary probability distribution $d^{\pi}(s|\theta)$, and the marginal rewards $Q^{\pi}(s,a)$.

Instead of computing $d^{\pi}(s|\theta)$ directly, we instead simply simulate the policy, depending on the fact that over a long simulation, we will visit each state with probability $d^{\pi}(s|\theta)$. Thus, for large enough T, we can compute

$$\nabla_{\theta} F^{\pi}(\theta) \approx \frac{1}{T} \sum_{t=1}^{T} \sum_{a} \frac{\partial \bar{p}^{\pi}(a|s_t,\theta)}{\partial \theta} Q^{\pi}(s_t,a), \tag{12.25}$$

where we simulate according to a known transition function $s_{t+1} = S^M(s_t, a, W_{t+1})$. We may simulate the process from a known transition function and a model of the exogenous information process W_t (if this is present), or we may simply observe the policy in action over a period of time.

This then leaves us with $Q^{\pi}(s_t, a)$. We are going to approximate this with estimates that we call $\bar{Q}_t^{\pi}(S_t|\theta)$, which we will compute by running a simulation starting at time t until T (or some horizon $t + H$). This requires running a different simulation that can be called a roll-out simulation, or a lookahead simulation. To avoid confusion, we are going to let $\tilde{S}_{tt'}$ be the state variable at time t' in a roll-out simulation that is initiated at time t. We let $\tilde{W}_{tt'}$ be the simulated random information between $t' - 1$ and t' for a simulation that is initiated at time t. Recognizing that $\tilde{S}_{tt} = S_t$, we can write

$$\bar{Q}_t^\pi(S_t|\theta) = \mathbb{E}_W \frac{1}{T-t} \sum_{t'=t}^{T-1} r(\tilde{S}_{tt'}, A^\pi(\tilde{S}_{tt'}|\theta)),$$

where $\tilde{S}_{t,t'+1} = S^M(\tilde{S}_{tt'}, A^\pi(\tilde{S}_{tt'}|\theta), \tilde{W}_{t,t'+1})$ represents the transitions in our lookahead simulation. Of course, we cannot compute the expectation, so instead we use the simulated estimate

$$\bar{Q}_t^\pi(S_t|\theta) \approx \frac{1}{T-t} \sum_{t'=t}^{T-1} r(\tilde{S}_{tt'}, A^\pi(\tilde{S}_{tt'}|\theta)). \tag{12.26}$$

We note that while we write this lookahead simulation as spanning the period from t to T, this is not necessary. We might run these lookahead simulations over a fixed interval $(t, t+H)$, and adjust the averaging accordingly.

We now have a computable estimate of $F^\pi(\theta)$ which we obtain from (12.26) by replacing $Q_t^\pi(S_t|\theta)$ with $\bar{Q}_t^\pi(S_t|\theta)$, giving us a sampled estimate of policy π using

$$F^\pi(\theta) \approx \sum_{t=0}^{T-1} \hat{Q}_t^\pi(S_t|\theta).$$

The final step is actually computing the derivative $\nabla_\theta F^\pi(\theta)$. For this, we are going to turn to numerical derivatives. Assume the lookahead simulations are fairly easy to compute. We can then obtain estimates of $\nabla_\theta \hat{Q}_t^\pi(S_t|\theta)$ using the finite difference. We can do this by perturbing each element of θ. If θ is a scalar, we might use

$$\nabla_\theta \hat{Q}_t^\pi(S_t|\theta) = \frac{\hat{Q}_t^\pi(S_t|\theta + \delta) - \hat{Q}_t^\pi(S_t|\theta - \delta)}{2\delta}. \tag{12.27}$$

If θ is a vector, we might do finite differences for each dimension, or turn to simultaneous perturbation stochastic approximation (SPSA) (see section 5.4.3 for more details).

This strategy was first introduced under the name of the REINFORCE algorithm. It has the nice advantage of capturing the downstream impact of changing θ on later states, but in a very brute force manner. This is actually a form of direct lookahead policy which we cover in depth in chapter 19.

Phew! Now you see why we marked this section with a **!

12.9 Supervised Learning

An entirely different approach to developing PFAs is to take advantage of the presence (if available) of an external source of decisions which we call "the supervisor." This might be a domain expert (such as a doctor making medical decisions, radiologists interpreting X-rays, or drivers operating a car), or perhaps simply a different optimization-based policy such as a deterministic lookahead. Supervised learning for decision problems is exactly analogous to supervised learning for machine learning.

Imagine that we have a set of decisions x^n from an external source (human or computer). Let S^n be our state variable, representing information available when the nth decision was made. For the moment, assume that we have access to a dataset $(S^n, x^n)_{n=1}^{N}$ from past history. Now we face a classical machine learning problem of fitting a function (policy) to this data. Start by assuming that we are going to use a simple linear model of the form

$$X^\pi(S|\theta) = \sum_{f \in \mathcal{F}} \theta_f \phi_f(S),$$

where $(\phi_f(S))_{f \in \mathcal{F}}$ is a set of features designed by a human (there is a vast machinery of statistical learning tools we can bring to bear on this problem). We can use our batch dataset to estimate $X^\pi(S|\theta)$, although more often we can use the tools in chapter 3 to adapt to new data in an online fashion.

Several issues arise when pursuing this approach:

- Our policy is never better than our supervisor, although in many cases a policy that is as good as an experienced supervisor might be quite good.
- In a recursive setting, we need to design algorithms that allow the policy to adapt as more data becomes available. Using a neural network, for example, can result in significant overfitting, producing unexpected results as the function adapts to noisy data.
- If our supervisor is a human, we are going to be limited in the number of times we can query our domain expert, raising the problem of efficiently designing questions.

Supervised learning can be a powerful strategy for finding an initial policy, and then using policy search methods (derivative-based or derivative-free) to further improve the policy. However, we face the issue of collecting data from our supervisor. If we have an extensive database of decisions and the corresponding state variables that capture the information we would use to make a decision, then we simply have a nice statistical challenge (albeit, not necessarily an easy one). However, it is often the case that we have to work with data arriving sequentially in an online manner. We can approach our policy estimation in two ways:

Active policy search – Here we are actively involved in the operation of the process to design better policies. We can do this in two ways:

Active policy adjustment – This involves adjusting the parameters controlling the policy, as we described above with policy search.

Active state selection – We may choose the state that then determines the decision. This might be in the form of choosing hypothetical situations (e.g. patient characteristics) and then asking the expert for his/her decision.

Passive policy search – In this setting, we are following some policy, and then selectively using the results to update our policy.

Active state selection is similar to derivative-free stochastic search (chapter 7). Instead of choosing x to obtain a noisy observation of $F(x) = \mathbb{E}F(x, W)$, we are choosing a state S to get a (possibly noisy) observation of an action x from some source. Active state selection can only be done in an offline setting (we cannot choose the characteristics of a patient walking into the hospital, but we can pose the characteristics of a hypothetical patient), but we are limited in terms of how many questions we can pose to our supervisor, especially if it is human (but also if it is a time consuming optimization model).

Passive policy search is an approach where we use our policy $X^\pi(S_t)$ to make decisions x_t that are then used to update the policy. Of course, if all we did was feed our own decisions back into the same function that produced the decisions, then we would not learn anything. However, it is possible to perform a weighted statistical fit, where we put a higher weight on decisions that perform better.

12.10 Why Does it Work?

12.10.1 Derivation of the Policy Gradient Theorem

We are going to provide the detailed derivation of

$$\frac{\partial F^\pi(\theta)}{\partial \theta} \;=\; \sum_s d^\pi(s|\theta) \sum_a \frac{\partial \bar{p}^\pi(a|s, \theta)}{\partial \theta} Q^\pi(s, a). \tag{12.28}$$

that we started in section 12.8.3.

We begin by defining two important quantities:

$$Q^\pi(s, a|\theta) = \sum_{t=1}^{\infty} \mathbb{E}\{r(s_t, a_t) - F^\pi(\theta)|s_0 = s, a_0 = a\},$$

$$V^\pi(s|\theta) = \sum_{t=1}^{\infty} \mathbb{E}\{r(s_t, a_t) - F^\pi(\theta)|s_0 = s\},$$

$$= \sum_{a \in A} \bar{p}^\pi(a_0 = a|s, \theta) \sum_{t=1}^{\infty} \mathbb{E}\{r(s_t, a_t) - F^\pi(\theta)|s_0 = s, a_0 = a\},$$

$$= \sum_{a} \bar{p}^\pi(a|s, \theta)Q^\pi(s, a). \tag{12.29}$$

Note that $Q^\pi(s, a|\theta)$ is quite different from the quantities $\bar{Q}^\pi(s, a|\theta)$ used above for the Boltzmann policy (which is consistent with Q-learning, which we first saw in section 2.1.6). $Q^\pi(s, a|\theta)$ sums the difference between the reward each period and the steady state reward per period (a difference that goes to zero on average), given that we start in state s and initially take action a. $V^\pi(s|\theta)$ is simply the expectation over all initial actions actions a as specified by our probabilistic policy.

We next rewrite $Q^\pi(s, a)$ as the first term in the summation, plus the expected value of the remainder of the infinite sum using

$$Q^\pi(s, a) = \sum_{t=1}^{\infty} \mathbb{E}\{r_t - F^\pi(\theta)|s_0 = s, a_0 = a\},$$

$$= r(s, a) - F^\pi(\theta) + \sum_{s'} P(s'|s, a)V^\pi(s'), \quad \forall s, a, \tag{12.30}$$

where $P(s'|s, a)$ is the one-step transition matrix (recall that this does not depend on θ). Solving for $F^\pi(\theta)$ gives

$$F^\pi(\theta) = r(s, a) + \sum_{s'} P(s'|s, a)V^\pi(s') - Q^\pi(s, a). \tag{12.31}$$

Now, note that $F^\pi(\theta)$ is not a function of either s or a, even though they both appear in the right hand side of (12.31). Noting that since our policy must pick some action, $\sum_{a \in A} \bar{p}^\pi(a|s, \theta) = 1$, which means

$$\sum_{a \in A} \bar{p}^\pi(a|s, \theta)F^\pi(\theta) = F^\pi(\theta), \quad \forall a.$$

This means we can take the expectation of (12.31) over all actions, giving us

$$F^\pi(\theta) = \sum_{a} \bar{p}^\pi(a|s, \theta) \left(r(s, a) + \sum_{s'} P(s'|s, a)V^\pi(s') - Q^\pi(s, a) \right), \tag{12.32}$$

for all states s. Taking a deep breath, we can now take derivatives using the following steps (explanations follow the equations):

$$\frac{\partial F^\pi(\theta)}{\partial \theta} = \frac{\partial}{\partial \theta}\left(\sum_a \bar{p}^\pi(a|s,\theta)\Big(r(s,a) + \sum_{s'} P(s'|s,a)V^\pi(s') - Q^\pi(s,a)\Big)\right) \quad (12.33)$$

$$= \sum_a \frac{\partial \bar{p}^\pi(a|s,\theta)}{\partial \theta}r(s,a) + \sum_a \frac{\partial \bar{p}^\pi(a|s,\theta)}{\partial \theta}\sum_{s'} P(s'|s,a)V^\pi(s')$$

$$+ \sum_a \bar{p}^\pi(a|s,\theta)\sum_{s'} P(s'|s,a)\frac{\partial V^\pi(s')}{\partial \theta} - \frac{\partial}{\partial \theta}\left(\sum_a \bar{p}^\pi(a|s,\theta)Q^\pi(s,a)\right) \quad (12.34)$$

$$= \sum_a \frac{\partial \bar{p}^\pi(a|s,\theta)}{\partial \theta}\Big(r(s,a) + \sum_{s'} P(s'|s,a)V^\pi(s')\Big)$$

$$+ \sum_a \bar{p}^\pi(a|s,\theta)\sum_{s'} P(s'|s,a)\frac{\partial V^\pi(s')}{\partial \theta} - \frac{\partial V^\pi(s)}{\partial \theta} \quad (12.35)$$

$$= \sum_a \frac{\partial \bar{p}^\pi(a|s,\theta)}{\partial \theta}\Big(Q^\pi(s,a) + F^\pi(\theta)\Big)$$

$$+ \sum_a \bar{p}^\pi(a|s,\theta)\sum_{s'} P(s'|s,a)\frac{\partial V^\pi(s')}{\partial \theta} - \frac{\partial V^\pi(s)}{\partial \theta} \quad (12.36)$$

$$= \sum_a \frac{\partial \bar{p}^\pi(a|s,\theta)}{\partial \theta}Q^\pi(s,a) + \sum_a \bar{p}^\pi(a|s,\theta)\sum_{s'} P(s'|s,a)\frac{\partial V^\pi(s')}{\partial \theta} - \frac{\partial V^\pi(s)}{\partial \theta}. \quad (12.37)$$

Equation (12.33) is from (12.32); (12.34) is the direct expansion of (12.33), where two terms vanish because $r(s,a)$ and $P(s'|s,a)$ do not depend on the policy $\bar{p}^\pi(a|s,\theta)$; (12.33) uses (12.29) for the last term; (12.36) uses (12.30); (12.29) uses the fact $F^\pi(\theta)$ is constant over states and actions, and $\sum_a \bar{p}^\pi(a|s,\theta) = 1$. Finally, note that equation (12.37) is true for all states.

We proceed to write

$$\frac{\partial F^\pi(\theta)}{\partial \theta} = \sum_s d^\pi(s|\theta)\frac{\partial F^\pi(\theta)}{\partial \theta} \quad (12.38)$$

$$= \sum_s d^\pi(s|\theta)\left(\sum_a \frac{\partial \bar{p}^\pi(a|s,\theta)}{\partial \theta}Q^\pi(s,a)\right.$$

$$\left. + \sum_a \bar{p}^\pi(a|s,\theta)\sum_{s'} P(s'|s,a)\frac{\partial V^\pi(s')}{\partial \theta} - \frac{\partial V^\pi(s)}{\partial \theta}\right). \quad (12.39)$$

Expanding gives us

$$\frac{\partial F^\pi(\theta)}{\partial \theta} = \sum_s d^\pi(s|\theta)\sum_a \frac{\partial \bar{p}^\pi(a|s,\theta)}{\partial \theta}Q^\pi(s,a)$$

$$+ \sum_s d^\pi(s|\theta)\sum_a \bar{p}^\pi(a|s,\theta)\sum_{s'} P(s'|s,a)\frac{\partial V^\pi(s')}{\partial \theta}$$

$$-\sum_s d^\pi(s|\theta)\frac{\partial V^\pi(s)}{\partial \theta} \tag{12.40}$$

$$= \sum_s d^\pi(s|\theta)\sum_a \frac{\partial \bar{p}^\pi(a|s,\theta)}{\partial \theta}Q^\pi(s,a)$$

$$+\sum_s d^\pi(s|\theta)\frac{\partial V^\pi(s)}{\partial \theta} - \sum_s d^\pi(s|\theta)\frac{\partial V^\pi(s)}{\partial \theta} \tag{12.41}$$

$$= \sum_s d^\pi(s|\theta)\sum_a \frac{\partial \bar{p}^\pi(a|s,\theta)}{\partial \theta}Q^\pi(s,a). \tag{12.42}$$

Equation (12.38) uses $\sum_s d^\pi(s|\theta) = 1$; (12.39) uses the fact (12.37) holds for all s; (12.40) simply expands (12.39); (12.41) uses the property that since $d^\pi(s)$ is the stationary distribution, then $\sum_s d^\pi(s|\theta)P(s'|s,a) = d^\pi(s'|\theta)$ (after substituting this result, then just change the index from s' to s). Equation (12.42) is the policy gradient theorem we first presented in equation (12.28) (and equation (12.23) in the body of the chapter).

12.11 Bibliographic Notes

Section 12.1 – The idea of modeling stochastic search algorithms (whether it is derivative-based or derivative-free) was first done (to our knowledge) in Powell (2019).

Section 12.2 – The concept that the search over policy function approximations is over the same classes of functions as would take place in any machine learning exercise seems to be new.

Section 12.5–12.6 – The concept of optimizing parameterized policies, which has been described as "policy search," has been actively studied since the 1990s. It is the reason we named this class the "policy search" class. Our presentation of policy search using numerical derivatives, or the methods of derivative-free stochastic search (both of which depend purely on simulating a policy as a black box) is well known in the reinforcement learning community (see Sigaud and Stulp (2019) for a recent and thorough review). We note that this review is specifically for continuous actions, but a parameterized policy can be used for discrete actions, and optimized using the same methods.

Section 12.7 – Both sections 12.5 and 12.6 depend purely on function approximations to perform stochastic search. There is a large class of dynamic programs where the future state S_{t+1} is a continuous function of S_t and x_t. These include, for example, resource allocation problems for managing money, water, blood, inventory, and electric power, where inventories of resources R_t are being allocated through decisions x_t to produce updated inventories R_{t+1}. The core equations ((12.15)–(12.16)) are little more than elaborate exercises in the chain rule that have long been used in control problems and neural networks (where it is referred to as backpropagation). See any standard treatment of discrete time optimal control (such as Kirk (2012), Stengel (1986), Sontag (1998), and Lewis and Vrabie (2012)). Our adaptation for parameterized policies was derived here from first principles, but the approach is straightforward.

Section 12.8 – Policy gradient methods have received considerable attention in the reinforcement learning community for problems with discrete states and actions. This section describes a method for computing policy gradients for discrete dynamic programs using a concept that has become known as the "policy gradient method," introduced in Sutton et al. (2000), and described nicely in the second edition of their book Sutton and Barto (2018)[Chapter 13].

Exercises

Review questions

12.1 Policy search is a sequential decision problem. Write out the elements of a policy search algorithm using our modeling framework.

12.2 What is an "affine policy"? Write out a general form for an affine policy. Imagine that we are managing an inventory storage problem where the state $S_t = (R_t, p_t)$ depends on the inventory we are holding R_t and the price we can sell the inventory p_t. Let x_t be the amount of our inventory to sell at time t. If we write our policy as

$$X^\pi(S_t|\theta) = \theta_0 + \theta_1 R_t + \theta_2 R_T^2 + \theta_3 p_t + \theta_4 p_t^2 + \theta_5 R_t p_t,$$

is this an affine policy? Why?

12.3 To do policy search it is critical that you know how to write out the objective function that you use to evaluate the performance of the policy.

(a) What is the objective function if you are tuning your policy in a simulator? Carefully explain each source of uncertainty (or randomness).

(b) What is the objective function if you are tuning your policy in the field?

Modeling questions

12.4 Assume we are going to search for policies for a simple inventory problem where the inventory R_t evolves according to

$$R_{t+1} = \max\{0, R_t + x_{t-\tau} - \hat{D}_{t+1}\},$$

where the random demand $\hat{D}_{t+1}$ follows a discrete uniform distribution from 1 to 10 with equal probability. An order x_t arrives at time $t + \tau$, which we will specify below. Assume $R_0 = 10$, and use the contribution function

$$C(S_t, x_t) = p_t \min\{R_t + x_{t-\tau}, \hat{D}_{t+1}\} - 15x_t,$$

where the price p_t is drawn from a uniform distribution between 16 and 25 with equal probability.

We are going to place our orders according to the order-up-to policy

$$X^{Inv}(S_t|\theta) = \begin{cases} \theta^{max} & \text{if } R_t < \theta^{min}, \\ 0 & \text{otherwise.} \end{cases}$$

We want to choose θ to solve

$$\max_\theta F(\theta) = \mathbb{E}_W \left\{ \sum_{t=0}^{100} C(S_t, x_t) | S_0 \right\}, \tag{12.43}$$

where $W = (W_1, ..., W_{100})$ is the vector of realizations of prices and demands.

(a) What is the state variable S_t at time t?

(b) What is the decision variable at time t? Does it matter that the decision at time t does not have any impact on the system until τ time periods later?

(c) What are the elements of the exogenous information variable W_t?

(d) What is the transition function? Recall that you need an equation for each element of S_t.

(e) The objective function in (12.43) maximizes the cumulative reward, but we are optimizing the policy in an offline simulator, which means we want to optimize the final reward, not the cumulative reward. Make the argument that (12.43) is still the correct objective. [Hint: look at Table 9.3 and identify which of the four classes of objective functions that equation (12.43) falls in.]

12.5 Assume you are tuning the parameters θ of a policy $X^\pi(S^n|\theta)$ to find $x^{\pi,N}$ in N iterations to maximize $\mathbb{E}F(\theta, W)$ using a gradient-based search algorithm. This means you have access to the gradient $\nabla_\theta F(\theta, W)$.

(a) Write out the five elements of a sequential decision problem (state variables, decision variables, exogenous information, transition function, and objective function).

(b) What is the exogenous information for this problem?

(c) Recalling the menu of stepsize policies that we can draw from (see section 6.2.3), what is meant by searching over policies?

Computational exercises

The next two exercises will optimize the policy modeled in exercise 12.4 using derivative-based methods.

12.6 Implement the basic stochastic gradient algorithm based on finite differences (see section 5.4.3). Use the harmonic stepsize

$$\alpha_n = \frac{\theta^{step}}{\theta^{step} + n - 1},\tag{12.44}$$

which means we also have to tune θ^{step}. Start by assuming $\tau = 1$.

(a) Run the algorithm 100 iterations for $\theta^{step} = 1, 5, 10, 20$ (just one sample path each) and report which one works best, and the value of θ that the algorithm returns.

(b) Run the algorithm 100 iterations for $\theta^{step} = 10$, and plot the objective function over the iterations for each value of θ^{step}. Repeat this 20 times to demonstrate the range of sample paths the algorithm can take. How many samples do you think you would need to reliably estimate which value of θ^{step} works best?

(c) Using the best value of θ^{step}, find the best value of θ when $\tau = 1, 5, 10$.

12.7 You are going to optimize the policy modeled in exercise 12.4 using the SPSA algorithm. Assume $\tau = 1$.

(a) Implement the simultaneous perturbation stochastic approximation (SPSA) algorithm (see section 5.4.4). Use the harmonic stepsize (see equation (12.44)), which means we also have to tune the stepsize parameter θ^{step}. Use a mini-batch of 1 for computing the gradient. Run the algorithm 100 iterations for $\theta^{step} = 1, 5, 10, 20$, where you run 20 repetitions for each value of θ^{step} and average the results. Report which one works best.

(b) Run the algorithm using $\theta^{step} = 10$ and mini-batch sizes of 1, 5, 10, and 20, and compare the performance over 100 iterations.

The next two exercises will optimize the policy modeled in exercise 12.4 using derivative-free methods. For each method, enumerate a set Θ of possible values for the two-dimensional ordering policy θ by varying θ^{min} over the values $2, 4, \dots, 10$, and varying θ^{max} over the range $6, 8, \dots, 20$ while excluding any combination where $\theta^{min} \geq \theta^{max}$. Let Θ be the set of allowable combinations of θ. Assume $\tau = 1$ throughout.

12.8 Lookup table with correlated beliefs: After building the set Θ, do the following:

(a) Initialize your belief by running five simulations for five different values of $\theta \in \Theta$. Average these results and set $\bar{\mu}_\theta^0$ to this average for all $\theta \in \Theta$. Compute the variance $\sigma^{2,0}$ of these five observations, and initialize the precision of the belief at $\beta_\theta^0 = 1/\sigma^{2,0}$ for all $\theta \in \Theta$. Let

$$\bar{F}^0 = \max_{\theta \in \Theta} \bar{\mu}_\theta^0$$

and report $\bar{F}^0$ (of course, $\bar{\mu}_\theta^0$ is the same for all θ, so you can just pick any θ).

(b) Assume that the estimates $\bar{\mu}_\theta^0$ are related according to

$$Cov(\bar{\mu}_\theta^0, \bar{\mu}_{\theta'}^0) = \sigma^0 e^{-\rho|\theta - \theta'|}.$$

Compute $Cov(\bar{\mu}_\theta^0, \bar{\mu}_{\theta'}^0)$ by running 10 simulations for each combination of $\theta = (4,6),(4,8),(4,10),(4,12),(4,14)$. Now find the value of ρ that produces the best fit of $Cov(\bar{\mu}_\theta^0, \bar{\mu}_{\theta'}^0)$ using these five datapoints. Now, fill out the matrix Σ^0 where

$$\Sigma_{\theta,\theta'}^0 = \sigma^0 e^{-\rho|\theta - \theta'|}$$

for all $\theta, \theta' \in \Theta$, and using the value of ρ that you determined above.

(c) Write out the equations for updating $\bar{\mu}_\theta^n$ using correlated beliefs (see section 3.4.2).

(d) Now use the interval estimation policy

$$\Theta^\pi(S^n | \theta^{IE}) = \arg \max_{\theta \in \Theta} \left(\bar{\mu}_\theta^n + \theta^{IE} \bar{\sigma}_\theta^n \right)$$

where $\bar{\sigma}_\theta^n = \Sigma_{\theta,\theta}^n$. Of course, we have now introduced another tunable parameter θ^{IE} in our policy to tune the parameters in our ordering policy $X^\pi(S_t | \theta)$. Get used to it - this happens a lot. Using $\theta^{IE} = 2$, execute the policy $\Theta^\pi(S^n | \theta^{IE})$ for 100 iterations, and report the simulated performance of the objective (12.43) as you progress. On a two-dimensional graph showing all the combinations of Θ, report how many times you sample each of the combinations.

(e) Repeat your search for $\theta^{IE} = 0, .5, 1, 2, 3$. Prepare a graph showing the performance of each value of θ^{IE}.

12.9 Response surface methods: In this exercise we are going to optimize θ by creating a statistical model of the function $F(\theta)$. After building the set Θ, do the following:

(a) Randomly pick 10 elements of Θ, simulate the policy 20 times, and then use the simulated performance of the policy to fit the linear model

$$\bar{F}^0(\theta) = \rho_0^0 + \rho_1^0 \theta^{min} + \rho_2^0 (\theta^{min})^2 + \rho_3^0 \theta^{max} + \rho_4^0 (\theta^{max})^2 + \rho_5^0 \theta^{min}\theta^{max}.$$

Use the methods in section 3.7 to fit this model.

(b) At iteration n, find

$$\theta^n = \arg\max_\theta \bar{F}^n(\theta).$$

We then run the policy using $\theta = \theta^n$ to obtain $\hat{F}^{n+1}(\theta^n)$. Add $(\theta^n, \hat{F}^{n+1})$ to the data used to fit the approximation to obtain the updated approximation $\bar{F}^{n+1}(\theta)$, and repeat. Run this for 20 iterations, and repeat 10 times. Report the average and the spread.

(c) Repeat the algorithm, but this time replace the policy for computing θ^n with

$$\hat{\theta}^n = \arg\max_\theta \bar{F}^n(\theta),$$

$$\theta^n = \hat{\theta}^n + \delta^n,$$

where

$$\theta^n = \begin{pmatrix} \theta_1^n \\ \theta_2^n \end{pmatrix}$$

and

$$\delta^n = \begin{pmatrix} \delta_1^n \\ \delta_2^n \end{pmatrix}.$$

The vector δ is a perturbation of magnitude r where

$$\delta_1^n + \delta_2^n = 0,$$

$$\sqrt{(\delta_1^n)^2 + (\delta_2^n)^2} = r.$$

These equations imply that

$$\delta_1^n = -\delta_2^n = r/\sqrt{2},$$

or

$$\delta_2^n = -\delta_1^n = r/\sqrt{2}.$$

This algorithm exploits the property that it is better to sample points that are displaced from the optimum. As is often the case, this simple policy involves another tunable parameter, the perturbation radius r. Start with $r = 4$. Run this algorithm for 20 iterations, and then do a final evaluation with $\delta^n = 0$ to see the performance based on the value of θ that is best given our approximate function. Repeat for $r = 0, 2, 6, 8$ and report which performs the best.

Problem solving questions

12.10 Imagine we have an asset selling problem where the policy is given by

$$X^\pi(S_t|\theta) = \begin{cases} 1 = \text{``sell''} & \text{if } p_t \geq \theta, \\ 0 = \text{``hold''} & \text{if } p_t < \theta. \end{cases} \qquad (12.45)$$

(a) Is this an affine policy? Why or why not?
(b) Now imagine that we do not know that this might be the right structure of the policy, and you want to design an affine policy. What might this look like? Do you think your affine policy might work well?
(c) What is meant by a monotone policy? Is the policy in (12.45) monotone?
(d) Imagine that you believe that your policy is monotone in price, but other than this, you do not know the shape of the function. Suggest an approximation strategy you might propose that allows you to require that the function be monotone in p_t, and sketch a method for estimating this function.

Sequential decision analytics and modeling

These exercises are drawn from the online book *Sequential Decision Analytics and Modeling* available at http://tinyurl.com/sdaexamplesprint.

12.11 Review the asset selling problem in chapter 2 up through 2.4. Three policies are suggested, but in this exercise we are going to focus on the tracking policy, which involves tuning a single parameter. We

will be using the Python module "AssetSelling" at http://tinyurl.com/
sdagithub, which contains the code to simulate the tracking policy.
This exercise will focus on derivative-free methods for performing the
parameter search.

(a) Run 20 simulations of the pricing model and determine from these
runs the largest and smallest prices. Divide this range into 20
segments. Now implement an interval estimation policy

$$X^{IE}(S^n|\theta^{IE}) = \arg\max_{x \in \mathcal{X}}(\bar{\mu}_x^n + \theta^{IE}\bar{\sigma}_x^n). \tag{12.46}$$

where $\mathcal{X}$ is the 20 possible values of the tracking parameter, $\bar{\mu}_x^n$ is
our estimate of the performance of the tracking parameter when it
takes value $x \in \mathcal{X}$. For this exercise, set $\theta^{IE} = 2$ (although this is
a parameter that would also need tuning). Show your estimates $\bar{\mu}_x^N$
for each value of x when your experimentation budget is $N = 20$,
and then when $N = 100$.

(b) This time, we are going to create a quadratic belief model where

$$\bar{F}^n(x) = \bar{\theta}_0^n + \bar{\theta}_1^n x + \bar{\theta}_2^n x^2,$$

where x is still the value of the tracking parameter. Test three
policies for choosing x^n (these are all presented in chapter 7):
(i) A greedy policy where $x^n = \arg\max_x \bar{F}^n(x)$.
(ii) An excitation policy $x^n = \arg\max_x \bar{F}^n(x) + \varepsilon^{n+1}$ where $\varepsilon^{n+1} \sim$
$N(0, \sigma^2)$, where σ^2 is the noise in the exploration process, which
is a parameter that has to be tuned.
(iii) A parameterized knowledge gradient policy where $x^n = $
$\arg\max_x \bar{F}^n(x) + Z$ where $Z = \pm r$, where r is a parameter that
needs to be tuned.
Simulate each policy for 100 iterations, and compare the performance of each policy.

12.12 Review the asset selling problem in chapter 2 up through 2.4. Three
policies are suggested, but in this exercise we are going to focus on
the tracking policy, which involves tuning a single parameter. We
will be using the Python module "AssetSelling" at http://tinyurl.com/
sdagithub, which contains the code to simulate the tracking policy.
This exercise will focus on derivative-based methods for performing
the parameter search.

(a) Produce an estimate of a stochastic gradient by running a simulation where the tracking parameter is set at x, and then again at $x+\delta$
where $\delta = 1$. Use a harmonic stepsize

$$\alpha_n = \frac{\theta^{step}}{\theta^{step} + n - 1},$$

where we leave the tuning of θ^{step} to you. Run this algorithm for 100 iterations, and find θ^{step} to produce the best solution $x^{\pi,N}$.

(b) Repeat (a), but this time repeat the simulation using a mini-batch m for $m = 1, 5, 10, 20$. Note that the best value of θ^{step} is likely to depend on m. Run the stochastic gradient algorithm for each value of m for $N = 100$ iterations, and compare the results.

Diary problem

The diary problem is a single problem you chose (see chapter 1 for guidelines). Answer the following for your diary problem.

12.13 Pick a particular decision in your diary problem (if there is more than one) and try to design a policy function approximation to make the decision. This will typically involve a tunable parameter (if you state a PFA without a tunable parameter, try to introduce one). Then, show how to tune the policy in the following settings:

(a) Offline, in a simulator. Remember that you will have both the tuning of the parameter(s), followed by testing. Write out the objective function using final reward formulation (if you do not remember this by now, flip back to equation (7.2)). Explicitly describe any uncertainties in your initial state S^0, along with the exogenous information W_t and the testing random variable $\widehat{W}_t$.

(b) Online, in the field. This means optimizing using the cumulative reward formulation (see equation (7.3)). Again – clearly define all the random variables.

Bibliography

Kirk, D.E. (2012). *Optimal Control Theory: An introduction.* New York: Dover.

Lewis, F.L. and Vrabie, D. (2012). *Design Optimal Adaptive Controllers,* 3e. Hoboken, NJ: JohnWiley & Sons.

Powell, W.B. (2019). A unified framework for stochastic optimization. *European Journal of Operational Research* 275: (3): 795–821.

Sigaud, O. and Stulp, F. (2019). Policy search in continuous action domains: An overview. *Neural Networks* 113: 28–40.

Sontag, E. (1998). *Mathematical Control Theory,* 2e., 1–544. Springer.

Stengel, R.F. (1986). *Stochastic optimal control: theory and application.* Hoboken, NJ: John Wiley & Sons.

Sutton, R.S. and Barto, A.G. (2018). *Reinforcement Learning: An Introduction*, 2e. Cambridge, MA: MIT Press.

Sutton, R.S., McAllester, D., Singh, S.P., and Mansour, Y. (2000). Policy gradient methods for reinforcement learning with function approximation. *Advances in neural information processing systems* 12 (22): 1057–1063.

13

Cost Function Approximations

Parametric function approximations (chapter 12) can be a particularly powerful strategy for problems where there is a clear structure to the policy. For example, buying when the price is below θ^{min} and selling when it is above θ^{max} is an obvious structure for many buy/sell problems. But PFAs do not scale to larger, more complex problems such as, say, scheduling an airline or managing an international supply chain. PFAs cannot even help you plan the path you will take with your car.

The problem with PFAs is that you either have to be able to identify a simple structural form (which means some form of linear or nonlinear model), or you can specify a high-dimensional architecture (locally constant or linear, full non-parametric, or a deep neural network) which will require a substantial number of training iterations (possibly in millions or tens of millions). There are many problems, however, where the decisions are high-dimensional, which means that lots of variables interact, such as the location of pieces on a chessboard, or the effect of surplus blood inventories in one region on the allocation of blood around the country. Learning these interactions in the presence of noise is especially difficult.

CFAs are a form of parameterized optimization models. Imagine that you have a problem that suggests a natural approximation as a deterministic optimization problem. These may be myopic (assigning available drivers in a ride-sharing fleet to waiting customers), or they may involve optimizing a deterministic approximation of the future (technically a form of direct lookahead approximation, but a simple one). An example is the use of deterministic shortest path problems in navigation systems, or optimizing inventory decisions over a planning horizon given a point forecast of demands.

The optimization problem may be as complicated as scheduling an airline, or as simple as trying to pick a medical treatment $x \in \mathcal{X} = \{x_1, \dots, x_M\}$ that will treat a patient's high blood sugar. Let $\bar{\mu}_x^n$ be the estimated reduction in blood sugar from treatment x_m after we have run n different tests, and let $\bar{\sigma}_x^n$ be the standard deviation of our estimate $\bar{\mu}_x^n$. Assuming our beliefs are independent of each other, our current state (belief state) is given by $S^n = B^n = (\bar{\mu}_x^n, \bar{\sigma}_x^n), x \in \mathcal{X}$. A greedy ("pure exploitation") policy would use the policy

$$X^{Explt}(S^n) = \arg\max_x \bar{\mu}_x^n.$$

Such a policy uses the treatment that appears to be best, but fails to recognize that after choosing x^n and observing $\hat{F}^{n+1} = F(x^n, W^{n+1})$ we can use this information to update our belief state (captured by S^n). The problem is that we may have an estimate $\bar{\mu}_x^n$ that is too low that would discourage us from trying it again. One way to fix this (which we introduced in chapter 7 as interval estimation) is by using the modified policy

$$X^{IE}(S_t|\theta) = \arg\max_{x \in \mathcal{X}} \left(\bar{\mu}_x^n + \theta\bar{\sigma}_x^n \right), \tag{13.1}$$

where θ is a parameter that has to be tuned through our usual objective function

$$\max_\theta F(\theta) = \mathbb{E} \sum_{t=0}^{T} C(S_t, X^\pi(S_t|\theta)). \tag{13.2}$$

We have tweaked the pure exploitation policy by adding an "uncertainty bonus" in (13.1) which encourages trying alternatives where $\bar{\mu}_x$ might be lower, but where there is sufficient uncertainty that it might actually be higher. This is a purely heuristic way of enforcing a tradeoff between exploration and exploitation (but a heuristic that enjoys some nice theoretical properties).

While our interval estimation policy is limited to discrete action spaces, parametric CFAs can actually be extended to very large-scale problems. Once you introduce an "$\arg\max_x$" into the policy, you open the door to using solvers for large linear, integer, nonlinear, and even nonlinear-integer programs as we illustrate later in this chapter. Suddenly we can now allow x_t to be vectors with hundreds of thousands of variables (dimensions).

The idea of using a parameterized optimization model is a widely used engineering heuristic, but has been completely overlooked as a valid way of building a policy for solving stochastic sequential decision problems. The point of departure between an ad-hoc heuristic and a formal optimization model is equation (13.2). Normally we do our parameter tuning in a simulator (presumably with a final reward objective), or online in the field (presumably with a cumulative reward objective). Either way, we need to explicitly formulate the parameter tuning process as an explicit optimization problem (such as (13.2)); if this is not done, then what you are doing is, in fact, just an engineering heuristic.

While using a parameterized optimization model is quite common in practice, using equation (13.2) to tune the parameters is not. As with PFAs, there are three dimensions in the use of parametric CFAs:

(1) Designing the parameterization – This is the art of any parametric model (including statistical models). CFAs begin as some form of deterministic optimization model, where the parameterization should be chosen to improve what can be achieved with the original deterministic approximation.

(2) Evaluating a parametric CFA – The most common way to evaluate a policy is a simulator, but there are many settings where simulators are either too time consuming or expensive to develop, or because we simply cannot create a mathematical model of the problem, requiring evaluation to be done in the field.

(3) Tuning the parameters – As we have seen in the chapters on stochastic search (chapters 5 and 7) and policy search (chapter 12), tuning the parameters θ using the objective function (13.2) is not easy. For this reason, it is quite common in the industry for someone to use intuition to simply pick values for θ. While the performance of the resulting policy may be reasonable, this is not optimization.

The research community has largely dismissed parameterized deterministic models as an "industrial heuristic." We claim that a parameterized optimization model is a powerful strategy for solving certain classes of stochastic optimization problems, and is just as valid as using any PFA, or any of the strategies that we are going to present later in this book. It all boils down to exploiting problem structure and insights into how uncertainty affects the solution.

We need to pause and make an important observation: PFAs and CFAs both look like parameterized policies, but they tend to be different in a critical way, especially when the PFA uses a generic architecture such as a linear model or neural network. PFAs using a generic architecture will provide no guidance in terms of the scaling of the vector θ. By contrast, if we start with a deterministic approximation, it introduces a tremendous amount of structure, which has the effect of scaling the problem. This dramatically simplifies the parameter search process.

The remainder of this chapter will focus on illustrating different ways to create parametric CFAs. Section 13.1 sets up some general notation. Then, section 13.2 presents examples of parameterizing the objective function, followed by section 13.3 which presents examples of parameterized constraints.

13.1 General Formulation for Parametric CFA

There are two ways to parameterize an optimization problem: through the objective function, and through the constraints. To capture these changes we define

$$\bar{C}^{\pi}(S_t, x_t | \theta) = \text{the modified objective function as determined by the policy } \pi, \text{ where } \theta \text{ represents the tunable parameters,}$$

$$\mathcal{X}_t^{\pi}(\theta) = \text{the modified set of constraints (that is, the feasible region) determined by policy } \pi, \text{ with tunable parameters } \theta.$$

A parametric CFA can be written in its most general form as

$$X^{CFA}(S_t|\theta) = \arg \max_{x_t \in \mathcal{X}_t^{\pi}(\theta)} \bar{C}^{\pi}(S_t, x_t|\theta), \tag{13.3}$$

where $\bar{C}^{\pi}(S_t, x_t|\theta)$ is a parametrically modified cost function, subject to a (possibly modified) set of constraints $\mathcal{X}^{\pi}(\theta)$, where θ is the vector of tunable parameters.

We now have a tunable policy $X^{CFA}(S_t|\theta)$ where we face the same challenge of finding θ as we did with PFAs in chapter 12. Note that θ might be a scalar, or may have dozens, even hundreds or thousands, of dimensions. We anticipate that the most common search procedures will be those based on either derivative-based stochastic search using numerical derivatives such as the SPSA algorithm described in section 12.5 (or section 5.4.4) or derivative-free stochastic optimization such as the methods outlined in section 12.6. It is possible that we might apply the exact gradient described in section 12.7, but taking the derivative of the policy when the policy is an optimization problem is likely going to be daunting.

13.2 Objective-Modified CFAs

We begin by considering problems where we modify the problem through the objective function to achieve desired behaviors. Including bonuses and penalties is a widely used heuristic approach to getting cost-based optimization models to produce desired behaviors, such as balancing real costs against penalties for poor service. Not surprisingly, we can use this approach to also produce robust behaviors in the presence of uncertainty.

We begin by presenting a general way of including linear cost correction models in the objective function. We then present three application settings: a dynamic assignment problem for assigning drivers to loads, a stochastic, dynamic shortest path problem, and a financial trading problem.

13.2.1 Linear Cost Function Correction

Although we favor parameterizations that are guided by the structure of the problem, a general approach to improving the performance of an optimization-based policy is to add a linear term to the objective, which gives us

$$X^{CFA-cost}(S_t|\theta) = \arg\max_{x_t \in \mathcal{X}_t} \left(C(S_t, x_t) + \sum_{f \in \mathcal{F}} \theta_f \phi_f(S_t, x_t) \right). \tag{13.4}$$

where $(\phi_f(S, x))_{f \in \mathcal{F}}$ is a set of features that depend first and foremost on x, and possibly on the state S. If a feature does not depend on the decision, then it would not affect the choice of optimal solution.

Designing the features for equation (13.4) is no different than designing the features for a linear policy function approximation (or, for that matter, any linear statistical model which we introduced in chapter 3). It is always possible to simply construct a polynomial comprised of different combinations of elements of x_t and S_t with different transformations (linear, square, ...), but many problems have very specific structure.

13.2.2 CFAs for Dynamic Assignment Problems

The truckload trucking industry requires matching drivers to loads, just as ride-sharing companies match drivers to riders. The difference with truckload trucking is that the customer is a load of freight, and sometimes the load has to wait a while (possibly several hours) before being picked up.

To model our problem we begin with defining the sets of resources and tasks which make up the state variables

$$\mathcal{D}_t = \text{the set of all drivers (with tractors) available at time } t,$$
$$\mathcal{L}_t = \text{the set of all loads waiting to be moved at time } t,$$
$$S_t = (\mathcal{D}_t, \mathcal{L}_t) = \text{the state of our system at time } t.$$

Our decision variables and costs are given by

$$x_{td\ell} = 1 \text{ if we assign driver } d \text{ to load } \ell \text{ at time } t, 0 \text{ otherwise,}$$
$$c_{td\ell} = \text{the contribution of assigning driver } d \in \mathcal{D}_t \text{ to load } \ell \in \mathcal{L}_t$$
$$\text{at time } t, \text{ including the revenue generated by the load, the}$$
$$\text{cost of moving empty to the load, as well as penalties for late}$$
$$\text{pickup or delivery.}$$

Finally, we have the post-decision sets of loads and drivers which we represent using

$$\mathcal{L}_t^x \;\; = \;\; \text{set of loads that were served at time } t, \text{ which is to say all } \ell \text{ where } \sum d \in \mathcal{D}_t x_{td\ell} = 1,$$

$$\mathcal{D}_t^x \;\; = \;\; \text{set of drivers that were dispatched at time } t, \text{ which is to say all } d \text{ where } \sum \ell \in \mathcal{L}_t x_{td\ell} = 1.$$

A myopic policy for assigning drivers to loads would be formulated as

$$X^{Assign}(S_t) = \arg\max_{x_t} \sum_{d\in\mathcal{D}_t} \sum_{\ell\in\mathcal{L}_t} c_{td\ell} x_{td\ell}. \tag{13.5}$$

Once we dispatch a driver (that is, $x_{td\ell} = 1$ for some $\ell \in \mathcal{L}_t$), we assume the driver vanishes (this is purely for modeling simplification). We then model drivers becoming available as an exogenous stochastic process along with the new loads. This is modeled using

$$\hat{L}_{t+1} \;\; = \;\; \text{exogenous process describing random loads (complete with origins and destinations) that were called in between } t \text{ and } t+1,$$

$$\hat{D}_{t+1} \;\; = \;\; \text{exogenous process describing drivers calling in between } t \text{ and } t+1 \text{ to say they are available (along with location).}$$

In practice $\hat{D}_t$ will depend on prior decisions, but this simplified model will help us make the point. The transition function would be given by

$$\mathcal{L}_{t+1} \;\; = \;\; \mathcal{L}_t \setminus \mathcal{L}_t^x \cup \hat{L}_{t+1}, \tag{13.6}$$

$$\mathcal{D}_{t+1} \;\; = \;\; \mathcal{D}_t \setminus \mathcal{D}_t^x \cup \hat{D}_{t+1}, \tag{13.7}$$

where $\mathcal{A} \setminus \mathcal{B}$ means we subtract set $\mathcal{B}$ from set $\mathcal{A}$. In real settings, however, loads that have been waiting too long may drop out and look for another carrier, which means we lose the load (and the revenue). Our myopic policy simply is not taking the value of what might happen in future time periods into account.

One way to handle this is to put a positive bonus for moving loads that have been delayed. Let

$$\tau_{t\ell} \;\; = \;\; \text{the time that load } \ell \in \mathcal{L}_t \text{ has been delayed as of time } t.$$

Now consider the modified policy

$$X^{CFA-Assign}(S_t|\theta) = \arg\max_{x_t} \sum_{d\in\mathcal{D}_t} \sum_{\ell\in\mathcal{L}_t} (c_{td\ell} + \theta\tau_{t\ell}) x_{td\ell}. \tag{13.8}$$

Now we have a modified cost function (we use the term "cost function" even though we are maximizing) that is parameterized by θ which places a bonus (assuming $\theta > 0$) on loads that have been delayed. The next challenge is to tune θ: Too large, and we move long distances to pull loads that have been waiting; too small, and we end up losing loads that have to wait too long. Our optimization problem is given by

$$\max_{\theta} \mathbb{E} \sum_{t=0}^{T} C(S_t, X^{CFA-Assign}(S_t | \theta)), \tag{13.9}$$

where

$$C(S_t, x_t) = \sum_{d \in \mathcal{D}_t} \sum_{\ell \in \mathcal{L}_t} c_{td\ell} x_{td\ell}.$$

We now face the problem of tuning θ to maximize profits. We may also set a target on, say, the number of loads that have been delayed more than 4 hours.

This is a classical use of a parametric cost function approximation for finding robust policies for a very high-dimensional resource allocation problem. The delay penalty parameter θ can be tuned in a simulator that represents the objective (13.9) along with the dynamics (13.6) and (13.7). In real applications, this tuning is often done (albeit in an ad hoc way) in an online setting based on real observations.

13.2.3 Dynamic Shortest Paths

Consider the problem of finding the best path through a network over time, as illustrated in Figure 13.1. Our navigation system uses best estimates of the times for each link in the network to plan a path to the destination, but as we progress along the path, new information arrives and the path is updated. This is a form of direct lookahead policy (that we consider in depth in chapter 19) using forecasts of future travel times.

The idea of planning paths into the future using a deterministic forecast is so familiar to us that we do not even challenge it, but this is a fully sequential, stochastic decision problem, with rolling forecasts. Now imagine that our shortest path takes us over a toll bridge which has to be lifted periodically to allow taller boats to traverse underneath. When this happens, traffic can be stopped for up to 20 minutes. It will take you 40 minutes to get to this bridge, and if you are delayed, you will miss your appointment.

When link times have distributions with long tails, we may wish to consider, for example, the 90th percentile of the time to traverse each link rather than the expectation. This is a form of parametric cost function approximation where we use a modified objective function.

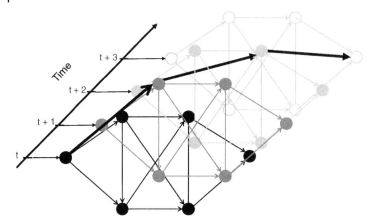

Figure 13.1 Illustration of a shortest path over a time-dependent network.

A sketch of the model using our standard framework is as follows:

State variables – We represent the location of a traveler by

$$R_t \quad = \quad \text{the next node where the traveler has to make a decision.}$$

Estimated travel costs are represented by

$$\tilde{c}_t \quad = \quad (\tilde{c}_{t,i,j})_{(i,j) \in \mathcal{N}},$$
$$= \quad \text{the vector of estimates of the cost to traverse link } (i, j)$$
$$\text{at time } t, \text{ given what is known at time } t.$$

We are also going to assume that we have a historical dataset that tells us the _distribution_ of travel costs. Since these distributions would be compiled based on many observations, we are going to assume that these are static (we would include these distributions in our initial state S_0, not in our dynamic state S_t).

The traveler's state S_t at time t is then

$$S_t = (R_t, \tilde{c}_t).$$

A common mistake is to assume that the state of our system is the location of the traveler. In a dynamic network, you have to include the estimates of the travel times on _every_ link of the network in the state variable, since these are being updated every time period.

Decision variables – The decision variables are given by

$$x_{tij} \quad = \quad \begin{cases} 1 & \text{if we traverse link } i \text{ to } j \text{ when we are at } i \text{ at time } t, \\ 0 & \text{otherwise.} \end{cases}$$

These are subject to constraints that ensure that from any node i, we have to go somewhere (until we reach our destination).

As always, we let $X^\pi(S_t|\theta)$ be our policy for determining which link (i, j) to traverse given that we are at node i.

Exogenous information – There are two types of exogenous information for this problem:

$$\hat{c}_{t+1,ij} \quad = \quad \text{This is the observed cost of traversing link } (i, j) \text{ after the traveler made the decision at time } t \text{ and traversed this link.}$$

The second type of new information is the updated estimates of the link costs. We are going to model the exogenous information as the change in the estimates:

$$\delta\tilde{c}_{t+1,ij} \quad = \quad \tilde{c}_{t+1,ij} - \tilde{c}_{t,ij},$$
$$\delta\tilde{c}_{t+1} \quad = \quad (\tilde{c}_{t+1,ij})_{(i,j)\in\mathcal{N}}.$$

Our exogenous information variable, then, is given by

$$W_{t+1} \quad = \quad (\hat{c}_{t+1}, \delta\tilde{c}_{t+1})$$

Transition function – The transition function for the forecasts evolves according to

$$\tilde{c}_{t+1,ij} = \tilde{c}_{t,ij} + \delta\tilde{c}_{t+1,ij}. \tag{13.10}$$

We update the physical state R_t using

$$R_{t+1} \quad = \quad \{j|x_{t,R_t,j} = 1\}. \tag{13.11}$$

In other words, if we are at node $i = R_t$ and we make the decision $x_{tij} = 1$ (which requires that we be at node i, since otherwise $x_{tij} = 0$), then $R_{t+1} = j$. Equations (13.10) and equation (13.11) make up our transition function:

$$S_{t+1} = S^M(S_t, X^\pi(S_t|\theta), W_{t+1}).$$

Objective function – We now write our objective function as

$$\min_\pi F^\pi(\theta) = \mathbb{E}\left\{ \sum_{t=0}^{T} \sum_{(i,j)\in\mathcal{N}} \hat{c}_{t+1,ij} X^\pi(S_t|\theta)|S_0 \right\}. \tag{13.12}$$

Note that our policy $X^\pi(S_t|\theta)$ is an indicator variable that is 1 if it specifies that the traveler should move over link (i, j) at time t, incurring the cost $\hat{c}_{t+1,ij}$.

Designing policies – There will always be some academic interest in solving the stochastic shortest path problem that we just sketched, but we are not aware

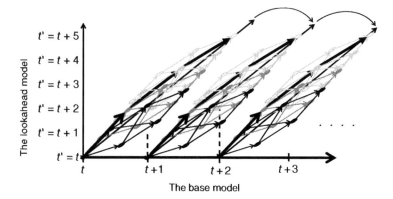

Figure 13.2 Illustration of rolling solution of deterministic shortest path problems using costs $\tilde{c}_t^\pi(\theta)$.

of any practical algorithms for solving, even approximately, the full dynamic shortest path problem that recognizes that the state variable captures the state of the entire graph.

For now, however, we are focusing on simple, practical solutions. The reality is that deterministic shortest path problems are exceptionally easy to solve (see our discussion in section 2.3.3). What we are going to propose is that instead of solving a deterministic shortest path problem using the estimates $\bar{c}_t$, we are going to use the θ-percentile of the distribution for each link. Let

$$\tilde{c}_{t,ij}^\pi(\theta) \quad = \quad \text{the } \theta\text{-percentile of the travel time for link } (i, j) \text{ given}$$
$$\text{our estimate at time } t.$$

We are going to solve a deterministic shortest path problem (as before), but using these modified link costs. Let $X^\pi(S_t|\theta)$ be the policy for choosing the next link based on solving the shortest path with these modified link costs.

Figure 13.2 demonstrates the process of solving shortest paths (which we illustrated in Figure 13.1) on a rolling basis. Each time we look ahead, we solve a deterministic shortest path using the θ-percentile costs $\tilde{c}_t^\pi(\theta)$. The solution to the shortest path problem at time t, when we are at a node i, simply tells us which node j to traverse to. By the time that we arrive at node j, the costs $\tilde{c}_t^\pi(\theta)$ would be updated, and we repeat the process.

All that remains is choosing θ. We do this by simulating our policy as we have done in the past where we have to estimate $F^\pi(\theta)$ in equation (13.12). Here we just have to apply our usual tools for stochastic search, recognizing that θ is a scalar, which means we just have a one-dimensional search. This would be fairly easy without the potentially high level of noise in the policy simulations.

13.2.4 Dynamic Trading Policy

We are going to describe a dynamic trading policy for determining which financial instruments to purchase that uses forecasts of stochastic prices that incorporate additional industrial statistics. The policy needs to balance risk with expected asset performance.

We briefly present a model of the problem using our standard framework. Of particular interest, however, is the policy that we suggest at the end that uses a modified objective function.

State variables – We represent the assets we may purchase using

$$\mathcal{J} \quad = \quad \text{the set of stocks we may hold a position in, with } i = 0 \text{ referring to cash,}$$

$$R_{ti} \quad = \quad \text{our position (in shares) in a particular stock } i \in \mathcal{J}, \text{ where } R_{ti} \text{ can be either positive (for a long position) or negative (for a short position), and where } R_{t,0} \text{ is the amount in cash,}$$

$$R_t \quad = \quad (R_{ti})_{i \in \mathcal{J}}.$$

Other information variables are

$$p_{ti} \quad = \quad \text{the price of stock } i,$$

$$p_t \quad = \quad (p_{ti})_{i \in \mathcal{J}},$$

$$f_{tt'i} \quad = \quad \text{the forecast, generated at time } t, \text{ of the price of stock } i \text{ at time } t' \text{ over a horizon } t' = t, \dots, t + H,$$

$$f_t \quad = \quad (f_{tt'i})_{i \in \mathcal{J}, t' = t, \dots, t+H}.$$

Our state variable is then

$$S_t = (R_t, p_t, f_t).$$

Decision variables – The decision variable is

$$x_{ti} \quad = \quad \text{the number of shares that we trade for each of the stocks. We use } x_{ti} > 0 \text{ to represent the number of shares we buy for stock } i, \text{ and } x_{ti} < 0 \text{ to represent a selling decision.}$$

The decision is constrained by the requirement that we have enough cash on hand to finance the purchasing decisions, given by

$$\sum_{i=1}^{M} x_{ti} p_{ti} \leq R_{t,0}.$$

We let $X^\pi(S_t | \theta)$ be the policy that determines x_t which satisfies this constraint.

Exogenous information – The exogenous information includes both the change in price and the change in forecasts given by

$$\hat{p}_{t+1,i} = \text{the change in the price of stock } i \text{ between } t \text{ and } t+1,$$

$$\hat{p}_t = (\hat{p}_{t+1,i})_{i \in \mathcal{I}}.$$

For the forecasts, the new information is contained in the new forecasts $f_{t+1,t',i}$. We would then write our exogenous information W_{t+1} as

$$W_{t+1} = (\hat{p}_{t+1}, f_{t+1}).$$

To simulate our process, we need to assume a probability model for $\hat{p}_{t+1,i}$. A simple model would be to assume that $\hat{p}_{t+1,i}$ is normally distributed with mean 0 and variance σ_i^2. Modeling these stochastic processes is important and can be quite challenging, but our interest right now is on the design of the policy.

Transition function – The transition equation for the position in a stock R_{ti} is given by

$$R_{t+1,i} = R_{ti} + x_{ti}. \tag{13.13}$$

The transition equation for the cash position $R_{t,0}$ is given by

$$R_{t+1,0} = R_{t0} - \sum_{i=1}^{M} x_{ti} p_{ti}. \tag{13.14}$$

The transition function for the price p_t would be given by

$$p_{t+1,i} = p_{ti} + \hat{p}_{t+1,i}. \tag{13.15}$$

Also, since the new forecasts are contained in the exogenous information, we can combine equations (13.13), (13.14), and (13.15) as

$$S_{t+1} = S^M(S_t, X^\pi(S_t | \theta), W_{t+1}), \tag{13.16}$$

where $X^\pi(S_t | \theta)$ denotes a policy that maps a state to a decision.

Objective function – Our single-period contribution function is given by

$$c^{trans} = \text{the transaction cost per dollar.}$$

The transaction cost per period is given by

$$C_t(S_t, x_t) = -c^{trans} \sum_{i=1}^{M} |x_{ti}| p_{ti}, \text{ for } t = 0, \dots, T-1,$$

where $|x_{ti}|$ is the absolute value of x_{ti}, which gives us the quantity of the trade (it does not matter whether we are buying or selling).

At the end of the day, we evaluate our risk using the quadratic function

$$\rho(R_T) = R_T' \Sigma R_T, \tag{13.17}$$

where Σ denotes the covariance matrix of the returns, which we assume we have estimated from historical data in advance. The final-period contribution function is then given by

$$C_T(S_T, x_T) = R_{T0} + \sum_{i=1}^{M} R_{Ti} p_{Ti} - \rho(R_T).$$

The objective function can now be written

$$\max_{\pi} \mathbb{E} \left\{ \sum_{t=0}^{T} C_t(S_t, X_t^{\pi}(S_t)) \Big| S_0 \right\}. \tag{13.18}$$

In practice, the expectation is approximated by using historical prices, which avoids the need to develop an underlying stochastic model.

Designing policies – We propose the following policy

$$X_t^{\pi}(S_t|\theta) = \arg\max_{x_t} \left(\sum_{i=1}^{M} \left((R_{ti} + x_{ti})(\tilde{f}_{ti}(\theta) - p_{ti}) - c^{trans}|x_{ti}|p_{ti}) - \rho(R_t + x_t) \right), \tag{13.19}$$

where $\tilde{f}_{ti}(\theta) = \sum_{s=1}^{H} \theta_s f_{t,t+s,i}$ represents an overall prediction of the future price using all available forecasts with different horizons and a tunable parameter vector $\theta = (\theta_1, \dots, \theta_H)$. This policy maximizes a utility function that balances the trade-off between return and risk. It can be seen that for the risk function (13.17), the policy can be computed efficiently by solving a convex optimization problem.

A popular approach for tuning policies in financial trading settings is to use historical prices, otherwise known as "back-testing." It is possible to tune the policy on a single, long series of prices pulled from history. As always, the danger is that the policy adapts to the vagaries of a particular price sequence from history that may not be replicated in the future. However, using a historical set of prices avoids the modeling approximations inherent in any mathematical model.

13.2.5 Discussion

Care has to be used if you want to use a stochastic gradient method for optimizing cost-modified CFAs since the objective function $F(\theta)$ (see equation 13.2) is generally not going to be differentiable with respect to θ. Small changes in θ

may produce sudden jumps, with intervals where there is no change at all. However, the expectation does help to smooth surfaces, so it all boils down to trying different methods to see which works the best.

13.3   Constraint-Modified CFAs

A particularly powerful approach to CFAs is to modify the constraints, since this provides the analyst with direct control over the solution. It helps if there is some intuition how uncertainty is likely to affect the final solution. While this is not always the case, it often is, and the idea of parametrically modifying constraints makes it possible to build this understanding into our solution.

The examples given here provide some illustrations:

■ **EXAMPLE 13.1**

Airlines routinely use deterministic scheduling models to plan the movements of aircraft. Such models have to be designed to represent the travel times between cities, which can be highly uncertain. To handle this, the airline uses travel times equal to the θ-percentile of the travel time distribution between each pair of cities (there may be different values of θ for different types of markets).

■ **EXAMPLE 13.2**

A retailer has to manage inventories for a long supply chain extending from the far East to North America. Uncertainties in production and shipping require that the retailer maintain buffer stocks. Let θ be the amount of buffer stock planned in the future (inventory is allowed to go to zero at the last minute), which enters the model through the constraints.

■ **EXAMPLE 13.3**

Independent system operators (ISOs) for the power grid have to plan how much energy to generate tomorrow based on a forecast of loads, as well as energy to be generated from wind and solar. They use a forecast factored by a vector θ with elements for each type of forecast.

We begin our discussion by describing how a set of linear constraints can be modified. We then present a study of a realistic, time-dependent energy storage problem in the presence of rolling forecasts of energy from wind.

13.3.1 General Formulation of Constraint-Modified CFAs

Constraint-modified CFAs can be written in the form

$$X^{Con-CFA}(S_t|\theta) = \arg \max_{x_t \in \mathcal{X}_t^\pi(\theta)} C(S_t, x_t), \tag{13.20}$$

where we are using a modified feasible region $\mathcal{X}_t^\pi(\theta)$ defined by

$$A_t^\pi(\theta^a)\tilde{x}_t \; = \; \theta^b \otimes b_t + \theta^c, \tag{13.21}$$

$$\tilde{x}_t \; \leq \; u_t - \theta^u, \tag{13.22}$$

$$\tilde{x}_t \; \geq \; 0 + \theta^\ell. \tag{13.23}$$

where $\theta^b \otimes b_t$ is the element by element product of the vector b with the similarly dimensioned vector of coefficients θ^b, plus a shift vector θ^c. The parameterization of the matrix $A_t^\pi(\theta^a)$ is how we would insert schedule slack for travel times, as well as any other adjustments that seem appropriate to the application. We then reduce the upper bounds u_t by a shift vector θ^u, and possibly raise the lower bounds by θ^ℓ. Our constraints are now parameterized by the (possibly high-dimensional) vector $\theta = (\theta^a, \theta^b, \theta^c, \theta^\ell, \theta^u)$.

The structure of the modified set of constraints hints at how we can expect to scale the vector θ. If our deterministic model closely matches what actually happens, then we would expect that $\theta^b \approx 1$, while $\theta^c, \theta^u, \theta^\ell \approx 0$. As uncertainty increases, we would expect θ^b to move away from 1 (but not too far), while we might expect $\theta^u, \theta^\ell \leq u_t$ while $\theta^c \leq b_t$. When you start doing stochastic search, you will appreciate that this type of scaling information is extremely valuable.

13.3.2 A Blood Management Problem

In section 8.3.2 we described a blood management problem where we have to manage eight types of blood, which can be only held for five weeks (the model works in one-week increments). Figure 13.3 provides all the ways that blood types can be substituted. Note that O-negative blood can be used for any blood type (this is the universal donor), but the supplies of O-negative do not come close to covering the entire demand for blood.

Our challenge is deciding which blood type to use for each patient, given the random demands for blood in the future. A mathematical model of this problem was already provided in section 8.3.2. Here, we provide an illustration of the model in Figure 13.4 which shows two time periods of a dynamic network, where all the different demands have been aggregated purely to streamline the graph and highlight decisions to use blood (if allowed) or to hold blood, where we have to keep track of aging. If we had perfect forecasts of blood demands, this would be a simple, time-dependent linear program.

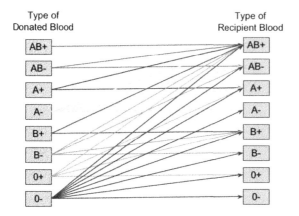

Figure 13.3 Allowable substitutions of different blood types.

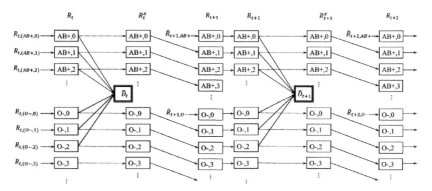

Figure 13.4 Multiperiod model of blood management, focusing on holding leftover blood over time.

We may assume that we will solve this problem once each week, using the forecasts for blood demands given by

$$f^D_{t,t'b} = \text{forecast for the demand for blood with attribute vector } b$$
$$\text{made at time } t, \text{ to serve a demand at time } t'.$$

If we use point forecasts (that is, assume that our forecasts $f^D_{tt'}$ are perfect), then we have a deterministic lookahead, just as we used for our dynamic shortest path problem in section 13.2.3. With the dynamic shortest path problem, we offered a solution for handling uncertainty in travel times by modifying the costs, using θ–percentiles instead of the means, which is a form of modified objective function.

With our blood management problem, ignoring the uncertainty in the forecasts might produce a solution where we use our entire inventory of O-negative blood. Intuition would say that we want to conserve our O-negative blood because it can be used to serve any form of random demand. One way to do this would be to inflate the demand for O-negative blood, which would encourage the model to maintain reserves of O-negative. To estimate the inflation, we might aggregate all the other blood types, and then take the difference between the mean and the $\theta-$percentile of the aggregate demand for the other blood types. This difference could then be added to the O-negative forecast.

With this modification, let $X_t^\pi(S_t|\theta)$ be the solution of how to allocate blood supplies at time t, given the modified demand for O-negative blood. We have to tune θ, ideally using a simulator, although it is not out of the question to experiment in the field (using, of course, a cumulative-reward objective).

13.3.3 An Energy Storage Example with Rolling Forecasts

Consider a general energy storage system depicted in Figure 13.5 which consists of energy from a wind farm, energy from the grid, a battery storage, and a load which could be a building, a university campus, or an entire city. The flows of energy have to be managed to meet a fairly consistent, if noisy, demand that depends on time of day (Figure 13.6(a)), which has to be planned in the presence of rolling forecasts of the energy from wind (Figure 13.6(b)). The demand follows familiar daily patterns, but the wind does not. In addition, the wind forecasts are not very accurate, and change quickly as the forecasts are updated.

We present our model in the usual five components: state variables, decision variables, exogenous information variables, transition function, and the objective function. We note that understanding the details of the model is not important. After presenting the model, we are going to present a policy that uses a deterministic lookahead which depends on forecasts of energy from the wind farm, as well as the demand for energy over the course of the day. We are going to parameterize these forecasts as a way of handling the uncertainty in the forecasts.

State variables – The planning of the system has to respond to the following information that is evolving over time:

$$D_t \quad = \quad \text{Demand (“load”) for power during hour } t.$$

$$E_t \quad = \quad \text{Energy generated from renewables (wind/solar) during hour } t.$$

$$R_t \quad = \quad \text{Amount of energy stored in the battery at time } t.$$

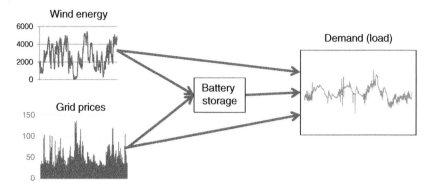

Figure 13.5 Energy storage system, including a renewable source (wind), energy from the grid at real-time prices, battery storage, and a load.

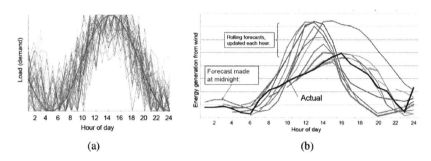

Figure 13.6 (a) Energy load by hour of day and (b) rolling forecast, updated hourly.

u_t = Limit on how much generation can be transmitted at time t (this is known in advance).

p_t = Price to be paid for energy drawn from the grid at time t.

We have access to rolling forecasts of the demand D_t and the energy from wind E_t, given by:

$f^D_{tt'}$ = Forecast of $D_{t'}$ made at time t.

$f^E_{tt'}$ = Forecast of $E_{t'}$ made at time t.

These variables make up our state variable:

$$S_t = (R_t, (f^D_{tt'})_{t' \geq t}, (f^E_{tt'})_{t' \geq t}).$$

Decision variables: – These are the flows between each of the elements of our energy system:

$$x_t = \text{Planned generation of energy during hour } t \text{ which consists of the following elements:}$$

$$x_t^{ED} = \text{flow of energy from wind to demand,}$$

$$x_t^{EB} = \text{flow of energy from wind to battery,}$$

$$x_t^{GD} = \text{flow of energy from grid to demand,}$$

$$x_t^{GB} = \text{flow of energy from grid to battery,}$$

$$x_t^{BD} = \text{flow of energy from battery to demand.}$$

We would normally write out the constraints that these flows have to satisfy. These consist of the flow conservation constraints, as well as upper bounds due to transmission constraints, as well as nonnegativity constraints on all the variables except x_t^{GB} since energy is allowed to flow both ways between the grid and the battery. For compactness, we are going to represent the constraints using

$$A_t x_t = R_t,$$
$$x_t \leq u_t,$$
$$x_t \geq 0.$$

Exogenous information – For the variables with forecasts (demand and wind energy), the exogenous information is the change in the forecast, or the deviation between forecast and actual:

$$\varepsilon_{t+1,\tau}^D = \text{Change in the forecast of demand (for } \tau > 1 \text{ periods in the future) that we first learn at time } t + 1, \text{ or the deviation between actual and forecast (for } \tau = 1).$$

$$\varepsilon_{t+1,\tau}^E = \text{Change in the forecast of wind energy (for } \tau > 1 \text{ periods in the future) that we first learn at time } t + 1, \text{ or the deviation between actual and forecast (for } \tau = 1).$$

We assume that prices evolve purely exogenously with deviations:

$$\hat{p}_{t+1} = \text{Change in grid prices between } t \text{ and } t + 1.$$

Our exogenous information is then

$$W_{t+1} = ((\varepsilon_{t+1,\tau}^D, \varepsilon_{t+1,\tau}^E)_{\tau \geq 1}, \hat{p}_{t+1}).$$

Transition function – The variables that evolve exogenously are

$$
\begin{aligned}
f^D_{t+1,t'} &= f^D_{tt'} + \varepsilon^D_{t+1,t'-t-1}, \quad t' = t+2, \dots, \\
D_{t+1} &= f^D_{t+1,t'} + \varepsilon^D_{t+1,1}, \\
f^E_{t+1,t'} &= f^E_{tt'} + \varepsilon^E_{t+1,t'-t-1}, \quad t' = t+2, \dots, \\
E_{t+1} &= f^E_{t+1,t'} + \varepsilon^E_{t+1,1}, \\
p_{t+1} &= p_t + \hat{p}_{t+1}.
\end{aligned}
$$

The energy in storage evolves according to

$$
R_{t+1,t'} = R_{tt'} + x^{EB}_{tt'} + x^{GB}_{tt'} - x^{BD}_{tt'}.
$$

The estimate $\tilde{R}_{t+1,t+1}$ becomes the actual energy in the battery as of time $t+1$, while $\tilde{R}_{t+1,t'}$ for $t' \geq t+2$ are projections that may change. These equations make up our transition function $S_{t+1} = S^M(S_t, x_t, W_{t+1})$.

Objective function – Our single-period contribution function is

$$
C(S_t, x_t) = p_t \left(x^{GB}_t + x^{GD}_t \right).
$$

Our objective function, then, would be

$$
\max_{\pi} F^\pi(\theta) = \mathbb{E} \left\{ \sum_{t=0}^{T} C(S_t, X^\pi(S_t|\theta)) | S_0 \right\}. \tag{13.24}
$$

As in the past, we can estimate this objective function by simulating our policy, which we present next.

Designing the policy – Given the complex interactions of time-dependent demands, time-varying energy from wind, and the constraints on transmission, we are going to develop a deterministic lookahead model (a form of DLA). Although we do not deal with DLAs in depth until chapter 19, a deterministic lookahead is fairly simple, and we are going to show how to parameterize the policy to handle the uncertainty in the forecasts.

We distinguish the decision we make at time t, x_t, and the *planned decisions* we make at time t over our planning horizon, which we indicate by $\tilde{x}_{tt'}$. Our planned decisions are given by

$$\tilde{x}_{tt'} \quad = \quad \text{planned generation of energy during hour } t' > t, \text{ where the plan is made at time } t, \text{ which is comprised of the following elements:}$$

$$\tilde{x}_{tt'}^{ED} \quad = \quad \text{flow of energy from renewables to demand,}$$

$$\tilde{x}_{tt'}^{EB} \quad = \quad \text{flow of energy from renewables to battery,}$$

$$\tilde{x}_{tt'}^{GD} \quad = \quad \text{flow of energy from grid to demand,}$$

$$\tilde{x}_{tt'}^{GB} \quad = \quad \text{flow of energy from grid to battery,}$$

$$\tilde{x}_{tt'}^{BD} \quad = \quad \text{flow of energy from battery to demand.}$$

We have to create projections of the energy in the battery over the horizon $t' > t$:

$$\tilde{R}_{t+1,t'} = \tilde{R}_{tt'} + \tilde{x}_{tt'}^{EB} + \tilde{x}_{tt'}^{GB} - \tilde{x}_{tt'}^{BD}.$$

The estimate $\tilde{R}_{t+1,t+1}$ becomes the actual energy in the battery as of time $t + 1$, while $\tilde{R}_{t+1,t'}$ for $t' \geq t + 2$ are projections that may change.

Our policy, then, is to optimize deterministically using point forecasts over a planning horizon $t, t + 1, \dots, t + H$:

$$X^{DLA}(S_t) = \arg\max_{x_t,(\tilde{x}_{tt'},t'=t+1,\dots,t+H)} \left(p_t(x_t^{GB} + x_t^{GD}) + \sum_{t'=t+1}^{t+H} \tilde{p}_{tt'}(\tilde{x}_{tt'}^{GB} + \tilde{x}_{tt'}^{GD}) \right) \tag{13.25}$$

subject to the following constraints: First, for time t we have

$$x_t^{BD} - x_t^{GB} - x_t^{EB} \quad \leq \quad R_t, \tag{13.26}$$

$$\tilde{R}_{t,t+1} - (x_t^{GB} + x_t^{EB} - x_t^{BD}) \quad = \quad R_t, \tag{13.27}$$

$$x_t^{ED} + x_t^{BD} + x_t^{GD} \quad = \quad D_t, \tag{13.28}$$

$$x_t^{EB} + x_t^{ED} \quad \leq \quad E_t, \tag{13.29}$$

$$x_t^{GD}, x_t^{EB}, x_t^{ED}, x_t^{BD} \quad \geq \quad 0. \tag{13.30}$$

Then, for $t' = t + 1, \dots, t + H$ we have

$$\tilde{x}_{tt'}^{BD} - \tilde{x}_{tt'}^{GB} - \tilde{x}_{tt'}^{EB} \quad \leq \quad \tilde{R}_{tt'}, \tag{13.31}$$

$$\tilde{R}_{t,t'+1} - (\tilde{x}_{tt'}^{GB} + \tilde{x}_{tt'}^{EB} - \tilde{x}_{tt'}^{BD}) \quad = \quad \tilde{R}_{tt'}, \tag{13.32}$$

$$\tilde{x}_{tt'}^{ED} + \tilde{x}_{tt'}^{BD} + \tilde{x}_{tt'}^{GD} \quad = \quad f_{tt'}^{D}, \tag{13.33}$$

$$\tilde{x}_{tt'}^{EB} + \tilde{x}_{tt'}^{ED} \quad \leq \quad f_{tt'}^{E}. \tag{13.34}$$

We are now going to focus on equations (13.33) and (13.34) since both depend on forecasts which are uncertain. In chapter 19 we are going to propose a general approach for creating lookahead policies that capture uncertainty. Here, we are going to do something simple (and very practical), which may even outperform the more complicated lookahead strategies we will describe later.

Our parameterized policy replaces equations (13.33) and (13.34) with

$$\tilde{x}_{tt'}^{ED} + \tilde{x}_{tt'}^{BD} + \tilde{x}_{tt'}^{GD} = \theta_{t'-t}^{D} f_{tt'}^{D}, \tag{13.35}$$

$$\tilde{x}_{tt'}^{EB} + \tilde{x}_{tt'}^{ED} \leq \theta_{t'-t}^{E} f_{tt'}^{E}. \tag{13.36}$$

Now let $X_t^{CFA}(S_t|\theta)$ be the policy that solves the optimization problem in (13.25) subject to the constraints (13.31)–(13.32) and (13.35)–(13.36). We have introduced the parameters $\theta = (\theta_\tau^E, \theta_\tau^D), \tau = 1, 2, \dots, H$ as a form of "discount factor" on the forecasts f_t^D and f_t^E.

We now face the problem of tuning θ, which means optimizing $F^\pi(\theta)$ in (13.24). For this we draw on our foundation of stochastic search. For this problem, we used the SPSA algorithm described in section 12.5 (see section 5.4.4 for a more detailed description) because it is well suited to handling multidimensional problems (θ has two 23-dimensional vectors).

We will not repeat any of the algorithmic steps (they have already been covered), but we share the following experiences with the numerical work:

- Simulations of the policy are relatively fast, requiring solving 24 relatively small linear programs (allowing us to perform an entire simulation in just a few seconds).
- Simulations of the policy are *very* noisy. It is necessary to average 1000 repetitions to get a reasonable estimate of the function (but always use whatever parallel computing capabilities you have available).
- This does not mean that we need to use a mini-batch with 1000 simulations in the SPSA calculation, but we did need mini-batches on the order of 20 to 40, which means we needed 40 to 80 function evaluations for each gradient.
- Do not forget the need to tune your stepsize formula (we used RMSProp from chapter 6). The tuning matters, and the tuning even depends on your choice of starting point.
- The problem is highly time-dependent, but our parameterized lookahead policy is completely stationary. For example, θ_τ depends on how many time periods into the future we are forecasting, but does not depend on the time t at which we are making the decision. This is the value of imbedding the forecast within the policy.

A nice property of the policy is that if the forecasts are perfect, then the optimal solution should be $\theta^* = 1$. Figure 13.7(a) tests this idea for a problem with perfect forecasts by setting $\theta_\tau = 1$ for all τ and then varying each θ_τ individually. The graph shows that $\theta_\tau^* = 1$ for each value of τ.

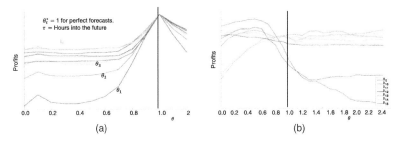

Figure 13.7 Objective vs. θ_τ for (a) perfect forecasts and (b) stochastic forecasts.

We ran the SPSA algorithm for a problem with imperfect (in fact, highly imperfect) forecasts. We then fixed θ_τ to the values produced by the SPSA algorithm, and repeated the exercise of varying θ_τ for individual values of τ. The results are shown in Figure 13.7(b), which shows that the optimum values have now moved well away from 1.0.

When doing stochastic search with any algorithm (derivative-based or derivative-free) is that it helps to understand the behavior of the surface $\mathbb{E}F^\pi(\theta, W)$. While the one-dimensional plots in Figure 13.7 hint at the behavior of the surface (for example, the function appears to have a single optimum in each dimension), but seeing the function in higher dimensions contributes to our understanding.

Figure 13.8 shows four sets of two-dimensional heatmaps, where darker red reflects higher values. Each heatmap shows the two values of θ_τ between 0 and 2, so the center is $\theta_\tau = 1$, which was optimal for the deterministic problem. Note the ridges in 13.8(a) and (b), which will cause problems for a gradient-based algorithm. These ridges would also create challenges for derivative-free search methods.

Figure 13.9 shows how much the profits improved by optimizing θ using the SPSA algorithm compared to the performance using $\theta = 1$. The runs were performed for different starting points θ^0, drawn from four different regions:

(1) The first region started from $\theta^0 = 1$.
(2) The second region was $\theta^0 \in [0, 1]$.
(3) The third region was $\theta^0 \in [.5, 1.5]$.
(4) The fourth region was $\theta^0 \in [1.0, 2.0]$.

We can draw several conclusions from this graph:

- The optimized CFA outperforms the basic deterministic lookahead (with $\theta = 1$) by 20 to 50 percent, which we consider significant.

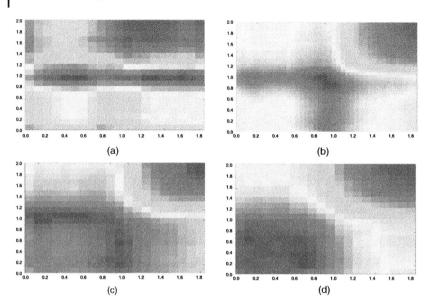

Figure 13.8 2-d heatmaps of the objective function for four different pairs of (θ_i, θ_j). Each dimension of each plot ranges from 0 to 2.

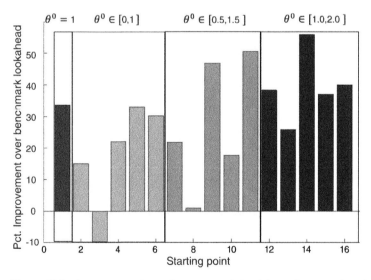

Figure 13.9 Improvement in profits using optimized θ over base results with $\theta = 1$, using starting point θ^0 drawn from each of four ranges: $\theta^0 = 1$, $\theta^0 \in [0,1]$, $\theta^0 \in [.5, 1.5]$, and $\theta^0 \in [1,2]$.

- The performance can vary widely as we randomize the starting points. However, starting with $\theta^0 = 1$ produced results that are comparable to or better than 12 out of 15 runs, but noticeably underperformed 3 of the runs. It is very nice to have a natural starting point, but more experiments are needed to understand the robustness of the optimized solutions.

- Not shown is the effect of tuning the stepsize policy, which was significant. Stepsize tuned for one starting region $[0, 1]$ but used for another starting region $[1, 2]$ could produce optimized values of θ that underperformed $\theta = 1$.

- The tuning process requires serious algorithmic work, but the resulting policy is no more complicated than a basic deterministic lookahead (that is, with $\theta = 1$).

- This is a highly nonstationary problem, with a dynamic, rolling forecast. However, our parametric CFA policy with an imbedded forecast is stationary (none of the parameters depend on time of day), which is very valuable for a problem which has strong time-of-day behavior. By imbedding the forecast, we turn a highly nonstationary problem into a stationary one that responds immediately to evolving forecasts.

There is one particularly important point about parametric CFAs in general (and parameterized lookaheads in particular):

> Many problems have complex dynamics, such as the presence of rolling forecasts for this energy problem. It is typically impossible to build these dynamics into the policies in the lookahead classes that we cover starting in chapter 14 (but especially chapter 19 on direct lookaheads), but it is quite easy to capture them in the simulation of the base model. For this reason, a carefully designed parametric CFA, tuned using the full base model which captures these dynamics, may outperform a much more complex stochastic lookahead policies that require approximations.

As with any parametric model (in optimization or statistics), there is always the question of the robustness of the model when it is implemented in new environments. These questions remain with both PFAs and CFAs. What seems to be most important is that the deterministic optimization model should capture important structural properties of the policy, which means that the tuning is just helping the policy to handle uncertainty.

13.4 Bibliographic Notes

Section 13.1 – The term "cost function approximation" was first proposed in Powell (2014). Powell and Meisel (2016) compared four classes of policies for an energy storage problem, with one being a simple version of a parameterized optimization model. The first paper to the idea of a CFA formally is Ghadimi et al. (2020). We note that the concept of parameterized optimization models is a widely used industry heuristic, but without the proper statement of an objective function.

Section 13.2 – The dynamic trading policy (section 13.2.4) was described by a graduate student based on his summer internship.

Section 13.2 – The energy storage problem with rolling forecasts (section 13.3.3) was first presented in Powell (2021). The model and algorithmic work was given in Ghadimi et al. (2020).

Exercises

Review questions

13.1 What are the two ways of parameterizing an optimization-based policy?

13.2 The dynamic assignment problem and the dynamic shortest path problem both parameterize the objective function, but motivated by completely different objectives. What are they?

13.3 What is the complete state variable of a dynamic shortest path problem?

Modeling questions

13.4 Using the model from section 8.3.2, write a model for the blood management problem in section 13.3.2 capturing the uncertainty in the forecasts for blood. Section 13.3.2 suggests a simple idea of inflating the demand for O-negative blood to have an adequate reserve in case we run short in our supply of other blood types. Of course, this ignores the ability to substitute across other blood types. You are going to develop a more general model for this problem based on a parameterized lookahead.

 (a) Write out the full, multiperiod model with random demands, including all five dimensions of a dynamic model.
 (b) Now introduce reserves θ_a for *each* blood type and write out this modified lookahead policy.
 (c) Write out the objective function for evaluating this policy.
 (d) The policy you have designed in (b) uses an additive adjustment. Now suggest a multiplicative adjustment as we used in section 13.3.3. How does this change the scaling of θ?

(e) Since our tunable parameter θ is now a vector with eight dimensions, sketch the calculations required to estimate a gradient using the SPSA algorithm.

13.5 The energy storage problem in section 13.3.3 has to manage a highly time-dependent demand (with consistent peaks and valleys), along with rolling forecasts that can exhibit highs and lows at any time of day. Given these characteristics,

(a) What does it mean to say that a "policy is stationary?"
(b) Is the policy defined by (13.25) with constraints (13.26)–(13.32), (13.35), and (13.36) stationary? What allows you to make this determination?
(c) Each element of the vector of parameters θ_τ was found to fall in the range [0,2]. In fact, if the forecasts were perfect then we know that $\theta_\tau = 1$. This is a very nice property. How is it that this CFA policy is so nicely scaled?

Computational exercises

13.6 From the supplementary materials page https://castlelab.princeton.edu/rlso_supplementary/, download the Python module (under Software) for the dynamic assignment problem. This software has modeled the dynamic assignment problem with the θ–percent costs $\bar{c}^\pi_{t,i,j}(\theta)$. Using this software, do the following:

(a) Simulate the performance of the policy using $\theta = 1$. Repeat this 20 times and estimate the mean and standard deviation of the performance of the policy, and report the results. Normally we would use an initial experiment like this to determine how many times we need to run the simulation, but for now we evaluate a policy by averaging across 20 simulations.
(b) Simulate the performance of the θ–percentile policies using $\theta = 0, .2, .4, .6, .8, .9$, and report which value produces the best results.

Theory questions

13.7 Show that the objective function for the policy defined by the optimization problem (13.25)–(13.32), along with the modified constraints (13.35)–(13.36), is concave in θ. Note: this requires a background in linear programming.

13.8   Argue why the performance of the policy $F(\theta)$ produced by simulating the policy $X^{\pi}(S_t|\theta)$ given by (13.25), subject to constraints (13.26)–(13.32), is *not* concave in θ.

Problem solving questions

13.9   You would like to purchase a laptop. Price is a concern, but so is reliability, as well as service. There are some retail chains that offer service on the models they sell. You have found that that you buy laptops every two years. You do some research to develop a sense about reliability, but you will also learn from your own experience. Let

$$\mathcal{J} \ = \ \text{The set of channels you can purchase the laptop from (retail outlets, websites),}$$

$$Q_i \ = \ \text{1 if channel } i \text{ offers repair service, 0 otherwise,}$$

$$\bar{\mu}_{ti} \ = \ \text{estimated probability that the laptop purchased from channel } i \text{ will need service, given the experience as of time } t,$$

$$p_{ti} \ = \ \text{price of a laptop purchased from channel } i \text{ at time } t,$$

$$R_{ti} \ = \ \text{1 if you are holding a laptop purchased from channel } i \text{ as of time } t, \text{ 0 otherwise,}$$

$$z_{ti} \ = \ \text{1 if you purchase a laptop from channel } i \text{ at time } t, \text{ 0 otherwise,}$$

$$\hat{F}_{ti} \ = \ \text{if a laptop purchased from channel } i \text{ needs a repair at time } i.$$

Use this notation to answer the following:

(a) Define the state variable S_t.
(b) Identify the decision variable and exogenous information variable. Create the notation for the policy for making the decision (we will design this).
(c) Give the equations for the transition function. Assume you are going to use exponential smoothing with parameter α to update your estimate of $\bar{\mu}_{ti}$.
(d) You want to minimize how much you spend, and you put a weight ρ^{serv} on the value of purchasing the laptop from a channel that offers service. Finally, you would like to limit the probability of needing service to less than 0.05. Use these guidelines to create an objective function for evaluating your policy.

Diary problem

The diary problem is a single problem you chose (see chapter 1 for guidelines).
Answer the following for your diary problem.

13.10   Do one of the following:

(a) Pick a decision in your problem that lends itself to being made by
solving a deterministic approximation over some horizon. Think
about how uncertainty might affect the quality of this solution,
and what you think should be done differently in the presence of
uncertainty. Try to suggest a parametrization that would make the
deterministic lookahead work better.

(b) Pick a decision in your problem where a myopic optimization is a
reasonable starting point. Now, think about how considering the
downstream impact of the decision might affect the decision you
are making now. Try to introduce a parametrization that would
make the myopic model work better.

Bibliography

Ghadimi, S., Perkins, R., and Powell, W.B. (2020). Reinforcement Learning via
Parametric Cost Function Approximation for Multistage Stochastic
Programming. https://arxiv.org/abs/2001.00831.

Powell, W.B. (2014). Clearing the jungle of stochastic optimization. *Informs
TutORials in Operations Research 2014*.

Powell, W.B. (2021). From reinforcement learning to optimal control: A unified
framework for sequential decisions. *Handbook on Reinforcement Learning and
Optimal Control, Studies in Systems, Decision and Control*, 29–74.

Powell, W.B. and Meisel, S. (2016). Tutorial on stochastic optimization in energy
Part II: An energy storage illustration. *IEEE Transactions on Power Systems*.

Part V – Lookahead Policies

Lookahead policies are based on estimates of the impact of a decision on the future. There are two broad strategies for doing this:

Value function approximations If we are in a state S_t and take an action x_t, then we observe new information W_{t+1} (which is random at time t) which takes us to a new state S_{t+1}, we might be able to approximate the value of being in state S_{t+1}. We can then use this to help us make a better decision x_t now if we can do a good job of approximating the value of being in state.

Direct lookahead approximations Here we explicitly plan decisions now, x_t, and into the future, $x_{t+1}, \dots, x_{t+H}$, to help us make the best decision x_t to implement now. The problem in stochastic models is that the decisions $x_{tt'}$ for $t' > t$ depend on future information, so they are random.

The choice between using value functions versus direct lookaheads boils down to a single equation which gives the optimal policy at time t when we are in state S_t:

$$
X_t^{\pi^*}(S_t) = \arg\max_{x_t \in \mathcal{X}_t} \left(C(S_t, x_t) + \mathbb{E}\left\{ \underbrace{ \max_{\pi \in \Pi} \mathbb{E}\left\{ \sum_{t'=t+1}^{T} C(S_{t'}, X_{t'}^{\pi}(S_{t'})) | S_{t+1} \right\} }_{\text{future contributions}} | S_t, x_t \right\} \right). \quad (13.37)
$$

The challenge is balancing the contributions now, given by $C(S_t, x_t)$, against future contributions. If we could compute the future contributions, this would be an optimal policy. However, computing future contributions in the presence of a (random) sequential information process is almost always computationally intractable.

There are problems where we can create reasonable approximations of the future contributions. When we do this around the post-decision state S_t^x (we can also write this as (S_t, x_t)), this would be called the post-decision value function that we write as $\overline{V}_t^x(S_t^x|\theta)$, and allows us to write our policy as

$$X_t^{VFA}(S_t|\theta) = \arg\max_{x_t \in \mathcal{X}_t}(C(S_t, x_t) + \overline{V}_t^x(S_t^x|\theta)). \tag{13.38}$$

Needless to say, the VFA policy in equation (13.38) looks a lot friendlier than the full DLA policy using equation (13.37). The challenge is creating a reasonably accurate approximation $\overline{V}_t^x(S_t^x|\theta)$ where

$$\overline{V}_t^x(S_t^x|\theta) \approx \mathbb{E}\left\{\max_{\pi \in \Pi} \mathbb{E}\left\{\sum_{t'=t+1}^{T} C(S_{t'}, X_{t'}^{\pi}(S_{t'}))S_{t+1}\right\} |S_t, x_t\right\}.$$

This begs the question: Can we create a sufficiently accurate approximation $\overline{V}_t^x(S_t^x|\theta)$? The answer is ... sometimes. It really depends on the problem.

Policies based on value functions have attracted considerable attention over the years from the academic community. In fact, terms like "dynamic programming" and "optimal control" are basically synonymous with value functions (or cost-to-go functions, as they are known in control theory). There are very small classes of problems where these can be computed exactly, hence the interest in fields that go by names like "approximate dynamic programming," "adaptive dynamic programming," or "reinforcement learning," although reinforcement learning has evolved to refer to an entire spectrum of policies that span, in the language of this book, all four classes of policies.

There is a wide range of strategies for approximating value functions which we have reviewed in chapter 3, all with their own strengths and weaknesses. The richness of these strategies explains why our coverage of VFA policies spans the following chapters:

Chapter 14: Exact dynamic programming – This chapter focus on a handful of sequential decision problems that can be solved exactly, which is to say, we can find provably optimal policies. Most of this presentation is centered on a field known as discrete Markov decision processes, which originated in the 1950s, and focuses on problems where there is a (not too large) set of discrete states, a (not too large) set of discrete actions, and random information W_{t+1} which allows us to take expectations. If these conditions are satisfied, these problems can be solved using a strategy that involves stepping backward through time computing the value of being in each state (this is often known as "backward dynamic programming"). The theory is very elegant, but it is rarely computable. However, the ideas lay the foundation for a variety of approximation strategies. We also touch on a special problem in optimal control called linear quadratic regulation which is a foundational result of the very large field of optimal control, with many applications in control of robots and aircraft.

Chapter 15: Backward approximate dynamic programming – This chapter describes how to do backward dynamic programming approximately for multidimensional (and even continuous) states, multidimensional (and even continuous) decisions, and complex, multidimensional exogenous information processes.

Chapter 16: Forward ADP I: The value of a policy – This chapter describes the fundamentals of approximating value functions for a fixed policy using forward methods, where we simulate forward in time. Forward methods create a natural mechanism for sampling states (pay attention to how we sample states in chapter 15).

Chapter 17: Forward ADP II: Policy optimization – This chapter extends the previous one by showing how to simultaneously learn and optimize over policies. The interaction between learning a value function while also searching for policies introduces a significant level of complexity that explains why this field is so rich.

Chapter 18: Forward ADP III: Convex functions – This chapter adapts the forward ADP methods to the context of convex problems, specifically motivated by resource allocation problems, which represents a massive problem class. Convexity (concavity when maximizing) makes it possible for us to handle very high-dimensional problems.

By contrast, we have a single chapter, chapter 19, on direct lookahead (DLA) policies for solving equation (13.37). Our core strategy for solving equation (13.37) will be to replace the base model with an approximate lookahead model that is easier to solve, while continuing to capture the most important elements of the problem.

Our approximate lookahead model might be deterministic or stochastic. If it is deterministic, we are going to assume that algorithms are available for solving the lookahead model. If it is stochastic, then we are faced with solving a stochastic optimization within the policy for our stochastic optimization problem, albeit a simplified one. Entire fields have been dedicated to specific strategies for approximating and solving lookahead models, but these methods basically draw on all the tools of the rest of the book. For this reason, chapter 19 focuses more on strategies for creating the lookahead model, since the entire rest of the book covers the methods for solving the lookahead model.

A brief history of approximate dynamic programming

Since the 1950s the standard approach for solving sequential decision problems (dynamic programs, optimal control problems) has been to start by stating Bellman's equation (or equivalently, the Hamilton-Jacobi equation) which characterizes an optimal policy. However, almost invariably these cannot actually be computed, so the natural approach has been to solve these equations approximately. By now, as you can see from our presentation in chapter 11, and then the discussion of PFAs and CFAs in chapters 12 and 13, we feel a more balanced perspective is needed.

Approximate dynamic programming has a long history of re-invention by different communities. The first attempt was in 1959 by Bellman himself when he realized that his use of discrete states would explode when there were multiple state variables, a behavior that became widely known as the "curse of dimensionality." Computational work in the core Markov decision process community largely died at that point, with subsequent work focusing more on the theory that is summarized in chapter 14.

In 1974, Paul Werbos showed how to derive estimates of value functions for control problems using a method he called "backpropagation," which initiated a long line of research in the controls community, primarily for continuous, deterministic problems, that continues today. In fact it was this community that initiated the use of neural networks for approximating what they called "cost to go" functions (value functions in this book).

Then, in the 1980s, Rich Sutton and his adviser Andy Barto were experimenting with learning algorithms in psychology, using the setting of describing how a mouse would learn to navigate a maze. Psychology has a long history, dating to 1897 with the research by Ivan Pavlov into training dogs to associate a particular signal, or cue (in this case ringing a bell), to elicit a response (salivating) at which point the dog would receive a treat. Through repeated trials, the dog could be trained to associate the ringing of a bell with receiving a treat that would cause the dog to salivate. The repeated trials reinforced the relationship between the bell and receiving a treat (and then salivating). This became known as "cue learning" (where the ringing of the bell is the "cue"), and the process of associating the cue with the reward became known as "reinforcement learning," terms that became popular in the 1940s and 1950s.

Sutton and Barto applied this same idea in the context of a maze, where a reward is not received until the mouse learns to find a path to a particular exit where there is a reward. As a result, an action does not immediately return a reward; instead, it just takes the mouse to a downstream state, which may eventually lead to a reward. This means the value of the action (turning left or right) depends on the state. They designed an algorithm that would, through many repetitions, learn "Q" factors, where $Q(s, a)$ is the value of taking an action a

when the mouse is in state s. The algorithm has two basic steps:

$$\hat{q}^{n+1}(s^n, a^n) = r(s^n, a^n) + \lambda \max_{a'} \bar{Q}^n(s^{n+1}, a'), \tag{13.39}$$

$$\bar{Q}^{n+1}(s^n, a^n) = (1 - \alpha_n)\bar{Q}^n(s^n, a^n) + \alpha_n \hat{q}^{n+1}(s^n, a^n). \tag{13.40}$$

The variables s^n and a^n are a current state and action (chosen according to rules that have to be designed). λ plays the role of a discount factor, but this has nothing to do with the time value of money. s^{n+1} is either observed from a physical system, or simulated from a known transition function given s^n and a^n. α_n is known variously as a stepsize or learning rate.

Equations (13.39) and (13.40) can be rewritten

$$\bar{Q}^{n+1}(s^n, a^n) = \bar{Q}^n(s^n, a^n) + \alpha_n(r(s^n, a^n)$$
$$+ \lambda \max_{a'} \bar{Q}^n(s^{n+1}, a') - -\bar{Q}^n(s^n, a^n)). \tag{13.41}$$

The quantity

$$(r(s^n, a^n) + \lambda \max_{a'} \bar{Q}^n(s^{n+1}, a') - -\bar{Q}^n(s^n, a^n))$$

became known in the reinforcement learning literature as a "temporal difference" with parameter λ, and as a result the update became known as "TD(λ)" (pronounced tee-dee-lambda). The parameter λ looks like a discount factor, but it is an *algorithmic discount factor* which has nothing to do with the time value of money (we use γ for this purpose).

At some point in the 1980s the connection between equations (13.39) and (13.40) and the field of discrete Markov decision processes was made, but it was not until 1992 that John Tsitsiklis bridged the updating equations (13.39) and (13.40) with the work on stochastic approximation methods (these are the stochastic gradient methods that we covered in chapter 5) that provided the basis for a convergence proof.

Equation (13.41) should look familiar: it is basically a stochastic gradient, with the difference that $\bar{Q}^n(s^{n+1}, a')$ and $\bar{Q}^n(s^n, a^n)$ are biased estimates of the true function. Tsitsiklis extended the theory on stochastic approximation methods to handle this. This work provided the spark for the landmark book *Neuro-Dynamic Programming* by Bertsekas and Tsitsiklis in 1996 which laid the theoretical foundation for convergence theory in the entire field of VFA-based policies.

For a number of years, and to some extent still today, equations (13.39) and (13.40) are most closely associated with the term "reinforcement learning." What this community has found, along with everyone doing research based on approximating value functions, is that value function approximations are effective on a fairly limited set of problems where using machine learning to approximate the value of being in a state produces effective policies (with reasonable effort – an often overlooked issue).

Today, the second edition of Sutton and Barto's highly popular *Reinforcement Learning: An Introduction* includes policies from all four classes of policies that we have introduced in this book. The "discovery" of strategies from the different classes of policies is a pattern that has been repeated across communities that work on sequential decision problems: fields including stochastic search, simulation-optimization, optimal control, and the multi-armed bandit community have all evolved the use of policies from the different classes of policies.

As you take the plunge into the rich set of strategies of approximating value functions, make sure that you have exhausted the simpler PFAs and CFAs, as well as the DLAs that we will cover in chapter 19. Keep in mind that sequential decision problems are ubiquitous, and that all decisions have to be made with some method. Now think about how many decisions are made by solving Bellman's equation (even approximately).

14

Exact Dynamic Programming

There are very specific classes of sequential decision problems that can be solved exactly, producing optimal policies. The most general class of problems fall under the umbrella known as discrete Markov decision processes, which are characterized by a (not too large) set of discrete states $\mathcal{S}$, and a (not too large) set of discrete actions $\mathcal{A}$. We deviate from our standard notation of using x for decisions to acknowledge the long history in this field of using a for action, where a is discrete (it could be an integer, a discretized continuous variable, or a categorical quantity such as color, medical treatment, or product recommendation). This is the notation that has been adopted by the reinforcement learning community.

It turns out that there is a wide range of applications with discrete actions, where the number of actions is "not too large," but the requirement that the state space is "not too large" is far more restrictive in practice. However, despite this limitation (which is severe), the study of this problem class has helped to establish the theory of sequential decision problems, and has laid the foundation for different algorithmic strategies even when the assumption of small state and action spaces does not apply.

The investigation of discrete Markov decision processes attracted a mathematically sophisticated community which has largely defined the work in this field up through the 1990s. A number of the equations in this chapter, while quite elegant (and sometimes quite sophisticated), are not computable for anything other than toy problems. This style sharply contrasts with the entire rest of the book. However, the algorithms in this chapter laid the foundation for entire classes of algorithms described in chapters 15–18 which scale to much larger (and in some cases extremely large) problems. We note that the foundation for this material is laid in sections 14.1–14.3.

Although this chapter is primarily focused on discrete dynamic programs (discrete states and actions), we pause first in section 14.4 to demonstrate how

Reinforcement Learning and Stochastic Optimization: A Unified Framework for Sequential Decisions, First Edition. Warren B. Powell.
© 2022 John Wiley & Sons, Inc. Published 2022 John Wiley & Sons, Inc.

the same equations can be used to solve certain continuous problems analytically. This section should be viewed as an exercise that illustrates the key ideas of sections 14.1–14.3 using a toy problem that can be solved using the same tools and concepts, but without any need for numerical computation. We then close with section 14.11 that presents the foundation of a very large field known as optimal control, where we can find optimal solutions to an important problem class known as linear quadratic regulation which has many applications in engineering.

14.1 Discrete Dynamic Programming

To understand the power of the Markov decision process framework, it is useful to return to the idea of a decision tree, illustrated in Figure 14.1. We enumerate the decisions out of each decision node (squares), and the random outcomes out of each outcome node (circles). If there are 10 possible decisions and 10 possible random outcomes, our tree is 100 times bigger after one sequence of decisions and random information. If we step forward 10 steps (10 decisions followed by

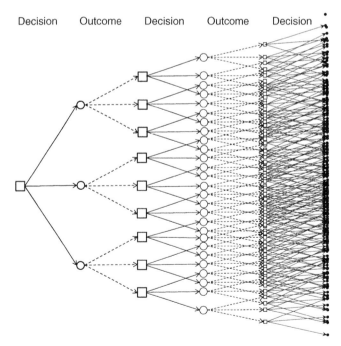

Figure 14.1 Decision tree illustrating the sequence of decisions and new information, illustrating the explosive growth of decision trees.

random information), our tree would have 100^{10} ending nodes. And this is not even a large problem (it is easy to find problems with far larger numbers of actions and outcomes). The explosive growth in the size of the decision trees is illustrated in Figure 14.1, where the number of decisions and outcomes is quite small.

The breakthrough of Markov decision processes (by Richard Bellman in the 1950s) was the recognition that each decision node corresponds to a state of a dynamic system. In the classical representation of a decision tree, decision nodes correspond to the entire history of the process up to that point in time. However, there are many settings where we may not need the entire history to make a decision.

Assume that the relevant information we need to make a decision can be represented by a state s that falls in a discrete set $S = (1, 2, \ldots, |S|)$, where S is small enough to enumerate. For example, S_t might be the number of units of blood in a hospital inventory. In this case, the number of decision nodes does not grow exponentially. Furthermore, we only need to know the inventory, and not the history of how we got there.

When we can exploit this more compact structure, our decision tree collapses into the diagram shown in Figure 14.2, where the number of states in each period is fixed. Note that the number of outcome nodes is potentially quite large (possibly infinite). For example, our random information may be continuous or multidimensional; this would be the second of the three curses of dimensionality we first introduced in section 2.1.3. (For a reminder of how complicated state variables can be, flip back to the energy storage illustrations in section 9.9.)

There are many problems where states are continuous, or the state variable is a vector producing a state space that is far too large to enumerate. In addition, the one-step transition matrix $p_t(S_{t+1}|S_t, a_t)$ can also be difficult or impossible to compute. So why cover material that is widely acknowledged to work only on small or highly specialized problems? There are (at least) four reasons:

(1) Some problems have small state and action spaces and can be solved with these techniques. In fact, it is often the case that the tools of Markov decision processes offers the only path to finding the *optimal* policy.

(2) We can use optimal policies, which are limited to fairly small problems, to evaluate approximation algorithms that can be scaled to larger problems.

(3) The theory of Markov decision processes can be used to identify structural properties that can help us identify properties of optimal policies that we can exploit in policy search algorithms.

(4) This material provides the intellectual foundation for approximation algorithms that can be scaled to far more complex problems, such as optimizing the locomotives for a major railroad, or optimizing a network of hydroelectric reservoirs.

Decision   Outcome   Decision   Outcome   Decision   Outcome

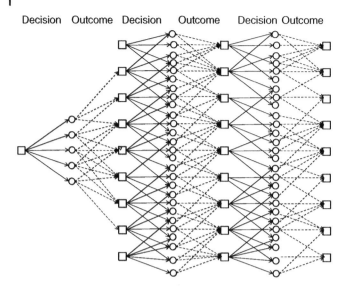

Figure 14.2 Collapsed version of the decision tree, when states do not capture entire history.

As with most of the chapters in the book, the body of this chapter focuses on the algorithms. Some of the elegant theory that has been developed for this field is presented in the "Why does it work" section (section 14.12). The intent is to allow the presentation of results to flow more naturally, but serious students of dynamic programming are encouraged to delve into these proofs, which are quite elegant. This is partly to develop a deeper appreciation of the properties of the problem as well as to develop an understanding of the proof techniques that are used in this field.

14.2 The Optimality Equations

In the last chapter, we illustrated a number of stochastic optimization models that involve solving the following objective function

$$\max_{\pi} \mathbb{E} \left\{ \sum_{t=0}^{T} \gamma^t C(S_t, A_t^{\pi}(S_t)) | S_0 \right\}. \tag{14.1}$$

The most important contribution of the material in this chapter is that it provides a path to optimal policies. In practice, optimal policies are rare, so even with the computational limitations, at least having a framework for characterizing optimal policies is exceptionally valuable.

14.2.1 Bellman's Equations

With a little thought, we realize that we do not have to solve this entire problem at once. Assume that we are solving a deterministic shortest path problem where S_t is the index of the node in the network where we have to make a decision. If we are in state $S_t = i$ (that is, we are at node i in our network) and take action $a_t = j$ (that is, we wish to traverse the link from i to j), our transition function will tell us that we are going to land in some state $S_{t+1} = S^M(S_t, a_t)$ (in this case, node j).

What if we had a function $V_{t+1}(S_{t+1})$ that told us the value of being in state S_{t+1} (giving us the value of the path from node j to the destination)? We could evaluate each possible action a_t and simply choose the action a_t that has the largest one-period contribution, $C_t(S_t, a_t)$, plus the value of landing in state $S_{t+1} = S^M(S_t, a_t)$ which we represent using $V_{t+1}(S_{t+1})$. Since this value represents the money we receive one time period in the future, we might discount this by a factor γ. In other words, we have to solve

$$a_t^*(S_t) = \arg\max_{a_t \in \mathcal{A}_t} \left(C_t(S_t, a_t) + \gamma V_{t+1}(S_{t+1}) \right),$$

where "arg max" means that we want to choose the action a_t that maximizes the expression in parentheses. We also note that S_{t+1} is a function of S_t and a_t, meaning that we could write it as $S_{t+1}(S_t, a_t)$. Both forms are fine. It is common to write S_{t+1} by itself, but the dependence on S_t and a_t needs to be understood.

The value of being in state S_t is the value of using the optimal decision $a_t^*(S_t)$. That is

$$
\begin{aligned}
V_t(S_t) &= \max_{a_t \in \mathcal{A}_t} \left(C_t(S_t, a_t) + \gamma V_{t+1}(S_{t+1}(S_t, a_t)) \right) \\
&= C_t(S_t, a_t^*(S_t)) + \gamma V_{t+1}(S_{t+1}(S_t, a_t^*(S_t))).
\end{aligned}
\tag{14.2}
$$

Equation (14.2) is the optimality equation for deterministic problems.

When we are solving stochastic problems, we have to model the fact that new information becomes available after we make the decision a_t. The result can be uncertainty in both the contribution earned, and in the determination of the next state we visit, S_{t+1}. For example, consider the problem of managing oil inventories for a refinery. Let the state S_t be the inventory in thousands of barrels of oil at time t (we require S_t to be integer). Let a_t be the amount of oil ordered at time t that will be available for use between t and $t + 1$, and let $\hat{D}_{t+1}$ be the demand for oil between t and $t + 1$. The state variable is governed by the simple inventory equation

$$S_{t+1}(S_t, a_t, \hat{D}_{t+1}) = \max\{0, S_t + a_t - \hat{D}_{t+1}\}.$$

We have written the state S_{t+1} using $S_{t+1}(S_t, a_t, \hat{D}_{t+1})$ to express the dependence on S_t, a_t, and $\hat{D}_{t+1}$, but it is common to simply write S_{t+1} and let the dependence on S_t, a_t, and $\hat{D}_{t+1}$ be implicit. Since $\hat{D}_{t+1}$ is random at time t when we have to choose a_t, we do not know S_{t+1}. But if we know the probability distribution of the demand $\hat{D}_{t+1}$, we can work out the probability that S_{t+1} will take on a particular value.

If $\mathbb{P}^D(d) = \mathbb{P}[\hat{D} = d]$ is our probability distribution, then we can find the probability distribution for S_{t+1} using

$$Prob(S_{t+1} = s') = \begin{cases} 0 & \text{if } s' > S_t + a_t, \\ \mathbb{P}^D(S_t + a_t - s') & \text{if } 0 < s' \leq S_t + a_t, \\ \sum_{d=S_t+a_t}^{\infty} \mathbb{P}^D(d) & \text{if } s' = 0. \end{cases}$$

These probabilities depend on S_t and a_t, so we write the probability distribution as

$$\mathbb{P}(S_{t+1}|S_t, a_t) = \text{the probability of } S_{t+1} \text{ given } S_t \text{ and } a_t.$$

We can then modify the deterministic optimality equation in (14.2) by summing over the probability of each possible value of S_{t+1} (which is the same as taking an expectation), giving us

$$V_t(S_t) \quad = \quad \max_{a_t \in \mathcal{A}_t} \left(C_t(S_t, a_t) + \gamma \sum_{s' \in \mathcal{S}} \mathbb{P}(S_{t+1} = s'|S_t, a_t) V_{t+1}(s') \right). \quad (14.3)$$

We refer to this as the *standard form* of Bellman's equations, since this is the version that is used by virtually every textbook on Markov decision processes. An equivalent form is to write

$$V_t(S_t) \quad = \quad \max_{a_t \in \mathcal{A}_t} \left(C_t(S_t, a_t) + \gamma \mathbb{E}\{V_{t+1}(S_{t+1}(S_t, a_t, W_{t+1}))|S_t\} \right), \quad (14.4)$$

where we simply use an expectation instead of summing over probabilities. We refer to this equation as the *expectation form* of Bellman's equation. This version is the standard style that we use in this book.

Equation (14.4) is generally written in the more compact form

$$V_t(S_t) \quad = \quad \max_{a_t \in \mathcal{A}_t} \left(C_t(S_t, a_t) + \gamma \mathbb{E}\{V_{t+1}(S_{t+1})|S_t\} \right), \quad (14.5)$$

where the functional relationship $S_{t+1} = S^M(S_t, a_t, W_{t+1})$ is implicit. At this point, however, we have to deal with some subtleties of mathematical notation. In equation (14.4) we have captured the *functional dependence* of S_{t+1} on S_t and a_t, but the expectation is actually over the random variable W_{t+1}, which may be completely independent of the state of the system, but there may be *conditional dependence* of W_{t+1} on the state S_t and/or the action a_t. For this reason, we will often write

$$V_t(S_t) = \max_{a_t \in A_t} \left(C_t(S_t, a_t) + \gamma \mathbb{E}\{V_{t+1}(S_{t+1})|S_t, a_t\} \right). \tag{14.6}$$

The standard form of Bellman's equation (14.3) has been popular in the research community since it lends itself to elegant algebraic manipulation when we assume we know the transition matrix. It is common to write it in a more compact form. Recall that a policy π is a rule that specifies the action a_t given the state S_t. In this chapter, it is easiest if we always think of a policy in terms of a rule "when we are in state s we take action a." This is a form of "lookup-table" representation of a policy that is very clumsy for most real problems, but it will serve our purposes here.

The probability that we transition from state $S_t = s$ to $S_{t+1} = s'$ can be written as

$$p_{ss'}(a) = \mathbb{P}(S_{t+1} = s'|S_t = s, a_t = a).$$

We would say that "$p_{ss'}(a)$ is the probability that we end up in state s' if we start in state s at time t when we are taking action a." Now assume that we have a function $A_t^\pi(s)$ that determines the action a we should take when in state s. It is common to write the transition probability $p_{ss'}(a)$ in the form

$$p_{ss'}^\pi = \mathbb{P}(S_{t+1} = s'|S_t = s, A_t^\pi(s) = a).$$

We can now write this in matrix form

$$P_t^\pi = \text{the one-step transition matrix under policy } \pi,$$

where $p_{ss'}^\pi$ is the element in row s and column s'. There is a different matrix P^π for each policy (decision rule) π.

Now let c_t^π be a column vector with element $c_t^\pi(s) = C_t(s, A_t^\pi(s))$, and let v_{t+1} be a column vector with element $V_{t+1}(s)$. Then (14.3) is equivalent to

$$\begin{bmatrix} \vdots \\ v_t(s) \\ \vdots \end{bmatrix} = \max_{\pi} \left(\begin{bmatrix} \vdots \\ c_t^\pi(s) \\ \vdots \end{bmatrix} + \gamma \begin{bmatrix} \ddots & & \\ & p_{ss'}^\pi & \\ & & \ddots \end{bmatrix} \begin{bmatrix} \vdots \\ v_{t+1}(s') \\ \vdots \end{bmatrix} \right), \tag{14.7}$$

where the maximization is performed for each element (state) in the vector. In matrix/vector form, equation (14.7) can be written

$$v_t = \max_{\pi} \left(c_t^\pi + \gamma P_t^\pi v_{t+1} \right). \tag{14.8}$$

Here, we maximize over policies because we want to find the best action for *each* state. The vector v_t is known widely as the *value function* (the value of being in each state). In control theory, it is known as the *cost-to-go function*, where it is typically denoted as *J*.

Equation (14.8) can be solved by finding a_t for each state *s*. The result is a decision vector $a_t^* = (a_t^*(s))_{s \in \mathcal{S}}$, which is equivalent to determining the best policy. This is easiest to envision when a_t is a scalar (how much to buy, whether to sell), but in many applications $a_t(s)$ is itself a vector. For example, assume our problem is to assign individual programmers to different programming tasks, where our state S_t captures the availability of programmers and the different tasks that need to be completed. Of course, computing a vector a_t for each state S_t which is itself a vector is much easier to write than to implement.

It is very easy to lose sight of the relationship between Bellman's equation and the original objective function that we stated in equation (14.1). To bring this out, we begin by writing the expected profits using policy π from time *t* onward

$$
F_t^\pi(S_t) \quad = \quad \mathbb{E}\left\{ \sum_{t'=t}^{T-1} C_{t'}(S_{t'}, A_{t'}^\pi(S_{t'})) + C_T(S_T)|S_t \right\}.
$$

$F_t^\pi(S_t)$ is the expected total contribution if we are in state S_t in time *t*, and follow policy π from time *t* onward. If $F_t^\pi(S_t)$ were easy to calculate, we would probably not need dynamic programming. Instead, it seems much more natural to calculate V_t^π recursively using

$$
V_t^\pi(S_t) \quad = \quad C_t(S_t, A_t^\pi(S_t)) + \mathbb{E}\left\{ V_{t+1}^\pi(S_{t+1})|S_t \right\}.
$$

It is not hard to show (by stepping backward in time) that

$$
F_t^\pi(S_t) = V_t^\pi(S_t).
$$

The proof, given in section 14.12.1, uses a proof by induction: assume it is true for V_{t+1}^π, and then show that it is true for V_t^π (not surprisingly, inductive proofs are very popular in dynamic programming).

With this result in hand, we can then establish the following key result. Let $V_t(S_t)$ be a solution to equation (14.4) (or (14.3)). Then

$$
\begin{aligned}
F_t^* \quad &= \quad \max_{\pi \in \Pi} F_t^\pi(S_t) \\
&= \quad V_t(S_t). \quad\quad\quad (14.9)
\end{aligned}
$$

Equation (14.9) establishes the equivalence between (a) the value of being in state S_t and following the optimal policy and (b) the optimal value function at state S_t. While these are indeed equivalent, the equivalence is the result of a theorem (established in section 14.12.1). However, it is not unusual to find people who lose sight of the original objective function. Later, we have to solve these equations approximately, and we will need to use the original objective function to evaluate the quality of a solution.

14.2.2 Computing the Transition Matrix

It is very common in stochastic, dynamic programming (more precisely, Markov decision processes) to assume that the one-step transition matrix P^π is given as data (remember that there is a different matrix for each policy π). In practice, we generally can assume we know the transition function $S^M(S_t, a_t, W_{t+1})$ from which we have to derive the one-step transition matrix.

Assume that the random information W_{t+1} that arrives between t and $t+1$ is independent of all prior information. Let Ω be the set of possible outcomes of our stochastic process, and let $w_{t+1} = W_{t+1}(\omega)$ be a particular realization (for simplicity, we assume that Ω is discrete, as in a set of sampled observations), where $\mathbb{P}(W_{t+1} = w_{t+1} = W_{t+1}(\omega))$ is the probability of outcome $W_{t+1} = w_t$. Also define the indicator function

$$\mathbb{1}_{\{X\}} = \begin{cases} 1 & \text{if the statement ``}X\text{'' is true.} \\ 0 & \text{otherwise.} \end{cases}$$

Here, "X" represents a logical condition (such as, "is $S_t = 6$?"). We now observe that the one-step transition probability $\mathbb{P}_t(S_{t+1}|S_t, a_t)$ can be written

$$\begin{aligned} \mathbb{P}_t(S_{t+1}|S_t, a_t) &= \mathbb{E}\mathbb{1}_{\{s'=S^M(S_t, a_t, W_{t+1})\}} \\ &= \sum_{\omega \in \Omega} \mathbb{P}(W_{t+1} = w_{t+1})\mathbb{1}_{\{s'=S^M(S_t, a_t, w_{t+1})\}}. \end{aligned}$$

So, finding the one-step transition matrix means that all we have to do is to sum over all possible outcomes of the information W_{t+1} and add up the probabilities that take us from a particular state-action pair (S_t, a_t) to a particular state $S_{t+1} = s'$. Sounds easy.

In some cases, this calculation is straightforward (consider our oil inventory example earlier in the section). But in other cases, this calculation is impossible. For example, W_{t+1} might be a vector of prices or demands. In this case, the set of outcomes Ω can be much too large to enumerate (this is the third curse of dimensionality), but we can work with a sampled set of outcomes, as we indicated in section 10.3.3.

While we can estimate the transition matrix statistically, our standard approach is to simulate the transition *function*, rather than compute (or even approximate) the one step transition *matrix*. We will first see this in an ADP setting in chapter 15. For the remainder of this chapter, we assume the one-step transition matrix is available.

14.2.3 Random Contributions

In many applications, the one-period contribution function is a deterministic function of S_t and a_t, and hence we routinely write the contribution as the deterministic function $C_t(S_t, a_t)$. However, this is not always the case. For example, a car traveling over a stochastic network may choose to traverse the link from node i to node j, and only learn the cost of the movement after making the decision. For such cases, the contribution function is random, and we might write it as

$$\hat{C}_{t+1}(S_t, a_t, W_{t+1}) \quad = \quad \text{the contribution received in period } t+1 \text{ given}$$
$$\text{the state } S_t \text{ and decision } a_t, \text{ as well as the new}$$
$$\text{information } W_{t+1} \text{ that arrives in period } t+1.$$

In this case, we simply bring the expectation in front, giving us

$$V_t(S_t) \quad = \quad \max_{a_t} \mathbb{E}\{\hat{C}_{t+1}(S_t, a_t, W_{t+1}) + \gamma V_{t+1}(S_{t+1})|S_t\}. \tag{14.10}$$

Now let

$$C_t(S_t, a_t) = \mathbb{E}\{\hat{C}_{t+1}(S_t, a_t, W_{t+1})|S_t\}.$$

Thus, we may view $C_t(S_t, a_t)$ as the expected contribution given that we are in state S_t and take action a_t.

14.2.4 Bellman's Equation Using Operator Notation*

The vector form of Bellman's equation in (14.8) can be written even more compactly using operator notation. Let $\mathcal{M}$ be the "max" (or "min") operator in (14.8) that can be viewed as acting on the vector v_{t+1} to produce the vector v_t. If we have a given policy π, we can write

$$\mathcal{M}^\pi v(s) = C_t(s, A^\pi(s)) + \gamma \sum_{s' \in \mathcal{S}} \mathbb{P}_t(s'|s, A^\pi(s))v_{t+1}(s').$$

Alternatively, we can find the best action, which we represent using

$$\mathcal{M}v(s) = \max_a \left(C_t(s, a) + \gamma \sum_{s' \in \mathcal{S}} \mathbb{P}_t(s'|s, a)v_{t+1}(s')\right).$$

Here, $\mathcal{M}v$ produces a vector, and $\mathcal{M}v(s)$ refers to element s of this vector. In vector form, we would write

$$\mathcal{M}v = \max_{\pi} \left(c_t^{\pi} + \gamma P_t^{\pi} v_{t+1} \right).$$

Now let V be the space of value functions. Then, $\mathcal{M}$ is a mapping

$$\mathcal{M} : V \to V.$$

We may also define the operator $\mathcal{M}^{\pi}$ for a particular policy π using

$$\mathcal{M}^{\pi}(v) = c_t^{\pi} + \gamma P^{\pi} v \tag{14.11}$$

for some vector $v \in V$. $\mathcal{M}^{\pi}$ is known as a *linear operator* since the operations that it performs on v are additive and multiplicative. In mathematics, the function $c_t^{\pi} + \gamma P^{\pi} v$ is known as an *affine function*. This notation is particularly useful in mathematical proofs (see some of the proofs in section 14.12), but we will not use this notation when we describe models and algorithms.

We see later in the chapter that we can exploit the properties of this operator to derive some very elegant results for Markov decision processes. These proofs provide insights into the behavior of these systems, which can guide the design of algorithms. For this reason, it is relatively immaterial that the actual computation of these equations may be intractable for many problems; the insights still apply.

14.3 Finite Horizon Problems

Finite horizon problems tend to arise in two settings. First, some problems have a very specific horizon. For example, we might be interested in the value of an American option where we are allowed to sell an asset at any time $t \le T$ where T is the exercise date. Another problem is to determine how many seats to sell at different prices for a particular flight departing at some point in the future. In the same class are problems that require reaching some goal (but not at a particular point in time). Examples include driving to a destination, selling a house, or winning a game.

A second class of problems is actually infinite horizon, but where the goal is to determine what to do right now given a particular state of the system. For example, a transportation company might want to know what drivers should be assigned to a particular set of loads right now. Of course, these decisions need to consider the downstream impact, so models have to extend into the future, but we simply do not need to optimize over an infinite horizon. For this reason, we might model the problem over a horizon T which, when solved, yields a decision of what to do now. This is known as a direct lookahead policy which we cover in chapter 19, but a DLA policy can involve solving a Markov decision process.

When we encounter a finite horizon problem, we assume that we are given the function $V_T(S_T)$ as data. Often, we simply use $V_T(S_T) = 0$ because we are primarily interested in what to do now, given by a_0, or in projected activities over some horizon $t = 0, 1, \dots, H$, where H is the length of a planning horizon. If we set T sufficiently larger than H, then we may be able to assume that the decisions $a_0, a_1, \dots, a_H$ are of sufficiently high quality to be useful.

Solving a finite horizon problem, in principle, is straightforward. The optimality equations give us

$$
\begin{aligned}
V_t(S_t) &= \max_{a_t \in \mathcal{A}} \mathbb{E}\{C_t(S_t, a_t) + \gamma V_{t+1}(S_{t+1}) | S_t\} \\
&= \max_{a_t \in \mathcal{A}} (C_t(S_t, a_t) + \gamma \mathbb{E}\{V_{t+1}(S_{t+1}) | S_t\}) \\
&= \max_{a_t \in \mathcal{A}} \left(C_t(S_t, a_t) + \gamma \sum_{s'} V_{t+1}(s') P(S_{t+1} = s' | S_t, a_t) \right), \qquad (14.12)
\end{aligned}
$$

where $P(s'|S_t, a_t)$ is the one-step transition matrix. If we could compute the one-step transition matrix (which this community typically assumes), then all we have to do is to execute equation (14.12) starting at the last time period T (where we might assume $V_T(S_T) = 0$ or some other ending value), and then stepping backward in time (the reason why this is called "backward dynamic programming").

It is important to realize that equation (14.12) was considered a major breakthrough when it was first discovered by Richard Bellman in the 1950s. Keep in mind that prior to this work, people approached these problems as decision trees as illustrated in Figure 14.1, which exploded in size extremely quickly. In effect, solving sequential stochastic optimization problems was considered completely intractable.

The implementation of backward dynamic programming is outlined in Figure 14.3. The algorithm is disarmingly simple; so simple, in fact, that it is likely that this is the reason that this field has focused primarily on steady state problems, which we will address shortly. What is overlooked, however, is that the one-step transition matrix is rarely computable, as it suffers from what is known as the three curses of dimensionality:

The state space – If the state variable S_t is an L-vector, where each dimension can take on one of K values, then the state space has K^L values, which grows very quickly with L.

The action space – The standard assumption in Markov decision processes is that a_t can take on a finite (say M, where M is not too large) values. While

there are many applications that fit this assumption, there are applications
where our decision is a vector (which is the reason that we use x_t for decisions in this book), and possibly a very high-dimensional vector. Problems
where x_t has ten thousand to hundred thousand dimensions arise frequently
in resource allocation problems, which we illustrated in section 8.3.

The outcome space – The exogenous information W_t may also be a vector, often
with continuous elements (examples of this are also found in section 8.3).
The size of the outcome space grows quickly with the dimensionality of W_t.

The one-step transition matrix as $|\mathcal{S}| \times |\mathcal{S}| \times |\mathcal{A}|$ elements, and each of these
elements requires an expectation over Ω. In other words, computing $P(s'|S_t, a_t)$
is the bottleneck.

We first saw backward dynamic programming in section 2.1.2 (and then
again in section 14.1) when we described a simple decision tree problem. The
only difference between the backward dynamic programming algorithm in
Figure 14.3 and our solution of the decision tree problem is primarily notational. Decision trees are visual and tend to be easier to understand, whereas

---

Step 0.   Initialization:

        Initialize the terminal contribution $V_T(S_T)$.
        Set $t = T - 1$.

Step 1a.   Step backward in time $t = T, T - 1, \dots, 0$:

    Step 2a.   Loop over states $s \in \mathcal{S} = \{1, \dots, |\mathcal{S}|\}$:
    Step 2b.   Initialize $V_t(s) = -M$ (where M is very large).

        Step 3a.   Loop over each action $a \in \mathcal{A}(s)$:

            Step 4a   Initialize $Q(s, a) = 0$.
            Step 4b.   Find the expected value of being in state s and taking
                  action a:
            Step 4c.   Compute $Q_t(s, a) = \sum_{w \in W} \mathbb{P}(w|s, a)V_{t+1}(s' = s^M(s, a, w))$.
            Step 4c.   If $Q_t(s, a) > V_t(s)$ then

                Step 3b.   Store the best value $V_t(s) = Q_t(s, a)$.
                Step 3c.   Store the best action $A_t(s) = a$.

Step 1b.   Return the value $V_t(s)$ and policy $A_t(s)$ for all $s \in \mathcal{S}$ and $t = 0, \dots, T$.

Figure 14.3   A backward dynamic programming algorithm.

in this section the methods are described using notation. However, decision tree problems are typically presented in the context of problems with relatively small numbers of states and actions: What job should I take? Should the United States put a blockade around Cuba? Should the shuttle launch have been canceled due to cold weather?

Another popular illustration of dynamic programming is the discrete asset acquisition problem. Assume that you order a quantity a_t at each time period to be used in the next time period to satisfy a demand $\hat{D}_{t+1}$. Any unused product is held over to the following time period. For this, our state variable S_t is the quantity of inventory left over at the end of the period after demands are satisfied. The transition equation is given by $S_{t+1} = [S_t + a_t - \hat{D}_{t+1}]^+$ where $[x]^+ = \max(x,0)$. The cost function (which we seek to minimize) is given by $\hat{C}_{t+1}(S_t, a_t) = c^h S_t + c^o \mathbb{1}_{\{a_t > 0\}}$, where $\mathbb{1}_{\{X\}} = 1$ if X is true and 0 otherwise. Note that the cost function is nonconvex. This does not create problems if we solve our minimization problem by searching over different (discrete) values of a_t. Since all of our quantities are scalar, there is no difficulty finding $C_t(S_t, a_t)$.

To compute the one-step transition matrix, let Ω be the set of possible outcomes of $\hat{D}_t$, and let $\mathbb{P}(\hat{D}_t = \omega)$ be the probability that $\hat{D}_t = \omega$. The one-step transition matrix is computed using

$$\mathbb{P}(s'|s, a) = \sum_{\omega \in \Omega} \mathbb{P}(\hat{D}_{t+1} = \omega) \mathbb{1}_{\{s' = [s+a-\omega]^+\}}$$

where Ω is the set of (discrete) outcomes of the demand $\hat{D}_{t+1}$.

Another example is the shortest path problem with random arc costs. Assume that you are trying to get from origin node q to destination node r in the shortest time possible. As you reach each intermediate node i, you are able to observe the time required to traverse each arc out of node i. Let V_j be the expected shortest path time from j to the destination node r. At node i, you see the link time $\hat{\tau}_{ij}$ which represents a random observation of the travel time. Now we choose to traverse arc (i, j^*) where j^* solves $\min_j(\hat{\tau}_{ij} + V_j)$. The choice of downstream node j^* is random since the travel time $\hat{\tau}_{ij}$ is random. We would then compute the value of being at node i using $V_i = \mathbb{E}\{\min_j(\hat{\tau}_{ij} + V_j)\}$.

14.4 Continuous Problems with Exact Solutions

There is a rich history in the study of Markov decision processes of specialized problems which yield exact solutions, especially in settings with continuous states and actions. In this section we illustrate two classic problems: the gambling problem, where we derive an optimal policy for determining how much to bet, and a continuous budgeting problem. These applications nicely illustrate the core principles without hiding behind the veil of computation.

14.4.1 The Gambling Problem

A gambler has to determine how much of his capital he should bet on each round of a game, where he will play a total of N rounds. He will win a bet with probability p and lose with probability $q = 1 - p$ (assume $q < p$). Let S^n be his total capital after n plays, $n = 0, 1, \ldots, N$, with S^0 being his initial capital. For this problem, S^n is the state of our system (his available capital) after n plays. Let x^n be the (discrete) amount he bets in round $n + 1$, where we require that $x^n \leq S^n$. He wants to maximize $\ln S^N$, which provides a strong penalty for ending up with a small amount of money at the end and a declining marginal value for higher amounts.

Let

$$
W^n = \begin{cases} 1 & \text{if the gambler wins the } n^{th} \text{ game,} \\ 0 & \text{otherwise.} \end{cases}
$$

The system evolves according to

$$S^{n+1} = S^n + x^n W^{n+1} - x^n(1 - W^{n+1}).$$

Let $V^n(S^n)$ be the value of having S^n dollars at the end of the nth game. The value of being in state S^n at the end of the nth round can be written as

$$
\begin{aligned}
V^n(S^n) &= \max_{0 \leq x^n \leq S^n} \mathbb{E}\{V^{n+1}(S^{n+1})|S^n\} \\
&= \max_{0 \leq x^n \leq S^n} \mathbb{E}\{V^{n+1}(S^n + x^n W^{n+1} - x^n(1 - W^{n+1}))|S^n\}.
\end{aligned}
$$

Here, we claim that the value of being in state S^n is found by choosing the decision that maximizes the expected value of being in state S^{n+1} given what we know at the end of the nth round.

We solve this by starting at the end of the Nth trial, and assuming that we have finished with S^N dollars, which means our ending value is

$$V^N(S^N) = \ln S^N.$$

Now step back to $n = N - 1$, where we may write

$$
\begin{aligned}
V^{N-1}(S^{N-1}) &= \max_{0 \leq x^{N-1} \leq S^{N-1}} \mathbb{E}\{V^N(S^{N-1} + x^{N-1}W^N - x^{N-1}(1 - W^N))|S^{N-1}\} \\
&= \max_{0 \leq x^{N-1} \leq S^{N-1}} \left[p\ln(S^{N-1} + x^{N-1}) + (1 - p)\ln(S^{N-1} - x^{N-1})\right].
\end{aligned}
$$

$$(14.13)$$

Let $V^{N-1}(S^{N-1}, x^{N-1})$ be the value within the max operator. We can find x^{N-1} by differentiating $V^{N-1}(S^{N-1}, x^{N-1})$ with respect to x^{N-1}, giving

$$\frac{\partial V^{N-1}(S^{N-1}, x^{N-1})}{\partial x^{N-1}} = \frac{p}{S^{N-1} + x^{N-1}} - \frac{1-p}{S^{N-1} - x^{N-1}}$$

$$= \frac{2S^{N-1}p - S^{N-1} - x^{N-1}}{(S^{N-1})^2 - (x^{N-1})^2}.$$

Setting this equal to zero and solving for x^{N-1} gives

$$x^{N-1} = (2p - 1)S^{N-1}.$$

The next step is to plug this back into (14.13) to find $V^{N-1}(s^{N-1})$ using

$$
\begin{aligned}
V^{N-1}(S^{N-1}) &= p\ln(S^{N-1} + S^{N-1}(2p-1)) + (1-p)\ln(S^{N-1} - S^{N-1}(2p-1)) \\
&= p\ln(S^{N-1}2p) + (1-p)\ln(S^{N-1}2(1-p)) \\
&= p\ln S^{N-1} + (1-p)\ln S^{N-1} + \underbrace{p\ln(2p) + (1-p)\ln(2(1-p))}_{K} \\
&= \ln S^{N-1} + K,
\end{aligned}
$$

where K is a constant with respect to S^{N-1}. Since the additive constant does not change our decision, we may ignore it and use $V^{N-1}(S^{N-1}) = \ln S^{N-1}$ as our value function for $N-1$, which is the same as our value function for N. Not surprisingly, we can keep applying this same logic backward in time and obtain

$$V^n(S^n) = \ln S^n \; (+K^n)$$

for all n, where again, K^n is some constant that can be ignored. This means that for all n, our optimal solution is

$$x^n = (2p - 1)S^n.$$

The optimal strategy at each iteration is to bet a fraction $\beta = (2p - 1)$ of our current money on hand. Of course, this requires that $p > .5$.

There is a long tradition in the study of Markov decision processes of deriving the structure of optimal policies. In some cases, such as this gambling problem, we can find the optimal solution (or optimal policy). In others, we can find the structure of the policy, such as showing that a "buy low, sell high" policy is optimum, leaving us with just the problem of finding the buy and sell points.

14.4.2 The Continuous Budgeting Problem

Assume that the resources we are allocating are continuous (for example, how much money to assign to various activities), which means that R_t is continuous, as is the decision of how much to budget. We are going to assume that the contribution from allocating x_t dollars to task t is given by

$$C_t(x_t) = \sqrt{x_t}.$$

This function assumes that there are diminishing returns from allocating additional resources to a task, as is common in many applications. We can solve this problem exactly using dynamic programming. We first note that if we have R_T dollars left for the last task, the value of being in this state is

$$V_T(R_T) = \max_{x_T \le R_T} \sqrt{x_T}.$$

Since the contribution increases monotonically with x_T, the optimal solution is $x_T = R_T$, which means that $V_T(R_T) = \sqrt{R_T}$. Now consider the problem at time $t = T - 1$. The value of being in state R_{T-1} would be

$$V_{T-1}(R_{T-1}) = \max_{x_{T-1} \le R_{T-1}} \left(\sqrt{x_{T-1}} + V_T(R_T(x_{T-1})) \right) \tag{14.14}$$

where $R_T(x_{T-1}) = R_{T-1} - x_{T-1}$ is the money left over from time period $T - 1$. Since we know $V_T(R_T)$ we can rewrite (14.14) as

$$V_{T-1}(R_{T-1}) = \max_{x_{T-1} \le R_{T-1}} \left(\sqrt{x_{T-1}} + \sqrt{R_{T-1} - x_{T-1}} \right). \tag{14.15}$$

We solve (14.15) by differentiating with respect to x_{T-1} and setting the derivative equal to zero (we are taking advantage of the fact that we are maximizing a continuously differentiable, concave function). Let

$$F_{T-1}(R_{T-1}, x_{T-1}) = \sqrt{x_{T-1}} + \sqrt{R_{T-1} - x_{T-1}}.$$

Differentiating $F_{T-1}(R_{T-1}, x_{T-1})$ and setting this equal to zero gives

$$\frac{\partial F_{T-1}(R_{T-1}, x_{T-1})}{\partial x_{T-1}} = \frac{1}{2}(x_{T-1})^{-\frac{1}{2}} - \frac{1}{2}(R_{T-1} - x_{T-1})^{-\frac{1}{2}}$$
$$= 0.$$

This implies

$$x_{T-1} = R_{T-1} - x_{T-1},$$

which gives

$$x_{T-1}^* = \frac{1}{2}R_{T-1}.$$

We now have to find V_{T-1}. Substituting x_{T-1}^* back into (14.15) gives

$$
\begin{aligned}
V_{T-1}(R_{T-1}) &= \sqrt{R_{T-1}/2} + \sqrt{R_{T-1}/2} \\
&= 2\sqrt{R_{T-1}/2}.
\end{aligned}
$$

We can continue this exercise, but there seems to be a bit of a pattern forming (this is a common trick when trying to solve dynamic programs analytically). It seems that a general formula might be

$$
V_{T-t+1}(R_{T-t+1}) = t\sqrt{R_{T-t+1}/t}, \tag{14.16}
$$

or, equivalently,

$$
V_t(R_t) = (T - t + 1)\sqrt{R_t/(T - t + 1)}. \tag{14.17}
$$

How do we determine if this guess is correct? We use a technique known as proof by induction. We assume that (14.16) is true for $V_{T-t+1}(R_{T-t+1})$ and then show that we get the same structure for $V_{T-t}(R_{T-t})$. Since we have already shown that it is true for V_T and V_{T-1}, this result would allow us to show that it is true for all t.

Finally, we can determine the optimal solution using the value function in equation (14.17). The optimal value of x_t is found by solving

$$
\max_{x_t} \left(\sqrt{x_t} + (T - t)\sqrt{(R_t - x_t)/(T - t)} \right). \tag{14.18}
$$

Differentiating and setting the result equal to zero gives

$$
\frac{1}{2}(x_t)^{-\frac{1}{2}} - \frac{1}{2}\left(\frac{R_t - x_t}{T - t}\right)^{-\frac{1}{2}} = 0.
$$

This implies that

$$
x_t = (R_t - x_t)/(T - t).
$$

Solving for x_t gives

$$
x_t^* = R_t/(T - t + 1).
$$

This gives us the very intuitive result that we want to evenly divide the available budget among all remaining tasks. This is what we would expect since all the tasks produce the same contribution.

14.5   Infinite Horizon Problems*

Most of this book focuses on finite horizon problems, which tends to be most useful for practical problems. The history of research in Markov decision processes has been to focus on infinite horizon problems. We speculate that if you assume that you are given the one-step transition matrix $P(S_{t+1} = s'|S_t = s, a)$, solving finite horizon problems become, well, trivial. Needless to say, this is far from the truth.

By contrast, infinite horizon problems are challenging with genuinely elegant mathematics, as we will see. We typically use infinite horizon formulations whenever we wish to study a problem where the parameters of the contribution function, transition function, and the process governing the exogenous information process do not vary over time. More importantly, infinite horizon problems provide a number of insights into the properties of problems and algorithms, drawing off an elegant theory that has evolved around this problem class. Even students who wish to solve complex, nonstationary problems will benefit from an understanding of this problem class.

We start with the finite-horizon version of Bellman's equation, which we saw earlier but repeat here

$$V_t(S_t) \quad = \quad \max_{a_t \in \mathcal{A}} \mathbb{E}\{C_t(S_t, a_t) + \gamma V_{t+1}(S_{t+1})|S_t\}. \tag{14.19}$$

We can think of a steady-state problem as one without the time dimension. Letting $V(s) = \lim_{t \to \infty} V_t(S_t)$ (and assuming the limit exists), we obtain the steady-state optimality equations

$$V(s) \quad = \quad \max_{a \in \mathcal{A}} \left\{ C(s, a) + \gamma \sum_{s' \in \mathcal{S}} \mathbb{P}(s'|s, a)V(s') \right\}. \tag{14.20}$$

The functions $V(s)$ can be shown (as we do later) to be equivalent to solving the infinite horizon problem

$$\max_{\pi \in \Pi} \mathbb{E}\left\{ \sum_{t=0}^{\infty} \gamma^t C_t(S_t, A_t^\pi(S_t)) \right\}. \tag{14.21}$$

Now define

$$P^{\pi,t} \quad = \quad t\text{-step transition matrix, over periods } 0, 1, \dots, t-1,$$
$$\text{given policy } \pi$$
$$= \quad \Pi_{t'=0}^{t-1} P_{t'}^\pi. \tag{14.22}$$

We further define $P^{\pi,0}$ to be the identity matrix. As before, let c_t^π be the column vector of the expected cost of being in each state given that we choose the action a_t described by policy π, where the element for state s is $c_t^\pi(s) = C_t(s, A^\pi(s))$.

The infinite horizon, discounted value of a policy π starting at time t is given by

$$v_t^\pi \;=\; \sum_{t'=t}^{\infty} \gamma^{t'-t} P^{\pi,t'-t} c_{t'}^\pi. \tag{14.23}$$

Assume that after following policy π_0 we follow policy $\pi_1 = \pi_2 = \dots = \pi$. In this case, equation (14.23) can now be written as (starting at $t = 0$)

$$v^{\pi_0} \;=\; c^{\pi_0} + \sum_{t'=1}^{\infty} \gamma^{t'} P^{\pi,t'} c_{t'}^\pi \tag{14.24}$$

$$= \; c^{\pi_0} + \sum_{t'=1}^{\infty} \gamma^{t'} \left(\Pi_{t''=0}^{t'-1} P_{t''}^\pi \right) c_{t'}^\pi \tag{14.25}$$

$$= \; c^{\pi_0} + \gamma P^{\pi_0} \sum_{t'=1}^{\infty} \gamma^{t'-1} \left(\Pi_{t''=1}^{t'-1} P_{t''}^\pi \right) c_{t'}^\pi \tag{14.26}$$

$$= \; c^{\pi_0} + \gamma P^{\pi_0} v^\pi. \tag{14.27}$$

Equation (14.27) shows us that the value of a policy is the single period reward plus a discounted final reward that is the same as the value of a policy starting at time 1. If our decision rule is stationary, then $\pi_0 = \pi_1 = \dots = \pi_t = \pi$, which allows us to rewrite (14.27) as

$$v^\pi \;=\; c^\pi + \gamma P^\pi v^\pi. \tag{14.28}$$

This allows us to solve for the stationary reward explicitly (as long as $0 \le \gamma < 1$), giving us

$$v^\pi \;=\; (I - \gamma P^\pi)^{-1} c^\pi.$$

We can also write an infinite horizon version of the optimality equations using our operator notation. Letting $\mathcal{M}$ be the "max" (or "min") operator (also known as the Bellman operator), the infinite horizon version of equation (14.11) would be written

$$\mathcal{M}^\pi(v) = c^\pi + \gamma P^\pi v. \tag{14.29}$$

There are several algorithmic strategies for solving infinite horizon problems. The first, value iteration, is the most widely used method. It involves iteratively estimating the value function. At each iteration, the estimate of the value function determines which decisions we will make and as a result, defines a policy. The second strategy is *policy iteration*. At every iteration, we define a policy (literally, the rule for determining decisions) and then determine the value function for that policy.

Careful examination of value and policy iteration reveals that these are closely related strategies that can be viewed as special cases of a general strategy that uses value and policy iteration. Finally, the third major algorithmic strategy exploits the observation that the value function can be viewed as the solution to a specially structured linear programming problem.

14.6 Value Iteration for Infinite Horizon Problems*

Value iteration is perhaps the most widely used algorithm in dynamic programming for infinite horizon problems because it is the simplest to implement and, as a result, often tends to be the most natural way of solving many problems. It is virtually identical to backward dynamic programming for finite horizon problems. In addition, most of our work in approximate dynamic programming is based on value iteration.

Value iteration comes in several flavors. The basic version of the value iteration algorithm is given in Figure 14.4. The proof of convergence (see section 14.12.2) is quite elegant for students who enjoy mathematics. The algorithm also has several nice properties that we explore shortly.

It is easy to see that the value iteration algorithm is similar to the backward dynamic programming algorithm. Rather than using a subscript t, which we decrement from T back to 0, we use an iteration counter n that starts at 0 and increases until we satisfy a convergence criterion. Here, we stop the algorithm when

$$\|v^n - v^{n-1}\| < \epsilon(1 - \gamma)/2\gamma,$$

Step 0. Initialization:

Set $v^0(s) = 0 \ \forall s \in \mathcal{S}$.
Fix a tolerance parameter $\epsilon > 0$.
Set $n = 1$.

Step 1. For each $s \in \mathcal{S}$ compute:

$$v^n(s) \quad = \quad \max_{a \in \mathcal{A}} \left(C(s, a) + \gamma \sum_{s' \in \mathcal{S}} \mathbb{P}(s'|s, a) v^{n-1}(s') \right). \tag{14.30}$$

Step 2. If $\|v^n - v^{n-1}\| < \epsilon(1 - \gamma)/2\gamma$, let π^ϵ be the resulting policy that solves (14.30), and let $v^\epsilon = v^n$ and stop; else set $n = n + 1$ and go to step 1.

Figure 14.4 The value iteration algorithm for infinite horizon optimization.

Replace Step 1 with

Step 1'. For each $s \in \mathcal{S}$ compute

$$v^n(s) = \max_{a \in \mathcal{A}} \left\{ C(s, a) + \gamma \left(\sum_{s' < s} \mathbb{P}(s'|s, a)v^n(s') + \sum_{s' \geq s} \mathbb{P}(s'|s, a)v^{n-1}(s') \right) \right\}$$

Figure 14.5 The Gauss-Seidel variation of value iteration.

where $\|v\|$ is the max-norm defined by

$$\|v\| = \max_s |v(s)|.$$

Thus, $\|v\|$ is the largest absolute value of a vector of elements. Thus, we stop if the largest change in the value of being in any state is less than $\epsilon(1 - \gamma)/2\gamma$ where ϵ is a specified error tolerance.

We next describe a Gauss-Seidel variant which is a useful method for accelerating value iteration, and a version known as relative value iteration.

14.6.1 A Gauss-Seidel Variation

A slight variant of the value iteration algorithm provides a faster rate of convergence. In this version (typically called the Gauss-Seidel variant), we take advantage of the fact that when we are computing the expectation of the value of the future, we have to loop over all the states s' to compute $\sum_{s'} \mathbb{P}(s'|s, a)v^n(s')$. For a particular state s, we would have already computed $v^{n+1}(\hat{s})$ for $\hat{s} = 1, 2, ..., s-1$. By simply replacing $v^n(\hat{s})$ with $v^{n+1}(\hat{s})$ for the states we have already visited, we obtain an algorithm that typically exhibits a noticeably faster rate of convergence. The algorithm requires a change to step 1 of the value iteration, as shown in Figure 14.5.

14.6.2 Relative Value Iteration

Another version of value iteration is called *relative value iteration*, which is useful in problems that do not have a discount factor or where the optimal policy converges much more quickly than the value function, which may grow steadily for many iterations. The relative value iteration algorithm is shown in Figure 14.6.

In relative value iteration, we focus on the fact that we may be more interested in the convergence of the difference $|v(s)-v(s')|$ than we are in the values of $v(s)$ and $v(s')$. This would be the case if we are interested in the best policy rather than the value function itself (this is not always the case). What often

Step 0. Initialization:

- Choose some $v^0 \in \mathcal{V}$.
- Choose a base state s^* and a tolerance ε.
- Let $w^0 = v^0 - v^0(s^*)e$ where e is a vector of ones.
- Set $n = 1$.

Step 1. Set

$$
\begin{aligned}
v^n &= \mathcal{M}w^{n-1}, \\
w^n &= v^n - v^n(s^*)e.
\end{aligned}
$$

Step 2. If $sp(v^n - v^{n-1}) < (1-\gamma)\varepsilon/\gamma$, go to step 3; otherwise, go to step 1.
Step 3. Set $a^\varepsilon = \arg\max_{a \in A} (C(a) + \gamma P^\pi v^n)$.

Figure 14.6   Relative value iteration.

happens is that, especially toward the limit, all the values $v(s)$ start increasing by the same rate. For this reason, we can pick any state (denoted s^* in the algorithm) and subtract its value from all the other states.

To provide a bit of formalism for our algorithm, we define the *span* of a vector v as follows:

$$
sp(v) = \max_{s \in \mathcal{S}} v(s) - \min_{s \in \mathcal{S}} v(s).
$$

Note that our use of "span" is different than the way it is normally used in linear algebra. Here and throughout this section, we define the norm of a vector as

$$
\|v\| = \max_{s \in \mathcal{S}} v(s).
$$

Note that the span has the following six properties:

(1) $sp(v) \geq 0$.
(2) $sp(u + v) \leq sp(u) + sp(v)$.
(3) $sp(kv) = |k| sp(v)$.
(4) $sp(v + ke) = sp(v)$.
(5) $sp(v) = sp(-v)$.
(6) $sp(v) \leq 2\|v\|$.

Property (4) implies that $sp(v) = 0$ does not mean that $v = 0$ and therefore it does not satisfy the properties of a norm. For this reason, it is called a *seminorm*.

The relative value iteration algorithm is simply subtracting a constant from the value vector at each iteration. Obviously, this does not change the optimal decision, but it does change the value itself. If we are only interested in the optimal policy, relative value iteration often offers much faster convergence, but it may not yield accurate estimates of the value of being in each state.

14.6.3 Bounds and Rates of Convergence

One important property of value iteration algorithms is that if our initial estimate is too low, the algorithm will rise to the correct value from below. Similarly, if our initial estimate is too high, the algorithm will approach the correct value from above. This property is formalized in the following theorem:

Theorem 14.6.1. For a vector $v \in \mathcal{V}$:

(a) If v satisfies $v \geq \mathcal{M}v$, then $v \geq v^*$.
(b) If v satisfies $v \leq \mathcal{M}v$, then $v \leq v^*$.
(c) If v satisfies $v = \mathcal{M}v$, then v is the unique solution to this system of equations and $v = v^*$.

The proof is given in section 14.12.3. It is a nice property because it provides some valuable information on the nature of the convergence path. In practice, we generally do not know the true value function, which makes it hard to know if we are starting from above or below (although some problems have natural bounds, such as nonnegativity).

The proof of the monotonicity property also provides us with a nice corollary. If $V(s) = \mathcal{M}V(s)$ for all s, then $V(s)$ is the unique solution to this system of equations, which must also be the optimal solution.

This result raises the question: What if some of our estimates of the value of being in some states are too high, while others are too low? This means the values may cycle above and below the optimal solution, although at some point we may find that all the values have increased (decreased) from one iteration to the next. If this happens, then it means that the values are all equal to or below (above) the limiting value.

Value iteration also provides a nice bound on the quality of the solution. Recall that when we use the value iteration algorithm, we stop when

$$\|v^{n+1} - v^n\| < \epsilon(1-\gamma)/2\gamma \tag{14.31}$$

where γ is our discount factor and ϵ is a specified error tolerance. It is possible that we have found the optimal policy when we stop, but it is very unlikely that we have found the optimal value functions. We can, however, provide a bound on the gap between the solution v^n and the optimal values v^* by using the following theorem:

Theorem 14.6.2. If we apply the value iteration algorithm with stopping parameter ϵ and the algorithm terminates at iteration n with value function v^{n+1}, then

$$\|v^{n+1} - v^*\| \le \epsilon/2. \tag{14.32}$$

Let π^ϵ be the policy that we terminate with, and let v^{π^ϵ} be the value of this policy. Then

$$\|v^{\pi^\epsilon} - v^*\| \le \epsilon.$$

The proof is given in section 14.12.4. While it is nice that we can bound the error, the bad news is that the bound can be quite poor. More important is what the bound teaches us about the role of the discount factor.

We can provide some additional insights into the bound, as well as the rate of convergence, by considering a trivial dynamic program. In this problem, we receive a constant reward c at every iteration. There are no decisions, and there is no randomness. The value of this "game" is quickly seen to be

$$v^* = \sum_{n=0}^{\infty} \gamma^n c$$

$$= \frac{1}{1 - \gamma} c. \tag{14.33}$$

Consider what happens when we solve this problem using value iteration. Starting with $v^0 = 0$, we would use the iteration

$$v^n = c + \gamma v^{n-1}.$$

After we have repeated this n times, we have

$$v^n = \sum_{m=0}^{n-1} \gamma^n c$$

$$= \frac{1 - \gamma^n}{1 - \gamma} c. \tag{14.34}$$

Comparing equations (14.33) and (14.34), we see that

$$v^n - v^* = -\frac{\gamma^n}{1 - \gamma} c. \tag{14.35}$$

Similarly, the change in the value from one iteration to the next is given by

$$\|v^{n+1} - v^n\| = \left| \frac{\gamma^{n+1}}{1 - \gamma} - \frac{\gamma^n}{1 - \gamma} \right| c$$

$$= \gamma^n \left| \frac{\gamma}{1 - \gamma} - \frac{1}{1 - \gamma} \right| c$$

$$= \gamma^n \left| \frac{\gamma - 1}{1 - \gamma} \right| c$$

$$= \gamma^n c.$$

If we stop at iteration $n + 1$, then it means that

$$\gamma^n c \leq \epsilon/2 \left(\frac{1 - \gamma}{\gamma} \right). \tag{14.36}$$

If we choose ϵ so that (14.36) holds with equality, then our error bound (from 14.32) is

$$\begin{aligned} \|v^{n+1} - v^*\| &\leq \epsilon/2 \\ &= \frac{\gamma^{n+1}}{1 - \gamma} c. \end{aligned}$$

From (14.35), we know that the distance to the optimal solution is

$$|v^{n+1} - v^*| = \frac{\gamma^{n+1}}{1 - \gamma} c,$$

which matches our bound.

This little exercise confirms that our bound on the error may be tight. It also shows that the error decreases geometrically at a rate determined by the discount factor. For this problem, the error arises because we are approximating an infinite sum with a finite one. For more realistic dynamic programs, we also have the effect of trying to find the optimal policy. When the values are close enough that we have, in fact, found the optimal policy, then we have only a Markov reward process (a Markov chain where we earn rewards for each transition). Once our Markov reward process has reached steady state, it will behave just like the simple problem we have just solved, where c is the expected reward from each transition.

14.7 Policy Iteration for Infinite Horizon Problems*

In policy iteration, we choose a policy and then find the infinite horizon, discounted value of the policy. This value is then used to choose a new policy. The general algorithm is described in Figure 14.7. Policy iteration is popular for infinite horizon problems because of the ease with which we can find the value of a policy. As we showed in section 14.5, the value of following policy π is given by

$$v^\pi = (I - \gamma P^\pi)^{-1} c^\pi. \tag{14.37}$$

While computing the inverse can be problematic as the state space grows, it is, at a minimum, a very convenient formula.

It is useful to illustrate the policy iteration algorithm in different settings. In the first, consider a batch replenishment problem where we have to replenish

Step 0. Initialization:

    Step 0a. Select a policy π^0.
    Step 0b. Set $n = 1$.

Step 1. Given a policy π^{n-1}:

    Step 1a. Compute the one-step transition matrix $P^{\pi^{n-1}}$.
    Step 1b. Compute the contribution vector $c^{\pi^{n-1}}$ where the element for state s is given
        by $c^{\pi^{n-1}}(s) = C(s, A^{\pi^{n-1}})$.

Step 2. Let $v^{\pi,n}$ be the solution to

$$(I - \gamma P^{\pi^{n-1}})v \;=\; c^{\pi^{n-1}}.$$

Step 3. Find a policy π^n defined by

$$a^n(s) \;=\; \arg\max_{a \in A} \left(C(a) + \gamma P^\pi v^n \right).$$

    This requires that we compute an action for each state s.

Step 4. If $a^n(s) = a^{n-1}(s)$ for all states s, then set $a^* = a^n$; otherwise, set $n = n + 1$ and go
    to step 1.

Figure 14.7   Policy iteration.

resources (raising capital, exploring for oil to expand known reserves, hiring people) where there are economies from ordering larger quantities. We might use a simple policy where if our level of resources $R_t < q$ for some lower limit q, we order a quantity $a_t = Q - R_t$. This policy is parameterized by (q, Q) and is written

$$A^\pi(R_t) = \begin{cases} 0, & R_t \geq q, \\ Q - R_t, & R_t < q. \end{cases} \tag{14.38}$$

For a given set of parameters $\pi = (q, Q)$, we can compute a one-step transition matrix P^π and a contribution vector c^π.

Policies come in many forms. For the moment, we simply view a policy as a rule that tells us what decision to make when we are in a particular state. In later chapters, we introduce policies in different forms since they create different challenges for finding the best policy.

Given a transition matrix P^π and contribution vector c^π, we can use equation (14.37) to find v^π, where $v^\pi(s)$ is the discounted value of started in state s and following policy π. From this vector, we can infer a new policy by solving

$$a^n(s) \;=\; \arg\max_{a\in\mathcal{A}} \left(C(a) + \gamma P^\pi v^n\right) \tag{14.39}$$

for each state s. For our batch replenishment example, it turns out that we can show that $a^n(s)$ will have the same structure as that shown in (14.38). So, we can either store $a^n(s)$ for each s, or simply determine the parameters (q, Q) that correspond to the decisions produced by (14.39). The complete policy iteration algorithm is described in Figure 14.7.

The policy iteration algorithm is simple to implement and has fast convergence when measured in terms of the number of iterations. However, solving equation (14.37) is quite hard if the number of states is large. If the state space is small, we can use $v^\pi \;=\; (I - \gamma P^\pi)^{-1} c^\pi$, but the matrix inversion can be computationally expensive. For this reason, we may use a hybrid algorithm that combines the features of policy iteration and value iteration.

14.8   Hybrid Value-Policy Iteration*

Value iteration is basically an algorithm that updates the value at each iteration and then determines a new policy given the new estimate of the value function. At any iteration, the value function is not the true, steady-state value of the policy. By contrast, policy iteration picks a policy and then determines the true, steady-state value of being in each state given the policy. Given this value, a new policy is chosen.

It is perhaps not surprising that policy iteration converges faster in terms of the number of iterations because it is doing a lot more work in each iteration (determining the true, steady-state value of being in each state under a policy). Value iteration is much faster per iteration, but it is determining a policy given an approximation of a value function and then performing a very simple updating of the value function, which may be far from the true value function.

A hybrid strategy that combines features of both methods is to perform a somewhat more complete update of the value function before performing an update of the policy. Figure 14.8 outlines the procedure where the steady-state evaluation of the value function in equation (14.37) is replaced with a much easier iterative procedure (step 2 in Figure 14.8). This step is run for M iterations, where M is a user-controlled parameter that allows the exploration of the value of a better estimate of the value function. Not surprisingly, it will generally be the case that M should decline with the number of iterations as the overall process converges.

Step 0. Initialization:

- Set $n = 1$.
- Select a tolerance parameter ε and inner iteration limit M.
- Select some $v^0 \in \mathcal{V}$.

Step 1. Find a decision $a^n(s)$ for each s that satisfies

$$a^n(s) \quad = \quad \arg\max_{a \in \mathcal{A}} \left\{ C(s, a) + \gamma \sum_{s' \in \mathcal{S}} \mathbb{P}(s'|s, a) v^{n-1}(s') \right\},$$

which we represent as policy π^n.

Step 2. Partial policy evaluation.

(a) Set $m = 0$ and let: $u^n(0) = c^\pi + \gamma P^{\pi^n} v^{n-1}$.
(b) If $\|u^n(0) - v^{n-1}\| < \varepsilon(1 - \gamma)/2\gamma$, go to step 3. Else:
(c) While $m < M$ do the following:
  (i) $u^n(m + 1) = c^{\pi^n} + \gamma P^{\pi^n} u^n(m) = \mathcal{M}^\pi u^n(m)$.
  (ii) Set $m = m + 1$ and repeat (i).
(d) Set $v^n = u^n(M), n = n + 1$ and return to step 1.

Step 3. Set $a^\varepsilon = a^{n+1}$ and stop.

Figure 14.8 Hybrid value/policy iteration.

14.9 Average Reward Dynamic Programming*

There are settings where the natural objective function is to maximize the *average* contribution per unit time. Assume we start in state s. Then, the average reward from starting in state s and following policy π is given by

$$\max_\pi F^\pi(s) = \max_\pi \lim_{T \to \infty} \frac{1}{T} \mathbb{E} \sum_{t=0}^{T} C(S_t, A^\pi(S_t)). \tag{14.40}$$

Here, $F^\pi(s)$ is the expected reward *per time period*. In matrix form, the total value of following a policy π over a horizon T can be written as

$$V_T^\pi = \sum_{t=0}^{T} (P^\pi)^t c^\pi,$$

where V_T^π is a column vector with element $V_T^\pi(s)$ giving the expected contribution over T time periods when starting in state s. We can get a sense of how $V_T^\pi(s)$ behaves by watching what happens as T becomes large. Assuming that our underlying Markov chain is ergodic (which means you can eventually get from any state to any other state with positive probability), we know that $(P^\pi)^T \to P^*$ where the rows of P^* are all the same.

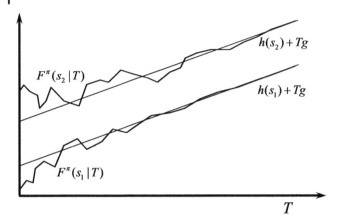

Figure 14.9 Cumulative contribution over a horizon T when starting in states s_1 and s_2, showing growth approaching a rate that is independent of the starting state.

Now define a column vector g given by

$$g^\pi = P^* c^\pi.$$

Since the rows of P^* are all the same, all the elements of g^π are the same, and each element gives the average contribution per time period using the steady state probability of being in each state. For finite T, each element of the column vector V_T^π is not the same, since the contributions we earn in the first few time periods depends on our starting state. But it is not hard to see that as T grows large, we can write

$$V_T^\pi \to h^\pi + Tg^\pi,$$

where the vector h^π captures the state-dependent differences in the total contribution, while g^π is the state-independent average contribution in the limit. Figure 14.9 illustrates the growth in V_T^π toward a linear function.

If we wish to find the policy that performs the best as $T \to \infty$, then clearly the contribution of h^π vanishes, and we want to focus on maximizing g^π, which we can now treat as a scalar.

14.10 The Linear Programming Method for Dynamic Programs**

Theorem 14.6.1 showed us that if

$$v \geq \max_a \left(C(s, a) + \gamma \sum_{s' \in \mathcal{S}} \mathbb{P}(s'|s, a) v(s') \right),$$

then v is an upper bound (actually, a vector of upper bounds) on the value of being in each state. This means that the optimal solution, which satisfies $v^* = c + \gamma P v^*$, is the smallest value of v that satisfies this inequality. We can use this insight to formulate the problem of finding the optimal values as a linear program. Let β be a vector with elements $\beta_s > 0$, $\forall s \in \mathcal{S}$. The optimal value function can be found by solving the following linear program

$$\min_{v} \sum_{s \in \mathcal{S}} \beta_s v(s) \tag{14.41}$$

subject to

$$v(s) \geq C(s, a) + \gamma \sum_{s' \in \mathcal{S}} \mathbb{P}(s'|s, a)v(s') \quad \text{for all } s \text{ and } a. \tag{14.42}$$

The linear program has a $|\mathcal{S}|$-dimensional decision vector (the value of being in each state), with $|\mathcal{S}| \times |\mathcal{A}|$ inequality constraints (equation (14.42)).

This formulation was viewed as primarily a theoretical result for many years, since it requires formulating a linear program where the number of constraints is equal to the number of states times the number of actions. While even today this limits the size of problems it can solve, modern linear programming solvers can handle problems with tens of thousands of constraints without difficulty. This size is greatly expanded with the use of specialized algorithmic strategies which are an active area of research as of this writing.

The advantage of the LP method over value iteration is that it avoids the need for iterative learning with the geometric convergence exhibited by value iteration. Given the dramatic strides in the speed of linear programming solvers over the last decade, the relative performance of value iteration over the linear programming method is an unresolved question. However, this question only arises for problems with relatively small state and action spaces. While a linear program with 50,000 constraints is considered large, dynamic programs with 50,000 states and actions tend to be relatively small problems.

14.11 Linear Quadratic Regulation

Easily the best known problem in optimal control is the problem known as *linear quadratic regulation*. This is a problem known only to the controls community, and for this reason, we are going to revert to classical controls notation, which is the only way it will ever appear in any popular presentation (except that we are still going to use time t in contrast with the more typical "k" used by the controls community). As you read this model, it is best to think of it in the context of a problem such as managing a robot or rocket where

x_t = the state vector, giving the location (in two or three dimensions), and velocity (again, in two or three dimensions),

u_t = the control vector, giving the force applied to each of the two (or three) dimensions.

The state evolves according to linear equations

$$x_{t+1} = A_t x_t + B_t u_t, \tag{14.43}$$

which captures the effect of force on location and velocity. Our goal is to find controls $u_1, \dots, u_T$ to minimize the costs given by

$$C_t(x_t, u_t) = \frac{1}{2} x_t^T Q_t x_t + u_t^T R_T u_t, \quad t = 1, \dots, T-1,$$

$$C_T(x_T) = \frac{1}{2} x_T^T S_T x_T,$$

where Q_t, R_t and S_T are symmetric, positive semidefinite matrices. The objective function is then given by

$$J = \frac{1}{2} \sum_{t=1}^{T} C_t(x_t, u_t) + \frac{1}{2} C_T(x_T),$$

subject to the system dynamics given by (14.43). Note that x_t and u_t are unconstrained. We are going to use the principle of *Lagrangian relaxation* and relax (14.43) and add the deviation to the objective function, giving us the *Lagrangian* (this is a standard optimization technique)

$$L(u_1, \dots, u_T, \lambda) = \frac{1}{2} \sum_{t=1}^{T} \left(C_t(x_t, u_t) + \lambda_{t+1}(A_t x_t + B_t u_t - x_{t+1}) \right) + \frac{1}{2} C_T(x_T). \tag{14.44}$$

The controls community then defines a portion of the Lagrangian, called the Hamiltonian, which is

$$H_t = \frac{1}{2} x_t^T Q_t x_t + u_t^T R_t u_t + \lambda_{t+1}(A_t x_t + B_t u_t).$$

Differentiating (14.44) with respect to λ_{t+1} and setting the derivative equal to zero (which would be true at optimality) returns our transition equation by setting

$$\frac{\partial L(u_1, \dots, u_T, \lambda)}{\partial \lambda_{t+1}} = (A_t x_t + B_t u_t - x_{t+1}) = 0,$$

from which we regain (14.43) (known as the *state equations*). Students of linear programming will recognize that λ_t as a dual variable.

Then, we obtain the *costate equations* by differentiating H_t with respect to x_t giving us

$$\lambda_t = \frac{\partial H_t}{\partial x_t} = Q_t x_t + A_t^T \lambda_{t+1}, \tag{14.45}$$

which essentially gives us our dual variables.

This system is solved essentially using the derivatives given in section 12.7. From these, we can derive the following feedback equations that are solved backward in time starting from a given S_T (typically determined by where we want our device to end up):

$$S_t = A_t^T \left[S_{t+1} - S_{t+1} B_t (B_t^T S_{t+1} B_t + R_t)^{-1} B_t^T S_{t+1} \right] A_t + Q_t. \tag{14.46}$$

We can then compute

$$K_t = \left(B_t^T S_{t+1} B_t + R_t \right)^{-1} B_t^T S_{t+1} A_t. \tag{14.47}$$

Our optimal control is then given by

$$u_t^* = -K_t x_t. \tag{14.48}$$

We note that these derivations have all been done in the context of a deterministic problem. One way to introduce uncertainty is with additive noise

$$x_{t+1} = A_t x_t + B_t u_t + w_t, \tag{14.49}$$

where w_t is random at time t (this is the classical style of the optimal control community – we use W_{t+1} elsewhere in this book). Additive noise can enter, for example, when exogenous forces (such as wind) interfere with the evolution of the system over time.

Adding additive noise as we do in equation (14.49) does not change our solution. When it is added to the Hamiltonian, we take expectations and since we will assume that $\mathbb{E} w_t = 0$, the noise term just drops out.

This is again a rare case of a truly optimal policy, which is very dependent on the characteristics of this problem:

- The quadratic form of the cost function (quadratic in both the state x_t and control u_t).
- The fact that it is completely unconstrained.

What is especially important about the optimal control given by (14.48) is that it is *linear* in the controls u_t. This suggests a starting point for problems that do not satisfy all of these conditions. One strategy that has been successfully applied is to assume that the policy is locally linear, which is that it is linear in the controls, but with coefficients that are defined only over specific regions.

14.12 Why Does it Work?**

The theory of Markov decision processes is especially elegant for students who enjoy probabilistic mathematics. While not needed for computational work, an understanding of why they work will provide a deeper appreciation of the properties of these problems.

Section 14.12.1 provides a proof that the optimal value function satisfies the optimality equations. Section 14.12.2 proves convergence of the value iteration algorithm. Section 14.12.3 then proves conditions under which value iteration increases or decreases monotonically to the optimal solution. Then, section 14.12.4 proves the bound on the error when value iteration satisfies the termination criterion given in section 14.6.3. Section 14.12.5 closes with a discussion of deterministic and randomized policies, along with a proof that deterministic policies are always at least as good as a randomized policy.

14.12.1 The Optimality Equations

Until now, we have been presenting the optimality equations as though they were a fundamental law of some sort. To be sure, they can easily look as though they were intuitively obvious, but it is still important to establish the relationship between the original optimization problem and the optimality equations. Since these equations are the foundation of dynamic programming, it seems beholden on us to work through the steps of proving that they are actually true.

We start by remembering the original optimization problem

$$
F_t^\pi(S_t) = \mathbb{E}\left\{\sum_{t'=t}^{T-1} C_{t'}(S_{t'}, A_{t'}^\pi(S_{t'})) + C_T(S_T)|S_t\right\}. \tag{14.50}
$$

Since (14.50) is, in general, exceptionally difficult to solve, we resort to the optimality equations

$$
V_t^\pi(S_t) = C_t(S_t, A_t^\pi(S_t)) + \mathbb{E}\{V_{t+1}^\pi(S_{t+1})|S_t\}. \tag{14.51}
$$

Our challenge is to show that these are the same. In order to establish this result, it is going to help if we first prove the following:

Lemma 14.12.1. Let S_t be a state variable that captures the relevant history up to time t, and let $F_{t'}(S_{t+1})$ be some function measured at time $t' \geq t + 1$ conditioned on the random variable S_{t+1}. Then

$$
\mathbb{E}\left[\mathbb{E}\{F_{t'}|S_{t+1}\}|S_t\right] = \mathbb{E}\left[F_{t'}|S_t\right]. \tag{14.52}
$$

Proof: This lemma is variously known as the law of iterated expectations or the tower property. Assume, for simplicity, that $F_{t'}$ is a discrete, finite random variable that takes outcomes in $\mathcal{F}$. We start by writing

$$\mathbb{E}\{F_{t'}|S_{t+1}\} = \sum_{f \in \mathcal{F}} f\mathbb{P}(F_{t'} = f|S_{t+1}). \qquad (14.53)$$

Recognizing that S_{t+1} is a random variable, we may take the expectation of both sides of (14.53), conditioned on S_t as follows:

$$\mathbb{E}\left[\mathbb{E}\{F_{t'}|S_{t+1}\}|S_t\right] = \sum_{S_{t+1} \in \mathcal{S}} \sum_{f \in \mathcal{F}} f\mathbb{P}(F_{t'} = f|S_{t+1}, S_t)\mathbb{P}(S_{t+1} = S_{t+1}|S_t). \quad (14.54)$$

First, we observe that we may write $\mathbb{P}(F_{t'} = f|S_{t+1}, S_t) = \mathbb{P}(F_{t'} = f|S_{t+1})$, because conditioning on S_{t+1} makes all prior history irrelevant. Next, we can reverse the summations on the right-hand side of (14.54) (some technical conditions have to be satisfied to do this, but these are satisfied if the random variables are discrete and finite). This means

$$
\begin{aligned}
\mathbb{E}\left[\mathbb{E}\{F_{t'}|S_{t+1} = S_{t+1}\}|S_t\right] &= \sum_{f \in \mathcal{F}} \sum_{S_{t+1} \in \mathcal{S}} f\mathbb{P}(F_{t'} = f|S_{t+1}, S_t)\mathbb{P}(S_{t+1} = S_{t+1}|S_t) \\
&= \sum_{f \in \mathcal{F}} f \sum_{S_{t+1} \in \mathcal{S}} \mathbb{P}(F_{t'} = f, S_{t+1}|S_t) \\
&= \sum_{f \in \mathcal{F}} f\mathbb{P}(F_{t'} = f|S_t) \\
&= \mathbb{E}\left[F_{t'}|S_t\right],
\end{aligned}
$$

which proves our result. Note that the essential step in the proof occurs in the first step when we add S_t to the conditioning. □

We are now ready to show the following:

Proposition 14.12.1. $F_t^{\pi}(S_t) = V_t^{\pi}(S_t)$.

Proof: To prove that (14.50) and (14.51) are equal, we use a standard trick in dynamic programming: proof by induction. Clearly, $F_T^{\pi}(S_T) = V_T^{\pi}(S_T) = C_T(S_T)$. Next, assume that it holds for $t + 1, t + 2, \dots, T$. We want to show that it is true for t. This means that we can write

$$V_t^{\pi}(S_t) = C_t(S_t, A_t^{\pi}(S_t)) + \mathbb{E}\left[\underbrace{\mathbb{E}\left\{\sum_{t'=t+1}^{T-1} C_{t'}(S_{t'}, A_{t'}^{\pi}(S_{t'})) + C_t(S_T(\omega))\Big|S_{t+1}\right\}}_{F_{t+1}^{\pi}(S_{t+1})}\Big|S_t\right].$$

We then use lemma 14.12.1 to write $\mathbb{E}\left[\mathbb{E}\{... |S_{t+1}\}|S_t\right] = \mathbb{E}\left[... |S_t\right]$. Hence,

$$V_t^\pi(S_t) \;=\; C_t(S_t, A_t^\pi(S_t)) + \mathbb{E}\left[\sum_{t'=t+1}^{T-1} C_{t'}(S_{t'}, A_{t'}^\pi(S_{t'})) + C_t(S_T)|S_t\right].$$

When we condition on S_t, $A_t^\pi(S_t)$ (and therefore $C_t(S_t, A_t^\pi(S_t))$) is deterministic, so we can pull the expectation out to the front giving

$$V_t^\pi(S_t) \;=\; \mathbb{E}\left[\sum_{t'=t}^{T-1} C_{t'}(S_{t'}, y_{t'}(S_{t'})) + C_t(S_T)|S_t\right]$$

$$=\; F_t^\pi(S_t),$$

which proves our result. □

Using equation (14.51), we have a backward recursion for calculating $V_t^\pi(S_t)$ for a given policy π. Now that we can find the expected reward for a given π, we would like to find the best π. That is, we want to find

$$F_t^*(S_t) \;=\; \max_{\pi \in \Pi} F_t^\pi(S_t).$$

If the set Π is infinite, we replace the "max" with "sup." We solve this problem by solving the optimality equations. These are

$$V_t(S_t) \;=\; \max_{a \in \mathcal{A}}\left(C_t(S_t, a) + \sum_{s' \in \mathcal{S}} p_t(s'|S_t, a)V_{t+1}(s')\right). \tag{14.55}$$

We are claiming that if we find the set of $V's$ that solves (14.55), then we have found the policy that optimizes F_t^π. We state this claim formally as:

Theorem 14.12.1. Let $V_t(S_t)$ be a solution to equation (14.55). Then

$$F_t^* \;=\; V_t(S_t)$$

$$=\; \max_{\pi \in \Pi} F_t^\pi(S_t).$$

Proof: The proof is in two parts. First, we show by induction that $V_t(S_t) \geq F_t^*(S_t)$ for all $S_t \in \mathcal{S}$ and $t = 0, 1, ..., T - 1$. Then, we show that the reverse inequality is true, which gives us the result.

Part 1:

We resort again to our proof by induction. Since $V_T(S_T) = C_t(S_T) = F_T^\pi(S_T)$ for all S_T and all $\pi \in \Pi$, we get that $V_T(S_T) = F_T^*(S_T)$.

Assume that $V_{t'}(S_{t'}) \geq F_{t'}^*(S_{t'})$ for $t' = t + 1, t + 2, \ldots, T$, and let π be an arbitrary policy. For $t' = t$, the optimality equation tells us

$$V_t(S_t) = \max_{a \in \mathcal{A}} \left(C_t(S_t, a) + \sum_{s' \in \mathcal{S}} p_t(s'|S_t, a) V_{t+1}(s') \right).$$

By the induction hypothesis, $F_{t+1}^*(s) \leq V_{t+1}(s)$, so we get

$$V_t(S_t) \geq \max_{a \in \mathcal{A}} \left(C_t(S_t, a) + \sum_{s' \in \mathcal{S}} p_t(s'|S_t, a) F_{t+1}^*(s') \right).$$

Of course, we have that $F_{t+1}^*(s) \geq F_{t+1}^\pi(s)$ for an arbitrary π. Also let $A^\pi(S_t)$ be the decision that would be chosen by policy π when in state S_t. Then

$$V_t(S_t) \geq \max_{a \in \mathcal{A}} \left(C_t(S_t, a) + \sum_{s' \in \mathcal{S}} p_t(s'|S_t, a) F_{t+1}^\pi(s') \right)$$

$$\geq C_t(S_t, A^\pi(S_t)) + \sum_{s' \in \mathcal{S}} p_t(s'|S_t, A^\pi(S_t)) F_{t+1}^\pi(s')$$

$$= F_t^\pi(S_t).$$

This means

$$V_t(S_t) \geq F_t^\pi(S_t) \quad \text{for all } \pi \in \Pi,$$

which proves part 1.

Part 2:

Now we are going to prove the inequality from the other side. Specifically, we want to show that for any $\epsilon > 0$ there exists a policy π that satisfies

$$F_t^\pi(S_t) + (T - t)\epsilon \geq V_t(S_t). \tag{14.56}$$

To do this, we start with the definition

$$V_t(S_t) = \max_{a \in \mathcal{A}} \left(C_t(S_t, a) + \sum_{s' \in \mathcal{S}} p_t(s'|S_t, a) V_{t+1}(s') \right). \tag{14.57}$$

We may let $a_t(S_t)$ be the decision rule that solves (14.57). This rule corresponds to the policy π. In general, the set $\mathcal{A}$ may be infinite, whereupon we have to replace the "max" with a "sup" and handle the case where an optimal decision may not exist. For this case, we know that we can design a decision rule $a_t(S_t)$ that returns a decision a that satisfies

$$V_t(S_t) \leq C_t(S_t, a) + \sum_{s' \in \mathcal{S}} p_t(s'|S_t, a) V_{t+1}(s') + \epsilon. \tag{14.58}$$

We can prove (14.56) by induction. We first note that (14.56) is true for $t = T$ since $F_T^\pi(S_t) = V_T(S_T)$. Now assume that it is true for $t' = t + 1, t + 2, \ldots, T$. We already know that

$$F_t^\pi(S_t) = C_t(S_t, A^\pi(S_t)) + \sum_{s' \in \mathcal{S}} p_t(s'|S_t, A^\pi(S_t)) F_{t+1}^\pi(s').$$

We can use our induction hypothesis which says $F_{t+1}^\pi(s') \geq V_{t+1}(s') - (T - (t + 1))\epsilon$ to get

$$F_t^\pi(S_t) \geq C_t(S_t, A^\pi(S_t)) + \sum_{s' \in \mathcal{S}} p_t(s'|S_t, A^\pi(S_t))[V_{t+1}(s') - (T - (t + 1))\epsilon]$$

$$= C_t(S_t, A^\pi(S_t)) + \sum_{s' \in \mathcal{S}} p_t(s'|S_t, A^\pi(S_t))V_{t+1}(s')$$

$$- \sum_{s' \in \mathcal{S}} p_t(s'|S_t, A^\pi(S_t))[(T - t - 1)\epsilon]$$

$$= \left\{ C_t(S_t, A^\pi(S_t)) + \sum_{s' \in \mathcal{S}} p_t(s'|S_t, A^\pi(S_t))V_{t+1}(s') + \epsilon \right\} - (T - t)\epsilon.$$

Now, using equation (14.58), we replace the term in brackets with the smaller $V_t(S_t)$ (equation (14.58)):

$$F_t^\pi(S_t) \geq V_t(S_t) - (T - t)\epsilon,$$

which proves the induction hypothesis. We have shown that

$$F_t^*(S_t) + (T - t)\epsilon \geq F_t^\pi(S_t) + (T - t)\epsilon \geq V_t(S_t) \geq F_t^*(S_t).$$

This proves the result. □

Now we know that solving the optimality equations also gives us the optimal value function. This is our most powerful result because we can solve the optimality equations for many problems that cannot be solved any other way.

14.12.2   Convergence of Value Iteration

We now undertake the proof that the basic value function iteration converges to the optimal solution. This is not only an important result, it is also an elegant one that brings some powerful theorems into play. The proof is also quite short. However, we will need some mathematical preliminaries:

Definition 14.12.1. Let $\mathcal{V}$ be a set of (bounded, real-valued) functions and define the norm of v by:

$$\|v\| = \sup_{s \in \mathcal{S}} v(s)$$

where we replace the "*sup*" with a "*max*" when the state space is finite. Since $\mathcal{V}$ is closed under addition and scalar multiplication and has a norm, it is a **normed linear space**.

Definition 14.12.2. $T : \mathcal{V} \rightarrow \mathcal{V}$ is a **contraction mapping** if there exists a γ, $0 \leq \gamma < 1$ such that:

$$\|Tv - Tu\| \leq \gamma\|v - u\|.$$

Definition 14.12.3. A sequence $v^n \in \mathcal{V}$, $n = 1, 2, \ldots$ is said to be a **Cauchy sequence** if for all $\epsilon > 0$, there exists N such that for all $n, m \geq N$:

$$\|v^n - v^m\| < \epsilon.$$

Definition 14.12.4. A normed linear space is **complete** if every Cauchy sequence contains a limit point in that space.

Definition 14.12.5. A **Banach space** is a complete normed linear space.

Definition 14.12.6. We define the norm of a matrix Q as

$$\|Q\| = \max_{s \in \mathcal{S}} \sum_{j \in \mathcal{S}} |q(j|s)|,$$

that is, the largest row sum of the matrix. If Q is a one-step transition matrix, then $\|Q\| = 1$.

Definition 14.12.7. The **triangle inequality** means that given two vectors $a, b \in \mathfrak{R}^n$:

$$\|a + b\| \leq \|a\| + \|b\|.$$

The triangle inequality is commonly used in proofs because it helps us establish bounds between two solutions (and in particular, between a solution and the optimum).

We now state and prove one of the famous theorems in applied mathematics and then use it immediately to prove convergence of the value iteration algorithm.

Theorem 14.12.2. (Banach Fixed-Point Theorem) Let $\mathcal{V}$ be a Banach space, and let $T : \mathcal{V} \rightarrow \mathcal{V}$ be a contraction mapping. Then:

(a) There exists a unique $v^* \in \mathcal{V}$ such that $Tv^* = v^*$.
(b) For an arbitrary $v^0 \in \mathcal{V}$, the sequence v^n defined by: $v^{n+1} = Tv^n = T^{n+1}v^0$ converges to v^*.

Proof: We start by showing that the distance between two vectors v^n and v^{n+m} goes to zero for sufficiently large n and by writing the difference $v^{n+m} - v^n$ using

$$v^{n+m} - v^n = v^{n+m} - v^{n+m-1} + v^{n+m-1} - \cdots - v^{n+1} + v^{n+1} - v^n$$

$$= \sum_{k=0}^{m-1} (v^{n+k+1} - v^{n+k}).$$

Taking norms of both sides and invoking the triangle inequality gives

$$
\begin{aligned}
\|v^{n+m} - v^n\| &= \left\| \sum_{k=0}^{m-1} (v^{n+k+1} - v^{n+k}) \right\| \\
&\leq \sum_{k=0}^{m-1} \|(v^{n+k+1} - v^{n+k})\| \\
&= \sum_{k=0}^{m-1} \|(T^{n+k}v^1 - T^{n+k}v^0)\| \\
&\leq \sum_{k=0}^{m-1} \gamma^{n+k} \|v^1 - v^0\| \\
&= \frac{\gamma^n(1-\gamma^m)}{(1-\gamma)} \|v^1 - v^0\|.
\end{aligned}
\tag{14.59}
$$

Since $\gamma < 1$, for sufficiently large n the right-hand side of (14.59) can be made arbitrarily small, which means that v^n is a Cauchy sequence. Since V is *complete*, it must be that v^n has a limit point v^*. From this we conclude

$$
\lim_{n \to \infty} v^n \to v^*.
\tag{14.60}
$$

We now want to show that v^* is a fixed point of the mapping T. To show this, we observe

$$
\begin{aligned}
0 &\leq \|Tv^* - v^*\| & (14.61) \\
&= \|Tv^* - v^n + v^n - v^*\| & (14.62) \\
&\leq \|Tv^* - v^n\| + \|v^n - v^*\| & (14.63) \\
&= \|Tv^* - Tv^{n-1}\| + \|v^n - v^*\| & (14.64) \\
&\leq \gamma\|v^* - v^{n-1}\| + \|v^n - v^*\|. & (14.65)
\end{aligned}
$$

Equation (14.61) comes from the properties of a norm. We play our standard trick in (14.62) of adding and subtracting a quantity (in this case, v^n), which sets up the triangle inequality in (14.63). Using $v^n = Tv^{n-1}$ gives us (14.64). The inequality in (14.65) is based on the assumption of the theorem that T is a contraction mapping. From (14.60), we know that

$$
\lim_{n \to \infty} \|v^* - v^{n-1}\| = \lim_{n \to \infty} \|v^n - v^*\| = 0.
\tag{14.66}
$$

Combining (14.61), (14.65), and (14.66) gives

$$
0 \leq \|Tv^* - v^*\| \leq 0,
$$

from which we conclude

$$\|Tv^* - v^*\| = 0,$$

which means that $Tv^* = v^*$.

We can prove uniqueness by contradiction. Assume that there are two limit points that we represent as v^* and u^*. The assumption that T is a contraction mapping requires that

$$\|Tv^* - Tu^*\| \leq \gamma\|v^* - u^*\|.$$

But, if v^* and u^* are limit points, then $Tv^* = v^*$ and $Tu^* = u^*$, which means

$$\|v^* - u^*\| \leq \gamma\|v^* - u^*\|.$$

Since $\gamma < 1$, this is a contradiction, which means that it must be true that $v^* = u^*$. □

We can now show that the value iteration algorithm converges to the optimal solution if we can establish that $\mathcal{M}$ is a contraction mapping. So we need to show the following:

Proposition 14.12.2. If $0 \leq \gamma < 1$, then $\mathcal{M}$ is a contraction mapping on $\mathcal{V}$.

Proof: Let $u, v \in \mathcal{V}$ and assume that $\mathcal{M}v \geq \mathcal{M}u$ where the inequality is applied elementwise. For a particular state s let

$$a_s^*(v) \in \arg\max_{a \in \mathcal{A}} \left(C(s,a) + \gamma \sum_{s' \in \mathcal{S}} \mathbb{P}(s'|s,a)v(s') \right)$$

where we assume that a solution exists. Then

$$0 \leq \mathcal{M}v(s) - \mathcal{M}u(s) \tag{14.67}$$

$$= C(s, a_s^*(v)) + \gamma \sum_{s' \in \mathcal{S}} \mathbb{P}(s'|s, a_s^*(v))v(s')$$

$$- \left(C(s, a_s^*(u)) + \gamma \sum_{s' \in \mathcal{S}} \mathbb{P}(s'|s, a_s^*(u))u(s') \right) \tag{14.68}$$

$$\leq C(s, a_s^*(v)) + \gamma \sum_{s' \in \mathcal{S}} \mathbb{P}(s'|s, a_s^*(v))v(s')$$

$$- \left(C(s, a_s^*(v)) + \gamma \sum_{s' \in \mathcal{S}} \mathbb{P}(s'|s, a_s^*(v))u(s') \right) \tag{14.69}$$

$$= \gamma \sum_{s' \in \mathcal{S}} \mathbb{P}(s'|s, a_s^*(v))[v(s') - u(s')] \tag{14.70}$$

$$\leq \gamma \sum_{s' \in \mathcal{S}} \mathbb{P}(s'|s, a_s^*(v)) \|v - u\| \tag{14.71}$$

$$= \gamma \|v - u\| \sum_{s' \in \mathcal{S}} \mathbb{P}(s'|s, a_s^*(v)) \tag{14.72}$$

$$= \gamma \|v - u\|. \tag{14.73}$$

Equation (14.67) is true by assumption, while (14.68) holds by definition. The inequality in (14.69) holds because $a_s^*(v)$ is not optimal when the value function is u, giving a reduced value in the second set of parentheses. Equation (14.70) is a simple reduction of (14.69). Equation (14.71) forms an upper bound because the definition of $\|v - u\|$ is to replace all the elements $[v(s) - u(s)]$ with the largest element of this vector. Since this is now a vector of constants, we can pull it outside of the summation, giving us (14.72), which then easily reduces to (14.73) because the probabilities add up to one.

This result states that if $\mathcal{M}v(s) \geq \mathcal{M}u(s)$, then $\mathcal{M}v(s) - \mathcal{M}u(s) \leq \gamma |v(s) - u(s)|$. If we start by assuming that $\mathcal{M}v(s) \leq \mathcal{M}u(s)$, then the same reasoning produces $\mathcal{M}v(s) - \mathcal{M}u(s) \geq -\gamma |v(s) - u(s)|$. This means that we have

$$|\mathcal{M}v(s) - \mathcal{M}u(s)| \leq \gamma |v(s) - u(s)| \tag{14.74}$$

for *all* states $s \in \mathcal{S}$. From the definition of our norm, we can write

$$\sup_{s \in \mathcal{S}} |\mathcal{M}v(s) - \mathcal{M}u(s)| = \|\mathcal{M}v - \mathcal{M}u\|$$

$$\leq \gamma \|v - u\|.$$

This means that $\mathcal{M}$ is a contraction mapping, which means that the sequence v^n generated by $v^{n+1} = \mathcal{M}v^n$ converges to a unique limit point v^* that satisfies the optimality equations. □

14.12.3 Monotonicity of Value Iteration

Infinite horizon dynamic programming provides a compact way to study the theoretical properties of these algorithms. The insights gained here are applicable to problems even when we cannot apply this model, or these algorithms, directly.

We assume throughout our discussion of infinite horizon problems that the reward function is bounded over the domain of the state space. This assumption is virtually always satisfied in practice, but notable exceptions exist. For example, the assumption is violated if we are maximizing a utility function that depends on the log of the resources we have at hand (the resources may be bounded, but the function is unbounded if the resources are allowed to hit zero).

Our first result establishes a monotonicity property that can be exploited in the design of an algorithm.

Theorem 14.12.3. For a vector $v \in \mathcal{V}$:

(a) If v satisfies $v \geq \mathcal{M}v$, then $v \geq v^*$.
(b) If v satisfies $v \leq \mathcal{M}v$, then $v \leq v^*$.
(c) If v satisfies $v = \mathcal{M}v$, then v is the unique solution to this system of equations and $v = v^*$.

Proof: Part (a) requires that

$$v \geq \max_{\pi \in \Pi}\{c^\pi + \gamma P^\pi v\} \tag{14.75}$$

$$\geq c^{\pi_0} + \gamma P^{\pi_0} v \tag{14.76}$$

$$\geq c^{\pi_0} + \gamma P^{\pi_0} (c^{\pi_1} + \gamma P^{\pi_1} v) \tag{14.77}$$

$$= c^{\pi_0} + \gamma P^{\pi_0} c^{\pi_1} + \gamma^2 P^{\pi_0} P^{\pi_1} v.$$

Equation (14.75) is true by assumption (part (a) of the theorem) and equation (14.76) is true because π_0 is some policy that is not necessarily optimal for the vector v. Using similar reasoning, equation (14.77) is true because π_1 is another policy which, again, is not necessarily optimal. Using $P^{\pi,(t)} = P^{\pi_0} P^{\pi_1} \cdots P^{\pi_t}$, we obtain by induction

$$v \geq c^{\pi_0} + \gamma P^{\pi_0} c^{\pi_1} + \cdots + \gamma^{t-1} P^{\pi_0} P^{\pi_1} \cdots P^{\pi_{t-1}} c^{\pi_t} + \gamma^t P^{\pi,(t)} v. \tag{14.78}$$

Recall that

$$v^\pi = \sum_{t=0}^{\infty} \gamma^t P^{\pi,(t)} c^{\pi_t}. \tag{14.79}$$

Breaking the sum in (14.79) into two parts allows us to rewrite the expansion in (14.78) as

$$v \geq v^\pi - \sum_{t'=t+1}^{\infty} \gamma^{t'} P^{\pi,(t')} c^{\pi_{t'+1}} + \gamma^t P^{\pi,(t)} v. \tag{14.80}$$

Taking the limit of both sides of (14.80) as $t \to \infty$ gives us

$$v \geq \lim_{t \to \infty} v^\pi - \sum_{t'=t+1}^{\infty} \gamma^{t'} P^{\pi,(t')} c^{\pi_{t'+1}} + \gamma^t P^{\pi,(t)} v \tag{14.81}$$

$$\geq v^\pi \quad \forall \pi \in \Pi. \tag{14.82}$$

The limit in (14.81) exists as long as the reward function c^π is bounded and $\gamma < 1$. Because (14.82) is true for all $\pi \in \Pi$, it is also true for the optimal policy, which means that

$$
\begin{aligned}
v &\geq v^{\pi*} \\
&= v^*,
\end{aligned}
$$

which proves part (a) of the theorem. Part (b) can be proved in an analogous way. Parts (a) and (b) mean that $v \geq v^*$ and $v \leq v^*$. If $v = \mathcal{M}v$, then we satisfy the preconditions of both parts (a) and (b), which means they are both true and therefore we must have $v = v^*$. □

This result means that if we start with a vector that is higher than the optimal vector, then we will decline monotonically to the optimal solution (almost – we have not quite proven that we actually get to the optimal). Alternatively, if we start below the optimal vector, we will rise to it. Note that it is not always easy to find a vector v that satisfies either condition (a) or (b) of the theorem. In problems where the rewards can be positive and negative, this can be tricky.

14.12.4 Bounding the Error from Value Iteration

We now wish to establish a bound on our error from value iteration, which will establish our stopping rule. We propose two bounds: one on the value function estimate that we terminate with and one for the long-run value of the decision rule that we terminate with. To define the latter, let π^ϵ be the policy that satisfies our stopping rule, and let v^{π^ϵ} be the infinite horizon value of following policy π^ϵ.

Theorem 14.12.4. If we apply the value iteration algorithm with stopping parameter ϵ and the algorithm terminates at iteration n with value function v^{n+1}, then

$$
\|v^{n+1} - v^*\| \leq \epsilon/2, \tag{14.83}
$$

and

$$
\|v^{\pi^\epsilon} - v^*\| \leq \epsilon. \tag{14.84}
$$

Proof: We start by writing

$$
\begin{aligned}
\|v^{\pi^\epsilon} - v^*\| &= \|v^{\pi^\epsilon} - v^{n+1} + v^{n+1} - v^*\| \\
&\leq \|v^{\pi^\epsilon} - v^{n+1}\| + \|v^{n+1} - v^*\|. \tag{14.85}
\end{aligned}
$$

Recall that π^ϵ is the policy that solves $\mathcal{M}v^{n+1}$, which means that $\mathcal{M}^{\pi^\epsilon}v^{n+1} = \mathcal{M}v^{n+1}$. This allows us to rewrite the first term on the right-hand side of (14.85) as

$$
\begin{aligned}
\|v^{\pi^\epsilon} - v^{n+1}\| &= \|\mathcal{M}^{\pi^\epsilon} v^{\pi^\epsilon} - \mathcal{M}v^{n+1} + \mathcal{M}v^{n+1} - v^{n+1}\| \\
&\leq \|\mathcal{M}^{\pi^\epsilon} v^{\pi^\epsilon} - \mathcal{M}v^{n+1}\| + \|\mathcal{M}v^{n+1} - v^{n+1}\| \\
&= \|\mathcal{M}^{\pi^\epsilon} v^{\pi^\epsilon} - \mathcal{M}^{\pi^\epsilon} v^{n+1}\| + \|\mathcal{M}v^{n+1} - \mathcal{M}v^n\| \\
&\leq \gamma \|v^{\pi^\epsilon} - v^{n+1}\| + \gamma \|v^{n+1} - v^n\|.
\end{aligned}
$$

Solving for $\|v^{\pi^\epsilon} - v^{n+1}\|$ gives

$$
\|v^{\pi^\epsilon} - v^{n+1}\| \leq \frac{\gamma}{1-\gamma} \|v^{n+1} - v^n\|.
$$

We can use similar reasoning applied to the second term in equation (14.85) to show that

$$
\|v^{n+1} - v^*\| \leq \frac{\gamma}{1-\gamma} \|v^{n+1} - v^n\|. \tag{14.86}
$$

The value iteration algorithm stops when $\|v^{n+1} - v^n\| \leq \epsilon(1-\gamma)/2\gamma$. Substituting this in (14.86) gives

$$
\|v^{n+1} - v^*\| \leq \frac{\epsilon}{2}. \tag{14.87}
$$

Recognizing that the same bound applies to $\|v^{\pi^\epsilon} - v^{n+1}\|$ and combining these with (14.85) gives us

$$
\|v^{\pi^\epsilon} - v^*\| \leq \epsilon,
$$

which completes our proof. □

14.12.5 Randomized Policies

We have implicitly assumed that for each state, we want a single action. An alternative would be to choose a policy probabilistically from a family of policies. If a state produces a single action, we say that we are using a *deterministic policy*. If we are randomly choosing an action from a set of actions probabilistically, we say we are using a *randomized policy*.

Randomized policies may arise because of the nature of the problem. For example, you wish to purchase something at an auction, but you are unable to attend yourself. You may have a simple rule ("purchase it as long as the price is under a specific amount") but you cannot assume that your representative will apply the same rule. You can choose a representative, and in doing so you are effectively choosing the probability distribution from which the action will be chosen.

Behaving randomly also plays a role in two-player games. If you make the same decision each time in a particular state, your opponent may be able to predict your behavior and gain an advantage. For example, as an institutional investor you may tell a bank that you are not willing to pay any more than $14 for a new offering of stock, while in fact you are willing to pay up to $18. If you always bias your initial prices by $4, the bank will be able to guess what you are willing to pay.

When we can only influence the likelihood of an action, then we have an instance of a randomized MDP. Let

$q_t^\pi(a|S_t)$ = The probability that decision a will be taken at time t given state S_t and policy π (more precisely, decision rule A^π).

In this case, our optimality equations look like

$$V_t^*(S_t) = \max_{\pi \in \Pi^{MR}} \sum_{a \in \mathcal{A}} \left[q_t^\pi(a|S_t) \left(C_t(S_t, a) + \sum_{s' \in \mathcal{S}} p_t(s'|S_t, a) V_{t+1}^*(s') \right) \right]. \quad (14.88)$$

Now let us consider the single best action that we could take. Calling this a^*, we can find it using

$$a^* = \arg\max_{a \in \mathcal{A}} \left[C_t(S_t, a) + \sum_{s' \in \mathcal{S}} p_t(s'|S_t, a) V_{t+1}^*(s') \right].$$

This means that

$$C_t(S_t, a^*) + \sum_{s' \in \mathcal{S}} p_t(s'|S_t, a^*) V_{t+1}^*(s') \geq C_t(S_t, a) + \sum_{s' \in \mathcal{S}} p_t(s'|S_t, a) V_{t+1}^*(s') \quad (14.89)$$

for all $a \in \mathcal{A}$. Substituting (14.89) back into (14.88) gives us

$$
\begin{aligned}
V_t^*(S_t) &= \max_{\pi \in \Pi^{MR}} \sum_{a \in \mathcal{A}} \left[q_t^\pi(a|S_t) \left(C_t(S_t, a) + \sum_{s' \in \mathcal{S}} p_t(s'|S_t, a) V_{t+1}^*(s') \right) \right] \\
&\leq \max_{\pi \in \Pi^{MR}} \sum_{a \in \mathcal{A}} \left[q_t^\pi(a|S_t) \left(C_t(S_t, a^*) + \sum_{s' \in \mathcal{S}} p_t(s'|S_t, a^*) V_{t+1}^*(s') \right) \right] \\
&= C_t(S_t, a^*) + \sum_{s' \in \mathcal{S}} p_t(s'|S_t, a^*) V_{t+1}^*(s').
\end{aligned}
$$

What this means is that if you have a choice between picking exactly the action you want versus picking a probability distribution over potentially optimal and nonoptimal actions, you would always prefer to pick exactly the best action. Clearly, this is not a surprising result.

The value of randomized policies arise primarily in two-person games, where one player tries to anticipate the actions of the other player. In such situations, part of the state variable is the estimate of what the other play will do when the game is in a particular state. By randomizing his behavior, a player reduces the ability of the other player to anticipate his moves.

14.13   Bibliographic Notes

This chapter presents the classic view of Markov decision processes, for which the literature is extensive. Beginning with the seminal text of Bellman (Bellman (1957)), there have been numerous, significant textbooks on the subject, including Howard (1960), Nemhauser (1966), White (1969), Derman (1970), Bellman (1971), Dreyfus and Law (1977), Dynkin and Yushkevich (1979), Denardo (1982), Ross (1983), and Heyman and Sobel (1984). As of this writing, the current high-water mark for textbooks in this area is the landmark volume by Puterman (2005). Most of this chapter is based on Puterman (2005), modified to our notational style.

Section 14.10 – The linear programming method was first proposed in Manne (1960) (see subsequent discussions in Derman (1962) and Puterman (2005)). The so-called linear programming method was ignored for many years because of the large size of the linear programs that were produced, but the method has seen a resurgence of interest using approximation techniques. Recent research into algorithms for solving problems using this method are discussed in section 17.10.

Section 14.11 – This section was adapted from Lewis and Vrabie (2012), section 2.2.

Exercises

Review questions

14.1   Discrete Markov decision processes have been studied since the 1950's as a way of solving stochastic, dynamic programs. Yet, in chapter 4, this is used as an example of a stochastic optimization problem that can be solved deterministically. Explain.

14.2   A classical inventory problem works as follows: Assume that our state variable R_t is the amount of product on hand at the end of time period t and that D_t is a random variable giving the demand during time interval $(t-1, t)$ with distribution $p_d = \mathbb{P}(D_t = d)$. The demand in time interval t must be satisfied with the product on hand at the beginning of the period. We can then order a quantity x_t at the end of period t that can be used to replenish the inventory in period $t + 1$.

   (a) Give the transition function that relates R_{t+1} to R_t if the order quantity is x_t (where x_t is fixed for all R_t).

    (b) Give an algebraic version of the one-step transition matrix $P^\pi = \{p_{ij}^\pi\}$ where $p_{ij}^\pi = \mathbb{P}(R_{t+1} = j | R_t = i, A^\pi = x_t)$.

14.3 Repeat the previous exercise, but now assume that we have adopted a policy π that says we should order a quantity $x_t = 0$ if $R_t \geq s$ and $x_t = Q - R_t$ if $R_t < q$ (we assume that $R_t \leq Q$). Your expression for the transition matrix will now depend on our policy π (which describes both the structure of the policy and the control parameter s).

Modeling questions

14.4 Every day, a salesman visits N customers in order to sell the R identical items he has in his van. Each customer is visited exactly once and each customer buys zero or one item. Upon arrival at a customer location, the salesman quotes one of the prices $0 < p_1 \leq p_2 \leq \dots \leq p_m$. Given that the quoted price is p_i, a customer buys an item with probability r_i. Naturally, r_i is decreasing in i. The salesman is interested in maximizing the total expected revenue for the day. Show that if $r_i p_i$ is increasing in i, then it is always optimal to quote the highest price p_m.

14.5 You need to decide when to replace your car. If you own a car of age y years, then the cost of maintaining the car that year will be $c(y)$. Purchasing a new car (in constant dollars) costs P dollars. If the car breaks down, which it will do with probability $b(y)$ (the breakdown probability), it will cost you an additional K dollars to repair it, after which you immediately sell the car and purchase a new one. At the same time, you express your enjoyment with owning a new car as a negative cost $-r(y)$ where $r(y)$ is a declining function with age. At the beginning of each year, you may choose to purchase a new car ($z = 1$) or to hold onto your old one ($z = 0$). You anticipate that you will actively drive a car for another T years.

    (a) Identify all the elements of a Markov decision process for this problem.

    (b) Write out the objective function which will allow you to find an optimal decision rule.

    (c) Write out the one-step transition matrix.

    (d) Write out the optimality equations that will allow you to solve the problem.

14.6 Describe the gambling problem in section 14.4.1 as a decision tree, assuming that we can gamble only 0, 1, or 2 dollars in each round (this is just to keep the decision tree from growing too large).

14.7 You are trying to find the best parking space to use that minimizes the time needed to get to your restaurant. There are 50 parking spaces, and you see spaces 1, 2, ... , 50 in order. As you approach each parking space, you see whether it is full or empty. We assume, somewhat heroically, that the probability that each space is occupied follows an independent Bernoulli process, which is to say that each space will be occupied with probability p, but will be free with probability $1 - p$, and that each outcome is independent of the other.

It takes 2 seconds to drive past each parking space and it takes 8 seconds to walk past. That is, if we park in space n, it will require $8(50 - n)$ seconds to walk to the restaurant. Furthermore, it would have taken you $2n$ seconds to get to this space. If you get to the last space without finding an opening, then you will have to drive into a special lot down the block, adding 30 seconds to your trip.

We want to find an optimal strategy for accepting or rejecting a parking space.

(a) Give the sets of state and action spaces and the set of decision epochs.

(b) Give the expected reward function for each time period and the expected final reward function.

(c) Give a formal statement of the objective function.

(d) Give the optimality equations for solving this problem.

(e) You have just looked at space 45, which was empty. There are five more spaces remaining (46 through 50). What should you do? Using $p = 0.6$, find the optimal policy by solving your optimality equations for parking spaces 46 through 50.

(f) Give the optimal value of the objective function in part (e) corresponding to your optimal solution.

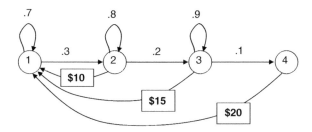

Computational exercises

14.8 We are going to use a very simple Markov decision process to illustrate how the initial estimate of the value function can affect convergence behavior. In fact, we are going to use a Markov reward process to illustrate the behavior because our process does not have any decisions. Assume we have a two-stage Markov chain with one-step transition matrix

$$P = \begin{bmatrix} 0.7 & 0.3 \\ 0.05 & 0.95 \end{bmatrix}.$$

The contribution from each transition from state $i \in \{1, 2\}$ to state $j \in \{1, 2\}$ is given by the matrix

$$\begin{bmatrix} 10 & 30 \\ 30 & 5 \end{bmatrix}.$$

That is, a transition from state 1 to state 2 returns a contribution of 30. Apply the value iteration algorithm for an infinite horizon problem (note that you are not choosing a decision so there is no maximization step). The calculation of the value of being in each state will depend on your previous estimate of the value of being in each state. The calculations can be easily implemented in a spreadsheet. Assume that your discount factor is .8.

(a) Plot the value of being in state 1 as a function of the number of iterations if your initial estimate of the value of being in each state is 0. Show the graph for 50 iterations of the algorithm.
(b) Repeat this calculation using initial estimates of 100.
(c) Repeat the calculation using an initial estimate of the value of being in state 1 of 100, and use 0 for the value of being in state 2. Contrast the behavior with the first two starting points.

14.9 Apply policy iteration to the problem given in exercise 14.8. Plot the average value function (that is, average the value of being in each state) after each iteration alongside the average value function found using value iteration after each iteration (for value iteration, initialize the value function to zero). Compare the computation time for one iteration of value iteration and one iteration of policy iteration.

14.10 Now apply the hybrid value-policy iteration algorithm to the problem given in exercise 14.8. Show the average value function after each major iteration (update of n) with $M = 1, 2, 3, 5, 10$. Compare the convergence rate to policy iteration and value iteration.

14.11 We have a four-state process (shown in the figure). In state 1, we will remain in the state with probability 0.7 and will make a transition to state 2 with probability 0.3. In states 2 and 3, we may choose between two policies: Remain in the state waiting for an upward transition or make the decision to return to state 1 and receive the indicated reward. In state 4, we return to state 1 immediately and receive $20. We wish to find an optimal long run policy using a discount factor $\gamma = .8$. Set up and solve the optimality equations for this problem.

14.12 Assume that you have been applying value iteration to a four-state Markov decision process, and that you have obtained the values over iterations 8 through 12 shown in the following table (assume a discount factor of 0.90). Assume you stop after iteration 12. Give the tightest possible (valid) bounds on the optimal value of being in each state.

	Iteration				
State	8	9	10	11	12
1	7.42	8.85	9.84	10.54	11.03
2	4.56	6.32	7.55	8.41	9.01
3	11.83	13.46	14.59	15.39	15.95
4	8.13	9.73	10.85	11.63	12.18

14.13 Assume that a control limit policy exists for our shuttle problem in exercise 2 that allows us to write the optimal dispatch rule as a function of s, as in $z^\pi(s)$. We may write $r(s, z)$ as a function of one variable, the state s.

(a) Illustrate the shape of $r(s, z(s))$ by plotting it over the range $0 < s < 3M$ (since we are allowing there to be more customers than can fill one vehicle, assume that we are allowed to send $z = 0, 1, 2, \dots$ vehicles in a single time period).

(b) Let $c = 10$, $h = 2$, and $M = 5$, and assume that $A_t = 1$ with probability 0.6 and is 0 with probability 0.4. Set up and solve a system of linear equations for the optimal value function for this problem in steady state.

Theory questions

14.14 Show that $\mathbb{P}(S_{t+\tau}|S_t)$, given that we are following a policy π (for stationary problems), is given by (14.22). [Hint: first show it for $\tau =$

1,2 and then use inductive reasoning to show that it is true for general τ.]

14.15 Repeat the derivation in section 14.4.2 assuming that the reward for task t is $c_t\sqrt{x_t}$.

14.16 Repeat the derivation in section 14.4.2 assuming that the reward for task t is given by $\ln(x)$.

14.17 Repeat the derivation in section 14.4.2 one more time, but now assume that all you know is that the reward is continuously differentiable, monotonically increasing and concave.

14.18 What happens to the answer to the budget allocation problem in section 14.4.2 if the contribution is convex instead of concave (for example, $C_t(x_t) = x_t^2$)?

14.19 In the proof of theorem 14.12.3 we showed that if $v \geq \mathcal{M}v$, then $v \geq v^*$. Go through the steps of proving the converse, that if $v \leq \mathcal{M}v$, then $v \leq v^*$.

14.20 Theorem 14.12.3 states that if $v \leq \mathcal{M}v$, then $v \leq v^*$. Show that if $v^n \leq v^{n+1} = \mathcal{M}v^n$, then $v^{m+1} \geq v^m$ for all $m \geq n$.

14.21 Consider a finite-horizon MDP with the following properties:

- $\mathcal{S} \in \mathfrak{R}^n$, the action space $\mathcal{A}$ is a compact subset of $\mathfrak{R}^n$, $\mathcal{X}(s) = \mathcal{X}$ for all $s \in \mathcal{S}$.
- $C_t(S_t, x_t) = c_t S_t + g_t(x_t)$, where $g_t(\cdot)$ is a known scalar function, and $C_T(S_T) = c_T S_T$.
- If decision x_t is chosen when the state is S_t at time t, the next state is

$$S_{t+1} = A_t S_t + f_t(x_t) + \omega_{t+1},$$

where $f_t(\cdot)$ is scalar function, and A_t and ω_t are respectively $n \times n$ and $n \times 1$-dimensional random variables whose distributions are independent of the history of the process prior to t.

(a) Show that the optimal value function is linear in the state variable.

(b) Show that there exists an optimal policy $\pi^* = (x_1^*, ..., x_{T-1}^*)$ composed of constant decision functions. That is, $A_t^{\pi^*}(s) = A_t^*$ for all $s \in \mathcal{S}$ for some constant A_t^*.

14.22 Assume that you have invested R_0 dollars in the stock market which evolves according to the equation

$$R_t = \gamma R_{t-1} + \varepsilon_t$$

where ε_t is a discrete, positive random variable that is independent and identically distributed and where $0 < \gamma < 1$. If you sell the stock at the end of period t, it will earn a riskless return r until time T, which means it will evolve according to

$$R_t = (1 + r)R_{t-1}.$$

You have to sell the stock, all on the same day, some time before T.

(a) Write a dynamic programming recursion to solve the problem.
(b) Show that there exists a point in time τ such that it is optimal to sell for $t \geq \tau$, and optimal to hold for $t < \tau$.
(c) How does your answer to (b) change if you are allowed to sell only a portion of the assets in a given period? That is, if you have R_t dollars in your account, you are allowed to sell $x_t \leq R_t$ at time t.

14.23 Show that the matrix H^n in the recursive updating formula from equation (3.68)

$$\bar{\theta}^n = \bar{\theta}^{n-1} - H^n x^n \hat{\varepsilon}^n$$

reduces to $H^n = 1/n$ for the case of a single parameter (which means we are using Y =constant, with no independent variables).

14.24 A dispatcher controls a finite capacity shuttle that works as follows: In each time period, a random number A_t arrives. After the arrivals occur, the dispatcher must decide whether to call the shuttle to remove up to M customers. The cost of dispatching the shuttle is c, which is independent of the number of customers on the shuttle. Each time period that a customer waits costs h. If we let $z = 1$ if the shuttle departs and 0 otherwise, then our one-period reward function is given by

$$c_t(s, z) = cz + h[s - Mz]^+,$$

where M is the capacity of the shuttle. Show that $c_t(s, a)$ is submodular where we would like to minimize r. Note that we are representing the state of the system after the customers arrive.

14.25 Assume that a control limit policy exists for our shuttle problem in exercise 2 that allows us to write the optimal dispatch rule as a function of s, as in $z^\pi(s)$. We may write $r(s, z)$ as a function of one variable, the state s.

(a) Illustrate the shape of $r(s, z(s))$ by plotting it over the range $0 < s < 3M$ (since we are allowing there to be more customers than can fill one vehicle, assume that we are allowed to send $z = 0, 1, 2, ...$ vehicles in a single time period).

(b) Let $c = 10$, $h = 2$, and $M = 5$, and assume that $A_t = 1$ with probability 0.6 and is 0 with probability 0.4. Set up and solve a system of linear equations for the optimal value function for this problem in steady state.

14.26 Show that the matrix H^n in the recursive updating formula from equation (3.68)

$$\bar{\theta}^n = \bar{\theta}^{n-1} - H^n x^n \bar{\varepsilon}^n$$

reduces to $H^n = 1/n$ for the case of a single parameter (which means we are using $Y =$ constant, with no independent variables).

Problem solving questions

14.27 You have to send a set of questionnaires to each of N population segments. The size of each population segment is given by w_i. You have a budget of B questionnaires to allocate among the population segments. If you send x_i questionnaires to segment i, you will have a sampling error proportional to

$$f(x_i) = 1/\sqrt{x_i}.$$

You want to minimize the weighted sum of sampling errors, given by

$$F(x) = \sum_{i=1}^{N} w_i f(x_i)$$

You wish to find the allocation x that minimizes $F(x)$ subject to the budget constraint $\sum_{i=1}^{N} x_i \leq B$. Set up the optimality equations to solve this problem as a dynamic program (needless to say, we are only interested in integer solutions).

14.28 An oil company will order tankers to fill a group of large storage tanks. One full tanker is required to fill an entire storage tank. Orders are placed at the beginning of each four week accounting period but do not arrive until the end of the accounting period. During this period, the company may be able to sell 0, 1, or 2 tanks of oil to one of the regional chemical companies (orders are conveniently made in units of storage tanks). The probability of a demand of 0, 1, or 2 is 0.40, 0.40, and 0.20, respectively.

A tank of oil costs $1.6 million (M) to purchase and sells for $2M. It costs $0.020M to store a tank of oil during each period (oil ordered in period t, which cannot be sold until period $t + 1$, is not charged any holding cost in period t). Storage is only charged on oil that is in the tank at the beginning of the period and remains unsold during the period. It is possible to order more oil than can be stored. For example, the company may have two full storage tanks, order three more, and then only sell one. This means that at the end of the period, they will have four tanks of oil. Whenever they have more than two tanks of oil, the company must sell the oil directly from the ship for a price of $0.70M. There is no penalty for unsatisfied demand.

An order placed in time period t must be paid for in time period t even though the order does not arrive until $t + 1$. The company uses an interest rate of 20 percent per accounting period (that is, a discount factor of 0.80).

(a) Give an expression for the one-period reward function $r(s, d)$ for being in state s and making decision d. Compute the reward function for all possible states (0, 1, 2) and all possible decisions (0, 1, 2).

(b) Find the one-step probability transition matrix when your action is to order one or two tanks of oil. The transition matrix when you order zero is given by

From-To	0	1	2
0	1	0	0
1	0.6	0.4	0
2	0.2	0.4	0.4

(c) Write out the general form of the optimality equations and solve this problem in steady state.

(d) Solve the optimality equations using the value iteration algorithm, starting with $V(s) = 0$ for $s = 0, 1$, and 2. You may use a programming environment, but the problem can be solved in a spreadsheet. Run the algorithm for 20 iterations. Plot $V^n(s)$ for $s = 0, 1, 2$, and give the optimal action for each state at each iteration.

(e) Give a bound on the value function after each iteration.

Sequential decision analytics and modeling

These exercises are drawn from the online book *Sequential Decision Analytics and Modeling* available at http://tinyurl.com/sdaexamplesprint.

14.29 We are going to perform experiments for an energy storage problem that we can solve exactly using backward dynamic programming. Download the code "EnergyStorage_I" from http://tinyurl.com/
sdagithub.

(a) Using the Python implementation of the basic model, run a grid
search for the parameter vector $\theta = (\theta^{buy}, \theta^{sell})$ by varying θ^{sell} over
the range from $20 to $60 in increments of $1 for prices, and vary-
ing θ^{buy} over the range from $20 to θ^{sell}, also in increments of $1.
Assume that the price process evolves according to

$$p_{t+1} = \min\{100, \max\{0, p_t + \varepsilon_{t+1}\}\}$$

where ε_{t+1} follows a discrete uniform distribution given by

$$\varepsilon_{t+1} = \begin{cases} -2 & \text{with prob. } 1/5 \\ -1 & \text{with prob. } 1/5 \\ 0 & \text{with prob. } 1/5 \\ +1 & \text{with prob. } 1/5 \\ +2 & \text{with prob. } 1/5 \end{cases}$$

Assume that $p_0 = \$50$.

(b) Now solve for an optimal policy by using the backward dynamic
programming strategy in section 14.3 of the text (the algorithm has
already been implemented in the Python module).

(i) Run the algorithm where prices are discretized in increments
of $1, then $0.50 and finally $0.25. Compute the size of the state
space for each of the three levels of discretization, and plot the
run times against the size of the state space.

(ii) Using the optimal value function for the discretization of $1,
compare the performance against the best buy-sell policy you
found in part (a).

(c) Repeat (b), but now assume that the price process evolves according
to

$$p_{t+1} = .5p_t + .5p_{t-1} + \varepsilon_{t+1}$$

where ε_{t+1} follows the distribution in part (1). You have to modify
the code to handle an extra dimension of the state variable. Compare the run times using the price models assumed in part (a) and
part (b) using the single discretization of $1.

(d) Section 8.3.1 of the sequential decision analytics notes introduces a time series model where

$$p_{t+1} \;=\; \bar{\theta}_{t0}p_t + \bar{\theta}_{t1}p_{t-1} + \bar{\theta}_{t2}p_{t-2} + \varepsilon_{t+1}. \qquad (14.90)$$

The section also provides the updating equations for $\bar{\theta}_t$.

(i) For this variation, present the full model of the problem using our canonical framework (states, decisions, exogenous information, transition function, objective function).

(ii) How many dimensions does the state variable have? Estimate how long it might take to solve this using Bellman's equation given your experience in parts (b) and (c).

(iii) Now consider optimizing the buy-sell policy of part (a). What effect does the more complex price model have on the design of this policy? In particular, how does your policy reflect the value of p_{t-1}?

Diary problem

The diary problem is a single problem you chose (see chapter 1 for guidelines). Answer the following for your diary problem.

14.30 Use your sequential model to write your problem as a dynamic program, and write out Bellman's equation for solving it. Note that you will have to write out the state variables, and then show mathematically how to compute the one-step transition matrix. It is unlikely that you would be able to solve this, so discuss the computational complexity of each of the elements that you would need to solve Bellman's equation. Note that if you have continuous elements in your state variable, you just have to treat the transition matrix as a function that you integrate over, rather than using discrete sums.

Bibliography

Bellman, R.E. (1957). *Dynamic Programming*. Princeton, N.J.: Princeton University Press.

Bellman, R.E. (1971). *Introduction to the Mathematical Theory of Control Processes*, Vol. II, New York: Academic Press.

Denardo, E.V. (1982). *Dynamic Programming*. Englewood Cliffs, NJ: PrenticeHall.

Derman, C. (1962). On sequential decisions and Markov chains. *Management Science* 9 (1): 16–24.

Derman, C. (1970). *Finite State Markovian Decision Processes*. New York: Academic Press.

Dreyfus, S. and Law, A. M. (1977). *The Art and Theory of Dynamic Programming*. New York: Academic Press.

Dynkin, E.B. and Yushkevich, A.A. (1979). Controlled Markov processes. *in volume Grundlehren der mathematischen Wissenschaften 235 of A Series of Comprehensive Studies in Mathematics*. New York: SpringerVerlag.

Heyman, D.P. and Sobel, M. (1984). *Stochastic Models in Operations Research, Volume II: Stochastic Optimization*. New York: McGraw Hill.

Howard, R.A. (1960). *Dynamic programming and Markov processes*. Cambridge, MA: MIT Press.

Lewis, F.L. and Vrabie, D. (2012). *Design Optimal Adaptive Controllers*, 3e. Hoboken, NJ: JohnWiley & Sons.

Manne, A.S. (1960). Linear programming and sequential decisions. *Management Science* 6 (3): 259–267.

Nemhauser, G.L. (1966). *Introduction to Dynamic Programming*. New York: JohnWiley & Sons.

Puterman, M.L. (2005). *Markov Decision Processes*, 2e. Hoboken, NJ: John Wiley and Sons.

Ross, S.M. (1983). *Introduction to Stochastic Dynamic Programming*. New York: Academic Press.

White, D.J. (1969). *Dynamic Programming*. San Francisco: HoldenDay.

15

Backward Approximate Dynamic Programming

Chapter 14 presented the most classical solution methods from discrete Markov decision processes, which are often referred to as "backward dynamic programming" since it is necessary to step backward in time, using the value $V_{t+1}(S_{t+1})$ to compute $V_t(S_t)$. While we can occasionally apply this strategy to problems with continuous states and decisions (as we did in section 14.4), most often this is used for problems with discrete states and decisions, and where the one-step transition matrix $P(S_{t+1} = s'|S_t = s, a)$ is known (that is, computable).

The field of discrete Markov decision processes has enjoyed a rich theoretical history, largely because of the elegance of discrete states and actions, and the assumption that we can compute expectations over W_{t+1}. This theory seems to have been self-perpetuating, since it is not supported by a class of well-motivated applications. However, as we see in this and later chapters, it has provided the foundation for powerful and practical approximation strategies.

The basic backward dynamic programming strategy used for discrete dynamic programming suffers from what we have identified as the three curses of dimensionality:

(1) State variables – As the state variable grows past three or four dimensions, the number of states tends to become too large to enumerate. In particular, there are many applications where some (or all) of the dimensions of the state variable are continuous.
(2) Decision variables – Enumerating all possible decisions tends to become intractable if there are more than three or four dimensions, unless it is possible to significantly prune the number of decision using constraints. Problems with more than three or four dimensions tend to require special structure such as convexity. For this reason, we adopted the classical notation of discrete actions a in chapter 14.4, but for reasons we make clear

Reinforcement Learning and Stochastic Optimization: A Unified Framework for Sequential Decisions, First Edition. Warren B. Powell.
© 2022 John Wiley & Sons, Inc. Published 2022 John Wiley & Sons, Inc.

shortly, this chapter reverts back to our standard notation x for decisions, where we are going to allow x to be multidimensional and continuous.

(3) Exogenous information variables – We assume that our exogenous information $W_t \in \mathcal{W} = \{w_1, ..., w_L\}$ and let

$$p_t^W(w|s, x) = \mathbb{P}[W_t = w|s, x].$$

As we pointed out in section 9.7 finding the one-step transition matrix requires computing the expectation

$$
\begin{aligned}
\mathbb{P}(s'|S_t^x = (s, x)) &= \mathbb{E}_{W_{t+1}}\{\mathbb{1}_{\{s'=S^M(s,x,W_{t+1})\}}|S_t = s, x_t = x\} \\
&= \sum_{w\in\mathcal{W}} p_{t+1}^W(W_{t+1} = w|s, x)\mathbb{1}_{\{s'=S^M(s,x,w)\}}. \quad (15.1)
\end{aligned}
$$

However, if W_{t+1} is a vector or continuous (instead of the discrete outcomes in $\mathcal{W}$), this becomes computationally intractable.

These computational issues have motivated the development of fields with names like "approximate dynamic programming," "heuristic dynamic programming" (an older term used in engineering), "adaptive dynamic programming," (a term adopted in engineering after 2010), "neuro-dynamic programming," or "reinforcement learning," (the highly popular field that evolved within computer science). All of these approaches are effectively a form of "forward approximate dynamic programming" since they are all based on the principle of stepping forward in time. Many authors (including this author) have assumed that if you cannot do "backward dynamic programming" (that is, the method described in section 14.3), then you need to turn to "approximate dynamic programming" (which means forward approximate dynamic programming). This chapter challenges this notion.

This chapter presents a strategy known as *backward approximate dynamic programming*, which has the notable feature that it can handle multidimensional (and continuous) state variables and exogenous information variables. In addition, under the right conditions, it can also handle multidimensional (and continuous) decision variables. In other words, backward approximate dynamic programming overcomes all three curses of dimensionality. However, it still struggles with the same challenge of any method based on approximating the value function: The quality of the policy depends heavily on how well we can approximate the value function, and there are many problems where high quality approximations are simply not possible. At the end of this chapter, we are going to present some strong empirical evidence supporting its effectiveness.

15.1 Backward Approximate Dynamic Programming for Finite Horizon Problems

We are going to start by illustrating backward approximate dynamic programming for finite horizon problems, which parallels backward dynamic programming that we introduced in chapter 14. We begin using classical lookup tables for the value functions, and then transition to continuous approximations.

While we will see that forward ADP methods can be quite powerful, we are going to first present the idea of backward approximate dynamic programming, which has received comparatively little attention in the research literature. Backward ADP can be viewed as an implementation of classical backward dynamic programming (see the algorithm in Figure 14.3) that uses sampling of states and exogenous information to avoid enumerating state spaces and information spaces. We still need to optimize over decisions, but this opens up the potential of exploiting structure such as concavity (convexity if minimizing) to use solvers for high-dimensional decisions.

In addition to scaling nicely to complex problems, we are going to close by presenting some empirical evidence supporting the use of backward ADP. However, as with any approximation method, we cannot make any broad statements about the performance of backward ADP over forward ADP methods (or any of the other classes of policies). It should be viewed as a powerful tool in the toolbox of any sequential decision scientist.

15.1.1 Some Preliminaries

We start by writing Bellman's equation broken into two steps: from pre-decision state S_t to post-decision state S_t^x, and then from post-decision state S_t^x to the next pre-decision state S_{t+1}:

$$V_t(S_t) = \max_{x_t} \left(C(S_t, x_t) + V_t^x(S_t^x) \right), \tag{15.2}$$

$$V_t^x(S_t^x) = \mathbb{E}_{W_{t+1}} \left\{ V_{t+1}(S_{t+1}) | S_t^x \right\}, \tag{15.3}$$

where

$$S_t^x = S^{M,x}(S_t, x_t),$$

$$S_{t+1} = S^{M,W}(S_t^x, W_{t+1}).$$

These steps are illustrated in Figure 15.1.

The computational challenges associated with these equations include:

- Computing $V_t(S_t)$ for each (presumably discrete) pre-decision state S_t in equation (15.2).

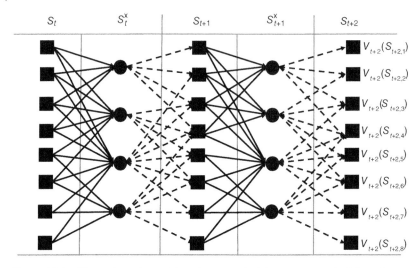

Figure 15.1 Illustration of transitions from pre-decision S_t to post-decision S_t^x to pre-decision S_{t+1} and so on.

- Optimizing over x_t if x_t is a vector in equation (15.2).
- Computing $V_t^x(S_t^x)$ for each post-decision S_t^x in equation (15.3).
- Computing the expectation $\mathbb{E}_{W_{t+1}}$ over the random variable W_{t+1} in equation (15.3).

We are going to break down these computational challenges one step at a time, as follows:

(1) Sampled states with lookup tables – The core idea of backward ADP is to avoid enumerating the entire state space by using a sampled set of states instead. In this first stage, we will still use a lookup table representation of the value functions, and we will also assume we can do full expectations, and maximize over all decisions (which generally means a not-too-large set of discrete decisions).

(2) Sampled expectations – Here we are going to replace the exact expectation over W_{t+1} with a sampled approximation.

(3) Parametric approximations of the value function – Here we replace the lookup table representation of the value function with a parametric (or nonparametric) approximation which helps with both the computation of value function approximations.

(4) Decisions – There are two strategies for handling multidimensional (possibly high dimensional) decisions:

(a) We can replace the maximization over decisions with a maximization over a sampled set.

(b) If we use a parametric approximation for $V_t^x(S_t^x)$, we may be able to solve equation (15.2) using classical optimization methods (linear, nonlinear, or integer programming).

We are going to start by describing backward ADP using lookup table models for the value function, and then we are going to transition to using continuous approximations.

15.1.2 Backward ADP Using Lookup Tables

The basic idea of backward approximate dynamic programming is to perform classical backward dynamic programming, using equations (15.2)–(15.3), but instead of enumerating all the states $\mathcal{S}$, we work with a sampled set $\hat{\mathcal{S}}$. We begin by illustrating the strategy using lookup table approximations for the value function approximations. This closely parallels classical backward dynamic programming (see, for example, equation (14.3)).

For now we are going to make the assumption (true for some, but hardly all, applications) that the post-decision state space $\mathcal{S}^x$ is "not too large." By contrast, we are going to allow the pre-decision state space $\mathcal{S}$ to be arbitrarily large. This situation arises frequently when there is information needed to make a decision, but which is no longer needed once a decision has been made. Some examples where this arises are:

■ **EXAMPLE 15.1**

As a car traverses from node i to node j on a transportation network, it incurs random costs $\hat{c}_{ij}$ which it learns when it first arrives at node i. The (pre-decision) state when it arrives at node i is then $S = (i, (\hat{c}_{ij})_j)$. After making the decision to traverse from i to some node j' (but before moving to j'), the post-decision state is $S^x = (j)$, since we no longer need the realization of the costs $(\hat{c}_{ij})_j$.

■ **EXAMPLE 15.2**

A truck driver arrives in city i and learns a set $\mathcal{L}_i$ of loads that need to be moved to other cities. This means when it arrives at i that the state of our driver is $S = (i, \mathcal{L}_i)$. Once the driver chooses a load $\ell \in \mathcal{L}_i$, but before moving to the destination of load ℓ, the (post-decision) state is $S^x = (\ell)$ (or we might use the destination of load ℓ).

■ **EXAMPLE 15.3**

A cement truck is given a set of orders to deliver set to a set of work sites. Let R_t be the inventory of cement, and let $\mathcal{D}_t$ be the set of construction sites needing deliveries (the set includes how much cement is needed by each site). The decision that needs to be made by the cement plant is how much cement to make to replenish inventory. The pre-decision state is $S_t = (R_t, \mathcal{D}_t)$, while the post-decision state is $S_t^x = R_t^x$ which is the amount of inventory left over after making all the deliveries.

In each of these examples, the number of pre-decision states may be extremely large. Instead of looping over all states in $\mathcal{S}$ (as we had to do in Figure 15.1), we are going to take a sample $\hat{\mathcal{S}}$ which is of manageable size. We see the power of Monte Carlo simulation in that the state variables can be both continuous and high-dimensional, since we control the number of samples in $\hat{\mathcal{S}}$. The only caveat is that we have to pre-specify a sampling region, which means we have to know something about the range of values of each dimension of S_t.

In addition to enumerating the post-decision states, we also assume (for now):

- There is a discrete set of decisions $x_t \in \{x_1, x_2, \dots, x_K\}$ that we can search over.
- There are discrete outcomes $W_{t+1} \in \{w_1, \dots, w_L\}$.
- We know the probability $p_t^W(w_\ell) = \mathbb{P}(W_{t+1} = w_\ell | S_t^x)$.

The steps of the algorithm are described in detail in Figure 15.3, but we refer to Figure 15.2 to explain the idea. The pre-decision states are depicted as squares while post-decision states are circles. We represent the states in our sampled set $\hat{\mathcal{S}}$ using the black squares. Assuming we know $\overline{V}_{t+2}(s)$ for states $s \in \hat{\mathcal{S}}$, we compute the value $\overline{V}_{t+1}^x(s)$ for each post-decision state s in $\mathcal{S}^x$ by taking the expectation over all random outcomes that take us to states in our sampled set $\hat{\mathcal{S}}$, given by the equation

$$V_{t+1}^x(S_{t+1}^x) = \frac{\sum_{\ell=1}^{L} p_{t+2}^W(w_\ell)\overline{V}_{t+2}(S_{t+2}(w_\ell))\mathbb{1}_{\{S_{t+2}(w_\ell)\in\hat{\mathcal{S}}\}}}{\sum_{\ell=1}^{L} p_{t+2}^W(w_\ell)\mathbb{1}_{\{S_{t+2}(w_\ell)\in\hat{\mathcal{S}}\}}}, \qquad (15.4)$$

where $S_{t+2}(w) = S^M(S_{t+1}^x, w)$. Note that equation (15.4) only includes transitions to values of S_{t+2} in the sampled set $\hat{\mathcal{S}}$, which means that we have to normalize the probabilities so that the probabilities of the outcomes that transition to states in $\hat{\mathcal{S}}$ sum to one.

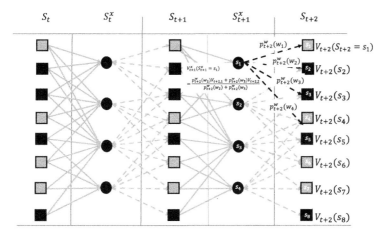

Figure 15.2 Calculation of the value of the post-decision state S_{t+1}^x using full expectation.

This quickly raises a potential problem. What if none of the random outcomes take us to states in $\hat{S}$? When this happens, we choose a subset of random outcomes from a post-decision state, find the pre-decision states that these outcomes take us to, and then add these states to the sampled set $\hat{S}$. We then repeat the calculation.

Once we have the value of being in each post-decision state, we then step back to find the value of being in each sampled pre-decision state, which is depicted in Figure 15.4. Since we assume we have computed the value of being in each post-decision state, finding the value of being in any pre-decision state involves simply searching over all decisions and finding the decision with the highest one-period reward plus downstream value.

15.1.3 Backward ADP Algorithm with Continuous Approximations

Now that we have sketched the basic idea of backward ADP, we are going to outline a fully scalable algorithm that can handle multidimensional and continuous state variables (pre-decision S_t and post-decision S_t^x), decisions x_t, and exogenous information W_{t+1}. We do this by using appropriately designed continuous approximations of the value function around the post-decision state variable.

A sketch of the algorithm is given in Figure 15.5. This algorithm has some nice features:

- Both the pre- and post-decision states S_t and S_t^x can be multidimensional and continuous.

Step 0. Initialization:

> **0a.** Initialize the terminal contribution $V_T(S_T)$.
> **0b.** Create a sampled set of pre-decision states $\hat{S}$ (we assume we can use this same sample each time period).
> **0c.** Create a full set of post-decision states S^x (presumably a manageable size).
> **0d.** Set $t = T - 1$.

Step 1a. Step backward in time $t = T, T - 1, \dots, 0$:

Compute the value of each post-decision state:

> **Step 2a.** Initialize pre-decision value function approximation $\overline{V}_t(s) = -M$.
> **Step 2b.** Loop over the sampled set of pre-decision states $s \in \hat{S}$.
> **Step 2c.** Loop over each decision $x \in \mathcal{X}(s)$:
>
> > **Step 3a.** Compute $Q_t(s, x) = C(s, x) + \overline{V}_t^x(s' = S^{M,x}(s, x))$.
> > **Step 3b.** If $Q_t(s, x) > \overline{V}_t(s)$ then set $\overline{V}_t(s) = Q_t(s, x)$.

Compute the value of each sampled pre-decision state:

> **Step 4a.** Initialize post-decision value function approximation $\overline{V}_t^x(s^x) = -M$.
> **Step 4b.** Loop over the full set of post-decision states $s^x \in S^x$.
> **Step 4c.** Step back in time: $t = t - 1$.
>
> > **Step 5a.** Initialize $Q(s, x) = 0$.
> > **Step 5b.** Initialize total probability $\rho = 0$.
> > **Step 5c.** Loop over each $w \in W$:
> > **Step 5d.** If $\rho > 0$ then (we have to normalize $Q_t(s, x)$ in case $\rho < 1$):
> >
> > > **Step 6a.** Compute $Q_t(s, x) = Q_t(s, x) + \mathbb{P}(w|s, x)\overline{V}_{t+1}(s' = S^M(s, x, w))$.
> > > **Step 6b.** $\rho = \rho + \mathbb{P}(w|s, x)$.
> > > **Step 6c.** $Q_t(s, x) = Q_t(s, x)/\rho$
> >
> > **Else:** Get here if $\rho = 0$, which means there were no random transitions to states in $\hat{S}$:
> >
> > > **Step 6d.** Choose a sample of outcomes $\hat{w}$ (at least one), find the downstream pre-decision state $\hat{s} = S^{M,W}(s, \hat{w})$, and add each $\hat{s}$ to $\hat{S}$.
> > > **Step 6e.** Return to step 4a.

Step 1b. Return the values $\overline{V}_t(s)$ for all $s \in S$ and $t = 0, \dots, T$.

Figure 15.3 A backward dynamic programming algorithm using lookup tables.

- The exogenous information W_t can also be multidimensional and continuous, as long as we have some mechanism for sampling the random variable. This may come from an underlying mathematical model, or it may come from historical observations.

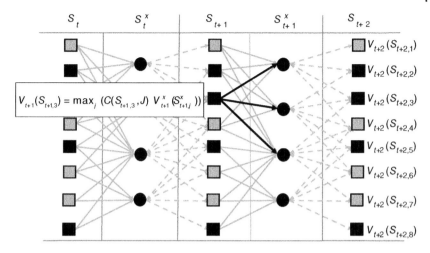

Figure 15.4 Calculation of value of pre-decision state S_{t+1} using full maximization.

---

(0)   Assume we have a value function approximation $\overline{V}_T(s)$.
(1)   Perform N samples for $n = 1, ..., N$:

  (1a)   Randomly sample a post-decision state $\hat{s}_t^{x,n}$ from the set $|Scalhat^x$.
  (1b)   Find a sample realization of $\hat{w}_{t+1}^n$ of the random variable W_{t+1} given that we are in state $\hat{s}_t^x$.
  (1c)   Simulate our way from $\hat{s}_t^{x,n}$ to $\hat{s}_{t+1}^n$ using $\hat{s}_{t+1}^n = S^{M,W}(\hat{s}_t^{x,n}, \hat{w}_{t+1}^n)$.
  (1d)   Compute a sample estimate $\hat{v}_{t+1}^n$ of the value of being in pre-decision state $\hat{s}_{t+1}^n$ using

  $$\hat{v}_{t+1}^n = \max_x \left(C(\hat{s}_{t+1}^n, x) + \overline{V}_{t+1}^x(\hat{s}_{t+1}^{n,x}) \right) \tag{15.5}$$

  where $\hat{s}_{t+1}^{n,x} = S^{M,x}(\hat{s}_t^n, x_t^n)$, and where x_t^n is the value of x that optimizes equation (15.5). We are now going to associate the value $\hat{v}_{t+1}^n$ with the previous post-decision state $\hat{s}_t^{x,n}$.

(2)   From step 1, we compile a set of observations $(\hat{s}_t^n, \hat{v}_{t+1}^n), n = 1, ..., N$.
(3)   Use the dataset $(\hat{s}_t^n, \hat{v}_{t+1}^n), n = 1, ..., N$ to fit a statistical model $\overline{V}_t^x(s)$ using any of the statistical methods in chapter 3 (but here we are doing batch learning). Some of the methods that have proven successful in this context are described in section 15.3.
(4)   Step back one time period and repeat until $t = 0$.

Figure 15.5 A backward dynamic programming algorithm for multidimensional applications.

- The decision x_t may be multidimensional and continuous (or discrete), but algorithms for solving multidimensional decision problems typically require concavity of $\left(C(\hat{s}_{t+1}^n, x) + \overline{V}_{t+1}^x(\hat{s}_{t+1}^{n,x}) \right)$ (convexity if minimizing). This is where some care might have to be put into the choice of architecture for the value function approximation.

An open question is: how well does the method work? The approximation for time t depends on the approximation for $t + 1$, which means the errors in the approximation for $t + 1$ propagate backward to t and, in fact, accumulate. Section 15.4 reports on three sets of empirical benchmarking experiments that support the accuracy and efficiency of backward approximate dynamic programming. However, we can obtain stronger results when we apply these ideas in the context of a stationary (steady state) problem, an idea that has evolved in the literature under the name "fitted value iteration."

15.2   Fitted Value Iteration for Infinite Horizon Problems

Most of this book focuses on finite horizon problems, since these represent the problems most often encountered in practice. However, the literature on Markov decision processes, as can be seen in the presentation in chapter 14, has emphasized the steady state version of Bellman's equation which is written:

$$V(s) \quad = \quad \max_{x \in \mathcal{X}} \left(C(s, x) + \gamma \mathbb{E}_W \{ V(S^M(s, x, W)) | s \} \right).$$

where $s' = S^M(s, x, W)$ is the state we land in given that we are now in state s, make decision x, and then observe W. We write the value function explicitly as a function of the transition function $S^M(s, x, W)$ to make the dependence on W explicit. Needless to say, computing this expectation is problematic, especially inside a max operator. Instead, we used a sampled estimate by choosing a random sample $W = \{w_1, w_2, \dots, w_L\}$.

The basic idea follows the steps of backward ADP. We choose a sample of states $\hat{S} = \{\hat{s}_1, \dots, \hat{s}_m, \dots, \hat{s}_M\}$. Assume we have an approximate value function $\overline{V}^{n-1}(s)$. Then, given $\overline{V}^{n-1}(s)$, we sample $\hat{s}_m \in \hat{S}$ and compute

$$\hat{v}_m^n = \max_{x \in \mathcal{X}} \left(C(\hat{s}_m, x) + \gamma \frac{1}{L} \sum_{\ell=1}^{L} \left(\overline{V}^{n-1}(S^M(\hat{s}_m, x, w_\ell)) | s \right) \right). \tag{15.6}$$

Repeat equation (15.6) for $m = 1, \dots, M$ until we have compiled a dataset $(\hat{s}_m, \hat{v}_m^n)$ for $m = 1, \dots, M$. Note that we index the value function approximation $\overline{V}^n(s)$ by iteration n, but the sampled states $\hat{s} \in \hat{S}$ are the same from one iteration to the next.

The next step is to use the dataset $(\hat{s}_m, \hat{v}_m^n)_{m=1}^M$ to create an updated value function approximation $\overline{V}^n(s)$, using any of the approximation architectures in chapter 3. Of course, solving equation (15.6) is more difficult than solving for $\hat{v}^n$ in equation (15.5) because we have chosen to illustrate fitted value iteration using value functions that depend on the pre-decision state, forcing us to use the sampled representation of the expectation. We could use the same strategy as we did in the finite-horizon case and compute the value function around the post-decision state. Similarly, we could use the sampled representation of the expectation illustrated in this section in the finite-horizon setting. We have decided to illustrate both methods, but either can be used in either setting.

The only real difference between the finite and infinite horizon versions is that the finite horizon algorithm involves a single backward pass over the horizon. There is no notion of convergence. By contrast, we can repeat our process for updating $\overline{V}^n(s)$ in the infinite horizon case for as many iterations as we like, opening the door to questions about convergence. Recall that we could obtain strict bounds on the error when we were using lookup table representations and assuming that we could compute the one-step transition matrix (see section 14.12.2).

We made the point in chapter 4 that classical discrete Markov decision processes, where we assume that the one-step transition matrix is known, is actually a deterministic problem (see section 4.2.5), as is any stochastic problem where the expectation can be computed exactly. In fact, in section 4.3 we made the point that replacing the expectation with a sampled approximation, as we are doing in equation (15.6), is simply replacing the original expectation that we could not compute, with an approximate expectation that we can compute. Once we do so, we are effectively turning our exact "deterministic" problem into an approximate "deterministic" problem. But if we continue to use a lookup table representation, we still suffer from the curse of dimensionality in the state space.

It is possible to show convergence results similar to those for the exact, discrete dynamic programming methods, but it requires an approximating architecture that is sufficiently flexible to allow arbitrarily accurate fits at the sampled states. This would not be possible if we were to use a low-dimensional parametric architecture (such as a quadratic fit). Gaussian process regression, kernel regression and neural networks are all approximation methods that can produce very accurate approximations, but any time you use these high-dimensional architectures, you run the risk of overfitting to noisy observations unless you have exceptionally large samples. So, pick your poison.

15.3 Value Function Approximation Strategies

We illustrated the basic idea of backward approximate dynamic programming using a standard lookup table representation for the value function, but this would quickly cause problems if we have a multidimensional state (the classic curse of dimensionality). In this section, we suggest three strategies for approximating value functions that mitigate this problem to some degree.

15.3.1   Linear Models

Arguably the most natural strategy for approximating the value function is to fit a statistical model, where the most natural starting point is a linear model of the form

$$\overline{V}_t(S_t|\theta_t) = \sum_{f\in\mathcal{F}} \theta_{tf}\phi_f(S_t).$$

Here, $\phi_f(S_t)$ are a set of appropriately chosen features. For example, if S_t is a continuous scalar (such as price), we might use $\phi_1(S_t) = S_t$ and $\phi_2(S_t) = S_t^2$.

The idea is very simple. For each $\hat{s}$ in our sampled set of pre-decision states $\hat{\mathcal{S}}$, compute a sampled estimate $\hat{v}_t^n$ of the value of being in a state s^n

$$\hat{v}_t^n = \arg\max_x \left(C(\hat{s}^n, x) + \mathbb{E}\{\overline{V}_{t+1}(S_{t+1})|\hat{s}^n\} \right),$$

where $S_{t+1} = S^M(\hat{s}^n, x, W_{t+1})$.

Now we have a set of data $(\hat{s}^n, \hat{v}_t^n)$ for $n = 1, \dots, |\hat{\mathcal{S}}|$. We can use this dataset to estimate any statistical model $\overline{V}_t(S_t|\theta_t)$ which gives us an estimate of the value of being in every state, not just the sampled states. For example, assume we have a linear model (remember this means linear in the parameters)

$$\begin{aligned} \overline{V}_t(S_t|\bar{\theta}_t) &= \bar{\theta}_{t1}\phi_1(S_t) + \bar{\theta}_{t2}\phi_2(S_t) + \bar{\theta}_{t3}\phi_3(S_t) + \dots, \\ &= \sum_{f\in\mathcal{F}} \theta_{tf}\phi_f(S_t), \end{aligned}$$

where $\phi_f(S_t)$ is some feature of the state. This might be the inventory R_t (money in the bank, units of blood), or R_t^2, or $\ln(R_t)$. Create the (column) vector ϕ^n using

$$\phi^n = \begin{pmatrix} \phi_1^n \\ \phi_2^n \\ \vdots \\ \phi_F^n \end{pmatrix}$$

where $\phi_f^n = \phi_f(S_t^n)$.

Let $\hat{v}_t^n$ be computed using (15.7), which we can think of as a sample realization of the estimate $\overline{V}_t^{n-1}(S_t)$. We can think of

$$\hat{\varepsilon}_t^n = \overline{V}_t^{n-1}(S_t) - \hat{v}_t^n$$

as the "error" in our estimate. Using the methods we first introduced in section 3.8.1, we can update our estimates of the parameter vector $\bar{\theta}_t^{n-1}$ using

$$\bar{\theta}_t^n = \bar{\theta}_t^{n-1} - H_t^n \phi_t^n \hat{\varepsilon}_t^n, \tag{15.7}$$

where H_t^n is a matrix computed using

$$H_t^n = \frac{1}{\gamma^n} M_t^{n-1},$$ (15.8)

where M_t^{n-1} is an $|\mathcal{F}|$ by $|\mathcal{F}|$ matrix which is updated recursively using

$$M_t^n = M_t^{n-1} - \frac{1}{\gamma_t^n}(M_t^{n-1}\phi_t^n(\phi_t^n)^T M_t^{n-1}).$$ (15.9)

γ_t^n is a scalar computed using

$$\gamma_t^n = 1 + (\phi_t^n)^T M_t^{n-1}\phi_t^n.$$ (15.10)

Parametric approximations are particularly attractive because we get an estimate of the value of being in *every* state from a small sample. The price we pay for this generality is the errors introduced by our parametric approximation.

15.3.2 Monotone Functions

There are a number of sequential decision problems where the state variable has three to six or seven dimensions, which tend to be the range where the state space is too large to estimate value functions using lookup tables. There are, however, a number of applications where the value function is monotone in each dimension, which is to say that as the state variable increases in each dimension, so does the value of being in the state. Some examples include:

- Optimal replacement of parts and equipment tend to exhibit value functions which are monotone in variables describing the age and/or condition of the parts.
- The problem of controlling the number of patients enrolled in clinical trials produces value functions that are monotone in variables such as the number of enrolled patients, the efficacy of the drug, and the rate at which patients drop out of the study.
- Initiation of drug treatments (statins for cholesterol, metformin for lowering blood sugar) result in value functions that are monotone in health metrics such as cholesterol or blood sugar, the age of a patient, and their weight.
- Economic models of expenditures tend to be monotone in the resources available (e.g. personal savings), and other indices such as stock market, interest rates, and unemployment.

Monotonicity can be exploited when we are using a lookup table representation of a value function. Assume that a state s consists of four dimensions $(s_{t1}, s_{t2}, s_{t3}, s_{t4})$, where each dimension takes on one of a set of discrete values, such as $S_{t2} \in \{S_{t2,1}, S_{t2,2}, S_{t2,3}, \ldots, S_{t2,J_2}\}$. Assume we have a sampled estimate of the value of being in state $\hat{s}^n$, which we might compute using

$$\hat{v}_t^n(\hat{s}^n) = \max_x \left(C(\hat{s}^n, x) + \mathbb{E}_{W_{t+1}} \{\overline{V}_t^{n-1}(S_{t+1}) | \hat{s}^n\} \right),$$

where $S_{t+1} = S^M(\hat{s}^n, x, W_{t+1})$. We might then use our sampled estimate (regardless of how it is found) to update the value function approximation at state $\hat{s}^n$ using

$$\overline{V}_t^n(\hat{s}^n) = (1 - \alpha_n)\overline{V}_t^{n-1}(\hat{s}^n) + \alpha_n \hat{v}_t^n(\hat{s}^n).$$

We assume that $\overline{V}_t^{n-1}(s)$ is monotone in s before the update. Assume that $s' \succ s$ means that each element $s'_{ij} \geq s_{ij}$. Then if $\overline{V}_t^{n-1}(s)$ is monotone in s, then $s' \succ s$ means that $\overline{V}_t^{n-1}(s') \geq \overline{V}_t^{n-1}(s)$. However, we cannot assume that this is true of $\overline{V}_t^n(s)$ just after we have done an update for state s_t^n. We can quickly check if $\overline{V}_t^n(s) \leq \overline{V}_t^n(s')$ for each s' with at least one element that is larger than the corresponding element of s.

The idea is illustrated in Figure 15.6. Starting with the upper left corner, we start with an initial value function $\overline{V}(s) = 0$, and make an observation (the blue dot) of 10 in the middle. We then use the monotone structure to make all points to the right and above of this point to equal 10. We then make an observation of 5, and use this observation to update all the points to the left and below the last observation.

Figure 15.7 shows snapshots from a video where monotonicity is being used to update a two-dimensional function. Again starting from the upper right, the first three screenshots were from the first 20 iterations, while the last one (lower right) was at the end, long after the function had stopped changing.

Monotonicity is an important structural property. When it holds, it dramatically speeds the process of learning the value functions. We have used this idea for matrices with as many as seven dimensions, although at that point a lookup representation of a seven-dimensional function becomes quite large.

There will be situations where a value function is monotone in some dimensions, but not in others. This can be handled (somewhat clumsily) but imposing monotonicity over the subset of states where monotonicity holds. For the remaining states, we have to resort to brute force lookup table methods. If $\tilde{s}$ is the set of states where the value function is not monotone, while $\breve{s}$ is the states over which the value function is monotone (of course, $s = (\tilde{s}, \breve{s})$), then we can think of a value function $\overline{V}(\tilde{s}, \breve{s})$ where we have a value function $\overline{V}(\tilde{s}, \breve{s})$ that is monotone in $\breve{s}$ for each state $\tilde{s}$ (we hope there are not too many of these).

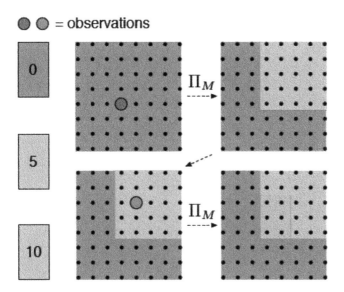

Figure 15.6 Illustration of the use of monotonicity. Starting from upper left: (1) Initial value function all 0, with observation (blue dot) of 10; (2) using observation to update all points to the right and above to 10; (3) new observation (pink dot) of 5; (4) updating all points to the left and below to 5. Modified from Jiang and Powell (2015)

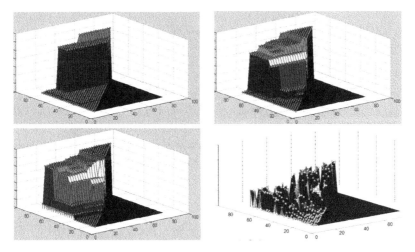

Figure 15.7 Video snapshot of use of monotonicity for a two-dimensional function for three updates; fourth snapshot (lower right) is a value function where monotonicity was not used.

15.3.3 Other Approximation Models

We encourage readers to experiment with other methods from chapter 3 (or your favorite book in statistics or machine learning). We note that approximation errors will accumulate with backward ADP, so you should not have much confidence that $\overline{V}_t(S_t)$ is actually a good approximation of the value of being in state S_t. However, we have found that even when there is a significant difference between $\overline{V}_t(S_t)$ and the true value function $V_t(S_t)$ (when we can find this), the approximation $\overline{V}_t(S_t)$ may still provide a high quality policy, but there are no guarantees.

15.4 Computational Observations

As of the writing of this book, backward approximate dynamic programming is a relatively new algorithmic strategy, which is surprising given that it is the approximate analog of classical backward dynamic programming (from chapter 14). The first reference in the literature appears to be 2013. For this reason, we begin with a presentation of several projects we have been directly involved with which produced some form of benchmarking of the solutions produced by backward approximate dynamic programming. We then share some notes on the methodology.

15.4.1 Experimental Benchmarking of Backward ADP

In this section we report on the empirical benchmarking of backward ADP in three very different settings. The first uses comparisons against the exact, optimal solution computed using the techniques of chapter 14. The second two examples are more complex, making exact solutions impossible. Instead, we compare backward ADP against policies that are already being used, one for the optimization of a battery storage system and the second for the allocation of resources in Africa by the International Monetary Fund.

Optimization of clinical trials

The problem faced by companies running clinical trials is to make the following decision at each point in time: the drug works (go to market, typically by selling the patent), the drug does not work (cancel the clinical trial), or continue testing. The state variable has three dimensions:

- The number of patients we have tested.
- A two-dimensional belief state, capturing the mean and variance of our estimate of the probability that the drug works.

This means our state variable has a single discrete dimension and two continuous dimensions. If we are willing to live with a discretized version of the continuous dimensions, this is a problem that can be solved optimally using the backward dynamic programming methods we presented in chapter 14. Optimal benchmarks for real problems are quite rare.

The results can be stated very simply:

- The optimal solution required 268 hours on a modern laptop.
- Backward approximate dynamic programming required 20 minutes, and the solution was within 1.2 percent of the performance of the policy produced by the optimal solution.

Optimizing a complex energy storage problem

We were given the problem of optimizing an energy storage device where we had to balance two revenue streams:

- We can use the battery to buy and sell energy from/to the grid. Electricity prices (known as LMPs, or "locational marginal prices") are updated every five minutes and can vary dramatically. Prices that might average $20/megawatt-hour can spike to $1000 or even $10,000.
- Grid operators will pay battery operators to help them with a process called "frequency regulation." Power over the grid fluctuates as a result of the random variations of loads placed on the grid. A grid operator might pay $30 per megawatt of power (each battery has a rated power rating, which gives how *fast* power moves into and out of the battery), but these prices also vary, and can increase to $500 per megawatt per hour or more.

When the grid operator is paying a battery to perform frequency regulation, the grid will send a signal every two seconds whether it wants the battery to charge or discharge (at some percentage of the battery's power rating), or do nothing. The grid never asks the battery to charge (or discharge) for extended periods, so these batteries do not have to be very large. Frequency regulation is purely for short-term smoothing of power variations.

When a battery operator is being paid to perform frequency regulation, then the expectation is that it will comply with the signals from the grid operator (these are known as "RegD" signals in the US). In practice, limitations of the device (it is not just batteries that perform this function – any generator, from natural gas turbines to coal plants may perform frequency regulation) mean that the device providing frequency regulation may not have perfect compliance with the RegD signal. For this reason, there are penalties for non-compliance. Figure 15.8 shows a plot of LMP and RegD prices over a period of

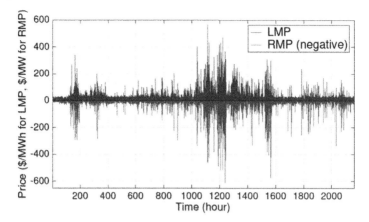

Figure 15.8 Real-time energy prices (blue), and regulation prices (red), January–March, 2015, revealing the high correlation between the two.

several months. It indicates the degree of volatility, along with the correlations between the two prices. This is an example of a problem where the modeling of the exogenous information processes is particularly important.

This raised the question: What if a battery operator (batteries typically have perfect compliance) occasionally disobeyed the RegD signal? In particular, what if the grid operator is asking the battery to buy electricity at a time when electricity prices are very high? The battery operator might wish to go against the RegD signal, sell into a high-priced market (this may last for just five minutes), but paying the penalty for noncompliance.

The challenge here is that following a RegD signal is trivial; the battery operator simply follows the RegD signal from the grid operator which specifies when to charge, discharge or do nothing. However, choosing between simply doing what the RegD signal tells us to do, or running against the signal to take advantage of price spikes on the grid, requires optimization-based logic. For example, the grid price may rise, but has it risen enough to suffer the consequences of noncompliance with the RegD signal?

The distinguishing characteristic of this problem is the number of time periods. Decisions are made every 2 seconds, which means there are 43,200 time periods in a day. Standard backward dynamic programming was prohibitive because of the size of the state space as well as the number of time periods. Forward ADP methods, which we will introduce in chapters 16–18 (stay tuned for these presentations) require iterative learning which would simply be too slow. In fact, this was the problem that motivated our first use of backward ADP.

Table 15.1 Comparison of revenues generated from backward ADP, combining revenues from frequency regulation and power purchase, to revenues from a pure frequency regulation policy.

Month	Backward ADP revenue	Pure RegD policy	Pct. improvement
January	22052	19131	10.27
February	51282	46331	10.68
March	36518	32329	12.95
April	24121	22272	8.3
May	31861	30232	5.39
June	18975	17999	5.42
July	18463	17152	7.64
August	15988	14750	8.39
September	22336	20462	9.16
October	17714	16553	7.01
November	15930	15033	5.97
December	15079	13901	8.47
Annual	290323	266151	9.08

This problem is too hard to compute optimal benchmarks, as we did in the clinical trials problem. However, we have a different benchmark which is extremely demanding: instead of optimizing over the two signals, we can just follow the RegD signal. This is very difficult competition, since we anticipate that the benefits of optimizing across both revenue streams will be modest. This means that we are not in a position to tolerate suboptimal performance since this would threaten the larger revenue stream from just following the RegD signal.

The results are shown in Table 15.1. These results show consistent, if modest, improvements from the combined signal produced using backward approximate dynamic programming. Again, we emphasize the challenge of competing against a pure frequency regulation policy which produces over 90 percent of the revenue using a very simple rule that is easy to follow.

Resource allocation in Africa

Our last demonstration involves a complex resource allocation problem faced by the International Monetary Fund (IMF) among projects within Africa. The widely used approach for solving this problem is a single-period linear program

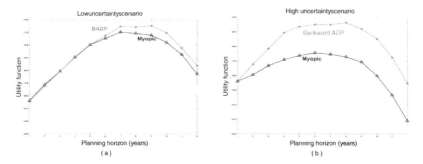

Figure 15.9 Performance of the policy produced by backward ADP using (a) low uncertainty and (b) high uncertainty in future forecasts.

that optimized a complex utility function for capturing the state of a country over the course of a year. The utility function would capture metrics about the economy, social metrics such as poverty, investments in infrastructure, and measures of instability (such as assassinations). The decisions were how much to invest in different projects, such as roads, education, health, and power generators. Given that these decisions cover resources being allocated across all countries in Africa, and all projects, it is a high-dimensional decision, with a very high-dimensional (and largely continuous) state vector.

The state of the art for this problem is the use of a linear program that optimizes the benefits within a single year, although it was clear that some investments had multiyear horizons. This was also a problem with tremendous uncertainties. In any given year insurgencies could arise and challenge the stability of a country. The emergence of diseases, or discoveries of natural resources, were frequent examples of high-impact sources of uncertainty.

This problem was solved using backward approximate dynamic programming, and compared to a myopic policy that is widely used in practice. The results are shown in Figure 15.9, which reports on two sets of simulations. Figure 15.9(a) shows the results of backward ADP for a simulation with relatively low noise, while Figure 15.9(b) shows the results for a setting with significant sources of uncertainty. Backward ADP outperformed the standard myopic policy for both the low noise and high noise situations. It did particularly well in the high noise environment, which is precisely the conditions where someone might say "we have so much uncertainty about the future, why plan for it?"

This application is a nice demonstration of backward ADP in a complex, high-dimensional resource allocation problem. In fact, it is a problem which clearly needs a policy in the lookahead class, but where direct lookahead policies (which we introduced in chapter 11, and cover in much more detail in chapter 19) are not an obvious approach.

15.4.2 Computational Notes

Some thoughts to keep in mind while designing and testing algorithms using backward approximate dynamic programming:

Approximation architectures – It is possible to use any of the statistical learning methods described in chapter 3 (or your favorite book on statistics/machine learning). We note that most of the methods in this book involve adaptive learning (this is the focus of chapter 3), but with backward ADP, we actually return to the more familiar setting (in the statistical learning community) of batch learning. Following standard advice in the specification of any statistical model, make sure that the dimensionality of the model (measured by the number of parameters) is much smaller than the number of observations to avoid overfitting.

Tuning – Virtually all adaptive learning algorithms have tunable parameters, and this is the Achilles heel of this entire approach to solving stochastic optimization problem. In chapter 9, section 9.11 summarizes four problem classes (see Table 9.3), where classes (1) and (4) are posed as finding the best learning policy. These "learning policies" represent the process of finding the best search algorithm, which includes tuning the parameters that govern a particular class of algorithm. In practice, this search for the best learning policy (or equivalently, the search for the best search algorithm) is typically done in an ad hoc way. There are thousands of papers which will prove asymptotic convergence, but the actual design of an algorithm depends on ad hoc testing.

Validating – A major challenge with any approximation strategy, backward ADP included, is validation. Backward ADP can work extremely well on problems where the value function is a fairly good approximation of the true value function, but there are no guarantees. It helps to have a good benchmark (in this case, the widely accepted myopic policy served this role) for comparison.

Performance – We have obtained exceptionally good performance on some problem classes, including energy storage problems with thousands of time periods. In comparisons against optimal policies (obtained using the methods from chapter 14 for low-dimensional problem instances), we have obtained solutions that were over 95 percent of optimality, but on occasions the performance was as low as 70 percent when we did a poor job with the approximations.

15.5 Bibliographic Notes

Section 15.1 – The first use of the term "backward approximate dynamic programming" in the published literature is in Senn et al. (2014), which is based on Senn's Ph.D. dissertation (in German), which appeared in 2013. This work was for a finite-horizon deterministic control problem. Cheng et al. (2018a) used backward ADP for a stochastic energy storage problem using the idea of a low-rank approximation for the value function. Cheng et al. (2018b) used a simpler linear architecture for an energy storage problem and showed that it was quite effective.

Section 15.2 – Fitted value iteration is basically backward approximate dynamic programming for infinite horizon problems. Szepesvári and Munos (2005) and Munos and Szepesv (2008) were the earliest papers that use the term "fitted value iteration." Fitted value iteration is a form of approximate value iteration which we consider in depth in chapter 17, which focuses on forward algorithms.

Section 15.4.1 – The work on backward ADP for clinical trials is taken from Tian et al. (2021). The experimental work for energy storage is taken from Cheng et al. (2018b). The work on allocating aid in Africa is taken from Aboagye and Powell (2018), which extended the seminal paper by Collier and Dollar (2002) which proposed the myopic policy for the same problem.

Exercises

Review questions

15.1 Contrast backward approximate dynamic programming for finite horizon problems versus infinite horizon problems in terms of the concept of "convergence" for each one.

Computational exercises

15.2 We are going to solve the continuous budgeting problem presented in section 14.4.2 using backward approximate dynamic programming. The problem starts with R_0 resources which are then allocated over periods 0 to T. Let x_t be the amount allocated in period t with contribution

$$C_t(x_t) = \sqrt{x_t}.$$

Assume that $T = 20$ time periods.

(a) Use the results of section 14.4.2 to solve this problem optimally. Evaluate your simulation by simulating your optimal policy 1000 times.

(b) Use the backward ADP algorithm described in Figure 15.5 to obtain the value function approximations using

$$\overline{V}_t(R_t) = \theta_{t0} + \theta_{t1}\sqrt{x_t}.$$

Use linear regression (either the methods in section 3.7.1, or a package) to fit $\overline{V}_t(R_t)$. Then, simulate this policy 1000 times (ideally using the same sample paths as you used for part (a)). How do you think θ_{t0} and θ_{t1} should behave?

(c) Use the backward ADP algorithm described in Figure 15.5 to obtain the value function approximations using

$$\overline{V}_t(R_t) = \theta_{t0} + \theta_{t1}R_t^x + \theta_{t2}(R_t^x)^2,$$

where R_t^x is the post-decision resource state $R_t^x = R_t - x_t$ (which is the same as R_{t+1} since transitions are deterministic).

Use linear regression (either the methods in section 3.7.1, or a package) to fit $\overline{V}_t(R_t)$. Then, simulate this policy 1000 times (ideally using the same sample paths as you used for part (a)).

15.3 Repeat exercise 15.2, but this time use

$$C(x_t) = \ln(x_t).$$

For part (b), use

$$\overline{V}_t(R_t) = \theta_{t0} + \theta_{t1}\ln(x_t).$$

15.4 In this exercise you are going to solve a simple inventory problem using Bellman's equations, to obtain an optimal policy. Then, the exercises that follow will have you implement various backward ADP policies that you can compare against the optimal policy you obtain in this exercise. Your inventory problem will span T time periods, with an inventory equation governed by

$$R_{t+1} = \max\{0, R_t - \hat{D}_{t+1}\} + x_t.$$

Here we are assuming that product ordered at time t, x_t, arrive at $t + 1$. Assume that $\hat{D}_{t+1}$ is described by a discrete uniform distribution between 1 and 20.

Next assume that our contribution function is given by

$$C(S_t, x_t) = 50 \min\{R_t, \hat{D}_{t+1}\} - 10x_t.$$

(a) Find an optimal policy by solving this dynamic program exactly using classical backward dynamic programming methods from chapter 14 (specifically equation (14.3)). Note that your biggest challenge will be computing the one-step transition matrix. Simulate the optimal policy 1,000 times starting with $R_0 = 0$ and report the performance.

(b) Now solve the problem using backward ADP using a simple quadratic approximation for the value function approximation:

$$\overline{V}_t^x(R_t^x) = \theta_{t0} + \theta_{t1}R_t^x + \theta_{t2}(R_t^x)^2.$$

where R_t^x is the post-decision resource state which we might represent using

$$R_t^x = \max\{0, R_t - \mathbb{E}\{\hat{D}_{t+1}\}\} + x_t.$$

Having found $\overline{V}_t^x(R_t^x)$, simulate the resulting policy 1,000 times, and compare your results to your optimal policy.

Sequential decision analytics and modeling

These exercises are drawn from the online book *Sequential Decision Analytics and Modeling* available at `http://tinyurl.com/sdaexamplesprint`.

15.5 We are going to perform experiments for an energy storage problem that we can solve exactly using backward approximate dynamic programming. Download the code "EnergyStorage_I" from `http://tinyurl.com/sdagithub`. This code is set up to solve the problem exactly using backward dynamic programming, where we have to enumerate the state space. Here, you will be asked to create a version of the code that uses backward approximate dynamic programming.

Assume that the price process evolves according to

$$p_{t+1} = \min\{100, \max\{0, p_t + \varepsilon_{t+1}\}\}$$

where ε_{t+1} follows a discrete uniform distribution given by

$$\varepsilon_{t+1} = \begin{cases} -2 & \text{with prob. } 1/5 \\ -1 & \text{with prob. } 1/5 \\ 0 & \text{with prob. } 1/5 \\ +1 & \text{with prob. } 1/5 \\ +2 & \text{with prob. } 1/5 \end{cases}$$

Assume that $p_0 = \$50$.

(a) Solve for an optimal policy by using the backward dynamic programming strategy in section 14.3 of the text (the algorithm has
already been implemented in the Python module).
  (i) Run the algorithm where prices are discretized in increments
of $\$1$, then $\$0.50$ and finally $\$0.25$. Compute the size of the state
space for each of the three levels of discretization, and plot the
run times against the size of the state space.
  (ii) Using the optimal value function for the discretization of $\$1$,
simulate the policy for each level of discretization of the prices
using 100 forward simulations, and report the estimated objective functions.
(b) Modify the code to solve the problem using the approximate
dynamic programming with lookup tables given in Figure 15.3.
Simulate the resulting policy (for each of the three levels of price
discretization) and report the results.
(c) Modify the code to solve the problem using the approximate
dynamic programming using a continuous approximation given in
Figure 15.5. Simulate the resulting policy (for each of the three levels of price discretization) and report the results. Use the linear
model of the post-decision value function

$$\overline{V}_t^x(S_t^x) = \sum_{f \in \mathcal{F}} \theta_f \phi_f(S_t^x)$$

with features

$$\phi_0(S_t^x) = 1,$$
$$\phi_1(S_t^x) = R_t^x,$$
$$\phi_2(S_t^x) = (R_t^x)^2,$$
$$\phi_3(S_t^x) = p_t,$$
$$\phi_4(S_t^x) = p_t^2,$$
$$\phi_5(S_t^x) = R_t^x p_t.$$

Simulate the policy you obtain with your approximate value function (using 100 simulations) and compare the results to the optimal policy.

(d) Repeat (c), but now assume that the price process evolves according to

$$p_{t+1} = .5p_t + .5p_{t-1} + \varepsilon_{t+1}$$

where ε_{t+1} follows the distribution as shown. Remember that you now have to include p_{t-1} in your state variable. Just use the single price discretization of $1. Please do the following:

(i) First compute the optimal policy following your approach in part (a). You have to modify the code to handle an extra dimension of the state variable. Compare the run times using the price models assumed in part (a) and part (b). How did the more complex state variable affect the solution time for the optimal algorithm and the backward approximate dynamic programming algorithm?

(ii) Compare the performance of the optimal solution to the solution obtained using backward approximate dynamic programming.

Diary problem

The diary problem is a single problem you chose (see chapter 1 for guidelines). Answer the following for your diary problem.

15.6 Take your formulation of your diary problem that you developed for the diary problem exercise in chapter 14, and sketch a backward ADP algorithm for the problem. Specify a value function approximation that you think might work.

Bibliography

Aboagye, N.K. and Powell, W.B. (2018). Stochastic optimization of official development assistance allocation.

Cheng, B., Asamov, T., and Powell, W.B. (2018a). Low-rank value function approximation for co-optimization of battery storage. *IEEE Transactions on Smart Grid* 9 (6): 6590–6598.

Cheng, B., Member, S., and Powell, W.B. (2018b). Transactions on smart grid co-optimizing battery storage for the frequency regulation and energy arbitrage using multi-scale dynamic programming. *IEEE Transactions on the Smart Grid* 9 (3): 1997–2005.

Collier, P. and Dollar, D. (2002). Aid allocation and poverty reduction. *European Economic Review* 46 (8): 1475–1500.

Jiang, D. R., and Powell, W. B. (2015). An Approximate Dynamic Programming Algorithm for Monotone Value Functions. *Operations Research*, 63 (6), 1489–1511. doi:10.1287/opre.2015.1425.

Munos, R. and Szepesv, C. (2008). Finite-time bounds for fitted value iteration. *Journal of Machine Learning Research* 1: 815–857.

Senn, M., Link, N., Pollak, J., and Lee, J.H. (2014). Reducing the computational effort of optimal process controllers for continuous state spaces by using incremental learning and post-decision state formulations. *Journal of Process Control* 24: 133–143.

Szepesvári, C. and Munos, R. (2005). Finite time bounds for sampling based fitted value iteration. *Proceedings of the 22nd International Conference on Machine Learning ICML '05* . 880–887.

Tian, Z., Han, W. and Powell, W.B. (2021). Adaptive learning of drug quality and optimization of patient recruitment for clinical trials with dropouts. *Manufacturing & Service Operations Management*.

16

Forward ADP I: The Value of a Policy

Chapter 14 laid the foundation for finding optimal policies for problems with discrete states, assuming that states and decisions can be enumerated, and the one-step transition matrix can be calculated. The chapter presented classical backward dynamic programming for finite horizon problems, but most of the chapter focused on infinite horizon problems, where we introduced several methods for computing the value function $V(s)$. Of these, the most important is value iteration, since this is relatively easy to compute, and it is the foundation for a number of approximation strategies.

In chapter 15, we introduced the idea of backward approximate dynamic programming (for finite horizon problems), also known as fitted value iteration for infinite horizon problems. Backward approximate dynamic programming is, surprisingly, a relatively recent invention, and while fitted value iteration is somewhat older, the attention it has received is a small fraction compared to the methods that we are going to present in this chapter, and chapters 17 and 18, which are all based on the principle of *forward* approximate dynamic programming.

We suspect the reason for the relative popularity of forward approximate dynamic programming is that it captures the dynamics of an actual physical system, which moves forward in time. It has the immediate benefit of avoiding any semblance of enumerating states, which avoids "the" curse of dimensionality (which is most often associated with vector-valued states). It even avoids the need to determine how to sample the state space, as is required in backward dynamic programming.

It also avoids any need to compute the one-step transition matrix, since we are either simulating the exogenous information W_{t+1}, or we are simply observing transitions from S_t to S_{t+1} from a physical system. When we step forward in time, it seems as if there is a natural sampling mechanism, and while this is true to a degree, we will have to pay attention to how we choose the decisions

Reinforcement Learning and Stochastic Optimization: A Unified Framework for Sequential Decisions, First Edition. Warren B. Powell.
© 2022 John Wiley & Sons, Inc. Published 2022 John Wiley & Sons, Inc.

that determine (up to a point) the next state we visit. By contrast, backward ADP requires pure, random sampling, which means our choice of states is not guided at all by the physics of the problem beyond assumptions about the range of states.

Most of the work in this area still assumes discrete decisions, which enjoys a very wide set of applications. We cover vector-valued decisions, but not until chapter 18 where we limit our focus to problems with concave (convex if minimizing) contribution functions (which translates to concavity in the value function).

In this chapter, we focus primarily on the different ways of calculating $\hat{v}^n$, and then using this information to estimate a value function approximation, *for a fixed policy*. The reason we do this is to resolve the subtleties of estimating the value of a policy before we allow the policy to evolve with the iterations, which introduces a significant complication. To emphasize that we are computing values for a fixed policy, we index parameters such as the value function V^π by the policy π. After we establish the fundamentals for estimating the value of a policy, chapter 17 addresses the process of searching for good policies.

16.1   Sampling the Value of a Policy

At first glance, the problem of statistically estimating the value of a fixed policy should not be any different than estimating a function from noisy observations. In fact, this can be true, but it depends on how $\hat{v}^n$ is being calculated. In time (especially in chapter 17 when we are optimizing over policies), we will have to learn to live with the reality that $\hat{v}^n$ is almost always a biased sampled estimate of the value of being in a state.

Our normal style has been to model finite horizon problems without a discount factor. Of course, discounting is essential in infinite horizon problems, as we saw in chapter 14. In the text that follows, we are going to sometimes switch between finite and infinite horizon, so we retain a discount factor γ even for the finite horizon case.

16.1.1   Direct Policy Evaluation for Finite Horizon Problems

Imagine that we have a fixed policy $X^\pi(s)$ which may take any of the forms described in chapter 11. For iteration n, if we are in state S_t^n at time t, we then choose decision $x_t^n = X^\pi(S_t^n)$, after which we sample the exogenous information W_{t+1}^n. We sometimes say that we are following sample path ω^n from which we observe $W_{t+1}^n = W_{t+1}(\omega^n)$. The exogenous information W_{t+1}^n may depend on both S_t^n and the decision x_t^n. From this, we may compute our contribution from

Step 0. Initialization:

> **Step 0a.** Initialize $\overline{V}^0$.
> **Step 0b.** Initialize S^1.
> **Step 0c.** Set $n = 1$.

Step 1. Choose a sample path ω^n.
Step 2. Choose a starting state S_0^n.
Step 3. Do for $t = 0, 1, \ldots, T$:

> **Step 3a.** $x_t^n = X^\pi(S_t^n)$.
> **Step 3b.** $\hat{C}_t^n = C(S_t^n, x_t^n)$.
> **Step 3c.** $W_{t+1}^n = W_{t+1}(\omega^n)$.
> **Step 3d.** $S_{t+1}^n = S^M(S_t^n, x_t^n, W_{t+1}^n)$.

Step 4. Compute $\hat{v}_0^n = \sum_{t=0}^{T} \gamma^t \hat{C}_t^n$.
Step 5. Increment n. If $n \leq N$ go to Step 1.
Step 6. Use the sequence of state-value pairs $(S^i, \hat{v}^i)_{i=1}^N$ to fit a value function approximation $\overline{V}^\pi(s)$.

Figure 16.1 Basic policy evaluation procedure.

$$\hat{C}_t^n = C(S_t^n, x_t^n).$$

Finally, we compute our next state from our transition function

$$S_{t+1}^n = S^M(S_t^n, x_t^n, W_{t+1}^n).$$

This process continues until we reach the end of our horizon T. The basic algorithm is described in Figure 16.1. In step 6, we use a batch routine to fit a statistical model. It is often more natural to use some sort of recursive procedure and imbed the updating of the value function within the iterative loop. The type of recursive procedure depends on the nature of the value function approximation. Later in this chapter, we describe several recursive procedures if we are using linear regression.

Finite horizon problems are sometimes referred to as *episodic*, where an episode refers to a simulation of a policy until the end of the horizon (also known as trials). However, the term episodic can also be interpreted more broadly. For example, an emergency vehicle may repeatedly return to base where the system then restarts. Each cycle of starting from a home base and then returning to the home base can be viewed as an episode. As a result, if we are working with a finite horizon problem, we prefer to refer to these specifically as such.

Evaluating a fixed policy is mathematically equivalent to making unbiased observations of a noisy function. Fitting a functional approximation is precisely what the entire field of statistical learning has been trying to do for decades. If we are fitting a linear model, then there are some powerful recursive procedures that can be used. These are discussed in section 16.1.2.

16.1.2 Policy Evaluation for Infinite Horizon Problems

Not surprisingly, infinite horizon problems introduce a special complication, since we cannot obtain an unbiased observation in a finite number of measurements. We present some methods that have been used for infinite horizon applications.

Recurrent Visits

There are many problems which are infinite horizon, but where the system resets itself periodically. A simple example of this is a finite horizon problem, where hitting the end of the horizon and starting over (as would occur in a game) can be viewed as an episode. A different example is a queueing system, where perhaps we are trying to manage the admission of patients to an emergency room. From time to time the queue may become empty, at which point the system resets and starts over. For such systems, it makes sense to estimate the value of following a policy π when starting from this base state.

Even if we do not have such a renewal system, imagine that we find ourselves in a state s. Now follow a policy π until we re-enter state s again. Let $R^n(s)$ be the reward earned, and let $\tau^n(s)$ be the number of time periods required before re-entering state s. Here, n is counting the number of times we visit state s. An observation of the average reward earned when in state s and following policy π would be given by

$$\hat{v}^n(s) = \frac{R^n(s)}{\tau^n(s)}.$$

$\hat{v}^n(s)$ would be computed when we return to state s. We might then update the *average* value of being in state s using

$$\bar{v}^n(s) = (1 - \alpha_{n-1})\bar{v}^{n-1}(s) + \alpha_{n-1}\hat{v}^n(s).$$

Note that as we make each transition from some state s' to some state s'', we are accumulating rewards in $R(s)$ for every state s that we have visited prior to reaching state s'. Each time we arrive at some state s'', we stop accumulating rewards for s'', and compute $\hat{v}^n(s'')$, and then smooth this into the current estimate of $\bar{v}(s'')$. Note that we have presented this only for the case of computing the average reward per time period.

Partial Simulations

While we may not be able to simulate an infinite trajectory, we may simulate a long trajectory T, long enough to ensure that we are producing an estimate that is "good enough." When we are using discounting, we realize that eventually γ^t becomes small enough that a longer simulation does not really matter. This idea can be implemented in a relatively simple way.

Consider the algorithm in Figure 16.1, and insert the calculation in step 3:

$$\bar{c}_t = \frac{t-1}{t}\bar{c}_{t-1} + \frac{1}{t}\hat{C}^n_t.$$

$\bar{c}_t$ is an average over the time periods of the contribution per time period. As we follow our policy over progressively more time periods, $\bar{c}_t$ approaches an average contribution per time period. Over an infinite horizon, we would expect to find

$$\hat{v}^n_0 = \lim_{t \to \infty} \sum_{t=0}^{\infty} \gamma^t \hat{C}^n_t = \frac{1}{1-\gamma}\bar{c}_{\infty}.$$

Now assume that we only progress T time periods, and let $\bar{c}_T$ be our estimate of $\bar{c}_{\infty}$ at this point. We would expect that

$$\hat{v}^n_0(T) = \sum_{t=0}^{T} \gamma^t \hat{C}^n_t$$

$$\approx \frac{1 - \gamma^{T+1}}{1-\gamma}\bar{c}_T. \tag{16.1}$$

The error between our T-period estimate $\hat{v}^n_0(T)$ and the infinite horizon estimate $\hat{v}^n_0$ is given by

$$\delta^n_T = \frac{1}{1-\gamma}\bar{c}_{\infty} - \frac{1 - \gamma^{T+1}}{1-\gamma}\bar{c}_T$$

$$\approx \frac{1}{1-\gamma}\bar{c}_T - \frac{1 - \gamma^{T+1}}{1-\gamma}\bar{c}_T$$

$$= \frac{\gamma^{T+1}}{1-\gamma}\bar{c}_T.$$

Thus, we just have to find T to make δ_T small enough. This strategy is imbedded in some optimal algorithms, which only require that $\delta^n_T \to 0$ as $n \to \infty$ (meaning that we have to steadily allow T to grow).

Infinite Horizon Projection

We can easily see from (16.1) that if we stop after T time periods, we will under-estimate the infinite horizon contribution by a factor $1 - \gamma^{T+1}$. Assuming that T is reasonably large (say, $\gamma^{T+1} < 0.1$), we might introduce the correction

$$\hat{v}_0^n = \frac{1}{1 - \gamma^{T+1}} \hat{v}_0^n(T).$$

In essence we are taking a sample estimate of a T-period path, and projecting it out over an infinite horizon.

16.1.3   Temporal Difference Updates

Assume that we are in state S_t^n and we make decision x_t^n (using policy π), after which we observe the information W_{t+1} which puts us in state $S_{t+1}^n = S^M(S_t^n, x_t^n, W_{t+1}^n)$. The contribution from this transition is given by $C(S_t^n, x_t^n)$. Imagine now that we continue this until the end of our horizon T. For simplicity, we are going to drop discounting. In this case, the contribution along this path would be

$$\hat{v}_t^n = C(S_t^n, x_t^n) + C(S_{t+1}^n, x_{t+1}^n) + \dots + C(S_T^n, x_T^n). \tag{16.2}$$

This is the contribution from following the path produced by a combination of the information from outcome ω^n (this determines $W_{t+1}^n, W_{t+2}^n, \dots, W_T^n$) and policy π. $\hat{v}_t^n$ is an unbiased sample estimate of the value of being in state S_t and following policy π over sample path ω^n. We can use a stochastic gradient algorithm to estimate the value of being in state S_t using

$$\bar{V}_t^n(S_t^n) = \bar{V}_t^{n-1}(S_t^n) - \alpha_n \left(\bar{V}_t^{n-1}(S_t^n) - \hat{v}_t^n \right). \tag{16.3}$$

We can obtain a richer class of algorithms by breaking down our path cost in (16.2) by using

$$\hat{v}_t^n = \sum_{\tau=t}^{T} C(S_\tau^n, x_\tau^n)$$

$$- \underbrace{\left\{ \sum_{\tau=t}^{T} \left(\bar{V}_\tau^{n-1}(S_\tau) - \bar{V}_{\tau+1}^{n-1}(S_{\tau+1}) \right) \right\} + \left(\bar{V}_t^{n-1}(S_t) - \bar{V}_{T+1}^{n-1}(S_{T+1}) \right)}_{=0}.$$

We now use the fact that $\bar{V}_{T+1}^{n-1}(S_{T+1}) = 0$ (this is where our finite horizon model is useful). Rearranging gives

$$\hat{v}_t^n = \bar{V}_t^{n-1}(S_t) + \sum_{\tau=t}^{T} \left(C(S_\tau^n, x_\tau^n) + \bar{V}_{\tau+1}^{n-1}(S_{\tau+1}) - \bar{V}_\tau^{n-1}(S_\tau)\right).$$

Let

$$\delta_\tau = C(S_\tau^n, x_\tau^n) + \bar{V}_{\tau+1}^{n-1}(S_{\tau+1}^n) - \bar{V}_\tau^{n-1}(S_\tau^n). \tag{16.4}$$

The terms δ_τ are called *temporal differences*. If we were using a standard single-pass algorithm, then at time t, $\hat{v}_t^n = C(S_t^n, x_t^n) + \bar{V}_{t+1}^{n-1}(S_{t+1}^n)$ would be our sample observation of being in state S_t, while $\bar{V}_t^{n-1}(S_t)$ is our current estimate of the value of being in state S_t. This means that the temporal difference at time t, $\delta_t = \hat{v}_t^n - \bar{V}_t^{n-1}(S_t)$, is the difference in our estimate of the value of being in state S_t between our current estimate and the updated estimate. The temporal difference is also known as the *Bellman error*.

Using (16.4), we can write $\hat{v}_t^n$ in the more compact form

$$\hat{v}_t^n = \bar{V}_t^{n-1}(S_t) + \sum_{\tau=t}^{T} \delta_\tau. \tag{16.5}$$

Substituting (16.5) into (16.3) gives

$$\bar{V}_t^n(S_t) = \bar{V}_t^{n-1}(S_t) - \alpha_{n-1}\left[\bar{V}_t^{n-1}(S_t) - \left(\bar{V}_t^{n-1}(S_t) + \sum_{\tau=t}^{T} \delta_\tau\right)\right]$$

$$= \bar{V}_t^{n-1}(S_t) + \alpha_{n-1}\sum_{\tau=t}^{T-1} \delta_\tau. \tag{16.6}$$

We next use this bit of algebra to build an important class of updating mechanisms for estimating value functions.

16.1.4 TD(λ)

The temporal differences δ_τ are the errors in our estimates of the value of being in state S_τ. We can think of each term in (16.6) as a correction to the estimate of the value function. It makes sense that updates farther along the path should not be given as much weight as those earlier in the path. As a result, it is common to introduce an artificial discount factor λ, producing updates of the form

$$\bar{V}_t^n(S_t) \quad = \quad \bar{V}_t^{n-1}(S_t) + \alpha_{n-1} \sum_{\tau=t}^{T} \lambda^{\tau-t} \delta_\tau. \tag{16.7}$$

We derived this formula without a time discount factor. We leave as an exercise to the reader to show that if we have a time discount factor γ, then the temporal-difference update becomes

$$\bar{V}_t^n(S_t) \quad = \quad \bar{V}_t^{n-1}(S_t) + \alpha_{n-1} \sum_{\tau=t}^{T} (\gamma\lambda)^{\tau-t} \delta_\tau. \tag{16.8}$$

Equation (16.8) shows that the discount factor γ, which is typically viewed as capturing the time value of money, and the algorithmic discount λ, which is a purely algorithmic device, have exactly the same effect. Not surprisingly, modelers in operations research have often used a discount factor γ set to a much smaller number than would be required to capture the time-value of money. Artificial discounting allows us to look into the future, but then discount the results when we feel that the results are not perfectly accurate.

Updates of the form given in equation (16.7) produce an updating procedure that is known as TD(λ) (or, temporal difference learning with discount λ). Here, λ is introduced as a form of algorithmic discounting, since it has nothing to do with the traditional use of discounting to reflect the value of money. Algorithmic discounting is a heuristic way of limiting the effect of decisions we plan on making in the future, given that our model of the future is imperfect.

The updating formula in equation (16.7) requires that we step all the way to the end of the horizon before updating our estimates of the value. There is, however, another way of implementing the updates. The temporal differences δ_τ are computed as the algorithm steps forward in time. As a result, our updating formula can be implemented recursively. Assume we are at time t' in our simulation. We would simply execute

$$\bar{V}_t^n(S_t^n) := \bar{V}_t^n(S_t) + \alpha_{n-1}\lambda^{t'-t}\delta_{t'} \quad \text{for all } t \le t'. \tag{16.9}$$

Here, our notation ":=" means that we take the current value of $\bar{V}_t^n(S_t)$, add $\alpha_{n-1}\lambda^{t'-t}\delta_{t'}$ to it to obtain an updated value of $\bar{V}_t^n(S_t)$. When we reach time $t' = T$, our value functions would have undergone a complete update. We note that at time t', we need to update the value function for every $t \le t'$.

16.1.5 TD(0) and Approximate Value Iteration

An important special case of TD(λ) occurs when we use $\lambda = 0$. In this case,

$$\bar{V}_t^n(S_t^n) \quad = \quad \bar{V}_t^{n-1}(S_t^n) + \alpha_{n-1}\big(C(S_t^n, x_t^n) + \gamma\bar{V}_{t+1}^{n-1}(S^M(S_t^n, x_t^n, W_{t+1}^n)) - \bar{V}_t^{n-1}(S_t^n)\big). \tag{16.10}$$

Now consider value iteration. In chapter 14, when we did not have to deal with Monte Carlo samples and statistical noise, value iteration (for a fixed policy) looked like

$$V_t^n(s) = C(s, X^\pi(s)) + \gamma \sum_{s' \in \mathcal{S}} p^\pi(s'|s) V_{t+1}^n(s').$$

In steady state, we would write it as

$$V^n(s) = C(s, X^\pi(s)) + \gamma \sum_{s' \in \mathcal{S}} p^\pi(s'|s) V^{n-1}(s').$$

When we use approximate dynamic programming, we are following a sample path that puts us in state S_t^n, where we observe a sample realization of a contribution $\hat{C}_t^n$, after which we observe a sample realization of the next downstream state S_{t+1}^n (the decision is determined by our fixed policy). A sample observation of the value of being in state S_t^n would be computed using

$$\hat{v}_t^n = C(S_t^n, x_t^n) + \gamma \bar{V}_{t+1}^{n-1}(S_{t+1}^n).$$

We can then use this to update our estimate of the value of being in state S_t^n using

$$
\begin{aligned}
\bar{V}_t^n(S_t^n) &= (1 - \alpha_{n-1}) \bar{V}_t^{n-1}(S_t^n) + \alpha_{n-1} \hat{v}_t^n \\
&= (1 - \alpha_{n-1}) \bar{V}_t^{n-1}(S_t^n) + \\
&\quad \alpha_{n-1}\left(C(S_t^n, x_t^n) + \gamma \bar{V}^{n-1}(S^M(S_t^n, x_t^n, W_{t+1}^n))\right).
\end{aligned}
\qquad (16.11)
$$

It is not hard to see that (16.10) and (16.11) are the same. The idea is popular because it is particularly easy to implement. It is also well suited to high-dimensional decision vectors x, as we illustrate in chapter 18.

Temporal difference learning derives its name because $\bar{V}^{n-1}(S)$ is viewed as the "current" value of being in state S, while $C(S, x) + \bar{V}^{n-1}(S^M(S, x, W))$ is viewed as the updated value of being in state S. The difference $\bar{V}^{n-1}(S) - (C(S, x) + \bar{V}^{n-1}(S^M(S, x, W)))$ is the difference in these estimates across iterations (or time), hence the name. TD(0) is a form of statistical bootstrapping, because rather than simulate the full trajectory, it depends on the current estimate of the value $\bar{V}^{n-1}(S^M(S, x, W))$ of being in the downstream state $S^M(S, x, W)$.

While TD(0) can be very easy to implement, it can also produce very slow convergence. The effect is illustrated using the simple five-state Markov chain shown in Figure 16.2, where the contribution of the transitions out of states 0 through 4, denoted by $\hat{c}$, is always 0, and then we receive 1 when we make the final transition out of state 5. When we apply TD(0) updating to estimate the value of each state, we produce the set of numbers shown in Table 16.1. In this

Figure 16.2 Five-state Markov chain for illustrating backward learning.

Table 16.1 Effect of stepsize on backward learning.

Iteration	$\bar{V}_0$	$\hat{v}_1$	$\bar{V}_1$	$\hat{v}_2$	$\bar{V}_2$	$\hat{v}_3$	$\bar{V}_3$	$\hat{v}_4$	$\bar{V}_4$	$\hat{v}_5$
0	0.000		0.000		0.000		0.000		0.000	1
1	0.000	0.000	0.000	0.000	0.000	0.000	0.000	0.000	1.000	1
2	0.000	0.000	0.000	0.000	0.000	0.000	0.500	1.000	1.000	1
3	0.000	0.000	0.000	0.000	0.167	0.500	0.667	1.000	1.000	1
4	0.000	0.000	0.042	0.167	0.292	0.667	0.750	1.000	1.000	1
5	0.008	0.042	0.092	0.292	0.383	0.750	0.800	1.000	1.000	1
6	0.022	0.092	0.140	0.383	0.453	0.800	0.833	1.000	1.000	1
7	0.039	0.140	0.185	0.453	0.507	0.833	0.857	1.000	1.000	1
8	0.057	0.185	0.225	0.507	0.551	0.857	0.875	1.000	1.000	1
9	0.076	0.225	0.261	0.551	0.587	0.875	0.889	1.000	1.000	1
10	0.095	0.261	0.294	0.587	0.617	0.889	0.900	1.000	1.000	1

illustration, there are no decisions and the contribution is zero for every other time period. A stepsize of $1/n$ was used throughout.

Table 16.1 illustrates that the rate of convergence for $\bar{V}_0$ is dramatically slower than for $\bar{V}_4$. The reason is that as we smooth $\hat{v}_t$ into $\bar{V}_{t-1}$, the stepsize has a discounting effect. The problem is most pronounced when the value of being in a state at time t depends on contributions that are a number of steps into the future (imagine the challenge of training a value function to play the game of chess). For problems with long horizons, and in particular those where it takes many steps before receiving a reward, this bias can be so serious that it can appear that temporal differencing (and algorithms that use it) simply does not work. We can partially overcome the slow convergence by carefully choosing a stepsize rule. Stepsizes are discussed in depth in chapter 6. See in particular the OSAVI stepsize policy (section 6.4) which is designed specifically for estimating value functions.

16.1.6 TD Learning for Infinite Horizon Problems

We can perform updates using a general TD(λ) strategy as we did for finite horizon problems. However, there are some subtle differences. With finite horizon problems, it is common to assume that we are estimating a different function $\bar{V}_t$ for each time period t. As we step through time, we obtain information that can be used for a value function at a *specific* point in time. With stationary problems, each transition produces information that can be used to update the value function, which is then used in all future updates. By contrast, if we update $\bar{V}_t$ for a finite horizon problem, then this update is not used until the next forward pass through the states.

When we move to infinite horizon problems, we drop the indexing by t. Instead of stepping forward in time, we step through iterations, where at each iteration we generate a temporal difference

$$\delta^n = C(s^n, x^n) + \gamma \bar{V}^{n-1}(S^{M,x}(s^n, x^n)) - \bar{V}^{n-1}(s^n).$$

To do a proper update of the value function at each state, we would have to use an infinite series of the form

$$\bar{V}^n(s) = \bar{V}^{n-1}(s) + \alpha_n \sum_{m=0}^{\infty} (\gamma\lambda)^m \delta^{n+m}, \tag{16.12}$$

where we can use any initial starting state $s^0 = s$. Of course, we would use the same update for each state s^m that we visit, so we might write

$$\bar{V}^n(s^m) = \bar{V}^{n-1}(s^m) + \alpha_n \sum_{n=m}^{\infty} (\gamma\lambda)^{(n-m)} \delta^n. \tag{16.13}$$

Equations (16.12) and (16.13) both imply stepping forward in time (presumably a "large" number of iterations) and computing temporal differences before performing an update. A more natural way to run the algorithm is to do the updates incrementally. After we compute δ^n, we can update the value function at each of the previous states we visited. So, at iteration n, we would execute

$$\bar{V}^n(s^m) := \bar{V}^n(s^m) + \alpha_n (\gamma\lambda)^{n-m} \delta^m, \quad m = n, n-1, \dots, 1. \tag{16.14}$$

We can now use the temporal difference δ^n to update the estimate of the value function for every state we have visited up to iteration n.

Figure 16.3 outlines the basic structure of a TD(λ) algorithm for an infinite horizon problem. Step 1 begins by computing the first post-decision state, after which step 2 makes a single step forward. After computing the temporal-difference in step 3, we traverse previous states we have visited in step 4 to update their value functions.

Step 0. Initialization:

> **Step 0a.** Initialize $\overline{V}^0(S)$ for all S.
> **Step 0b.** Initialize the state S^0.
> **Step 0c.** Set $n = 1$.

Step 1. Choose ω^n.
Step 2. Solve

$$x^n = \arg\max_{x \in \mathcal{X}^n} \left(C(S^n, x) + \gamma \overline{V}^{n-1}(S^{M,x}(S^n, x))) \right). \tag{16.15}$$

Step 3. Compute the temporal difference for this step:

$$\delta^n = C(S^n, x^n) + \gamma \left(\overline{V}^{n-1}(S^{M,x}(S^n, x^n)) - \overline{V}^{n-1}(S^n) \right).$$

Step 4. Update $\overline{V}$ for $m = n, n-1, \ldots, 1$:

$$\overline{V}^n(S^m) = \overline{V}^{n-1}(S^m) + (\gamma\lambda)^{n-m}\delta^n. \tag{16.16}$$

Step 5. Compute $S^{n+1} = S^M(S^n, x^n, W(\omega^n))$.
Step 6. Let $n = n + 1$. If $n < N$, go to step 1.

Figure 16.3 A TD(λ) algorithm for infinite horizon problems.

In step 3, we update all the states $(S^m)_{m=1}^n$ that we have visited up to then. Thus, at iteration n, we would have simulated the partial update

$$\bar{V}^n(S^0) = \bar{V}^{n-1}(S^0) + \alpha_{n-1} \sum_{m=0}^{n} (\gamma\lambda)^m \delta^m. \tag{16.17}$$

This means that at any iteration n, we have updated our values using biased sample observations (as is generally the case in value iteration). We avoided this problem for finite horizon problems by extending out to the end of the horizon. We can obtain unbiased updates for infinite horizon problems by assuming that all policies eventually put the system into an "absorbing state." For example, if we are modeling the process of holding or selling an asset, we might be able to guarantee that we eventually sell the asset.

One subtle difference between temporal difference learning for finite horizon and infinite horizon problems is that in the infinite horizon case, we may be visiting the same state two or more times on the same sample path. For the finite horizon case, the states and value functions are all indexed by the time that we visit them. Since we step forward through time, we can never visit the same state at the same point in time twice in the same sample path. By contrast, it is quite easy in a steady-state problem to revisit the same state over and over again. For example, we could trace the path of our nomadic trucker (introduced in section 2.3.4.1), who might go back and forth between the same pair of locations in the same sample path. As a result, we are using the value function to determine what state to visit, but at the same time we are updating the value of being in these states.

16.2 Stochastic Approximation Methods

A central idea in recursive estimation is the use of stochastic approximation methods and stochastic gradients. We have already seen this in one setting in the chapter on derivative-based stochastic optimization in section 5.3.1. We review the idea again here, but in a different context. We begin with the same stochastic optimization problem, which we originally introduced as the problem

$$\min_x \mathbb{E}F(x, W).$$

Now assume that we are choosing a scalar value v to solve the problem

$$\min_v \mathbb{E}F(v, \hat{V}), \tag{16.18}$$

where

$$F(v, \hat{V}) = \frac{1}{2}(v - \hat{V})^2,$$

and where $\hat{V}$ is a random variable with unknown mean. We would like to use a series of sample realizations $\hat{v}^n$ to guide an algorithm that generates a sequence v^n that converges to the optimal solution v^* that solves (16.18). We use the same basic strategy as we introduced in section 5.3.1 where we update v^n using

$$
\begin{aligned}
v^n &= v^{n-1} - \alpha_{n-1}\nabla F(v^{n-1}, \hat{v}^n) \tag{16.19} \\
&= v^{n-1} - \alpha_{n-1}(v^{n-1} - \hat{v}^n).
\end{aligned}
$$

Now if we make the transition that instead of updating a scalar v^n, we are updating $\bar{V}_t^n(S_t^n)$. This produces the updating equation

$$\bar{V}_t^n(S_t^n) = \bar{V}_t^{n-1}(S_t^n) - \alpha_{n-1}(\bar{V}_t^{n-1}(S_t^n) - \hat{v}^n). \tag{16.20}$$

If we use $\hat{v}^n = C(S_t^n, x_t^n) + \gamma \bar{V}^{n-1}(S_{t+1}^n)$, we quickly see that the updating equation produced using our stochastic gradient algorithm (16.20) gives us the same update that we obtained using temporal difference learning (equation (16.10)) and approximate value iteration (equation (16.11)). In equation (16.19), α_n is called a stepsize, because it controls how far we go in the direction of

$\nabla F(v^{n-1}, \hat{v}^n)$, and for this reason this is the term that we adopt for α_n throughout this book. In contrast to our first use of this idea in section 5.3, where the stepsize had to serve a scaling function, in this setting the units of the variable being optimized, v^n, and the units of the gradient are the same. Indeed, we can expect that $0 < \alpha_n \leq 1$, which is a major simplification.

Now consider what happens when we replace the lookup table representation $\bar{V}(s)$ that we used earlier, with a linear regression $\bar{V}(s|\theta) = \theta^T \phi$. Now we want to find the best value of θ, which we can do by solving

$$\min_\theta \mathbb{E} \frac{1}{2} (\bar{V}(s|\theta) - \hat{v})^2.$$

Applying a stochastic gradient algorithm, we obtain the updating step

$$\theta^n = \theta^{n-1} - \alpha_{n-1} (\bar{V}(s|\theta^{n-1}) - \hat{v}^n) \nabla_\theta \bar{V}(s|\theta^n). \tag{16.21}$$

Since $\bar{V}(s|\theta^n) = \sum_{f \in \mathcal{F}} \theta_f^n \phi_f(s) = (\theta^n)^T \phi(s)$, the gradient with respect to θ is given by

$$\nabla_\theta \bar{V}(s|\theta^n) = \begin{pmatrix} \frac{\partial \bar{V}(s|\theta^n)}{\partial \theta_1} \\ \frac{\partial \bar{V}(s|\theta^n)}{\partial \theta_2} \\ \vdots \\ \frac{\partial \bar{V}(s|\theta^n)}{\partial \theta_F} \end{pmatrix} = \begin{pmatrix} \phi_1(s^n) \\ \phi_2(s^n) \\ \vdots \\ \phi_F(s^n) \end{pmatrix} = \phi(s^n).$$

Thus, the updating equation (16.21) is given by

$$\begin{aligned} \theta^n &= \theta^{n-1} - \alpha_{n-1} (\bar{V}(s|\theta^{n-1}) - \hat{v}^n) \phi(s^n) \\ &= \theta^{n-1} - \alpha_{n-1} (\bar{V}(s|\theta^{n-1}) - \hat{v}^n) \begin{pmatrix} \phi_1(s^n) \\ \phi_2(s^n) \\ \vdots \\ \phi_F(s^n) \end{pmatrix}. \end{aligned} \tag{16.22}$$

Using a stochastic gradient algorithm requires that we have some starting estimate θ^0 for the parameter vector, although $\theta^0 = 0$ is a common choice.

While this is a simple and elegant algorithm, we have reintroduced the problem of scaling. Just as we encountered in section 5.3, the units of θ^{n-1} and the units of $(\bar{V}(s|\theta^{n-1}) - \hat{v}^n) \phi(s^n)$ may be completely different. What we have learned about stepsizes still applies, except that we may need an initial stepsize that is quite different than 1.0 (our common starting point).

Our experimental work has suggested that the following policy works well: When you choose a stepsize formula, scale the first value of the stepsize so that the change in θ^n in the early iterations of the algorithm is approximately 20 to 50 percent (you will typically need to observe several iterations). You want to see individual elements of θ^n moving consistently in the same direction during the early iterations. If the stepsize is too large, the values can swing wildly, and the algorithm may not converge at all. If the changes are too small, the algorithm may simply stall out. It is very tempting to run the algorithm for a period of time and then conclude that it appears to have converged (presumably to a good solution). While it is important to see the individual elements moving in the same direction (consistently increasing or decreasing) in the early iterations, it is also important to see oscillatory behavior toward the end.

16.3 Bellman's Equation Using a Linear Model*

It is possible to solve Bellman's equation for infinite horizon problems by starting with the assumption that the value function is given by a linear model $V(s) = \theta^T \phi(s)$ where $\Phi(s)$ is a column vector of basis functions for a particular state s. Of course, we are still working with a single policy, so we are using Bellman's equation only as a method for finding the best linear approximation for the infinite horizon value of a fixed policy π.

We begin with a derivation based on matrix linear algebra, which is more advanced and which does not produce expressions that can be implemented in practice. We follow this discussion with a simulation-based algorithm which can be implemented fairly easily.

16.3.1 A Matrix-based Derivation**

In section 16.5.2, we provided a geometric view of basis functions, drawing on the elegance and obscurity of matrix linear algebra. We are going to continue this presentation and present a version of Bellman's equation assuming linear models. However, we are not yet ready to introduce the dimension of optimizing over policies, so we are still simply trying to approximate the value of being in a state. Also, we are only considering infinite horizon models, since we have already handled the finite horizon case. This presentation can be viewed as another method for handling infinite horizon models, while using a linear architecture to approximate the value function.

First recall that Bellman's equation (for a fixed policy) is written

$$V^\pi(s) = C(s, X^\pi(s)) + \gamma \sum_{s' \in \mathcal{S}} p(s'|s, X^\pi(s))V^\pi(s').$$

In vector-matrix form, we let V^π be a vector with element $V^\pi(s)$, we let c^π be a vector with element $C(s, X^\pi(s))$ and finally we let P^π be the one-step transition matrix with element $p(s'|s, X^\pi(s))$ at row s, column s'. Using this notation, Bellman's equation becomes

$$V^\pi = c^\pi + \gamma P^\pi V^\pi,$$

allowing us to solve for V^π using

$$V^\pi = (I - \gamma P^\pi)^{-1} c^\pi.$$

This works with a lookup-table representation (a value for each state). Now assume that we replace V^π with an approximation $\bar{V}^\pi = \Phi\theta$ where, Φ is a $|\mathcal{S}| \times |\mathcal{F}|$ matrix with element $\Phi_{s,f} = \phi_f(s)$. Also let

$$d_s^\pi \;=\; \text{the steady state probability of being in state } s \text{ while following policy } \pi,$$

$$D^\pi \;=\; \text{a } |\mathcal{S}| \times |\mathcal{S}| \text{ diagonal matrix where the state probabilities } (d_1^\pi, \dots, d_{|\mathcal{S}|}^\pi) \text{ make up the diagonal.}$$

We would like to choose θ to minimize the weighted sum of errors squared, where the error for state s is given by

$$\epsilon^n(s) = \sum_f \theta_f \phi_f(s) - \left(c^\pi(s) + \gamma \sum_{s' \in \mathcal{S}} p^\pi(s'|s, X^\pi) \sum_f \theta_f^n \phi_f(s') \right) \quad (16.23)$$

The first term on the right hand side of (16.23) is the predicted value of being in each state given θ, while the second term on the right hand side is the "predicted" value computed using the one-period contribution plus the expected value of the future which is computed using θ^n. The expected sum of errors squared is then given by

$$\min_\theta \sum_{s \in \mathcal{S}} d_s^\pi \left(\sum_f \theta_f \phi_f(s) - \left(c^\pi(s) + \gamma \sum_{s' \in \mathcal{S}} p^\pi(s'|s, X^\pi) \sum_f \theta_f^n \phi_f(s') \right) \right)^2.$$

In matrix form, this can be written as

$$\min_\theta (\Phi\theta - (c^\pi + \gamma P^\pi \Phi \theta^n))^T D^\pi (\Phi\theta - (c^\pi + \gamma P^\pi \Phi \theta^n)) \quad (16.24)$$

where D^π is a $|\mathcal{S}| \times |\mathcal{S}|$ diagonal matrix with elements d_s^π which serves a scaling role (we want to focus our attention on states we visit the most). We can find the optimal value of θ (given θ^n) by taking the derivative of the function being optimized in (16.24) with respect to θ and setting it equal to zero. Let θ^{n+1} be the optimal solution, which means we can write

$$\Phi^T D^\pi (\Phi \theta^{n+1} - (c^\pi + \gamma P^\pi \Phi \theta^n)) = 0, \quad (16.25)$$

We can find a fixed point $\lim_{n \to \infty} \theta^n = \lim_{n \to \infty} \theta^{n+1} = \theta^*$, which allows us to write equation (16.25) in the form

$$A\theta^* = b, \quad (16.26)$$

where $A = \Phi^T D^\pi (I - \gamma P^\pi) \Phi$ and $b = \Phi^T D^\pi c^\pi$. This allows us, in theory at least, to solve for θ^* using

$$\theta^* = A^{-1}b, \tag{16.27}$$

which can be viewed as a scaled version of the normal equations (equation 3.40). Equation (16.27) is very similar to our calculation of the steady state value of being in each state introduced in chapter 14, given by

$$V^\pi = (I - \gamma P^\pi)^{-1} c^\pi.$$

Equation (16.27) differs only in the scaling by the probability of being in each state (D^π) and then the transformation to the feature space by Φ.

We note that equation (16.25) can also be written in the form

$$A\theta - b = \Phi^T D^\pi \left(\Phi\theta - (c^\pi + \gamma P^\pi \Phi\theta) \right). \tag{16.28}$$

The term $\Phi\theta$ can be viewed as the approximate value of each state. The term $(c^\pi + \gamma P^\pi \Phi\theta)$ can be viewed as the one-period contribution plus the expected value of the state that you transition to under policy π, again computed for each state. Let δ^π be a column vector containing the temporal difference for each state when we choose a decision according to policy π. By tradition, the temporal difference has always been written in the form $C(S_t, x) + \bar{V}(S_{t+1}) - \bar{V}(S_t)$, which can be thought of as "estimated minus predicted." If we continue to let δ^π be the traditional definition of the temporal difference, it would be written

$$\delta^\pi = -\left(\Phi\theta - (c^\pi + \gamma P^\pi \Phi\theta) \right). \tag{16.29}$$

The pre-multiplication of δ^π by D^π in (16.28) has the effect of factoring each temporal difference by the probability that we are in each state. Then pre-multiplying $D^\pi \delta^\pi$ by Φ^T has the effect of transforming this scaled temporal difference for each state into the feature space.

The goal is to find the value θ that produces $A\theta - b = 0$, which means we are trying to find the value θ that produces a scaled version of $\Phi\theta - (c^\pi + \gamma P^\pi \Phi\theta) = 0$, but transformed to the feature space.

Linear algebra offers a compact elegance, but at the same time can be hard to parse, and for this reason we encourage the reader to stop and think about the relationships. One useful exercise is to think of a set of basis functions where we have a "feature" for each state, with $\phi_f(s) = 1$ if feature f corresponds to state s. In this case, Φ is the identity matrix. D^π, the diagonal matrix with diagonal elements d_s^π giving the probability of being in state s, can be viewed as scaling quantities for each state by the probability of being in a state. If Φ is the identity matrix, then $A = D^\pi - \gamma D^\pi P^\pi$ where $D^\pi P^\pi$ is the matrix of *joint* probabilities of being in state s *and* then transitioning to state s'. The vector b becomes the vector of the cost of being in each state (and then taking the a corresponding to policy π) times the probability of being in the state.

When we have a smaller set of basis functions, then multiplying c^π or $D^\pi(I - \gamma P^\pi)$ times Φ has the effect of scaling quantities that are indexed by the state into the feature space, which also transforms an $|\mathcal{S}|$-dimensional space into an $|\mathcal{F}|$-dimensional space.

16.3.2 A Simulation-based Implementation

No one actually computes expressions such as those given in section 16.3.1. In practice, we simulate everything.

We start by simulating a trajectory of states, decisions, and information,

$$(S^0, x^0, W^1, S^1, x^1, W^2, \dots, S^n, x^n, W^{n+1}).$$

Recall that $\phi(s)$ is a column vector with an element $\phi_f(s)$ for each feature $f \in \mathcal{F}$. Using our simulation shown earlier, we also obtain a sequence of column vectors $\phi(s^i)$ and contributions $C(S^i, x^i)$. We can create a sample estimate of the $|\mathcal{F}|$ by $|\mathcal{F}|$ matrix A in section 16.3.1 using

$$A^n = \frac{1}{n} \sum_{i=0}^{n-1} \phi(S^i)(\phi(S^i) - \gamma\phi(S^{i+1}))^T. \tag{16.30}$$

We can also create a sample estimate of the vector b using

$$b^n = \frac{1}{n} \sum_{i=0}^{n-1} \phi(S^i)C(S^i, x^i). \tag{16.31}$$

To gain some intuition, again stop and assume that there is a feature for every state, which means that $\phi(S^i)$ is a vector of 0's with a 1 corresponding to the element for state S^i, which means it is a kind of indicator variable telling us what state we are in. The term $(\phi(S^i) - \gamma\phi(S^{i+1}))$ is then a simulated version of $D^\pi(I - \gamma P^\pi)$, weighted by the probability that we are in a particular state, where we replace the probability of being in a state with a sampled realization of actually being in a particular state.

We are going to use this foundation to introduce two important algorithms for infinite horizon problems when using linear models to approximate value function approximations. These are known as *least squares temporal difference learning* (LSTD), and *least squares policy evaluation* (LSPE).

16.3.3 Least Squares Temporal Difference Learning (LSTD)

As long as A^n is invertible (which is not guaranteed), we can compute a sample estimate of θ using

$$\theta^n = (A^n)^{-1} b^n. \tag{16.32}$$

This algorithm is known in the literature as *least squares temporal difference learning*. As long as the number of features is not too large (as is typically the case), the inverse is not too hard to compute. LSTD can be viewed as a batch algorithm which operates by collecting a sample of temporal differences, and then using least squares regression to find the best linear fit.

We can see the role of temporal differences more clearly by doing a little algebra. We use equations (16.30) and (16.31) to write

$$
\begin{aligned}
A^n\theta^n - b^n &= \frac{1}{n}\sum_{i=0}^{n-1}\left(\phi(S^i)(\phi(S^i) - \gamma\phi(S^{i+1}))^T\theta^n - \phi(S^i)C(S^i, x^i)\right) \\
&= \frac{1}{n}\sum_{i=0}^{n-1}\phi(S^i)\left(\phi(S^i)^T\theta^n - (c^{\pi} + \alpha\phi(S^{i+1})^T\theta^n)\right) \\
&= \frac{1}{n}\sum_{i=0}^{n-1}\phi(S^i)\delta^i(\theta^n),
\end{aligned}
$$

where $\delta^i(\theta^n) = \phi(S^i)^T\theta^n - (c^{\pi} + \alpha\phi(S^{i+1})^T\theta^n)$ is the ith temporal difference given the parameter vector θ^n. Thus, we are doing a least squares regression so that the sum of the temporal differences over the simulation (which approximations the expectation) is equal to zero. We would, of course, like it if θ could be chosen so that $\delta^i(\theta) = 0$ for all i. However, when working with sample realizations the best we can expect is that the average across the observations of $\delta^i(\theta)$ tends to zero.

16.3.4 Least Squares Policy Evaluation

LSTD is basically a batch algorithm, which requires collecting a sample of n observations and then using regression to fit a model. An alternative strategy, known as *least squares policy evaluation* (or LSPE), uses a stochastic gradient algorithm which successively updates estimates of θ. The basic updating equation is

$$\theta^n = \theta^{n-1} - \frac{\alpha}{n}G^n\sum_{i=0}^{n-1}\phi(S^i)\delta^i(n), \tag{16.33}$$

where G^n is a scaling matrix. Although there are different strategies for computing G^n, the most natural is a simulation-based estimate of $(\Phi^T D^{\pi}\Phi)^{-1}$ which can be computed using

$$G^n = \left(\frac{1}{n+1} \sum_{i=0}^{n} \phi(S^i)\phi(S^i)^T \right)^{-1}.$$

To visualize G^n, return again to the assumption that there is a feature for every state. In this case, $\phi(S^i)\phi(S^i)^T$ is an $|S|$ by $|S|$ matrix with a 1 on the diagonal for row S^i and column S^i. As n approaches infinity, the matrix

$$\left(\frac{1}{n+1} \sum_{i=0}^{n} \phi(S^i)\phi(S^i)^T \right)$$

approaches the matrix D^π of the probability of visiting each state, stored in elements along the diagonal.

16.4 Analysis of TD(0), LSTD, and LSPE Using a Single State*

A useful exercise to understand the behavior of recursive least squares, LSTD and LSPE is to consider what happens when they are applied to a trivial dynamic program with a single state and a single decision. Obviously, we are interested in the policy that chooses the single decision. This dynamic program is equivalent to computing the sum

$$F = \mathbb{E} \sum_{i=0}^{\infty} \gamma^i \hat{C}^i, \tag{16.34}$$

where $\hat{C}^i$ is a random variable giving the ith contribution. If we let $\bar{c} = \mathbb{E}\hat{C}^i$, then clearly $F = \frac{1}{1-\gamma}c$. But let's pretend that we do not know this, and we are using these various algorithms to compute the expectation.

16.4.1 Recursive Least Squares and TD(0)

Let $\hat{v}^n$ be an estimate of the value of being in state S^n. We continue to assume that the value function is approximated using

$$\bar{V}(s) = \sum_{f \in \mathcal{F}} \theta_f \phi_f(s).$$

We wish to choose θ by solving

$$\min_{\theta} \sum_{i=1}^{n} \left(\hat{v}^i - \left(\sum_{f \in \mathcal{F}} \theta_f \phi_f(S^i) \right) \right)^2.$$

Let θ^n be the optimal solution. We can determine this recursively using the techniques presented earlier in this chapter which gives us the updating equation

$$\theta^n = \theta^{n-1} - \frac{1}{1 + (x^n)^T M^{n-1} x^n} M^{n-1} x^n (\bar{V}^{n-1}(S^n) - \hat{v}^n) \tag{16.35}$$

where $x^n = (\phi_1(S^n), \ldots, \phi_f(S^n), \ldots, \phi_F(S^n))$, and the matrix M^n is computed using

$$M^n = M^{n-1} - \frac{1}{1 + (x^n)^T M^{n-1} x^n} \left(M^{n-1} x^n (x^n)^T M^{n-1} \right).$$

If we have only one state and one decision, we only have one basis function $\phi(s) = 1$ and one parameter $\theta^n = \bar{V}^n(s)$. Now the matrix M^n is a scalar and equation (16.35) reduces to

$$\begin{aligned}
v^n &= v^{n-1} - \frac{M^{n-1}}{1 + M^{n-1}} (v^{n-1} - \hat{v}^n) \\
&= \left(1 - \frac{M^{n-1}}{1 + M^{n-1}} \right) v^{n-1} + \frac{M^{n-1}}{1 + M^{n-1}}.
\end{aligned}$$

If $M^0 = 1$, then $M^{n-1} = 1/n$, giving us

$$v^n = \frac{n-1}{n} v^{n-1} + \frac{1}{n} \hat{v}^n.$$

Imagine now that we are using TD(0) where $\hat{v}^n = \hat{C}^n + \gamma v^{n-1}$. In this case, we obtain

$$v^n = \left(1 - (1-\gamma)\frac{1}{n} \right) v^{n-1} + \frac{1}{n} \hat{C}^n. \tag{16.36}$$

Equation (16.36) can be viewed as an algorithm for finding

$$v = \sum_{n=0}^{\infty} \gamma^n \hat{C}^n,$$

where the solution is $v^* = \frac{1}{1-\gamma} \mathbb{E}\hat{C}$.

Equation (16.36) shows us that recursive least squares, when $\hat{v}^n$ is computed using temporal difference learning, has the effect of successively adding sample realizations of costs, with a "discount factor" of $1/n$. The factor $1/n$ arises directly as a result of the need to smooth out the noise in $\hat{C}^n$. For example, if $\hat{C} = c$ is a known constant, we could use standard value iteration, which would give us

$$v^n = c + \gamma v^{n-1}.$$ (16.37)

It is easy to see that v^n in (16.37) will rise much more quickly toward v^* than the algorithm in equation (16.36). We return to this topic in some depth in chapter 6.

16.4.2 Least Squares Policy Evaluation

LSPE requires that we first generate a sequence of states S^i and contributions $\hat{C}^i$ for $i = 1, \ldots, n$. We then compute θ by solving the regression problem

$$\theta^n = \arg\min_{\theta} \sum_{i=1}^{n} \left(\sum_{f} \theta_f \phi_f(S^i) - (\hat{C}^i + \gamma \bar{V}^{n-1}(S^{i+1})) \right)^2.$$

For a problem with one state where $\theta^n = v^n$, this reduces to

$$v^n = \arg\min_{\theta} \sum_{i=1}^{n} \left(\theta - (\hat{C}^i + \gamma v^{n-1}) \right)^2.$$

This problem can be solved in closed form, giving us

$$v^n = \left(\frac{1}{n} \sum_{i=1}^{n} \hat{C}^i \right) + \gamma v^{n-1}.$$

16.4.3 Least Squares Temporal Difference Learning

Finally, we showed that the LSTD procedure finds θ by solving the system of equations

$$\sum_{i=1}^{n} \phi_f(S^i)(\phi_f(S^i) - \gamma \phi_f(S^{i+1}))^T \theta^n = \sum_{i=1}^{n} \phi_f(S^i)\hat{C}^i,$$

for each $f \in \mathcal{F}$. Again, since we have only one basis function $\phi(s) = 1$ for our single state problem, this reduces to finding the scalar $\theta^n = v^n$ using

$$v^n = \frac{1}{1-\gamma} \left(\frac{1}{n} \sum_{i=1}^{n} \hat{C}^n \right).$$

16.4.4 Discussion

This presentation illustrates three different styles for estimating an infinite horizon sum. In recursive least squares, equation (16.35) demonstrates the successive smoothing of the previous estimate v^n and the latest estimate $\hat{v}^n$.

We are, at the same time, adding contributions over time while also trying to smooth out the noise.

LSPE, by contrast, separates the estimation of the mean of the single period contribution, and the process of summing contributions over time. At each iteration, we improve our estimate of $\mathbb{E}\hat{C}$, and then accumulate our latest estimate in a telescoping sum.

LSTD, finally, updates its estimate of $\mathbb{E}\hat{C}$, and then projects this over the infinite horizon by factoring the result by $1/(1 - \gamma)$.

16.5 Gradient-based Methods for Approximate Value Iteration*

There has been a strong desire for approximation algorithms with the following features:

(1) Off-policy learning.
(2) Temporal-difference learning.
(3) Linear models for value function approximation.
(4) Complexity (in memory and computation) that is linear in the number of features.

The last requirement is primarily of interest in specialized applications which require thousands or even millions of features. Off-policy learning is desirable because it provides an important degree of control over exploration. Temporal-difference learning is useful because it is so simple, as are the use of linear models, which make it possible to provide an estimate of the entire value function with a small number of measurements.

Off-policy, temporal-difference learning was first introduced in the form of Q-learning using a lookup table representation, where it is known to converge. But we lose this property if we introduce value function approximations that are linear in the parameters. In fact, Q-learning can be shown to diverge for any positive stepsize. The reason is that there is no guarantee that our linear model is accurate, which can introduce significant instabilities in the learning process.

We begin by describing how to estimate linear value functions using approximate value iteration. Then section 16.5.2 provides a geometric view of linear models.

16.5.1 Approximate Value Iteration with Linear Models**

Q-learning and temporal difference learning can be viewed as forms of stochastic gradient algorithms, but the problem with earlier algorithms when we use linear value function approximations can be traced to the choice of objective function. For example, if we wish to find the best linear approximation $\bar{V}(s|\theta)$, a hypothetical objective function would be to minimize the expected mean squared difference between $\bar{V}(s|\theta)$ and the true value function $V(s)$. If d_s^π is the probability of being in state s, this objective would be written

$$MSE(\theta) = \frac{1}{2} \sum_s d_s^\pi (\bar{V}(s|\theta) - V(s))^2.$$

If we are using approximate value iteration, a more natural objective function is to minimize the mean-squared Bellman error. We use the Bellman operator $\mathcal{M}^\pi$ (as we did in chapter 14) for policy π to represent

$$\mathcal{M}^\pi v = c^\pi + \gamma P^\pi v,$$

where v is a column vector giving the value of being in state s, and c^π is the column vector of contributions $C(s, X^\pi(s))$ if we are in state s and choose a decision x according to policy π. This allows us to define

$$MSBE(\theta) = \frac{1}{2} \sum_s d_s^\pi \left(\bar{V}(s|\theta) - (c^\pi(s) + \gamma \sum_{s'} p^\pi(s'|s) \bar{V}(s'|\theta)) \right)^2$$

$$= \|\bar{V}(\theta) - \mathcal{M}\bar{V}(\theta)\|_D^2.$$

We can minimize $MSBE(\theta)$ by generating a sequence of states $(S^1, \dots, S^i, S^{i+1}, \dots)$ and then computing a stochastic gradient

$$\nabla_\theta MSBE(\theta) = \delta^{\pi,i}(\phi(S^i) - \gamma\phi(S^{i+1})),$$

where $\phi(S^i)$ is a column vector of basis functions evaluated at state S^i. The scalar $\delta^{\pi,i}$ is the temporal difference given by

$$\delta^{\pi,i} = \bar{V}(S^i|\theta) - (c^\pi(S^i) + \gamma\bar{V}(S^{i+1}|\theta)).$$

We note that $\delta^{\pi,i}$ depends on the policy π which affects both the single period contribution and the likelihood of transitioning to state S^{i+1}. To emphasize that we are working with a fixed policy, we carry the superscript π throughout.

For this section, we are defining the temporal difference as

$$\delta^{\pi,i} = \bar{V}(S^i|\theta) - (c^\pi(S^i) + \gamma\bar{V}(S^{i+1}|\theta))$$

because it is a natural byproduct when deriving algorithms based on stochastic gradient methods. Earlier in this chapter, we defined the temporal difference as $\delta_\tau = C(S_\tau^n, x_\tau^n) + \bar{V}_{\tau+1}^{n-1}(S_{\tau+1}^n) - \bar{V}_\tau^{n-1}(S_\tau^n)$ (see equation (16.4)), which is more natural when used to represent telescoping sums (for example, see equation

(16.5)). A stochastic gradient algorithm, then, would seek to optimize θ using

$$
\begin{aligned}
\theta^{n+1} &= \theta^n - \alpha_n \nabla_\theta MSBE(\theta) & (16.38) \\
&= \theta^n - \alpha_n \delta^{\pi,n}(\phi(S^n) - \gamma\phi(S^{n+1})). & (16.39)
\end{aligned}
$$

Were we to use the more traditional definition of a temporal difference, our equation would be written

$$
\theta^{n+1} = \theta^n + \alpha_n \delta^{\pi,n}(\phi(S^n) - \gamma\phi(S^{n+1})),
$$

which runs counter to the classical statement of a stochastic gradient algorithm (given in equation (16.38)) for minimization problems.

A variant of this basic algorithm, called the generalized TD(0) (or, GTD(0)) algorithm, is given by

$$
\theta^{n+1} = \theta^n - \alpha_n(\phi(S^n) - \gamma\phi(S^{n+1}))\phi(S^n)^T u^n, \qquad (16.40)
$$

where

$$
u^{n+1} = u^n - \beta_n(u^n - \delta^{\pi,n}\phi(S^n)). \qquad (16.41)
$$

α_n and β_n are both stepsizes. u^n is a smoothed estimate of the product $\delta^{\pi,n}\phi(S^n)$.

Gradient descent methods based on temporal differences will not minimize $MSBE(\theta)$ because there does not exist a value of θ that would allow $\hat{v}(s) = c^\pi(s) + \gamma\bar{V}(s|\theta)$ to be represented as $\bar{V}(s|\theta)$. We can fix this using the mean squared projected Bellman error ($MSPBE(\theta)$) which we compute as follows. It is more compact to do this development using matrix-vector notation. We first recall the projection operator Π given by

$$
\Pi = \Phi(\Phi^T D^\pi \Phi)^{-1}\Phi^T D^\pi.
$$

(See section 16.5.2 for a derivation of this operator.) If V is a vector giving the value of being in each state, ΠV is the nearest projection of V on the space generated by $\theta\phi(s)$. We are trying to find $\bar{V}(\theta)$ that will match the one-step lookahead given by $\mathcal{M}^\pi\bar{V}(\theta)$, but this produces a column vector that cannot be represented directly as $\Phi\theta$, where Φ is the $|\mathcal{S}| \times |\mathcal{F}|$ matrix of feature vectors ϕ. We accomplish this by pre-multiplying $\mathcal{M}^\pi V(\theta)$ by the projection operator Π. This allows us to form the mean squared projected Bellman error using

$$
\begin{aligned}
MSPBE(\theta) &= \frac{1}{2}\|\bar{V}(\theta) - \Pi\mathcal{M}^\pi\bar{V}(\theta)\|_D^2 & (16.42) \\
&= \frac{1}{2}\left(\bar{V}(\theta) - \Pi\mathcal{M}^\pi\bar{V}(\theta)\right)^T D\left(\bar{V}(\theta) - \Pi\mathcal{M}^\pi\bar{V}(\theta)\right). & (16.43)
\end{aligned}
$$

We can now use this new objective function as the basis of an optimization algorithm to find θ. Recall that D^π is a $|\mathcal{S}| \times |\mathcal{S}|$ diagonal matrix with elements d_s^π, giving us the probability that we are in state s while following policy π. We use D^π as a scaling matrix to give us the probability that we are in state s. We start by noting the identities

$$
\begin{aligned}
\mathbb{E}[\phi\phi^T] &= \sum_{s\in\mathcal{S}} d_s^\pi \phi_s \phi_s^T \\
&= \Phi^T D^\pi \Phi.
\end{aligned}
$$

$$
\begin{aligned}
\mathbb{E}[\delta^\pi \phi] &= \sum_{s\in\mathcal{S}} d_s^\pi \phi_s \left(c^\pi(s) + \gamma \sum_{s'\in\mathcal{S}} p^\pi(s'|s)\bar{V}(s'|\theta) - \bar{V}(s|\theta) \right) \\
&= \Phi^T D^\pi (\mathcal{M}^\pi \bar{V}(\theta) - \bar{V}(\theta)).
\end{aligned}
$$

The derivations here and in the following lines make extensive use of matrices, which can be difficult to parse. A useful exercise is to write out the matrices assuming that there is a feature $\phi_f(s)$ for each state s, so that $\phi_f(s) = 1$ if feature f corresponds to state s. See exercise 16.12.

We see that the role of the scaling matrix D^π is to enable us to take the expectation of the quantities $\phi\phi^T$ and $\delta^\pi \phi$. We are going to simulate these quantities, where a state will occur with probability d_s^π. We also use

$$
\begin{aligned}
\Pi^T D^\pi \Pi &= (\Phi(\Phi^T D^\pi \Phi)^{-1}\Phi^T D^\pi)^T D^\pi (\Phi(\Phi^T D^\pi \Phi)^{-1}\Phi^T D^\pi) \\
&= (D^\pi)^T \Phi(\Phi^T D^\pi \Phi)^{-1}\Phi^T D^\pi \Phi(\Phi^T D^\pi \Phi)^{-1}\Phi^T D^\pi \\
&= (D^\pi)^T \Phi(\Phi^T D^\pi \Phi)^{-1}\Phi^T D^\pi.
\end{aligned}
$$

We have one last painful piece of linear algebra that gives us a more compact form for MSPBE(θ). Pulling the $1/2$ to the left hand side (this will later vanish when we take the derivative), we can write

$$
\begin{aligned}
2MSPBE(\theta) &= \|\bar{V}(\theta) - \Pi\mathcal{M}^\pi \bar{V}(\theta)\|_D^2 \\
&= \|\Pi(\bar{V}(\theta) - \mathcal{M}^\pi \bar{V}(\theta))\|_D^2 \\
&= (\Pi(\bar{V}(\theta) - \mathcal{M}^\pi \bar{V}(\theta)))^T D^\pi (\Pi(\bar{V}(\theta) - \mathcal{M}^\pi \bar{V}(\theta))) \\
&= (\bar{V}(\theta) - \mathcal{M}^\pi \bar{V}(\theta))^T \Pi^T D^\pi \Pi(\bar{V}(\theta) - \mathcal{M}^\pi \bar{V}(\theta)) \\
&= (\bar{V}(\theta) - \mathcal{M}^\pi \bar{V}(\theta))^T (D^\pi)^T \Phi(\Phi^T (D^\pi)\Phi)^{-1}\Phi^T D^\pi (\bar{V}(\theta) - \mathcal{M}^\pi \bar{V}(\theta)) \\
&= (\Phi^T D^\pi (\mathcal{M}^\pi \bar{V}(\theta) - \bar{V}(\theta)))^T (\Phi^T D^\pi \Phi)^{-1}\Phi^T D^\pi (\mathcal{M}\bar{V}(\theta) - \bar{V}(\theta)) \\
&= \mathbb{E}[\delta^\pi \phi]^T \mathbb{E}[\phi\phi^T]^{-1}\mathbb{E}[\delta^\pi \phi]. \quad (16.44)
\end{aligned}
$$

We next need to estimate the gradient of this error $\nabla_\theta MSPBE(\theta)$. Keep in mind that $\delta^\pi = c^\pi + \gamma P^\pi \Phi\theta - \Phi\theta$. If ϕ is the column vector with element $\phi(s)$, assume that s' occurs with probability $p^\pi(s'|s)$ under policy π, and let ϕ' be the corresponding column vector. Differentiating (16.44) gives

$$
\begin{aligned}
\nabla_\theta MSPBE(\theta) &= \mathbb{E}[(\gamma\phi' - \phi)\phi^T]\mathbb{E}[\phi\phi^T]^{-1}\mathbb{E}[\delta^\pi \phi] \\
&= -\mathbb{E}[(\phi - \gamma\phi')\phi^T]\mathbb{E}[\phi\phi^T]^{-1}\mathbb{E}[\delta^\pi \phi].
\end{aligned}
$$

We are going to use a standard stochastic gradient updating algorithm for minimizing the error given by $MSPBE(\theta)$, which is given by

$$\theta^{n+1} = \theta^n - \alpha_n \nabla_\theta MSPBE(\theta) \tag{16.45}$$

$$= \theta^n + \alpha_n \mathbb{E}[(\phi - \gamma\phi')\phi^T]\mathbb{E}[\phi\phi^T]^{-1}\mathbb{E}[\delta^\pi\phi]. \tag{16.46}$$

We can create a linear predictor which approximates

$$w \approx \mathbb{E}[\phi\phi^T]^{-1}\mathbb{E}[\delta^\pi\phi].$$

where w is approximated using

$$w^{n+1} = w^n + \beta_n(\delta^{\pi,n} - (\phi^n)^T w^n)\phi^n.$$

This allows us to write the gradient

$$\nabla_\theta MSPBE(\theta) = -\mathbb{E}[(\phi - \gamma\phi')\phi^T]\mathbb{E}[\phi\phi^T]^{-1}\mathbb{E}[\delta^\pi\phi]$$

$$\approx -\mathbb{E}[(\phi - \gamma\phi')\phi^T]w.$$

We have now created the basis for two algorithms. The first is called generalized temporal difference 2 (GTD2), given by

$$\theta^{n+1} = \theta^n + \alpha_n(\phi^n - \gamma\phi^{n+1})((\phi^n)^T w^n). \tag{16.47}$$

Here, ϕ^n is the column vector of basis functions when we are in state S^n, while ϕ^{n+1} is the column vector of basis functions for the next state S^{n+1}. Note that if equation (16.47) is executed right to left, all calculations are linear in the number of features F.

An important feature of the algorithm, especially for applications with large number of features, is that the algorithm is linear in the number of features.

A variant, called TDC (temporal difference with gradient corrector) is derived by using a slightly modified calculation of the gradient

$$\nabla_\theta MSPBE(\theta) = -\mathbb{E}[(\phi - \gamma\phi')\phi^T]\mathbb{E}[\phi\phi^T]^{-1}\mathbb{E}[\delta^\pi\phi]$$

$$= -\left(\mathbb{E}[\phi\phi^T] - \gamma\mathbb{E}[\phi'\phi^T]\right)\mathbb{E}[\phi\phi^T]^{-1}\mathbb{E}[\delta^\pi\phi]$$

$$= -\left(\mathbb{E}[\delta^\pi\phi] - \gamma\mathbb{E}[\phi'\phi^T]\mathbb{E}[\phi\phi^T]^{-1}\mathbb{E}[\delta^\pi\phi]\right)$$

$$\approx -\left(\mathbb{E}[\delta^\pi\phi] - \gamma\mathbb{E}[\phi'\phi^T]w\right).$$

This gives us the TDC algorithm

$$\theta^{n+1} = \theta^n + \alpha_n\left(\delta^{\pi,n}\phi^n - \gamma\phi^{n'}((\phi^n)^T w^n)\right). \tag{16.48}$$

GTD2 and TDC are both proven to converge to the optimal value of θ for a fixed implementation policy $X^\pi(s)$ which may be different than the learning (behavior) policy. That is, after updating θ^n where the temporal difference $\delta^{\pi,n}$ is computed assuming we are in state S^n and follow policy π, we are allowed to follow the learning policy to determine S^{n+1}. This allows us to directly control the states that we visit, rather than depending on the decisions made by the implementation policy.

16.5.2 A Geometric View of Linear Models*

For readers comfortable with linear algebra, we can obtain an elegant perspective on the geometry of basis functions. In section 3.7.1, we found the parameter vector θ for a regression model by minimizing the expected square of the errors between our model and a set of observations. Assume now that we have a "true" value function $V(s)$ which gives the value of being in state s, and let $p(s)$ be the probability of visiting state s. We wish to find the approximate value function that best fits $V(s)$ using a given set of basis functions $(\phi_f(s))_{f \in \mathcal{F}}$. If we minimize the expected square of the errors between our approximate model and the true value function, we would want to solve

$$\min_{\theta} F(\theta) = \sum_{s \in \mathcal{S}} p(s) \left(V(s) - \sum_{f \in \mathcal{F}} \theta_f \phi_f(s) \right)^2, \tag{16.49}$$

where we have weighted the error for state s by the probability of actually being in state s. Our parameter vector θ is unconstrained, so we can find the optimal value by taking the derivative and setting this equal to zero. Differentiating with respect to $\theta_{f'}$ gives

$$\frac{\partial F(\theta)}{\partial \theta_{f'}} = -2 \sum_{s \in \mathcal{S}} p(s) \left(V(s) - \sum_{f \in \mathcal{F}} \theta_f \phi_f(s) \right) \phi_{f'}(s).$$

Setting the derivative equal to zero and rearranging gives

$$\sum_{s \in \mathcal{S}} p(s) V(s) \phi_{f'}(s) = \sum_{s \in \mathcal{S}} p(s) \sum_{f \in \mathcal{F}} \theta_f \phi_f(s) \phi_{f'}(s). \tag{16.50}$$

At this point, it is much more elegant to revert to matrix notation. Define an $|\mathcal{S}| \times |\mathcal{S}|$ diagonal matrix D where the diagonal elements are the state probabilities $p(s)$, as follows

$$D = \begin{pmatrix} p(1) & 0 & & 0 \\ 0 & p(2) & & 0 \\ \vdots & 0 & \cdots & \vdots \\ 0 & \vdots & & p(|\mathcal{S}|) \end{pmatrix}.$$

Let V be the column vector giving the value of being in each state

$$V = \begin{pmatrix} V(1) \\ V(2) \\ \vdots \\ V(|\mathcal{S}|) \end{pmatrix}.$$

Finally, let Φ be an $|\mathcal{S}| \times |\mathcal{F}|$ matrix of the basis functions given by

$$\Phi = \begin{pmatrix} \phi_1(1) & \phi_2(1) & & \phi_{|\mathcal{F}|}(1) \\ \phi_1(2) & \phi_2(2) & \cdots & \phi_{|\mathcal{F}|}(2) \\ \vdots & \vdots & & \vdots \\ \phi_1(|\mathcal{S}|) & \phi_2(|\mathcal{S}|) & & \phi_{|\mathcal{F}|}(|\mathcal{S}|) \end{pmatrix}.$$

Recognizing that equation (16.50) is for a particular feature f', with some care it is possible to see that equation (16.50) for all features is given by the matrix equation

$$\Phi^T DV = \Phi^T D\Phi\theta. \tag{16.51}$$

It helps to keep in mind that Φ is an $|\mathcal{S}| \times |\mathcal{F}|$ matrix, D is an $|\mathcal{S}| \times |\mathcal{S}|$ diagonal matrix, V is an $|\mathcal{S}| \times 1$ column vector, and θ is an $|\mathcal{F}| \times 1$ column vector. The reader should carefully verify that (16.51) is the same as (16.50).

Now, pre-multiply both sides of (16.51) by $(\Phi^T D\Phi)^{-1}$. This gives us the optimal value of θ as

$$\theta = (\Phi^T D\Phi)^{-1}\Phi^T DV. \tag{16.52}$$

This equation is closely analogous to the normal equations of linear regression, given by equation (3.37), with the only difference being the introduction of the scaling matrix D which captures the probability that we are going to visit a state.

Now, pre-multiply both sides of (16.52) by Φ, which gives

$$\Phi\theta = \bar{V} = \Phi(\Phi^T D\Phi)^{-1}\Phi^T DV.$$

$\Phi\theta$ is, of course, our approximation of the value function, which we have denoted by $\bar{V}$. This, however, is the best possible value function given the set of functions $\phi = (\phi_f)_{f \in \mathcal{F}}$. If the vector ϕ formed a complete basis over the space formed by the value function $V(s)$ and the state space $\mathcal{S}$, then we would obtain $\Phi\theta = \bar{V} = V$. Since this is generally not the case, we can view $\bar{V}$ as the nearest point projection (where "nearest" is defined as a weighted measure using the state probabilities $p(s)$) onto the space formed by the basis functions. In fact, we can form a projection operator Π defined by

$$\Pi = \Phi(\Phi^T D\Phi)^{-1}\Phi^T D$$

so that $\bar{V} = \Pi V$ is the value function closest to V that can be produced by the set of basis functions.

This discussion brings out the geometric view of basis functions (and at the same time, the reason why we use the term "basis function"). There is an extensive literature on basis functions that has evolved in the approximation literature.

16.6 Value Function Approximations Based on Bayesian Learning*

A different strategy for updating value functions is one based on Bayesian learning. Assume that we start with a prior $V^0(s)$ of the value of being in state s, and we assume that we have a known covariance function $Cov(s, s')$ that captures the relationship in our belief about $V(s)$ and $V(s')$. A good example where this function would be known might be a function where s is continuous (or a discretization of a continuous surface), where we might use

$$Cov(s, s') \propto e^{-\frac{\|s-s'\|^2}{b}} \tag{16.53}$$

where b is a bandwidth. This function captures the intuitive behavior that if two states are close to each other, their covariance is higher. So, if we make an observation that raises our belief about $V(s)$, then our belief about $V(s')$ will increase also, and will increase more if s and s' are close to each other. We also assume that we have a variance function $\lambda(s)$ that captures the noise in a measurement $\hat{v}(s)$ of the function at state s.

Our Bayesian updating model is designed for applications where we have access to observations $\hat{v}^n$ of our true function $V(s)$ which we can view as coming from our prior distribution of belief. This assumption effectively precludes using updating algorithms based on approximate value iteration, Q-learning, and least squares policy evaluation. We cannot eliminate the bias, but we describe how to minimize it. We then describe Bayesian updating using lookup tables and parametric models.

16.6.1 Minimizing Bias for Infinite Horizon Problems

We would very much like to have observations $\hat{v}^n(s)$ which we can view as an unbiased observation of $V(s)$. One way to do this is to build on the methods described in section 16.1.

To illustrate, assume that we have a policy π that determines the decision x_t we take when in state S_t, generating a contribution $\hat{C}_t^n$. Assume we simulate this policy for T time periods using

$$\hat{v}^n(T) = \sum_{t=0}^{T} \gamma^t \hat{C}_t.$$

If we have a finite horizon problem and T is the end of our horizon, then we are done. If our problem has an infinite horizon, we can project the infinite horizon value of our policy by first approximating the one-period contribution using

$$\bar{c}_T^n = \frac{1}{T} \sum_{t=0}^{T} \hat{C}_t^n.$$

Now assume this estimates the average contribution per period starting at time $T + 1$. Our infinite-horizon estimate would be

$$\hat{v}^n = \hat{v}_0(T) + \gamma^{T+1} \frac{1}{1-\gamma} \bar{c}_T^n.$$

Finally, we use $\hat{v}^n$ to update our value function approximation $\bar{V}^{n-1}$ to obtain $\bar{V}^n$.

We next illustrate the Bayesian updating formulas for lookup tables and parametric models.

16.6.2 Lookup Tables with Correlated Beliefs

Up until now when we used a lookup table model for $\bar{V}^n(s)$, updating $\bar{V}^n(s)$ for some state s would not affect the estimates $\bar{V}^n(s')$ for other states $s' \neq s$. With our Bayesian model, we can do much more if we have access to a covariance function such as the one we illustrated in equation (16.53).

Assume that we have discrete states, and assume that we have a covariance function $Cov(s, s')$ in the form of a covariance matrix Σ where $Cov(s, s') = \Sigma(s, s')$. Let V^n be our vector of beliefs about the value $V(s)$ of being in each state (we use V^n to represent our Bayesian beliefs, so that $\bar{V}^n$ can represent our frequentist estimates). Also let Σ^n be the covariance matrix of our belief about the vector V. If $\hat{v}^n(S^n)$ is an (approximately) unbiased sample observation of $V(s)$, the Bayesian formula for updating V^n is given by

$$\bar{V}^{n+1}(s) = V^n(s) + \frac{\hat{v}^n(S^n) - V^n(s)}{\lambda(S^n) + \Sigma^n(S^n, S^n)} \Sigma^n(s, S^n).$$

This has to be computed for each s (or at least each s where $\Sigma^n(s, S^n) > 0$). We update the covariance matrix using

$$\Sigma^{n+1}(s, s') = \Sigma^n(s, s') - \frac{\Sigma^n(s, S^n)\Sigma^n(S^n, s')}{\lambda(S^n) + \Sigma^n(S^n, S^n)}.$$

16.6.3 Parametric Models

For most applications, a parametric model (specifically, a linear model) is going to be much more practical. Our frequentist updating equations for our regression vector θ^n were given as

$$\theta^n = \theta^{n-1} - \frac{1}{\gamma^n} M^{n-1}\phi^n \hat{\varepsilon}^n, \tag{16.54}$$

$$M^n = M^{n-1} - \frac{1}{\gamma^n}(M^{n-1}\phi^n(\phi^n)^T M^{n-1}), \tag{16.55}$$

$$\gamma^n = 1 + (\phi^n)^T M^{n-1}\phi^n, \tag{16.56}$$

where $\hat{\varepsilon}^n = \bar{V}(\theta^{n-1})(S^n) - \hat{v}^n$ is the difference between our current estimate $\bar{V}(\theta^{n-1})(S^n)$ of the value function at our observed state S^n and our most recent observation $\hat{v}^n$. The adaptation for a Bayesian model is quite minor. The matrix M^n represents

$$M^n = [(X^n)^T X^n]^{-1}.$$

It is possible to show that the covariance matrix Σ^θ (which is dimensioned by the number of basis functions) is given by

$$\Sigma^\theta = M^n \lambda.$$

In our Bayesian model, λ is the variance of the difference between our observation $\hat{v}^n$ and the true value function $v(S^n)$, where we assume λ is known. This variance may depend on the state that we have observed, in which case we would write it as $\lambda(s)$, but in practice, since we do not know the function $V(s)$, it is hard to believe that we would be able to specify $\lambda(s)$. We replace M^n with $\Sigma^{\theta, n}$ and rescale γ^n to create the following set of updating equations

$$\theta^n = \theta^{n-1} - \frac{1}{\gamma^n}\Sigma^{\theta, n-1}\phi^n \hat{\varepsilon}^n, \tag{16.57}$$

$$\Sigma^{\theta, n} = \Sigma^{\theta, n-1} - \frac{1}{\gamma^n}(\Sigma^{\theta, n-1}\phi^n(\phi^n)^T\Sigma^{\theta, n-1}), \tag{16.58}$$

$$\gamma^n = \lambda + (\phi^n)^T\Sigma^{\theta, n-1}\phi^n. \tag{16.59}$$

16.6.4   Creating the Prior

Approximate dynamic programming has been approached from a Bayesian perspective in the research literature, but otherwise has apparently received very little attention. We suspect that while there exist many applications in stochastic search where it is valuable to use a prior distribution of belief, it is much harder to build a prior on a value function.

Lacking any specific structural knowledge of the value function, we anticipate that the easiest strategy will be to start with $V^0(s) = v^0$, which is a constant across all states. There are several strategies we might use to estimate v^0. We might sample a state S^i at random, and find the best contribution $\hat{C}^i = \max_a C(S^i, a)$. Repeat this n times and compute

$$\bar{c} = \frac{1}{n} \sum_{i=1}^{n} \hat{C}^i.$$

Finally, let $v^0 = \frac{1}{1-\gamma} \bar{c}$ if we have an infinite horizon problem. The hard part is that the variance λ has to capture the variance of the difference between v^0 and the true $V(s)$. This requires having some sense of the degree to which v^0 differs from $V(s)$. We recommend being very conservative, which is to say choose a variance λ such that $v^0 + 2\sqrt{\lambda}$ easily covers what $V(s)$ might be. Of course, this also requires some judgment about the likelihood of visiting different states.

16.7   Learning Algorithms and Atepsizes

A useful exercise to understand the behavior of recursive least squares, LSTD and LSPE is to consider what happens when they are applied to a trivial dynamic program with a single state and a single decision. Obviously, we are interested in the policy that chooses the single decision. This dynamic program is equivalent to computing the sum

$$F = \mathbb{E} \sum_{i=0}^{\infty} \gamma^i \hat{C}^i, \qquad (16.60)$$

where $\hat{C}^i$ is a random variable giving the ith contribution. If we let $\bar{c} = \mathbb{E}\hat{C}^i$, then clearly $F = \frac{1}{1-\gamma} \bar{c}$. But let's pretend that we do not know this, and we are using these various algorithms to compute the expectation.

We first used the single-state problem in section 16.4, but did not focus on the implications for stepsizes. Here, we use our ability to derive analytical solutions for the optimal value function for least squares temporal differences (LSTD), least squares policy evaluation (LSPE), and recursive least squares and temporal differences. These expressions allow us to understand the types of behaviors we would like to see in a stepsize formula.

In the remainder of this section, we start by assuming that the value function is approximated using a linear model

$$\bar{V}(s) = \sum_{f \in \mathcal{F}} \theta_f \phi_f(s).$$

However, we are going to then transition to a problem with a single state, and a single basis function $\phi(s) = 1$. We assume that $\hat{v}$ is a sampled estimate of the value of being in the single state.

16.7.1 Least Squares Temporal Differences

In section 16.3 we showed that the LSTD method, when using a linear architecture, applied to infinite horizon problems required solving

$$\sum_{i=1}^{n} \phi_f(S^i)(\phi_f(S^i) - \gamma \phi_f(S^{i+1}))^T \theta = \sum_{i=1}^{n} \phi_f(S^i)\hat{C}^i,$$

for each $f \in \mathcal{F}$. Let θ^n be the optimal solution. Again, since we have only one basis function $\phi(s) = 1$ for our single state problem, this reduces to finding $v^n = \theta^n$

$$v^n = \frac{1}{1 - \gamma} \left(\frac{1}{n} \sum_{i=1}^{n} \hat{C}^n \right). \tag{16.61}$$

Equation (16.61) shows that we are trying to estimate $\mathbb{E}\hat{C}$ using a simple average. If we let $\bar{C}^n$ be the average over n observations, we can write this recursively using

$$\bar{C}^n = \left(1 - \frac{1}{n} \right) \bar{C}^{n-1} + \frac{1}{n}\hat{C}^n.$$

For the single state (and single decision) problem, the sequence $\hat{C}^n$ comes from a stationary sequence. In this case a simple average is the best possible estimator. In a dynamic programming setting with multiple states, and where we are trying to optimize over policies, v^n would depend on the state. Also, because the policy that determines the decision we take when we are in a state is changing over the iterations, the observations $\hat{C}^n$, even when we fix a state, would be nonstationary. In this setting, simple averaging is no longer the best. Instead, it is better to use

$$\bar{C}^n = (1 - \alpha_{n-1})\bar{C}^{n-1} + \alpha_{n-1}\hat{C}^n, \tag{16.62}$$

and use one of the stepsizes described in section 6.1, 6.2, or 6.3. As a general rule, these stepsize rules do not decline as quickly as $1/n$.

16.7.2 Least Squares Policy Evaluation

Least squares policy evaluation, which is developed using basis functions for infinite horizon applications, finds the regression vector θ by solving

$$\theta^n = \arg\min_\theta \sum_{i=1}^n \left(\sum_f \theta_f \phi_f(S^i) - \left(\hat{C}^i + \gamma \bar{V}^{n-1}(S^{i+1})\right) \right)^2 .$$

When we have one state, the value of being in the single state is given by $v^n = \theta^n$ which we can write as

$$v^n = \arg\min_\theta \sum_{i=1}^n \left(\theta - \left(\hat{C}^i + \gamma v^{n-1}\right) \right)^2 .$$

This problem can be solved in closed form, giving us

$$v^n = \left(\frac{1}{n} \sum_{i=1}^n \hat{C}^i \right) + \gamma v^{n-1} .$$

Similar to LSTD, LSPE works to estimate $\mathbb{E}\hat{C}$. For a problem with a single state and decision (and therefore only one policy), the best estimate of $\mathbb{E}\hat{C}$ is a simple average. However, as we already argued with LSTD, if we have multiple states and are searching for the best policy, the observation $\hat{C}$ for a particular state will come from a nonstationary series. For such problems, we should again adopt the updating formula in (16.62) and use one of the stepsize rules described section 6.1, 6.2, or 6.3.

16.7.3 Recursive Least Squares

Using our linear model, we start by using the following standard least squares model to fit our approximation

$$\min_\theta \sum_{i=1}^n \left(\hat{v}^i - \left(\sum_{f \in \mathcal{F}} \theta_f \phi_f(S^i) \right) \right)^2 .$$

As we have already discussed in chapter 3, we can fit the parameter vector θ using least squares, which can be computed recursively using

$$\theta^n = \theta^{n-1} - \frac{1}{1 + (x^n)^T B^{n-1} x^n} B^{n-1} x^n (\bar{V}^{n-1}(S^n) - \hat{v}^n)$$

where $x^n = (\phi_1(S^n), \ldots, \phi_f(S^n), \ldots, \phi_F(S^n))$, and the matrix B^n is computed using

$$B^n = B^{n-1} - \frac{1}{1 + (x^n)^T B^{n-1} x^n} \left(B^{n-1} x^n (x^n)^T B^{n-1} \right).$$

For the special case of a single state, we use the fact that we have only one basis function $\phi(s) = 1$ and one parameter $\theta^n = \bar{V}^n(s) = v^n$. In this case, the matrix B^n is a scalar, and the updating equation for θ^n (now v^n), becomes

$$
\begin{aligned}
v^n &= v^{n-1} - \frac{B^{n-1}}{1 + B^{n-1}} (v^{n-1} - \hat{v}^n) \\
&= \left(1 - \frac{B^{n-1}}{1 + B^{n-1}}\right) v^{n-1} + \frac{B^{n-1}}{1 + B^{n-1}} \hat{v}^n.
\end{aligned}
$$

If $B^0 = 1$, $B^{n-1} = 1/n$, giving us

$$v^n = \left(1 - \frac{1}{n}\right) v^{n-1} + \frac{1}{n} \hat{v}^n. \tag{16.63}$$

Now imagine we are using approximate value iteration. In this case, $\hat{v}^n = \hat{C}^n + \gamma v^n$. Substituting this into equation (16.63) gives us

$$
\begin{aligned}
v^n &= \left(1 - \frac{1}{n}\right) v^{n-1} + \frac{1}{n} (\hat{C}^n + \gamma \hat{v}^n) \\
&= \left(1 - \frac{1}{n}(1 - \gamma)\right) v^{n-1} + \frac{1}{n} \hat{C}^n. \tag{16.64}
\end{aligned}
$$

Recursive least squares has the behavior of averaging the observations of $\hat{v}$. The problem is that $\hat{v}^n = \hat{C}^n + \gamma v^n$, since $\hat{v}^n$ is also trying to be a discounted accumulation of the costs. Assume that the contribution was deterministic, where $\hat{C} = c$. If we were doing classical approximate value iteration, we would write

$$v^n = c + \gamma v^{n-1}. \tag{16.65}$$

Comparing (16.64) and (16.65), we see that the one-period contribution carries a coefficient of $1/n$ in (16.64) and a coefficient of 1 in (16.64). We can view equation (16.64) as a steepest ascent update with a stepsize of $1/n$. If we change the stepsize to 1, we obtain (16.65).

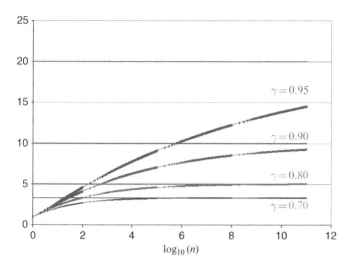

Figure 16.4 $\bar{v}^n$ plotted against $\log_{10}(n)$ when we use a $1/n$ stepsize rule for updating.

16.7.4 Bounding $1/n$ Convergence for Approximate value Iteration

It is well known that a $1/n$ stepsize will produce a provably convergent algorithm when used with approximate value iteration. Experimentalists know that the rate of convergence can be quite slow, but people new to the field can sometimes be found using this stepsize rule. In this section, we hope to present evidence that the $1/n$ stepsize should never be used with approximate value iteration or its variants.

Figure 16.4 is a plot of v^n computed using equation (16.64) as a function of $\log_{10}(n)$ for $\gamma = 0.7, 0.8, 0.9$, and 0.95, where we have set $\hat{C} = 1$. For $\gamma = 0.90$, we need 10^{10} iterations to get $\bar{v}^n = 9$, which means we are still 10 percent from the optimal. For $\gamma = 0.95$, we are not even close to converging after 100 billion iterations.

It is possible to derive compact bounds, $v^L(n)$ and $v^U(n)$ for $\bar{v}^n$ where

$$v^L(n) < v^n < v^U(n).$$

These are given by

$$v^L(n) = \frac{c}{1-\gamma}\left(1 - \left(\frac{1}{1+n}\right)^{1-\gamma}\right), \tag{16.66}$$

$$v^U(n) = \frac{c}{1-\gamma}\left(1 - \frac{1-\gamma}{\gamma n} - \frac{1}{\gamma n^{1-\gamma}}\left(\gamma^2 + \gamma - 1\right)\right). \tag{16.67}$$

Using the formula for the lower bound (which is fairly tight when n is large enough that v^n is close to v^*), we can derive the number of iterations to achieve a particular degree of accuracy. Let $\hat{C} = 1$, which means that $v^* = 1/(1 - \gamma)$. For a value of $v < 1/(1 - \gamma)$, we would need at least $n(v)$ to achieve $\bar{v}^* = v$, where $n(v)$ is found (from (16.66)) to be

$$n(v) \geq [1 - (1 - \gamma)v]^{-1/(1-\gamma)}. \tag{16.68}$$

If $\gamma = 0.9$, we would need $n(v) = 10^{20}$ iterations to reach a value of $v = 9.9$, which gives us a one percent error. On a 3-GHz chip, assuming we can perform one iteration per clock cycle (that is, 3×10^9 iterations per second), it would take 1,000 years to achieve this result.

16.7.5 Discussion

We can now see the challenge of choosing stepsizes for approximate value iteration, temporal-difference learning and Q-learning, compared to algorithms such as LSPE, LSTD, and approximate policy iteration (the finite horizon version of LSPE). If we observe $\hat{C}$ with noise, and if the discount factor $\gamma = 0$ (which means we are not trying to accumulate contributions over time), then a stepsize of $1/n$ is ideal. We are just averaging contributions to find the average value. As the noise in $\hat{C}$ diminishes, and as γ increases, we would like a stepsize that approaches 1. In general, we have to strike a balance between accumulating contributions over time (which is more important as γ increases) and averaging the observations of contributions (for which a stepsize of $1/n$ is ideal).

By contrast, LSPE, LSTD, and approximate policy iteration are all trying to estimate the average contribution per period for each state. The values $\hat{C}(s, x)$ are nonstationary because the policy that chooses the decision is changing, making the sequence $\hat{C}(s^n, x^n)$ nonstationary. But these algorithms are not trying to simultaneously accumulate contributions over time.

16.8 Bibliographic Notes

Section 16.1 – This section reviews a number of classical methods for estimating the value of a policy drawn from the reinforcement learning community. The best overall reference for this is Sutton and Barto (2018). Least-squares temporal differencing is due to Bradtke and Barto (1996).

Section 16.2 – Tsitsiklis (1994) and Jaakkola et al. (1994) were the first to make the connection between emerging algorithms in approximate dynamic programming (Q-learning, temporal difference learning) and the field of stochastic approximation theory (Robbins and Monro (1951), Blum (1954), Kushner and Yin (2003)).

Section 16.3 – The development of Bellman's equation using linear models is based on Tsitsiklis and Van Roy (1997), Lagoudakis and Parr (2003) and Bertsekas (2017). Tsitsiklis and Van Roy (1997) highlights the central role of the D-norm used in this section, which also plays a central role in the design of a simulation-based version of the algorithm. Ljung and Soderstrom (1983) and Young (1984) provide nice treatments of recursive statistics. Precup et al. (2001) gives the first convergent algorithm for off-policy temporal-difference learning using basis functions by using an adjustment which based on the relative probabilities of choosing an action from the target and behavioral policies. Lagoudakis et al. (2002) and Bradtke and Barto (1996) present least squares methods in the context of reinforcement learning. Van Roy and Choi (2006) uses the Kalman filter to perform scaling for stochastic gradient updates, avoiding the scaling problems inherent in stochastic gradient updates such as equation (16.22). Nedic and Bertsekas (2003) describes the use of least squares equation with a linear (in the parameters) value function approximation using policy iteration and proves convergence for TD(λ) with general λ. Bertsekas et al. (2004) presents a scaled method for estimating linear value function approximations within a temporal differencing algorithm.

Section 16.4 – The analysis of dynamic programs with a single state is based on Ryzhov et al. (2015).

Section 16.5 – Baird (1995) provides a nice example showing that approximate value iteration may diverge when using a linear architecture, even when the linear model may fit the true value function perfectly. Tsitsiklis and Van Roy (1997) establishes the importance of using Bellman errors weighted by the probability of being in a state. de Farias and Van Roy (2000) shows that there does not necessarily exist a fixed point to the projected form of Bellman's equation $\Phi\theta = \Pi \mathcal{M}\Phi\theta$ where $\mathcal{M}$ is the max operator. This paper also shows that a fixed point does exist for a projection operator Π_D defined with respect to the norm $\|\cdot\|_D$ which weights a state s with the probability d_s of being in this state. This result is first shown for a fixed policy, and then for a class of randomized policies. GTD2 and TDC are due to Sutton et al. (2009), with material from Sutton et al. (2008).

Section 16.6 – Dearden et al. (1998b) introduces the idea of using Bayesian updating for Q-learning. Dearden et al. (2013) then considers model-based Bayesian learning. Our presentation is based on Ryzhov and Powell (2010) which introduces the idea of correlated beliefs.

Exercises

Review questions

16.1 Describe in words (no mathematics) the difference between implementing TD(0) and TD(1).

16.2 Describe in words, with only necessary mathematics, the essential differences between LSTD and LSPE.

16.3 Show that updating the value of being in a state using, for example, temporal difference updates (section 16.1.3) are basically stochastic gradient updates (see section 16.2). This means that temporal difference updates are solving a particular optimization problem. What is the optimization problem?

Computational exercises

16.4 We are going to again try to use approximate dynamic programming to estimate a discounted sum of random variables:

$$F^T = \mathbb{E} \sum_{t=0}^{T} \gamma^t R_t,$$

where R_t is a random variable that is uniformly distributed between 0 and 100 (you can use this information to randomly generate outcomes, but otherwise you cannot use this information). This time we are going to use a discount factor of $\gamma = .95$. We assume that R_t is independent of prior history. We can think of this as a single state Markov decision process with no decisions.

(a) Using the fact that $\mathbb{E}R_t = 50$, give the exact value for F^{100}.

(b) Propose an approximate dynamic programming algorithm to estimate F^T. Give the value function updating equation, using a stepsize $\alpha_t = 1/t$.

(c) Perform 100 iterations of the approximate dynamic programming algorithm to produce an estimate of F^{100}. How does this compare to the true value?

(d) Compare the performance of the following stepsize rules: Kesten's rule, the stochastic gradient adaptive stepsize rule (use $\nu = .001$), $1/n^\beta$ with $\beta = .85$, the Kalman filter rule, and the optimal stepsize rule. For each one, find both the estimate of the sum and the variance of the estimate.

16.5 Figure 16.2 shows a five-state Markov chain where we transition from state 0 to 1 to 2 until transition out of state 5, earning contributions of 0 from each transition until we earn 1 when we transition from state 5, at which point we terminate. Table 16.1 shows the value of being in each state after each iteration of a TD(0) learning algorithm (otherwise known as approximate value iteration). Repeat the calculations in Table 16.1 using a fixed stepsizes of

    (a) $\alpha = 1.0$.
    (b) $\alpha = 0.5$.
    (c) $\alpha = 0.1$.
    (d) $\alpha = 0.05$.
    (e) Compare the rates of convergence. Why wouldn't we always use $\alpha = 1.0$?

16.6 Consider a Markov decision process with a single state and single action. Assume that we do not know the expected value of the contribution $\hat{C}$, but each time it is sampled, draw a sample realization from the uniform distribution between 0 and 20. Also assume a discount factor of $\gamma = 0.90$. Let $V = \sum_{t=0}^{\infty} \gamma^t \hat{C}_t$. The exercises that follow can be formed in a spreadsheet. Estimate V using LSTD using 100 iterations.

16.7 Repeat exercise 16.6, estimating V with LSPE using 100 iterations.

16.8 Repeat exercise 16.6, estimating V using recursive least squares, executing the algorithm for 100 iterations.

16.9 Repeat exercise 16.6, estimating V using temporal differencing (approximate value iteration) and a stepsize of $1/n^{.7}$.

16.10 Repeat exercise 16.6, estimating V using temporal differencing (approximate value iteration) and a stepsize of $5/(5 + n - 1)$.

16.11 Repeat exercise 16.10 using a discount factor of 0.95.

Theory questions

16.12 We are going to walk through the derivation of the equations in section 16.5 assuming that there is a feature for each state, where $\phi_f(s) = 1$ if feature f corresponds to state s, and 0 otherwise. When asked for a sample of a vector or matrix, assume there are three states and three features. As we did in section 16.5, let d_s^π be the probability of being in state s under policy π, and let D^π be the diagonal matrix consisting of the elements d_s^π.

    (a) What is the column vector ϕ if $s = 1$? What does $\phi\phi^T$ look like?

    (b) If d_s^π is the probability of being in state s under policy π, write out $\mathbb{E}[\phi\phi^T]$.

    (c) Write out the matrix Φ.

    (d) What is the projection matrix Π?

    (e) Write out equation (16.44) for $MSPBE(\theta)$.

16.13   Write out all the equations in section 16.5 for a problem where the state s is an integer quantity $\{0, 1, 2, \ldots, S\}$, and where

$$\bar{V}(s|\theta) = \theta_0 + \theta_1 s.$$

16.14   Write out all the equations in section 16.5 for a problem where there is a feature $\phi_f(s)$ for each state, where $\phi_f(s) = 1$ if $f = s$.

Diary problem

The diary problem is a single problem you chose (see chapter 1 for guidelines). Answer the following for your diary problem.

16.15   Using the policy that you designed in exercise 12.13, sketch the steps for estimating the value of the policy using the following methods:

    (a) TD(0) – Temporal differencing with $\lambda = 0$.

    (b) TD(1) – Temporal differencing with $\lambda = 1$.

    (c) For your diary problem, discuss what appear to be the strengths and weaknesses of TD(0) and TD(1).

Bibliography

Baird, L.C. (1995). Residual algorithms: Reinforcement learning with function approximation. *In Proceedings of the Twelfth International Conference on Machine Learning*, 30–37.

Bertsekas, D., Borkar, V.S., and Nedic, A. (2004). Improved temporal difference methods with linear function approximation. In: *Handbook of Learning and Approximate Dynamic Programming* (eds. J. Si, A. G. Barto, W. B. Powell and D. Wunsch), 233–257. New York: IEEE Press.

Bertsekas, D.P. (2017). *Dynamic Programming and Optimal Control: Approximate Dynamic Programming*, 4e. Belmont, MA: Athena Scientific.

Blum, J. (1954). Multidimensional stochastic approximation methods. *Annals of Mathematical Statistics* 25: 737–744.

Bradtke, S.J. and Barto, A.G. (1996). Linear least-squares algorithms for temporal difference learning. *Machine Learning* 22 (1): 33–57.

de Farias, D. P. and Van Roy, B. (2000). On the existence of fixed points for approximate value iteration and temporal-difference learning. *Journal of Optimization Theory and Applications* 105 (3): 589–608.

Dearden, R., Friedman, N., and Andre, D. (2013). Model-based Bayesian exploration. arXiv, https://arxiv.org/abs/1301.6690.

Dearden, R., Friedman, N., and Russell, S. (1998b). Bayesian Q-Learning. *Proceedings of the National Conference on Artificial Intelligence*. 761–768.

Jaakkola, T., Jordan, M.I., and Singh, S.P. (1994). On the convergence of stochastic iterative dynamic programming algorithms. *Neural Computation* 6 (6): 1185–1201.

Kushner, H.J. and Yin, G.G. (2003). *Stochastic Approximation and Recursive Algorithms and Applications*. New York: Springer.

Ljung, L. and Soderstrom, T. (1983). *Theory and Practice of Recursive Identification*. Cambridge, MA: MIT Press.

Lagoudakis, M. and Parr, R. (2003). Least-squares policy iteration. *Journal of Machine Learning Research* 4: 1107–1149.

Lagoudakis, M., Parr, R., and Littman, M. (2002). Least squares methods in reinforcement learning for control. *Methods and Applications of Artificial Intelligence*, 752–752.

Nedic, A., Bertsekas, D.P. (2003). Least squares policy evaluation algorithms with linear function approximation. *Discrete Event Dynamic Systems* 13 (1): 79–110.

Precup, D., Sutton, R.S., and Dasgupta, S. (2001). Off-policy temporal-difference learning with function approximation. In: *19th International Conference on Machine Learning*, 417–424.

Robbins, H. and Monro, S. (1951). A stochastic approximation method. *The Annals of Mathematical Statistics* 22 (3): 400–407.

Ryzhov, I.O. and Powell, W.B. (2010). Approximate dynamic programming with correlated bayesian beliefs. In: *Forty-Eighth Annual Allerton Conference on Communication, Control, and Computing*. Monticello, IL.

Ryzhov, I.O., Frazier, P.I., and Powell, W.B. (2015). A new optimal stepsize for approximate dynamic programming. *IEEE Transactions on Automatic Control* 60 (3): 743–758.

Sutton, R.S. and Barto, A.G. (2018). *Reinforcement Learning: An Introduction*, 2e. Cambridge, MA: MIT Press.

Sutton, R.S., Maei, H.R., Precup, D., Bhatnagar, S., Silver, D., Szepesvári, C., and Wiewiora, E. (2009). Fast gradient-descent methods for temporal-difference learning with linear function approximation. *Proceedings of the 26th Annual International Conference on Machine Learning ICML '09.* 1–8.

Sutton, R.S., Szepesvari, C., and Maei, H.R. (2008). A convergent O (n) algorithm for off-policy temporal-difference learning with linear function approximation. In: *Proceedings of the Neuro Information Processing Society.* Vancouver, 1–8.

Tsitsiklis, J.N. (1994). Asynchronous stochastic approximation and q-learning. *Machine Learning* 16: 185–202.

Tsitsiklis, J.N. and Van Roy, B. (1997). An analysis of temporal-difference learning with function approximation. *IEEE Transactions on Automatic Control* 42 (5): 674–690.

Van Roy, B. and Choi, D.P. (2006). A generalized Kalman filter for fixed point approximation and efficient temporal-difference learning. *Discrete Event Dynamic Systems* 16: 207–239.

Young, P. (1984). *Recursive Estimation and TimeSeries Analysis.* Berlin, Heidelberg: SpringerVerlag.

17

Forward ADP II: Policy Optimization

We are now ready to tackle the problem of searching for good policies while simultaneously trying to produce good value function approximations. The guiding principle in this chapter is that we can find good policies if we can find good value function approximations. The problem is that finding good value function approximations requires that we be simulating "good" policies (using the methods of chapter 16). It is the interaction between the two that creates all the complications.

The algorithmic strategies presented in this chapter are all based on algorithms we first presented in chapter 14, with two notable exceptions:

- We never take expectations – Random variables are always handled through either Monte Carlo simulation, historical trajectories, or direct field observations.
- We use machine learning to approximate functions – This means we have to deal with estimation errors due to noise, errors due to biased observations, and structural errors from the chosen approximating architecture.

The statistical tools presented in chapter 3 focused on finding the best statistical fit of a function that we can only observe with noise, but where we assumed that the observations are unbiased. In chapter 16, we saw that the sampled estimate $\hat{v}_t^n$ of the value of being in state S_t^n could be biased for several reasons:

- If we are using approximate value iteration, the value functions have to steadily accumulate downstream values (recall the slow convergence illustrated in Table 16.1).
- The sampled $\hat{v}_t^n$ might depend on downstream value function approximations, which might produce structural biases (e.g. if we use a linear approximation of a nonlinear function).

Reinforcement Learning and Stochastic Optimization: A Unified Framework for Sequential Decisions, First Edition. Warren B. Powell.
© 2022 John Wiley & Sons, Inc. Published 2022 John Wiley & Sons, Inc.

- $\hat{v}_t^n$ depends on the policies that are being used to make decisions in the future which in turn depend on value function approximations which are (a) incorrect and (b) changing over the iterations.

In all three cases, our observations of $\hat{v}_t^n$ are biased, but in a way that is also changing over iterations as we search for better policies.

When we write our generic optimization problem

$$\max_\pi \mathbb{E}\left\{\sum_{t=0}^{T} \gamma^t C(S_t, X_t^\pi(S_t))|S_0\right\}, \tag{17.1}$$

the maximization over policies can mean choosing one of the approximation strategies for $\overline{V}_t(S_t)$ from chapter 3, and choosing the parameters that control the approximation. A useful way to express this search is to let $f \in \mathcal{F}$ be the set of architectures (functions), and let $\theta \in \Theta^f$ be any tunable parameters for functions in class f, which means our policy π is an element of $(f \in \mathcal{F}, \theta \in \Theta^f)$. Our search over policies is then the same as

$$\max_{\pi=(f\in\mathcal{F},\theta\in\Theta^f)} \mathbb{E}\left\{\sum_{t=0}^{T} \gamma^t C(S_t, X_t^\pi(S_t))|S_0\right\}.$$

For example, we might be choosing between a myopic policy, or perhaps a simple linear architecture with one basis function

$$\overline{V}_t(S_t) = \theta_0 + \theta_1 S_t, \tag{17.2}$$

or perhaps a linear architecture with two basis functions,

$$\overline{V}_t(S_t) = \theta_0 + \theta_1 S_t + \theta_t S^2. \tag{17.3}$$

We might even use a nonlinear architecture such as

$$\overline{V}_t(S_t) = \frac{e^{\theta_0+\theta_1 S}}{1 + e^{\theta_0+\theta_1 S}}.$$

We can try estimating value functions with each of these architectures (which still requires searching for θ for each function class), and then compare the performance of the resulting policies using the objective function in (17.1), which is how we would actually perform the search over function classes (admittedly this is ad hoc).

We begin our presentation with an overview of the basic algorithmic strategies that we cover in this chapter.

17.1 Overview of Algorithmic Strategies

The algorithmic strategies that we examine in this chapter are based on the principles of value iteration and policy iteration, first introduced in chapter 14. We continue to adapt our algorithms to finite and infinite horizons.

Basic value iteration for finite horizon problems work by solving

$$V_t(S_t) = \max_{x_t} \left(C(S_t, x_t) + \gamma \mathbb{E}\{V_{t+1}(S_{t+1})|S_t, x_t\} \right). \tag{17.4}$$

Equation (17.4) works by stepping backward in time, where $V_t(S_t)$ is computed for each (presumably discrete) state S_t. This is classical "backward" dynamic programming which suffers from the well-known curse of dimensionality, because we typically are unable to "loop over all the states."

Approximate dynamic programming approaches finite horizon problems by solving problems of the form

$$\hat{v}_t^n = \max_{x_t} \left(C(S_t^n, x_t) + \gamma \overline{V}_{t+1}^{x,n-1}(S^{M,x}(S_t^n, x_t)) \right). \tag{17.5}$$

Here, we have formed the value function approximation around the post-decision state. We execute the equations by stepping forward in time which creates a natural state sampling procedure known in the reinforcement literature as *trajectory following*. If x_t^n is the decision that optimizes (17.5), then we compute our next state using $S_{t+1}^n = S^M(S_t^n, x_t^n, W_{t+1}^n)$ where W_{t+1}^n is sampled from some distribution. The process runs until we reach the end of our horizon, at which point we return to the beginning of the horizon and repeat the process.

Classical value iteration for infinite horizon problems is centered on the basic iteration

$$V^n(S) = \max_{x} \left(C(S, x) + \gamma \mathbb{E}\{V^{n-1}(S')|S\} \right). \tag{17.6}$$

Again, equation (17.6) has to be executed for each state S. After each iteration, the new estimate V^n replaces the old estimate V^{n-1} on the right, after which n is incremented.

When we use approximate methods, we might observe an estimate of the value of being in a state using

$$\hat{v}^n = \max_{x} \left(C(S^n, x) + \gamma \overline{V}^{x,n-1}(S^{M,x}(S^n, x^n)) \right). \tag{17.7}$$

We then use the observed state-value pair $(S^n, \hat{v}^n)$ to update the value function approximation using whatever architecture we have chosen.

Using $\hat{v}^n$ to update the value function approximation can introduce a significant level of noise, that is then translated to the behavior of the policy producing unpredictable effects (this is well known to experimentalists in the ADP community). One strategy for mitigating this noise is to imbed a policy approximation loop within an outer loop where policies are updated. Assume we fix our policy using

$$X^{\pi,n}(S) = \arg\max_{x \in \mathcal{X}} \left(C(S, x) + \gamma \overline{V}^{x,n-1}(S^{M,x}(S, x)) \right). \tag{17.8}$$

Now perform the loop over $m = 1, \dots, M$,

$$\hat{v}^{n,m} = \max_{x \in \mathcal{X}} \left(C(S^{n,m}, x) + \gamma \overline{V}^{x,n-1}(S^{M,x}(S^{n,m}, x)) \right)$$

where $S^{n+1,m} = S^M(S^{n,m}, x^{n,m}, W^{n+1,m})$. Note that the value function $\overline{V}^{x,n-1}(s)$ remains constant within this inner loop. After executing this loop, we take the series of observations $\hat{v}^{n,1}, \dots, \hat{v}^{n,M}$ and use them to update $\overline{V}^{x,n-1}(s)$ to obtain $\overline{V}^{x,n}(s)$.

Typically, $\overline{V}^{x,n}(s)$ does not depend on $\overline{V}^{x,n-1}(s)$, other than to influence the calculation of $\hat{v}^{n,m}$. If M is large enough, $\overline{V}^{x,n}(s)$ will represent an accurate approximation of the value of being in state s while following the policy in equation (17.8). In fact, it is specifically because of this ability to approximate a policy that approximate policy iteration is emerging as a powerful algorithmic strategy for approximate dynamic programming. However, the cost of using the inner policy evaluation loop can be significant, and for this reason approximate value iteration and its variants remain popular.

Repeated evaluations of a policy helps reduce the noise, but does not eliminate the errors in the approximation itself, possibly due to the choice of architecture, or possibly due to the reality that our observations $\hat{v}^n$ are based on approximations which means that our policy is suboptimal, biasing the estimates $\hat{v}^n$. In other words, there is a lot going on that distorts the trajectory of the algorithm.

The remainder of the chapter is organized around covering the following strategies:

Approximate value iteration – These are policies that iteratively update the value function approximation, and then immediately update the policy (by using the updated value function approximation). We strive to find a value function approximation that estimates the value of being in each state while following a (near) optimal policy, but only in the limit. We intermingle the treatment of finite and infinite horizon problems. Variations include:

- Lookup table representations – Here we introduce three major strategies that reflect the use of the pre-decision state, state-decision pairs, and the post-decision state:
  - AVI for pre-decision state – Approximate value iteration using the classical pre-decision state variable.
  - *Q*-learning – Estimating the value of state-decision pairs.
  - AVI for the post-decision state – Approximate value iteration where value function approximations are approximated around the post-decision state.
- Parametric architectures – We summarize some of the extensive literature which depends on linear models (basis functions), and touch on nonlinear models.

Approximate policy iteration – These are policies that attempt to explicitly approximate the value of a policy to some level of accuracy within an inner loop, within which the policy is held fixed.

- API using lookup tables – We use this setting to present the basic idea.
- API using linear models – This strategy continues to attract attention because of its simplicity.
- API using nonparametric models – Nonparametric models offer significantly greater flexibility, but the price is that they are less stable (they can respond much more quickly to random variations) and require considerably more observations.

The linear programming method – The linear programming method, first introduced in chapter 14, can be adapted to exploit value function approximations.

17.2   Approximate Value Iteration and *Q*-Learning Using Lookup Tables

Arguably the most natural and elementary approach for approximate dynamic programming uses approximate value iteration. In this section we explore the following topics related to this important algorithmic strategy:

- Value iteration using a pre-decision state variable.
- *Q*-learning.
- Value iteration using a post-decision state variable.
- Value iteration using a backward pass.

17.2.1 Value Iteration Using a Pre-Decision State Variable

Classical value iteration (for a finite-horizon problem) estimates the value of being in a specific state S_t^n

$$\hat{v}_t^n = \max_{x_t} \left(C(S_t^n, x_t) + \gamma \mathbb{E}\{V_{t+1}(S_{t+1}) | S_t^n\} \right), \tag{17.9}$$

where $S_{t+1} = S^M(S_t^n, x_t, W_{t+1}^n)$, and S_t^n is the state that we are in at time t, iteration n. We assume that we are following a sample path ω^n, where we compute $W_{t+1}^n = W_{t+1}(\omega^n)$. After computing $\hat{v}_t^n$, we update the value function using the standard equation

$$\overline{V}_t^n(S_t^n) = (1 - \alpha_{n-1})\overline{V}_t^{n-1}(S_t^n) + \alpha_{n-1}\hat{v}_t^n. \tag{17.10}$$

If we sample states at random (rather than following the trajectory) and repeat equations (17.9) and (17.10), we will eventually converge to the correct value of being in each state. Note that we are assuming a finite-horizon model, and that we can compute the expectation exactly. When we can compute the expectation exactly, this is very close to classical value iteration, with the only exception that we are not looping over all the states at every iteration.

One reason to use the pre-decision state variable is that for some problems, computing the expectation is easy. For example, W_{t+1} might be a binomial random variable (did a customer arrive, did a component fail) which makes the expectation especially easy. If this is not the case, then we have to approximate the expectation. For example, we might use

$$\hat{v}_t^n = \max_{x_t} \left(C(S_t^n, x_t) + \gamma \sum_{\hat{\omega} \in \hat{\Omega}^n} p^n(\hat{\omega}) \overline{V}_{t+1}^{n-1}(S^M(S_t^n, x_t, W_{t+1}(\hat{\omega}))) \right). \tag{17.11}$$

Either way, using a lookup table representation we can update the value of being in state S_t^n using

$$\overline{V}_t^n(S_t^n) = (1 - \alpha_{n-1})\overline{V}_t^{n-1}(S_t^n) + \alpha_{n-1}\hat{v}_t^n.$$

Keep in mind that if we can compute an expectation (or if we approximate it using a large sample $\hat{\Omega}$), then the stepsize should be much larger than when we are using a single sample realization (as we did with the post-decision formulation). An outline of the overall algorithm is given in Figure 17.1.

At this point a reasonable question to ask is: Does this algorithm work? The answer is ... possibly, but not in general. Before we get an algorithm that will work (at least in theory), we need to deal with what is known as the exploration-exploitation problem, which we address in section 17.5.

Step 0. Initialization:

    Step 0a. Initialize $\overline{V}_t^0$, $t \in \mathcal{T}$.
    Step 0b. Set $n = 1$.
    Step 0c. Initialize S^0.

Step 1. Sample ω^n.

    Step 2. Do for $t = 0, 1, \dots, T$:

        Step 2a. Choose $\hat{\Omega}^n \subseteq \Omega$ and solve:

$$\hat{v}_t^n = \max_{a_t} \left(C_t(S_t^{n-1}, x_t) + \gamma \sum_{\hat{\omega} \in \hat{\Omega}^n} p^n(\hat{\omega}) \overline{V}_{t+1}^{n-1} (S^M(S_t^{n-1}, x_t, W_{t+1}(\hat{\omega}))) \right)$$

        and let x_t^n be the value of x_t that solves the maximization problem.
        Step 2b. Compute:

$$S_{t+1}^n = S^M(S_t^n, x_t^n, W_{t+1}(\omega^n)).$$

        Step 2c. Update the value function:

$$\overline{V}_t^n \leftarrow U^V(\overline{V}_t^{n-1}, S_t^n, \hat{v}_t^n)$$

Step 3. Increment n. If $n \leq N$, go to Step 1.
Step 4. Return the value functions $(\overline{V}_t^n)_{t=1}^T$.

Figure 17.1 Approximate dynamic programming using a pre-decision state variable.

17.2.2 Q-Learning

One of the earliest and most widely studied algorithms in the reinforcement learning literature is known as Q-learning. The name is derived simply from the notation used in the algorithm, and appears to have initiated the tradition of naming algorithms after the notation.

To motivate Q-learning, return for the moment to the classical way of making decisions using dynamic programming. Normally we would want to solve

$$x_t^n = \arg \max_{x_t \in \mathcal{X}_t^n} \left(C_t(S_t^n, x_t) + \gamma \mathbb{E} \left\{ \overline{V}_{t+1}^{n-1} (S_{t+1}(S_t^n, x_t, W_{t+1})) \mid S_t^n, x_t \right\} \right). \quad (17.12)$$

Solving equation (17.12) can be problematic for two different reasons. The first is that we may not be able to compute the expectation because it is computationally too complex (the second curse of dimensionality). The second is that we may simply not have the information we need to compute the expectation. This might happen if (a) we do not know the probability distribution of the random information or (b) we may not know the transition function. In either of these cases, we say that we do not "know the model" and we need to use a "model-free" formulation.

When we can compute the expectation, which means we have the transition function and we know the probability distribution, then we are using what is known as a "model-based" formulation. Many authors equate "model-based" with knowing the one-step transition matrix, but this ignores the many problems where we know the transition function, we know the probability law for the exogenous information, but we simply cannot compute the transition function either because the state space is too large (or continuous), or the exogenous information is multidimensional.

Earlier, we circumvented this problem by approximating the expectation by using a subset of outcomes (see equation (17.11)), but this can be computationally clumsy for many problems. One thought is to solve the problem for a single sample realization

$$x_t^n = \arg\max_{x_t \in \mathcal{X}_t^n} \left(C_t(S_t^n, x_t) + \gamma \overline{V}_{t+1}^{n-1}(S_{t+1}(S_t^n, x_t, W_{t+1}(\omega^n))) \right). \quad (17.13)$$

The problem is that this means we are choosing x_t for a particular realization of the future information $W_{t+1}(\omega^n)$. If we use the same sample realization of $W_{t+1}(\omega^n)$ to make the decision that will actually happen (when we step forward in time), then this is what is known as cheating (peeking into the future), which can seriously distort the behavior of the system. If we use a single sample realization for $W_{t+1}(\omega)$ that is different than the one we use when we simulate forward, then this is simply unlikely to produce good results (imagine computing averages based on a single observation).

What if we instead choose the decision x_t^n first, then observe W_{t+1}^n (so we are not using this information when we choose our decision) and then compute the cost? Let the resulting cost be computed using

$$\hat{q}_t^n(S_t, x_t) = C(S_t, x_t) + \gamma \overline{V}_{t+1}^{n-1}(S^M(S_t^n, x_t, W_{t+1}(\omega^n))). \quad (17.14)$$

We could now smooth these values to obtain

$$\bar{Q}_t^n(S_t, x_t) = (1 - \alpha_{n-1})\bar{Q}_t^{n-1}(S_t^n, x_t^n) + \alpha_{n-1}\hat{q}_t^n(S_t, x_t).$$

Not surprisingly, we can compute the value of being in a state from the Q-factors using

$$\overline{V}_t^n(S_t) = \max_x \bar{Q}_t^n(S_t, x). \quad (17.15)$$

If we combine (17.15) and (17.14), we obtain

$$\hat{q}_t^n = C(S_t, x_t) + \gamma \max_{x_{t+1}} \bar{Q}^{n-1}(S_{t+1}, x_{t+1}),$$

where $S_{t+1} = S^M(S_t^n, x_t, W_{t+1}(\omega^n))$ is the next state resulting from the decision x_t and the sampled information $W_{t+1}(\omega^n)$.

The functions $Q_t(S_t, x_t)$ are known as *Q-factors* and they capture the value of being in a state and taking a particular decision. Recall from section 9.4.5 that a state-decision pair (S_t, x_t) is a form of post-decision state, although it is typically the least-compact form for representing a post-decision state.

We can now choose a decision by solving

$$x_t^n = \arg \max_{x_t \in \mathcal{X}_t^n} \bar{Q}_t^{n-1}(S_t^n, x_t). \tag{17.16}$$

Note that once we know the Q-factors, we can choose a decision without knowing anything else, which is one reason why Q-learning is often described as a method for problems where we can observe a process (such as doctors making decisions) and learn decisions without having a transition function or a model for rewards or uncertainties (also known as model-free dynamic programming).

The complete algorithm is summarized in Figure 17.2.

A variation of Q-learning is known as "Sarsa" which stands for "state, action, reward, state, action" (the computer science community has a culture of naming its algorithms around its notation). Imagine that we start in a state s and make decision x. After this, we observe a reward r and the next state s'. Finally, use some policy to choose the next decision x'.

17.2.3 Value Iteration Using a Post-Decision State Variable

For the many applications that lend themselves to a compact post-decision state variable, it is possible to adapt approximate value iteration to value functions estimated around the post-decision state variable. At the heart of the algorithm we choose decisions (and estimate the value of being in state S_t^n) using

$$\hat{v}_t^n = \arg \max_{x_t \in \mathcal{X}_t} \left(C(S_t^n, x_t) + \gamma \bar{V}_t^{n-1}(S^{M,x}(S_t^n, x_t)) \right).$$

The distinguishing feature when we use the post-decision state variable is that the maximization problem is now deterministic. The key step is how we update the value function approximation. Instead of using $\hat{v}_t^n$ to update a pre-decision value function approximation $\bar{V}^{n-1}(S_t^n)$, we use $\hat{v}_t^n$ to update a post-decision value function approximation around the *previous* post-decision state $S_{t-1}^{x,n}$. This is done using

Step 0. Initialization:

> **Step 0a.** Initialize an approximation for the value function $\bar{Q}_t^0(S_t, x_t)$ for all states S_t and decisions $x_t \in \mathcal{X}_t, t = \{0, 1, \dots, T\}$.
> **Step 0b.** Set $n = 1$.
> **Step 0c.** Initialize S_0^1.

Step 1. Choose a sample path ω^n.

> **Step 2.** Do for $t = 0, 1, \dots, T$:

>> **Step 2a:** Determine the decision using ϵ-greedy. With probability ϵ, choose a decision x^n at random from $\mathcal{X}$. With probability $1 - \epsilon$, choose a^n using

$$x_t^n \quad = \quad \arg\max_{x_t \in \mathcal{X}_t} \bar{Q}_t^{n-1}(S_t^n, x_t).$$

>> **Step 2b.** Sample $W_{t+1}^n = W_{t+1}(\omega^n)$ and compute the next state $S_{t+1}^n = S^M(S_t^n, x_t^n, W_{t+1}^n)$.
>> **Step 2c.** Compute

$$\hat{q}_t^n \quad = \quad C(S_t^n, x_t^n) + \gamma \max_{x_{t+1} \in \mathcal{X}_{t+1}} \bar{Q}_{t+1}^{n-1}(S_{t+1}^n, x_{t+1}).$$

>> **Step 2d.** Update $\bar{Q}_t^{n-1}$ and $\bar{V}_t^{n-1}$ using.

$$\bar{Q}_t^n(S_t^n, x_t^n) \quad = \quad (1 - \alpha_{n-1})\bar{Q}_t^{n-1}(S_t^n, x_t^n) + \alpha_{n-1}\hat{q}_t^n$$

Step 3. Increment n. If $n \leq N$ go to Step 1.
Step 4. Return the Q-factors $(\bar{Q}_t^n)_{t=1}^T$.

Figure 17.2 A Q-learning algorithm.

$$\bar{V}_{t-1}^n(S_{t-1}^{x,n}) \quad = \quad (1 - \alpha_{n-1})\bar{V}_{t-1}^{n-1}(S_{t-1}^{x,n}) + \alpha_{n-1}\hat{v}_t^n.$$

The post-decision state not only allows us to solve deterministic optimization problems, there are many applications where the post-decision state has either the same dimensionality as the pre-decision state, or, for some applications, a much lower dimensionality.

A complete summary of the algorithm is given in Figure 17.3.

Q-learning shares certain similarities with dynamic programming using a post-decision value function. In particular, both require the solution of a deterministic optimization problem to make a decision. However, Q-learning accomplishes this goal by creating a post-decision state given by the state/decision pair (S, x) (we first introduced this form of post-decision state in section 9.4.5). We then have to learn the value of being in (S, x), rather than the value of being in state S alone (which is already very hard for most problems).

Step 0. Initialization:

    Step 0a. Initialize an approximation for the value function $\overline{V}_t^0(S_t^x)$ for all post-decision states S_t^x, $t = \{0, 1, \ldots, T\}$.
    Step 0b. Set $n = 1$.
    Step 0c. Initialize $S_0^{x,1}$.

Step 1. Choose a sample path ω^n.

    Step 2. Do for $t = 0, 1, \ldots, T$:

        Step 2a: Determine the decision using ε-greedy. With probability ε, choose a decision x^n at random from $\mathcal{X}$. With probability $1 - \varepsilon$, choose a^n using

$$\hat{v}_t^n \;\; = \;\; \arg\max_{x_t \in \mathcal{X}_t} \left(C(S_t^n, x_t) + \gamma \overline{V}_t^{n-1} (S^{M,x}(S_t^n, x_t)) \right).$$

        Let x_t^n be the decision that solves the maximization problem.
        Step 2b. Update $\overline{V}_{t-1}^{n-1}$ using:

$$\overline{V}_{t-1}^n (S_{t-1}^{x,n}) \;\; = \;\; (1 - \alpha_{n-1})\overline{V}_{t-1}^{n-1} (S_{t-1}^{x,n}) + \alpha_{n-1}\hat{v}_t^n.$$

        Step 2c. Sample $W_{t+1}^n = W_{t+1}(\omega^n)$ and compute the next state $S_{t+1}^n = S^M(S_t^n, x_t^n, W_{t+1}^n)$.

Step 3. Increment n. If $n \leq N$ go to Step 1.
Step 4. Return the value functions $(\overline{V}_t^n)_{t=1}^T$.

Figure 17.3   Approximate value iteration for finite horizon problems using the post-decision state variable.

If we compute the value function approximation $\overline{V}^n(S^x)$ around the post-decision state $S^x = S^{M,x}(S, x)$, we can create Q-factors directly from the contribution function and the post-decision value function using

$$\bar{Q}^n(S, x) = C(S, x) + \gamma \overline{V}_t^n (S^{M,x}(S, x)).$$

Viewed this way, approximate value iteration using value functions estimated around a post-decision state variable is equivalent to Q-learning. However, if the post-decision state is compact, then estimating $\overline{V}(S^x)$ is much easier than estimating $\bar{Q}(S, x)$.

17.2.4   Value Iteration Using a Backward Pass

Classical approximate value iteration, which is equivalent to temporal difference learning with $\lambda = 0$ (also known as TD(0)), can be implemented using a pure forward pass, which enhances its simplicity. However, there are problems where it is useful to simulate decisions moving forward in time, and then updating value functions moving backward in time. This is also known as temporal difference learning with $\lambda = 1$, but we find "backward pass" to be more descriptive. The algorithm is depicted in Figure 17.4.

In this algorithm, we step forward through time creating a trajectory of states, decisions, and outcomes. We then step backward through time, updating the

---

Step 0. Initialization:

Step 0a. Initialize $\overline{V}_t^0$, $t \in \mathcal{T}$.
Step 0b. Initialize S_0^1.
Step 0c. Choose an initial policy $X^{\pi,0}$.
Step 0d. Set $n = 1$.

Step 1. Choose a sample path ω^n.
Step 2: Do for $t = 0, 1, 2, \ldots, T$:

Step 2a: Find

$$x_t^n = X_t^{\pi,n-1}(S_t^n)$$

Step 2b: Update the state variable

$$S_{t+1}^n = S^M(S_t^n, x_t^n, W_{t+1}(\omega^n)).$$

Step 3: Set $\hat{v}_{T+1}^n = 0$ and do for $t = T, T-1, \ldots, 1$:

Step 3a: Update $\hat{v}_t^n$ using

$$\hat{v}_t^n = C(S_t^n, x_t^n) + \gamma \hat{v}_{t+1}^n.$$

Step 3b: Update the value function approximation $\overline{V}_t^n$ using

$$\overline{V}_t^n \leftarrow U^V(\overline{V}_t^{n-1}, S_t^{x,n}, \hat{v}_t^n).$$

Step 3c. Update the policy

$$X_t^{\pi,n}(S) = \arg\max_{x \in \mathcal{X}} \left(C(S_t^n, x) + \gamma \overline{V}_t^n(S^{M,x}(S_t^n, x)) \right).$$

Step 4. Increment n. If $n \leq N$ go to Step 1.
Step 5. Return the value functions $(\overline{V}_t^N)_{t=1}^T$.

Figure 17.4 Double-pass version of the approximate dynamic programming algorithm for a finite horizon problem.

value of being in a state using information from the same trajectory in the future. We are going to use this algorithm to also illustrate ADP for a time-dependent, finite horizon problem. In addition, we are going to illustrate a form of policy evaluation. Pay careful attention to how variables are indexed.

The idea of stepping backward through time to produce an estimate of the value of being in a state was first introduced in the control theory community under the name of *backpropagation through time* (BTT). The result of our backward pass is $\hat{v}_t^n$, which is the contribution from the sample path ω^n and a particular policy. Our policy is, quite literally, the set of decisions produced by the value function approximation $\overline{V}^{n-1}$. Unlike our forward-pass algorithm (where $\hat{v}_t^n$ depends on the approximation $\overline{V}_t^{n-1}(S_t^x)$), $\hat{v}_t^n$ is a valid, unbiased estimate of the value of being in state S_t^n at time t and following the policy produced by $\overline{V}^{n-1}$.

We introduce an inner loop so that rather than updating the value function approximation with a single $\hat{v}_0^n$, we average across a set of samples to create a more stable estimate, $\bar{v}_0^n$.

These two strategies are easily illustrated using our simple asset selling problem. For this illustration, we are going to slightly simplify the model we provided earlier, where we assumed that the change in price, $\hat{p}_t$, was the exogenous information. If we use this model, we have to retain the price p_t in our state variable (even the post-decision state variable). For our illustration, we are going to assume that the exogenous information is the price itself, so that $p_t = \hat{p}_t$. We further assume that $\hat{p}_t$ is independent of all previous prices (a pretty strong assumption). For this model, the pre-decision state is $S_t = (R_t, p_t)$ while the post-decision state variable is simply $S_t^x = R_t^x = R_t - x_t$ which indicates whether we are holding the asset or not. Further, $S_{t+1} = S_t^x$ since the resource transition function is deterministic.

With this model, a single-pass algorithm (approximate value iteration) is performed by stepping forward through time, $t = 1, 2, \ldots, T$. At time t, we first sample $\hat{p}_t$ and we find

$$\hat{v}_t^n = \max_{x_t \in \{0,1\}} \left(\hat{p}_t^n x_t + (1 - x_t)(-c_t + \bar{v}_t^{n-1}) \right). \tag{17.17}$$

Assume that the holding cost $c_t = 2$ for all time periods.

Table 17.1 illustrates three iterations of a single-pass algorithm for a three-period problem. We initialize $\bar{v}_t^0 = 0$ for $t = 0, 1, 2, 3$. Our first decision is x_1 after we see $\hat{p}_1$. The first column shows the iteration counter, while the second shows the stepsize $\alpha_{n-1} = 1/n$. For the first iteration, we always choose to sell because $\bar{v}_t^0 = 0$, which means that $\hat{v}_t^1 = \hat{p}_t^1$. Since our stepsize is 1.0, this produces $\bar{v}_{t-1}^1 = \hat{p}_t^1$ for each time period.

Table 17.1 Illustration of a single-pass algorithm.

Iteration	α_{n-1}	t=0				t=1				t=2				t=3	
		$\bar{v}_0$	$\hat{v}_1$	$\hat{p}_1$	x_1	$\bar{v}_1$	$\hat{v}_2$	$\hat{p}_2$	x_2	$\bar{v}_2$	$\hat{v}_3$	$\hat{p}_3$	x_3	$\bar{v}_3$
0		0				0				0				0
1	1	30	30	30	1	34	34	34	1	31	31	31	1	0
2	0.50	31	32	24	0	31.5	29	21	0	29.5	30	30	1	0
3	0.3	32.3	35	35	1	30.2	27.5	24	0	30.7	33	33	1	0

In the second iteration, our first decision problem is

$$\hat{v}_1^2 = \max\{\hat{p}_1^2, -c_1 + \bar{v}_1^1\}$$
$$= \max\{24, -2 + 34\}$$
$$= 32,$$

which means $x_1^2 = 0$ (since we are holding). We then use $\hat{v}_1^2$ to update $\bar{v}_0^2$ using

$$\bar{v}_0^2 = (1 - \alpha_1)\bar{v}_0^1 + \alpha_1 \hat{v}_1^1$$
$$= (0.5)30.0 + (0.5)32.0$$
$$= 31.0.$$

Repeating this logic, we hold again for $t = 2$ but we always sell at $t = 3$ since this is the last time period. In the third pass, we again sell in the first time period, but hold for the second time period.

It is important to realize that this problem is quite simple, and we do not have to deal with exploration issues. If we sell, we are no longer holding the asset and the forward pass should stop (more precisely, we should continue to simulate the process given that we have sold the asset). Instead, even if we sell the asset, we step forward in time and continue to evaluate the state that we are holding the asset (the value of the state where we are not holding the asset is, of course, zero). Normally, we evaluate only the states that we transition to (see Step 2b), but for this problem, we are actually visiting all the states (since there is, in fact, only one state that we really need to evaluate).

Now consider a double-pass algorithm. Table 17.2 illustrates the forward pass, followed by the backward pass, where for simplicity we are going to use only a single inner iteration ($M = 1$). Each line of the table only shows the numbers determined during the forward or backward pass. In the first pass, we always sell (since the value of the future is zero), which means that at each time period the value of holding the asset is the price in that period.

Table 17.2 Illustration of a double-pass algorithm.

		t=0	t=1				t=2				t=3			
Iteration	**Pass**	$\bar{v}_0$	$\hat{v}_1$	$\hat{p}_1$	x_1	$\bar{v}_1$	$\hat{v}_2$	$\hat{p}_2$	x_2	$\bar{v}_2$	$\hat{v}_3$	$\hat{p}_3$	x_3	$\bar{v}_3$
0		0				0				0				0
1	Forward	$\rightarrow$	$\rightarrow$	30	1	$\rightarrow$	$\rightarrow$	34	1	$\rightarrow$	$\rightarrow$	31	1	
1	Back	30	30	$\leftarrow$	$\leftarrow$	34	34	$\leftarrow$	$\leftarrow$	31	31	$\leftarrow$	$\leftarrow$	0
2	Forward	$\rightarrow$	$\rightarrow$	24	0	$\rightarrow$	$\rightarrow$	21	0	$\rightarrow$	$\rightarrow$	27	1	
2	Back	26.5	23	$\leftarrow$	$\leftarrow$	29.5	25	$\leftarrow$	$\leftarrow$	29	27	$\leftarrow$	$\leftarrow$	0

In the second pass, it is optimal to hold for two periods until we sell in the last period. The value $\hat{v}_t^2$ for each time period is the contribution of the rest of the trajectory which, in this case, is the price we receive in the last time period. So, since $a_1 = a_2 = 0$ followed by $a_3 = 1$, the value of holding the asset at time 3 is the $27 price we receive for selling in that time period. The value of holding the asset at time $t = 2$ is the holding cost of -2 plus $\hat{v}_3^2$, giving $\hat{v}_2^2 = -2 + \hat{v}_3^2 = -2 + 27 = 25$. Similarly, holding the asset at time 1 means $\hat{v}_1^2 = -2 + \hat{v}_2^2 = -2 + 25 = 23$. The smoothing of $\hat{v}_t^n$ with $\bar{v}_{t-1}^{n-1}$ to produce $\bar{v}_{t-1}^n$ is the same as for the single pass algorithm.

The value of implementing the double-pass algorithm depends on the problem. For example, imagine that our asset is an expensive piece of replacement equipment for a jet aircraft. We hold the part in inventory until it is needed, which could literally be years for certain parts. This means there could be hundreds of time periods (if each time period is a day) where we are holding the part. Estimating the value of the part now (which would determine whether we order the part to hold in inventory) using a single-pass algorithm could produce extremely slow convergence. A double-pass algorithm would work dramatically better. But if the part is used frequently, staying in inventory for only a few days, then the single-pass algorithm will work fine.

17.3 Styles of Learning

At this point it is useful to pause and discuss the different styles in which we can use the ideas from section 17.2 and chapter 16. In this section, we contrast three settings in which we might apply these ideas:

- The basic offline learning problem that we have been solving up to now using a simulator to train value functions.

- An online learning problem that would arise if we were optimizing a system while it operates in the field.
- An approximate lookahead policy where we apply these methods purely to make a decision x_t at time t.

17.3.1 Offline Learning

The algorithms presented in chapter 16 and section 17.2 are written in the context of running a simulator to approximate the expectation

$$F^\pi = \mathbb{E} \sum_{t=0}^{T} C(S_t, X^\pi(S_t)), \tag{17.18}$$

where, if we are simulating a sample path ω^n, we would write the results of a single simulation as

$$\hat{F}^\pi(\omega^n) = \sum_{t=0}^{T} C(S_t(\omega^n), X_t^\pi(S_t(\omega^n))),$$

where our transitions evolve according to

$$S_{t+1}(\omega^n) = S^M(S_t(\omega^n), X_t^\pi(S_t(\omega^n)), W_{t+1}(\omega^n))$$

for a sequence of exogenous inputs $(W_1(\omega^n), \ldots, W_T(\omega))$. We have been using this base model (where we use "base model" as it was introduced in chapter 9) with a policy

$$X_t^\pi(S_t) = \arg\max_x \left(C(S_t, x) + \overline{V}_t^{x,n-1}(S_t^x) \right), \tag{17.19}$$

where S_t^x is our post-decision state, and $\overline{V}_t^{x,n-1}(S_t^x)$ is our post-decision value function approximation learned after $n - 1$ updates. We may use TD(0), TD(1), or the general TD(λ) updates for using a sampled estimate $\hat{v}_t^n$ to udpate $\overline{V}^{x,n-1}(S_t)$ to obtain $\overline{V}_t^{x,n}(S_t)$ using any approximation architecture. The ultimate goal is to solve the problem

$$\max_\pi F^\pi$$

using specific classes of value function approximations (assume we are restricting ourselves to VFA-based policies).

This whole approach assumes we are doing offline learning in a simulator, where we assume we have access to the transition function $S_{t+1} = S^M(S_t, x_t, W_{t+1})$ and a way of sampling $(W_1, \ldots, W_T)$. We use this setting to do repeated training iterations, and it is particularly important when we use TD(λ)

for $\lambda > 0$ since this requires the backward communication of updates described in section 16.1.4 (see in particular equation (16.13)).

We remind the reader not to confuse offline learning with off policy learning. Offline learning means we are (typically) learning in a simulator where we do not care how well we are doing, while we are learning the value functions. We just care how well our final policy works after we have estimated our value functions.

17.3.2 From Offline to Online

Now imagine that we are trying to design our VFA-based policy without a simulator. Instead, we have an actual physical system we are trying to learn from and control. In this setting, we are no longer going to depend on knowing the transition function or observing the exogenous information W_t; instead we are simply going to make a decision x_t and then observe the next state S_{t+1} (classic model-free dynamic programming). Although not critical for this discussion, we can assume that decisions are being made with our VFA-based policy that is being updated as we go, but how are these updates happening?

First, it does not make sense to be learning a time-dependent policy $X_t^\pi(S_t)$ since once we pass time t, we are no longer interested in $X_t^\pi(S_t)$. So let's start by assuming that we are going to estimate a stationary policy $X^\pi(S_t)$ and a stationary value function approximation $\overline{V}^{x,n}(S_t)$. Remember that in our offline setting, n counted how many times we had simulated our process $W_1, \ldots, W_T$. We see that in our online setting, $n = t$ because we update our value function approximation (which we label with n) once per time period (indexed by t).

Next, we can certainly apply classical $TD(0)$ updates as we step forward, and this can work perfectly well for some problem classes. If this is the case, then we can step forward from state S_t, execute action $x_t = X^\pi(S_t)$ using $\overline{V}^{x,n-1}$. We then get our updated estimate of the value of being in state S_t given by $\hat{v}_t^n$, which we use to update our value function approximation to obtain $\overline{V}^{x,n-1}$.

While $TD(0)$ works very well in some problem classes, there are many problems where $TD(\lambda)$, possibly using $\lambda = 1$, can work much better. If you need any convincing, flip back to Table 16.1 and the discussion around those calculations to remind yourself how slow $TD(0)$ can be. So we have to ask, if we transition to online learning, have we lost this powerful algorithmic strategy?

Fortunately, the answer is no, but we have to do some extra work. As we progress forward in time, we need to retain at least some history of states $S_{t'}$, decisions $x_{t'}$, states $S_{t'}$ (or, for our illustration, post-decision states $S_{t'}^x$), and contributions $c_{t'} = C(S_{t'}, x_{t'})$ for $t' = t - 1, t - 2, \ldots, t - H$. For convenience we compile this sequence into a history that allows us to trace backward in time.

Now recall how we did our TD(λ) updates for our discounted infinite-horizon problem in equation (16.12), but now we are going to first adapt it to an undiscounted, finite-horizon setting using

$$\overline{V}^n(s) \;=\; \overline{V}^{n-1}(s) + \alpha_n \sum_{m=0}^{H} (\lambda)^m \delta^{n+m}, \tag{17.20}$$

where δ^n is our usual temporal-difference update

$$\delta^n \;=\; C(s^n, x^n) + \overline{V}^{n-1}(S^{M,x}(s^n, x^n)) - \overline{V}^{n-1}(s^n).$$

We are going to execute equation (17.20) adaptively, going backward in time. To make the logic as clear as possible, we are going to assume a lookup table value function, and we are going to start by indexing the value function by the time t' when we visit state $S_{t'}$ just so we can keep track of the incremental updating. For this reason, we begin by defining

$$\overline{V}^x_{t',t'}(S_{t'}) \;=\; \overline{V}^x_{t'}(s) = \text{the starting value of the estimate of } \overline{V}_{t'}(S_{t'})$$
$$\text{as of time } t',$$

$$\overline{V}^x_{t',t}(s) \;=\; \text{the partial update of } \overline{V}^x_{t'}(s) \text{ that has occurred by}$$
$$\text{time } t \geq t'.$$

Assume that $\overline{V}^x_{t'}(S_{t'})$ is the approximate value of being in state $S_{t'}$ when we visited it at time t'. By time $t > t'$, we would have a partially updated estimate $\overline{V}^x_{t',t}(S_{t'})$ of the value of being in state $S_{t'}$ given by

$$\overline{V}^x_{t',t}(S_{t'}) \;=\; \overline{V}^x_{t'}(S_{t'}) + \alpha_{t'} \sum_{\tau=t'}^{t} \lambda^{\tau-t'} \delta_\tau. \tag{17.21}$$

This means that our update by time $t+1$ would be

$$\overline{V}^x_{t',t+1}(S_{t'}) \;=\; \overline{V}^x_{t'}(S_{t'}) + \alpha_{t'} \sum_{\tau=t'}^{t+1} \lambda^{\tau-t'} \delta_\tau,$$

$$=\; \overline{V}^x_{t',t}(S_{t'}) + \lambda^{t+1-t'} \delta_{t+1}. \tag{17.22}$$

This means that as we step forward to time $t+1$, we have to run backward through history adding $\lambda^{t+1-t'} \delta_{t+1}$ to each $\overline{V}_{t'}(S_{t'})$ for $t' = t, t-1, t-2, ...,$ until $\lambda^{t+1-t'}$ is small enough that we can stop.

As a final step, we drop the time index because we are updating a stationary policy.

17.3.3 Evaluating Offline and Online Learning Policies

Almost completely overlooked in the research literature is the recognition that if you are learning online, you need to use a cumulative reward objective. Offline (which is how most algorithms are tested), you should be using a final reward objective, which means the class 4 objective in Table 9.3 in section 9.11, given by

$$\max_{\pi^{lrn}} \mathbb{E}\{C(S, X^{\pi^{imp}}(S|\theta^{imp}), \widehat{W})|S^0\} =$$

$$\mathbb{E}_{S^0} \mathbb{E}^{\pi^{lrn}}_{W^1,\dots,W^N|S^0} \mathbb{E}^{\pi^{imp}}_{S|S^0} \mathbb{E}_{\widehat{W}|S^0} C(S, X^{\pi^{imp}}(S|\theta^{imp}), \widehat{W}). \tag{17.23}$$

Note that we are evaluating the learning policy π^{lrn}, but this may be the same as (or closely related to) the implementation policy. If we are using a perturbed implementation policy (for example, adding in a noise term as is done in an excitation policy), then the $\max_{\pi^{lrn}}$ really means maximizing over the noise in the excitation policy.

In section 9.12 we show that you can simulate this (otherwise intimidating) expression. Let ω be a single sample path of the training observations $W_1(\omega), \dots, W_T(\omega)$, and let ψ be a single observation of the testing random variable $\widehat{W}(\psi)$. An estimate of the value of the learning policy π^{lrn} is then

$$F^\pi(\theta^{lrn}|\omega, \psi) = \frac{1}{T} \sum_{t=0}^{T} C(S_t(\psi), X^{\pi^{imp}}(S_t(\psi)|\theta^{imp}, \omega), \widehat{W}_{t+1}(\psi)). \tag{17.24}$$

We finally average over a set of K samples of ω, and L samples of ψ, giving us

$$\bar{F}^\pi(\theta^{lrn}) = \frac{1}{K}\frac{1}{L} \sum_{k=1}^{K} \sum_{\ell=1}^{L} F^\pi(\theta^{lrn}|\omega^k, \psi^\ell). \tag{17.25}$$

In plain English, this means training $\overline{V}_t(S_t)$, then fixing $\overline{V}_t(S_t)$, and running simulations to see how it performs. It is when we are simulating the policy (holding $\overline{V}_t(S_t)$ fixed) that we are approximating the expectation $\mathbb{E}^{\pi^{imp}}_{S|S^0}$ in equation (17.23).

17.3.4 Lookahead Policies

Another perspective of approximate dynamic programming is in the context of a lookahead policy. This is an idea that we are going to revisit in more depth in chapter 19 which focuses on lookahead policies, but for completeness we are going to hint at what we would do right now for comparison.

Imagine that we feel that to make a good decision now, we have to plan into the future using our best estimates, say, of forecasts of various activities. We may have a situation such as planning inventories for a complex supply chain where a stationary policy would not work. Also, as we discuss in chapter 19, lookahead policies have the feature of imbedding a lot of information in the form of latent variables, which is information that affects the modeling as we project into the future, but without adding to the complexity of the state variable as we project into the future.

This idea requires that we set up an approximate model that is then solved with approximate dynamic programming, using any of the algorithms discussed so far. We end up solving a problem at some time t with the same structure as the one behind (17.18), but it starts at time t. Also, because it is in a lookahead model, it can be simpler, so we use modified states, decisions, and exogenous information which we introduce in more detail in section 19.2:

$$X_t^\pi(S_t) = \arg\max_{x_t} \tilde{E} \left\{ \sum_{t'=t}^{t+H} C(\tilde{S}_{tt'}, \tilde{X}_{t'}^\pi(\tilde{S}_{tt'})) \right\}. \tag{17.26}$$

In other words, our policy will be to solve an approximate lookahead model, and use the decision $x_t = \tilde{X}_t^\pi(\tilde{S}_{tt})$ that looks best right now. We note that this has to be re-optimized (possibly from scratch, but not necessarily) each time period. Also, while this idea is computing value functions to obtain good policies, the primary interest is in the decision of what to do at time t.

17.4 Approximate Value Iteration Using Linear Models

Approximate value iteration, Q-learning, and temporal difference learning (with $\lambda = 0$) are clearly the simplest methods for updating an estimate of the value of being in a state. Linear models are the simplest methods for approximating a value function. Not surprisingly, then, there has been considerable interest in putting these two strategies together.

Figure 17.5 depicts a basic adaptation of linear models updated using recursive least squares in an approximate value iteration. However, not only are there no convergence proofs for this algorithm, there are examples that show that it may not converge, even for problems where the linear approximation has the potential for identifying the correct value function. This said, the method is popular because of its relative simplicity, and because it seems to work for many applications (recall that we used linear architectures for the benchmarking studies for backward approximate dynamic programming in section 15.4.1 with very good results).

The most important step whenever a linear model is used, regardless of the setting, is to choose the basis functions carefully so that the linear model has a chance of representing the true value function over the widest range of states.

The biggest strength of a linear model is also its biggest weakness. A large error can distort the update of θ^n which then impacts the accuracy of the entire approximation. Since the value function approximation determines the policy (see Step 1), a poor approximation leads to poor policies, which then distorts the observations $\hat{v}^n$. This can be a vicious circle from which the algorithm may never recover.

A second step is in the specific choice of recursive least squares updating. Figure 17.5 refers to the classic recursive least squares updating formulas in equations (3.41)–(3.45) in chapter 3. However, buried in these formulas is the implicit use of a stepsize rule of $1/n$. We show in chapter 6 that a stepsize $1/n$ is particularly bad for approximate value iteration (as well as Q-learning and TD(0) learning). While this stepsize can work well (indeed, it is optimal) for stationary data, it is very poorly suited for the backward learning that arises in approximate value iteration. Fortunately, the problem is easily fixed if we replace the updating equations for M^n and γ, which are given as

$$
M^n = M^{n-1} - \frac{1}{\gamma^n}(M^{n-1}\phi^n(\phi^n)^T M^{n-1}),
$$

$$
\gamma^n = 1 + (\phi^n)^T M^{n-1}\phi^n,
$$

Step 0. Initialization:

    Step 0a. Initialize $\overline{V}^0$.
    Step 0b. Initialize S^1.
    Step 0c. Set $n = 1$.

Step 1. Solve

$$
\hat{v}^n = \max_{x \in \mathcal{X}^n} \left(C(S^n, x) + \gamma \sum_f \theta_f^{n-1} \phi_f(S^{M,x}(S^n, x)) \right) \tag{17.27}
$$

    and let x^n be the value of x that solves (17.27).
Step 2. Update the value function recursively using equations (3.41)–(3.45) from chapter 3 to obtain θ^n.
Step 3. Choose a sample $W^{n+1} = W(\omega^{n+1})$ and determine the next state using some policy such as

$$
S^n = S^M(S^n, x^n, W^{n+1}).
$$

Step 4. Increment n. If $n \leq N$ go to Step 1.
Step 5. Return the value functions $\overline{V}^N$.

Figure 17.5 Approximate value iteration using a linear model.

in equations (3.44) and (3.45) with

$$M^n = \frac{1}{\lambda}\left(M^{n-1} - \frac{1}{\gamma^n}(M^{n-1}\phi^n(\phi^n)^T M^{n-1})\right),$$

$$\gamma^n = \lambda + (\phi^n)^T M^{n-1}\phi^n,$$

in equations (3.47) and (3.48). Here, λ discounts older errors. $\lambda = 1$ produces the original recursive formulas. When used with approximate value iteration, it is important to use $\lambda < 1$. In section 3.8.2, we argue that if you choose a stepsize rule for α_n such as $\alpha_n = \theta^{step}/(\theta^{step} + n - 1)$, you should set λ_n at iteration n using

$$\lambda_n = \alpha_{n-1}\left(\frac{1 - \alpha_n}{\alpha_n}\right).$$

Approximate value iteration using a linear architecture has to be used with care. Provable convergence results are rare, and there are examples of divergence. As with all policies (whether they use value function approximations or not), the performance of any particular policy is very problem dependent. It is particularly valuable to design some sort of benchmark. If you are using value functions, then your problem likely falls in a class that requires a policy that estimates the downstream impact of a decision made now. This means that some form of direct lookahead approximation (described in chapter 19) might be a natural benchmark.

17.5 On-policy vs. off-policy learning and the exploration–exploitation problem

One of the most difficult challenges in approximate dynamic programming is managing the exploration of the state space to ensure that we get a good approximation of $V_t(S_t)$ over the set of states S_t that we are likely to visit. We have to deal with the following problems:

- We do not know in advance the set of states that we are most likely to visit. At iteration n, we have an approximation $\overline{V}^n(S)$. If we stopped now, our policy would be given by

$$x_t^n = \arg\max_{x_t \in \mathcal{X}_t}\left(C(S_t^n, x_t) + \mathbb{E}\{\overline{V}_{t+1}^n(S_{t+1})|S_t^n, x_t\}\right). \tag{17.28}$$

This would then lead us to state $S_{t+1}^n = S^M(S_t^n, x_t^n, W_{t+1}^n)$. Moving to state S_{t+1}^n means we are using *trajectory following*, and it suggests that S_{t+1}^n is a reasonable state to visit. However, it depends on our current value function approximation $\overline{V}_{t+1}^n(S_{t+1})$ which might be quite poor.

- For a stochastic problem where W_{t+1} is chosen from a probability distribution, the sampled value $\hat{v}_t^n$ of being in a state, calculated using

$$\hat{v}_t^n = \max_{x_t \in \mathcal{X}_t} \left(C(S_t^n, x_t) + \mathbb{E}\{\overline{V}_{t+1}^n(S_{t+1})|S_t^n, x_t\} \right).$$

is a random variable (and this can be a very noisy random variable), which means that our value function approximations $\overline{V}_t^n(S_t)$ are themselves random variables. If $\overline{V}_t^n(S_t)$ overestimates the value of being in a state, our system will be attracted to that state and visit it more often than it should. Similarly, if we have underestimated $\overline{V}_t^n(S_t)$, the system will avoid decisions that take us to S_t, limited our ability to fix the error.
- The estimate $\hat{v}_t^n$ depends on $\overline{V}_{t+1}^n(S_{t+1})$ which means that $\hat{v}_t^n$ is biased.
- While the noise in $\hat{v}_t^n$ due to W_{t+1} can create errors in our estimate of $\overline{V}_t^n(S_t)$, we may also introduce structural errors if we use any form of parametric or locally parametric belief model.

We start our discussion with some terminology. We then transition to discussing the issues associated with lookup table representations, and then to the use of generalized learning methods.

17.5.1 Terminology

We begin our discussion by establishing a few terms:

The implementation policy $X^{\pi^{imp}}(S_t)$ – If $\overline{V}^n(S_t)$ is our value function approximation after n training iterations for time t, then the implementation policy is the policy we obtain from using these value function approximations, which means

$$X^{\pi^{imp},n}(S_t) = \arg\max_{x_t \in \mathcal{X}_t} \left(C(S_t, x_t) + \mathbb{E}\{\overline{V}_{t+1}^n(S_{t+1})|S_t, x_t\} \right). \tag{17.29}$$

The implementation policy would be $X^{VFA,N}(S_t)$ after we have exhausted our training iterations. The implementation policy is referred to as the *target policy* in computer science.

The learning policy $X^{\pi^{lrn}}(S_t)$ – This is the policy we use while we are learning the value function approximations. We may choose to use our implementation policy, which is to say we are using equation (17.28) to determine the decision x_t^n we make now to determine the state $S_{t+1}^n = S^M(S_t^n, x_t^n, W_{t+1}^n)$ we visit next (during iteration n). The learning policy is known as the *behavior policy* in computer science. Other learning policies might include:

- Random – Choose x_t^n randomly from the set (or region) $\mathcal{X}_t$.

- Epsilon-greedy – Choose x_t^n at random from $\mathcal{X}_t$ with probability ϵ, and use the implementation policy $x_t^n = X^{VFA,n}(S_t)$ with probability $1 - \epsilon$.
- Interval estimation – Choose x_t^n from

$$X^{IE}(S_t|\theta^{IE}) = \arg\max_{x_t \in \mathcal{X}_t}\left((C(S_t, x_t) + \mathbb{E}\{\overline{V}_{t+1}^n(S_{t+1})|S_t, x_t\}) + \theta^{IE}\bar{\sigma}_t^n(S_t) \right)$$

where $\bar{\sigma}_t^n(S_t)$ is the standard deviation of our estimate $\overline{V}_{t+1}^n(S_{t+1})$.

- Perturbed implementation policy (for continuous decisions):

$$X^{\pi^{lrn}}(S_t) = X^{\pi^{imp}}(S_t) + \varepsilon_{t+1}, \tag{17.30}$$

where $\varepsilon_{t+1} \sim N(0, \sigma_\varepsilon^2)$.

We could tap any of the learning policies from chapter 7, but there is a strong bias toward policies that are simple and easy to compute.

As a general rule, we only use the learning policy to determine a state to visit. If we use our learning policy to choose x_t^n, we would not use $\hat{v}_t^n = C(S_t^n, x_t^n) + \mathbb{E}\{\overline{V}_{t+1}^n(S_{t+1})|S_t^n, x_t^n\}$ to update the estimate of our value function.

On policy learning – This is when we use our implementation policy $X^{\pi^{imp}}(S_t)$ to guide the choice of decision from which we do our learning.

Off policy learning – This is when we use our learning policy $X^{\pi^{lrn}}(S_t)$ to guide the choice of the next state.

Policies like the perturbed implementation policy in (17.30) are attractive (where applicable) because they are well suited to serve as an implementation policy that pays a small price to continue learning.

17.5.2 Learning with Lookup Tables

A considerable amount of work in approximate dynamic programming started in computer science and operations research using lookup table representations of value functions. Lookup tables offer the attraction that in the limit, they can provide a perfect fit. The downside is that straightforward implementations mean that visit state s teaches us nothing about state s'. Most of the literature on the exploration-exploitation problem is focused on lookup table representations.

Consider the two-state dynamic program illustrated in Figure 17.6. Assume we start in state 1, and further assume that we initialize the value of being in each of the two states to $\overline{V}^0(1) = \overline{V}^0(2) = 0$. We see a negative contribution of -$5 to move from state 1 to 2, but a contribution of $0 to stay in state 1. We do not see the contribution of $20 to move from state 2 back to state 1, to it appears to be best to stay in state 1. This is where we need a learning policy to perform forced exploration.

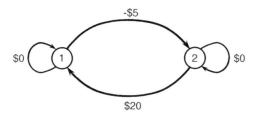

Figure 17.6 Two-state dynamic program, with transition contributions.

A more realistic version of this issue can be illustrated with our nomadic trucker problem that we introduced in section 2.3.4.1. Assume we dispatcher our truck driver using the tractory-following implementation policy. We would obtain the results shown in Figure 17.7a, where the circles are proportional to the value function approximations. We see from the figure that the trucker ended up visiting just seven cities after 500 dispatches.

An alternative strategy is to start with optimistic estimates of the value of being in each city to encourage exploration, while still using just the implementation policy. This produces the value functions depicted in Figure 17.7b, which shows that the trucker is visiting far more cities. Note that this is not an ideal solution, as it is effectively suggesting that the trucker should visit any city he has not yet visited. Of course, we can tune our optimistic estimate (presumably in a simulator) so that we pick a "high enough" initial estimate.

17.5.3 Learning with Generalized Belief Models

Exploration policies depend heavily on how we are approximating the value function. With lookup tables, visiting a state s teaches us nothing about other states, which makes exploration exceptionally important. The argument that we would make is that the vast majority of real problems have exceptionally large (frequently infinite) state spaces, which limits the value of lookup table representations. Just skim the state variables in the simple inventory problems we reviewed in section 9.9 (which grew to 42 dimensions) to remind yourself how quickly state spaces can grow even on simple problems.

Chapter 3 offers a variety of strategies for some form of generalized learning, where visiting a state s teaches us about the value of many other states. Examples include:

Lookup tables with correlated beliefs – We can often express a relationship between pairs of states through a covariance matrix Σ. Using the tools of section 3.4.2, we can visit one state and then update many other states through the relationship captured in Σ. It may be the only property we have is smoothness, but this can still be a powerful property.

(a) Low initial estimate of the value function.

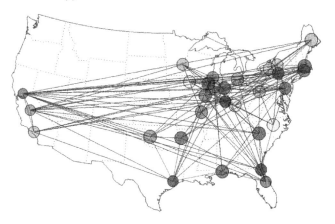

(b) High initial estimate of the value function.

Figure 17.7 The effect of value function initialization on search process. Case (a) uses a low initial estimate and produces limited exploration; case (b) uses a high initial estimate, which forces exploration of the entire state space.

Monotonicity – If the value of $V(s)$ increases (or decreases) in each dimension (or a subset of dimensions), we can use this property to update many states from a single observation.

Linear models – The simplest belief model, and as a result the one that has attracted the most attention, is a linear model (remember this means linear in the parameters) with the general form

$$\overline{V}_t(S_t|\theta_t) = \sum_{f \in \mathcal{F}} \theta_{tf} \phi_f(S_t).$$

Note that we have used our standard time-indexed form (this is not standard in communities such as computer science), but it is relatively easy to estimate time-dependent VFAs.

Linear models are widely used, but this does not mean they work well, and we revisit this in our presentation below. Linear models are used because they are simple and provide an answer, but often users have no idea how good the answer is. It is unlikely, for example, that a single linear model would accurately represent a value function over the entire range of states that we actually visit.

Nonlinear models – Nonlinear parametric models offer the same generalized learning that linear models do, although it introduces other issues that we discuss later in the chapter.

Convex approximations – In chapter 18, we are going to see that convexity is a powerful property that allows us to estimate accurate value function approximations without the need for any explicit exploration logic.

Locally linear approximations – Here we are creating linear models for a set of regions that we can represent as a series of state spaces $(\mathcal{S}_1, \dots, \mathcal{S}_I)$. Note that once we introduce the idea of local approximations, we re-introduce the need to visit states within the different regions $\mathcal{S}_i$. Presumably the number of regions will be dramatically smaller than the number of states, so this is an improvement.

Neural networks – Neural networks are such highly flexible architectures that we almost return to the same situation we were with lookup tables, but without the intuition of learning locally.

Linear models are easily the most popular form of value function approximation, and as with all parametric models, offer the power of generalized learning where a model with K parameters (where K is typically on the order of 10 to 100 parameters, but may be in the 1000s). This means that with relatively few training iterations, we will at least get an estimate of the value of being in *every* state.

We still have the same issues with bias (in the values of $\hat{v}_t^n$) and noise, but we also have to deal with structural error, since we cannot expect a linear model to be globally accurate. Adding more features (which increases K) is not a cure-all, since it can introduce unwanted variability. For example, we may be fitting a smooth, concave function (envision a nice grassy hilltop) where a quadratic will do a good, if not perfect, job of capturing the general shape. Higher-order functions could introduce unwanted undulations.

Exploration with a parametric model has a completely different behavior than using a lookup table (potentially even nonparametric models). The classical intuition about exploration-exploitation is entirely different with parametric models. A good example is the problem of learning a linear demand curve that we showed in Figure 12.2 in section 12.6.2. While this is not a value function, it nicely illustrates how learning a linear function requires observations that are removed from the center of mass of other observations. Learning a linear function (in any setting) is best done by observing extreme points. The problem (which arose in Figure 12.2) is that visiting extreme points may be expensive if you are learning in an offline setting.

When using parametric value function approximations, it has been our experience that the most popular strategy used in practice is to use a learning policy that consists of a perturbed implementation policy of the sort introduced in section 12.6.2. These are most naturally implemented using continuous decisions and states, but it is possible to use a soft-max (Boltzmann) policy to choose among categorical alternatives.

17.6 Applications

There are many problems where we can exploit structure in the state variable, allowing us to propose functions characterized by a small number of parameters which have to be estimated statistically. Section 3.6.3 represented one version where we had a parameter for each (possibly aggregated) state. The only structure we assumed was implicit in the ability to specify a series of one or more aggregation functions.

The remainder of this section illustrates the use of regression models in specific applications which include pricing an American option and playing lose tic-tac-toe, followed by a brief discussion of deterministic problems that arise in engineering control problems and games.

17.6.1 Pricing an American Option

Consider the problem of determining the value of an American-style put option which gives us the right to sell an asset (or contract) at a specified price at any of a set of discrete time periods. For example, we might be able to exercise the option on the last day of the month over the next 12 months.

Assume we have an option that allows us to sell an asset at $1.20 at any of four time periods. We assume a discount factor of 0.95 to capture the time value of money. If we wait until time period 4, we must exercise the option, receiving zero if the price is over $1.20. At intermediate periods, however, we may choose to hold the option even if the price is below $1.20 (of course, exercising it if the price is above $1.20 does not make sense). Our problem is to determine whether to hold or exercise the option at the intermediate points.

Table 17.3 Ten sample realizations of prices over four time periods.

	Stock prices			
	Time period			
Outcome	1	2	3	4
1	1.21	1.08	1.17	1.15
2	1.09	1.12	1.17	1.13
3	1.15	1.08	1.22	1.35
4	1.17	1.12	1.18	1.15
5	1.08	1.15	1.10	1.27
6	1.12	1.22	1.23	1.17
7	1.16	1.14	1.13	1.19
8	1.22	1.18	1.21	1.28
9	1.08	1.11	1.09	1.10
10	1.15	1.14	1.18	1.22

From history, we have found 10 samples of price trajectories which are shown in Table 17.3.

If we wait until time period 4, our payoff is shown in Table 17.4, which is zero if the price is above 1.20, and $1.20 - p_4$ for prices below \$1.20.

At time $t = 3$, we have access to the price history (p_1, p_2, p_3). Since we may not be able to assume that the prices are independent or even Markovian (where p_3 depends only on p_2), the entire price history represents our state variable, along with an indicator that tells us if we are still holding the asset. We wish to predict the value of holding the option at time $t = 4$. Let $V_4(S_4)$ be the value of the option if we are holding it at time 4, given the state (which includes the price p_4) at time 4. Now let the conditional expectation at time 3 be

$$\overline{V}_3(S_3) = \mathbb{E}\{V_4(S_4)|S_3\}.$$

Our goal is to approximate $\overline{V}_3(S_3)$ using information we know at time 3. We propose a linear regression of the form

$$Y = \theta_0 + \theta_1 X_1 + \theta_2 X_2 + \theta_3 X_3,$$

where

$$Y = V_4,$$
$$X_1 = p_2,$$

Table 17.4 The payout at time 4 if we are still holding the option.

| | Option value at $t = 4$ | | | |
| | | Time period | | |
Outcome	1	2	3	4
1	-	-	-	0.05
2	-	-	-	0.07
3	-	-	-	0.00
4	-	-	-	0.05
5	-	-	-	0.00
6	-	-	-	0.03
7	-	-	-	0.01
8	-	-	-	0.00
9	-	-	-	0.10
10	-	-	-	0.00

$$X_2 = p_3,$$
$$X_3 = (p_3)^2.$$

The variables X_1, X_2, and X_3 are our basis functions. Keep in mind that it is important that our explanatory variables X_i must be a function of information we have at time $t = 3$, whereas we are trying to predict what will happen at time $t = 4$ (the payoff). We would then set up the data matrix given in Table 17.5.

We may now run a regression on this data to determine the parameters $(\theta_i)_{i=0}^3$. It makes sense to consider only the paths which produce a positive value in the fourth time period, since these represent the sample paths where we are most likely to still be holding the asset at the end. The linear regression is only an approximation, and it is best to fit the approximation in the region of prices which are the most interesting (we could use the same reasoning to include some "near misses"). We only use the value function to estimate the value of holding the asset, so it is this part of the function we wish to estimate. For our illustration, however, we use all 10 observations, which produces the equation

$$\overline{V}_3 \approx 0.0056 - 0.1234p_2 + 0.6011p_3 - 0.3903(p_3)^2.$$

$\overline{V}_3$ is an approximation of the expected value of the price we would receive if we hold the option until time period 4. We can now use this approximation to help us decide what to do at time $t = 3$. Table 17.6 compares the value of

Table 17.5 The data table for our regression at time 3.

	Regression data			
	Independent variables			Dependent variable
Outcome	X_1	X_2	X_3	Y
1	1.08	1.17	1.3689	0.05
2	1.12	1.17	1.3689	0.07
3	1.08	1.22	1.4884	0.00
4	1.12	1.18	1.3924	0.05
5	1.15	1.10	1.2100	0.00
6	1.22	1.23	1.5129	0.03
7	1.44	1.13	1.2769	0.01
8	1.18	1.21	1.4641	0.00
9	1.11	1.09	1.1881	0.10
10	1.14	1.18	1.3924	0.00

exercising the option at time 3 against holding the option until time 4, computed as $\gamma \overline{V}_3(S_3)$. Taking the larger of the two payouts, we find, for example, that we would hold the option given samples 1-4, 6, 8, and 10, but would sell given samples 5, 7, and 9.

We can repeat the exercise to estimate $\overline{V}_2(S_t)$. This time, our dependent variable "Y" can be calculated two different ways. The simplest is to take the larger of the two columns from Table 17.6 (marked in bold). So, for sample path 1, we would have $Y_1 = \max\{.03, 0.03947\} = 0.03947$. This means that our observed value is actually based on our approximate value function $\overline{V}_3(S_3)$.

An alternative way of computing the observed value of holding the option in time 3 is to use the approximate value function to determine the decision, but then use the actual price we receive when we eventually exercise the option. Using this method, we receive 0.05 for the first sample path because we decide to hold the asset at time 3 (based on our approximate value function) after which the price of the option turns out to be worth 0.05. Discounted, this is worth 0.0475. For sample path 2, the option proves to be worth 0.07 which discounts back to 0.0665 (we decided to hold at time 3, and the option was worth 0.07 at time 4). For sample path 5 the option is worth 0.10 because we decided to exercise at time 3.

Regardless of which way we compute the value of the problem at time 3, the remainder of the procedure is the same. We have to construct the independent variables "Y" and regress them against our observations of the value of the

Table 17.6 The payout if we exercise at time 3, and the expected value of holding based on our approximation. The best decision is indicated in bold.

	Rewards	
		Decision
Outcome	Exercise	Hold
1	0.03	$0.04155 \times .95 = \mathbf{0.03947}$
2	0.03	$0.03662 \times .95 = \mathbf{0.03479}$
3	0.00	$0.02397 \times .95 = \mathbf{0.02372}$
4	0.02	$0.03346 \times .95 = \mathbf{0.03178}$
5	**0.10**	$0.05285 \times .95 = 0.05021$
6	0.00	$0.00414 \times .95 = \mathbf{0.00394}$
7	**0.07**	$0.00899 \times .95 = 0.00854$
8	0.00	$0.01610 \times .95 = \mathbf{0.01530}$
9	**0.11**	$0.06032 \times .95 = 0.05731$
10	0.02	$0.03099 \times .95 = \mathbf{0.02944}$

option at time 3 using the price history (p_1, p_2). Our only change in methodology would occur at time 1 where we would have to use a different model (because we do not have a price at time 0).

17.6.2 Playing "Lose Tic-Tac-Toe"

The game of "lose tic-tac-toe" is the same as the familiar game of tic-tac-toe, with the exception that now you are trying to make the other person get three in a row. This nice twist on the popular children's game provides the setting for our next use of regression methods in approximate dynamic programming.

Unlike our exercise in pricing options, representing a tic-tac-toe board requires capturing a discrete state. Assume the cells in the board are numbered left to right, top to bottom as shown in Figure 17.8a. Now consider the board in Figure 17.8b. We can represent the state of the board after the tth play using

$$S_{ti} = \begin{cases} 1 & \text{if cell } i \text{ contains an "X,"} \\ 0 & \text{if cell } i \text{ is blank,} \\ -1 & \text{if cell } i \text{ contains an "O,"} \end{cases}$$

Figure 17.8 Some tic-tac-toe boards. (17.8a) Our indexing scheme. (17.8b) Sample board.

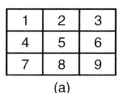

(a)   (b)

$$S_t = (S_{ti})_{i=1}^{9}.$$

We see that this simple problem has up to $3^9 = 19,683$ states. While many of these states will never be visited, the number of possibilities is still quite large, and seems to overstate the complexity of the game.

We quickly realize that what is important about a game board is not the status of every cell as we have represented it. For example, rotating the board does not change a thing, but it does represent a different state. Also, we tend to focus on strategies (early in the game when it is more interesting) such as winning the center of the board or a corner. We might start defining variables (basis functions) such as

$$\phi_1(S_t) = \text{1 if there is an "X" in the center of the board, 0 otherwise,}$$

$$\phi_2(S_t) = \text{the number of corner cells with an "X,"}$$

$$\phi_3(S_t) = \text{the number of instances of adjacent cells with an "X"}$$
$$\text{(horizontally, vertically, or diagonally).}$$

There are, of course, numerous such functions we can devise, but it is unlikely that we could come up with more than a few dozen (if that) which appeared to be useful. It is important to realize that we do not need a value function to tell us to make obvious moves.

Once we form our basis functions, our value function approximation is given by

$$\overline{V}_t(S_t) = \sum_{f \in \mathcal{F}} \theta_{tf} \phi_f(S_t).$$

We note that we have indexed the parameters by time (the number of plays) since this might play a role in determining the value of the feature being measured by a basis function, but it is reasonable to try fitting a model where $\theta_{tf} = \theta_f$. We estimate the parameters θ by playing the game (and following some policy) after which we see if we won or lost. We let $Y^n = 1$ if we won the nth game, 0 otherwise. This also means that the value function is trying to approximate the probability of winning if we are in a particular state.

We may play the game by using our value functions to help determine a policy. Another strategy, however, is simply to allow two people (ideally, experts) to play the game and use this to collect observations of states and game outcomes. This is an example of . If we lack a "supervisor" then we have to depend on simple strategies combined with the use of slowly learned value function approximations. In this case, we also have to recognize that in the early iterations, we are not going to have enough information to reliably estimate the coefficients for a large number of basis functions.

17.6.3 Approximate Dynamic Programming for Deterministic Problems

There has been considerable interest in applying ADP to two classes of deterministic problems:

- Engineering control problems – Imagine making decisions about how to control a drone or robot, where we have to apply a multidimensional force vector u_t to the device (using the notation of control theory) to minimize some performance metric.
- Playing games – Reinforcement learning/approximate dynamic programming has attracted attention for games such as computer Go, chess and an array of video games.

Neural networks have proven to be very popular in both settings, with reports of considerable success (although the techniques for computer games tend to require a hybrid policy). As we pointed out when we first introduced neural networks in section 3.9.3, the high-dimensionality of neural networks tends to make them sensitive to noise. However, for deterministic problems this is not an issue, and the ability of neural networks to represent complex functions without the struggle of identifying reasonable architectures can be particularly powerful.

It is beyond the scope of this volume to describe developments in these two rich fields in any depth. We encourage readers interested in either of these problem classes to look for more specialized presentations.

17.7 Approximate Policy Iteration

One of the most important tools in the toolbox for approximate dynamic programming is approximate policy iteration. This algorithm is neither simpler nor more elegant than approximate value iteration, but it can offer stronger convergence guarantees if the policy is evaluated within a specified tolerance.

In this section we review several flavors of approximate policy iteration, including:

(a) Finite horizon problems using lookup tables.
(b) Finite horizon problems using linear models.
(c) Infinite horizon problems using linear models.

Finite horizon problems allow us to obtain Monte Carlo estimates of the value of a policy by simulating the policy until the end of the horizon. Note that a "policy" here always refers to decisions that are determined by value function approximations. We use the finite horizon setting to illustrate approximating value function approximations using lookup tables and basis functions, which allows us to highlight the strengths and weaknesses of the transition to basis functions.

We then present an algorithm based on least squares temporal differences (LSTD) and contrast the steps required for finite horizon and infinite horizon problems when using linear models.

17.7.1 Finite Horizon Problems Using Lookup Tables

A fairly general purpose version of an approximate policy iteration algorithm is given in Figure 17.9 for an infinite horizon problem. This algorithm helps to illustrate the choices that can be made when designing a policy iteration algorithm in an approximate setting.

The algorithm features three nested loops. The innermost loop steps forward and backward in time from an initial state $S^{n,0}$. The purpose of this loop is to obtain an estimate of the value of a path. Normally, we would choose T large enough so that γ^T is quite small (thereby approximating an infinite path).

The next outer loop repeats this process M times to obtain a statistically reliable estimate of the value of a policy (determined by $\overline{V}^{\pi,n}$). The third loop, representing the outer loop, performs policy updates (in the form of updating the value function). In a more practical implementation, we might choose states at random rather than looping over all states.

Readers should note that we have tried to index variables in a way that shows how they are changing (do they change with outer iteration n? inner iteration m? the forward look-ahead counter t?). This does not mean that it is necessary to store, for example, each state or decision for every n, m, and t. In an actual implementation, the software should be designed to store only what is necessary.

We can create different variations of approximate policy iteration by our choice of parameters. First, if we let $T \to \infty$, we are evaluating a true infinite horizon policy. If we simultaneously let $M \to \infty$, then $\bar{v}^n$ approaches the exact, infinite horizon value of the policy π determined by $\overline{V}^{\pi,n}$. Thus, for $M = T = \infty$, we have a Monte Carlo-based version of exact policy iteration.

Step 0. Initialization:

Step 0a. Initialize $\overline{V}^{\pi,0}$.
Step 0b. Set a look-ahead parameter T and inner iteration counter M.
Step 0c. Set $n = 1$.

Step 1. Sample a state S_0^n and then do:
Step 2. Do for $m = 1, 2, \dots, M$:

Step 3. Choose a sample path ω^m (a sample realization over the lookahead horizon T).
Step 4. Do for $t = 0, 1, \dots, T$:

Step 4a. Compute

$$x_t^{n,m} \quad = \quad \arg \max_{x_t \in \mathcal{X}_t^{n,m}} \left(C(S_t^{n,m}, x_t) + \gamma \overline{V}^{\pi,n-1}(S^{M,x}(S_t^{n,m}, x_t)) \right).$$

Step 4b. Compute

$$S_{t+1}^{n,m} \quad = \quad S^M(S_t^{n,m}, x_t^{n,m}, W_{t+1}(\omega^m)).$$

Step 5. Initialize $\hat{v}_{T+1}^{n,m} = 0$.
Step 6. Do for $t = T, T - 1, \dots, 0$:

Step 6a. Accumulate $\hat{v}^{n,m}$:

$$\hat{v}_t^{n,m} \quad = \quad C(S_t^{n,m}, x_t^{n,m}) + \gamma \hat{v}_{t+1}^{n,m}.$$

Step 6b. Update the approximate value of the policy:

$$\bar{v}^{n,m} = \left(\frac{m-1}{m}\right)\bar{v}^{n,m-1} + \frac{1}{m}\hat{v}_0^{n,m}.$$

Step 7. Update the value function at S^n:

$$\overline{V}^{\pi,n} = (1 - \alpha_{n-1})\bar{v}^{n-1} + \alpha_{n-1}\hat{v}_0^{n,M}.$$

Step 8. Set $n = n + 1$. If $n < N$, go to Step 1.
Step 9. Return the value functions $(\overline{V}^{\pi,N})$.

Figure 17.9 A policy iteration algorithm for infinite horizon problems.

We can choose a finite value of T that produces values $\hat{v}^{n,m}$ that are close to the infinite horizon results. We can also choose finite values of M, including $M = 1$. When we use finite values of M, this means that we are updating the policy before we have fully evaluated the policy. This variant is known in the literature as *optimistic policy iteration* because rather than wait until we have a true estimate of the value of the policy, we update the policy after each sample (presumably, although not necessarily, producing a better policy). We may also think of this as a form of partial policy evaluation, not unlike the hybrid value/policy iteration described in section 14.8.

17.7.2 Finite Horizon Problems Using Linear Models

The simplest demonstration of approximate policy iteration using linear models is in the setting of a finite horizon problem. Figure 17.10 provides an adaption of the algorithm using lookup tables when we are using linear models. There is an outer loop over n where we fix the policy using

$$X_t^\pi(S_t) = \arg\max_{x_t}\left(C(S_t, x_t) + \gamma \sum_f \theta_{tf}^{\pi,n}\phi_f(S_t, x_t)\right). \tag{17.31}$$

We are assuming that the basis functions are not themselves time-dependent, although they depend on the state variable S_t (and decision x) which, of course, is time dependent. The policy is determined by the parameters $\theta_{tf}^{\pi,n}$.

We update the policy $X_t^\pi(s)$ by performing repeated simulations of the policy in an inner loop that runs $m = 1, \dots, M$. Within this inner loop, we use recursive least squares to update a parameter vector $\theta_{tf}^{n,m}$. This step replaces Step 6b in Figure 17.9.

If we let $M \to \infty$, then the parameter vector $\theta_t^{n,M}$ approaches the best possible fit for the policy $X_t^\pi(s)$ determined by $\theta^{\pi,n-1}$. However, it is very important to realize that this is not equivalent to performing a perfect evaluation of a policy using a lookup table representation. The problem is that (for discrete states), lookup tables have the *potential* for perfectly approximating a policy, whereas this is not generally true when we use basis functions. If we have a poor choice of basis functions, we may be able to find the best possible value of $\theta^{n,m}$ as m goes to infinity, but we may still have a terrible approximation of the policy produced by $\theta^{\pi,n-1}$.

17.7.3 LSTD for Infinite Horizon Problems Using Linear Models

We have built the foundation for approximate policy iteration using lookup tables and basis functions for finite horizon problems. We now make the transition to infinite horizon problems using linear models, where we introduce the dimension of projecting contributions over an infinite horizon. There are several ways of accomplishing this (see section 16.1.2). We use least squares temporal differencing, since it represents the most natural extension of classical policy iteration for infinite horizon problems.

Step 0. Initialization:

Step 0a. Fix the basis functions $\phi_f(s)$.
Step 0b. Initialize $\theta_{tf}^{\pi,0}$ for all t. This determines the policy we simulate in the inner loop.
Step 0c. Set $n = 1$.

Step 1. Sample an initial starting state S_0^n:

Step 2. Initialize $\theta^{n,0}$ (if $n > 1$, use $\theta^{n,0} = \theta^{n-1}$), which is used to estimate the value of policy π produced by $\theta^{pi,n}$. $\theta^{n,0}$ is used to approximate the value of following policy π determined by $\theta^{\pi,n}$.
Step 3. Do for $m = 1, 2, \dots, M$:

Step 4. Choose a sample path ω^m.
Step 5. Do for $t = 0, 1, \dots, T$:

Step 5a. Compute

$$x_t^{n,m} \;=\; \arg\max_{x_t \in \mathfrak{X}_t^{n,m}} \left(C(S_t^{n,m}, x_t) + \gamma \sum_f \theta_{tf}^{\pi,n-1} \phi_f(S^{M,x}(S_t^{n,m}, x_t)) \right)$$

Step 5b. Compute

$$S_{t+1}^{n,m} \;=\; S^M(S_t^{n,m}, x_t^{n,m}, W_{t+1}(\omega^m)).$$

Step 6. Initialize $\hat{v}_{T+1}^{n,m} = 0$.
Step 7. Do for $t = T, T-1, \dots, 0$:

$$\hat{v}_t^{n,m} \;=\; C(S_t^{n,m}, x_t^{n,m}) + \gamma \hat{v}_{t+1}^{n,m}.$$

Step 8. Update $\theta_t^{n,m-1}$ using recursive least squares to obtain $\theta_t^{n,m}$ (see section 3.8).

Step 9. Set $n = n + 1$. If $n < N$, go to Step 1.
Step 10. Return the value functions $(\overline{V}^{\pi,N})$.

Figure 17.10 A policy iteration algorithm for finite horizon problems using linear models.

To begin, we let a sample realization of a one-period contribution, given state S^m and decision x^m be given by

$$\hat{C}^m = C(S^m, x^m).$$

As in the past, we let $\phi^m = \phi(S^m)$ be the column vector of basis functions evaluated at state S^m. We next fix a policy which chooses decisions greedily based on a value function approximation given by $\overline{V}^n(s) = \sum_f \theta_f^n \phi_f(s)$ (see equation

(17.31)). Imagine that we have simulated this policy over a set of iterations $i = (0, 1, \ldots, m)$, giving us a sequence of contributions $\hat{C}^i$, $i = 1, \ldots, m$. Drawing on the foundation provided in section 16.3, we can use standard linear regression to estimate θ^m using

$$\theta^m = \left[\frac{1}{1+m} \sum_{i=0}^{m} \phi_i(\phi^i - \gamma\phi^{i+1})^T \right]^{-1} \left[\frac{1}{1+m} \sum_{i=1}^{m} \phi^i \hat{C}^i \phi^i \right]. \tag{17.32}$$

As a reminder, the term $\phi^i - \gamma\phi^{i+1}$ can be viewed as a simulated, sample realization of $I - \gamma P^\pi$, projected into the feature space. Just as we would use $(I - \gamma P^\pi)^{-1}$ in our basic policy iteration to project the infinite-horizon value of a policy π (for a review, see section 14.7), we are using the term

$$\left[\frac{1}{1+m} \sum_{i=0}^{m} \phi_i(\phi^i - \gamma\phi^{i+1})^T \right]^{-1}$$

to produce an infinite-horizon estimate of the feature-projected contribution

$$\left[\frac{1}{1+m} \sum_{i=1}^{m} \phi^i \hat{C}^i \phi^i \right].$$

Equation (17.32) requires solving a matrix inverse for every observation. It is much more efficient to use recursive least squares, which is done by using

$$\epsilon^m = \hat{C}^m - (\phi^m - \gamma\phi^{m+1})^T \theta^{m-1}, \tag{17.33}$$

$$M^m = M^{m-1} - \frac{M^{m-1}\phi^m(\phi^m - \gamma\phi^{m+1})^T M^{m-1}}{1 + (\phi^m - \gamma\phi^{m+1})^T M^{m-1}\phi^m}, \tag{17.34}$$

$$\theta^m = \theta^{m-1} + \frac{\epsilon^m M^{m-1}\phi^m}{1 + (\phi^m - \gamma\phi^{m+1})^T M^{m-1}\phi^m}. \tag{17.35}$$

Figure 17.11 provides a detailed summary of the complete algorithm. The algorithm has some nice properties if we are willing to assume that there is a vector θ^* such that the true value function $V(s) = \sum_{f \in \mathcal{F}} \theta_f^* \phi_f(s)$ (admittedly, a pretty strong assumption). First, if the inner iteration limit M increases as a function of n so that the quality of the approximation of the policy gets better and better, then the overall algorithm will converge to the true optimal policy. Of course, this means letting $M \to \infty$, but from a practical perspective, it means that the algorithm can find a policy arbitrarily close to the optimal policy.

Second, the algorithm can be used with vector-valued and continuous decisions. There are several features of the algorithm that allow this. First, computing the policy $X^\pi(s|\theta^n)$ requires solving a deterministic optimization problem. If we are using discrete decisions, it means simply enumerating the decisions and choosing the best one. If we have continuous decisions, we need to solve

Step 0. Initialization:

> **Step 0a.** Initialize θ^0.
> **Step 0b.** Set the initial policy:

$$A^\pi(s|\theta^0) = \arg\max_{a \in \mathcal{A}} \left(C(s, x) + \gamma \phi(S^M(s, x))^T \theta^0 \right).$$

> **Step 0c.** Set $n = 1$.

Step 1. Do for $n = 1, \dots, N$.

> **Step 2.** Initialize S_0^n.
> **Step 3.** Do for $m = 0, 1, \dots, M$:

> > **Step 4.** Initialize $\theta^{n,m}$.
> > **Step 5.** Sample W^{m+1}.
> > **Step 6.** Do the following:

> > > **Step 6a.** Computing the decision $x^{n,m} = X^\pi(S^m|\theta^{n-1})$.
> > > **Step 6b.** Compute the post-decision state $S^{x,m} = S^{M,x}(S^{n,m}, x^{n,m})$.
> > > **Step 6c.** Compute the next pre-decision state $S^{n,m+1} = S^M(S^{n,m}, x^{n,m}, W^{m+1})$.
> > > **Step 6d.** Compute the input variable $\phi(S^{n,m}) - \gamma\phi(S^{n,m+1})$ for equation (17.32).

> > **Step 7.** Do the following:

> > > **Step 7a.** Compute the response variable $\hat{C}^m = C(S^{n,m}, x^{n,m}, W^{m+1})$.
> > > **Step 7b.** Compute $\theta^{n,m}$ using equation (17.32).

> **Step 8.** Update θ^n and the policy:

$$\theta^{n+1} = \theta^{n,m}$$
$$X^{\pi,n+1}(s) = \arg\max_{x \in \mathcal{X}} \left(C(s, x) + \gamma\phi(S^M(s, x))\theta^{n+1} \right).$$

Step 9. Return the $X^\pi(s|\theta^N)$ and parameter θ^N.

Figure 17.11 Approximate policy iteration for infinite horizon problems using least squares temporal differencing.

a nonlinear programming problem. The only practical issue is that we may not be able to guarantee that the objective function is concave (or convex if we are minimizing). Second, note that we are using trajectory following (also known as on-policy training) in Step 6c, without an explicit exploration step. It can be very difficult implementing an exploration step for multidimensional decision vectors.

We can avoid exploration as long as there is enough variation in the states we visit that allows us to compute θ^m in equation (17.32). When we use lookup tables, we require exploration to guarantee that we eventually will visit every state infinitely often. When we use basis functions, we only need to visit states with sufficient diversity that we can estimate the parameter vector θ^m. In the language of statistics, the issue is one of *identification* (that is, the ability to estimate θ) rather than exploration. This is a much easier requirement to satisfy, and one of the major advantages of parametric models.

17.8 The Actor–Critic Paradigm

It is very popular in some communities to view approximate dynamic programming in terms of an "actor" and a "critic." Simply put, the actor is a policy that chooses the decision, and the critic is the value function that evaluates the action produced by the policy. In engineering control applications, where states and controls are continuous, it is common to represent both the policy and the approximate value function using (typically shallow) neural networks, and hence some authors refer to "actor nets" and "critic nets." Note that in this setting, the actor is a form of policy function approximation.

The policy iteration algorithm in Figure 17.12 provides one illustration of the actor–critic paradigm. The decision function is equation (17.36), where $V^{\pi,n-1}$ determines the policy (in this case). This is the actor. Equation (17.37), where we update our estimate of the value of the policy, is the critic. We fix the actor (that is, we fix the value function approximation used by the actor) for a period of time and perform repeated iterations where we try to estimate value functions given a particular actor (policy). From time to time, we stop and use our value function to modify our behavior (something critics like to do). In this case, we update the behavior by replacing V^{π} with our current $\overline{V}$.

In other settings, the policy is a policy function approximation of some form that maps the state to a decision. For example, if we are driving through a transportation network (or traversing a graph) the policy might be of the form "when at node i, go next to node j," which would be a form of lookup table policy. As we update the value function, we may decide the right policy at node i is to traverse to node k. Once we have updated our policy, the policy itself does not directly depend on a value function.

Another example might arise when determining how much of a resource we should have on hand. We might solve the problem by maximizing a function of the form $f(x) = \beta_0 - \beta_1(x - \beta_2)^2$. Of course, β_0 does not affect the optimal quantity. We might use the value function to update β_0 and β_1. Once these are determined, we have a function that does not itself directly depend on a value function.

Step 0. Initialization:

Step 0a. Initialize $V_t^{\pi,0}$, $t \in \mathcal{T}$.
Step 0b. Set $n = 1$.
Step 0c. Initialize S_0^1.

Step 1. Do for $n = 1, 2, \ldots, N$:

Step 2. Do for $m = 1, 2, \ldots, M$:

Step 3. Choose a sample path ω^m.
Step 4. Initialize $\hat{v}^m = 0$
Step 5. Do for $t = 0, 1, \ldots, T$:

Step 5a. Solve:

$$x_t^{n,m} = \arg \max_{x_t \in \mathcal{X}_t^{n,m}} \left(C_t(S_t^{n,m}, x_t) + \gamma V_t^{\pi,n-1}(S^{M,x}(S_t^{n,m}, x_t)) \right) \tag{17.36}$$

Step 5b. Compute:

$$S_t^{x,n,m} = S^{M,x}(S_t^{n,m}, x_t^{n,m}),$$
$$S_{t+1}^{n,m} = S^{M,W}(S_t^{x,n,m}, W_{t+1}(\omega^m)).$$

Step 6. Do for $t = T - 1, \ldots, 0$:

Step 6a. Accumulate the path cost (with $\hat{v}_T^m = 0$)

$$\hat{v}_t^m = C_t(S_t^{n,m}, x_t^m) + \gamma \hat{v}_{t+1}^m.$$

Step 6b. Update approximate value of the policy starting at time t:

$$\overline{V}_{t-1}^{n,m} \leftarrow U^V(\overline{V}_{t-1}^{n,m-1}, S_{t-1}^{x,n,m}, \hat{v}_t^m) \tag{17.37}$$

where we typically use $\alpha_{m-1} = 1/m$.

Step 7. Update the policy value function

$$V_t^{\pi,n}(S_t^x) = \overline{V}_t^{n,M}(S_t^x) \quad \forall t = 0, 1, \ldots, T.$$

Step 8. Return the value functions $(V_t^{\pi,N})_{t=1}^T$.

Figure 17.12 Approximate policy iteration using value function-based policies.

17.9   Statistical Bias in the Max Operator★

A subtle type of bias arises when we are optimizing because we are taking the maximum over a set of random variables. In algorithms such as Q-learning or approximate value iteration, we are computing $\hat{q}_t^n$ by choosing the best of a set of decisions which depend on $\bar{Q}^{n-1}(S, x)$. The problem is that the estimates $\bar{Q}^{n-1}(S, x)$ are random variables. In the best of circumstances, assume that $\bar{Q}^{n-1}(S, x)$ is an unbiased estimate of the true value $V_t(S^x)$ of being in (post-decision) state S^x. Because it is still a statistical estimate with some degree of variation, some of the estimates will be too high while others will be too low. If a particular decision takes us to a state where the estimate just happens to be too high (due to statistical variation), then we are more likely to choose this as the best decision and use it to compute $\hat{q}^n$.

To illustrate, assume we have to choose a decision $x \in \mathcal{X}$, where $C(S, x)$ is the contribution earned by using decision x (given that we are in state S) which then takes us to (post-decision) state $S^{M,x}(S, x)$ where we receive an estimated value $\overline{V}(S^{M,x}(S, x))$. Normally, we would update the value of being in state S by computing

$$\hat{v}^n = \max_{x \in \mathcal{X}} \left(C(S, x) + \overline{V}^{x,n-1}(S^{M,x}(S, x)) \right).$$

We would then update the value of being in state S using our standard update formula

$$\overline{V}^n(S) = (1 - \alpha_{n-1})\overline{V}^{n-1}(S) + \alpha_{n-1}\hat{v}^n.$$

Since $\overline{V}^{n-1}(S^{M,x}(S, x))$ is a random variable, sometimes it will overestimate the true value of being in state $S^{M,x}(S, x)$ while other times it will underestimate the true value. Of course, we are more likely to choose a decision that takes us to a state where we have overestimated the value.

We can quantify the error due to statistical bias as follows. Fix the iteration counter n (so that we can ignore it), and let

$$U_x = C(S, x) + \overline{V}(S^{M,x}(S, x))$$

be the estimated value of using decision x. The statistical error, which we represent as β, is given by

$$\beta = \mathbb{E}\{\max_{x \in \mathcal{X}} U_x\} - \max_{x \in \mathcal{X}} \mathbb{E}U_x. \tag{17.38}$$

The first term on the right-hand side of (17.38) is the expected value of $\overline{V}(S)$, which is computed based on the best observed value. The second term is the correct answer (which we can only find if we know the true mean). We can get an estimate of the difference by using a statistical technique known as the "plug-in principle." We assume that $\mathbb{E}U_x = \overline{V}(S^{M,x}(S, x))$, which means that we assume that the estimates $\overline{V}(S^{M,x}(S, x))$ are correct, and then try to estimate $\mathbb{E}\{\max_{x \in \mathcal{X}} U_x\}$. Thus, computing the second term in (17.38) is easy.

The challenge is computing $\mathbb{E}\{\max_{x \in \mathcal{X}} U_x\}$. We assume that while we have been computing $\overline{V}(S^{M,x}(S, x))$, we have also been computing $\bar{\sigma}^2(x) = Var(U_x) = Var(\overline{V}(S^{M,x}(S, x)))$. Using the plug-in principle, we are going to assume that the estimates $\bar{\sigma}^2(x)$ represent the true variances of the value function approximations. Computing $\mathbb{E}\{\max_{x \in \mathcal{X}} U_x\}$ for more than a few decisions is computationally intractable, but we can use a technique called the Clark approximation to provide an estimate. This strategy finds the exact mean and variance of the maximum of two normally distributed random variables, and then assumes that this maximum is also normally distributed. Assume the decisions can be ordered so that $\mathcal{X} = \{1, 2, \dots, |\mathcal{X}|\}$. Now let

$$\bar{U}_2 = \max\{U_1, U_2\}.$$

We can compute the mean and variance of $\bar{U}_2$ as follows. First, we temporarily define α using

$$\alpha^2 = \sigma_1^2 + \sigma_2^2 - 2\sigma_1\sigma_2\rho_{12}$$

where $\sigma_1^2 = Var(U_1)$, $\sigma_2^2 = Var(U_2)$, and ρ_{12} is the correlation coefficient between U_1 and U_2 (we allow the random variables to be correlated, but shortly we are going to approximate them as being independent). Next find

$$z = \frac{\mu_1 - \mu_2}{\alpha}.$$

where $\mu_1 = \mathbb{E}U_1$ and $\mu_2 = \mathbb{E}U_2$. Now let $\Phi(z)$ be the cumulative standard normal distribution (that is, $\Phi(z) = \mathbb{P}[Z \le z]$ where Z is normally distributed with mean 0 and variance 1), and let $\phi(z)$ be the standard normal density function. If we assume that U_1 and U_2 are normally distributed (a reasonable assumption when they represent sample estimates of the value of being in a state), then it is a straightforward exercise to show that

$$\mathbb{E}\bar{U}_2 = \mu_1 \Phi(z) + \mu_2 \Phi(-z) + \alpha\phi(z), \tag{17.39}$$

$$Var(\bar{U}_2) = [(\mu_1^2 + \sigma_1^2)\Phi(z) + (\mu_1^2 + \sigma_2^2)\Phi(-z) + (\mu_1 + \mu_2)\alpha\phi(z)]$$
$$- (\mathbb{E}\bar{U}_2)^2. \tag{17.40}$$

Now assume that we have a third random variable, U_3, where we wish to find $\mathbb{E}\max\{U_1, U_2, U_3\}$. The Clark approximation solves this by using

$$\bar{U}_3 = \mathbb{E}\max\{U_1, U_2, U_3\}$$
$$\approx \mathbb{E}\max\{U_3, \bar{U}_2\},$$

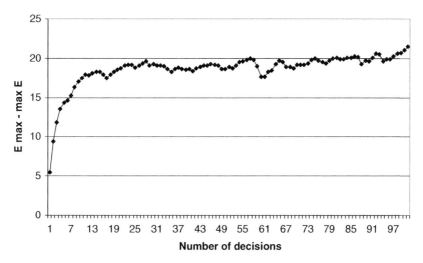

Figure 17.13 $\mathbb{E} \max_x U_x - \max_x \mathbb{E} U_x$ for 100 decisions, averaged over 30 sample realizations. The standard deviation of all sample realizations was 20.

where we assume that $\bar{U}_2$ is normally distributed with mean given by (17.39) and variance given by (17.40). For our setting, it is unlikely that we would be able to estimate the correlation coefficient ρ_{12} (or ρ_{23}), so we are going to assume that the random estimates are independent. This idea can be repeated for large numbers of decisions by using

$$\bar{U}_x = \mathbb{E} \max\{U_1, U_2, \dots, U_x\}$$
$$\approx \mathbb{E} \max\{U_x, \bar{U}_{x-1}\}.$$

We can apply this repeatedly until we find the mean of $\bar{U}_{|\mathcal{X}|}$, which is an approximation of $\mathbb{E}\{\max_{x \in \mathcal{X}} U_x\}$. This, in turn, allows us to compute an estimate of the statistical bias β given by equation (17.38).

Figure 17.13 plots $\beta = \mathbb{E} \max_x U_x - \max_x \mathbb{E} U_x$ as it is being computed for 100 decisions, averaged over 30 sample realizations. The standard deviation of each U_x was fixed at $\sigma = 20$. The plot shows that the error increases steadily until the set $\mathcal{X}$ reaches about 20 or 25 decisions, after which it grows much more slowly. Of course, in an approximate dynamic programming application, each U_x would have its own standard deviation which would tend to decrease as we sample a decision repeatedly (a behavior that the approximation above captures nicely).

This brief analysis suggests that the statistical bias in the max operator can be significant. However, it is highly data dependent. If there is a single dominant decision, then the error will be negligible. The problem only arises when there are many (as in 10 or more) decisions that are competitive, and where the standard deviation of the estimates is not small relative to the differences between the means. Unfortunately, this is likely to be the case in most large-scale applications (if a single decision is dominant, then it suggests that the solution is probably obvious).

The relative magnitudes of value iteration bias over statistical bias will depend on the nature of the problem. If we are using a pure forward pass (TD(0)), and if the value of being in a state at time t reflects rewards earned over many periods into the future, then the value iteration bias can be substantial (especially if the stepsize is too small).

Value iteration bias has long been recognized in the dynamic programming community. By contrast, statistical bias appears to have received almost no attention, and as a result we are not aware of any research addressing this problem. We suspect that statistical bias is likely to inflate value function approximations fairly uniformly, which means that the impact on the policy may be quite small. However, if the goal is to obtain the value function itself (for example, to estimate the value of an asset or a contract), then the bias can distort the results.

17.10 The Linear Programming Method Using Linear Models*

In section 14.10, we showed that the determination of the value of being in each state can be found by solving the following linear program

$$\min_{v} \sum_{s \in \mathcal{S}} \beta_s v(s) \tag{17.41}$$

subject to

$$v(s) \geq C(s, x) + \gamma \sum_{s' \in \mathcal{S}} p(s'|s, x) v(s') \quad \text{for all } s \text{ and } x. \tag{17.42}$$

The problem with this formulation arises because it requires that we enumerate the state space to create the value function vector $(v(s))_{s \in \mathcal{S}}$. Furthermore, we have a constraint for each state-decision pair, a set that will be huge even for relatively small problems.

We can partially solve this problem by replacing the discrete value function with a regression function such as

$$\overline{V}(s|\theta) = \sum_{f \in \mathcal{F}} \theta_f \phi_f(s).$$

where $(\phi_f)_{f\in\mathcal{F}}$ is an appropriately designed set of basis functions. This produces a revised linear programming formulation

$$\min_{\theta} \sum_{s\in\mathcal{S}} \beta_s \sum_{f\in\mathcal{F}} \theta_f \phi_f(s)$$

subject to:

$$v(s) \ \geq\ C(s,x) + \gamma \sum_{s'\in\mathcal{S}} p(s'|s,x) \sum_{f\in\mathcal{F}} \theta_f \phi_f(s') \quad \text{for all } s \text{ and } x.$$

This is still a linear program, but now the decision variables are $(\theta_f)_{f\in\mathcal{F}}$ instead of $(v(s))_{s\in\mathcal{S}}$. Note that rather than use a stochastic iterative algorithm, we obtain θ directly by solving the linear program.

We still have a problem with a huge number of constraints. Since we no longer have to determine $|\mathcal{S}|$ decision variables (in (17.41)–(17.42) the parameter vector $(v(s))_{s\in\mathcal{S}}$ represents our decision variables), it is not surprising that we do not actually need all the constraints. One strategy that has been proposed is to simply choose a random sample of states and decisions. Given a state space $\mathcal{S}$ and set of decisions $\mathcal{X}$, we can randomly choose states and decisions to create a smaller set of constraints.

Some care needs to be exercised when generating this sample. In particular, it is important to generate states roughly in proportion to the probability that they will actually be visited. Then, for each state that is generated, we need to randomly sample one or more decisions. The best strategy for doing this is going to be problem-dependent.

This technique has been applied to the problem of managing a network of queues. Figure 17.14 shows a queueing network with three servers and eight queues. A server can serve only one queue at a time. For example, server A might be a machine that paints components one of three colors (say, red, green, and blue). It is best to paint a series of parts red before switching over to blue. There are customers arriving exogenously (denoted by the arrival rates λ_1 and λ_2). Other customers arrive from other queues (for example, departures from queue 1 become arrivals to queue 2). The problem is to determine which queue a server should handle after each service completion.

If we assume that customers arrive according to a Poisson process and that all servers have negative exponential service times (which means that all processes are memoryless), then the state of the system is given by

$$S_t = R_t = (R_{ti})_{i=1}^{8},$$

where R_{ti} is the number of customers in queue i. Let $\mathcal{K} = \{1,2,3\}$ be our set of servers, and let a_t be the attribute vector of a server given by $a_t = (k, q_t)$, where k is the identity of the server and q_t is the queue being served at time t.

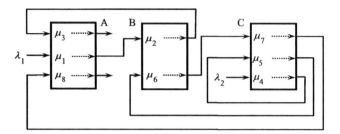

Figure 17.14 Queueing network with three servers serving a total of eight queues, two with exogenous arrivals (λ) and six with arrivals from other queues. Adapted from de Farias and Van Roy (2003).

Each server can only serve a subset of queues (as shown in Figure 17.14). Let $\mathcal{D} = \{1, 2, \dots, 8\}$ represent a decision to serve a particular queue, and let $\mathcal{D}_a$ be the decisions that can be used for a server with attribute a. Finally, let $x_{tad} = 1$ if we decide to assign a server with attribute a to serve queue $d \in \mathcal{D}_a$.

The state space is effectively infinite (that is, too large to enumerate). But we can still sample states at random. Research has shown that it is important to sample states roughly in proportion to the probability they are visited. We do not know the probability a state will be visited, but it is known that the probability of having a queue with r customers (when there are Poisson arrivals and negative exponential servers) follows a geometric distribution. For this reason, it was chosen to sample a state with $r = \sum_i R_{ti}$ customers with probability $(1 - \gamma)\gamma^r$, where γ is a discount factor (a value of 0.95 was used).

Further complicating this problem class is that we also have to sample decisions. Let $\mathcal{X}$ be the set of all feasible values of the decision vector x. The number of possible decisions for each server is equal to the number of queues it serves, so the total number of values for the vector x is $3 \times 2 \times 3 = 18$. In the experiments for this illustration, only 5,000 states were sampled (in portion to $(1 - \gamma)\gamma^r$), but all the decisions were sampled for each state, producing 90,000 constraints.

Once the value function is approximated, it is possible to simulate the policy produced by this value function approximation. The results were compared against two myopic policies: serving the longest queue, and first-in, first-out (that is, serve the customer who had arrived first). The costs produced by each policy are given in Table 17.7, showing that the ADP-based strategy significantly outperforms these other policies.

Considerably more numerical work is needed to test this strategy on more realistic systems. For example, for systems that do not exhibit Poisson arrivals or negative exponential service times, it is still possible that sampling states based on geometric distributions may work quite well. More problematic is the rapid growth in the feasible region $\mathcal{X}$ as the number of servers, and queues per server, increases.

Table 17.7 Average cost estimated using simulation. Data from de Farias and Van Roy (2003).

Policy	Cost
ADP	33.37
Longest	45.04
FIFO	45.71

An alternative to using constraint sampling is an advanced technique known as column generation. Instead of generating a full linear program which enumerates all decisions (that is, $v(s)$ for each state), and all constraints (equation (17.42)), it is possible to generate sequences of larger and larger linear programs, adding rows (constraints) and columns (decisions) as needed. These techniques are beyond the scope of our presentation, but readers need to be aware of the range of techniques available for this problem class.

17.11 Finite Horizon Approximations for Steady-State Applications

It is easy to assume that if we have a problem with stationary data (that is, all random information is coming from a distribution that is not changing over time), then we can solve the problem as an infinite horizon problem, and use the resulting value function to produce a policy that tells us what to do in any state. If we can, in fact, find the optimal value function for every state, this is true.

There are many applications of infinite horizon models to answer policy questions. Do we have enough doctors? What if we increase the buffer space for holding customers in a queue? What is the impact of lowering transaction costs on the amount of money a mutual fund holds in cash? What happens if a car rental company changes the rules allowing rental offices to give customers a better car if they run out of the type of car that a customer reserved?

These are all dynamic programs controlled by a constraint (the size of a buffer or the number of doctors), a parameter (the transaction cost), or the rules governing the physics of the problem (the ability to substitute cars). We may be interested in understanding the behavior of such a system as these variables are adjusted. For infinite horizon problems that are too complex to solve exactly, ADP offers a way to approximate these solutions.

Infinite horizon models also have applications in operational settings. Assume that we have a problem governed by stationary processes. We could solve the steady-state version of the problem, and use the resulting value function to define a policy that would work from any starting state. This works if we have, in fact, found at least a close approximation of the optimal value function for any starting state. However, if you have made it this far in this book, then that means you are interested in working on problems where the optimal value function cannot be found for all states. Typically, we are forced to approximate the value function, and it is always the case that we do the best job of fitting the value function around states that we visit most of the time.

When we are working in an operational setting, then we start with some known initial state S_0. From this state, there are a range of "good" decisions, followed by random information, that will take us to a set of states S_1 that is typically heavily influenced by our starting state. Figure 17.15 illustrates the phenomenon. Assume that our true, steady-state value function approximation looks like the sine function. At time $t = 1$, the probability distribution of the state S_t that we can reach is shown as the shaded area. Assume that we have chosen to fit a quadratic function of the value function, using observations of S_t that we generate through Monte Carlo sampling. We might obtain the dotted curve labeled as $\overline{V}_1(S_1)$, which closely fits the true value function around the states S_1 that we have observed.

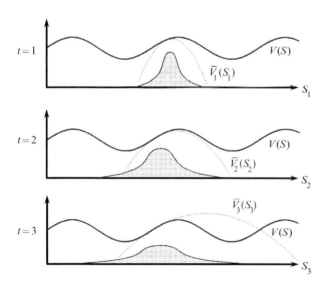

Figure 17.15 Exact value function (sine curve) and value function approximations for $t = 1, 2, 3$, which change with the probability distribution of the states that we can reach from S_0.

For times $t = 2$ and $t = 3$, the distribution of states S_2 and S_3 that we actually observe grows wider and wider. As a result, the best fit of a quadratic function spreads as well. So, even though we have a steady-state problem, the best value function approximation depends on the initial state S_0 and how many time periods into the future that we are projecting. Such problems are best modeled as finite horizon problems, but only because we are forced to approximate the problem.

17.12   Bibliographic Notes

Section 17.2 – Approximate value iteration using lookup tables encompasses the family of algorithms that depend on an approximation of the value of a future state to estimate the value of being in a state now, which includes Q-learning and temporal-difference learning. These methods represent the foundation of approximate dynamic programming and reinforcement learning.

Section 17.4 – The problems with the use of linear models in the context of approximate value iteration (TD learning) are well known in the research literature. Good discussions of these issues are found in Bertsekas and Tsitsiklis (1996), Tsitsiklis and Van Roy (1997), Baird (1995), and Precup et al. (2001), to name a few.

Section 17.7 – Bradtke and Barto (1996) first introduced least squares temporal differencing, which is a way of approximating the one-period contribution using a linear model, and then projecting the infinite horizon performance. Lagoudakis and Parr (2003) describes the least squares policy iteration algorithm (LSPI) which uses a linear model to approximate the Q-factors, which is then imbedded in a model-free algorithm.

Section 17.8 – There is a long history of referring to policies as "actors" and value functions as "critics" (see, for example, Barto et al. (1983), Williams and Baird (1990), Bertsekas and Tsitsiklis (1996), and Sutton and Barto (2018)). Borkar and Konda (1997) and Konda and Borkar (1999) analyze actor–critic algorithms as an updating process with two time-scales, one for the inner iteration to evaluate a policy, and one for the outer iteration where the policy is updated. Konda and Tsitsiklis (2003) discusses actor–critic algorithms using linear models to represent both the actor and the critic, using bootstrapping for the critic. Bhatnagar et al. (2009) suggest several new variations of actor–critic algorithms, and proves convergence when both the actor and the critic use bootstrapping.

Section 17.10 – Schweitzer and Seidmann (1985) describes the use of basis functions in the context of the linear programming method. The idea is further developed in de Farias and Van Roy (2003) which also develops performance guarantees. Farias and Roy (2001) investigates the use of constraint sampling and proves results on the number of samples that are needed.

Exercises

Review questions

17.1 Explain the difference between on-policy and off-policy learning.

17.2 Contrast, using only necessary notation (but you will need some) the essential differences between ADP using a pre-decision state, a post-decision state, and Q-learning.

17.3 Contrast ADP using a post-decision state versus Q-learning, given that (S, x) is a form of post-decision state. Are these equivalent?

17.4 Discuss using Q-learning where x is a vector.

17.5 Explain in words the difference between the single-pass and double-pass versions of forward ADP. Can you give an example of a problem where you would need to use a backward pass?

17.6 Contrast a backward pass with backward approximate dynamic programming. Are these equivalent? If not, how are they different?

17.7 Use notation to explain what is meant by the "actor" and the "critic" in the actor–critic paradigm.

Modeling questions

17.8 The most common strategy for using approximate dynamic programming is to train value function approximations offline using a simulator. Using the language introduced in section 9.11, where we classified problems based on (a) whether they were state-independent or state-dependent, and (b) whether we were optimizing the final reward or cumulative reward, training VFAs would fall in the class of state-dependent problems, where we are maximizing the final reward. In its most compact form, this objective can be written (see, for example, Table 9.3)

$$\max_{\pi^{lrn}} \mathbb{E}\{C(S, X^{\pi^{imp}}(S|\theta^{imp}), W)|S_0\}. \tag{17.43}$$

In equation (9.43), we expanded the expectations to make the underlying random variables explicit, which produced the equivalent expression

$$\max_{\pi^{lrn}} \mathbb{E}\{C(S, X^{\pi^{imp}}(S|\theta^{imp}), \widehat{W})|S^0\} =$$

$$\mathbb{E}_{S^0} \mathbb{E}^{\pi^{lrn}}_{W^1,\dots,W^N|S^0} \mathbb{E}^{\pi^{imp}}_{S|S^0} \mathbb{E}_{\widehat{W}|S^0} C(S, X^{\pi^{imp}}(S|\theta^{imp}), \widehat{W}). \qquad (17.44)$$

Using the context of the forward ADP algorithms presented in this chapter, answer the following:

(a) When optimizing over policies (such as the learning policies π^{lrn}), we have to search over classes of policies $f \in \mathcal{F}^{lrn}$, and any tunable parameters $\theta \in \Theta^f$ within that class. Give two examples of "policy classes" and an example of a tunable parameter for each class.

(b) Throughout the book, we have used as our default objective for dynamic programming the function

$$\max_{\pi} \mathbb{E}\left\{\sum_{t=0}^{T} C(S_t, X^{\pi}(S_t))|S_0\right\}. \qquad (17.45)$$

This chapter (and we could include the backward ADP methods of chapter 15) presents different methods for training VFAs, after which we would run simulations to test the effectiveness by simulating the objective in (17.45). Explain what is meant by π^{lrn} and π^{imp} in equation (17.44).

(c) In section 9.11, we identified equation (17.43) as the objective for optimizing final reward and we showed (in equation (9.44)) that this could be simulated using

$$\max_{\pi^{lrn}} \mathbb{E}_{S^0} \mathbb{E}^{\pi^{imp}}_{((W^n_t)^T_{t=0})^N_{n=0}|S^0} \left(\mathbb{E}^{\pi^{imp}}_{(\widehat{W}_t)^T_{t=0}|S^0} \frac{1}{T} \sum_{t=0}^{T-1} C(S_t, X^{\pi^{imp}}(S_t|\theta^{imp}), \widehat{W}_{t+1}) \right). \qquad (17.46)$$

Make the case that when I am designing algorithms to solve the cumulative reward objective in (17.45) that I am actually solving the final-reward optimization problem given in (17.46).

Computational exercises

17.9 We are going to revisit exercise 15.4 using forward ADP, which we repeat here. In this exercise you are going to solve a simple inventory problem using Bellman's equations, to obtain an optimal policy. Then, the exercises that follow will have you implement various backward ADP policies that you can compare against the optimal policy you obtain in this exercise. Your inventory problem will span T time periods, with an inventory equation governed by

$$R_{t+1} = \max\{0, R_t - \hat{D}_{t+1}\} + x_t.$$

Here we are assuming that product ordered at time t, x_t, arrive at $t + 1$. Assume that $\hat{D}_{t+1}$ is described by a discrete uniform distribution between 1 and 20.

Next assume that our contribution function is given by

$$C(S_t, x_t) = 50 \min\{R_t, \hat{D}_{t+1}\} - 10x_t.$$

(a) Find an optimal policy by solving this dynamic program exactly using classical backward dynamic programming methods from chapter 14 (specifically equation (14.3)). Note that your biggest challenge will be computing the one-step transition matrix. Simulate the optimal policy 1,000 times starting with $R_0 = 0$ and report the performance.

(b) Now solve the problem using forward ADP using a simple quadratic approximation for the value function approximation:

$$\overline{V}_t^x(R_t^x) = \theta_{t0} + \theta_{t1}R_t^x + \theta_{t2}(R_t^x)^2$$

where R_t^x is the post-decision resource state which we might represent using

$$R_t^x = \max\{0, R_t - \mathbb{E}\{\hat{D}_{t+1}\}\} + x_t.$$

Use 100 forward passes to estimate $\overline{V}_t(S_t)$ using the algorithm in Figure 17.3.

(c) Having found $\overline{V}_t^x(R_t^x)$, simulate the resulting policy 1,000 times, and compare your results to your optimal policy.

(d) Repeat (b) and (c) but this time use a value function approximation that is only linear in R_t^x:

$$\overline{V}_t^x(R_t^x) = \theta_{t0} + \theta_{t1}R_t^x.$$

How does the resulting policy compare your results from part (c)?

17.10 We are going to revisit exercise 15.2 using forward ADP, which we repeat here. We are going to solve the continuous budgeting problem presented in section 14.4.2 using backward approximate dynamic programming. The problem starts with R_0 resources which are then allocated over periods 0 to T. Let x_t be the amount allocated in period t with contribution

$$C_t(x_t) = \sqrt{x_t}.$$

Assume that $T = 20$ time periods.

(a) Use the results of section 14.4.2 to solve this problem optimally. Evaluate your simulation by simulating your optimal policy 1000 times.

(b) Use the forward ADP algorithm described in Figure 17.3 to obtain the value function approximations using

$$\overline{V}_t(R_t) = \theta_{t0} + \theta_{t1}\sqrt{x_t}.$$

Use 100 forward passes to estimate $\overline{V}_t(R_t)$. Use linear regression (either the methods in section 3.7.1, or a package) to fit $\overline{V}_t(R_t)$. Then, simulate this policy 1000 times (ideally using the same sample paths as you used for part (a)). How do you think θ_{t0} and θ_{t1} should behave?

(c) Use the forward ADP algorithm described in Figure 15.5 to obtain the value function approximations using

$$\overline{V}_t(R_t) = \theta_{t0} + \theta_{t1}R_t^x + \theta_{t2}(R_t^x)^2,$$

where R_t^x is the post-decision resource state $R_t^x = R_t - x_t$ (which is the same as R_{t+1} since transitions are deterministic).

Use linear regression (either the methods in section 3.7.1, or a package) to fit $\overline{V}_t(R_t)$. Then, simulate this policy 1000 times (ideally using the same sample paths as you used for part (a)).

17.11 Repeat exercise 7.10, but this time use

$$C(x_t) = \ln(x_t).$$

For part (b), use

$$\overline{V}_t(R_t) = \theta_{t0} + \theta_{t1}\ln(x_t).$$

Theory questions

17.12 Prove that the newsvendor objective function

$$F(x) = \mathbb{E}\{p\min\{x, W\} - cx\}$$

is concave in x as long as $p \geq c$.

Problem-solving questions

17.13 We are going to try again to solve our asset selling problem, We assume we are holding a real asset and we are responding to a series of offers. Let $\hat{p}_t$ be the tth offer, which is uniformly distributed between 500 and 600 (all prices are in thousands of dollars). We also assume that each offer is independent of all prior offers. You are willing to consider up to 10 offers, and your goal is to get the highest possible price. If you have not accepted the first nine offers, you must accept the 10th offer.

(a) Write out the decision function you would use in a dynamic programming algorithm in terms of a Monte Carlo sample of the latest price and a current estimate of the value function.

(b) Write out the updating equations (for the value function) you would use after solving the decision problem for the tth offer.

(c) Implement an approximate dynamic programming algorithm using *synchronous* state sampling. Using 1000 iterations, write out your estimates of the value of being in each state immediately after each offer. For this exercise, you will need to discretize prices for the purpose of approximating the value function. Discretize the value function in units of 5 dollars.

(d) From your value functions, infer a decision rule of the form "sell if the price is greater than $\bar{p}_t$."

17.14 We wish to use Q-learning to solve the problem of deciding whether to continue playing a game where you win \$1 if you flip a coin and see heads, and lose \$1 if you see tails. Using a stepsize $\alpha = \dfrac{\theta}{\theta+n}$, implement the Q-learning algorithm in equations (11.18) and (11.19). Initialize your estimates $\bar{Q}(s, a) = 0$, and run 1000 of the algorithm using $\theta = 1$, 10, 100, and 1000. Plot Q^n for each of the three values of θ, and discuss the choice you would make if your budget was $N = 50$, 100, or 1000.

Sequential decision analytics and modeling

These exercises are drawn from the online book *Sequential Decision Analytics and Modeling* available at http://tinyurl.com/sdaexamplesprint.

17.15 Review chapter 5, sections 5.1–5.6, on stochastic shortest path problems. We are going to focus on the extension in section 5.6, where costs $\hat{c}_{ij}$ are random, and a traveler gets to see the costs $\hat{c}_{ij}$ out of node i when the traveler arrives at node i, and before she has to make a decision which link to move over. Software for this problem is available at http://tinyurl.com/sdagithub – download the module "StochasticShortestPath_Dynamic."

(a) Write out the pre- and post-decision state variables for this problem.

(b) Given a value function approximation $\overline{V}_t^{x,n}(S_t^x)$ around the post-decision state S_t^x, describe the steps for updating $\overline{V}_t^{x,n}(S_t^x)$ to obtain $\overline{V}_t^{x,n+1}(S_t^x)$.

(c) Using the Python module, compare the performance using the following stepsize formulas:

(i) $\alpha_n = 0.10$.

(ii) $\alpha_n = \dfrac{1}{n}$.

(iii) $\alpha_n = \dfrac{\theta^{step}}{\theta^{step}+n-1}$ with $\theta^{step} = 10$.

Run the algorithm for 10, 20, 50, and 100 training iterations, and then simulate the resulting policy. Report the performance of the policy resulting from each stepsize formula, given the number of training iterations.

17.16 Review chapter 13, sections 13.1–13.4, on the blood management problem. Software for this problem is available at http://tinyurl.com/sdagithub – download the module "BloodManagement."

(a) Write out the pre- and post-decision state variables for this problem.

(b) Given a value function approximation $\overline{V}_t^{x,n}(S_t^x)$ around the post-decision state S_t^x, describe the steps for updating $\overline{V}_t^{x,n}(S_t^x)$ to obtain $\overline{V}_t^{x,n+1}(S_t^x)$. Note that $\overline{V}_t^{x,n}(S_t^x)$ is piecewise linear and separable.

(c) Using the Python module, compare the performance using the following stepsize formulas:

(i) $\alpha_n = 0.10$.

(ii) $\alpha_n = \dfrac{1}{n}$.

(iii) $\alpha_n = \dfrac{\theta^{step}}{\theta^{step}+n-1}$ with $\theta^{step} = 10$.

Run the algorithm for 10, 20, 50, and 100 training iterations, and then simulate the resulting policy. Report the performance of the policy resulting from each stepsize formula, given the number of training iterations.

Diary problem

The diary problem is a single problem you chose (see chapter 1 for guidelines). Answer the following for your diary problem.

17.17 For your diary problem, compare the pure forward pass algorithm in Figure 17.3 to the two-pass algorithm in Figure 17.4 in terms of both computational complexity and likely performance.

Bibliography

Baird, L.C. (1995). Residual algorithms: Reinforcement learning with function approximation. In: *Proceedings of the Twelfth International Conference on Machine Learning.* 30–37.

Barto, A.G., Sutton, R.S., and Anderson, C.W. (1983). Neuron-like elements that can solve difficult learning control problems. *IEEE Transactions on Systems, Man and Cybernetics* 13 (5): 834–846.

Bertsekas, D.P. and Tsitsiklis, J.N. (1996). *Neuro-Dynamic Programming*, Belmont, MA: Athena Scientific.

Bhatnagar, S., Sutton, R.S., Ghavamzadeh, M., and Lee, M. (2009). 'Natural actor{critic algorithms. *Automatica* 45 (11): 2471–2482.

Borkar, V. and Konda, V.R. (1997). The actor-critic algorithm as multi-time-scale stochastic approximation. *Sadhana* 22 (4): 525–543.

Bradtke, S.J. and Barto, A.G. (1996). Linear least-squares algorithms for temporal difference learning. *Machine Learning* 22 (1): 33–57.

de Farias, D.P. and Van Roy, B. (2003). The linear programming approach to approximate dynamic programming. *Operations Research* 51: 850–865.

Farias, D. and Roy, B. (2001). On constraint sampling for the linear programming approach to approximate dynamic. *Mathematics of Operations Research* 29 (3): 462–478.

Konda, V.R. and Borkar, V.S. (1999). Actor-critic–type learning algorithms for Markov decision processes. *SIAM Journal on Control and Optimization* 38: 94.

Konda, V.R. and Tsitsiklis, J.N. (2003). On actor-critic algorithms. *SIAM Journal on Control and Optimization* 42 (4): 1143–1166.

Lagoudakis, M. and Parr, R. (2003). Least-squares policy iteration. *Journal of Machine Learning Research* 4: 1107–1149.

Precup, D., Sutton, R.S., and Dasgupta, S. (2001). Off-policy temporal-difference learning with function approximation. In: *19th International Conference on Machine Learning*, 417–424.

Schweitzer, P. and Seidmann, A. (1985). Generalized polynomial approximations in Markovian decision processes. *Journal of Mathematical Analysis and Applications* 110 (6): 568–582.

Sutton, R.S. and Barto, A.G. (2018). *Reinforcement Learning: An Introduction*, 2e. Cambridge, MA: MIT Press.

Tsitsiklis, J.N. and Van Roy, B. (1997). An analysis of temporal-difference learning with function approximation. *IEEE Transactions on Automatic Control* 42 (5): 674–690.

Williams, R.J. and Baird, L.C. (1990). A mathematical analysis of actor-critic architectures for learning optimal controls through incremental dynamic programming. In: *Sixth Yale Workshop on Adaotive and Learning Systems.*, 96–101. New Haven.

18

Forward ADP III: Convex Resource Allocation Problems

In chapter 3, we introduced general purpose approximation tools for approximating functions without assuming any special structural properties. In this chapter, we focus on approximating value functions that arise in dynamic resource allocation problems where contribution functions (and, as a byproduct, value functions) tend to be convex (concave if maximizing) in the resource dimension. It is standard practice in the optimization community to refer to these problems as "convex" since minimization is standard, but we will stick to our standard practice of maximizing.

For example, if R is the amount of resource available (water, oil, money, or vaccines) and $V(R)$ is the value of having R units of our resource, we often find that $V(R)$ will be concave in R (where R is often a vector). Often, it is piecewise linear, whether R is discrete (e.g. inventories of trucks or units of blood) or continuous (as would arise if we are managing energy or money). Value functions with this structure yield to special approximation strategies, and some of the issues we encountered in the previous two chapters (notably the exploration–exploitation problem) vanish.

There is a genuinely vast range of problems that can be broadly described as dynamic resource allocation. Table 18.1 provides just a hint of the diversity of application settings in this domain. Almost all of these settings involve multidimensional decisions, as we manage different resources (doctors, truck trailers, blood, etc.), different types of resources (physician specialties, trailer types, blood types), any of which may be spatially distributed.

We are going to begin with a simple scalar problem where R_t is the amount of resource (energy in a battery, cash on hand, inventory of parts, etc.) on hand at time t. We then transition to vector-valued problems. These arise in many settings, but we are going to use the context of spatially distributed problems as our motivating application, where we define

Reinforcement Learning and Stochastic Optimization: A Unified Framework for Sequential Decisions, First Edition. Warren B. Powell.
© 2022 John Wiley & Sons, Inc. Published 2022 John Wiley & Sons, Inc.

Table 18.1 Sample list of resource allocation problems arising in different problem domains.

Major field	Problem	Resource
Energy	Grid operations	Energy generators
	Grid operations	Natural gas supplies
	Grid operations	Energy from wind
	Battery storage	Storage capacity
	Battery storage	Energy in the battery
	Building management	Building temperature
Health	Public health	COVID tests
	Public health	Vaccines
	Public health	Nurses
	Public health	Blood inventories
	Hospitals	ICU capacity
	Hospitals	Physicians
	Hospitals	Nurses
	Hospitals	Medications
	Hospitals	Blood supplies
Logistics	Inventory management	On-hand inventory
	Inventory management	Material handling
	Manufacturing	Stamping machines
	Manufacturing	Robots
	Supply chain	Suppliers
	Supply chain	Raw materials
Freight transportation	Truck operations	Drivers
	Truck operations	Loads
	Truck operations	Trailers
	Rail operations	Locomotives
	Rail operations	Freight cars
	Ocean	Vessels
	Ocean	Port handling capacity
Finance	Trading	Investments
	Trading	Cash
	Trading	Risk exposure
Laboratory sciences	Equipment	Microscopes
	Equipment	Scanners
	Equipment	Computers
	Materials	Oxygen
	Materials	Metals
	People	Scientists
	People	Technicians

$$
\begin{aligned}
R_{ti} &= \text{quantity of resource available at location } i \in \mathcal{I} \text{ at time } t, \\
R_t &= \text{the resource state vector,} \\
&= (R_{ti})_{i \in \mathcal{I}}.
\end{aligned}
$$

Depending on the underlying problem, the spatially distributed problem may be spread over tens, hundreds, or many thousands of locations, creating potentially a very high dimensional problem. We will use the spatially distributed setting to motivate vector-valued resource state variables, but vector-valued resource allocation problems arise in a variety of settings:

$$
\begin{aligned}
R_{tk} &= \text{Quantity of resource of type } k \in \mathcal{K} \text{ (type of shirt, color)} \\
&\quad \text{which can be substituted (at a cost) to satisfy a demand } D_{t\ell} \\
&\quad \text{for products of type } \ell .
\end{aligned}
$$

$$
\begin{aligned}
R_{tt'} &= \text{Resources that we know about at time } t \text{ that will be} \\
&\quad \text{available to be used at time } t'.
\end{aligned}
$$

$$
\begin{aligned}
R_{ta} &= \text{Resources (such as people or complex equipment) with} \\
&\quad \text{attribute vector } a = (a_1, a_2, \dots, a_M) \in \mathcal{A}.
\end{aligned}
$$

The notation R_{ta} is the most general, but opens up the door to a potentially very high dimensional resource vector if the attribute vector a has more than two or three dimensions.

We consider a series of strategies for approximating the value function using increasing sophistication:

Piecewise linear, concave – We start with this for a simple, scalar inventory problem to demonstrate the power of concavity.

Separable, piecewise linear, concave – These functions are especially useful when we are interested in integer solutions. Separable functions are relatively easy to estimate and offer special structural properties when solving the optimality equations.

General nonlinear regression equations – Here, we bring the full range of tools available from the field of statistics.

Cutting planes – This is a technique for approximating multidimensional, piecewise linear functions that has proven to be particularly powerful for multistage linear programs such those that arise in dynamic resource allocation problems.

Linear approximations – There are problems where value functions that are linear in the resources can be quite useful, especially for very high-dimensional problems, where the number of resources, say, with attribute vector a is typically 0 or 1.

Resource allocation with an exogenous state variable – All of the approximations up to now consist purely of a resource vector R_t in the state variable. There are problems where we need to capture other information, that we identify by I_t, giving us a state variable $S_t = (R_t, I_t)$, and where we do not enjoy the structure of concavity (or convexity) in I_t.

An important dimension of this chapter will be our use of derivatives to estimate value functions, rather than just the value of being in a state. When we want to determine how much oil should be sent to a storage facility, what matters most is the marginal value of additional oil. For some problem classes, this is a particularly powerful device that dramatically improves convergence.

This chapter will expect the reader has a background in linear programming. We will assume some understanding with the tools for solving linear programs (although no working knowledge of the algorithms is needed). More important will be an understanding of dual variables, which we use for estimating value functions.

18.1 Resource Allocation Problems

In chapter 8 we presented a number of problems that could be described as resource allocation problems. In this chapter, we are going to use three to illustrate different algorithmic strategies: our familiar newsvendor problem, a two-stage resource allocation problem with substitution, and finally a very general, multiperiod resource allocation problem.

18.1.1 The Newsvendor Problem

Perhaps the most elementary resource allocation problem is known as the newsvendor problem, which we first introduced in section 2.3.1. Here, we first allocate a quantity of a resource ("newspapers") x paying a unit cost c per newspaper, then observe a demand D, where we sell the smaller of x and D at a price p.

Newsvendor problems arise throughout stochastic resource allocation problems. For example, a transportation company (a railroad, an airline, or a shipping company) often has to place orders for equipment a year or more in advance. The company hopes that all the equipment will be used, and will be enough to satisfy demand. If the company orders too much, it faces an overage situation. If the company has ordered too few, then it is in an underage situation.

In our notation, we would define

$$
\begin{aligned}
x &= \text{the order quantity that can be used to satisfy upcoming} \\
&\quad \text{demands (not yet revealed),} \\
D &= \text{the demand that arises during time interval 1,} \\
c &= \text{the unit purchase cost of assets,} \\
p &= \text{the price for each unit of demand that is satisfied.}
\end{aligned}
$$

Our contribution function is given by

$$F(x) = \mathbb{E}F(x, D) = \mathbb{E}\{p \min[x, D] - cx\}. \tag{18.1}$$

We assume (as occurs in the real newsvendor problem) that unused assets have no value (as would happen if we were actually managing newspapers). Each time period is a new problem.

Figure 18.1(a) shows the shape of $F(x, D)$ for different values of D assuming, of course, that the price p is greater than the cost of the inventory c, where profits are maximized at $x = D$. Figure 18.1(b) gives a probability distribution for the random variable D, and finally Figure 18.1(c) is the expected profits given that we order a quantity x, and then observe the random demand D. This figure illustrates the fundamental concave shape for the newsvendor problem, which is behind the concave shape of many resource allocation problems where we are trying to match a supply against a random demand. This behavior persists even for much more complex resource allocation problems, as long as revenues are linear in p, and costs are linear in c.

18.1.2 Two-Stage Resource Allocation Problems

In the newsvendor problem, we assume there is a single type of resource being used to satisfy a single type of demand. There are a number of settings where we have to allocate different types of resources now, after which we see the demand, and then we get to make a final decision of which resources should satisfy which demand.

■ **EXAMPLE 18.1**

An electric power utility needs to purchase expensive components that cost millions of dollars and require a year or more to order. The industry needs to maintain a supply of these in case of a failure. The problem is to determine how many units to purchase, when to purchase them, what

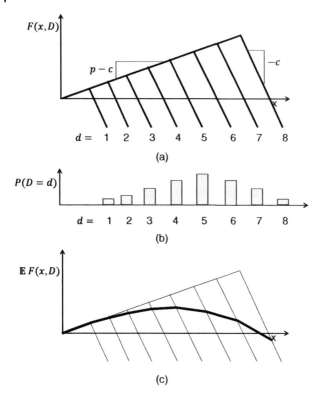

Figure 18.1 (a) The shape of the newsvendor problem for different values of the demand D; (b) The probability $P(D = d)$ of each outcome of demand; (c) The expected profits as a function of x.

features they should have, and where they should be stored. When a failure occurs, the company will find the closest unit that has the features required for a particular situation.

■ **EXAMPLE 18.2**

An investment bank needs to allocate funds to various investments (long-term, high risk investments, real estate, stocks, index funds, bonds, money markets, CD's). As opportunities arise, the bank will move money from one investment to another, but these transactions can take time to execute and cost money (for example, it is easiest and fastest to move money out of a money market fund).

■ **EXAMPLE 18.3**

An online bookseller prides itself in fast delivery, but this requires holding books in inventory. If orders arrive when there is no inventory, the seller may have to delay filling the order (and risk losing it) or purchase the books at a higher cost from the publisher. If the inventory is too high, the company has to choose between holding the books in inventory (tying up space and capital), discounting the book to increase sales, or selling inventory to another distributor (at a substantial discount).

■ **EXAMPLE 18.4**

An automotive manufacturer has to decide what models to design and build, and with what features. Given a three year design and build cycle, they have to create cars that will respond to an uncertain marketplace in the future. Once the models are built, customers have to adjust and purchase models that are closest to their wishes.

In all of these problems, we make an initial allocation decision. This could be the decision to purchase a type of equipment, build a particular model of car, or stock inventories of different types of product. Once the initial decision is made, we see information about the demand for the asset as well as the prices/costs derived from satisfying a demand (which can also be random). After this information is revealed, we may make new decisions. The goal is to make the best initial decisions given the potential downstream decisions that might be made. These problems are illustrated in Figure 18.2.

This problem combines "what to do" (what type of product, where to store it) with "how much." It is a basic building block for much more complex, fully sequential resource allocation problems which we present next.

18.1.3 A General Multiperiod Resource Allocation Model*

The insights behind the newsvendor problem and the two-state resource allocation model can be leveraged into a fairly general model for dynamic resource allocation problems. In this model, we are managing a "resource" (people, equipment, blood, money, etc.) to serve "demands" (tasks, customers, and jobs). We note that this model is quite general, and can be used for some fairly complex resource allocation problems.

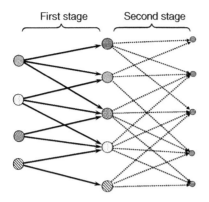

First stage   Second stage

Figure 18.2 A two-stage allocation problem. The first-stage decisions have to be made before the second stage information becomes known. When this information is revealed, it is possible to re-allocate resources.

We describe the resources and demands using:

$$R_{ta} \quad = \quad \text{The number of resources with attribute } a \in \mathcal{A} \text{ in the system at time } t.$$

$$R_t \quad = \quad (R_{ta})_{a \in \mathcal{A}}.$$

$$D_{tb} \quad = \quad \text{The number of demands of type } b \in \mathcal{B} \text{ in the system at time } t.$$

$$D_t \quad = \quad (D_{tb})_{b \in \mathcal{B}}.$$

Both a and b are vectors of attributes of resources and demands. The state of our system is given by

$$S_t \quad = \quad (R_t, D_t).$$

New information is represented as exogenous changes to the resource and demand vectors, as well as to other parameters that govern the problem. These are modeled using:

$$\hat{R}_{t+1,a} \quad = \quad \text{Exogenous changes to } R_{ta} \text{ from information that arrives during time interval } t \text{ (between } t \text{ and } t+1).$$

$$\hat{D}_{t+1,b} \quad = \quad \text{Exogenous changes to } D_{tb} \text{ from information that arrives during time interval } t \text{ (between } t \text{ and } t+1).$$

Our information process, then, is given by

$$W_{t+1} = (\hat{R}_{t+1}, \hat{D}_{t+1}).$$

In a blood management problem, $\hat{R}_{t+1}$ included blood donations. In a model of complex equipment such as aircraft or locomotives, $\hat{R}_{t+1}$ would also capture equipment failures or delays. In a product inventory setting, $\hat{R}_{t+1}$ could represent theft of product. $\hat{D}_{t+1}$ usually represents new customer demands, but can also represent changes to an existing demand or cancelations of orders.

Decisions are modeled using:

$$\mathcal{D}^D = \text{Decision to satisfy a demand with attribute } b \text{ (each decision } d \in \mathcal{D}^D \text{ corresponds to a demand attribute } b_d \in \mathcal{B}).$$

$$\mathcal{D}^M = \text{Decision to modify a resource (each decision } d \in \mathcal{D}^M \text{ has the effect of modifying the attributes of the resource). } \mathcal{D}^M \text{ includes the decision to "do nothing."}$$

$$\mathcal{D} = \mathcal{D}^D \cup \mathcal{D}^M.$$

$$x_{tad} = \text{The number of resources that initially have attribute } a \text{ that we act on with a decision of type } d \in \mathcal{D}.$$

$$x_t = (x_{tad})_{a \in \mathcal{A}, d \in \mathcal{D}}.$$

The decisions have to satisfy constraints such as

$$\sum_{d \in \mathcal{D}} x_{tad} = R_{ta}, \tag{18.2}$$

$$\sum_{a \in \mathcal{A}} x_{tad} \leq D_{tb_d}, \quad d \in \mathcal{D}^D, \tag{18.3}$$

$$x_{tad} \geq 0. \tag{18.4}$$

We let $\mathcal{X}_t$ be the set of x_t that satisfy (18.2)–(18.4). As before, we assume that decisions are determined by a class of decision functions

$$X_t^\pi(S_t) = \text{a function that returns a decision vector } x_t \in \mathcal{X}_t, \text{ where } \pi \in \Pi \text{ is an element of the set of functions (policies) } \Pi.$$

The transition function is given generically by

$$S_{t+1} = S^M(S_t, x_t, W_{t+1}).$$

We now have to deal with each dimension of our state variable. The most difficult, not surprisingly, is the resource vector R_t. This is handled primarily through the attribute transition function

$$a_t^x = a^{M,x}(a_t, d),$$

where a_t^x is the post-decision attribute (the attribute produced by action of type a before any new information has become available). For algebraic purposes, we define the indicator function

$$\delta_{a'}(a, d) = \begin{cases} 1 & \text{if } a' = a_t^x = a^{M,x}(a_t, d), \\ 0 & \text{otherwise.} \end{cases}$$

Using matrix notation, we can write the post-decision resource vector R_t^x using

$$R_t^x = \Delta R_t,$$

where Δ is a matrix in which $\delta_{a'}(a, d)$ is the element in row a' and column (a, d). We emphasize that the function $\delta_{a'}(a, d)$ and matrix Δ are used purely for notational convenience; in a real implementation, we just work with the transition function $a^{M,x}(a_t, d_t)$. The pre-decision resource state vector is given by

$$R_{t+1} = R_t^x + \hat{R}_{t+1}.$$

We model demands in a simple way. If a resource is assigned to a demand, then it is "served" and then it vanishes from the system. Otherwise, it is held to the next time period. Let

$$
\begin{aligned}
\delta D_{tb_d}(x) &= \text{the number of demands of type } b_d \text{ that are served} \\
&\quad\ \text{at time } t, \\
&= \sum_{a \in \mathcal{A}} x_{tad} \quad d \in \mathcal{D}^D, \\
\delta D_t &= (\delta D_{tb})_{b \in \mathcal{B}}.
\end{aligned}
$$

The demand transition function can be written

$$
\begin{aligned}
D_t^x &= D_t - \delta D_t(x), \\
D_{t+1} &= D_t^x + \hat{D}_t.
\end{aligned}
$$

The last dimension of our model is the objective function. For our resource allocation problem, we define a contribution for each decision given by

$$
\begin{aligned}
c_{ad} &= \text{contribution earned (negative if it is a cost) from using} \\
&\quad\ \text{decision } d \text{ acting on resources with attribute } a.
\end{aligned}
$$

The contribution function for time period t is assumed to be linear, given by

$$C(S_t, x_t) = \sum_{a \in \mathcal{A}} \sum_{d \in \mathcal{D}} c_{ad} x_{tad}.$$

The objective function is now given by

$$\max_{\pi \in \Pi} \mathbb{E} \left\{ \sum_{t=0}^{T} C(S_t, X_t^{\pi}(S_t)) | S_0 \right\}.$$

18.2   Values Versus Marginal Values

It is common in dynamic programming to talk about the problem of estimating the value of being in a state. When we are working on resource allocation problems, we have to make the transition from using the value of being in a state (which is extremely high dimensional for these problems) to using the *marginal* value of an additional resource R_{ta} with attribute a (if we are using our multiattribute notation). So, instead of finding a single value $V_t(R_t)$ for a high-dimensional vector R_t, we are going to compute values $\hat{v}_{ta}$ for each $a \in \mathcal{A}$ giving the marginal value of increasing R_{ta}. This means we are going to compute a vector of marginal values $\hat{v}_t = (\hat{v}_{ta})_{a \in \mathcal{A}}$ instead of a single $V_t(R_t)$.

We are going to use the context of resource allocation problems to illustrate the power of using the gradient. In principal, the challenge of estimating the slope of a function is the same as that of estimating the function itself (the slope is simply a different function). However, there can be important, practical advantages to estimating slopes. First, we may be able to approximate $V_t(R_t)$ using a linear approximation, or a piecewise linear, separable approximation.

A second and equally important difference is that if we estimate the value of being in a state, we get one estimate of the value of being in a state when we visit that state. When we estimate a gradient, we get an estimate of a derivative for *each* type of resource *all at the same time*. For example, if $R_t = (R_{ta})_{a \in \mathcal{A}}$ is our resource vector and $V_t(R_t)$ is our value function, then the gradient of the value function with respect to R_t would look like

$$\nabla_{R_t} V_t(R_t) = \begin{pmatrix} \hat{v}_{ta_1} \\ \hat{v}_{ta_2} \\ \vdots \\ \hat{v}_{ta_{|\mathcal{A}|}} \end{pmatrix},$$

where

$$\hat{v}_{ta_i} = \frac{\partial V_t(R_t)}{\partial R_{ta_i}}.$$

There may be additional work required to obtain each element of the gradient, but the incremental work can be far less than the work required to get the value function itself. This is particularly true when the optimization problem naturally returns these gradients (for example, dual variables from a linear program), but this can even be true when we have to resort to numerical derivatives. Once we have all the calculations to solve a problem, solving small perturbations can be relatively inexpensive.

There is one important problem class where finding the value of being in a state is equivalent to finding the derivative. That is the case of managing a single resource. In this case, the state of our system (the resource) is the attribute vector a, and we are interested in estimating the value $V(a)$ of our resource being in state a. Alternatively, we can represent the state of our system using the vector R_t, where $R_{ta} = 1$ indicates that our resource has attribute a (we assume that $\sum_{a \in \mathcal{A}} R_{ta} = 1$). In this case, the value function can be written

$$V_t(R_t) = \sum_{a \in \mathcal{A}} v_{ta} R_{ta}.$$

Here, the coefficient v_{ta} is the derivative of $V_t(R_t)$ with respect to R_{ta}.

In a typical implementation of an approximate dynamic programming algorithm, we would only estimate the value of a resource when it is in a particular state (given by the attribute vector a). This is equivalent to finding the derivative $\hat{v}_a$ only for the value of a where $R_{ta} = 1$. By contrast, computing the gradient $\nabla_{R_t} V_t(R_t)$ implicitly assumes that we are computing $\hat{v}_a$ for each $a \in \mathcal{A}$. There are some algorithmic strategies (we will describe an example of this in section 18.6) where this assumption is implicit in the algorithm. Computing $\hat{v}_a$ for all $a \in \mathcal{A}$ is reasonable if the attribute state space is not too large (for example, if a is a physical location among a set of several hundred locations). If a is a vector, then enumerating the attribute space can be prohibitive (it is, in effect, the "curse of dimensionality" revisited).

Given these issues, it is critical to first determine whether it is necessary to estimate the slope of the value function, or the value function itself. The result can have a significant impact on the algorithmic strategy.

18.3 Piecewise Linear Approximations for Scalar Functions

There are many problems where we have to estimate the value of having a quantity R of some resource (where R is a scalar). We might want to know the value of having R dollars in a budget, R pieces of equipment, or R units of some inventory. R may be discrete or continuous, but we are going to focus on problems where R is either discrete or is easily discretized.

Assume we have a function that is monotonically decreasing, which means that while we do not know the value function exactly, we know that $V(R+1) \leq V(R)$ (for scalar R). If our function is piecewise linear concave, then we will assume that $V(R)$ refers to the slope at R (more precisely, to the right of R). Assume our current approximation $\overline{V}^{n-1}(R)$ satisfies this property, and that at iteration n, we have a sample observation of $V(R)$ for $R = R^n$. If our function is piecewise linear concave, then $\hat{v}^n$ would be a sample realization of a derivative of the function. If we use our standard updating algorithm, we would write

$$\overline{V}^{n}(R^n) = (1 - \alpha_{n-1})\overline{V}^{n-1}(R^n) + \alpha_{n-1}\hat{v}^n.$$

After the update, it is quite possible that our updated approximation no longer satisfies our monotonicity property. We review two strategies for maintaining monotonicity:

The leveling algorithm – A simple method that imposes monotonicity by simply forcing elements of the series which violate monotonicity to a larger or smaller value so that monotonicity is restored.

The CAVE algorithm – If there is a monotonicity violation after an update, CAVE simply expands the range of the function over which the update is applied.

18.3.1 The Leveling Algorithm

The leveling algorithm uses a simple updating logic that can be written as follows:

$$\overline{V}^{n}(y) = \begin{cases} (1 - \alpha_{n-1})\overline{V}^{n-1}(R^n) + \alpha_{n-1}\hat{v}^n & \text{if } y = R^n, \\ \overline{V}^{n}(y) \vee \left\{ (1 - \alpha_{n-1})\overline{V}^{n-1}(R^n) + \alpha_{n-1}\hat{v}^n \right\} & \text{if } y > R^n, \\ \overline{V}^{n}(y) \wedge \left\{ (1 - \alpha_{n-1})\overline{V}^{n-1}(R^n) + \alpha_{n-1}\hat{v}^n \right\} & \text{if } y < R^n, \end{cases} \quad (18.5)$$

where $x \wedge y = \max\{x, y\}$, and $x \vee y = \min\{x, y\}$. Equation (18.5) starts by updating the slope $\overline{V}^{n}(y)$ for $y = R^n$. We then want to make sure that the slopes are declining. So, if we find a slope to the right that is larger, we simply bring it down to our estimated slope for $y = R^n$. Similarly, if there is a slope to the left that is smaller, we simply raise it to the slope for $y = R^n$. The steps are illustrated in Figure 18.3.

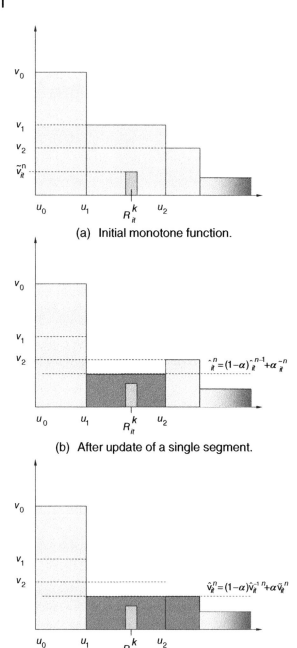

(a) Initial monotone function.

$$\hat{v}_{it}^{n}=(1-\alpha)\hat{v}_{it}^{n-1}+\alpha\tilde{v}_{it}^{n}$$

(b) After update of a single segment.

$$\hat{v}_{it}^{n}=(1-\alpha)\hat{v}_{it}^{n-1}+\alpha\tilde{v}_{it}^{n}$$

(c) After leveling operation.

Figure 18.3 Steps of the leveling algorithm. Figure 18.3a shows the initial monotone function, with the observed R and observed value of the function $\hat{v}$. Figure 18.3b shows the function after updating the single segment, producing a non-monotone function. Figure 18.3c shows the function after monotonicity restored by leveling the function.

18.3.2 The CAVE Algorithm

A particularly useful variation is to perform an initial update (when we compute $\bar{y}$) over a wider interval than just $y = R^n$. Assume we are given a parameter δ^0 which has been chosen so that it is approximately 20 to 50 percent of the maximum value that R^n might take. Now compute $\overline{V}^n(y)$ using

$$
\overline{V}^n(y) = \begin{cases} (1 - \alpha_{n-1})\overline{V}^{n-1}(y) + \alpha_{n-1}\hat{v}^n, & R^n - \delta^n \le y \le R^n + \delta^n, \\ \overline{V}^{n-1}(y) & \text{otherwise.} \end{cases}
$$

Here, we are using $\hat{v}^n$ to update a wider range of the interval. We then apply the same logic for maintaining monotonicity (concavity if these are slopes). We start with the interval $R^n \pm \delta^0$, but we have to periodically reduce δ^0. We might, for example, track the objective function (call it F^n), and update the range using

$$
\delta^n = \begin{cases} \delta^{n-1} & \text{if } F^n \ge F^{n-1} - \epsilon, \\ \max\{1, .5\delta^{n-1}\} & \text{otherwise.} \end{cases}
$$

While the rules for reducing δ^n are generally ad hoc, we have found that this is critical for fast convergence. The key is that we have to pick δ^0 so that it plays a critical scaling role, since it has to be set to be roughly on the order of the maximum value that R^n can take.

The CAVE algorithm, properly tuned, is likely to be the better of the two methods, but tuning is important and introduces an additional step. We suggest using CAVE if you anticipate that you are going to be doing quite a bit of work with a particular problem class.

18.4 Regression Methods

As in chapter 3 we can create regression models where the are manipulations of the number of resources of each type. For example, we might use

$$
\overline{V}(R) = \theta_0 + \sum_{a \in A} \theta_{1a} R_a + \sum_{a \in A} \theta_{2a} R_a^2, \tag{18.6}
$$

where $\theta = (\theta_0, (\theta_{1r})_{r \in R}, (\theta_{2r})_{r \in R})$ is a vector of parameters that are to be determined. The choice of explanatory terms in our approximation will generally reflect an understanding of the properties of our problem. For example, equation (18.6) assumes that we can use a mixture of linear and separable quadratic terms. A more general representation is to assume that we have developed a family $\mathcal{F}$ of basis functions $(\phi_f(R))_{f \in \mathcal{F}}$. Examples of a basis function are

$$\phi_f(R) = R_{a_f}^2,$$

$$\phi_f(R) = \left(\sum_{a \in \mathcal{A}_f} R_a \right)^2 \quad \text{for some subset } \mathcal{R}_f,$$

$$\phi_f(R) = (R_{a_1} - R_{a_2})^2,$$

$$\phi_f(R) = |R_{a_1} - R_{a_2}|.$$

A common strategy is to capture the number of resources at some level of aggregation. For example, if we are purchasing emergency equipment, we may care about how many pieces we have in each region of the country, and we may also care about how many pieces of a type of equipment we have (regardless of location). These issues can be captured using a family of aggregation functions G_f, $f \in \mathcal{F}$, where $G_f(a)$ aggregates an attribute vector a into a space $\mathcal{R}^{(f)}$ where for every basis function f there is an element $a_f \in \mathcal{R}^{(f)}$. Our basis function might then be expressed using

$$\phi_f(R) = \sum_{a \in \mathcal{A}} 1_{\{G_f(a)=a_f\}} R_a.$$

We have written our basis functions purely in terms of the resource vector, but it is possible for them to be written in terms of other parameters in a more complex state vector, such as asset prices.

Given a set of basis functions, we can write our value function approximation as

$$\overline{V}(R|\theta) = \sum_{f \in \mathcal{F}} \theta_f \phi_f(R). \tag{18.7}$$

It is important to keep in mind that $\overline{V}(R|\theta)$ (or more generally, $\overline{V}(S|\theta)$), is any functional form that approximates the value function as a function of the state vector parameterized by θ. Equation (18.7) is a classic linear-in-the-parameters function. We are not constrained to this form, but it is the simplest and offers some algorithmic shortcuts.

The issues that we encounter in formulating and estimating $\overline{V}(R|\theta)$ are the same that any student of statistical regression would face when modeling a complex problem. The major difference is that our data arrives over time (iterations), and we have to update our formulas recursively. Also, it is typically the case that our observations are nonstationary. This is particularly true when an update of a value function depends on an approximation of the value function in the future (as occurs with value iteration or any of the TD(λ) classes of algorithms). When we are estimating parameters from nonstationary data, we do not want to equally weight all observations.

The problem of finding θ can be posed in terms of solving the following stochastic optimization problem

$$\min_{\theta} \mathbb{E} \frac{1}{2} (\overline{V}(R|\theta) - \hat{V})^2.$$

We can solve this using a stochastic gradient algorithm, which produces updates of the form

$$
\begin{aligned}
\bar{\theta}^n &= \bar{\theta}^{n-1} - \alpha_{n-1}(\overline{V}(R^n|\bar{\theta}^{n-1}) - \hat{V}(\omega^n))\nabla_{\theta}\overline{V}(R^n|\theta^n) \\
&= \bar{\theta}^{n-1} - \alpha_{n-1}(\overline{V}(R^n|\bar{\theta}^{n-1}) - \hat{V}(\omega^n)) \begin{pmatrix} \phi_1(R^n) \\ \phi_2(R^n) \\ \vdots \\ \phi_F(R^n) \end{pmatrix}.
\end{aligned}
$$

If our value function is linear in R_t, we would write

$$\overline{V}(R|\theta) = \sum_{a \in \mathcal{A}} \theta_a R_a.$$

In this case, our number of parameters has shrunk from the number of possible realizations of the entire vector R_t to the size of the attribute space (which, for some problems, can still be large, but nowhere near as large as the original state space). For this problem, $\phi(R^n) = R^n$.

It is not necessarily the case that we will always want to use a linear-in-the-parameters model. We may consider a model where the value increases with the number of resources, but at a declining rate that we do not know. Such a model could be captured with the representation

$$\overline{V}(R|\theta) = \sum_{a \in \mathcal{A}} \theta_{1a} R_a^{\theta_{2a}},$$

where we expect $\theta_2 < 1$ to produce a concave function. Now, our updating formula will look like

$$
\begin{aligned}
\theta_1^n &= \theta_1^{n-1} - \alpha_{n-1}(\overline{V}(R^n|\bar{\theta}^{n-1}) - \hat{V}(\omega^n))(R^n)^{\theta_2}, \\
\theta_2^n &= \theta_2^{n-1} - \alpha_{n-1}(\overline{V}(R^n|\bar{\theta}^{n-1}) - \hat{V}(\omega^n))(R^n)^{\theta_2} \ln R^n
\end{aligned}
$$

where we assume the exponentiation operator in $(R^n)^{\theta_2}$ is performed componentwise.

We can put this updating strategy in terms of temporal differencing. As before, the temporal difference is given by

$$\delta_\tau \;=\; C_\tau(R_\tau, x_{\tau+1}) + \overline{V}_{\tau+1}^{n-1}(R_{\tau+1}) - \overline{V}_\tau^{n-1}(R_\tau).$$

The original parameter updating formula (equation (16.7)) when we had one parameter per state now becomes

$$\bar{\theta}^n \;=\; \bar{\theta}_t^{n-1} + \alpha_{n-1} \sum_{\tau=t}^{T} \lambda^{\tau-t} \delta_\tau \nabla_\theta \overline{V}(R^n | \bar{\theta}^n).$$

It is important to note that in contrast with most of our other applications of stochastic gradients, updating the parameter vector using gradients of the objective function requires mixing the units of θ with the units of the value function. In these applications, the stepsize α_{n-1} has to also perform a scaling role.

18.5 Separable Piecewise Linear Approximations

Scalar, piecewise linear functions have proven to be an exceptionally powerful way of solving high dimensional stochastic resource allocation problems. We can describe the algorithm with a minimum of technical details using what is known as a "plant-warehouse-customer" model, which we presented in section 18.1.2. Imagine that we have the problem depicted in Figure 18.4a. We start by shipping "product" out of the four "plant" nodes on the left, and we have to decide how much to send to each of the five "warehouse" nodes in the middle. After making this decision, we then observe the demands at the five "customer" nodes on the right.

We can solve this problem using separable, piecewise linear value function approximations. Assume we have an initial estimate of a piecewise linear value function for resources at the warehouses (setting these equal to zero is fine). This gives us the network shown in Figure 18.4b, which is a small linear program, even when we have hundreds (or thousands) of plant and warehouse nodes. Solving this problem gives us a solution of how much to send to each node.

We then use the solution to the first stage (which gives us the resources available at each warehouse node), take a Monte Carlo sample of each of the demands, and solve a second linear program that sends product from each warehouse to each customer. What we want from this stage is the dual variable for each warehouse node, which gives us an estimate of the marginal value of resources at each node. Note that some care needs to be used here, because these dual variables are not actually estimates of the value of one more resource, but rather are subgradients, which means that they may be the value of the last resource or the next resource, or something in between.

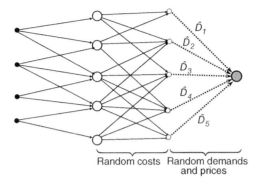

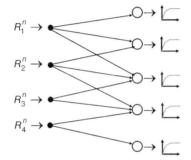

Random costs **Random demands and prices**

(a) The two-stage problem with stochastic second-stage data.

(b) Solving the first stage using a separable, piecewise linear approximation of the second stage.

Duals:

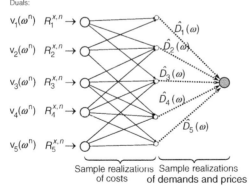

Sample realizations **Sample realizations**
of costs **of demands and prices**

(c) Solving a Monte Carlo realization of the second stage and obtaining dual variables.

Figure 18.4 Steps in estimating separable, piecewise-linear approximations for two-stage stochastic programs.

Finally, we use these dual variables to update the piecewise linear value functions using the methods described earlier. This process is repeated until the solution no longer seems to be improving.

Although we have described this algorithm in the context of a two-stage problem, the same basic strategy can be applied for problems with many time periods. Using approximate value iteration (TD(0)), we would step forward in time, and after solving each linear program we would stop and use the dual variables for the constraints (18.2) to update the value functions from the previous time period (more specifically, around the previous post-decision state). For a finite horizon problem, we would proceed until the last time period, then repeat the entire process until the solution seems to be converging.

With more work, we can implement a backward pass (TD(1)) by avoiding any value function updates until we reach the final time period, but we would have to retain information about the effect of incrementing the resources at each node by one unit (this is best done with a numerical derivative). We would then need to step back in time, computing the marginal value of one more resource at time t using information about the value of one more resource at time $t + 1$. These marginal values would be used to update the value function approximations.

This algorithmic strategy has some nice features:

- This is a very general model with applications that span equipment, people, product, money, energy, and vaccines. It is ideally suited for "single layer" resource allocation problems (one type of resource, rather than pairs such as pilots and aircraft, blood and patients, or trucks and deliveries), although many two-layer problems can be reasonably approximated as single-layer problems.
- The methodology scales to very large problems, with hundreds or thousands of nodes, and tens of thousands of dimensions in the decision vector.
- We do not need to solve the exploration-exploitation problem. A pure exploitation strategy works fine. The reason has to do with the concavity of the value function approximations, which has the effect of pushing suboptimal value functions toward the correct solution.
- Piecewise linear value function approximations are quite robust, and avoid making any simplifying assumptions about the shapes of the value functions.

18.6 Benders Decomposition for Nonseparable Approximations**

While the use of separable, piecewise linear approximations has proven effective (especially for discrete problems where flows need to be integer), the use of a separable approximation will inevitably introduce errors. It is possible to create a nonseparable approximation using an approach called *Benders decomposition* which approximates the value function by minimizing over a set of linear hyperplanes, known as cutting planes.

We are going to begin by presenting the idea of Benders decomposition for a simple two-stage resource allocation problem.

18.6.1 Benders' Decomposition for Two-Stage Problems

Cutting planes represent a powerful strategy for representing concave (or convex if we are minimizing), piecewise-linear functions for multidimensional problems. This method evolved originally as a technique in the 1970s for solving complex integer programs which benefited from separating decision variables into two classes (say, optimizing warehouse locations, and then allocating demands to warehouses). The method was then adapted to the types of sequential decision problems in the early 1990s that arise in two-stage and multistage stochastic resource allocation problem.

Historically, dynamic programming has been viewed as a technique for small, discrete optimization problems, while stochastic programming has been the field that handles uncertainty within math programs (which are typically characterized by high-dimensional decision vectors and large numbers of constraints). The connections between stochastic programming and dynamic programming, historically viewed as diametrically competing frameworks, have been largely overlooked. This section is designed to bridge the gap between stochastic programming and approximate dynamic programming. Our presentation is facilitated by notational decisions (such as our use of x_t for decisions), and our use of the post-decision state variable, which eliminates the expectation from within the maximization problem for each period.

In this section, we are going to put our sampling in the context of an iterative algorithm, where we choose sample ω^n at the nth iteration. This contrasts with our previous style of choosing a fixed sample $w_1, \dots, w_K$. We just want to emphasize that the change in notation reflects a change in the context of how sampling is done. We do this because we may have to choose a sample $w_1^n, \dots, w_K^n$ at the nth iteration.

For example, let R_t be the vector of inventories of product at each of the fulfillment centers, and let x_t be the replenishment decisions that will arrive at time $t + 1$. The decisions x_t have to satisfy a set of constraints that we represent generically as

$$A_t x_t = R_t.$$

These inventories have to be used to satisfy random demands D_{t+1} (which have the same dimensionality as R_t). The inventories R_{t+1} are then given by

$$R_{t+1} = B_t x_t + \hat{R}_{t+1},$$

where x_t is the vector of flows from one facility to the next, and the matrix B_t sums the flows into each facility. Here, we have added in some noise, $\hat{R}_{t+1}$, that might account for damaged or delayed shipments. We would also have observed the demands D_{t+1}, and updated transportation costs c_{t+1}, which are random because of the need to move by for-hire trucking companies. We also note that the matrices A_t and B_t capture travel times; if these are random, then at time t the matrices A_{t+1} and B_{t+1} are also random.

This means the information that is revealed by time $t + 1$ is

$$W_{t+1} = (A_{t+1}, B_{t+1}, c_{t+1}, D_{t+1}, \hat{R}_{t+1}),$$

which in turn gives us our (pre-decision) state at time $t + 1$ as

$$S_{t+1} = (R_{t+1}, A_{t+1}, B_{t+1}, c_{t+1}, D_{t+1}).$$

We are going to simplify our presentation by assuming that $A_{t+1}, B_{t+1}, c_{t+1}$ and D_{t+1} are independent of any previous information (a property known as *inter-stage independence* in the stochastic programming literature), which means that our post-decision state is

$$S_t^x = R_t^x = B_t x_t.$$

Since S_t^x is determined by x_t, the stochastic programming literature writes this state variable as x_t which, while mathematically accurate, is of much higher dimensionality than R_t^x (which could be a scalar if we have a single warehouse that holds inventories). Either representation works fine for what we are going to do.

If we use the pre-decision state, the problem to find x_t at time t is given by

$$\max_{x_t} \left(c_t x_t + \mathbb{E}_{W_{t+1}} V_{t+1}(R_{t+1}, W_{t+1}) \right). \tag{18.8}$$

Note that the information vector W_{t+1} is extremely high dimensional, which would complicate both taking the expectation $\mathbb{E}_{W_{t+1}}$ as well as approximating $V_{t+1}(R_{t+1}, W_{t+1})$. But if we use the post-decision state, we get the much simpler problem

$$\max_{x_t} \left(c_t x_t + V_t^x(R_t^x) \right). \tag{18.9}$$

This problem is solved subject to the constraints,

$$A_t x_t = R_t, \tag{18.10}$$

$$x_t \geq 0, \tag{18.11}$$

where (18.10) represents constraints on how much inventory can be placed in each fulfillment center (captured by R_t).

We then solve the second stage problem (at time $t+1$) to determine x_{t+1}, given the first stage decisions. Assume that we observe outcome ω for the random variable W. We get to see the new information $W_{t+1}(\omega)$ before we compute x_{t+1}, so we capture this by writing $x_{t+1}(\omega)$. The resulting problem would be written

$$V_{t+1}(x_t, W_{t+1}(\omega)) = \max_{x_{t+1}(\omega)} c_{t+1}(\omega) x_{t+1}(\omega), \tag{18.12}$$

subject to, for all $\omega \in \Omega$,

$$A_{t+1}(\omega) x_{t+1}(\omega) \leq R_{t+1}(\omega), \tag{18.13}$$

$$B_{t+1}(\omega) x_{t+1}(\omega) \leq D_{t+1}(\omega), \tag{18.14}$$

$$x_{t+1}(\omega) \geq 0. \tag{18.15}$$

Equation (18.13) imposes flow conservation on the flows of inventories. Equation (18.14) represents the demand constraints, where we assume our contribution vector c_{t+1} is designed to give a high incentive to meet demand. Let $\beta_{t+1}(\omega)$ be the dual variable of the resource constraint (18.13) which reflects the effect of the time t decision x_t on time period $t+1$.

Our strategy will be to replace $V_t^x(x_t)$ with an approximation that is created by generating a series of hyperplanes, and then taking the minimum across these hyperplanes as our approximation. This "approximation" will, in the limit, produce an exact representation of $V_t^x(x_t)$.

The value function $V_{t+1}(x_t, W_{t+1})$ is known in the stochastic programming literature as the *recourse function* since it allows us to respond to different outcomes using the *recourse variables* $x_{t+1}(\omega)$ which are chosen after choosing x_t and observing $W_{t+1}(\omega)$. Thus, we might want to satisfy demand in Texas from a nearby fulfillment center in Houston, but if that center does not have sufficient inventory, our *recourse* is to satisfy demand from a more distant center in Chicago.

We face the challenge of approximating the function $V_{t+1}(x_t) = \mathbb{E} V_{t+1}(x_t, W_{t+1})$ so that we can solve the initial problem for x_t in equation (18.9). It would also be nice if we could do this in a way so that we can solve the first stage problem as a linear program, which makes it easy to handle the vector x_t. There are several strategies we can draw on, but here we are going to illustrate a powerful idea known as Benders decomposition. In a nutshell, our second stage function $V_{t+1}(x_t, W_{t+1})$ is a linear program, which means that it is concave in the right hand side constraint $B_1 x_0$ (because we are maximizing).

We illustrate Benders decomposition in the context of solving a sampled version of the problem. We do this by replacing our original full sample space Ω (over which the original expectation $\mathbb{E}$ is defined) with a sampled set of outcomes $W = (\omega^1, \dots, \omega^N)$. For each solution, we would obtain the optimal value $\hat{V}_{t+1}(x_t, w)$, and the corresponding dual variable $\beta(w)$ for $w \in W$. We then average over the outcomes to create an approximation of the post-decision value function $V_t^x(x_t)$ which we denote $\overline{V}_t^x(x_t)$, given by

$$\overline{V}_t^x(x_t) = \frac{1}{N} \sum_{n=1}^N \hat{V}_{t+1}(x_t, \omega^n).$$

Benders decomposition iteratively builds up an approximation of $V_{t+1}(x_t)$ by constructing a series of *supporting hyperplanes* (see Figure 18.5) derived by solving the second stage linear program for individual samples w of the random vector W_{t+1}. We do this by solving equations (18.12)–(18.15) for the sample realizations $\Omega = \{\omega^n, n = 1, \dots, N\}$, and obtain

$$\alpha_{t+1}^n(\omega^n) = V_{t+1}(x_t, W_{t+1}(\omega^n)),$$
$$\beta_{t+1}^n = \beta_{t+1}(\omega^n).$$

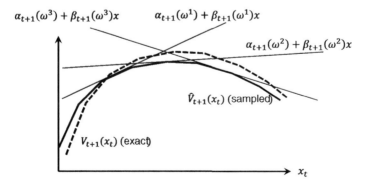

Figure 18.5 Illustration of Benders cuts shown next to exact $(V_{t+1}(x_t))$ and sampled $(\hat{V}_{t+1}(x_t))$ recourse functions.

where $\beta_{t+1}(\omega^n)$ is the dual variable for constraint (18.13). We then solve

$$x_t^* = \arg\max_{x_t, z_t} (c_t x_t + z_t), \tag{18.16}$$

subject to (18.10)–(18.11) and

$$z_t \leq \alpha_{t+1}^n(\omega^n) + \beta_{t+1}^n(\omega^n) x_t, \quad n = 1, \dots, N. \tag{18.17}$$

Equation (18.17) creates a multidimensional envelop, depicted in Figure 18.5, which depicts the sampled function $\hat{V}_{t+1}(x_t)$ and the original true function $V_{t+1}(x_t)$. Note that the hyperplanes touch the sampled function $\hat{V}_{t+1}(x_t)$, but only approximate the true function $V_{t+1}(x_t)$.

Our indexing of time deserves a bit of explanation. The coefficients $\alpha_{t+1}^n(\omega^n)$ and $\beta_{t+1}^n(\omega^n)$ are indexed by $t+1$ because they depend on the specific sampled observation $W_{t+1}(\omega)$ of the new information that becomes known by $t+1$. However, z_t works like an expectation; equation 18.17 is taking the minimum across all these cuts, creating z_t which does not depend on a single realization ω.

The steps of the algorithm implementing this method are shown in Figure 18.6.

We close by noting that this is one way of solving convex problems, but it requires assuming that the sampled approximation will provide a good solution. This has opened a body of literature focusing on the design of good samples, which is challenging in the high dimensional settings of linear programs.

It would be easy to conclude that using multidimensional Benders cuts would be better than using separable, piecewise linear approximations. The separable, piecewise linear approximations are particularly useful when managing discrete resources (trucks, locomotives) since it is much easier to obtain integer solutions when using a piecewise linear approximation where the kinks occur on integer values of R_{ta}.

We compared the two approaches in the setting of managing energy for a fleet of batteries connected on the grid, where the amount of energy being stored in each battery is continuous. Also, because it is easy to move energy between any pair of locations on the grid, we would expect the problem to be highly nonseparable. Figure 18.7 shows the performance of Benders cuts (see the upper bound performance) against separable, piecewise linear value function approximations for grids with 25 batteries (a) and 50 batteries (b).

The results show that the SPWL approximation has slightly faster convergence for the case with 25 batteries. For the grid with 50 batteries, the separable approximation seems to show much faster convergence. We suspect a reason that the separable approximation works so well is that the updates are more efficient; Benders cuts in high dimensions are less efficient because the cuts do

---

Step 0. Initialization:

Step 0a. Initialize V_t^0
Step 0b. Set $n = 1$.

Step 1. Solve

$$x_t^n = \arg\max_{x_t, z_t} (c_t x_t + z_t),$$

subject to

$$z_t \leq \alpha_{t+1}^m(\omega^m) + \beta_{t+1}^m(\omega^m)x_t, \quad m = 1, \dots, n-1.$$

Step 2. For $k = 1, \dots, K$:

$$\hat{V}_{t+1}(x_t^n, W_{t+1}(\omega^k)) = \max_{x_{t+1}(\omega^k)} c_{t+1}(\omega^k)x_{t+1}(\omega^k),$$

subject to (18.13)-(18.15). Obtain dual $\beta_{t+1}^n(\omega^k)$ for equation (18.14) for each ω^k.

Step 3. Compute:

$$\alpha_t^n = \frac{1}{K} \sum_{k=1}^K \hat{V}_{t+1}(x_t^n, \omega^k),$$

$$\beta_t^n = \frac{1}{K} \sum_{k=1}^K \beta_{t+1}^n(\omega^k).$$

Step 4. Increment n. If $n \leq N$ go to Step 1.
Step 5. Return solution x_t^N.

Figure 18.6 Illustration of Benders decomposition for two-stage stochastic optimization using sampled model.

not contribute to the quality of the marginal value of *each* battery, whereas this is not the case with the separable approximations, where each VFA is updated every iteration.

18.6.2 Asymptotic Analysis of Benders with Regularization**

The previous section described the basic idea of Benders decomposition using a fixed sample to represent the uncertainty of the second stage. Here, we present an asymptotic version of Benders that is in the theme of the other iterative algorithms presented in this chapter. This version was first introduced as *stochastic decomposition*. We begin by introducing the basic algorithm, followed by a variant known as regularization that has been found to stabilize performance.

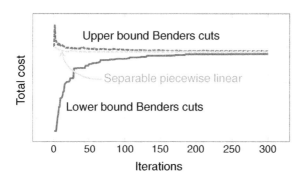

(a) Benders (upper bound) vs. separable, piecewise linear for grid with 25 batteries.

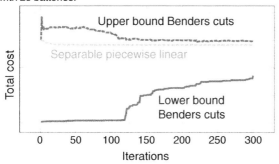

(b) Benders (upper bound) vs. separable, piecewise linear for grid with 50 batteries.

Figure 18.7 Comparison of Benders cuts vs. separable, piecewise linear value function approximations for allocating energy between a fleet of batteries over a power grid.

The basic algorithm

We begin by presenting the two-stage stochastic programming model we first presented in section 18.6:

$$\max_{x_0} \left(c_0 x_0 + \mathbb{E} Q_1(x_0, W) \right), \tag{18.18}$$

subject to

$$A_0 x_0 = b, \tag{18.19}$$

$$x_0 \geq 0. \tag{18.20}$$

We are going to again solve the original problem (18.18) using a series of Benders cuts, but this time we are going to construct them somewhat different. The approximated problem still looks like

$$x^n = \arg\max_{x_0, z}(c_0 x_0 + z), \tag{18.21}$$

subject to (18.19)–(18.20) and

$$z \le \alpha_m^n + \beta_m^n x_0, \quad m = 1, \dots, n - 1. \tag{18.22}$$

Of course, for iteration $n = 1$ we do not have any cuts.

The second stage problem which is solved for a given value $W(\omega)$ which specifies the costs and the demand D_1. In our iterative algorithm, we solve the problem for ω^n, using the solution x_0^n from the first stage

$$Q_1(x_0^n, \omega^n) = \max_{x_1(\omega^n)} c_1(\omega^n) x_1(\omega^n), \tag{18.23}$$

subject to:

$$A_1 x_1(\omega^n) \le B_1 x_0^n, \tag{18.24}$$

$$B_1 x_1(\omega^n) \le D_1(\omega^n), \tag{18.25}$$

$$x_1(\omega^n) \ge 0. \tag{18.26}$$

As before, we let $\hat{\beta}^n$ be the dual variable for the resource constraint (18.24) when we solve the problem using sample ω^n. Then let

$$\alpha_n^n = \frac{1}{n} \sum_{m=1}^{n} Q_1(x_0^m, \omega^m),$$

$$\beta_n^n = \frac{1}{n} \sum_{m=1}^{n} \hat{\beta}^m.$$

Thus, we compute α_n^n by averaging all the prior objective functions for the second stage, and then we compute β_n^n by averaging all the prior dual variables. We finally update all prior α_m^n and β_m^n for $m < n$ using

$$\alpha_m^n = \frac{n-1}{n} \alpha_m^{n-1}, \quad m = 1, \dots, n-1,$$

$$\beta_m^n = \frac{n-1}{n} \beta_m^{n-1}, \quad m = 1, \dots, n-1.$$

Aside from the differences in how the Benders cuts are computed, the major difference between this implementation and our earlier sampled solution given in section 18.6 is that in this recursive formulation, the samples ω are drawn from the full sample space Ω rather than a sampled one. When we solve the sampled version of the problem, we solve it exactly in a finite number of iterations, but we only obtain an optimal solution to a sampled problem. Here, we have an algorithm that will asymptotically converge to the optimal solution of the original problem.

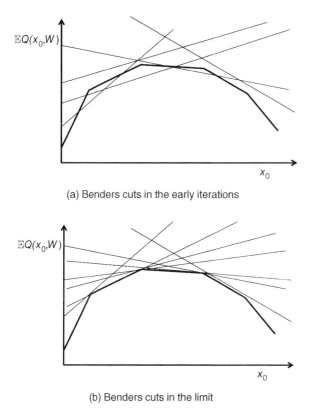

(a) Benders cuts in the early iterations

(b) Benders cuts in the limit

Figure 18.8 Illustration of cuts generated using stochastic decomposition (a) in the early iterations and (b) in the limit.

Figure 18.8 illustrates the cuts generated using stochastic decomposition. It is useful to compare the cuts generated using stochastic decomposition to those generated when we used a sampled version of the problem in section 18.6 as depicted in Figure 18.5. When we were solving our sampled version, we could compute the expectation exactly, which is why the cuts were tight. Here, we are sampling from the full probability space, and as a result we get cuts that approximate the function, but nothing more. However, in the limit as $n \to \infty$, the cuts will converge to the true function in the vicinity of the optimum.

Which is better? Hard to say. While it is nice to have an algorithm that is asymptotically optimal, we can only run a finite number of iterations. The sampled problem will be more stable due to the averaging that takes place in every iteration, but we then have to solve a linear program for every ω in the sampled problem, a step that involves much more computational overhead than the recursive version.

18.6.3 Benders with Regularization

Regularization is a tool that will come up repeatedly when estimating functions from data. The same is true with Benders decomposition. Regularization is handled through a minor modification of the approximate optimization problem (18.21) which becomes

$$x^n = \arg\max_x \left(c_0 x_0 + z + \rho^n (x - \bar{x}^{n-1})^2\right), \tag{18.27}$$

which is solved subject to (18.19)–(18.20) and the Benders cut constraints (18.22). The parameter ρ^n is a decreasing sequence that needs to be scaled to handle the difference in the units between the costs and the term $(x - \bar{x}^{n-1})^2$. $\bar{x}^{n-1}$ is the regularization term which is updated each iteration; the idea with regularization is to keep x^n from straying too far from a previous solution, especially in the early iterations.

The use of the squared deviation $(x - \bar{x}^{n-1})^2$ is known as L_2 regularization, which might be written as $\|x - \bar{x}^{n-1}\|_2^2$. An alternative is L_1 regularization which minimizes the absolute value of the deviation, which would be written as $|x - \bar{x}^{n-1}|$.

There are different ways of setting the regularization term, but the simplest one just uses $\bar{x}^{n-1} = x^{n-1}$. Other ideas involve smoothing several previous iterations. The regularization coefficient is any declining sequence such as

$$\rho^k = r\rho^{k-1}$$

for some factor $r < 1$, starting with an initial ρ^0 that has to be chosen to handle the scaling.

Properly implemented, regularization offers not only theoretical guarantees, but has also been found to accelerate convergence and stabilize the performance of the algorithm.

18.7 Linear Approximations for High-Dimensional Applications

Imagine now that we are going to use basis functions that are linear in the resource variables (not to be confused with linear models that are linear in the parameters). A linear approximation for our value function approximation is given by

$$\overline{V}^n(R_t) = \sum_{a \in \mathcal{A}} \bar{v}_{ta}^n R_{ta},$$

where $\bar{v}_{ta}^n$ is our estimate of the marginal value of resources with attribute a after n iterations. We can estimate these slopes by using the methods just idescribed to obtain $\hat{v}_{ta}^n$ which is our sampled estimate of the marginal value of R_{ta} at iteration n. We then update our estimate of the linear approximation using

$$\bar{v}_{ta}^n = (1 - \alpha_n)\bar{v}_{ta}^{n-1} + \alpha_n\hat{v}_{ta}^n,$$

where α_n is our stepsize (see chapter 6) which might depend on how many times we have observed $\hat{v}_a^n$ for attribute a.

Linear value function approximations can be particularly useful for high-dimensional problems where the space of attributes $\mathcal{A}$ is so large that the elements of R_{ta} are likely to be quite small (mostly zero, sometimes 1). However, this creates the problem that we may have trouble computing $\hat{v}_{ta}^n$ for every $a \in \mathcal{A}$.

Consider, for example, the problem of assigning truck drivers to loads, a is for the attributes of the truck driver. The attribute vector a might include

$$a_1 \;\; = \;\; \text{current location of the driver (or location where he is headed),}$$

$$a_2 \;\; = \;\; \text{the location of where he lives,}$$

$$a_3 \;\; = \;\; \text{how many days he has been driving since he was last at home,}$$

$$a_4 \;\; = \;\; \text{the type of truck trailer he is pulling (for example, a dry van or refrigerated trailer),}$$

$$a_5 \;\; = \;\; \text{the driver's nationality.}$$

The size of this attribute space could easily be in the millions, while we may be optimizing a fleet of 100 or 1000 drivers. Computing $\hat{v}_{ta}^n$, which might require reoptimizing the time t optimization problem to get the marginal value of R_{ta} for each $a \in \mathcal{A}$, would be computationally expensive.

The way to circumvent this "curse of dimensionality" is to use the power of hierarchical aggregation introduced in section 3.6.1. The idea is to create a family of reduced attribute vectors $a^{(g)}$ where $a^{(0)}$ is the original full attribute vector. Then, create a series of more compact vectors $a^{(1)}, a^{(2)}, \ldots, a^{(G)}$, where $a^{(G)}$ might have a single attribute with a small number of values. Hierarchical aggregation maintains a set of weights $w_a^{(g),n}$ that depend on the level of aggregation, as well as the attribute vector a, where

$$\sum_{g=0}^{G} w_a^{(g),n} = 1.$$

These weights are given by one over the variance plus square of the bias, normalized so they sum to 1 (see section 3.6.1 for details).

We then maintain a value of averaged estimates $\bar{v}_a^{(g),n}$, recognizing that we may not have any observations for some aggregation levels g for some (in fact many) attribute vectors a. When this is the case, we set $w_a^{(g),n} = 0$. Finally, we obtain our estimate of the marginal value of R_{ta} using

$$\bar{v}_a^n = \sum_{g=0}^{G} w_a^{(g),n} \bar{v}_a^{(g),n}.$$

18.8 Resource Allocation with Exogenous Information State

The methods in this chapter have demonstrated how we can find effective value function approximations for resource allocation problems, even when the resource vector R_t is very high-dimensional. All of this work assumed that R_t was the *only* state variable. The message here is simple: high-dimensional problems are not hard, as long as we can exploit structure such as concavity (or convexity), or linearity.

The problem is that there are many resource allocation problems where the resource vector is not the only information in the state variable. We might have additional information covering a host of activities: temperature, weather forecasts, market prices, the humidity in a laboratory, competitor behavior ..., the list can be endless. This is information that we would put in our information state variable I_t (see section 9.4).

We need to emphasize that we would only include information in I_t if it is changing over time and, of course, if it affects the behavior of our system. It does not matter if our decisions do or do not affect the trajectory of I_t. When we have an information state variable, then our system state variable is given by (R_t, I_t). The difficulty is that while the value function may be concave (or convex) in R_t, this property typically does not translate to I_t. In particular, we handle I_t as if it is just additional dimensions to R_t, since I_t typically affects how *each* element of R_t affects the value function. So, if we were using our linear value function approximation with slope $\bar{v}_{ta}^n$, we would now want to write $\bar{v}_{ta}^n(I_t)$ to express the dependence on the information state I_t.

Imagine, for example, that we can express our information state as a (not too large) set $\mathcal{J} = \{i_1, i_2, ..., i_{|\mathcal{J}|}\}$. Instead of estimating a single value $\bar{v}_{ta}^n$ for each attribute a, we now have to compute $\bar{v}_{ta}^n(i)$ for $i \in \mathcal{J}$. If there are 10 information states, then we just made our problem 10 times harder. However, I_t could be a multidimensional (and possibly continuous) vector.

There are special cases we can handle. The most important arises when I_{t+1} is independent of I_t (or R_t). For example, I_t might be the attributes of a patient, where we are comfortable assuming that the attributes of the patient arriving at $t + 1$ has nothing to do with the patient that arrived at time t. In this case, the post-decision state $S_t^x = R_t^x$, which means we forget the information state. This is important because we are typically estimating a value function approximation around the post-decision state.

The property that I_{t+1} does not depend on I_t is referred to as "interstage independence" in the stochastic programming community. While convenient, it does not happen very often. Not surprisingly, this issue arises frequently in machine learning. We saw this earlier in section 7.13.6 for a class of active learning problem (also known as a multiarmed bandit problem) known as the *contextual bandit* problem. This discussion offered a novel perspective of this problem, but otherwise did not offer a solution.

A potential solution approach is to draw a page from our work on hierarchical aggregation. Assume we can create a family of information state variables $(I_t^{(0)}, I_t^{(1)}, ... , I_t^{(G)})$ where $I_t^{(0)}$ is our original complete information state, while $I_t^{(g)}$ is a series of successively more aggregate variables for $g = 1, 2, ... G$. Assume that each of these variables can be discretized into a set $\mathcal{I}^{(g)}$ of decreasing size. Finally assume that the set $\mathcal{I}^{(G)}$ is relatively small, meaning that we have no difficult creating estimates for each value in $\mathcal{I}^{(G)}$. We simulate I_t and the identify the corresponding elements in each set $\mathcal{I}^{(g)}$. We can then apply our methods of hierarchical aggregation to create a weighted estimate.

18.9 Closing Notes

This chapter has highlighted a variety of complex resource allocation problems, but our examples are all limited to what are known as *single layer* resource allocation problems. For example, we are managing water in reservoirs, money, blood (of different types), and trucks. These problems lend themselves to the kinds of convex approximations described in this chapter.

Imagine now that we want to manage blood, and we have to serve two types of patients: those requiring emergency surgeries that have to be satisfied now, and elective surgeries that can be delayed. In the case of elective surgeries, we have two classes of resources: the blood and the (elective) patients. Without the presence of elective patients, our post-decision state variables would consist only of the different types of blood.

It is *much* harder to approximate the value of the blood resource vector R_t^{blood} when we have the ability to delay elective surgeries, since now the marginal value of an extra unit of blood with attribute a depends on the set of elective surgeries. Separable approximations are unlikely to work, and as interactions become more complex, we need substantially more Benders cuts to capture these. Requiring integer solutions adds additional complexity.

When the future becomes sufficiently complicated, we often have to turn to direct lookahead policies, which we introduce next in chapter 19.

18.10   Bibliographic Notes

Section 18.1.1 – In operations research, the newsvendor problem (previously known as the "single period inventory problem" or "the newsboy problem"), arises throughout stochastic resource allocation problems. It is a simple problem that makes it useful for illustrating concepts in stochastic optimization. Qing et al. (2011) provides a good general review; Petruzzi and Dada (1999) provides a somewhat older review (but much of the research was done decades ago). There is still continued interest in specialized topics; for example, DeYong (2020) reviews the newsvendor literature related to price-setting research.

Section 18.1.2 – There is an extensive literature exploiting the natural convexity of $Q(x_0, W_1)$ in x_0, starting with Van Slyke and Wets (1969), followed by the seminal papers on stochastic decomposition (Higle and Sen, 1991) and the stochastic dual dynamic programming (SDDP) (Pereira and Pinto, 1991). A substantial literature has unfolded around this work, including Shapiro (2011) who provides a careful analysis of SDDP, and its extension to handle risk measures (Shapiro et al. (2013), Philpott et al. (2013)). A number of papers have been written on convergence proofs for Benders-based solution methods, but the best is Girardeau et al. (2014). Kall and Wallace (2009) and Birge and Louveaux (2011) are excellent introductions to the field of stochastic programming. King and Wallace (2012) is a nice presentation on the process of modeling problems as stochastic programs. A modern overview of the field is given by Shapiro et al. (2014).

Section 18.1.3 – The notation in this section was developed in Powell et al. (2001), and applied in a number of papers, including Simao et al. (2009) (for truckload trucking) and Bouzaiene-Ayari et al. (2016) (for a locomotive management problem).

Section 18.2 – The decision of whether to estimate the value function or its derivative is often overlooked in the dynamic programming literature, especially within the operations research community. In the controls community, use of gradients is sometimes referred to as *dual heuristic dynamic programming* (see Werbos (1992) and Venayagamoorthy and Harley (2002)). The operations research community is very familiar with the idea of using marginal values (see, for example, the methods cited in sections 18.5 and 18.6), while the computer science community (among others), works almost exclusively with problems (such as those defined over graphs) where we need the value of being in a state (rather than the marginal value).

Section 18.3 – The CAVE algorithm was first proposed in Godfrey and Powell (2001) for the newsvendor problem, and then extended to spatial resource allocation problems in fleet management in Powell and Godfrey (2002) and Godfrey and Powell (2002). The theory behind the projective SPAR algorithm is given in Powell et al. (2004). A proof of convergence of the leveling algorithm is given in Topaloglu and Powell (2003). A convergence proof for a version of the piecewise linear, separable approximation is given in Zhou et al. (2020).

Section 18.6 – The first paper to formulate a math program with uncertainty appears to be Dantzig and Ferguson (1956). For a broad introduction to the field of stochastic optimization, see Ermoliev (1988) and Pflug (1996). For complete treatments of the field of stochastic programming, see Shapiro (2003), Birge and Louveaux (2011), Kall and Mayer (2005), and Shapiro et al. (2014). For an easy tutorial on the subject, see Sen and Higle (1999). A very thorough introduction to stochastic programming is given in Ruszczyński and Shapiro (2003). Mayer (1998) provides a detailed presentation of computational work for stochastic programming. There has been special interest in the types of network problems we have considered (see Wallace (1986), S.W. and Wallace (1987), and Birge et al. (1988)). Rockafellar and Wets (1991) presents specialized algorithms for stochastic programs formulated using scenarios. This modeling framework has been of particular interest in the are of financial portfolios (Mulvey and Ruszczyński (1995)). Benders' decomposition for two-stage stochastic programs was first proposed by Van Slyke and Wets (1969) as the "L-shaped" method. Higle and Sen (1991) introduce stochastic decomposition, which is a Monte-Carlo based algorithm that is most similar in spirit to approximate dynamic programming. Chen and Powell (1999) present a variation of Benders that falls between stochastic decomposition and the L-shaped method. The relationship between Benders' decomposition and dynamic programming is often overlooked. A notable exception is Pereira and Pinto (1991), which uses Benders to solve a resource allocation problem arising in the management of reservoirs. This paper presents Benders as a method for avoiding the curse of dimensionality of dynamic programming. For an excellent review of Benders' decomposition for multistage problems, see Ruszczyński (2003). Benders has been extended to multistage problems in Birge (1985), Ruszczyński (1993), and Chen and Powell (1999), which can be viewed as a form of approximate dynamic programming using cuts for value function approximations.

Section 18.7 – High-dimensional applications arise when, for example, we need to estimate $\bar{v}_a$ where $a \in \mathcal{A}$ is a multidimensional vector where the set $\mathcal{A}$ can have a number of elements that far exceeds our budget for observations. This section used hierarchical learning which was developed in George et al. (2008) VFA paper (and described in section 3.6. These methods were used in Simao et al. (2009) a is for the attributes of a truck driver.

Section 18.8 – The stochastic optimization literature has long realized that it is relatively easy to approximate a function $V(R)$ which is concave (convex if minimizing) in R, and where R may be high dimensional. However, there are many problems that involve managing resource allocation problems that combine a resource vector R_t with an exogenously evolving information state I_t, which means the state of the system is $S_t = (R_t, I_t)$. I_t is often relatively unstructured data such as weather, prices, forecasts, the humidity in a laboratory, and so on. The stochastic programming community often assumes "interstage independence" which means that the post-decision state $S_t^x = R_t^x$ (that is, it does not depend on I_t); see Morton (1996), Queiroza and Morton (2013). Asamov and Powell (2018) presents a regularization algorithm that assumes that I_t can take on a "finite" (that is, not too large) number of discrete values $I_1, I_2, \dots, I_K$.

Exercises

Review questions

18.1 Give three examples of resource allocation problems not listed in Table 18.1. Describe the type of resource (or resources) and the decisions that need to be made.

18.2 What is meant by a "two-stage" resource allocation problem? Given an example.

18.3 Equations (18.12)–(18.15) use variables such as $x_{t+1}(\omega)$, $A_{t+1}(\omega)$ and $b_{t+1}(\omega)$? What is meant by ω, and what do we mean when we write it as an argument as we did with these three variables?

18.4 Imagine that we have to solve a newsvendor problem with dynamically varying costs and prices:

$$\max_{x \leq R_t} F(x) = \mathbb{E}\{p_t \max\{x, W_{t+1}\} - c_t x | S_t\}$$

What is the resource state variable? What are the "exogenous information state" variables?

Modeling questions

18.5 Following the general modeling style of section 18.1.3, create your own model of a fleet of autonomous electric vehicles, where the goal is to simulate the dispatching process over the course of the day. The following are some general guidelines to follow when creating your model:

- Assume that you are modeling a region (say a state) that has been divided into a set of zones $z \in \mathcal{Z}$. There may be anywhere between 100 and 10,000 zones depending on the size of the region and the size of the zones.
- Assume you are modeling time in 15 minute increments, over an entire day.
- You will need to model a fleet of vehicles $i \in \mathcal{I}$.
- Let b_{ti} be the battery charge level in vehicle i at time t. You may assume that all vehicles are fully charged at the beginning of the day. Let η^{move} be the rate that each car consumes energy while moving, and η^{idle} for the energy consumption rate while sitting idle.
- Let a_{ti} be the characteristics of vehicle i at time t, which will include current location (if idle), or the location it is heading to if in the middle of a trip; the time period it is expected to arrive (if moving); and the battery charge level b_{ti}.
- Let $\hat{D}_{t+1,zz'}$ be the number of new requests for trips that arrive between t and $t + 1$ to travel from zone z to zone z'. Let $x_{tdi} = 1$ if we choose to implement decision $d \in \mathcal{D}$ for vehicle i.
- You will need to introduce a set of decisions $\mathcal{D}$ where $d \in \mathcal{D}$ can be: move to pick up a customer, move empty, do nothing, move to a recharging station to be recharged (introduce notation for recharging stations).

Set up all five elements of a dynamic model, using $X^{\pi}(S_t)$ as your policy. Then, suggest two policies that you think might work, one from the policy search class, and one from the lookahead class.

Computational exercises

18.6 Consider a newsvendor problem where we solve

$$\max_{x} \mathbb{E}F(x, \hat{D}),$$

where

$$F(x, \hat{D}) = p \min(x, \hat{D}) - cx.$$

We have to choose a quantity x before observing a random demand $\hat{D}$. For our problem, assume that $c = 1$, $p = 2$, and that $\hat{D}$ follows a discrete uniform distribution between 1 and 10 (that is, $\hat{D} = d, d = 1, 2, \dots, 10$ with probability 0.10). Approximate $\mathbb{E}F(x, \hat{D})$ as a piecewise linear function using the methods described in section 18.3, using a stepsize $\alpha_{n-1} = 1/n$. Note that you are using derivatives of $F(x, \hat{D})$ to estimate the slopes of the function. At each iteration, randomly choose x between 1 and 10. Use sample realizations of the gradient to estimate your function. Compute the exact function and compare your approximation to the exact function.

18.7 Repeat exercise 18.6, but this time approximate $\mathbb{E}F(x, \hat{D})$ using a linear approximation:

$$\bar{F}(x) = \theta x.$$

Compare the solution you obtain with a linear approximation to what you obtained using a piecewise-linear approximation. Now repeat the exercise using demands that are uniformly distributed between 500 and 1000. Compare the behavior of a linear approximation for the two different problems.

18.8 Repeat exercise 18.6, but this time approximate $\mathbb{E}F(x, \hat{D})$ using the Leveling algorithm. Start with an initial approximation given by

$$\bar{F}^0(x) = \theta_0(x - \theta_1)^2.$$

Use the recursive regression methods of sections 18.4 and 3.8 to fit the parameters. Justify your choice of stepsize rule. Compute the exact function and compare your approximation to the exact function.

18.9 Repeat exercise 18.6, but this time approximate $\mathbb{E}F(x, \hat{D})$ using the regression function given by

$$\bar{F}(x) = \theta_0 + \theta_1 x + \theta_2 x^2.$$

Use the recursive regression methods of sections 18.4 and 3.8 to fit the parameters. Justify your choice of stepsize rule. Compute the exact function and compare your approximation to the exact function. Estimate your value function approximation using two methods:

(a) Use observations of $F(x, \hat{D})$ to update your regression function.
(b) Use observations of the derivative of $F(x, \hat{D})$, so that $\bar{F}(x)$ becomes an approximation of the derivative of $\mathbb{E}F(x, \hat{D})$.

18.10 Approximate the function $\mathbb{E}F(x, \hat{D})$ in exercise 18.6, but now assume that the random variable $\hat{D} = 1$ (that is, it is deterministic). Using the following approximation strategies:

(a) Use a piecewise linear value function approximation. Try using both left and right derivatives to update your function.

(b) Use the regression $\bar{F}(x) = \theta_0 + \theta_1 x + \theta_2 x^2$.

18.11 We are going to solve the basic asset acquisition problem (section 8.2.1) where we purchase assets (at a price p^p) at time t to be used in time interval $t + 1$. We sell assets at a price p^s to satisfy the demand $\hat{D}_t$ that arises during time interval t. The problem is to be solved over a finite time horizon T. Assume that the initial inventory is 0 and that demands follow a discrete uniform distribution over the range $[0, D^{max}]$. The problem parameters are given by

$$
\begin{aligned}
\gamma &= 0.8, \\
D^{max} &= 10, \\
T &= 20, \\
p^p &= 5, \\
p^s &= 8.
\end{aligned}
$$

Solve this problem by estimating a piecewise linear value function approximation (section 18.3). Choose $\alpha_{n+1} = a/(a + n)$ as your step-size rule, and experiment with different values of a (such as 1, 5, 10, and 20). Use a single-pass algorithm, and report your profits (summed over all time periods) after each iteration. Compare your performance for different stepsize rules. Run 1000 iterations and try to determine how many iterations are needed to produce a good solution (the answer may be substantially less than 1000).

18.12 Repeat exercise 18.11, but this time use the Leveling algorithm to approximate the value function. Use as your initial value function approximation the function

$$
\bar{V}_t^0(R_t) = \theta_0(R_t - \theta_2)^2.
$$

For each of the exercises that follow, you may have to tweak your step-size rule. Try to find a rule that works well for you (we suggest sticking with a basic $a/(a + n)$ strategy). Determine an appropriate number of training iterations, and then evaluate your performance by averaging results over 100 iterations (testing iterations) where the value function is not changed.

(a) Solve the problem using $\theta_0 = 1, \theta_1 = 5$.

(b) Solve the problem using $\theta_0 = 1, \theta_1 = 50$.

(c) Solve the problem using $\theta_0 = 0.1, \theta_1 = 5$.

(d) Solve the problem using $\theta_0 = 10, \theta_1 = 5$.

(e) Summarize the behavior of the algorithm with these different parameters.

18.13 Repeat exercise 18.11, but this time assume that your value function approximation is given by

$$\overline{V}_t^0(R_t) = \theta_0 + \theta_1 R_t + \theta_2 R_t^2.$$

Use the recursive regression techniques of sections 18.4 and 3.8 to determine the values for the parameter vector θ.

18.14 Repeat exercise 18.11, but this time assume you are solving an infinite horizon problem (which means you only have one value function approximation).

18.15 Repeat exercise 18.13, but this time assume an infinite horizon.

18.16 Repeat exercise 18.11, but now assume the following problem parameters:

$$\gamma = 0.99,$$
$$T = 200,$$
$$p^p = 5,$$
$$p^s = 20.$$

For the demand distribution, assume that $\hat{D}_t = 0$ with probability 0.95, and that $\hat{D}_t = 1$ with probability 0.05. This is an example of a problem with low demands, where we have to hold inventory for a fairly long time.

Sequential decision analytics and modeling

These exercises are drawn from the online book *Sequential Decision Analytics and Modeling* available at http://tinyurl.com/sdaexamplesprint.

18.17 Read sections 13.1–13.4 on the blood management problem. An approximate dynamic programming algorithm has been implemented in Python, which can be downloaded from http://tinyurl.com/sdagithub using the module "BloodManagement."

(a) Use the ADP algorithm to create piecewise linear value function approximations, and simulate the resulting policy. Report the objective function you obtain from simulating the policy.

(b) Now set all the VFA's equal to zero and simulate the resulting myopic policy. Compare the performance of the myopic policy to the VFA-based policy.

(c) Increase the supply of blood by 50 percent (multiply all input supplies by 1.5), and repeat the comparison of parts a and b. How did the relative value of the VFA-based policy change in the presence of inflated supplies of blood?

(d) Repeat part (c), but this time multiply the supplies by 0.80. How does this affect the results?

Diary problem

The diary problem is a single problem you chose (see chapter 1 for guidelines). Answer the following for your diary problem.

18.18 This chapter is designed for problems that enjoy the property of concavity (convexity if minimizing). This often arises in the context of resource allocation problems, but may arise in other settings as well. Identify any state variables where you believe the value function would be concave (or convex) with respect to this (these) state variables, and if these are present in your problem, describe an approximation architecture that exploits this property.

Bibliography

Asamov, T. and Powell, W.B. (2018). Regularized decomposition of high dimensional multistage stochastic programs with Markov uncertainty. *SIAM Journal on Optimization* 28 (1): 575–595.

Birge, J.R. (1985). Decomposition and partitioning techniques for multistage stochastic linear programs. *Mathematical Programming* 33: 25–41.

Birge, J. R. and Louveaux, F. (2011). *Introduction to Stochastic Programming*, 2e. New York: Springer.

Birge, J.R., Wallace, S.W. (1988). A separable piecewise linear upper bound for stochastic linear programs. *SIAM Journal of Control and Optimization* 26: 1–14.

Bouzaiene-Ayari, B., Cheng, C., Das, S., Fiorillo, R., and Powell, W.B. (2016). From single commodity to multiattribute models for locomotive optimization: A comparison of optimal integer programming and approximate dynamic programming. *Transportation Science* 50 (2): 1–24.

Chen, Z.-L., and Powell, W.B. (1999). A convergent cutting-plane and partial-sampling algorithm for multistage linear programs with recourse. *Journal of Optimization Theory and Applications* 103 (3): 497–524.

Dantzig, G.B. and Ferguson, A. (1956). The allocation of aircrafts to routes: An example of linear programming under uncertain demand. *Management Science* 3: 45–73.

DeYong, G. (2020). The price-setting newsvendor: Review and extensions. *International Journal of Production Research* 58 (6): 1776–1804.

Ermoliev, Y. (1988). Stochastic quasigradient methods. In: *Numerical Techniques for Stochastic Optimization* (eds. Y. Ermoliev and R. Wets). Berlin: Springer-Verlag.

George, A., Powell, W.B., and Kulkarni, S. (2008). Value function approximation using multiple aggregation for multiattribute resource management. *Journal of Machine Learning Research*. 2079–2111.

Girardeau, P., Leclere, V., and Philpott, A.B. (2014). On the convergence of decomposition methods for multistage stochastic convex programs. *Mathematics of Operations Research* 40 (1): 130–145.

Godfrey, G.A. and Powell, W.B. (2001). An Adaptive, distribution-free algorithm for the newsvendor problem with censored demands, with applications to inventory and distribution. *Management Science* 47 (8): 1101–1112.

Godfrey, G.A. and Powell, W.B. (2002). An adaptive dynamic programming algorithm for dynamic fleet management, II: Multiperiod travel times. *Transportation Science* 36 (1): 40–54.

Higle, J.L. and Sen, S. (1991). Stochastic decomposition: An algorithm for two-stage linear programs with recourse. *Mathematics of Operations Research* 16 (3): 650–669.

Kall, P. and Mayer, J. (2005). *Stochastic Linear Programming: Models, Theory, and Computation*. New York: Springer.

Kall, P. and Wallace, S.W. (2009). *Stochastic Programming*, Vol. 10, Hoboken, NJ: John Wiley & Sons.

King, A.J. and Wallace, S.W. (2012). *Modeling with Stochastic Programming*. New York: Springer Verlag.

Mayer, J. (1998). *Stochastic Linear Programming Algorithms: A Comparison Based on a Model Management System*. Springer.

Morton, D.P. (1996). An enhanced decomposition algorithm for multistage stochastic hydroelectric scheduling. *Annals of Operations Research* 64: 211–235.

Mulvey, J.M. and Ruszczyński, A. (1995). A new scenario decomposition method for large scale stochastic optimization. *Operations Research* 43: 477–490.

Pereira, M.F. and Pinto, L.M.V.G. (1991). Multistage stochastic optimization applied to energy planning. *Mathematical Programming* 52: 359–375.

Petruzzi, N. C. and Dada, M. (1999). Pricing and the Newsvendor Problem: A Review with Extensions. *Operations Research* 47: 183–194.

Pflug, G. (1996). *Optimization of Stochastic Models: The Interface Between Simulation and Optimization, Kluwer International Series in Engineering and Computer Science: Discrete Event Dynamic Systems*, Boston: Kluwer Academic Publishers.

Philpott, A.B., De Matos, V., and Finardi, E. (2013). On solving multistage stochastic programs with coherent risk measures. *Operations Research* 51 (4): 957–970.

Powell, W.B. and Godfrey, G.A. (2002). An adaptive dynamic programming algorithm for dynamic fleet management, I: Single period travel times. *Transportation Science* 36 (1): 40–54.

Powell, W.B., Ruszczyński, A., and Topaloglu, H. (2004). Learning algorithms for separable approximations of discrete stochastic optimization problems. *Mathematics of Operations Research* 29 (4): 814–836.

Powell, W.B., Simao, H.P., and Shapiro, J.A. (2001). A representational paradigm for dynamic resource transformation problems. In: *Annals of Operations Research* (eds. F.C. Coullard and J. Owens), 231–279. J. C. Baltzer AG.

Queiroza, A.R. and Morton, D.P. (2013). Sharing cuts under aggregated forecasts when decomposing multi-stage stochastic programs. *Operations Research Letters* 41 (3): 311–316.

Rockafellar, R.T. and Wets, R.J.B. (1991). Scenarios and policy aggregation in optimization under uncertainty. *Mathematics of Operations Research* 16 (1): 119–147.

Ruszczyński, A. (1993). Parallel decomposition of multistage stochastic programming problems. *Mathematical Programming* 58: 201–228.

Ruszczyński, A. (2003). *Decomposition Methods*. Amsterdam: Elsevier.

Ruszczyński, A. and Shapiro, A. (2003). *Handbooks in Operations Research and Management Science: Stochastic Programming*, Vol. 10, Amsterdam: Elsevier.

Sen, S. and Higle, J.L. (1999). An introductory tutorial on stochastic linear programming models. *Interfaces* (April): 33–61.

Shapiro, A. (2003). *Stochastic Programming*, Vol. 10, Chichester, U.K.: John Wiley & Sons.

Shapiro, A. (2011), Analysis of stochastic dual dynamic programming method. *European Journal of Operational Research* 209 (1): 63–72.

Shapiro, A., Dentcheva, D., and Ruszczyński, A. (2014). *Lectures on Stochastic Programming: Modeling and Theory*, 2e. Philadelphia: SIAM.

Shapiro, A., Tekaya, W., Da Costa, J.P., and Soares, M.P. (2013). Risk neutral and risk averse Stochastic Dual Dynamic Programming method. *European Journal of Operational Research* 224 (2): 375–391.

Simao, H.P., Day, J., George, A.P., Gifford, T., Powell, W.B., and Nienow, J. (2009). An approximate dynamic programming algorithm for large-scale fleet management: A case application. *Transportation Science* 43 (2): 178–197.

Wallace, S.W. (1987). A piecewise linear upper bound on the network recourse function. *Mathematical Programming* 38: 133–146.

Topaloglu, H. and Powell, W.B. (2003). An algorithm for approximating piecewise linear concave functions from sample gradients. *Operations Research Letters* 31: 66–76.

Van Slyke, R.M. and Wets, R.J.B. (1969). L-shaped linear programs with applications to optimal control and stochastic programming. *SIAM Journal of Applied Mathematics* 17: 638–663.

Venayagamoorthy, G. and Harley, R. (2002). Comparison of heuristic dynamic programming and dual heuristic programming adaptive critics for neurocontrol of a turbogenerator. *Neural Networks, IEEE* 13 (3): 764–773.

Wallace, S.W. (1986). Solving stochastic programs with network recourse. *Networks* 16: 295–317.

Werbos, P.J. (1992). Approximate dynamic programming for real-time control and neural modelling. In: *Handbook of Intelligent Control: Neural, Fuzzy, and Adaptive Approaches* (eds. D.J. White and D.A. Sofge).

Qing, Y., Wang, R., Vakharia, A.J., Chen, Y., Seref, M.M. (2011). The newsvendor problem: Review and directions for future research. *European Journal of Operational Research* 213: 361–374.

Zhou, S., Wang, F., Shi, N., and Xu, Z. (2020). A new convergent hybrid learning algorithm for two-stage stochastic programs. *European Journal of Operational Research* 283 (1): 33–46.

19

Direct Lookahead Policies

Up to now we have considered three classes of policies: policy function approximations (PFAs), parametric cost function approximations (CFAs), and policies that depend on value function approximations (VFAs) which approximate the impact of a decision on the future through the state variable. All three of these policies depend on approximating some function, which means we are limited by our ability to create approximations that work well in practice.

Not surprisingly, we cannot always develop sufficiently accurate functional approximations. Policy function approximations have been most successful when decisions are simple decisions (think of buy low, sell high policies) or low-dimensional continuous controls that can be approximated using parametric or nonparametric functions (these might range from a linear function to a neural network). Cost function approximations require a deterministic model that provides a reasonable approximation. Value function approximations work well when the value function exhibits structure that can be exploited using the family of approximating architectures we presented in chapter 3 or chapter 18.

When all else fails (and it often does), we have to resort to direct lookahead policies (DLAs), which optimize over some horizon to help capture the impact of decisions made now on activities in the future, from which we can extract the decision we would make now. A few examples of problems which are likely going to require a full direct lookahead policy are:

■ **EXAMPLE 19.1**

Knowing whether to turn left or right at an intersection requires planning your path all the way to the destination.

Reinforcement Learning and Stochastic Optimization: A Unified Framework for Sequential Decisions, First Edition. Warren B. Powell.
© 2022 John Wiley & Sons, Inc. Published 2022 John Wiley & Sons, Inc.

■ **EXAMPLE 19.2**

Imagine a hurricane moving through a region, as often happens in the southeastern US as well as southeast Asia. It is necessary to evacuate regions, but given the presence of network bottlenecks, regions need to be evacuated in a coordinated way. It is important to plan an evacuation for each zone z at each point in time to identify the zones where it is critical to evacuate now given the current state of the hurricane.

■ **EXAMPLE 19.3**

Knowing whether you can use your valuable supply of O-negative blood (which everyone can use) depends on the planned flow of donations and surgeries, as well as the age of your current O-negative inventories.

■ **EXAMPLE 19.4**

A financial planner planning for your retirement needs to estimate the risk that your portfolio might not hit a target for you to retire comfortably. This assessment will determine what investments to make now.

Direct lookahead policies represent a much more brute force approach for making decisions and, not surprisingly, are typically very hard computationally. As a result, the challenge here is introducing approximations that make this problem tractable. We divide these approximations into two broad categories:

Deterministic lookahead – This is the most common approach in practice when we need to use a direct lookahead policy. There are problems where this works quite well (think of navigation systems helping you find a good path over a dynamically varying transportation network), but there are settings where deterministic approximations simply will not work.

Stochastic lookahead – When we need a direct lookahead policy but where a deterministic lookahead would ignore critical issues, then we have to explore the world of stochastic lookahead models, which will be the focus of most of this chapter.

Our presentation is organized as follows:

Part I: Foundational material – This consists of the following general topics for DLA policies:

Section 19.1 – Creating the optimal direct lookahead policy. This serves as the foundation for any direct lookahead approximation.

Section 19.2 – We present the notation we use for our approximate lookahead model, and discuss the different approximation strategies.

Section 19.3 – We discuss the idea of using objectives for the lookahead model that are different than the base model.

Section 19.4 – We return to the familiar territory of evaluating a policy, a dimension that is often overlooked in the context of DLA policies.

Section 19.5 – We close Part I of this chapter with a discussion of why a DLA policy might be appropriate.

Part II: Deterministic DLA models – Section 19.6 discusses the simple but popular idea of using deterministic lookahead models.

Part III: Stochastic DLA models – We divide this substantial topic as follows:

Section 19.7 – We start with a quick tour through the four classes of policies, but discussed in the context of using them within a DLA model, where the tradeoff between computational cost and solution quality changes from when they are used in the base model:

Section 19.7.1 – Lookahead PFAs tend to be popular when available because they are computationally the easiest, but they are not the easiest to design.

Section 19.7.2 – Lookahead CFAs can be easier to tune if a deterministic approximation is available.

Section 19.7.3 – We describe strategies for using approximate dynamic programming, and particularly backward ADP, in the context of a lookahead model.

Section 19.7.4 – Using a DLA policy within a lookahead DLA model will be computationally demanding, but may be a necessary fallback.

Given this tour, we now present a few specialized results that have attracted considerable attention from different communities.

Section 19.8 – This is a thorough presentation of the popular idea of Monte Carlo tree search for problems with discrete actions. We cover both classical (pessimistic) MCTS and optimistic MCTS.

Section 19.9 – Next we cover two-stage stochastic programming which is widely used in the research literature for vector-valued decisions.

19.1 Optimal Policies Using Lookahead Models

Direct lookahead policies are best described by restating our objective function

$$F(S_0) = V_0(S_0) = \max_{\pi \in \Pi} \mathbb{E} \left\{ \sum_{t'=0}^{T} C(S_{t'}, X_{t'}^{\pi}(S_{t'})) | S_0 \right\}. \tag{19.1}$$

We remind the reader of our habit of always conditioning the expectation on the starting state S_0; if you change the starting state, it can have an impact on the optimal policy. This means that we should technically be writing our optimal policy as a function of S_0, as in $\pi^*(S_0)$. Up to now we have typically left this dependence implicit, but as we progress, we will see that we need to remind ourselves of this dependence.

Now imagine solving equation (19.1) starting at time t,

$$V_t(S_t) = \max_{\pi \in \Pi} \mathbb{E} \left\{ \sum_{t'=t}^{T} C(S_{t'}, X_{t'}^{\pi}(S_{t'})) | S_t \right\}, \tag{19.2}$$

which we can also write as

$$V_t(S_t) = \max_{x_t \in \mathcal{X}_t} \Bigg(C(S_t, x_t) +$$

$$\mathbb{E}_{S_t} \mathbb{E}_{W_{t+1}|S_t} \left\{ \max_{\pi \in \Pi} \mathbb{E}_{S_{t+1}} \mathbb{E}_{W_{t+1}\cdots W_T|S_{t+1}} \left\{ \sum_{t'=t+1}^{T} C(S_{t'}, X_{t'}^{\pi}(S_{t'})) | S_{t+1} \right\} | S_t, x_t \right\} \Bigg), \tag{19.3}$$

where $\mathcal{X}_t$ are the constraints given the state S_t (which is implicit by indexing the constraint set by time t). We have written the expectations in (19.4) in the full expanded form. The expectations over S_t (or the imbedded S_{t+1}) would also handle any belief states. It is easy to see looking at (19.4) why we rarely use the expanded form, and instead we just write

$$V_t(S_t) = \max_{x_t \in \mathcal{X}_t} \Bigg(C(S_t, x_t) + \mathbb{E} \left\{ \max_{\pi \in \Pi} \mathbb{E} \left\{ \sum_{t'=t+1}^{T} C(S_{t'}, X_{t'}^{\pi}(S_{t'})) | S_{t+1} \right\} | S_t, x_t \right\} \Bigg). \tag{19.4}$$

We just caution the reader that when you see equation (19.4), we mean the expression in equation (19.3).

We can now write this as a policy for making the decision x_t at time t:

$$X_t^{DLA}(S_t) = \arg\max_{x_t} \Bigg(C(S_t, x_t) + \mathbb{E} \left\{ \underbrace{\max_{\pi \in \Pi} \mathbb{E} \left\{ \sum_{t'=t+1}^{T} C(S_{t'}, X_{t'}^{\pi}(S_{t'})) | S_{t+1} \right\}}_{V_{t+1}(S_{t+1})} | S_t, x_t \right\} \Bigg). \tag{19.5}$$

We note in passing that if we could compute (19.5), the policy $X_t^{DLA}(S_t)$ would be the optimal policy.

If we were able to compute the value function $V_{t+1}(S_{t+1})$ in equation (19.4), we would be able to write our policy as

$$X_t^{DLA}(S_t) = \arg\max_{x_t} \left(C(S_t, x_t) + \mathbb{E}\{V_{t+1}(S_{t+1})|S_t, x_t\} \right). \qquad (19.6)$$

or, using the post-decision value function $V_t^x(S_t^x)$ which we introduced in chapter 15,

$$X_t^{DLA}(S_t) = \arg\max_{x_t} \left(C(S_t, x_t) + V_t^x(S_t^x) \right). \qquad (19.7)$$

Remember that the optimization problem using the post-decision state is a deterministic problem (that is, no imbedded expectation), which opens the door to problems where x_t is a vector (possibly even a very high-dimensional vector), as long as we approximate $V_t^x(S_t^x)$ with a function that is linear or concave in x_t.

Of course, the versions of the policies using value functions are attractive because they are so compact, but if we could compute them, or even develop reasonable approximations, we would be drawing on the techniques we introduced in chapters 14–18. The reason we have direct lookahead policies is specifically for the many problems where value functions are simply not effective. Some examples of problems where the value of the future is not easy to approximate include:

- Problems with complex interactions – Imagine a stochastic scheduling problem (routing vehicles, scheduling machines, scheduling doctors, etc.) which involve complex interactions in the future between different types of resources (vehicles and loads; machines and jobs; doctors, nurses, and patients). To make a decision now (for example, to commit to serve a job or patient in the future), it is necessary to explicitly plan the schedule in the future.
- Problems with forecasts – Consider a problem of managing inventories of products over a holiday, where we have a forecast $f_t = (f_{tt'})_{t' \geq t}$ for the demands. Since forecasts evolve over time, they should be a part of the state variable, but this is never done. The most natural approach is to model forecasts as latent variables (which means we ignore that the forecasts themselves are changing), but this requires that we model the problem over a planning horizon.
- Multilayer resource allocation problems – Value functions can be effective when modeling single layer resource allocation problems (managing water,

blood, money, trucks, etc.), but many problems involve multiple layers (jobs and machines, trucks and packages, blood and patients). It is very hard to capture the value of machines (for example) when it depends on the number of jobs.

The policy in equation (19.5) can look daunting, but this is the starting point for the rest of this chapter. Figure 19.1 helps to explain it by drawing parallels with a decision tree which we first saw back in section 2.1.2. Figure 19.1 uses the convention that square nodes represent pre-decision states S_t (or $S_{t'}$ for $t' > t$) where decisions are made. Solid lines are decisions; dashed lines represent random outcomes, over which we have to take expectations. Keep in mind that we would never use a decision tree if, for example, x_t was a vector. In this chapter, we are going to keep open the possibility that x_t is a vector, which means the idea of enumerating all possible values of x_t is by itself intractable.

Whenever we see expectations, we generally assume that we cannot compute it since random variables can be continuous and/or vector-valued. However, Monte Carlo simulation is such a powerful tool, it is safe to assume that we are going to turn to Monte Carlo methods to approximate any expectation.

Figure 19.1 draws a line from the policy π that we are maximizing over to each decision node starting at time $t + 1$. This reflects the property that the policy π in the lookahead model has to specify the decision for *each* decision node, from time $t + 1$ onward. Designing this policy is easily the most difficult aspect

$$X_t^*(S_t) = argmax_x \left(C(S_t, x_t) + \mathbb{E} \left\{ \max_\pi \mathbb{E} \left\{ \sum_{t'=t+1}^{T} C\left(S_t, X_{t'}^\pi(S_{t'})\right) | S_{t+1} \right\} | S_t, x_t \right\} \right)$$

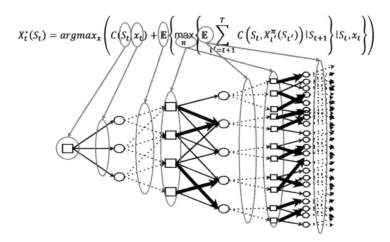

Figure 19.1 Relationship between the lookahead policy in equation (19.5) and a decision tree.

of our lookahead policy in equation (19.5). We refer to this policy as the *lookahead policy*, but we also sometimes call it the "policy within a policy." Here, not only do we face the problem of searching over policies (a challenge we have been addressing since chapter 11), we have to do this *within the expectation!* This means we do this for *every* x_t and *every* outcome of W_{t+1} (more specifically, for every state S_{t+1})!! Needless to say we are going to need to design some shortcuts.

It should not be lost on us that the problem of finding a lookahead policy starting at time $t + 1$ is really the same problem we are facing of designing a policy for time t (or time 0). All we have done is to push the problem out one time period. However, we also need to recognize that we are never going to actually implement the decisions produced by the lookahead policy; these are just to help us make a better decision now. We are going to exploit this observation by making the argument that we can better tolerate a suboptimal policy in the lookahead model than we can in the base model.

There are entire fields of research that pursue different strategies for approximating the lookahead model. For example:

Rolling horizon procedure – Also known as a "receding horizon procedure," it refers to the process of optimizing over an interval $(t, t + H)$, implementing decisions for time t, rolling to $t + 1$ (and sampling/observing new information), and then solving over the interval $(t + 1, t + H + 1)$ (hence the name "rolling horizon"). Most of the time rolling horizon procedures use deterministic approximations of the future.

Model predictive control – This is the term used in the engineering-controls community, and refers to the fact that if we create a lookahead model, we need an explicit model of the problem, which means this is just another name for "direct lookahead policy." The controls literature in engineering focuses mostly on deterministic problems, and as a result, MPC (the standard abbreviation for model predictive control) is typically associated with deterministic models of the future, and MPC is most often associated with deterministic lookaheads.

Stochastic programming – The stochastic programming community simplifies the lookahead model in several ways, but the two most important are (a) we represent future uncertainties with a sampled set of scenarios, and (b) we simplify the future by assuming that we first see the entire future,and then optimize given that we are allowed to see the entire future (this is known as a "two-stage" approximation, reviewed shortly).

Monte Carlo tree search – MCTS is an algorithmic strategy (which we review shortly) for adaptively searching in a forward manner a stochastic search tree for problems with discrete action spaces. Asymptotically it will explore the entire search tree, which means it can be used to solve the base problem, but in practice it is a partial search, which means it is a direct lookahead policy.

Rollout policies – Here we assume that the lookahead policy is replaced with a simple function that is easy to compute, but technically this is just another word for any approximate policy.

Robust optimization – The field of robust optimization replaces the model of uncertainty with an "uncertainty set" where future outcomes $W_{t+1}, \dots, W_{t+H}$ falls within a bounded region, and we then maximize $x_{t+1}, \dots, x_{t+H}$ for a particular realization $w_{t+1}, \dots, w_{t+H}$ which represents the worst outcome given the decisions.

Approximate dynamic programming – There are many applications of approximate dynamic programming which are actually implemented in the context of a stochastic lookahead model (but the authors forgot to tell you). Often, it is specifically the approximations that we are going to introduce next for creating approximate lookahead models that makes ADP possible.

Our entire presentation is centered on the idea of replacing the lookahead model with an approximation that is easier to solve, although we do present approaches that will solve the lookahead model to optimality, at least asymptotically. We begin by providing a notational system for modeling an approximate lookahead model, and then present different approximation strategies. The remainder of the chapter is then dedicated to describing these strategies in greater depth.

19.2 Creating an Approximate Lookahead Model

The art of direct lookahead policies is centered on replacing the true model (that is, our base model) with an approximate lookahead model that captures the most important aspects of our problem, while introducing simplifications that make the lookahead model tractable.

There are many research papers solving "stochastic, dynamic models" which are actually stochastic lookahead models with variables that are being held constant, which means they are not even discussed. These are not always obvious. In fact, whether a stochastic, dynamic model is a lookahead model or a base model often depends on how it is used. If we just want to implement the decision in the first period, after which new information arrives and the entire process repeats, it is a lookahead model. However, we might use our stochastic, dynamic model to test the effect of changes in input parameters over the entire horizon. In this case, the model is a base model being used as a simulator.

Two more examples illustrate this issue:

■ **EXAMPLE 19.5**

Brazil uses stochastic optimization models to plan their use of hydroelectric reservoirs. The plan optimizes over a 10 year period, and can be used in one of two ways. The first is to determine the flow of water between reservoirs over the upcoming week. This process is repeated each week. Used in this way, it is a stochastic lookahead model. However, the same model can be used to test changing the capacity of pumps for moving water between reservoirs. In this setting, the model is used as a simulator (that is, a base model).

■ **EXAMPLE 19.6**

A stochastic model for optimizing the movement of trucks between locations can be used to determine how trucks should be dispatched. This model, which optimizes flows over a week, can then be updated every hour with new forecasts. Used in this way, it is a lookahead model. However, the same model can be used to simulate the effect of different fleet sizes, in which case the model is a base model used for strategic planning.

We now introduce notation for the lookahead model, followed by a discussion of the different classes of approximations that we can introduce.

19.2.1 Modeling the Lookahead Model

We begin by observing that a DLA policy is solving a model, called the lookahead model, that is distinct from the base model that we are trying to solve by designing an effective policy. This means we need notation that is specific to the lookahead model.

We begin by noting that a lookahead model has to be indexed by the time t at which it is being formulated. Since it extends over a horizon from t to $\min\{t + H, T\}$, we index every variable by t (which fixes the information content of the model) and t', which is the time period within the lookahead horizon.

Then, we suggest using the same variables as in the base model, but with a "tilde." Thus, $\tilde{S}_{tt'}$ would be the state in our lookahead model at time t' within the lookahead horizon, for a model being solved at time t. $\tilde{S}_{tt'}$ might have fewer variables than S_t (or $S_{t'}$), and we might also use a different level of aggregation. Similarly, $\tilde{x}_{tt'}$ would be the decision we make at time t' using our lookahead policy $\tilde{\pi}$, given by the function $\tilde{X}_t^{\tilde{\pi}}(\tilde{S}_{tt'})$, with exogenous information $\widetilde{W}_{tt'}$ being the simulated information in our lookahead model that we first observe at time t'. With this notation, our sequence of states, decisions and "exogenous" information, for a lookahead model generated at time t, would look like

$$(\tilde{S}_{tt}, \tilde{x}_{tt}, \tilde{W}_{t,t+1}, \tilde{S}_{t,t+1}, \tilde{x}_{t,t+1}, \tilde{W}_{t,t+2}, \dots, \tilde{S}_{tt'}, \tilde{x}_{tt'}, \tilde{W}_{t,t'+1}, \dots).$$

Note that in a base model, we may be running a process online which means that after we make a decision x_t at time t, the exogenous information W_{t+1} would be observed from a physical process. In a lookahead model, $\tilde{W}_{t,t+1}$ must come from a model.

Using this notation, our direct lookahead policy would be written

$$X_t^{DLA}(S_t) = \arg\max_{x_t} \left(C(S_t, x_t) + \tilde{\mathbb{E}} \left\{ \max_{\tilde{\pi} \in \tilde{\Pi}} \tilde{\mathbb{E}} \left\{ \sum_{t'=t+1}^{t+H} C(\tilde{S}_{tt'}, \tilde{X}_{tt'}^{\tilde{\pi}}(\tilde{S}_{tt'})) | \tilde{S}_{t,t+1} \right\} | S_t, x_t \right\} \right) (19.8)$$

where $\tilde{S}_{t,t'+1} = \tilde{S}^M(\tilde{S}_{tt'}, \tilde{X}_t^{\tilde{\pi}}(\tilde{S}_{tt'}), \tilde{W}_{t,t'+1})$ describes the dynamics within our lookahead model, and where $\tilde{X}_t^{\tilde{\pi}}(\tilde{S}_{tt'})$ is the lookahead policy corresponding to $\tilde{\pi}$.

Here, we write $\tilde{\Pi}$ as a modified set of policies, and $\tilde{\mathbb{E}}$ as a the expectation over a modified set of random outcomes. In fact, later we are going to introduce the possibility of using a different uncertainty operator just for the lookahead model, such as one that evaluates extreme events to capture risk. We might even be modeling time differently (e.g. hourly time steps instead of 5 minutes), but we are going to keep the same time notation for simplicity.

19.2.2 Strategies for Approximating the Lookahead Model

There are a variety of strategies that we can use for approximating the lookahead model to make solving (19.5) computationally tractable. These include:

(1) **Horizon truncation** – We may reduce the horizon from (t, T) to $(t, t+H)$, where H is a suitable, short horizon that is chosen to capture important behaviors. For example, we might want to model water reservoir management over a 10 year period, but a lookahead policy that extends one year (capturing a full cycle of seasons) might be enough to produce high quality decisions. We can then simulate our policy to produce forecasts of flows over all 10 years.

(2) **Outcome aggregation or sampling** – Instead of using the full set of outcomes Ω (which is often infinite), we can use Monte Carlo sampling to choose a small set of possible outcomes that start at time t (assuming we are in state S_t^n during the nth simulation through the horizon) through the end of our horizon $t + H$. We refer to this as $\tilde{\Omega}_t^n$ to capture that it is constructed for the decision problem at time t while in state S_t^n. The simplest

model in this class is a deterministic lookahead, which uses a single point estimate.

(3) **Discretization** – Time, states, and decisions may all be discretized in a way that makes the resulting model computationally tractable. In some cases, this may result in a Markov decision process that may be solved exactly using backward dynamic programming (which we introduced in chapter 14). Because the discretization generally depends on the current state S_t, this model will have to be solved all over again after we make the transition from t to $t + 1$.

(4) **Stage aggregation** – A stage represents the process of revealing information followed by the need to make a decision. We may approximate the future by aggregating stages to reduce the growth of the problem.

(5) **Latent variables** – We may ignore some variables in our lookahead model as a form of simplification. For example, a forecast of weather or future prices can add a number of dimensions to the state variable (we demonstrated this in our energy storage example in section 9.9). While we have to track these in the base model (including the evolution of these forecasts), we can hold them fixed in the lookahead model, and then ignore them in the state variable (these become *latent variables*).

(6) **Policy approximation** – The lookahead model still requires that we design a lookahead policy, which means we have to find a "policy-within-a-policy." While we may have chosen to use a lookahead policy for the base model, we would typically choose something simpler as the policy within the lookahead model.

The remainder of this chapter describes different strategies that have been used for approximating lookahead models. The most complex approximation strategy is the design of the lookahead policy for stochastic lookaheads, since this is where we have to recognize that while we are trying to design a policy for our original base model, using a stochastic lookahead model still requires that we solve a stochastic optimization problem starting at time $t + 1$ onward.

Below we provide a brief discussion of each of these strategies. The strategy of policy approximation is so rich that, beyond a brief sketch, we defer a more complete treatment of this to later in the chapter.

Horizon truncation

Horizon truncation is easily the simplest approximation that is almost always used in lookahead models. The general idea is to choose a horizon H that is long enough to capture activities in the future that we feel will affect decisions now. In an energy storage problem, this might require planning at least a day (sometimes two) into the future to capture daily cycles. In seasonal inventory planning, we might need a horizon that extends past peak periods such as a major holiday. A longer horizon usually means better results (but not always).

Outcome aggregation or sampling

We typically approximate the expectations in the lookahead model using samples. The first expectation is over the random variables in $\tilde{W}_{t+1}$, while the second expectation is over entire sample paths of the sequence $\tilde{W}_{t+2}, \tilde{W}_{t+3}, \dots, \tilde{W}_{t+H}$. We might write the first set of samples as $\tilde{\Omega}_{t+1}$, while the second set of samples might be represented using the set $\tilde{\Omega}_{[t+2,t+H]}$.

As with the horizon, the more samples you have, the better your results will be, but the marginal improvement to the policy tends to decline while the computational cost increases. How much it increases depends on how you are making decisions. If we use a rollout policy, we have to repeat this for every sample, so the computational cost increases linearly with the size of the sample. However, there is a method called two-stage stochastic programming (described next) where we optimize over all scenarios, over all time periods, all at once. These problems can become quite large, and the CPU times can increase much faster than linearly with the sample size.

It is important to address the issue of sampling (which is always necessary) in conjunction with the design of the lookahead policy which we address shortly.

Stage aggregation

A common approximation is a two-stage formulation (see Figure 19.2(a)), where we make a decision x_t, then observe all future events (until $t + H$), and then make all remaining decisions. By contrast, a multistage formulation would explicitly model the sequence: decision, information, decision, information, and so on. Figure 19.2(b) illustrates the many possible paths in a multistage formulation, where we have highlighted a single history $h_{tt'}$ from t to t', followed by a set of outcomes after t' that share the common history $h_{tt'}$.

There is a substantial literature using the two-stage representation (Figure 19.2(a)) for solving stochastic resource allocation problems, dating back to the 1950s. Section 19.9 provides a more detailed summary of this approach.

Latent variables

One of the most powerful, and subtle, approximation strategies involves simplifying the lookahead model by simply holding some variables constant, which means that they are treated as latent variables within the lookahead model. Some examples are:

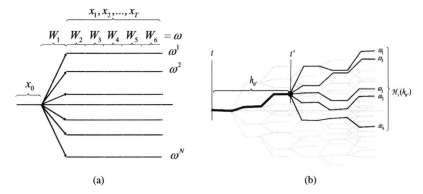

Figure 19.2 Illustration of (a) a two-stage scenario tree and (b) a multistage scenario tree showing a common history $h_{tt'}$ up to time t', with different sample paths after t'.

- Forecasts – Easily one of the most common variables that are treated as latent variables is a forecast $f^W_{tt'}$ of a quantity $W_{t'}$ made at time t. As we step from time t to $t+1$ in our base model, we would obtain updated forecasts $f^W_{t+1,t'}$. However, when we transition from time t' to $t'+1$ in the lookahead model, we do not update our vector of forecasts. As a result, the vector $(f^W_{tt'})_{t' \geq t}$ is held constant in the lookahead model, and as such are treated as latent variables.
- Beliefs – There are many settings where we have a belief B_t about a parameter, such as how a patient responds to a drug, such as one of two dozen drugs for reducing blood sugar. If we make a decision (such as prescribing medication) and then observe how the patient responds, then we learn from this response and update our belief. We may decide to try a medication to reduce blood sugar just to see how the patient responds. We then observe the response and update the belief. We may choose to make decisions in the lookahead model without updating beliefs, primarily because this can be computationally expensive. But we would update beliefs once we make our decision x_t and then observe the response W_{t+1} in the base model (which can be the field).
- We may be optimizing the movement of a utility truck cleaning up after a storm. Two forms of information are phone calls from customers complaining about power outages, and then the observations made by the utility truck observing where there are outages. To make the decision of how to route the truck, we certainly want to model the observations of the truck driver in the lookahead model, but we may wish to ignore new phone calls from customers in the lookahead model (remember that these would just be simulated phone calls).

There is no simple formula for deciding which dynamic state variables can be held fixed in the lookahead model. The key is to think about how the dynamic information would affect decisions in the future, and the extent to which this would affect the decision x_t that is implemented in the base model.

Policy approximation

The most challenging problem when creating a lookahead model is the design of the lookahead policy (or the policy-within-the-policy) given by equation (19.8). Just as most of this book is on the design of good (ideally optimal) policies, the lookahead policy has its own optimization over policies imbedded within the lookahead. This sounds like a circular problem, since we are designing the lookahead policy as part of creating our DLA policy for the base model. A natural question is: if we have already decided we need a direct lookahead policy, shouldn't we just use a DLA policy (within our DLA policy)?

The first problem is that DLA policies (for stochastic lookahead models) can be computationally expensive, as this chapter will demonstrate. The lookahead DLA policy has to be computed many times, so using a DLA policy in the lookahead model may not be computationally feasible (although a deterministic lookahead is not out of the question). The good news is that since the lookahead policy is just being used to approximate decisions (we are not actually implementing the lookahead policy), we can get away with simpler policies that are "good enough" and computationally simpler.

This said, it is important to think about all four classes of policies when designing a lookahead policy. It can be worthwhile to flip back to our original discussion of how to design a policy in chapter 11 (in particular the discussion in section 11.10). The important differences between choosing a policy for the base model and choosing a policy within a lookahead model are:

- Computation – The policy within the DLA typically has to be simulated many times, so computational demands are much more important when designing a lookahead policy compared to the base model, where the policy only has to be computed once per time period.
- Effect of suboptimality – Suboptimality of a policy for the base model translates directly to decisions that are being implemented. By contrast, errors in a lookahead policy have a less direct impact on the decisions that are actually being implemented.

In other words, the emphasis in the lookahead policy shifts more to minimizing computational cost and less to creating the highest quality decisions.

We delve into each of the four classes in more depth, but for now we make the following comments:

PFAs – Policy function approximations are easily the simplest to compute, but we have to find the function first, which involves its own estimation problem. Fitting a PFA is the most difficult among the four classes. Later we introduce some short cuts.

CFAs – Cost function approximations involve an imbedded optimization, which means it is capturing some problem structure. As a result, the tuning becomes easier, but the function will be more expensive to compute.

VFAs – There are inherent parallels between PFAs and VFAs, but VFAs tend to be easier to approximate (compared to PFAs and pure CFAs) since it is incorporating information about the downstream effect of a decision, whereas PFAs and CFAs have to slowly learn which parameter settings work best over time through repeated simulations.

DLAs – More than any of the other policy classes, DLAs are directly optimizing over a horizon, so they do the most to incorporate problem structure directly within the policy. Deterministic DLAS are, of course, relatively easy to compute (but as a rule, harder than any of the other three classes), but will typically involve tunable parameters to compensate for the errors introduced by the deterministic model. By contrast, stochastic DLAs are easily the hardest to compute of all the policies, but minimize tuning (which may still be required as a result of the approximations we introduce).

19.3   Modified Objectives in Lookahead Models

The earlier discussion described ways of creating approximate lookahead models, where the approximations are introduced explicitly to simplify the process of computing the lookahead model.

There are, however, reasons to change the lookahead model to create a policy with desired behaviors. We begin our discussion with the important special case of risk-adjusted policies, where we replace the expectation operator (which simply averages over outcomes) with a risk operator that puts different weights on specific events, typically related to the tails of distributions. We then discuss other types of modifications that we have encountered in our own work.

19.3.1   Managing Risk

When uncertainty is involved (as it is throughout this volume), it is very common that some events are viewed as particularly undesirable. Some examples are:

- Arriving late to an appointment.
- Not meeting the energy requirements of a system.
- Overdosing a patient, or using a drug that produces a bad reaction.
- Losing money on an investment, or losing an amount of money that exceeds the reserves (such as an insurance policy) that is used to cover these events.
- Losing a competition by choosing the wrong players.
- Running out of time before completing enough experiments to find the right material or design.

There is an extensive (and mathematically deep) field of research devoted to the handling of risk. We focus on the issues associated with modeling and computation. We begin by introducing an uncertainty operator $\rho(\cdot)$ that we use instead of an expectation that allows us to manipulate a sequence of contributions in any way. We follow this with three different ways of creating risk-adjusted behaviors, followed by a discussion on evaluating risk-adjusted policies.

Uncertainty operators and risk-adjusted performance

It is not unusual to need to incorporate risk into the design of a policy. There are several ways in which we can consider risk:

- Uncertainty in the contribution – We may be trying to maximize our total contribution (or minimize a cost) where we are worried about the possibility of low contributions.
- Risk that we violate a constraint, such as not arriving on time, or not meeting demand.
- Uncertainty in a separate performance metric – We may be trying to minimize the cost of delivering goods to customer, but we also wish to ensure that a driver gets home on time.

Concern about the uncertainty in the objective can be handled by replacing the expectation operator with a more general uncertainty operator $\rho(\cdot)$. We can define our operator $\rho(F_t^\pi)$ in different ways:

- Value at risk – We would use the distribution of F_t^π to compute the α-percentile, where we might use $\alpha = 0.10$ to capture the lower tail. We would write our uncertainty operator as $\rho_\alpha(F_t^\pi)$ to give us the α-percentile.
- Conditional value at risk – Widely known as CVar, which is also known under other names (such as average value at risk), CVar uses the average of the values $\hat{F}_t^\pi$ that fall below the α-percentile (assuming we are worried about low values).
- Risk adjusted performance – If $\bar{F}_t^\pi$ is the average of the outcomes and $\bar{\sigma}_t^\pi$ is the standard deviation, our risk adjusted performance might be

$$\rho(F_t^\pi|\theta) = \bar{F}_t^\pi - \theta\bar{\sigma}_t^\pi,$$

where we might choose $\theta \approx 2$ to capture the lower tail.

- Intra-horizon deviations – Rather than focusing on aggregate statistics over the entire lookahead horizon, we may wish to focus on, say, the worst performance in an individual time period, or the number of times the contribution is above or below some target. In this case, we could write our risk metric using

$$\rho(C(\tilde{S}_{t,t+1}, \tilde{X}^\pi(\tilde{S}_{t,t+1})), C(\tilde{S}_{t,t+2}, \tilde{X}^\pi(\tilde{S}_{t,t+2})), \dots, C(\tilde{S}_{t,t+H}, \tilde{X}^\pi(\tilde{S}_{t,t+H}))).$$

These risk metrics are all defined in terms of the variability of actual performance over different sample outcomes (often referred to as scenarios).

Typically we would use a modified contribution function such as

$$\bar{C}^{risk}(S_t, x_t) = \mathbb{E}_{W_{t+1}} C(S_t, x_t, W_{t+1}) + \eta\rho(S_t, x_t, W_{t+1})$$

which combines an expected contribution $\mathbb{E}_{W_{t+1}} C(S_t, x_t, W_{t+1})$ with our risk metric $\rho(S_t, x_t, W_{t+1})$ which we assume has access to current state information S_t, the decision x_t, as well as the random information (stock returns, patient performance) W_{t+1}. The parameter η governs the tradeoff between the importance of the expected contribution versus the outcome of tail events. Of course, we are applying this risk-adjusted contribution in the context of our lookahead model, which means we will use our lookahead state variable $\tilde{S}_{tt'}$, decision $\tilde{x}_{tt'} = \tilde{X}^\pi(\tilde{S}_{tt'})$, and our modeled exogenous information $\tilde{W}_{t,t'+1}$.

For example, let the performance of the lookahead model following sample path $\tilde{\omega}$ be given by

$$\tilde{F}_t^\pi(S_t, x_t|\tilde{\omega}) = C(S_t, x_t) + \sum_{t'=t+1}^{t+H} C(\tilde{S}_{tt'}(\tilde{\omega}), \tilde{X}^\pi(\tilde{S}_{tt'}(\tilde{\omega})), \tilde{W}_{t,t'+1}(\tilde{\omega}))$$

where $\tilde{S}_{t,t'+1}(\tilde{\omega}) = \tilde{S}^M(\tilde{S}_{tt'}(\tilde{\omega}), \tilde{x}_{tt'} = \tilde{X}^\pi(\tilde{S}_{tt'}(\tilde{\omega})), \tilde{W}_{t,t'+1}(\tilde{\omega}))$ is the state at time $t'+1$ while following sample path $\tilde{\omega}$. $\tilde{F}_t^\pi(S_t, x_t|\tilde{\omega})$ is the total contribution from following lookahead policy $\tilde{X}^\pi(\tilde{S}_{tt'})$ along sample path $\tilde{\omega}$. We can then define $\tilde{F}_t^\pi(S_t, x_t)$ to be a random variable with outcomes $\tilde{F}_t^\pi(\tilde{\omega}_i)$ for sample path $\tilde{\omega}_i \in \tilde{\Omega}_{[t+2,t+H]}$.

We could, then, compute a risk-adjusted version of this total performance

$$\tilde{F}_t^{risk-sum,\pi}(S_t, x_t) = \mathbb{E}\tilde{F}_t^\pi + \eta\rho(\tilde{F}_t^\pi). \tag{19.9}$$

Alternatively, we could have summed the risk-adjusted contributions

$$\bar{F}_t^{sum-risk,\pi}(S_t, x_t) = C(S_t, x_t) + \sum_{t'=t+1}^{t+H} \mathbb{E}C(\tilde{S}_{tt'}, \tilde{x}_{tt'}, \tilde{W}_{t,t+1}) +$$

$$\eta\rho(\tilde{S}_{tt'}, \tilde{x}_{tt'}, \tilde{W}_{t,t+1}). \tag{19.10}$$

If we were not considering risk, these two metrics would be the same, since the expectation of the sum of a set of random variables is equal to the sum of the expectations. This is not the case when we introduce risk measures. For this reason

$$\bar{F}_t^{risk-sum,\pi}(S_t, x_t) \neq \bar{F}_t^{sum-risk,\pi}(S_t, x_t),$$

and it is not even clear what the relationship would be. Even more important is that we are making decisions $\tilde{x}_{tt'} = \tilde{X}^\pi(\tilde{S}_{tt'})$ over time within the lookahead, so we have to consider the evaluation of the entire range of partial sums of contributions. The design of this lookahead policy has actually attracted considerable interest the research literature (see the bibliographic notes for some additional readings). This topic is beyond the scope of this book, but this discussion serves as a brief introduction.

To capture risk in a lookahead policy, it helps to describe the entire information flow in the lookahead model using

$$\tilde{h}_{[t,t']}^\pi = (\tilde{S}_{tt}, \tilde{x}_{tt}, \tilde{W}_{t,t+1}, \dots, \tilde{S}_{t,t'}).$$

We call $\tilde{h}_{[t,t']}^\pi$ a "history" realizing that we are actually looking forward in time, and this is the history relative to time t' in the future. The history $\tilde{h}_{[t,t']}^\pi$ is a random vector which produces the realization $\tilde{h}_{[t,t']}^\pi(\omega)$ when following sample path ω, using policy π.

Given this history, we can compile any activity metrics we need. Let

$$\rho^{LA}(\tilde{h}_{[t,t']}^\pi|\theta) = \text{performance metric derived from the history}$$
$$\tilde{h}_{[t,t']}^\pi \text{ in the lookahead model.}$$

The risk operator $\rho^{LA}(h_{[t,t+H]}|\theta)$ can capture virtually any statistic from the future history, which is computed from the family of sample paths. So, just as an expectation averages across sample paths, the risk operator $\rho^{LA}(h_{[t,t+H]}|\theta)$ can be used to take a worst case, or the 10th or 90th percentile. The key is that we are simulating its value, so it is readily computable from a simulated lookahead model.

Risk-adjusted policies

Ultimately, however we handle risk, it is still just a lookahead policy, a topic that is easy to overlook. For example, we might create our risk-adjusted lookahead policy by using

$$X_t^{risk}(S_t|\theta) = \arg\max_{x_t \in \mathcal{X}_t} \bar{F}_t^{sum-risk,\pi}(S_t, x_t). \tag{19.11}$$

If x_t is discrete, which is to say that $\mathcal{X}_t = \{x_1, \ldots, x_M\}$, computing this policy is fairly straightforward. If it is continuous, and possibly vector-valued, we would need to draw on the tools of derivative-based (chapter 5) or derivative-free (chapter 7) stochastic search.

There is actually a wide range of applications where risk is an important issue. How risk is captured is highly problem dependent. Financial applications tend to work with tail deviations; energy applications will balance expected costs with penalties for outages; applications in health will focus on the possibility of bad health outcomes; design of buildings and bridges tend to emphasize the possibility of failure under extreme wind or earthquakes; inventory problems typically focus on avoiding running out of the resource being supplied; the control of electric vehicles might focus on the possibility of the battery running low.

Chance-constrained policies

The policy in equation (19.11) was designed to maximize (or minimize) a risk adjusted objective function. An alternative approach is to retain our original objective that we presented in equation (19.5)

$$X_t^{CC}(S_t|\theta) = \arg\max_{x_t} \left(C(S_t, x_t) + \mathbb{E}\left\{ \max_{\pi \in \Pi} \mathbb{E}\left\{ \sum_{t'=t+1}^{T} C(S_{t'}, X_{t'}^{\pi}(S_{t'}))|S_{t+1} \right\} |S_t, x_t \right\} \right) \tag{19.12}$$

but now we introduce a constraint that our risk operator $\rho^{LA}(\tilde{h}_{[t,t+H]}^{\pi})$ (which is a random variable since $\tilde{h}_{[t,t+H]}^{\pi}$ is random) meets some target θ^{target} with a desired probability P^{target}. This means we wish to maximize expected contributions over our lookahead model, subject to the probabilistic constraint

$$\mathbb{P}[\rho^{LA}(\tilde{h}_{[t,t+H]}^{\pi}) \le \theta^{target}] \ge P^{target}. \tag{19.13}$$

The constraint (19.13) is known as a *chance constraint*. Although computing the probability $\mathbb{P}[\rho^{LA}(\tilde{h}_{[t,t+H]}^{\pi}) \le \theta^{target}]$ can be challenging, a natural approach is to approximate it using a sample, just as we have been using to approximate expectations in the lookahead policy (or to evaluate the policy).

Robust optimization

Robust optimization first emerged as a way of solving engineering design problems that arise in areas such as civil and mechanical engineering. The problem is to design buildings subject to stresses from wind. The challenge is to find the least cost design (a minimization problem) under the worst wind condition, given by speed and direction. A mechanical engineer would face a similar problem designing the wing of a commercial aircraft, where the goal is to minimize weight while handling the most extreme stresses from wind.

Let $F(x, w)$ be the cost of design x given wind conditions w, where we might set $F(x, w)$ to a very large number of our design fails. Then let W be the space of "wind" outcomes, with element $w \in W$. Let x capture all the design choices. We want to solve the optimization problem

$$\min_{x \in X} \max_{w \in W} F(x, w). \tag{19.14}$$

This formulation is known as a *robust optimization problem* since our solution x has to do well under the worst conditions. The feasible region X is understood to be governed by physical constraints that are well defined. By contrast, the design of the set W is not as well defined, since we have to address the issue of the likelihood of extreme events. Let θ be a parameter that governs the likelihood of events that we want to include in W. Then, we would write this set as $W(\theta)$ to reflect the dependence of the set W on this parameter.

There are several strategies that we can use to construction $W(\theta)$:

- Box constraints – We can represent $W(\theta)$ as a set of constraints of the form

  $$w^{min} \leq w \leq w^{max}.$$

  If w is a vector (say, wind speed and direction), we would have w^{min} and w^{max} for each dimension. These limits might be chosen so that the probability that each dimension falls outside of this range is θ.

  The problem with this strategy is that the model would likely pick the most extreme values of each dimension of w. So, we might have high wind, and separately a direction that produces the most stress. This logic would likely choose w that is the most extreme of both values.

- Joint constraints – Here we construct a region such that the probability that all dimensions fall outside of the boundary with some probability θ. Joint regions are harder to create and work with, especially when we adapt this idea to sequential problems.

This strategy has been adapted for fully sequential problems by formulating a direct lookahead model using the same principles, with the following changes:

- Replace x with $(\tilde{x}_{tt}, \tilde{x}_{t,t+1}, \ldots, \tilde{x}_{t,t+H})$.
- Let w be a specific realization of $(\tilde{W}_{t,t+1}, \ldots, \tilde{W}_{t,t+H})$, and let $\mathcal{W}(\theta)$ be the set of all possible values of w that we are willing to consider. We let $w = (\tilde{w}_{t,t+1}, \ldots, \tilde{w}_{t,t+H})$ be a specific sequence of realizations.

Switching to our more standard style of maximizing over x, our multiperiod robust optimization problem would be written

$$\max_{(\tilde{x}_{tt}, \ldots, \tilde{x}_{t,t+H}) \in \mathcal{X}} \quad \min_{(\tilde{w}_{tt}, \ldots, \tilde{w}_{t,t+H}) \in \mathcal{W}(\theta)} F(\tilde{x}_t, \tilde{w}_t). \tag{19.15}$$

Note that the optimization problem stated by (19.15) is a deterministic optimization problem. Our problem is to pick a single vector $\tilde{x}_t = (\tilde{x}_{tt}, \ldots, \tilde{x}_{t,t+H})$ and a single $\tilde{w}_t = (\tilde{w}_{tt}, \ldots, \tilde{w}_{t,t+H})$. Understanding that we are only interested in $\tilde{x}_{tt}$ (as is the case with all our direct lookahead policies), we can now write our robust optimization model as a policy

$$X_t^{RO}(S_t|\theta) = \arg\min_{(\tilde{x}_{tt}, \ldots, \tilde{x}_{t,t+H}) \in \mathcal{X}} \quad \max_{(\tilde{w}_{tt}, \ldots, \tilde{w}_{t,t+H}) \in \mathcal{W}(\theta)} F(\tilde{x}_t, \tilde{w}_t). \tag{19.16}$$

Although we are optimizing over the entire vector $(\tilde{x}_{tt}, \ldots, \tilde{x}_{t,t+H})$, the policy only implements $\tilde{x}_{tt}$.

Robust optimization has been promoted by some as a way of avoiding the need to create an underlying probability distribution, although this is a bit misleading. Creating an uncertainty set $\mathcal{W}(\theta)$ that produces desired behaviors is the most difficult challenge, both computationally (since the boundaries of $\mathcal{W}(\theta)$ have to consider the joint likelihood of all the random events) and in terms of modeling the underlying problem. Thus, while the policy (19.16) does not have an explicit probability calculation, the underlying probability model is imbedded in the creation of $\mathcal{W}(\theta)$.

Often overlooked by the robust optimization community is that the formulation (19.16) is inherently "two-stage" and ignores the ability to make new decisions as information unfolds. How important this is depends on the problem setting.

19.3.2 Utility Functions for Multiobjective Problems

There are many settings where we evaluate a policy using well-defined metrics such as profits, costs, or the time to complete a task. However, we need policies that understand that the real world is more complex. Some examples are:

- A ride hailing service or trucking company needs to minimize the miles that a driver moves empty, but the company also has to recognize that drivers need to get home on time.

- Similarly, while minimizing empty miles we want to make sure that customers are served on time.
- We wish to schedule machinery to maximize throughput, but machines have to be maintained, and it helps to have some slack in the schedule for maintenance or to absorb delays when a machine had to be stopped for mechanical reasons.
- We want to find the shortest path over a dynamic network, but without too many turns.
- We wish to optimize our bids in a market, but have to be careful not to become predictable to our competitors.

A simple and practical way of handling these issues is to introduce bonuses and penalties in the lookahead model, which are ignored when we are evaluating policy (which might occur in the field). The use of bonuses and penalties to handle different objectives is a widely used heuristic. Often, these bonuses and penalties are introduced to find a solution that meets target performance metrics other than minimizing costs (for example), which means we are typically trying to meet a chance constraint. Tuning these bonuses and penalties closely parallels policy search, and uses the same tools.

19.3.3 Model Discounting

When we introduced approximations in our lookahead model, it can make sense to discount decisions being simulated farther in the future under the rationale that the cumulative effect of our approximations make these decisions less important to the process of determining the best decision now. A discounted lookahead policy would be written

$$X_t^{DLA}(S_t) = \arg\max_{x_t}\left(C(S_t, x_t) + \tilde{\mathbb{E}}\left\{\max_{\tilde{\pi}\in\tilde{\Pi}}\tilde{\mathbb{E}}\left\{\sum_{t'=t+1}^{t+H}\lambda^{t'-t}C(\tilde{S}_{tt'}, \tilde{X}_{tt'}^{\tilde{\pi}}(\tilde{S}_{tt'}))|\tilde{S}_{t,t+1}\right\}|S_t, x_t\right\}\right). \quad (19.17)$$

The parameter λ serves the same role as λ in TD(λ) (see section 16.1.4), in that it plays the role of an *algorithmic* discount factor.

19.4 Evaluating DLA Policies

There is a surprisingly substantial tradition in the literature focusing on stochastic lookaheads to focus on solving the stochastic lookahead without recognizing that the solution is just the computation of a policy that still needs to be evaluated. Often, so much work is put into solving a stochastic lookahead policy, that the idea of simulating what is an otherwise cumbersome policy can seem impractical.

The methods for evaluating stochastic lookahead policies are the same we would use for any policy. There are two fundamental strategies we can use:

- Offline, using a simulator – This is the best way to do a comprehensive comparison of policies, including tuning policies within a class. This requires building a simulator, which can be a significant project in many settings such as energy, transportation and logistics, health, and finance. A simulator requires both a model of the dynamics of the physical system (which is captured in the transition function $S^M(S_t, x_t, W_{t+1})$) and a model of the exogenous information process $W_1, W_2, \ldots, W_t, \ldots$. There are two ways to simulate the information process:

  - Using a mathematical model – This requires creating mathematical models of the random variables in W_{t+1} given the current state S_t and most recent decision x_t. Mathematical models can be quite sophisticated (chapter 10 provides an introduction to a field known as uncertainty quantification), but provide the ability to perform repeated simulations, including simulations of physical processes that did not exist in history (such as major investments into wind and solar farms for energy generation).
  - Using history – This is easily the most common approach used in finance, where backtesting is a standard method. This involves creating sample paths of $W_1, W_2, \ldots, W_t, \ldots$ from different periods of history.

  Offline simulators enjoy the advantages of controlled experiments, the ability to test in parallel, and in some cases the ability to approximate derivatives, opening the door to derivative-based stochastic search.

- Online, in the field – Simulators can be expensive to build (and have their own errors), so the alternative is often to observe how policies work in the field. Online evaluation means experimenting with a real physical system, which means that you have to actually live with the performance of a policy for a period of time. While field evaluations avoid modeling errors, they are slow (it takes a day to simulate a day), and there is no way to approximate derivatives. However, there are many settings where building a simulator is just not possible or appropriate, so this is not a strategy to be discarded.

We need to make the point that simulating policies can be quite noisy. While this is not universally true, we recommend that methods for policy evaluation and tuning be designed with the possibility that this may be true, and that some initial experiments be run to evaluate the level of uncertainty.

While we will note in closing that all policies need to be evaluated, simulating a PFA or CFA policy is fundamental to its effectiveness, while the same is not true of a carefully designed stochastic DLA policy.

19.4.1 Evaluating Policies in a Simulator

Assume (as we will do moving forward) that we are going to pick a lookahead policy $\tilde{X}^{\tilde{\pi}}_t(\tilde{S}_t|\theta)$ which may be from any of the four classes. This means we can write the policy in (19.8) without the imbedded $\max_{\tilde{\pi}}$ operator:

$$
X^{DLA}_t(S_t|\theta) \;=\; \arg\max_{x_t}\left(C(S_t, x_t) + \bar{\mathbb{E}}\left\{ \bar{\mathbb{E}}\left\{ \sum_{t'=t+1}^{t+H} C(\tilde{S}_{tt'}, \tilde{X}^{\tilde{\pi}}_{tt'}(\tilde{S}_{tt'}|\theta))|\tilde{S}_{t,t+1} \right\} |S_t, x_t \right\} \right) \tag{19.18}
$$

where θ captures any parameters needed to control our policy (and by this, we mean the entire DLA policy, not just the lookahead policy $\tilde{X}^{\tilde{\pi}}_{tt'}(\tilde{S}_{tt'}|\theta)$). We approximate the first expectation in (19.18) using a set of samples of $\tilde{W}_{t+1}$ that we represent using the set $\tilde{\Omega}_{t+1}$. We then approximate the second expectation by creating a series of sample paths $\omega \in \tilde{\Omega}_{[t+2,t+H]}$ where we use our pre-determined policy $\tilde{X}^{\tilde{\pi}}_t(\tilde{S}_t|\theta)$ to run a simulation along the sample path ω, creating the sequence

$$
(\tilde{S}_{tt}, \tilde{x}_{tt}, \tilde{W}_{t,t+1}(\omega), \tilde{S}_{t,t+1}, \tilde{x}_{t,t+1}, \tilde{W}_{t,t+2}(\omega), \dots, \tilde{S}_{tt'}, \tilde{x}_{tt'}, \tilde{W}_{t,t'+1}(\omega), \dots),
$$

where $\tilde{x}_{tt'} = \tilde{X}^{\pi}(\tilde{S}_{tt'}|\theta)$ and where $\tilde{S}_{t,t'+1} = \tilde{S}^M(\tilde{S}_{tt'}, \tilde{x}_{tt'}, \tilde{W}_{t,t'+1})$. From this sequence we can accumulate contributions from each decision using $C(\tilde{S}_{tt'}, \tilde{x}_{tt'})$.

The steps of the procedure are given in Figure 19.3 which returns an approximation $\bar{F}_t(x_t)$ for the performance of the policy starting at time t in state S_t.

The procedure in Figure 19.3 can, of course, be used to produce the estimate

$$
\bar{F}_0(x_0|S_0) \approx \mathbb{E}\{F(x^{\pi,N}, W)|S_0\}.
$$

We can think of $\bar{F}_0(x_0|S_0)$ as a sampled estimate of $\mathbb{E}\{F(x^{\pi,N}, W)|S_0\}$, which can be used in any of our algorithms for derivative-free stochastic search. Since we are using a simulator, we would use the objective for offline learning given (in chapter 7) by

$$
\max_{\pi} \mathbb{E}\{F(x^{\pi,N}, W)|S_0\}.
$$

Any of the four classes of policies may be used here. The choice of learning policy should be guided by the speed with which simulations can be run, and the variability from one run of the simulator to the next.

19.4.2 Evaluating Risk-Adjusted Policies

Surprisingly, a common strategy for evaluating a risk-adjusted policy is to simulate it N times and take an average. Let $X^{RA}_t(S_t|\theta)$ be any of our risk-adjusted policies: $X^{risk}_t(S_t|\theta)$ (equation (19.11)), $X^{CC}_t(S_t|\theta)$ (equations (19.12)–(19.13)),

Step 0. Initialization:

    Step 0a. Set initial state $\tilde{S}_t$.
    Step 0b. Choose the sample sets $\tilde{\Omega}_{t+1}$ for $\tilde{W}_{t+1}$, and $\tilde{\Omega}_{[t+2,t+H]}$ for the entire sequence $\tilde{W}_{t+2}, \dots, \tilde{W}_{t+H}$.

Step 1. Do for $x_t = x_1, x_2, \dots, x_M$:

    Step 2. Compute $C_t(x_t) = C(\tilde{S}_t, x_t)$:
    Step 3. Do for $\tilde{w}_{t+1} = \{\tilde{w}_1, \tilde{w}_2, \dots, \tilde{w}_M\} \in \tilde{\Omega}_{t+1}$:

        Step 4a. Find state $\tilde{S}_{t+1} = \tilde{S}^M(\tilde{S}_t, x_t, \tilde{w}_{t+1})$.
        Step 4b. Choose policy parameters $(\theta^{min}, \theta^{max})$ given $\tilde{S}_{t+1}$ [We don't actually do this in practice]
        Step 4c. Do for each sample path $\omega \in \tilde{\Omega}_{[t+2,t+H]}$: [These simulations should be done in parallel]

            Step 5a. Simulate $(\theta^{min}, \theta^{max})$ policy over $t' = t + 1, t + 2, \dots, t + H$ [Works for any parameterized policy]

                Step 6a. Find $\tilde{x}_{t'} = \tilde{X}^\pi(\tilde{S}_{t'}|\theta^\pi = (s, S))$.
                Step 6b. Find contribution $C_{t'}(\tilde{x}_{t'}, \omega) = C(\tilde{S}_{t'}, \tilde{x}_{t'})$.
                Step 6c. Find $\tilde{W}_{t'+1}(\omega)$.
                Step 6d. Find next state $\tilde{S}_{t'+1} = \tilde{S}^M(\tilde{S}_{t'}, \tilde{x}_{t'}, \tilde{W}_{t'+1})$.

            Step 5b. Accumulate total contribution $F_t^\pi(x_t, \omega) = C_t(x_t) + \sum_{t'=t+1}^{t+H} C_{t'}(\tilde{x}_{t'}, \omega)$.

    Step 7. Find $\bar{F}_t^\pi(x_t|S_t) = \frac{1}{M} \sum_{i=1}^{M} F_t^\pi(x_t, \omega_i)$.

Step 8. Find $x_t^* = \arg\max_{x_t} \bar{F}(x_t|S_t)$ and return x_t^* and the function $\bar{F}_t^\pi(x_t|S_t)$.

Figure 19.3 Simulation of a lookahead policy.

or $X_t^{RO}(S_t|\theta)$ (equation (19.16)). One way to evaluate these policies is to simulate a sample path ω to obtain

$$F^{RA}(\omega|\theta) = \sum_{t=0}^{T} C(S_t(\omega), X^{RA}(S_t(\omega)|\theta)). \tag{19.19}$$

We might do several simulations and take an average

$$\bar{F}^{RA}(\theta) = \frac{1}{N} \sum_{n=1}^{N} F^{RA}(\omega^n|\theta), \tag{19.20}$$

which means we are evaluating our policy using a standard expectation-based metric, which we have written previously as

$$F^{RA}(\theta) = \mathbb{E}\left\{\sum_{t=0}^{T} C(S_t, X^{RA}(S_t|\theta))|S_0\right\}. \tag{19.21}$$

Now we just have to search for the best θ using our standard stochastic search methods.

The attentive reader might raise the question: we are using one objective in the lookahead model for the policy, and a different objective to evaluate the policy. Is this right? We first note that top professionals in the field have been known to do this (see the bibliographic notes for this section at the end of the chapter). Second, it can be argued that our risk-adjusted policy can be viewed just as we would any other policy. If it is executed frequently, it may make sense to evaluate its performance over time (which means taking an average).

This said, there is a logical inconsistency between using one objective in the lookahead model and a different objective for evaluating the policy. Think about making a decision at time 0: The lookahead model in a risk-adjusted policy is using one metric, and then we evaluate the policy using a different method. We can immediately create a policy that would work better by simply switching to an expectation-based lookahead model.

Evaluating a policy using risk metrics is fairly straightforward when using sampling-based methods. The operator $\rho^{LA}(h_{[t,t+H]}|\theta)$ is quite general, since virtually any metric can be calculated from a simulated history of contributions and other statistics. The only issue when evaluating a policy would arise when using a chance-constrained formulation, since we may find that we are evaluating the chance constraint. This would have to be handled by adding a penalty term, which is equivalent to dropping the chance constraint and moving these violations into the objective function.

19.4.3 Evaluating Policies in the Field

Evaluating a policy in the field means that we are going to be using the tools of derivative-free stochastic search, which we reviewed in chapter 7. In addition, it means that we have to use the objective which optimizes cumulative reward, given by

$$\max_{\pi} \mathbb{E}_{S_0}\mathbb{E}_{W_1,...,W_T|S_0}\left\{\sum_{t=0}^{T} C(S_t, X^{\pi}(S_t))|S_0\right\},$$

since we have to live with our performance while learning. Since learning is inherently quite slow, we suggest one of the lookahead policies in chapter 7

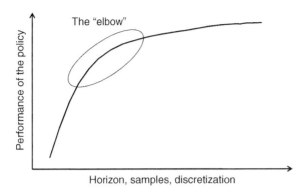

Figure 19.4   The performance elbow, which shows the improvement as we increase the horizon, number of samples or discretization intervals.

such as the knowledge gradient or one of its variants, which tend to be best suited for learning with limited budgets.

19.4.4   Tuning Direct Lookahead Policies

The first three approximate strategies (the horizon, outcome sampling, and discretization intervals) are the most straightforward, since they fall into the "more is better" categories. More time periods (long horizons), more samples, and a finer discretization is always better. The problem is the tradeoff between solution quality and computational cost.

Finding the best value for each of these modeling choices tends to involve creating the plot depicted in Figure 19.4, which illustrates the "performance elbow" which is the point where increasing the model parameter (horizon, outcomes, discretization intervals) produces diminishing returns. The point where CPU times become impractical depends entirely on the situation.

Whether we are searching over discrete choices (such as different classes of policies, or different types of approximation architectures), or continuous parameters, policy search is an exercise in stochastic search. Most frequently we are approaching the problem using the tools of derivative-free stochastic search (see chapter 7), but we may be able to use exact or numerical derivatives for continuous parameters, and apply the methods of chapter 5. Either way, the discussion on policy search in chapter 12 is worth reviewing.

19.5   Why Use a DLA?

Given the complexity of designing and computing a direct lookahead policy, it is fair to ask: what are the benefits? It is helpful to contrast two classes of policies for one of the most classical sequential decision problems: inventory planning. The simplest inventory problem features orders that arrive instantly, so the inventory equation is given by

$$R_{t+1} = \max\{0, R_t + x_t - \hat{D}_{t+1}\},$$

where $\hat{D}_{t+1}$ is the demand arriving between t and $t+1$ that is not known when we place our ordering decision x_t (which arrives instantly). For this basic problem, our state variable is $S_t = R_t$. A natural policy (in fact, one that is known to be optimal for this problem) is a simple PFA known as an order-up-to policy written as

$$X^\pi(S_t|\theta) = \begin{cases} \theta^{max} - R_t & \text{if } R_t < \theta^{min}, \\ 0 & \text{otherwise.} \end{cases} \qquad (19.22)$$

where we let $\theta = (\theta^{min}, \theta^{max})$.

This simple inventory problem is one of those cases where a reasonable policy is given by a known, analytical function. Now imagine that we are faced with the following series of variation that might easily arise for our inventory problem:

(1) Start by introducing lead times, where an order x_t placed at time t arrives τ time periods later. In fact, imagine we are in North America ordering product from China, with a lead time of 100 days, although this can vary over a range of 50 days.
(2) Next imagine that we recognize that at time 0, we have a ship enroute carrying an order from China that will arrive in 10 days.
(3) Instead of one cargo ship arriving in 10 days, we have three ships arriving in 5, 15, and 40 days.
(4) We are given information about a storm moving through the Pacific that could delay ships that intersect the path of the storm by 5 to 10 days. Alternatively, we have to plan for the event that a storm *might* arrive during the next 50 days, in which case it would delay our shipment.
(5) Finally, we are given the option of placing a rush order via air freight that will arrive in 10 days.

We then pose the question: how do each of these variations change our decision?

Capturing these additional problem characteristics using an order-up-to policy is hard, because the only parameters we can control is the vector $\theta = (\theta^{min}, \theta^{max})$. This means we have to make θ a function of the state S_t, which we would write as $\theta(S_t)$. Information about ships arriving in the future and weather means that S_t is becoming fairly high dimensional (and remember, we are allowing ourselves to know that the starting policy is of the order-up-to variety). Even with this information, it would be quite difficult to design the function $\theta(S_t)$.

Now consider what happens when we use a direct lookahead. All of the information about ships arriving and weather (including uncertainty in the weather and the option of placing a rush order using air freight) would be captured in the lookahead model. As we simulate into the future, we would capture the ships arriving (this schedule is in the state variable), the random effect of the storm (this would be modeled in the random variables $\tilde{W}_{tt'}$ in the lookahead model), and the ability to place a rush order using air freight (this would be a decision $\tilde{x}_{tt'}$ in the lookahead model, determined by some reasonable lookahead policy). Of course, this simulation has to be run for each ordering decision x_t now, but remember that all this can be done in parallel.

So, while we pay a computational cost for using a stochastic direct lookahead policy, we obtain a policy that not only captures all the dimensions of a very complex state variable S_t, we can also capture our ability to make decisions in the future. If we change the schedule of inbound ships, this may change our order decision x_t now. Similarly, if we introduce new decisions that we may make in the future in our lookahead model that help us respond to random events, that may change which x_t now is the best decision.

In other words, our DLA policy captures our highly complex state variable (the schedule of ships arriving), random future events (such as the storm), as well as options that help us to mitigate these random events (such as the rush order of inventory). This is a very responsive policy, without the need to do sophisticated machine learning.

This finishes Part I of this chapter, where we have discussed the general setting of DLA policies and the different strategies for overcoming the computational issues inherent in stochastic lookahead models. We next turn our attention first to the use of deterministic lookahead policies, which are easily the most popular class of policies for solving lookahead problems. Then, the rest of the chapter covers methods for approximating stochastic lookahead policies.

19.6 Deterministic Lookaheads

The most widely used approximation used when designing a lookahead model is to assume that the problem is deterministic. This eliminates both expectations, which then means that instead of optimizing over policies, we just optimize over the decisions as with any deterministic model. The policy would then be written as

$$X_t^{DLA-Det}(S_t) \;=\; \arg\max_{x_t}\left(C(S_t, x_t) + \max_{x_{t+1},\ldots,x_{t+H}} \sum_{t'=t+1}^{T} C(S_{t'}, x_{t'}) \right),$$

$$= \arg\max_{x_t,\dots,x_{t+H}} \left(C(S_t, x_t) + \sum_{t'=t+1}^{T} C(S_{t'}, x_{t'}) \right),$$

$$= \arg\max_{x_t,\dots,x_{t+H}} \sum_{t'=t}^{T} C(S_{t'}, x_{t'}). \tag{19.23}$$

subject to the necessary constraints.

This is one policy where it does not matter whether x_t is a scalar (continuous or discrete) or a vector. Even if x_t is a scalar, the optimization problem in (19.23) still has to optimize over the vector $x_t, \dots, x_{t+H}$.

There are a few myths that we have to address about deterministic lookahead policies:

Myth 1 – *Deterministic models are easy to solve.* While this can be true, there are complex problems where optimizing over long horizons can be very expensive. For example, optimizing a fleet of truck drivers or locomotives over horizons of a week or more can be computationally completely intractable. We are now going to suggest stochastic lookahead models that are easier to solve than a deterministic lookahead for some problems.

Figure 19.5 shows the increase in CPU times for a transportation application that involves managing locomotives over time. If we increase the horizon to four days, which is not unreasonable in this setting, the CPU time to solve a single instance of a lookahead model grows to 50 hours (this work was done around year 2010 using Cplex and a large memory computer).

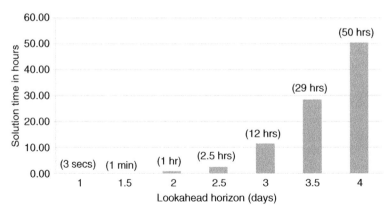

Figure 19.5 Growth in CPU times as we increase the horizon in a deterministic lookahead model.

Myth 2 – *Since we are obtaining an optimal solution, "it" is optimal.* This is a surprisingly common misconception. Often people do not appreciate that optimizing over a horizon to determine what to do now is a policy to solve a problem, not the problem itself, and an optimal solution to an approximate lookahead model is *not* an optimal policy! For example, think about planning a path using a navigation system, which finds the best path given point estimates of the time required to traverse each link of the network. This is likely to be a good policy, but because we are ignoring the uncertainty in those link times, this is not an optimal policy.

Myth 3 – *Even though the policy may not be optimal, it is a good starting point.* While there are many problems for which this is true, imagine using a deterministic lookahead to create a policy for buying and selling assets in the face of stochastically varying prices. The best estimate of future prices is the current price, so the forecast of future prices is constant. A natural policy is to buy when the price goes below some point, and sell when it goes above another point. A deterministic lookahead will not even produce a reasonable approximation of this policy. This said, it is entirely possible that a deterministic lookahead *may* be a good starting point, as we demonstrated in chapter 13 when we used a deterministic lookahead for our energy storage problem, with parametric modifications that had to be tuned.

We would still say that a deterministic lookahead should be considered as a possible starting point, but given the diversity of application settings, you have to think about whether a deterministic lookahead would capture the right behaviors for your problem. You cannot just adopt a deterministic lookahead (or any policy) blindly.

19.6.1   A Deterministic Lookahead: Shortest Path Problems

The best way to illustrate a lookahead policy is the process of finding the best path over a transportation network where travel times are evolving randomly as traffic moves. Imagine that we are trying to get from an origin q to a destination r, and assume that we are at an intermediate node i (trying to get to r). Our navigation system will recommend that we go from i to some node j by first finding the shortest path from i to r, and then using this path to determine what to do now.

This problem is solved as a deterministic (but time-dependent) dynamic programming problem. To simplify the notation, we are going to assume that each movement over a link (i, j) takes one time period. We represent our traveler only when he is at a node, since this is the only time when there is a real decision. Imagine that it is time t, and that we are at node q heading to r. Define

c_{tij} = the estimated cost, made at time t, of traversing link (i, j),

x_{tij} = the flow that we plan, at time t, on traversing link (i, j) (typically at some time in the future).

In our shortest path problem, the flow x_{tij} is either 1, meaning that link (i, j) is in the shortest path from q to r, and 0 otherwise.

Assume that we are sure we can arrive by time T, but if we arrive earlier, then we do nothing at node r until time T. We can write our problem as

$$\min_{x_t, \dots, x_T} \sum_{t'=t}^{T} \sum_i \sum_j c_{tij} x_{tij} \tag{19.24}$$

subject to flow conservation constraints:

$$\sum_j x_{tqj} = 1, \tag{19.25}$$

$$\sum_k x_{t',ki} - \sum_j x_{t'+1,ij} = 0, \quad t' = t, \dots, T-1, \forall i, \tag{19.26}$$

$$\sum_i x_{T-1,ir} = 1. \tag{19.27}$$

The optimization model (19.24)–(19.27) is a lookahead model that is optimizing the problem from time t until the end of the horizon T. Constraint (19.25) specifies that one unit of flow has to go out of origin node q at time t. Constraints (19.26) ensure that flow into each intermediate node is equal to the flow out. Finally, constraint (19.27) ensures that there is one unit of flow into node r at time T.

Shortest path problems are always solved as (highly specialized) deterministic dynamic programs. When combined with some careful software engineering, problem (19.24)–(19.27) can be easily solved using Bellman's equation for deterministic problems:

$$V_{t'i} = \min_j \left(c_{t'ij} + V_{t'+1,j} \right), \tag{19.28}$$

for $t' = t, \dots, T$ and all nodes i.

Both the linear program (19.24)–(19.27) and the deterministic dynamic program (19.28) represent deterministic lookahead models. When we solve the linear program, all we use is the decision x_t that tells us what to do at time t. Similarly, we use the decision from our dynamic program

$$x_t^* = \arg\min_j \left(c_{tqj} + V_{t+1,j} \right)$$

which tells us which node j we should go to. If $x_{tqj} = 1$, we will reoptimize when we arrive at node j at time $t+1$, at which time the costs may have changed.

This problem is a nice illustration of a complex stochastic network problem where the state variable for the original problem is not just the location of the traveler, but the current estimate of the travel times on each link in the entire network. Needless to say, this is an exceptionally high-dimensional stochastic optimization problem for which there is no known algorithm that could solve it to optimality. By virtue of our choice of a deterministic lookahead model, we can solve the lookahead model optimally as a dynamic program, but this does not make it an optimal policy.

Shortest path problems represent a familiar application (we use this any time we use a navigation system) where a deterministic lookahead policy seems to provide useful guidance. If only all problems were this easy, as we demonstrate in the next section.

19.6.2 Parameterized Lookaheads

One of the most powerful strategies for designing DLAs is to use a parameterized deterministic lookahead, as we did in section 13.3.3 for a stochastic, time-dependent energy storage problem. Given how important this strategy is for DLA policies, we are going to briefly summarize this strategy here.

Recall that we were optimizing over the following flows connecting a wind farm, the grid, a battery storage device, all serving a demand:

$$
\begin{aligned}
\tilde{x}_{tt'} \quad &= \quad \text{Planned generation of energy during hour } t' > t, \text{ where} \\
&\qquad \text{the plan is made at time } t, \text{ which is comprised of the} \\
&\qquad \text{following elements:} \\[4pt]
\tilde{x}_{tt'}^{ED} \quad &= \quad \text{flow of energy from renewables to demand,} \\[4pt]
\tilde{x}_{tt'}^{EB} \quad &= \quad \text{flow of energy from renewables to battery,} \\[4pt]
\tilde{x}_{tt'}^{GD} \quad &= \quad \text{flow of energy from grid to demand,} \\[4pt]
\tilde{x}_{tt'}^{GB} \quad &= \quad \text{flow of energy from grid to battery,} \\[4pt]
\tilde{x}_{tt'}^{BD} \quad &= \quad \text{flow of energy from battery to demand.}
\end{aligned}
$$

The goal was to optimize the available energy E_t from the wind farm, energy from the grid that can be purchased at price p_t, to serve a demand D_t. Given the time-varying nature of demands, capacity constraints on storage and transmission, and the highly stochastic nature of wind energy, we need to plan into the future, where we use forecasts of demands D_t and energy from the wind farm E_t that we represent using

$$f_{tt'}^{D} = \text{forecast of demand } D_{t'} \text{ made at time } t,$$

$$f_{tt'}^{E} = \text{forecast of wind energy } E_{t'} \text{ made at time } t.$$

We then created a deterministic lookahead policy where we optimize over a horizon $t, \dots, t+H$ which is given by

$$X^{DLA}(S_t|\theta) = \arg\max_{x_t, (\tilde{x}_{tt'}, t'=t+1,\dots,t+H)} \left(p_t(x_t^{GB} + x_t^{GD}) + \sum_{t'=t+1}^{t+H} \tilde{p}_{tt'}(\tilde{x}_{tt'}^{GB} + \tilde{x}_{tt'}^{GD}) \right) \quad (19.29)$$

subject to the following constraints: First, for time t we have:

$$x_t^{BD} - x_t^{GB} - x_t^{EB} \leq R_t, \quad (19.30)$$

$$\tilde{R}_{t,t+1} - (x_t^{GB} + x_t^{EB} - x_t^{BD}) = R_t, \quad (19.31)$$

$$x_t^{ED} + x_t^{BD} + x_t^{GD} = D_t, \quad (19.32)$$

$$x_t^{EB} + x_t^{ED} \leq E_t, \quad (19.33)$$

$$x_t^{GD}, x_t^{EB}, x_t^{ED}, x_t^{BD} \geq 0. \quad (19.34)$$

Then, for $t' = t+1, \dots, t+H$ we have:

$$\tilde{x}_{tt'}^{BD} - \tilde{x}_{tt'}^{GB} - \tilde{x}_{tt'}^{EB} \leq \tilde{R}_{tt'}, \quad (19.35)$$

$$\tilde{R}_{t,t'+1} - (\tilde{x}_{tt'}^{GB} + \tilde{x}_{tt'}^{EB} - \tilde{x}_{tt'}^{BD}) = \tilde{R}_{tt'}, \quad (19.36)$$

$$\tilde{x}_{tt'}^{ED} + \tilde{x}_{tt'}^{BD} + \tilde{x}_{tt'}^{GD} = \theta_{t'-t}^{D} f_{tt'}^{D}, \quad (19.37)$$

$$\tilde{x}_{tt'}^{EB} + \tilde{x}_{tt'}^{ED} \leq \theta_{t'-t}^{E} f_{tt'}^{E}. \quad (19.38)$$

The two key constraints are equations (19.37) and (19.38) which use the forecasts $f_{tt'}^{D}$ and $f_{tt'}^{E}$. Given the uncertainty in these forecasts, we multiply both by coefficients $\theta_{t'-t}^{D}$ and $\theta_{t'-t}^{E}$. These coefficients give us a parameterized policy $X^{DLA}(S_t|\theta)$, where we now face the challenge of tuning θ. We now have to tune θ by optimizing

$$\max_{\theta} F^{\pi}(\theta) = \mathbb{E}\left\{ \sum_{t=0}^{T} C(S_t, X^{\pi}(S_t|\theta))|S_0 \right\}. \quad (19.39)$$

A detailed summary of the tuning process for this problem is given in section 13.3.3. This application brings out the general strategy of using the concept of parametric cost function approximations in the setting of deterministic lookaheads. It is our belief that this strategy is actually widely used in practice, but in an ad hoc manner. What is missing is the formal process of tuning the parameters, which means solving the optimization problem given by (19.39), whether it is done in an online or offline setting.

We emphasize that no strategy is a panacea, but tuned deterministic lookaheads represent a practical and powerful strategy that has been largely ignored by the academic community. We note that one major strength of this approach over stochastic lookaheads is that we can tune θ in a very realistic simulator (or the real world), avoiding the array of approximations that are typically needed for stochastic lookaheads, as we describe next.

As with any parametric model, designing the parameterization is an art that requires some intuition into the structure of the problem and how uncertainty would affect a deterministic solution. We encourage readers to return to chapter 13 for the presentation on parameterized deterministic models.

19.7 A Tour of Stochastic Lookahead Policies

We are now going to review each of the four classes of policies as potential candidates for the lookahead policy. While this tour parallels the path of this entire book that focuses on covering all four classes of policies, when viewed as candidates for a lookahead policy inside a direct lookahead policy shifts the emphasis to computation, with relaxed emphasis on solution quality.

As of this writing, there is very little research addressing the effect of suboptimal lookahead policies on the performance of a DLA policy. While there is some theoretical research, we anticipate that this evaluation is always going to be problem dependent, and will require testing in a simulator.

19.7.1 Lookahead PFAs

Our lookahead policy $X_t^{DLA}(S_t)$, with the imbedded lookahead policy $\tilde{X}_{tt'}^{\tilde{\pi}}(\tilde{S}_{tt'})$, is given by

$$X_t^{DLA}(S_t) = \arg\max_{x_t} \left(C(S_t, x_t) + \tilde{E}\left\{ \max_{\tilde{\pi} \in \tilde{\Pi}} \tilde{E}\left\{ \sum_{t'=t+1}^{T} C(S_{t'}, \tilde{X}_{tt'}^{\tilde{\pi}}(\tilde{S}_{tt'})) | \tilde{S}_{t+1} \right\} | S_t, x_t \right\} \right). \quad (19.40)$$

To break this down, assume for illustration that we are solving an inventory replenishment problem, and that our lookahead policy $\tilde{X}_{tt'}^{\tilde{\pi}}(\tilde{S}_{tt'})$ follows an order-up-to policy (see equation (19.22)), where we trigger an order when the inventory R_t falls below θ^{min}, at which time we order $\tilde{X}_{tt'}^{\tilde{\pi}}(\tilde{S}_{tt'}|\tilde{\theta}) = (\theta^{max} - R_t)$. Let $\tilde{\theta} = (\theta^{min}, \theta^{max})$ be our tunable parameters. If we fix this as our lookahead policy, we would replace the inner $\max_{\tilde{\pi}}$ with $\max_{\tilde{\theta}}$, giving us

$$X_t^{DLA}(S_t) = \arg\max_{x_t} \left(C(S_t, x_t) + \tilde{E}\left\{ \max_{\tilde{\theta}} \tilde{E}\left\{ \sum_{t'=t+1}^{T} C(S_{t'}, \tilde{X}_{tt'}^{\tilde{\pi}}(\tilde{S}_{tt'}|\tilde{\theta})) | \tilde{S}_{t+1} \right\} | S_t, x_t \right\} \right) \quad (19.41)$$

Written this way, we see that the optimal $\tilde{\theta}^*$ depends on the state S_{t+1}, just as our optimal policy $X^*(S_t)$ from solving equation (19.1) should be written

$X^*(S_t|S_0)$. We already discussed the challenges of creating a function $\theta^*(S_t)$ in section 19.5. We either need a general function $\theta^*(S_t)$ so we can compute $\tilde{\theta} = \theta^*(S_{t+1})$, or we have to find the optimal $\tilde{\theta}$ for each state S_{t+1} in (19.41). To be completely honest, neither of these seem feasible.

A more realistic alternative is that we are going to pick a single $\theta = (\theta^{min}, \theta^{max})$ for our DLA policy $X_t^{DLA}(S_t|\theta)$. Now our policy looks like

$$X_t^{DLA}(S_t|\theta) = \arg\max_{x_t}\left(C(S_t, x_t) + \tilde{E}\left\{\tilde{E}\left\{\sum_{t'=t+1}^{T} C(S_{t'}, \tilde{X}_{t'}^\pi(\tilde{S}_{tt'}|\tilde{\theta}))|\tilde{S}_{t+1}\right\}|S_t, x_t\right\}\right).$$

We still have to tune θ just as we would tune any parameter in a policy (as we have done for PFAs and CFAs), but now this only has to be done once (offline), which makes our policy much easier to compute.

Of course, we have traded off solution quality for computational simplicity, but this is an example of prioritizing computation in a DLA policy, recognizing that a suboptimal policy does not have as much of an adverse impact since we are not actually implementing the decisions.

Figure 19.6 illustrates the calculations behind our lookahead policy in equation (19.41). We first have to loop over the possible values of x_t (so, x_t better be a discrete scalar). We then have to approximate the expectation over the random variable $\tilde{W}_{t+1}$ using a Monte Carlo sample. Then, we create a series of sample paths, where one sample path ω is a sequence of observations of $\tilde{W}_{t+2}, \tilde{W}_{t+3}, \dots, \tilde{W}_{t+H}$. These sample paths are used to simulate our policy for the remainder of the horizon.

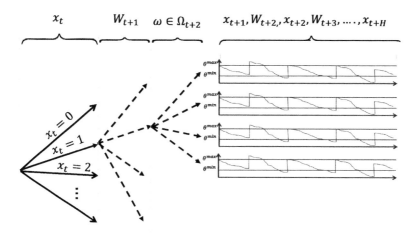

Figure 19.6 Illustration of the simulation of a direct lookahead inventory policy with an imbedded order-up-to policy within the lookahead model.

This particular solution to the lookahead policy is unique to this application. However, the idea of adopting a policy in the lookahead model that is easier to compute, while trading off solution quality, is a general strategy that can be applied to any problem. The challenge is identifying that easier-to-compute and good-enough lookahead policy combines art and science.

19.7.2 Lookahead CFAs

From the perspective of choosing lookahead policies, PFAs and CFAs are both parameterized policies that can be approached the same way. There are two major differences when handling CFAs:

- CFAs are computationally more expensive than PFAs, since there is always an imbedded optimization (which may be as simple as a sort, or as complex as an integer program). Imbedded in a DLA, it may be necessary to compute the CFA policy hundreds or thousands of times.
- The search for the best parameter vector θ for a CFA tends to be simpler than that for a PFA, since the CFA incorporates a considerable amount of problem structure in the deterministic optimization model.

Other than these issues, evaluating CFAs parallels the evaluation of PFAs.

19.7.3 Lookahead VFAs for the Lookahead Model

VFA-based policies tend to be easier to approximate than PFAs if we do not know the structure of the policy. Another benefit is that it is easier to create time-dependent policies using VFAs than with PFAs. This said, any strategy for solving, even approximately, a dynamic program for each state $\tilde{S}_{tt'}$ to determine $\tilde{x}_{tt'}$ would be hopelessly impractical. Just the same, we have to realize that for problems where a direct lookahead is appropriate, a backward ADP policy (as described in chapter 15), would not be able to capture the complex interactions in the future.

There is another approach that exploits the power of VFA-based policies within a direct lookahead. It consists of two steps:

Step 1: Use backward approximate programming (from chapter 15) once on the lookahead model, which means we are limited to the information in the approximate state $\tilde{S}_{tt'}$ rather than the base state S_t, as we did in chapter 15. From this, we obtain value function approximations $\overline{V}_{tt'}^x(\tilde{S}_{tt'})$ for each time period $t' = t, t+1, \ldots, t+H$ in the lookahead model.

Step 2: Now use these VFAs as the basis for the policy

$$\tilde{X}_{tt'}^{VFA}(\tilde{S}_{tt'}) = \arg\max_{\tilde{x}_{tt'}} \left(C(\tilde{S}_{tt'}, \tilde{x}_{tt'}) + \overline{V}_{tt'}^{x}(\tilde{S}_{tt'}^{x}) \right),$$

that we then use in step (6a) of the algorithm in Figure 19.3. The policy $\tilde{X}_{tt'}^{VFA}(\tilde{S}_{tt'})$ is used to simulate the downstream impact of the decision x_t.

This approach combines the power of a VFA-based policy with the higher level of detail we obtain in a DLA-based policy, since there may be information in the state S_t in the base model that is captured in the lookahead model as latent variables, but not in the lookahead state $\tilde{S}_{tt'}$.

19.7.4 Lookahead DLAs for the Lookahead Model

So now we return to the idea of using a direct lookahead policy as the lookahead policy in our lookahead model. This is easily going to be the most computationally demanding strategy that we could use for our lookahead policy. We have to first remember that we have to revisit all of the approximation strategies we used to formulate our original lookahead model. We anticipate two in particular will make a DLA policy in the lookahead computationally possible (if not attractive):

- The planning horizon – Let $\tilde{H}$ be the planning horizon within the lookahead DLA model. We would expect $\tilde{H} < H$, but this is clearly a parameter that can be tuned. However, if we aggressively shorten the horizon, we may obtain a practical policy. For example, imagine that we are planning the dispatch of driverless electric vehicles, which means we own the vehicle and need to plan its activities into the future when we make dispatch decisions now. We may decide that we want to plan until the end of the day before making a decision now. However, our policy within the DLA might myopically optimize just one dispatch into the future.
- A deterministic lookahead policy – We might use a deterministic lookahead for the DLA policy within a stochastic model. Using our ride hailing example, we could sample demands into the future and then solve this optimally, which would give us an optimistic estimate of the value of being in a future state $\tilde{S}_{tt'}$.

Remember that both of these policies have to be solved for each simulated state $\tilde{S}_{tt'}$ that we encounter as we simulate into the future for our DLA policy. Even if these prove to be too expensive, they may serve as benchmarks for other policies.

19.7.5 Discussion

It is not possible to say which of these strategies would be best for a particular problem. Even a single problem class can come in different flavors that require different types of policies. Our goal is to challenge the reader to always consider all four classes of policies, and to use all the tools available. The range of sequential decision problems is vast, and we are trying to make the entire span of the toolbox that the field can offer.

Next, we are going to consider two widely studied (and used) lookahead policies that have attracted considerable attention in the academic literature. The two strategies are:

- Monte Carlo tree search for discrete decisions – This approach has become wildly popular in computer science, largely motivated by its use in games, where direct lookaheads are particularly useful.
- Two-stage stochastic programming for vector decisions – This policy was first developed in the 1950s for a special problem class (make decision, see information, make decision, stop) but has been primarily used as a policy for fully sequential problems.

A third strategy, robust optimization, is one that we introduced earlier in section 19.3 as an example of a lookahead policy with a different objective. This also has attracted considerable attention in the academic literature, although it is unclear how widely it is used in practice. Unfortunately, there is often a big gap between the methods that attract academic attention, and those that are actually used in practice.

19.8 Monte Carlo Tree Search for Discrete Decisions

For problems where the number of decisions per state is not too large (but where the set of random outcomes may be quite large, even infinite), we may replace the explicit enumeration of the entire tree with a heuristic policy to evaluate what might happen after we reach a state. Imagine that we are trying to evaluate if we should choose decision $\tilde{x}_{tt}$ which takes us to state $\tilde{S}_t^x$ (the post-decision state), after which we choose at random an outcome of $\tilde{W}_{t,t+1}$, which puts us in the next (pre-decision) state $\tilde{S}_{t,t+1}$. If we repeat this process for each decision $\tilde{x}_{tt}$, and then for each decision $\tilde{x}_{t,t+1}$ out of each of the downstream states $\tilde{S}_{t,t+1}$, the tree would explode in size.

Monte Carlo tree search offers a method that samples the tree in an intelligent way. Given an infinite budget, MCTS will ultimately learn the entire tree, but the hope is that it produces a high quality (perhaps near optimal) initial decision that we can implement in our base model at a reasonable computational cost. As a direct lookahead, MCTS offers a framework where we can draw on many of the other tools we have introduced, but this section will provide a basic introduction to MCTS that is well suited to problems with discrete sets of decisions, and where the set of decisions out of each state is not too large.

19.8.1 Basic Idea

Monte Carlo tree search is a technique that is very popular in computer science (although the roots of the method are in operations research), where it has been primarily used for deterministic problems. MCTS (as it is widely known) proceeds by applying a simple test to each decision out of a node, and then uses this test to choose one decision to explore. This may result in a traversal to a state we have visited before, at which point we simply repeat the process, or we may find ourselves at a new state. We then call a *rollout policy* which is some policy (literally, any of the policies we have discussed up to now) for making decisions that depends on the problem at hand. The rollout policy gives us an estimate of the value of being in this new state. If the state is attractive enough, it is added to the tree.

Each node (state) in the tree is described by four quantities:

(1) The pre-decision value function $\tilde{V}_{tt'}(\tilde{S}_{tt'})$, the post-decision value function $\tilde{V}^x_{tt'}(\tilde{S}^x_{tt'})$, and the contribution $C(\tilde{S}_{tt'}, \tilde{x}_{tt'})$ from being in state $\tilde{S}_{tt'}$ and taking decision $\tilde{x}_{tt'}$.
(2) The visit count, $N(\tilde{S}_{tt'})$, which counts the number of times we have performed rollouts (explained next) from state $\tilde{S}_{tt'}$.
(3) The decision counter, $N(\tilde{S}_{tt'}, \tilde{x}_{tt'})$, which counts the number of times we have taken decision $\tilde{x}_{tt'}$ from state $\tilde{S}_{tt'}$.
(4) The set of decisions $\mathcal{X}_s$ from each state s and the random outcomes $\tilde{\Omega}_{t,t'+1}(\tilde{S}^x_{tt'})$ that might happen when in post-decision state $\tilde{S}^x_{tt'}$.

19.8.2 The Steps of MCTS

Monte Carlo tree search progresses in four steps which are illustrated in Figure 19.7, where the detailed steps of the MCTS are described in a series of procedures. Note that, as before, we let $\tilde{S}_{tt'}$ be the pre-decision state, which is the node that precedes a decision. A deterministic function $S^{M,x}(\tilde{S}_{tt'}, \tilde{x}_{tt'})$ takes us to a post-decision state $\tilde{S}^x_{tt'}$, after which a Monte Carlo sample of the exogenous information takes us to the next pre-decision state $\tilde{S}_{t,t'+1}$.

(1) Selection – There are two steps in the selection phase. The first (and most difficult) requires choosing a decision, while the second involves taking a Monte Carlo sample of any random information.

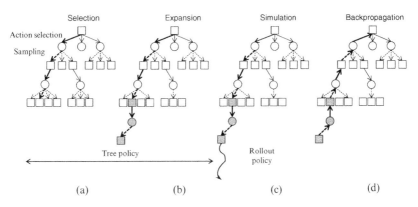

Figure 19.7 Illustration of Monte Carlo tree search, illustrating (left to right): selection, expansion, simulation, and backpropagation.

(1a) Choosing the decision – The first step from any node (already generated) is to choose a decision (see Figure 19.7(a) and the algorithm in Figure 19.8). The most popular policy for choosing a decision is to use a type of upper confidence bound (recall we introduced UCB policies in section 7.5) adapted for trees, hence its name, Upper Confidence bounding for Trees (UCT). We could simply choose the decision that appears to be best, but we could get stuck in a solution were we avoid decisions that do not look attractive. Since our estimates are only approximations, we have to recognize that we may not have explored them enough (the classic exploration-exploitation tradeoff). In this setting, the UCT policy is given by

$$X_{tt'}^{UCT}(\tilde{S}_{tt'}|\theta^{UCT}) = \arg\max_{\tilde{x} \in \tilde{X}_{tt'}} \left((C(\tilde{S}_{tt'}, \tilde{x}) + \overline{V}_{tt'}^x(\tilde{S}_{tt'}^x)) + \theta^{UCT}\sqrt{\frac{\ln N(\tilde{S}_{tt'})}{N(\tilde{S}_{tt'}, \tilde{x}_{tt'})}} \right).$$

The parameter θ^{UCT} has to be tuned, just as we would tune any policy. As with UCB policies, the square root term is designed to encourage exploration, by putting a bonus for decisions that have not been explored as often. A nice feature of UCT policies is that they are very easy to compute, which is important in an MCTS setting where we need to quickly evaluate many decisions.

(1b) Sampling the outcome – Here we assume that we can simply take a Monte Carlo sample of any random information (see section 10.4). There are settings where simple Monte Carlo sampling is not very efficient, such as when the random outcome might be a success or failure, where one or the other dominates.

(2) Expansion – If the decision we choose earlier is one we have chosen before, then we progress to the next post-decision state (the solid line connecting

---

function $MCTS(S_t)$
Step 0. Create root note $\tilde{S}_{tt} = S_t$; set iteration counter $n = 0$.
Step 1. while $n < n^{thr}$

    Step 1.1 $\tilde{S}_{tt'} \leftarrow TreePolicy(\tilde{S}_{tt})$
    Step 1.2 $\bar{V}_{tt'}(\tilde{S}_{tt'}) \leftarrow SimPolicy(\tilde{S}_{tt'})$
    Step 1.3 $Backup(\tilde{S}_{tt'}, \bar{V}_{tt'}(\tilde{S}_{tt'}))$
    Step 1.4 $n \leftarrow n + 1$

Step 2. $\tilde{x}_t^* = \arg\max_{\tilde{x}_{tt} \in \tilde{x}_{tt}(\tilde{S}_{tt})} \bar{C}(\tilde{S}_{tt}, \tilde{x}_{tt}) + \bar{V}_{tt}^x(\tilde{S}_{tt}^x)$
Step 3. return x_t^*.

Figure 19.8 Sampled MCTS algorithm.

the square node to the round node in Figure 19.7(b)) at which point we then sample another random outcome which brings us to a new pre-decision state (see the algorithm in Figure 19.9). But if we have not chosen this decision before, then we expand our tree by first adding the link associated with the decision to the post-decision state node, followed by a Monte Carlo sample which takes us to the subsequent pre-decision state. At this point we have to deal with the fact that we would not have an estimate of the value of being in this state (which we need for our UCT policy). To overcome this, we call our simulation policy, which is a form of roll-out policy (discussed next).

(3) Simulation – The simulation step assumes we have access to some policy which is easy to execute that allows us to obtain a quick and reasonable estimate of the value of being in a state (see Figure 19.7(c) and the algorithm in Figure 19.10). Of course, this is very problem dependent. Some strategies include:

- A myopic policy, which greedily makes choices. There are problems where myopic policies are reasonable starting estimates (of course they are suboptimal). However, such a greedy policy can be extremely poor (imagine finding the shortest path through a network by always choosing the shortest link out of a node).

- A parameterized policy with reasonable estimates of the parameters. We might have a rule for selling an asset if its price rises by some percentage. Such a rule will not be ideal, but it will be reasonable.

- A posterior bound. We might sample all future information, and then make the best decision assuming that this future information comes true.

(4) Backpropagation – After simulating forward using our rollout policy to obtain an initial estimate of the value of being in our newly generated state,

function $TreePolicy(\tilde{S}_{tt'})$
Step 0. $t' \leftarrow t$
Step 1. while $\tilde{S}_{tt'}$ is non-terminal **do**

    Step 2. if $|\tilde{\mathcal{A}}_{tt'}(\tilde{S}_{tt'})| < d^{thr}$ **do** (Expanding a decision out of a pre-decision state)

        Step 2.1 Choose decision $\tilde{x}_{tt'}^{*}$ by optimizing on the basis of the contribution of the decision $\tilde{C}(\tilde{S}_{tt'}, \tilde{x}_{tt'})$, then taking a Monte Carlo sample to the next pre-decision state $\tilde{S}_{t,t'+1}$, and then finally using the rollout policy to approximate the value of being in state $\tilde{S}_{t,t'+1}$.
        Step 2.2 $\tilde{S}_{tt'}^{x} = S^{M}(\tilde{S}_{tt'}, \tilde{x}_{tt'}^{*})$ (Expansion step)
        Step 2.3 $\tilde{\mathcal{X}}_{tt'}(\tilde{S}_{tt'}) \leftarrow \tilde{\mathcal{X}}_{tt'}(\tilde{S}_{tt'}) \bigcup \{\tilde{x}_{tt'}^{*}\}$
        Step 2.4 $\tilde{\mathcal{X}}_{tt'}^{u}(\tilde{S}_{tt'}) \leftarrow \tilde{\mathcal{X}}_{tt'}^{u}(\tilde{S}_{tt'}) - \{\tilde{x}_{tt'}^{*}\}$

        else Step 2.5 $\tilde{x}_{tt'}^{*} = \arg\max_{\tilde{x}_{tt'} \in \tilde{\mathcal{X}}_{tt'}(\tilde{S}_{tt'})} \left(\left(\tilde{C}(\tilde{S}_{tt'}, \tilde{x}_{tt'}) + \bar{V}_{tt'}^{x}(\tilde{S}_{tt'}^{x}) \right) + \theta^{UCT} \sqrt{\frac{\ln N(\tilde{S}_{tt'})}{N(\tilde{S}_{tt'}, \tilde{x}_{tt'})}} \right)$

        Step 2.6 $\tilde{S}_{tt'}^{x} = S^{M}(\tilde{S}_{tt'}, \tilde{x}_{tt'}^{*})$

    end if
    Step 3 if $|\tilde{\Omega}_{t,t'+1}(\tilde{S}_{tt'}^{x})| < e^{thr}$ **do** (Expanding an exogenous outcome out of a post-decision state)
        Step 3.1 Choose exogenous event $\tilde{W}_{t,t'+1}$,
        Step 3.2 $\tilde{S}_{t,t'+1} = S^{M,x}(\tilde{S}_{tt'}^{x}, \tilde{W}_{t,t'+1})$ (Expansion step)
        Step 3.3 $\tilde{\Omega}_{t,t'+1}(\tilde{S}_{tt'}^{x}) \leftarrow \tilde{\Omega}_{t,t'+1}(\tilde{S}_{tt'}^{x}) \bigcup \{\tilde{W}_{t,t'+1}\}$
        Step 3.4 $\tilde{\Omega}_{t,t'+1}^{u}(\tilde{S}_{tt'}^{x}) \leftarrow \tilde{\Omega}_{t,t'+1}^{u}(\tilde{S}_{tt'}^{x}) - \{\tilde{W}_{t,t'+1}\}$
        Step 3.5 $t' \leftarrow t' + 1$
        return $\tilde{S}_{tt'}$ (stops execution of **while** loop)

        else Step 3.6 Choose exogenous event $\tilde{W}_{t,t'+1}$,
        Step 3.7 $\tilde{S}_{t,t'+1} = S^{M,x}(\tilde{S}_{tt'}^{x}, \tilde{W}_{t,t'+1})$
        Step 3.8 $t' \leftarrow t' + 1$
        end if
    end while

Figure 19.9 The tree policy.

function $SimPolicy(\tilde{S}_{tt'})$
Step 0. Choose a sample path $\tilde{\omega} \in \tilde{\Omega}_{tt'}$
Step 1. while $\tilde{S}_{tt'}$ is non-terminal
    Step 2.1 Choose $\tilde{x}_{tt'} \leftarrow \tilde{\pi}(\tilde{S}_{tt'})$ where $\tilde{\pi}$ is the rollout policy.
    Step 2.2 $\tilde{S}_{t,t'+1} \leftarrow S^{M}(\tilde{S}_{tt'}, \tilde{x}_{tt'}(\tilde{\omega}))$
    Step 2.3 $t' \leftarrow t' + 1$
end while
return $\bar{V}_{tt'}(\tilde{S}_{tt'})$ (Value function of $\tilde{S}_{tt'}$)

Figure 19.10 This function simulates the policy.

---

function $Backup(\tilde{S}_{tt'}, \mathcal{V}_{tt'}^x(\tilde{S}_{tt'}))$
while $\tilde{S}_{tt'}$ is not null **do**

    Step 1.1 $N(\tilde{S}_{tt'}) \leftarrow N(\tilde{S}_{tt'}) + 1$
    Step 1.2 $t^* \leftarrow t'\text{-}1.$
    Step 1.3 $N(\tilde{S}_{t,t^*-1}, \tilde{x}_{t,t^*-1}) \leftarrow N(\tilde{S}_{t,t^*-1}, \tilde{x}_{t,t^*-1}) + 1$
    Step 1.4 $\mathcal{V}_{t,t^*-1}^x(\tilde{S}_{t,t^*-1}^x) \leftarrow \dfrac{1}{\sum_{\tilde{\omega}_{t,t^*+1} \in \tilde{\Omega}_{t,t^*+1}(\tilde{S}_{tt^*}^x)} p(\tilde{\omega}_{t,t^*+1})}\cdot$
        $E_g[p(\tilde{W}_{t,t^*+1})/g(\tilde{W}_{t,t^*+1})\mathcal{V}_{tt^*}^x(S^{M,x}(\tilde{S}_{tt^*}^x, \tilde{W}_{t,t^*+1}))]$
    Step 1.5 $\tilde{S}_{tt^*} \leftarrow$ predecessor of $\tilde{S}_{tt^*}^x$
    Step 1.6 $\Delta \leftarrow \bar{C}(\tilde{S}_{tt^*}, \tilde{x}_{tt^*}) + \mathcal{V}_{tt^*}^x(\tilde{S}_{tt^*}^x)$
    Step 1.7 $\mathcal{V}_{tt^*}(\tilde{S}_{tt^*}) \leftarrow \mathcal{V}_{tt^*}(\tilde{S}_{tt^*}) + \dfrac{\Delta - \mathcal{V}_{tt^*}(\tilde{S}_{tt^*})}{N(\tilde{S}_{tt^*})}$
    Step 1.8 $t' \leftarrow t^*$

end while

Figure 19.11 Backup process which updates the value of each decision node in the tree.

we now backtrack and obtain updated estimates of the value of each of the states on the path to the newly generated state (see Figure 19.7(d) and the algorithm in Figure 19.11).

Figure 19.12 shows a tree produced by an MCTS algorithm, which illustrates the varying degrees to which MCTS explores the tree. An indication that MCTS is adding value is the presence of narrow sections of the tree which are explored at much greater depth than other portions of the tree. If the tree is fairly balanced, then it means that MCTS is not pruning decisions which means it is basically enumerating the tree. Of course, the real question is how well the resulting tree works as a policy to solve the base model.

19.8.3 Discussion

MCTS is a true mixture – it is a DLA that uses a UCB policy (a form of CFA – see section 7.5), a rollout policy (typically a PFA or CFA), and VFAs. Perhaps the biggest weakness is the rollout policy used to obtain an initial estimate of the value of being at a node. Since we have no guarantees regarding the quality of the rollout policy, the initial estimate will underestimate the true value (on average).

MCTS has been proven to be asymptotically optimal. That is, given an infinite search budget, MCTS will ultimately sample each decision from each state (and

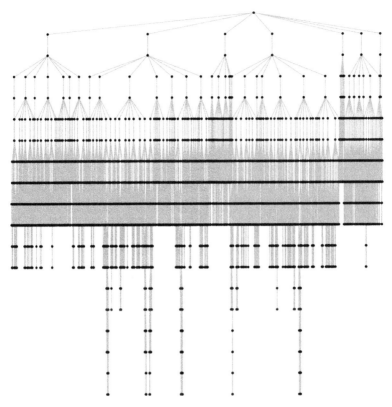

Figure 19.12 Sample of a tree produced by Monte Carlo tree search, illustrating the variable depth produced by an MCTS algorithm.

all the states will be enumerated) infinitely often. The reason is the UCT policy given by

$$X_{tt'}^{UCT}(\tilde{S}_{tt'}|\theta^{UCT}) = \arg\max_{\tilde{x} \in \tilde{X}_{tt'}} \left((C(\tilde{S}_{tt'}, \tilde{x}) + \bar{V}_{tt'}^{x}(\tilde{S}_{tt'}^{x})) + \theta^{UCT} \sqrt{\frac{\ln N(\tilde{S}_{tt'})}{N(\tilde{S}_{tt'}, \tilde{x}_{tt'})}} \right).$$

The key is the term

$$\sqrt{\frac{\ln N(\tilde{S}_{tt'})}{N(\tilde{S}_{tt'}, \tilde{x}_{tt'})}},$$

where $N(\tilde{S}_{tt'})$ is the number of times we visit state $\tilde{S}_{tt'}$ and $N(\tilde{S}_{tt'}, \tilde{x}_{tt'})$ is the number of times we visit state $\tilde{S}_{tt'}$ and choose decision $\tilde{x}_{tt'}$. This term goes to infinity as we continue to search the tree. If we do not try decision $\tilde{x}_{tt'}$, then the numerator grows while the denominator holds constant. If we have never tried the decision $\tilde{x}_{tt'}$, then the denominator is zero which forces us to choose among the decisions $\tilde{x}_{tt'}$ that have never been tried until all decisions have been tried at least once (which forces us to explore all the downstream states).

In short, this "uncertainty bonus" term (as it is often known in the bandit community where the logic originated) is the secret that forces us to eventually try everything. This means that even if our rollout policy produces a poor estimate of $\tilde{V}_{tt'}^x(\tilde{S}_{tt'}^x)$, we will still try every decision $\tilde{x}_{tt'}$, that will eventually force us to visit every state $\tilde{S}_{tt'}$.

Asymptotic convergence proofs need to be taken with a grain of salt (and this includes the proofs in section 5.10). We never run algorithms to the limit, which means what we care about is how well it works within our computation budget. MCTS is no exception. How well it works in practice depends on the computation budget, which depends on choices such as the lookahead policy, as well as the details of how it is coded. For example, there are many opportunities to use parallel computation (we might explore hundreds of decisions all at the same time), which will affect how deep we can explore tree (and how thoroughly) within our budget.

19.8.4 Optimistic Monte Carlo Tree Search

A variant of MCTS is one that is called *optimistic* MCTS, and it works with one minor modification. Instead of simulating a rollout policy starting at a node $\tilde{S}_{t,t''}$ until the end of the horizon, we generate the sample path of exogenous information $\tilde{W}_{t,t''}, \tilde{W}_{t,t''+1}, \ldots, \tilde{W}_{t,t+H}$. Then, we assume that this entire sample path is known to us, and solve to optimality the deterministic optimization problem that determines the decisions $\tilde{x}_{t,t''}, \tilde{x}_{t,t''+1}, \ldots, \tilde{x}_{t,t+H}$ starting at node $\tilde{S}_{t,t''}$. We then compute the contribution over this path and use it to initialize the value at node $\tilde{S}_{t,t''}$.

By optimizing over the entire sample path, we are making decisions that are allowed to "see" into the future. This produces an optimistic estimate of the value of being at node $\tilde{S}_{t,t''}$ rather than the pessimistic estimate from traditional MCTS. This is a powerful result because it allows us to prove asymptotic optimality *without* requiring that we visit every state, or even test every decision. This is an important feature for problems with a large number of decisions, but it comes at a price. Solving the deterministic optimization problem over the sample path can be quite expensive.

There are two ways to run optimistic MCTS. The least expensive is the logic sketched earlier where we optimize over the entire sample path, using information that would not otherwise be available. This is known as *information relaxation*. A more advanced algorithm uses penalties that try to enforce

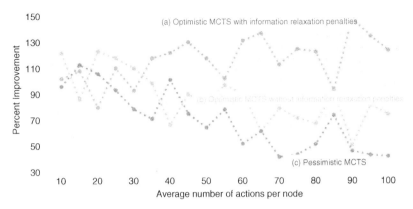

Figure 19.13 Performance of (a) optimistic MCTS with information relaxation penalties, (b) optimistic MCTS without information relaxation penalties, and (c) pessimistic MCTS.

the requirement that the decision $\tilde{x}_{tt''}$ be independent of information arriving after t''. Needless to say, this idea is even more expensive.

Figure 19.13 shows the improvement for the following policies as a function of the number of decisions per node:

(a) Optimistic MCTS where penalties have been added to try to encourage decisions $\tilde{x}_{t,t''}$ that do not reflect future events (that is, $\tilde{w}_{t,t''+1}$ and later).
(b) Optimistic MCTS without the information relaxation penalties.
(c) The classical "pessimistic" MCTS.

The calculation of the information penalties for (a) involves advanced methods that are beyond the scope of this book, but comparing these three strategies provides some insights into both MCTS as well as two-stage stochastic programming which we describe now.

The results show that optimistic MCTS, with the information relaxation penalties, performs significantly better than pessimistic MCTS as the number of decisions per node exceed 30. Optimistic MCTS without the information relaxation penalties also outperforms pessimistic MCTS, but the difference is not as dramatic. What this figure does not show is the difference in computation cost. Optimistic MCTS, even without information relaxation penalties, is computationally more expensive than pessimistic MCTS. The computation of the information relaxation penalties imposes an even greater burden. However, this issue has to be addressed in the context of modern computing environments where massive parallelism may be possible.

This is an area that requires more research. It is well known that MCTS is limited to problems where the number of decisions per node is "not too large," but the idea of using optimistic estimates of the value of the future may expand the size of problems (that is, in terms of the number of decisions) that it can be used for.

The idea of making decisions such as $\tilde{x}_{t,t''}$ using information from a (simulated) future is used in other methods, as we will see next with two-stage stochastic optimization. Figure 19.13 is an initial indication of the effect of using information relaxation in a lookahead policy.

19.9 Two-Stage Stochastic Programming for Vector Decisions*

After deterministic linear programming was invented, researchers quickly realized that there were uncertainties in many applications, leading to an effort (initiated by none other than George Dantzig, who invented the simplex method that started the math programming revolution) to incorporate uncertainty into a linear program.

Not surprisingly, the introduction of uncertainty into mathematical programs opened a Pandora's box of modeling and computational issues that challenges the research community today. For decades the community focused on what became known as the two-stage stochastic programming problem, which consists of "make decision, see information, make one more decision." The two-stage stochastic programming formulation remains a foundational tool for approximating fully sequential problems. We introduce the basic two-stage stochastic programming problem, and then show how it can be used to build a lookahead policy for the fully sequential inventory problem we introduced earlier. We then show how this can be used as an approximate lookahead model for fully sequential problems.

As a historical note: the field of stochastic programming emerged in the 1950's alongside the development of the field of Markov decision processes by Richard Bellman, with stochastic programming focusing on vector-valued decisions and Markov decision processes working with discrete decisions. These communities evolved in parallel with distinctly different notational systems, modeling frameworks, and relative emphasis on theory vs. computation.

19.9.1 The Basic Two-Stage Stochastic Program

In section 18.6, we introduced what is known as the *two stage stochastic program* where we make an initial decision x_0 (such as where to locate warehouses), after which we see information $W_1 = W_1(\omega)$ (which might be the demands for product), and then we make a second set of decisions $x_1(\omega)$ which depend on this information (the decisions x_1 are known as the *recourse variables*).

As we did in section 18.6, the two-stage stochastic programming problem is written

$$\max_{x_0} \left(c_0 x_0 + \mathbb{E}V_1(x_0, W_1) \right), \tag{19.42}$$

subject to the constraints,

$$A_0 x_0 \;\; = \;\; b_0, \tag{19.43}$$

$$x_0 \;\; \geq \;\; 0. \tag{19.44}$$

The initial decisions x_0 (which determines the inventories in the warehouses) then impacts the decisions that can be made after the information (the demand) becomes known, producing the second stage problem

$$V_1(x_0, \omega) = \max_{x_1(\omega)} c_1(\omega) x_1(\omega), \tag{19.45}$$

subject to, for all $\omega \in \Omega$,

$$A_1 x_1(\omega) \;\; \leq \;\; B_1 x_0, \tag{19.46}$$

$$B_1 x_1(\omega) \;\; \leq \;\; D_1(\omega), \tag{19.47}$$

$$x_1(\omega) \;\; \geq \;\; 0. \tag{19.48}$$

Note that we write the value function for the second state, $V_1(x_0, W_1)$ in terms of x_0, although x_0 is not a proper state variable. For example, we could write

$$R_1 = B_1 x_0,$$

and then write the first stage objective as

$$\max_{x_0} \left(c_0 x_0 + \mathbb{E}V_1(R_1, W_1) \right).$$

In fact, for many applications R_1 is lower dimensional (and possibly much lower dimensional) than x_0. For example, R_1 might be a vector of inventories, where R_{1i} is the amount of inventory at location i, while x_0 is a high dimensional vector with elements x_{0ij}. Our version shown earlier represents standard convention in this community.

We cannot actually compute this model if Ω represents all the potential outcomes, so we have to use the idea of a sampled model that we first introduced in section 4.3. We note that even when Ω is carefully designed and "not too big," the two-stage problem can still be hard to solve (if the decision vectors are large enough).

There are several computational strategies that have been used to solve problem (19.42)-(19.48):

The "deterministic equivalent" method – The problem (19.42)–(19.48), when formulated using a sampled set of observations Ω, is basically a single (potentially large) deterministic linear program, leading some to refer to this problem as the "deterministic equivalent." Modern solvers can handle problems with hundreds of thousands of variables with minimal training (specialists have worked on problems with millions of variables).

Relaxation – If we replace x_0 with $x_0(\omega)$, this means that we are allowing x_0 to see the future. We can fix this with a *nonanticipativity constraint* that looks like

$$x_0(\omega) = x_0, \text{ for all } \omega \in \hat{\Omega}. \tag{19.49}$$

This formulation allows us to design algorithms that relax this constraint, allowing us to solve $|\Omega|$ independent problems with logic that penalizes deviations of (19.49). This methodology has become known under the name of *progressive hedging* which has made it possible to approach two-stage stochastic programs that would otherwise be too large.

Benders decomposition – In section 18.6 we introduced the idea of using Benders decomposition, where the function $\mathbb{E}V_1(x_0, W_1)$ is replaced with a series of cuts. This is really a form of approximate dynamic programming, which we showed in chapter 18.

We are now going to transition to using this two-stage model as a policy for fully sequential problems.

19.9.2 Two-Stage Approximation of a Sequential Problem

While there are true two-stage stochastic programming problems, there is a vast range of fully sequential problems of the types that we have been pursuing in this book that involve vector-valued decisions in the presence of uncertainty. A widely used strategy is to solve these problems by approximating them as two-stage problems, where there is a decision to be made now at time t, represented by x_t, after which we then pretend that we observe all the future information over the rest of our horizon $(t + 1, t + H)$. After this information is revealed, we then make all remaining decisions $\tilde{x}_{tt'}$ for $t' = t + 1, \dots, t + H$, which represents (along with the revealed information) the second stage of our decision problem.

We are going to illustrate these ideas using the energy storage problem we first addressed in section 13.3.3, where we are pulling energy from the grid (at a random price), a wind farm (with random supply), to meet a time-dependent load, with a storage device to smooth out the different sources of variability. It may be helpful to flip back to this section to review the model, but we will repeat the definitions of a few variables. Our state variables were given by:

$$D_t \quad = \quad \text{Demand (``load'') for power during hour } t.$$

$$E_t \quad = \quad \text{Energy generated from renewables (wind/solar) during hour } t.$$

$$R_t \quad = \quad \text{Amount of energy stored in the battery at time } t.$$

$$u_t \quad = \quad \text{Limit on how much generation can be transmitted at time } t \text{ (this is known in advance).}$$

$$p_t \quad = \quad \text{Price to be paid for energy drawn from the grid at time } t.$$

The decision variables were given by:

$$x_t^{ED} \quad = \quad \text{Flow of energy from wind to demand.}$$

$$x_t^{EB} \quad = \quad \text{Flow of energy from wind to battery.}$$

$$x_t^{GD} \quad = \quad \text{Flow of energy from grid to demand.}$$

$$x_t^{GB} \quad = \quad \text{Flow of energy from grid to battery.}$$

$$x_t^{BD} \quad = \quad \text{Flow of energy from battery to demand.}$$

Let x_t be the vector of all five flows, and let $\tilde{x}_{tt}, \dots, \tilde{x}_{t,t+H}$ be the set of vectors over our planning horizon. As we did in section 13.3.3, we want to plan over a horizon where we no longer use point forecasts of the demands $D_{t'}$ and energy from wind or solar, $E_{t'}$. Instead, we want to explicitly model the uncertainty in $D_{t'}, E_{t'}$ and $p_{t'}$ for $t' > t$.

First, we are going to create a set of M sample paths that we call $\tilde{\Omega}_t$ (indexed by t because it is generated at time t), where each $\tilde{\omega}_t \in \tilde{\Omega}_t$ represents a full sequence of the random variables $W_{t+1}, W_{t+2}, \dots, W_{t+H}$. Given $\tilde{\omega}_t \in \tilde{\Omega}_t$ (that is, given the rest of the future), we index all future decisions with the same outcome $\tilde{\omega}_t$, giving us the vector $\tilde{x}_{tt'}(\tilde{\omega}_t)$, for $t' = t + 1, t + 2, \dots, t + H$. The outcomes $\tilde{\omega}_t$ are known as *scenarios* in the language of stochastic programming.

Next we are going to make decisions $\tilde{x}_{t,t+1}(\tilde{\omega}_t), \tilde{x}_{t,t+2}(\tilde{\omega}_t), \dots, \tilde{x}_{t,t+H}(\tilde{\omega}_t)$ for each sample path $\tilde{\omega}_t$. Since $\tilde{\omega}_t$ indexes the entire sample path, that means each $\tilde{x}_{tt'}(\tilde{\omega}_t)$ is allowed to "see" the entire future. However, we only allow this for decisions $\tilde{x}_{tt'}$ in the future (that is, $t' > t$). The current decision that we would implement, $\tilde{x}_{tt}$, is not indexed by $\tilde{\omega}_t$.

We now have the following sets of variables:

- Variables indexed by (tt) (that is, known or determined now), which include:
  - Parameters or quantities, including costs and constraints.
  - Decisions $\tilde{x}_{tt}$ which will be implemented now.
- Variables indexed by (tt') for $t' > t$, which include:
  - Parameters or quantities in the future which are not known now.
  - Decisions $\tilde{x}_{tt'}$ that we plan to help us make $\tilde{x}_{tt}$, but which will not be implemented.

There are two ways to model $\tilde{x}_{tt}$. The first is to treat it as a single vector that must be determined now. The second is to allow it to also be indexed by $\tilde{\omega}_t$, creating a family of variables $\tilde{x}_{tt}(\tilde{\omega}_t)$, each of which is allowed to see the entire sample path (which means it is allowed to see into the future). This raises the question: which one to implement? We need just a single decision to be implemented, which we do by imposing the constraint

$$\tilde{x}_{tt}(\tilde{\omega}_t) = x_t. \tag{19.50}$$

Equation (19.50) is known in the stochastic programming literature as a *nonanticipativity constraint*, which we first saw in chapter 2 (see equation (2.25)). The first question that a reader should ask is: why would we even use the notation $x_{tt}(\omega)$? The reason is computational. Imagine that we write the time t decision as $x_{tt}(\omega)$ and temporarily ignore equation (19.50). In this case, the problem would decompose into a series of problems, one for each $\tilde{\omega}_t \in \tilde{\Omega}_t$. These are much smaller problems than a single problem where we have to deal with all the different scenarios at the same time, but it means that we can get a different answer $\tilde{x}_{tt}(\tilde{\omega}_t)$ at time t, which is a problem. However, there are algorithmic strategies that take advantage of this problem structure.

We can model the first period decision as $\tilde{x}_{tt}(\tilde{\omega}_t)$ and impose a nonanticipativity constraint (19.50), or we can simply use x_t and just let all the future decisions $\tilde{x}_{tt'}(\tilde{\omega}_t)$ for $t' > t$ depend on the scenario. If we use the latter formulation (which is simpler to write, but may be harder to solve), we can write our policy at time t as

$$X_t^{SP}(S_t|\theta) = \arg\min_{x_t} \left(p_t \left(x_t^{GB} + x_t^{GD} \right) + \right.$$

$$\left. \min_{(\tilde{x}_{tt'}(\tilde{\omega}_t))_{t'=t+1}^{t+H}, \tilde{\omega}_t \in \tilde{\Omega}_t} \sum_{\tilde{\omega}_t \in \tilde{\Omega}_t} P(\tilde{\omega}_t) \sum_{t'=t+1}^{t+H} \left(\tilde{p}_{tt'}(\tilde{\omega}_t) \left(\tilde{x}_{tt'}^{GB}(\tilde{\omega}_t) + \tilde{x}_{tt'}^{GD}(\tilde{\omega}_t) \right) - \theta^{pen} \tilde{x}_{tt'}^{slack}(\tilde{\omega}_t) \right) \right) \tag{19.51}$$

subject to the constraints

$$\tilde{R}_{t'+1}(\tilde{\omega}_t) - \left(\tilde{x}_{tt'}^{GB}(\tilde{\omega}_t) + \tilde{x}_{tt'}^{EB}(\tilde{\omega}_t) - \tilde{x}_{tt'}^{BD}(\tilde{\omega}_t) \right) = \tilde{R}_{t'}(\tilde{\omega}_t), \tag{19.52}$$

$$\tilde{x}_{tt'}^{ED}(\tilde{\omega}_t) + \tilde{x}_{tt'}^{EB}(\tilde{\omega}_t) \leq \bar{h}_{t'}(\tilde{\omega}_t), \tag{19.53}$$

$$\tilde{x}_{tt'}^{BD}(\tilde{\omega}_t) + \tilde{x}_{tt'}^{GD}(\tilde{\omega}_t) + \tilde{x}_{tt'}^{ED}(\tilde{\omega}_t) + \tilde{x}_{tt'}^{slack}(\tilde{\omega}_t) = \bar{D}_{t'}(\tilde{\omega}_t), \tag{19.54}$$

$$\tilde{x}_{tt'}^{GB}(\tilde{\omega}_t), \tilde{x}_{tt'}^{EB}(\tilde{\omega}_t), \tilde{x}_{tt'}^{BD}(\tilde{\omega}_t), \tilde{x}_{tt'}^{ED}(\tilde{\omega}_t), \tilde{x}_{tt'}^{slack}(\tilde{\omega}_t) \geq 0. \tag{19.55}$$

Equation (19.52) is the flow conservation constraint for our battery, while equation (19.53) limits the power available from the wind or solar farm. Equation (19.54) limits how much we can deliver to the customer, where $\tilde{x}_{tt'}^{slack}$ is the slack variable that captures how much demand is not satisfied, which carries a penalty θ^{pen} in the objective function.

This formulation allows us to make a decision at time t while modeling the stochastic variability in future time periods. In the stochastic programming community, the outcomes $\tilde{\omega}_t$ are often referred to as *scenarios*. If we model 20 scenarios, then our optimization problem becomes roughly 20 times larger, so the introduction of uncertainty in the future comes at a significant computational cost. However, this formulation allows us to capture the variability in future outcomes, which can be a significant advantage over using simple forecasts. Furthermore, while the problem is certainly much larger, we can approach it using one of the three algorithmic strategies described in section 19.9.1.

Although the two-stage approximation is dramatically smaller than a full multistage model, even two-stage approximations can be quite challenging. The problem is that there are many applications where even a deterministic lookahead model can be hard to solve. Fortunately, modern solvers for linear, integer, and convex optimization problems have improved dramatically over the years, along with the speed of computers and the ability to support parallelization for optimization solvers.

19.9.3 Discussion

The stochastic programming community almost universally views the two-stage lookahead policy as "the problem" that they are solving. Keep in mind that there are many applications (often with integer variables) where the two-stage stochastic program is quite challenging to solve. Research teams have sometimes spent years developing these models. Overlooked is the reality that the problem defined by (19.51)–(19.55) is actually a policy to solve our base model (such as that given in equation (19.1)). As with all approximate lookahead models (and there are a number of approximations in this model), an optimal solution of an approximate lookahead model is never an optimal policy.

Most users of two-stage stochastic programming will focus on the approximation of using a sample of scenarios. Often overlooked, surprisingly, are the errors from the two-stage approximation, where decisions

$\tilde{x}_{t,t+1}(\tilde{\omega}_t), \tilde{x}_{t,t+2}(\tilde{\omega}_t), \ldots, \tilde{x}_{t,t+H}(\tilde{\omega}_t)$ are allowed to see the entire sequence of exogenous random variables of prices, demands, and energy from renewables. The effect of this approximation is, of course, highly problem dependent, but this is where we refer back to the simulations reported in Figure 19.13 for MCTS which indicates that the two-stage approximation may in fact produce a significant reduction in the quality of the policy when simulated in the base model.

19.10 Observations on DLA Policies

A number of comments are in order regarding the use of lookahead policies:

- It is common when using lookahead policies to form a model, and then assume that this is the model we have to solve. The real problem is the base model, while the lookahead policy is just one way of creating a policy for the base model.
- It is sometimes hard to know if a model is a lookahead model or a base model. There are many instances of stochastic dynamic programs which are base models, but these might also be lookahead models. We provide some guidance in the text that follows.
- Optimal solutions of stochastic lookahead models can be quite difficult to find, but an optimal solution of an approximate lookahead model is not an optimal policy for the base model, which is the real model being solved.
- While simulating a policy (any policy) is typically the best way to tune and compare policies, this is critical when using policy search (policy function approximations, or cost function approximations). By contrast, it is less important when using lookahead models. Given the complexity of building simulators, along with the approximations inherent in any simulation, there are many situations where lookahead policies are just tested in the field. When this process is used, it is important to recognize that this is happening, and to collect performance metrics to evaluate the performance of the policy, and to assess changes made in the policy.

As of this writing, the vocabulary of "base models" and "lookahead models" has not entered the language of modeling in stochastic optimization. As a result, it can be difficult to identify whether a stochastic optimization model is a base model or lookahead model. Some guidelines include:

- How is the model being used? If the primary output of a model is the decision we make in the first time period, this is almost always a lookahead model. On the other hand, it is possible the model is being used to answer strategic questions, where we are using optimization to simulate good decisions. In this case, the model is a base model. Note that the same model can serve as either a lookahead model or a base model.

- If we are using a deterministic lookahead model, or a stochastic lookahead using scenario trees, to make decisions in a setting that is changing dynamically, then this is a lookahead model. The future decisions in the lookahead model (that is, $\tilde{x}_{tt'}$ for $t' > t$) are not going to be implementable in a stochastic setting.

Perhaps the most important take-away of this chapter is the need to clearly distinguish the lookahead model from the base model, and to remember that the problem we are trying to solve is the base model. Then, the challenge is to design a lookahead model that strikes a balance between realism and computational tractability, especially in the design of the lookahead policy. Our goal has been to provide a menu of choices along both dimensions, with the hope that some combination will work for a specific application.

19.11   Bibliographic Notes

Section 19.1 – Lookahead models fall in the broad class of policies known in the controls community as "model predictive control" (or MPC), although MPC is often equated with deterministic lookaheads. For a deterministic base model, a deterministic lookahead, which looks all the way to the end of the horizon, is an optimal policy (see Morari et al. (2002) and Camacho and Bordons (2003)). For stochastic problems, a full stochastic lookahead is equivalent to Bellman's optimality principle (see Puterman (2005)).

Section 19.2 – Our method of developing the approximate lookahead, including the identification of the different types of approximations and the notation, was first presented in Powell (2014).

Section 19.3 – The idea of using modified objectives in a lookahead policy, while still using expectations to evaluate the policy, has been used in an ad-hoc manner. For example, Ben-Tal et al. (2005) uses a robust optimization objective to solve an inventory problem, but then states (p. 262):

> "To evaluate the actual outcomes that might result from employing this model, we ran hundreds of simulations with different data sets and compared the mean performance vis-à-vis the mean PH [planning horizon] outcome."

In other words, they are simulating the robust policy hundreds of time and taking an average (that is, approximating an expectation). We believe that simulating policies that are computed using various risk metrics, is probably quite common, but it is done (as was the case in Ben-Tal et al. (2005)) without an explicit model of the policy.

Section 19.3.1 – The subject of risk in sequential decision problems under uncertainty is exceptionally rich, with intellectually deep issues and mathematics. At the heart of the issue is the (typical) lack of additivity of risk across time periods, something we have taken for granted when working with expectations. It was the work of Andrzej Ruszczynski on dynamic risk measures (starting with Ruszczyński (2010)) that largely launched the study of risk measures in dynamic programs. For an introduction to this line of research, we recommend starting with his tutorial Ruszczyński (2014). Philpott and De Matos (2012) provides a very nice summary of nested risk measures in a direct lookahead policy for a water reservoir planning model for New Zealand which is solved using SDDP (from chapter 18). The paper then illustrates the subsequent simulation of the policy, where both the average performance (which determines the cost of operation) and the risk of power outages are considered. Shapiro et al. (2013) illustrates these issues for a hydroelectric planning problem for Brazil. Maceira et al. (2014) addresses risk in setting spot prices for the Brazilian system. Collado et al. (2017) illustrate risk modeling for a risk-averse stochastic path detection problem.

Section 19.6 – Deterministic lookaheads are a widely used heuristic for making sequential decisions in dynamic environments. This is known as "model predictive control" in the optimal control literature, which is the only community to have seriously studied the properties of deterministic lookahead policies, although this is all in the setting of deterministic problems (see Morari et al. (2002) and Camacho and Bordons (2003)). Secomandi (2008) studies the effect of reoptimization on rolling horizon procedures as they adapt to new information.

Section 19.7 – The idea of using any of the four classes of policies (or a hybrid) as the lookahead policy is new, although MCTS is an inherently hybrid approach.

Section 19.8 – The foundation for the ideas behind MCTS are based on Chang et al. (2005). The first use of "Monte Carlo tree search" appeared in Coulom (2006). A review of Monte Carlo tree search is given in Browne et al. (2012), although this is primarily for deterministic problems. Other recent reviews include Auger et al. (2013) and Munos (2014). For a recent review of MCTS and its application in games such as computer Go, see Fu (2017). A nice tutorial on MCTS is given in Fu (2018).

Central to the success of MCTS is having an effective rollout policy to get an initial approximation of the value of being in a leaf node. Rollout policies were originally introduced and analyzed in Bertsekas et al. (1997) and Bertsekas and Castanon (1999).

Section 19.8.4 – Optimistic MCTS is based on the idea of using a deterministic sample in the estimation of the value of a node, and then optimizing using the complete future, which is a form of information relaxation, producing an optimistic estimate of the value of being at a node. Jiang et al. (2020) presents an asymptotic proof of convergence of optimistic MCTS.

Section 19.9 – Two-stage resource allocation problems are widely used to illustrate general two-stage stochastic programming, first introduced by Dantzig (1955), which launched the field of stochastic programming (see Birge and Louveaux (2011), Kall and Wallace (2009), Shapiro et al. (2014)). We are not aware of any analysis of the performance of two-stage stochastic programming models when used as policies in sequential resource allocation problems.

Exercises

Review questions

19.1 Describe at least three examples of problems where (a) it seems apparent that you need to anticipate the downstream impact of a decision now and (b) you do not think you can reasonably approximate this using a value function approximation. Feel free to use settings from your own personal experiences.

19.2 The variables in the lookahead model, $\tilde{S}_{tt'}$, $\tilde{x}_{tt'}$, and $\tilde{W}_{tt'}$, are all indexed by two time indices (t and t'). Explain the purpose of each time index.

19.3 List the six classes of approximation strategies that can be applied to develop an approximate lookahead model. Briefly explain each.

19.4 List three examples (not given in section 19.3.1) of risk. These may be drawn from any setting.

19.5 Since you have already decided to use a direct lookahead policy (at least, that is why you are reading this chapter), why don't you use a DLA for your choice of lookahead policy. What are the major concerns when designing the lookahead policy?

Problem solving questions

19.6 If we are using a PFA (parameterized by θ) as our lookahead policy (see equation (19.40)), we need to optimize θ. What is the objective function that we should, in theory, use to optimize θ? What does your optimal θ depend on?

19.7 We are trying to find the best path through a dynamic transportation network, so that we travel from origin r to destination s, and arrive at the destination node s by a specific time t^{target}. We want to:

(a) Begin by modeling the deterministic shortest path problem as a lookahead model, but modified so that you capture how late you are relative to t^{target} when you arrive at the destination node s. What is the state variable that you need for this problem?

(b) Give the Bellman equation for this deterministic approximation.

(c) Assume that you are going to solve this problem using updated estimates of travel costs as you arrive to each node. Model the state variable for this sequential decision problem.

(d) Give the expression for simulating the performance of this policy for a sample path of updated costs.

19.8 In our presentation of the dynamic shortest path problem (see section 13.2.3), we noted that the state variable (assuming the traveler takes one time period to traverse each link in the network) was

$$S_t = (R_t, \tilde{c}_t).$$

where R_t captures the node where the traveler is currently located (which means they have to make a decision), and $\tilde{c}_t = (\tilde{c}_{t,k\ell})_{k,\ell \in \mathcal{N}}$ is the vector of the current estimates of the travel time on each link (k, ℓ), which is updated according to

$$\tilde{c}_{t+1,k\ell} = \tilde{c}_{t,k\ell} + \delta\tilde{c}_{t+1,k\ell}.$$

The standard way of solving this problem is to create a lookahead policy that requires solving a deterministic shortest path problem.

(a) What approximations are being made when we replace the original problem with the deterministic shortest path problem. Be specific (it is not enough to just say "it is deterministic").

(b) What is the state variable of the deterministic problem?

(c) One approach for incorporating uncertainty is to replace the point estimate (as of time t) of $\tilde{c}_{t,k\ell}$ with the $\tilde{c}^\pi_{t,i,j}(\theta)$ which is defined by

$$\tilde{c}^\pi_{t,i,j}(\theta) \quad = \quad \text{the } \theta\text{-percentile of the travel time for link } (i,j)$$
$$\text{given our estimate at time } t.$$

Rewrite the policy in (19.5) using the parameterized lookahead policy and describe how we are supposed to go about optimizing θ. Explain why the optimal value of θ is a function, and describe what it is a function of.

(d) Rewrite the policy in equation (19.5) if we fix define θ to be a scalar. Give the objective function for optimizing θ.

19.9 You are tasked with purchasing futures contracts for natural gas. Let $x_{tt'}$ be the quantity of natural gas that you wish to purchasing now for delivery at time t', at a price $p_{tt'}$. The prices are seasonal, but with a fair amount of noise. If you purchase more than you need, you have the ability to store the excess. Let u be the capacity of the storage reservoir for natural gas. Assume you initially have R_0.

Create a lookahead policy for helping to make the set of decisions $x_t = (x_{tt'})_{t' \geq t}$ that need to be implemented now. Remember that you will have to model decisions $\tilde{x}_{t,t',t''}$ in the lookahead model, which represent contracts that we *plan* now, but where we do not actually make the decision to place the purchase until time t' for deliveries at time t''. Let $\tilde{p}_{t,t',t''}$ be the forecast of the forward price for natural gas that might be in place at time t' for delivery at time t''. Note that these forecasts evolve continuously according to the MMFE model (see, for example, section 9.4). Do not worry about computing the lookahead policy.

19.10 Using your model from exercise 19.9, create a deterministic version of your lookahead model.

(a) In view of the seasonal nature of the demand for natural gas (highest in the winter for heating), and the capacity constraint on storage, common on the strengths of a deterministic lookahead mode.

(b) Now consider the potential price variations (which can be substantial) but which would never be forecasted (price spikes occur for very transient reasons), common on the behaviors that you would like in a policy, but which a deterministic lookahead model might miss.

19.11 Using your model from exercise 19.9, and your deterministic lookahead from exercise 19.10, design a parameterized lookahead that will produce a policy that will work better given the uncertainty of price spikes. Your parameters should take on values such as 0 (if additive) or 1 (if multiplicative) in the event that there are no price variations. Show how your parameterized policy can be tuned so that it will be better positioned to handle price spikes (assume that these happen at random, and cannot be forecasted).

Sequential decision analytics and modeling

These exercises are drawn from the online book *Sequential Decision Analytics and Modeling* available at `http://tinyurl.com/sdaexamplesprint`.

19.12 Read sections 6.1–6.4 on the dynamic shortest path problem. The method describes using dynamic programming to solve a deterministic approximation of the network, and uses this as a deterministic lookahead policy. The algorithm has been implemented in Python, which can be downloaded from `http://tinyurl.com/sdagithub` using the module "StochasticShortestPath_Dynamic."

(a) Use the deterministic lookahead model as a policy (which is already coded) and simulate the policy on the network that is provided. Report the performance of the policy in terms of the time required to complete the path.

(b) Now, modify the costs on the links so that you are using the θ-percentile rather than the mean. Use the assumed mean and standard deviation for the random costs $\hat{c}_{ij}$ to create a θ-percentile cost $\bar{c}(\theta)_{ij}$. This is a new deterministic shortest path problem. Modify the code to simulate the performance of this policy for $\theta = 0.5, 0.7, 0.9$. Simulate each policy over 20 samples, and report the mean and standard deviation for each value of θ.

Diary problem

The diary problem is a single problem you chose (see chapter 1 for guidelines). Answer the following for your diary problem.

19.13 Choose a decision in your diary problem that might reasonably be made using a direct lookahead policy (recall that *any* decision can, in principle, be made using a direct lookahead policy). Now answer the following:

(a) Describe the decision and how the decision impacts the future.

(b) Assess the pros and cons of using a deterministic lookahead model for a policy. Do you feel that this would be a good approximation?

(c) Suggest an approximate lookahead model, and describe at least one candidate for the "policy within a policy."

Bibliography

Auger, D., Couëtoux, A., and Teytaud, O. (2013). Continuous upper confidence trees with polynomial exploration Consistency. In: *Joint European Conference on Machine Learning and Knowledge Discovery in Databases,* 194–209. Springer.

Ben-Tal, A., Golany, B., Nemirovski, A., and Vial, J.-P. (2005). Retailer-supplier flexible commitments contracts: A robust optimization approach. *Manufacturing & Service Operations Management* 7 (3): 248–271.

Bertsekas, D.P. and Castanon, D.A. (1999). Rollout algorithms for stochastic scheduling problems. *Journal of Heuristics* 5: 89–108.

Bertsekas, D.P., Tsitsiklis, J.N., and Wu, C. (1997). Rollout algorithms for combinatorial optimization. *Journal of Heuristics* 3 (3): 245–262.

Birge, J.R. and Louveaux, F. (2011). *Introduction to Stochastic Programming*, 2e. New York: Springer.

Browne, C.B., Powley, E., Whitehouse, D., Lucas, S.M., Cowling, P.I., Rohlfshagen, P., Tavener, S., Perez, D., Samothrakis, S., and Colton, S. (2012). A survey of Monte Carlo tree search methods. *IEEE Transactions on Computational Intelligence and AI in Games* 4 (1): 1–49.

Camacho, E. and Bordons, C. (2003), *Model Predictive Control.* London: Springer.

Chang, H. ., Lee, H.-G., Fu, M. C., and Marcus, S.I. (2005). Evolutionary policy iteration for solving Markov Decision processes. *IEEE Transactions on Automatic Control* 50 (11): 1804–1808.

Collado, R., Meisel, S., and Priekule, L. (2017). Risk-averse stochastic path detection. *European Journal of Operational Research* 260 (1): 195–211.

Coulom, R. (2006). Efficient selectivity and backup operators in Monte-Carlo tree search. In: *Lecture Notes in Computer Science (including subseries Lecture Notes in Artificial Intelligence and Lecture Notes in Bioinformatics)*, 72–83. New York: Springer-Verlag.

Dantzig, G.B. (1955). Linear programming with uncertainty. *Management Science* 1: 197–206.

Fu, M.C. (2017). Markov Decision Processes, AlphaGo, and Monte Carlo Tree Search: Back to the Future. In: *TutORials in Operations Research*, 68–88.

Fu, M.C. (2018). Monte Carlo tree search: A tutorial. In: *Winter Simulation Conference*, 222–236.

Jiang, D.R., Al-Kanj, L., and Powell, W.B. (2020). Optimistic Monte Carlo Tree Search with Sampled Information Relaxation Dual Bounds Optimistic Monte Carlo Tree Search with Sampled Information Relaxation Dual Bounds, (September), arXiv, https://arxiv.org/abs/1605.05711.

Kall, P. and Wallace, S.W. (2009). *Stochastic Programming*, Vol. 10. Hoboken, NJ: John Wiley & Sons.

Maceira, M.E., Marzano, L.G., Penna, D.D., Diniz, A.L., and Justino, T.C. (2014). Application of CVaR risk aversion approach in the expansion and operation planning and for setting the spot price in the Brazilian hydrothermal interconnected system. *Proceedings 2014 Power Systems Computation Conference, PSCC 2014.*

Morari, M., Lee, J.H., and Garc, C.E. (2002). *Model Predictive Control*. New York: Springer-Verlag.

Munos, R. (2014). *From Bandits to MonteCarlo Tree Search: The Optimistic Principle Applied to Optimization and Planning*, Vol. 7.

Philpott, A.B. and De Matos, V.L. (2012). Dynamic sampling algorithms for multistage stochastic programs with risk aversion. *European Journal of Operational Research* 218 (2): 470–483.

Powell, W.B. (2014). Clearing the jungle of stochastic optimization. *Informs TutORials in Operations Research 2014.*

Puterman, M.L. (2005). *Markov Decision Processes*, 2e. Hoboken, NJ: John Wiley and Sons.

Ruszczyński, A. (2010). Riskaverse dynamic programming for Markov decision processes. *Mathematical Programming* 125: 235–261.

Ruszczyński, A. (2014). Advances in risk-averse optimization. In: *Informs Tutorials in Operations Research*, 168–190. Baltimore, MD: Informs. chapter 9.

Secomandi, N. (2008), An analysis of the control-algorithm re-solving issue in inventory and revenue management. *Manufacturing & Service Operations Management* 10 (3): 468–483.

Shapiro, A., Dentcheva, D., and Ruszczyński, A. (2014). *Lectures on Stochastic Programming: Modeling and theory*, 2e. Philadelphia: SIAM.

Shapiro, A., Tekaya, W., Da Costa, J.P., and Soares, M.P. (2013). Risk neutral and risk averse stochastic dual dynamic programming method. *European Journal of Operational Research* 224 (2): 375–391.

Part VI – Multiagent Systems

Part VI of our book consists of a single chapter on multiagent systems, but this chapter opens up an entirely new line of thinking. This chapter builds entirely on our universal framework, since each agent can be modeled using the same framework we have developed earlier in the book. Decisions made by each agent will draw on the same classes of policies.

We begin by revisiting basic learning problems, but now these are presented using a two-agent model: an environment agent, and a controlling agent. We contrast the resulting model to the approach used by a substantial and mature literature known as "partially observable Markov decision processes" (or POMDPs). We will show that using our approach produces models that are more practical and scalable than those developed in the POMDP literature. We also feel that our approach fixes a fundamental error made in the POMDP literature regarding knowledge of the transition function.

We then transition to systems with multiple controlling agents, where we use different policies to achieve different behaviors. We also introduce the idea that we can model different levels of beliefs about other agents, which spans beliefs about what another agent knows, to beliefs about how they behave. This is a modeling choice rather than a comparison of algorithms to solve a specific model. Multiagent systems open up an entirely new approach for modeling and controlling complex systems.

There is an extensive literature on multiagent systems. The vast majority of applications use fairly trivial policies. At the other extreme are very sophisticated modeling papers that use the POMDP modeling framework and work to achieve some form of optimality as a system. These approaches tend to be limited to relatively small systems.

Our approach is to model each agent using an extended version of our universal modeling framework that accommodates a new dimension, which is communication. We make decisions using our four classes of policies to find the best policy that can be computed in reasonable time, given the data available to the controlling agent.

We note that our universal modeling framework provides for belief states alongside deterministically known physical and informational parameters and quantities. We have illustrated many applications where belief states were not present. In multiagent systems, belief states are always present since there is always unknown information about other agents. Of course, we may choose to ignore belief state variables, but this will be an explicit modeling choice. However, the ability to learn from decisions will almost always be present, if we wish to take advantage of what we learn.

20

Multiagent Modeling and Learning

There is a host of problems that are best approached as multiagent systems, where the use of multiple agents allows us to capture a division of knowledge. The simplest example is any learning problem where there is a truth (which we can model as known only to an "environment agent") that needs to be learned by an agent that is making decisions (which we will call a "controlling agent"). However, this is just the beginning of the variety of systems that can be captured by exploiting the concept of multiple agents.

In this chapter, we are going to introduce the fundamental elements of a multiagent model, motivated by applications of increasing complexity. We start with an overview of multiagent systems, where we summarize the dimensions of a multiagent system, outline how to generalize our modeling framework to the multiagent environment, and then cover the area of communication which arises purely because of the presence of multiple agents.

The remainder of the chapter is divided between two-agent systems, and systems with multiple (possibly many) agents.

We begin by showing how to model pure learning problems as two-agent systems, with an environment agent that contains the ground truth, and a controlling agent that has to learn the environment to make decisions. We use the setting of mitigating the flu in a population, and develop a spectrum of models which we then use to illustrate the use of different classes of policies. We contrast our modeling strategy for pure learning problems with an established field known as partially observable Markov decision processes (or POMDPs) where learning problems are modeled and solved as a single system.

We then introduce a two-agent version of the newsvendor problem to illustrate two agents that are ostensibly cooperative, but with different objectives. This application provides a nice setting for learning the behavior of other controlling agents.

Reinforcement Learning and Stochastic Optimization: A Unified Framework for Sequential Decisions, First Edition. Warren B. Powell.
© 2022 John Wiley & Sons, Inc. Published 2022 John Wiley & Sons, Inc.

The second half of the chapter moves to multiagent systems, beginning with a classical system involving hundreds of independent agents that represent thermostats for apartments in a large building. The chapter closes with a cooperative system involving different hospitals managing, and sharing, blood supplies.

Multiagent systems are a rich problem class, and a single chapter will not be able to cover all the dimensions of multiagent systems. Our goal, rather, is to illustrate how to apply our universal policy, and to illustrate how the four classes of policies can be applied in this context.

20.1 Overview of Multiagent Systems

We begin by describing the dimensions of a multiagent system, followed by a presentation of how to model communication, which is the modeling element not present in our original (single-agent) framework. We then describe how to model a multiagent system, and discuss controlling architectures.

20.1.1 Dimensions of a Multiagent System

There are a number of dimensions to multiagent systems. A sample includes:

(1) The agents – We start with the list of agents and their capabilities.
(2) Learning – This includes:
  • Learning the environment.
  • Learning about other agents, which includes:
    – Learning what they know.
    – Learning about their behavior (specifically, how they make decisions).
(3) Communication – Communication consists the following:
  • We have to describe which agents can send information to, and receive information from, other agents.
  • The speed and capacity of communication between each pair of agents.
  • The accuracy of the communication between each pair of agents. Accuracy may be a result of technology (communication errors) or a choice (one agent providing biased information to another).
(4) Coordination – This describes any mechanisms used to coordinate the behavior of multiple agents to achieve a common goal.
(5) Reward structure – How agents interact depends on how they are rewarded. Competition (or cooperation) is a matter of degree, and may range from zero-sum games (e.g. competition for an open resource), to agents who might be on the same team but with different objectives. Agents with the same reward structure should learn cooperative behaviors.

(6) Resources – Agents often have to manage resources. A common example is energy, but an agent may be distributing vaccines, medical supplies, ammunition, food, water, parts, etc. We would need to specify:
  • Which resources does an agent need to manage?
  • How much of the resource the agent can store?
  • How much is consumed by the agent itself?
  • What are the exogenous demands that have to be satisfied (and how are these learned)?
  • How is it replenished (does it return to a home base, refilling stations, or can it be visited by replenishment agents)?

While individual agents can have a wide range of capabilities in the context of an application domain, the capabilities that relate directly to the control of a multiagent system include:

• The environment agent – This agent cannot make any decisions, or perform any learning (that is, anything that implies intelligence). This is the agent that would know the truth about the environment, which includes unknown parameters that we are trying to learn, or which performs the modeling of physical systems that are being observed by other agents. Controlling agents are, however, able to change the environment.
• Controlling agents – These are agents that make decisions that act on other agents, or the ground truth agent (acting as the environment). Controlling agents may communicate information to other controlling and/or learning agents. Controlling agents may also change the environment, or the state (physical, informational, and/or belief) of other controlling agents.
• Learning agents – These agents do not make any decisions, but can observe and perform learning (about the ground truth and/or other controlling agents), and communicate beliefs to other agents.
• Replenishment agents – These are agents with the ability to replenish resources. They may be stationary or mobile, where a mobile replenishment agent might be replenished by a stationary replenishment agent. These agents can perform learning and make decisions, which make them similar to controlling agents, but with a narrow set of activities.

There can be many types of agents who make decisions, communicate information, and/or perform learning. For example, we might list:

• Single stationary device – Examples include robotic arms used in manufacturing, field cameras, as well as machinery such as a heating and air conditioning system for a building.
• Single mobile device – These can include land robots, flying drones, and underwater vehicles. In time this will also include driverless vehicles.

- Fleets of devices – We may have a group of robots, drones, and underwater vehicles, and in time may include fleets of autonomous electric vehicles.
- Individual people in the field – This could be a medical technician making decisions about how to test and/or treat people for disease, a policeman working a neighborhood, or a soldier acting alone. It could even be individuals in a population making decisions about whether protect themselves from exposure to a virus or to get vaccinated.
- Teams of people in the field – This might be a group of people under the control of a single person, as might arise in military operations, or medical personnel responding to the outbreak of disease.
- Individual people managing a set of resources for a company – This could be someone assigning locomotives to trains for a specific train yard (or region of the country), a manager making manufacturing and inventory decisions for a single supplier in a supply chain.
- Senior managers making decisions that guide lower level managers – For example, a senior manager might set productivity targets that are used to evaluate field managers who actually assign people to tasks. The term "senior manager" can apply to any decision-maker within a company making decisions about budgets, pricing of products, or marketing.

We note that devices and people can both be agents, but they are very different types of agents, since devices will struggle to develop skills that we take for granted in any human. People can develop much more complex behaviors than devices, which introduces a much more complex learning challenge.

20.1.2 Communication

Communication is a characteristic of multiagent systems that does not exist in any form with our basic single agent model. There are a number of dimensions to modeling communication between agents. These include:

- Communication architecture – We have to decide who can communicate what to whom. It will generally not be the case in more complex systems that any agent can (or would) send everything in their state vector S_{tq} to every other agent. We may have coordinating agents that communicate with everyone, make decisions and then send these decisions (in some form) to other agents. We introduced the sets Q_q^+ and Q_q^- that capture the sets of agents that agent q can act on, or those which can act on q, with any type of decision. We can capture just the information architecture using

$$\mathcal{I}_q^+ \quad = \quad \text{the agents to whom agent } q \text{ can send information to,}$$
$$\mathcal{I}_q^- \quad = \quad \text{the agents that can send information to agent } q.$$

- Active observations – We may choose to observe the environment or other agents, by running a test (e.g. to see who is infected with the flu) or with sensors such as radar. We may observe the location of an agent, resources under the agent's control, and decisions the agent may make. For example, a navy ship has to make the decision whether to turn on its radar to observe another ship, which simultaneously reveals the location of the ship sending the radar.

- Receiving information – If information is sent from q' to q, the agent q has to update their own beliefs, which has to reflect the confidence that agent q has in the information coming from agent q'. Assume that we have just obtained from agent q' updated information $\widehat{W}_{q'qi}$ about the ith element of the state variable S_{qi} for agent q. Let

$$\beta_{qi} \quad = \quad \text{the precision (one over the variance) of our belief}$$
$$\text{in the estimate } S_{qi},$$

$$\beta_{q'q}^W \quad = \quad \text{the precision in the information flowing from } q' \text{ to } q$$
$$\text{(we could make this depend on } i \text{ as well),}$$

$$\delta_{q'q} \quad = \quad \text{the bias introduced by agent } q' \text{ when sending}$$
$$\text{information to agent } q.$$

If we receive the information $\widehat{W}_{q'qi}$, we update our estimate S_{qi} and its precision β_{qi} using the formulas we first introduced in section 3.4.1:

$$S_{qi} \quad \leftarrow \quad \frac{\beta_{qi} S_{qi} + \beta_{q'q}^W \widehat{W}_{q'qi}}{\beta_{qi} + \beta_{q'q}^W}, \tag{20.1}$$

$$\beta_{qi} \quad \leftarrow \quad \beta_{qi} + \beta_{q'q}^W. \tag{20.2}$$

Note that our precision $\beta_{q'q}$ depends on both the sending and receiving agents, which means it depends on relationships, rather than just the reliability of either agent.

- Sending information – We have to model the act of sending information in S_{tq} to another agent q'. The information may be sent accurately, or with some combination of noise and bias.

- Signal distortion – We have to capture the presence of noise which reflects the difference between what is being sent, and what is received. Signal distortion can arise in two ways:

  - Passive distortion – This is a byproduct of technology (the communication channel may introduce noise) and environment (weather or exposure to magnetic fields can introduce noise).

  - Active distortion – This is where the sending agent intentionally distorts the information being sent. There are different types of active distortion:

* Active noise – This is where the sending agent adds a zero-mean noise term to hide the true mean.
* Active bias – The sending agent may intentionally bias a signal for any of a variety of reasons.

20.1.3 Modeling a Multiagent System

To model our system, we begin by using our standard vocabulary from our universal framework and simply add the index q (which we use to index agents). However, we start with the architecture that describes how they communicate and interact, which is a new element to our modeling framework that only arises with multiagent systems.

The agent architecture:
We begin by describing the set of agents which we model using:

$$\mathcal{Q} \;=\; \text{The set of agents.}$$

$$\mathcal{Q}_q^+ \;=\; \text{The set of agents } q' \text{ that agent } q \text{ can affect with a decision}$$
$$x_{tqq'}, \text{ where these decisions can represent sending}$$
$$\text{information, money, or physical resources.}$$

We will use our conventional notation for capturing constraints on the flow of physical resources. The flow of information, on the other hand, is described by bandwidth constraints (say, bits per second), as well as the reliability of the information. For now, since information is used to update our beliefs B_{tq}, we are going to introduce the vector $\zeta_{qq'}$:

$$\zeta_{qq'i} \;=\; \text{the vector of parameters governing the speed, capacity}$$
$$\text{and reliability of the information about data element } i$$
$$\text{that is communicated from } q \text{ to any agent } q' \in \mathcal{Q}_q^+,$$

$$\;=\; (\beta_{qq'i}, \delta_{qq'i}, \eta_{qq'i}), \text{ where:}$$

$$\beta_{qq'i} \;=\; \text{the precision (inverse of the variance) of the reliability}$$
$$\text{of the information sent from } q \text{ to } q' \text{ in terms of its accuracy}$$
$$\text{of describing some data element } i,$$

$$\delta_{qq'i} \;=\; \text{the bias in the information sent about data element } i$$
$$\text{sent from } q \text{ to } q',$$

$$\eta_{qq'i} \;=\; \text{the energy required to send information about data}$$
$$\text{element } i \text{ from } q' \text{ to } q.$$

Communication is an entirely new dimension to our modeling framework, since this does not arise at all in the context of a single agent system.

State variables:

As with single agent systems, each agent has a state variable with the same three classes of information:

$$R_{tq} \quad = \quad \text{The state of resources controlled by agent } q \text{ at time } t.$$

$$I_{tq} \quad = \quad \text{Any other information known to agent } q \text{ at time } t.$$

$$B_{tq} \quad = \quad \text{The beliefs of agent } q \text{ about anything known to any other agent (and therefore not known to agent } q). \text{ This covers parameters in the ground truth, anything known by any other agent (for example, the resources that an agent } q' \text{ might be controlling), and finally, beliefs about how other agents make decisions.}$$

Belief states are the richest and most challenging dimension of multiagent systems, especially when there is more than one controlling agent, as would occur in competitive games.

Decision variables:

Since communication and learning is such a fundamental component of multiagent systems, we are going to introduce new decision variables specifically for the purpose of capturing information sent from an agent q to another agent q':

$$z_{tqq'i} \quad = \quad \text{Information about data element } i \text{ sent from agent } q \text{ at time } t \text{ to agent } q' \text{ arriving at time } t+1.$$

$$z_{tqi} \quad = \quad (z_{tqq'i})_{q' \in Q_q^+}.$$

The reliability, bias and energy required to send information are described by $\zeta_{qq'i}$.

In addition, we are still going to have our traditional "x" variables that will capture the movement of physical resources, and will still continue to control (or influence) observations that we make, as we have done previously:

$$x_{tq} \quad = \quad \text{Decisions (scalar or vector) of decisions made by agent } q, \text{ which might include:}$$

$$x_{tqq'} \quad = \quad \text{Decisions made by agent } q \text{ that act on agent } q'.$$

Whereas we would require that $x_t \in \mathcal{X}_t$ for our single-agent notation, we now use $x_{tq} \in \mathcal{X}_{tq}$ which might simply require that x_{tq} be drawn from a finite set (say, of drugs, or choice of people), or it can represent flow conservation constraints that we can write using matrix notation as

$$A_{tq}x_{tq} = R_{tq},$$
$$x_{tq} \leq u_{tq},$$
$$x_{tq} \geq 0.$$

Decisions are made by the policy $X_q^\pi(S_{tq})$.

If $\mathcal{X}_t$ is our feasible region on our decisions x_{tq}, how do we represent constraints on the flow of information? The simple reality is: there are no constraints on the flow of information. Normally, we would expect that $z_{tqq'}$ would contain at least a noisy estimate of one or more of the variables in our state S_{tq} (remember that S_{tq} is our state of knowledge), but this is not necessary. Agents can transmit information that is simply incorrect, and this happens in adversarial settings where an agent may wish to misinform an adversary. Of course, we see this happening all the time on social media!

A new dimension in the formulation of decisions that is introduced in the study of multiagent systems is the decision to communicate. We have already seen decisions to observe in chapter 7, where we choose x^n and then observe a function $\hat{F}^n = F(x^n, W^{n+1})$. However, now we have the ability to choose from what we know (in our state S_t at time t) and then communicate it with some level of precision to another agent.

Exogenous information variables:
Exogenous information to an agent may come from outside of our system, or from another agent:

$$W_{tq} = \text{Exogenous information arriving to agent } q \text{ from any}$$
$$\text{exogenous source, which may include:}$$
$$W_{tqq'} = \text{Information arriving to agent } q \text{ from agent } q'.$$

The information in $W_{tqq'}$ will typically be a byproduct of a decision $x_{tqq'}$. It is important to be clear about the timing of when the information from a decision $x_{tqq'}$ made at time t arrives to agent q'. We are going to assume that it arrives at time $t + 1$.

Transition function:
This is simply

$$S_{t+1,q} = S_q^M(S_{tq}, x_{tq}, W_{t+1,q}),$$

which we have used throughout the book. Now, we have a transition function for each agent q.

Objective function:

We assume that each agent has a performance metric that we write as

$$C_q(S_{tq}, x_{tq}) = \text{the contribution earned by agent } q \text{ from being in}$$
$$\text{state } S_{tq} \text{ and making decision } x_{tq},$$

where $x_{tq} = X_q^\pi(S_{tq})$.

We then face the problem of optimizing across policies. This is not as straightforward as it is with single agents. Two possible optimization mechanisms include:

- Each agent optimizes their own policy – While this fits in the framework we have covered in the rest of the book, we still encounter the reality that all the agents are presumably searching for policies at the same time. Since the policy of an agent q' can influence the exogenous information $W_{t+1,q'q}$ that arrives to agent q, it means that the exogenous information processes are evolving as part of the policy search process (whether it is online or offline).
- A single optimizing process (not necessarily an agent) may be managing the process of searching over all the policies in search of a globally optimizing set of policies. Keep in mind that the policy used by each agent q can only depend on what agent q knows, which is captured by their state variable S_{tq}.

Since there are so many mechanisms for searching for policies, we are simply going to introduce the problem of searching for policies. Keep in mind that while we may, of course, be searching across policy classes, we think it is more common that the class of policy will have been chosen for each agent, leaving the optimization to consist of any tunable parameters.

There is a tendency in the literature on multiagent systems to work with a "system state" $S_t = (S_{tq})_{q \in Q}$. We take the position that this is meaningless, since no agent ever sees all this information, including a central controlling agent, which is not allowed to see the states S_{tq} of the individual agents (although there can be information sharing). We would approach the modeling of each agent as its own system, with the understanding that a challenge of any intelligent agent is to develop models that help the agent to forecast the exogenous information process W_{tq}. Of course, this depends on the policy being used by the agent, and the anticipated behaviors of other agents.

20.1.4 Controlling Architectures

Now that we have a model, it is worth discussing designing policies. In our single agent problems up to now, we assume that we have a single performance metric that we are optimizing. Multiagent systems are more complex. We still assume that each agent has a single objective, but it has the ability to influence the behavior of other agents, which means an agent can change the environment it works in to improve its own performance. The structure of a system influences how each agent behaves.

Given the diversity of types of agents, it should not be surprising that there is a wide range of multiagent systems. For this reason, we list a few examples of systems that illustrate interesting multiagent settings:

- Learning systems (controlling agent and environment agent) – A controlling agent can learn an environment, but can also modify the environment to help its own performance. We illustrate this in section 20.2 using the context of mitigating the spread of the flu in a population.
- Two-agent adversarial systems (games) – These often describe zero-sum games, but may include semi-cooperative settings where two agents interact with different objectives (but not necessarily completely opposite). We illustrate this using a semi-cooperative game we call the two-agent newsvendor problem (see section 20.4).
- Oligopolistic systems – These often arise in markets where there is a small number of players (say, three to five) which are more complex than two-agent systems, but small enough that the effect of individual players on a market can be discerned.
- Multiple independent agents – This describes systems where each agent is behaving completely independently of every other agent, but using a policy that has been chosen to achieve a system-wide goal. We illustrate this in section 20.5 using a collection of thermostats in a building.
- Multiple cooperating agents – This can describe teams of people working together, or groups of suppliers who make up the supply chain for a product. We use the setting of coordinating the blood supplies managed by different hospitals in section 20.6.
- Hierarchical systems – These would describe central (or centralized) managers controlling field agents, or even layers of managers controlling the next lower layer, as arises in the military or large companies. In this context, agents typically have to balance their local performance (which can include the performance of the agents below them) with following the guidance of higher level agents (ideally these are aligned, but as we all know ...).

20.2 A Learning Problem – Flu Mitigation

We are going to use the problem of protecting a population against the flu as an illustrative example of a learning problem. It will start as a learning problem with an unknown but controllable parameter, which is the prevalence of the flu in the population. We will use this to illustrate different classes of policies, after which we will propose several extensions.

20.2.1 Model 1: A Static Model

Let μ be the prevalence of the flu in the population (that is, the fraction of the population that has come down with the flu). In a static problem where we have an unknown parameter μ, we make observations using

$$W_{t+1} = \mu + \varepsilon_{t+1}, \tag{20.3}$$

where the noise $\varepsilon_{t+1} \sim N(0, \sigma_W^2)$ is what keeps us from observing μ perfectly.

We express our belief about μ by assuming that $\mu \sim N(\bar{\mu}_t, \bar{\sigma}_t^2)$. Since we fix the assumption of normality, we express our belief about μ as $B_t = (\bar{\mu}_t, \bar{\sigma}_t^2)$. We are again going to express uncertainty using $\beta_t = 1/\bar{\sigma}_t^2$ which is the precision of our estimate of μ, and $\beta^W = 1/\sigma_W^2$ is the precision of our observation noise ε_{t+1}.

We need to estimate the number of people with the disease by running tests, which produces the noisy estimate W_{t+1}. We represent the decision to run a test by the decision variable x_t^{obs} where

$$x_t^{obs} = \begin{cases} 1 & \text{if we observe the process and obtain } W_{t+1}, \\ 0 & \text{if no observation is made.} \end{cases}$$

If $x_t^{obs} = 1$, then we observe W_{t+1} which we can use to update our belief about μ using

$$\bar{\mu}_{t+1} = \frac{\beta_t \bar{\mu}_t + \beta^W W_{t+1}}{\beta_t + \beta^W}, \tag{20.4}$$

$$\beta_{t+1} = \beta_t + \beta^W. \tag{20.5}$$

If $x_t^{obs} = 0$, then $\bar{\mu}_{t+1} = \bar{\mu}_t$ and $\beta_{t+1} = \beta_t$.

For this problem our state variable is our belief about μ, which we write

$$S_t = B_t = (\bar{\mu}_t, \beta_t).$$

If this was our problem, it would be an instance of a one-armed bandit. We might assess a cost for making an observation, along with a cost of uncertainty. For example, assume we have the following costs:

$$
\begin{aligned}
c^{obs} &= \text{The cost of sampling the population to estimate the} \\
&\quad\text{number of people infected with the flu.} \\
C^{unc}(S_t) &= \text{The cost of uncertainty,} \\
&= c^{unc}\bar{\sigma}_t. \\
C(S_t, x_t) &= c^{obs}x_t^{obs} + C^{unc}(S_t).
\end{aligned}
$$

Using this information, we can put this model in our canonical framework as follows:

State variables – $S_t = (\bar{\mu}_t, \beta_t)$.
Decision variables – $x_t = x_t^{obs}$ determined by our policy $X^{obs}(S_t)$ (to be determined later).
Exogenous information – W_{t+1} which is our noisy estimate of how many people have the flu from equation (20.3) (and we only obtain this if $x^{obs} = 1$).
Transition function – Equations (20.4) and (20.5).
Objective function – We would write our objective as

$$
\max_{\pi} \mathbb{E}\left\{\sum_{t=0}^{T} C(S_t, x_t)|S_0\right\}. \tag{20.6}
$$

We now need a policy $X^{obs}(S_t)$ to determine x_t^{obs}. We can use any of the four classes of policies described in section 7.3 or chapter 11. We sketch examples of policies in section 20.2.5.

20.2.2 Variations of Our Flu Model

We are going to present a series of variations of our flu model to bring out different modeling issues. These variations produce the following models:

Model 2: A time-varying model.
Model 3: A time-varying model with drift.
Model 4: A dynamic model with a controllable truth.
Model 5: A flu model with a resource constraint and exogenous state.
Model 6: A spatial model.

These variations are designed to bring out the modeling issues that arise when we have an evolving truth (with known dynamics), an evolving truth with unknown dynamics (the drift), an unknown truth that we can control (or influence), followed by problems that introduce the dimension of having a known and controllable physical state.

(2) A time-varying model

If the true prevalence of the flu is evolving exogenously (as we would expect in this application), then we would write the true parameter as depending on time, μ_t, which might evolve according to

$$\mu_{t+1} = \max\{0, \mu_t + \varepsilon^\mu_{t+1}\}, \tag{20.7}$$

where $\varepsilon^\mu_{t+1} \sim N(0, \sigma^{\mu,2})$ describes how our truth is evolving. If the truth evolves with zero mean and known variance $\sigma^{\varepsilon,2}$, our belief state is the same as it was with a static truth (that is, $S_t = (\bar{\mu}_t, \beta_t)$). What does change is the transition function which now has to reflect both the noise of an observation ε_{t+1} as well as the uncertainty in the evolution of the truth, captured by ε^μ_{t+1}.

Remark: When μ was a constant, we did not have a problem referring to it as a parameter, whereas the state of our system is the belief which evolves over time (state variables should only include information that changes over time). When μ is changing over time, in which case we write it as μ_t, then it is more natural to think of the value of μ_t as the state of the system, but not observable to the controller. For this reason, many authors would refer to μ_t as a *hidden state*. However, we still have the belief about μ_t, which creates some confusion: What is the state variable? We are going to resolve this confusion.

(3) A time-varying model with drift

Now assume that

$$\varepsilon^\mu_{t+1} \sim N(\delta, \sigma^{\varepsilon,2}).$$

If $\delta \neq 0$, then it means that μ_t is drifting higher or lower (for the moment, we are going to assume that δ is a constant). We do not know δ, so we would assign a belief such as

$$\delta \sim N(\bar{\delta}_t, \bar{\sigma}^{\delta,2}_t).$$

Again let the precision be given by $\beta^\delta_t = 1/\bar{\sigma}^{\delta,2}_t$.

We might update our estimate of our belief about δ using

$$\hat{\delta}_{t+1} = W_{t+1} - W_t.$$

Now we can update our estimate of the mean and variance of our belief about δ using

$$\bar{\delta}_{t+1} = \frac{\beta^\delta_t \bar{\delta}_t + \beta^W \hat{\delta}_{t+1}}{\beta^\delta_t + \beta^W}, \tag{20.8}$$

$$\beta^\delta_{t+1} = \beta^\delta_t + \beta^W. \tag{20.9}$$

In this case, our state variable becomes

$$S_t = B_t = \left((\bar{\mu}_t, \beta_t), (\bar{\delta}_t, \beta_t^\delta)\right).$$

Here, we are modeling only the belief about μ_t, while μ_t itself is just a dynamically varying parameter. This changes in the next example.

(4) A dynamic model with a controllable truth
Now consider what happens when our decisions might actually change the truth μ_t. Let

$$x_t^{vac} = \text{the number of vaccination shots we administer in the region.}$$

We assume that the vaccination shots reduce the presence of the disease by θ^{vac} for each vaccinated patient, which is x_t^{vac}. We are going to assume that the decision made at time t is not implemented until time $t + 1$. This gives us the following equation for the truth

$$\mu_{t+1} = \max\{0, \mu_t - \theta^{vac} x_{t-1}^{vac} + \varepsilon_{t+1}^\mu\}. \tag{20.10}$$

We express our belief about the presence of the disease by assuming that it is Gaussian where $\mu_t \sim N(\bar{\mu}_t, \sigma_t^2)$. Again letting the precision be $\beta_t = 1/\sigma_t^2$, our belief state is $B_t = (\bar{\mu}_t, \beta_t)$, with transition equations similar to those given in equations (7.26) and (7.27) but adjusted by our belief about what our decision is doing. If we make an observation (that is, if $x_t^{obs} = 1$), then

$$\bar{\mu}_{t+1} = \frac{\beta_t(\bar{\mu}_t - \theta^{vac} x_{t-1}^{vac}) + \beta^W W_{t+1}}{\beta_t + \beta^W}, \tag{20.11}$$

$$\beta_{t+1} = \beta_t + \beta^W. \tag{20.12}$$

If $x_t^{obs} = 0$, then $\bar{\mu}_{t+1} = \bar{\mu}_t - \theta^{vac} x_{t-1}^{vac}$, and $\beta_{t+1} = \beta_t$.

This setting introduces a modeling challenge: Is the state μ_t? Or is it the belief $(\bar{\mu}_t, \beta_t)$? When μ_t was static or evolved exogenously, it seemed clear that the state was our belief about μ_t. However, now that we can control μ_t, it seems more natural to view μ_t as the state. This problem is an instance of a partially observable Markov decision problem. Later we are going to review how the POMDP community models these problems, and offer a different approach.

This problem has an unobservable state that is controllable. The next two problems will introduce the dimension of combining both observable and unobservable states that are both controllable.

(5) A flu model with a resource constrained and exogenous state
Now imagine that we have a limited number of vaccinations that we can administer. Let R_0 be the number of vaccinations we have available. Our vaccinations

x_t^{vac} have to be drawn from this inventory. We might also introduce a decision x_t^{inv} to add to our inventory (at a cost). This means our inventory evolves according to

$$R_{t+1} = R_t + x_{t-1}^{inv} - x_{t-1}^{vac},$$

where we require $x_{t-1}^{vac} \leq R_t$. We still have our decision of whether to observe the environment x_t^{obs}, so our decision variables are

$$x_t = (x_t^{inv}, x_t^{vac}, x_t^{obs}).$$

While we are at it, we might as well include information about the weather such as temperature I_t^{temp} and humidity I_t^{hum} which can contribute to the spread of the flu. We would model these in our "other information" variable

$$I_t = (I_t^{temp}, I_t^{hum}).$$

Our state variable becomes

$$S_t = \left(R_t, (I_t^{temp}, I_t^{hum}), (\bar{\mu}_t, \beta_t) \right). \tag{20.13}$$

We now have a combination of a controllable physical state R_t that we can observe perfectly, exogenous environmental information $I_t = (I_t^{temp}, I_t^{hum})$, and the belief state $B_t = (\bar{\mu}_t, \beta_t)$ which captures our distribution of belief about the controllable state μ_t that we cannot observe.

Note how quickly our solvable two-dimensional problem just became a much larger five-dimensional problem. This is a big issue if we are trying to use Bellman's equation, but only if we are using a lookup table representation of the value function (otherwise, we do not care).

(6) A spatial model

Imagine that we have to allocate our supply of flu vaccines over a set of regions $\mathcal{J}$. For this problem, we have a truth μ_{ti} and belief $(\bar{\mu}_{ti}, \beta_{ti})$ for each region $i \in \mathcal{J}$. Next assume that x_{ti}^{vac} is the number of vaccines allocated to region i, which is subject to the constraint

$$\sum_{i \in \mathcal{J}} x_{ti}^{vac} \leq R_t. \tag{20.14}$$

Our inventory R_t now evolves according to

$$R_{t+1} = R_t + x_t^{inv} - \sum_{i \in \mathcal{J}} x_{ti}^{vac}.$$

The coupling constraint (20.14) prevents us from solving for each region independently. This produces the state variable

$$S_t = \left(R_t, (\bar{\mu}_{ti}, \beta_{ti})_{i \in \mathcal{J}} \right). \tag{20.15}$$

What we have done with this extension is to create a state variable that is potentially very high dimensional, since spatial problems may easily range from hundreds to thousands of regions.

20.2.3 Two-Agent Learning Models

There are two perspectives that we can take in any learning problem: one from the perspective of the environment, and one from the perspective of the controller that makes decisions:

The environment perspective – The environment (sometimes called the "ground truth") knows μ_t, but cannot make any decisions (nor does it do any learning).

The controller perspective – The controller makes decisions that affect the environment, but is not able to see μ_t. Instead, the controller only has access to the belief about μ_t.

The model of the environment agent is given in Figure 20.1. The model of the controlling agent is given in Figure 20.2.

It is best to think of the two perspectives as agents, each working in their own world. There is the "environment agent" which does not make decisions, and the "controlling agent" which makes decisions, and performs learning about the environment that cannot be observed (such as μ_t). Once we have identified our two agents, we need to define what is known by each agent. This begins with who knows what about parameters such as μ_t, but it does not stop there.

Table 20.1 shows the environmental state and controlling state variables for each of the variations of our flu problem that we presented in section 20.2.2. A few observations are useful:

- The two-agent perspective means we have two systems. The environment agent is a simple system with no decisions, but with access to μ_t and the dynamics of how vaccinations affect μ_t. The state of the system for the environment agent is S_t^{env} which includes μ_t. The state for the system for the

---

State variables $S_t^{env} = (\mu_t, \delta)$ (we include the drift δ, even if it is not changing).

Decision variables There are no decisions.

Exogenous information $W_{t+1}^{env} = \varepsilon_{t+1}^{\mu}$.

Transition function $S^{env} = S^{M.env}(S_t^{env}, , W_{t+1}^{env})$, which includes equation (20.7) describing the evolution of μ_t.

Objective function Since there are no decisions, we do not have an objective function.

Figure 20.1   The canonical model of the environment.

State variables $S_t^{cont} = \left((\bar{\mu}_t, \beta_t), (\bar{\delta}_t, \beta_t^\delta)\right)$.

Decision variables $x_t = (x_t^{vac}, x_t^{obs})$.

Exogenous information W_{t+1}^{cont} which is our noisy estimate of how many people have the flu (and we only obtain this if $x_t^{obs} = 1$).

Transition function $S_{t+1}^{cont} = S^{M,cont}(S_t^{cont}, x_t, W_{t+1}^{cont})$, which consist of equations (20.11) and (20.12).

Objective function We can write this in different ways. Assuming we are implementing this in a field situation, we want to optimize cumulative reward. Let:

$$c^{obs} = \text{The unit cost of sampling the population to estimate the number of people infected with the flu.}$$

$$C^{vac}(\bar{\mu}_t) = \text{The cost we assess when we think that the number of infected people is } \bar{\mu}_t.$$

Now let $C^{cont}(S_t, x_t) = c^{obs} x_t^{obs} + C^{vac}(\bar{\mu}_t)$ be the cost at time t when we are in state S_t and make decision x_t (note that x_t^{vac} impacts S_{t+1}). Finally, we want to optimize

$$\max_\pi \mathbb{E} \left\{ \sum_{t=0}^{T} C^{cont}(S_t, X_t^\pi(S_t)) | S_0 \right\}. \tag{20.16}$$

Figure 20.2 The canonical model of the controlling agent.

Table 20.1 Environmental state variables and controller state variables for different models.

Model	S_t^{env}	S_t^{cont}	Description
(1)	(μ_t)	$(\bar{\mu}_t, \beta_t)$	Static, unknown truth
(2)	$((\mu_t), (I_t^{temp}, I_t^{hum}))$	$(R_t, (I_t^{temp}, I_t^{hum}), (\bar{\mu}_t, \beta_t))$	Resource constrained with exogenous information
(3)	(μ_t, δ)	$((\bar{\mu}_t, \beta_t), (\bar{\delta}_t, \beta_t^\delta))$	Dynamic model with uncertain drift
(4)	$(\mu_t, x_{t-1}^{vac}, \theta^{vac})$	$(\bar{\mu}_t, \beta_t)$	Dynamic model with a controllable truth
(5)	$((\mu_{ti})_{i\in\mathcal{I}}, x_{t-1}^{vac}, \theta^{vac})$	$(R_t, (\bar{\mu}_t, \beta_t))$	Resource constrained model
(6)	$((\mu_{ti})_{i\in\mathcal{I}}, x_{t-1}^{vac}, \theta^{vac})$	$(R_t, (\bar{\mu}_{ti}, \beta_{ti})_{i\in\mathcal{I}})$	Spatially distributed model

---

controlling agent, S_t^{cont}, is the belief about μ_t, along with any other information known to the controlling agent such as R_t. The two systems are completely distinct, beyond the ability to communicate.

- In model 2, we model the temperature I_t^{temp} and humidity I_t^{hum} as state variables for both the environment, which presumably would control changes to these variables, and the controlling agent, since we have assumed that the controlling agent is able to observe these perfectly. We could, of course, insist that the controller can only observe these through imperfect instruments, in which case they would be handled in the same way we handle μ_t.

- Normally a state variable S_t should only include information that changes over time (otherwise the information would go in the initial state S_0). For this presentation, we included information such as the drift δ (model 3) and the effect of vaccinations on the prevalence of the flu θ^{vac} (model 4) in the environmental state variables to indicate information known to the environment but not to the controlling agent.

- In model 4, we include the decision x^{vac}_{t-1} in the state variable for the environment. We assume that the controlling agent makes the decision to vaccinate x^{vac}_{t-1} at time $t-1$ which is then communicated to the environment (which is how it gets added to S^{env}_t) and is then implemented during time period t. The information arrives to the environment through the exogenous information variable W^{env}_t.

- For models 5 and 6, we see how quickly we can go from two or three dimensions, to hundreds or thousands of dimensions. The spatially distributed model cannot be solved using standard discrete representations of state spaces, but approximate dynamic programming has been used for very high-dimensional resource allocation problems (see chapter 18).

In addition to modeling what each agent knows, we have to model communication. This will become an important issue when we model multiple controlling agents which we address in section 20.5. For our problem with a single controlling agent and a passive environment, there are only two types of communication: (1) the ability of the controlling agent to observe the environment (with noise) and (2) the communication of the decision x^{vac}_t to the environment.

It is not hard to see that *any* learning problem can (and we claim should) be presented using this "two-agent" perspective.

20.2.4   Transition Functions for Two-Agent Model

Our two-agent model has focused on what each agent knows (the state variable), but there is another dimension that deserves a closer look, which is the transition function itself. Assume that the true model describing the evolution of μ_t (known only to the environment) is

$$
\begin{aligned}
\mu_{t+1} = {} & \theta^\mu_0 \mu_t + \theta^\mu_{24} \mu_{t-24} + (\theta^{temp}_0 U_t + \theta^{temp}_1 U_{t-1} + \theta^{temp}_2 U_{t-2}) \\
& - (\theta^{vac}_1 x^{vac}_{t-1} + \theta^{vac}_2 (x^{vac}_{t-1})^2) + \varepsilon^\mu_{t+1}.
\end{aligned} \tag{20.17}
$$

where

$$U_t = \left(\max\{0, I_t^{temp} - I_t^{threshold}\} \right)^2$$

where $I^{threshold}$ is a threshold temperature (say, 25 degrees F) below which colds and sneezing begins to spread the flu. The inclusion of temperature over the current and two previous time periods captures lag in the onset of the flu due to cold temperatures.

For certain classes of policies, the controlling agent needs to develop its own model of the evolution of the flu. The controlling agent would not know the true dynamics in equation (20.17) and might instead use the following time-series model for the observed number of flu cases W_t:

$$W_{t+1} = \theta_0^W W_t + \theta_1^W W_{t-1} + \theta_2^W W_{t-2} - \theta^{vac} x_{t-1}^{vac} + \varepsilon_{t+1}^W. \tag{20.18}$$

Our model in equation (20.18) is a reasonable time-series model for the sequence of observations $(W_1, \dots, W_t)$ to predict W_{t+1}. There are, however, several errors in this model:

- The controlling agent is using observations W_t, W_{t-1} and W_{t-2} while the environment uses μ_t, which is not observable to the controller.
- The controlling agent did not realize there was a 24-hour lag in the development of the flu.
- The controlling agent is ignoring the effect of temperature.
- The controlling agent is not properly capturing the effect of vaccinations on infections.

Just the same, ignorance is bliss and our controlling agent moves forward with their best effort at modeling the evolution of the flu. Assume that the time-series model (20.18) is a reasonable fit of the data. We suspect that a careful examination of the errors (they should be independent and identically distributed) might fail a proper statistical test, but it is also possible that we cannot reject the hypothesis that the errors do satisfy the appropriate conditions. This does not mean that the model is correct; it just means that we do not have the data to reject it.

Now imagine that a graduate student is writing a simulator for the flu model, and assume that there is only one person writing the code (which is typically what happens in practice). Our erstwhile graduate student will create the true transition equation (20.17). When she goes to create the transition model used by the controller, she would create the best approximation possible given the information she was allowed to use, but she would know immediately that there are a number of errors in her approximation. This would allow her to declare that this model is "non-Markovian," but it is only because she is using her knowledge of the true model.

We offer the observation that almost all statistical models (such as equation (20.18)) are just approximations, which means given enough data, we would be able to argue that there is some violation (typically that the errors are independent and identically distributed). Someone developing this model might insist, without any data, that the approximate transition function (20.18) is "non-Markovian" by simply comparing the model to the "true" model (20.17), which is unknown to the controller (but known to the modeler). In effect, the modeler is cheating by using her knowledge of the true model, which simply would not happen in practice (this is a true story).

20.2.5 Designing Policies for the Flu Problem

Once we formulate our models of each agent, we need to design policies for the controlling agent. The creation of effective, high quality policies can be major projects. What we want to do is to sketch examples of each of the four classes of policies to help reinforce why it is important to understand all four classes.

Policy function approximations
Policy function approximations are analytical functions that map states to actions. Of the four classes of policies, this is the only class that does not involve an imbedded optimization problem.

For our flu problem, it is common to use the structure of the problem to identify simple functions for making decisions. For example, we might use the following rule for determining whether to make an observation of the environment:

$$X^{pfa-obs}(S_t|\theta^{obs}) = \begin{cases} 1 & \bar{\sigma}_t/\bar{\mu}_t \geq \theta^{obs}, \\ 0 & \text{otherwise}. \end{cases} \tag{20.19}$$

The policy captures the intuition that we want to make an observation when the level of uncertainty (captured by the standard deviation of our estimate of the true prevalence), relative to the mean, is over some number. The parameter θ^{obs} has to be tuned, which we do using the objective function (20.6). A nice feature of the tunable parameter is that it is unitless.

To determine x_t^{vac}, we might set μ^{vac} as a target infection level, and then vaccinate at a level that we believe (or hope?) that we get down to the target. To do this, first compute

$$\zeta_t(\theta^{vac}) = \frac{1}{\theta^{vac}} \max\{0, (\bar{\mu}_t - \mu^{vac})\}.$$

We can view ζ_t as the distance to our goal μ^{vac}. This calculation ignores the uncertainty in our estimate $\bar{\mu}_t$, so instead we might want to use

$$\zeta_t(\theta^\zeta) = \max\{0, (\bar{\mu}_t + \theta^\zeta \bar{\sigma}_t - \mu^{vac})\}.$$

This policy is saying that μ_t *might* be as large as $\bar{\mu}_t + \theta^\zeta \bar{\sigma}_t$, where θ^ζ is a tunable parameter. Now our policy for x^{vac} would be be

$$X^{pfa-vac}(S_t|\theta^{vac}, \theta^\zeta) = \frac{1}{\theta^{vac}}\zeta_t(\theta^\zeta). \tag{20.20}$$

Using our policy $X^{obs}(S_t)$, we can write our policy for $x_t = (x_t^{vac}, x_t^{objs})$ as

$$X^{PFA}(S_t|\theta) = \left(X^{pfa-vac}(S_t|\theta^{vac}, \theta^\zeta), X^{pfa-obs}(S_t|\theta^{obs})\right),$$

where $\theta = (\theta^{vac}, \theta^{obs}, \theta^\zeta)$. This policy would have to be tuned in the objective function (20.16). This policy could then be compared to that obtained by approximating Bellman's equation.

An alternative approach for designing a policy function approximation is to assume that it is represented by a linear model

$$X^{PFA}(S_t|\theta) = \sum_{f \in \mathcal{F}} \theta_f \phi_f(S_t).$$

Parametric functions are easy to estimate, but they require that we have some intuition into the structure of the policy. An alternative is to use a neural network, where θ is the weights on the links in the graph of the neural network. It is important to keep in mind that neural networks tend to be very high dimensional (θ may have thousands, even hundreds of thousands, of dimensions), and they may not replicate obvious properties. Either way, we would tune θ using the objective function in (20.16).

Cost function approximations

We are going to illustrate CFAs using the spatially distributed flu vaccination problem, where we assume we are allowed to observe just one region $x \in \mathcal{J}$ at a time (we have just one inspection team). Assume that we can only treat one region at a time, where we are always going to treat the region that has the highest estimated prevalence of the flu.

We do not know μ_{tx}, but at time t assume that we have an estimate $\bar{\mu}_{tx}$ for the prevalence of the flu in region $x \in \mathcal{J}$, where we assume that $\mu_x \sim N(\bar{\mu}_{tx}, \bar{\sigma}_{tx}^2)$. We use this belief to decide which region to vaccinate, which we describe using the policy

$$X^{vac}(S_t) = \arg\max_{x \in \mathcal{J}} \bar{\mu}_{tx}.$$

We then have to decide which region to observe. We can approach this problem as a multiarmed bandit problem, where we have to decide which region ("arm" in bandit lingo) to observe. The most popular class of policies for learning problems in the computer science community is known as upper confidence bounding for multiarmed bandit problems. A class of UCB policy is interval estimation which would choose the region x that solves

$$X^{obs-IE}(S_t|\theta^{IE}) = \arg\max_{x \in \mathcal{J}} \left(\bar{\mu}_{tx} + \theta^{IE}\bar{\sigma}_{tx}\right) \tag{20.21}$$

where $\bar{\sigma}_{tx}$ is the standard deviation of the estimate $\bar{\mu}_{tx}$.

The policy $X^{obs-IE}(S_t|\theta^{IE})$ is a form of parametric cost function approximation; it requires solving an imbedded optimization problem, and there is no explicit effort to approximate the impact of a decision now on the future. It is easy to compute, but θ^{IE} has to be tuned. To do this, we need an objective function. Note that we are going to tune the policy in a simulator, which means we have access to μ_{tx} for all $x \in \mathcal{J}$.

Let $x_t^{obs} = X^{obs-IE}(S_t|\theta^{IE})$ be the region we choose to observe given what we know in S_t. This gives us the observation

$$W_{t+1,x_t^{obs}} = \mu_{t,x_t^{obs}} + \varepsilon_{t+1},$$

where $\varepsilon \sim N(0, \sigma_W^2)$. We would then use this observation to update the estimates $\bar{\mu}_{t,x_t^{obs}}$ using the Bayesian updating equations for the mean (7.26) and precision (7.27) (see equations (20.11) and (20.12)).

It is important to remember that the true prevalence μ_{tx} is changing over time as a result of our policy of observation and vaccination, so we are going to refer to it as $\bar{\mu}_{tx}^{\pi}(\theta^{IE})$, where the observation policy is parameterized by θ^{IE}.

We are learning in the field, which means we want to minimize the prevalence of the flu over all regions, over time. Since we are using a simulator to evaluate policies, we would evaluate our performance using the true level of flu prevalence, given by

$$F^{\pi}(\theta^{IE}) = \mathbb{E}_{S_0} \left\{ \sum_{t=0}^{T} \sum_{x \in \mathcal{J}} \bar{\mu}_{tx}^{\pi}(\theta^{IE})|S^0 \right\}. \tag{20.22}$$

We then need to tune our policy by solving

$$\min_{\theta^{IE}} F^{\pi}(\theta^{IE}).$$

Policies based on value functions

Any sequential decision problem with a properly defined state variable can be solved using Bellman's equation

$$V_t(S_t) = \max_x \left(C(S_t, x_t) + \mathbb{E}\{V_{t+1}(S_{t+1})|S_t, x_t\} \right),$$

which gives us the policy

$$X^{VFA}(S_t) = \arg\max_x \left(C(S_t, x_t) + \mathbb{E}\{V_{t+1}(S_{t+1})|S_t, x_t\} \right).$$

In practice we cannot compute $V_t(S_t)$, so we resort to methods that approximate the value function (see chapters 15, 16, and 17).

The use of approximate value functions has been recognized for a wide range of dynamic programming and stochastic control problems. However, it has been largely overlooked for problems with a belief state, with the notable exception of the literature on Gittins indices (see section 7.6), which reduces high-dimensional belief states (the beliefs across an entire set of arms) down to one dynamic program per arm.

In principle, approximate dynamic programming can be applied to even high-dimensional problems, including those with belief states, by replacing the value function $V_t(S_t)$ with a statistical model such as

$$V_t(S_t) \approx \overline{V}_t(S_t|\theta) = \sum_{f \in \mathcal{F}} \theta_f \phi_f(S_t),$$

where $(\phi_f(S_t))_{f \in \mathcal{F}}$ is a set of features. Alternatively, we might approximate $\overline{V}_t(S_t)$ using a neural network.

We note that we might write our policy as

$$X^{VFA}(S_t|\theta) = \arg\max_x \left(C(S_t, x) + \sum_{f \in \mathcal{F}} \theta_f \phi_f(S_t, x) \right), \tag{20.23}$$

where $(\phi_f(S_t, x_t)_{f \in \mathcal{F}}$ is a set of features involving both S_t and x_t. For example, we might design something like

$$X^{VFA}(S_t|\theta) = \arg\max_x \left(C(S_t, x) + \left(\theta_{t0} + \theta_{t1}\bar{\mu}_t + \theta_{t2}\bar{\mu}_t^2 + \theta_{t3}\bar{\sigma}_t + \theta_{t4}\beta_t\bar{\sigma}_t \right) \right).$$

There are a variety of strategies for fitting θ that have been developed, as we reviewed in chapters 3 and 16.

Direct lookahead policy

Direct lookahead policies involve solving an approximate lookahead model that we previously gave in equation (11.24), but repeat it here for convenience:

$$X_t^{DLA}(S_t) = \arg\max_{x_t} \left(C(S_t, x_t) + \tilde{E}\left\{ \max_{\tilde{\pi}} \tilde{E}\left\{ \sum_{t'=t+1}^{T} C(\tilde{S}_{tt'}, \tilde{X}^{\tilde{\pi}}(\tilde{S}_{tt'}))|\tilde{S}_{t,t+1}\right\} |S_t, x_t\right\} \right). \tag{20.24}$$

The problem with direct lookahead policies is that it requires solving a stochastic optimization problem (to solve the original stochastic optimization problem). To make it tractable, we can introduce various approximations. Some that are relevant for our problem setting could be:

(1) Use a deterministic approximation. These are effective for pure resource allocation problems (google maps using a deterministic lookahead to find

the best path to the destination over a stochastic graph), but seem unlikely to work well for learning problems.

(2) Use a parameterized policy for $\tilde{\pi}$. We could use any of the policies suggested earlier as our lookahead policy. We would then also have to use Monte Carlo sampling to approximate the expectations.

(3) We can solve a simplified Markov decision process.

(4) We could approximate the lookahead using Monte Carlo tree search.

We are going to illustrate the third approach. We start with model 3 of the flu problem, which requires the state variable

$$S_t^{cont} = \left((\bar{\mu}_t, \beta_t), (\bar{\delta}_t, \beta_t^{\delta}) \right).$$

We might be able to do a reasonable job of solving a dynamic program with a two-dimensional state variable (using discretization), but not a four-dimensional state. One approximation strategy is to fix the belief about the drift δ by holding $(\bar{\delta}_t, \beta_t^{\delta})$ constant. This means that we continue to model the true δ with uncertainty, but we ignore the fact that we can continue to learn and update the belief. This means the state variable $\tilde{S}_{tt'}$ in the lookahead model is given by

$$\tilde{S}_{tt'} = (\tilde{\bar{\mu}}_{tt'}, \tilde{\beta}_{tt'}).$$

Assuming that we can discretize the two-dimensional state, we can solve the lookahead model using classical backward dynamic programming on this approximate model (we could do this in steady state, or over a finite horizon, which makes more sense). Solving this model will give us exact value functions $\tilde{V}_{tt'}(\tilde{S}_{tt'})$ for our approximate lookahead model, from which we can then find the decision to make now given by

$$X_t^{\pi}(S_t) = \arg \max_x \left(C(S_t, x) + \mathbb{E}\{\tilde{V}_{t,t+1}(\tilde{S}_{t,t+1}) | S_t\} \right). \tag{20.25}$$

Then, we implement $x_t = X_t^{\pi}(S_t)$, step forward to $t + 1$, observe W_{t+1}, update to state S_{t+1} and repeat the process.

A hybrid policy

We have two types of decisions: whether to observe x_t^{obs}, and how many to vaccinate x_t^{vac}. We can combine them into a single, two-dimensional decision $x_t = (x_t^{obs}, x_t^{vac})$ and then think of enumerating all possible actions. However, we can also use hybrids. For example, we could use the policy function in equation (20.19), but then turn to any of the other four classes of policies for x_t^{vac}. This not only reduces the dimensionality of the problem, but might help if we feel that we have confidence in the function for x_t^{obs} (perhaps a UCB-style policy?) but are less confident designing a function for the more complex x_t^{vac} since it is managing physical resources.

20.3 The POMDP Perspective*

The POMDP community approaches the controllable version of our flu problem by viewing it as a dynamic program with state μ_t and action x_t that controls (or at least influences) this state. Viewed from this perspective, μ_t is *the* state of the system. Any reference to "the state" refers to the environment state μ_t in our two-agent model. In our resource constrained system, we would add R_t to the state variable giving us $S_t = (R_t, \mu_t)$, but for now we are going to focus on the unconstrained problem.

The community then shifts to the idea of modeling the belief about μ_t, and then introduces the "belief MDP" where the belief is the state (instead of μ_t).

The problem with these two versions of a Markov decision process is that there is not a clear model of who knows what. There is also the confusion of a "state" s (sometimes called the physical state), and our belief $b(s)$ giving the probability that we are in state s, where $b(s)$ is its own state variable!! This issue arises not just in who has access to the value of μ_t, but also information about the transition function. We resolved this confusion with our two-agent model.

To help us present the POMDP perspective, we are going to make the following assumptions:

(A1) We are solving the problem in steady state.
(A2) Our state space is discrete, which means we can write $S_t \in \mathcal{S} = \{s_1, ..., s_K\}$. For example, our state may be the blood sugar of a patient (discretized). We are not able to observe the state perfectly (our observation of blood sugar comes with sampling error and the natural variations of blood sugar in the patient).
(A3) We act on the (unknown) state with a decision x_t, which might be the choice of diet to control blood sugar (or a choice of medication, including the type of drug and its dosage).
(A4) We can compute the one-step transition matrix $p(s'|s, x)$ which is the probability that we transition to $S = s'$ given that we are in state s and take action x. It is important to remember that $p(s'|s, x)$ is computed using

$$p(s'|s, x) = \mathbb{E}_S \mathbb{E}_{W|S} \{\mathbb{1}_{\{S_{t+1} = s' = S^M(s, x, W)\}} | S_t = s\}.$$

The first expectation $\mathbb{E}_S$ captures our uncertainty about the state S (such as the actual blood sugar), while the second expectation $\mathbb{E}_{W|S}$ captures the noise in the observation of S. Computing the one-step transition matrix

Table 20.2 Table of notation for POMDPs.

Physical (unobservable) system	
$S_t = s$	Physical (unobservable) state s
x_t	Decision (made by the controller) that acts on S_t
W_{t+1}	Exogenous information impacting the physical state S_t
$S^M(s, x, w)$	Transition function for physical state $S_t = s$ given $x_t = x$ and $W_{t+1} = w$
$p(s'|s, x)$	$Prob[S_{t+1} = s'|S_t = s, x_t = x]$
Controller system	
$b(s)$	Probability (belief) we are in physical (unobservable) state s
W^{obs}	Noisy observation of physical state $S_t = s$
$\mathcal{W}^{obs}$	Space of outcomes of W^{obs}
$p^{obs}(w|s)$	Probability of observing $W^{obs} = w$ given $S_t = s$

$p(s'|s, x)$ means we need to know the transition function $S^M(s, x, W)$, which may not be known (we may not know how a patient responds to a diet or drug). In addition we need to know the probability distribution for W.

Table 20.2 presents the notation we use in our model.

With these assumptions, we can formulate the familiar form of Bellman's equations for discrete states and actions

$$V(s) = \max_x \left(C(s, x) + \sum_{s' \in \mathcal{S}} p(s'|s, x)V(s') \right). \tag{20.26}$$

The problem with solving Bellman's equation (20.26) to determine actions is that the controller determining x is not able to see the state s. The POMDP community addresses this by creating a belief $b(s)$ for each state $s \in \mathcal{S}$. At any point in time, we can only be in one state, which means

$$\sum_{s \in \mathcal{S}} b(s) = 1.$$

The POMDP literature then creates what is known as the *belief MDP* in terms of the belief state vector $b = (b(s_1), ..., b(s_K)) = (b_1, ..., b_K)$ where b_k is the probability (our belief) that $\mu = s_k$. This is a dynamic program whose state is given by the continuous vector b. We next introduce the transition function for the belief vector b given by

$$B^M(b, x, W) = \text{the transition function that gives the probability}$$

vector $b' = B^M(b, x, W)$ when the current belief vector (the prior) is b, we make decision x and then observe the random variable W, which is a noisy observation of μ_t if we have chosen to make an observation.

The function $B^M(b, x, W)$ returns a vector b' that has an element $b'(s)$ for each physical state s. To be clear, if $b_t(s)$ is our belief that we are in state s, then b_{t+1} is the vector of beliefs that we are in each state s' which is given by

$$b_{t+1} = B^M(b_t, x_t, W_{t+1}).$$

We will write $B^M(b, x, W)(s)$ to refer to element s of the vector returned by $B^M(b, x, W)$.

The derivation of the belief transition function is a moderately difficult exercise in the use of Bayes theorem, which we defer until section 20.8.1. However, we can show that Bellman's equation (20.26) can be written as

$$V(b_t) = \max_x \left(C(b_t, x) + \sum_{s \in \mathcal{S}} b_t(s) \sum_{s' \in \mathcal{S}} p(s'|s, x) \times \right.$$

$$\left. \sum_{w^{obs} \in \mathcal{W}^{obs}} P^{obs}(w^{obs}|s', x)V(B^M(b_t, x, w^{obs})) \right) \qquad (20.27)$$

If equation (20.27) can be solved, then the policy for making decisions for the controller is given by

$$X^*(b_t) = \arg \max_x \left(C(b_t, x) + \sum_{s \in \mathcal{S}} b_t(s) \sum_{s' \in \mathcal{S}} p(s'|s, x) \right.$$

$$\left. \sum_{w^{obs} \in \mathcal{W}^{obs}} P^{obs}(w^{obs}|s', x)V(B^M(b_t, x, W^{obs})) \right).$$

As if the computations behind these equations were not daunting enough, we need to also realize that we are combining the decisions of the controller with a knowledge of the dynamics (captured by the transition matrix) of the physical system. The one-step transition matrix requires knowledge of both the transition function $S^M(s, x, W)$, which will not always be known to the controller. In section 20.2.4, we presented a setting where the controlling agent did not know the true transition function.

A challenge here is that even through we have discretized the unobservable state, the vector b is continuous. However, Bellman's equation using the state b has some nice properties that the research community has exploited. Just the same, it is still limited to problems where the state space $\mathcal{S}$ of the unobservable system is relatively small. Keep in mind that small problems can easily produce state spaces of 10,000, and we could never execute these equations with a state space that large (think of the size of our state space when we have a belief for each of 50 states in the US, or 3,000 counties). For this reason, the POMDP community has developed a variety of approximation strategies.

By contrast, our two-agent model avoids the assumption that the controller knows the transition function, and opens the door to using any of the four classes of policies which, as we have shown, can be designed to scale to very high dimensional problems. Just as important, it opens the door to simple PFAs and CFAs that are likely to be much easier to explain and implement.

20.4 The Two-Agent Newsvendor Problem

In section 2.3.1 we introduced the newsvendor problem, which is a basic problem that arises in many applications involving the allocation of resources under uncertainty. In this problem, we allocate a quantity x_t, then observe a demand $\hat{R}_{t+1}$. We are going to use the cost minimizing version, where there is an underage cost c^u for each unit of unsatisfied demand, and an overage cost c^o for each unit of excess inventory (that is still not held until the next time period). This produces a cost function $F(x, D)$ given by

$$F(x, D) = c^u \max\{0, D - x\} + c^o \max\{x - D\}. \tag{20.28}$$

There are many applications where a "field agent" (that we designate "q") perceives a need to provide resources to meet an estimated demand, but has to ask for the resources from a "central agent" (q') who may not fill the entire request. The field agent typically has a higher cost of running out than having too much (that is, $c_q^u > c_q^o$). Imagine running out of food, blood, or ammunition, compared to having excess quantities of those three resources.

Let

$$
\begin{aligned}
\hat{R}_{t+1} &= \text{actual demand for resources at time } t + 1, \\
\hat{R}_t^e &= \text{estimate of the demand } \hat{R}_{t+1} \text{ made at time } t, \\
&= \mathbb{E}\hat{R}_{t+1}.
\end{aligned}
$$

In other words, $\hat{R}_t^e$ is an unbiased estimate of $\hat{R}_{t+1}$, information that we were not provided in our original newsvendor problem.

We can find an optimal solution to the problem in equation (20.28) if we knew the distribution of $\hat{R}_{t+1}$ given the estimate R_t^e, but let's assume we do not know the distribution. However, we can still say that the optimal solution would be given by

$$X^\pi(S_t|\theta_t) = R_t^e + \theta_t, \tag{20.29}$$

where S_t is what is known by the decision-maker at time t. We might start by assuming that $S_t = R_t^e$, but the state variable will also depend on the process for adaptively updating θ_t.

The real challenge that our field agent faces is that he has to ask for the resources from a "central agent," who has her own objective. This is a twist (that arises often in practice) that we did not face in our original newsvendor problem. Let

$$
\begin{aligned}
x_{tqq'} &= \text{the request made by field agent } q \text{ to the central agent } q' \\
&\quad\ \text{for resources,} \\
&= X_q^\pi(S_{tq}|\theta_{tq}), \\
x_{tq'q} &= \text{the amount that the central agent } q' \text{ decides to give to the} \\
&\quad\ \text{field agent } q, \\
&= X_{q'}^\pi(S_{tq'}|\theta_{tq'}).
\end{aligned}
$$

This produces a cost function $F_q(x, \hat{R})$ for our field agent q given by

$$F_q(x_{tqq'}, \hat{R}_{t+1}) = c_q^u \max\{0, \hat{R}_{t+1} - x_{tq'q}\} + c_q^o \max\{0, x_{tq'q} - \hat{R}_{t+1}\}, \tag{20.30}$$

where the amount that our central agent provides to the field agent, $x_{tq'q}$, depends on the original request $x_{tqq'}$ made by the field agent to the central agent, although the relationship is something that the field agent has to try to estimate.

To understand how the central agent might make her decision $x_{tq'q}$, we have to formulate her objective function, which is given by

$$F_{q'}(x_{tq'q}, \hat{R}_{t+1}) = c_{q'}^u \max\{0, \hat{R}_{t+1} - x_{tq'q}\} + c_{q'}^o \max\{0, x_{tq'q} - \hat{R}_{t+1}\}. \tag{20.31}$$

The only difference between the objective for agent q and agent q' is that q uses the costs (c_q^u, c_q^o) while agent q' uses $(c_{q'}^u, c_{q'}^o)$. Typically, $c_q^u > c_{q'}^u$ and $c_q^o < c_{q'}^o$, reflecting the likelihood that field agents have a much stronger desire not to run out of resources.

Note that the performance of the central agent still depends on satisfying the unknown demand $\hat{R}_{t+1}$, but it might be reasonable to assume that the under-age and overage costs for the central agent might satisfy $c_{q'}^u = c_{q'}^o$, which means she really wants the field agent to match expected demand as closely as possible, while the field agent wants to guard against being under. This creates a tension (which happens frequently in practice) where the field agent wants higher resource levels than the central agent, who is less sensitive to under-age and more sensitive to the cost of the excess inventory. This highlights how two agents, presumably working together (but with different goals) can end up behaving competitively.

Recall that the field agent is given an unbiased estimate R_t^e of $\hat{R}_{t+1}$, but since $c_q^u < c_q^o$, it is to be expected that the optimal value of θ_{tq} in equation (20.29) means that $\theta_{tq} > 0$, which means that the request $x_{tqq'}$ made by agent q to agent q', despite being based on the unbiased estimate R_t^e, is likely to be a biased estimate of $\hat{R}_{t+1}$.

We are going to borrow from the original newsvendor policy in equation (20.29) and propose a policy for agent q of the same form

$$X_q^\pi(S_{tq}|\theta_{tq}) = x_{tqq'} = R_t^e + \theta_{tq}. \tag{20.32}$$

In this case, we now have two reasons to believe that $\theta_{tq} > 0$:

- Since $c_q^u > c_q^o$, we will want to order a quantity greater than our estimate R_t^e.
- For reasons we present now, we are going to expect that our central agent is going to give the field agent *less* than he asks, which means we expect $x_{tq'q} < x_{tqq'}$. The field agent will know this, and use this as a reason to further inflate his request.

Since it is likely that $x_{tqq'}$ is biased upward, it makes sense to propose a policy for the central agent of the form

$$X_{q'}^\pi(S_{tq}|\theta_{tq}) = x_{tqq'} - \theta_{tq'}. \tag{20.33}$$

We have subtracted the correction term $\theta_{tq'}$ so that we can expect $\theta_{tq'} > 0$.

The policy in (20.33) uses only the request from agent q to guide the decision of the central agent. This makes the central agent relatively easy to manipulate. For example, while the central agent will learn that the request $x_{tqq'}$ is inflated and make adjustments (as shown in (20.33)), the field agent could keep inflating $x_{tqq'}$.

A more plausible model would be to assume that the central agent balances the request $x_{tqq'}$ from the field agent against an independent source of knowledge. One idea would be to estimate the bias and variance of the request from the field agent, and the bias and variance of an independent source of knowledge, using the methods described in section 3.5. Let $w_{tqq'}$ and $w_{t \cdot q'}$ be the weights given to each source of information, which are computed by taking the inverse of the variance of each estimate plus the square of the bias, normalized so the two weights sum to one (see section 3.6.3). We then obtain a blended estimate of the need using

$$x_{tq'}^{blend} = w_{tqq'} x_{tqq'} + w_{t \cdot q'} x_{t \cdot q'}. \tag{20.34}$$

This mechanism also encourages a level of truthfulness from the field agent, since significant bias or noise would have the effect of reducing $w_{tqq'}$.

If we adopt the policies (20.32) and (20.33) for agents q and q', we then need to design policies for adjusting θ_{tq} and $\theta_{tq'}$ which we represent using

$$\theta_{t+1,q} = \Theta_q^\pi(S_{tq}),$$

$$\theta_{t+1,q'} = \Theta_{q'}^\pi(S_{tq'}).$$

Now we face the challenge of designing the adaptive learning policies $\Theta_q^\pi(S_{tq})$ and $\Theta_{q'}^\pi(S_{tq'})$.

Figure 20.3 depicts the flow of information in a graphical format, similar to that used by engineers to draw circuits. This figure shows where noise might enter the different forms of communication. It shows the field agent making decisions based on the input from the field, and then sending a request to the central agent. It then shows the central agent combining the request from the field agent with its own information about the environment to make its own recommendation.

We need to recognize that there is not a "right" way to design these policies, since this has to do with modeling human behavior. For example, it is possible that the central agent (who has to manage hundreds of field agents) uses a simple rule

$$\Theta_{q'}^\pi(S_{tq'}|\rho_{q'}) = (1 - \rho_{q'})\theta_{tq'} + \rho_{q'}(x_{tqq'} - \hat{R}_{t+1}). \tag{20.35}$$

Note that with this policy, our state variable for q' would be $S_{tq'} = (x_{tqq'}, \theta_{tq'})$.

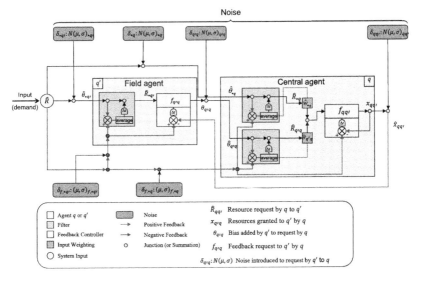

Figure 20.3 Information flow diagram for a two-agent newsvendor problem.

The policy in (20.35) is a simple reactive policy (a form of PFA) with a smoothing parameter $\rho_{q'}$ that will need to be tuned. Agent q' does not know the initial estimate R_t^e (this information is private to the field agent), but we assume that q' is able to learn the actual demand $\hat{R}_{t+1}$ at the end of the period (note that this is not always true in practice). Agent q' would like to know how much the field agent is biasing his request, but her best estimate of R_t^e is $\hat{R}_{t+1}$.

Similarly, we might pose a policy for agent q

$$\Theta_q^\pi(S_{tq}|\rho_q) = (1 - \rho_q)\theta_{tq} + \rho_q(x_{tqq'} - x_{tq'q}). \tag{20.36}$$

Again, this is a simple reactive policy that exploits the ability of the field agent to use the difference between what the field agent asked for, $x_{tqq'}$, and what the central agent provided, $x_{tq'q}$.

While there is not a "right" policy for either agent, we can pose the question of designing the best (or optimal) policy for one agent (say the field agent q), given an assumed policy $X_{q'}^\pi(S_{tq'})$ for the central agent. In practice, we should be considering all four classes of policies, although PFAs such as the policies (20.32) and (20.33) are going to be the most popular in a setting like this. This means that each agent faces a derivative-free stochastic search problem, so the methods of chapter 7 come into play, with one twist. Since the other agent is likely to be constantly adjusting their policy, each agent faces a nonstationary learning problem.

This very simple model has tremendous richness in terms of designing policies which nicely illustrate some of the challenges of multiagent settings. Some strategies that we can consider using include:

- If each agent uses a simple adjustment policy such as those depicted in equations (20.32) and (20.33), they each run the risk of the other agent learning their policy, and then making adjustments. A simple strategy to hide a policy is to introduce some noise. For example, the central agent might use the policy

$$X_{q'}^\pi(S_{tq}|\theta_{tq}) = x_{tqq'} - \theta_{tq'} + \varepsilon_{q'},$$

where $\varepsilon_{q'} \sim N(0, \sigma_{q'}^2)$ is a zero mean noise term. The challenge of the central agent is to choose the right value of the variance $\sigma_{q'}^2$; too low, and the field agent will be able to learn her policy; too high, and the central agent is responding with resource levels that are highly suboptimal for the central agent, and may simply result in the field agent further inflating his requests.

- The field agent may anticipate the policy $\Theta_{q'}^\pi(S_{tq'})$ for updating their adjustment, and build this response mechanism into their own choice. This means approximating $\partial\Theta_{q'}^\pi(S_{tq'})/\partial\theta_{tq}$.

- Each agent can turn the problem of optimizing their adjustment as an active learning problem. That is, consider discretizing θ_{tq} (from the perspective of the field agent), and then turning this into an active learning problem, which can be approached using the methods of chapter 7.
- A common strategy is to guess at the other agent's state variable, which means understanding what information they are using to make their decisions, and then trying to manipulate it (without their knowledge, of course).

One theme that should emerge from all of these policies is that learning the behavior (that is, the policy) of other agents requires looking at their decisions.

20.5 Multiple Independent Agents – An HVAC Controller Model

Consider the problem of controlling the air conditioning in each apartment of a large building with two goals in mind:

- The temperature of each apartment needs to stay within the temperature range E^{min} and E^{max}, with specific financial penalties (in the form of rent discounts) when the temperature falls outside of this range.
- The building pays real-time grid prices that change every 5 minutes, and the building operator would like to minimize what it pays for electricity.

The air conditioning is controlled by a thermostat in each apartment, which we are going to model as an agent. The thermostats do not communicate.

20.5.1 Model

We model the system for each thermostat q using our standard framework as follows:

State variables:

$$
\begin{aligned}
E_{tq}^{in} &= \text{Current temperature of apartment } q. \\
E_t^{out} &= \text{Current outdoor temperature.} \\
H_{tq} &= \begin{cases} 1 & \text{If the air conditioner in apartment } q \text{ is currently on.} \\ 0 & \text{If the air conditioner is currently off.} \end{cases} \\
S_{tq} &= \text{State of the air conditioning system in apartment } q \\
&= (E_{tq}^{in}, E_t^{out}, H_{tq}).
\end{aligned}
$$

Decision variables:

$$
x_{tq} = \begin{cases} +1 & \text{If we wish to turn the AC on (requires } H_{tq} = 0). \\ 0 & \text{If we wish to leave it unchanged.} \\ -1 & \text{If we wish to turn the AC off (requires } H_{tq} = 1). \end{cases}
$$

We let $X_q^\pi(S_{tq})$ be the policy that determines x_{tq} given the information in S_{tq}.

Exogenous information variables:

There are two ways to model our system:

- Model-free – We use this approach if we do not wish to assume that we know anything about the dynamics about how the temperature E_{tq}^{in} evolves over time. In this case, we assume only that we observe $E_{t+1,q}^{in}$, which means that this is the exogenous information, so $W_{t+1,q} = E_{t+1,q}^{in}$. We know $H_{t+1,q}$ since we know H_{tq} and x_{tq} which then determines $H_{t+1,q}$, which means that we could write $W_{t+1,q} = S_{t+1,q}$ (that is, our exogenous information is the state variable).

- Model-based – This opens the door to a much richer model. Assume that we have access to the external temperature (call this E_t^{out}), and that we have access to a dynamic model based on the thermodynamic properties of the apartment. In this case, our exogenous information would be $W_{t+1,q} = E_{t+1}^{out}$ (assuming the thermostat has access to the outside temperature).

Transition function:

If we use our model-free formulation, then $S_{t+1,q} = W_{t+1,q}$, which is to say that we simply observe the state at time $t + 1$ rather than calculating how we reach this state given x_{tq} and observing $EW_{t+1,q}$.

A model-based representation might allow us to observe the external temperature E_{t+1}^{out} and then use a dynamic equation such as

$$
E_{t+1,q}^{in} = E_{tq}^{in} + \rho_q\left(E_{t+1}^{out} - E^0\right) + \varepsilon_{t+1,q}, \tag{20.37}
$$

where E^0 is a base temperature (around 65 degrees) and ρ_q is a coefficient that reflects the thermal transfer for apartment q.

Objective function:

A natural way to evaluate our performance is to measure deviations outside of our range $(\theta^{min}, \theta^{max})$, as in

$$
C_q(S_{tq}, x_{tq}) = c^u \max\{\theta^{min} - E_t, 0\} + c^o \max\{E_t - \theta^{max}, 0\}.
$$

The coefficients c^u and c^o would have to be chosen to reflect the discomfort of a building that is too cold (in the summer time) versus too warm. We could, if we wish, also include operating cost, giving us

$$C(S_t, x_t) = c^u \max\{\theta^{min} - E_t, 0\} + c^o \max\{E_t - \theta^{max}, 0\} + c^{oper} H_t.$$

Since c^{oper} reflects an actual operating cost, c^u and c^o would then have to be scaled to capture the relative importance of operating cost, versus the discomfort of being cold or hot. This is a nice example of a multiobjective cost function.

20.5.2 Designing Policies

Stationary controllers are, of course, the simplest, but even with our basic temperature-controlling HVAC controller, it may still make sense to consider all four classes of policies:

Policy function approximation – These are easily the most widely used policies in practice. A natural policy (for air conditioning) would be to use

$$X_q^{AC}(S_t | \theta) = \begin{cases} 0 & \text{if } E_t < \theta^{min}, \\ +1 & \text{if } E_t > \theta^{max}, \\ H_t & \theta^{min} \leq E_t \leq \theta^{max}. \end{cases}$$

We intentionally left θ constant across all the apartments. The building manager, who is not strictly an agent (since she plays no role other than setting θ), might set this policy by solving

$$\min_{\theta} \sum_{t=0}^{T} \sum_{q \in \mathcal{Q}} C_q(S_{tq}, X_{tq}^{PFA}(S_{tq}|\theta)). \tag{20.38}$$

There are several ways that this simple policy can be generalized:

- θ can depend on each agent to capture different thermal transfer rates among apartments.
- θ can be made time-dependent to capture time-of-day effects.

A major limitation of PFAs can arise if we can anticipate changes in the temperature (as might happen every morning in winter) that produce changes in our internal temperature E_{tq}^{in} that are larger than we can compensate for using our HVAC system. It may be necessary to begin pre-heating (in winter) or pre-cooling (in summer) to compensate for peak periods.

Cost function approximation – CFAs open the door to using deterministic rolling horizon procedures, as we did in our energy example in section 13.3.3. This just leaves the need for performing tuning. We can do this offline as we did in the example in section 13.3.3, but a nice challenge would be to do this online, in the field, which means we would be optimizing cumulative performance. We can apply the techniques of derivative-free stochastic search in chapter 7.

Value function approximation – It is natural in this type of application, with clear time-of-day patterns, to expect that we would need a policy that depends on time of day. We hinted at this when we suggested a time-dependent control parameter θ in our PFA, but this turned a two-dimensional search problem into what might be a much higher dimensional search (especially if we make θ depend on 5-minute increments). VFA-based policies, on the other hand, tend to be naturally time-dependent. We can use any of the approximate dynamic programming algorithms described in either chapter 15 (backward approximate dynamic programming) or chapters 16–17 (forward approximate dynamic programming).

Direct lookahead – Time-dependent applications are natural candidates for lookahead policies, as we saw with the energy storage problem in section 13.3.3, which is actually a combination DLA-CFA, since the lookahead was parameterized. Imagine that we are given a forecast of daily temperatures that allows us to optimize the process of tuning the air conditioning on and off to maintain proper temperatures. This would involve solving an integer program over a rolling horizon. While this is not a particularly difficult problem using modern integer programming solvers, it is introducing substantially greater computations compared to the other policies. If this is done using a central server, this approach could be possible, but if the calculations have to be performed on each thermostat in each apartment, it would be out of the question.

20.6 Cooperative Agents – A Spatially Distributed Blood Management Problem

In section 8.3.2 we introduced the problem of managing inventories of blood, where we captured the eight different types of blood, the age (0 to 5 weeks), and modeled the ability to substitute different types of blood. However, the model did not capture location. Now assume that each hospital is managing its own inventories of blood, but has the ability to send blood from one location to another. Further assume (for the moment) that hospitals are altruistic – a life is a life and each hospital wants to make sure its blood inventories are best used, whether it is at their own or another hospital.

We are going to use the strategy of separable, piecewise linear value function approximations that we introduced in section 18.5. We could use Benders cuts (described in section 18.6) but the use of separable piecewiselinear approximations can be accomplished with much simpler communication between agents.

The mathematical model has already been described (in section 8.3.2), as has the logic for doing separable piecewise linear approximations (in section 18.5). For this reason, we are going to review the model and approximation methodology graphically. Figure 20.4(a) shows the network model that we would use for a single hospital q, where we have a node for each blood type and age, with arcs to demands for each blood type (if substitution is allowed).

Figure 20.4(b) cleans up the figure by sweeping all the demands into a single node, making it easier to see the blood that is held, first flowing into the node for the post-decision state (the leftover blood before new demands become known), and then into the pre-decision state of the next week (at which point the blood "ages" by one week). The policy for a single agent using separable, piecewise linear value functions would be written

$$X^{VFA-PWL}(S_t) = \arg\max_{x_t \in \mathcal{X}_t} \left(C(S_t, x_t) + \sum_{a \in \mathcal{A}} \overline{V}^x_{ta}(R^x_{ta}(x_t)) \right), \qquad (20.39)$$

where $R^x_{ta}(x_t)$ is the number of units of blood with attribute a at the end of week t (that is, the post-decision resource state), where $\overline{V}^x_{ta}(R^x_{ta}(x_t))$ is the piecewise linear value function for $R^x_{ta}(x_t)$ units of blood.

Finally, Figure 20.4(c) illustrates piecewise linear value functions being attached to each post-decision node, producing a deterministic linear program that is easy to solve, assuming that each hospital has access to this type of computing resource (it is easier to make this assumption for a hospital than a thermostat!).

These value function approximations are only for the value of blood held at the same hospital for the following week. We compute these piecewise linear approximations by taking the marginal value of an additional unit of blood of each type and age, and use this with algorithms such as the CAVE or Leveling algorithms (reviewed in section 18.3) to create concave (if maximizing) approximations of the value of additional units of blood (for each type and age).

Now imagine that we are applying this logic for hospital q. The piecewise linear value function approximations require only that we obtain the marginal value $\hat{v}_{tqa}$ for an additional unit of blood with attribute a (this would include blood type and age) for blood stored at hospital q. This is used to update the piecewise linear value functions for the value function approximations $\overline{V}_{t-1,qa}(R_{t-1,qa})$ as we step back in time.

In addition to making blood choice decisions at hospital q, we can also decide to move blood to another hospital q' (or from another hospital q' to hospital q). To help this process, we not only need the value of blood stored at hospital q, but

Figure 20.4 (a) Blood network showing assignments of blood supplies to demands of different blood types. (b) Blood network with demand arcs aggregated, and showing holding to post-decision R_t^x and the following pre-decision R_{t+1}. (c) Blood network showing piecewise linear value functions attached to post-decision resource nodes.

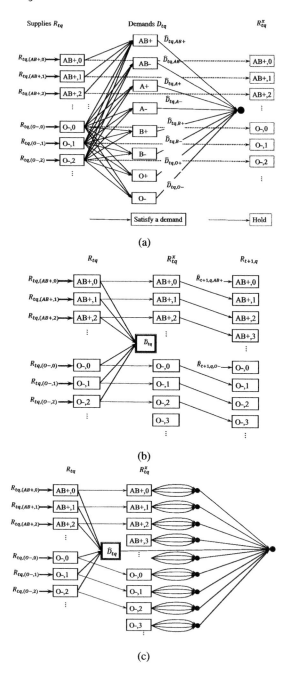

also blood stored at other hospitals q'. This means that hospital q, for example, needs estimates $\overline{V}^x_{tq'a}(R^x_{tq'a})$ for each hospital q' to help decide if blood should be moved to q'.

We do not have to solve this using a global optimization over all hospitals (there are thousands in the US). Instead, we can use a multiagent formulation and solve a single small model for each hospital, communicating marginal values between hospitals.

When we allow interhospital transfers, we produce the network depicted in Figure 20.5 where hospital q is now allowed to move each unit of blood (identified by type and age) to other hospitals that are represented only as the piecewise linear value functions. Mathematically, we would write the policy for hospital q as

$$X^{VFA-PWL}_q(S_t) = \arg\max_{x_{tq}\in\mathcal{X}_{tq}}\left(C(S_{tq}, x_{tq}) + \sum_{q'\in\mathcal{Q}}\sum_{a\in\mathcal{A}}\overline{V}^x_{tq'a}(R^x_{tq'a}(x_t))\right). \tag{20.40}$$

Comparing the single agent policy in equation (20.39) to the multiagent policy in (20.40), we see they are basically the same, aside from making the decisions for each hospital separately at time t.

We have to emphasize that using this approach requires that the hospitals be willing to share their marginal values $\hat{v}_{tqa}$ for each unit of blood with each other. This approach could easily be scaled over thousands of hospitals if necessary.

This logic, carefully implemented, can produce a near-optimal solution if we are trying to optimize across all the hospitals. But what if we do not believe

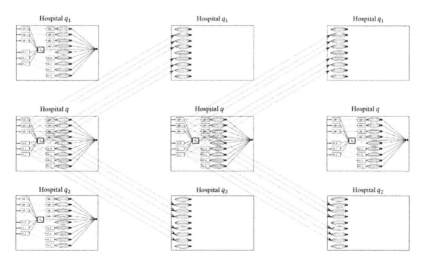

Figure 20.5 Blood management for multiple hospitals as a multiagent system.

that our hospitals are completely altruistic? We can achieve greedy behaviors by simply discounting the value of blood at hospitals other than our own. If the discount factor is set all the way down to zero, then we are back to completely greedy behaviors.

20.7 Closing Notes

There is an extensive literature on multiagent systems. Much of this is centered on devices (drones, robots, unmanned aerial vehicles) where belief formation about the environment is the central challenge. These issues are largely outside of the scope of this book.

There is an equally substantial literature that focuses on specific contextual domains, where the behavior of the agents (that is, the policies) is designed for the specific problem domain such as modeling teams of robotic soccer players, drones investigating an urban environment, flocks of birds and swarms of insects. Often these models combine the design of policies with the detailed control of the physics of the device (accelerating/decelarating, turning, taking off, and landing).

Our brief presentation here is designed to provide some basic notation and sketch some of the issues that arise in multiagent control. The goal here was primarily to demonstrate how to extend our modeling framework, to highlight the potential for using all four classes of policies.

20.8 Why Does it Work?

20.8.1 Derivation of the POMDP Belief Transition Function

This section derives Bellman's equation given by equation (20.27).

The belief transition function is an exercise in Bayes' theorem. Let $b_t(s)$ be the probability we are in state $S_t = s'$ (this is our prior) at time t. We assume we have access to the distribution

$$P^{obs}(w^{obs}|s) = \text{the probability that we observe } W^{obs} = w^{obs}$$
$$\text{if we are in state } s.$$

Assume we are in state $S_t = s$ and take action x_t and observe $W^{obs}_{t+1} = w^{obs}$. The updated belief distribution $b_{t+1}(s')$ would then be

$$b_{t+1} \quad (s'|b_t, x_t, W^{obs}_{t+1} = w^{obs}) = B^M(b_t, x, W^{obs}_{t+1} = w^{obs})(s')$$
$$= Prob[S_{t+1} = s'|b_t, x_t, W^{obs}_{t+1} = w^{obs}]$$

$$= \frac{Prob[W^{obs}_{t+1} = w^{obs}|b_t, x_t = x, S_{t+1} = s']Prob[S_{t+1} = s'|b_t, x_t]}{Prob[W^{obs} = w^{obs}|b_t, x_t]} \quad (20.41)$$

$$= \frac{Prob[W^{obs}_{t+1} = w^{obs}|b_t, x_t, S_{t+1} = s']\sum_{s \in \mathcal{S}} Prob[S_{t+1} = s'|S_t = s, b_t, x_t]Prob[S_t = s|b_t, x_t]}{Prob[W^{obs}_{t+1} = w^{obs}|b_t, x_t]} \quad (20.42)$$

$$= \frac{Prob[W^{obs}_{t+1} = w^{obs}|S_{t+1} = s']\sum_{s \in \mathcal{S}} Prob[S_{t+1} = s'|S_t = s, x_t]b_t(s)}{Prob[W^{obs}_{t+1} = w^{obs}|b_t, x_t]} \quad (20.43)$$

$$= \frac{P^{obs}(w^{obs}|s')\sum_{s \in \mathcal{S}} P(s'|s, x_t)b_t(s)}{P^{obs}(w^{obs}|b_t, x_t)} \quad (20.44)$$

Equation (20.41) is a straightforward application of Bayes' theorem, where all probabilities are conditioned on the decision x_t and the prior b_t (which has the effect of capturing history). Equation (20.42) handles the transition from conditioning on the belief $b(s)$ that $S_t = s$, to the state $S_{t+1} = s'$ from which observations are made. The remaining equations reduce (20.42) by recognizing when conditioning on $b_t(s)$ does not matter, and substituting in the names of the variables for the differenet probabilities.

We compute the denominator in equation (20.41) using

$$Prob[W^{obs}_{t+1} = w^{obs}|b_t, x_t] = \sum_{s' \in \mathcal{S}} Prob[W^{obs}_{t+1} = w^{obs}|S_{t+1} = s']$$

$$\cdot \sum_{s \in \mathcal{S}} Prob[S_{t+1} = s'|S_t = s, x_t]Prob[S_t = s|b_t, x_t] \quad (20.45)$$

$$= \sum_{s' \in \mathcal{S}} P^{obs}(w^{obs}|s')\sum_{s \in \mathcal{S}} P(s'|s, x_t)b_t(s). \quad (20.46)$$

Equation (20.44) is fairly straightforward to compute as long as the state space $\mathcal{S}$ is not too large (actually, it has to be fairly small), the observation probability distribution $P^{obs}(w^{obs}|S = s, x)$ is known, and the one-step transition matrix $P(s'|s, x)$ is known. Knowledge of $P^{obs}(w^{obs}|S = s, x)$ requires an understanding of the structure of the process of observing the unknown system. For example, if we are sampling the population to learn about who has the flu, we might use a binomial sampling distribution to capture the probability that we sample someone with the flu. Knowledge of the one-step transition matrix $P(s'|s, x)$, of course, requires an understanding of the underlying dynamics of the physical system.

This said, note that we have three summations over the state space to compute a single value of $b_{t+1}(s')$. This has to be repeated for each $s' \in \mathcal{S}$, and it has to be computed for each action x_t and observation W^{obs}_{t+1}. That is a lot of nested loops. The problem is that we are modeling two transitions: the evolution of the state S_t, and the evolution of the belief vector $b_t(s)$. This would not be an issue if we were just simulating the two systems. Equations (20.44) and (20.46) require computing expectations to find the transition probabilities for both the physical state S_t and the belief state b_t.

The POMDP community then approaches solving this dynamic program through Bellman's equation (for steady state problems) that can be written

$$V(b_t) = \max_x \left(C(b_t, x) + \mathbb{E}\{V(B^M(b_t, x, W^{obs}_{t+1}))|b_t, x\} \right).$$

It is better to expand the expectation operator over the actual random variables that are involved. Assume that we are in physical state S_t with belief vector b_t. The expectation would then be written

$$V(b_t) = \max_x \left(C(b_t, x) + \mathbb{E}_{S_t|b_t} \mathbb{E}_{S_{t+1}|S_t} \right.$$

$$\mathbb{E}_{W^{obs}_{t+1}|S_{t+1}} \{V(B^M(b_t, x, W^{obs}))|b_t, x\}). \tag{20.47}$$

Here, $\mathbb{E}_{S_t|b_t}$ integrates over the state space for S_t using the belief distribution $b_t(s)$. $\mathbb{E}_{S_{t+1}|S_t}$ takes the expectation over S_{t+1} given S_t. Finally, $\mathbb{E}_{W^{obs}_{t+1}|S_{t+1}}$ integrates over the space of observations given we are in state S_{t+1}. These expectations would be computed using

$$V(b_t) = \max_x \left(C(b_t, x) + \sum_{s \in S} b_t(s) \sum_{s' \in S} p(s'|s, x) \right.$$

$$\left. \sum_{w^{obs} \in W^{obs}} P^{obs}(w^{obs}|s', x)V(B^M(b_t, x, w^{obs})) \right). \tag{20.48}$$

20.9 Bibliographic Notes

Section 20.1 – The idea of modeling control as being distributed among multiple agents has an endless array of applications, including traffic and transportation (Chen et al. (2010)), building control systems (Zhao et al. (2013)), animal science (Tang and Bennett (2010)), agriculture (Tang and Bennett (2010)), and energy (González-briones et al. (2018)), to name just a few. There are many books on the topic (Tecuci (1998) and D'Inverno and Luck (2001) are two examples), and a growing number of tutorial articles using the umbrella of "multi-agent reinforcement learning" (Chen et al. (2010), Busoniu et al. (2011) and Dorri et al. (2018)). Abara et al. (2017) provides an introduction to agent-based modeling and software.

The presentation of multiagent systems in this book follows a different style, first sketched in Powell (2021), where each agent is modeled using our universal framework. We augment our universal framework, which is inherently single-agent, with the dimension of communication. We distinguish major classes of agents, where controlling agents can use any of the four classes of policies (which is new).

Section 20.2 – The flu mitigation model is taken from Powell (2020).

Section 20.3 – This presentation of POMDPs, and the observation that the classical POMDP framework makes the assumption that the controlling agent knows the transition function for the environment, is taken from Powell (2020).

Section 20.4 – Figure 20.3 was prepared by Gunter Schemmann and has been used in a course, ORF 411, taught by Warren Powell for many years. There is an extensive literature on the newsvendor problem, and yet virtually no papers that reference "two-agent newsvendor." This section is based on work by Brian Cheung while working as a Ph.D. student at Princeton.

Section 20.5 – While the idea of modeling a HVAC controller as an agent is quite common (see, for example, Dorri et al. (2018) and Zhao et al. (2013)), our presentation uses the framework in this book, which is new.

Section 20.6 – The blood management model uses the nonlinear value functions first presented in Godfrey and Powell (2001) and adapted for fleet management problems in Powell and Godfrey (2002). This approximation method has then been used in a spatial decomposition approach in Shapiro and Powell (2006), where the spatial decomposition can be viewed as a multiagent decomposition. This idea was then applied to a real application for managing locomotives (a system installed in 2006 and still running as of the writing of this book), described in Bouzaiene-Ayari et al. (2016). The adaptation to the blood management problem here is new, and was used because the problem is much simpler.

Exercises

Review questions

20.1 What types of agents are there?

20.2 We can use our universal modeling framework to model each agent, but multiagent systems do introduce new elements that only arise when we have two or more agents. At a high level, what are these new elements?

20.3 Give the state variables for agents S_{tq} and $S_{tq'}$ for the two-agent newsvendor problem in section 20.4.

Modeling questions

20.4 What has to be specified to fully describe the communication architecture of a multiagent system?

20.5 What observable information might agent q use to learn about the behavior of an agent q'? By "behavior" we are referring to the policy $X_{q'}^\pi(S_{tq'})$ used by agent q'.

20.6 Multiagent systems require that agents develop beliefs about other agents. Think of all the dimensions of our universal framework that we would use to describe an agent. Using the elements of this framework as a starting point, list every belief that one controlling agent might form about another controlling agent.

20.7 Model all five elements of each agent of the two-agent newsvendor problem in section 20.4 as sequential decision problems. Label the agents A and B, and put these labels in the subscripts of the variables associated with each agent. You do not have to specify the policy.

Problem solving questions

20.8 Assume there are two agents that we will designate A and B, who are trying to mitigate flu in their respective communities (perhaps neighboring states). We could model agents A and B using the models in section 20.2 if there were no interactions, but the populations of the two countries travel between each other, spreading infections. Let

$$P_{ti}^H = \text{the true number of healthy people in state } i \text{ at time } t,$$

$$P_{ti}^I = \text{the true number of infected people in state } i \text{ at time } t,$$

$$P_{ti}^V = \text{the number of vaccinated people in state } i \text{ at time } t \text{ (this would be known to agent } i),$$

$$\mu_{ti} = \text{the true fraction of the population of state } i \text{ that has the flu at time } t, \text{ for } i = A, B,$$

$$\rho_{ij} = \text{fraction of customers who move from state } i \text{ to state } j \text{ each time period, which applies equally to the entire population,}$$

$$x_{ti}^{vac} = \text{number of vaccinations administered in state } i \text{ at time } t,$$

$$x_{tij}^{share} = \text{the number of vaccinations that state } i \text{ sends to state } j \text{ at time } t,$$

$$x_{ti}^{test} = \text{the number of people who are tested in state } i$$
$$\text{between } t \text{ and } t+1, \text{ producing noisy estimates}$$
$$\hat{P}_{t+1,i}^{H} \text{ of } P_{ti}^{H} \text{ and } \hat{P}_{t+1,i}^{I} \text{ of } P_{ti}^{I} \text{ (deciding to test the}$$
$$\text{population at time } t \text{ produces observations by}$$
$$\text{time } t+1),$$

$$\bar{P}_{ti}^{H} = \text{the estimated number of healthy people in state } i$$
$$\text{at time } t,$$

$$\bar{P}_{ti}^{I} = \text{the estimated number of infected people in state } i$$
$$\text{at time } t.$$

Assume initially that there is no information sharing between the two agents (e.g. of the estimates $\bar{P}_{t}^{H}$ or $\bar{P}_{t}^{I}$).

People are vaccinated without regard to their status as healthy or infected. Healthy people are protected from further infections, but infected people remain infected if the vaccination comes after their initial infection. Create three agents: environment (that covers both states), and each controlling agent. For each agent, do the following:

(a) Define the state variables, decision variables, exogenous information variables and transition function (assume ρ_{ij} is known to agent i (and ρ_{ji} is known to agent j).

(b) Assume the objective of each controlling agent is to minimize the number of infected people in their state. Write out the objective function for each controlling agent.

(c) Keep in mind that the populations might be quite different, which means that if $P_{ti}^{H} \gg P_{tij}^{H}$, state j may have an incentive to send vaccinations to state i, given that a fraction ρ of the people in state i will travel to state j. Given this, design a simple PFA to determine x_{ti}^{vac} and x_{tij}^{share} for each state.

20.9 Assume that each agent in our flu problem is willing to share their observations $\hat{P}_{t}^{H}$ and $\hat{P}_{t}^{I}$. Further assume that the transition rates ρ_{ij} and ρ_{ji} are unknown to both agents. Describe how this affects the formulation of the state variables, exogenous information variables, and the transition functions. Note that each agent will have to create their own estimate of ρ_{ij}.

20.10 The central agent for the two-agent newsvendor problem can blend requests from the field agent with external sources of knowledge as is done in equation (20.34). This process (which we believe is quite common when dealing with people) raises a number of questions:

(a) It would seem that a field agent could always do better by constantly raising his requests relative to the estimate R_t^e to stay one step ahead of the estimates of the central agent. Describe how the blending process can discourage this.

(b) Describe in words how being too predictable can hurt the field agent.

(c) If being too predictable can hurt the field agent, describe a noise and biasing strategy that could help the field agent. Contrast the long run effectiveness of bias versus noise.

20.11 Assume that the central agent in the two-agent newsvendor problem (section 20.4) is not allowed to see $\hat{R}_{t+1}$, but does see total cost (combining overage costs and underage costs), as might be typical in many organizations.

(a) Update your models for the central agent.

(b) Suggest a learning policy for the central agent given the information available to her.

20.12 For the two-agent newsvendor problem (section 20.4) we are going to introduce two changes: First, we now assume the field agent can hold excess inventory to a later time period and second, the cost of underage c^u now varies randomly over time, so we will model it as c_t^u. This means that there will be some time periods where c_t^u may be much higher than other time periods. This variation sets up an incentive for the field agent to *hoard* resources during periods where the cost of underage is low so that they might be available when the underage cost is high.

(a) Update your model for the field agent. Assume that the central agent is unaware of resources being held for future time periods.

(b) Suggest a policy (for the field agent) for holding inventory. Note that you may hold resources for future periods while not meeting demand in the current period.

20.13 Assume that the field agent in our two-agent newsvendor problem (section 20.4) is trying to anticipate the behavior of the central agent (there is an initial discussion of this in the text, but without details). Your goal in this exercise is to fill in the details.

(a) Suggest a belief model for the field agent of the behavior of the central agent.

(b) Incorporate this belief model into your state variable, and outline the five elements of a sequential decision system for the field agent. Be sure to include the updating equations for all the state variables in your transition function, including any that are needed for your belief model of the central agent.

(c) Design an ordering policy for the field agent. If your policy required introducing additional state variables, be sure to update your model in part (b).

20.14 Assume that the central agent in our two-agent newsvendor problem (section 20.4) is trying to anticipate the behavior of the field agent, without any assumptions that the field is trying to anticipate the central agent (as we did in exercise 20.13). Your goal in this exercise is to fill in the details.

(a) Suggest a belief model for the central agent of the behavior of the field agent.

(b) Incorporate this belief model into your state variable, and outline the five elements of a sequential decision system for the central agent. Be sure to include the updating equations for all the state variables in your transition function, including any that are needed for your belief model of the field agent.

(c) Design an ordering policy for the central agent. If your policy required introducing additional state variables, be sure to update your model in part (b).

20.15 (Advanced) Repeat exercise 20.14 assuming that you have already done exercise 20.13, which means you are modeling the central agent while anticipating that the field agent is trying to anticipate the behavior of the central agent.

Sequential decision analytics and modeling

These exercises are drawn from the online book *Sequential Decision Analytics and Modeling* available at http://tinyurl.com/sdaexamplesprint.

20.16 This exercise will focus on the multiagent problems in chapters 10 and 11, so please begin by reviewing this material. This exercise will use the Python module "TwoNewsvendor" available at http://tinyurl.com/sdagithub.

(a) Run the code with the basic learning model and see what the two players settle for in the end (the biases that get chosen more often).

(b) Fix the central command bias at -4. Make the code run the problem with the field agent treating it as a learning process with several values for the UCB parameter (1, 2, 4, 8, 16, 32) and make a graph round vs choice to see how the learning policy selects what biases to use.

(c) Using code from the first two-newsvendor model and code from the modules for the learning approach, write your own module where the field agent treats the problem as a learning problem, each round choosing a bias in [0, 10], and the central command treats the problem using the strategy from the first model (computes bias, adds his own bias plus some noise).

(d) Consider now a two-newsvendor problem where the central command also has some external information about the demand. What he has is a much noisier estimate of the demand (say the noise is, for our spreadsheet data where the demand is always between 20 and 40, three times bigger than the noise the noise from the source communicating with the field agent). Redefine the bias from the central command as the quantity that he adds to the estimate he gets. Try a learning approach where the bias he selects is chosen in the interval $[-5, 5]$. Run the program and compare the results with the old learning process.

(e) (Punishing strategy 1) Consider the case where the field agent is using a learning approach and the central command is using a punishing strategy. Since he knows the field gets a bigger penalty for providing less than the demand, the central command will compute the previous field bias (for time $t - 1$) and if it is positive, it will apply a bias twice as big in magnitude and of opposite sign to the field's request. Run this experiment and see what the field's prevalent bias will be after the 4000 rounds.

Diary problem

The diary problem is a single problem you chose (see chapter 1 for guidelines). Answer the following for your diary problem.

20.17 If your diary problem has either the dimension of learning (such as learning an unknown environment), or multiple decision makers, you can use the framework in this chapter. Given this, do the following:

(a) Describe each agent, and characterize each as either an environment agent, a controlling agent (which may or may not have learning), and possibly a pure learning agent.

(b) For each controlling agent, identify belief variables that capture beliefs that the controlling agent would have to create about unknown parameters, quantities. If there is more than one controlling agent, it will be necessary for each agent to develop beliefs about other controlling agents, but leave this for part (c). While it is important to introduce notation, start by describing the beliefs and unknown quantities in words.

(c) If there is more than one controlling agent, it will be necessary to create beliefs about other controlling agents.

(d) For each agent, create a full model including all five elements (decisions and objective functions will be missing for agents other than controlling agents).

Bibliography

Abara, S., Theodoropoulos, G.K., Lemarinier, P., and O'Hared, G.M. (2017). Agent based modelling and simulation tools: A review of the state-of-art software. *Computer Science Review* 24: 13–33.

Bouzaiene-Ayari, B., Cheng, C., Das, S., Fiorillo, R., and Powell, W.B. (2016). From single commodity to multiattribute models for locomotive optimization: A comparison of optimal integer programming and approximate dynamic programming. *Transportation Science* 50 (2): 1–24.

Busoniu, L., Babuska, R., and Schutter, B.D. (2011). A comprehensive survey of multiagent reinforcement learning. *IEEE Transactions on Systems, Man and Cybernetics Part C: Applications and Reviews* 38 (2): 156–172.

Chen, B., Cheng, H.H., and Member, S. (2010). A review of the applications of agent technology in traffic and transportation systems. *IEEE Transactions on Intelligent Transportation Systems* 11 (2): 485–497.

D'Inverno, M. and Luck, M. (2001). *Understanding Agent Systems*. New York: Springer.

Dorri, A.L.I., Member, S., Kanhere, S.S., and Member, S. (2018). Multi-agent systems : A survey. *IEEE Transactions on Industry Applications* 49 (1): 28573–28593.

Godfrey, G.A. and Powell, W.B. (2001). An adaptive, distribution-free algorithm for the newsvendor problem with censored demands, with applications to inventory and distribution. *Management Science* 47 (8): 1101–1112.

González-briones, A., Prieta, F.D.L., Omatu, S., and Corchado, J.M. (2018). Multi-agent systems applications in energy optimization problems : A state-of-the-art review. *Energies* 11: 1–28.

Powell, W.B. (2020). On state variables, bandit problems and POMDPs, arXiv, https://arxiv.org/abs/2002.06238.

Powell, W.B. (2021). From reinforcement learning to optimal control: A unified framework for sequential decisions. In: *Handbook on Reinforcement Learning and Optimal Control, Studies in Systems, Decision and Control*, 29–74.

Powell, W.B. and Godfrey, G.A. (2002). An adaptive dynamic programming algorithm for dynamic fleet management, I: Single period travel times. *Transportation Science* 36 (1): 40–54.

Shapiro, J.A. and Powell, W.B. (2006). A metastrategy for large-scale resource management based on informational decomposition. *INFORMS Journal on Computing* 18 (1): 43–60.

Tang, W. and Bennett, D.A. (2010). Agent-based modeling of animal movement : A review. *Geography Compass* 7: 682–700.

Tecuci, G. (1998). *Building Intelligent Agents*. Academic Press.

Zhao, P., Member, S., Suryanarayanan, S., Member, S., Simões, M. G., and Member, S. (2013). An energy management system for building structures using a multi-agent decision-making control methodology. *IEEE Transactions on Industry Applications* 49 (1): 322–330.

Index

a

Active learning, 61
Actor-critic algorithm, 907
Ad-clicks, 459
Adaptive learning algorithms, 202
    as sequential decision, 202
    designing policies, 209
    learning problems, 202
Affine function, 747
Aggregation, modeling, 125
Aleatoric uncertainty, 564
Algorithm
    approximate value iteration
        linear model, 886
    Benders for two-stage, 951
    indifference zone selection, 383
    optimal computing budget
        allocation, 383
    temporal-difference learning
        infinite horizon, 833
    actor-critic, 907
    ADP with pre-decision state, 872
    approximate policy iteration,
        901, 907
        linear model, 903
        LSTD, 905
    backward ADP
        lookup tables, 800
        parametric model, 800

backward dynamic
    programming, 749
bias-adjusted Kalman filter
    stepsize, 297
direct lookahead policy
    simulator, 994
double-pass ADP, 876
Gauss-Seidel variation, 758
hybrid value/policy iteration, 764
least squares temporal
    differencing, 905
MCTS algorithm
    backup step, 1014
    roll-out policy, 1011
    simulation step, 1012
    tree policy, 1012
policy evaluation, 825
policy iteration, 763
Q-learning
    finite horizon, 875, 876
relative value iteration, 758
shortest path algorithm, 76
value iteration, 757
American option, 894
Anytime problems, 41
Apparent convergence, 300
Applications
    ad-clicks, 320, 370, 459
    aid in Africa, 813

Reinforcement Learning and Stochastic Optimization: A Unified Framework for Sequential Decisions, First Edition. Warren B. Powell.
© 2022 John Wiley & Sons, Inc. Published 2022 John Wiley & Sons, Inc.

American option, 55, 894
asset selling, 54, 662
asset valuation, 437
bidding, 992
blood management, 450, 715,
    927, 1070
blood sugar, 662
budgeting problem, 752
building management, 1067
business, 5
carbon monoxide, 567
cash management, 186
chemical diffusion, 570
clinical trials, 55, 810
computer simulations, 320
cost-minimizing newsvendor, 226
COVID pandemic, 575
currency exchange, 513
demand estimation, 567, 569
diabetes medication, 376, 490
diseases, 114
drones, 580, 611
dynamic assignment, 447
dynamic resource allocation, 927
dynamic shortest path, 707
e-commerce, 5, 227
economics, 5
electric utility, 932
electricity contracts, 509
electricity price forecasting, 565
electricity prices, 589
energy, 927
energy forecasting, 489
energy investments, 186
energy modeling
    crossing times, 593
    jump-diffusion, 590
    mean reversion, 590
    quantile distributions, 591
    regime shifting, 592
energy planning, 986

energy storage, 471, 523, 611,
    662, 811
    active learning, 526
    passive learning, 525
    rolling forecast, 717
    rolling forecasts, 526
    time series price, 525
energy storage policies, 627
    CFA, 628
    comparison, 629
    DLA, 629
    hybrid, 629
    PFA, 628
    VFA, 628
engineering, 5
engineering design, 185
European options, 55
field experiments, 320
finance, 5
financial trading, 611, 711, 986
fleet management, 78
    multiple drivers, 80
    nomadic trucker, 79
flu mitigation, 1044
    controllable truth, 1048
    designing policies, 1054
    resource constrained, 1049
    spatial model, 1049
    static model, 1045
    time-varying, 1047
    two-agent learning, 1050
    with drift, 1047
freight transportation, 5
gambling problem, 751
games, 514
graph problems, 433
grid model, 570
homeland security, 55
HVAC controller, 1067
    model, 1067
    policies, 1069
information collection, 5

inventory, 6, 13, 73, 226, 439, 611
investment banking, 932
laboratory experimentation, 458
load matching, 705
logistics, 800
machine replacement, 55
machine scheduling, 992
manufacturing, 328, 933
materials science, 5, 320, 376, 986
medical, 514
medical decisions, 81, 114, 320,
     406, 457, 580, 986
medical diagnostics, 567
medical research, 5
money, 927
nested newsvendor, 226
newsvendor, 70, 189, 406, 930
nomadic trucker, 124, 434
oil inventories, 580
online bookseller, 933
police patrolling, 610
power grid, 567, 572
power transformers, 509
pricing, 80, 114, 572
public health, 5, 114
rat diet, 572
resource allocation, 446
     multiperiod, 933
     two-stage, 931
ride hailing, 991, 1008
rockets, 47
scientific exploration, 82
shortest path, 76, 189, 992, 1001
     deterministic, 76
     dynamic, 78
     robust, 78
     stochastic, 77
shuttle bus, 662
SpaceX, 47
sports, 114, 320
stochastic shortest path, 433
stopping problems, 54

supply chain, 5, 6
tic-tac-toe, 898
transformer replacement, 435
transportation, 320, 799
trucking, 489, 799, 927, 991
two-agent newsvendor, 1062
water reservoir, 611
Approximate dynamic
     programming, 50
single state, 855
     LSPE, 857
     LSTD, 856
     recursive least squares, 857
Approximate linear programming
     method
     linear model, 912
Approximate policy iteration, 900
     linear model, 903
     lookup tables, 901
Approximate value iteration, 828
     backward pass, 876
     bounding $1/n$ convergence, 859
     geometric view, 850
     gradient-based methods, 845
     linear models, 845, 886
     lookup tables, 871
     post-decision state, 875
     pre-decision state, 872
     slow backward learning, 832
     TD(0), 830
Approximation architectures, 165
Asset valuation, 437

_b_
Backgammon, 492
Backpropagation, 879
Backward ADP
     benchmarking, 810
          aid in Africa, 813
          clinical trials, 810
          energy storage, 811
     finite horizon, 797
     lookup table, 800

parametric model, 800
VFA approximation, 805
  linear model, 806
  monotone function, 807
Backward dynamic
  programming, 749
Base model, 528
Base models vs. lookahead
  models, 528
Basis functions, 941
  American option, 896
  approximate linear
    programming, 913
  tic-tac-toe, 899
Bayesian prior, 112
Bayesian updating
  correlated normal beliefs, 113
  independent beliefs, 112
  normal-normal, 112
  precision, 113
Behavior policy, 208, 890
Belief MDP, 1060
Belief state, 408, 495
Bellman error, 829
Bellman's equation, 741
  curse of dimensionality, 47
  deterministic, 741
  expectation form, 742
  linear model, 837
  linear operator, 747
  matrix form, 743
  operator form, 746
  standard form, 742
Bellman's optimality equation, 46
Benders decomposition, 950
  asymptotic analysis, 952
  two-stage, 947
Benders' decomposition
  regularization, 952
Beta distribution, 587
Bias and variance, 118
Bias in max operator, 909

Blood management problem, 450
Boltzmann exploration, 337, 627
Budgeting problem, 752

C

Chance constrained
  programming, 61
Classification of problems, 529
  state-dependent, cumulative
    reward, 531
  state-dependent, final reward, 532
  state-independent, cumulative
    reward, 530
  state-independent, final
    reward, 530
Competitive analysis, 391
Conditional value at risk, 523
Contextual bandit, 406
Control law, 49
Correlated beliefs, 113
Cost function approximation
  constraint-modified, 714
    blood management, 715
    dynamic trading, 711
    energy storage, 717
      parameterized policy, 722, 1004
      SPSA, 722
      tuning, 723
    general form, 703
    heat maps, 723
    interval estimation, 702
      uncertainty bonus, 702
  objective-modified, 704
    cost function correction, 705
    dynamic assignment, 705
  pure exploitation, 702
  shortest path, 707
Cost-to-go function, 48
COVID pandemic, 575
Crossing times, 593
Cumulative reward, 42
  adaptive policy, 401
  bandit problems, 323

cumulative loss, 393
examples, 59
expectation
  state-dependent, 530
expensive experiments, 349
knowledge gradient, 362, 369,
    370, 400
learning, 206
multiarmed bandit, 57
newsvendor, 71
policy evaluation, 641, 996
simulated
  state-dependent, 533
  state-independent, 533
state-independent, 530
stochastic gradient algorithm,
    224, 319, 327
stochastic search, 187
two-agent, 1051
Curse of dimensionality, 47, 162
Cutting planes, 946

d

Data driven, 70, 82
Decision tree, 44, 738
Decision variables, 65
  constraints, 504
  execution decisions, 503
  policies, 505
  strategic decisions, 503
  tactical decisions, 503
  types, 502
Decisions, 500
Deep neural networks, 154
Derivative-free stochastic search
  CFA policies, 335
    Bayes greedy, 336, 337
    greedy policy, 335
    Interval estimation, 337
    interval estimation, 337
    UCB policy, 337
  designing policies, 394
  direct lookahead, 348

multiperiod deterministic, 355
multiperiod stochastic, 357
restricted multiperiod, 353
single period lookahead, 350
exogenous state information, 405
large choice sets, 403
learning in batches, 380
policy function
  approximation, 333
scaling, 396
tuning, 398
VFA policy, 338
  backward ADP, 342
  Gittins indices, 343
VFA-policy
  Beta-Bernoulli, 340
Designing a policy, 631
Diagnostic uncertainty, 561
Direct lookahead
    approximations, 616
Direct lookahead model
  approximation strategies, 980
    horizon truncation, 981
    latent variables, 982
    policy approximation, 984
    sampling, 982
    stage aggregation, 982
  lookahead objectives, 985
    chance-constrained, 989
    managing risk, 985
    model discounting, 992
    risk measures, 986
    risk-adjusted policies, 989
    robust optimization, 990
    utility functions, 991
  notation, 980
Direct lookahead policy
  approximate dynamic
      programming, 978
  approximate lookahead, 980
  backward ADP lookahead
      policy, 1007

creating lookahead model, 978
deterministic, 1000
discounted, 992
DLA lookahead policy, 1008
fixed lookahead policy, 994
lookahead CFA, 1007
lookahead PFA, 1005
lookahead policy, 977
MCTS, 978, 1009
model predictive control, 977
optimal, 974
policy-within-a-policy, 977
robust optimization, 978
rolling horizon procedure, 977
rollout policy, 978
shortest path, 1001
stochastic programming, 977
why use, 997
Distributional uncertainty, 587
Double-pass ADP algorithm, 876
Dynamic assignment problem, 447
Dynamic program
budgeting problem, 752
gambling problem, 751
Dynamic programming
model-free, 516, 875

e
Endogenous learning, 101, 105
Energy storage model, 523
active learning, 526
passive learning, 525
rolling forecasts, 526
time-series prices, 525
Energy storage policies, 627
CFA, 628
DLA, 629
Hybrid, 629
PFA, 628
VFA, 628
Episodic, 825
Epistemic uncertainty, 564
Epsilon-greedy, 626

Evaluating policies, 385
competitive analysis, 391
dynamic regret, 390
empirical performance, 386
indifference zone, 392
opportunity cost, 391
probability of correction
selection, 393
quantiles, 386
static regret, 387, 388
subset selection, 394
Excitation policy, 627, 675
Exogenous information, 65, 506
adversarial processes, 513
deterministic, 514
lagged, 510
modeling, 506, 511
scenarios, 509
supervisory processe, 514
Exogenous learning, 104
Expectation
compact form, 40
expanded form, 40
Expected improvement, 351
Experimental uncertainty, 561
Exploration
Boltzmann exploration, 627
epsilon-greedy, 626
excitation, 627
Thompson sampling, 627
Exploration vs. exploitation, 626
Exponential smoothing, 109, 274

f
Final reward, 42, 187, 206, 329
asymptotic convergence, 392
bandit problems, 323
cost function approximation, 702
expectation
state-dependent, 530, 531
state-independent, 529
expensive experiments, 349
knowledge gradient, 351, 362, 364

multiarmed bandit, 60
newsvendor, 70
policy search, 668
simulated
  state-dependent, 534, 885
  state-independent, 533
stochastic gradient algorithm, 235
stochastic search, 224
Finite-horizon approximations
  steady state, 915
Fitted value iteration, 804, 823
Fleet management problem, 78
  fleet, 80
  nomadic trucker, 79
Forecasts
  rolling, 75, 497, 526, 635, 707, 717

g
Gambling problem, 751
Games
  backgammon, 492
  chess, 514, 900
  computer Go, 514, 900
  maze, 52
  video, 154
  video games, 514
Gaussian process regression, 117
Gittins indices, 343
  normally distributed rewards, 346

h
Hamiltonian, 48
Hierarchical aggregation, 121
  correlations, 169
  estimation, 129
  estimation, 125
Hyperparameter, 229, 587, 589

i
Implementation policy, 208, 337
Inferential uncertainty, 561
Infinite horizon, 755
  policy iteration, 901

temporal-difference learning, 832
Information acquisition, 456
Interval estimation, 337
Inventory problems, 73
  basic, 439
  batch replenishment, 444
  forecasts, 75
  general, 440
  lagged, 443
  with lags, 75
  without lags, 73
Inverse optimization, 105

j
Jump diffusion, 590

k
k-nearest neighbor, 150
Kalman filter, 293
Kernel regression, 151
Knowledge gradient, 362
  correlated beliefs, 375
  cumulative reward, 369
  final reward, 351, 364
  posterior reshaping, 354
  restricted lookahead, 353
  sampled belief, 370

l
Laboratory experimentation, 458
Lagged information, 510
Lagrangian relaxation, 768
Lasso, 134
Latent variable, 496, 563
Learning
  approximation strategies, 106
  Bayesian
    state variable, 113
  bias and variance, 118
  deep neural networks, 154
  endogenous, 101, 105
  exogenous, 101, 104
  exponential smoothing, 109

frequentist
  state variable, 111
Gaussian process regression, 117
inverse optimization, 105
k-nearest neighbor, 150
kernel regression, 151
linear regression, 132
local polynomial regrerssion, 153
lookup tables, 106
  Bayesian, 111
  frequentist, 110
maximum likelihood
    estimation, 141
model-based, 52
model-free, 52
nonlinear parametric models, 140
nonparametric definition, 149
nonparametric models, 107
nonstationary, 159
  learning process, 161
  transient truth, 160
  truth, 159
normal equations, 133
parametric models, 106
  linear, 106
  nonlinear, 107
sampled belief models, 141
Sherman-Morrison
  derivation, 168
supervised, 101
support vector machines, 155
Learning policy, 208, 337, 890
Learning process, 161
Learning rate, 235
Least squares policy evaluation,
    841, 857
Least squares temporal difference,
    840, 841, 856, 903, 905
Leveling algorithm, 939
Linear model
  basis functions, 131
  geometric view, 850

Lasso, 134
  recursive least squares, 136
  sparse additive, 134
Linear models, 131
Linear programming method,
    766, 767
Linear quadratic regulation, 48, 767
Linear regression, 132
  Longstaff and Schwartz, 894
Local polynomial regression, 153
Longstaff and Schwartz, 894
  basis functions, 896
Lookahead dynamic
    programming, 885
Lookahead model, 528
Lookup tables
  aggregation, 121
  Bayesian updating, 111
    state variable, 113
  frequentist updating, 110
    state variable, 111
  hierarchical aggregation, 121, 169
LSPE, 841
  single state, 844
LSTD, 840, 841
  single state, 844

m
Markov decision process, 45, 737
  average reward, 765
  backward dynamic
      programming, 749
  Bellman's equation, 46, 741
  cost-to-go function, 744
  curse of dimensionality, 47
  finite horizon, 747
  infinite horizon, 755
  linear programming method, 767
  objective function, 740
  optimality equation, 741
  partially observable, 488
  random contributions, 746
  state transitions, 739

transition matrix, 745
value function, 744
value iteration
    stopping rule, 757
Martingale model of forecast
    evolution, 498
Maximum likelihood
    estimation, 141
MCTS
    backpropagation step, 1014
    expansion step, 1012
    selection step, 1010
    simulation step, 1012
    steps, 1010
Mean reversion, 590
Measure-theoretic view of
    information, 535
Medical decision making, 81, 457
MMFE, 498
Model
    base model, 528
    lookahead, 528
    nominal, 528
    true, 528
    universal, 467
Model first, then solve, 4, 7, 11, 27,
    433, 470, 474, 505, 603
Model predictive control, 62
Model-free dynamic programming,
    516, 875
Modeling frameworks, 68
Modeling time, 478
Monte Carlo, 582
Monte Carlo sampling, 582
Monte Carlo simulation, 560
    inverse cumulative method, 585
Monte Carlo tree search, 1009, 1010
    optimistic, 1016
Monte-Carlo simulation, 581
Multiagent systems
    blood management, 1070
        multiple agent, 1073

communication, 1038
    active observations, 1039
    architecture, 1038
    receiving information, 1039
    sending information, 1039
    signal distortion, 1039
controlling architecture, 1043
dimensions, 1036
    communication, 1036
    coordination, 1036
    learning, 1036
    the agents, 1036
flu mitigation, 1044
    controllable truth, 1048
    designing policies, 1054
    multiarmed bandit, 1056
    neural network policy, 1055
    resource-constrained, 1049
    spatial model, 1049
    static model, 1045
    time-varying model, 1047
    time-varying with drift, 1047
HVAC controller, 1067
    model, 1067
    policies, 1069
modeling, 1040
two-agent learning model, 1050
    transition functions, 1052
two-agent newsvendor, 1062
types
    controlling, 1037
    environment, 1037
    learning, 1037
    replenishment, 1037
Multiarmed bandit, 57
    active learning, 61, 325
    backward ADP, 342
    best hitter, 59
    best medication, 59
    best path, 59
    beta-bernoulli belief, 340
    Boltzmann exploration, 337

computer science, 321
contextual, 405, 406
derivative-free stochastic opt., 325
dynamic programming, 338
flu mitigation, 1056
Gittins, 323
Gittins indices, 343
  normal rewards, 346
history, 60
interval estimation, 337
interval estimation policy, 59
nonstationary, 399
objective function, 58
one-armed, 321
sequential decision problem, 325
state variable, 58
story, 321
Thompson sampling, 337
transient learning, 401
two-armed, 321
uncertainty bonus, 1016
upper confidence bounding,
    323, 337
variations, 323

n

Neural network, 106, 107
actor-critic, 907
batch learning, 109
classification, 107, 243
deep, 53, 364, 655
drone, 611
estimation, 242, 658, 1055
four layer, 269
games, 632, 900
learning, 335
limitations, 663, 672, 701, 805
medical, 514
parametric, 143
policy, 331, 666, 971
properties, 246
robot, 632, 656, 900
stepsizes, 285

value function, 50, 734, 907,
    971, 1057
value functions, 608, 624
Newsvendor
final reward, 70
Newsvendor problem, 70
contextual, 71
cumulative reward, 71
final reward, 70
multidimensional, 72
Nomadic trucker, 79, 434, 491
Nonanticipativity constraint, 57, 201
Nonparametric model
definition, 149
Nonparametric models, 149
k-nearest neighbor, 150
kernel regression, 151
local polynomial regression, 153
Nonparametric statistics
tree regression, 156
Nonstationary learning, 159
learning process, 161
martingale truth, 159
transient truth, 160
Normal equations, 133

o

Objective function, 66
conditional value at risk, 523
cumulative reward, 206
final reward, 206
performance metrics, 518
probability of correct
    selection, 393
regret, 208
robust optimization, 523
static regret, 388
value at risk, 523
Observational uncertainty, 560
Off-policy learning, 890
Offline learning, 882
On-policy learning, 890
Online learning, 883

Opportunity cost, 391
Optimal computing budget
    allocation, 383
Optimal control, 47
  control law, 49
  cost-to-go function, 48
  Hamiltonion, 48
  Lagrangian relaxation, 768
  linear decision rule, 769
  linear quadratic regulation,
    48, 767
  state equation, 768
  stochastic, 49
  transition function, 47
Optimal stopping, 54
Optimality equations
  proof, 770
Optimistic policy iteration, 903
Optimum-deviation policy, 674

p

Parametric models
  linear, 131
  nonlinear, 140
Partial policy evaluation, 903
Partially observable Markov
    decision process, 563, 1059
Partially observable states, 495
Performance elbow, 997
Policy
  affine, 612
  behavior, 208, 667, 890
  Boltzmann, 627
  choosing policy class, 638
  complexity tradeoffs, 636
  cost function approximation, 613
  definition, 604
  designing, 631
  direct lookahead, 616, 974
  energy storage problem, 627
    CFA, 628
    DFA, 629
    Hybrid, 629

    PFA, 628
    VFA, 628
  epsilon-greedy, 626
  evaluation, 532, 641
  examples, 604
  excitation, 627
  four classes, 209
  four classes of policies, 606
  hybrid, 620
  implementation, 208, 337,
    644, 667
  learning, 208, 337, 644, 667, 890
  linear decision rule, 612
  measurability, 538
  optimistic, 903
  partial policy evaluation, 903
  performance-based, 668
  policy function
    approximation, 610
  randomized, 626
  roll-out heuristic, 1009
  sampling the shoulders, 673
  soft issues, 644
  supervised, 668
  target, 208, 667, 890
  Thompson sampling, 627
  tree search, 1009
  tuning, 642
  value function
    approximation, 614
Policy evaluation, 641
  infinite horizon, 826
    infinite horizon projection, 828
    partial simulations, 827
    recurrent visits, 826
    temporal difference update, 828
Policy function approximation, 659
  affine, 660
  affine policy, 660
  Boltzmann, 659
  contextual, 665
  limitations, 632

locally linear, 663
monotone policies, 661
nonlinear, 662
Policy gradient theorem, 683, 684
  computation, 684
Policy iteration, 763
  hybrid value/policy
    iteration, 764
  infinite horizon, 901
Policy search
  correlated knowledge
    gradient, 677
  derivative-free methods, 670
  exact derivatives
    continuous, 677
    discrete, 680
    policy gradient theorem, 683
  excitation, 675
  interval estimation, 675
  knowledge gradient, 677
  numerical derivatives, 669
  optimum-deviation, 674
  sampled belief, 676
  supervised learning, 686
Policy-within-a-policy, 620
POMDP, 488, 563, 1059
  belief MDP, 1060
Post-decision state, 79, 490
Pricing, 80
Prognostic uncertainty, 560

q

Q-learning, 875
  algorithm, 875, 876
  augmented state, 493
  deep neural network, 53
  expected rewards, 688
  linear model, 886
  lookup tables, 873
  off-policy, 845
  recursive least squares, 887

temporal-difference
    updating, 51, 735, 845
  updating, 51, 277, 615, 735
Quantile distribution, 591

r

Random number seed, 582
Randomized policies, 626, 781
Receding horizon procedure, 63
Recursive least squares, 136
  derivation, 166
  multiple observations, 139
  nonstationary data, 138
  stationary data, 136
Regret, 208
Reinforcement learning, 50
  Q-learning, 51
Resource allocation, 446, 930
  blood management, 450
  dynamic assignment, 447
  exogenous information state, 958
  general resource allocation, 933
  information acquisition, 456
  newsvendor, 930
  two-stage resource
    allocation, 931
Response surface method, 335
Risk
  robust optimization, 523
Risk measures, 523
  conditional value at risk, 523
  value at risk, 523
Risk operator, 987
Robust optimization, 63, 523
  as a policy, 991
  optimization problem, 991
  robust policy, 64
  uncertainty set, 64
    box constraints, 990
Roll-out heuristic, 1009
Rolling forecast, 635

Rolling forecasts, 497
Rolling horizon procedure, 63

S
Sample average approximation, 194
  convergence rate, 197
Sampled belief models, 141
Sampled models, 193, 595
Sampling the shoulders, 673
SARSA, 875
Scientific exploration, 82
Sequential kriging, 352
Sherman-Morrison formula,
      115, 168
Shortest path
  information collecting, 459
  stochastic, 433
Shortest path problem, 76
  dynamic, 78
  robust, 78
  stochastic, 77
  deterministic, 76
Simulation optimization, 60, 382
  indifference zone algorithm, 383
  OCBA, 60, 383
Simultaneous perturbation
      stochastic approximation
  see SPSA, 240
Soft max, 627
Software, 469
Specification error, 569
SPSA, 240
  mini-batch, 240
  tuning, 722
State variables, 65
  asset pricing, 437
  belief, 408
  belief state, 485, 495
  definition, 481
  dynamic assignment problem, 450
  factored, 498
  flat representation, 498
  gambling problem, 751

information state, 485
lagged, 490
latent variable, 496
physical state, 408, 485
post-decision state, 79, 490, 491
pre-decision state, 491
rolling forecasts, 497
shortest path, 493
three types, 485
Stepsize policies
  apparent convergence, 301
  BAKF proof, 303
  bias-adjusted Kalman filter, 297
  convergence conditions, 276
  deterministic, 276
    constant, 278
    harmonic, 279
    McClain, 280
    polynomial learning rate, 280
    search-then-converge, 281
  guidelines, 301
  infinite horizon
    bounds, 859
  introduction, 233
  optimal
    approx. value iteration, 297
    BAKF, 295
    nonstationary - I, 293
    nonstationary - II, 294
    stationary, 291
  optimal stepsizes, 289
  stochastic, 282
    AdaGrad, 287
    ADAM, 287
    convergence conditions, 283
    Kesten's rule, 285
    RMSProp, 288
    stochastic gradient adaptive
      stepsize, 286
    Trigg, 286

Stochastic approximation procedure
- see stochastic gradient
algorithm, 233
Stochastic decomposition, 950, 952
Stochastic gradient algorithm, 225,
233, 835
as a sequential decision
problem, 247
finite differences, 238
for neural networks, 242
gradient smoothing, 237
martingale proof, 256
mini-batches, 240
older proof, 252
second order methods, 237
SPSA, 240
Stochastic modeling
crossing times, 593
electricity prices, 589
energy illustration
crossing times, 593
jump diffusion, 590
mean reversion, 590
quantile distribution, 591
regime shifting, 592
quantile distribution, 591
regime shifting, 592
Stochastic optimal control, 49
Stochastic optimization
asymptotic formulation, 42
cumulative reward, 42
final reward, 42
Stochastic programming, 56
Benders decomposition, 946
interstage independence, 948, 959
lookahead policy, 57
nonanticipativity, 57
nonanticipativity constraint, 57
recourse variables, 56
two-stage, 56
Stochastic search, 40
adaptive learning, 184

cumulative reward, 187, 224
derivative-based, 42
derivative-free, 43
deterministic methods, 184, 188
chance-constrained, 190
Markov decision process, 192
newsvendor, 189
optimal control, 191
shortest path, 189
final reward, 187, 224
newsvendor, 183
sampled approximations, 184
sampled models, 193
chance-constrained, 195
convergence, 197
linear program, 194
parametric model, 196
state-dependent, 187
state-independent, 187
Stopping time, 54
Supervised learning, 101, 900
Support vector machines, 155

t

Target policy, 208, 890
TD(λ), 829
TD(0)
recursive least squares, 842
TD(0) updating
slow backward learning, 832
Temporal difference learning, 829
Temporal-difference learning, 828
infinite horizon, 832
Thompson sampling, 337, 627
Three curses of dimensionality,
47, 748
Tic-tac-toe, 898
Time, 478
Trajectory following, 890
Transformer replacement, 435
Transient learning model, 401
knowledge gradient, 402
Transition function, 65, 515

batch replenishment, 445
Transition matrix, 745
Tree regression, 156
Trucking, 491
Two-stage stochastic
    programming, 1018
    Benders' decomposition, 947

u

Uncertainty
    adversarial, 564
    aleatoric, 564
    algorithmic, 573
    algorithmic uncertainty, 561
    coarse-grained, 564
    communication, 572
    communication uncertainty, 561
    control, 571
    control uncertainty, 561
    diagnostic, 561, 567
    distributional, 564
    epistemic, 564
    experimental, 561, 568
    fine-grained, 564
    goal, 574
    goal uncertainty, 561
    implementation, 571
    inferential, 561, 567
    model, 569
    model specification, 569
    model uncertainty, 561
    observational, 560, 562
    political, 574
    political uncertainty, 562
    prognostic, 560, 564
    regulatory, 574
    transitional, 571
    transitional uncertainty, 561
Uncertainty quantification, 560
Universal modeling framework, 10,
    64, 65, 467, 469
    base model, 528
    compact formulation, 68

decision variables, 65, 500
    energy storage illustration, 523
    exogenous information, 506
    expanded formulation, 65
    lookahead model, 528
    objective function, 66, 518
        risk measures, 523
        uncertainty operators, 523
    problem classifications, 529
    software, 469
    state variable, 481
    state variables, 65
    transition function, 65, 515
Upper confidence bounding, 337

v

Value at risk, 523
Value function approximation, 80
    Bayesian learning, 852
        correlated beliefs, 853
        creating the prior, 855
        parametric models, 854
    cutting planes, 946
    leveling, 939
    regression methods, 941
        separable piecewise linear, 944
    tic-tac-toe, 898
Value function approximations
    linear approximation, 956
    piecewise linear, 938
Value iteration, 757
    algorithmic update, 869
    bound, 761
    bounds, 760
    error bound, 780
    Gauss-Seidel variation, 758
    monotone convergence, 778
    monotonic behavior, 760
    pre-decision state, 872
    proof of convergence, 774
    relative value iteration, 758
    stopping rule, 757

Printed and bound by CPI Group (UK) Ltd, Croydon, CR0 4YY

27/10/2024

14580674-0001